Petroleum Refining Design and Applications Handbook
Volume 5

Scrivener Publishing
100 Cummings Center, Suite 541J
Beverly, MA 01915-6106

Publishers at Scrivener
Martin Scrivener (martin@scrivenerpublishing.com)
Phillip Carmical (pcarmical@scrivenerpublishing.com)

Petroleum Refining Design and Applications Handbook

Volume 5

- Pressure Relieving Devices and Emergency Relief System Design
- Process Safety and Energy Management
- Product Blending
- Cost Estimation and Economic Evaluation
- Sustainability in Petroleum Refining
- Process Safety Incidents

A. Kayode Coker

Scrivener Publishing

WILEY

This edition first published 2023 by John Wiley & Sons, Inc., 111 River Street, Hoboken, NJ 07030, USA and Scrivener Publishing LLC, 100 Cummings Center, Suite 541J, Beverly, MA 01915, USA
© 2023 Scrivener Publishing LLC
For more information about Scrivener publications please visit www.scrivenerpublishing.com.

All rights reserved. No part of this publication may be reproduced, stored in a retrieval system, or transmitted, in any form or by any means, electronic, mechanical, photocopying, recording, or otherwise, except as permitted by law. Advice on how to obtain permission to reuse material from this title is available at http://www.wiley.com/go/permissions.

Wiley Global Headquarters
111 River Street, Hoboken, NJ 07030, USA

For details of our global editorial offices, customer services, and more information about Wiley products visit us at www.wiley.com.

Limit of Liability/Disclaimer of Warranty
While the publisher and authors have used their best efforts in preparing this work, they make no representations or warranties with respect to the accuracy or completeness of the contents of this work and specifically disclaim all warranties, including without limitation any implied warranties of merchant-ability or fitness for a particular purpose. No warranty may be created or extended by sales representatives, written sales materials, or promotional statements for this work. The fact that an organization, website, or product is referred to in this work as a citation and/or potential source of further information does not mean that the publisher and authors endorse the information or services the organization, website, or product may provide or recommendations it may make. This work is sold with the understanding that the publisher is not engaged in rendering professional services. The advice and strategies contained herein may not be suitable for your situation. You should consult with a specialist where appropriate. Neither the publisher nor authors shall be liable for any loss of profit or any other commercial damages, including but not limited to special, incidental, consequential, or other damages. Further, readers should be aware that websites listed in this work may have changed or disappeared between when this work was written and when it is read.

Library of Congress Cataloging-in-Publication Data

ISBN 9781394206988

Cover image: Refinery I Christian Lagerek | Dreamstime.com
Cover design by Kris Hackerott

Set in size of 11pt and Minion Pro by Manila Typesetting Company, Makati, Philippines

Printed in the USA

10 9 8 7 6 5 4 3 2 1

Companion Web Page

This multi-volume set includes access to its companion web page, from which can be downloaded useful software, spreadsheets, and other value-added products related to the books. To access it, follow the instructions below:

1. Go to https://scrivenerpublishing.com/coker_volume_five/
2. Enter your email in the username field
3. Enter "Refining" in the password field

In Loving Memory of

My Parents

Gabriel Shodipọ Coker and Modupe Ajibikẹ Coker

and

Dr. Soni Oyekan (A Great Chemical Engineer)

Finally

Without Him, I am nothing. Life is in Almighty God the Father (Creator) alone.

In Him alone is the energy that lies in Life.

> "God wills that His Laws working in Creation should be quite familiar to man, so that he can adjust himself accordingly, and with their help can complete and fulfill his course through the world more easily and without ignorantly going astray."
>
> *Abd-ru-shin*
> *(In the Light of Truth)*
>
> **The Laws of Creation**
>
> The Law of Motion
> The Law of the Attraction of Homogeneous Species
> The Law of Gravitation
> The Law of Reciprocal Action

"What is Truth?"

Pilate (John 18, 38)

"Only the truth is simple."

Sebastian Haffner

"Woe to the people to whom the truth is no longer sacred!"

Friedrich Christoph Schlosser

"Truth does not conform to us, dear son but we have to conform with it."

Matthias Claudius

"Nothing will give safety except truth. Nothing will give peace except the serious search for truth."

Blaise Pascal

"Truth is the summit of being; justice is the application of it to affairs."
Ralph Waldo Emerson

"The ideals which have lighted my way, and time after time have given me new courage to face life cheerfully, have been Kindness, Beauty and Truth."
Albert Einstein

"It irritates people that the truth is so simple."
Johann Wolfgang von Goethe

"I know that this plainness of speech makes them hate me; and what is this hatred but a proof that I am speaking the truth? – this is the occasion and reason of their slander of me, and you will find out in this or in any future inquiry."
Socrates

"Aglow with the Light of the Divine, I surrender my whole attention to the Presence of Truth that guides my path."
Michael Bernard Beckwith

"Truth means the congruence of a concept with its reality."
G.W. Friedrich Hegel

"Truth is the revealing gloss of reality."
Simone Well

"We are the Multi-dimensional Universe becoming aware of Itself. Live in this One Truth – That God is Real As your very Life!"
Michael Bernard Beckwith

"Truth is a torch, but a tremendous one. That is why we hurry past it, shielding our eyes, even terrified of getting burnt."
Johann Wolfgang von Goethe

"Truth is the spirit's sun."
Marquis de Vauvenargues

"You will recognise the Truth, and the truth will set you free."
John, 8:32

"Truth is the Eternal – Unchangeable! Which never changes in its form, but is as it has been eternally and will ever remain, as it is now. Which can therefore never be subjected to any development either, because it has been perfect from the very beginning. Truth is real, it is 'being'! Only being is true life. The entire Universe is "supported" by this Truth!"
Abd-ru-shin

Truth

To honour God in all things and to perform everything solely to the glory of God

Abd-ru-shin

(In the Light of Truth)

Awake!

Keep the heart of your thoughts pure, by so doing you will bring peace and be happy.

Love thy neighbour, which means honour him as such!

Therein lies the adamantine command: You must never consciously harm him, either in his body or in his soul, either in his earthly possessions or in his reputation!

He who does not keep this commandment and acts otherwise, serves not God but the darkness, to which he gives himself as a tool!

Honour be to God Who only sows Love! Love also in the The Law of the destruction of the darkness!

Abd-ru-shin

(In the Light of Truth)

Love & Gratitude

Crystal Images © Office Masaru Emoto, LLC

Contents

Preface		xxiv
Acknowledgments		xxvii
23	**Pressure Relieving Devices and Emergency Relief System Design**	1
	23.0 Introduction	1
	23.1 Types of Positive Pressure Relieving Devices (See Manufacturers' Catalogs for Design Details)	2
	23.2 Types of Valves/Relief Devices	6
	Conventional Safety Relief Valve	6
	Balanced Safety Relief Valve	7
	Special Valves	7
	Rupture Disk	7
	Example 23.1	15
	23.3 Materials of Construction	18
	Safety and Relief Valves: Pressure-Vacuum Relief Values	18
	Rupture Disks	19
	23.4 General Code Requirements [1]	20
	23.5 Relief Mechanisms	20
	Reclosing Devices, Spring Loaded	20
	Non-Reclosing Pressure Relieving Devices	21
	23.6 Pressure Settings and Design Basis	21
	23.7 Unfired Pressure Vessels Only, But Not Fired or Unfired Steam Boilers	24
	Non-Fire Exposure	24
	External Fire or Heat Exposure Only and Process Relief	24
	23.8 Relieving Capacity of Combinations of Safety Relief Valves and Rupture Disks or Non-Reclosure Devices (Reference ASME Code, Par. UG-127, U-132)	24
	Primary Relief	24
	Rupture Disk Devices, [44] Par UG-127	25
	Footnotes to ASME Code	26
	23.9 Establishing Relieving or Set Pressures	27
	Safety and Safety Relief Valves for Steam Service	28
	23.10 Selection and Application	28
	Causes of System Overpressure	28
	23.11 Capacity Requirements Evaluation for Process Operation (Non-Fire)	29
	Installation	34
	23.12 Piping Design	37
	Pressure Drops	37
	Line Sizing	37
	23.13 Selection Features: Safety, Safety-Relief Valves, and Rupture Disks	44
	23.14 Calculations of Relieving Areas: Safety and Relief Valves	46
	23.15 Standard Pressure Relief Valves Relief Area Discharge Openings	46

23.16	Sizing Safety Relief Type Devices for Required Flow Area at Time of Relief	47
23.17	Effects of Two-Phase Vapor-Liquid Mixture on Relief Valve Capacity	47
23.18	Sizing for Gases or Vapors or Liquids for Conventional Valves with Constant Backpressure Only	47
	Procedure	48
	Establish Critical Flow for Gases and Vapors	48
	Example 23.2: Flow through Sharp Edged Vent Orifice (Adapted after [41])	54
23.19	Orifice Area Calculations [42]	54
23.20	Sizing Valves for Liquid Relief: Pressure-Relief Valves Requiring Capacity Certification [5D]	60
23.21	Sizing Valves For Liquid Relief: Pressure Relief Valves Not Requiring Capacity Certification [5D]	61
23.22	Reaction Forces	66
	Example 23.3	67
	Solution	67
	Example 23.4	69
	Solution	70
23.23	Calculations of Orifice Flow Area using Pressure Relieving Balanced Bellows Valves, with Variable or Constant Backpressure	72
23.24	Sizing Valves for Liquid Expansion (Hydraulic Expansion of Liquid Filled Systems/Equipment/Piping)	80
23.25	Sizing Valves for Subcritical Flow: Gas or Vapor But Not Steam [5d]	81
23.26	Emergency Pressure Relief: Fires and Explosions Rupture Disks	84
23.27	External Fires	84
23.28	Set Pressures for External Fires	85
23.29	Heat Absorbed	85
	The Severe Case	85
23.30	Surface Area Exposed to Fire	86
23.31	Relief Capacity for Fire Exposure	87
23.32	Code Requirements for External Fire Conditions	87
23.33	Design Procedure	88
	Example 23.5	88
	Solution	88
23.34	Pressure Relief Valve Orifice Areas on Vessels Containing Only Gas, Unwetted Surface	92
23.35	Rupture Disk Sizing Design and Specification	93
23.36	Specifications to Manufacturer	93
23.37	Size Selection	94
23.38	Calculation of Relieving Areas: Rupture Disks for Non-Explosive Service	94
23.39	The Manufacturing Range (MR)	95
23.40	Selection of Burst Pressure for Disk, P_b (Table 23.3)	95
	Example 23.6: Rupture Disk Selection	98
23.41	Effects of Temperature on Disk	98
23.42	Rupture Disk Assembly Pressure Drop	101
23.43	Gases and Vapors: Rupture Disks [5a, Par, 4.8]	101
	Volumetric Flow: scfm Standard Conditions (1.4.7 psia and 60°F)	102
	Steam: Rupture Disk Sonic Flow; Critical Pressure = 0.55 and P_2/P_1 is Less Than Critical Pressure Ratio of 0.55	103
23.44	API for Subsonic Flow: Gas or Vapor (Not Steam)	103
23.45	Liquids: Rupture Disk	104
23.46	Sizing for Combination of Rupture Disk and Pressure Relief Valve in Series Combination	105
	Example 23.7: Safety Relief Valve for Process Overpressure	106
	Example 23.8: Rupture Disk External Fire Condition	106
	Solution	107
	Heat Input	107

	Total Heat Input (from Figure 23.30a)	107
	Quantity of Vapor Released	107
	Critical Flow Pressure	107
	Disk Area	108
	Example 23.9: Rupture Disk for Vapors or Gases; Non-Fire Condition	108
	Solution	108
	Example 23.10: Liquids Rupture Disk	109
	Example 23.11: Liquid Overpressure, Figure 23.34	110
23.47	Pressure-Vacuum Relief for Low-Pressure Storage Tanks	110
23.48	Basic Venting For Low-Pressure Storage Vessels	111
23.49	Non-Refrigerated Above Ground Tanks; API-Std. 2000	112
23.50	Boiling Liquid Expanding Vapor Explosions (BLEVEs)	113
	Ignition of Flammable Mixtures	116
23.51	Managing Runaway Reactions	116
	Hydroprocessing Units	117
	Acid/Base Reactions	118
	Methanation	118
	Alkylation Unit Acid Runaway	118
	23.51.1 Runaway Reactions: DIERS	118
23.52	Hazard Evaluation in the Chemical Process Industries	120
23.53	Hazard Assessment Procedures	121
	Exotherms	122
	Accumulation	122
23.54	Thermal Runaway Chemical Reaction Hazards	122
	Heat Consumed Heating the Vessel. The ϕ-Factor	123
	Onset Temperature	124
	Time-To-Maximum Rate	125
	Maximum Reaction Temperature	125
	Vent Sizing Package (VSP)	126
	Vent Sizing Package 2™ (VSP2™)	127
	Advanced Reactive System Screening Tool (ARSST)	128
23.55	Two-Phase Flow Relief Sizing for Runaway Reaction	128
	Runaway Reactions	131
	Vapor Pressure Systems	132
	Gassy Systems	132
	Hybrid Systems	132
	Simplified Nomograph Method	134
	Vent Sizing Methods	138
	Vapor Pressure Systems	138
	Fauske's Method	140
	Gassy Systems	142
	Homogeneous Two-Phase Venting Until Disengagement	143
	Two-Phase Flow Through an Orifice	144
	Conditions of Use	145
23.56	Discharge System	145
	Design of The Vent Pipe	145
	Safe Discharge	146
	Direct Discharge to The Atmosphere	147
	Example 23.12	147
	Tempered Reaction	147
	Solution	147

	Example 23.13	149
	Solution	149
	Example 23.14	150
	Solution	151
	Example 23.15	152
	Solution	152
	DIERS Final Reports	155
23.57	Sizing for Two-Phase Fluids	155
	Example 23.16	161
	Solution	162
	Example 23.17	164
	Solution	164
	Example 23.18	172
	Example 23.19	177
	Solution	178
	Type 3 Integral Method [5]	179
	Example 23.20 [76]	180
	Solution	181
23.58	Flares/Flare Stacks	182
	Flares	184
	Sizing	184
	Flame Length [5c]	186
	Flame Distortion [5c] Caused by Wind Velocity	187
	Flare Stack Height	189
	Flaring Toxic Gases	194
	Purging of Flare Stacks and Vessels/Piping	195
	Pressure Purging	195
	Example 23.21: Purge Vessel by Pressurization Following the Method of [41]	195
23.59	Compressible Flow for Discharge Piping	197
	Design Equations for Compressible Fluid Flow for Discharge Piping	197
	Critical Pressure, P_{crit}	200
	Compressibility Factor Z	201
	Friction factor, f	202
	Discharge Line Sizing	203
23.60	Vent Piping	204
	Discharge Reactive Force	204
	Example 23.22	205
	Solution	206
	Example 23.23: Flare and Relief Blowdon System	208
	Solution	208
	A Rapid Solution for Sizing Depressuring Lines [5c]	208
	Codes and Standards	212
	Discharge Locations	213
	Process Safety Incidents with Relief Valve Failures and Flarestacks	214
	A Case Study on Williams Geismar Olefins Plant, Geismar, Louisiana [95]	214
	Process Flow of the Olefins	214
	The Incident	216
	Technical Analysis	219
	Key Lessons	222
	Explosions in Flarestacks	225
	Relief Valves	227

		Location	228
		Relief Valve Registers	228
		Relief Valve Faults [92]	229
		Tailpipes [92]	230
		GLOSSARY	230
		Acronyms and Abbreviations	239
		Nomenclature	240
		Subscripts	244
		Greek Symbols	244
	References		245
		World Wide Web on Two-Phase Relief Systems	247

24 Process Safety and Energy Management in Petroleum Refinery — 249

24.1	Introduction		249
24.2	Process Safety		250
	24.2.1	Process Safety Information	253
	24.2.2	Conduct of Operations (COO) and Operational Discipline (OD)	254
		Process Safety Culture: BP Refinery Explosion, Texas City, 2005	257
		Detailed Description	257
		Causes	258
		Key Lessons	260
		Process Safety Culture	260
		Selected CSB Findings	260
		Selected Baker Panel Finding	261
		Process Knowledge Management	261
		Training and Performance Assurance	261
		Management of Change (MOC)	261
		Asset Integrity and Reliability	261
	24.2.3	Process Hazard Analysis	262
		Safe Operating Limits	263
		Impact on Other Process Safety Elements	264
24.3	General Process Safety Hazards in a Refinery		265
		Desalters	266
		Critical Operating Parameters Impacting Process Safety	266
		The Quality of Aqueous Effluent from Desalters	267
		Desalter Water Supply	267
		Vibration within Relief Valve (RV) Pipework	267
		Example of Process Safety Incidents and Hazards	267
		Hydrotreating [2]	267
24.4	Example of Process Safety Incidents and Hazards		267
		Catalytic Cracking [2]	270
24.5	Process Safety Hazards		270
		Reforming	271
		Alkylation [2]	271
		Hydrotreating Units	271
	24.5.1	Examples of Process Safety Incidents and Hazards	272
		HF release, Texas City, TX, 1987 [2]	272
		HF release, Corpus Christi, TX, 2009	272
		HF release at Philadelphia Energy Solutions Refining and Marketing LLC (PES), Philadelphia 2019	273
		Post-Incident Activities	276

		Coking [2]	277
		Equilon Anacortes Refinery Coking Plant Accident, 1998	277
		Design Considerations	278
24.6	Hazards Relating to Equipment Failure		278
24.7	Columns and Other Process Pressure Vessels and Piping		279
		Corrosion	279
		Corrosion Inhibitors	280
24.8	Inadequate Design and Construction		290
		Corrosion within "dead legs"	290
24.9	Inadequate Material of Construction Specification		290
24.10	Material Failures and Process Safety Prevention Programs		291
		Piping Repair Incident at Tosco Avon Refinery, CA, USA	291
		Lessons Learned from this accident	297
24.11	Hazard and Operability Studies (HAZOP)		297
		Study Co-ordination	303
	24.11.1	HAZOP Documentation Requirements	303
	24.11.2	The Basic Concept of HAZOP	304
	24.11.3	Division into Sections	304
		Use of Guidewords	304
	24.11.4	Conducting a HAZOP Study	305
		Define Objective and Scope	306
		Prepare for the Study	307
		Record the Results	307
	24.11.5	Hazop Case Study [8]	307
	24.11.6	HAZOP of a Batch Process	308
		Limitations of HAZOP Studies	315
		Conclusions	315
24.12	HAZAN		315
24.13	Fault Tree Analysis		317
24.14	Failure Mode and Effect Analysis (FMEA)		318
		Methodology of FMEA	318
		Definition of System to be Evaluated	318
		Level of Analysis	318
		Analysis of Failures	318
24.15	The Swiss Cheese Model		319
24.16	Bowtie Analysis		320
		Validity Rules for Barriers	320
		Example	322
		Process Safety Isolation Practices in Petroleum Refinery and Chemical Process Industries	322
24.17	Inherently Safer Plant Design		325
		Inherently Safer Plant Design in Reactor Systems	327
24.18	Energy Management in Petroleum Refinery		330
		Total cost of energy	331
		Energy Policy	331
		Crude Distillation Unit	332
		Heat Exchangers	332
		Steam Traps	333
		Optimization of Refinery Steam/Power System	333
		Reducing fouling/surface cleaning/surface coating in heat exchanger/furnace	333
		Pumping System	333
		Electric Drives	334

			Furnace System	334
			Compressed Air	335
			Flare System	335
		24.18.1	Environmental Impact of Flaring	336
		24.18.2	Environmental Impact of Petroleum Industry	337
		24.18.3	Environmental Impact Assessment (EIA)	339
		24.18.4	Pollution Control Strategies in Petroleum Refinery	340
		24.18.5	Energy Management and Co_2 Emissions in Refinery	345
	24.19	Benchmarking in Refinery		345
Glossary				346
Acronyms and Abbreviations				354
References				354
25	**Product Blending**			**357**
	25.0	Introduction		357
	25.1	Blending Processes		360
		25.1.1	Gasoline Blending	361
	25.2	Ternary Diagram of Crude Oils		361
		25.2.1	Elemental Analysis and Ternary Classification of Crude Oils	361
		25.2.2	Reading a Ternary Diagram	363
			Solution	364
			Example 25.1	364
			Solution	365
			Solution	365
			Example 25.2	368
			Solution	368
	25.3	Reid Vapor Pressure Blending		369
			Example 25.4	370
			Solution	370
			Example 25.5	370
			Solution	370
			Example 25.6	374
		25.3.1	Reid Vapor Pressure Blending for Gasolines and Naphthas	376
			Example 25.7	376
			Solution	376
	25.4	Flash Point Blending		377
			Example 25.8	378
			Solution	378
	25.5	Alternative Methods for Determining the Blend Flash Point		380
			Example 25.9	380
			Solution	380
			Example	382
			Example 25.10	382
			Solution	382
			Example 25.11	386
	25.6	Pour Point Blending		386
			Example 25.12	387
			Solution	387
			Example 25.13 [2]	388
			Solution	388
			Example 25.14 [3]	391

		Example 25.15	394
25.7	Cloud Point Blending		395
		Example 25.16	396
		Solution	396
25.8	Aniline Point Blending		396
		Example 25.17	397
		Solution	397
	25.8.1	Alternative Aniline Point Blending	397
		Example 25.18	398
		Solution	400
25.9	Smoke Point Blending		401
		Example 25.19	401
		Solution	401
	25.9.1	Smoke Point of Kerosenes	402
25.10	Viscosity Blending		403
		Example 25.20	403
		Solution	403
25.11	Regular Gasoline		404
25.12	Product Blending		405
	25.12.1 Premium Gasoline		405
25.13	Viscosity Prediction From the Crude Assay		405
25.14	Gasoline Octane Number Blending		407
		Example 25.21	408
		Solution	409
		Example 25.22	410
		Solution	410
		Example 25.23	411
		Solution	412
25.15	Other Blending Correlations		414
		Cetane Index	414
		Diesel Index	414
		U.S. Bureau of Mines Correlation Index (BMCI)	414
		Aromaticity Factor	414
25.16	Fluidity of Residual Fuel Oils		415
	25.16.1 Fluidity Test		415
	25.16.2 Fluidity Blending		416
		Example 25.24	416
		Solution	416
25.17	Conversion of Kinematic Viscosity to Saybolt Universal		416
	Viscosity or Saybolt Furol Viscosity		416
	25.17.1 Conversion to Saybolt Universal Viscosity		416
		Example 25.25	417
		Solution	417
	25.17.2 Conversion to Saybolt Furol Viscosity		417
		Example 25.26	418
		Solution	418
	25.17.3 Refractive Index of Petroleum Fractions		418
		Example 25.27	419
		Solution	419
25.18	Determination of Molecular-Type Composition		419
		Example 25.28	421

		Solution	421
		Example 25.29	422
		Solution	422
	25.19	Determination of Viscosity From Viscosity/Temperature Data at Two Points	423
		Example 25.30	424
		Solution	424
	25.20	Linear Programming (LP) for Blending	426
		LP Software	428
		The Excel Solver	428
	25.20.1	Mathematical Formulation	430
	25.20.2	Problem Solution	433
		Notation	434
		Example 25.31	434
		Solution	435
		Example 25.32	435
		Solution	435
		Example 25.33	438
		Solution	438
		Example 25.34	438
		Solution	438
		Example 25.35	442
		Solution	442
		Example 25.36	449
		Solution	449
		A Case Study	450
		Solution	453
	25.21	Environmental Concern of Gasoline Blending	455
	25.21.1	Operation of Catalytic Converter	457
	25.21.2	Effectiveness of Catalytic Converters	460
	References		464
	Bibliography		466

26 Cost Estimation and Economic Evaluation — 467

26.1	Introduction		467
26.2	Refinery Operating Cost		468
	26.2.1	Theoretical Sales Realization Valuation Method	470
		Example 26.1	470
	26.2.2	Cost Allocation for Actual Usage	471
26.3	Capital Cost Estimation		471
26.4	Equipment Cost Estimations by Capacity Ratio Exponents		472
26.5	Yearly Cost Indices		476
		Example 26.2	477
		Solution	477
		Example 26.3	480
		Solution	480
26.6	Factored Cost Estimate		480
26.7	Detailed Factorial Cost Estimates		481
		Zevnik and Buchanan's Method	483
		Timm's Method	484
		Bridgwater's Method	484
26.8	Bare Module Cost for Equipment		487

26.9	Summary of the Factorial Method	488
26.10	Computer Cost Estimating	488
26.11	Project Evaluation	491
	Introduction	491
26.12	Cash Flows	493
	Return on Investment (ROI)	493
	Accounting Coordination	494
	Payback Period (PBP)	494
	Example 26.5	495
	Present Worth (or Present Value)	496
	Net Present Value (NPV)	497
	The Profitability Index (PI)	502
	Example 26.6	503
	Discounted Cash Flow Rate of Return (DCFRR)	503
	Relationship between PBP and DCFRR	503
	Example 26.7	507
	Solution	507
	Example 26.8	509
	Solution	509
26.12.1	Incremental Criteria	510
	Depreciation	512
26.12.2	Profitability	515
	Example 26.9	516
	Solution	516
	Example 26.10	516
	Solution	520
	Economic Analysis	521
	Example 26.11	524
	Solution	524
26.12.3	Inflation	532
26.12.4	Sensitivity Analysis	533
26.13	Refining Economics	533
	Crude Slates	533
	Refinery Configuration	534
	Product Slates	534
	Refinery Utilization	534
	Environmental Initiatives	535
26.13.1	Refinery Margin Definitions	535
	Example 26.12	536
	Solution	536
	Example 26.13	537
26.13.2	Refinery Complexity	537
	Example 26.14	538
	Solution	538
26.13.3	Supply and Demand Balance	539
	Product Quality	539
	Standard Density	539
	Blending Components	539
	Constraining Properties	539
	Quality Premiums/Discounts	539
	A Case Study [44]	540

		Problem Statement	540
		Process Description	542
		Catalytic Reformer	542
		Naphtha Desulfurizer	544
		Summary of Investment and Utilities Costs	545
		Calculation of Direct Annual Operating Costs	545
		On-Stream Time	546
		Water Makeup	546
		Power	546
		Fuel	546
		Royalties	547
		Catalyst Consumption	548
		Insurance	548
		Local Taxes	548
		Maintenance	548
		Miscellaneous Supplies	548
		Plant Staff and Operators	548
		Calculations of Income before Income Tax	549
		Summary of Direct Annual Operating Costs	549
		Calculation of ROI	550
26.14	Global Effects on Refining Economy		552
	26.14.1	Carbon Tax	557
26.15	Economic Terminologies on Sustainability		558
		Carbon footprint	558
		Global Warming Potential (GWP)	558
		An Improved Method of Using GWPs	560
		Solution	562
		Carbon Dioxide Equivalent	565
		Carbon Credit	566
		Carbon Offset	566
		Carbon Price	567
		Nomenclature	567
References			568
Bibliography			569

27 Sustainability in Engineering, Petroleum Refining and Alternative Fuels — 571

27.0	Introduction		571
27.1	Impacts on the Overall Greenhouse Effect		576
27.2	Carbon Capture and Storage in Refineries		578
27.3	Sustainability in the Refinery Industries		580
27.4	Sustainability in Engineering Design Principles		582
27.5	Alternative Fuels (Biofuels)		587
27.6	Process Intensification (PI) in Biodiesel		589
27.7	Biofuel from Green Diesel		592
		Analysis	592
		Processing of Biodiesel	592
	27.7.1	Specifications of Biodiesel	596
		Advantages	597
		Disadvantages	597
	27.7.2	Bioethanol	597
	27.7.3	Biodiesel Production	601

		Application	601
		Process	602
		Reaction Chemistry	603
		Economics	603
	27.7.4	An Alternative Process of Manufacturing Biodiesel	604
		Reaction Chemistry	607
	27.7.5	Biofuel from Algae	607
	27.7.6	Economic Viability of Algae	608
27.8	Fast Pyrolysis		609
	27.8.1	Fast Pyrolysis Principle	609
	27.8.2	Fast Pyrolysis Technologies	610
	27.8.3	Minerals of Biomass	611
	27.8.4	Applications of Fast Pyrolysis Liquid	611
		Heat and Power	611
	27.8.5	Chemicals and Materials	613
	27.8.6	Bio-Fuels-Fast Pyrolysis Bio-Oil (FPBO) from Biomass Residues	613
		Feedstocks	614
	27.8.7	Properties of Pyrolysis Oil	615
		Main advantages	616
27.9	Acid Gas Removal		617
		Chemical Solvent Processes	617
		Physical Solvent Processes	617
	27.9.1	Process Description of Amine Gas Treating	618
		Chemical Reactions	618
		For hydrogen sulfide H_2S removal:	618
		For carbon dioxide (CO_2) removal	618
		Amines Used [48]	621
	27.9.2	Equilibrium Data for Amine–Sour Gas Systems	625
	27.9.3	Emerging Technologies [48]	625
		Chemistry	627
	27.9.4	Advanced Amine Based Solvents	627
		Chemistry	628
		Disadvantages of Amine Solvents	628
	27.10	Alkaline Salt Process (Hot Carbonate)	629
		Split Flow Process of Potassium Carbonate Process	630
		Two Stage Process	630
27.11	Ionic Liquids		632
		Disadvantages	632
		Viscosity	633
		Tunability	633
		A Case Study of Acid Gas Sweetening with DEA (Schlumberger and Honeywell UniSim® Design Suite R470 Technology)	634
		Learning Objectives	634
		Building the Simulation	636
		Defining the Simulation Basis	636
		Amines Property Package	636
		Column Overview	636
		Contactor	636
		Adding the Basics	636
		Adding the feed streams	636
		Physical Unit Operations	638

	Separator Operation	638
	Contactor Operation	639
	Valve Operation	641
	Separator Operation	641
	Heat Exchanger Operation	642
	Regenerator Operation	643
	Mixer Operation	644
	Cooler Operation	646
	Pump Operation	646
	Adding Logical Unit Operations	647
	Set Operation	647
	Recycle Operation	648
	Save your case	649
	Analyzing the Results	649
27.12	Advanced Modeling	650
27.13	Carbon Capture and Storage (CCS)	652
27.14	Risk Management	655
27.15	The Institution of Chemical Engineers (IChemE, U.K.) Position on Climate Change	655
	27.15.1 Net Zero Carbon Emissions	656
	Emissions Reduction must Start NOW	656
	27.15.2 Guided by UN Sustainable Development Goals	656
	Systems Thinking	657
	Global Mechanisms	657
	Best Available Techniques	657
	Innovation	657
	27.15.3 Training and Application of Skills	657
	Education	657
27.16	Oil & Gas and Petrochemical Companies with Zero Carbon Emissions Targets by 2050	658
	Hydroflex™ Technology	659
	Evonik and Siemens Energy Partnership	661
27.17	Offshore Petroleum Regulator for Environment and Decommissioning (OPRED), UK	661
	Wood Plc, UK	663
	Tata Chemicals Europe (TCE)	663
	Hengli Petrochemical (Dalian) Co. Ltd. (HPDC)	663
	Saudi Aramco	663
	Processing	664
27.18	Gas Heated Reformer (GHR)	668
27.19	Pressure Swing Adsorption (PSA)	672
27.20	Distribution and Storage	675
	Applications	677
27.21	Steam Methane Reforming (SMR) for Fuel Cells	678
27.22	New Technologies of Carbon Capture Storage	680
27.23	Carbon Clean Process Design (CC)	681
	Advantages	692
	Advantages	693
	27.23.1 Cyclone Carbon Clean Technology	693
	27.23.2 CycloneCC Technology	698
27.24	Electrochemically Mediated Amine Regeneration (EMAR)	700
	Mechanism	701
27.25	Refinery of the Future	704
27.26	The Crude Oil to Chemical Strategy (COC)	710

27.27 Available Crude to Chemicals Processing Routes 720
27.28 Chemical Looping 721
27.29 Conclusions 722
Glossary 723
References 729
Bibliography 732

Appendix D 733

Glossary of Petroleum and Petrochemical Technical Terminologies 809

About the Author 937

Index 939

Preface

Petroleum refining is a complex industry that worldwide produces more than $10 billion worth of refined products. Improvements in the design and operation of these facilities can deliver large economic value for refiners. Furthermore, economic, regulatory and environmental concerns impose significant pressure on refiners to provide safe working conditions and at the same time optimize the refining process. Refiners have considered alternative processing units and feedstocks by investing in new technologies.

The United States, Europe and countries elsewhere in the world are now embarking on the full electrification of automobiles. Furthermore, the pandemic of the coronavirus with lockdowns in many countries has restricted the movement of people, resulting in less use of aviation fuel and motor gasoline. This has resulted in crude being sold at $42.0 per barrel, presenting problems to oil producers and refiners. The venture of electrification still poses inherent problems of resolving rechargeable batteries and fuel cells and providing charging stations along various highways and routes. Oil and natural gas will for the foreseeable future form an important part of everyday life. Their availability has changed the whole economy of the world by providing basic needs for mankind in the form of fuel, petrochemicals and feedstocks for fertilizer plants and energy for the power sector.

The average crude oil price of Brent crude has varied between $41.96/bbl. in 2020 to $82.5/bbl. in 2023, and Western Texas Intermediate (WTI) crude oil between $39.7/bbl. in 2020 to $75.7/bbl. in 2023 respectively, caused by various factors such as the war between Russia and Ukraine. Furthermore, global oil demand is set to rise by 1.9 mb/d in 2023 to a record 101.7 md/d, with nearly half the gain from China following the lifting of its Covid restrictions. Refining capacity is the maximum volume of crude oil that refineries can produce in a day. Setting the US as a benchmark for the world, it had 135 operable petroleum refineries and a total of refining capacity of 19 million barrels per day in 2020 to 128 operable refineries with a total crude distillation capacity of 17.9 million barrels per day, a loss of 1.1 million barrels. In the same period of time the world lost a total of 3.3 million barrels of daily refining capacity, and about 1/3 of these losses occurred in the US.

With this realignment, and planned refining openings and capacity expansions in Asia, trade press reports suggest China will overtake the US as the country with the most refining capacity by the end of 2023.

World Economic Situation

Russia's war in Ukraine is further expected to advance economic consensus's expectations for higher price inflation and slower global GDP growth.

On Oil

As solid global oil demands is expected to reach record highs in 2023 per International U.S. Energy Information Administration (EIA), supply challenges have persisted for oil and natural gas production. The global demand of 98.8 million barrels per day (mb/d) in Q2, 2022 is projected to grow to a record high 102.7 mb/d in December 2023 per (EIA).

Uncertainties: This results in effective Russian production losses and potential OPEC and US growth.

US petroleum net exports reached a record high as 94.3%, 7.3% above its 5-year average. Presently US refineries are operating at or near maximum utilization and about 1/3 of recent refining capacity loss are due to conversions to biofuels plants (e.g., renewable fuels and oxygenated plants).

US refining is a long-cycle business. Refiners could bring more refining capacity online despite these challenges, and the result could be higher demand and higher costs for crude oil.

Presently, the world economy runs on oil and natural gas, and the processing of these feedstocks for producing fuels, and value-added products has become an essential activity in modern society. The availability of liquefied natural gas (LNG) has enhanced the environment, and recent development in the technology of natural gas to liquids (GTL) has further improved the availability of fuel to transportation and other sectors.

The complex processing of petroleum refining has created a need for environmental, health, and safety management procedures and safe work practices. These procedures are established to ensure compliance with applicable regulations and standards such as hazard communications (PHA, HAZOPS, MoC, and so on), emissions, waste management (pollution that includes volatile organic compounds (VOC), carbon monoxide, sulfur oxides (SO_x), nitrogen oxides (NO_x), particulates, ammonia (NH_3), hydrogen sulfide (H_2S), and toxic organic compounds) and waste minimization. These pollutants are often discharged as air emissions, wastewater or solid wastes. Furthermore, concerns over issues such as the depletion of the ozone layer that results in global warming is increasingly having a significant impact on earth's nature and mankind, and carbon dioxide (CO_2) is known to be the major culprit of global warming. Other emissions such as CH_4, H_2S, NO_x, and SO_x from petroleum refining have adversely impacted the environment, and agencies such as Occupational Safety and Health Administration (OSHA), and Environmental Protection Agency (EPA), Health and Safety Executive (U.K. HSE) have imposed limits on the emissions of these compounds upon refiners.

Flaring has become more complicated and concerns about its efficiency have been increasing and discussed by experts. The OSHA, EPA and HSE have imposed tighter regulations on both safety and emission control, which have resulted in higher levels of involvement in safety, pollution, emissions and so on.

Petroleum refining is one of the important sectors of the world economy, and it's playing a crucial and pivotal role in industrialization, urbanization, and meeting the basic needs of mankind by supplying energy for industrial and domestic transportation, feedstock for petrochemical products as plastics, polymers, agrochemicals, paints, and so on. Globally, it processes more materials than any other industry, and with a projected increase in population to around 8.1 billion by 2025, increasing demand for fuels, electricity and various consumer products made from the petrochemical route is expected via the petroleum and refining process.

Petroleum Refining Design and Applications Handbook Volume Five, is a continuation of the previous volumes, comprising of five chapters including extensive case studies of process safety incidents in the refineries, a revised glossary of petroleum and technical terminology, process data sheets, appendices, Excel spreadsheet programs, computer developed programs, UniSim–Design simulation software, cases studies and a Conversion Table. Chapter 23 describes pressure relieving device types, rupture disks, materials of construction, general code requirements (ASME), pressure settings and design basis, safety and safety relief valves for various services and sizing of the valves for steam, air, liquid, vapor and external fires conditions; piping design involving pressure drops to the relief valve and tail pipe, the effects of two-phase, vapor-liquid mixture on relief valve capacity, runaway reactions involving gassy, vapor pressure and hybrid (gas + vapor) systems using Design Institute for Emergency Relief Systems (DIERS) and simplified nomograph methodology, hazard evaluation in the chemical process industries, sizing for two-phase fluids using types 1 and 2 (Omega) and 3 (Integral) methods, flares and flare stacks and sizing, compressible flow for discharge piping, and case studies. Chapter 24 describes process safety and energy management in petroleum refinery involving the BP Texas refinery incident in 2005, which killed 15 people and injured over 170. The chapter reviews 14 elements of Occupational Safety and Hazard Administration (OSHA) process safety management program with relevance to the safety culture at BP Texas refinery. It further reviews process hazard analysis, general process safety hazards in a refinery, and examples of process safety incidents and hazards, hazards relating to equipment failure, classification of corrosion, and corrosion inhibitors, case studies, hazard and operability studies (Hazop), checklist for a productive hazop study, Hazop of a batch process, limitations of Hazop studies, Hazard analysis (Hazan), energy policy, flare system, environmental impact of flaring, environmental impact of petroleum industry, environmental impact assessment (EIA), pollution control strategies in petroleum refinery, benchmarking in refinery, and glossary. Chapter 25 reviews petroleum product blending processes, elemental analysis and ternary classification of crude oils, Reid vapor pressure blending, flash point blending, pour point blending, cloud point blending, aniline point blending, smoke point blending, viscosity blending, gasoline octane number blending, diesel index, US Bureau of Mines Correlation index (BMCI), aromaticity factor, fluidity of residual fuel oils, fluidity test, fluidity blending, refractive index of petroleum fractions, and linear programming (LP) for blending. It further reviews the environmental concerns of gasoline blending; a gasoline blend stock specifically formulated with ethanol referred to as a blend stock for oxygenated blending

(BOD), the operation and effectiveness of catalytic converters. Chapter 26 reviews cost estimation and economic evaluation in the refinery. This involves refinery operating costs, theoretical sales realization valuation method, and cost allocation for actual usage, capital cost estimation, equipment cost estimations by capacity ratio exponents, factored cost estimate, project evaluation using annual return on investment (ROI), payback period (PBP), net present value (NPV), the average rate of return (ARR), present value ratio (PVR) or the internal rate of return (IRR), discounted cash flow rate on return (DCFRR), effect of inflation on NPV, sensitivity analysis, refining economics, refinery margin definitions, global effects on refining economy, carbon tax, carbon footprint, global warming potential (GWP), carbon dioxide equivalent, carbon credit, carbon offset and carbon price. Chapter 27 describes sustainability in engineering, petroleum refining and alternative fuels. The chapter reviews the impacts on the overall greenhouse effect, carbon capture and storage in refineries, sustainability in engineering design principles, principles of green engineering, alternative fuels (biofuels), process intensification in biofuels, biofuel from green diesel, specifications of biofuels, bioethanol, biodiesel production, biofuel from algae, fast pyrolysis principle and technologies, acid-gas removal using the amine solutions (MEA, DEA, DGA, MDEA, TEA, DIPA), reviews of emerging technologies, a case study of acid gas sweetening with DEA, carbon capture and storage, net zero carbon emissions, Oil & Gas and petrochemical companies with zero carbon emissions targets by 2050, gas heated reformer (GHR), pressure swing adsorption (PSA), methane steam reforming for fuel cells, new technologies for carbon capture storage, and the refinery of the future.

Finally, there are case studies of process safety incidents in these volumes, which the author hopes will spur readers to process safety management, investigating the root causes and near misses of incidents in the refinery plants, finding ways in mitigating these incidents in the future thereby saving lives of personnel in the refinery facilities and chemical process industries worldwide.

The US Chemical Safety and Hazard Investigation Board (www.csb.gov) has provided case studies of process safety incidents with animations, and recently the U.K. Institution of Chemical Engineers (IChemE) (www.icheme.org) from IChemE Safety and Loss Prevention Special Interest Group (SIG) has published case studies of lessons learned database (LLD). It provides major process safety incident vs. root cause map matrix in a quick reference guide (https://lnkd.in/dm3t5VPe) in process safety incidents. Readers are advised to view these websites and will find them educational and informative.

Finally, the volume provides a glossary of petroleum and technical terminology, process datasheets, and a conversion table, developed Microsoft Excel spreadsheet programs and developed programs including UniSim design software programs that can be readily accessed from the publisher's website using a dedicated password.

A Kayode Coker (www.akctechnology.com)

Acknowledgments

This project is the culmination of nine years of research, collating relevant materials from organizations, institutions, companies and publishers, developing Excel spreadsheet programs and computer programs; using Honeywell's UniSim Design steady state simulation programs and providing the majority of the drawings in the text.

Sincere gratitude to Honeywell Process Solutions for granting permission to incorporate the use of UniSim Design simulation and many other suites of software programs in the book. I express my thanks to Dr. Jamie Barber of Honeywell Process Solutions for his friendship and help over many years of using the UniSim Design simulation software. The late Dr. Soni Oyekan and I had numerous fruitful discussions on petroleum blending products and also on renewable fuels for transportation and other uses and oxygenated plants, and would like to record my appreciation of our friendship and his help in this final volume. To Mr. Ahmed Mutawa formerly of SASREF Co., Saudi Arabia for developing the Conversion Table program for the book.

Many organizations, institutions and companies as Gas Processor Suppliers Association (GPSA), USA, Honeywell Process Solutions, Saudi Aramco Shell Refinery Co. (SASREF), Absoft Corporation, USA, American Institute of Chemical Engineers, The Institution of Chemical Engineers, U.K., Chemical Engineering magazine by Access Intelligence, USA, Hydrocarbon Processing magazine have readily given permission for the use of materials and their release for publication. I greatly acknowledge and express my deepest gratitude to these organizations.

I have been privileged to have met with Phil Carmical, Publisher at Scrivener Publishing Co., some twenty years ago. Phil initiated the well-known Ludwig's project at the time during his tenure at Gulf Publishing Co., and Elsevier, respectively. His suggestions in collaborating on these important works some nine years ago were timely to the engineering community, as I hope that these works will be greatly beneficial to this community world-wide. I'm deeply grateful to Phil for agreeing to collaborate with me, his suggestions and assistance since. I believe that upon completing this aspect of the project that the book will save lives in the refinery industry.

I also wish to express my thanks to the Wiley-Scrivener team: Kris Hackerott – Graphics Designer, Bryan Aubrey – Copy editor, Myrna Ting – Typesetter and her colleagues. I am truly grateful for your professionalism, assistance and help in the production of this volume.

Finally,

Bow down in humility before the Greatness of God, Whose Love is never-ending, and who sends us his help at all times.
He alone is Life and the Power and the Glory for ever and ever.

A. Kayode Coker

23

Pressure Relieving Devices and Emergency Relief System Design

23.0 Introduction

The subject of process safety is so broad in scope that this chapter must be limited to the application, design, rating, and specifications for process over-pressure relieving devices for flammable vapors, process explosions and external fires on equipment; and the venting or flaring of emergency or excess discharge of gases to a vent flare stack. The subject of fire protection cannot be adequately covered; however, the engineer is referred to texts dealing with the subject in a thorough manner [1–6].

The possibilities for development of excess pressure exist in nearly every process plant. Due to the rapidly changing and improved data, codes, regulations, recommendations, and design methods, it is recommended that reference be made to the latest editions of the literature listed in this chapter. While attempting to be reliable in the information presented, I am not responsible or liable for interpretation or the handling of the information by experienced or inexperienced engineers. This chapter's subject matter is vital to the safety of plants' personnel and facilities.

It is important to understand how the overpressure can develop (source) and what might be the eventual results. The mere solving of a formula to obtain an orifice area is secondary to an analysis and understanding of the pressure system. Excess pressure can develop from explosion, chemical reaction, reciprocating pumps or compressors, external fire around equipment, and an endless list of related and unrelated situations. In addition to the possible injury to personnel, the loss of equipment can be serious and an economic setback.

Most states have laws specifying the requirements regarding application of pressure-relieving devices in refineries, process, and steam power plants. In essentially every instance, at least part of the reference includes the *A.S.M.E. Boiler and Pressure Vessel Code*, Section VIII, Division 1 (Pressure Vessels) and/or Division 2 [1]; and Section VII, *Recommended Rules for Care of Power Boilers* [7]. In addition, the publications of the American Petroleum Institute are helpful in evaluation and design. These are API-RP-520 [8], *Design and Installation of Pressure-Relieving Systems in Refineries; Part I-Design; Part II-Installation*; and API-RP-521 [9], *Guide for Pressure Relief and Depressurizing Systems, ANSI/ ASME B31.1 Power Piping*; B16.34; and NFPA-1; [10], Sections 30, 68, and 69.

The ASME Code requires that all pressure vessels be protected by a pressure-relieving device that shall prevent the internal pressure from increasing more than 10% above the maximum allowable working pressures of the vessel (MAWP) to be discussed later. Except where multiple relieving devices are used, the pressure shall not increase more than 16% above the MAWP or, where additional pressure hazard is created by the vessel being exposed to external heat (not process related) or fire, supplemented pressure relieving devices must be installed to prevent the internal pressure from rising more than 21% above the MAWP. (See [1] sections U-125 and UG-126.) The best practice in industrial design recommends that (a) all pressure vessels of any pressure be designed, fabricated, tested and code-stamped per the applicable ASME code [1] or API Codes and Standards [5] and (b) that pressure relieving devices be installed for pressure relief and venting per codes [1, 5, 8 and 9].

Although not specifically recognized in the titles of the codes, the rupture disk as a relieving device is, nevertheless, included in the requirements as an acceptable device.

Usual practice is to use the terms "safety valve" or "relief valve" to indicate a relieving valve for system overpressure and this will be generally followed here. When specific types of valves are significant, they will be emphasized.

This chapter reviews runaway chemical reactions in process equipment and further presents design methodologies for sizing vents involving two-phase flow and compressible flow for discharge piping. The chapter provides a case study of process safety incident in a refinery; describes sizing methodology of a flare stack and explosions in flare stacks.

23.1 Types of Positive Pressure Relieving Devices (See Manufacturers' Catalogs for Design Details)

Relief Valve: A relief valve is an automatic spring-loaded pressure-relieving device actuated by the static pressure upstream of the valve, and which opens further with increase in pressure over the opening pressure. It is used primarily for liquid service [1, 8] (Figures 23.1a and 23.1b). The rated capacity is usually attained at 25% overpressure.

Safety Valve: This is an automatic pressure-relieving device actuated by the static pressure upstream of the valve and characterized by rapid full opening or "pop" action upon opening [1, 8], but does not reseat. It is used for steam or air service (Figure 23.2). The rated capacity is reached at 3, 10 or 20% overpressure, depending upon applicable code.

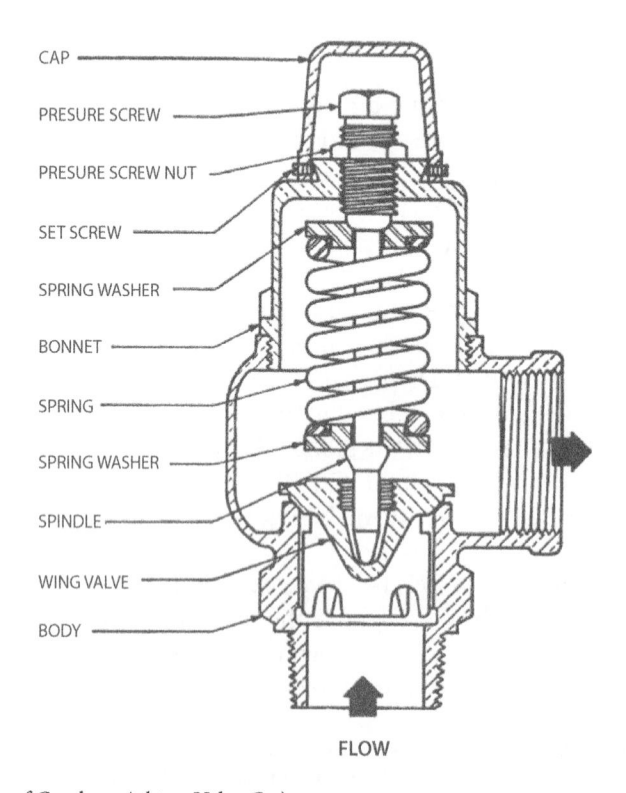

Figure 23.1a Relief valve. (Courtesy of Crosby – Ashton Valve Co.)

Pressure-Relieving Devices

Figure 23.1b Accessories for all types of safety relieving valves. (Courtesy of Crosby – Ashton valve Co.)

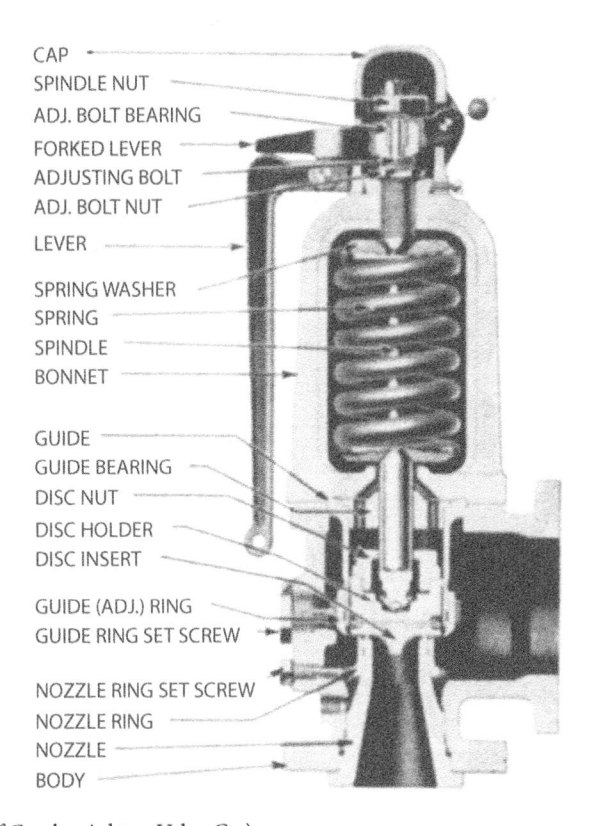

Figure 23.2 Safety valve. (Courtesy of Crosby-Ashton Valve Co.)

4 Petroleum Refining Design and Applications Handbook Volume 5

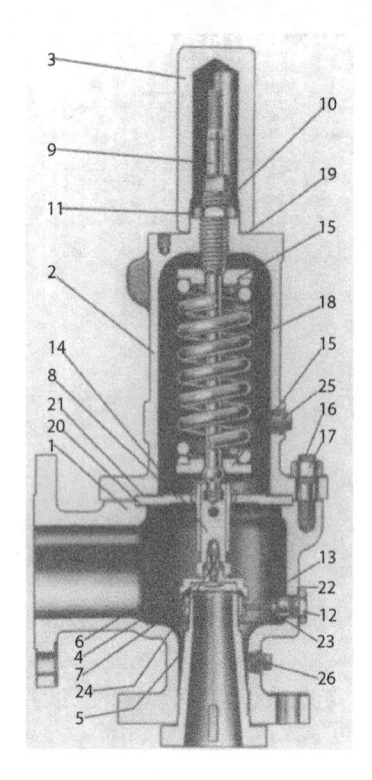

Bill of Materials-Conventional

ITEM	PART NAME		MATERIAL
1	Body	28( )A10 thru 26( )A26	SA-216 GR. WCB, Carbon Steel
		26( )A32 thru 26( )A36	SA-217 GR. WC6, Alloy St. (1¼ CR-½ Moly)
2	Bonnet	26( )A10 thru 26( )A36	SA-216 GR. WCB, Carbon Steel
		26( )A32 thru 26( )A36	SA-217 GR. WC6, Alloy St. (1¼ CR-½ Moly)
3	Cap. Plain Screwed		Carbon Steel
4	Disc		Stainless Steel
5	Nozzle		316 St. St.
6	Disc Holder		300 Series St. St.
7	Blow Down Ring		300 Series St. St.
8	Steeve Guide		300 Series St. St.
9	Stem		Stainless Steel
10	Spring Adjusting Screw		Stainless Steel
11	Jam Nut (Spr. Adj. Scr.)		Stainless Steel
12	Lack Screw (B.D.R)		Stainless Steel
13	Lack Screw Stud		Stainless Steel
14	Stem Retainer		Stainless Steel
15	Spring Button		Carbon Steel Rust Proofed
16	Body Stud		ASTM A193 Gr. B7, Alloy St.
17	Hex Nut (Body)		ASTM A194 Gr. 2H, Alloy St.
18	Spring	26( )A10 thru 26( )A16	Carbon Steel Rust Proofed
		26( )A20 thru 26( )A36	High Temp. Alloy Rust Proofed
19	Cap Gasket		Soft Iron or Steel
20	Body Gasket		Soft Iron or Steel
21	Bonnet Gasket		Soft Iron or Steel
22	Lock Screw Gasket		Soft Iron or Steel
23	Hex Nut (B.D.R.L.S.)		Stainless Steel
24	Lock Screw (D.H.)		Stainless Steel
25	Pipe Plug (Bonnet)		Steel
26	Pipe Plug (Body)		Steel

Also suitable for liquid service where ASME Code certification is not required.

Figure 23.3a Conventional or unbalanced nozzle safety relief valve. (By permission from Teledyne Farris Engineering Co.)

Figure 23.3b Safety relief valve with rubber or plastic seats. (By permission from Anderson, Greenwood and Co.)

Bill of Materials-BalanSeal

ITEM	PART NAME		MATERIAL
1	Body	26( )B10 thru 26( )B26	SA-216 GR. WCB, Carbon Steel
		26( )B32 thru 26( )B36	SA-217 GR. WC6, Alloy St. (1¼ CR–½ Moly)
2	Bonnet	26( )B10 thru 26( )B36	SA-216 GR. WCB, Carbon Steel
		26( )B32 thru 26( )B36	SA-217 GR. WC6, Alloy St. (1¼ CR–½ Moly)
3	Cap, Plain Screwed		Carbon Steel
4	Disc		Stainless Steel
5	Nozzle		316 St. St.
6	Disc Holder		300 Series St. St.
7	Blow Down Ring		300 Series St. St.
8	Sleeve Guide		300 Series St. St.
9	Stem		Stainless Steel
10	Spring Adjusting Screw		Stainless Steel
11	Jam Nut (Spr. Adj. Scr.)		Stainless Steel
12	Lock Screw (B.D.R)		Stainless Steel
13	Lock Screw Stud		Stainless Steel
14	Stem Retainer		Stainless Steel
15	Bellows		316L St. St.
16	Bellows Gasket		Flexible Graphite
17	Spring Buttom		Carbon Steel Rust Proofed
18	Body Stud		ASTM A193 Gr. B7, Alloy St.
19	Hex Nut (Body)		ASTM A194 Gr. 2H, Alloy St.
20	Spring	26( )B10 thru 26( )B16	Carbon Steel Rust Proofed
		26( )B20 thru 26( )B36	High Temp. Alloy Rust Proofed
21	Cap Gasket		Soft Iron or Steel
22	Body Gasket		Soft Iron or Steel
23	Bonnet Gasket		Soft Iron or Steel
24	Lock Screw (D.H.)		Soft Iron or Steel
25	Hex Nut (B.D.R.L.S.)		Stainless Steel
26	Lock Screw (D.H.)		Stainless Steel
27	Pipe Plug (Body)		Steel

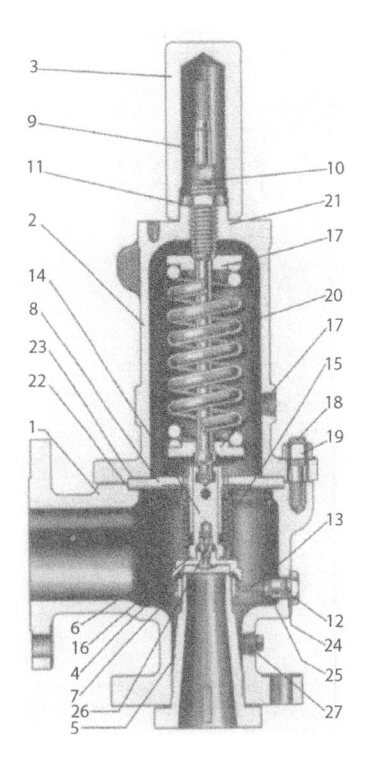

Also suitable for liquid service where ASME Code certification is not required.

Figure 23.4 Balanced nozzle safety relief valve, Balanseal®. (By permission from Teledyne Farris Engineering Co.)

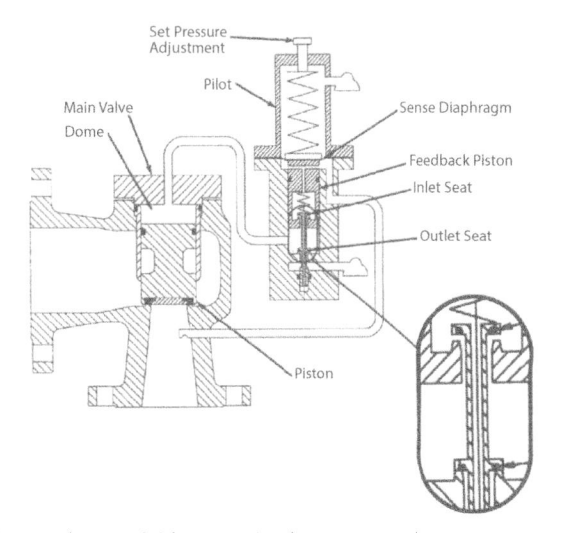

With no system pressure, the pilot inlet seat is open and outlet seat is closed. As pressure is admitted to the main valve inlet, it enters the pilot through a filter screen and is transmitted through passages in the feedback piston, past the pilot seat, into the main valve dome to close the main valve piston.

As system pressure increase and approaches valve set pressure, it acts upward on the sense diaphragm, with the feedback piston moving upward to close the inlet seat, thus sealing in the main valve dome pressure, as the outlet seat is also closed. A small, futher increase in system pressure opens the outlet seat, venting the main valve dome pressure. This reduced dome pressure acts on the unbalanced feedback piston to reduce feedback piston lift, tending to "lock in" the dome pressure. Thus, at any stable inlet pressure there will be no pilot flow (i.e. zero leakage).

As inlet pressure rises above set pressure, dome pressure reduction will be such as to provide modulating action of the main valve piston proportional to the process upset. The spool/feedback piston combination will move, responding to system pressure, to alternately allow pressure in the main valve dome to increase or decrease, thus moving the main valve piston to the exact lift that will keep system pressure constant at the required flow. Full main valve lift, and therefore full capacity, is achieved with 5% overpressure. As system pressure decreases below set pressure, the feedback piston moves downward and opens the inlet seat to admit system pressure to the dome, closing the main valve.

Due to the extremely small pilot flow, the pilot on gas/vapor valves normally discharge to atmosphere through a weather and bug-proof fitting. Pilots liquid service valves have their discharge piped to the main valve outlet.

Figure 23.5a Pilot – operated safety relief valve. (By permission from Anderson, Greenwood and Co.)

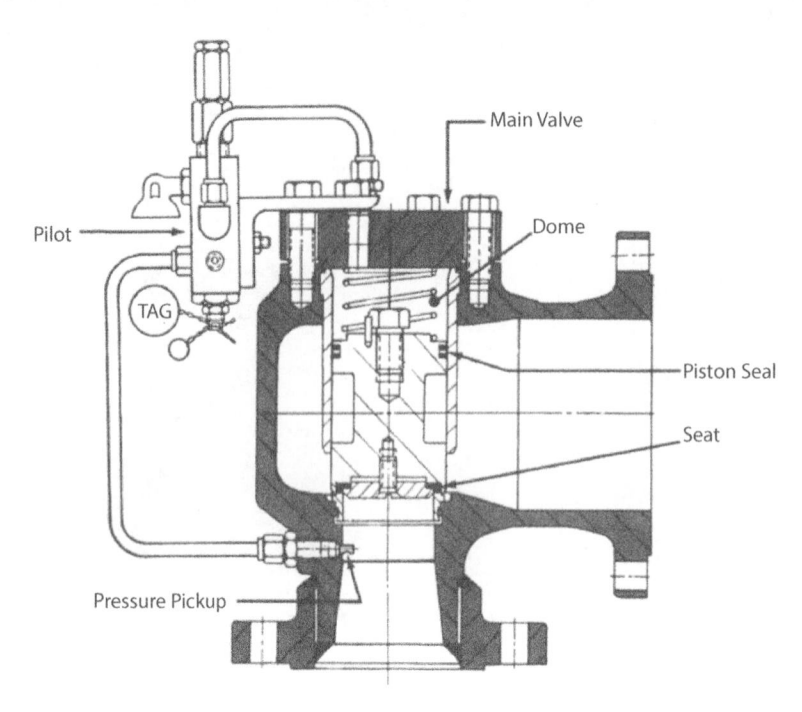

Figure 23.5b Safety relief valve mechanism as connected to a non-flow (zero flow) pilot safety valve (By permission from Anderson Greenwood and Co.)

Safety-Relief Valve: This is an automatic pressure-relieving device actuated by the static pressure upstream of the valve and characterized by an adjustment to allow reclosure, either a "pop" or a "non-pop" action, and a nozzle type entrance; and it reseats as pressure drops. It is used on steam, gas, vapor, and liquid (with adjustments), and is probably the most general type of valve in refineries, petrochemical, and chemical plants (Figures 23.3a, 23.3b, and 23.4). The rated capacity is reached at 3 or 10% overpressure, depending upon code and/or process conditions. It is suitable for use either as a safety or as a relief valve [1, 8]. It opens in proportion to an increase in internal pressure.

Pressure-Relief Valve: The term "Pressure-relief valve" applies to relief valves, safety valves, or safety-relief valves [8].

Pilot Operated Safety Valves: When properly designed, this type of valve arrangement conforms to the ASME code. It is a pilot operated pressure relief valve in which the major relieving device is combined with and is controlled by a self-activating auxiliary pressure relief valve (see Figures 23.5a and b).

23.2 Types of Valves/Relief Devices

There are many design features and styles of safety relief valves, such as flanged ends, screwed ends, valves fitted internally for corrosive service, high-temperature service, cryogenic service/low temperatures, with bonnet or without, nozzle entrance or orifice entrance, and resistance to discharge piping strains on body. Yet most of these variations have little, if anything to do with the actual performance to relieve overpressure in a system/vessel.

A few designs are important to the system arrangement and relief performance. They are as follows.

Conventional Safety Relief Valve

This valve design has the spring housing vented to the discharge side of the valve. The performance of the valve upon relieving overpressures is directly affected by any changes in the backpressure on the valve (opening pressure, closing pressure, relieving capacity referenced to opening pressure) (see Figures 23.3, 23.6a, and 23.6b) [11]. When connected to a multiple relief valve manifold, the performance of the valve can be somewhat unpredictable from a relieving capacity standpoint due to the varying backpressure in the system.

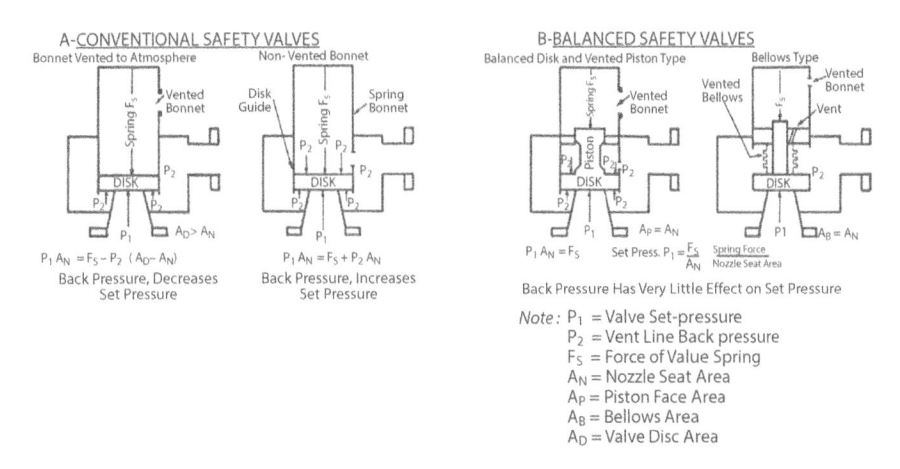

Figure 23.6a Effect of back pressure on set pressure of safety or safety relief valves. (By permission from Recommended Practice for Design and Construction of Pressure – Relieving Systems in Refineries, API RP – 520, 5th. Ed., American Petroleum Institute (1990).)

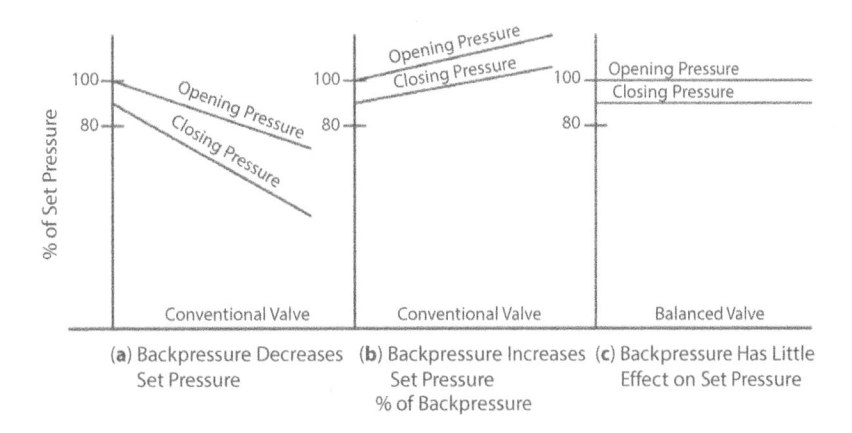

Figure 23.6b Diagram of approximate effects of back pressure on safety relief valve operation. (Adapted by permission from Teledyne Farris Engineering Co.)

Balanced Safety Relief Valve

This valve provides an internal design (usually bellows) above/on the seating disk in the huddling chamber that minimizes the effect of backpressure on the performance of the valve (opening pressure, closing pressure, and relieving capacity) [11] (see Figures 23.4, 23.6, and 23.6a, and 23.6b).

Special Valves

   a) internal spring safety relief valve
   b) power-actuated pressure relief valve
   c) temperature-actuated pressure relief valve

These last three are special valves from the viewpoint of refineries, chemical, and petrochemical plant applications, but they can be designed by the major manufacturers and instrumentation manufacturers as these are associated with instrumentation controls (Figure 23.7a). Care must be taken in the system design to make certain it meets all ASME code requirements.

Rupture Disk

A rupture disk is a non-reclosing thin diaphragm (metal, plastic, carbon/graphite (non-metallic)) held between flanges and designed to burst at a predetermined internal pressure. Each bursting requires the installation of a new

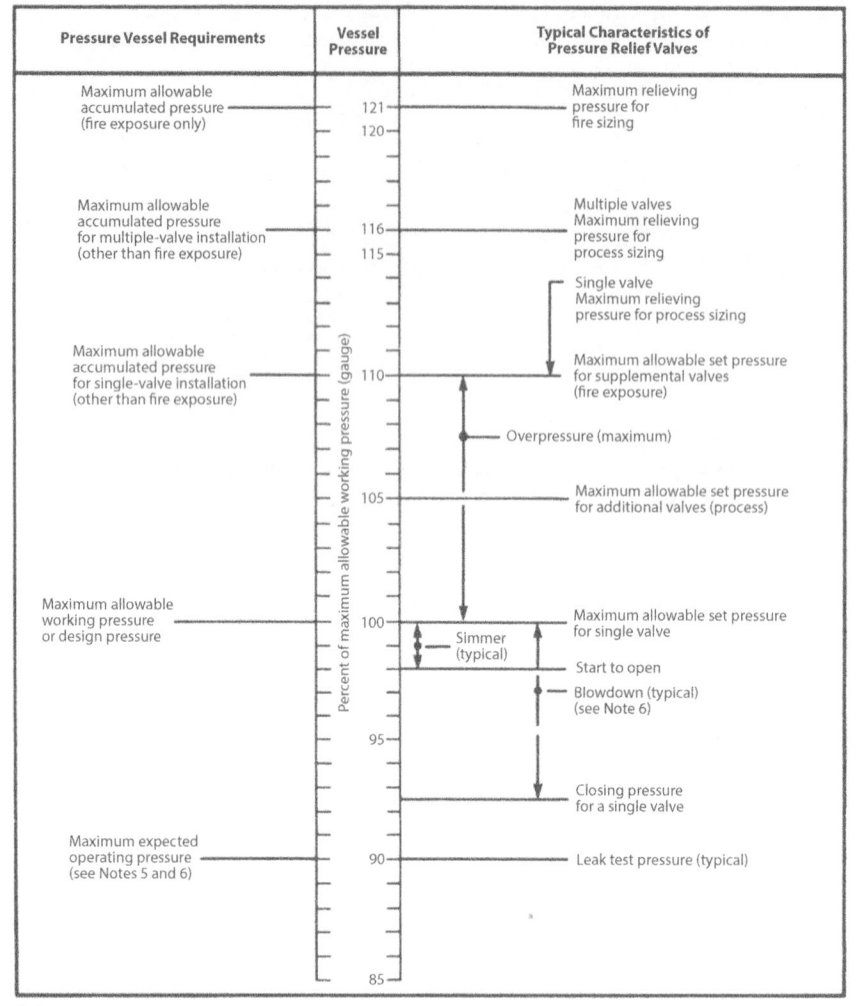

Figure 23.7a Pressure level relationship for pressure – relief valve installed on a pressure vessel (vapor phase). Single valve (or more) used for process or supplemental valves for external fire (see labelling on chart). (Reprinted by permission from Sizing, Selection and Installation of Pressure Relieving Devices in Refineries, Part 1 "Sizing and Selection", API RP- 520, 5th ed., Jul 1990, American Petroleum Institute.)

disk. It is used in corrosive service, toxic, or "leak-proof" applications, and for required bursting pressures not easily accommodated by the conventional valve such as explosions. It is applicable to steam, gas, vapor, and liquid systems (Figures 23.7b, 23.8a-k, and 23.9a-f). There are at least four basic types of styles of disks, and each requires specific design selection attention.

An explosion rupture disc is a special disc (or disk) designed to rupture at high rates of pressure rise, such as runaway reactions. It requires special attention from the manufacturer [11].

Other rupture devices suitable for certain applications are as follows [11]:

- breaking pin device
- shear pin device
- fusible plug device

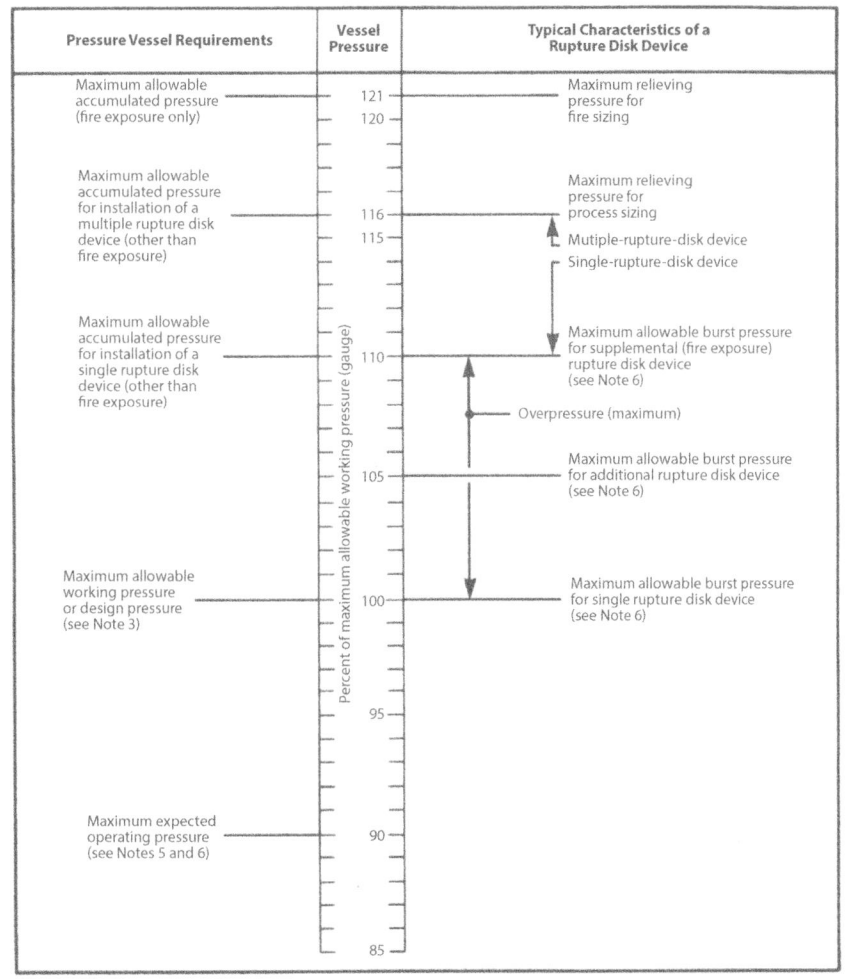

PRESSURE RELIEVING DEVICES AND EMERGENCY RELIEF SYSTEM DESIGN   9

Figure 23.7b Pressure level relationships for rupture disk devices. (Reprinted by permission from Sizing, Selection and Installation of Pressure Relieving Devices in Refineries, Part 1 "Sizing and Selection", API RP- 520, 5th ed., Jul 1990, American Petroleum Institute.)

Set Pressure: The set pressure, in pounds per square inch gauge (barg), is the inlet pressure at which the safety or relief valve is adjusted to open [8, 9]. This pressure is set regardless of any backpressure on the discharge of the valve, and is not to be confused with a manufacturer's spring setting.

Overpressure: Pressure increase over the set pressure of the primary relieving device is overpressure. It is the same as accumulation when the relieving device is set at the maximum allowable working pressure (MAWP) of the vessel [8].

Accumulation: Pressure increase over the MAWP of the vessel during discharge through the safety or relief valve, expressed as a percent of that pressure, pounds per square inch (bar), is called accumulation [8].

Blowdown: Blowdown is the difference between the set pressure and the reseating pressure of a safety or relief valve, expressed as a percent of the set pressure, or pounds per square inch (bar) [8].

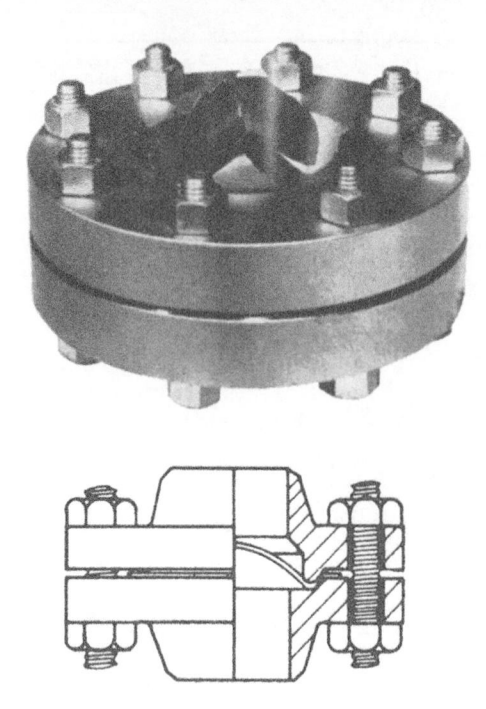

Figure 23.8a Metal type frangible disk (above) with cross section (below). (Courtesy of Black, Sivalis and Bryson Safety Systems, Inc.)

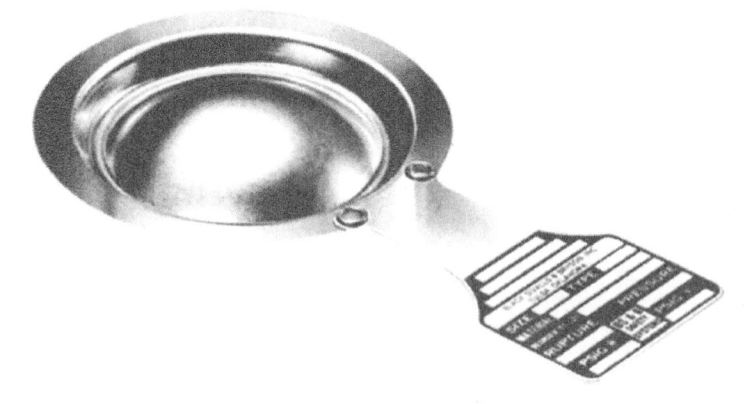

Figure 23.8b Standard rupture disk. A prebulged rupture disk available in a broad range of sizes, pressures, and metals. (By permission from B.S. & B. Safety Systems, Inc.)

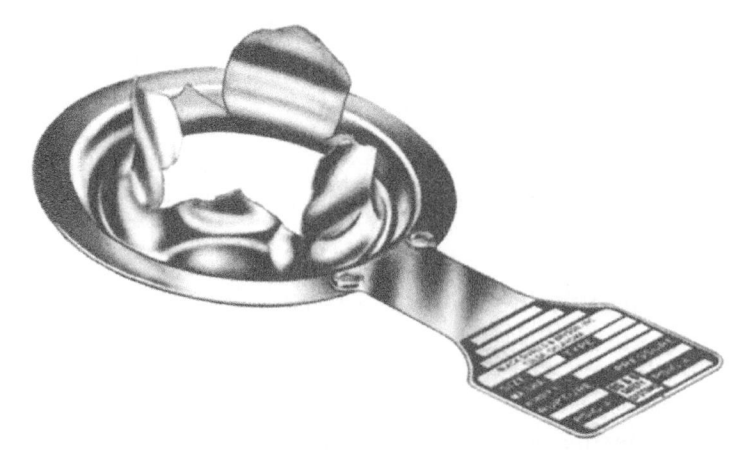

Figure 23.8c Disk of Figure 23.8b after rupture. Note 30o angular seating in holder is standard for prebulged solid metal disk. (By permission from B.S. & B. Safety Systems, Inc.)

Figure 23.8d Disk of Figure 23.8b with an attached (underside) vacuum support to prevent premature rupture in service with possible less than atmospheric pressure on underside and/or pulsation service. (By permission from B.S. & B. Safety Systems, Inc.)

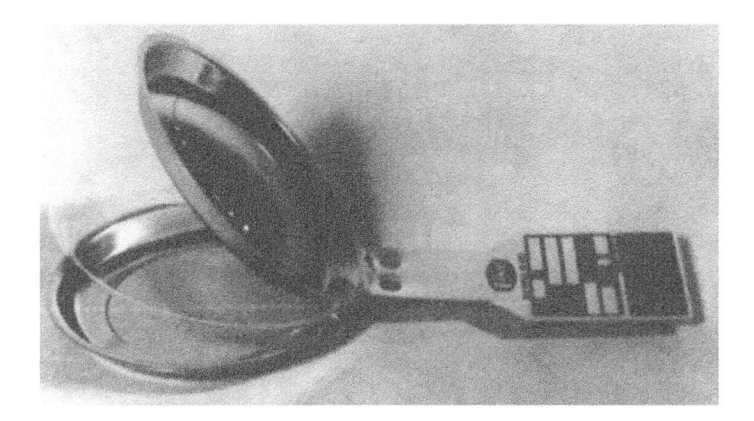

Figure 23.8e Rupture disk (top) with Teflon® or other corrosion-resistance film/sheet seal, using an open retaining ring. For positive pressure only. (By permission from Fike Metal Products Div., Fike Corporation, Blue Springs, MO.)

Figure 23.8f(a) Rupture disk (top), similar to Figure 23.8e, except a metal vacuum support is added (see Figure 23.8f(b)). (By permission from Fike Metal Products Div., Fike Corporation, Blue Springs, MO.)

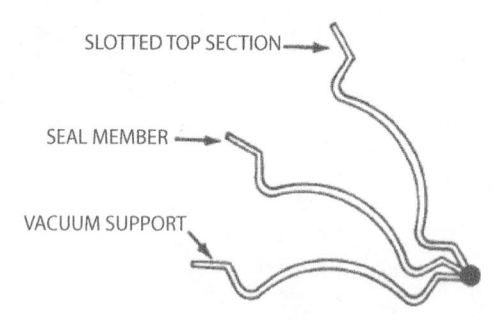

Figure 23.8f(b) Cross section of disk assembly for Figure 23.8f(a). (By permission from Fike Metal Products Div., Fike Corporation, Blue Springs, MO.)

Figure 23.8g(a) Reverse buckling® disk, showing top holder with knife blades (underside) that cut the disk at time of rupture. (By permission, B. S. & B Safety Systems, Inc.)

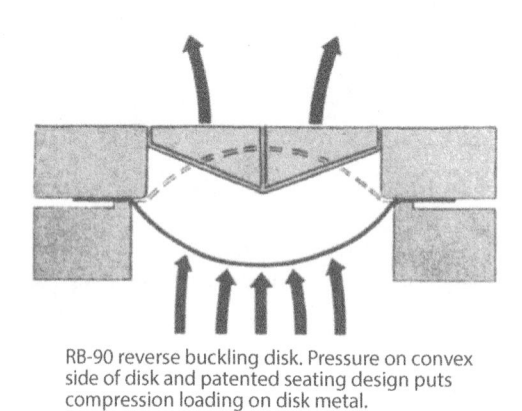

Figure 23.8g(b) Reverse buckling® disk. Pressure on convex side of disk and patented seating design puts compression loading on disk metal. (By permission, B. S. & B Safety Systems, Inc.)

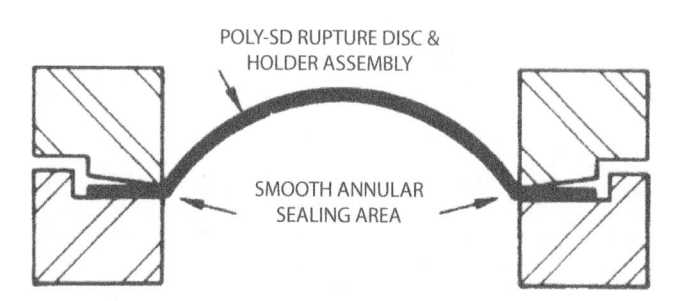

Figure 23.8h Special metal disk holder for polymer systems using a smooth disk surface to reduce polymer adherence and a smooth annular sealing area. Usually thick to avoid need for vacuum support and to allow for corrosion attack. (By permission from Fike Metal Products Co. Div., Fike Corporation, Inc.)

Figure 23.8i Flat disk used for low pressure and for isolation of corrosive environments. Usual pressure range is 2–15 psig with ± 1 psi tolerance. Stainless steel disk with Teflon® seal usually standard. (By permission from Fike Metal Products Div., Fike Corporation. Catalog 73877-1, p. 35.)

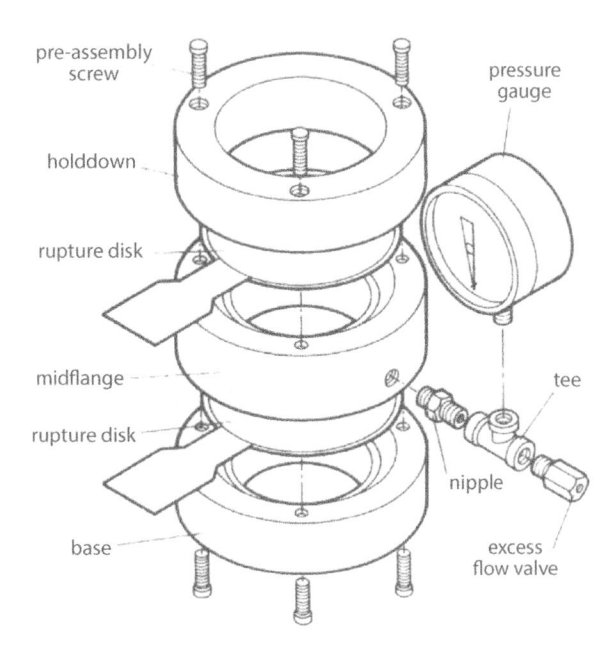

Figure 23.8j Exploded view of double disk assembly. Usually burst pressure is same for each disk. Used for corrosive/toxic conditions to avoid premature loss of process and at remote locations. Note the use of tell-tale between disks. (By permission from Fike Metal Products Div., Fike Corporation, Inc.)

Back Pressure: This is the pressure existing at the outlet or discharge connection of the pressure-relieving device, resulting from the pressure in the discharge system of the installed device [11]. This pressure may be only atmospheric if discharge is directly to atmosphere, or it may be some positive pressure due to pressure drops of flow of discharging vapors/gases (or liquids where applicable) in the pipe collection system or a common header, which in turn may be connected to a blowdown or flare system with definite backpressure conditions during flow, psig (gauge). The pressure drop during flow discharge from the safety relief valve is termed "built-up backpressure."

Burst Pressure: This is the inlet static pressure at which a rupture disk pressure-relieving device functions or opens to release internal pressure.

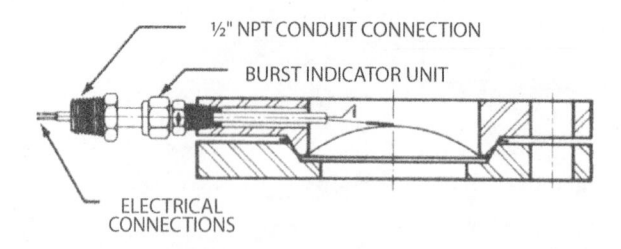

Figure 23.8k Rupture disk with burst indicator. Several other techniques available. (By permission from Fike Metal Products Div., Fike Corporation, Inc.)

Figure 23.9a Non-metal frangible disk. Ruptured disk showing complete breakout of membrane. (Courtesy of Frails Industries, Inc.)

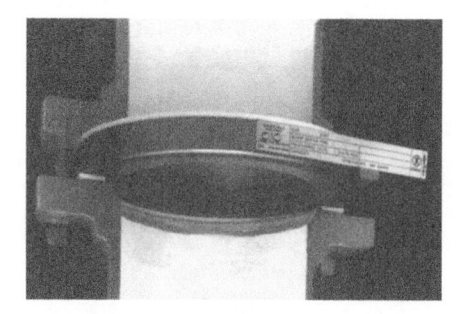

Figure 23.9b Standard non-metal frangible disk (graphite); Teflon® coatings or linings are available on entire disk. (By permission from Zook enterprises.)

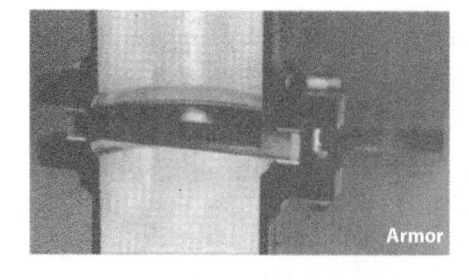

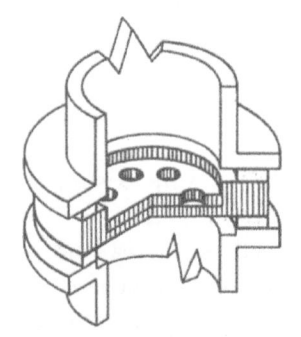

Figure 23.9c Armored graphite disk. Note: steel ring bonded to circumference of disk to increase safety in toxic or flammable services and improve reliability by preventing unequal piping stresses from reaching the pressure membrane. Teflon® coatings or linings are available on the entire disk. (By permission from Zook Enterprises.)

Vent side (lower pressure side)

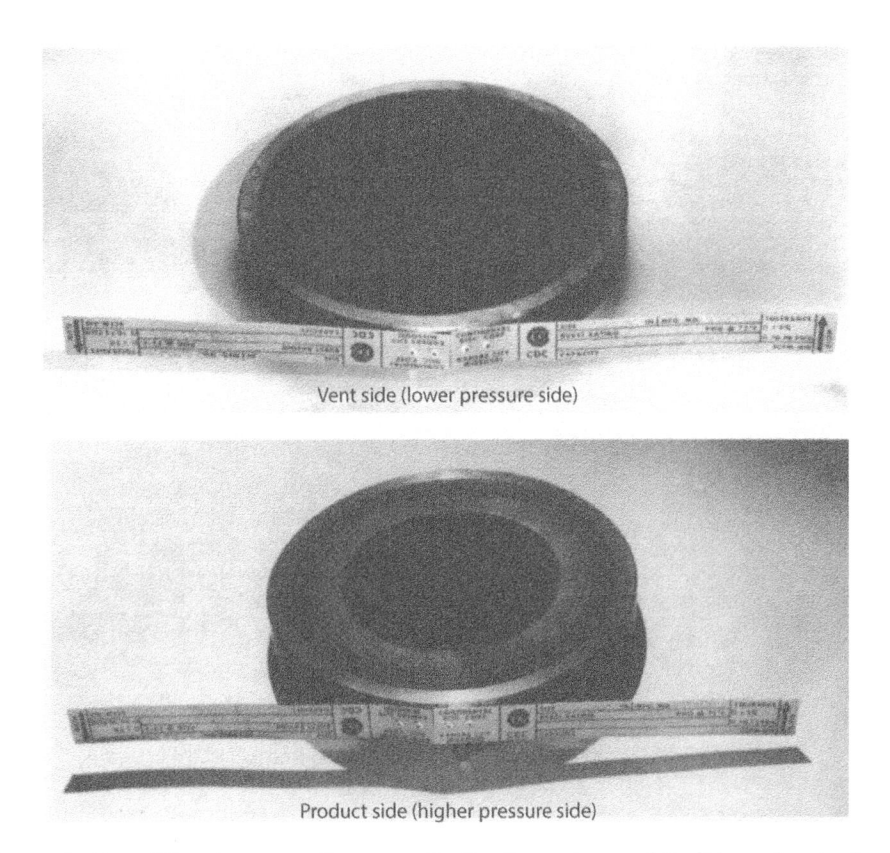

Product side (higher pressure side)

Figure 23.9d Protection against two different pressures from opposite directions using graphite disks, such as in closed storage tanks; particularly API – type to guard against failure of primary breathers, conservation vents, and so on. These require a differential of at least 10 psig between the two burst ratings, depending on diameters of disks. (By permission from Continental Disk Corporation.)

Design Pressure: This is the pressure used in the vessel design to establish the minimum code permissible thickness for containing the pressure in pounds per square inch gauge (barg).

Maximum Allowable Working Pressure (MAWP): This is the maximum pressure in pounds per square inch gauge (barg) permissible at the top of a completed vessel in its operating position for a specific designated temperature. This pressure is calculated in accordance with the ASME code (Par. UG-98) [1] for all parts or elements of the vessel using closest next larger to calculated value nominal thickness (closest standard for steel plate) (see Par. UG-A22) but exclusive of any corrosion allowance or other thickness allowances for loadings (see ASME Par.-UG-22) on vessels other than pressure (e.g., extreme wind loadings for tall vessels).

The MAWP is calculated using nominal standard steel plates (but could be other metal-use code stresses) thickness, using maximum vessel operating temperature for metal stress determinations. See Ref. [1] Par. UG-98.

Example 23.1

Hypothetical vessel design, carbon steel grade A-285, Gr C
Normal operating: 45 psig at 600°F
Design pressure: 65 psig at 700°F corresponding to the 65 psig.
Assume calculated thickness per ASME code Par. UG- 27: 0.43 in.
Closest standard plate thickness to fabricate vessel is 0.50 in. with -0.01 in. and +0.02 in. tolerances at mill.
Then

1. Using 0.50 in. - 0.01 in. (tolerance) = 0.49 in. min. thickness.
2. Using 0.50 in. + 0.02 in. (tolerance) = 0.52 in. max. thickness.

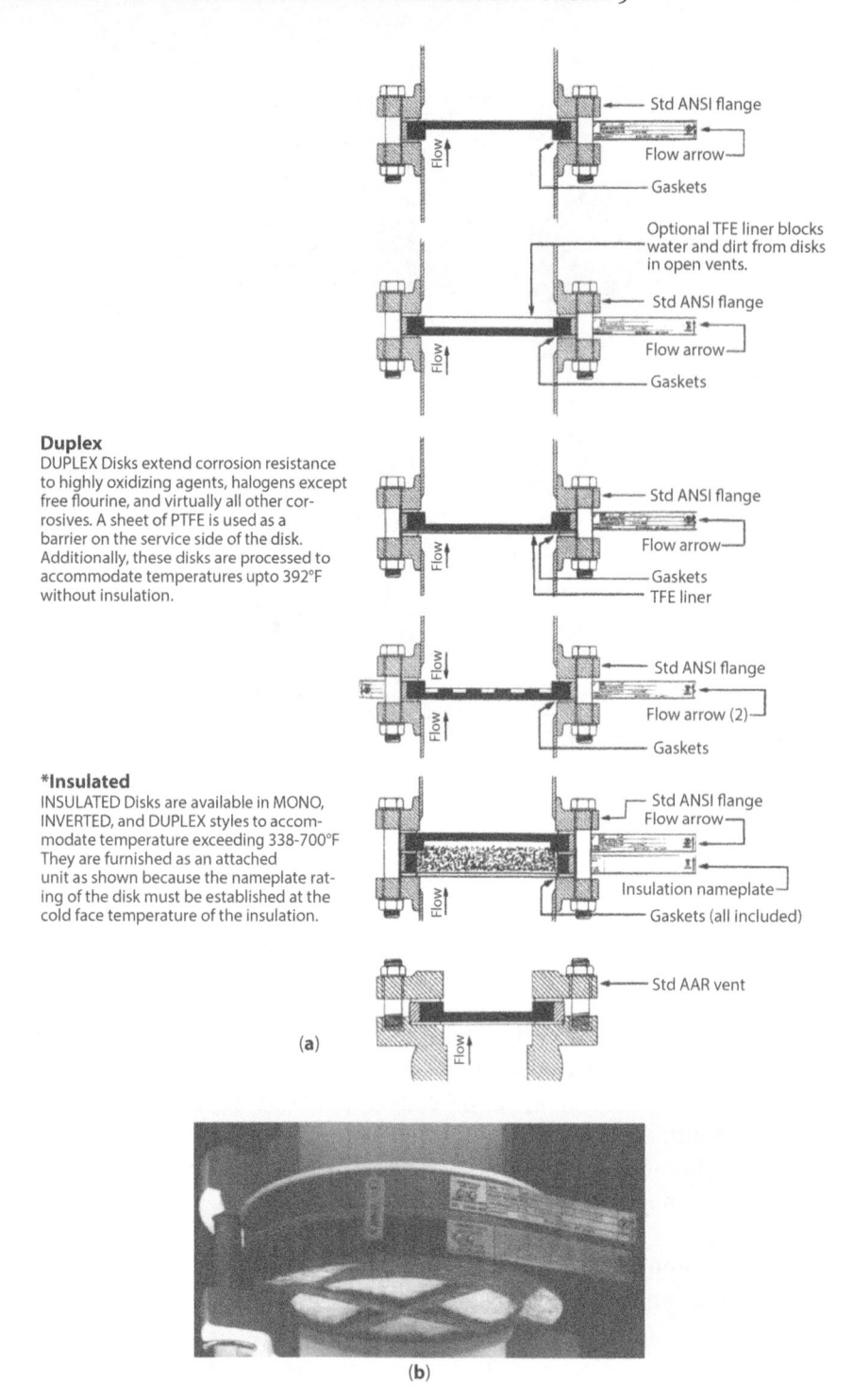

16 Petroleum Refining Design and Applications Handbook Volume 5

Duplex
DUPLEX Disks extend corrosion resistance to highly oxidizing agents, halogens except free flourine, and virtually all other corrosives. A sheet of PTFE is used as a barrier on the service side of the disk. Additionally, these disks are processed to accommodate temperatures upto 392°F without insulation.

*Insulated
INSULATED Disks are available in MONO, INVERTED, and DUPLEX styles to accommodate temperature exceeding 338-700°F They are furnished as an attached unit as shown because the nameplate rating of the disk must be established at the cold face temperature of the insulation.

Figure 23.9e (a) & (b) Duplex and insulated disks. (By permission from Zook Enterprises.)

Generally, for design purposes, with this type of tolerance, nominal thickness = 0.50 in. can be used for calculations.

Now, using Par. UG-27, 0.50 in. thickness and ASME code stress at 750°F (estimated or extrapolated) per Par. UCS-23 at 750°F, the maximum allowable stress in tension is 12,100 psi.

Recalculate pressure (MAWP) using Par. UG-27 [1]

For cylindrical shells under internal pressure:

(1) Circumferential stress (longitudinal joint).

Figure 23.9f For pressure ratings of 15 psig or lower, subject to internal vacuum conditions, a vacuum support is required that is an integral part of the rupture disk and cannot be added in the field. (By permission from Falls Industries.)

$$P_d = SEt/(R_i + 0.6t), \text{ psi} = \text{psig} \tag{23.1}$$

$$t = PR/[SE - 0.6P] \tag{23.2}$$

where

- t = minimum actual plate thickness of shell, no corrosion, = 0.50″
- P_d = design pressure, for this example equals the MAWP, psi
- R_i = inside radius of vessel, no corrosion allowance added, in.
- S = maximum allowable stress, psi, from Table UCS-23
- E = joint efficiency for welded vessel joint, plate to plate to heads. See ASME Par. UW-12, nominal = 85% = 0.85
- t = required thickness of shell, exclusive of corrosion allowance, inches

(2) Longitudinal stress (circumferential joints).

$$P_d = 2SEt/(R - 0.4t) \tag{23.3}$$

$$t = PR/[2SE + 0.4P] \tag{23.4}$$

The vessel shell wall thickness shall be the greater of Eq. (23.2) or (23.4), or the pressure shall be the lower of Eq. (23.1) or (23.3) [1].

For the above example, assume calculated MAWP (above) = 80 psig. *This is the maximum pressure that any safety relief valve can be set to open.*

For pressure levels for pressure relief valves referenced to this MAWP, see Figures 23.7a and b.

Operating Pressure: This is the pressure, psig (barg), to which the vessel is expected to be subjected during normal or the maximum probable pressure during upset operations. There is a difference between a pressure generated internally due to controlled rising vapor pressure (and corresponding temperature) and that generated due to an unexpected runaway reaction, where reliance must depend on the sudden release of pressure at a code conformance pressure/temperature. In this latter case, careful examination of the possible conditions for a runaway reaction should be made. This examination is usually without backup data or a firm basis for calculating possible maximum internal vessel pressure to establish a maximum operating pressure and, from this, a design pressure.

Design pressure of a vessel: This is the pressure established as a nominal maximum above the expected process maximum operating pressure. This design pressure can be established based on experience/practice and suggests a percentage increase of the vessel design pressure above the expected maximum process operating pressure level. There is no code requirement for establishing the design pressure. Good judgment is important in selecting each of these pressures. See operating pressure description in above paragraph. Depending on the actual operating pressure level, the increase usually varies from a minimum of 10% higher or 25 psi, whichever is greater, to much higher increases. For instance, if the maximum expected operating pressure in a vessel is 150 psig, then experience might suggest that the design pressure be set for 187–200 psig. Other factors known regarding the possibility of a runaway reaction might suggest setting it at 275 psig. A good deal of thought needs to enter into this pressure level selection. (Also see section on explosions and DIERS technology in this chapter [12, 13].)

Relieving Pressure: This is the pressure-relief device's set pressure plus accumulation or overpressure (see Figures 23.7a and b). For example, at a set pressure equal to the maximum allowable at the MAWP of the vessel of 100 psig, and for process internal vessel pressure, the pressure-relief device would begin relieving at nominal 100 psig (actually begins to open at 98 psig; see Figures 23.7a and b) and the device (valve) would be relieving at its maximum conditions at 110 psig (the 10 psig is termed the "accumulation pressure") for a single valve installation, or 116 psig, for a multiple valve installation on the same vessel. These are all process situations, which do not have an external fire around the vessel (see External Fire discussion later in this chapter and Figures 23.7b, 23.31a, b for these allowable pressure levels) and in no case do the figures apply to a sudden explosion internally.

Reseating pressure: This is the pressure level after valve opening under pressure that the internal static pressure falls to when there is no further leakage through the pressure relief valve (see Figure 23.7a).

Closing pressure: This is the pressure established as decreasing inlet pressures when the disk of the valve seats and there is no further tendency to open or close.

Simmer: This is the audible or visible escape of fluid between the seat and disk of a pressure-relieving valve at an inlet static pressure below the popping pressure, but at no measurable capacity of flow (for compressible fluid service).

Popping pressure: This is the pressure at which the internal pressure in a vessel rises to a value that causes the inlet valve seat to begin to open and to continue in the opening direction to relieve the internal overpressure greater than the set pressure of the device (for compressible fluid service).

23.3 Materials of Construction

Safety and Relief Valves: Pressure-Vacuum Relief Values

For most process applications, the materials of construction can be accommodated to fit both the corrosive-erosive and mechanical strength requirements. Manufacturers have established standard materials, which will fit a large percentage of the applications, and often only a few parts need to be changed to adapt the valve to a corrosive service. Typical standard parts are (see Figures 23.3, 23.3a, and 23.4) as follows:

	Option 1 (typical only)
Body	carbon steel, SA 216, gr. WCB
Nozzle	316 stainless steel
Disc/Seat	stainless steel
BlowDown Ring	300 Ser. stainless steel
Stem or Spindle	stainless steel
Spring	C.S. rust proof or high temp. alloy, rust proof
Bonnet	SA-216, Gr. WCB carbon steel
Bellows	316L stainless steel
	Option 2 (typical only)
Body	316 stainless steel
Nozzle	17-4 stainless steel or 316 stainless steel
Disc/Seat	Teflon, Kel-F, Vespel or Buna-N
Blow Down Ring	316 stainless steel
Stem or Spindle	17-4 stainless steel or 316 SS
Spring	316 stainless steel
Bonnet	316 stainless steel
Bellows	-

For pressure and temperature ratings, the manufacturers' catalogs must be consulted. In high pressure and/or temperature, the materials are adjusted to the service.

For chemical service, the necessary parts are available in 3.5% nickel steel; Monel; Hastelloy C; Stainless Type 316, 304, etc.; plastic-coated bellows; nickel; silver; nickel-plated springs and other workable materials.

The designer must examine the specific valve selected for a service and evaluate the materials of construction in contact with the process as well as in contact or exposed to the vent or discharge system. Sometimes the corrosive nature of the materials in the vent system presents a serious corrosion and fouling problem on the back or discharge side of the valve while it is closed.

For these special situations, properly designed rupture disks using corrosion-resistant materials can be installed both before the valve inlet as well as on the valve discharge. For these cases, refer to both the valve manufacturer and the rupture disk manufacturers (see Section 23.4 for this condition).

Rupture Disks

Rupture disks are available in:

1. practically all metals that can be worked into thin sheets, including lead, Monel, nickel, aluminum, silver, Inconel, 18-8 stainless steel, platinum, copper, Hastelloy, and others.

2. plastic-coated metals, lead-lined aluminum, lead-lined copper.
3. plastic seals of polyethylene, Kel-F®, and Teflon®.
4. graphite, impregnated graphite or carbon.

The selection of the material suitable for the service depends upon the corrosive nature of the fluid and its bursting characteristics in the pressure range under consideration. For low pressure, a single standard disk of some materials would be too thin to handle and maintain its shape, as well as give a reasonable service life from the corrosion and fatigue standpoints (see Section 23.9).

23.4 General Code Requirements [1]

It is essential that the ASME code requirements be understood by the designer and individual rating and specifying the installation details of the safety device. It is not sufficient to merely establish an orifice diameter, since process considerations that might cause overpressure must be thoroughly explored in order to establish the maximum relieving conditions.

An abbreviated listing of the key rating provisions is given in paragraphs UG-125 - 135 of the ASME code, Section 8, Div. 1, for unfired pressure vessels [1].

1. All pressure vessels covered by Division 1 or 2 of Section VIII are to be provided with protective over pressure devices. There are exceptions covered by paragraph U-l of the code, and in order to omit a protective device, this paragraph should be examined carefully. For example, vessels designed for above 3000 psi are not covered; also vessels with <120 gallons of water, vessels with inside diameter not over 6 in. (at any pressure), vessels having internal or external operating pressures not over 15 psig (regardless of size), and a few other conditions may not be subject to this code.
2. Unfired steam boilers must be protected.
3. Pressure relief must be adequate to prevent internal pressures from rising over 10% above the MAWP, except when the excess pressure is developed by external fire or other unforeseen heat source (see design details in later paragraph). Papa [14] proposes an improved technique for relief sizing (also see Figures 23.7a and b).
4. When a pressure vessel is exposed to external heat or fire, supplemental pressure-relieving devices are required for this excessive pressure. These devices must have capacity to limit the overpressure to not more than 21% above the MAWP of the vessel (see Figures 23.7a and b). A single relieving device may be used to handle the capacities of paragraph UG-125 of the code, provided it meets the requirements of both conditions described.
5. Rupture disks may be used to satisfy the requirements of the code for conditions such as corrosion and polymer formations, which might make the safety/relief valve inoperative, or where small leakage by a safety valve cannot be tolerated. They are particularly helpful for internal explosion pressure release.
6. Liquid relief valves should be used for vessels that operate full of liquid.

23.5 Relief Mechanisms

Reclosing Devices, Spring Loaded

Safety and relief valves must be the direct spring-loaded type, and for pressure ranges noted below, the code [1] requires the following:

Set pressure	Maximum spring reset referenced to set pressure*
≤250 psig	±10%
≥ 250 psig	±5%

*Marked on value

The set pressure tolerances of pressure-relief valves are not to exceed ±2 psi for pressures up to and including 70 psig and ±3% for pressures above 70 psig. Indirect operation of safety valves, for example, by pilot valve, is not acceptable unless the primary unloading valve will automatically open and will operate fully in accordance with design relieving capacity conditions if some essential part of the pilot or auxiliary device should fail [1].

The pilot valve is a self-actuated pressure relief valve that controls the main valve opening.

Non-Reclosing Pressure Relieving Devices

Rupture disks must have a specified bursting pressure at a specified temperature. There must be complete identification of the metallurgy (if metal) or other properties if graphite or plastic, and the disk must be guaranteed by the manufacturer to burst within 5% (±) of the specified bursting pressure at the rated temperature.

The connection nozzle holding the disk must have a net cross-sectional area no less than that required for the design rated conditions of the disk.

The certification of disk performance is to be based on actual bursting tests of two or more disks from a lot of the same material of the same size as the disk to be sold by the manufacturer. The holder for the test disks must be identical to the design, dimensions, and so on for the disk being certified (for details, see ASME code, Par. UG-127 [1]).

23.6 Pressure Settings and Design Basis

Unfired steam boilers, that is, nominally termed "waste heat boilers" or "heat exchangers," which generate steam by heat interchange with other fluids (see ASME code [1] Par. U-1 (g)), should be equipped with pressure-relieving devices required by the ASME Code, Section 1, as far as applicable; otherwise, use Par. UG-125. Vessels, which per Par. U-1(g), follow Par. UG-125ff are as follows:

1. Evaporators or heat exchangers.
2. "Vessels in which steam is generated by the use of heat resulting from operation of a processing system containing a number of pressure vessels such as used in the manufacturer of refineries, chemical and petrochemical products" [1].
3. Par. U-1 (h) "Pressure vessels or parts subject to direct firing from the combustion of any fuel, which are not within the scope of Sections I, III or IV, may be constructed in accordance with the rules of Section VIII, Div. I, Par. UW-2 (d)" [1].

To meet code requirements, the relieving device must be directly open to the system to be relieved; see Figures 23.10, 23.11, and 23.12. For Figures 23.10, 23.11, and 23.12, the rupture disk and the relief valve must be designed to handle the relieving capacity at the relieving temperature without allowing more than a 10% pressure buildup above the MAWP of the unfired pressure vessel (or corresponding overpressure for other code requirements). Figure 23.11 requires that the rupture disk be designed the same as for Figure 23.10, 23.13a, and b, and Figure 23.12 requires that the relief valve be the primary device and meets the process relief requirements; it may have additional capacity to accommodate such conditions as external fire, or this additional requirement may be installed in a separate relief valve or rupture disk as

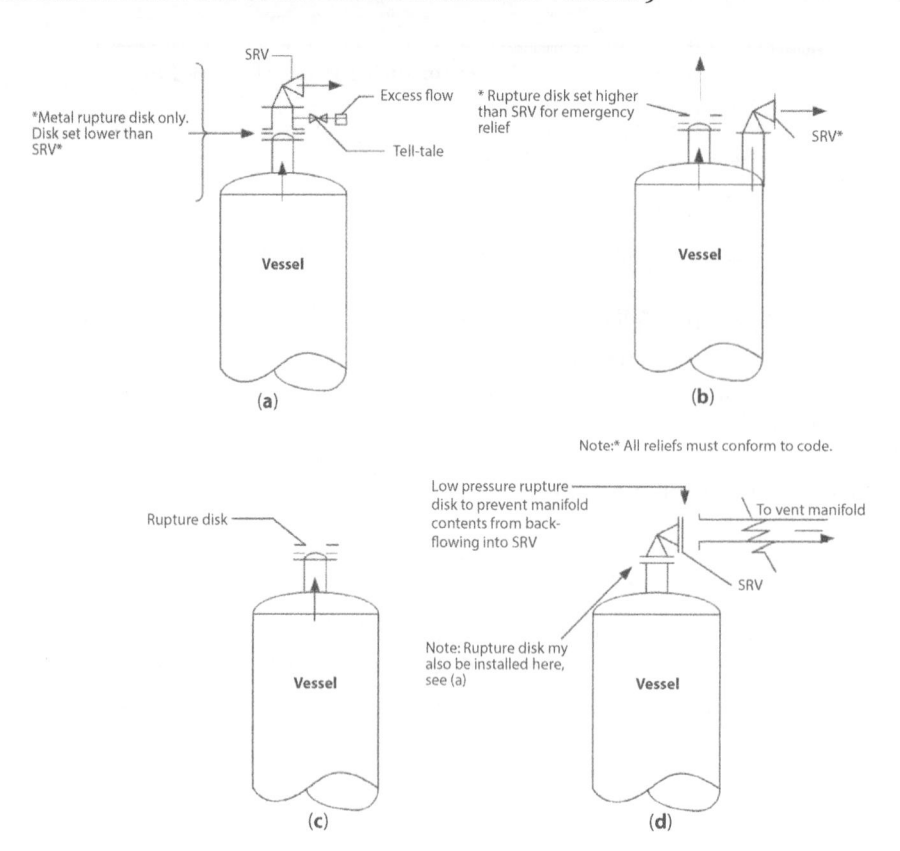

Figure 23.10 Rupture disk installations.

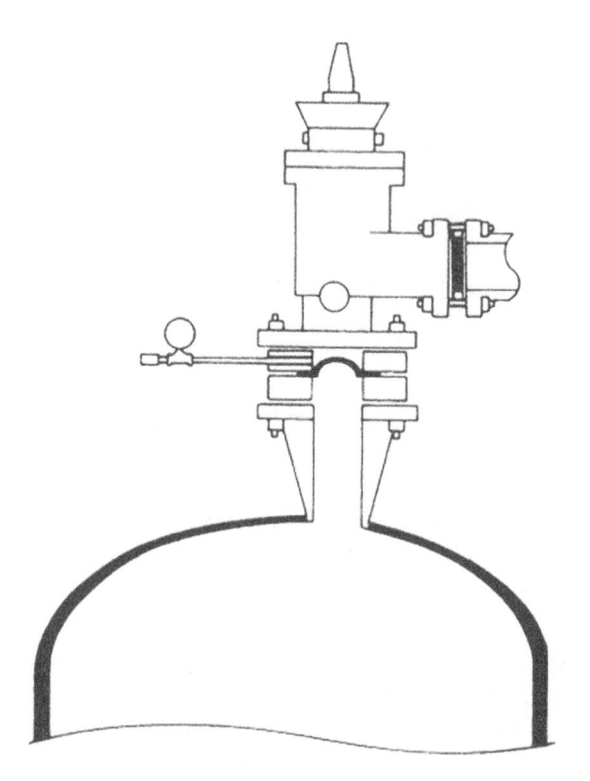

Figure 23.11 Safety valve and rupture disk installation using pressure rupturing disk on inlet to safety relief valve, and low pressure disk on valve discharge to protect against backflow/corrosion of fluid on valve discharge side, possibly discharge manifold. (By permission from Fike Metal Products Div., Fike Corporation, Inc.)

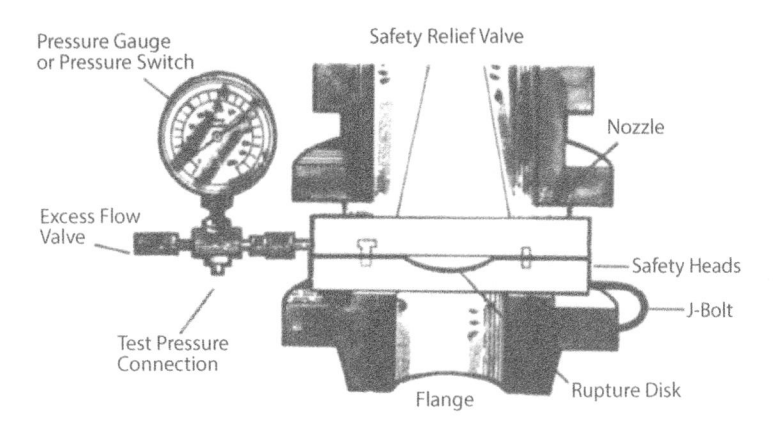

Figure 23.12 Rupture disk mounted beneath a pressure – relieving spring – loaded valve. A reverse buckling® disk arrangement is often recommended here. (By permission from B. S. & B. Safety Systems, Inc.)

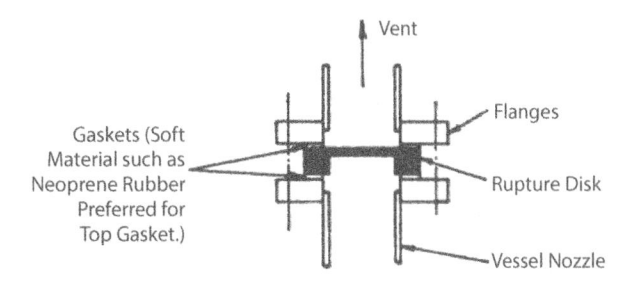

Install Disk with Pressure Membrane up. When inverted, the Disk Bursts at about 65% Increase in Pressure. Disk Must be Positioned True Center of Vent Line and Nozzle. If Eccentric, Burst Characteristics Might Not Hold True.

Figure 23.13a Installation of graphite rupture disk. (Adapted by permission from Falls Industries, Inc.)

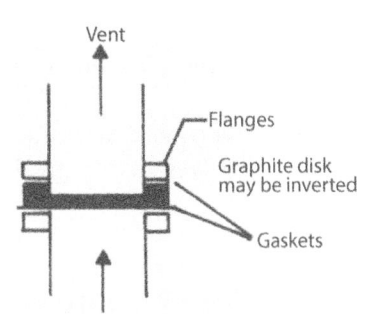

Inverted graphite disk bursts at higher pressure than with flat surface on top.

Figure 23.13b Inverted graphite disk bursts at higher pressure than with flat surface on top. (Adapted by permission from Falls Industries, Inc.)

shown. Also, the separate rupture disk may be in a secondary function not covered by the code for such conditions as runaway reactions and internal explosion. For these conditions, the setting of the rupture disk is left up to the designer and may be higher than that for the usual relief. Of course, it should be set sufficiently below the rupture condition for the vessel or component in order to avoid a hazardous condition and meet code requirements.

23.7 Unfired Pressure Vessels Only, But Not Fired or Unfired Steam Boilers

Non-Fire Exposure

Single pressure-relief valve installation must be set to operate at a pressure not exceeding the MAWP of the vessel, [1] Par. UG-134, but may be set to operate at pressures below the MAWP. The device must prevent the internal pressure from rising more than 10% above the MAWP.

For multiple pressure-relief valves installation, if the required capacity is provided using more than one pressure relieving device, (i) only one device must be set at or below the MAWP of the vessel, and (ii) the additional device(s) may be set to open at higher pressures, but in no case at a pressure any higher than 105% of the MAWP. The combination of relieving valves must prevent the pressure from rising more than 16% above the MAWP. See ASME Ref [1] Par. UG-125C and G-1 and Par. UG-134a.

External Fire or Heat Exposure Only and Process Relief

Valves to protect against excessive internal pressures must be set to operate at a pressure not in excess of 110% of the MAWP of the vessel (ASME Par. UG-134b).

When valves are used to meet the requirements of both Par. UG-125(c) and UG-125c-(2), that is, both internal process pressure and external fire/heat requirements, the valve(s) must "be set to operate not over the MAWP of the vessel. For these conditions of the additional hazard of extreme fire or heat, supplemental" pressure-relieving devices must be installed to protect the vessel. The supplemental devices must be capable of preventing the pressures from rising more than 21% above the MAWP (*Note*: this is not the setting). The same pressure-relieving devices may be used to satisfy the capacity requirements of Par. UG-125c or C(1) and Par. UG-125c-(2) provided the pressure setting requirements of Par. UG-134(a) are met. See Par. (A) 1 and 2 above and see Figure 23.7a.

When pressure-relief devices are intended primarily for protection against overpressure due to external fire or heat, have no permanent supply connection, and are used for storage at ambient temperature of non-refrigerated liquefied compressed gases, they are excluded from requirements of Par. UG-125c (1) and C (2), with specific provisions. See ASME code [1] for detailed references and conditions.

- Vessels operating completely filled with liquid must be equipped with liquid relief valves, unless otherwise protected (Par. UG-125-3(g)).
- Safety and safety relief valves for steam service should meet the requirements of ASME Par. UG-131 (b), [1]. Note that the requirements for these valves are slightly different than for process type valves.

23.8 Relieving Capacity of Combinations of Safety Relief Valves and Rupture Disks or Non-Reclosure Devices (Reference ASME Code, Par. UG-127, U-132)

Primary Relief

A single rupture disk can be used as the only overpressure protection on a vessel or system (Figure 23.10). The disk must be stamped by the manufacturer with the guaranteed bursting pressure at a specific temperature. The disk must rupture within ±5% of its stamped bursting pressure at its specified burst temperature of operation. The expected burst temperature may need to be determined by calculation or extrapolation to be consistent with the selected pressure.

The set burst pressure should be selected to permit a sufficiently wide margin between it and the vessels used or design operating pressure and temperature to avoid premature failure due to fatigue or creep of metal or plastic coatings.

Selected Portions of ASME Pressure Vessel Code, quoted by permission [1].

Section VIII, Division I Superscript = Footnote reference July 1, 1989 Edition in Code Figure No., for this text.

Rupture Disk Devices, [44] Par UG-127

1. General
   a. Every rupture disk shall have a stamped bursting pressure within a manufacturing design range [1] at a specified disk temperature [16] and shall be marked with a lot number, and shall be guaranteed by its manufacturer to burst within 5% (plus or minus) of its stamped bursting pressure at the coincident disk temperature.
2. Capacity Rating
   a. The calculated capacity rating of a rupture disk device shall not exceed a value based on the applicable theoretical formulas (see Par. UG-131) for the various media multiplied by K = coefficient = 0.62. The area A (square inches) in the theoretical formula shall be the minimum net area existing after burst [17].
3. Application of Rupture Disks
   a. A rupture disk device may be used as the sole pressure relieving device on a vessel. *Note:* When rupture disk devices are used, it is recommended that the design pressure of the vessel be sufficiently above the intended operating pressure to provide sufficient margin between operating pressure and rupture disk bursting pressure to prevent premature failure of the rupture disk due to fatigue or creep. Application of rupture disk devices to liquid service should be carefully evaluated to assure that the design of the rupture disk device and the dynamic energy of the system on which it is installed will result in sufficient opening of the disk.
   b. A rupture disk device may be installed between a pressure relief valve [18] and the vessel provided (see Figure 23.10).
      i. The combination of the spring loaded safety or safety relief valve and the rupture disk device is ample in capacity to meet the requirements of UG-133 (a) and (b).
      ii. The stamped capacity of a spring loaded safety or safety relief valve (nozzle type) when installed with a rupture disk device between the inlet of the valve and the vessel shall be multiplied by a factor of 0.80 of the rated relieving capacity of the valve alone, or alternatively, the capacity of such a combination shall be established in accordance with Par. 3 below.
      iii. The capacity of the combination of the rupture disk device and the spring loaded safety or safety relief valve may be established in accordance with the appropriate paragraphs of UG-132, Certification of Capacity of Safety Relief Valves in Combination with Non-reclosing Pressure Relief Devices.
      iv. The space between a rupture disk device and a safety or safety relief valve shall be provided with a pressure gauge, a try cock, free vent, or suitable telltale indicator. This arrangement permits detection of disk rupture or leakage [19].
      v. The opening [20] provided through the disk, after burst, is sufficient to permit a flow equal to the capacity of the valve (Par. 2 and 3 above), and there is no chance of interference with proper functioning of the valve; but, in no case shall this area be less than 80% of the area of the inlet of the valve unless the capacity and functioning of the specific combination of rupture disk and valve have been established by test in accordance with UG-132.

Note that in lieu of testing, Par (b) 2 and (b) 3 above allow the use of a capacity factor of 0.80 as a multiplier on the stamped capacity of the spring loaded safety relief valve (nozzle type). Some manufacturers test specific valve/rupture disk combinations and determine the actual capacity factor for the combination, and then use this for the net capacity determination. See Figures 23.10, 23.11, 23.12, 23.13a and b.

    c. A rupture disk device may be installed on the outlet side [21] of a spring loaded safety relief valve, which is opened by direct action of the pressure in the vessel provided (Figure 23.12).

1. The valve is so designed that it will not fail to open at its proper pressure setting regardless of any backpressure that can accumulate between the valve disk and the rupture disk. The space between the valve disk and rupture disk shall be vented or drained to prevent accumulation of pressure due to a small amount of leakage from the valve [22].
2. The valve is ample in capacity to meet the requirements of UG-133 (a) and (b).
3. The stamped bursting pressure of the rupture disk at the coincident disk temperature plus any pressure in the outlet piping shall not exceed the design pressure of the outlet portion of the safety or safety relief valve and any pipe or fitting between the valve and the rupture disk device. However, in no case shall the stamped bursting pressure of the rupture disk at the coincident operating temperature plus any pressure in the outlet piping exceed the maximum allowable working pressure of the safety or safety relief valve.
4. The opening provided through the rupture disk device after breakage is sufficient to permit a flow equal to the rated capacity of the attached safety or safety relief valve without exceeding the allowable overpressure.
5. Any piping beyond the rupture disk cannot be obstructed by the rupture disk or fragment.
6. The contents of the vessel are clean fluids, free from gumming or clogging matter, so that accumulation in the space between the valve inlet and the rupture disk (or in any other outlet that may be provided) will not clog the outlet.
7. The bonnet of the safety relief valve shall be vented to prevent accumulation of pressure.

Footnotes to ASME Code

47. The minimum net flow area is the calculated net area after a complete burst of the disk with appropriate allowance for any structural members, which may reduce the net flow through the rupture disk device. The net flow area for sizing purposes shall not exceed the nominal pipe size area of the rupture disk device.
48. Use of a rupture disk device in combination with a safety or safety relief valve shall be carefully evaluated to ensure that the media being handled and the valve operational characteristics will result in pop action of the valve coincident with the bursting of the rupture disk.
49. Users are warned that a rupture disk will not burst at its design pressure if backpressure builds up in the space between the disk and the safety or safety relief valve, which will occur should leakage develop in the rupture disk due to corrosion or other cause.
50. This use of a rupture disk device in series with the safety or safety relief valve is permitted to minimize the loss by leakage through the valve of valuable or of noxious or otherwise hazardous materials and where a rupture disk alone or disk located on the inlet side of the valve is impracticable, or to prevent corrosive gases from a common discharge line from reaching the valve internals.
51. Users are warned that an ordinary spring loaded safety relief valve will not open at its set pressure if backpressure builds up in the space between the valve and rupture disk. A specially designed valve is required, such as a diaphragm valve or a valve equipped with a bellows above the disk.

(Source: Reprinted with ASME permission. ASME Pressure Vessel Code, Section VIII, Division I, UG-127, 1989 Edition, pp. 86–88.)

23.9 Establishing Relieving or Set Pressures

The pressure at which the valve is expected to open (set pressure) is usually selected as high as possible consistent with the effect of possible high pressure on the process as well as the containing vessel. Some reactions have a rapid increase in temperature when pressure increases, and this may fix the maximum allowable process pressure. In other situations, the pressure rise above operating must be kept to some differential, and the safety valve must relieve at the peak value. A set pressure at the maximum value (whether MAWP of vessel, or other, but insuring protection to the weakest part of the system) requires the smallest valve. Consult manufacturers for set pressure compensation (valve related) for temperatures >200°F (> 93.3 °C).

When the pressure rise in a system is gradual and not "explosive" in nature, a safety or safety relief valve is the proper device, but when it is critical to completely depressurize a system or the rate of pressure increase might be expected to be rapid, then a rupture disk is the proper device. Properly designed, a pilot operated valve may be selected after checking its performance with the manufacturer.

Often a system (a group of vessels not capable of being isolated from each other by block valves, or containing restriction to flow and release of pressure) may need a relief valve set reasonably close, set +15-20% when system is below 1000 psig; above, typically use 7-15% above as set criteria related to normal operating pressure to catch any pressure upswing. Then this may have a backup valve set higher (but within code) to handle further pressure increase. Or, the second device may be a rupture disk. It is not unusual to have two relief devices on the same equipment set at different pressures.

For situations where explosion may involve chemical liquid, vapor, or dust, it is generally advisable to obtain rate of pressure rise data and peak explosion pressure data in order to intelligently establish the design parameters. Such data are available [23–30]; however, it is important to evaluate whether the conditions are comparable between the systems when selecting the values for design. In general, the lower the setting for pressure relief, the lower will be the final internal peak pressure in the vessel. It is extremely important to realize that the higher the system pressure before relief, the higher will be the peak pressure attained in the vessel. In some difficult cases, it may be advisable to set relief devices at two pressures, one lower than the other. Each must be designed for the conditions expected when it relieves, and one or all must satisfy code requirements or be more conservative than the code.

For pulsating service, the set pressure is usually set greater than the nominal 10% or 25 psig above the average operating pressure of the system in order to avoid unnecessary releases caused by surging pressure peaks, but still not exceeding the MAWP of the vessel/system. Careful analysis must be made of the proper set condition.

Safety relief valves are available for relieving or set pressures as low as 2, 10, and 20 psig, as well as higher pressures. Lower pressures are available on special order. Usually a more accurate relief is obtained from the higher pressures.

Safety relief valves are normally tested in the shop, or even on the equipment at atmospheric temperature. The set tolerances on the valves as manufactured are established by the code as discussed earlier. In order to recognize the difference between the test temperature and the actual operating temperature at actual relief, the corrections shown in Tables 23.1a and b are applied. An increase in temperature above design causes a reduction in valve set pressure due to the effects of temperature on the spring and body.

Testing of pressure relieving spring loaded valves at atmospheric temperature requires an adjustment in set pressure at ambient conditions to compensate for higher operating temperatures. For process services, see Table 23.1a and for saturated steam, use Table 23.1b.

Table 23.1a Compensation Factors for Safety Relief Valves between Atmospheric Test Temperatures and Actual Operating Temperature [31].

Operating temperature °F	Percent increase in set pressure at atmospheric temperature
-450 - 200	None
201 - 450	2
451 - 900	3
901 - 1200	4

By permission, Teledyne Farris Engineering Corp., Cat. FE-316, p. 12.

Table 23.1b Set Pressure Compensation for Saturated Steam Service Safety-Relief Valves between Atmospheric Test Temperature and Actual Operating Temperature.

Saturated steam pressure set pressure (psig)	% Increase in spring settling
10 - 100	2
101 - 300	3
301 - 1000	4
1001 - 3000	5

By permission, Teledyne Farris Engineering Corp., Cat. FE-316, p. 12.

Safety and Safety Relief Valves for Steam Service

Pressure relieving devices in process plants for process and utility steam systems must conform to the requirements of ASME [1] Par. UG-131b. This is not necessarily satisfactory to meet the ASME Power Boiler Code for applications on power generating equipment.

Vessels or other pressure containing equipment that operates filled with liquid must be provided with liquid relief valves, unless protected otherwise [1]. Any liquid relief valve must be at least 0.5 in. in pipe size, [1] Par UG-128 (see [32]).

23.10 Selection and Application

Causes of System Overpressure

Figure 23.14 shows the steps in sizing relief-device; Figure 23.14a, Operational Check Sheet [33], lists 16 possible causes of overpressure in a process system, and Figure 23.14b shows a typical relief valve process data sheet. There are many others, and each system should be reviewed for its peculiarities. System evaluation is the heart of a realistic,

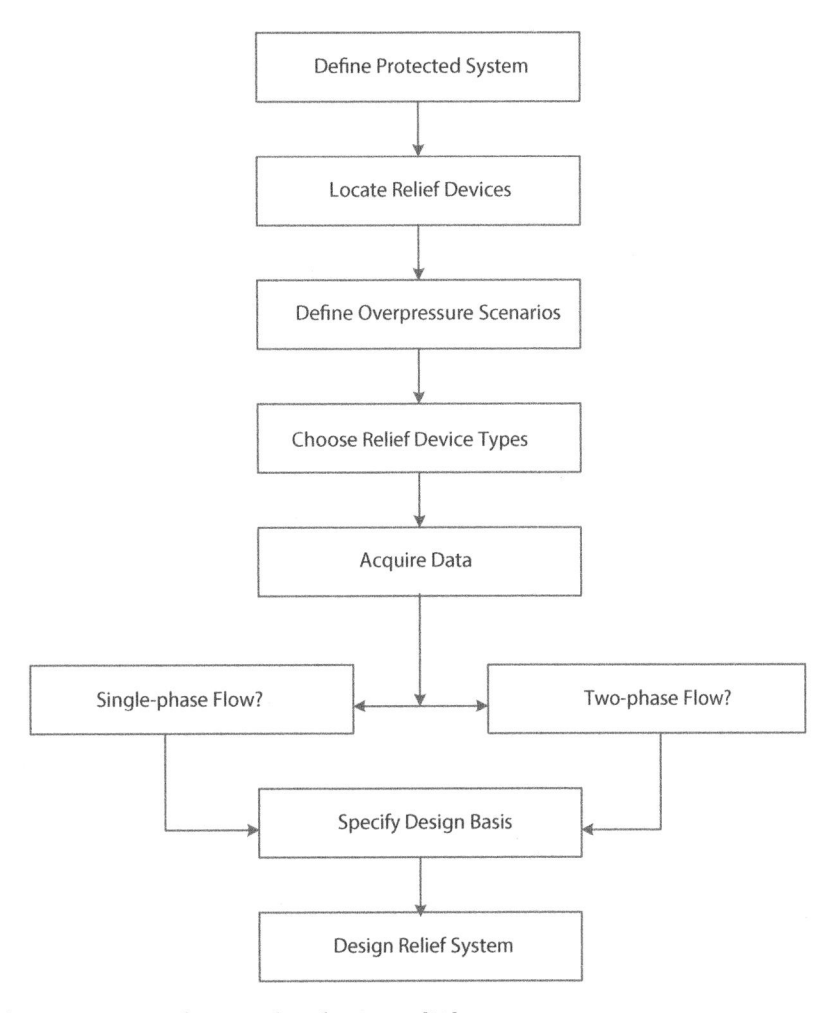

Figure 24.14 The relief-device sizing procedures involves three steps [90].

safe, and yet economical overpressure protection installation on any single equipment or any group of equipment. Solving formulas with the wrong basis and/or data can be disastrous. The following should be reviewed:

1. The sources of possible overpressure.
2. Maximum overpressure possible from all sources.
3. Maximum rate of volume increase at the burst pressure, and temperature at this condition.
4. Length of duration of overpressure.

23.11 Capacity Requirements Evaluation for Process Operation (Non-Fire)

A major requirement in the design and engineering of a plant or system is to ensure safe equipment operation. A great effort in this respect is directed to determine the pressure limits of equipment and to protect that equipment from dangerous overpressure conditions. Pressure relief valves are employed for this protection service, although under certain condition, bursting disks (rupture disks) may be used.

The cost of providing facilities to relieve all possible emergencies simultaneously can be high. Every emergency situation arises from a specific scenario. The simultaneous occurrence of two or more emergencies or contingencies is not common. However, problems such as utility failures, must be considered on a plant or area-wide basis and may create simultaneous relief requirements that set relief system sizes and back pressures. Electrical power failures are

	SAFETY SALVE DESIGN OPERATIONAL CHECK SHEET						
Date:					Job No.:		
Checked:					By:		
	Vessel or System:	Process Evaporator					
	Design Pressure:	75 psig					
	Allowable Pressure for Capacity Relief:		75 + (75+ 10%) = 82.5 psig				
	Operating Conditions:	Fluid:	PDC			Mol Wt.	113.5
		sp.gr.	1.16			Temp.	Oper.
		Latent Heat	125 BTU/lb.			Corrosive	No
	Physical Conditions:	Vessel Dia.	5 feet		× length	6 feet	
		Insulation:	Yes, 2"				
		Fire Control Measures:		No Sprinkler Sys.			
	Cause of Overpressure		Capacity Requirement, lb/hr.				
1.	Failure Cool Water/Elect/Mechanical		----------------------------------				
2.	Reflux and/or Condensing Failure		----------------------------------				
3.	Entrance of a Highly Volatile Fuel		----------------------------------				
4.	Vapor generation, external fire		----2,920----				
5.	Excessive Operating Heat Inputs		----22,500----				
6.	Accumulation of Non-Condensibles		----------------------------------				
7.	Closed Outlets		----22,500 or less----				
8.	Failure of Automatic Controls/Instr.		----22,500 or less----				
9.	Internal Explosions (Use Rupture Disk)		----------------------------------				
10.	Chemical Reaction/Run-a-Way		----------------------------------				
11.	Two Phase Flow Conditions		----------------------------------				
12.	Inadvertent opening valve into system		----------------------------------				
13.	Check Valve failure		----------------------------------				
14.	Cooling Fans failure		----------------------------------				
15.	Heat Exchanger Tube rupture/failure		----------------------------------				
16.	Circulating Pump failure		----------------------------------				
	Causes that may occur simultaneously:						
	---------- Any of the four considered may occur simultaneously -------						
	Probability of occurence: --Closed outlet with a failure of automatic control resulting in excessive heat input ----						
	Allowance to made:---------None---------						
	Relief capacity used for sizing valve(s)--------------22,500 lb./hr.						
	Auxiliaries		Cause of overpressure		Capacity Requirement		
1.	Exchangers		Split tube(s)		1482 lb. hr.		
			Thermal Vaporization		----------------------------------		
2.	Pumps		Discharge Restriction		----------------------------------		
3.	Length of Line		Thermal Vaporization		----------------------------------		

Figure 23.14a Safety valve design operational check sheet. (Adapted and added to by permission from N.E. Sylvander and D. L. Klatz, Design and Construction of Pressure Relieving Systems, Univ. of Michigan, Ann Arbor (1948). Six items of overpressure list above by E. E. Ludwig [34] and from API Rec. Practice 521 (1982)).

known to cause large relief loads. Unless a failure can cascade into multiple types of failure, an emergency which can arise from two or more types of failure (e.g., the simultaneous failure of a control valve and cooling water) is not usually considered when sizing safety equipment.

Each system and item of equipment should be examined for operational safety as set forth by specific plant area (and process fluids) requirements and the codes previously cited. The codes particularly [5a -d, 8 -10] establish guides based on wide experience, and are sound requirements for design. Relief capacity is based on the most severe requirement of a system, including possible two-phase flow [13]. A system is generally equipment or groups of equipment that is isolated by shut-off valves. Within these isolated systems, a careful examination of the probable causes of overpressure is made [35]. Figures 23.15, 23.16, and 23.17 are suggested guides [33].

Capacities are calculated for conditions of temperature and pressure at actual state of discharge. Final discharge pressure is the set pressure plus overpressure.

It *must be emphasized* that the determination of the anticipated maximum overpressure volume at a specified pressure and temperature is vital to a proper protection of the process system. The safety relief calculations should be

ΣΩΣ AJC TECHNOLOGY	RELIEF DEVICE PHILOSOPHY SHEET EQUIPMENT No.: DATE:		DOCUMENT/ITEM REFERENCE SHEET No.: OF			
CHECKED BY:	MADE BY:					1
						2
DESIGN CODES: VESSELS	EXCHANGERS		LINERS			
						4
OTHER REQUIREMENTS						5
						6
BASIS FOR CALCULATION:						7
						8
SET PRESSURE, psig: :MAX. BACK PRESSURE	(a) BEFORE RELIEVING		(b) WHILE RELIEVING			9
NORMAL CONDITIONS UNDER RELIEF DEVICE:			Calculated			10
STATE: TEMPERATURE, °F: PRESSURE, psig:	POSSIBLE CAUSE?	FLUID RELIEVED	RELIEF RATE, lb/h	ORIFICE AREA, in².		11
						12
HAZARDS CONSIDERED						13
1. Outlets blocked						14
2. Control Valve malfunction						15
3. Machine trip/overspeed/density change						16
4. Exchanger tube rupture						17
5. Power failure/ Voltage dip						18
6. Instrument air failure						19
7. Cooling failure						20
8. Reflux failure						21
9. Abnormal entry of volatile liquid						22
10. Loss of liquid level						23
11. Abnormal chemical reaction						24
12. Boxed in thermal expansion						25
13. External fire						26
14. (specify)						27
15. (specify)						28
16. (specify)						29
SELECTED DESIGN CASE:						30
						31
RELIEVED FLUID: STATE DENSITY / MW:	TEMPERATURE:		Cp./Cv:			32
COMPOSITION:			FLASHING:			33
RELIEF RATE REQUIRED, lb/h: ORIFICE SELECTED:	AREA, in²:		TYPE:			34
ACTUAL CAPACITY, lb/h						35
REMARKS/SKETCH						36
						37–46
Issue No:	1	Date	2	Date	3 Date	47
Made/Revised by						48
Checked by						49
Approved- Process						50
Approved- by						51

Figure 23.14b Relief valve process data sheet.

performed at the actual worst conditions of the system, for example, at the allowable accumulated pressure and its corresponding process temperature. These can be tedious and perhaps time-consuming calculations, but they must not be "glossed" over but developed in a manner that accounts for the seriousness of the effort. *They must be documented carefully and preserved permanently.*

The situation is just as critical, if not more so, for runaway reactions or reaction conditions that are not adequately known. They should be researched or investigated by laboratory testing for possible runaway conditions and then the kinetic and heat/pressure rise calculations should be performed, even if some assumptions must be made to establish a basis. Refer to later paragraphs and the American Institute of Chemical Engineers Design Institute for Emergency Relief (AIChE/DIERS) [13]. At the time of a vessel or pressure/vacuum system failure, the calculations for the effected pressure-relief devices are always reviewed by plant management and the Occupational Safety and

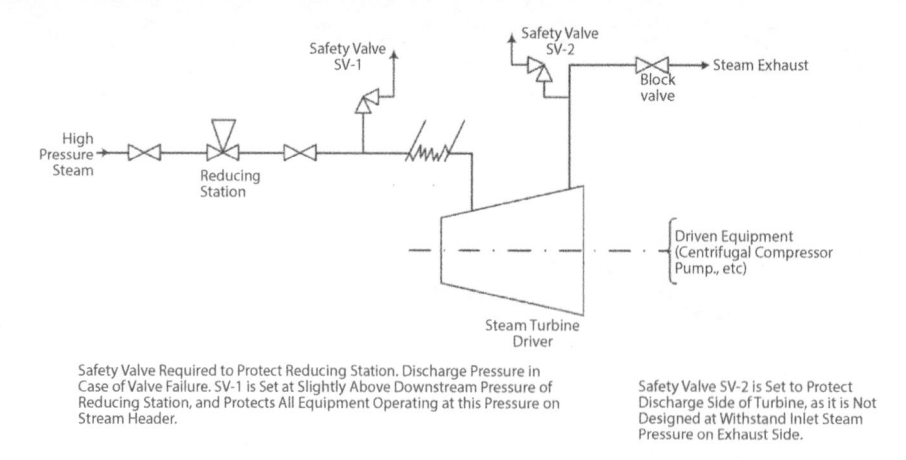

Figure 23.15 Safety valve protecting specific equipment operating.

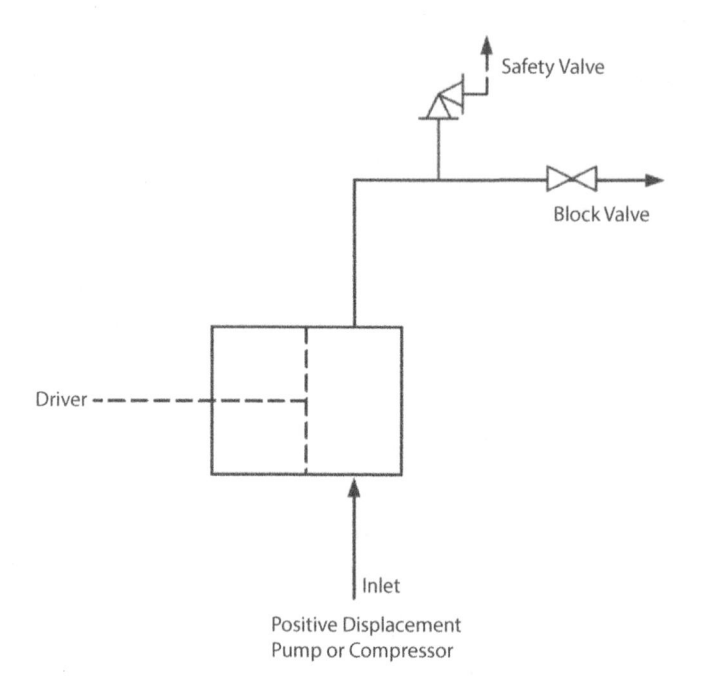

Figure 23.16 Safety valve in positive displacement system.

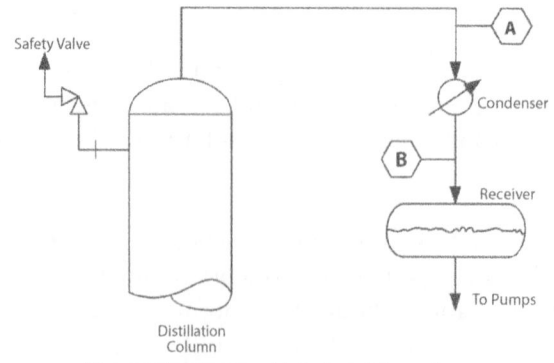

Figure 23.17 System protected by safety valve on a distillation column.

Health Administration (OSHA) inspectors. A few notes on causes of process system failures are noted below, with additional comments in API-521 [5a - d] [9].

Failure of Cooling Water: Assume all cooling media fail; determine relief capacity for the total vapors entering the vessel, including recycle streams (see [36] and [8]).

Reflux Failure: (a) At the top of a distillation column, the capacity is total overhead vapor [8]; (b) when source of heat is in feed stream, the capacity is vapor quantity calculated in immediate feed zone [36]; (c) when reboilers supply heat to system, the capacity is feed plus reboil vapors [36]. Each situation must be examined carefully.

Blocked Outlets on Vessels: (a) For liquid, the capacity is the maximum pump-in rate. (b) For liquid-vapor system, the capacity is total entering vapor plus any generated in the vessel [8].

Blocked Outlets and Inlets: For systems, lines or vessels capable of being filled with liquid and heated by the sun or process heat require thermal relief to accommodate the liquid expansion (assuming vaporization is negligible).

Instrument Failure: Assume instrument control valves freeze or fail in open position (or closed, whichever is worse), determine the capacity for relief based on flows, temperatures, or pressures possible under these circumstances. The judicious selection of instrument failure sequence may eliminate or greatly reduce relief valve requirements.

Equipment Failure: Pumps, tubes in heat exchangers and furnaces, turbine drivers and governor, compressor cylinder valves are examples of equipment that might fail and cause overpressure in the process. If an exchanger tube splits or develops a leak, a high-pressure fluid will enter the low side, overpressuring either the shell or the channels and associated system as the case may be.

Vacuum: (a) Removal of liquid or vapor at greater rate than entering the vessel, the capacity is determined by the volume displaced. (b) Injecting cold liquid into hot (steamed out) vessel, the condensing steam will create vacuum and must be relieved. Capacity is equivalent to vapor condensed.

Other guidelines that provide advice on where relief devices are required are as follows [90]:

- All pressure vessels require overpressure protection.
- All low-pressure storage tanks require pressure and vacuum relief for normal operation (e.g., pumping in and out, tank breathing caused by temperature changes). Tanks must also be protected from any emergency events that could create an abnormally high venting load (e.g., fire exposure, procedural failure during line blowing, etc.).
- Positive-displacement pumps, compressors, and turbines require relief devices on the discharge side for deadhead protection.
- Segments of liquid-filled piping that have a high risk of overpressure due to thermal expansion (e.g., unloading lines) should have relief devices.
- Piping that can be overpressured due to process control failure (e.g., high-pressure steam letdown control into a low-pressure steam header) needs relief devices.
- A vessel jacket is usually considered a distinctly separate pressure vessel and requires its own overpressure protection.

In-breathing and Out-breathing Pump In and Out: See section on Pressure-Vacuum Relief for Low Pressure Storage Tanks.

Electrical power failure:

- May be local or refinery – wide event.
- May be total or partial.

- Increases superimposed back pressure on a relief device.
- May cascade into instrument air, steam, and other utility failures.
- Often sets the relief sizing and flare size.
- May be necessary to look hard at timing of different reliefs.
- May require emergency depressuring of some units (e.g., hydrocrackers), which would impose a large back pressure.

Other failures to consider:

- Mal-operation
- Temperature runaway – depressuring
- Mechanical failure of equipment (e.g., reboiler steam valve fails to open)
- Abnormal heat input.
- Split exchanger tube.
- Automatic control failure.
- Instrument air failure.
- Valve left open
- Steam-out
- Thermal (blocked-in line or equipment)
- Any other type of failure that can create a relief condition.

Installation

Never place a block valve on the discharge side of a pressure relief device of any kind, except see [1] Par. U-135 (e).

Never place a block valve on the inlet side of a pressure relief device of any kind, unless it conforms to the code practice for rupture disks or locking devices. See [1] Par. UG-135 (e) and Appendix M, ASME code.

Note that the intent of the ASME Code is to ensure that under those circumstances where a pressure relieving device can be isolated by a block valve from its pressure, or its discharge, a responsible individual locks and unlocks the block valve to the safe open position and that this individual remains at the block valve the entire time that the block valve is closed.

Safety (relief, or safety-relief) valves are used for set pressures from 10 to 10,000 psig (0.69–690 bar) and even higher. At the low pressures, the sensitivity to relieving pressure is not always as good as is required for some processes, and for this reason, most valve installations start at 15 to 20 psig (1.03–1.38 bar).

Figures 23.10 and 23.18 illustrate a few typical safety valve installations. Care must be shown in designing any manifold discharge headers collecting the vents from several valves. Sharp bends are to be avoided. Often two or more collection systems are used in order to avoid discharging a high-pressure valve into the same header with a low-pressure valve. The simultaneous discharge of both valves might create too great a backpressure on the low pressure valve, unless adequate arrangement has been made in the valve design and selection. The balanced safety relief valve can overcome most of the problems of this system.

Whenever possible, the individual installation of valves is preferred, and these should be connected directly to the vessel or pipeline [1, 37]. If a block-type valve is considered necessary for a single valve installation, it must be of the full open type and locked open with the key in responsible hands, as stated earlier.

Dual installations are frequently made in continuous processes to allow switching from one valve to another without shutdown of the pressure system. A special three-way plug of full open type is installed directly on the vessel, and the safety valves are attached to it with short piping (Figure 23.18). The three-way valve ensures that one side of the safety valve pair is always connected to the vessel, as this pattern valve does not have a blind point during switching

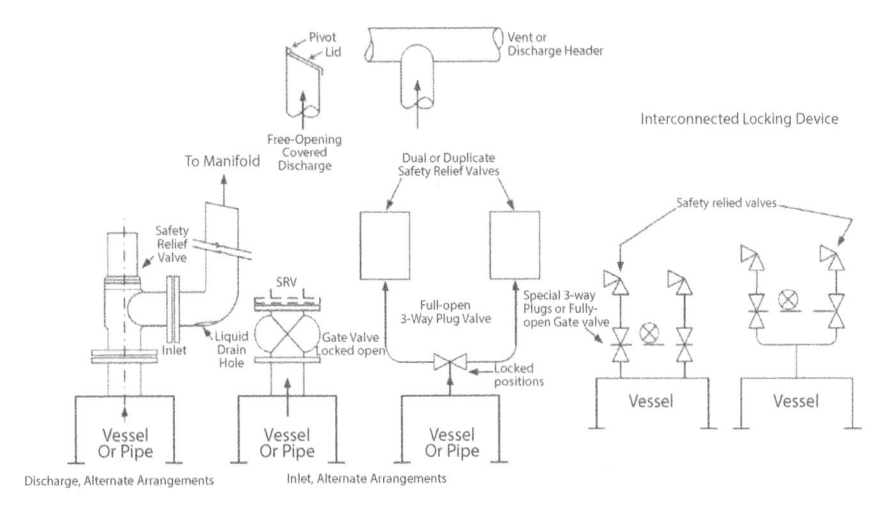

Figure 23.18 Safety relief valve installations.

(Figure 23.19). One of the important justifications for this dual arrangement is that safety-relief valves may leak on reseating after discharging. This leak may be caused by a solid particle lodged on the seat. This valve can be removed for repair and cleaning after the process has been switched to the second valve. Each valve must be capable of relieving the full process requirements. Multiple valves may also be individually installed separately on a vessel.

Figure 23.19 illustrates a newer approach at simplifying the dual safety relief valve installation, ASME Sect. VIII, Div. 1, UG-135(b) [1] and API RP-520, Part II Conformance [5]. Note that the safety relief valves (SRV) are mounted on top of each of one dual vertical connection and are bubble tight. Also see cross-section view. The flow C_v values for each size device are available from the manufacturer.

Relief valves should be installed in a vertical position, with the spring acting towards the ground. This is because they are designed to be installed in this orientation, and if they are horizontal, they may not open and close as designed. The majority of relief valves are installed in the correct position, but there are examples where space constraints or line Δps have led engineers to install them horizontally. Figure 23.19a shows one such valve that is installed horizontally on an air compressor. Vibrations caused the spring to erode the metalwork to such an extent that it protrudes through the bonnet, rendering it useless [87].

The Anderson Greenwood & Co. (AGCO) Safety Selector Valve body houses a uniquely designed switching mechanism. The internal rotor smoothly diverts flow to either safety relief valve. Conventional direct spring-operated valves or pilot operated valves may be used. The inactive valve is totally isolated by external adjustment. To begin the switchover, the retraction bushing is rotated to its stop. This separates the isolation disk from the standby valve channel and temporarily "floats" it in the main valve cavity. The index shaft is then rotated 180° to the alternate channel. The retraction bushing is then returned to its original position, securely seating the isolation disk beneath the valve taken out of service. A red pointer indicates which valve is in service, and double padlocking provisions allow the safety selector valve to be locked in either safety relief valve position. The padlocks or car seals can only be installed with the internals in the proper position. No special tools are necessary for switching.

The alternate concept that has been in use for many years is to fabricate or purchase a Tee connection upon which the two-safety relief valves can be mounted on top of their full-port plug or gate valve with required locking lugs.

Rupture disks are often used in conjunction with safety valves as shown in Figures 23.10, 23.11, 23.12, and 23.18.

Inlet piping is held to a minimum, with the safety device preferably mounted directly on the equipment and with the total system pressure drop loss to pressure relief valve inlet not exceeding 3% of the set pressure in psig, of maximum

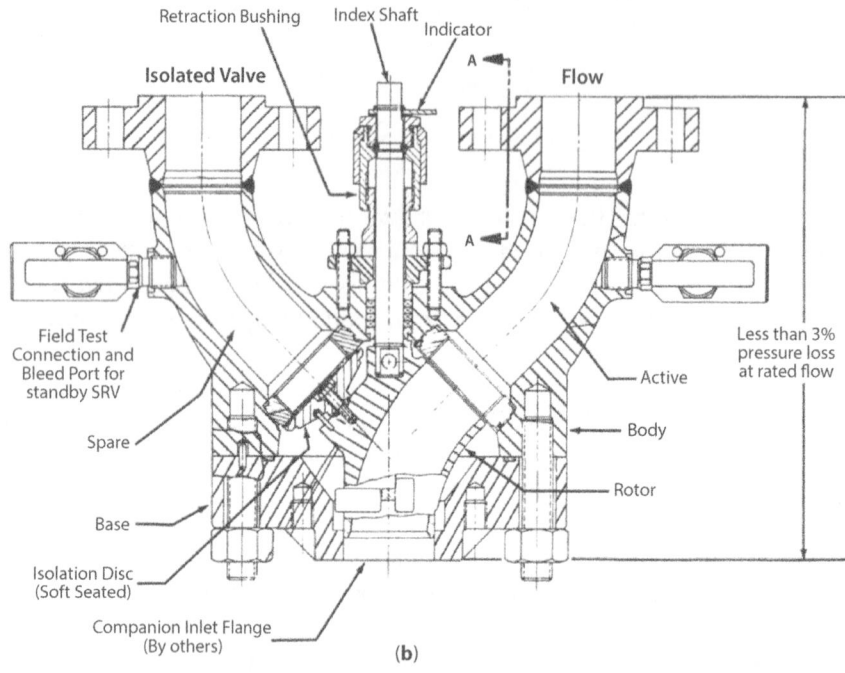

36 PETROLEUM REFINING DESIGN AND APPLICATIONS HANDBOOK VOLUME 5

Figure 23.19 Safety selector valve for dual relief valve installation with switching. (By permission from Anderson, Greenwood and Co.® AGGO.)

Figure 23.19a Relief valve installed horizontally. The markings on the bonnet result from the spring working its way out. (Source: Chris Flower and Adam Wills, The Chemical Engineer, U.K., pp. 24-31, June 2017.)

relief flowing conditions [8]. To conform to the code (see ASME code, Sect. VIII, Div. 1-UG-127 [1]), avoid high inlet pressure drop and possible valve chatter:

1. Never make pipe connection smaller than valve or disk inlet.
2. Keep friction pressure drop very low, not over 1-2% of allowable pressure for capacity relief [1, 5, 8, 33, 37].
3. Velocity head loss should be low, not over 2% of allowable pressure for capacity relief [33].

23.12 Piping Design

Pressure Drops

There are Δp limitations around relief valves, typically 3% of the set pressure on the inlet and 10% of the set pressure on the outlet. The main reason for this is stability and reliability of the valve in operation. A relief valve works by a simple force balance across the seat of the relief valve. When the valve is closed, the "downwards" or closing force is provided by the spring and any pressure in the body of the valve. The "upwards" or opening force is provided by the pressure of the process fluid acting over the nozzle area (i.e., force = pressure x area). When the valve opens, two effects can alter the force balance and cause the valve to reseat.

The first is the loss of inlet pressure; the fluid is now flowing and frictional losses will reduce the pressure so that the opening force is decreased. The second is increased pressure in the body of the valve, thus increasing the closing force. Relief valves are designed to cope with these effects: the seat area is larger than the nozzle area so that the flowing fluid acts over a larger area, providing more upwards force when opened compared to when closed. The designer should be cautious not to use 3% and 10% for the Δp acceptance criteria. These are common and typical numbers as there are relief valves on the market that are different. For example, some valves will only accept 5% backpressure [87].

Line Sizing

As well as Δp criteria, there are also design rules that the engineer is required to comply with. The nominal pipe diameter of the inlet piping should be no less than the valve inlet size (unlike that shown in Figure 23.20g). This includes all elements of the pipework, e.g., if an isolation valve is installed, it should be full bore. The criteria become more noticeable, if other piping components are used that are not full bore. For example, where there is a bursting disc in the line with a vacuum support, here the free flow area through these devices should be the same or larger than the free flow area into the inlet of the valve. There are two reasons: firstly meeting the 3% inlet Δp criteria and secondly to ensure that the valve is determining the capacity of the relief system rather than another piping component [87].

For the discharge line, API does not provide a similar line size requirement. Appendix M of ASME VIII—Rules for Construction of Pressure Vessels—does state that the discharge pipe shall be at least of the same size as the valve outlet, which matches European codes that explicitly state this requirement. Generally, it is very difficult to get a backpressure of less than 10% with pipework smaller than the valve outlet. When engineers are performing preliminary designs of a relief system, it is recommended going one size up on the pipework at either side of a relief valve in order to meet the Δp criteria for both the present and potential future modifications.

Discharge piping must be sized for low-pressure drop at the maximum flow not only from anyone valve but also for the combined flow possibilities in the discharge collection manifold all the way to the vent release point, whether it be a flare, incinerator, absorber, or other arrangement [9] (Figures 23.20a–h).

Conventional safety relief valves, as usually installed, produce unsatisfactory performance when variable backpressure exists [5, 8] (see Figure 23.6a). Additionally, the same variable backpressure forces affect the set pressure release. At low backpressures, the valve flow falls rapidly as compared with the flow for a theoretical nozzle (see Figures 23.19 and 23.20 [5a]).

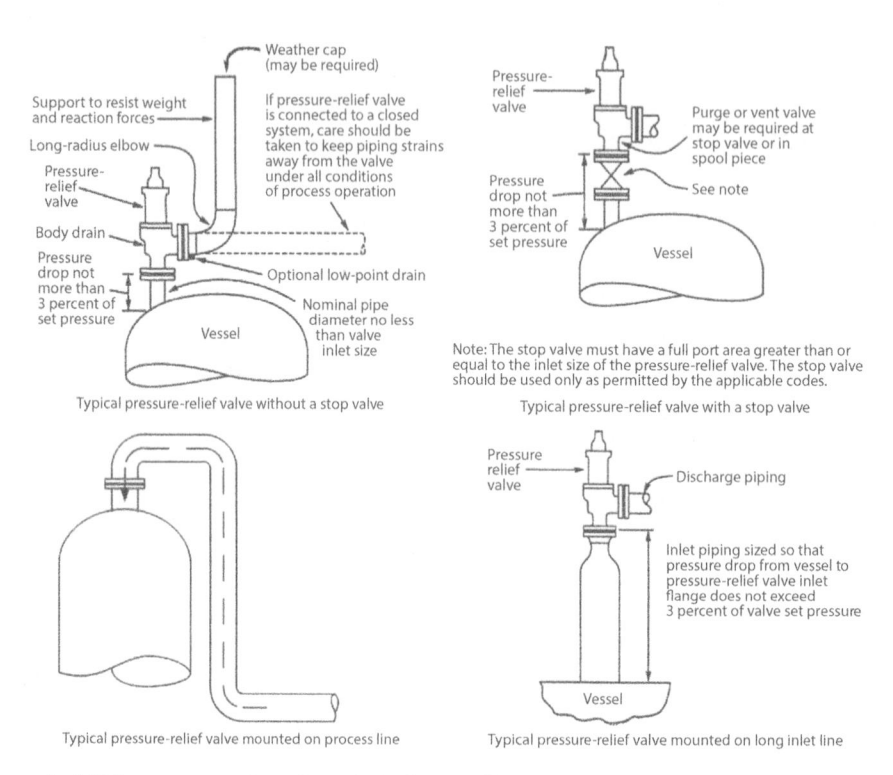

Figure 23.20a Recommended API – 520 piping for safety relief valve installations. (Reprinted by permission from American Petroleum Institute, Sizing, Selection and Installation of Pressure Relieving Devices in Refineries, Part II – Installations, API RP – 520, 3rd. ed., Nov. 1988.)

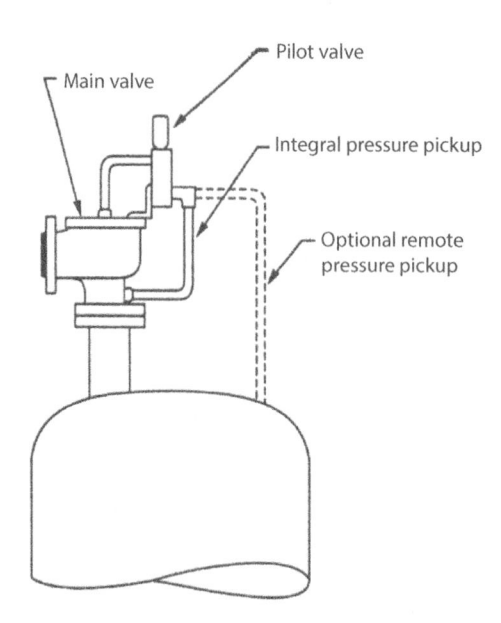

Figure 23.20b Typical pilot – operated pressure relief valve installation.

Pressure Relieving Devices and Emergency Relief System Design 39

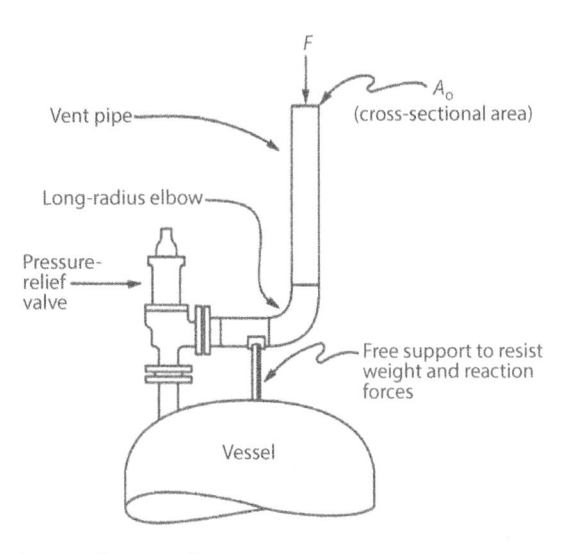

Figure 23.20c Typical pressure – relief valve installation with vent pipe.

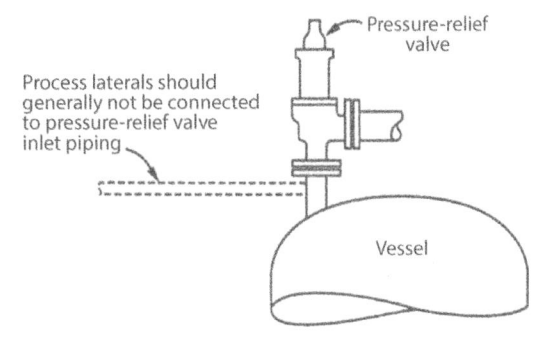

Figure 23.20d Typical installation avoiding process laterals connected to pressure-relief valve inlet piping.

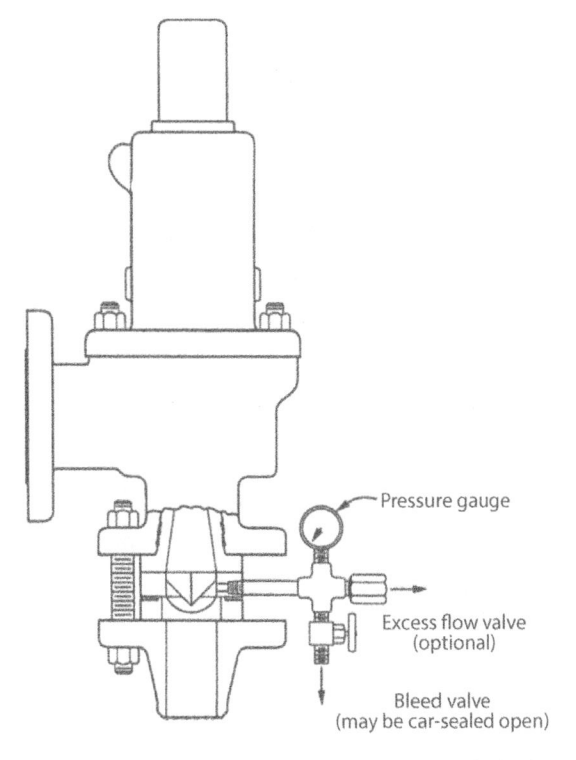

Figure 23.20e Typical rupture disk assembly installed in combination with a pressure – relief valve.

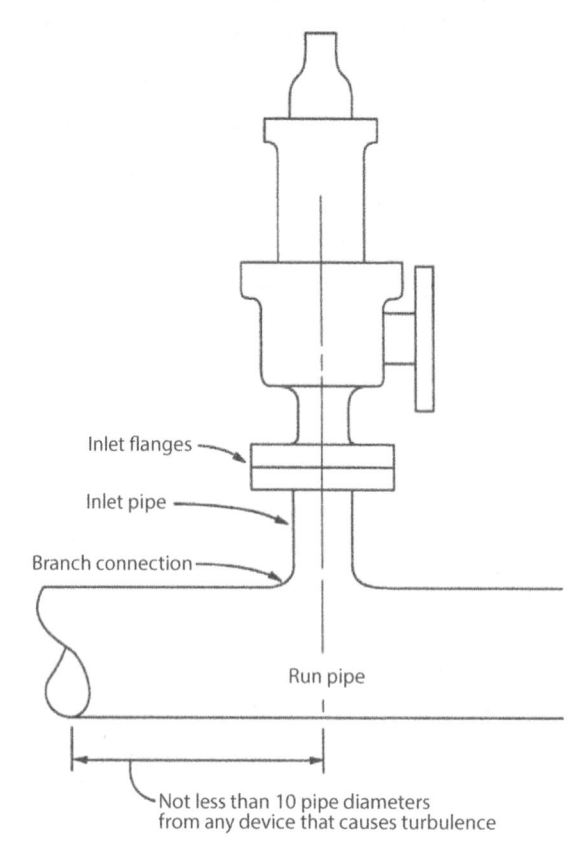

Figure 23.20f Typical installation avoiding excessive turbulence at pressure – relief valve inlet.

Figure 23.20g Inlet piping expanding into the relief valve. Incorrect installation may lead to valves chattering. (Source: Chris Flower and Adam Wills, The Chemical Engineer, U.K., pp. 24-31, June 2017.)

Figure 23.20h An unrestrained metal cover placed over an outlet pipe could become a missile. (Source: Chris Flower and Adam Wills, The Chemical Engineer, U.K., pp. 24-31, June 2017).

For conventional valves, pressure drop or variations in backpressure should not exceed 10% of set pressure. Because most process safety valves are sized for critical pressure conditions, the piping must accommodate the capacity required for valve relief and not have the pressure at the end of vent or manifold exceed the critical pressure. Designing for pressure 30-40% of critical with balanced valves yields smaller pipes yet allows proper functioning of the valve. The discharge line size must not be smaller than the valve discharge. Check the manufacturer for valve performance under particular conditions, especially with balanced valves that can handle 70-80% of set pressure as backpressure.

For non-critical flow, the maximum backpressure must be set and pressure drop calculated by the usual friction equations.

When process conditions permit, the low-pressure range is handled by bursting disks, which will relieve down to 2 psig. These disks are also used up to 100,000 psig and above. The rupture pressures and manufacturing ranges of metal disks are given in Tables 23.2 and 23.3. For non-metallic materials such as graphite, bursting pressures are available from the manufacturers. From these manufacturing tolerances, it can be seen that the relation of disk bursting pressure to the required relieving pressure must be carefully considered. Manufacturing practice is to furnish a disk, which will burst within a range of pressures and tolerances, and whose rated pressure is the result of bursting tests of representative sample disks, which burst within the range specified. The engineer should specify only ASME code certified disks. It is not possible to obtain a disk for the usual process application set to burst at a given pressure, as is the relieving pressure of a safety valve. An increase in temperature above the disk rating temperature 72°F (22°C) decreases the bursting pressure 70-90% depending upon the metal and temperature (Tables 23.4a and b).

The minimum rupture pressure of disks of various metals and combinations varies so widely that individual manufacturer must be consulted.

For the usual installation, the rupture disk is installed as a single item between special flanges, which hold the edges securely and prevent pulling and leakage. If the system is subjected to vacuum or pressure surges, a vacuum support must be added to prevent collapse of the sealing disk. The flanges that hold the disk may be slip-on, weld neck, and so on. Disks to fit screwed and union-type connections are also available (see Figures 23.8 and 23.9).

The service life of a rupture disk is difficult to predict, since corrosion, cycling pressures, temperature, and other process conditions can all affect the useful life and cause premature failure. A graphite-type disk is shown in Figure 23.9. In some processes, it is safer to replace disks on a schedule after the life factor has been established, as a planned shutdown is certainly less costly than an emergency one.

Table 23.2 Typical Prebulged Solid Metal Disk Manufacturing Ranges and Tolerances at 72°F.

Specified burst pressure rating, psig	Manufacturing range (%)		Rated (stamped) burst tolerance %
	Under	Over	
2–5	–40	+40	± 25
6–8	–40	+40	± 20
9–12	–30	+30	± 15
13–14	–10	+20	± 10
15–19	–10	+20	± 5
20–50	–4	+14	± 5
51–100	–4	+10	± 5
101–500	–4	+7	± 5
501–up	–3	+6	± 5

Note:

1. Special reduced manufacturing ranges can be obtained for the STD prebulged metal disk, ¼, ½, and ¾ ranges are available upon request. Please consult your representative or the factory for additional information.
2. Burst tolerances are the maximum expected variation from the disk's rated (stamped) burst pressure.
3. Standard-type rupture disk comply with ASME code requirements.

Manufacturing Range
The manufacturing range is defined as the allowable pressure range within which a rupture disk is rated. It is based upon the customer specified burst pressure. The manufacturing ranges for Continental's standard rupture disk are outlined in this table.

Burst Tolerance
After the disk has been manufactured and tested, it is stamped with the rated burst pressure. The rated (stamped) burst pressure is established by bursting a minimum of two disks and averaging the pressures at which the disks burst. This average is the rated (stamped) burst pressure of the disk. Standard rupture disks above 15 psig at 72°F are provided with a burst tolerance of ±5% of the rated (stamped) burst pressure. This is in accordance with the ASME code. Burst tolerances for disks below 15 psig at 72°F are outlined in this table. Burst tolerance applies only to the rated (stamped) burst pressure of the disk. Burst certificates are provided with each disk lot.
By permission, Continental Disk Corporation, Catalog STD-1184.

Rupture disks are often placed below a safety valve to prevent corrosive, tarring, or other material from entering the valve nozzle. Only disks that do not disintegrate when they burst (Figures 23.10, 23.11, 23.12, and 23.18) can be used below a safety valve, as foreign pieces that enter the valve might render it useless. This is acceptable to certain code applications [1]. These disks are also used to provide secondary relief when in parallel with safety valves set at lower pressures. They can also be installed on the discharge of a safety valve to prevent loss of hazardous vapors, but caution should be used in any serious situation.

Table 23.3 Typical Metal Disk (Single) Bursting Pressures at 72°F using Different Metals.

Size (in.)	Disk minimum burst pressure (psig) (without liners)					
	Alum	Silver	Nickel	Monel	Inconel	316SS
¼	160	450	600	700	1120	1550
½	65	220	300	350	560	760
1	29	120	150	180	250	420
1½	22	80	100	116	160	275
2	13	48	60	70	110	150
3	10	35	45	50	80	117
4	7	26	35	40	70	90
6	5	20	25	30	47	62
8	4	15	20	23	34	51
10	4	-	16	17	30	43
12	3	-	13	15	25	36
14	3	-	11	13	21	31
16	3	-	10	12	19	28
18	3	-	9	11	17	24
20	3	-	8	9	16	22
24	3	-	-	-	-	-
30	-	-	-	-	-	-
36	-	-	-	-	-	-

(-) = Consult factory

*Special designs of some manufacturers may exceed 150,000 psi for small sizes. The pressures listed are generally typical but certainly not the only ones available for the size shown.

Note:

1. Maximum burst pressure depends upon disk size and application temperature. Pressures to 80,000* psig are available.
2. Other materials and sizes are available upon request.
3. Other liner materials are available upon request. Minimum burst pressures will change with change in liner material.
4. For larger sizes or sizes not shown, consult your representative, or the factory. Courtesy, Continental Disk Corp., Bul. 1184, p. 4-5.

Table 23.4a Typical Recommended Maximum Temperatures for Metals Used in Disks.

Metals	Maximum temperature, (°F)
Aluminum	250 – 260
Silver	250 – 260
Nickel	750 – 800
Monel	800
Inconel	900 – 1000
316 Stainless Steel	900

Source: Various manufacturers' technical catalogs.

Table 23.4b Typical Recommended Maximum Temperatures for Linings and Coatings with Metals Used with Disks.

Metal	Maximum Temperature (°F)
Teflon®FEP Plastic	400
Polyvinylchloride	150 – 180
Lead	250

Source: Various manufacturers' technical catalogs.

23.13 Selection Features: Safety, Safety-Relief Valves, and Rupture Disks

Referring to the description and definitions in the introduction for this chapter, it is important to recognize that in order to accomplish the required pressure relief, the proper selection and application of device type is essential.

Safety Valve: This is normally used for steam service, but suitable for gases or vapors. When used in steam generation and process steam service, the valves conform to the *ASME Power Boiler Code* as well as the *ASME Pressure Vessel Code*, Section VIII, and are tested at capacity by the National Board of Boiler and Pressure Vessel Inspectors. This type of valve characteristically "pops" full open and remains open as long as the overpressure exists.

Relief Valve: This is normally selected for liquid relief service such as hydraulic systems, fire and liquid pumps, marine services, liquefied gases, and other total liquid applications. The valve characteristically opens on overpressure to relieve its rated capacity and then reseats.

Safety-Relief Valve: This is normally selected for vapors and gases as may be found in all types of industrial processes. Characteristically, this valve will open only enough to allow the pressure to drop below the set pressure, and then it will reseat until additional overpressure develops. If the pressure persists or increases, then the valve will remain open or increase its opening up to the maximum design, but as the pressure falls, the valve follows by closing down until it is fully reseated. However, as in any installation of any "safety" type valve, the valve may not reseat completely gas tight. In such cases, it may be necessary to switch to a stand-by valve and remove the leaking valve for repair (see Figures 23.7a and b).

Special Valves: Because of the difficult and special sealing requirements of some fluids such as chlorine and Dowtherm, special valves have been developed to handle the requirements.

Vacuum Relief and Combined Pressure-Vacuum Relief for Low-Pressure Conditions: This is normally used for low pressures such as 1 ounce water to 1.5 psig above atmospheric by special spring or dead weight loading, and for vacuum protection such as 0.5 psi below atmospheric. Usually these conditions are encountered in large process- that is, processing of crude oil, ammonia, storage tanks and so on.

Rupture Disks: This is used for low- as well as high-pressure protection of vessels and pipelines where sudden and total release of overpressure is required. Once the disk has ruptured, the process system is exposed to the environment of the backpressure of the discharge system, whether atmospheric or other. The process system is depressurized and the disk must be replaced before the process can be restarted. Typically, the types of disks available are:

1. Solid Metal Rupture Disk (Figure 23.8). This is the original type of rupture device, available in various metals and non-metals. It should normally not be used for operating pressures greater than 70% of rupture pressure in a non-corrosive environment. The metal disks are designed with a domed or hemispherical shape, with pressure on the concave side. As the pressure internally increases, the metal wall thins as the metal stretches to achieve a smaller radius of curvature. After the wall has thinned sufficiently, it will burst to relieve the pressure and tension loading on the metal. The accuracy of metal disks is ±5%, except for the Reverse Buckling Disk Assembly, which is ±2%. The usual recommended maximum operating temperatures for metal rupture disks is given in Tables 23.4a and b (also see Figures 23.8g and n).
   a) **Solid metal disk with vacuum support**: When vacuum can occur internally in the system, or when external pressure on the convex side of the disk can be greater than the pressure on the concave side of the disk, a vacuum support is necessary to prevent reversal of the disk (Figure 23.8d).
   b) **Solid metal disk with rating near minimum for size**: With a rating pressure near the minimum available for the size and material of construction, a special thin disk is attached to the lower and possibly the upper sides of the rupture disk to ensure freedom from deformity caused by the condition of the disk holding surfaces (Figure 23.8e). There are several versions of what to include under such conditions; therefore, it is advisable to clearly explain the installation conditions and application to the manufacturer.
   c) **Composite rupture disk**: This type consists of a metal disk (not necessarily solid, it may have slots) protected by an inner and/or an outer membrane seal (Figure 23.8e). There are several possible arrangements, including vacuum support, as for the styles of paragraph (b) above. This general class has the same use-rating limitation as for the solid disk.
   d) **Reverse acting or buckling disk assembly**: This design allows the disk to be operated in a system at up to 90% of its rated burst pressure. The pressure is operating on the convex side of the disk, and when bursting pressure is reached, the disk being in compression reverses with a snap action at which time the four knife edges (Figure 23.8g) cut the metal and it clearly folds back without fragmentation. There is another version of the same concept of reverse buckling, but it uses a pre-scored disk and thereby omits the knife blades. These types of disks do not need vacuum supports, unless there is unusually high differential pressure across the disk.
2. Graphite Rupture Disk (Figure 23.9). There are special designs of disks and disk assemblies for specific applications, and the manufacturer should be consulted for his recommendation. Disks are available for pressure service, pressure-vacuum applications, high-temperature conditions, and close tolerance bursting conditions. The bursting accuracy of most designs is ±5% for rated pressures above 15 psig and 0.75% for rated pressures 14 psig (0.96 barg) and below. It should be noted that these ratings are not affected by temperature up to 300°F (149°C).

A new concept in graphite disks includes addition of a fluorocarbon film barrier between the process and the disk, and is termed a "duplex disk." These disks are suitable for temperatures to 392°F (200°C), with accuracies as just mentioned.

Graphite disks are normally used in corrosive services and/or high-temperature situations where metal wall thickness and corrosion rates make the metal units impractical because of unpredictable life cycles. The disks are available down to 1 psig ±0.75 psig and are not affected by fatigue cracking. An interesting feature is the use of standard ASA (ANSI) flanges rather than special flanges (see Figure 23.9c).

It is important to recognize here also that once the disk bursts, the system is depressed, and there will be fragments of graphite blown out with the venting system. Special discharge designs are often used to prevent plugging of discharge pipe and fragments from being sprayed into the surrounding environment.

23.14 Calculations of Relieving Areas: Safety and Relief Valves

References to the ASME Code [1] and the API Code [5, 8] are recommended in order that the design engineer may be thoroughly aware of the many details and special situations that must be recognized in the final sizing and selection of a pressure relieving device. All details of these codes cannot be repeated here; however, the usually important requirements are included for the typical chemical and petrochemical application for the guidance of the engineer.

Before performing sizing/design of relief valves calculations, a thorough examination of the possible causes and flow conditions of temperature and pressure should be determined. From this list, select the most probable and perhaps the worst-case possibility and establish it as a design basis (Figures 23.14a and b [38]).

When the possibilities of internal explosion or a runaway chemical reaction exist, or are even suspected, they must also be rigorously examined and calculations performed to establish the magnitude of the flow, pressure, and temperature problems. Select the worst condition and plan to provide for its proper release to prevent rupture of equipment. This latter situation can only be handled by application of rupture disks and/or remote sensing and predetermined rupture of the disks (see Figures 23.5a, 23.8k, and 23.8l) or remote sensing and application of quenching of the reaction/developing explosive condition by automatic process action and/or commercial application of quenching medium.

23.15 Standard Pressure Relief Valves Relief Area Discharge Openings

The "orifice" area of these devices (see table below) is at the outlet end of the SRV nozzle through which the discharging vapor/gases/liquids must pass. These values are identified in industry as: (valve body inlet size in.) × (orifice letter) × (valve body outlet size, in.). For example, a valve would be designated 3E4.

The standard orifice area designations are (also refer to mechanical illustrations of valves, previously shown in this chapter):

Orifice letter	D	E	F	G	H	J
Area, in^2.	0.11	0.196	0.307	0.503	0.785	1.287
Area, cm^2	0.71	1.27	1.98	3.25	5.06	8.30
Orifice letter	K	L	M	N	P	Q
Area, in^2.	1.838	2.853	3.600	4.340	6.380	11.05
Area, cm^2	11.85	18.4	23.2	28.0	41.20	71.30

Orifice letter	R	T	V*	W	W2*	X*
Area, in².	16.0	26.0	42.19	57.26	93.6	101.8
Area, cm²	103.20	167.70	272.19	369.42	603.87	656.77
Orifice letter	Y*	Z*	Z_2^*	AA	BB	BB2
Area, in².	128.8	159.0				
Area, cm²	830.97	1025.8				

*Note: These letters and orifice areas are not consistent for these large orifices between various manufacturers. Some sizes go to 185 in², which is a very large valve. When two valves are shown, they represent two different published values by manufacturers.

23.16 Sizing Safety Relief Type Devices for Required Flow Area at Time of Relief

Before initiating any calculations, it is necessary to establish the general category of the pressure relief valve being considered. This section covers conventional and balanced spring-loaded types.

Given the rate of fluid flow to be relieved, the usual procedure is to first calculate the minimum area required in the valve orifice for the conditions contained in one of the following equations. In the case of steam, air, or water, the selection of an orifice may be made directly from the capacity tables if so desired. In either case, the second step is to select the specific type of valve that meets the pressure and temperature requirements.

General equations are given first to identify the basic terms that correlate with ASME Pressure Vessel Code, Section VIII.

It is recommended that computations of relieving loads avoid cascading of safety factors or multiple contingencies beyond the reasonable flow required to protect the pressure vessel.

23.17 Effects of Two-Phase Vapor-Liquid Mixture on Relief Valve Capacity

Many process systems where conditions for safety relief valve discharge are not single phase of all liquid (through the valve) or all vapor, but a mixture either inside the "containing" vessel or quite often as the fluid passes through the valve orifice; the liquid flashes to partial vapor, or the flashing starts just ahead of the orifice. Here, a mixture attempts to pass through the orifice, and the size must be sufficient or a restriction will exist and pressure will build up in the vessel due to inadequate relief. This problem was of considerable concern to the Design Institute for Emergency Relief of the AIChE during their studies [13]. As a result, considerable research was performed leading to design techniques to handle this problem. A review of two-phase relief shall be presented; meanwhile the designer is referred to the references in the bibliography of this chapter. Also see Leung [39] for detailed procedure and additional references.

23.18 Sizing for Gases or Vapors or Liquids for Conventional Valves with Constant Backpressure Only

This type of valve may be used when the variations in backpressure on the valve discharge connection do not exceed 10% of the valve set pressure, and provided this backpressure variation does not adversely affect the set pressure.

Procedure

1. For a new installation, establish pressure vessel normal maximum operating pressure, and temperature, and then the safe increment above this for vessel design conditions and determine the MAWP of the new vessel. (Have qualified fabricator or designer establish this.)
2. Establish the maximum *set* pressure for the pressure relieving valves as the MAWP, or lower, but *never* higher.
3. Establish actual *relieving* pressure (and corresponding temperature) from Figure 23.7a (at 110% of set pressure for non-fire and non-explosive conditions). Explosive conditions may require total separate evaluation of the set pressure (*never above the MAWP*), which should be lower or staged; or, most likely, will not be satisfied by a standard SRV due to the extremely rapid response needed. The capacity for flow through the valve is established by these conditions.
4. For *existing vessel* and re-evaluation of pressure relieving requirements, start with the known MAWP for the vessel, recorded on the vessel drawings and on its ASME certification papers. Then follow steps 2 and 3 above.

Establish Critical Flow for Gases and Vapors

Critical or sonic flow will usually exist for most (compressible) gases or vapors discharging through the nozzle orifice of a pressure relieving valve. The rate of discharge of a gas from a nozzle will increase for a decrease in the absolute pressure ratio P_2/P_1 (exit/inlet) until the linear velocity in the throat of the nozzle reaches the speed of sound in the gas at that location. Thus, the critical or sonic velocity or critical pressures are those conditions that exist when the gas velocity reaches the speed of sound. At that condition, the actual pressure in the throat will not fall below P_1/r_c even if a much lower pressure exists downstream [16]. The maximum velocity at the outlet end (or restriction) in a pipe or nozzle is sonic or critical velocity. This is expressed as [40]:

English Engineering units

$$v_s = \sqrt{kgRT} = \sqrt{kg(144)P'\overline{V}} = 68.1\sqrt{kP'\overline{V}}$$
$$= 223\left(\frac{kT}{M}\right)^{0.5} \tag{23.5}$$

In metric units

$$v_s = \sqrt{\gamma RT} = \sqrt{\gamma P'\overline{V}} = 316.2\sqrt{\gamma p'\overline{V}}$$
$$= 91.2\left(\frac{\gamma T}{M}\right)^{0.5} \tag{23.5a}$$

where

k = ratio of specific heats at constant pressure/constant volume, C_p/C_v (see Table 23.5).
v_s = sonic velocity of gas, ft/s (m/s)
g = acceleration of gravity, 32.2 ft/s²
R = individual gas constant = (MR/M) = 1545/M

In metric units

$$R = R_o/M \quad J/kg\ K$$

Table 23.5 Properties of Gases and Vapors.

Gases and vapors	Hydrocarbons reference symbols	Chemical formula	Mol. wt.	R = 1545/mol. wt.	Critical conditions Pressure (Psia)	Critical conditions Temperature (°R)	Boiling point (F) @ 14.7 Psia	Specific volume Cu ft/lb @ 14.7 Psia & 60F (Z factor accounted For)	Latent heat of vaporization (Btu/lb @ 14.7 Psia)	Specific heat constant pressure (C_p @ 60F)	Specific heat constant volume (C_v @ 60F)	Specific heat ratio K = C_p/C_v
1. Acetylene	C_2	C_2H_2	26.04	59.5	905	557	-118.7	14.37	256.0	0.397	0.320	1.24
2. Air		N_2+O_2	28.29	53.3	547	239	-317.7	13.09	91.8	0.240	0.171	1.40
3. Ammonia		NH_3	17.03	90.8	1657	731	-28.1	22.10	590.0	0.523	0.399	1.31
4. Argon		Ar	39.94	38.7	705	272	-30.3	9.50	71.7	0.125	0.075	1.66
5. Benzene		C_6H_6	78.11	19.8	714	1013	176.2	*	169.3	0.240	0.215	1.12
6. Iso-Butane	iC_4	C_4H_{10}	58.12	26.6	529	735	10.9	6.26	157.8	0.387	0.352	1.10
7. n-Butane	nC_4	C_4H_8	58.12	26.6	551	766	31.1	6.25	165.9	0.397	0.363	1.09
8. Iso-Butylene	iC_4	C_4H_{10}	56.10	27.5	580	753	19.6	6.54	169.5	0.368	0.333	1.10
9. Butylene	nC_4	C_4H_8	56.10	27.5	583	756	20.7	6.54	167.9	0.327	0.292	1.11
10. Carbon Dioxide		CO_2	44.01	35.1	1073	548	-109.3	8.53	248.8[1]	0.199	0.153	1.30
11. Carbon Monoxide		CO	28.01	55.1	514	242	-313.6	13.55	91.0	0.248	0.177	1.40
12. Carbureted Water Gas (3)		-	19.48	79.5	454	235	-	19.60	-	0.281	0.208	1.35
13. Chlorine		Cl_2	70.91	21.8	1119	751	-29.6	5.25	123.8	0.115	0.084	1.36
14. Coke Oven Gas (3)		-	11.16	138.5	407	197	-	34.10	-	0.679	0.514	1.32
15. n-Decane	nC_{10}	$C_{10}H_{22}$	142.28	10.9	312	1115	345.2	*	120.0	0.401	0.387	1.03
16. Ethane	C_2	C_2H_6	30.07	51.5	708	550	-127.5	12.52	210.7	0.410	0.343	1.19
17. Ethyl Alcohol		C_2H_5OH	46.07	33.5	927	930	172.9	*	368.0	0.370	0.328	1.13
18. Ethyl Chloride		C_2H_4Cl	64.52	23.9	764	829	54.4	5.59	168.5	0.274	0.230	1.19

(Continued)

Table 23.5 Properties of Gases and Vapors. (Continued)

Gases and vapors	Hydrocarbons reference symbols	Chemical formula	Mol. wt.	R = 1545/mol. wt.	Critical conditions Pressure (Psia)	Critical conditions Temperature (°R)	Boiling point (F) @ 14.7 Psia	Specific volume Cu ft/lb @ 14.7 Psia & 60F (Z factor accounted For)	Latent heat of vaporization (Btu/lb @ 14.7 Psia)	Specific heat constant pressure (C_p @ 60F)	Specific heat constant volume (C_v @ 60F)	Specific heat ratio K = C_p/C_v
19. Ethylene	C_2	C_2H_4	28.05	55.1	749	510	-154.7	13.40	207.6	0.361	0.291	1.24
20. Flue Gas (2)		-	30.00	51.5	563	265	-	12.63	-	0.240	0.174	1.38
21. Helium		He	4.00	386.0	33	9	-450.0	94.91	9.9	1.24	0.748	1.66
22. n-Heptane	nC_7	C_7H_{16}	100.20	15.4	397	973	209.2	*	136.2	0.399	0.379	1.05
23. n-Hexane	nC_6	C_6H_{14}	86.17	17.9	434	915	155.7	*	144.8	0.398	0.375	1.06
24. Hydrogen		H_2	2.02	765.0	188	60	-423.0	187.80	194.0	3.41	2.42	1.41
25. Hydrogen Sulfide		H_2S	34.08	45.3	1306	673	-76.5	11.00	236.0	0.254	0.192	1.32
26. Methane	C	CH_4	16.04	96.4	673	344	-258.8	23.50	219.7	0.526	0.402	1.31
27. Methyl Alcohol		CH_3OH	32.04	48.3	1157	924	148.1	*	473.0	0.330	0.275	1.20
28. Methyl Chloride		CH_3Cl	50.49	30.6	968	750	-10.8	6.26	184.2	0.200	0.167	1.20
29. Natural Gas (3)		-	18.82	82.1	675	379	-	20.00	-	0.485	0.382	1.27
30. Nitrogen		N_2	28.02	55.1	492	228	-320.0	13.53	85.8	0.248	0.177	1.40
31. n-Nonane	nC_9	C_9H_{20}	128.25	12.0	335	1073	303.4	*	125.7	0.400	0.385	1.04
32. Iso-Pentane	iC_5	C_5H_{12}	72.15	21.4	483	830	82.1	*	145.7	0.388	0.361	1.08
33. n-Pentane	nC_5	C_5H_{12}	72.15	21.4	485	847	96.9	*	153.8	0.397	0.370	1.07
34. Pentylene	C_5	C_5H_{10}	70.13	22.0	586	854	86.0	*	149.0	0.382	0.353	1.08
35. n-Octane	nC_8	C_8H_{18}	114.22	13.5	362	1025	258.2	*	131.7	0.400	0.382	1.05
36. Oxygen		O_2	32.00	48.3	730	278	-297.4	11.85	92.0	0.219	0.156	1.40

(Continued)

Table 23.5 Properties of Gases and Vapors. (Continued)

Gases and vapors	Hydrocarbons reference symbols	Chemical formula	Mol. wt.	R = 1545/mol. wt.	Critical conditions Pressure (Psia)	Critical conditions Temperature (°R)	Boiling point (F) @ 14.7 Psia	Specific volume Cu ft/lb @ 14.7 Psia & 60F (Z factor accounted For)	Latent heat of vaporization (Btu/lb @ 14.7 Psia)	Specific heat constant pressure (C_p @ 60F)	Specific heat constant volume (C_v @ 60F)	Specific heat ratio K = C_p/C_v
37. Propane	C_3	C_3H_8	44.09	35.1	617	666	-43.7	8.45	183.5	0.388	0.342	1.13
38. Propylene	C_3	C_3H_6	42.08	36.7	668	658	-53.9	8.86	188.2	0.354	0.307	1.15
39. Refinery Gas (High Paraffin) (4)		-	28.83	53.6	674	515	-	13.20	-	0.395	0.33	1.20
40. Refinery Gas (High Olefin) (4)		-	26.26	58.8	639	456	-	14.40	-	0.397	0.33	1.20
41. Sulphur Dioxide		SO_2	64.06	24.1	1142	775	14.0	5.80	168	0.147	0.118	1.24
42. Water Vapor		H_2O	18.02	85.8	3208	1166	212.0	*	970.3	0.445	0.332	1.33

By permission, Elliott Turbomachinery Co., Inc.

*These substances are not in a vapor state at 14.7 psia and 60°F and therefore sp. Vol. values are not listed.

NOTES: Most values taken from Natural Gasoline Supply Men's Association Engineering Data Book, 1951 – 6th.Ed.

1. Heat of Sublimation.
2. Flue gas – Approximate values of based on 80.5% N_2, 16% CO_2, 3.5% O_2. Actual properties depend on exact composition. Reference: Mark's Engineering Handbook.
3. Carbureted Water Gas, Coke Oven Gas and Natural Gas. Based on average compositions. Actual properties will differ depending on exact compositions. Reference: Perry's Handbook (3rd Edition).
4. Refinery gas (High Paraffin) – Has a greater mol. Percent of saturated hydrocarbons (example C_2H_6)

Refinery gas (High Olefins) – Has a greater mol. Percent of unsaturated hydrocarbons (example C_2H_4)
Reference: Perry's Handbook (3rd ed.)

where R_o = 8314 J/kg mol K, M = molecular weight of the gas).

MR = universal gas constant = 8314 J/k mol K
M = molecular weight
T = upstream absolute temperature, °R = °F + 460 (K = °C + 273.15)
$\overline{V}$ = specific volume of fluid, ft³/lb (m³/kg)
$P_1 = P'$ = upstream pressure, psia (N/m² absolute (pascal))
p' = pressure, bar
d = pipe inside diameter, in.
W = gas rate, lb/h
Z = gas compressibility factor
$P_c = P_{cri}$ = critical pressure, psia
γ = ratio of specific heat at constant pressure to specific heat at constant volume
 = C_p/C_v.

The critical pressure at a pipe outlet is [5c]:

$$P_{crit} = \left[\frac{W}{408d^2}\right]\left(\frac{ZT}{M}\right)^{0.5}, \text{psia} \tag{23.6}$$

The velocity v_s will occur at the outlet end or in a restricted area [40] when the pressure drop is sufficiently high. The conditions of temperature, pressure, and specific volume are those occurring at the point in question.

Critical pressure will normally be found between 53 and 60% of the upstream pressure, P', at the time of relief from overpressure, including accumulation pressure in psia. That is, P' represents the actual pressure at which the relief device is "blowing" or relieving, which is normally above the set pressure by the amount of the accumulation pressure (see Figure 23.7a).

Thus, if the downstream or backpressure on the valve is less than 53–60% (should be calculated) of the values of P', note above, critical (sonic) flow will usually exist. If the downstream pressure is over approximately 50% of the relief pressure, P', the actual critical pressure should be calculated to determine the proper condition. Calculation of critical pressure [41]:

$$P_c = P_1\left[\frac{2}{(k+1)}\right]^{\frac{k}{(k-1)}} \tag{23.7}$$

$$\frac{P_c}{P_1} = r_c = \left[\frac{2}{(k+1)}\right]^{\frac{k}{(k-1)}} \tag{23.8}$$

For critical flow conditions at β ≤ 0.2. This equation is conventionally solved by Figure 23.21.

At critical conditions, the maximum flow through the nozzle or orifice is [41]

$$W_{max(critical\ flow)} = C_o A P_o \sqrt{\frac{kg_c M}{R_g T_o}\left(\frac{2}{k+1}\right)^{\frac{(k+1)}{(k-1)}}} \tag{23.9}$$

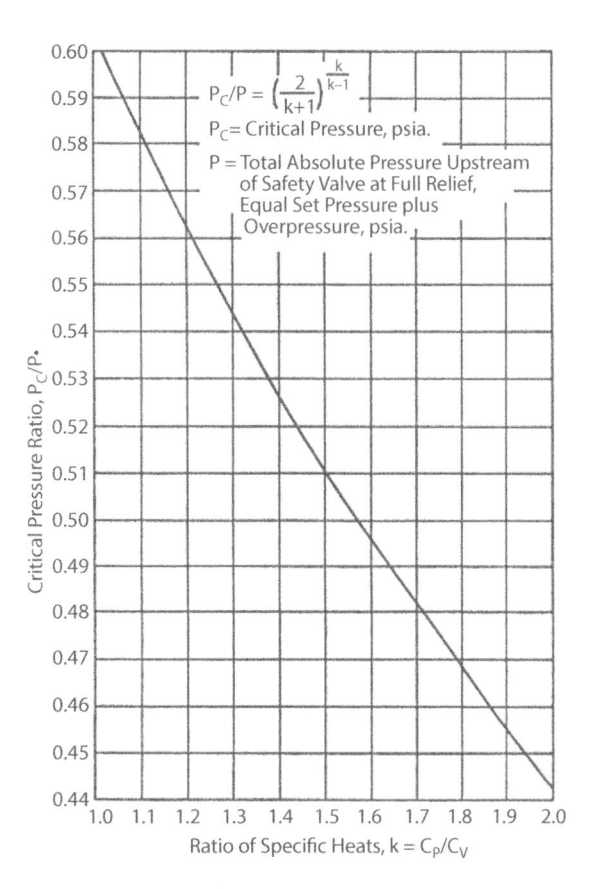

Figure 23.21 Critical back pressure ratio for vapors and gases.

where

- M = molecular weight of vapor or gas, lb_m/lb mol.
- T_o = temperature of service, °R = (°F + 460).
- R_g = ideal universal gas constant = 1545, ft-lb_f/lb mol °R, also = MR.
- C_o = discharge coefficient for sharp-edged orifice.
  - = 0.61 for Reynolds number > 30,000 and not sonic.
  - = 1.0 for sonic flow, C_o increases from 0.61 to 1.0. (use 1.0 to be conservative [16, 41], 5th ed.)
- A = area of opening, orifice, or hole, or nozzle, ft^2.
- P_1 = P_o = upstream pressure, lb_f/ft^2, abs (psfa).
- g_c = conversion factor, 32.174 lb_m/lb_f (ft/s^2)
- β = ratio of orifice diameter/pipe diameter (or nozzle inlet diameter).
- W_{max} = maximum mass flow at critical or choked conditions, lb/s.
- $P_c = P_{crit}$ = critical flow throat pressure, psia = sonic = choked pressure
  - = maximum downstream pressure producing maximum flow.

If the downstream pressure exceeds the critical flow pressure, then sub-critical pressure will occur and the equations for sub-critical flow should be used.

When the downstream pressure is less than (or below) the critical or choked pressure, the velocity and fluid flow rate at a restriction or throat will not/cannot be increased by lowering the downstream pressure further, and the fluid velocity at the restriction or throat is the velocity of sound at the conditions [41].

The critical or sonic ratio is conveniently shown on Figure 23.21, but this does not eliminate the need for calculating the P_c/P_1 ratio for a more accurate result.

Example 23.2: Flow through Sharp Edged Vent Orifice (Adapted after [41])

A small hole has been deliberately placed in a vessel near the top to provide a controlled vent for a nitrogen purge/blanket. The hole is 0.2 in. diameter with the vessel operating at 150 psig at 100°F. Determine the flow through this vent hole. Assume it acts as a sharp-edged orifice.

k (for nitrogen) = 1.4
From Eq. 23.8,

$$P_c = (150+14.7)\left[\frac{2}{(1.4+1)}\right]^{1.4/(1.4-1)} = 87.0 \text{ psia, critical pressure}$$

Hole area = A = $\pi d^2/4$ = $\pi (0.2)^2/4$ = 0.0314 in^2 = 0.0002182 ft^2
Discharge coef. C_o = assumed = 1.0 (Note, could calculate Re to verify)
Inside pressure = 150 psig + 14.7 = 164.7 psia
T_o = 100 + 460 = 560°R

$$W_{max} = \left[1.0(0.0002182)(164.7)\left(144 \text{ in}^2/\text{ft}^2.\right)\right]\left[\sqrt{\frac{(1.4)(32.174)(28)}{(1545)(560)}\left[\frac{2}{1.4+1}\right]^{(1.4+1)/(1.4-1)}}\right]$$

$$= 0.1143 \text{ lb/s}.$$

critical flow rate, W_{max} = 0.1143 lb/s.

23.19 Orifice Area Calculations [42]

Calculations of orifice flow area for conventional pressure-relieving valves and flow are critical (sonic) through part of relieving system, i.e., backpressure is less than 55% of the absolute relieving pressure (including set pressure plus accumulation). See Figure 23.7a; use K_b = 1.0 (Figure 23.26), constant backpressure with variation not to exceed 10% of the set pressure.

  a) **for vapors and gases**, in lb/h; K_b = 1.0; "C" from Figure 23.25, P is the relieving pressure absolute, psia

$$A = \frac{W\sqrt{TZ}}{CK_d P_1 K_b \sqrt{M}}, \text{in}^2. \qquad (23.10)$$

(Effective net discharge area)

where

  C = gas or vapor flow constant
  K_b = 1 when backpressure is below 55% of absolute relieving pressure
  K_d = coefficient of discharge (0.953)
  M = molecular weight of gas or vapor, lb_m/lb mole.
  P_1 = relieving pressure, psia = set pressure + overpressure + 14.7

W = required vapor or gas capacity, lb/h.
T = inlet temperature, °R = °F + 460.
Z = compressibility factor corresponding to T and P.

Metric units in kg/h

$$A = \frac{1.317 \, W \sqrt{TZ}}{C \, K_d \, P_1 \, K_b \sqrt{M}}, \text{cm}^2 \qquad (23.10a)$$

where

A = required orifice area in cm²
C = gas or vapor flow constant
W = required vapor capacity, kg/h
T = inlet temperature, K = (°C + 273.15)
K_b = vapor or gas flow correction factor for constant backpressure above critical pressure,
 = 1 when backpressure is below 55% of absolute relieving pressure.
K_d = coefficient of discharge (0.953 for vapors, gases)
P_1 = relieving pressure in bar abs = set pressure + overpressure + 1.013
Z = compressibility factor corresponding to T and P (Z = 1.0)

    b) **For vapors and gases**, in scfm, K_b = 1.0

$$A = \frac{V \sqrt{GTZ}}{1.175 \, C P_1 \, K_d \, K_b}, \text{in}^2. \qquad (23.11)$$

where

G = specific gravity of gas (Molecular weight of gas/Molecular weight of air
 = 29.0)
T = inlet temperature, °R = °F + 460
P_1 = relieving pressure, psia = set pressure + overpressure + 14.7
V = required gas capacity, scfm
Z = compressibility factor corresponding to T and P (Z = 1.0)

Metric units in Normal m³/h:

$$A = \frac{V \sqrt{GTZ}}{3.344 \, C K_d \, P_1 \, K_b}, \text{cm}^2 \qquad (23.11a)$$

where

A = required orifice area in cm²
C = gas or vapor flow constant
G = specific gravity of gas (Molecular weight of gas /Molecular weight of air = 29.0)
V = required gas capacity, m³/h
T = inlet temperature, K = (°C + 273.15)
K_d = coefficient of discharge (0.953 for vapors, gases)
K_b = vapor or gas flow correction factor for constant backpressure above critical pressure (K_b = 1.0)
P_1 = relieving pressure in bar abs = set pressure + overpressure + 1.013
Z = compressibility factor corresponding to T and P (Z = 1.0)

Pressure relief devices in gas or vapor service that operate at critical conditions may be sized using the following equations:

US Customary units:

$$A = \frac{W}{CK_d P_1 K_b K_c} \sqrt{\frac{TZ}{M}} \quad (23.11b)$$

or

$$A = \frac{V\sqrt{TZM}}{6.32 CK_d P_1 K_b K_c} \quad (23.11c)$$

or

$$A = \frac{V\sqrt{TZG}}{1.175 CK_d P_1 K_b K_c} \quad (23.11d)$$

SI Units:

$$A = \frac{13,160\,W}{CK_d P_1 K_b K_c} \sqrt{\frac{TZ}{M}} \quad (23.11e)$$

or

$$A = \frac{35,250\,V\sqrt{TZM}}{CK_d P_1 K_b K_c} \quad (23.11f)$$

or

$$A = \frac{189,750\,V\sqrt{TZG}}{CK_d P_1 K_b K_c} \quad (23.11g)$$

where

- A = required effective discharge area of the device, in² (mm²)
- W = required flow through the device, lb/h, (kg/h)
- C = coefficient determined from an expression of the ratio of the specific heats ($k = C_p/C_v$) of the gas or vapor at inlet relieving conditions (see Figure 23.25 or Table 23.6) or

$$C = 520\sqrt{k\left(\frac{2}{k+1}\right)^{(k+1)/(k-1)}} \quad (23.11h)$$

Table 23.6 Values of Coefficient C.

k	C	k	C	k	C	k	C
1.00	315°	1.30	347	1.60	372	1.90	394
1.01	317	1.31	348	1.61	373	1.91	395
1.02	318	1.32	349	1.62	374	1.92	395
1.03	319	1.33	350	1.63	375	1.93	396
1.04	320	1.34	351	1.64	376	1.94	397
1.05	321	1.35	352	1.65	376	1.95	397
1.06	322	1.36	353	1.66	377	1.96	398
1.07	323	1.37	353	1.67	378	1.97	398
1.08	325	1.38	354	1.68	379	1.98	399
1.09	326	1.39	355	1.69	379	1.99	400
1.10	327	1.40	356	1.70	380	2.00	400
1.11	328	1.41	357	1.71	381	—	—
1.12	329	1.42	358	1.72	382	—	—
1.13	330	1.43	359	1.73	382	—	—
1.14	331	1.44	360	1.74	383	—	—
1.15	332	1.45	360	1.75	384	—	—
1.16	333	1.46	361	1.76	384	—	—
1.17	334	1.47	362	1.77	385	—	—
1.18	335	1.48	363	1.78	386	—	—
1.19	336	1.49	364	1.79	386	—	—
1.20	337	1.50	365	1.80	387	—	—
1.21	338	1.51	365	1.81	388	—	—
1.22	339	1.52	366	1.82	389	—	—
1.23	340	1.53	367	1.83	389	—	—
1.24	341	1.54	368	1.84	390	—	—
1.25	342	1.55	369	1.85	391	—	—
1.26	343	1.56	369	1.86	391	—	—
1.27	344	1.57	370	1.87	392	—	—

(*Continued*)

Table 23.6 Values of Coefficient C. (*Continued*)

k	C	k	C	k	C	k	C
1.28	345	1.58	371	1.88	393	—	—
1.29	346	1.59	372	1.89	393	—	—
1.30	347	1.60	373	1.90	394	—	—

^aThe limit of C, as k approaches 1.00 is 315.

(Source: Sizing, Selection and Installation of Pressure – Relieving Devices in Refineries, Part 1 – Sizing and Selection, API RP 520, 7th ed., January 2000).

Where k cannot be determined, it is suggested that a value of C equal to 315 be used. The units for C are $\dfrac{\sqrt{lb_m \times lb_{mole} \times {}^\circ R}}{lb_f \times h}$

K_d = effective coefficient of discharge. For preliminary sizing, use the following values:

  (i) 0.975 when a pressure relief valve is installed with or without a rupture disk in combination.
  (ii) 0.62 when a pressure-relief valve is not installed and sizing is for a rupture disk.

P_1 = upstream relieving pressure, psia (kPaa). This is the set pressure plus the allowable overpressure plus atmospheric pressure.
K_b = capacity correction factor due to backpressure. The backpressure correction factor applies to balanced bellows valves only. For conventional and pilot operated valves, $K_b = 1.0$
K_c = combination correction factor for installation with a rupture disk upstream of the pressure relief valve
  = 1.0 when a rupture disk is not installed.
  = 0.9 when a rupture disk is installed in combination with a pressure relief valve and the combination does not have a published value.
T = relieving temperature of the inlet gas or vapor, °R = °F + 460 (K = °C + 273)
Z = compressibility factor for the deviation of the actual gas from a perfect gas, a ratio evaluated at inlet relieving conditions.
M = molecular weight of the gas or vapor at inlet relieving conditions.
V = required flow through the device, scfm at 14.7 psia and 60°F (Nm³/min at 0°C and 101.3 kPaa)
G = specific gravity of gas at standard conditions referred to air at standard conditions (normal conditions).
G = 1.00 for air at 14.7 psia and 60°F (101.3 kPaa and 0°C).

c) **For steam, in lb/h;** $K_b = 1.0$ and $K_{sh} = 1.0$ for saturated steam when backpressure is below 55% of absolute relieving pressure [5d].

$$A = \dfrac{W_s}{51.5\, P_1\, K_d\, K_b\, K_c\, K_n\, K_{sh}}, in^2. \tag{23.12}$$

SI units:

$$A = \frac{190.4 \times W_s}{P_1 K_d K_b K_c K_n K_{sh}}, \text{mm}^2 \qquad (23.12a)$$

where

W_s = required steam capacity in lb/h (kg/h).
K_d = effective coefficient of discharge. For preliminary sizing, use the following values:
  (i) 0.975 when a pressure-relief valve is installed with or without a rupture disk in combination.
  (ii) 0.62 when a pressure-relief valve is not installed and sizing is for a rupture disk.
K_b = 1 when backpressure is below 55% of absolute relieving pressure. The backpressure correction factor applies to balanced bellows valves only. For conventional valves, use a value for K_b = 1.0
K_c = combination correction factor for installation with a rupture disk upstream of the pressure relief valve.
  = 1.0 when a rupture disk is not installed.
  = 0.9 when a rupture disk is installed in combination with a pressure-relief valve and the combination do not have a published value.
K_n = Napier steam correction factor for set pressures between 1500 and 2900 psig.
  = 1 when $P_1 \leq 1500$ psia (10,339 kPaa)

$$= \frac{0.1906 \times P_1 - 1000}{0.2292 \times P_1 - 1061} \text{ (US customary units)}$$

$$= \frac{0.02764 \times P_1 - 1000}{0.03324 \times P_1 - 1061} \text{ (SI units)}$$

where $P_1 \geq 1500$ psia (10,339 kPaa) and ≤ 3200 psia (22,057 kPaa).

K_{hs} = 1 for saturated steam.
P_1 = relieving pressure, psia = set pressure + overpressure + 14.7
  = relieving pressure, kPaa = set pressure + overpressure + 101.3

Metric units in kg/h

$$A = \frac{W_s}{52.49 \, P_1 K_d K_b K_{sh}}, \text{cm}^2 \qquad (23.12b)$$

where

W_s = required steam capacity, kg/h.
K_b = 1 when backpressure is below 55% of absolute relieving pressure.
K_d = coefficient of discharge (K_d = 0.953).
K_{hs} = 1 for saturated steam.
P_1 = relieving pressure, bara = set pressure + overpressure + 1.013.

  d) For air, in scfm, K_b = 1.0 when the backpressure is below 55% of absolute relieving pressure

$$A = \frac{V_a \sqrt{T}}{418\, K_d\, P_1\, K_b}, \text{in}^2 \qquad (23.13)$$

where

V_a = relieving air capacity, scfm
K_b = 1 when backpressure is below 55% of absolute relieving pressure.
K_d = coefficient of discharge (K_d = 0.953)
P_1 = relieving pressure, psia = set pressure + overpressure + 14.7.
T = inlet temperature, °R = °F + 460.

Metric units in m³/h

$$A = \frac{V_a \sqrt{T}}{1189.3\, K_d\, P\, K_b}, \text{cm}^2 \qquad (23.13a)$$

where

V_a = required air capacity in m³/h.
K_b = 1 when backpressure is below 55% of absolute relieving pressure.
K_d = coefficient of discharge (K_d = 0.953).
P = relieving pressure, bara = set pressure + overpressure + 1.013
T = inlet temperature, K = (°C + 273.15)

e) **For liquids, US gpm**

K_p = 1.0 at 10% overpressure
K_u = 1.0 at normal viscosities
$\Delta P = P_1 - P_2$ = upstream pressure, psig (set + overpressure) – total backpressure, psig.

23.20 Sizing Valves for Liquid Relief: Pressure-Relief Valves Requiring Capacity Certification [5D]

Section VIII, Division I, of the ASME Code requires that capacity certification be obtained for pressure-relief valves designed for liquid service. The procedure for obtaining capacity certification includes testing to determine the rated coefficient of discharge for the liquid relief valves at 10 % overpressure.

ASME Code valves: Board Certified for liquids only [5d].

$$A = \frac{V_L \sqrt{G}}{38\, K_d\, K_w\, K_c\, K_u \sqrt{P_1 - P_2}}, \text{in}^2 \qquad (23.14)$$

S.I. units

$$A = \frac{11.78 \times V_L \sqrt{G}}{K_d\, K_w\, K_c\, K_u \sqrt{P_1 - P_2}}, \text{mm}^2 \qquad (23.14a)$$

where

- A = required effective discharge area, in² (mm²)
- G = specific gravity of the liquid at the flowing temperature referred to water at standard conditions (density of liquid/ density of water ≈ 62.3 lb/ft³)
- V_L = required liquid capacity, U.S. gpm (l/min)
- K_d = rated coefficient of discharge that should be obtained from the valve manufacturer.

For a preliminary sizing,

- = 0.65 when a pressure relief valve is installed with or without a rupture disk in combination.
- = 0.62 when a pressure relief valve is not installed and sizing is for a rupture disk.
- K_w = variable or constant backpressure sizing factor, balanced valves, liquid only (Figure 23.28). If the backpressure is atmospheric, K_w = 1.0. Balanced bellows valves in backpressure service will require the correction factor. Conventional and pilot operated valves require no special correction.
- K_c = combination correction factor for installation with a rupture disk upstream of the pressure relief valve.
- = 1.0 when a ruptured disk is not installed.
- = 0.9 when a rupture disk is installed in combination with a pressure relief valve and the combination does not have a published value.
- K_u = viscosity correction factor (K_u = 1 at normal viscosities).

$$= \left(0.9935 + \frac{2.878}{Re^{0.5}} + \frac{342.75}{Re^{1.5}}\right)^{-1.0}$$

- P_1 = upstream relieving pressure. This is the set pressure plus allowable overpressure, psig (kPag).
- P_2 = backpressure at outlet, psig (kPag).

23.21 Sizing Valves For Liquid Relief: Pressure Relief Valves Not Requiring Capacity Certification [5D]

Before the ASME code incorporated requirements for capacity certification, valves were generally sized for liquid service, which assumes an effective coefficient of discharge, K_d = 0.62, and 25 % overpressure.

This method will typically result in an oversized design where a liquid valve is used for an application with 10% overpressure. A K_p correction factor of 0.6 is used for this situation [5d].

US Customary Units

$$A = \frac{V_L \sqrt{G}}{38 K_d K_w K_c K_p K_u \sqrt{1.25 P_1 - P_2}}, \text{in}^2 \qquad (23.15)$$

S.I. Units

$$A = \frac{11.78 \times V_L \sqrt{G}}{K_d K_w K_c K_u K_p \sqrt{1.25 P_1 - P_2}}, \text{mm}^2 \qquad (23.15a)$$

where

- A = required effective discharge area, in² (mm²)
- G = specific gravity of liquid at flowing temperature referred to water at standard conditions (density of liquid/ density of water ≈ 62.3 lb_m/ft³).
- V_L = required liquid capacity, U.S. gpm (l/min).

K_d = rated coefficient of discharge that should be obtained from the valve for a preliminary sizing, an effective discharge coefficient can be used as follows:
= 0.65 when a pressure relief valve is installed with or without a rupture disk in combination,
= 0.62 when a pressure relief valve is not installed and sizing is for a rupture disk.

K_w = variable or constant backpressure sizing factor, balanced valves, liquid only (Figure 23.28). If the backpressure is atmospheric, K_w = 1.0. Balanced bellows valves in backpressure service will require the correction factor.

Conventional and pilot operated valves require no special correction.

K_c = combination correction factor for installation with a rupture disk upstream of the pressure-relief valve.
= 1.0 when a ruptured disk is not installed.
= 0.9 when a rupture disk is installed in combination with a pressure-relief valve and the combination does not have a published value.

K_p = liquid capacity correction factor for overpressures. At 25% overpressure, K_p = 1.0. For overpressure other than 25%, K_p is determined from Figure 23.22

K_u = viscosity correction factor (K_u = 1 at normal viscosities).

$$= \left(0.9935 + \frac{2.878}{Re^{0.5}} + \frac{342.75}{Re^{1.5}}\right)^{-1.0}$$

P_1 = set pressure at inlet, psig (kpag)
P_2 = total backpressure at outlet, psig (kPag).

Metric units in dm³/min

$$A = \frac{V_L \sqrt{G}}{84.89 \, K_d \, K_p \, K_u \sqrt{1.25 P_1 - P_2}}, \text{cm}^2 \tag{23.15a}$$

where

A = required effective discharge area, cm²
G = specific gravity of liquid at flowing temperature referred to water at standard conditions.

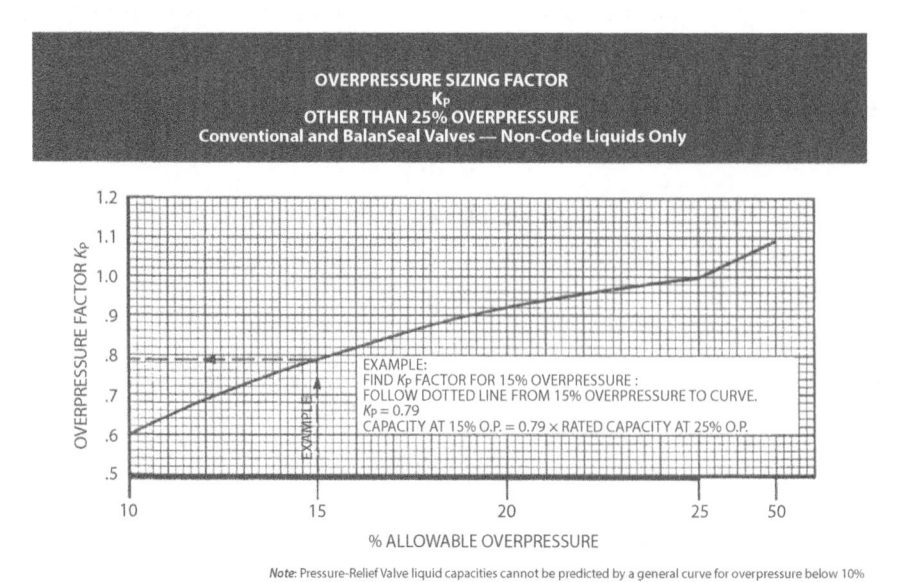

Figure 23.22 Liquids overpressure sizing factor, K_p, for other than 25% overpressure. Applies to non-code liquids only using conventional and balanced valves. (By permission from Teledyne Farris Engineering Co.)

V_L = required liquid capacity, US gpm (l/min).

K_d = 0.62 for a preliminary sizing estimation; otherwise, rated coefficient of discharge should be obtained from the valve manufacturer.

K_p = liquid capacity correction factor for overpressures. At 25%. Overpressure, K_p = 1.0. For overpressure other than 25%, K_p is determined from Figure 23.22

K_u = viscosity correction factor (K_u = 1 at normal viscosities).

P_1 = set pressure at inlet, barg.

P_2 = backpressure at outlet, barg.

When sizing a relief valve for viscous liquid service, the orifice area is first calculated for non-viscous service application (i.e., K_u = 1.0) to obtain a preliminary required discharge area, A, and can be obtained from Eq. 25.5a. From API Std 526 standard orifice sizes, the next orifice size larger than A is used in determining the Reynolds number (Re).

To apply the viscosity correction K_u, a preliminary or trial calculation should be made for the areas required using the equation of paragraph (e) above or the modified equation (still ASME conformance [5] but not capacity certified). A simplified equation based on the ASME Pressure Vessel Code equations, Section VIII, Div. 1, Mandatory Appendix XI uses K coefficient of discharge in the equations, where K is defined as 90% of the average K_d of certified tests with compressible or incompressible fluids, see [42], pg 40.

For first trial, assume K_u for viscosity = 1.0

For final calculation use K_u from Figures 23.23 and 23.23a or a correlation equation defined by

$$K_u = -2.38878 + 1.2502(\ln Re) - 0.17510(\ln Re)^2 + 0.01087(\ln Re)^3 - 0.00025(\ln Re)^4 \qquad (23.15c)$$

K_u is then substituted in Equations 23.15 and 23.15a (S.I. units). Determine the needed Reynolds number, Re, using the next size larger orifice. Area is determined from that made in the first trial calculation [5].

$$Re = V_L \frac{(2800\,G)}{\mu\sqrt{A}}, \text{ or} \qquad (23.16)$$

or

$$Re = \frac{12,700\,V_L}{(U\sqrt{A})}, \text{(do not use when U < 100 SSU)} \qquad (23.17)$$

SI units

$$Re = V_L \frac{(18,800\,G)}{\mu\sqrt{A}} \qquad (23.16a)$$

or

$$Re = \frac{(85,220\,V_L)}{U\sqrt{A}} \qquad (23.17a)$$

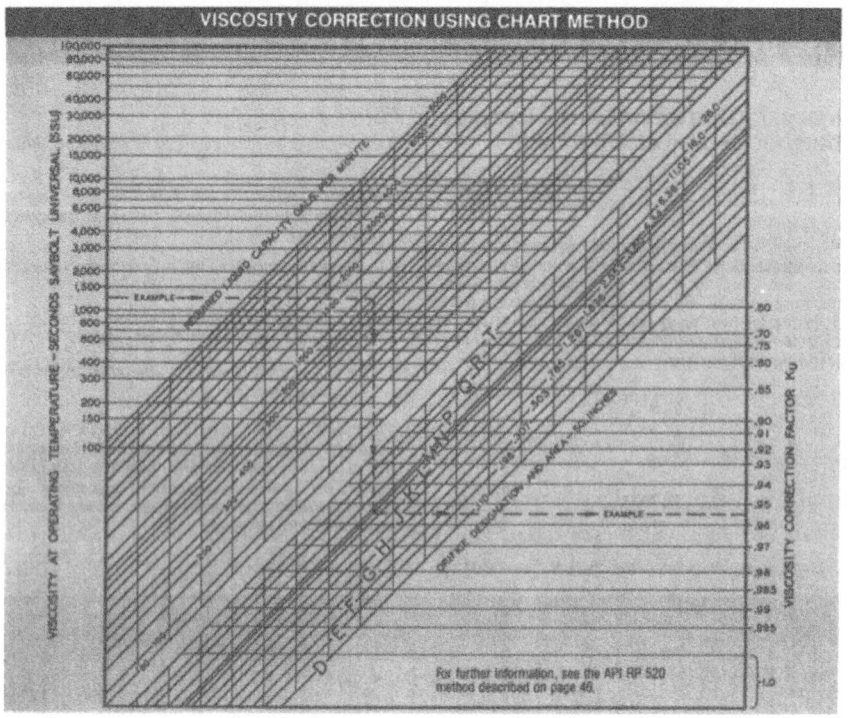

Figure 23.23 Liquids viscosity correction using chart method for Ku. (By permission from Teledyne Farris Engineering Co.)

where

Re = Reynolds number
μ = absolute viscosity at the flowing temperature, cP
P_1 = set pressure, psig
U = viscosity at the flowing temperature, Secs. Saybolt Universal (SSU).
P_2 = backpressure, psig.

Note: Equation 25.17a is not recommended for viscosities less than 100 Secs. Saybolt Universal.

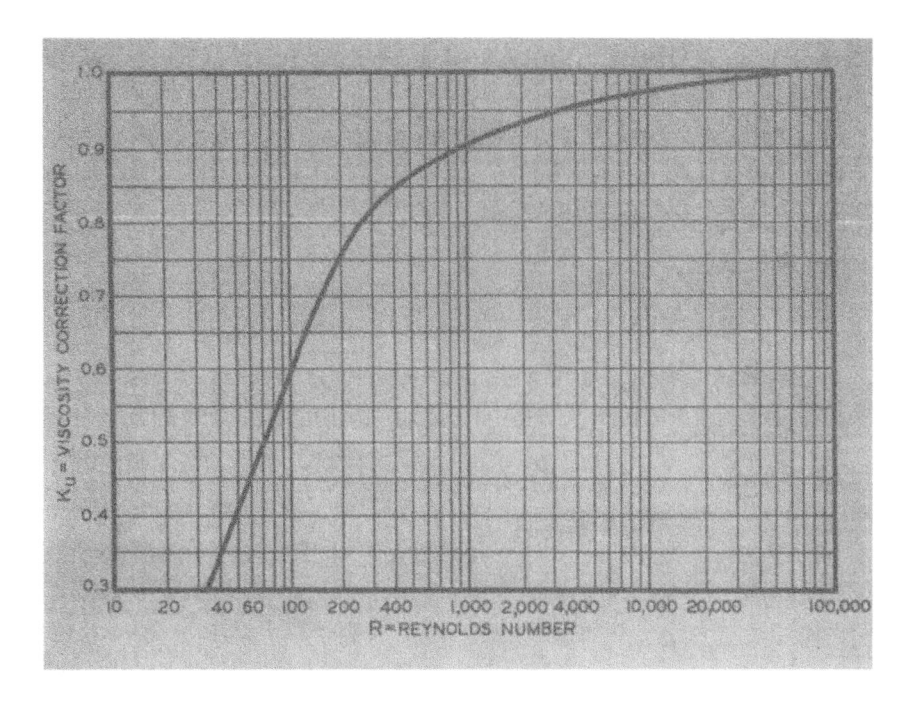

Viscosity Correction Using Reynold's Number Method of API RP520

As an alternate to Figure 7-23, you may wish to use the method given in API RP 520 for sizing viscous liquids.

When a relief valve is sized for viscous liquid service, it is suggested that it be sized first as for nonviscous type application in order to obtain a preliminary required discharge area, A. From manufacturer's standard orifice sizes, the next larger orifice size should be used in determining the Reynold's number R, from either of the following relationships:

$$R = \frac{V_L(2{,}800\,G)}{\mu\sqrt{A}}$$

or

$${}^*R = \frac{12{,}700\,V_L}{U\sqrt{A}}$$

*Use of this equation is not recommended for viscosities less than 100 SSU

Where:

V_L = flow rate at the flowing temperature in U.S. gallons per minute.

G = specific gravity of the liquid at the following temperature referred to water = 1.00 at 70 degrees Fahrenheit.

μ = absolute viscosity at the flowing temperature, in centipoises.

A = effective discharge area, in square inches (from manufacturers' standard orifice areas).

U = viscosity at the flowing temperature, in Saybolt Universal seconds.

After the value of R is determined, the factor K_V† is obtained from the graph. Factor K_V is applied to correct the "preliminary required discharge area." If the corrected area exceeds the "chosen standard orifice area", the above calculations should be repeated using the next larger standard orifice size.

†K_v of API = K_u of Teledyne Farris

Figure 23.23a Viscous liquid valve sizing using the method of API RP – 520. (Reprinted by permission from Teledyne Farris Engineering Co., and Sizing, Selection and Installation of Pressure Relieving Devices in Refineries, Part I "Sizing and Selection", API RP – 520, 5th ed., July 1990, American Petroleum Institute.)

Table 23.7 shows a list of actual K_d values from a variety of manufacturers and models of relief valves. The table shows the variations of K_d and some closer to API 520 values than other. One of the reasons for the variation is the lift of the relief valve, which is the distance between the nozzle and the valve seat. When relief valves are sized, it is the orifice area that is determined, which is shown in green in Figure 23.24. For high lift valves (high K_d value valves), the orifice is the minimum flow area and defines the flow. However, for low lift valves, it is the curtain area (shown in purple in Figure 23.24 (b)), which is the minimum flow area. Valves are specified using orifice area, not flow areas, so the valve shown in Figures 23.23 and 23.24 (a) and (b) will have a low K_d, and if the engineer has used a value of 0.975 in their calculations, then the valve will be undersized by a significant factor.

Table 23.7 Actual K_d Values of a Selection of Relief Valves.

Gas duty valves		Liquid duty valves	
Manufacturer and model	Actual K_d	Manufacturer and model	Actual K_d
Brody 3500	0.957	Farris 1850	0.724
Farris 1850	0.724	Lesser 488	0.524
Lesser 488	0.801	Safety Systems WB100	0.653
Safety Systems WB400	0.975	Crosby 900 Series	0.735
Crosby JOS-E	0.961		

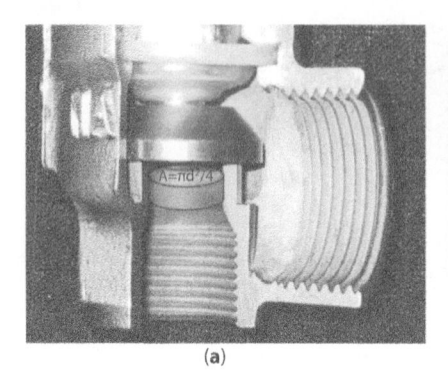

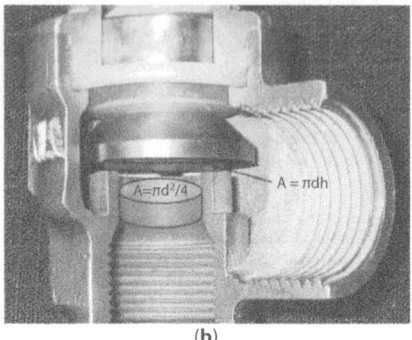

Figure 23.24 (a) Orifice area in a relief valve (green); (b) Orifice area in a relief valve (green), Curtain area (purple). This valve will have a low discharge coefficient as the curtain area is lower than the orifice area. (Source: Chris Flower and Adam Wills, The Chemical Engineer, U.K., pp. 24- 31, June 2017.)

23.22 Reaction Forces

The discharge of a pressure relief valve with unsupported discharge piping will impose a reactive load on the inlet of the valve as a result of the reaction force of the flowing fluid. This is particularly essential where piping discharging to atmosphere includes a 90° turn and has no support for the outlet piping. All reactive loading due to the operation of the valve is then transmitted to the valve and inlet piping.

The following formula is based on a condition of critical steady state flow of a compressible fluid that discharges to the atmosphere through an elbow and a vertical discharge pipe. The reaction force (F) includes the effects of both momentum and static pressure [5d]. The formula is applicable for any gas, vapor, or steam and is expressed by

In US customary Units

$$F = \frac{W}{366}\sqrt{\frac{kT}{(k+1)M}} + (AP) \tag{23.17b}$$

In metric units

$$F = 129\,W\sqrt{\frac{kT}{(k+1)M}} + 0.1(AP) \tag{23.17c}$$

where

- A = area of the outlet at the point of discharge, in² (mm²)
- C_p = specific heat at constant pressure.
- C_v = specific heat at constant volume.
- F = reaction force at the point of discharge to the atmosphere, lb_f (N)
- k = ratio of specific heats (C_p/C_v) at the outlet conditions.
- M = molecular weight of the process fluid
- P = static pressure within the outlet at the point of discharge, psig (barg)
- T = temperature at the outlet, °R (K)
- W = flow of any gas or vapor, lb_m/h (kg/s)

Example 23.3

In a process plant, ammonia is used to control the pH during a production process to obtain maximum yield of a product. For this purpose, liquid ammonia is vaporized by passing steam through a coil in a vaporizer. A relief valve from the vaporizer is set at a pressure of 290 psig. If the relieving rate of ammonia vapor is 620 lb/h, determine the size of the relief valve required to relieve ammonia vapor during an emergency.

Design data:

Ratio of specific heat capacities (C_p/C_v) = 1.33
Molecular weight of ammonia = 17.03
Critical pressure, P_c = 111.3 atm.
Critical temperature, T_c = 405.6 K
Constants in Antoine Equation:

A = 16.9481
B = 213.25
C = -32.98

Solution

From Antoine's equation

$$\ln P = A - \frac{B}{T+C}$$

Relieving pressure = mm Hg.
Relieving temperature = K
Relieving pressure = set pressure + over pressure + atmospheric pressure
  = 290 + (290 x 0.1) + 14.7
  = 333.7 psia.
Conversion 1 psi = 51.715 mm Hg.
  1 atm = 14.7 psi
Relieving pressure = 17,257.3 mm Hg.
Relieving temperature from Antoine's equation

$$\ln P = A - \frac{B}{T+C}$$

$$\ln(17,257.296) = 16.9481 - \frac{2132.5}{(T-32.98)}$$

$$-7.192(T-32.98) = -2132.5$$

$$-7.192T + 237.19 = -2132.5$$

$$T = \frac{2369.69}{7.192}$$

$$= 329.5 \text{ K} (593.45°\text{R})$$

Critical pressure = 1636.1 psia
Critical temperature = 703.41 °R
Since compressibility factor Z is a function of temperature T_r and P_r, then

$$T_r = T/T_c = 593.45/730.41$$
$$= 0.812$$
$$P_r = P/P_c = 333.71/1636.11$$
$$= 0.204$$

From the Nelson and Obert chart, Z = 0.837.
Coefficient "C" for gas related to specific heats.
Substituting the value for k in Eq. 23.11h,

$$C = 520 \left[k \left\{ \frac{2}{(k+1)} \right\}^{\frac{(k+1)}{(k-1)}} \right]^{0.5}$$

$$= 520 \left[(1.33) \left\{ \frac{2}{1.33+1} \right\}^{\frac{(1.33+1)}{(1.33-1)}} \right]^{0.5}$$

$$= 350$$

For vapor, the orifice area (from Eq. 23.10) is

$$A = \frac{W\sqrt{TZ}}{C K_d P K_b \sqrt{M}}, \text{in}^2.$$

$$A = \frac{620\sqrt{(593.45)(0.812)}}{(350)(0.953)(333.7)(1)\sqrt{17.03}}$$

$$= 0.0296 \text{ in}^2$$

The nearest standard orifice area is 0.110 in², having an inlet and outlet valve body sizes of 1 and 2 in respectively. The designation is 1D2 pressure relief valve.

Table 23.8 Results of Relief Valve Sizing for Vapor Flow of Example 23.3.

Results of relief valve sizing for vapor flow		
Flow Type	Vapor	
Set Pressure	290	psig
Relieving Pressure	333.7	psia
Inlet Temperature	132.5	°F
Inlet Temperature	592.5	°R
Ratio of Specific heat capacities, k	1.33	
Constant Coefficient, C	349.8	
Gas Compressibility factor Z	0.812	
Relieving flow rate	620	lb/h
$K_b =$	1	
$K_d =$	0.953	
Molecular weight of vapor, M_w	17.03	
Calculated orifice area	0.03	in²
Nearest orifice area	0.11	in²
Orifice Size	D	
Maximum vapor rate	2302	lb/h

The maximum vapor rate with the nearest standard orifice area is

$$W = A C K_d P K_b \sqrt{M/(TZ)} \sqrt{M/(TZ)}$$
$$= (0.11)(350)(0.953)(333.7)(1)\sqrt{17.03/(593.45 \times 0.81)}$$
$$= 2305 \text{ lb/h}.$$

The Excel spreadsheet program Example 23.3.xlsx shows the calculations of Example 23.3 for vapor relief valve sizing, and Table 23.8 shows the results of Example 23.3.

Example 23.4

A bellows type pressure relief valve is required to protect a vessel containing an organic liquid. The required relieving capacity is 310 U.S. gpm. The inlet temperature is 170°F and the set pressure is 100 psig. Allowable overpressure is 25% with a built-up backpressure is 25 psig. The fluid's physical properties such as the specific gravity and viscosity are 1.45 and 3200 cP, respectively. Determine the orifice size of the valve.

Take the correction factors

$$K_d = 0.62, K_w = 0.92, K_c = K_p = K_u = 1$$

Solution

Non-ASME Code Liquid Valves [5c] non-board certified for liquids, but code acceptable for other services. K_p from Figure 23.22, $K_d = 0.62$, and 25% overpressure.

Substituting the values in Eq. 23.15,

$$A = \frac{V_L \sqrt{G}}{38 K_d K_w K_c K_p K_u \sqrt{1.25 P_1 - P_2}}, \text{in}^2$$

$$A = \frac{(310)\sqrt{1.45}}{38(0.62)(0.92)(1.0)(1.0)(1.0)\sqrt{1.25(100) - 25}}$$

$$= 1.722 \text{ in}^2$$

The next standard orifice area is 1.838 in² having an inlet and outlet valve body sizes of 3 and 4 in respectively. The designation is 3K4 pressure relief valve.

The maximum flow rate with the standard orifice area is

$$V_L = A(38)(K_d)(K_w)(K_c)(K_p)(K_u)\sqrt{(1.25 P_1 - P_2)/G}$$

$$= 1.838(38)(0.62)(0.92)(1.0)(1.0)(1.0)\sqrt{(1.25(100) - 25)/1.45}$$

$$= 330.85 \text{ gpm.}$$

Next, calculate the Reynolds number using the manufacturer's orifice area from Eq. 23.16:

$$Re = V_L \frac{(2800 G)}{\mu \sqrt{A}}, \text{ or}$$

$$Re = \frac{(310)(2800 \times 1.45)}{(3200)\sqrt{1.838}}$$

$$= 290.0$$

Determine the viscosity correction factor K_u from Eq. 23.15c,

$$K_u = -2.38878 + 1.2502(\ln Re) - 0.17510(\ln Re)^2$$
$$+ 0.01087(\ln Re)^3 - 0.00025(\ln Re)^4$$

$$K_u = -2.38878 + 1.2502(\ln 290) - 0.17510(\ln 290)^2$$
$$+ 0.01087(\ln 290)^3 - 0.00025(\ln 290)^4$$

$$= 0.794$$

The required orifice area with calculated $K_u = 0.794$ is

$$A = \frac{(310)\sqrt{1.45}}{38(0.62)(0.92)(1.0)(1.0)(0.794)\sqrt{1.25(100) - 25}}$$

$$= 2.17 \text{ in}^2.$$

The nearest standard orifice area is 2.853 in², with L designation having preferred valve body sizes, 3–4 in. or 4–6 in.

The maximum liquid flow rate with the standard orifice area is

$$V_L = A(38)(K_d)(K_w)(K_c)(K_p)(K_u)\sqrt{(1.25P_1 - P_2)/G}$$
$$= 2.853(38)(0.62)(0.92)(1.0)(0.794)(1.0)\sqrt{(1.25(100) - 25)/1.45}$$
$$= 407.76 \text{ gpm.}$$

Substituting the values in Eq. 23.14,

ASME Code valves: Board Certified for liquids only.

$$A = \frac{V_L\sqrt{G}}{38 K_d K_w K_c K_u \sqrt{P_1 - P_2}}$$

$$A = \frac{(310)\sqrt{1.45}}{38(0.62)(0.92)(1)(1)\sqrt{100 - 25}}$$

$$= 1.989 \text{ in}^2.$$

The error in the orifice is between the ASME code and Non-ASME code is

$$\varepsilon = \frac{(1.989 - 1.722)}{1.722} \times 100\% = 15.5\%$$

Percentage deviation in the calculated orifice areas between the ASME code valves and Non-ASME code Liquid Valves [5a] non-board certified for liquids is 15.5%. This is because the Non-ASME code uses a 25% overpressure, whereas the ASME code formula is based on only 10%. It should be noted that Eq. 23.15 is applicable only to relief valves not requiring capacity certification (Section 4.6 of API 520). The Excel spreadsheet Example 23.4.xlsx shows the calculations for liquid relief valve sizing and Table 23.9 shows the results of Example 23.4.

Table 23.9 Results of Relief Valve Sizing for Liquid Flow of Example 23.4.

Results of relief valve sizing for liquid flow		
Flow Type	Liquid	
Set Pressure at inlet	100	psig
Back Pressure at outlet	25	psig
Specific gravity of liquid	1.45	
Viscosity of Liquid	3200	cP
Relieving flow rate	310	US gal.
K_d	0.62	

(Continued)

Table 23.9 Results of Relief Valve Sizing for Liquid Flow of Example 23.4. (*Continued*)

Results of relief valve sizing for liquid flow		
K_w	0.92	
K_c	1	
K_p	1	
K_u	1	
Calculated orifice area	1.722	in²
Nearest orifice area	1.838	in²
Orifice Size	K	
Maximum Liquid flow rate	330.8	US gal.
Reynolds number	290	
Calculated Ku	0.794	
Calculated orifice area	2.17	in²
Nearest orifice area	2.853	in²
Orifice Size	L	
Maximum Liquid flow rate	407.6	US gal.

23.23 Calculations of Orifice Flow Area using Pressure Relieving Balanced Bellows Valves, with Variable or Constant Backpressure

Must be used when backpressure variation exceeds 10% of the set pressure of the valve. Flow may be critical or non-critical for balanced valves. All orifice areas, A, in² [42]. The sizing procedure is the same as for conventional valves listed above (Eq. 23.10), but uses equations given below incorporating the correction factors K_v and K_w. With variable backpressure, use maximum value for P_2 [5a, 42].

a) For vapors or gases, lb/h

$$A = \frac{W\sqrt{TZ}}{CK_d P_1 K_v \sqrt{M}}, in^2 \tag{23.18}$$

Metric units in kg/h

$$A = \frac{1.317\, W\sqrt{TZ}}{CK_d P_1 K_v \sqrt{M}}, cm^2 \tag{23.18a}$$

b) For vapors or gases, scfm

$$A = \frac{V\sqrt{GTZ}}{1.175\, CK_d P_1 K_v}, in^2 \tag{23.19}$$

In metric units in Normal m³/h

For vapor or gases, Normal m³/h

$$A = \frac{V\sqrt{GTZ}}{3.344\,C\,K_d\,P_1\,K_v}, \text{cm}^2 \quad (23.19a)$$

c) For steam, lb/h

$$A = \frac{W_s}{51.5\,K_d\,K_v\,K_{sh}\,K_n\,P_1}, \text{in}^2. \quad (23.20)$$

Metric units, kg/h

$$A = \frac{W_s}{52.49\,K_d\,K_v\,K_{sh}\,K_n\,P_1}, \text{cm}^2 \quad (23.20a)$$

d) For air, scfm

$$A = \frac{V_a\sqrt{T}}{418\,K_d\,P_1\,K_v}, \text{in}^2. \quad (23.21)$$

Metric units, Normal m³/h

$$A = \frac{V_a\sqrt{T}}{1189.3\,K_d\,P_1\,K_v}, \text{cm}^2 \quad (23.21a)$$

e) For liquids, gpm; ASME Code valve

$$A = \frac{V_L\sqrt{G}}{38.0\,K_d\,K_w\,K_u\sqrt{P}}, \text{in}^2. \quad (23.22)$$

f) For liquids, gpm, non-ASME Code valve

$$A = \frac{V_L\sqrt{G}}{38.0\,K_d\,K_p\,K_w\,K_u\sqrt{(1.25P_1 - P_2)}}, \text{in}^2. \quad (23.23)$$

Metric unit in dm³/min.

$$A = \frac{V_L\sqrt{G}}{84.89\,K_d\,K_p\,K_w\,K_u\sqrt{(1.25P_1 - P_2)}}, \text{cm}^2 \quad (23.23a)$$

When the backpressure, P_2, is variable, use the maximum value.

where (Courtesy of Teledyne Farris Engineering Co. [42]):

- A = required orifice area in in². This is as defined in the ASME code and ANSI/API Std 526.
- W = required vapor capacity in lb/h.
- W_s = required steam capacity in lb/h.
- V = required gas capacity in scfm.
- V_a = required air capacity in scfm.
- V_L = required liquid capacity, gpm (dm³/min).
- G = specific gravity of gas (air = 1.0) or specific gravity of liquid (water = 1.0) at actual discharge temperature. A specific gravity at any lower temperature will obtain a safe valve size.
- M = average molecular weight of vapor.
- P_1 = relieving pressure in psia = [set pressure, psig + over pressure, psig + 14.7] psia. Minimum overpressure = 3 psi.
- P_1 = set pressure at inlet, psig.
- P_2 = backpressure at outlet, psig.
- ΔP = Set pressure + over pressure, psig - backpressure, psig. At 10% overpressure $\Delta P = 1.1 P_1 - P_2$. Below 30 psig set pressure, $\Delta P = P_1 + 3 - P_2$.
- T = inlet temperature of absolute, °R = (°F + 460).
- Z = compressibility factor corresponding to T and P. If this factor is not available, compressibility correction can be safely ignored by using a value of Z = 1.0.
- C = gas or vapor flow constant; see Figure 23.25.
- k = ratio of specific heats, C_p/C_v. If a value is not known the use of k = 1.001, C = 315 will result in a safe valve size. Isentropic coefficient, n, may be used instead of k (Table 23.5).
- K_p = liquid capacity correction factor for overpressures lower than 25% for non-code liquids equations only (see Figure 23.22).
- K_b = vapor or gas flow correction factor for constant backpressures above critical pressure (see Figure 23.26).

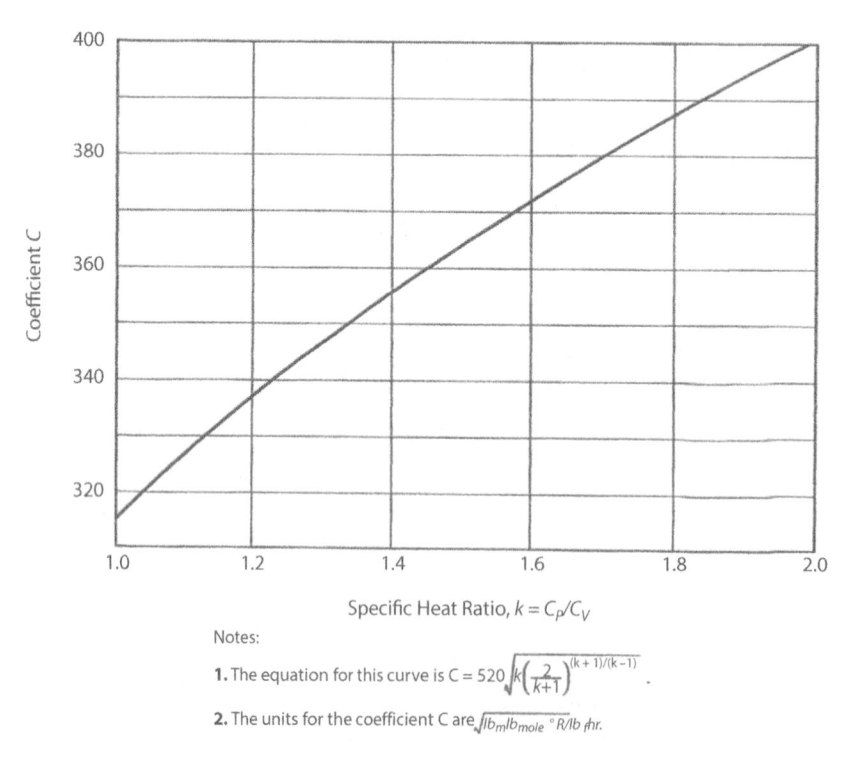

Notes:
1. The equation for this curve is $C = 520 \sqrt{k \left(\frac{2}{k+1}\right)^{(k+1)/(k-1)}}$.
2. The units for the coefficient C are $\sqrt{lb_m lb_{mole} \, °R/lb \, hr}$.

Figure 23.25 Constant "C" for gas or vapor related to specific heats. (By permission from Sizing, Selection and Installation of Pressure-Relieving Devices in Refineries, Part I "Sizing and Selection", API RP – 520, 5th ed., July 1990.)

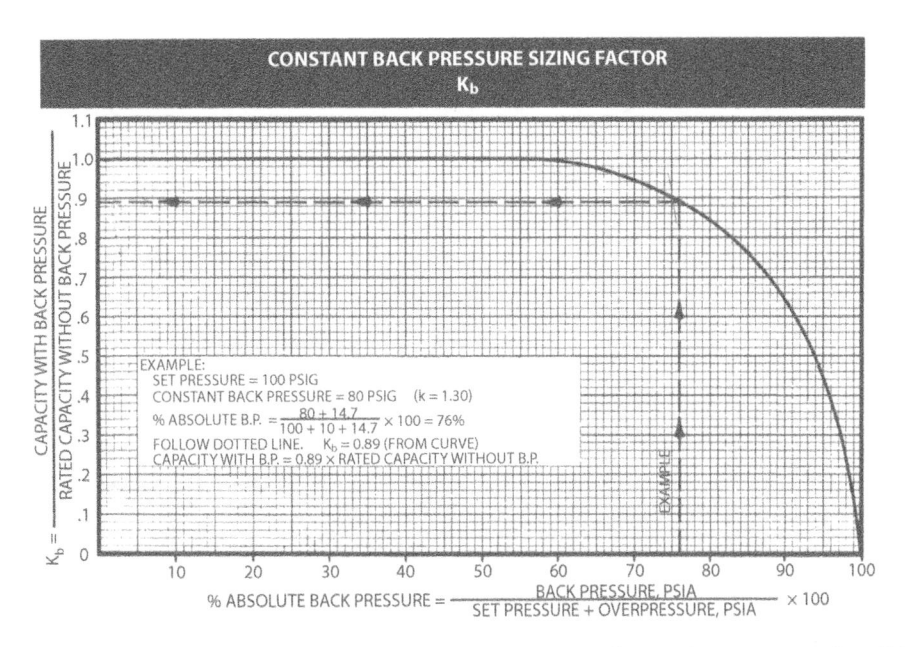

Figure 23.26 Constant back pressure sizing factor, K_b, conventional valves – vapors and gases. (By permission from Teledyne Farris Engineering Co.)

K_v = vapor or gas flow factor for variable backpressures for balanced seal valves only (see Figures 23.27a and 23.27b).

K_w = liquid flow factor for variable backpressures for balanced seal valves only (see Figure 23.28). For atmos., $K_w = 1.0$.

K_u = liquid viscosity correction factor (see Figures 23.23 or 23.23a).

K_{sh} = steam superheat correction factor (see Table 23.10).

K_n = Napier steam correction factor for set pressures between 1500 and 2900 psig (see Table 23.11).

K_d = coefficient of discharge [42]:*
  0.953 for air, steam, vapors and gases[†]
  0.724 for ASME Code liquids
  0.64 for non-ASME Code liquids
  0.62 for rupture disks and non-reclosing spring loaded devices ASME [1], Par. UG – 127

*0.975 per API RP-520, balanced valve.

[†] Some manufacturers' National Board-Certified Tests will have different values for some of their valves. Be sure to obtain the manufacturer's certified coefficient for the valve you select.

To convert flow capacity from scfm to lb/h use

$$W = \frac{(M)(V)}{6.32}$$

where

  M = molecular weight of flowing media.
  V = flow capacity, sfcm
  W = flow capacity, lb/h

Where the pressure relief valve is used in series with a rupture disk, a combination capacity of 0.8 must be applied to the denominator of the referenced equations. Refer to a later section of this text or to specific manufacturers.

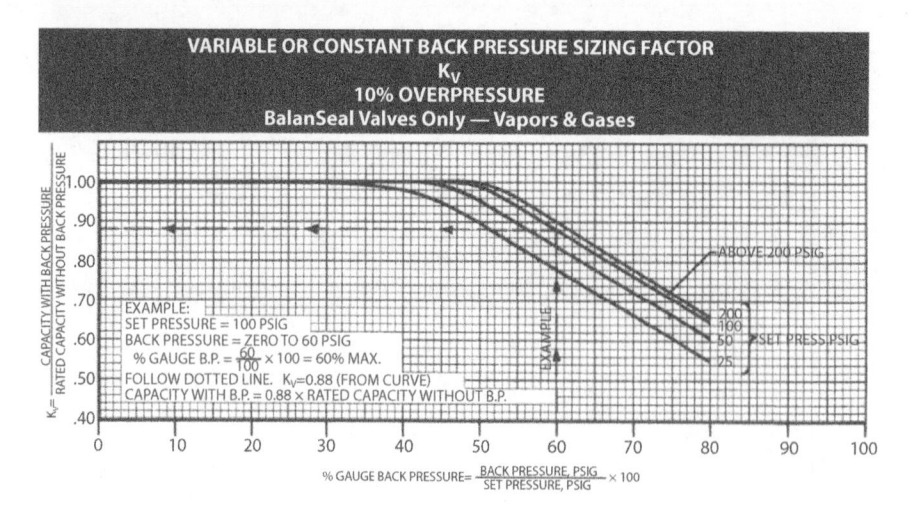

Figure 23.27a Variable or constant back pressure sizing factor K_v, at 10% overpressure. BalanSeal® valves only – vapors and gases. (By permission from Teledyne Farris Engineering Co.)

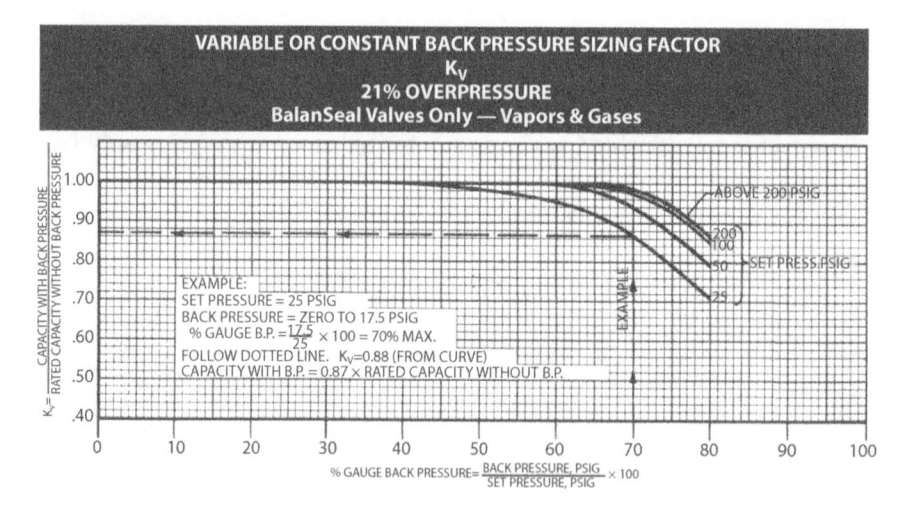

Figure 23.27b Variable or constant back pressure sizing factor, K_v, at 21% overpressure, BalanSeal® valves only – vapors and gases. (By permission from Teledyne Farris Engineering Co.)

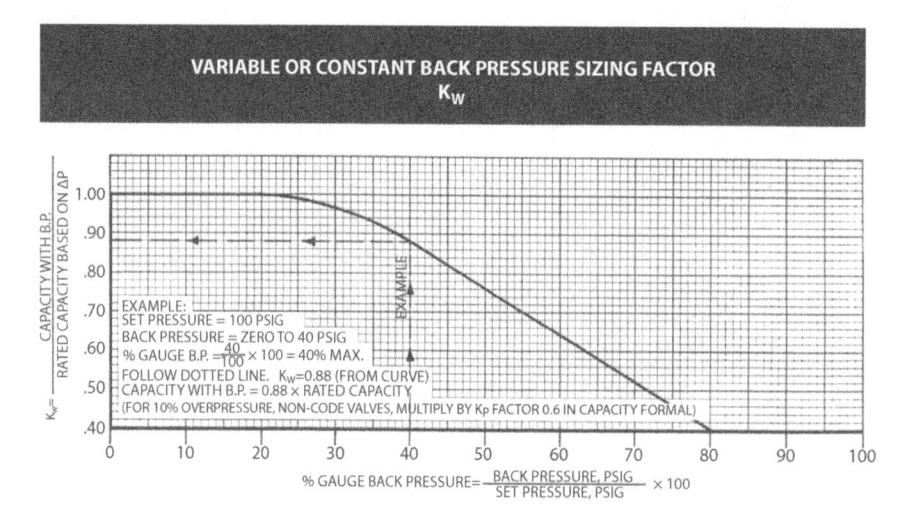

Figure 23.28 Variable or constant back pressure factor, K_w, for liquids only, BalanSeal® valves. Use this factor as a divisor to results of constant back pressure equations or tables. (By permission from Teledyne Farris Engineering Co.)

PRESSURE RELIEVING DEVICES AND EMERGENCY RELIEF SYSTEM DESIGN 77

Table 23.10 Steam Superheat Correction Factors, K_{sh}.

Set pressure psig	Saturated steam temperature °F.	Total temperature in degrees fahrenheit																																					
		280	300	320	340	360	380	400	420	440	460	480	500	520	540	560	580	600	620	640	660	680	700	720	740	760	780	800	820	840	860	880	900	920	940	960	980	1000
15	250	100	100	100	99	99	98	98	97	96	95	94	93	92	91	90	89	88	87	86	86	85	84	83	83	82	81	81	80	79	79	78	78	77	76	76	75	75
20	259	100	100	100	99	99	98	98	97	96	95	94	93	92	91	90	89	88	87	86	86	85	84	83	83	82	81	81	80	79	79	78	78	77	77	76	75	75
40	278	-	100	100	100	99	98	98	97	96	95	94	93	92	91	90	89	88	87	87	86	85	84	84	83	82	82	81	80	80	79	79	78	77	77	76	75	75
60	308	-	-	100	100	99	99	98	98	96	95	94	93	92	91	90	89	88	87	87	86	85	84	84	83	82	82	81	80	80	79	79	78	77	77	76	75	75
80	324	-	-	-	100	99	99	99	87	96	95	94	93	92	91	90	89	89	88	87	86	85	84	84	83	82	82	81	80	80	79	79	78	77	77	76	76	75
100	338	-	-	-	-	100	100	99	98	96	96	94	93	92	91	90	90	89	88	87	86	85	85	84	83	82	82	81	80	80	79	79	78	77	77	76	76	75
120	350	-	-	-	-	100	100	99	98	97	96	95	94	93	92	91	90	89	88	87	86	85	85	84	83	82	82	81	80	80	79	79	78	77	77	76	76	75
140	361	-	-	-	-	-	100	100	98	97	96	95	94	93	92	91	90	89	88	87	86	86	85	84	83	82	82	81	80	80	79	79	78	78	77	76	76	75
160	371	-	-	-	-	-	100	100	99	98	97	95	94	93	92	91	90	89	88	87	86	86	85	84	83	82	82	81	80	80	79	79	78	78	77	76	76	75
180	380	-	-	-	-	-	-	100	99	98	97	96	95	93	92	91	90	89	88	87	86	86	85	84	83	82	82	81	80	80	79	79	78	78	77	77	76	75
200	388	-	-	-	-	-	-	100	99	98	97	96	95	93	92	91	90	89	88	87	86	86	85	84	83	82	82	81	81	80	79	79	78	78	77	77	76	75
220	395	-	-	-	-	-	-	100	99	98	97	96	95	94	92	91	90	89	88	87	87	86	85	84	83	82	82	81	81	80	79	79	78	78	77	77	76	75
240	403	-	-	-	-	-	-	-	100	99	98	97	95	94	93	92	90	89	88	87	87	86	85	84	83	82	82	81	81	80	79	79	78	78	77	77	76	76
260	409	-	-	-	-	-	-	-	100	99	98	97	96	94	93	92	91	90	89	88	87	86	85	84	84	83	82	82	81	80	80	79	79	78	78	77	77	76
280	416	-	-	-	-	-	-	-	100	99	98	97	96	95	93	92	91	90	89	88	87	86	85	85	84	83	82	82	81	80	80	79	79	78	78	77	77	76
300	422	-	-	-	-	-	-	-	-	100	99	98	96	95	93	92	91	90	89	88	87	86	85	85	84	83	82	82	81	80	80	79	79	78	78	77	77	76
350	436	-	-	-	-	-	-	-	-	100	100	98	97	96	94	93	91	90	89	88	87	86	86	85	84	83	83	82	81	81	80	80	79	79	78	78	77	76
400	448	-	-	-	-	-	-	-	-	-	100	99	98	96	95	93	92	91	90	89	88	87	86	85	84	83	83	82	81	81	80	80	79	79	78	78	77	76
450	460	-	-	-	-	-	-	-	-	-	100	99	98	97	96	94	92	91	90	89	88	87	86	85	85	84	83	82	82	81	80	80	79	79	78	78	77	76
500	470	-	-	-	-	-	-	-	-	-	-	100	99	98	96	94	93	92	91	89	88	87	87	86	85	84	83	82	82	81	80	80	79	79	78	78	77	76
550	480	-	-	-	-	-	-	-	-	-	-	100	99	98	97	95	94	92	91	90	89	88	87	86	85	84	83	83	82	81	81	80	80	79	79	78	77	76
600	489	-	-	-	-	-	-	-	-	-	-	-	100	99	97	96	94	93	91	90	89	88	87	86	85	84	84	83	82	81	81	80	80	79	79	78	77	76
650	497	-	-	-	-	-	-	-	-	-	-	-	100	99	98	96	95	93	92	91	89	88	87	86	85	84	84	83	82	81	81	80	80	79	79	78	77	76
700	506	-	-	-	-	-	-	-	-	-	-	-	-	100	99	97	96	94	93	91	90	89	88	87	86	85	84	83	82	82	81	80	80	79	79	78	77	76
750	513	-	-	-	-	-	-	-	-	-	-	-	-	100	99	98	96	95	93	92	90	89	88	87	86	85	84	83	83	82	81	81	80	79	79	78	77	76
800	520	-	-	-	-	-	-	-	-	-	-	-	-	-	100	99	97	95	94	92	91	90	88	87	86	85	84	84	83	82	81	81	80	79	78	78	77	76
850	527	-	-	-	-	-	-	-	-	-	-	-	-	-	100	99	98	96	94	93	92	90	89	88	87	86	85	84	83	82	81	81	80	79	78	78	77	76

(Continued)

Table 23.10 Steam Superheat Correction Factors, K_{sh}. (Continued)

Set pressure psig	Saturated steam temperature °F.	Total temperature in degrees fahrenheit																																				
		280	300	320	340	360	380	400	420	440	460	480	500	520	540	560	580	600	620	640	660	680	700	720	740	760	780	800	820	840	860	880	900	920	940	960	980	1000
900	533	-	-	-	-	-	-	-	-	-	-	-	-	-	100	100	99	97	95	93	92	90	89	88	87	86	85	84	83	82	81	81	80	79	79	78	77	77
950	540	-	-	-	-	-	-	-	-	-	-	-	-	-	-	100	99	97	95	94	92	91	89	88	87	86	85	84	83	82	82	81	80	79	79	78	77	77
1000	546	-	-	-	-	-	-	-	-	-	-	-	-	-	-	100	99	98	96	94	93	91	90	89	87	86	85	84	83	83	82	81	80	79	79	78	77	77
1050	552	-	-	-	-	-	-	-	-	-	-	-	-	-	-	100	100	99	97	95	93	92	90	89	88	87	86	85	84	83	82	81	80	80	79	78	77	77
1100	558	-	-	-	-	-	-	-	-	-	-	-	-	-	-	100	100	99	98	96	94	92	91	89	88	87	86	85	84	83	82	81	80	80	79	78	78	77
1150	563	-	-	-	-	-	-	-	-	-	-	-	-	-	-	-	100	99	98	97	95	93	91	90	88	87	86	85	84	83	82	81	81	80	79	78	78	77
1200	569	-	-	-	-	-	-	-	-	-	-	-	-	-	-	-	100	99	99	97	95	93	91	90	89	87	86	85	84	83	82	81	81	80	79	78	78	77
1250	574	-	-	-	-	-	-	-	-	-	-	-	-	-	-	-	-	100	99	98	96	94	92	91	89	88	87	85	85	84	83	82	81	80	79	79	78	77
1300	579	-	-	-	-	-	-	-	-	-	-	-	-	-	-	-	-	100	99	98	97	95	93	91	89	88	87	86	85	84	83	82	81	80	80	79	78	78
1350	584	-	-	-	-	-	-	-	-	-	-	-	-	-	-	-	-	-	100	99	97	95	93	92	90	89	88	86	85	84	83	82	81	81	80	79	78	78
1400	588	-	-	-	-	-	-	-	-	-	-	-	-	-	-	-	-	-	100	99	97	96	94	92	91	89	88	86	85	84	83	82	81	81	80	79	78	78
1450	593	-	-	-	-	-	-	-	-	-	-	-	-	-	-	-	-	-	100	99	98	96	94	93	91	89	88	87	86	84	83	82	82	81	80	79	78	78
1500	597	-	-	-	-	-	-	-	-	-	-	-	-	-	-	-	-	-	-	100	98	97	95	93	91	90	88	87	86	84	83	83	82	81	80	79	78	77
1600	606	-	-	-	-	-	-	-	-	-	-	-	-	-	-	-	-	-	-	100	100	98	96	94	92	90	88	87	86	85	84	83	82	81	80	79	78	77
1700	615	-	-	-	-	-	-	-	-	-	-	-	-	-	-	-	-	-	-	-	100	98	97	95	92	91	89	87	86	85	84	83	82	81	80	79	77	76
1800	622	-	-	-	-	-	-	-	-	-	-	-	-	-	-	-	-	-	-	-	100	98	97	95	93	91	89	87	86	85	84	83	82	81	80	79	77	76
1900	630	-	-	-	-	-	-	-	-	-	-	-	-	-	-	-	-	-	-	-	-	100	97	96	93	91	89	87	86	84	83	82	81	80	79	78	77	76
2000	636	-	-	-	-	-	-	-	-	-	-	-	-	-	-	-	-	-	-	-	-	100	98	96	94	91	89	87	86	84	83	82	81	80	79	78	76	75
2100	644	-	-	-	-	-	-	-	-	-	-	-	-	-	-	-	-	-	-	-	-	100	98	96	94	92	89	87	86	84	83	82	80	79	78	77	76	75
2200	650	-	-	-	-	-	-	-	-	-	-	-	-	-	-	-	-	-	-	-	-	100	98	96	94	92	90	87	86	84	83	82	80	79	78	77	76	75
2300	658	-	-	-	-	-	-	-	-	-	-	-	-	-	-	-	-	-	-	-	-	100	98	96	94	92	90	87	86	84	83	82	80	79	78	77	76	75
2400	663	-	-	-	-	-	-	-	-	-	-	-	-	-	-	-	-	-	-	-	-	100	98	96	94	92	90	87	86	84	83	82	80	79	78	77	76	75
2500	669	-	-	-	-	-	-	-	-	-	-	-	-	-	-	-	-	-	-	-	-	100	98	96	94	92	90	87	86	84	83	82	80	79	78	77	76	75
2600	675	-	-	-	-	-	-	-	-	-	-	-	-	-	-	-	-	-	-	-	-	-	100	96	93	92	90	87	86	84	83	82	80	79	78	77	76	75
2700	680	-	-	-	-	-	-	-	-	-	-	-	-	-	-	-	-	-	-	-	-	-	100	95	93	91	90	87	86	84	83	82	80	79	78	77	76	75
2800	686	-	-	-	-	-	-	-	-	-	-	-	-	-	-	-	-	-	-	-	-	-	-	95	93	90	89	87	86	84	83	82	80	79	78	77	76	75
2900	691	-	-	-	-	-	-	-	-	-	-	-	-	-	-	-	-	-	-	-	-	-	-	-	-	87	86	84	83	82	80	79	78	77	76	75	74	73

Courtesy of Teledyne-Farris Engineering Co., Cat. 187C.

Table 23.11 Napier Steam Correction Factors, K_n, for Set Pressures between 1500 and 2900 psig.

Sizing factors for steam

K_n Napier correction factor for set pressures between 1500 and 2900 at 10% overpressure

Equation:

$$K_n = \frac{0.1906P - 1000}{0.2292P - 1061} \text{ where}$$

P = relieving pressure, psia

Set Pres. psig	K_n	Set Pres. psig	K_n	Set Pres. psig	K_n	Set Pres. psig	K_n	Set Pres. psig	K_n	Set Pres. psig	K_n	Set Pres. psig	K_n
1500	1.005	1640	1.014	1780	1.025	1920	1.037	2060	1.050	2200	1.066	2340	1.083
1510	1.005	1650	1.015	1790	1.026	1930	1.038	2070	1.051	2210	1.067	2350	1.085
1520	1.006	1660	1.016	1800	1.026	1940	1.039	2080	1.052	2220	1.068	2360	1.086
1530	1.007	1670	1.016	1810	1.027	1950	1.040	2090	1.053	2230	1.069	2370	1.087
1540	1.007	1680	1.017	1820	1.028	1960	1.040	2100	1.054	2240	1.070	2380	1.089
1550	1.008	1690	1.018	1830	1.029	1970	1.041	2110	1.055	2250	1.072	2390	1.090
1560	1.009	1700	1.019	1840	1.030	1980	1.042	2120	1.057	2260	1.073	2400	1.092
1570	1.009	1710	1.019	1850	1.031	1990	1.043	2130	1.058	2270	1.074	2410	1.093
1580	1.010	1720	1.020	1860	1.031	2000	1.044	2140	1.059	2280	1.075	2420	1.095
1590	1.011	1730	1.021	1870	1.032	2010	1.045	2150	1.060	2290	1.077	2430	1.096
1600	1.011	1740	1.021	1880	1.033	2020	1.046	2160	1.061	2300	1.078	2440	1.098
1610	1.012	1750	1.023	1890	1.034	2030	1.047	2170	1.062	2310	1.079	2450	1.099
1620	1.013	1760	1.023	1900	1.035	2040	1.048	2180	1.063	2320	1.081	2460	1.101
1630	1.014	1770	1.024	1910	1.036	2050	1.049	2190	1.049	2330	1.082	2470	1.102

Set Pres. psig	K_n	Set Pres. psig	K_n	Set Pres. psig	K_n
2480	1.104	2620	1.128	2760	1.157
2490	1.105	2630	1.130	2770	1.159
2500	1.107	2640	1.132	2780	1.161
2510	1.109	2650	1.134	2790	1.164
2520	1.110	2660	1.136	2800	1.166
2530	1.112	2670	1.138	2810	1.169
2540	1.114	2680	1.140	2820	1.171
2550	1.115	2690	1.142	2830	1.174
2560	1.117	2700	1.144	2840	1.176
2570	1.119	2710	1.146	2850	1.179
2580	1.121	2720	1.148	2860	1.181
2590	1.122	2730	1.150	2870	1.184
2600	1.124	2740	1.152	2880	1.187
2610	1.126	2750	1.155	2890	1.189
				2900	1.192

Courtesy Teledyne-Farris Engineering Co., Cat. 187C.

23.24 Sizing Valves for Liquid Expansion (Hydraulic Expansion of Liquid Filled Systems/Equipment/Piping)

The API Code RP-520 [5a] suggests the following to determine the liquid expansion rate to protect liquid filled (full) systems or locations where liquid could be trapped in parts of a system or an area could be subjected to blockage by process or operational accident. When thermal input from any source can/could cause thermal expansion of the enclosed liquid:

$$\text{gpm} = BH/(500\,G\,C_h) \qquad (23.24)$$

This relation can be converted to solve for the required orifice area at 25% overpressure for non-viscous liquids discharging to atmosphere [31]

$$A = BH/(13600)(P_1 G)^{0.5},\,\text{in}^2. \qquad (23.24a)$$

where

- gpm = flow rate at the flowing temperature, US gal/min.
- B = cubical expansion coefficient per °F for the liquid at the expected temperature. This information is obtained from the process design data; however, for values, see specific liquid data or see table below for typical values.
- H = total heat transfer rate, Btu/h. For heat exchangers, Ref [5a] recommends that this value be taken as the maximum exchanger duty during operation.
- G = specific gravity referred to water =1.0 at 60°F. Ignore liquid compressibility.
- C_h = specific heat of the trapped fluid, Btu/lb/°F
- A = required orifice area, in².
- P_1 = set pressure of valve, psig

Reference [5a] in Appendix D cautions that if the vapor pressure of the fluid at the temperature is greater than the relief set/design pressure, then the valve must be capable of handling the rate of vapor generation. Other situations should be examined as the thermal relief by itself may be insufficient for total relief.

Typical Values for Cubic Expansion Coeffcient*

Liquid	Value
3-34.9 °API gravity	0.0004
35-50.9 °API gravity	0.0005
51-63.9 °API gravity	0.0006
64-78.9 °API gravity	0.0007
79-88.9 °API gravity	0.0008
89-93.9 °API gravity	0.00085
94-100 °API gravity and lighter	0.0009
Water	0.0001

(Source: By permission API, Ref [5a])

Note: Reference of the API gravity values to refinery and petrochemical plant fluids will show that they correspond to many common hydrocarbons.

The method provided is only for short-term protection in some cases. If the blocked-in liquid has a vapor pressure higher than the relief design pressure, the pressure relief device should be capable of handling the vapor generation rate. However, if discovery and correction before liquid boiling are expected, then vaporization will not be accounted for sizing the pressure-relief device.

23.25 Sizing Valves for Subcritical Flow: Gas or Vapor But Not Steam [5d]

If the ratio of backpressure to inlet pressure to valve exceeds the critical pressure ratio, P_c/P_1

$$\left(\frac{P_c}{P_1}\right) = \left[\frac{2}{(k+1)}\right]^{k/(k-1)}$$

the flow through the valve is subcritical. The required area (net, free unobstructed) is calculated for a conventional relief valve, including sizing a pilot-operated relief valve [5d]:

$$A = \frac{W}{735 \times F_2 K_d K_c} \sqrt{\frac{ZT}{MP_1(P_1 - P_2)}}, \text{in}^2. \qquad (23.25a)$$

or

$$A = \frac{V}{4645 \times F_2 K_d K_c} \sqrt{\frac{ZTM}{P_1(P_1 - P_2)}}, \text{in}^2. \qquad (23.25b)$$

or

$$A = \frac{V}{864 \times F_2 K_d K_c} \sqrt{\frac{ZTG}{P_1(P_1 - P_2)}}, \text{in}^2. \qquad (23.25c)$$

In S.I. Units:

$$A = \frac{17.9 \times W}{F_2 K_d K_c} \sqrt{\frac{ZT}{MP_1(P_1 - P_2)}}, \text{mm}^2 \qquad (23.26a)$$

or

$$A = \frac{47.95 \times V}{F_2 K_d K_c} \sqrt{\frac{ZTM}{P_1(P_1 - P_2)}}, \text{mm}^2 \qquad (23.26b)$$

or

$$A = \frac{258 \times V}{F_2 K_d K_c} \sqrt{\frac{ZTG}{P_1(P_1 - P_2)}}, mm^2 \qquad (23.26c)$$

When using a balanced/bellows relief valve in the sub-critical, use Equations 23.18 through 23.22; however, the backpressure correction factor for this condition should be supplied by the valve manufacturer [5d]. For sub-critical, conventional valve:

$$F_2 = \sqrt{\left[\frac{k}{k-1}\right](r)^{2/k}\left[\frac{1-(r)^{(k-1)/k}}{1-r}\right]} \qquad (23.27)$$

where

A = required effective discharge area of the device, in² (mm²)
W = required flow through the device, lb/h (kg/h)
F_2 = coefficient of subcritical flow = $\sqrt{\left(\frac{k}{k-1}\right)(r)^{2/k}\left[\frac{1-r^{(k-1)/k}}{1-r}\right]}$ (Figure 23.29);
r = ratio of backpressure to upstream relieving pressure, P_2/P_1
K_d = effective coefficient of discharge. For preliminary sizing
    (i) 0.975 when a pressure relief valve is installed with or without a rupture disk in combination.
    (ii) 0.62 when a pressure relief valve is not installed and sizing is for a rupture disk.
K_c = combination correction factor for installation with a rupture disk upstream of the pressure relief valve.
   = 1.0 when a rupture disk is not installed.
   = 0.9 when a rupture disk is installed in combination with a pressure relief valve and the combination does not have a published value.
Z = compressibility factor for the deviation of the actual gas from a perfect gas, evaluated at relieving inlet conditions.

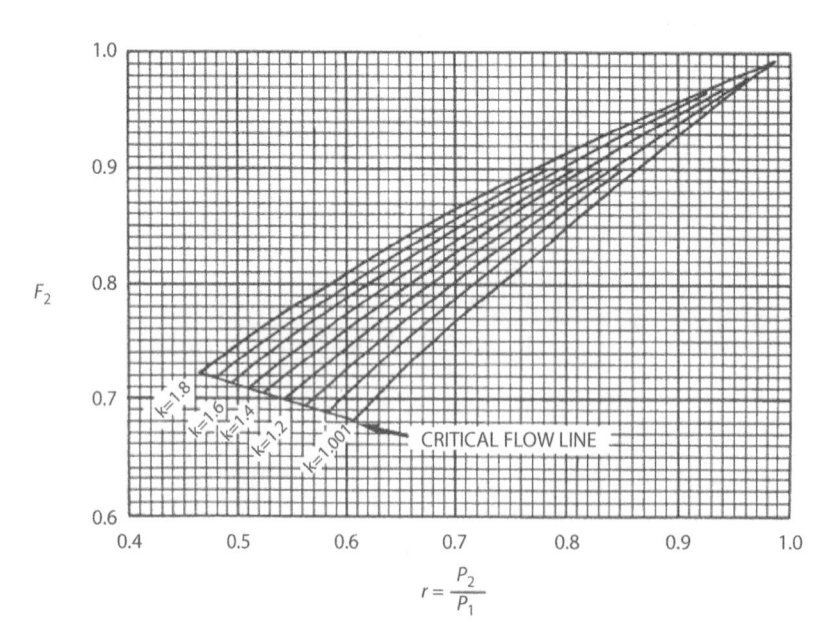

Figure 23.29 Values of F_2 for subcritical flow of gases and vapors. (By permission from Sizing, Selection and Installation of Pressure-Relieving Devices in Refineries, Part I "Sizing and Selection", API RP – 520, 5th ed., July 1990, American Petroleum Institute).

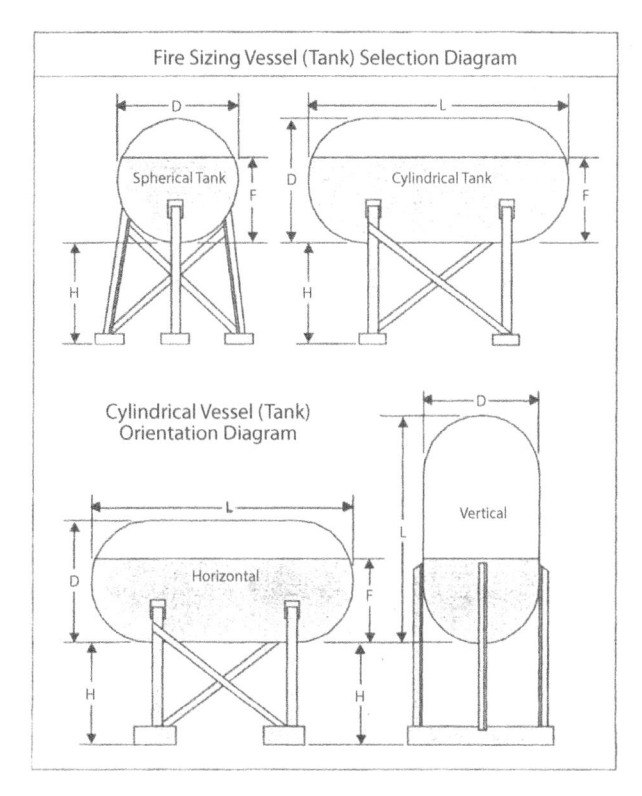

Figure 23.30 Vessel types for external fires.

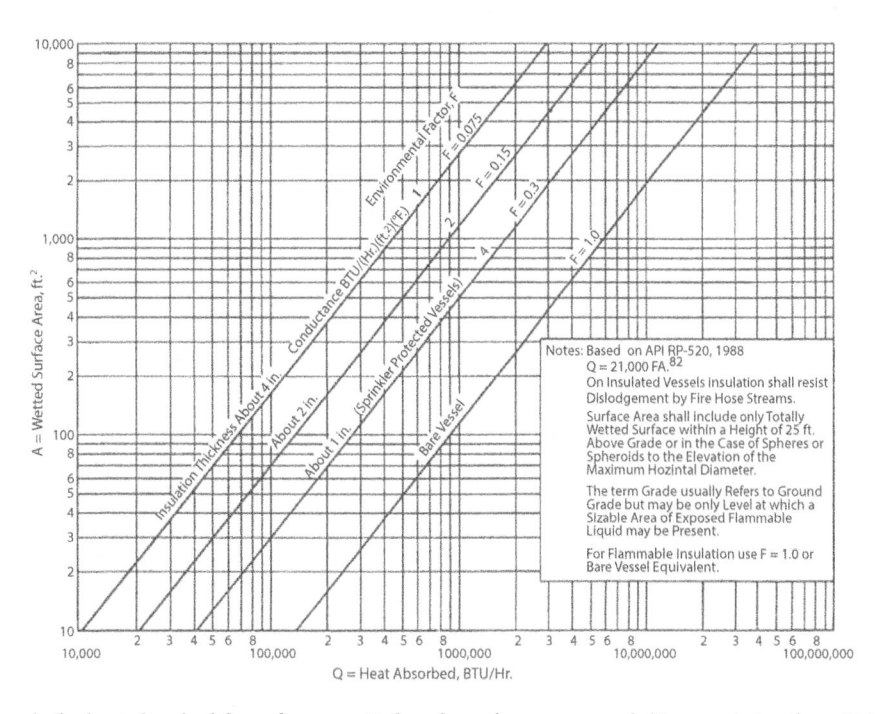

Figure 23.30a API formula for heat absorbed from fire on wetted surface of pressure vessel. (By permission from Sizing, Selection and Installation of Pressure-Relieving Devices in Refineries, Part I "Sizing and Selection", API RP – 520, 5th ed., July 1990).

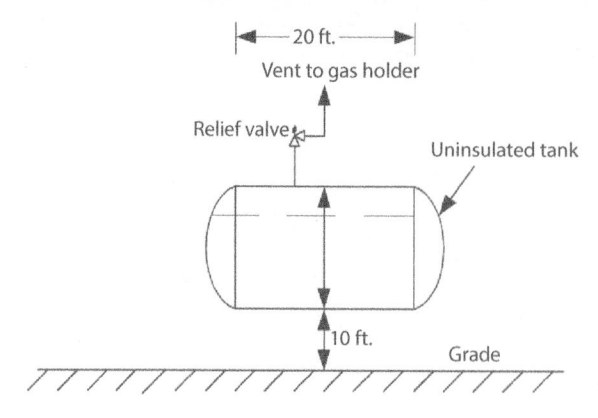

Figure 23.30b A horizontal, non-insulated storage tank of Example 23.5.

T = relieving temperature of the inlet gas or vapor, °R = °F + 460 (K = °C + 273)
M = molecular weight of the gas or vapor.
P_1 = upstream relieving pressure, psia (kPaa). This is the set pressure plus the allowable overpressure plus atmospheric pressure.
P_2 = backpressure, psia (kPaa).
V = required flow through the device, scfm at 14.7 psia and 60°F (Nm³/min at 101.325 kPaa and 0°C)
G = specific gravity of gas at standard conditions referred to air at standard conditions (normal conditions). G = 1 for air at 14.7 psia and 60°F (101.325 kPaa and 0°C)

23.26 Emergency Pressure Relief: Fires and Explosions Rupture Disks

Process systems can develop pressure conditions that cannot timely or adequately be relieved by pressure relieving valves as described earlier. These conditions are primarily considered to be (1) internal process explosions due to runaway reactions (see Design Institute for Emergency Relief Systems (DIERS) [13] in pressure vessels or similar containers such as an atmospheric grain storage silo (dust explosion typically) or storage bin; (2) external fires developed around, under, or encompassing a single process vessel or a system of process equipment, or an entire plant; and (3) other conditions in which rapid/instantaneous release of developed pressure and large volumes of vapor/liquid mixture is vital to preserve the integrity of the equipment. For these conditions, a rupture disk may perform a vital safety-relief function. Sometimes the combination of a rupture disk and pressure-relieving valve will satisfy a prescribed situation, but the valve cannot be relied on for instantaneous release (response time lag of usually a few seconds).

The ASME Pressure Vessel Code [1] and the API codes or recommended procedures [5, 8 and 9] recognize and set regulations and procedures for capacity design, manufacture, and installation of rupture disks, once the user has established the basis of capacity requirements.

23.27 External Fires

There have been at least six different formulas proposed and used to determine the proper and adequate size of rupture disk openings for a specific relieving condition. The earlier studies of Sylvander and Katz [33] led to the development of the ASME and API recommendations. This approach assumes that a fire exists under or around the various vessels in a process. This fire may have started from static discharge around flammable vapors, flammable liquids released into an area drainage ditch, combustible gas/vapors released through flange leaks or ruptures, overpressure, or many other potential hazards. The codes [9] suggest typical lists of potential hazards and some approach to determine the types of process conditions that can cause a fire. A suggested extended list (Figure 23.14a) is presented earlier in this chapter and in the following paragraphs. There are no formulas to establish a code capacity for the volume of vapors to be released under anyone of the possible plant "failure" conditions. Therefore, the codes assume, based on evaluation of test data, that fire is under or around the various vessels and that absorbed heat vaporizes the contained fluid. The information presented is taken from API-RP-520/521 latest editions [5, 8 and 9] and the ASME code [1]. The designer should be familiar with the details in these codes. Figure 23.30 shows various vessel types with dimensions for an external fire sizing.

23.28 Set Pressures for External Fires

The MAWP (discussed earlier) for the vessel or each vessel should be the maximum set pressure for the rupture disk. Furthermore, estimated flame temperatures of usually 2500–3500°F (1371–1927°C) should be used to establish the reduced vessel metal wall temperatures (recognizing the benefits of code recommended fireproof insulation if properly applied to prevent dislodging by fire water hose pressures impacting on the insulation). The MAWP should then be re-established by calculation using the metal wall code allowable stresses at the new estimated reduced metal temperature. This should be the maximum set pressure for the rupture disk provided the new lower value does not cause it to be below or too close to the usual expected process operating temperature. In such a case, the set pressure should be 25% above the operating condition, exclusive of fire, not exceeding the MAWP values [34].

When a rupture disk relieves/blows/ruptures, it creates a rapid depressuring of the process system and a likely discharge of some or all vapor/liquid in the vessel(s), discharging to the properly designed disposal system. Therefore, great care should be given to setting the rupture disk pressure because it does not have an accumulation factor, but bursts at the prescribed pressure of the disk, taking into account the code allowed manufacturing tolerances. Two-phase flow will most likely occur when the disc (or safety-relief valve) blows (see later references to explosions and DIERS work [13, 22]).

For unexpected runaway or process overpressure not subject to external fire, the rupture disk set pressure, which is the bursting pressure, should be sufficiently higher than the expected "under acceptable control" conditions for the operation to avoid the frequent burst and shutdown of the process. Usually this is found to be about 20–30% above the maximum expected peak operation pressure. Again, recognize the requirements for the relieving device set pressure not to exceed the actual vessel MAWP at the expected relieving temperature (by calculation or pilot plant test data).

23.29 Heat Absorbed

The amount of heat absorbed by a vessel exposed to an open fire is markedly affected by the size and character of the installation and by the environment. These conditions are evaluated by the following equivalent formulas, in which the effect of size on the heat input is shown by the exponent of A_w, the vessel wetted area, and the effect of other conditions, including vessel external insulation, is included in a factor F [5]:

$$q = 21,000 \, F \, A_w^{-0.18} \qquad (23.28)$$

$$Q = 21,000 \, F \, A_w^{+0.82} \qquad (23.29)$$

The Severe Case

Where the facility does not have prompt fire fighting equipment and inadequate drainage of flammable materials away from the vessel:

$$Q = 34,500\, F\, A_w^{+0.82} \tag{23.30}$$

where

q = average unit heat absorption, in Btu/h. ft² of wetted surface.
Q = total heat absorption (input) to the wetted surface, in Btu/h.
A_w = total wetted surface, ft².

The expression $A_w^{-0.18}$ is the area exposure factor or ratio. This recognizes that large vessels are less likely to be completely exposed to the flame of an open fire than small vessels. It is recommended that the total wetted surface (A in the foregoing formulas) be limited to that wetted surface included within a height of 25 feet above "grade" or, in the case of spheres and spheroids, to the elevation of the maximum horizontal diameter or a height of 25 feet, whichever is greater. (A more conservative approach is recommended.) The term "grade" usually refers to ground grade, but may be any level at which a sizable area of exposed flammable liquid could be present (Figure 23.30) [5, 8].

F = environment factor, values of which are shown in Table 23.10 for various types of installation.

Surface areas of vessel elliptical heads can be estimated by 1.15 × cross-sectional area of vessel.

These are the basic formulas for the usual installation, with good drainage and available fire-fighting equipment. These formulas are plotted on Figure 23.30a showing curves for Q for various values of factor F. The approximate amount of insulation corresponding to the factors is indicated.

Referring to the wetted surface, A_w, the surface areas of ASME flanged and dished head, ASME elliptical heads, hemispherical heads, and so on, are often the end assemblies on a cylindrical vessel. If a formula is not available to accurately estimate the wetted surface, or the blank diameters used for fabrication, which would give a close approximation of the inside surface of the head, use an estimated area for the dished or elliptical heads as 1.2 × cross-section area of the vessel based on its diameter.

23.30 Surface Area Exposed to Fire

The surface area of a vessel exposed to fire, which is effective in generating vapor, is that area wetted by its internal liquid contents. The liquid contents under variable level conditions should ordinarily be taken at the average inventory (e.g., see note below).

1. *Liquid-full vessels* (such as treaters) operate liquid full. Therefore, the wetted surface would be the total vessel surface within the height limitation.
2. *Surge drums* (vessels) usually operate about half full. Therefore, the wetted surface would be calculated at 50% of the total vessel surface, but higher if design is based on greater figure.
3. *Knockout drums* (vessels) usually operate with only a small amount of liquid. Therefore, the wetted surface would be in proportion, but to maximum design liquid level.
4. *Fractionating columns* usually operate with a normal liquid level in the bottom of the column and a level of liquid on each tray. It is reasonable to assume that the wetted surface be based on the total liquid within the height limitation—both on the trays and in the bottom.
5. *Working storage tanks'* wetted surface is usually calculated on the average inventory, but at least 25 ft height, unless liquid level can reasonably be established as higher; if established as higher, then use higher value. This should be satisfactory not only because it conforms to a probability, but also because it provides a factor of safety in the time needed to raise the usually large volume of the liquid's sensible heat to its boiling point.

It is recommended that the wetted area be at least to the height as defined in the definition of area, A_w.

Note: E.E. Ludwig [34] suggested that determining A_w values may be more conservative and not conform exactly to code [5a, c, and d] recommendations. The Code [5a], Part 1, Sect D, Par. D.4] reads, "to determine vapor generation, only that portion of the vessel that is wetted by its internal liquid and is equal to or less than 25 feet above the source of flame needs to be recognized."

> 6. *E. E. Ludwig's experience* [34] in investigating many industrial fires and explosions, it is suggested that the height limit of 25 ft above "grade" or fire source level is too low for many process plants, and therefore, the effect of a large external fire around equipment can reach to 100 ft with 75 ft perhaps being acceptably conservative. Ludwig expressed concern in using the 25-ft limit, for example, for a horizontal butane storage "bullet" tank, 15 ft diameter and raised 15 ft off grade to its bottom. Furthermore, the fact that any fire will engulf the entire vessel should be considered and the wetted surface should be the entire vessel. The same concern applies to a vertical distillation column over 25 ft high. He suggested that the wetted surface should be at least 80% of the vessel height and further recognized that the tray liquid will wet the walls and be evaporated only as long as there is liquid to drain off the trays. But for a conservative approach, he assumed that there are always wetted walls in the column.

For packed columns, the wetted walls are at least to the top of the packing, with some entrainment height above that. Therefore, a vertical packed column of 60 ft of packing above the liquid level in the sump, which has a sump of 10 ft and a skirt of 10 ft, would be considered 70 ft of vertical height of wetted perimeter, not counting the skirt. If the wetted area reached a total height of 80 ft from grade, minus the unwetted area of the skirt height, because no liquid is in the skirt space, the 80 ft would not be used to establish the vertical height for fire exposure, but use 60 ft + 10 ft, or 70 ft wetted height. This approach is more conservative than the code [5a, c and d], which is based on Ludwig's experience in investigating the damage levels from fires [34].

Each situation must be evaluated on its own merits or conditions and operating situation, and even its environment with respect to the plant flammable processing equipment.

23.31 Relief Capacity for Fire Exposure

In calculating the relief capacity to take care of external fire, the following equation is used:

$$W = Q/L \tag{23.31}$$

where

W = weight rate of flow of vapors, lb/h
L = latent heat at allowable pressure, Btu/lb.
Q = total heat absorption from external fire, Btu/h

23.32 Code Requirements for External Fire Conditions

Paragraph UG-125 (3) of the ASME code [1] requires that supplemental relieving capacity be available for an unfired pressure vessel subject to external accidental fire or other unexpected source of heat. For this condition, relieving devices must be installed to prevent the pressure from rising more than 21% [9] above the MAWP of the vessel. The set pressure should not exceed the vessel MAWP. A single relief device may be used for this capacity as long as it also meets the normal overpressure design for other possible causes of 10%. If desirable, multiple separate devices can be installed to satisfy both potential overpressure situations.

For this condition, the API-RP-521 code [9] (Figure 23.7a) shows an allowable 16% maximum accumulation relieving pressure above the set pressure. For external fire conditions on a vessel, the maximum allowable accumulation pressure is 21% above the set pressure [9] for both single or multiple relieving devices (Figure 23.7a).

23.33 Design Procedure

The usual procedure for determining relief area requirements is

1. Determine the external surface area exposed to fire, as set forth by

$$Q = 21,000 \, F \, A_w^{+0.82} \tag{23.29}$$

and Table 23.12 and Section 23.26

2. Determine the heat absorbed, Q, from Figure 23.30a.
3. Calculate the rate of vaporization of liquid from

$$W = \frac{Q}{L} \tag{23.31}$$

4. Verify critical pressure from Equation 23.7 and establish actual backpressure for relieving device.
5. Calculate relieving area by applicable equation for critical or non-critical flow, using the flow rate determined in (3) above. (See Equation 23.10 and following.) The area actually selected for orifice of safety type valve must have orifice equal to or greater than calculated requirements. For a rupture disk application, the full free open cross-sectional area of pipe connections in inlet and exit sides must be equal to or greater than the calculated area.
6. Select a valve or rupture disk to accommodate the service application.
7. To provide some external protection against the damage that an external fire can do to a pressure relief valve or rupture disk, Ludwig [34] recommends that these devices be insulated after installation in such a manner as not to restrict their action but to provide some measure of reliable performance, even if the vessel is not insulated.

Example 23.5

Calculate the required area for the relief valve for a horizontal non-insulated storage tank, exposed to fire and containing liquid vinyl chloride monomer ($CH_2 = CHCl$). The tank dimensions are shown in Figure 23.30b. The design pressure is 100 psig, and the discharge from the relief valve will be vented to a gas holder operating at 0.5 psig. A 20% accumulation (over pressure) is assumed over the design pressure. The average inventory of the tank contents will equal 75% of the vessel's inside diameter.

Solution

The wetted surface for the vessel equals the wetted area of the two heads plus that of the cylindrical section.

Thus:

$$A_w = \frac{2\pi (yD_1)^2}{4} + D_2 L \pi$$

Table 23.12 Environmental Factor, F.

Type of installation		Factor*
Bare vessel		1.0
Insulated vessel† (These arbitrary insulation conductance values are shown as examples and are in British Thermal Units per hour per square foot per °F)		
(a) 4.0 Btu/ hr/ sq ft/°F	(1 in thick)	0.3
(b) 2.0 Btu/ hr/ sq ft/°F	(2 in thick)	0.15
(c) 1.0 Btu/ hr/ sq ft/°F	(4 in thick)	0.075
(d) 0.67 Btu/ hr/ sq ft/°F		0.05
(e) 0.5 Btu/ hr/ sq ft/°F		0.0376
(f) 0.4 Btu/ hr/ sq ft/°F		0.03
(g) 0.32 Btu/ hr/ sq ft/°F		0.026
Water-application facilities, on bare vessel**		1.0
Depressurizing and emptying facilities††		1.0
Underground storage		0.0
Earth-covered storage above grade		0.03
*	These are suggested values for the conditions assumed in code [33] Par D 5.21. When these conditions do not exist, engineering judgment should be exercised either in selecting a higher factor in providing means of protecting vessels from fire exposure as suggested in [33], par. D.8.	
†	Insulation shall resist dislodgement by fire hose streams. For example, a temperature difference of 1600°F was used. These conductance values are based on insulation having conductivity of 48TS/hr-ft-°F per inch at 1600°F and correspond to various thicknesses of insulation between 1 and 12 inches.	
**	See code for recommendations regarding water application and insulation.	
††	Depressurizing will provide a lower factor if done promptly, but no credit is to be taken when safety valves are being sized for fire exposure. See [33], Part I, par. D. 8.2.	
	By permission, API-RP-520, American Petroleum Institute, Div. of Refining (1967) and adapted for this current edition by this author from later editions of the code (197) and (1990). Items d, e, and f above from API-RP-520, 5th Ed. (1990). For complete reference, see the latest code cited in its entirety.	

where

y = the fraction of the vessel's internal diameter (ID) that is equivalent to average liquid inventory.
D_1 = diameter of circular blank from which head is shaped (D_1 will depend on the type of head).
D_2 = y times the ID of the tank.
L = tangent-to-tangent length of the cylindrical section.

D_1 = 11 ft.

$$A_w = \frac{2\pi[(0.75)(11)]^2}{4} + (0.75)(10)(20)\pi$$

$$= 578.15 \text{ sq. ft.}$$

The heat input to the wetted surface of the vessel is given by Eq. (23.29),

$$Q = 21{,}000 \, F \, A_w^{+0.82}$$

where

F = environmental factor, F = 1.0 (Bare vessel).

$$Q = 21{,}000(1)(578.15)^{0.82}$$

$$= 3.864 \times 10^6 \text{ Btu/h.}$$

The mass flow rate of vinyl chloride m, lb/h is

$$m = Q/\lambda = 3.864 \times 10^6 / 116$$

$$= 33{,}314.6 \text{ lb/h.}$$

The relieving temperature at the set pressure plus 20% over pressure plus 14.7 is 135°F. At this temperature, the latent heat of vaporization is 116 Btu/lb. Molecular weight of vinyl chloride is = 62.5.

Ratio of specific heat capacities k = 1.17.
The relieving pressure, P
P = set pressure + overpressure + atmospheric pressure
 = 100 + (100 x 0.2) + 14.7
 = 134.7 psia.

The critical properties of vinyl chloride are
P_c = 809 psia.
T_c = 313.7°F

$$P_r = P/P_c = 134.7/809 = 0.167$$

$$T_r = T/T_r = (135+460)/(313.7+460) = 0.77$$

From the compressibility factor chart,

$$Z = 0.86$$

The value of C with k = 1.17 is obtained by substituting the value of k in Eq. 23.11h:

$$C = 520 \sqrt{k \left(\frac{2}{k+1}\right)^{(k+1)/(k-1)}}$$

$$C = 520\sqrt{1.17\left(\frac{2}{1.17+1}\right)^{(1.17+1)/(1.17-1)}}$$

$$C = 334.17$$

For vapors and gases, in lb/h; $K_b = 1.0$; "C" from Figure 23.25, P is the relieving pressure absolute, psia. Substituting the values in Eq. 23.10,

$$A = \frac{W\sqrt{TW}}{CK_d P K_b \sqrt{M}}, \text{ in}^2$$

$$A = \frac{(33314.5)\sqrt{(595)(0.86)}}{(334.17)(0.953)(134.7)(1)\sqrt{62.5}}$$
$$= 2.22 \text{ in}^2$$

The nearest standard orifice area is 2.853 in². The pressure relief valve designation is L and the preferred valve body size is 3–4 in. or 4–6 in.

The maximum vapor rate with the nearest standard orifice area is

$$W = ACK_d PK_b \sqrt{M/(TZ)}$$
$$= (2.853)(334.17)(0.953)(134.7)(1)\sqrt{62.5/(595 \times 0.86)}$$
$$= 42,772.2 \text{ lb/h}.$$

The Excel spreadsheet Example 23.5.xlsx shows the calculations for fire relief valve sizing, and Table 23.13 shows the results for fire relief condition for Example 23.5.

Table 23.13 Results of Relief Valve Sizing for Fire Flow.

Results of Relief Valve Sizing for Fire flow		
Relief Type	Fire	
Set Pressure at inlet	100	psig
Relieving Pressure	134.7	psig
Relieving Temperature	135	°F
Molecular weight, M_w	62.5	
Compressibility factor, Z	0.86	
K_d	0.953	
K_b	1	

Ratio of specific heat capacities, k=C_p/C_v	1.17	
Environmental factor, F	1	
Latent heat of vaporization, λ	116	Btu/lb
Calculated heat input, Q	3864493	Btu/h
Calculated mass flow rate of liquid, m	33314.6	lb/h
Calculated constant, C	334.17	
Calculated area, A	2.22	in²
Nearest orifice area	2.853	in²
Orifice Size	L	
Maximum Liquid flow rate	42772.2	lb/h

23.34 Pressure Relief Valve Orifice Areas on Vessels Containing Only Gas, Unwetted Surface

Due to gas expansion from external fire, the API code [8] provides for calculation of the pressure-relief valve orifice area for a gas containing vessel exposed to external fire on the unwetted surface:

$$A = F' A_3 / \sqrt{P_1}, \text{in}^2 \tag{23.32}$$

Based on air and perfect gas laws, vessel is uninsulated, and it will not reach rupture conditions. Review [5a] for specific design situations

where

    A = effective discharge area of valve, in²
    A_3 = exposed surface area of vessel, ft²
    F' = operating environment factor, minimum value recommended = 0.01
        when the minimum value is unknown, use F' = 0.045, which can be calculated by [5c]

$$F' = \frac{0.1406}{CK_d} \left[\frac{(T_\omega - T_1)^{1.25}}{T_1^{0.6506}} \right] \tag{23.33}$$

    P_1 = upstream relieving pressure, in psia. This is the set pressure plus the allowable overpressure plus the atmospheric pressure, psia.

where

    C = coefficient determined by the ratio of the specific heats of the gas at standard conditions. This can be obtained from Equation 2 [5a] in 4.3.2.1 of API Recommended Practice 520, Part I, or Figure 23.25.
    K_d = coefficient of discharge (obtainable from the valve manufacturer). K_d is equal to 0.975 for sizing relief valves.
    T_w = vessel wall temperature, in degrees Rankine (°R=°F + 460).
    T_1 = gas temperature, absolute, in degrees Rankine (°R=°F + 460), at the upstream pressure, determined from the following relationship:

$$T_1 = \frac{P_1}{P_\eta} T_\eta$$

where

P_η = normal operating gas pressure, psia
T_η = normal operating gas temperature, in degrees Rankine (°R=°F + 460).

The recommended maximum vessel wall temperature for the usual carbon steel plate materials is 1100°F. Where vessels are fabricated from alloy materials, the value for T_w should be changed to a more appropriate recommended maximum [5a].

The relief load can be calculated directly, in pounds per hour [5a]:

$$W = 0.1406\sqrt{MP_1}\left[A_3 \frac{(T_\omega - T_1)^{1.25}}{T_1^{1.1506}}\right], \text{lb/h} \qquad (23.34)$$

23.35 Rupture Disk Sizing Design and Specification

Rupture or burst pressure of the metal disks must be specified at least 25–40% greater than the normal non-pulsing operating pressure of the vessel or system being protected. For low pressures less than 5 to 10 psig operating, the differences between operating pressure and set pressure of a valve or disk may need to be greater than that just cited.

For mild pulsations, the disk bursting pressure should be 1.75 times the operating pressure; and for strong pulsations, use two times the operating pressure [43]. Non-metallic impregnated graphite disks may be used to burst at 1.34 times operating pressure as these are less subject to fatigue. The bursting pressure must never be greater than the MAWP of the vessel, and proper allowance must be made for the possible pressure variations, plus and minus, due to the manufacturer's rupture pressure range. See Table 23.2 and specific manufacturers' literature. The ASME Code [1] Par. UG-127 requires disks to burst within 5%± of the stamped bursting pressure at a specified disk temperature at time of burst.

23.36 Specifications to Manufacturer

When ordering rupture disks, the following information and specifications should be given to the manufacturer.

1. Net inside diameter of opening leading to the flange or holding arrangement for the disk, inches; or the cubic feet of vapor at stated conditions of burst pressure, or both.
2. Preferred material of construction, if known; otherwise state service to obtain recommendation.
3. Type of hold-down arrangement: flanged (slip on, weld neck, screwed, stud) union, screwed, or special.
4. Material of construction for hold-down (flange, screwed) arrangement. Usually forged carbon steel is satisfactory, although aluminum or other material may be required to match vessel and/or atmosphere surrounding the disk assembly.
5. Temperature for (a) continuous operation and (b) at burst pressure.
6. Required relief or burst pressure in vessel and the backpressure on the disk, if any.
7. Disks to be ASME Code certified.

When the flow capacity for relief can be given to the disk manufacturer, together with the conditions at bursting pressure (including temperature), the manufacturer can check against a selected size and verify the ability of the disk to relieve the required flow.

23.37 Size Selection

Rupture disks are used for the same purpose as safety valves and, in addition, serve to relieve internal explosions in many applications. If the pressure rise can be anticipated, then the volume change corresponding to this change can be calculated by simple gas laws, and the capacity of the disk at the relieving pressure is known. The system must be examined and the possible causes of overpressure and their respective relief capacities identified before a reliable size can be determined. See Figure 23.14a.

23.38 Calculation of Relieving Areas: Rupture Disks for Non-Explosive Service

The vessel nozzle diameter (inside) or net free area for relief of vapors through a rupture disk for the usual process applications is calculated in the same manner as for a safety relief valve, except that the nozzle coefficient is 0.62 for vapors and liquids. Most applications in this category are derived from predictable situations where the flow rates, pressures, and temperatures can be established with a reasonable degree of certainty.

For rupture disk sizing the downstream pressure is assumed to reach the critical flow pressure, although the downstream pressure initially may be much lower. Under these conditions, the flow through the "orifice" that the disk produces on rupture is considered to be at critical flow. The assumptions of critical pressure do not apply where a fixed downstream side pressure into which the disk must relieve is greater than the critical pressure.

The coefficient of discharge, K_d, is the actual flow divided by the theoretical flow and must be determined by tests for each type or style and size of rupture disk as well as pressure-relieving valve. For rupture disks, the minimum net flow area is the calculated net area after a complete burst of the disk, making allowance for any structural members that could reduce the net flow area of the disk. For sizing, the net flow area must not exceed the nominal pipe size area of the rupture disk assembly [1].

The bursting pressure, P_b, of the conventional tension loaded disk is a function of the material of which the disk is fabricated, as well as its thickness and diameter; the temperature at which the disk is expected to burst, and not just the temperature corresponding to the disk set pressure [17]. This type disk is best suited to be set on P_b at least 30% above the system operating pressure. The reverse-buckling disk with knives to aid the bursting is compression loaded because the dome of the disk faces the internal vessel pressure. The bursting pressure, P_b, of this disk is dependent on the dome's geometrical shape and the characteristics of the knife blades, but it is essentially independent of thickness and generally does not need a vacuum support [17]. It is often used when the operating pressure is as high as $0.9P_b$. There are some potential problems or even hazards with this design if the knives fail, come loose, or corrode, and the use must be examined carefully. This disk like any other should never be installed upside down from its original design position. It should not be used in partial or total liquid service.

The reverse-buckling disk, without knives but with a pre-scored disk surface, offers some features that do not depend on the knives being in place because the thickness of the metal disk dome along the score line determines the bursting pressure of the disk.

23.39 The Manufacturing Range (MR)

The ASME code [1] requires that a ruptured disk must be stamped with a bursting pressure that falls within the manufacturing range (MR). This range identifies the allowable range of variation from a specified burst pressure to the actual burst pressure provided by the manufacturer and as agreed upon with the disk user. The stamped burst pressure of a lot of rupture disks is the average burst pressure of all the destructive tests performed per code requirements. The average of the tests must fall within the manufacturing range (see Table 23.2).

The thickness of one material for manufacture of a disk, along with the specific disk type, is the key factor in establishing at what pressure range a disk of a specified bursting diameter will actually burst on test and then in actual service.

To specify a rupture disk:

- identify the desired bursting pressure, P_b
- list the required MR

For example, a system requiring a bursting pressure, P_b, of 150 psig at 400°F would have an MR range of -4% to +7% at the operating temperature with a burst pressure tolerance of ±5%. The disk supplied by a specific manufacturer (MR varies with manufacturer and pressure ranges) could have a bursting pressure as low as -(0.04)(150) = -6 psi, or 144 psig; or as high as (+0.07)(150) = +10.5 or 160.5 psig. If the disk is stamped at the operating temperature at 144 psig, it could burst at ±5% or 7.2 psi or 136.8 psig or 151.2 psig. On the other hand, if the disk were stamped at operating temperature to burst at 160.5 psig (its highest), then it could actually burst ±5% of this, or ±8.03 psi, giving an actual burst pressure of 168.5 psig to 152.5 psig. Adapted from [17] by permission.

The code requires that the disks be burst on test by one of three methods using four sample disks, but not less than 5% from each lot. Figure 23.31 illustrates test results for burst pressure versus temperature of a disk design, all fabricated from the same material, and of the same diameter.

It is critical that such an examination be made to be certain that the bursting or set pressure at this temperature does not exceed the MAWP of the vessel at the operating temperature per the ASME code [1] (see Figures 23.32a and b).

As allowed by the code [1], the average of the manufacturer's disks burst tests could be stamped, for example, (144 + 160.5)/2 = 152.3 psig with an actual ±5% of 152.3 psig allowed for actual burst pressure of any disk *at the operating temperature*.

23.40 Selection of Burst Pressure for Disk, P_b (Table 23.3)

It is essential to select the type or style of rupture disk before making the final determination of the final burst pressure, and even this selection must recognize the pressure relationships between the disk's manufacturing range and the vessel's MAWP. (Also see Figures 23.32a and b.)

Table 23.14 summarizes the usual recommended relationship between the operating pressure of a process (should be the maximum expected upper range level) and the set pressure of the rupture disk. Recognize that the set pressure of the disk must not exceed the MAWP of the vessel (see Figures 23.32a and b). The burst pressure, P_b, can now be defined. The use of the MR discussed earlier applied to the burst pressure will establish the most probable maximum P_b.

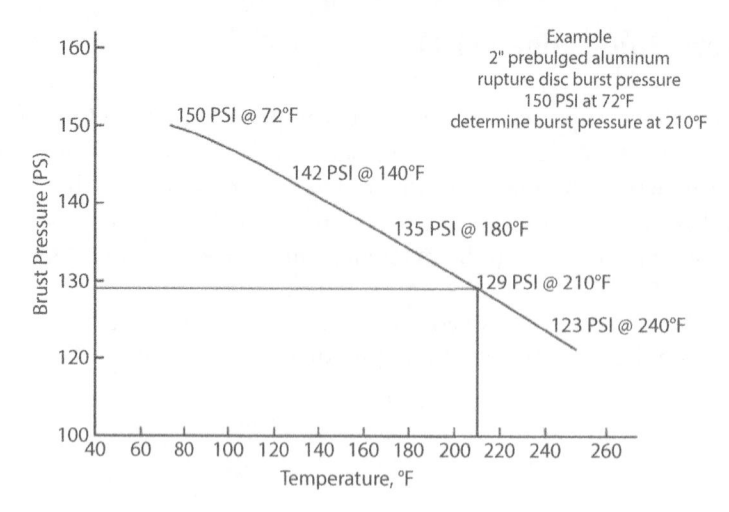

Figure 23.31 Establishing stamped rupture disk bursting pressure at coincident temperature, Method 2. (By permission from Fike Metal Products Div., Fike Corporation.)

The burst pressure maximum cannot exceed the MAWP of the vessel. Depending on the situation, it may be necessary to work backward to the operating pressure maximum to see if this is usable. Table 23.14 summarizes typical rupture disk characteristics noting that the maximum normal operating pressure of the system is shown as a function of the rupture disk bursting pressure, P_b.

From this guide to determine the bursting pressure of the rupture disk, it is apparent that some thought must be given to the process and the equipment design and the ultimate MAWP. It is not an arbitrary selection. When changing services on a vessel, the MAWP and the operating pressures of the process must be established and then the bursting specifications of the disk determined. When reordering rupture disks for a specific service to repeat a

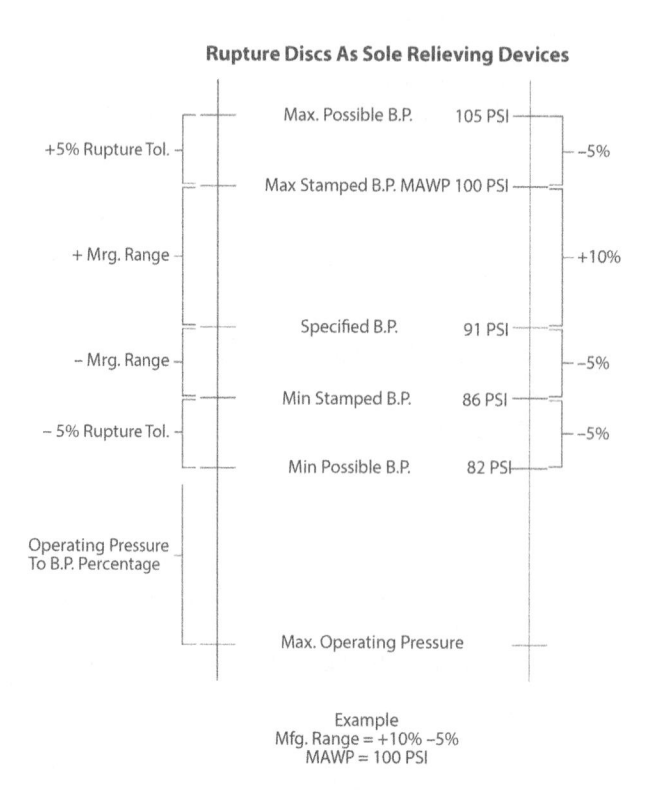

Figure 23.32a Rupture disks as sole relieving devices. (By permission from Fike Metal Products Div., Fike Corporation.)

Rupture Discs Used in Multiple or For Additional Hazard Due to External Heat

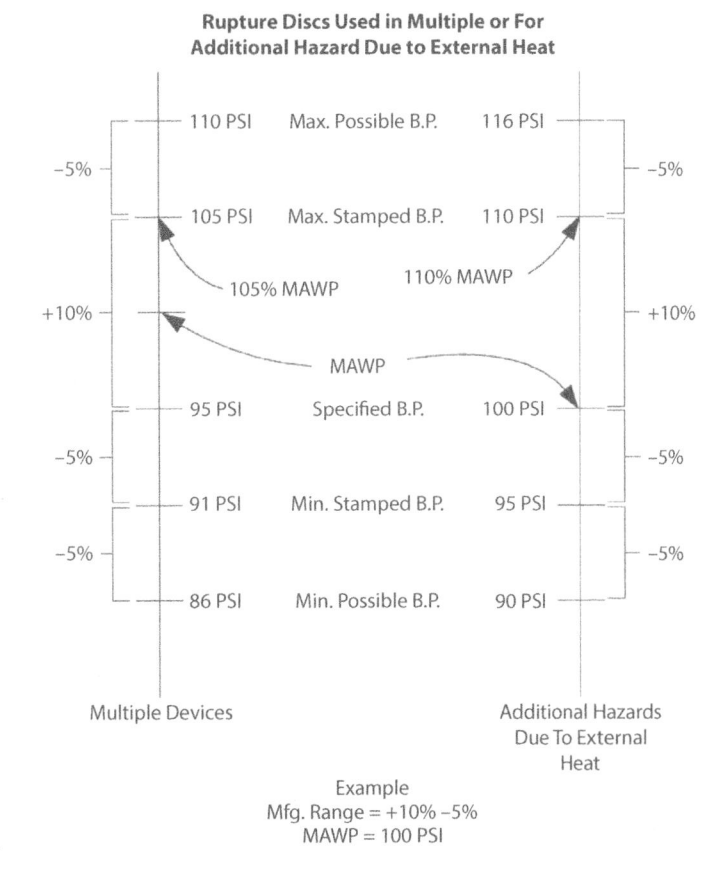

```
                110 PSI    Max. Possible B.P.    116 PSI
   -5%                                                       -5%
                105 PSI    Max. Stamped B.P.     110 PSI
                           105% MAWP    110% MAWP
   +10%                            MAWP                      +10%
                 95 PSI    Specified B.P.        100 PSI
   -5%                                                       -5%
                 91 PSI    Min. Stamped B.P.      95 PSI
   -5%                                                       -5%
                 86 PSI    Min. Possible B.P.     90 PSI

           Multiple Devices              Additional Hazards
                                         Due To External Heat
```

Example
Mfg. Range = +10% −5%
MAWP = 100 PSI

Figure 23.32b Rupture disks used in multiple or for additional hazard due to external heat. (By permission from Fike Metal Products Div., Fike Corporation.)

Table 23.14 Summary of Rupture-Disk Characteristics.

Type of disk	Vacuum support required?	Fragment upon rupture?	Gas service?	Liquid or partially-liquid service?	Maximum normal operating pressure
Conventional	Sometimes	*Yes/No*	Yes	Yes	0.7 P_b
Pre-scored					
Tension-loaded	No	No	Yes	Yes	0.85 P_b
Composite	Sometimes	Yes	Yes	Yes	0.8 P_b
Reverse-					
Buckling with					
Knife blades	No	No	Yes	No	0.9 P_b
Pre-scored					
Reverse-buckling	No	No	Yes	No	0.9 P_b

By permission, Nazario, F.N., Chem. Eng., June 20, 1988, p. 86 [17].
* Depends on manufacturer's specifications (E.E. Ludwig [34]).

performance of a ruptured disk, it is important to set the bursting specifications and the MR exactly the same as the original order; otherwise, the maximum stamped P_b could be too high for the vessel and the system not properly protected against overpressure.

Example 23.6: Rupture Disk Selection

Examination of the temperature control ranges of a process reactor reveals that the normal controls are to maintain a pressure of the reacting mixture of 80 psig, while the upper extreme could be 105 psig, which would be defined as the normal maximum operating pressure.

Select a conventional rupture disk; then from Table 23.14, $P_{max\ op} = 0.7\ P_b$.

Thus,

$$105 = 0.7\ P_b$$

$P_{b\text{-min}} = 105/0.7 = 150$ psig min. rupture disk burst pressure

From manufacturing range table for this type of disk, MR = +10/-5% at 150 psig rupture pressure minimum. The maximum rupture pressure = 150 + 10% = 165 psig, plus/minus the disk tolerance of 5%, allowing a final maximum burst pressure of 173.2 psig.

The minimum stamped bursting pressure of the disk would be 150 psig + 5% tolerance = 157.5 psig.

With a -5% MR, the specified burst pressure would be 157.5 + 5% = 165.4 psig.

With a +10% upper MR, the maximum stamped burst pressure of the disk could be 165.4 + 10% = 181.9 psig.

Using the code allowed tolerance for burst, the disk could burst at 181.9 + 5% = 190.9 psig. The MAWP for the vessel cannot be less than 181.9 psig.

23.41 Effects of Temperature on Disk

The temperature at the burst pressure must be specified to the manufacturer, as this is essential in specifying the metal or composite temperature stresses for the disk finally supplied. Higher temperatures reduce the allowable working stress of the disk materials. Reference [17] shows that temperature has an effect on metals in decreasing order, with the least effect on the lowest listed metal:

- aluminum
- stainless steel (changes after 400°F)
- nickel
- Inconel

When specifying the material at the disk temperature, the heat loss at the disk/disk holder as well as in a flowing pipe must be recognized and the assembly may need to be insulated. This is important as it relates to the actual temperature at the bursting pressure. Establishing this burst temperature is an essential part of the system safety and must not be guessed at or taken lightly.

Table 23.15 presents a temperature conversion table for various metals from one manufacturer for conventional pre-bulged, tension-loaded disks with pressure on concave side (not prescored) as an illustration of the effect of lower or elevated temperatures referenced to 72°F on the burst pressure of a stamped disk. For other types of disk designs and from other manufacturers, the specific data for the style disk must be used to make the appropriate temperature correction.

Table 23.15 Temperature Conversion Table for Conventional Rupture Disks Only.

Temperature correction factor in % from rupture pressure at 72° F											
Disk temp. (°F)	Rupture disk metals						Disk temp. (°F)	Rupture disk metals			
	Alum	Silver	Nickel	Monel	Inconel	316 S.S.		Nickel	Monel	Inconel	316 S.S.
−423	170	164	165	155	132	200	300	93	87	94	84
−320	152	152	144	140	126	181	310	92	87	94	84
−225	140	141	126	129	120	165	320	92	86	94	83
−200	136	138	122	126	118	160	330	92	86	94	83
−150	129	130	116	123	115	150	340	92	86	94	83
−130	127	126	116	121	114	145	350	91	85	93	82
−110	122	123	115	120	113	141	360	91	85	93	82
−100	120	122	115	119	112	139	370	91	85	93	82
−90	120	121	114	118	112	136	380	91	85	93	82
−80	120	120	114	118	112	136	390	90	84	93	81
−70	119	120	113	116	110	132	400	90	84	93	81
−60	119	119	112	115	110	130	410	90	84	93	81
−50	119	118	112	114	109	128	420	90	84	93	81
−40	118	117	111	113	108	125	430	89	84	93	81
−30	117	115	110	112	108	123	440	89	83	93	80
−20	116	112	109	111	107	121	450	89	83	93	80
−10	115	110	108	110	106	118	460	88	83	93	80
0	114	108	107	109	105	116	470	88	83	93	80
10	113	107	106	108	105	114	480	87	83	93	80
20	111	105	105	106	104	112	490	87	82	94	80
30	110	104	104	105	103	110	500	86	82	94	79
40	108	103	103	104	102	107	520	85	82	94	79
50	106	102	102	103	102	105	540	84	82	94	79
60	103	101	101	101	101	103	560	83	81	94	79
72	100	100	100	100	100	100	580	82	81	94	78
80	100	100	100	99	100	99	600	81	81	94	78
90	99	99	99	98	99	98	620	79	80	94	77

(Continued)

Table 23.15 Temperature Conversion Table for Conventional Rupture Disks Only. (*Continued*)

Temperature correction factor in % from rupture pressure at 72° F											
Disk temp. (°F)	Rupture disk metals						Disk temp. (°F)	Rupture disk metals			
	Alum	Silver	Nickel	Monel	Inconel	316 S.S.		Nickel	Monel	Inconel	316 S.S.
100	98	99	99	97	99	98	640	78	80	94	77
110	97	98	98	96	99	95	660	77	79	93	77
120	97	98	98	95	98	94	680	76	79	93	76
130	96	97	97	95	98	93	700	75	78	93	76
140	95	96	97	94	98	92	720	73	77	93	76
150	94	95	96	93	97	91	740	72	77	93	76
160	93	94	96	93	97	90	760	–	76	93	75
170	92	93	96	92	97	90	780	–	76	93	75
180	90	92	95	92	96	89	800	–	75	92	75
190	89	91	95	91	96	89	820	–	–	92	75
200	88	90	95	91	95	88	840	–	–	92	75
210	87	89	94	90	95	88	860	–	–	92	75
220	85	87	94	90	95	87	880	–	–	91	74
230	84	86	94	89	95	87	900	–	–	91	74
240	84	84	94	89	95	86					
250	81	84	93	89	95	86					
260			93	88	94	86					
270			93	88	94	85					
280			93	88	94	85					
290			93	87	94	84					
300			93	87	94	84					

By permission from B.S. and B. Safety Sytems, Inc.

Example:

What is rupture pressure at 500° F of a nickel disk rated 300 psi at 72° F?

1. Consult temperature conversion at 500° F of a nickel disk rated 300 psi at 72° F?
2. Multiply disk rating at 72° F by correction factor: 300 × 0.86 = 258

Rupture pressure of a nickel disk rated 300 psi at 72° F is therefore 258 psi at 500° F.
If you require a disk for a specific pressure at elevated or cold temperature and want to determine if it is a standard disk, convert the required pressure at elevated or cold temperature at 72° F.

Note: This conversion table does not apply to Type D or reverse buckling disks.

23.42 Rupture Disk Assembly Pressure Drop

The ruptured or burst disk on a vessel or pipe system presents a pressure drop to flow at that point, and it can be estimated by assuming the disk is a flat plate-orifice [17] with a discharge coefficient, K_d, of 0.62. As an alternate, the disk assembly can be assumed to be the equivalent of a section of pipe equal to 75 nominal disk diameters in length.

23.43 Gases and Vapors: Rupture Disks [5a, Par, 4.8]

The sizing is based on use of the ASME Code [44] flow coefficient:

K_d = 0.62 [1] (Par. UG-127) for standard metal disks, but use
K_d = 0.888 for graphite disks [45]
K_d = actual flow/theoretical flow = coefficient of discharge

To select the proper sizing equation, determine whether the flowing conditions are sonic or subsonic from the equations. When the absolute pressure downstream or exit of the throat is less than or equal to the critical flow pressure, P_c, then the flow is critical and the designated equations apply [5a]. When the downstream pressure exceeds the critical flow pressure, P_c, then sub-critical flow will occur, and the appropriate equations should be used [5a].

When P_1 is increased, the flow through an open disk increases and the pressure ratio, P_2/P_1, decreases when P_2 does not change, until a value of P_1 is reached, and there is no further increase in mass flow through the disk. The value of P_1 becomes equal to P_c, and the ratio is the critical pressure ratio, and the flow velocity is sonic (equals the speed of sound).

The maximum velocity (sonic) of a compressible fluid in pipes is [40]

$$v_s = \sqrt{kgRT} \text{ or} \tag{23.35}$$

$$v_s = \sqrt{kgP'\overline{V}(144)} \tag{23.36}$$

$$v_s = 68.1\sqrt{kP'\overline{V}} \tag{23.37}$$

where

- v_s = sonic or critical velocity of a gas, ft/s
- k = ratio of specific heats, C_p/C_v
- g = acceleration of gravity = 32.2 ft/s²
- R = individual gas constant = MR/M = 1545/M
- M = molecular weight
- MR = universal gas constant = 1545
- T = absolute temperature, °R = (460 + t°F)
- t = temperature, °F
- P' = pressure, psia, at outlet end or a restricted location in pipe when pressuredrop is sufficiently high
- $\overline{V}$ = specific volume of fluid, ft³/lb

For sonic flow [5a].

When actual pressure ratio, P_2/P_1, is less than critical pressure ratio, flow is sonic or critical. From Eq. 23.7, critical pressure ratio is

$$\frac{P_c}{P_1} = \left[\frac{2}{(k+1)}\right]^{k/(k-1)}$$

where

P_c = critical flow throat pressure, psia
P_b = stamped bursting pressure, psia = burst pressure + overpressure allowance (ASME Code of 10%) plus atmospheric pressure, psia

Important note: when actual system ratio, P_2/P_1, is less than critical pressure ratio calculated above by Equation 23.7, flow is sonic. When actual P_2/P_1 ratio is greater than critical pressure ratio, flow is subsonic.

P_2 = backpressure or exit pressure, psia
P_1 = upstream relieving pressure, psia

For sonic flow conditions [46]:

From Eq. 23.10,

$$A = \frac{W}{CK_d P_b K_b}\left(\frac{ZT}{M}\right)^{0.5}$$

where

A = minimum net required flow discharge area after complete burst of disk, in²
C = sonic flow constant for gas or vapor based on ratio of specific heats, k, Figure 23.25, when k is not known, use k = 1.001, or C = 315
W = required flow, lb/h
M = molecular weight
K_d = coefficient of discharge, K= 0.62 for rupture disks, except some coefficients are different. For example, the Zook graphite standard ASME disks when tested mono-style (Figures 23.9b and 23.13a) have a K_d of 0.888, and when inverted (Figure 23.13b) have a K_d of 0.779. Consult manufacturer for special disks.
T = flowing relieving temperature, °F + 460 = °R absolute
Z = compressibility factor for deviation from perfect gas if known; otherwise, use Z = 1.0 for pressures below 250 psia, at inlet conditions.
P_b = stamped bursting pressure plus overpressure allowance (ASME 10% or 3 psi whichever is greater) plus atmospheric pressure (14.7), psia

Volumetric Flow: scfm Standard Conditions (1.4.7 psia and 60°F)

$$A = \frac{Q_s(MTZ)^{0.5}}{6.326 CK_d P_b}, \text{in}^2 \qquad (23.38)$$

Q_s = required flow, ft³/min at standard conditions of 14.7 psia and 60°F, scfm

Actual flowing conditions, acfm:

$$A = \frac{5.596 Q_A}{CK_d} \sqrt{\frac{M}{TZ}} \qquad (23.39)$$

Q_A = required flow, ft³/min at actual conditions, acfm

Steam: Rupture Disk Sonic Flow; Critical Pressure = 0.55 and P_2/P_1 is Less Than Critical Pressure Ratio of 0.55

API reference [5a] [46] dry and saturated steam pressure up to 1500 psig:

$$A = \frac{W}{51.5 K_d P_b K_n K_{sh}}, \text{in}^2 \qquad (23.40)$$

where

W = flow, lb/h
K_d = coefficient of discharge = 0.62.
K_n = correction for Napier equation = 1.0 when $P_1 \leq 1515$ psig
   = $(0.1906 P_1 - 1000)/(0.2292 P_1 - 1061)$
      where $P_1 > 1515$ psia and ≤ 3215 psia, Table 23.10.
K_{sh} = superheat correction factor; see Table 23.10. For saturated steam at any pressure, $K_{sh} = 1.0$
P_b = stamped bursting pressure, psia

23.44 API for Subsonic Flow: Gas or Vapor (Not Steam)

For rupture disks, pressure ratio is greater than critical pressure; mass flow: lb/h

$$P_c/P > \text{critical pressure ratio} [2/(k+1)]^{k/(k-1)}$$

$$A = \frac{W}{735 C_2 K_d} \sqrt{\frac{ZT}{MP_1(P_1 - P_2)}} \qquad (23.41)$$

where

C_2 = subsonic flow conditions based on ratio of specific heats (see Table 23.15 and equation for C_2)

Volumetric flow, scfm conditions:

$$A = \frac{Q_s}{4645.2 C_2 K_d} \sqrt{\frac{ZTM}{P_1(P_1 - P_2)}} \qquad (23.42)$$

Actual flowing conditions, acfm:

$$A = \frac{Q_A}{131.43 C_2 K_d} \sqrt{\frac{P_1 M}{ZT(P_1 - P_2)}} \qquad (23.43)$$

Converting actual flow conditions to standard conditions of 60°F and 14.7 psia [47]:

$$Q_s = \left[\left(\frac{520}{14.7}\right)\left(\frac{P_{act}}{T_{act}}\right)\right](Q_{act}) = \text{scfm, } 60°F \text{ and } 14.7 \text{ psia} \qquad (23.44)$$

where

P_{act} = pressure actual, psia
T_{act} = temperature actual, °R = °F + 460°F
Q_{act} = flow at actual conditions: acfm, at actual flowing conditions
Q_s = required flow, ft³/min at standard conditions (14.7 psia at 60°F), (60°F + 460 = 520°R)

23.45 Liquids: Rupture Disk

The test for critical or non-critical does not apply. These equations apply to single-phase (at inlet) liquids, non-flashing to vapor on venting; fluid viscosity is less than or equal to water [46].

ASME mass flow:

$$A = \frac{W}{2407 K_d \sqrt{(P_b - P_2)\rho_1}}, \text{ in}^2 \qquad (23.45)$$

where

W = flow, lb/h
ρ = fluid density, lb/ft³
K_d = coefficient of discharge = 0.62
P_b = stamped bursting disk pressure plus accumulation of 10% plus atmospheric pressure, psia
P_2 = pressure on outlet side of rupture disk, psia

$$\text{Volumetric flow: } A = \frac{W_L \sqrt{SpGr}}{38 K_d \sqrt{(P_b - P_2)}}, \text{ in}^2 \qquad (23.46)$$

where

$SpGr$ = fluid specific gravity relative to water = 1.0 at 60°F
W_L = liquid flow, gpm

23.46 Sizing for Combination of Rupture Disk and Pressure Relief Valve in Series Combination

When the rupture disk is installed on the inlet side of the pressure relief valve (see Figures 23.10, 23.11, and 23.12), the ASME code requires that for untested disk–valve combinations that the relieving capacity of the combination be reduced to 80% of the rated relieving capacity of the pressure relief valve [1].

For flow tested combinations, see a few typical data in Table 23.16. Note, for example, that using a Continental disk reverse acting knife blade rupture disc with a Crosby JOS/JBS pressure relief valve, the combined effect is to multiply the rated capacity of the Crosby valve by a multiplier of 0.985 for a set pressure in the 60–74 psig range using a 1.5 in.

Table 23.16 Constant, C_2, for Gas or Vapor for Subsonic Flow Conditions.

k	P_2/P_1 0.95	0.90	0.85	0.80	0.75	0.70	0.65	0.60	0.55	0.50	0.45
1.001	158.1	214.7	251.9	277.8	295.7	307.4	313.7				
1.05	158.4	215.5	253.3	280.0	298.7	311.2	318.4	321.0			
1.10	158.7	216.3	254.7	282.0	301.5	314.8	322.9	326.4			
1.15	158.9	216.9	255.9	283.9	304.1	318.2	327.1	331.4			
1.20	159.2	217.6	257.0	285.6	306.5	321.3	331.0	336.0			
1.25	159.4	218.1	258.1	287.2	308.7	324.2	334.5	340.4			
1.30	159.6	218.7	259.1	288.7	310.8	326.9	337.9	344.4	346.8		
1.40	159.9	219.2	260.0	290.1	312.7	329.4	341.1	348.2	351.3		
1.45	160.0	220.1	261.6	292.6	316.2	334.0	346.8	355.2	359.5		
1.50	160.2	220.5	262.3	293.7	317.8	336.0	349.4	358.3	363.3		
1.55	160.3	220.8	263.0	294.8	319.3	338.0	351.8	361.3	366.8		
1.60	160.4	221.2	263.7	295.8	320.7	339.8	354.1	364.2	370.2	372.3	
1.65	160.6	221.5	264.3	296.7	322.0	341.6	356.3	366.87	373.4	376.1	
1.70	160.7	221.8	264.8	297.6	323.2	343.2	358.4	369.4	376.4	379.6	
1.75	160.8	222.1	265.4	298.5	324.4	344.8	360.4	371.8	379.3	383.0	
1.80	160.9	222.4	265.9	299.3	325.5	346.2	362.3	374.1	382.1	386.3	
1.90	161.0	222.9	266.9	300.7	327.6	349.0	365.7	378.4	387.2	392.3	
2.00	161.2	233.4	267.7	302.1	329.5	351.5	368.9	382.3	391.8	397.8	400.1
2.10	161.4	223.8	268.5	303.3	331.2	353.7	371.8	385.8	396.1	402.8	405.9
2.20	161.5	224.2	269.2	304.4	332.8	355.8	374.4	389.1	400.1	407.5	411.4
2.30	161.6	224.5	269.9	305.4	334.2	357.7	376.9	392.1	403.7	411.8	416.4

$$C_2 = 735\sqrt{\frac{k}{k-1}\left[\left(\frac{P_2}{P_1}\right)^{\frac{2}{k}} - \left(\frac{P_2}{P_1}\right)^{\frac{k+1}{k}}\right]}$$

By permission Continental Disk Corp., Cat. 1-1110, p. 4.

disc with Monel metal. Other disk and valve manufacturers have their own combination data, which, when available, avoids the requirement of derating the capacity to 80% of the rated capacity of the pressure relief valve. Disks (metal or graphite) that fragment should never be used because these may become potential problems for the safety valve performance. Therefore, a non-fragmenting disk should be selected, such as a reverse acting/reverse buckling preferably pre-scored design, but knife blades are a viable alternate.

Example 23.7: Safety Relief Valve for Process Overpressure

The conditions set forth on the Operational Checklist, Figure 23.14a, are used in the example specified on the specification form (Figure 23.33).

```
                                                    Spec. Dwg. No.
Job No. _____                      A. _____
                                                    Page ___ of ___ Pages
B/M No. _____                      Unit Price
                                                    No. Units

                SAFETY VALVE SPECIFICATIONS    Item No.
                        DESCRIPTION
Make _____  Model _____  Standard Type-Back Pressure  Standard
Size (Inlet × Orifice No. × Outlet)   4  ×  L  ×  6    Phase  Vapor
Set Press¹   75   (psig)  @  347  °F   Full Flow Back Press.  0  psig
Req'd Orifice Area   2.02   sq. in.: Selected Orifice Area  2.853  sq. in.
Accessories²   Screwed Cap, No Test Gag; No Lifting Gear
Inlet Nozzle: Press. Class   250 # –4"              Facing  Raised
Outlet Nozzle: Press. Class  125 # –6"              Facing  Raised
Mfgr's Rating   250 #   psig (Max) @   450  °F:  125 # (outlet) PSIG (Min) @  300  °F
¹Refers to Initial relief pressure.   ²Refers to cap type, lifting, gag, etc.
                        MATERIALS
Body and Bonnet    Cast Iron                Trim     Bronze
Nozzle and Disc    Bronze                   Bellows  None
Spring    Carbon Steel, cadium plated, or equal
                         REMARKS
Use three-way valve (Yes) (No)   No

                    PROCESS RATING DATA
Location   On Evaporator Shell
Fluid   PDC      Flow Based On  Operations   Set pressure Based On  Vessel Max.
System Oper. Press.  60   psig @  347  °F: MW*  113.5
Req'd Cap.  22,500  lbs/hr. @  347  °F  Phase: ___  Liquid Dens.* ___  lbs/cu. ft.
Liq. Viscosity* ___  SSU:XA overpressure  10 %   %:  Back Press. Corr. Factor ___
*Refers to properties at Set Pressure
                      CALCULATIONS
Use: A = W(Z T/M)^(1/2) / C K_d P K_b
Since data for c_p/c_v not readily available, use k = 1.001 and C = 315, Z = 1.0, K_b = 1.0

P (at relieving condition) = (75)(1.1 + 14.7 = 97.2 psia
T = 347 + 460 = 807R
A = 22,500 [(1)(807)/113.5]^(1/2) / (315)(0.953)(1.0)
A = 2.056 sq. in.
Note that critical flow conditions apply, since relieving pressure of 97.2 psia is over twice backpressure
of 14.7 psia.

By _____  Chk'd _____  App. _____  Rev. _____  Rev. _____  Rev. _____
```

Figure 23.33 Safety relief valve specification for process overpressure example.

Example 23.8: Rupture Disk External Fire Condition

An uninsulated 12 x 36-ft horizontal storage tank containing CCl_4 is to be protected from overpressure due to external fire by means of a rupture disk. The tank does not have a sprinkler system. Storage pressure is 5 psig and should not exceed 10 psig. Tank is assumed to be full. k = 1.13. Disk burst pressure to be 10 psig. MAWP of vessel is 50 psig. Discharge line backpressure is 1 psig.

Solution

Heat Input

Using ASME flanged and dished heads (F&D) from Appendix Tables of Blanks, the circle size is 152 in. for a 12 ft diameter tank. Then add 3 in. straight flange, which becomes 158 in., which is 158/12 = 13.166 ft diameter. Area of this diameter for surface area of head = 136.14 ft² equivalent surface area of one head. For a horizontal vessel, there are two heads possibly exposed to fire.

External Surface Area: cylindrical area + surface area heads (2)

$$= \pi(12)(36) + \left(\frac{158}{12}\right)^2 \left(\frac{\pi}{4}\right) 2$$

$$= 1357 + 272 = 1629 \text{ ft}^2. \text{ (approximate)}$$

Total Heat Input (from Figure 23.30a)

$Q = 900 \times 10^6$ Btu/h at area of 1629 ft²

Solving equation: $Q = 21{,}000 \, F \, A^{0.82}$

$Q = 21{,}000 \, (1.0) \, (1629)^{0.82} = 9{,}036{,}382$ Btu/h

Quantity of Vapor Released

Latent heat of CCl_4 = 85 Btu/lb

$$W = \frac{9{,}000{,}000}{85} = 106{,}000 \frac{\text{lb}}{\text{h}} \text{ (rounded)}$$

Critical Flow Pressure

$$P = 10 + 14.7 = 24.7 \text{ psia}$$

$$P_c = P\left[\frac{2}{(k+1)}\right]^{k/(k-1)} = 24.7\left[\frac{2}{(1.13+1)}\right]^{1.13/(1.13-1)}$$

$$P_c = 24.7 \, (0.578) = 14.3 \text{ psia, or}$$

$$\frac{P_2}{P_1} \text{actual} = \left[\frac{14.7 + 1.0}{(10)(1.1) + 14.7}\right] = 0.610$$

$$\frac{P_c}{P_1} = \left[\frac{2}{(1.13+1)}\right]^{1.13/(1.13-1)} = 0.578$$

$$\text{Flow is subcritical, since } \left(\frac{P_2}{P_1}\right) > \frac{P_c}{P_1}$$

Disk Area

$$A = \frac{W}{735 F_2 K_d} \sqrt{\frac{ZT}{MP_b(P_b - P_2)}}$$

where

W = 106,000 lb/h
K_d = 0.62
P_b = (10) (1.10 see Figure 23.11a) + 14.7 = 25.7 psia
P_2 = 1.0 + 14.7 = 15.7 psia
M = 154
T = 460 + 202 = 662°R (B.pt of CCl_4 at 10 psig)
k = 1.13
F_2 = 0.716 interpolated from Figure 23.29
Z = 1.0, compressibility factor

Area calculated substituting in above relation:

$$A = \frac{W}{735 F_2 K_d} \sqrt{\frac{ZT}{MP_b(P_b - P_2)}}$$

$$= \frac{106,000}{(735)(0.716)(0.62)} \sqrt{\frac{(1.0)(662)}{(154)(25.7)(25.7 - 15.7)}}$$

$$A = 42.016 \, in^2$$

Diameter is

$$d = \sqrt{\frac{4 \, Area}{\pi}}, in.$$

$$d = \sqrt{\frac{(4)(42.016)}{\pi}}$$

$$= 7.3 \, in.$$

Next standard size: Choose an 8 in. Sch. 40 pipe, has a cross-sectional area of 50.0 in^2. So an 8-inch frangible rupture disk will be satisfactory. Disk material to be lead or lead covered aluminum.

Example 23.9: Rupture Disk for Vapors or Gases; Non-Fire Condition

Determine the rupture disk size required to relieve the pressure in a process vessel with the following conditions: k = 1.4.

Vessel: MAWP = 85 psig; also = disk set pressure
Vapor flow to relieve = 12,000 std. ft^3/min at 60°F and 14.7 psia
Flowing temperature = 385°F
Vapor Mol. Wt = 28
Backpressure on discharge of disk= 30 psig

Solution

Determine if conditions on rupture are critical or non-critical, based on 10% overpressure for primary relief.

$$\frac{P_2}{P_1} = \left[\frac{30+14.7}{(85)(1.10)+14.7}\right] = 0.413$$

Critical pressure ratio: $\dfrac{P_c}{P_1} = \left[\dfrac{2}{(1.4+1)}\right]^{1.4/(1.4-1)} = 0.528$

Since actual $P_2/P_1 <$ critical ratio 0.528, the flow is sonic.
Critical pressure = P_{cr} = 108.2 (0.528) = 57.12 psia
Flow area required to relieve: [5a]

$$A = \frac{Q_s (MTZ)^{0.5}}{6.32 C K_d P_b} \tag{23.47}$$

where

Q_s = 12,000 scfm
M = 28
C = 356, Figure 23.25
K_d = 0.62 (rupture disk)
P_b = 108.2 psia
Z = 1.0
T = 385 + 460 = 845°R

$$A = \frac{1200[(28)(845)(1.0)]^{0.5}}{6.32(356)(0.62)(108.2)}$$

A = 12.23 in².

Diameter is

$$d = \sqrt{\frac{4\,\text{Area}}{\pi}}, \text{in.}$$

$$d = \sqrt{\frac{(4)(12.23)}{\pi}}$$

$$= 3.9 \text{ in.}$$

Choose next standard size: a 4 in. Sch. 40 pipe with an area of 12.6 in². This should be adequate.

Example 23.10: Liquids Rupture Disk

Determine the rupture disk size required to relieve the following operating condition:
Backpressure = 0 psig
Toluene flow = 1800 gpm, SpGr = 0.90
Pressure vessel: MAWP = 25 psig
Relieving pressure: set pressure; use 25 psig
Actual relief pressure = 25 + 10% = 27.5 psig

$$A = \frac{W_L}{38K_d}\sqrt{\frac{SpGr}{(P_1 - P_2)}}$$

Fluid: Toluene,
Flow = 1800 gpm
SpGr = 0.90
K_d = 0.62 per API [5a], or from manufacturer
P_1 = 27.5 + 14.7 = 42.2 psia
P_2 = 0 + 14.7 = 14.7 psia

The required area,

$$A = \frac{1800}{38(0.62)}\sqrt{\frac{0.90}{(42.2 - 14.7)}}$$
$$= 13.82 \text{ in}^2$$

Diameter is

$$d = \sqrt{\frac{4\,\text{Area}}{\pi}}, \text{ in.}$$

$$d = \sqrt{\frac{(4)(13.82)}{\pi}}$$

$$= 4.19 \text{ in.}$$

Use Standard pipe size: 5 in. (cross sectional area = 19.6 in²) disk, or check manufacturer. Use inlet and discharge pipe size = 6 in. std.

Example 23.11: Liquid Overpressure, Figure 23.34

A check of possible overpressure on a heat exchanger handling 95% aqua ammonia in the tubes indicates that tube failure is probably the condition requiring maximum relieving capacity. The aqua is being pumped by a positive displacement pump; 30 psia steam on shell side heats (not vaporizes) the aqua. In the case of tube failure, the aqua would flow into the shell and soon keep the steam from entering. The relief valve must prevent the shell from failing. The shell is designed for a working pressure of 210 psig.

The calculations are shown on the specification sheet (Figure 23.34).

23.47 Pressure-Vacuum Relief for Low-Pressure Storage Tanks

In order to accommodate "breathing" of tanks and other equipment due to temperature changes, pumping in and out, internal vapor condensation, and other situations, adequate safety vacuum relief must be provided. In many cases, both pressure and vacuum relief are needed (see Figure 23.35). For the average product storage tank, the API Guide For Tank Venting RP 2000 [48] serves to set the minimum venting quantities for various tank capacities. In addition to these tabulated values, calculations are made to satisfy each condition of operation to ensure that there is no situation requiring more than this vent capacity. Emergency vent capacity is also required to supplement the normal requirement in case of external fire or other unusual condition.

The normal venting to be provided must not allow pressure or vacuum conditions to develop, which could cause physical damage to the tanks [48].

```
                                                    Spec. Dwg. No.
Job No. _____                             A. _____
                                                    Page ___ of ___ Pages
B/M No. _____                             Unit Price _____
                                                    No. Units _____
```

SAFETY VALVE SPECIFICATIONS   Item No.
DESCRIPTION
Make _____ Model _____   Standard Type-Back Pressure Standard
Size (Inlet × Orifice No. × Outlet) ___ × ___ × ___   Phase Liquid
Set Press¹ __157__ (psig) @ __100__ °F   Full Flow Back Press. __0__ psig
Req'd Orifice Area __0.0951__ sq. in.: Selected Orifice Area __0.110__ sq. in.
Accessories² No lift gear, No finned bonnet, No test gag
Inlet Nozzle: Press. Class 1 1/2", 150 #   Facing Raised
Outlet Nozzle: Press. Class 2", 150 #   Facing Raised
Mfgr's Rating __230__ psig (Max) @ __100__ °F: __160__ PSIG (Min) @ __450__ °F
¹Refers to Initial relief pressure.   ²Refers to cap type, lifting, gag, etc.
MATERIALS
Body and Bonnet Cast carbon steel   Trim Stainless steel
Nozzle and Disc Orged stainless steel   Bellows None
Spring Carbon steel
REMARKS
Use threeway valve (Yes) (No) No
Plant standards do not accept 1" flanged connections, therefore use 1½"
PROCESS RATING DATA
Location Shell side of heater
Fluid 95% Aqua Ammonia Flow Based On Lube failure and pump capacity Set pressure Based Max Working On Press.
System Oper. Press. __30(shell)__ psig @ __275__ °F: MW* __17__
Req'd Cap. __40 gpm__ lbs/hr. @ __100__ °F Phase: Liquid Liquid Dens.* __36__ lbs/cu. ft.
Liq. Viscosity* __0.05 cp__ SSU:XA overpressure __25__ %: Back Press. Corr. Factor __none__
*Refers to properties at Set Pressure
CALCULATIONS
Max Working Pressure (shell side) = 210 psig
* For 25% accumulation, set pressure = (0.75)(210) = 157 psig
* Relieving pressure @ 10% overpressure = [157 + 10%] + (14.7) = 187.4 psia
For Liquid Relief: per ASME Code
$A = \text{gpm} (\sqrt{sp\ gr})/38.0 K_d \sqrt{\Delta P(K_u)}$
$K_d = 0.64$; $K_u = 1.0$ @ normal viscosity; gpm = 40
$\Delta P = 157 + 10\%(157) - 0 = 172.7$ psig
$A = (40) (\sqrt{0.578})/38.0 (0.64) \sqrt{172.7} = 0.095$ in².
Select 1" std. pipe size with cross-sectional area = 0.864 in².
By _____	Chk'd _____	App. _____	Rev. _____	Rev. _____	Rev. _____

Figure 23.34 Safety relief valve specification for liquid overpressure for Example 23.11.

23.48 Basic Venting For Low-Pressure Storage Vessels

Usually reference to low pressure venting is associated with large storage tanks of several thousand gallons capacity (ranging from a few thousand to a million); however, small low-pressure tanks can be handled in the same manner. The usual operating pressure range for the typical tank is 0.5 oz./in.² to about 1.5 psig. Since low-pressure vessels have pressure ratings expressed in various units, Table 23.17 can be useful for conversion.

Typical large storage vessels are illustrated in Figure 23.36. Usual operating/design pressure (max) is shown below. Operations associated with storage tanks should be carefully analyzed, since there are several factors that can significantly influence the safety relieving requirements. Usually, these are [48]

1. Normal operation
    a) Outbreathing or pressure relief
    b) Inbreathing or vacuum relief
2. Emergency conditions
    a) Pressure venting
3. For tank design per API Standard 650 with weak roof to shell designs (roof lifts up), the venting requirements of API-Std-2000 do not apply for emergency venting to atmosphere or elsewhere.

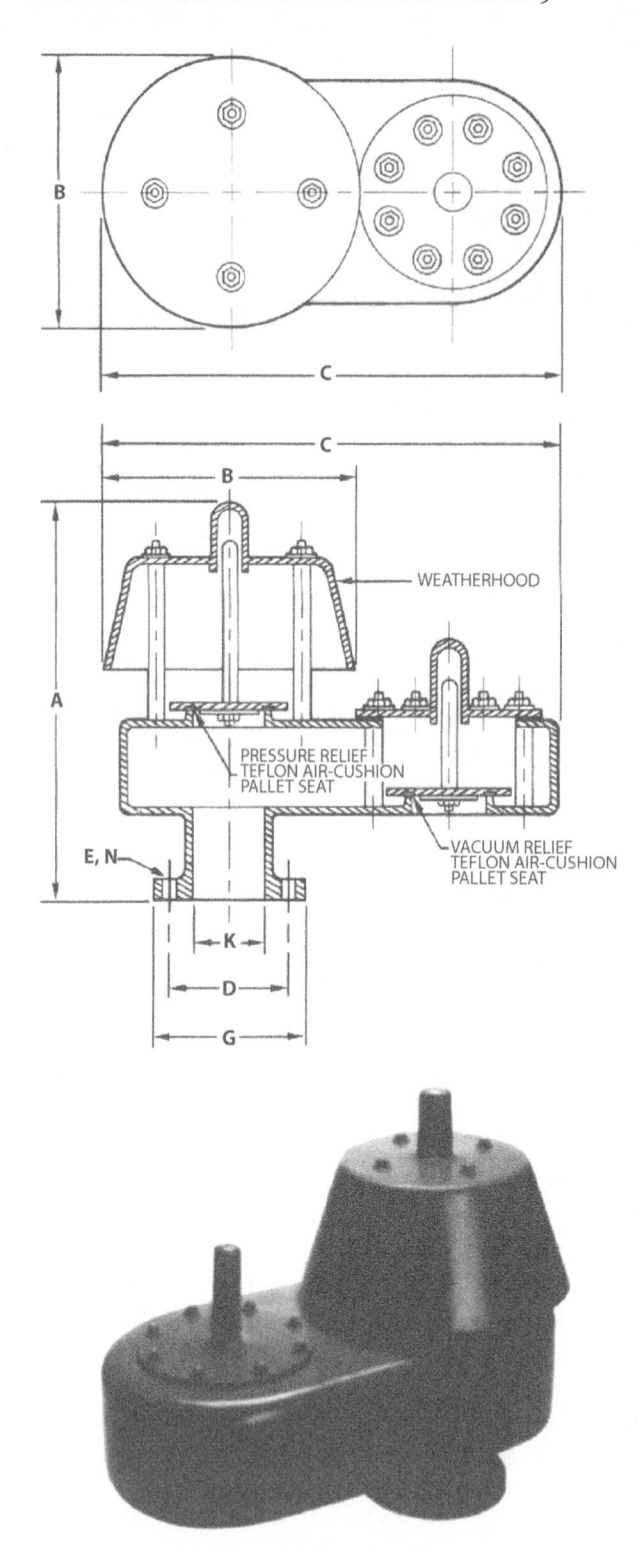

Figure 23.35 Dead weight type pressure and vacuum – relief valve for low pressure storage vessels. (By permission from the Protectoseal Co.)

23.49 Non-Refrigerated Above Ground Tanks; API-Std. 2000

Normal operations are to be established within the design parameters for the tank, thereby avoiding conditions that would damage it. Normal venting capacity should be at least the sum of venting required for oil/fluids movement and thermal effect. Required capacity can be reduced when liquid volatility is such that vapor generation or condensation in the allowable operating range of vessel pressure will provide all or part of the venting requirements.

Table 23.17 Rupture Disk/Relief Valve Combination Capacity Factors.

Disc type	Disk size	Disk material	Set pressure psig	Teledyne ferris 2600 & 4500	Dresser 1900 1900-30 1900-35	Crosby JOS/ JBS	Crosby JB	Crosby JO	Kunkle 5000 thru 5999	Lonergan D & DB
ZAP	1.5	Monel	60-74	0.982	-	0.985	0.983	0.980	-	-
(contd)	1.5	Monel	75 Plus	0.982	-	0.985	0.990	0.986	-	-
	1.5	Nickel	30-49	-	-	0.965	-	0.975	-	-
	1.5	Nickel	50-59	0.989	0.984	0.965	-	0.975	0.988	0.966
	1.5	Nickel	60 Plus	0.985	0.984	0.992	-	0.994	0.988	0.996
ZAP	3.0	Stainless Steel	15-29	0.963	0.966	-	-	-	-	-
	3.0	Stainless Steel	30-34	0.993	0.966	0.955	-	-	-	-
	3.0	Stainless Steel	35-49	0.993	0.966	0.970	-	-	-	-
	3.0	Stainless Steel	50-59	0.993	0.966	0.970	0.976	0.981	-	-

Extracted by permission:
Continental Disc Corp., Bul. #1-1111, pg. 4. Only portion of original tables presented for illustration.

Note:
ZAP is a reverse-acting rupture disk using replaceable knife blades. Patented. Other disk and valve manufacturer.

Outbreathing conditions are usually established when (a) the tank is being filled and the vapor space is being displaced with liquid, (b) thermal expansion and evaporation of the liquid, and (c) external fire on the vessel creating additional heat input to the contents.

The standard [48] specifies venting capacity of

1. Twelve hundred cubic foot of free air per hour for every 100 barrels (4200 gal/h) of maximum filling rate, for liquids with flash points below 100°F.
2. Six hundred cubic foot of free air per hour for each 100 barrels (4200 gal/h) of maximum filling rate, for liquids with flash points 100° F and above.
3. Thermal outbreathing or venting requirements, including thermal evaporation for a fluid (the code refers to oil) with a flash point of 100°F or below; use at least the figures of column 4 in Table 23.18.
4. Thermal outbreathing or venting requirements, including thermal evaporation for a fluid (the code uses oil) with a flash point of 100°F or above; use at least the figures of column 3 in Table 23.19.
5. To attain the total venting (outbreathing) requirements for a tank, refer to the appropriate flash point column and add the outbreathing plus the thermal venting flows.

Tank vent equipment ratings are expressed as free air capacity at 14.7 psia and 60°F, and in order to handle vapors from liquids of the chemical and petrochemical industry, corrections must be made. Likewise, corrections are required to recognize temperatures other than 60°F as in Table 23.20.

23.50 Boiling Liquid Expanding Vapor Explosions (BLEVEs)

This particular type of explosion is less known and understood, but nevertheless is an important type for damage considerations. This is a type of pressure release explosion and there are several descriptions.

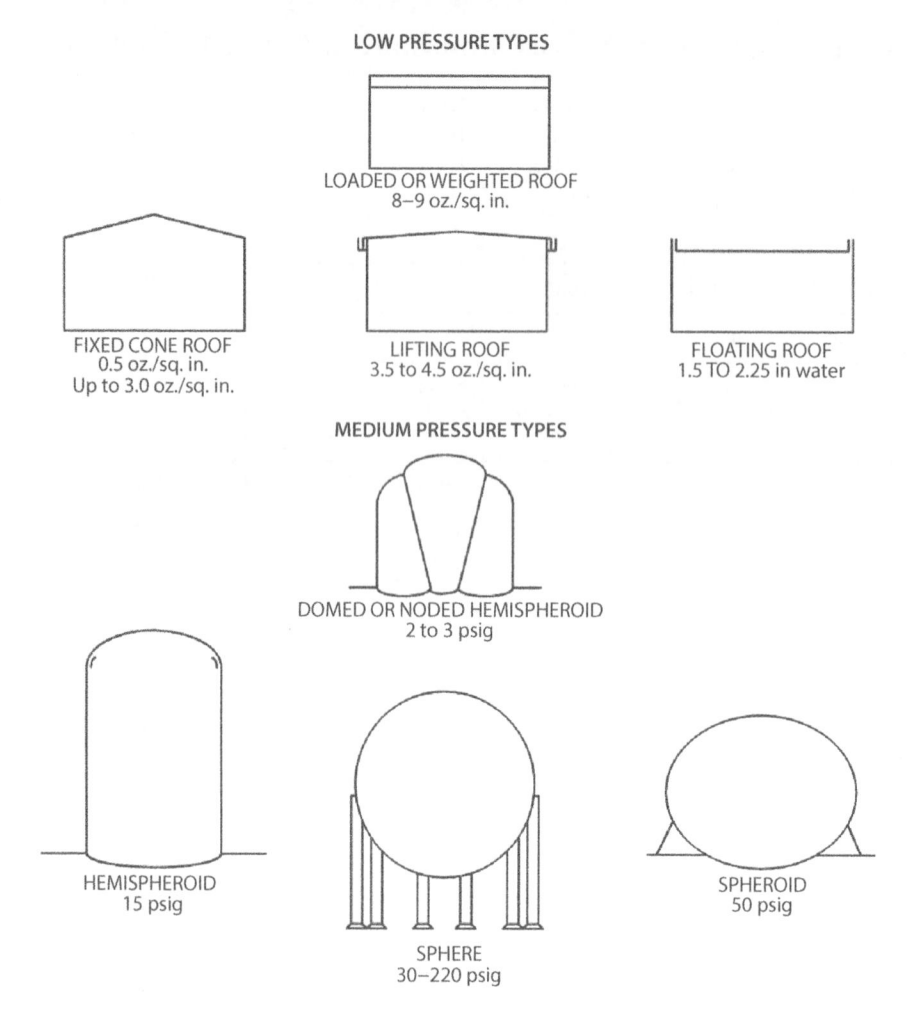

Figure 23.36 Representative configurations of large storage tanks.

Kirkwood [2] describes BLEVEs referenced to flammable liquids as occurring when a confined liquid is heated above its atmospheric boiling point by an external source of heat or fire and is suddenly released by the rupture of the container due to over-pressurization by the expanding liquid. A portion of the superheated liquid immediately flashes to vapor and is ignited by the external heat source.

The NFPA [6] contains extensive descriptions of BLEVEs and describes them in summary as paraphrased here with permission: liquefied gases stored in containers at temperatures above their boiling points at NTP will remain under pressure only as long as the container remains closed to the atmosphere. If the pressure is suddenly released to atmosphere due to failure from metal over-stress (by external fire or heat, corrosion penetration or external impact, the

Table 23.18 Convenient Pressure Conversions.

Oz./in.^{2}	lb./in.^{2}	in. Hg (0°C)	in. H$_2$O (4°C)
1	0.06250	0.1272513	1.730042
16	1	2.036021	27.68068
7.85846	0.4911541	1	13.59548
0.57802	0.0361262	0.07355387	1

Table 23.19 Total Rate of Emergency Venting Required for Fire Exposure vs. Wetted Surface Area (wetted area vs. ft.³ of free air/h, 14.7 psia, 60°F).

Wetted area[a] (square feet)	Venting requirement (cubic feet of free air[b] per hour)	Wetted area[a] (square feet)	Venting requirement (cubic feet of free air[b] per hour)
20	21,100	350	288,000
30	31,600	400	312,000
40	42,100	500	354,000
50	52,700	600	392,000
60	63,200	700	428,000
70	73,700	800	462,000
80	84,200	900	493,000
90	94,800	1000	524,000
100	105,000	1200	557,000
120	126,000	1400	587,000
140	147,000	1600	614,000
160	168,000	1800	639,000
180	190,000	2000	662,000
200	211,000	2400	704,000
250	239,000	2800	742,000
300	265,000	>2800[c]	-

By permission: API-Std-2000, 3rd Ed., 1982, reaffirmed Dec. 1987, American Petroleum Institute [48].

NOTE: Interpolation for intermediate values. The total surface area does not include the area of ground plates but does include roof areas less than 30 ft above grade.

[a]The wetted area of the tank or storage vessel shall be calculated as follows: For spheres and spheroids, the wetted area is equal to 55 % of the total surface area or the surface area to a height of 30 ft (9.14 meters) whichever is greater. For horizontal tanks, the wetted area is equal to 75 % of the total surface area. For vertical tanks, the wetted area is equal to the total surface area of the shell within a maximum height of 30 ft (9.14 meters) above grade.

[b]At 14.7 pounds per square inch absolute (1.014 bar) and 60°F (15.56°C).

[c]For wetted surfaces larger than 2800 ft² (260.1, m²), see 1.3.2.1, 1.3.2.2., and 1.3.2.4.

heat stored in the liquid generates very rapid vaporization of a portion of the liquid proportional to the temperature difference between that of the liquid at the instant the container fails and the normal boiling point of the liquid.

Often this can generate vapor from about one-third to one-half of the liquid in the container. The liquid vaporization is accompanied by a large liquid to vapor vaporization, which provides the energy for propagation of vessel cracks, propulsion of pieces of the container, rapid mixing of the air and vapor resulting in a characteristic fire-ball upon ignition by the external fire or other source that caused the BLEVE to develop in the first place, with atomization of the remaining cold liquid. Often the cold liquid from the vessel is broken into droplets that can burn as they fly out of the vessel. Often this cold liquid can escape ignition and may be propelled 1.5 mi or more from the initial state. In most BLEVEs, the failure originates in the vapor space above the liquid, and it is this space that is most subject to external overheating of the metal.

Table 23.20 Gravity and Temperature Correction Factors for Low Pressure Venting Calculations for Vapors.

Sp. Gr. of gas Air = 1.00	Sp. Gr. factors	Sp. Gr. of gas air = 1.00	Sp. Gr. factors	Temp. (°F)	Factor	Temp. (°F)	Factor	Temp. (°F)	Factor
0.20	0.447	1.10	1.050	5	1.0575	70	0.9905	200	0.8932
0.30	0.548	1.20	1.095	10	1.0518	80	0.9813	220	0.8745
0.40	0.632	1.30	1.141	15	1.0463	90	0.9732	240	0.8619
0.50	0.707	1.40	1.185	20	1.0408	100	0.9638	260	0.8498
0.60	0.775	1.50	1.223	25	1.0355	110	0.9551	280	0.8383
0.65	0.806	1.60	1.265	30	1.0302	120	0.9469	300	0.8272
0.70	0.837	1.70	1.305	35	1.0249	130	0.9388	320	0.8165
0.75	0.866	1.80	1.340	40	1.0198	140	0.9309	340	0.8063
0.80	0.894	1.90	1.380	45	1.0147	150	0.9233	360	0.7963
0.85	0.922	2.00	1.412	50	1.0098	160	0.9158	380	0.7868
0.90	0.949	2.50	1.581	55	1.0048	170	0.9084	400	0.7776
0.95	0.975	3.00	1.731	60	1.0000	180	0.9014	420	0.7687
1.00	1.000	3.50	1.870						
1.05	1.025	4.00	2.000						

By permission, Groth Equipment Corp., Tank Equipment Division.

The application of water externally to the vapor space of a vessel or application of insulation can often protect against BLEVEs.

A relief valve will not actually handle the vapor generated because its set pressure is usually higher than the boiling point pressure created by the hole or crack in the vessel; therefore it will not relieve at the lower pressure.

Lees [98] points out the effects of a BLEVE depend on whether the liquid in the vessel is flammable. The initial explosion may generate a blast wave and fragments from the vessel. For a flammable material, the conditions described in [6] above may results, and even a vapor cloud explosion (VCE) may result.

Ignition of Flammable Mixtures

Ignition can take place for any flammable mixture within the concentration ranges for the respective lower explosive limit (LEL) and upper explosive limit (UEL). The conditions for ignition may vary with the specific mixture, the type of oxidant (usually air or pure oxygen), the temperature and pressure of the system. Ignition may result from electrical spark, static spark and contact with hot surfaces (autoignition) [15].

Where Autoignition Temperature (AIT) = the minimum temperature occurs at which a material begins to self-heat at a high enough rate to result in combustion.

23.51 Managing Runaway Reactions

Many processes in petroleum refining are exothermic, that is they release heat. Combustion is a typical example, but the main processes are:

- Hydroprocessing (e.g., hydrocracking, hydrotreating, isomerization)
- Acid/base reactions - e.g., amine absorption.
- Methanation.
- Alkylation unit acid runaway.

The processes depend on various mechanisms to control the heat of reaction:

- Hydrocrackers, hydrotreaters, and isomerization units depend on the continuous flow of oil and hydrogen through the reactors along with injected cold quench gas to remove the heat away.
- Polymerization reactors may depend on coolant – jacketed or exchanger – style reactors or may have large recycle cooling streams with quenches.
- Amine absorption and other acid/base reactions depend on dilution of the heat in a flowing mass or a circulating cooling system.
- Methanators depend on flow and short residence times to carry heat away.
- An alkylation unit depends on maintaining proper chemistry to avoid acid runaway conditions.

Failures of the control mechanisms can lead to very high temperatures in the units. Hydrocrackers, for example, can exceed 1900°F (1038°C) in a runaway and those temperatures can be reached in only a few minutes or seconds.

Loss of containment can result when the reaction temperature exceeds the allowable metal temperature with the equipment still at pressure (i.e., you exceed the ultimate yield stress of the material under the process condition).

Process licensors and industry have developed specific methods to manage the affect processes in runaway temperature conditions. Some of these techniques are illustrated as follows [97].

Hydroprocessing Units

If temperatures in the hydroprocessing unit exceeds about 850°F (~ 454°C), the unit can enter a hydrocracking runaway condition in which catalytic and thermal cracking will drive temperatures up rapidly. Once started, the reaction is self-sustaining. A hydrocracking catalyst does not have to be present for this to occur. The reactor seldom fails in a runaway because the metal mass slows its heat up. The normal failure point is at the reactor outlet line or the first effluent exchanger. Fatalities have resulted from these failures.

Control of a true hydrocracking runaway requires immediate and aggressive depressuring of the unit to the flare. The depressuring point is usually the cold high-pressure separator; although if there is an amine absorber, depressuring may be downstream of it. Depressuring rate would typically be somewhere between 100–300 psi/min to start (this decays as the pressure drops). The unit should be at flare pressure within about 15 min. It should not be repressured until cooled, possibly using nitrogen gas. Licensors have specific guidelines for these procedures and the required equipment.

Not all units can get into a runaway, so analysis is needed to determine the requirements. Some specific scenarios to consider when determining the needs for a depressuring system are the following:

- Does the unit operate near 850°F at any point in its run?
- Does the unit have a hydrocracking catalyst?
- If flow stops (either treat or recycle gas or a single quench flow), will the bed get too hot?
- Can a sudden change in feedstock (like a lot more cracked feed) drive temperatures quickly to runaway conditions?

The depressuring system is normally manual activated for hydrotreaters. It may be automatically activated under defined conditions for hydrocrackers, with manual activation also available.

For units where runaway is only a remote possibility, the unit should still be capable of depressuring in a reasonable amount of time in case of a fire or other release scenarios. This is part of API Recommended Practices (API RP – 520 Parts I and II).

Another variable is temperature, which can be a challenge in these units. Temperatures are known at a few points in a reactor and possibly on the reactor shell. Maldistribution (from e.g., internal damage, previous runaway, or bad catalyst loading) can result in undetected hot spots. If a hot spot occurs near the reactor wall, this can be missed completely and thus careful consideration of the numbers and placement of temperature measuring points is required, both in the reactors and on their shells. Licensors have guidelines for temperature monitoring locations, and many companies have their own complete guidelines or requirements.

If an automatic depressuring system is prescribed, it normally will be a safety integrity, integrated or instrumented system (SIS), with appropriate security, redundancy, maintenance and testing requirements.

Acid/Base Reactions

For many acid/base reactions, temperature is controlled by a large circulating stream or a large base volume of cool material, with injection of acid or base into the much larger mass. If temperature gets too high, they can be controlled just by stopping the feed of one or more of the reactants.

A case in amine and other absorbers that depend on an acid/base reaction, the reaction generates a significant amount of heat, which was moderated by using a larger circulating mass of the absorbent solution. This acts as a flywheel to control the temperature. However, if the solution rate gets too low, the temperatures in the absorber can rise to the point where the solution actually begins regenerating in the absorber tower. This results in a phenomenon where the absorber accumulates hot liquid in the trays and periodically blows the hot liquid overhead. The process repeats on some frequency. This is remedied by increasing the absorbent rate to ensure the rich solution loading is at the correct level.

Methanation

Methanation is a highly exothermic reaction that is part of hydrogen plants and other units where carbon oxides in hydrogen may poison a process. A methanator reactor can get into a runaway condition if the upstream carbon oxide removal process fails or the incoming CO and CO_2 exceed the tolerable levels. These reactors generally have good thermocouple coverage. However, as they are in a vapor phase, maldistribution does not occur. The reactors have normally relatively thinner walls than hydroprocessing units, so they can heat up extremely quickly. Fast action is needed to avoid loss of containment.

If a runaway condition is detected in a methanator by any high reactor or outlet temperature, the reactor is automatically shut down by a feed block valve and the incoming gas is sent to flare until the problem is resolved. The auto-shutdown system will generally be a safety instrumented system (SIS) – type design with appropriate security, redundancy, maintenance, and testing requirements. There are provisions for manually activating the shutdown.

Alkylation Unit Acid Runaway

An acid runaway in an alkylation unit can result when the acid strength drops too low. The more dilute acid promotes rapid, exothermic polymerization of olefins over the reaction of an olefin and isobutane as desired. The polymers tend to dissolve in the acid, further diluting it and pushing more polymerization.

An acid runaway is controlled by stopping the olefinic feed, but leaving the paraffinic feed in. Fresh acid is brought in to restore the acid strength. Vendors have recommended approaches in much more detail for managing acid runaway.

23.51.1 Runaway Reactions: DIERS

One of the standardized methods for predicting or controlling runaway reaction that may lead to explosions (deflagration or detonations) is the Fauske approach (Figure 23.37). Others are presented elsewhere [49].

Accordingly, to emphasize the safety problems affecting all industrial process plants and laboratories, the AIChE established the industry-supported Design Institute for Emergency Relief Systems (DIERS). The purposes of the Institute are [22]

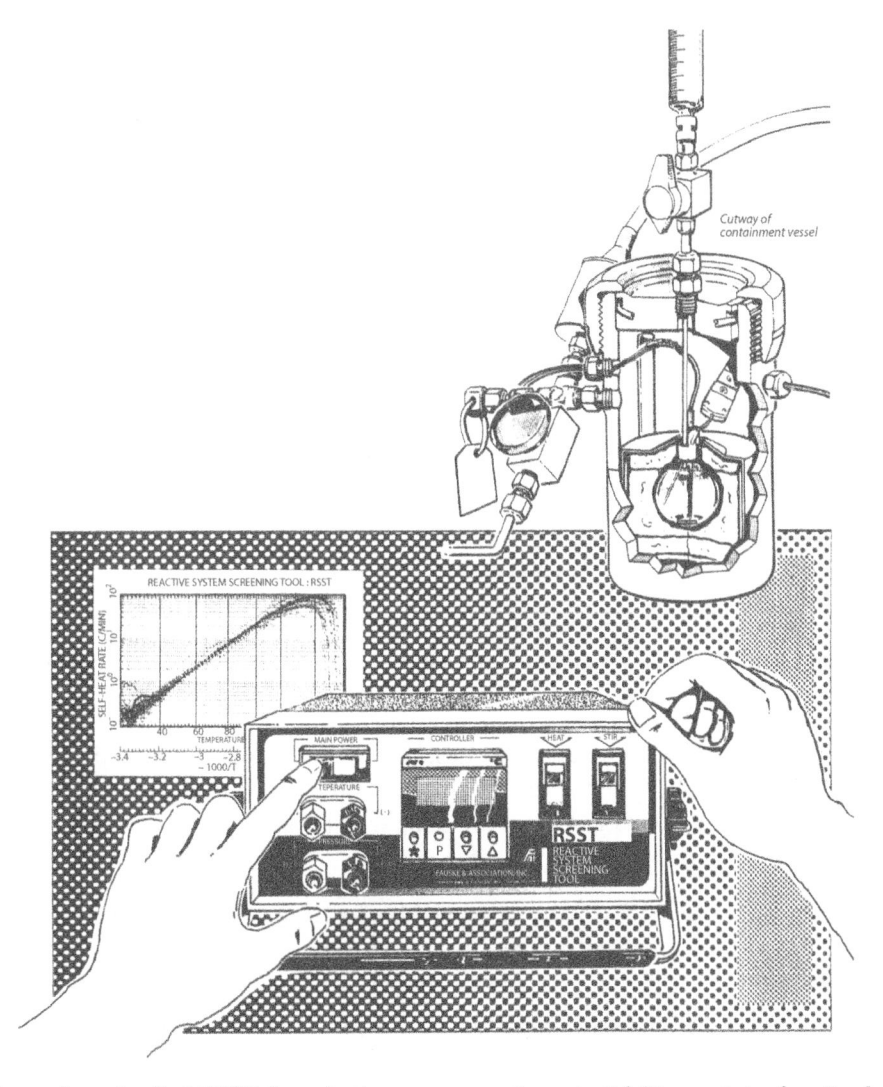

Figure 23.37 Reactive System Screening Tool (RSST) for evaluating runaway reaction potential. (By permission from Fauske and Associates, Inc.)

- Reduce the frequency, severity, and consequences of pressure producing accidents.
- Promote the development of new techniques that will improve the design of emergency relief systems.
- Understand runaway reactions.
- Study the impact of two-phase flow on pressure relieving device systems.

In the refineries and CPI, chemical manufacture especially in the fine, pharmaceutical, and specialty chemical industries involves the processing of reactive chemicals, toxic or flammable liquids, vapors, gases, and powders. The safety records of these industries have improved in recent years; however, fires, explosions, and incidents involving hazardous chemical reactions do still occur. A basis for good engineering practice in assessing chemical reaction hazards is essential, with the aim to help designers, engineers, and scientists responsible for testing and operating chemical plants to meet the statutory duties of safety imposed by governmental organizations (e.g., US. Environmental Protection Agency (EPA), Occupational Safety and Health Administration (OSHA), USA, Health & Safety Executive, UK, and others).

The control of chemical reactions (e.g., esterification, sulfonation, nitration, alkylation, polymerization, oxidation, reduction, and halogenation) and associated hazards are an essential aspect of chemical manufacture in the refineries and CPIs. The industries manufacture nearly all their products such as inorganic, organic, agricultural, polymers, and pharmaceuticals through the control of reactive chemicals. The reactions that occur are generally without incident. Barton and Nolan [50] examined exothermic runaway incidents and found that the principal causes were as follows.

- inadequate temperature control.
- inadequate agitation.
- inadequate maintenance.
- raw material quality.
- little or no study of the reaction chemistry and thermochemistry.
- human factors.

Other factors that are responsible for exothermic incidents are as follows.

- poor understanding of the reaction chemistry resulting in badly designed plant.
- under-rated control and safety back-up systems.
- inadequate procedures and training.

The research and technical evaluations have provided industry with extremely valuable information and design procedures, including, but not limited to, two-phase flow phenomena and runaway reactions during safety/over-pressure relief.

23.52 Hazard Evaluation in the Chemical Process Industries

The safe design and operation of chemical processing equipment require detailed attention to the hazards inherent in certain chemicals and processes. Chemical plant hazards can occur from many sources; the principal ones arise from

- fire and explosion hazards.
- thermal instability of reactants, reactant mixtures, isolated intermediates, and products.
- rapid gas evolution, which can pressurize and possibly rupture the plant.
- rapid exothermic reaction, which can raise the temperature or cause violent boiling of the reactants and also lead to pressurization.

Earlier reviews have been concerned with energy relationships for a particular chemical process, which are based upon two general classifications of chemical processes, conventional and hazardous [51]. The former is used to describe a non-explosive non-flammable reaction and the latter to describe an explosive and flammable type reaction. The division of reactions does not account for the varying degree of safety or hazards of a particular reaction, which may lie between these extremes, and consequently becomes too limiting in the design. Safety is most likely to be neglected with conventional reactions while overdesign may be the expected practice with hazardous reactions. These limitations can be avoided by introducing a third class of reactions, "special," which covers the intermediate area between conventional and hazardous where reactions are relatively safe. Shabica [51] has proposed some guidelines for this third classification, and Figure 23.38 shows the increasing degree of hazard of a particular process.

Chemical reaction hazards must be considered in assessing whether a process can be operated safely on the manufacturing scale. Furthermore, the effect of scale-up is particularly important. A reaction, which is innocuous on the

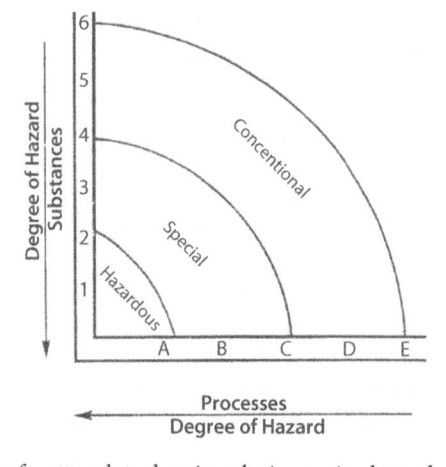

Figure 23.38 The increasing degree of hazard of matter plotted against the increasing hazard of a process. (Source: Shabica [66].)

laboratory or pilot plant scale, can be disastrous on a full-scale manufacturing plant. Therefore, the heat release from a highly exothermic process, for example, the reduction of an aromatic nitro compound, can be controlled easily in laboratory glassware. However, if the same reaction is carried out in a full-scale plant reactor with a smaller surface area to volume ratio, efficient cooling is essential; otherwise, a thermal runaway and violent decomposition may occur. Similarly, a large quantity of gas produced by the sudden decomposition of a diazonium compound can be easily vented on the laboratory scale, but the same decomposition on the large scale could pressurize and rupture a full-scale plant.

In addition, consequences of possible process maloperations, such as incorrect charging sequence, contamination of reactants, agitation failure, and poor temperature control, addition of reactants too quickly, omitting one of the reactants, and incorrect reactant concentration (recycling) must be considered. A number of parameters govern the reaction hazards associated with a process. These include the following:

- chemical components.
- process temperature.
- process pressure.
- thermochemical characteristics of the reaction.
- reaction rate.
- reaction ratios.
- solvent effects.
- batch size.
- operational procedure.

The assessment of the hazards of a particular process requires investigations of the effects of these parameters by experimental work, the interpretation of the results in relation to the plant situation, and the definition of a suitable basis for safe operation.

23.53 Hazard Assessment Procedures

Hazard assessments are essential and should be performed on all chemical processes. The reactors such as batch, semi-batch, and continuous can be employed for carrying out various operations. Many industrial reactions are exothermic (i.e., are accompanied by the evolution of heat) and therefore overheating can occur. In batch operations, all the reactants are added to the reactor at the start of the reaction. In semi-batch operations, one or more of the reactants are charged to the reactor at the start, and others are then metered in, giving some control over the rate of reaction, and thus the rate of heat production. Overheating often results in thermal runaway, which is characterized by progressive increases in the rate of heat generation, and hence temperature and pressure.

Thermal runaway is a particular problem in unsteady state batch reactions, where the rate of reaction and therefore the rate of heat production vary with time. The consequences of thermal runaway can sometimes be very severe as in the incidents at Seveso [52]. In this case, a bursting disk ruptured on a reactor. The reactor was used to manufacture trichlorophenol at a temperature of 170–185°C and it was heated by steam at 190°C. The batch had been dehydrated and left at a temperature of 158°C over the weekend. Ethylene glycol and caustic soda give exothermic secondary reactions producing sodium polyglycolates, sodium oxalate, sodium carbonate, and hydrogen thus exhibiting an autocatalytic behavior. These reactions caused a temperature rise allowing the production of tetrachlorodioxin and the subsequent vessel pressurization. The Bhopal fatal incident [53] (is not a thermal runaway) involved a toxic release of an intermediate, methyl isocyanate (MIC), which resulted in the fatality of over 2000 people. A runaway polymerization of highly toxic MIC occurred in a storage tank. The runaway polymerization caused an uncontrolled pressure rise, which caused a relief valve to lift and discharge a plume of toxic vapor over the city.

The task of specifying the design, operation and control of a reactor with stirrer, heating or cooling coils, reflux facilities, and emergency relief venting can pose a problem if all the time-dependent parameters are not considered. The use of batch processing techniques in the fine chemical industry is often characterized by

- multi-product plant (must have adaptable safety system).
- complex developing chemistry.
- high frequency of change.
- process control is simpler than continuous processes.

These factors are attributed to batch and semi-batch processes than continuous processes. However, the use of continuous processes on fine chemical manufacturing sites is limited. It is often preferable to use semi-batch mode as opposed to batch processes.

Exotherms

Temperature-induced runaways have many causes including the following:

- loss of cooling.
- loss of agitation.
- excessive heating.

During a temperature-induced upset, the reactor temperature rises above the normal operating target. When any of the temperature elements senses a high reactor temperature, the programmable logic controller (PLC) software shutdown system automatically puts the reactor on idle (isolates). This action doubles the cooling water flow to the reactor by opening a bypass valve. If these reactions are unsuccessful in terminating the temperature rise, the shutdown system opens the quench valves, dumping the reaction mass into the water-filled quench tank. If the dump valves fail to operate, the reactor contents soon reach a temperature where a violent self-accelerating decomposition occurs. The uncontrolled exotherm generates a large volume of gas and ejects the process material out of the reactor into the containment pot.

Accumulation

It is important to know how much heat of reaction can accumulate when assessing the hazards relating with an exothermic reaction. Accumulation in a batch or semi-batch process can be the result of

- adding a reactant too quickly.
- loss of agitation.
- carrying out the reaction at too low a temperature.
- inhibition of the reaction.
- delayed initiation of the desired reaction.

Reactants can accumulate when the chosen reaction temperature is too low, and as such the reaction continues even after the end of the addition. In such a case, a hazardous situation could occur if cooling were lost as exemplified [54].

Impurities or the delayed addition of a catalyst causes inhibition or delayed initiation resulting in accumulation in the reactors. The major hazard from accumulation of the reactants is due to a potentially rapid reaction and consequent high heat output that occurs when the reaction finally starts. If the heat output is greater than the cooling capacity of the plant, the reaction will runaway. The reaction might commence if an agitator is restarted after it has stopped, a catalyst is added suddenly, or because the desired reaction is slow to start.

23.54 Thermal Runaway Chemical Reaction Hazards

Thermal runaway reactions are the results of chemical reactions in batch or semi-batch reactors. A thermal runaway commences when the heat generated by a chemical reaction exceeds the heat, which can be removed to the surroundings as shown in Figure 23.39. The surplus heat increases the temperature of the reaction mass, which causes the reaction rate to increase, and subsequently accelerates the rate of heat production. Thermal runaway occurs as follows: as the temperature rises, the rate of heat loss to the surroundings increases approximately linearly with temperature. However, the rate of reaction, and thus the rate of heat generation, increases exponentially. If the energy release is large enough, high vapor pressures may result or gaseous products may be formed, which can finally lead to overpressurization, and possible mechanical destruction of the reactor or vessel (thermal explosion).

The energy released might result from the wanted reaction or from the reaction mass if the materials involved are thermodynamically unstable. The accumulation of the starting materials or intermediate products is an initial stage of a runaway reaction. Figure 23.40 illustrates the common causes of reactant accumulation. The energy release with the reactant accumulation can cause the batch temperature to rise to a critical level thereby triggering the secondary (unwanted) reactions. Thermal runaway starts slowly and then accelerates until finally it may lead to an explosion.

Heat Consumed Heating the Vessel. The φ-Factor

The fraction of heat required to heat up the vessel, rather than its contents, depends on the heat capacity of the vessel (i.e., how much energy is required to raise the temperature of the vessel with respect to the reaction mass). The heat loss to the vessel is known as the φ-factor [55]. The φ-factor is the ratio of the total heat capacity of sample and vessel to that of the sample alone and is defined by

$$\phi = \frac{M_s C_{ps}(\text{sample}) + M_v C_{pv}(\text{vessel})}{M_s C_{ps}(\text{sample})} \tag{23.48}$$

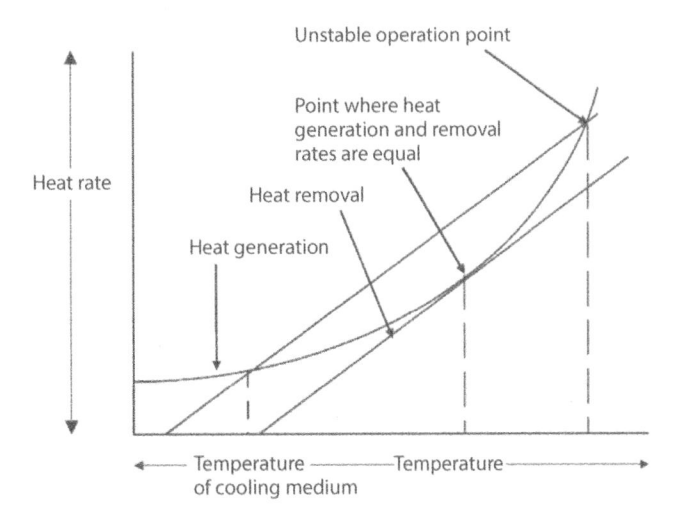

Figure 23.39 A typical curve of heat rate vs. temperature.

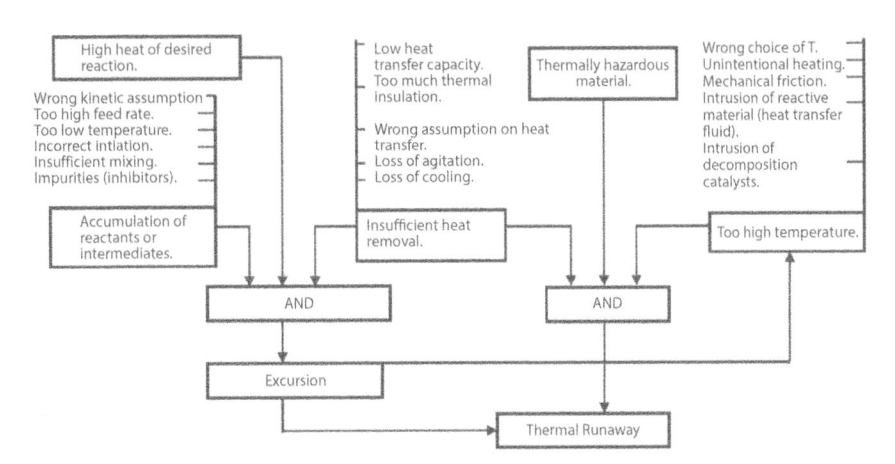

Figure 23.40 Causes of runaways in industrial reactors. (Source: W. Regenass, Safe Operation of Exothermic Reactions, Internal Ciba-Geigy Publication, 1984.)

where

M_s = mass of the sample (kg).
C_{ps} = specific heat capacity of sample (J/kg.K).
M_v = mass of the vessel (kg).
C_{pv} = specific heat capacity of vessel (J/kg.K).

The ϕ-factor does not account for the heat loss to the environment. It is used to adjust the self-heating rates as well as the observed adiabatic temperature rise.

As the scale of operation increases, the effect of the heat consumption by the plant typically reduces, and thus the extent to which the kinetics of the runaway reaction is influenced by the plant is reduced. For plant scale vessels, the ϕ-factor is usually low (i.e., 1.0–1.2) depending on the heat capacity of the sample and the vessel fill ratio. Laboratory testing for vent sizing must simulate these low ϕ-factors. If the laboratory ϕ-factor is high, several anomalies will occur.

- The rate of reaction will be reduced (i.e., giving incorrect reaction rate data for vent sizing).
- The magnitude of the exotherm will be smaller by a factor of ϕ.
- Measured pressure effects will be smaller than those that would occur on the plant.
- The induction period of a thermal runaway reaction will be increased compared to the plant.

The consequences of these erroneous anomalies will be

- inadequate safety design system.
- undersized emergency relief system.
- unknown decompositions may occur at elevated temperatures, which may not be realized in the laboratory.

Onset Temperature

The reaction onset temperature is that temperature where it is assumed that significant fuel consumption begins. The onset temperature is expressed by [56]

$$T_{onset} = \frac{B}{\ln\left[\dfrac{x_o \Delta H_{Rx} \cdot A}{C \cdot \phi \cdot T_{ex}^*}\right]} \qquad (23.49)$$

where

$$B = \frac{E_A}{R} \qquad (23.50)$$

A = Rate constant (Pre-exponential factor from Arrhenius equation
k = $A \exp(-E_A/RT)$, sec^{-1} (i.e. for a first order reaction)
B = Reduced activation energy, K
C = Liquid heat capacity of the product (J/kg.K)
E_A = Activation energy (J/mol)
R = Gas constant (8.314 J/mol.K)
T_{onset} = Onset temperature, K
X_o = Initial mass fraction
ΔH_{Rx} = Heat of decomposition, J/kg
ϕ = Dimensionless thermal inertial factor for the sample holder or product container (PHI-factor).

$T^{\bullet}_{ex}$ = Bulk heat-up rate driven by an external heat source (°C/sec).

Time-To-Maximum Rate

We may select an onset temperature based upon an arbitrary time-to-maximum-rate from the relation

$$t_{mr} = \frac{C.\phi.T^2 \exp(B/T_{onset})}{x_o.\Delta H_{Rx}.A.B} \quad (23.51)$$

Re-arranging Equation 23.51 yields

$$T_{onset} = \frac{B}{\ln\left[\dfrac{t_{mr}.x_o.\Delta H_{Rx}.A.B}{C.\phi.T^2}\right]} \quad (23.52)$$

Equation 23.52 gives an onset temperature T_{onset} that corresponds to a time-to-maximum rate t_{im} (min) using a successive substitution solution procedure. An initial guess of T = 350K in the RHS of Eq. 23.52 will give a solution value of T_{onset} on the LHS of Eq. 23.52 within 1% or on an absolute basis ±3°C. Convergence is reached within several successive substitution iterations.

Maximum Reaction Temperature

Once we have determined the onset temperature, we can then obtain the maximum reaction temperature by the adiabatic temperature rise and any contribution due to external heat input. The theoretical adiabatic temperature increase is

$$\Delta T_{adia} = \frac{x_o.\Delta H_{Rx}}{C.\phi} \quad (23.53)$$

The contribution to the overall temperature increase from external heat input is defined by

$$\Delta T_{ex} = \frac{T^{\bullet}_{ex}.t_{mr}}{2} \quad (23.54)$$

where

ΔT_{ex} = The temperature increase attributed to external heating effects, °C.
$T^{\bullet}_{ex}$ = Bulk heat-up rate due to external heating alone, °C/min
t_{im} = Time-to-maximum rate as determined by Eq. 23.52 with temperature T set equal to T_{onset}, min

From these expressions, the maximum reaction temperature T_{max} is defined by

$$T_{max} = T_{onset} + \Delta T_{adia} + \Delta T_{ex} \quad (23.55)$$

To obtain updated listing of the published information on this research, contact the AIChE office in New York. The work is original and conducted by thoroughly qualified researchers/engineers. The work on runaway reactions is the first systematized examination of the subject and is really the only design approach available, which requires careful study.

Two-phase flow is an important aspect of venting relief as well as of runaway reactions, and is a complicated topic when related to liquid flashing in a vessel as it discharges on pressure relief. It cannot be adequately covered by conventional fluid two-phase flow. The following briefly reviews different calorimetric methods employed for screening and testing two-phase relief system. Other techniques are illustrated elsewhere [49].

Vent Sizing Package (VSP)

The vent sizing package (VSP) was developed by Fauske & Associates Inc. The VSP and its latest version VSP2 employ the low thermal mass test cell stainless steel 304 and Hastelloy test cell with a volume of 120 ml contained in a 4 l high-pressure vessel as shown in Figure 23.41. The typical ϕ-factor is 1.05–1.08 for a test cell wall thickness of 0.127 to 0.178 mm. Measurements consist of sample temperature T_1 and pressure P_1, external guard temperature T_2, and containment pressure P_2. During the runaway or the self-heating period, the guard heater assembly serves to provide an adiabatic environment for the test sample by regulating T_2 close to T_1. For closed (non-vented) test cells, the containment vessel serves to prevent bursting of the test cell by regulating its own pressure P_2 to follow the test cell sample pressure P_1. This pressure-tracking feature makes possible the utilization of the thin wall (low ϕ-factor) test cell design. Vented or open tests, where the vapor or gas generated is vented either into the containment vessel or to an external container are unique capabilities of the VSP instruments. The typical onset sensitivity is 0.1°C/min for the VSP and 0.05°C/min for the VSP2.

The VSP experiments allow the comparison of various process versions, the direct determination of the wanted reaction adiabatic temperature rise, and the check of the possible initiations of secondary reactions. If no secondary reaction is initiated at the wanted reaction adiabatic final temperature, a further temperature scan allows the determination of the temperature difference between the wanted reaction adiabatic final temperature and the subsequent decomposition reaction onset temperature.

The VSP experiments are not suitable to measure the controlling reactant accumulation under normal process conditions. This is obtained using reaction calorimeters. The VSP experiments are suitable to assess the consequences of runaway decomposition reactions after their identification using other screening methods (Differential Thermal Analysis (DTA), Differential Scanning Calorimetry (DSC), and Accelerating Rate Calorimeter (ARC)). The decomposition is then initiated by a temperature scan or an isothermal exposure, depending on the known kinetic behavior

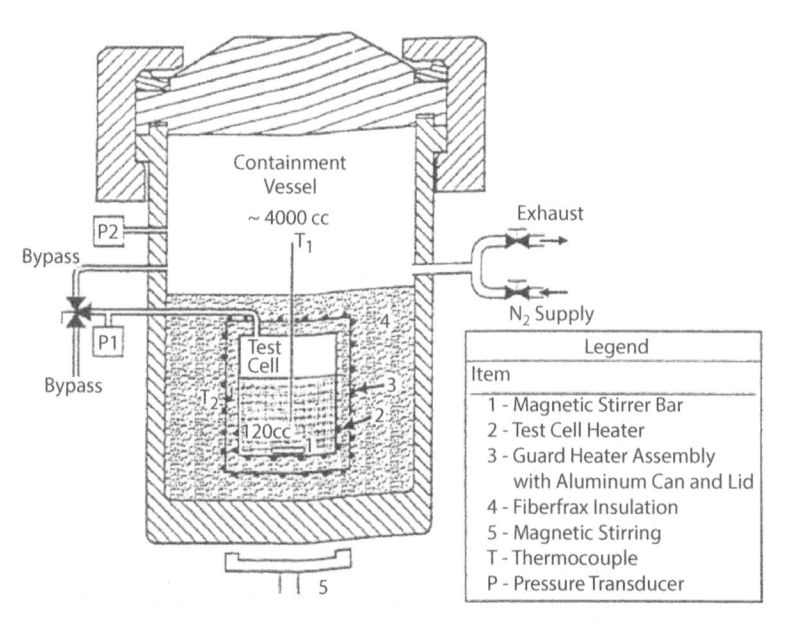

Figure 23.41 Vent sizing package (VSP) apparatus. (By permission from Fauske and Associates, Inc.)

of the reaction. In some instances where a reliable baseline cannot be determined on DTA thermograms, a VSP experiment enables a better and safer determination of the decomposition exotherm.

The VSP experiments provide both thermal information on runaway reactions and information on pressure effects. The type of pressurization, following the DIERS methodology (i.e., vapor pressure, production of non-condensable gases, or both) can be determined from VSP experiments. The following experimental conditions are readily achievable using the VSP [57, 58].

- Temperature up to 350°C–400°C under temperature scan conditions.
- Pressure up to 200 bar.
- Closed or open test cells. Open test cells are connected to a second containment vessel. Test cells of various materials including stainless steel, Hastelloy C276, and glass test cells. Glass test cells of various sizes and shape are suitable for testing fine chemical and pharmaceutical products/reaction conditions, when the process is in glass vessels only or when small samples only are available, or if the samples are very expensive.
- Mechanical stirring is recommended for testing polymerization reactions or viscous reaction mixtures. This requires the use of taller containment vessels than the original one to install the electric motor for the agitator.

Vent Sizing Package 2™ (VSP2™)

Figure 23.42 shows the Vent Sizing Package 2 system. It uses a patented low thermal mass temperature and pressure equalized 120 ml test cell configuration. The equipment is versatile and enables users to obtain accurate adiabatic temperature and pressure rate data for the fastest runaway reactions. The test data can directly be applied to process scale without performing tedious computations. VSP2 tests are employed to model upset conditions as

- loss of cooling.
- loss of stirring.
- mischarge of reagents.
- batch contamination.
- fire exposure heating.

The VSP2 provides vent sizing and thermal data under adiabatic runaway reaction conditions, which can be directly applied to process scale.

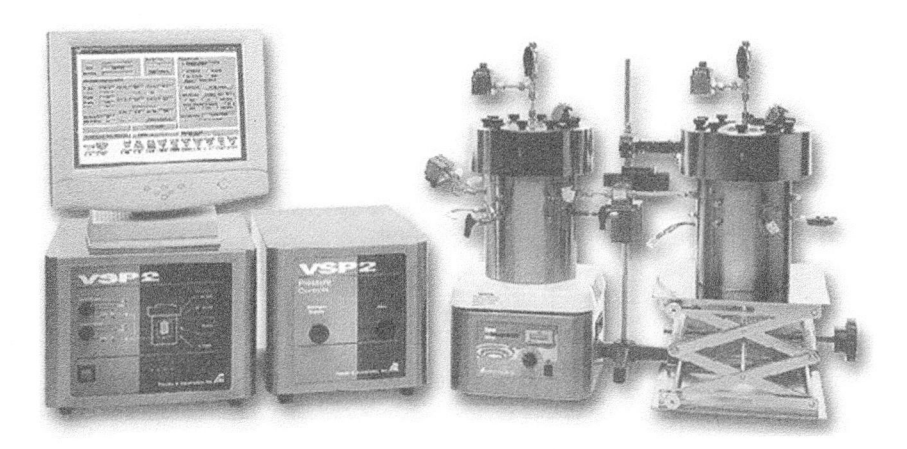

Figure 23.42 The Vent Sizing package 2TM (VSP 2TM) apparatus. (By permission from Fauske and Associates, Inc.)

Advanced Reactive System Screening Tool (ARSST)

Fauske & Associates Inc. developed the ARSST as an easy and inexpensive screening tool for characterizing chemical systems and acquiring relief system design data. It can safely identify potential chemical hazards in the process industry. The ARSST (Figures 23.43a, b, and c) consists of a spherical glass test cell and immersion heater. It has surrounding jacket heater and insulation, thermocouples, and a pressure transducer, a stainless steel containment vessel that serves as both a pressure simulator and safety vessel. The ARSST uses a small magnetic stir bar, which is placed in the test cell and driven by an external magnetic stirrer.

The sample cell volume is 10 ml and the containment volume is 350 ml. The apparatus has a low effective heat capacity relative to that of the sample whose value, expressed as the capacity ratio, is approximately 1.04 (i.e., quite adiabatic). This key feature allows the measured data to be directly applied to process scale.

The ARSST features a heat-wait-search (HWS) mode of operation that provides onset detection sensitivity as low as 0.1 °C/min and isothermal operation at elevated temperature. It can readily cope with endothermic behavior (phase change) and an optical flow regime detector enables the ARSST operator to distinguish between "foamy" and "non-foamy" runaway reactions (Figure 23.43c). The ARSST is computer controlled, which records time, temperatures, and pressure and heater power during a test. Figures 23.44–23.48 show typical plots generated by the equipment [59].

23.55 Two-Phase Flow Relief Sizing for Runaway Reaction

Many methods have been used to size relief systems: area/volume scaling, mathematical modeling using reaction parameters and flow theory, and empirical methods by the Factory Insurance Association (FIA). The DIERS of the AIChE has carried out studies of sizing reactors undergoing runaway reactions. Intricate laboratory instruments as described earlier have resulted in better vent sizes.

A selection of relief venting as the basis of safe operation is based upon the following considerations [60]:

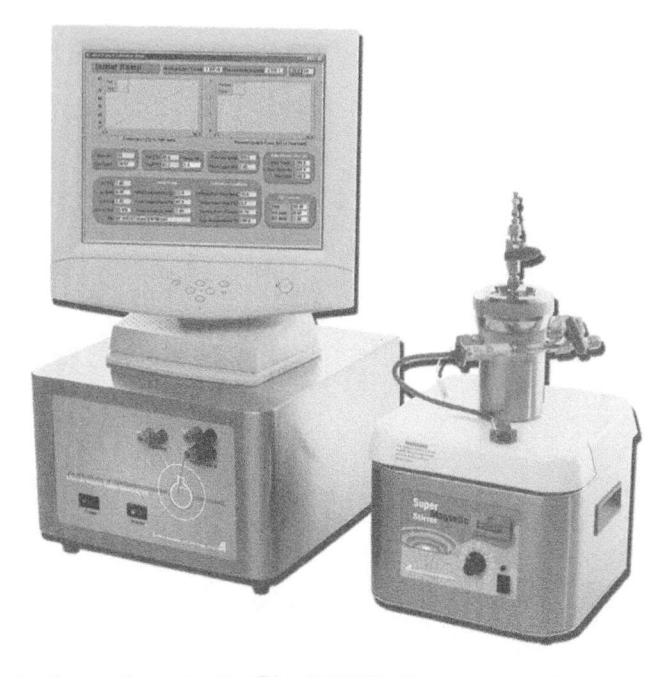

Figure 23.43a The Advanced Reactive System Screening Tool™ (ARSST™). (By permission from Fauske and Associates, Inc.)

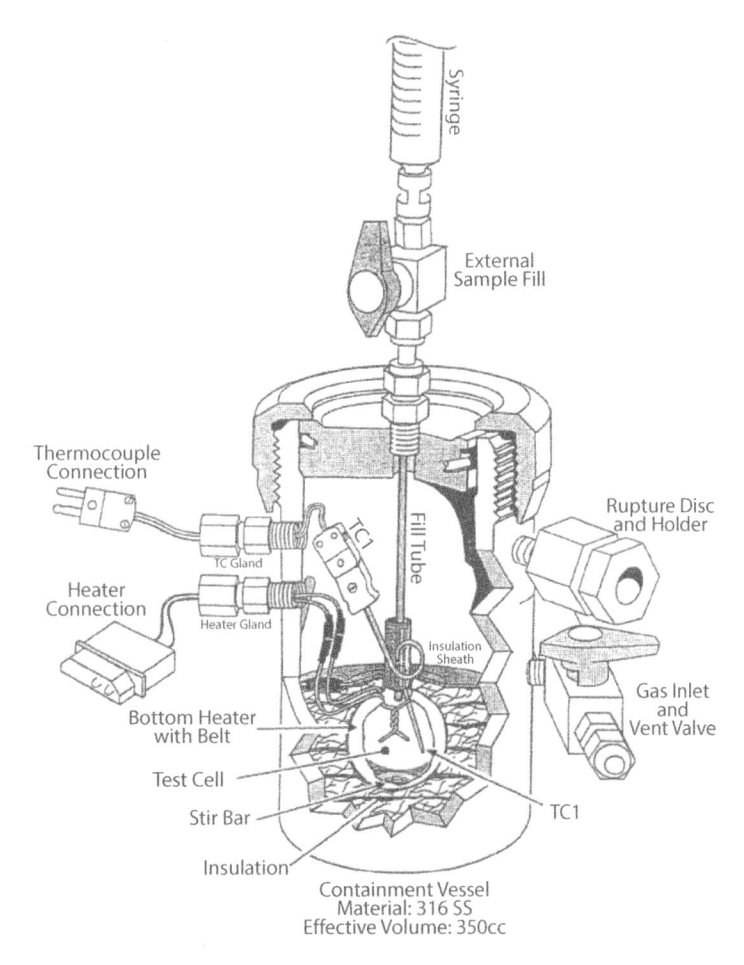

Figure 23.43b Schematic of the Advanced Reactive System Screening Tool™ (ARSS™) containment vessel and internals. (By permission from Fauske and Associates, Inc.)

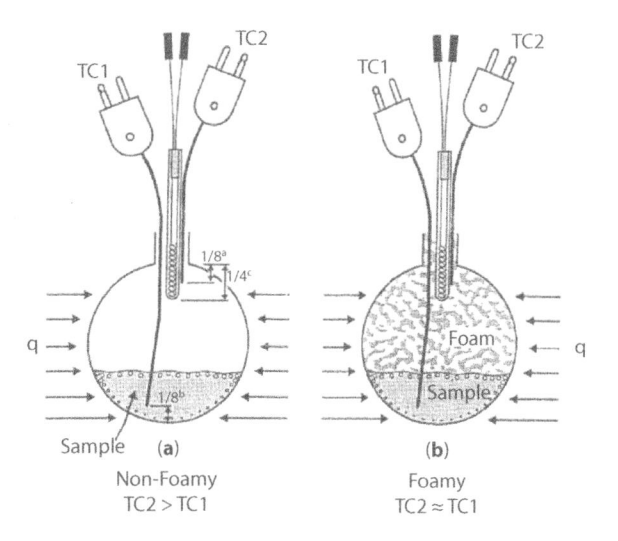

Figure 23.43c The Advanced Reactive System Screening Tool™ (ARSST™) flow regime detector. (By permission from Fauske and Associates, Inc.)

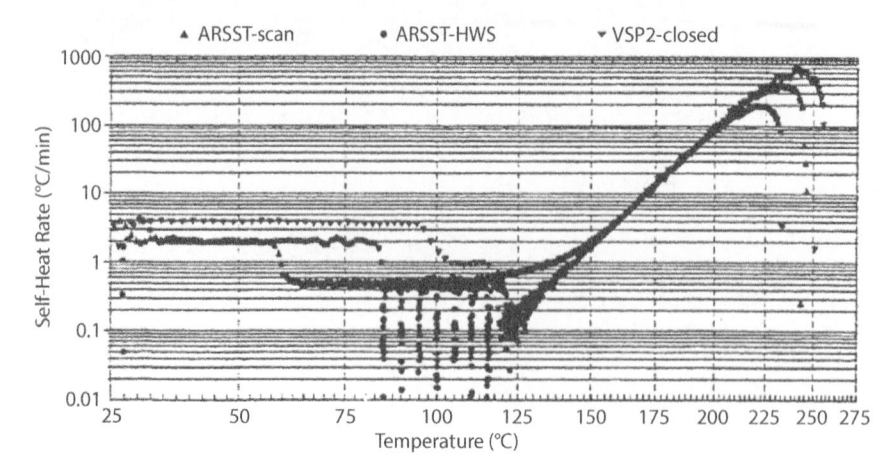

Figure 23.44 Self-heat rate data for 25% DTBP in Toluene. (By permission from Fauske and Associates, Inc.)

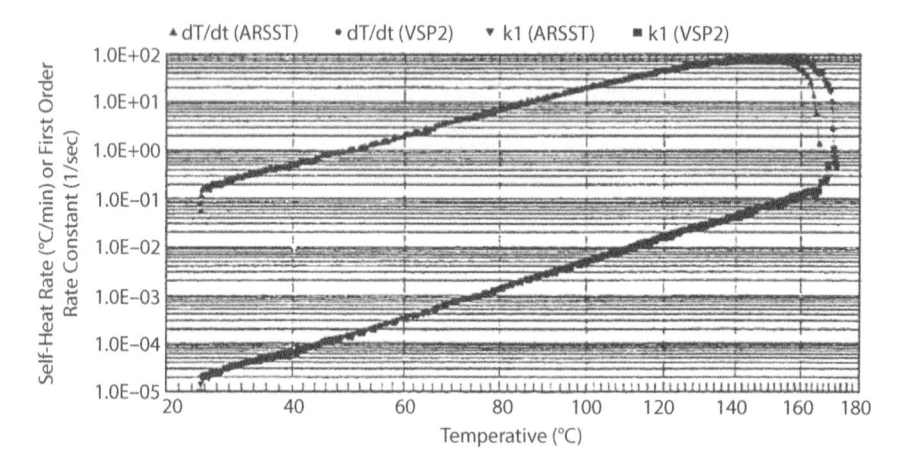

Figure 23.45 Self-heat rate data and first-order rate constant for Methanol/Acetic Anhydride. (By permission from Fauske and Associates, Inc.)

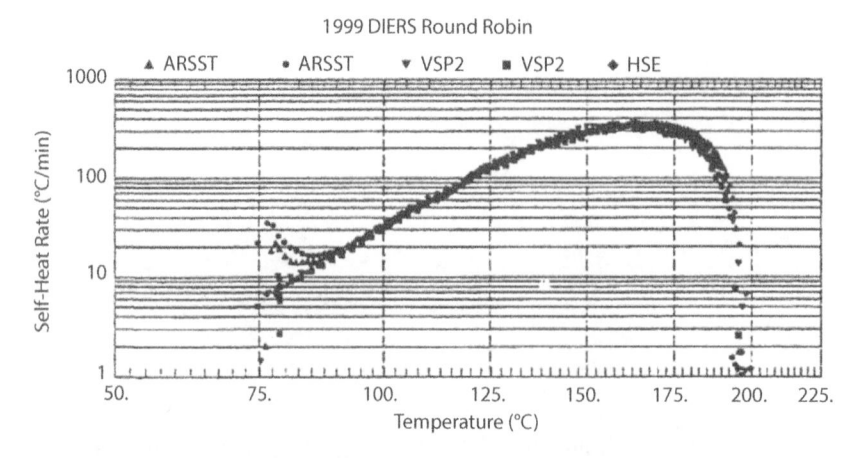

Figure 23.46 Composite self-heat plot of selected Round Robin data. (By permission from Fauske and Associates, Inc.)

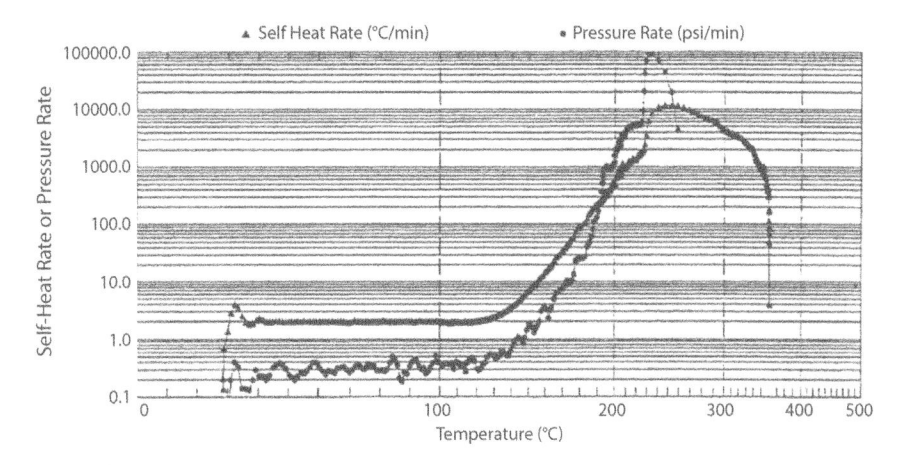

Figure 23.47 Self-heat rate and pressure rate data for neat DTBP at 300 psig in the ARSST. (By permission from Fauske and Associates, Inc.)

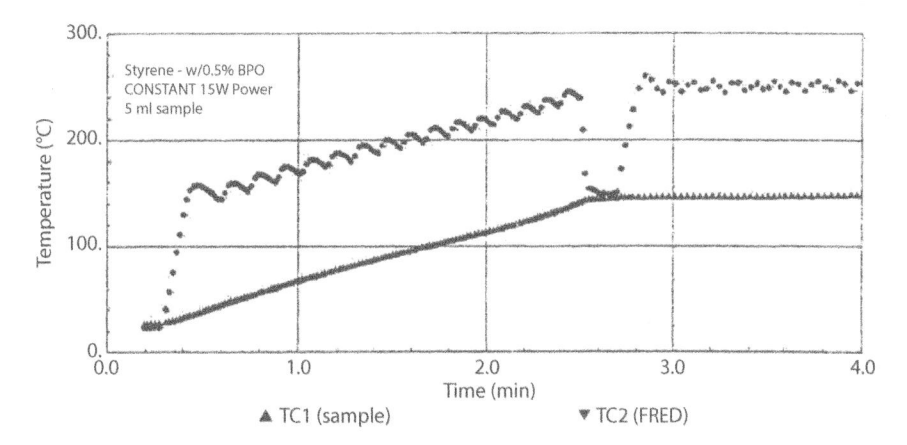

Figure 23.48 Example of flow regime detection with the ARSST. (By permission from Fauske and Associates, Inc.)

- compatibility of relief venting with the design and operation of the plant/process.
- identifying the worst scenarios.
- type of reaction.
- means of measuring the reaction parameters during the runaway reaction.
- relief sizing procedure.
- design of the relief system including discharge ducting and safe discharge area.

Runaway Reactions

A runaway reaction occurs when an exothermic system becomes uncontrollable. The reaction leads to a rapid increase in the temperature and pressure, which if not relieved can rupture the containing vessel. A runaway reaction happens because the rate of reaction, and therefore the rate of heat generation, increases exponentially with temperature. In contrast, the rate of cooling increases only linearly with temperature. Once the rate of heat generation exceeds available cooling, the rate of increase in temperature becomes progressively faster. Runaway reactions nearly always result in two-phase flow reliefs. In reactor venting, reactions essentially fall into three classifications:

1. Vapor pressure systems.
2. Hybrid (gas + vapor) systems.
3. Gassy systems.

The significance of these categories in terms of relief is that, once the vent has opened, both vapor pressure and hybrid reactions temper by losing enough heat through vaporization, to maintain temperature and pressure at an acceptable level. In a gas generating system, there is negligible or sometimes no control of temperature during venting, such that relief sizing is based upon the peak gas generation rate. Experimental studies conducted by the testing methods must not only be able to differentiate between the reaction types, but must also simulate large-scale process conditions. The results of the experimental studies should greatly enhance the design of the relief system. We shall review the reactor venting categories as follows.

Vapor Pressure Systems

In this type of reaction, no permanent gas is generated. The pressure generated by the reaction is due to the increasing vapor pressure of the reactants, products, and/or inert solvent as the temperature rises.

It is the rate of temperature increase (i.e., power output) between the set pressure and the maximum allowable pressure that determines the vent size and not the peak rate. Boiling is attained before potential gaseous decomposition (i.e., the heat of reaction is removed by the latent heat of vaporization). The reaction is tempered, and the total pressure in the reactor is equal to the vapor pressure. The principal parameter determining the vent size is the rate of the temperature rise at the relief set pressure.

Systems that behave in this manner obey the Antoine relationship between pressure P and temperature T as represented by

$$\ln P = A + B/T \tag{23.56}$$

where A and B are constants and T is the absolute temperature. An example is the methanol and acetic anhydride reaction.

Gassy Systems

Here, the system pressure is due entirely to the pressure of non-condensible gas rather than the vapor pressure of the liquid. The gas is the result of decomposition. The exothermic heat release is largely retained in the reaction mass since the cooling potential of volatile materials is not available. As such, both the maximum temperature and maximum gas generation rate can be attained during venting. Gaseous decomposition reactions occur without tempering. The total pressure in the reactor is equal to the gas pressure. The principal parameter determining the vent size is the maximum rate of pressure rise. Unlike the vapor-pressure systems, the pressure is controlled (and reduced) without cooling the reaction.

A survey within the Fine Chemical Manufacturing Organization of ICI has shown that gassy reaction systems predominate due to established processes such as nitrations, diazotizations, sulfonations, and many other types of reactions [61]. Very few vapor pressure systems have been identified, which also generated permanent gas, i.e., hybrid type.

Hybrid Systems

These are systems that have a significant vapor pressure and at the same time produce non-condensible gases. Gaseous decomposition reaction occurs before boiling: the reaction is still tempered by vapor stripping. The total pressure in the reactor is the summation of the gas partial pressure and the vapor pressure. The principal parameters determining the vent size are the rates of temperature and pressure rise corresponding to the tempering condition. A tempered reactor contains a volatile fluid that vaporizes or flashes during the relieving process. This vaporization removes energy via the heat of vaporization and tempers the rate of temperature rise due to the exothermic reaction.

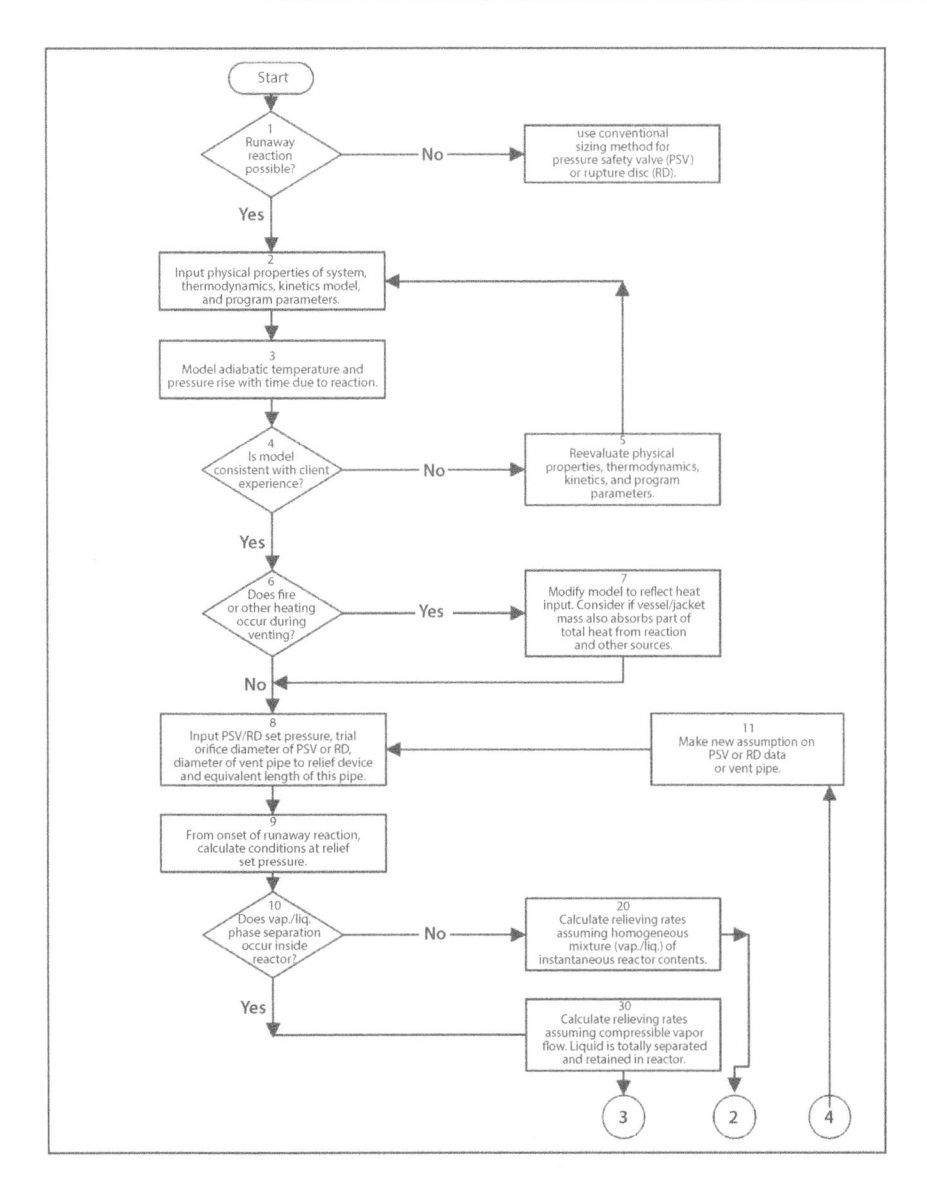

Figure 23.49 Block flow diagram showing the reaction kinetics and thermodynamics needed to create a reactor model. (Reproduced with the permission from the AIChE copyrights © 1996 AIChE, All rights reserved.)

In some hybrid systems, the vapor generation in a vented reaction is high enough to remove sufficient latent heat to moderate or "temper" the runaway, i.e., to maintain constant temperature. This subsequently gives a smaller vent size.

Richter and Turner [62] have provided systematic schemes for sizing batch reactor relief systems. They employed logic diagrams that outlined the various decisions to produce a model of the system. Figure 23.49 reviews the reaction kinetics and thermodynamics required to create a reactor model. Figure 23.50 shows a sequence of steps used to model flow from the reactor, assuming relief is occurring as a homogeneous vapor–liquid mixture. The DIERS program has supported the use of a homogeneous vapor–liquid mixture (froth) model, which relies on the assumption that the vapor phase is in equilibrium with the batch liquid phase [63]. Figure 23.51 shows the steps used to model compressible vapor venting from the reactor. Table 23.21 lists formulae for computing the area of the three systems.

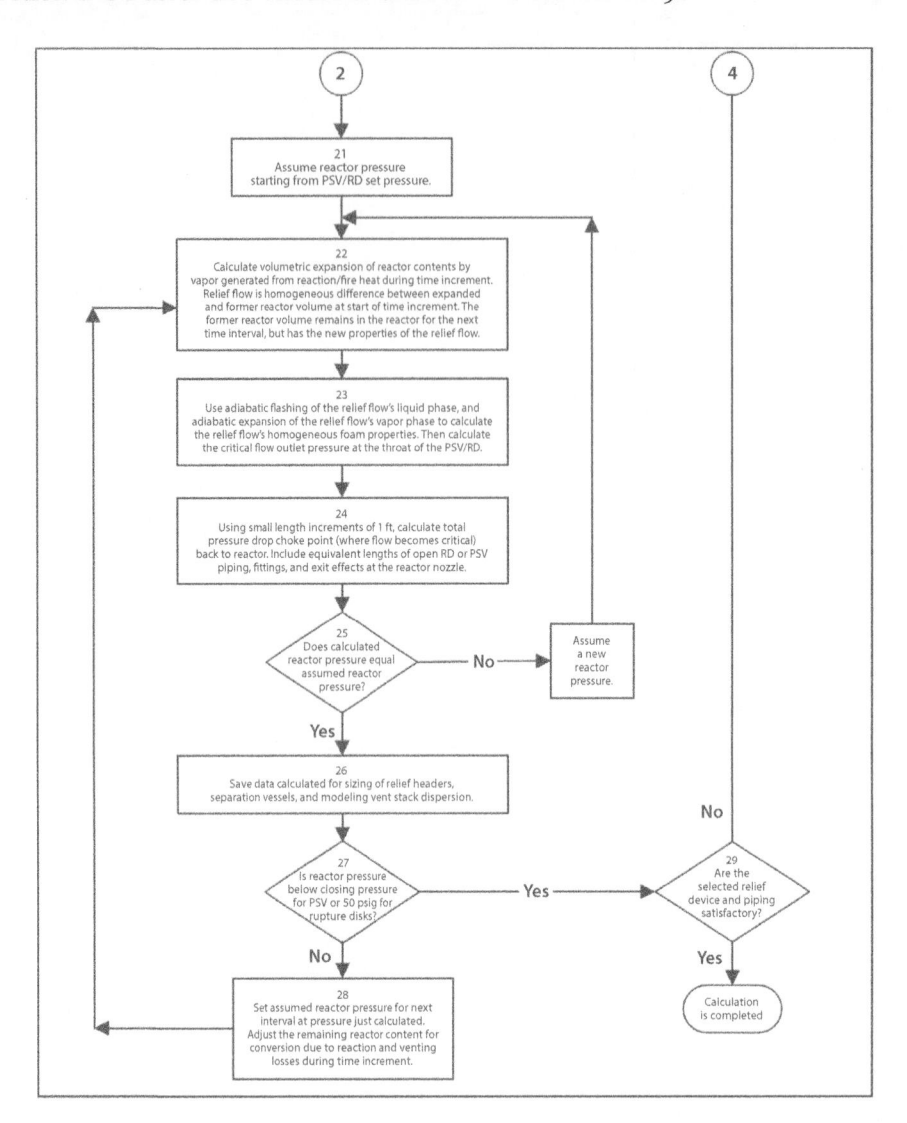

Figure 23.50 Steps used to model flow from the reactor, assuming relief is occurring as a homogeneous vapor – liquid mixture. (Reproduced with the permission from the AIChE copyrights © 1996 AIChE, All rights reserved.)

Simplified Nomograph Method

Boyle [67] and Huff [65] first accounted for two-phase flow with relief system design for runaway chemical reactions. Computer simulation approaches to vent sizing involve extensive thermokinetic and thermophysical characterization of the reaction system. Fisher [66] has provided an excellent review of emergency relief system design involving runaway reactions in reactors and vessels. The mass flux through the relief device representing choked two-phase flow through a hole is expressed as

$$G = \frac{Q_m}{A} = \frac{\Delta H_V}{\upsilon_{fg}} \left(\frac{g_c}{C_P T_S} \right)^{0.5} \qquad (23.57)$$

For two-phase flow through pipes, we apply an overall dimensionless discharge coefficient ψ. Equation 23.57 is referred to as the equilibrium rate model (ERM) for low-quality choked flow. Leung [67] indicated that Eq. 23.57 be multiplied by a factor of 0.9 to bring the value in line with the classic homogeneous equilibrium model (HEM). Equation 23.57 becomes

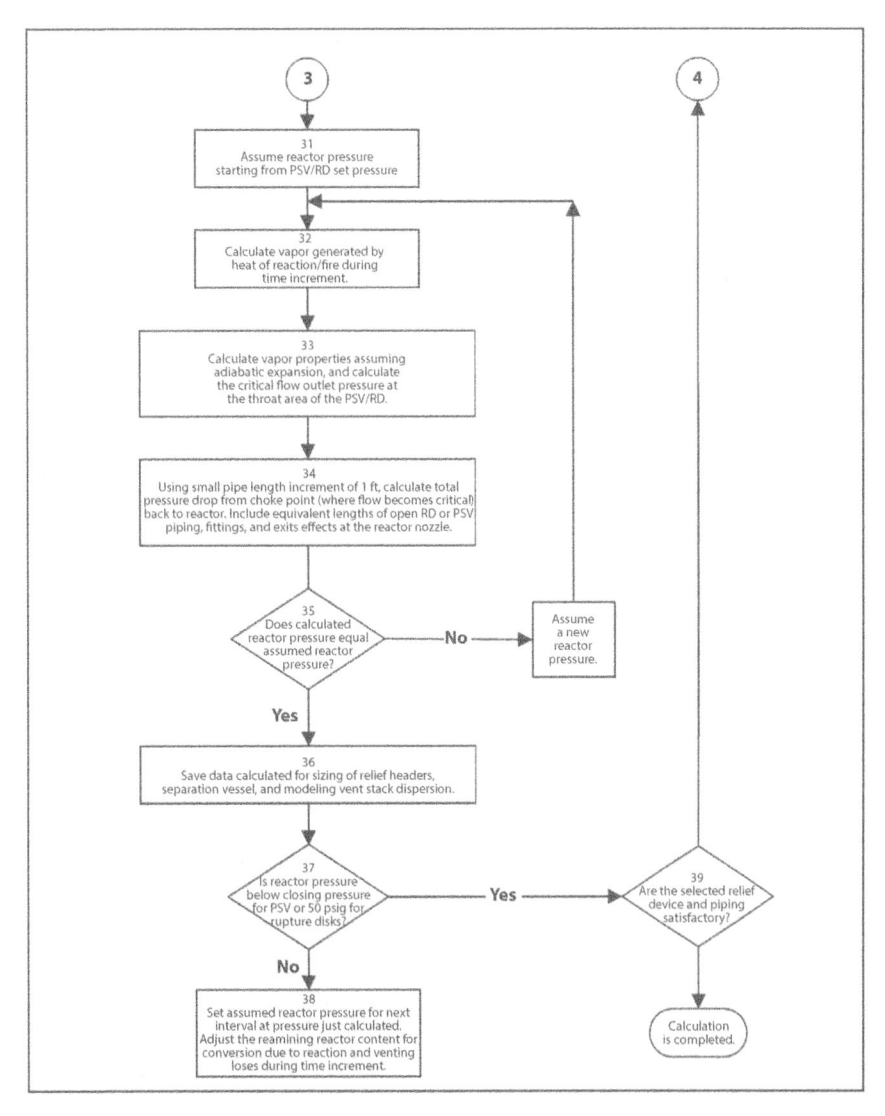

Figure 23.51 Steps used to model compressible vapor venting from the reactor. (Reproduced with the permission from the AIChE copyrights © 1996 AIChE, All rights reserved.)

$$G = \frac{Q_m}{A} = 0.9\psi \frac{\Delta H_V}{v_{fg}} \left(\frac{g_C}{C_P T_S} \right)^{0.5} \qquad (23.58)$$

where

- A = area of the hole, m².
- g_c = correction factor $1.0 \frac{\text{kg.m}}{\text{sec}^2} / \text{N}$.
- Q_m = mass flow through the relief, kg/s.
- ΔH_v = the heat of vaporization of the fluid, J/kg.
- v_{fg} = the change of specific volume of the flashing liquid, m³/kg.
- C_p = the heat capacity of the fluid, J/kg.K.
- T_s = the absolute saturation temperature of the fluid at the set pressure, K.

Table 23.21 Vent Areas and Diameters of the Three Systems.

Vapor system	Gassy system	Hybrid system
$A = 1.5 \times 10^{-5} \left(\dfrac{m_o \dfrac{dT}{dt}}{F.P_s} \right)$	$A = 3 \times 10^{-6} \left(\dfrac{1}{F} \right) \left(\dfrac{m_o}{m_t} \right) \left(\dfrac{\dfrac{dP}{dt}}{P_{MAP}^{1.5}} \right)$	$A = 5.6 \times 10^{-6} \left(\dfrac{1}{F} \right) \left(\dfrac{m_o}{m_t} \right) \left(\dfrac{\dfrac{dP}{dt}}{P_s^{1.5}} \right)$
$d = \left(\dfrac{4A}{\pi} \right)^{0.5}$	$d = \left(\dfrac{4A}{\pi} \right)^{0.5}$	$d = \left(\dfrac{4A}{\pi} \right)^{0.5}$
L/D=0, F=1.00	L/D=0 F=1.00	L/D=0, F=1.00
L/D=50, F=0.85	L/D=50, F=0.70	L/D=50, F=0.70
L/D=100, F=0.75	L/D=100, F=0.60	L/D=100, F=0.60
L/D=200, F=0.65	L/D=200, F=0.45	L/D=200, F=0.45
L/D=400, F=0.50	L/D=400, F=0.33	L/D=400, F=0.33

The result is applicable for homogeneous venting of a reactor (low quality, not restricted just to liquid inlet condition). Figure 23.52 gives the value of ψ for L/D ratio. For a pipe length zero, $\psi = 1$, as the pipe length increases, the value of ψ decreases. Equation 23.58 can be further rearranged in terms of a more convenient expression as follows:

$$\frac{\Delta H_V}{\upsilon_{fg}} = T_s \frac{dP}{dT} \tag{23.59}$$

Substituting Eq. 23.59 into Eq. 23.58 gives

$$G = 0.9\psi \frac{dP}{dT} \left(\frac{g_C T_S}{C_P} \right)^{0.5} \tag{23.60}$$

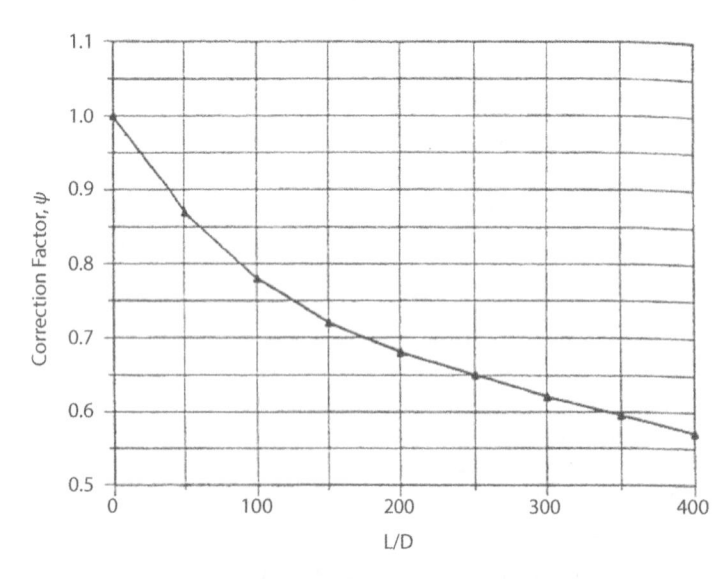

Figure 23.52 Correction factor versus L/D for two-phase flashing flow through pipes. (Source: J. C. Leung and M. A. Grolmes, "The Discharge of Two-Phase Flashing Flow in a Horizontal Duct", AIChE J., Vol. 33, No. 3, p. 524, 1987.)

The exact derivative is approximated by a finite difference derivative to yield

$$G \cong 0.9\psi \frac{\Delta P}{\Delta T}\left(\frac{g_C T_S}{C_P}\right)^{0.5} \tag{23.61}$$

where

ΔP = the overpressure.
ΔT = the temperature rise corresponding to the overpressure.

Fauske [68] has developed a simplified chart for the two-phase calculation. He expressed the relief area as

$$A = \frac{V\rho}{G\Delta t_V} \tag{23.62}$$

where

A = the relief vent area, m².
V = reactor volume, m³.
ρ = density of the reactants, kg/m³.
G = mass flux through the relief, kg/sec. m².
Δt_v = venting time, sec.

Boyle [64] developed Eq. 23.62 by defining the required area as that size that would empty the reactor before the pressure could rise above some allowable pressure for a given vessel. The mass flux G is given by Eq. 23.58 or Eq. 23.61, and the venting time is given by

$$\Delta t_V = \frac{(\Delta T)(C_P)}{q_S} \tag{23.63}$$

where

ΔT = the temperature rise corresponding to the overpressure ΔP.
T = the temperature.
C_p = the heat capacity.
q_s = the energy release rate per unit mass at the set pressure of the relief system.

Combining Equations 23.62, 23.63, and 23.57 yields

$$A = V\rho(g_C C_P T_S)^{-0.5}\frac{q_S}{\Delta P} \tag{23.64}$$

Equation 23.64 gives a conservative estimate of the vent area and the simple design method represents overpressure (ΔP) between 10 and 30%. For a 20% absolute overpressure, a liquid heat capacity of 2510 J/kg K for most organics, and considering that a saturated water relationship exists, the vent size area per 1,000 kg of reactants is given by

$$A = \left(\frac{m^2}{1{,}000\ kg}\right) = \frac{0.00208\left(\dfrac{dT}{dt}\right)}{P_s}\frac{°C/min.}{bar} \tag{23.65}$$

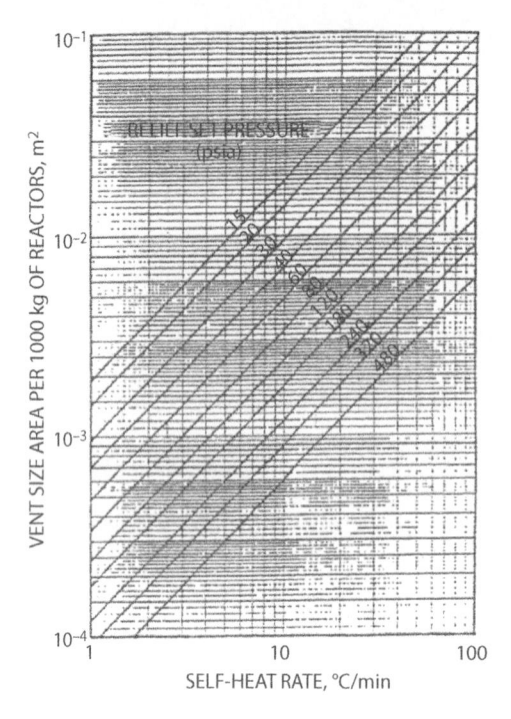

Figure 23.53 A vent sizing nomograph for tempered (high vapor-pressure) runaway chemical reactions. (Source: H. K. Fauske, "Generalized Vent Sizing Monogram for Runaway Chemical Reactions", Plant/Operations Prog., Vol. 3, No. 4, 1984. Reproduced with permission from the AIChE, Copyrights © 1984, All rights reserved.)

Figure 23.53 shows a nomograph for determining the vent size. The vent area is calculated from the heating rate, the set pressure, and the mass of reactants. The nomograph is used for obtaining quick vent sizes and checking the results of the more rigorous computation. Crowl and Louvar [41] have expressed that the nomograph data of Figure 23.53 applies for a discharge coefficient of $\psi = 0.5$, representing a discharge L/D of 400.0. However, use of the nomograph at other discharge pipe lengths and different ψ requires a suitable conversion.

Vent Sizing Methods

Vents are usually sized on the assumption that the vent flow is

- all vapor or gas
- all liquid and
- a two-phase mixture of liquid and vapor or gas.

The first two cases represent the smallest and largest vent sizes required for a given rate at increased pressure. Between these cases, there is a two-phase mixture of vapor and liquid. It is assumed that the mixture is homogeneous, that is, that no slip occurs between the vapor and liquid. Furthermore, the ratio of vapor to liquid determines whether the venting is closer to the all vapor or all liquid case. As most relief situations involve a liquid fraction of over 80%, the idea of homogeneous venting is closer to all liquid than all vapor. Table 9.40 shows vent area for different flow regimes.

Vapor Pressure Systems

These systems are called "tempering" (i.e., to prevent temperature rise after venting) systems as there is sufficient latent heat available to remove the heat of reaction, and to temper the reaction at the set pressure. The vent requirements for such systems are estimated from the Leung's Method [69, 70].

$$A = \frac{M_o q}{G\left[\left(\frac{V}{M_O} T_S \frac{dP}{dT}\right)^{0.5} + (C_V \Delta T)^{0.5}\right]^2} \qquad (23.66)$$

Alternatively, the vent area can be expressed as

$$A = \frac{M_o q}{G\left[\left(\frac{V}{M_O} \frac{\Delta H_V}{\upsilon_{fg}}\right)^{0.5} + (C_V \Delta T)^{0.5}\right]^2} \qquad (23.67)$$

where

- M_o = the total mass contained within the reactor vessel prior to relief, kg
- q = the exothermic heat release rate per unit mass, $\frac{J}{kg \cdot sec}$, $\frac{W}{kg}$
- V = the volume of the vessel, m³
- C_V = the liquid heat capacity at constant volume, J/kg.K
- ΔH_v = heat of vaporization of the fluid, J/kg
- υ_{fg} = change of specific volume of the flashing liquid, $(\upsilon_g - \upsilon_f)$, m³/kg

The heating rate q is defined by

$$q = \frac{1}{2} C_V \left[\left(\frac{dT}{dt}\right)_S + \left(\frac{dT}{dt}\right)_m\right] \qquad (23.68)$$

The first derivative Eq. 23.68, denoted by the subscript "s," corresponds to the heating rate at the set pressure, and the second derivative, denoted by subscript "m," corresponds to the temperature rise at the maximum turnaround pressure.

The above equations assume the following:

- Uniform froth or homogeneous vessel venting occurs.
- The mass flux, G, varies little during the relief.
- The reaction energy per unit mass, q, is treated as constant.
- Constant physical properties C_v, ΔH_v, and υ_{fg}.
- The system is a tempered reactor system. This applies to the majority of reaction systems.

We should take care of using consistent units in applying the above two-phase equations. The best procedure is to convert all energy units to their mechanical equivalents before solving for the relief area, especially when Imperial (English) units are used. To be consistent, use the SI unit.

The vapor pressure systems obey the Antoine relationships given by Eq. 23.56:

$$\ln P = A + \frac{B}{T}$$

Differentiating Equation 23.56 yields

$$\frac{1}{P}\frac{dP}{dT} = -\frac{B}{T^2} \qquad (23.69)$$

$$\frac{dP}{dT} = -\frac{B}{T^2}P \qquad (23.70)$$

An equation representing the relief behavior for a vent length L/D < 400 is given by [113]

$$M_O = \frac{(D_p)^2 (\Delta P_S)}{2.769 \left(\frac{dT}{dt}\right)_S} \left(\frac{T_S}{C_p}\right)^{0.5} \qquad (23.71)$$

where

- M_o = allowable mass of the reactor mixture charge (kg) to limit the venting overpressure to P_p (psig)
- D_p = rupture disk diameter, in.
- ΔP_s = the allowable venting overpressure (psi), that is the maximum venting pressure minus the relief device set pressure
- P_p = maximum venting pressure (psig)
- P_s = the relief device set pressure (psig). Note that the relief device set pressure can range from the vessel's MAWP to significantly below the MAWP
- T_s = the equilibrium temperature corresponding to the vapor pressure where the vapor pressure is the relief device set pressure (K)
- $\left(\frac{dT}{dt}\right)_S$ = the reactor mixture self-heat rate $\left(\frac{°C}{min}\right)$ at temperature T_s (K) as determined by a DIERS or equivalent test
- C_p = specific heat of the reactor mixture (cal/g-K or Btu/lb °F)

Equation (23.71) is a dimensional equation; therefore, the dimensions given in the parameters must be used.

Fauske's Method

Fauske [70] represented a nomograph for tempered reactions as shown in Figure 23.53. This accounts for turbulent flashing flow and requires information about the rate of temperature rise at the relief set pressure. This approach also accounts for vapor disengagement and frictional effects including laminar and turbulent flow conditions. For turbulent flow, the vent area is

$$A = \frac{1}{2} \frac{M_O \left(\frac{dT}{dt}\right)_S (\alpha_D - \alpha_O)}{F \left(\frac{T_S}{C_S}\right)^{0.5} \Delta P (1 - \alpha_O)} \quad \text{for } 0.1P_S \leq \Delta P \leq 0.3P_S \qquad (23.72)$$

where

- M_o = Initial mass of reactants, kg
- $\left(\frac{dT}{dt}\right)_S$ = Self-heat rate corresponding to the relief set pressure (K/s)

$\left(\dfrac{dT}{dt}\right)_m$ = Self-heat rate at turnaround temperature (K/s)

P_S = Relief set pressure (Pa)
α_D = Vessel void fraction corresponding to complete vapor disengagement.
α_o = Initial void fraction in vessel
T_S = Temperature corresponding to relief actuation, K
C_S = Liquid specific heat capacity, J/kg K
ΔP = Equilibrium overpressure corresponding to the actual temperature rise, Pa.
ΔT = Temperature rise following relief actuation, K
F = Flow reduction correction factor for turbulent flow
(L/D=0, F≈1.0; L/D=50, F≈0.85; L/D=100, F≈0.75;
L/D=200, F≈0.65; L/D=400, F≈0.55).
where L/D is the length-to-diameter ratio of the vent line.

Figure 23.54 shows a sketch of temperature profile for high vapor pressure systems.

The VSP bench scale apparatus can be employed to determine the information about the self-heat-rate and vapor disengagement when this is not readily available. In addition, the VSP equipment can be used for flashing flow characteristics using a special bottom vented test cell. Here, the flow rate G_o (kg/sm^2) is measured in a simulated vent line (same L/D ratio) of diameter D_o using the vent sizing package (VSP) apparatus. The following recommended scale-up approach in calculating the vent size is

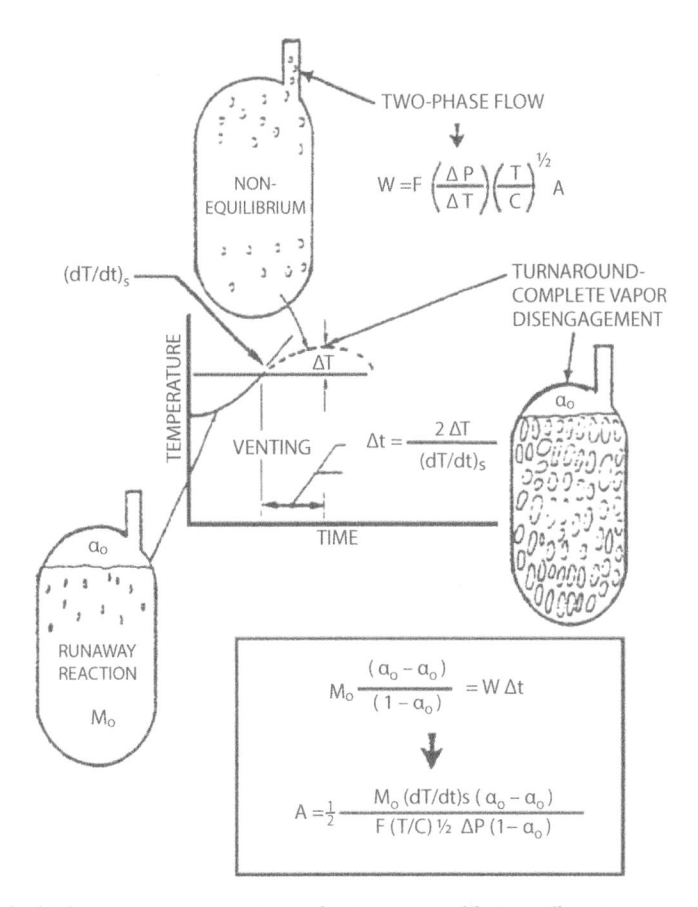

Figure 23.54 Vent sizing model for high vapor-pressure systems; due to non-equilibrium effects turnaround in temperature is assumed to coincide with the onset of complete vapor disengagement.

$$\text{If } G_O\left(\frac{D_T}{D_O}\right) \geq G_T \cong F\left(\frac{\Delta P}{\Delta T}\right)\left(\frac{T_S}{C_{P,S}}\right)^{0.5}$$

where D_T is the vent diameter required for turbulent flow and is expressed by

$$D_T \cong \frac{3}{2}\left[\frac{M_O \frac{dT}{dt}(\alpha_D - \alpha_O)}{FP_S(1-\alpha_O)}\right]^{0.5}\left(\frac{C_{P,S}}{T_S}\right)^{0.25}$$

If $G_O\left(\frac{D_T}{D_O}\right) \leq G_T \cong F\left(\frac{P}{T}\right)\left(\frac{T_S}{C_{P,S}}\right)^{0.5}$, the required vent diameter for laminar flow D_L is given by

$$D_L = \left(D_T^2 D_O \frac{G_T}{G_O}\right)^{0.33}$$

Gassy Systems

The major method of vent sizing for gassy system is two-phase venting to keep the pressure constant. This method was employed before DIERS with an appropriate safety factor [71]. The vent area is expressed by

$$A = \frac{Q_g(1-\alpha)\rho_f}{G_1} = \frac{Q_g M}{GV} \qquad (23.73)$$

where

- Q_g = volumetric gas generation rate at temperature and in reactor during relief, m³/sec
- M = mass of liquid in vessel, kg
- P_f = liquid density, kg/m³
- G, G_1 = mass vent capacity per unit area, kg/sec m²
- α = void fraction in vessel
- V = total vessel volume, m³

Unlike systems with vapor present, gassy systems do not have any latent heat to temper the reaction. The system pressure increases as the rate of gas generation with temperature increases, until it reaches the maximum value. The vent area could be underestimated, if sizing is dependent on the rate of gas generated at the set pressure. Therefore, it is more plausible to size the vent area on the maximum rate of gas generation. Homogeneous two-phase venting is assumed even if the discharge of liquid during venting could reduce the rate of gas generation even further. The vent area is defined by

$$A = 3.6 \times 10^{-3} Q_g \left(\frac{M_O}{VP_m}\right)^{0.5} \qquad (23.74)$$

The maximum rate of gas generation during a runaway reaction is proportional to the maximum value of $\frac{dP}{dt}$ and can be calculated from

$$Q_{g1} = \frac{M_O}{M_t}\left(\frac{V_t}{P_m}\frac{dP}{dt}\right) \tag{23.75}$$

An equation representing the relief behavior for a length L/D<400 as in the tempered system is given by [72]

$$M_O = \left[V_P \Delta P_P \left\{\frac{(D_P)^2(M_S)(P_P)}{(2.07)(T_T)(dP/dt)}\right\}^2\right]^{1/3} \tag{23.76}$$

where

V_p = vessel total volume (gal)
P_p = maximum allowable venting pressure (psia)
M_s = sample mass used in a DIERS test or equivalent test, g
$\frac{dP}{dt}$ = pressure rise in test (psi/s)
T_T = maximum temperature in test (K)
P_{amb} = ambient pressure at the end of the vent line, psia
$\Delta P_p = P_p - P_{amb}$, psi

Equation 23.76 is a dimensional equation and the dimensions given in the parameters must be used.

Homogeneous Two-Phase Venting Until Disengagement

ICI [71] developed a method for sizing a relief system that accounts for vapor/liquid disengagement. They proposed that homogeneous two-phase venting occurs that increases to the point of disengagement. Furthermore, they based their derivations upon the following assumptions:

- Vapor phase sensible heat terms may be neglected.
- Vapor phase mass is negligible.
- Heat evolution rate per unit mass of reactants is constant (or average value can be used).
- Mass vent rate per unit area is approximately constant (or safe value can be used).
- Physical properties can be approximated by average values.

The vent area is given by

$$A = \frac{qV(\alpha - \alpha_O)}{G\upsilon_f\left\{\dfrac{h_{fg}\upsilon_f(\alpha - \alpha_O)}{\upsilon_{fg}(1-\alpha_O)(1-\alpha)} + C\Delta T\right\}} \tag{23.77}$$

where

A = vent area, m^2
C = Liquid specific heat, J/kg K
G = Mass vent capacity per unit area, kg/m^{2}s
h_{fg} = Latent heat, J/kg

q = Self-heat rate, W/kg
V = Total vessel volume, m³
ΔT = Temperature rise corresponding to the overpressure, K
α_o = Initial void fraction
α = Void fraction in vessel
υ_f = Liquid specific volume, m³/kg
υ_{fg} = Difference between vapor and liquid specific volumes, m³/kg

It is valid if disengagement occurs before the pressure would have turned over during homogeneous venting; otherwise, Equation 23.77 gives an unsafe (too small) vent size. Therefore, it is necessary to verify that

$$q > \frac{GAh_{fg}\upsilon_f^2}{V\upsilon_{fg}(1-\alpha)^2} \qquad (23.78)$$

is satisfied at the point of disengagement.

Two-Phase Flow Through an Orifice

Sizing formulae for flashing two-phase flow through relief devices were obtained through DIERS. It is based upon Fauske's equilibrium rate model (ERM) and assumes frozen flow (non-flashing) forms a stagnant vessel to the relief device throat. This is followed by flashing to equilibrium in the throat. The orifice area is expressed by

$$A = \frac{W}{C_D}\left\{\left(\frac{xV_G}{kP_1}\right) + \left[\frac{(V_G - V_L)^2 C_L T_1}{\lambda^2}\right]\right\}^{0.5} \qquad (23.79)$$

where

A = Vent area, m²
C_D = Actual discharge coefficient
C_L = Average liquid specific heat, J/kg K
k = isentropic coefficient
P_1 = Pressure in the upstream vessel, N/m²
T_1 = Temperature corresponding to relief actuation, K
x = Mass fraction of vapor at inlet
V_G = Specific volume of gas, m³/kg
V_L = Specific volume of liquid, m³/kg
W = Required relief rate, kg/s
λ = Latent heat, J/kg

Theoretical rate is given by

$$W = A_f(2\Delta P \bullet \rho)^{0.5} \qquad (23.80)$$

For a simple sharp-edged orifice, the value of C_D is well established (about 0.6). For safety relief valves, its value depends upon the shape of the nozzle and other design features. In addition, the value of C_D varies with the conditions at the orifice. For saturated liquid at the inlet, Eq. 23.79 simplifies to yield

$$A = \frac{W(V_G - V_L)(C_L T_1)^{0.5}}{C_D \lambda} = \frac{W}{C_D \left(\dfrac{dP}{dT}\right)\left(\dfrac{T_1}{C_L}\right)^{0.5}} \qquad (23.81)$$

These equations are based upon the following assumptions:

- Vapor phase behaves as an ideal gas.
- Liquid phase is an incompressible fluid.
- Turbulent Newtonian flow.
- Critical flow. This is usually the case since critical pressure ratios of flashing liquid approach the value of 1.

It is recommended that a safety factor of at least 2.0 be used [71]. In certain cases, lower safety factors may be employed. The designer should consult the appropriate process safety section in the engineering department for advice.

Conditions of Use

- If Fauske's method yields a significantly different vent size, then the calculation should be reviewed.
- The answer obtained from the Leung's method should not be significantly smaller than that from Fauske's method.
- ICI recommends a safety factor of 1–2 on flow or area. The safety factor associated with the inaccuracies of the flow calculation will depend on the method used, the phase nature of the flow, and the pipe friction. For two-phase flow, use a safety factor of 2 to account for friction or static head.
- Choose the smaller of the vent size from the two methods.

A systematic evaluation of venting requires information on

- the reaction—vapor, gassy, or hybrid.
- flow regime—foamy or non-foamy.
- vent flow—laminar or turbulent.
- vent sizing parameters are dT/dt, $\Delta P/\Delta T$, and ΔT.

It is important to ensure that all factors (e.g., long vent lines) are accounted for, independent of the methods used. Designers should ascertain that a valid method is chosen rather than the most convenient or the best one. For ease of use, the Leung's method for vapor pressure systems is rapid and easy. Different methods should give vent sizes within a factor of 2. Nomographs give adequate vent sizes for long lines to L/D of 400, but sizes are divisible by 2 for nozzles. A computer software VENT has been developed to size two-phase relief for vapor, gassy, and hybrid systems.

23.56 Discharge System

Design of The Vent Pipe

The nature of the discharging fluid is necessary in determining the relief areas. DIERS and ICI techniques can analyze systems that exhibit "natural" surface active foaming and those that do not. The DIERS further found that small quantities of impurities can affect the flow regime in the reactor. In addition, a variation in impurity level could arise by changing the supplier of a particular raw material. Therefore, care is needed in sizing emergency relief on homogeneous vessel behavior, that is, two-phase flow. In certain instances, pressure relief during a runaway reaction can result in three-phase discharge, if solids are suspended in the reaction liquors. Solids can also be entrained by turbulence caused by boiling/gassing in the bulk of the liquid. Caution is required in sizing this type of relief system. Especially, where there is a significantly static head of fluid in the discharge pipe. Another aspect in the design of the relief system includes the possible blockage in the vent line. This could arise from the process material being solidified in cooler sections of the reactor. It is important to consider all discharge regimes when designing the discharge pipe work.

Safe Discharge

Reactors or storage vessels are fitted with overpressure protection vent directly to roof level. Such devices (e.g., relief valves) protect only against common process maloperations and not runaway reactions. The quantity of material ejected and the rate of discharge are low resulting in good dispersion. The increased use of rupture (bursting) discs can result in large quantities (95% of the reactor contents) being discharged for foaming systems.

The discharge of copious quantities of chemicals directly to atmosphere can give rise to secondary hazards, especially if the materials are toxic and can form a flammable atmosphere (e.g., vapor or mist) in air. In such cases, the provision of a knockout device (scrubber, dump tank) of adequate size to contain the aerated/foaming fluid will be required.

The regulatory authorities can impose restrictions on the discharge of the effluent from the standpoint of pollution. Therefore, reliefs are seldom vented to the atmosphere. In most cases, a relief is initially discharged to a knockout

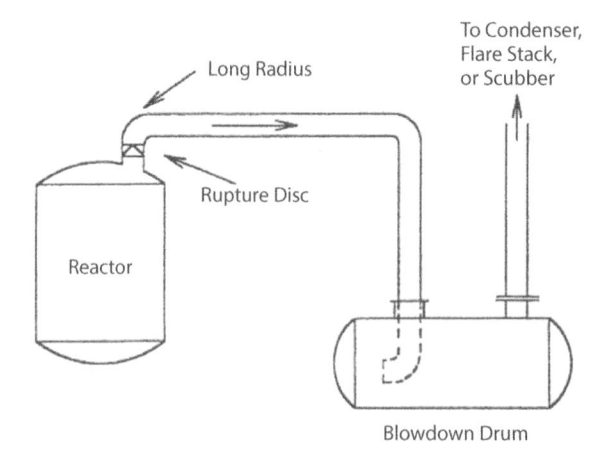

Figure 23.55 Relief containment system with blowdown drum. The blowdown drum separates the vapor from the liquid. (Source: Grossel [73].)

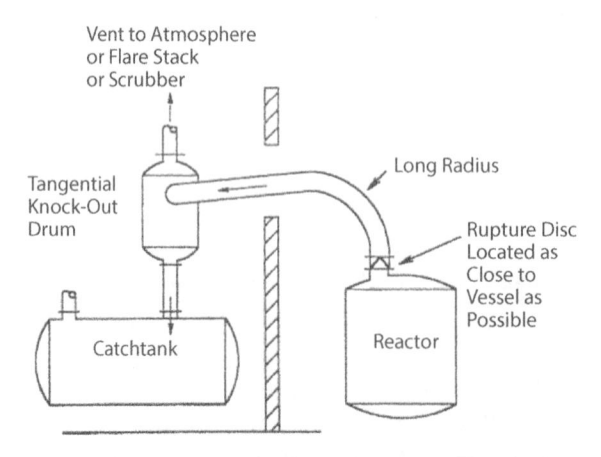

Figure 23.56 Tangential inlet knockout drum with separate liquid catch-tank. (Source: Grossel [73].)

system to separate the liquid from the vapor. The liquid is collected and the vapor is then discharged to another treatment unit. The vapor treatment unit depends upon the hazards of the vapor, which may include a vent condenser, scrubber, incinerator, flare, or a combination of these units. This type of system is referred to as the total containment system as shown in Figure 23.55. The knockout drum is sometimes called a catch tank or blowdown drum. The horizontal knockout serves both as a vapor–liquid separator as well as a holdup vessel for the disengaged liquid. These types are commonly used where there is greater space as in the petroleum refineries and petrochemical plants. The two-phase mixture enters at one end, and the vapor leaves at the opposite end. Inlets may be provided at each end with a vapor outlet at the center of the drum to minimize vapor velocities involving two-phase streams with very high vapor flow rates. When there is limited space in the plant, a tangential knockout drum is employed as illustrated in Figure 23.56. Coker [74] has given detailed design procedures of separating gas–liquid separators and in volume 3 of the volume series.

Direct Discharge to The Atmosphere

Careful consideration and analysis by management are essential before flammable or hazardous vapors are discharged to the atmosphere. Consideration and analysis must ensure that the discharge can be carried out without creating a potential hazard or causing environmental problems. The possible factors are as follows:

- exposure of plant personnel and/or the surrounding population to toxic vapors or corrosive chemicals.
- formation of flammable mixtures at ground level or elevated structures.
- ignition of vapors at the point of emission (blowdown drum vent nozzle).
- air pollution.

These factors and methods for evaluating their effects are elaborated in API publications (API RP521) [5c].

We should take special care in the design of the vapor vent stack to ensure that the tip faces straight up (i.e., no goose-neck) in order to achieve good dispersion. The stack should not be located near a building so as to avoid vapor drifting into the building. However, if the drum is near a building, the stack should extend at least 12 ft above the building floor. Grossel [75] has provided various descriptions of alternative methods of disposal.

Example 23.12

Tempered Reaction

An 800-gal reactor containing a styrene ($C_6H_5CH=CH_2$) mixture with a specific heat of 0.6 cal/gm °C has a 10-in. rupture disk and a vent line with equivalent length of 400. The vessel MAWP is 100 psig and the rupture disk set pressure is 20 psig. The styrene mixture had a self-heat rate of 60°C/min at 170°C as it is tempered in a DIERS venting test. What is the allowable reactor mixture charge to limit the overpressure to 10% over the set pressure?

Solution

Using Equation 23.71, we have

$$M_O = \frac{(D_P)^2 (\Delta P_S)}{2.769 \left(\dfrac{dT}{dt}\right)_S} \cdot \left(\frac{T_S}{C_P}\right)^{0.5}$$

where

M_o = allowable mass of the reactor mixture charge (kg) to limit the venting overpressure to P_p (psig).
D_p = 10 in.
P_s = the relief device set pressure = 20 psig

P_p = maximum venting pressure = 1.10(20) = 22.0 psig

ΔP_s = the allowable venting overpressure (psi), i.e., the maximum venting pressure minus the relief device set pressure.

(22.0 − 20) = 2 psi

T_s = the equilibrium temperature corresponding to the vapor pressure where the vapor pressure is the relief device set pressure (K) = 170 °C (170 + 273.15 = 443.15K)

$\left(\dfrac{dT}{dt}\right)_s$ = the reactor mixture self-heat rate $\left(\dfrac{°C}{min}\right)$ at temperature T_s (K) = 60 °C/min

C_p = specific heat of the reactor mixture (cal/g-K) = 0.6 cal/gm.K.

$$M_O = \dfrac{(10^2)(2.0)}{(2.769)(60)}\left(\dfrac{443.15}{0.6}\right)^{0.5}$$

$$M_O = 32.75 \text{ kg}$$

The density of styrene is 0.9 g/cm³. Therefore, the quantity charged in the reactor is

$$\dfrac{(32.72)\text{kg}\,(10^3)\text{g}}{1\text{ kg}} \times \dfrac{1\text{ cm}^3}{0.9\text{ g}} \times \dfrac{1\text{l}}{10^3\text{ cm}^3} \times \dfrac{1\text{ gal}}{3.785\text{l}}$$

$$= 9.6 \text{ gal} (\approx 10 \text{ gal}).$$

The reactor charge is quite small for an 800-gal reactor. If the pressure is allowed to rise to 10% above MAWP, then

$$\Delta P_s = 1.1(100) - 20$$

$$= 90 \text{ psi}.$$

The amount charged is

$$M_o = \dfrac{(10^2)(90)}{(2.769)(60)}\left(\dfrac{443.15}{0.6}\right)^{0.5}$$

$$M_O = 1472.2 \text{ kg}$$

$$= \dfrac{1472.2}{(0.9)(3.785)} \text{ gal}$$

$$= 432 \text{ gal}.$$

This shows that a much larger initial charge will be required. An Excel spreadsheet program (Example 23.12.xls) has been developed for this example and the results are shown below.

Calculations of Two-phase flow for a Tempered Reaction		
Size of Rupture disc, D_p	10	in.
Relief set pressure, P_s	20	psig
Specific heat capacity of the reactor mixture, C_p	0.6	J/kg.°C

Equilibrium temperature, T_s	170	°C
Equilibrium temperature, T_s	443.15	K
Density of liquid in the reactor	0.9	gm/cm³
Reactor mixture self-heat rate, $(dT/dt)_s$	60	°C/min.
Maximum Allowable Working Pressure, MAWP	100	psig
Maximum venting pressure, P_p	22	psig
Maximum allowable venting overpressure, ΔP_s	2	psi
Allowable mass of the reactor mixture, M_o	32.72	kg	The reactor charge is small for an 800-gal reactor
Quantity charged to the reactors	9.6	gal	For a rise in pressure to 10% above MAWP
Allowable venting overpressure	90	psi
The amount charged, M_o	1472.2	kg
The quantity charge to the reactor, M_o	432.2	gal

Example 23.13

A 700 gal reactor with a net volume of 850 gal containing an organic mixture has a 10 in. rupture disk and a vent line with an equivalent length L/D = 400. The vessel MAWP = 100 psig and the rupture disk set pressure is 20 psig. A venting test that was carried out showed that the reaction was "gassy." The test mass was 30 g, the peak rate of pressure rise was 550 psi/min, and the maximum test temperature was 300°C. Determine the allowable reactor mixture charge to limit the overpressure to 10% of the MAWP.

Solution

Using Equation 23.76 for "gassy" reaction, the amount charge is

$$M_O = \left[V_P \Delta P_P \left\{ \frac{(D_P)^2 (M_S)(P_P)}{(2.07)(T_T)(dP/dt)} \right\}^2 \right]^{1/3}$$

where

V_p = vessel total volume (gal) = 850 gal
P_p = maximum allowable venting pressure (psia)
     1.10 (100) + 14.7 = 124.7 psia.
M_s = sample mass used in a DIERS test or equivalent test = 30 g
$\frac{dP}{dt}$ = pressure rise in test (psi/sec). = 550 psi/min = 9.17 psi/sec
T_T = maximum temperature in test (K). (300°C = 573.15K)
ΔP_p = $P_p - P_{amb}$ = 124.7 - 14.7 = 110 psi
P_{amb} = ambient pressure at the end of the vent line = 14.7 psia

$$M_O = \left[(850)(110) \left\{ \frac{(10)^2 (30)(124.7)}{(2.07)(573.15)(9.17)} \right\}^2 \right]^{1/3}$$

M_o = 479.9 kg

The amount charge to the reactor is 480 kg. A Microsoft Excel spreadsheet (Example 23.13.xls) has been developed for this example.

Calculations of Two-phase flow for a Gassy Reaction				
Size of Rupture disc, D_p		10	in.	
Maximum Allowable Working Pressure, MAWP		100	psig	
Sample mass use in a DIERS test or equivalent test		30	g	
Maximum temperature in test		300	°C	
Maximum temperature in test		573.15	K	
Pressure rise in test,	(dP/dt)	550	psi/sec.	
Ambient pressure		14.7	psia	
Vessel total volume		850	gal	
Maximum allowable venting pressure, P_p		124.7	psig	
Differential pressure,	ΔP_p	110	psi	
Pressure rise in test,	(dP/dt)	9.17	psi/min.	
Allowable reactor mixture charge, M_o		479.94	kg	Reactor mixture charge to limit the over pressure to 10% over the set pressure

Example 23.14

Determine the vent size for a vapor pressure system using the following data and physical properties.
Reactor Parameters:

Volume, V = 10 m³.
Mass, M = 8,000 kg
Vent opening pressure = 15 bara. (217.5 psia)
Temperature at set pressure T_s = 170°C (443.15K)
Overpressure allowed above operating pressure = 10%
$\qquad$ = 1 bar (10^5 Pa).

Material Properties:

Specific heat, C_p = 3000 J/kg.K.
Slope of vapor pressure and temperature curve dP/dT = 20,000 Pa/K.
Rate of Reaction $\Delta T = dP/20,000 = 10^5/20,000 = 5K$.

Rate of set pressure, $(dT/td)_s$ = 6.0 K/min.
Rate of maximum pressure = 6.6 K/min.
Average rate = 6.3 K/min.
$\qquad$ = 0.105 K/s.

Solution

Using Fauske's nomograph of Figure 23.53 at a self-heat rate of 6.3 K/min and a set pressure of 217.5 psia, the corresponding vent is per 1,000 kg of reactants = 0.0008 m².

The vent area of 8,000 kg reactants = 0.0064 m².

The vent size

$$d = \left(\frac{4 \text{ Area}}{\pi}\right)^{0.5} = 90.27 \text{ mm } (3.6 \text{ in.}).$$

If Figure 23.53 is applicable for F = 0.5, then for F = 1.0, the area is

$$A = (0.0064 \text{ m}^2)\left(\frac{0.5}{1.0}\right) = 0.0032 \text{ m}^2$$

The area assumes a 20% absolute overpressure.

The result can be adjusted for other overpressures by multiplying the area by a ratio of 20/(new absolute percent overpressure).

Using the Leung's method:
Assuming L/D = 0, F = 1.0
Two-phase mass flux from Eq. 23.61 gives

$$G = (0.9)(1.0)(20,000)\left(\frac{1.0 \times 443.15}{3,000}\right)^{0.5}$$

$$= 6918.1 \frac{\text{kg}}{\text{ms}^2}.$$

Rate of heat generation: q

$$q = C_p \frac{dT}{dt}\left(\frac{J.K}{kg.K.s}\right) = (3,000)(0.105)\frac{W}{kg} = 315 \frac{W}{kg}.$$

From Eq. 23.66, the vent area A is

$$A = \frac{8,000 \times 315}{6918.1\left\{\left(\frac{10}{8,000} \times 443.15 \times 20,000\right)^{0.5} + (3,000 \times 5.0)^{0.5}\right\}^2}$$

$$A = 7.024 \times 10^{-3} \text{ m}^2.$$

$$d = 94.6 \text{ m } (3.27 \text{ in}).$$

The vent size is about 4.0 in.

Using the Fauske's method:
Assuming $\alpha = 1.0$, $\alpha_o = 0.0$, $F = 1.0$

$$\Delta P = 1 \text{bar} = 10^5 \text{ N/m}^2.$$

From Equation 23.72, the vent area is

$$A = \frac{1}{2} \cdot \left[\frac{(8,000)(0.105)(1-0)}{(1.0)\left(\frac{443.15}{3,000}\right)^{0.5}(10^5)(1-0)} \right]$$

$A = 10.927 \times 10^{-3} \text{m}^2$.
$d = 117.59\text{m} \ (4.64\text{in})$.

Example 23.15

A 3500-gallon reactor with styrene monomer undergoes adiabatic polymerization after being heated inadvertently to 70°C. The MAWP of the reactor is 5 bara. Determine the relief vent diameter required. Assume a set pressure of 4.5 bara and a maximum pressure of 5.4 bara. Other data and physical properties are given as follows:

Data

Volume (V)   = 13.16 m³, (3,500 gal)
Reaction mass (m_o), kg. = 9,500
Set temperature (T_s) = 209.4 °C = 482.5K

Data from VSP:

Maximum temperature (T_m)   219.5°C = 492.7K

$\left(\frac{dT}{dt}\right)_s$ 29.6°C/min = 0.493 K/s (sealed system)

$\left(\frac{dT}{dt}\right)_m$ 39.7°C/min = 0.662 K/s

Physical property data:

	4.5 bar set.	5.4 bar set.
V_f, m³/kg	0.001388	0.001414
V_g, m³/kg ; ideal gas assumed	0.08553	0.07278
C_p, kJ/kg.K	2.470	2.514
ΔH_v, kJ/kg.	310.6	302.3

Solution

The heating rate q is determined by Equation (23.68)

$$q = \frac{1}{2} C_V \left[\left(\frac{dT}{dt}\right)_S + \left(\frac{dT}{dt}\right)_m \right]$$

Assuming $C_v = C_p$

$$q = \frac{1}{2}(2.470 \text{ kJ/kg.K})[0.493+0.662](\text{K/s})$$

$$= 1.426 \frac{\text{kJ}}{\text{kg.s}}$$

The mass flux through the relief G is given by Equation 23.58, assuming $L/D = 0$ and $\psi = 1.0$

$$G = \frac{Q_m}{A} = 0.9\psi \frac{\Delta H_V}{\upsilon_{fg}} \left(\frac{g_C}{C_P T_S}\right)^{0.5}$$

$$G = \frac{(0.9)(1.0)(310660 \text{ J/kg})[1(\text{Nm})/\text{J}]}{(0.08553 - 0.001388)\text{m}^3/\text{kg}} \left\{ \frac{\left[1\left(\text{kg m/s}^2\right)/\text{N}\right]}{(2,470 \text{ J/kg K})(482.5 \text{ K})[1(\text{Nm})/\text{J}]} \right\}^{0.5}$$

$$= 3043.81 \frac{\text{kg}}{\text{m}^2\text{s}}$$

The relief area is determined by Equation 23.67

$$A = \frac{M_O q}{G\left[\left(\frac{V}{M_O} \frac{\Delta HV}{\upsilon_{fg}}\right)^{0.5} + (C_V \Delta T)^{0.5}\right]^2}$$

The change in temperature ΔT is $T_m - T_s$

$$\Delta T = 492.7 - 482.5 = 10.2\text{K}$$

$$A = \frac{(9500 \text{ kg})(1426 \text{ J/kg.s})[1(\text{Nm})/\text{J}]}{(3043.81 \text{ kg}/\text{m}^2\text{s}) \left[\left\{\left(\frac{13.16 \text{ m}^3}{9500 \text{ kg}}\right)\left(\frac{310660 \text{ J/kg}[1(\text{Nm})/\text{J}]}{0.08414 \text{ m}^3/\text{kg}}\right)\right\}^{0.5} + \left[\left(2470 \frac{\text{J}}{\text{kg.K}}\right)(10.2 \text{ K})[1(\text{Nm})/\text{J}]\right]^{0.5} \right]^2}$$

$$A = 0.0839 \text{m}^2.$$

The required relief diameter is

$$d = \left(\frac{4A}{\pi}\right)^{0.5}$$

$$= \left(\frac{(4)(0.0839 \text{ m}^2)}{3.14}\right)^{0.5} = 0.327 \text{ m}$$

$$d = 327 \text{ mm } (12.87 \text{ inches}).$$

We shall consider the situation that involves all vapor relief. The size of a vapor phase rupture disk required is determined by assuming that all of the heat energy is absorbed by the vaporization of the liquid. At the set temperature, the heat release rate q is given by

$$q = C_V\left(\frac{dT}{dt}\right)_s = \left(2.470 \frac{kJ}{kg.K}\right)\left(0.493 \frac{K}{s}\right)$$

$$q = 1.218 \frac{kJ}{kg.s}$$

The vapor mass flow through the relief is then

$$Q_m = \frac{q m_o}{H_v}$$

$$= \frac{(1218 \text{ J/kg})(9500 \text{ kg})}{(310660 \text{ J/kg})}$$

$$= 37.25 \cdot \frac{kg}{s}$$

The required relief area for vapor is determined by

$$A = \frac{Q_m}{C_o P}\left\{\frac{R_g T}{\gamma g_c M}\left(\frac{2}{\gamma+1}\right)^{\left(\frac{\gamma+1}{\gamma-1}\right)}\right\}^{0.5}$$

where

Q_m = discharge mass flow, kg/s
C_o = discharge coefficient
A = area of discharge
P = absolute upstream pressure
γ = heat capacity ratio for the gas
g_c = gravitational constant
M = molecular weight of the gas (M = 104 for styrene)
R_g = ideal gas constant $\left(8314 \text{ Pa m}^3/\text{kg molK}\right)$
T^s = absolute temperature of the discharge
Assuming $C_o = 1$ and $\gamma = 1.23$

Table 23.22 Vent Areas for Different Flow Regimes.

Type of flow	Required vent area as a multiple of all vapor vent area
All vapor	1
Two-phase:Churn turbulent	2-5
Bubbly	7
Homogeneous	8
All liquid	10

$$A = \frac{(37.25 \text{ kg/s})}{(1.0)(4.5 \text{ bar})(10^5 \text{ Pa/bar})\left[1(\text{N/m}^2)/\text{Pa}\right]} \left\{ \frac{(8314 \text{ Pa m}^3/\text{kg}-\text{mol})(482.5 \text{ K})\left[1(\text{N/m}^2)/\text{Pa}\right]}{(1.32)\left[1(\text{kg m/s}^2)/\text{N}\right](104 \text{ kg/kg}-\text{mol})} \right\}^{0.5}$$

$$\times \left\{ \left(\frac{2}{2.32}\right)^{\left(\frac{2.32}{-0.32}\right)} \right\}^{0.5}$$

$$A = 0.0242 \text{ m}^2.$$

The relief diameter = 0.176 m
d = 176mm (6.9 inches).

Thus, the size of the relief device is significantly smaller than for two-phase flow. Sizing for all vapor relief will undoubtedly give an incorrect result, and the reactor would be severely tested during this runaway occurrence. Table 23.22 shows the results of the VENT software program of Example 23.15.

Articles describing procedures related to the DIERS development of this entire subject have been published by some members of the DIERS group. These are referenced here and the detailed descriptions and illustrations in the noted articles can be most helpful to the potential user. A useful website for two-phase runaway flow system is www.fauske.com.

DIERS Final Reports

Refer to the bibliography at the end of this chapter for a listing of the final reports of this program [13].

23.57 Sizing for Two-Phase Fluids

Sizing for two-phase fluids is based upon the procedures outlined in API RP 520 [5]. In this design, the physical parameters are established carefully because a little change in the physical parameters may alter the results substantially. The design procedure uses the following definitions:

Noncondensible gas: This is a gas that is not easily condensed under normal pressure and temperature conditions. The common noncondensible gases are air, nitrogen (N_2), oxygen (O_2), hydrogen (H_2), carbon dioxide (CO_2), hydrogen sulfide (H_2S), and carbon monoxide (CO).

Highly subcooled liquid: It is a liquid that does not flash after passing through the PRV.

Subcooled liquid: It is a liquid that flashes after passing through the PRV.

The following three different types are possible for designing a two-phase PRV:

Type 1 (omega method): Subcooled (including saturated) liquid enters the PRV and flashes. No condensable vapor or noncondensable gas is present. This calculation follows the method defined in Section D 2.2 in API RP 520 [5].

Type 2 (omega method): All other cases. This calculation follows the method defined in Section D.2.3 in API RP 520 [5].

Type 3 (integration method): For all cases using the numerical integration method. This calculation follows the method defined in Section D2.1 in API RP 520 [5].

Type 1 (Omega Method)

This method is used for sizing pressure relief valves handling a subcooled (including saturated) liquid at the inlet. No condensable vapor or non-condensable gas should be present at the inlet. The subcooled liquid either flashes upstream or downstream of the pressure relief valve throat depending on which subcooling region the flow falls into. The equations also apply to all liquid scenarios. The following steps are as follows:

Step 1. Calculate the Saturated Omega Parameter, ω_s

For multicomponent systems with nominal boiling range* less than 150°F or single component systems, use either Eq. 23.82 or 23.83. If Eq. 23.82 is used, the fluid must be far from its thermodynamic critical point ($T_r \leq 0.9$ or $P_r \leq 0.5$)**

$$\omega_s = 0.185 \rho_{lo} C_p T_o P_s \left(\frac{\upsilon_{vls}}{h_{vls}} \right)^2 \tag{23.82}$$

where

ρ_{lo} = liquid density at the PRV inlet, lb/ft³
C_p = liquid specific heat at constant pressure at the PRV inlet (Btu/lb °R)
T_o = temperature at the PRV inlet °R = °F + 460
P_s = saturation (vapor) pressure corresponding to T_o (psia). For a multicomponent system, use the bubble point pressure corresponding to T_o.
υ_{vls} = difference between the vapor and liquid specific volumes at P_s (ft³/lb)
h_{vls} = latent heat of vaporization at P_s (Btu/lb). For multicomponent systems, h_{vls} is the difference between the vapor and liquid specific enthalpies at P_s

* The nominal boiling range is the difference in the atmospheric boiling points of the lightest and heaviest components in the system.

** Other assumptions that apply include: heat of vaporization and the heat capacity of the fluid are constant through the nozzle, behavior of the fluid vapor pressure with temperature follows the Clapeyron equation, and isenthalpic (constant enthalpy) flow process.

For multicomponent systems with nominal boiling range greater than 150°F or single component systems near the thermodynamic critical point, use Eq. 23.83.

$$\omega_s = 9 \left[\frac{\rho_{lo}}{\rho_9} - 1 \right] \tag{23.83}$$

where

ρ_o = liquid density at the PRV inlet, lb/ft³

ρ_9 = density evaluated at 90% of the saturation (vapor) pressure P_s corresponding to the PRV inlet temperature T_o, lb/ft³. For a multicomponent system, use the bubble point pressure corresponding to T_o for P_s. When determining ρ_9, the flash calculation should be carried out isentropically, but an isenthalpic (adiabatic) flash is sufficient.

Step 2. Determine the Subcooling Region

The subcooling region is determined as:

$$P_s > \eta_{st} P_o \Rightarrow \text{low subcooling region (flashing occurs upstream of throat)} \quad (23.84)$$

$$P_s < \eta_{st} P_o \Rightarrow \text{high subcooling region (flashing occurs at the throat).} \quad (23.85)$$

where

η_{st} = transition saturation pressure ratio

$$\eta_{st} = \frac{2\omega_s}{1 + 2\omega_s} \quad (23.86)$$

P_o = pressure at the PRV inlet (psia). This is the PRV set pressure (psig) plus the allowable overpressure (psi) plus atmospheric pressure.

Step 3. Determine if the Flow Is Critical or Subcritical

For the low subcooling region, the flow behavior is determined by

$$P_c > P_a \Rightarrow \text{critical flow} \quad (23.87)$$

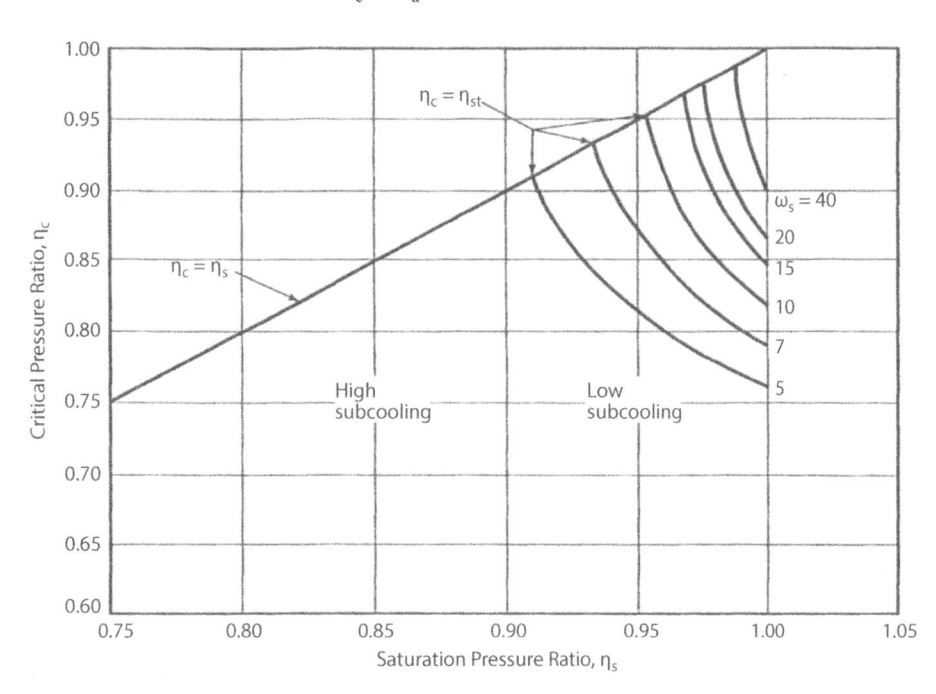

Figure 23.57 Correlation for nozzle critical flow of inlet subcooled liquids. (Source: API RP 520, Sizing Selection, and Installation of Pressure – Relieving Devices in Refineries, Part 1- Sizing and Selection, 7th ed., 2020.)

$$P_c < P_a \Rightarrow \text{subcritical flow} \qquad (23.88)$$

For the high subcooling region, the flow behavior is determined by

$$P_s > P_a \Rightarrow \text{critical flow} \qquad (23.89)$$

$$P_s < P_a \Rightarrow \text{subcritical flow (all-liquid flow)} \qquad (23.90)$$

where

P_c = critical pressure, psia

$$= \eta_c P_o \qquad (23.91)$$

η_c = critical pressure ratio and is estimated from Figure 23.57 using the value of η_s

where

$$\eta_s = \text{saturation pressure ratio} = \frac{P_s}{P_o} \qquad (23.92)$$

P_a = downstream backpressure (psia).

Step 4. Calculate the Mass Flux

In the **low subcooling region, use Eq. 23.93**. If the flow is critical, use η_c for η, and if the flow is subcritical, use η_a for η. **In the high subcooling region, use Eq. 23.94**. If the flow is critical, use P_s for P and if the flow is subcritical (all-liquid flow), use P_a for P

$$G = \frac{68.09\left(2(1-\eta_s) + 2\left[\omega_s \eta_s \ln\left(\frac{\eta_s}{\eta}\right) - (\omega_s - 1)(\eta_s - \eta)\right]\right)^{0.5}}{\omega_s\left(\frac{\eta_s}{\eta} - 1\right) + 1} \sqrt{P_o \rho_{lo}} \qquad (23.93)$$

$$G = 96.3\,[\rho_{lo}(P_o - P)]^{0.5} \qquad (23.94)$$

where

G = mass flux (lb/s ft²)

$$\eta_a = \text{backpressure ratio/subcritical pressure ratio} = \frac{P_a}{P_o} \qquad (23.95)$$

where

P_a = downstream backpressure (psia).

Step 5. Calculate the Required Area of the PRV

The following equation is applicable to turbulent flow systems, as most two-phase relief scenarios are within the turbulent flow regime.

$$A = 0.3208 \frac{Q \rho_{lo}}{K_d K_b K_c G} \qquad (23.96)$$

where

A = required effective discharge area, in².
Q = volumetric flow rate, gpm
K_d = discharge coefficient that should be obtained from the valve manufacturer. For a preliminary sizing estimation, a discharge coefficient 0.65 for subcooled liquids and 0.85 for saturated liquids.
K_b = backpressure correction factor for liquid that should be obtained from the valve manufacturer. For a preliminary sizing estimation, use Figure 23.58. The back pressure correction factor applies to balanced-bellows valves only.
K_c = combination correction factor for installations with a rupture disk upstream of the pressure relief valve.
   = 1.0 when a rupture disk is not installed.
   = 0.9 when a rupture disk is installed in combination with a pressure relief valve and the combination does not have a published value.

SI units

For $\eta_s \leq \eta_{st}$

$$\eta_c = \eta_{st} \qquad (23.97)$$

For $\eta_s \geq \eta_{st}$, the value of η_c is calculated using Eq. 23.98 or approximated using Eq. 23.99

$$\frac{\left(\omega_s + \frac{1}{\omega_s} - 2\right)}{2\eta_s} \eta_c^2 - 2(\omega_s - 1)\eta_s + \omega_s \eta_s \ln\left(\frac{\eta_c}{\eta_s}\right) + 1.5\omega_s \eta_s - 1 = 0 \qquad (23.98)$$

$$\eta_c = \eta_s \left(\frac{2\omega}{2\omega - 1}\right) \left[1 - \sqrt{1 - \frac{1}{\eta_s}\left(\frac{2\omega - 1}{2\omega}\right)}\right] \qquad (23.99)$$

where

η_s = saturation pressure ratio = P_s/P_o
η_a = subcritical pressure ratio = P_a/P_o

The mass flux for the low subcooling region is calculated by

$$G = \frac{\left(2(1-\eta_s) + 2\left[\omega_s \eta_s \ln\left(\frac{\eta_s}{\eta}\right) - (\omega_s - 1)(\eta_s - \eta)\right]\right)^{0.5}}{\omega_s \left(\frac{\eta_s}{\eta} - 1\right) + 1} \sqrt{P_o \rho_{lo}} \qquad (23.100)$$

In the above equation, if the flow is critical, use η_c for η, and if the flow is subcritical, use η_a for η.

The mass flux for the high subcooling region is calculated by

$$G = 1.414 \, [\rho_{lo} (P_o - P)]^{0.5} \qquad (23.101)$$

In the above equation, if the flow is critical, use P_s for P, and if the flow is subcritical, use P_a for P.

The PRV area is calculated by

$$A = 16.67 \frac{Q \rho_{lo}}{K_d K_b K_v G} \tag{23.102}$$

where

A = required effective discharge area, mm².
Q = volumetric flow rate, l/min
K_d = discharge coefficient that should be obtained from the valve manufacturer. For a preliminary sizing estimation, a discharge coefficient 0.65 for subcooled liquids and 0.85 for saturated liquids.
K_b = backpressure correction factor for liquid that should be obtained from the valve manufacturer. For a preliminary sizing estimation, use Figure 23.58
K_v = viscosity correction factor.
G = mass flux, kg/(s m²)
ω_s = saturated omega parameter

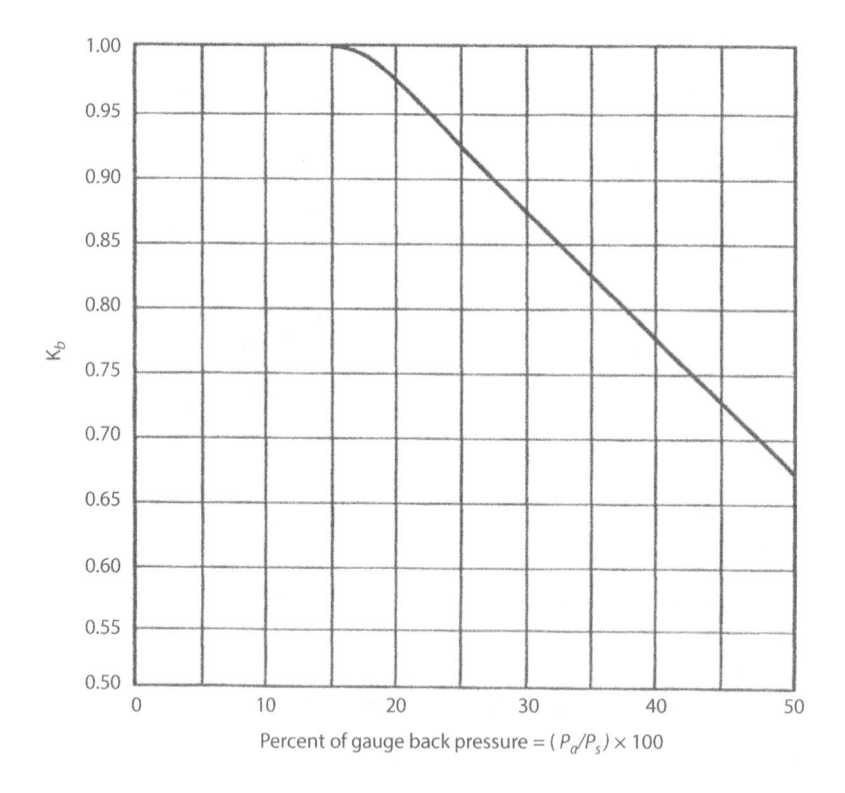

K_b = correction factor due to back pressure.
P_a = back pressure, in psig.
P_s = set pressure, in psig.

Note: The curve above represents values recommended by various manufacturers. This curve may be used when the manufacturer is not known. Otherwise, the manufacturer should be consulted for the applicable correction factor.

Figure 23.58 Back pressure correction factor, K_b for balanced – bellows pressure relief valves (Liquids). (Source: API RP 520, Sizing Selection, and Installation of Pressure – Relieving Devices in Refineries, Part 1- Sizing and Selection, 7th ed., 2020.)

ρ_{lo} = liquid density at the PRV inlet, kg/m³

ρ_{90} = density evaluated at 90% of the saturation (vapor) pressure, P_s, corresponding to the PRV inlet relieving temperature, T_o, kg/m³

P_s = saturation vapor pressure, Pa

P_o = PRV relieving pressure, Pa

P_c = critical pressure, Pa

P_a = downstream backpressure, Pa

T_o = relieving temperature, °C

η_a = subcritical pressure ratio/backpressure ratio = P_a/P_o

η_c = critical pressure ratio

η_s = saturation pressure ratio = P_s/P_o

η_{st} = transition saturation pressure ratio.

Example 23.16

The following relief requirements are given:

a.	Required propane volumetric flow rate caused by blocked in pump	= 100 gal/min
b.	Relief valve set at the design pressure of the equipment	= 260 psig
c.	Downstream total backpressure of (superimposed backpressure = 0 psig, built-up backpressure = 10 psig)	= 10 psig (24.7 psia)
d.	Inlet temperature at the PRV	= 60°F (519.7°R)
e.	Inlet liquid propane density at the PRV	= 31.92 lb/ft³
f.	Inlet liquid propane specific heat at constant pressure at the PRV	= 0.6365 Btu/lb°R
g.	Saturation pressure of propane corresponding to 60°F	107.6 psia
h.	Specific volume of propane liquid at the saturation pressure	0.03160 ft³/lb
i.	Specific volume of propane vapor at the saturation pressure	1.001 ft³/lb
j.	Latent heat of vaporization for propane at the saturation pressure	152.3 Btu/lb

The following data are derived:

a. Permitted accumulation is 10%
b. Relieving pressure of 1.10 × 260 = 286 psig (300.7 psia)
c. Percent of gauge backpressure = (10/260) × 100 = 3.8%. Since the downstream built-up backpressure is less than 10% of the set pressure, a conventional pressure relief valve may be used. Thus, the backpressure correction factor K_b = 1.0.
d. Since the propane is subcooled, a discharge coefficient K_d of 0.65 can be used.

Solution

Step 1. Calculate the saturated omega parameter, ω_s

Since the propane system is a single-component system far from its thermodynamic critical point, the saturated omega parameter is calculated from Eq. 23.82

$$\omega_s = 0.185(31.92)(0.6365)(519.67)(107.6)\left(\frac{1.001-0.0316}{152.3}\right)^2$$

$$= 8.515$$

Step 2. Determine the subcooling region.

The transition saturation pressure ratio η_{st} is calculated from Eq. 23.86:

$$\eta_{st} = \frac{2 \times 8.515}{1 + 2 \times 8.515}$$

$$= 0.9445$$

The liquid is determined to fall into the high subcooling region since $P_s < \eta_{st} P_o$.
i.e., $107.6 < 0.9445 \times 300.7 = 284.02$

Step 3. Determine if the flow is critical or subcritical

The flow is determined to be critical since $P_s > P_a$.
i.e., $107.6 > 24.7$

Step 4. Calculate the mass flux

Since the flow is critical, substitute P_s for P, and the mass flux from Eq. 23.94 is

$$G = 96.3 \, [(31.92)(300.7 - 107.6)]^{0.5}$$

$$= 7560.5 \text{ lb/s-ft}^2$$

Step 5. Calculate the required area of the PRV from Eq. 23.96:

$$A = 0.3208 \frac{(100)(31.92)}{(0.65)(1.0)(1.0)(7560.5)}$$

$$= 0.208 \text{ in}^2$$

The table below shows the orifice designation and effective area from the Crosby Catalog. Thus, the next standard orifice area is "F" orifice pressure relief valve = 0.307 in².

The maximum flow, W = (0.307) (0.65) (1) (1) (7560.5) (60)/(0.3208 × 7.4805)
$= 37721.5$ lb/h

Microsoft Excel spreadsheet (Example 23.16.xlsx) shows the calculations of Example 23.16, and the results of the calculation are shown below.

Results of Type 1 Omega Method (Subcooled) for liquid flow		
Flow Type	Type 1 Omega Method	
Volumetric flow rate of liquid	100	US gal./min
Set Pressure at inlet	260	psig
Downstream total back pressure	10	psig
Inlet temperature at the PRV	60	°F
Inlet liquid density at the PRV	31.92	lb/ft^3
Inlet liquid specific heat at constant pressure at the PRV	0.637	Btu/lb°R
Saturation pressure of propane corresponding to 60°F	107.6	psia
Specific volume of liquid propane at the saturation pressure	0.0316	ft^3/lb
Specific volume of propane vapor at the saturation pressure	1.001	ft^3/lb
Latent heat of vaporization for propane at the saturation pressure	152.3	Btu/lb
Relieving pressure	300.7	psia
K_b	1	
K_d	0.65	
K_c (rupture disk not installed)	1	
K_c (rupture disk installed)	0.9	
Saturated Omega Parameter, ω_s	8.5203	
Saturation pressure, P_{st}	284	psia
Flow Type	Critical flow	
Mass flux, G	7560	lb/(s.ft^2)
Effective required area, A	0.208	in^2
Next standard orifice area	0.307	in^2
Maximum flow rate with the next standard orifice size	37721	lb/h

The manufacturer's orifice effective area with designation is shown below.

Orifice Designation and Effective Area: Crosby Catalog		
Orifice Designation	**in^2**	**mm^2**
D	0.11	71
E	0.196	126
F	0.307	198
G	0.503	325

Orifice Designation and Effective Area: Crosby Catalog		
Orifice Designation	in²	mm²
H	0.785	506
J	1.287	830
K	1.838	1186
-	2.461	1588
L	2.853	1841
M	3.6	2323
N	4.34	2800
-	5.546	3578
P	6.379	4116
-	9.866	6365
Q	11.05	7129
R	16	10323
-	22.22	14335
T	26	16774
-	39.51	25490

Example 23.17

Design a pressure-relieving valve for a subcooled liquid that enters a PRV and flashes, using the following parameter:

Flow rate	= 15,000 kg/h
Liquid density at the PRV inlet	= 700 kg/m³
PRV set pressure	= 1000 kPaG
Bubble point pressure	= 850 kPaG
Density at 90% of bubble point pressure	= 600 kg/m³
PRV backpressure	= 50 kPaG
Allowable overpressure	= 10%
Type of PRV	= Conventional

Solution

1. The saturated omega parameter from Eq. 23.83 is

$$\omega_s = 9\left[\frac{\rho_{lo}}{\rho_9} - 1\right]$$

$$\omega_s = 9\left[\frac{700}{600} - 1\right]$$

$$= 1.5$$

2. The transition saturation pressure ratio η_{st} from Eq. 23.86 is

$$\eta_{st} = \frac{2 \times 1.5}{1 + 2 \times 1.5}$$

$$= 0.75$$

3. Determine the subcooling region.

The subcooling region is determined as

$$P_s > \eta_{st} P_o \Rightarrow \text{low subcooling region (flashing occurs upstream of throat)} \qquad (23.84)$$

$$P_s < \eta_{st} P_o \Rightarrow \text{high subcooling region (flashing occurs at the throat).} \qquad (23.85)$$

The saturation pressure P_{st} is

$$P_{st} = \eta_{st} \cdot P_o$$

$$= 0.75 \,(1.1 \times 1000 + 101.35)$$

$$= 901.0 \text{ kPaa}$$

Bubble point pressure $P_s = 850 + 101.35 = 951.35$ kPaa.
Since the bubble point pressure > saturation pressure, then it will be in the low subcooling region. i.e., 951.35 > 901.0

4. Saturation pressure ratio η_s from Eq. 23.92 is

$$\eta_s = \text{saturation pressure ratio} = \frac{P_s}{P_o} = \frac{951.35}{1201.35}$$

$$= 0.792$$

Since the saturated pressure ratio > the transition pressure ratio, i.e., $\eta_s > \eta_{st}$, 0792 > 0.75, then the critical pressure ratio η_c can be determined from Eq. 23.99

$$\eta_c = \eta_s \left(\frac{2\omega}{2\omega - 1}\right)\left[1 - \sqrt{1 - \frac{1}{\eta_s}\left(\frac{2\omega - 1}{2\omega}\right)}\right] \qquad (23.99)$$

$$\eta_c = 0.792\left(\frac{2\times 1.5}{2\times 1.5 -1}\right)\left[1-\sqrt{1-\frac{1}{0.792}\left(\frac{2\times 1.5 -1}{2\times 1.5}\right)}\right]$$

$$= 0.716$$

5. The critical pressure P_c is determined from Eq. 23.91

$$P_c = \eta_c P_o = 0.716 \times 1201.35$$

$$= 860.2 \text{ kPaa}$$

The backpressure P_a = 50 + 101.35 = 151.35 kPaa

Since the critical pressure > the backpressure, i.e., $P_c > P_a$, then the flow is critical.

6. The mass flux, G for the low subcooling region is calculated from Eq. 23.100, and for critical flow, use η_c for η

$$G = \frac{\left(2(1-\eta_s)+2\left[\omega_s\eta_s \ln\left(\frac{\eta_s}{\eta}\right)-(\omega_s-1)(\eta_s-\eta)\right]\right)^{0.5}}{\omega_s\left(\frac{\eta_s}{\eta}-1\right)+1}\sqrt{1000 P_o \rho_{lo}}$$

$$G = \frac{\left(2(1-0.792)+2\left[(1.5)(0.792)\ln\left(\frac{0.792}{0.716}\right)-(1.5-1)(0.792-0.716)\right]\right)^{0.5}}{1.5\left(\frac{0.792}{0.716}-1\right)+1}\sqrt{1201.35\times 700\times 1000}$$

$$= 19037.5 \text{ kg}/(\text{s m}^2)$$

7. The volumetric flow rate Q is

$$Q = \frac{W}{\rho_{lo}} = \frac{15{,}000}{700}\left\{\frac{\text{kg}}{\text{h}}\cdot\frac{\text{m}^3}{\text{kg}}\cdot\frac{10^3 \text{l}}{\text{m}^3}\cdot\frac{\text{h}}{60\text{min}}\right\},$$

$$= 357.14 \text{ l/min}.$$

Since the region is low subcooling, the discharge coefficient K_d is assumed as 0.85.

8. The effective PRV area is calculated from Eq. 23.102

$$A = 16.67\frac{Q\rho_{lo}}{K_d K_b K_v G}$$

$$A = 16.67\frac{(357.14)(700)}{(0.85)(1)(1)(19037.5)}$$

$$= 257.54 \text{ mm}^2$$

The next standard orifice size is G with area = 325 mm².

The maximum flow with the standard size is

$$W = (325)(0.85)(1)(1)(19037.5)(60)/(16.67 \times 1000.)$$

$$= 18929 \text{ kg/h}$$

Microsoft Excel spreadsheet (Example 23.17.xlsx) shows the calculations of Example 23.17 and the results are shown below.

Results of Type 1 Omega Method (Subcooled) for liquid flow		
Flow Type	SUBCOOLED	
Volumetric flow rate of liquid	357.14	l/min
Set Pressure at inlet	1000	kPaG
Downstream total back pressure	50	kPaG
Inlet liquid density at the PRV	700	kg/m³
Flow rate	15000	kg/h
Allowable overpressure	10	%
TYPE OF PRV	Conventional	
Relieving pressure of 1.1 x 1000 + 101.35	1201.4	kPaa
K_b	1	
K_d	0.85	
K_c (rupture disk not installed)	1	
K_c (rupture disk installed)	0.9	
K_v	1	
Saturated Omega Parameter, ω_s	1.5	
Saturation pressure, P_{st}	901.01	kPaa
Bubble point pressure	951.35	kPaa
Type of liquid phase	Low subcooling	
Saturation pressure, ratio, P_s/P_o	0.792	
Type of pressure phase	Critical	
Critical pressure ratio	1.1879	
Critical pressure	859.57	kPaa
Back pressure	151.35	kPaa
Mass flux, G	19050	kg/(s.m²)

Effective required area, A	257	mm²
Next standard orifice area	325	mm²
Maximum flow rate with the next standard orifice size	18941	kg/h

Step 2. (Omega Method): Sizing for Two-Phase Flashing Flow with a Noncondensable Gas Through a Pressure Relief Valve [5]

In this method, the term vapor (subscript v) will be used to refer to the condensable vapor present in the two-phase flow and the term gas (subscript g) will be used to refer to the non-condensable gas. The following procedure is as follows:

Step 1. Calculate the inlet void fraction, α_o

$$\alpha_o = \frac{x_o \upsilon_{vgo}}{\upsilon_o} \qquad (23.103)$$

where

x_o = gas or combined vapor and gas mass fraction (quality) at the PRV inlet
υ_{vgo} = specific volume of the gas or combined vapor and gas at the PRV inlet, ft³/lb
υ_o = specific volume of the two-phase system at the PRV inlet, ft³/lb

Step 2. Calculate the omega parameter, ω.

For systems that satisfy all of the following conditions, use Eq. 23.104

  a. Contains less than 0.1 weight % hydrogen
  b. Nominal boiling range* less than 150°F.
  c. Either P_{vo}/P_o less than 0.9 or P_{go}/P_o greater than 0.1
  d. Far from its thermodynamic critical points ($T_r \leq 0.9$ or $P_r \leq 0.5$)**

$$\omega = \frac{\alpha_o}{k} + 0.185(1-\alpha_o)\rho_{lo} C_p T_o P_{vo} \left(\frac{\upsilon_{lo}}{h_{vlo}}\right)^2 \qquad (23.104)$$

where

P_{vo} = saturation (vapor) pressure corresponding to the inlet temperature T_o (psia). For a multicomponent system, use the bubble point pressure corresponding to T_o.
P_o = pressure at the PRV inlet (psia). This is the PRV set pressure (psig) plus the allowable overpressure (psi) plus atmospheric pressure.
P_{go} = noncondensable gas partial pressure at the PRV inlet (psia).
k = ratio of specific heats of the gas or combined vapor and gas. If the specific heat ratio is unknown, a value of 1 can be used.
ρ_{lo} = liquid density at the PRV inlet, lb/ft³

C_p = liquid specific heat at constant pressure at the PRV inlet, Btu/lb - °R
T_o = temperature at the PRV inlet, °R = (°F + 460)
v_{vlo} = difference between the vapor† (not including any condensable gas present) and liquid specific volumes at the PRV inlet, ft³/lb
h_{vlo} = latent heat of vaporization at the PRV inlet, Btu/lb. For multicomponent systems, h_{vlo} is the difference between the vapor and liquid specific enthalpies.

Go to step 3 to determine if the flow is critical or subcritical. For systems that satisfy one of the following conditions, use Eq. 23.105

    a. Contains more than 0.1 weight % hydrogen
    b. Nominal boiling range greater than 150°F.
    c. Either P_{vo}/P_o greater than 0.9 or P_{go}/P_o less than 0.1
    d. Near its thermodynamic critical point ($T_r \geq 0.9$ or $P_r \geq 0.5$).

$$\omega = 9\left(\frac{v_9}{v_o} - 1\right) \qquad (23.105)$$

where

v_9 = specific volume evaluated at 90% of the PRV inlet pressure P_o (ft³/lb). When determining v_9, the flash calculation should be carried out isentropically, but an isenthalpic (adiabatic) flash is sufficient.

Go to step 4 to determine if the flow is critical or subcritical.

Step 3. Determine if the flow is critical or subcritical

$P_c > P_a \Rightarrow$ critical flow
$P_c < P_a \Rightarrow$ subcritical flow

where

P_c = critical pressure, psia
    = $[y_{go} \eta_{gc} + (1 - y_{go}) \eta_{vc}] P_o$
y_{go} = inlet gas mole fraction in the vapor phase. Can be determined using given mole composition information or the following equation = P_{go}/P_o
η_{gc} = nonflashing critical pressure ratio from (Figure 23.60) using the value of $\omega = \alpha_o/k$
η_{vc} = flashing critical pressure ratio from Figure 23.60 using the value of ω
P_a = downstream backpressure, psia

Go to step 5.

Step 4. Determine if the flow is critical or subcritical (Eq. 23.105)

$P_c > P_a \Rightarrow$ critical flow
$P_c < P_a \Rightarrow$ subcritical flow

where

$$P_c = \text{critical pressure, psia} = \eta_c P_o \quad (23.106)$$

η_c = critical pressure ratio from Figure 23.60. This ratio can also be obtained from the following expression:

$$\eta_c^2 + (\omega^2 - 2\omega)(1-\eta_c)^2 + 2\omega^2 \ln \eta_c + 2\omega^2(1-\eta_c) = 0 \quad (23.107)$$

P_a = downstream backpressure (psia).

Step 5. Calculate the mass flux (ω calculated from Eq. 23.104)

For critical flow, use Eq. 23.108

$$G = 68.09 \left[\frac{P_o}{v_o} \left(\frac{y_{go} \eta_{gc}^2 k}{\alpha_o} \right) + \frac{(1-y_{go})\eta_{vc}^2}{\omega} \right]^{1/2} \quad (23.108)$$

where

G = mass flux, lb/s – ft²

For subcritical flow, an iterative solution is required. Eqs. 23.109 and 23.110 are solved simultaneously for η_g and η_v:

$$\eta_a = y_{go} \eta_g + (1 - y_{go}) \eta_v \quad (23.109)$$

$$\frac{\alpha_o}{k}\left(\frac{1}{\eta_g} - 1\right) = \omega\left(\frac{1}{\eta_v} - 1\right) \quad (23.110)$$

where

η_g = nonflashing partial pressure ratio.
η_v = flashing partial pressure ratio.

Calculate the mass flux by

$$G = \left[y_{go} G_g^2 + (1-y_{go}) G_v^2 \right]^{0.5} \quad (23.111)$$

where

G_g = nonflashing mass flux, lb/s-ft²

$$G_g = \frac{68.09\left\{-2\left[\frac{\alpha_o}{k}\ln(\eta_g) + \left(\frac{\alpha_o}{k} - 1\right)(1-\eta_g)\right]\right\}^{0.5}}{\frac{\alpha_o}{k}\left(\frac{1}{\eta_g} - 1\right) + 1} \sqrt{\frac{P_o}{v_o}} \quad (23.112)$$

G_v = flashing mass flux, lb/s – ft²

$$G_v = \frac{68.09\{-2[\omega\ln(\eta_v)+(\omega-1)(1-\eta_v)]\}^{0.5}}{\omega\left(\frac{1}{\eta_v}-1\right)+1}\sqrt{\frac{P_o}{v_o}} \qquad (23.113)$$

Go to step 7.

Step 6. Calculate the mass flux (ω is calculated from Eq. 23.105).

For critical flow, use Eq. 23.114. For subcritical flow, use Eq. 23.115

$$G = 68.09\,\eta_c\left(\frac{P_o}{v_o\,\omega}\right)^{0.5} \qquad (23.114)$$

$$G = \frac{68.09\{-2[\omega\ln(\eta_a)+(\omega-1)(1-\eta_a)]\}^{0.5}}{\omega\left(\frac{1}{\eta_a}-1\right)+1}\sqrt{\frac{P_o}{v_o}} \qquad (23.115)$$

where

G = mass flux, lb/s-ft^2
η_a = backpressure ratio = $\dfrac{P_a}{P_o}$

Step 7. Calculate the required area of the PRV

$$A = \frac{0.04\,W}{K_d\,K_b\,K_c\,G} \qquad (23.116)$$

where

- A = required effective discharge area, in^2
- W = mass flow rate, lb/h
- K_d = discharge coefficient that should be obtained from the valve manufacturer. For a preliminary sizing estimation, a discharge coefficient of 0.85 can be used.
- K_b = backpressure correction factor for vapor that should be obtained from the valve manufacturer. For a preliminary sizing estimation, use Figure 23.59. The backpressure correction factor applies to balanced bellows valves only.
- K_c = combination correction factor for installations with a rupture disk upstream of the pressure relief valve.
  = 1.0 when a rupture disk is not installed.
  = 0.9 when a rupture disk is installed in combination with a pressure relief valve and the combination does not have a published value.

*The nominal boiling range is the difference in the atmospheric boiling points of the lightest and heaviest components in the system.

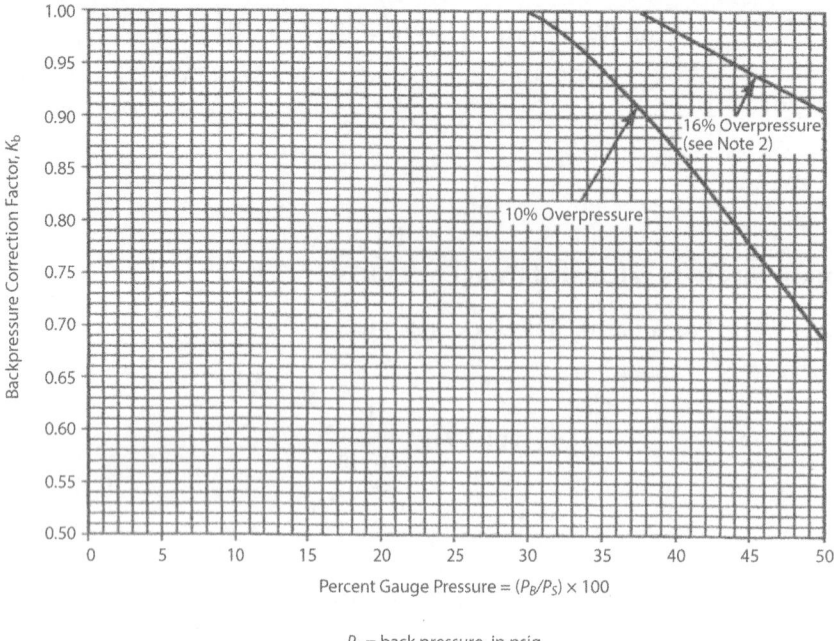

172 Petroleum Refining Design and Applications Handbook Volume 5

[Figure: Backpressure Correction Factor K_b vs Percent Gauge Pressure chart, showing curves for 10% Overpressure and 16% Overpressure (see Note 2). X-axis: Percent Gauge Pressure = $(P_B/P_S) \times 100$, 0 to 50. Y-axis: Backpressure Correction Factor, K_b, 0.50 to 1.00.]

P_B = back pressure, in psig.
P_S = set pressure, in psig.

Notes:
1. The curves above represent a compromise of the values recommended by a number of relief valve manufacturers and may be used when the make of the valve or the critical flow pressure point for the fluid is unknown. When the make of the valve is known, the manufacturer should be consulted for the correction factor. These curves for a given set pressure. For set pressures below 50 psig or subcritical flow, the manufacturer must be consulted for values of K_b–
2. See paragraph 3.3.3.
3. For 21% overpressure, K_b equals 1.0 up to P_B/P_S = 50%.

Figure 23.59 Back pressure correction factor, K_b for balanced bellows pressure relief valves (vapors and gases). (Source: API RP 520, Sizing Selection, and Installation of Pressure – Relieving Devices in Refineries, Part 1- Sizing and Selection, 7th ed., 2020.)

**Other assumptions that apply include: ideal gas behavior, heat of vaporization, and the heat capacity of the fluid are constant through out the nozzle, behavior of the fluid vapor pressure with temperature follows the Clapeyron equation, and isenthalpic (constant enthalpy) flow process.

†To obtain the vapor specific volume when a noncondensable gas is present at the PRV inlet, use the vapor partial pressure (from the mole composition) and the ideal gas law to calculate the volume.

Example 23.18

The following requirements are given as follows:

a.	Required gas oil hydrotreater (GOHT) flow rate caused by operational upset.	= 160,000 lb/h
b.	Inlet temperature at the PRV.	= 450°F (909.67°R)
c.	Relief valve set pressure design of the equipment.	= 600 psig
d.	Downstream total backpressure (superimposed backpressure = 0 pisg, built-up backpressure = 55 psig)	= 55 psig (69.9 psia)
e.	Two-phase specific volume at the inlet of PRV	= 0.1549 ft³ / lb
f.	Mass fraction of the vapor and gas at the inlet of PRV	= 0.5596

g.	Combined specific volume of the vapor and gas at the inlet of PRV	= 0.2462
h.	Inlet gas mole fraction in the vapor phase. Non condensable gases in the GOHT system include hydrogen, nitrogen, and hydrogen sulfide	= 0.4696
i.	Specific heat ratio, k	= 1

Here, the following data are as follows:

a. Overpressure of 10%.
b. Relieving pressure of 1.10 × 600 = 660 psig (674.7 psia)
c. Percent of gauge backpressure = (55/600) × 100 = 9.2%

Since the downstream backpressure is less than 10% of the set pressure, a conventional pressure relief valve is used. Thus, the backpressure correction factor K_b = 1.0

Set 1. Calculate the inlet void fraction.

The inlet void fraction α_o is determined by Eq. 23.103

$$\alpha_o = \frac{x_o \upsilon_{vgo}}{\upsilon_o}$$

$$= \frac{0.5596 \times 0.2462}{0.1549}$$

$$= 0.8894$$

Step 2. Calculate the omega parameter, ω.

Since the gas oil hydrotreater has a nominal boiling range greater than 150°F, Eq. 23.105 is used to calculate ω. The specific volume calculated at 0.9 × 674.7 = 607.2 psia using the results of an isenthalpic (i.e., adiabatic) flash calculation from a process simulator is 0.1737 ft³/lb. The ω parameter is determined by Eq. 23.105:

$$\omega = 9\left(\frac{\upsilon_9}{\upsilon_o} - 1\right)$$

$$= 9\left(\frac{0.1737}{0.1549} - 1\right)$$

$$= 1.0923$$

Step 4. Determine if the flow is critical or subcritical.

The critical pressure ratio η_c is 0.62 from Figure 23.60 with ω = 1.0923 or from Eq. 23.117.

$$\eta_c = [1 + (1.0446 - 0.0093431\, \omega^{0.5})\, \omega^{-0.56261}]^{(-0.70356 + 0.014685 \ln \omega)}$$

$$= 0.62$$

The critical pressure P_c is

$$P_c = \eta_c \times P_o = 0.62 \times 674.7$$

$$= 418.3 \text{ psia}$$

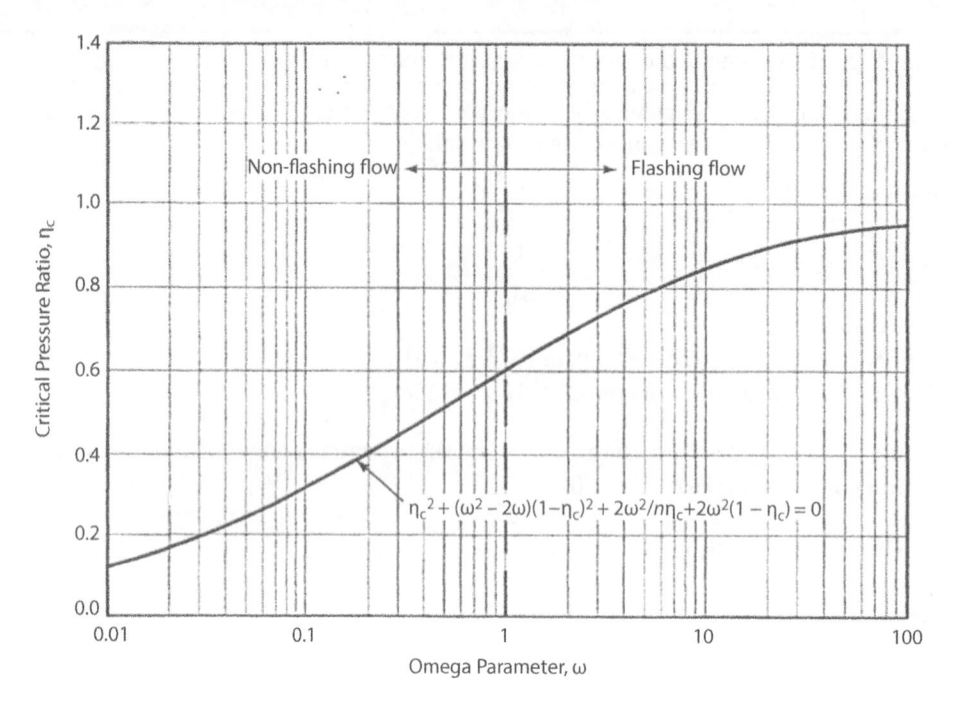

Figure 23.60 Correlation for nozzle critical flow of flashing and nonflashing systems. (Source: API RP 520, Sizing Selection, and Installation of Pressure – Relieving Devices in Refineries, Part 1- Sizing and Selection, 7th ed., 2020.)

The flow is determined to be critical since $P_c > P_a$, i.e., 418.3 > 69.7

Step 6. Calculate the mass flux G.

The mass flux G is calculated from Eq. 23.114

$$G = 68.09 \eta_c \left(\frac{P_o}{v_o \omega} \right)^{0.5}$$

$$G = 68.09(0.62)\left(\frac{674.7}{0.1549 \times 1.0923} \right)^{0.5}$$

$$= 2665.2 \text{ lb/s-ft}^2$$

Step 8. Calculate the required area of the PRV from Eq. 23.116

$$A = \frac{0.04\, W}{K_d\, K_b\, K_c\, G}$$

$$= \frac{0.04(160,000)}{(0.85)(1.0)(1.0)(2665.2)}$$

$$= 2.825 \text{ in}^2.$$

The next recommended standard orifice pressure relief size is "L" = 2.853 in^2.

The maximum flow with the standard orifice size is

$$W = (2.853)(0.85)(1.0)(1.0)(2665.2)/0.04$$

$$= 161{,}581 \text{ lb/h}$$

Microsoft Excel spreadsheet (Example 23.18.xlsx) shows the calculations of Example 23.18, and the results of the calculations are shown below.

Results of Type 2 Omega Method for two-phase flashing		
Flow Type	Two-phase flashing	
Flow rate	160000	lb/h
Set Pressure at inlet	600	psig
Downstream total back pressure	55	psig
Inlet temperature at the PRV	450	°F
Two-phase specific volume at the inlet of PRV	0.1549	ft³/lb
Specific volume at 90% at the inlet of the PRV	0.1737	ft³/lb
Mass fraction specific volume at the inlet of PRV	0.5596	
Combined specific volume of the vapor and gas at the inlet of PRV	0.2462	ft³/lb
Inlet gas mole fraction in the vapor phase	0.4696	
Specific heat ratio, k	0.1737	
Relieving pressure of 1.1 x 260 + 14.7	674.7	psia
K_b	1	
K_d	0.85	
K_c (rupture disk not installed)	1	
K_c (rupture disk installed)	0.9	
Calculate the inlet void fraction:	0.8894	
Saturated Omega Parameter, ω_s	1.092	
The critical pressure, P_c is:	416.89	psia
Total back pressure, P_a	69.7	psia
Flow Type	Critical	
Mass flux, G	2657.16	lb/(s.ft²)
Effective required area, A	2.834	in²
Next standard orifice area	2.853	in²
Maximum flow rate with the next standard orifice size	161093.6	lb/h

SI Units

The ω parameter is calculated from

$$\omega = 9\left(\frac{v_9}{v_o} - 1\right) \qquad (23.105)$$

where

v_9 = specific volume evaluated at 90% of the PRV inlet pressure, m³/kg
v_o = specific volume of the two-phase system at the PRV inlet, m³/kg

The flow condition is calculated as

$P_c > P_a \Rightarrow$ critical flow
$P_c < P_a \Rightarrow$ subcritical flow

where

$$P_c = \text{critical pressure, psia} = \eta_c P_o \qquad (23.106)$$

where

P_c is the critical pressure, Pa
$P_c = \eta_c P_o$

The critical pressure ratio η_c is obtained from Figure 23.70 or can be calculated by:

$$\eta_c^2 + (\omega^2 - 2\omega)(1-\eta_c)^2 + 2\omega^2 \ln \eta_c + 2\omega^2(1-\eta_c) = 0 \qquad (23.107)$$

Or

$$\eta_c = [1 + (1.0446 - 0.0093431\, \omega^{0.5})\, \omega^{-0.56261}]^{(-0.70356 + 0.014685 \ln \omega)} \qquad (23.117)$$

where

P_o = PRV relieving pressure, Pa
P_a = downstream backpressure, Pa

The mass flux is calculated as follows:

Critical flow:

$$G = \eta_c \left(\frac{P_o}{v_o \omega}\right)^{0.5} \qquad (23.118)$$

Subcritical flow:

$$G = \frac{\{-2[\omega \ln \eta_a + (\omega - 1)(1 - \eta_a)]\}^{0.5}}{\omega\left(\frac{1}{\eta_a} - 1\right) + 1} \sqrt{\frac{P_o}{v_o}} \qquad (23.119)$$

where

- G = mass flux, kg/s m²
- Po = PRV relieving pressure, Pa
- v_o = specific volume at the PRV inlet, m³/kg
- η_a = backpressure ratio, = P_a/P_o

The effective discharge area is calculated by

$$A = \frac{277.8\, W}{K_d\, K_b\, K_c\, K_v\, G} \qquad (23.120)$$

where

- A = effective discharge area, mm²
- W = mass flow rate, kg/h
- K_d = discharge coefficient, the preliminary value of 0.85.
- K_b = back-pressure correction factor.
- K_c = combination correction factor
- K_v = viscosity correction factor.

Example 23.19

Design a pressure relieving valve for a two-phase fluid at the inlet of the PRV. The following process parameters are to be used for the design:

Mass flow rate	= 10,000 kg/h
Specific volume at PRV relieving pressure	= 0.1 m³/kg
Specific volume at 90% of relieving pressure	= 0.15 m³/kg
PRV set pressure	= 1000 kPaG
PRV backpressure	= 150 kPaG
Discharge coefficient	= 0.85
Overpressure	= 10%
Type of PRV	= balanced bellow

Solution

Omega parameter, ω from Eq. 23.105:

$$\omega = 9\left(\frac{0.15}{0.1} - 1\right)$$
$$= 4.5$$

The critical pressure ratio η_c is obtained from Figure 23.70 or can be calculated from Eq. 23.107 or Eq. 23.117.

$$\eta_c^2 + (\omega^2 - 2\omega)(1-\eta_c)^2 + 2\omega^2 \ln \eta_c + 2\omega^2(1-\eta_c) = 0$$

Using the Solver from Excel spreadsheet for Eq. 23.107 gives

$$\eta_c = 0.779$$

Critical pressure ratio from Eq. 23.107,

$$P_c = \eta_c P_o$$
$$= 0.779 \times (1.1 \times 1000 + 101.35)$$
$$= 935.85 \text{ kPaa}$$

Since the critical pressure > backpressure, the flow is critical,

i.e., 935.85 > 251.35 kPa

The mass flux for critical flow is from Eq. 23.118 is

$$G = \eta_c \left(\frac{P_o}{v_o \omega}\right)^{0.5}$$

$$G = 0.779 \left(\frac{1201.350 \times 1000}{0.1 \times 4.5}\right)^{0.5}$$
$$= 1272.8 \text{ kg}/(\text{s m}^2)$$

The effective discharge area is calculated from Eq. 23.120:

$$A = \frac{277.8 \times 10{,}000}{(0.85)(1)(1)(1)(1272.8)}$$
$$= 2567.7 \text{ mm}^2$$

The next standard orifice size is N, area = 2800 mm².

The maximum flow rate with the standard orifice is

$$W = (2800)(0.85)(1)(1)(1)(1272.8)/277.8$$
$$= 10904.5 \text{ kg/h}$$

Microsoft Excel spreadsheet (Example 23.19.xlsx) shows the calculations of Example 23.19, and the results of the calculations are shown below.

Results of Type 2 Omega Method for two-phase flashing		
Noncondensable gas		
Flow Type	balanced below	
Flow rate	10000	kg/h
Set Pressure at inlet	1000	kPag
Downstream total back pressure	150	kPaG
Allowable overpressure	10	%
Specific volume at 90% of PRV relieving pressure	0.15	m³/kg
Relieving pressure of 1.1 x 1000 + 101.35	1201.4	kPaa
K_b	1	
K_d	0.85	
K_c (rupture disk not installed)	1	
K_c (rupture disk installed)	0.9	
Viscosity correction factor, K_v	1	
Saturated Omega Parameter,	4.5	
The critical pressure ratio,	0.78	
The critical pressure, P_c	937.02	kPaa
Total back pressure, P_a	251.35	kPaa
Flow is	Critical	
Mass flux, G	1274.4	kg/(sm²)
Effective required area, A	2564.5	mm²
Next standard orifice area	2800	mm²
Maximum flow rate with the next standard orifice size	10918	kg/h

Type 3 Integral Method [5]

In the integration, the mass flux is calculated by

$$G^2 = \left[\rho_1^2 \left(-2\int_{P_o}^{P} \frac{dP}{\rho}\right)\right]_{max} \quad (23.121)$$

The value of the integral can be approximated by

$$\int_{P_o}^{P_t} \frac{dP}{\rho} \cong \sum_{i=0}^{t} 2\left(\frac{P_{i+1} - P_i}{\rho_{i+1} - \rho_i}\right) \qquad (23.122)$$

And the overall mass density of the fluid is

$$\rho_m = \alpha \rho_v + (1 - \alpha) \rho_l \qquad (23.123)$$

where

- G = mass flux, kg/(s -m²)
- υ = specific volume of the fluid, m³/kg
- ρ = mass density of the fluid, kg/m³.
- ρ_m = mixed density, kg/m³
- ρ_l = density of liquid, kg/m³.
- ρ_v = density of vapor, kg/m³
- P = stagnation pressure of the fluid, Pa
- o = condition at the inlet of the nozzle.
- t = condition at the throat of the nozzle.

The effective orifice area is calculated from Eq. 23.120:

$$A = \frac{277.8\,W}{K_d\,K_b\,K_c\,K_v\,G}$$

where

- A = effective discharge area, mm²
- W = mass flow rate, kg/h
- K_d = discharge coefficient
  - = 0.85 for a two-phase fluid at the PRV inlet
  - = 0.65 for a single liquid phase
  - = 0.975 for a single vapor phase.
- K_b = backpressure correction factor (Figure 23.59)
- K_c = combination correction factor
- K_v = viscosity correction factor.

Example 23.20 [76]

Using the integration method, design a PRV for the following process parameters.

Type of fluid	= two-phase saturated
Mass flow rate	= 15,000 kg/h
PRV set pressure	= 1726.1 kPaG
End pressure for integration	= 288.7 kPaG
PRV backpressure	= 100 kPaG
Valve discharge coefficient	= 0.85
Overpressure	= 10%

The pressure–density relationship is given by

Pressure kPaa	Density kg/m³	Pressure kPaa	Density kg/m³
2000	15.281	1160	9.634
1930	14.826	1090	9.141
1860	14.368	1020	8.643
1790	13.908	950	8.137
1720	13.444	880	7.634
1650	12.98	810	7.634
1580	12.511	740	6.596
1510	12.041	670	6.068
1440	11.566	600	5.531
1370	11.088	530	4.985
1300	10.607	460	4.429
1230	10.122	390	3.858

Solution

Value of integral at 2000 kPaa = 0
  Value of integral at 1930 kPaa = - 2 × 70,000 / (15.281 + 14.826) = - 4650.1 m²/s²
  Mass flux t 1930 kPaa = 1429.7 kg/(s-m²)
  Value of integral at 1860 kPaa = - 4650.1 – 4795.5 = -9445.6 m²/s²
  Mass flux at 1860 kPaa = 1977.5 kg/(s-m²)
  This calculation continues till the maximum mass flux is achieved.
  The maximum mass flux achieved at 1160 kPaa = 3564.3 kg/(s-m²)
  The effective orifice area is calculated from Eq. 23.120

$$A = \frac{277.8\,W}{K_d\,K_b\,K_c\,K_v\,G}$$

$$A = \frac{277.8(15,000)}{(0.85)(1.0)(1.0)(1.0)(3564.3)}$$

$$= 1375.4 \text{ mm}^2$$

Arun Datta [76] has provided Excel Visual Basic software programs on the 3 Omega Parameter methods for sizing relief valves.

23.58 Flares/Flare Stacks

Flares are useful for the proper disposal of waste or emergency released gas/vapors and liquids. The effects on the environment and the thermal radiation from the flare must be recognized and designed for. Flares may be "ground" flares or they may be mounted on a tall stack to move the venting away from immediate plant areas. Figure 23.61 illustrates a plant flare stack system. The flow noted "from processes" could also include pressure relief valve discharges when properly designed for backpressure. This requires proper manifold design and, for safety, requires that the "worst case" volume condition be used particularly assuming that all relief devices discharge at the same time and any other process vents are also flowing. The piping systems sequence of entrance of the flare is important to backpressure determination for all respective relief devices.

The *knock-out drums or separator tanks/pots* can be designed using the techniques described elsewhere [74]. API-RP 521 [9] specifies 20–30 min holdup liquid capacity from relief devices plus a vapor space for dropout and a drain volume.

The unit should have backup instrumentation to ensure liquid level control to dispose of the waste recovered liquid.

The *seal tank/pot* is not a separator but a physical liquid seal (Figure 23.62) to prevent the possibilities of backflash from the flare from backing into the process manifolds. It is essential for every stack design.

The backpressure created by this drum is an additive to the pipe manifold pressure drops and the pressure loss through the separator. Therefore, it cannot be independently designed and not "integrated" into the backpressure system. The flow capacity of the relief valve(s) must not be reduced due to backpressure on the valves' discharge side (outlet). The total backpressure of the system must be limited to 10% of the set pressure of each pressure relief valve that may be relieving concurrently [5]. When balanced relief valves are used, the manifold backpressure can be higher, less than 30% of the valve's set pressure, psia [5].

The key detail of a seal drum is the liquid seal:

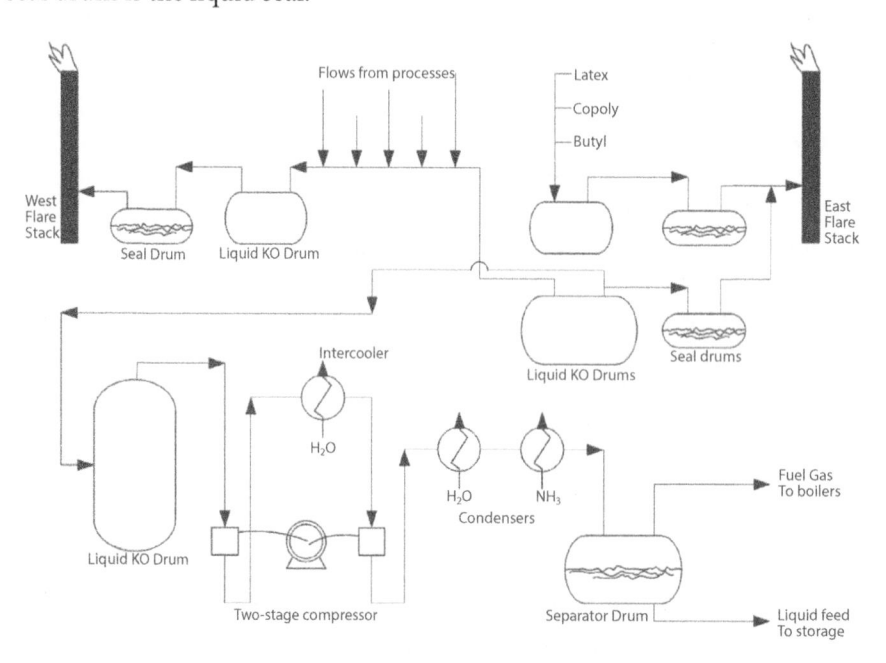

Figure 23.61 Illustration of one of many collection arrangements for process flow and/or relief valve discharge collections to relieve to one or more plant flare stacks. (By permission from Livingston, D. D., Oil Gas J., April 28, 1980.)

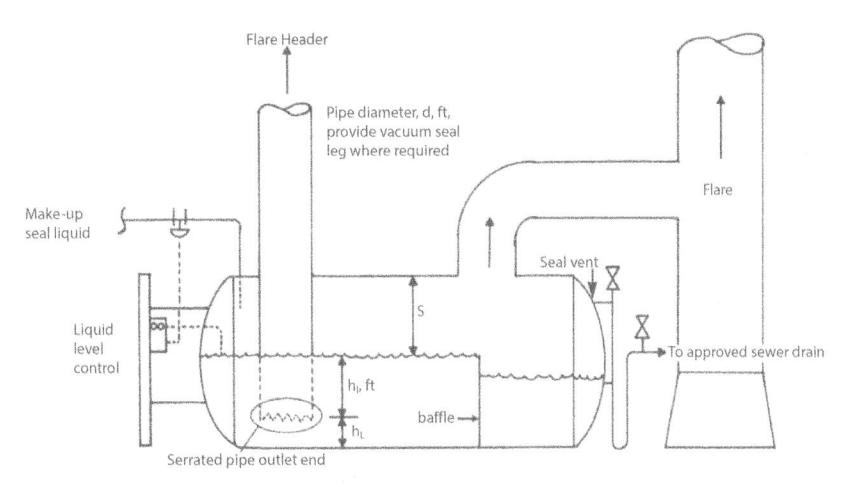

Figure 23.62 Suggested seal pot/drum for flare stack system (see API RP – 521, Fig. B-1, 3rd. ed., 1990). (Design adapted with permission from E. E. Ludwig [34] from API RP 521, 3rd. ed. (1990) American Petroleum Institute [5].)

$$h_1 = 144 P''/\rho; \text{ see Figure 23.62} \tag{23.124}$$

where h_1 = seal, submerged, ft
 P'' = maximum header exit pressure into seal, psig
 ρ = density of seal liquid, lb/ft^3

Calculate the cross section of the drum volume for vapor above the liquid level, establishing the level referenced to h_1, plus clearance to drum bottom, h_1, normally 12 to 18 in. This would be a segment (horizontal vessel) of a circle. Reference [5] recommends that the cross-section area of the vapor space above the liquid be at least equivalent to that of a circle diameter [(2) (inlet pipe)]. Thus, the cross-section area of vapor space A_s with equivalent diameter S should be at least [(2) (a_p)] where a_p is the cross-sectional area of the inlet pipe [5]. To avoid bubble burst slugging, Ludwig [34] suggests that the cross-sectional area of S be calculated to have a vapor velocity of less than the entrainment velocity for a mist sized liquid particle, or that the vapor area be approximately one-third the cross-section area of the diameter of the horizontal drum. For a vertical drum, Reference [5] recommends that the vapor disengaging height be 3 ft.

When vacuum can form in the system due to condensing/cooling hot vapor entering, the seal drum liquid volume and possibly the seal drum diameter/length must be adjusted to maintain a seal when/if the seal fluid is drawn up into the inlet piping. A vacuum seal leg should be provided on the inlet 1.2 times the expected equivalent vacuum height in order to maintain a seal.

The following design points should be considered:

- Provide liquid low-level alarms to prevent loss of liquid by evaporation, entrainment, leaks, or failure of the makeup liquid system.
- Use a sealing liquid that has a relatively low vapor pressure, and is not readily combustible, and will not readily freeze. Quite often, glycol or mixtures are used. In freezing conditions, the unit should receive personal inspection for condition of liquid.
- Provide overflow anti-siphon seal drains.
- Provide inlet vacuum seal legs.
- Some hydrocarbons may form gel clusters or layers with some sealant fluids; therefore, providing for cold weather heating and/or cleaning of the unit is necessary.

- Reference [5] suggests minimum design pressure for such a seal vessel of 50 psig, ASME code stamped (Ludwig [34]). Most flare seal drums operate at 0–5 psig pressure.
- Be extremely cautious and do not install lightweight gauge glass liquid level columns. Rather use the heavier shatterproof style.
- Provide reliable seal liquid makeup, using liquid level gauging and monitoring with recording to ensure good records of performance. The liquid level must be maintained; otherwise, the hazards of a bleed through or backflow can become serious.

Flares

Flares are an attempt to deliberately burn the flammable safety relief and/or process vents from a plant. The height of the stack is important to the safety of the surroundings and personnel, and the diameter is important to provide sufficient flow velocity to allow the vapors/gases to leave the top of the stack at sufficient velocities to provide good mixing and dilution after ignition at the flare tip by pilot flames.

API [5] discusses factors influencing flare design, including the importance of proper stack velocity to allow jet mixing. Stack gases must not be diluted below the flammable limit. The exit velocity must not be too low to allow flammable gases to fall to the ground and become ignited. The atmospheric dispersion calculations are important for the safety of the plant. Computer models can be used to evaluate the plume position when the flare leaves the stack under various atmospheric wind conditions. This should be examined under alternate possibilities of summer through winter conditions (also see [77]).

The velocities of the discharge of relief devices through a stack usually exceed 500 ft/s. Because this stream exits as a jet into the air, it is sufficient to cause turbulent mixing [5].

For a flare stack to function properly and to handle the capacity that may be required, the flows under emergency conditions from each of the potential sources must be carefully evaluated. These include, but may not be limited to, pressure relief valves and rupture disks, process blowdown for startup, shutdown, upset conditions, and plant fires creating the need to empty or blowdown all or parts of a system.

Sizing

Diameter: sizing based on stack velocity [5c], solve for "d"

$$\text{Mach} = (1.702)(10^{-5})\left(\frac{W}{P_t d^2}\right)\sqrt{\frac{T}{(kM)}} \tag{23.125}$$

In metric units

$$\text{Mach} = (11.61)(10^{-5})\left(\frac{W}{P_t d^2}\right)\sqrt{\frac{T}{(kM)}} \tag{23.126}$$

where

Mach = ratio of vapor velocity to sonic velocity in vapor, dimensionless. Mach = 0.5 for peak for short term flow, and 0.2 for more normal and frequent conditions [5c]
W = vapor relief rate to stack, lb/h (kg/s)
P_t = pressure of the vapor just inside flare tips (at top), psia. For atmospheric release, P_t = 14.7 psia (101.3 kPa abs)
d = flare tip diameter, ft (m), (end, or smallest diameter)
T = temperatures of vapors just inside flare tip, °R= °F + 460, (K = °C + 273)
k = ratio of specific heats, C_p/C_v for vapor being relieved
M = molecular weight of vapor

A peak velocity through the flare end (tip) of as much as 0.5 Mach is generally considered a peak, short term. A more normal steady state velocity of 0.2 Mach is for normal conditions and prevents flare/lift off [78]. Smokeless (with steam injection) flare should be sized for conditions of operating smokeless, which means vapor flow plus steam flow [5]. Pressure drops across the tip of the flare have been used satisfactorily up to 2 psi. It is important not to be too low and get flashback (without a molecular seal) or blow off where the flame blows off the tip (see [79]) (Figure 23.63).

Another similar equation yielding close results [80]:

$$d_t^2 = (W/1370)\sqrt{T/M}, \text{(generally, for smokeless flares)} \qquad (23.127)$$

based on Mach 0.2 limitation velocity, k = C_p/C_v = 0.2 and gas constant R = 1545 (ft-lb_f/(°R) (mole))

- d_t = flare tip diameter, in
- W = gas vent rate, lb/h
- T = gas temperature in stack, °R
- M = molecular weight of gas/vapor

For *non-smokeless* flares (no steam injection), about 30% higher capacity can be allowed [80]. Therefore, the diameter of a non-smokeless flare stack is approximately (0.85) (diameter of the smokeless flare stack).

The amount of steam injection required for smokeless flares is

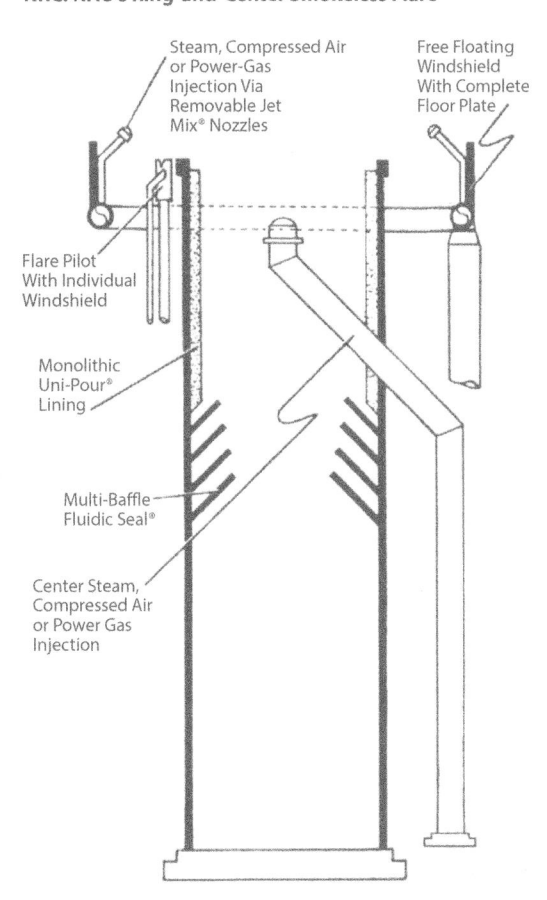

Figure 23.63 Flare stack arrangement for smokeless burning and back flash protection with Fluidic Seal® molecular seal. Steam can be injected into the flare to introduce air to the fuel by use of jets inside the stream and around the periphery. (By permission from Straitz, J. F., III, "Make the Flare Protect the Environment", Hydroc. Proc., p. 131, Oct. 1977.)

$$W_{steam} = W_{hc} (0.68 - 10.8/M) \qquad (23.128)$$

where

W_{steam} = steam injected, lb/h
W_{hc} = hydrocarbons to be flared, lb/h
M = molecular weight of hydrocarbons (average for mixture, hydrocarbons only)

For specific details, consult a flare system design manufacturer.
This calculation is based on a steam–CO_2 weight ratio of approximately 0.7 [33A, Par 5.4.3.2.1].
These should be sized for conditions under which they will operate smokelessly.

Flame Length [5c]

$$\text{Heat liberated or released by flame, } Q_f = (W)(H_c) \qquad (23.129)$$

where

Q_f = heat released by flame, Btu/h
$W_{hc} = W$ = gas/vapor flow rate, lb/h
H_c = heat of combustion of gas/vapor, Btu/lb

The heat liberated Q in Btu/h (kW) is calculated as follows (see API Recommended Practice 521, Figures 23.6a and b):

$$Q = (100,000)(21,500)$$

$$= 2.15 \times 10^9 \text{ Btu/h}.$$

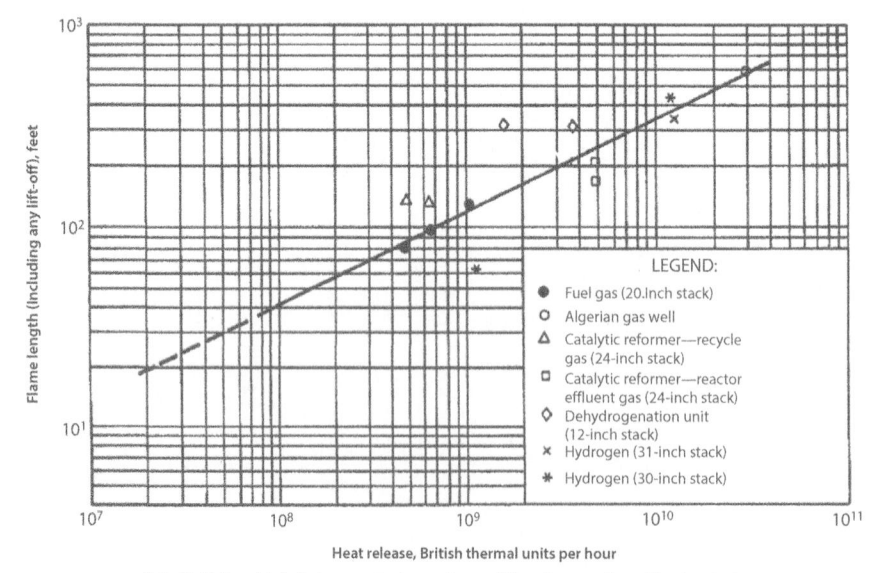

Figure 23.64 Flame length versus heat release: industrial sizes and releases (customary units). (Reprinted by permission from American Petroleum Institute, API RP 521, Guide for Pressure Relieving and Depressuring Systems, 3rd. Nov. 1990 [5].)

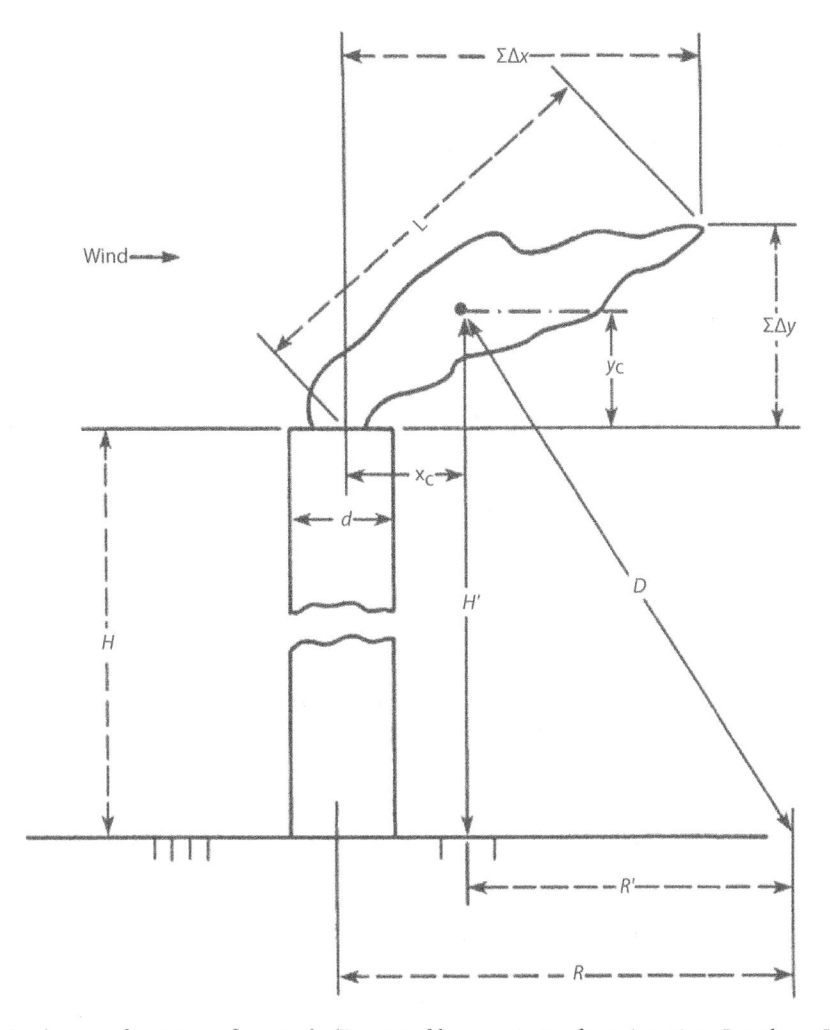

Figure 23.65 Dimensional references for sizing a flare stack. (Reprinted by permission from American Petroleum Institute, API RP – 521, Guide for Pressure Relieving and Depressuring Systems, 3rd. ed. Nov. 1990 [36].)

In metric units,

$$Q = (12.6)(50 \times 10^3)$$

$$= 6.3 \times 10^5 \text{ kW.}$$

Note: For many hydrocarbon–air mixtures, the value of H_c ranges from 20,000 to 22,000 Btu/lb. Referring to Figure 23.64 at the calculated heat release, H_c, read the flame length and refer to dimensional diagram for flame plume from a stack (Figure 23.65).

Flame Distortion [5c] Caused by Wind Velocity

Referring to Figure 23.66, the flame distortion is determined as $\frac{\Delta x}{L}$ or $\frac{\Delta y}{L}$.

Calculate U_j using the "d" determined for the selected Mach No. in earlier paragraph.

$$\frac{U_\infty}{U_j} = \frac{\text{wind velocity}}{\text{flare tip velocity}} \quad (23.130)$$

$$U_j = (\text{flow})/(\pi d^2/4), \text{ft/s.} \quad (23.131)$$

Flow, $F_1 = (W/3600)(379.1/MW)[460 + °F/520]$, ft³/s based on 60°F and 14.7 psia, and 359 ft³/mol
U_∞ = lateral wind velocity, ft/s
U_j = exit gas velocity from stack, ft/s
°F = flowing temperature

Kent [81] presents an alternate calculation method for flame distortion.

Read U_∞/U_j on Figure 23.66 and determine ratio:

$\Delta y/L_f = a$ (vertical)

$\Delta x/L_f = b$ (horizontal)

Then vertical: $\Delta y = L_f(a)$
horizontal: $\Delta x = L_f(b)$
L_f = length of flame, ft

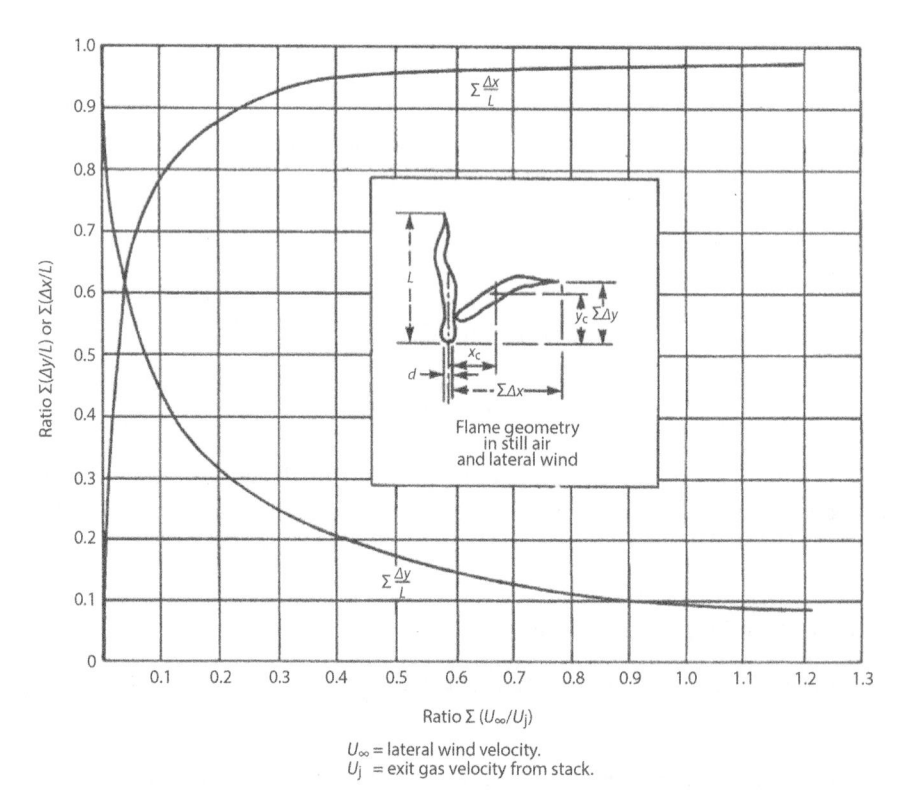

Figure 23.66 Approximate flame distortion due to lateral wind on jet velocity from flare stack. (Reprinted by permission from American Petroleum Institute, API RP – 521, Guide for Pressure Relieving and Depressuring Systems, 3rd. ed. Nov. 1990 [5].)

Flare Stack Height

The importance of the stack height (see Figure 23.65) is (a) to discharge the burning venting gases/vapors sufficiently high into the air so as to allow safe dispersion (b) to keep the flare flame of burning material sufficiently high to prevent the radiated heat from damaging equipment and facilities and from creating a life safety hazard to ground personnel. Figure 23.67 summarizes the accepted data for heat radiation related to human exposure time. Figure 23.68 summarizes the maximum radiation intensity related to a human escape time, allowing a 5-s reaction time to take action to escape, before the heat intensity injures the individual. Kent [81] suggests an escape velocity of 20 ft/s. The heat radiation is an important factor in locating/spacing of equipment with respect to one or more flares. The use of protective clothing and safety hard hats aids in extending the time of exposure when compared to bare skin.

The distance required between a flare stack venting and a point of exposure to thermal radiation is expressed [5c] [79] as

$$D_F = \sqrt{\tau F Q_r / (4\pi K)} \qquad (23.132)$$

where

D_F = minimum distance from the midpoint of a flame to the object, at ground level, ft (see Figure 23.65) (Note that this is not the flare stack height, but a part of calculation procedure)

F = fraction of heat radiated

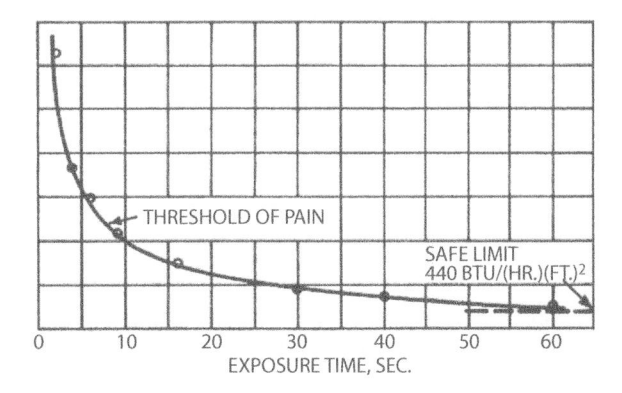

Figure 23.67 Heat radiation intensity vs. exposure time or bare skin at the threshold of pain. (By permission from Kent, Hydrocarbon Processing, Vol. 43, No. 8 (1964), p. 121 [81].)

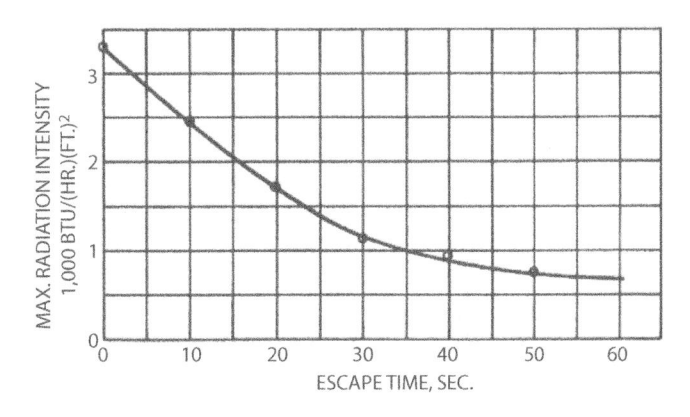

Figure 23.68 Maximum radiation intensity vs. escape time based on 5s reaction time. (By permission from Kent, Hydrocarbon Processing, Vol. 43, No. 8 (1964), p. 121 [81].)

This references to the total heat of combustion of a flame and selected values are [5c, 79] as follows:

Hydrocarbon	F range	F range average
Methane	0.10 to 0.20*	0.15
Natural Gas	0.19 to 0.23	0.21
Propane	-	0.33**
Butane	0.21 to 0.30	0.28
Hydrogen	0.10 to 0.17	0.15

*0.20 used for methane with carbon weight ratio of 0.333.
**With weight ratio of 0.222.

When in doubt, to be safe, use 0.4 [79] or 1.0 [5c]

τ = fraction heat intensity k transmitted through the atmosphere, usually assumed 1.0 (see later equation for modifying) [5c]
Q_r = heat release (lower heating valve), Btu/h

Kent [81] proposes total heat release:

$$Q_r = W \sum nh_c (379/M) \qquad (23.133)$$

or (59) $Q_n = 20,000\ W$

where

M = molecular weight
h_c = net calorific heat value, Btu/std.ft³
h_c = 50 M + 100 for hydrocarbons Btu/std.ft³ (LHV) at 14.7 psia and 60°F
h_c = Σnh_c for gas mixtures, Btu/std.ft³
n = mol fraction combustion compound(s)
f = fraction of radiated heat = 0.20 $[h_c/900]^{1/2}$
W = gas/vapors flow, lb/h
K = allowable radiation, Btu/h ft² (see Table 23.23)
   Select acceptable value for "conditions assumed"

Reasonable heat intensity K values are 1,500 Btu/h ft². When referred to Table 23.24

F = fraction of heat radiated
$\tau = 0.79\ (100/r)^{1/16} (100/D_F)^{1/16}$, from [5c].
r = relative humidity, %

When steam is injected at a rate of approximately 0.3 lb of steam per pound of flare gas, the fraction of heat radiated is decreased by 20%. τ is based on hydrocarbon flame at 2240°F, 80°F dry bulb air, relative humidity >10%, distance from flame between 100 and 500 ft, and is acceptable to estimate under wide conditions.

A slightly altered form of the D_F equation above [82, 83] for spherical radiation:

Table 23.23 Vent Sizing for Two-Phase (Runaway Reactions) Flow for a Tempered System of Example 23.15.

Mass of reactant in the vessel, kg:	9500.000
Reactor volume, m^3:	13.160
Slope of vapor pressure temp. curve: Nm^2 K:	7649.734
Latent heat of vaporization, KJ/kg:	.31060000E+03
Specific heat capacity of liquid, KJ/kg, K:	2.470
Set temperature, °C:	209.400
Maximum temperature, °C:	219.500
Specific volume of gas, m^3/kg:	0.085530
Specific volume of liquid, m^3/kg:	0.001388
Difference between gas and liquid specific volume m^3/kg:	0.084142
Self heat rate at set temperature, K/s:	0.4930
Self heat rate at maximum temperature, K/s:	0.6620
Mass flux per unit are, kg/m^2, s:	3043.067
Heat release rate per unit mass, kW/kg:	1.426
Vent area, m^2:	0.84579520E−01
Vent size, m:	0.32816150E+00
Vent size, mm:	0.32816150E+03
Vent size, in.:	0.12919750E+02

$$I = (\text{flow})(\text{NHV})(\varepsilon)/\left(4\pi D_F^2\right), \text{Btu/h ft}^2 \quad (23.134)$$

where

I = radiation intensity at point of object on ground level from midpoint of flame, Figures 23.66 and 23.68
Flow = gas flow rate, lb/h (or sft³/h)
NHV = net heating value of flare gas, Btu/lb, or (Btu/scf)
ε = emissivity

Height of stack for still air [81]:

$$H = \left(L^2 + \frac{f\,Q_r}{\pi\,q_M}\right)^{0.5} - L \quad (23.135)$$

The shortest stack exists when q_M = 3,300 Btu/h ft² (Figure 23.68). The limiting radial distance from the flame allowing for speed of escape of 20 ft/s is [81]

Table 23.24 Recommended Design Flare Radiation Levels Including Solar Radiation.

Permissible design level (K)		
British thermal units per hour per square foot	Kilowatts per square meter	Conditions
5000	15.77	Heat intensity on structures and in areas where operators are not likely to be performing duties and where shelter from radiant heat is available (for example, behind equipment).
3000	9.46	Value of K at design flare release at any location to which people have access (for example, at grade below the flare or a service platform of a nearby tower); exposure should be eliminated to a few seconds, sufficient for escape only.
2000	6.31	Heat intensity in areas where emergency actions lasting up to 1 min may be required by personnel without shielding but with appropriate clthing.
1500	4.73	Heat intensity in areas where emergency actions lasting several minutes may be required by personnel without shielding but with appropriate clothing.
500	1.58	Value of K design flare release at any location where personnel are continously exposed.

Note: On towers or other elevated structures where rapid escape is not possible ladders must provided on the side, away from the flare, so the structure can provide some shielding when K is greater than 2000Btu/h/ft² (631kW/m²). Reprinted by permission from API RP-52. Guide for Pressure Relieving and Depressuring Systems, 3rd ed. Nov 1990. American Petroleum Institute [5].

$$y = 20\, t_e = [x^2 - H(H+L)]^{0.5} \tag{23.136}$$

where

H = height of flare stack, ft, Figure 23.65
L = height of flame (length of flame from top of stack to flame tip), ft
D = X = radial distance from flame core (center) to grade, ft
R = y = radial distance from base of stack, ft, to grade intersection with D
t_e = time interval for escape, s
    Note: For safety, personnel and equipment should be outside the "y" distance.
R = distance from flame center to point X on ground (see Figure 23.69)

This has been shown to be quite accurate for distances as close to the flame as one flame length [82].

	Emissivity Values [83]
Carbon Monoxide	0.075
Hydrogen	0.075
Hydrogen Sulfide	0.070
Ammonia	0.070
Methane	0.10
Propane	0.11
Butane	0.12

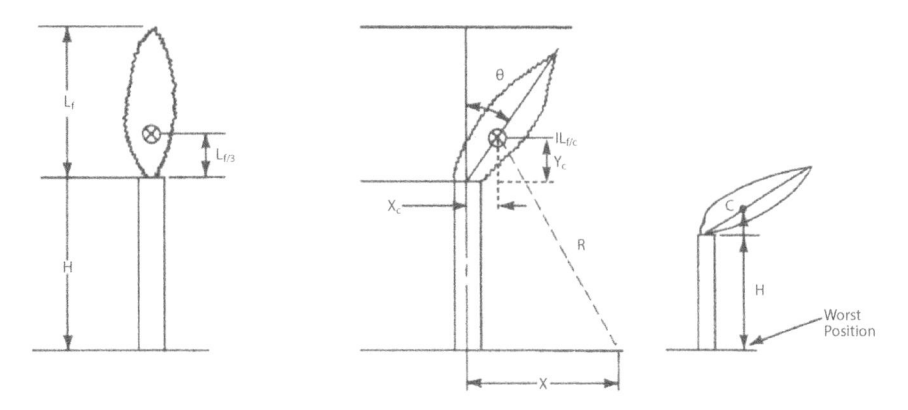

Figure 23.69 Diagrams of alternate flare stack design of Straitz. (By permission from Straitz, J. F., III and Altube, R. J., NAO, Inc. [83].)

Ethylene	0.12
Propylene	0.13
Maximum	0.13

Length of flame [83] (see Figure 25.69):

$$L_f = 10 \, (D) \, (\Delta P_t/55)^{1/2} \tag{23.137}$$

where

L_f = length of flame, ft
D = flare tip diameter, in
ΔP_t = pressure drop at the tip, in. of water

This gives flame length for conditions other than maximum flow.

The center of the flame is assumed to be located a distance of one-third the length of the flame from the tip, $L_f/3$ [83]. The flame angle is the vector addition of the wind velocity and the gas exit velocity.

$$V_{exit}, \text{gas exit velocity} = 550\sqrt{\Delta P_t/55}, \text{ft/s} \tag{23.138}$$

From Figure 23.69

$$X_c = (L_f/3)(\sin \theta)$$

$$Y_c = (L_f/3)(\cos \theta)$$

$$\text{Distance, } R = \sqrt{(X - X_c)^2 + (H + Y_c)^2} \tag{23.139}$$

For the worst condition of gas flow and wind velocities, vertically below flame center:

Then $R = H + Y_c$
 $H = R - (L_f/3)(\cos \theta)$
 $\theta = \tan^{-1}(V_{wind}/V_{gas\,exit})$

This assumes that the flame length stays the same for any wind velocity that is not rigidly true. With a wind greater than 60 mi/h, the flame tends to shorten. Straitz [83] suggests that practically this can be neglected.

Design values for radiation levels usually used [83] are as follows:

1. Equipment protection: 3,000 Btu/h.ft^2
2. Personnel, short time exposure: 1,500 Btu/h.ft^2
3. Personnel, continuous exposure: 440 Btu/h.ft^2
4. Solar radiation adds to the exposure, so on sunny days, continuous personnel exposure: 200–300 Btu/h.ft^2

Determine flare stack height above ground (grade):

Refer to Figure 23.65. Based on the Mach velocity of the vapor/gases leaving the top tip of the flare stack (see Eq. 23.125), determine the Mach number, e.g., 0.2; then from Figure 23.65

$$\text{where } H' = H + \tfrac{1}{2}(\Delta y) \tag{23.140}$$

$$\text{and } R' = R - \tfrac{1}{2}(\Delta x) \tag{23.141}$$

Δy and Δx from previous calculations under flame distortion.

Refer to Table 23.23 and select the "condition" for radiation level, K, and ground distance, R, from stack. Solve for H' and R' using the ground distance selected, R, from stack, and use the Δx previously calculated. Then, determined height of stack, H, by

$$D^2 = (R')^2 + (H')^2 \tag{23.142}$$

Substitute the previously calculated value of the distance from center of flame to grade, D, and also R'.
First solve for H'; then
H (height of stack) = H' – ½ (Δy)

$$\text{(previously calculated)} \tag{23.143}$$

Flaring Toxic Gases

The flaring of toxic gases involves special reviews. The Environmental Protection Agency (EPA) through the Chemical Manufacturers Association (CMA) provides test programs, and the destruction efficiency for certain combustible toxic material in a properly operated flare may be in the range of 98% [5c].

Depending upon the gases being flared and the flare style used, the minimum allowable net lower heating value should be in the range of 200–300 Btu/Scf. If the Btu/Scf value drops below this range, a special flare design may be required.

To ensure safe operation during periods when the flare may not have a flame present, ground level concentration calculations for hazardous components should be performed assuming the flare as a vent only. Other safeguards may be required to mitigate ground level exposure hazards. Reliable continuous pilot monitoring is essential when flaring toxic gases.

Purging of Flare Stacks and Vessels/Piping

- Vacuum cycle
- Pressure cycle
- Continuous, flow through

There are several different approaches to purging: Purging a system of flammable gas/vapor mixtures generally involves adding an inert gas such as nitrogen to the system. Sometimes the volumes of nitrogen are large, but it is still less expensive than most other nonflammable gas (even CO and CO_2 have to be used cautiously) and certainly air cannot be used because it introduces oxygen that could aggravate the flammability problem of flammability limits (also see [84]).

Pressure Purging

The inert gas is added under pressure to the system to be purged. This is then vented or purged to the atmosphere; usually more than one cycle of pressurization followed by venting is necessary to drop the concentration of a specific flammable or toxic component to a pre-established level 1.

To determine the number of purge cycles and achieve a specified component concentration after "j" purge cycles of pressure (or vacuum) and relief [41]:

$$y_j = y_o \left(n_L/n_H\right)^j - y_o \left(P_L/P_H\right)^j \tag{23.144}$$

Repeat the process as required to decrease the oxidant concentration to the desired level.

where

P_H = initial high pressures, mmHg
P_L = initial low pressure or vacuum, mm Hg
y_o = initial concentration of component (oxidant) under low pressure, mol fraction
n_H = number of mols at pressure condition
n_L = number of mols at atmospheric pressure or low pressure conditions
j = number of purge cycles (pressuring and relief)
y_j = specified component concentration after "j" purges.

Note: The above equation assumes pressure limits P_H and P_L are identical for each cycle and the total mols of nitrogen added for each cycle is constant [41].

Example 23.21: Purge Vessel by Pressurization Following the Method of [41]

A process vessel of 800-gal capacity is to have the oxygen content reduced from 21% oxygen (air). The system before process startup is at ambient conditions of 14.7 psia and 80°F. Determine the number of purges to reduce the oxygen content to 1 ppm (10^{-6} lb mol) using purchased nitrogen and used at 70 psig and 80°F to protect the strength of the vessel. How much nitrogen would be required?

Using Eq. 23.144:

y_o = initial mol fraction of oxygen. This is now the concentration of oxygen at end of the first pressuring cycle (not venting or purging).

At high pressure pressurization:

- y_o = 21 lb mol oxygen/100 total mols in vessel (initial)
- $y_o = (0.21)(P_o/P_H)$, composition for the high pressure condition
- P_o = beginning pressure in vessel, 14.7 psia
- P_H = high pressure of the purge nitrogen
- $y_o = (0.21)[14.7/(70 + 14.7)] = 0.03644$

The final oxygen concentration y_f is to be 1 ppm (10^{-6} lb mol/total mols)

$$y_f = y_o (P_L/P_H)^j \qquad (23.145)$$

$$10^{-6} = 0.21\,[14.7/(70 + 14.7)]^j$$

Solving by taking the natural logarithms:

$\ln[y_f/y_o] = j \ln[P_L/P_H]$
$\ln[10^{-6}/0.21] = j \ln[14.7/84.7]$
$j = 6.99$ cycles

Use seven minima, perhaps use eight, for assurance that purging is complete. Note that the above relationships hold for vacuum purging. Keep in mind the relationships between high and low pressure in the system and use mmHg for pressure if it is more convenient. For sweep-through purging, see [41].

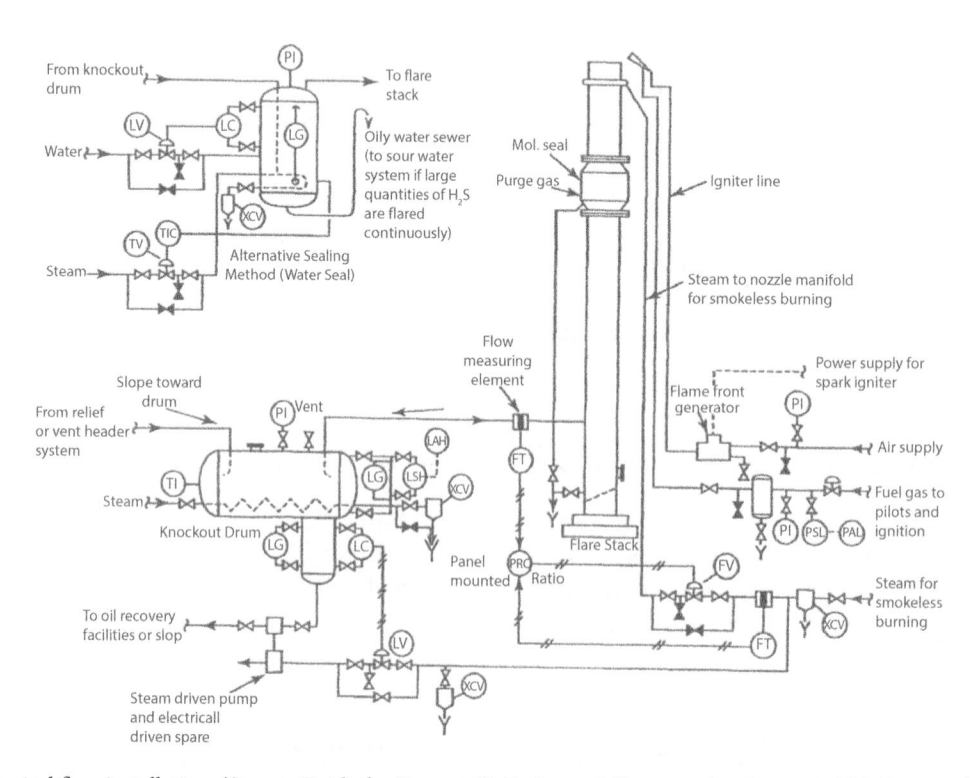

Figure 23.70 A typical flare installation. (Source: Guide for Pressure-Relieving and Depressuring Systems, API RP 521, 4th. Ed., 1997.)

Total mols nitrogen required [41]

$$n_{N_2} = j(P_H - P_L)\left[V/(R_g T)\right]$$
$$= 7.0(84.7 - 14.7)[(800/7.48)/10.3(80 + 460)] \quad (23.146)$$
$$= 8.98 \text{ mols nitrogen}$$

lb nitrogen = 8.98 (28) = 125.72 lb
V = 800 gal volume
R_g = 10.73 psi ft³/lb mol °R
T = nitrogen temperature, 80°F + 460 = 540 °R

Figure 23.70 shows a typical flare installation

23.59   Compressible Flow for Discharge Piping

The design of discharge piping or headers from relief valves for gases generally relates closely to isothermal conditions. Lapple [73] presented equations for compressible flow for both isothermal and adiabatic conditions. An important guideline for sizing discharge lines and headers from a relief valve using compressible fluids is to prevent the backpressure at relief valve outlets from reducing the fluid relieving capacity of the valve and header system. Sometimes designs that do that can also cause vibration in the discharge lines and unacceptable noise.

Conversely, if the backpressure is excessive, the relief valve may fail to lift at its set pressure. Conventional relief valves tolerate backpressures up to 10% of their set pressures, while balance bellows type can tolerate up to 30–50% of set pressure. The capacity of the valves is reduced above these tolerances.

The design of relief valves is governed by well-established guides such as API RP-520 [8], which employs a kinetic energy correction factor, and API RP-521 [5c], which relies on the Lapple chart. The limitation in successfully employing these methods is that they are based upon the unknown backpressure or header inlet pressure when the valve is discharging. Therefore, these methods often require a tedious trial-and-error solution. Figure 23.71 illustrates a typical discharge line (or tail pipe) from a safety relief valve.

Design Equations for Compressible Fluid Flow for Discharge Piping

The following equations are used to determine the pressure drop for compressible fluid flow.

The isothermal flow equation based on inlet pressure is

$$f_D \frac{L}{D} = \left(\frac{1}{M_1^2}\right)\left[1 - \left(\frac{P_2}{P_1}\right)^2\right] - \ln\left(\frac{P_1}{P_2}\right)^2 \quad (23.147)$$

where

$$r = P_1/P_2 \quad (23.148)$$

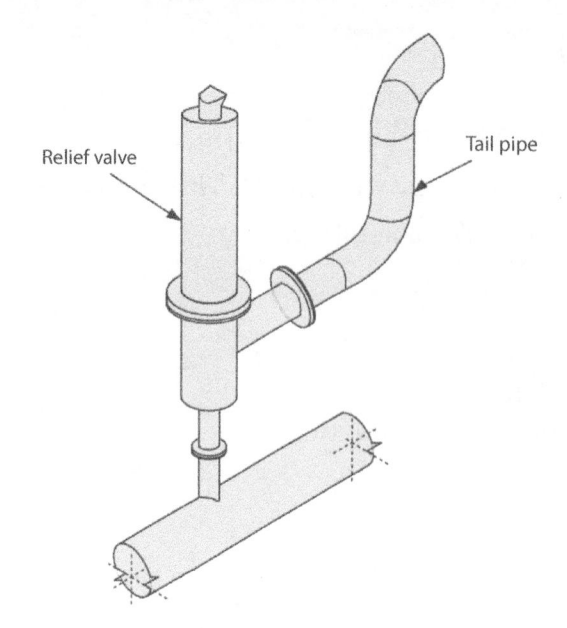

Figure 23.71 A typical relief valve and tail pipe.

Substituting Eq. 23.148 into Eq. 23.147 gives

$$f_D \frac{L}{D} = \frac{1}{M_1^2}\left[1 - \frac{1}{r^2}\right] - \ln r^2 \qquad (23.149)$$

Rearranging Equation 23.149 as a function of r gives

$$F(r) = M_1^2 \left(f_D \frac{L}{D}\right) - 1 + \frac{1}{r^2} + M_1^2 \ln r^2 \qquad (23.150)$$

The isothermal flow equation based on outlet pressure:

$$f_D \frac{L}{D} = \left(\frac{1}{M_2^2}\right)\left(\frac{P_1}{P_2}\right)^2 \left[1 - \left(\frac{P_2}{P_1}\right)^2\right] - \ln\left(\frac{P_1}{P_2}\right)^2 \qquad (23.151)$$

where

$$r = \frac{P_1}{P_2}$$

Substituting Eq. 23.148 into Eq. 23.151 gives

$$f_D \frac{L}{D} = \left(\frac{1}{M_2^2}\right) r^2 \left[1 - \frac{1}{r^2}\right] - \ln r^2 \qquad (23.152)$$

Rearranging Equation 23.152 as a function of r gives

$$F(r) = r^2 - 1 - M_2^2 \ln r^2 - M_2^2 f_D \frac{L}{D} \tag{23.153}$$

where

- D = header diameter, ft (meters)
- f_D = Moody (Darcy) friction factor
- L = header equivalent length, ft (meters)
- M_1 = Mach number of the inlet pipe
- M_2 = Mach number of the outlet pipe
- P_1, P_2 = inlet and outlet header pressures, psia (kilopascals absolute)

$$M = \frac{v_g}{v_s} \tag{23.154}$$

$$v_g = \frac{W}{\rho} \quad \text{or} \quad v_g = \frac{0.0509 W}{(\rho)(d^2)}$$

$$v_s = 223 \left(\frac{kT}{M_w}\right)^{0.5} \quad \text{or} \quad v_s = 68 \left(\frac{P_1}{\rho}\right)^{0.5}$$

$$v_g = \left(\frac{W}{A}\right)\left(\frac{ZRT}{PM_w}\right) \tag{23.155}$$

The Mach number can be determined by

$$M = \left(\frac{W}{A}\right)\left(\frac{ZRT}{PM_w}\right)\left(\frac{1}{233}\right)\left(\frac{M_w}{T}\right)^{0.5} \tag{23.156}$$

or

$$M = 0.00001336 \left(\frac{W}{PA}\right)\left(\frac{ZRT}{M_w}\right)^{0.5} \tag{23.157}$$

The outlet Mach number is given by

$$M_2 = 1.702 \times 10^{-5} \left(\frac{W}{P_2 D^2}\right)\left(\frac{ZT}{kM_w}\right)^{0.5} \tag{23.158}$$

In metric units:

$$M_2 = 3.23 \times 10^{-5} \left(\frac{W}{P_2 D^2}\right)\left(\frac{ZT}{kM_w}\right)^{0.5} \tag{23.159}$$

where

- A = pipe internal cross-sectional area, ft²
- W = gas flow rate, lb/hr (kg/h)
- Z = gas compressibility factor
- T = flowing temperature, °R = (°F + 460), K= (273.15 + °C)
- M_w = gas molecular weight
- P = P_1 or P_2, depending on input parameter, psia (kPa abs)
- R = Individual gas constant = MR/M_w = 1545/MW
- MR = Universal gas constant

In metric units:

- R = Individual gas constant = R_o/M_w, J/kg K
- R_o = Universal gas constant = 8314 J/kg mol K

Both graphical and computerized methods have been developed for solving Eqs. 23.147 and 23.151 and calculating pipe inlet pressure [85, 86]. Figure 23.72 gives a typical graphical representation of Eq. 23.147. The figure may be used to calculate the inlet pressure, P_1, for a line segment of constant diameter where the outlet pressure is known.

Critical Pressure, P_{crit}

If both high- and low-pressure relief valves need to relieve simultaneously, parallel high- and low-pressure headers terminating at the flare knockout drum are the economical choice. It is essential to check for critical flow at key points in the high-pressure header. The critical pressure at the pipe outlet can be determined by setting $M_2 = 1.0$ (sonic flow) in Eq. 23.152 as follows:

$$P_{critical} = 1.702 \times 10^{-5} \left(\frac{W}{D^2} \right) \left(\frac{ZT}{kM_w} \right)^{0.5} \tag{23.160}$$

or

Alternatively, the critical pressure can be expressed by

$$P_{crit} = \left(\frac{W}{408 d^2} \right) \left(\frac{ZT}{M_w} \right)^{0.5} \tag{23.161}$$

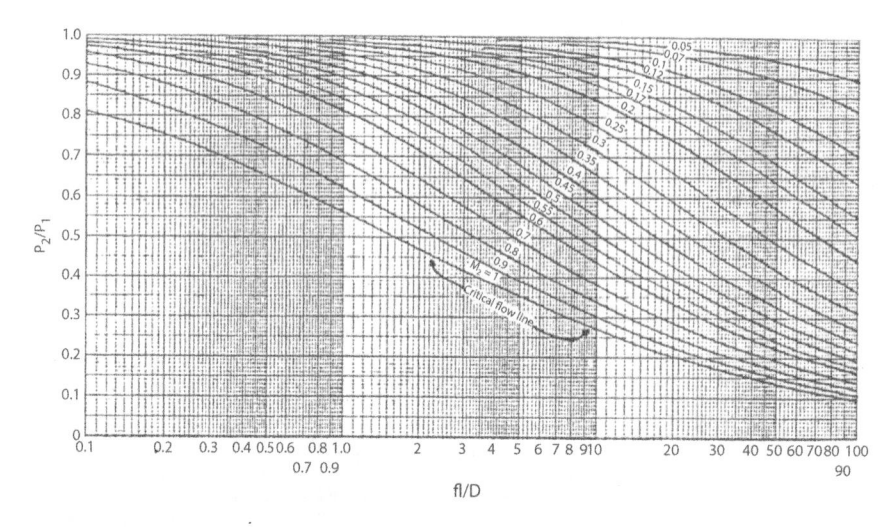

Figure 23.72 Isothermal flow chart. (Source: Mah, H. Y. [132].)

where

P_{crit} = critical pressure, psia
W = gas flow rate, lb/h
D = pipe internal diameter, in
Z = gas compressibility factor
T = gas temperature, °R
M_w = gas molecular weight

In metric units:

$$P_{critical} = 3.23 \times 10^{-5} \left(\frac{W}{D^2}\right)\left(\frac{ZT}{kM_w}\right)^{0.5} \qquad (23.162)$$

where

$P_{critical}$ is critical pressure, psia, (kPaa)

If the critical pressure is less than the pipe outlet pressure, the flow is subsonic. If the critical pressure is greater than the pipe outlet pressure, the flow is sonic and $M_2 = 1$. Therefore, the pipe inlet pressure P_1 is calculated from Eq, 23.147 with P_2 equal to the critical pressure.

The specific gravity is defined by

$$S_{60g} = \frac{\rho_{60g}}{\rho_{60g}} = \frac{M_g}{M_a} \qquad (23.163)$$

Compressibility Factor Z

Compressibility factors (Z) are available in charts or tables as a function of pseudo reduced temperatures and pressures, T_r, and P_r. Use of these charts is often time consuming and sometimes requires difficult calculations. The following equations allow the compressibility factor to be determined. This method gives a compressibility factor to within 5% for natural hydrocarbon gases with specific gravities between 0.5 and 0.8 and for pressures up to 5,000 psia. Eq. 23.164 gives the compressibility factor Z as

$$Z = F_1 \left\{ \frac{1}{\left[1 + \frac{\left(A_6 \bullet P \bullet 10^{[1.785 S_g]}\right)}{T^{3.825}}\right]} + F_2 \bullet F_3 \right\} + F_4 + F_5 \qquad (23.164)$$

where

$$F_4 = \{0.154 - 0.152 S_g\} P^{(3.18 S_g - 1.0)} e^{(-0..5P)} - 0.02$$

$$F_2 = 1.4 e^{\{-0.0054(T-460)\}}$$

$$F_3 = A_1 P^5 + A_2 P^4 + A_3 P^3 + A_4 P^2 + A_5 P$$

$$F_4 = \{0.154 - 0.152 S_g\} \, P^{(3.B \, Sg - 1.0)} \, e^{(-0.5P)} - 0.20$$

$$F_5 = 0.35 \, \{(0.6 \, S_g) \, e^{\, 1.039(P \, 18)}$$

The values of the constants A_1, A_2, A_3, A_4, A_5 and A_6 are

$A_1 = 0.001946$
$A_2 = -0.027635$
$A_3 = 0.136315$
$A_4 = -0.23849$
$A_5 = 0.1055168$
$A_6 = 3.44 \times 10^8$

The specific gravity of natural gas can be calculated from its density or molecular weight. This is expressed as the ratio of the gas density at 60°F and 1 atm (14.7 psia), $\rho_{gas, \, 60°F}$, to the density of air, $\rho_{air, \, 60°F}$, under the same conditions.

$$S_g = \frac{\text{density of gas}}{\text{density of air}}$$

$$= \frac{\rho_{gas, 60°F}}{\rho_{air, 60°F}}$$

Using the molecular weight of the gas, S_g can be expressed as

$$S_g = \frac{\text{molecular weight of gas}}{\text{moleculare weight of air}} = \frac{M_{w, gas}}{M_{w, air}}$$

Average molecular weight and viscosity:

$$M_w = \sum W \Big/ \sum (W/M_w) \tag{23.165}$$

$$T = \sum W_i T \Big/ \sum W_i \tag{23.166}$$

$$\mu = \sum x_i \mu_i (M_w)_i^{0.5} \Big/ \sum x_i (M_w)_i^{0.5} \tag{23.167}$$

Friction factor, f

The explicit equation for the friction factor (Chen friction factor, f_C) is expressed by

$$\frac{1}{\sqrt{f_C}} = -4 \log \left\{ \frac{\varepsilon}{3.7D} - \frac{5.02}{Re} \log A \right\} \tag{23.168}$$

where

$$A = \frac{\varepsilon/D}{3.7} + \left(\frac{6.7}{Re}\right)^{0.9}$$

and

ε = pipe roughness, ft.

The Darcy friction factor $f_D = 4 f_C$

The Newton-Raphson method is employed to solve the iterative process from the ratio of the inlet and outlet pressures. This is expressed in the form

$$X_{i+1} = X_i - \frac{F(X_i)}{F'(X_i)} \tag{23.169}$$

X_i is the guessed or assumed root of the equation given by $F(X) = 0$. $F(X_i)$ is the value of the objective function. $F'(X_i)$ is the value of the differential of the objective function. The i is the iteration counter, and i_{max} is the maximum iteration.

where i = 1, 2, 3, …..imax.

For isothermal flow equation based on the inlet and outlet Mach numbers, differentiating Eqs. 23.150 and 23.153 with respect to r gives

$$F'(r) = -\frac{2}{r^3} + \frac{2M_1^2}{r} \tag{23.170}$$

$$F'(r) = 2r - \frac{2M_2^2}{r} \tag{23.171}$$

A computer program (KAFLO [86]) has been developed to calculate the pressure drop (ΔP) and Mach number for compressible fluid flow in a network of connecting pipes from a pressure relief valve or valves. The program is based on the assumptions that the flow of gas through the discharge lines is isothermal, and that either the inlet or exit pressure is known. The Mach numbers are evaluated at both the inlet and outlet. The types of pipe fittings are incorporated in the program for selection. The program displays a message if the exit Mach number is greater than 0.7, signifying that the outlet gas velocity is too close to sonic velocity (i.e., the pipe size is too small). A larger pipe size is then required before the program proceeds to calculate the Reynolds number and the pressure drop in the pipe system.

Discharge Line Sizing

The following steps are used to size flare manifolds and relief valve blowdown systems:

1. The design starts at the flare tip where the outlet pressure is atmospheric. The calculation is worked back toward each relief valve in the system.
2. A size is assumed for each pipe section, and the maximum allowable velocity at each section inlet and outlet corresponds to a Mach number of 0.7. This criterion is applied to avoid pipe vibration and noise generation caused by excess velocity in the lines.
3. Properties in the common headers may be estimated from the relationships given in Eqs. 23.164–23.167. In those equations, i is the ith component.
4. The inlet pressure is calculated for each section of the line. At each downstream line, P_1 is taken as the outlet pressure of the upstream line, P_2, and a new upstream pressure P_1 is calculated. The operation is repeated, working back toward each relief valve.
5. The maximum allowable backpressure, MABP, is taken as 10% of the set pressure for conventional relief valves and 40% of the set pressure for balanced-bellows relief valves.
6. Check all relief valves against their MABP.

    - Case I. The calculated backpressure at the lowest set relief valve on a header is much smaller than its MABP. Reduce the header size.
    - Case II. The calculated backpressure at the lowest set relief valve on a header is close to and below its MABP. The header size is correct.
    - Case III. The calculated backpressure at the lowest set relief valve on a header is above its MABP. Increase the header size.

7. If there is a great difference between the calculated backpressure and the MABP, the longest header should be decreased in size until the calculated backpressure is close to the MABP.

23.60 Vent Piping

In general, the discharge piping or tail pipe should be as direct and as vertical as possible. Horizontal runs and elbows should be limited or at best avoided.

Vent lines must never have any pockets (traps), and valves should never be installed between the relieving device and the vessel or system it is protecting. Pipe fittings should be kept to a minimum.

If in doubt about pipe size, use a size larger.

Discharge Reactive Force

Reaction forces are made up of two elements. The first element follows Newton's third law that every action has an equal and opposite reaction. When a fluid is passing through the relief device and pipework, this results in an equal and opposite force on the pipework. The second element is pressure-related. If there is a choke point, that is, the gas or vapor is travelling at sonic velocity, then a pressure discontinuity will be created. This is a point where the fluid pressure suddenly changes, and since, for the same area, force is proportional to pressure, the forces on both sides of the pressure discontinuity will be unbalanced, leading to a resultant force on the pipework.

When a pressure relief valve discharges without supported discharge piping, its discharge will impose a reaction force due to the flowing fluid. This force will be transmitted into the valve structure, the mounting nozzle, and supporting vessel shell. All reactive loading and resulting stresses are dependent on the reaction force and the piping configuration. Therefore, the designer must ensure that the reaction forces and associated bending moments will not cause excessive stresses on the system's components. Furthermore, the location of an elbow and any support in the discharge system to direct the fluid up into a vent pipe must be considered in the analysis of the bending moments.

The following equation is based on the condition of critical steady state of a compressible fluid that discharges to the atmosphere through an elbow and a vertical discharge pipe. The reaction force F includes the effects of both momentum and static pressure. For any gas or vapor:

$$F = \frac{W\sqrt{\frac{kT}{(k+1)M}}}{366} + (A_o P_2) \tag{23.172}$$

where

F = reaction force at the point of discharge to the atmosphere, lb_f (Newtons)
W = flow of any gas or vapors, lb/h (kg/s)
k = ratio of specific heats (C_p/C_v)
C_p = specific heat at constant pressure
C_v = specific heat at constant volume
T = temperature at inlet °R = °F + 460 (K = °C + 273.15)
M = molecular weight of the process fluid
A_o = area of the outlet at the point of discharge, sq. in. (mm²)
P_2 = static pressure at the point of discharge, psig (barg)

Example 23.22

Size the tail pipe for steam flowing through a 6-in. Sch. 40 pipe from a pressure relief valve at a set pressure of 110.4 psig, under the following conditions:

Parameters	
Flow rate, lb/h	20,000
Temperature, °F	320.0
Outlet pressure, psia.	14.7
Viscosity, cP	0.0144
Ratio of specific heats (C_p/C_v)	1.3
Compressibility factor, Z	1
Molecular weight, M_w	18
Gas density, lb/ft³	0.0455
Pipe length, ft.	12

Pipe fittings	Number
90° Elbows (Long radius)	1
45° Elbows (Long radius)	1
Entrance	1
Exit	1

Solution

A computer program KAFLO for sizing the discharge line or tail pipe of a relief valve or valves for compressible, isothermal gas flow is employed to solve Example 23.22. The program uses the Newton-Raphson method (Eq. 23.169) to solve the iterative process of Eqs. 23.170 and 23.171, using $r = P_1/P_2 = 1.0$ as the starting default value. The KAFLO calculates the pressure drop, the inlet and outlet Mach numbers, Reynolds number, friction factor, the total length of pipe (using the 2-K method to determine the equivalent length of pipe from pipe fittings), and summing with the

Table 23.25 Computer Results of Example 23.17.

Line	
Normal size, in.	6
Schedule number:	40
Internal diameter: d, in.	6.065
Flow rate: W, lb/hr.	20,000
Compressibility factor: Z	1
Gas density: ρ, lb/ft³	0.042
Gas viscosity: μ cP	0.0144
Ratio of specific heats: k, (C_p/C_v)	1.32
Gas molecular weight: M_w, lb/lb. mole	18.0
Gas temperature: °F	320
Actual pipe length: L, ft.	12
$r = (P_1/P_2 = 1)$	
Equivalent length of pipe: L_{eq}, ft.	62.549
Gas Reynolds number: Re	6,961,150
Pipe roughness: ε, ft.	0.00015
Darcy friction factor: f_D	0.0154
Total length of pipe: ft.	74.549
Gas inlet pressure: P_1, psia.	19.5367
Gas outlet pressure: P_2, psia.	14.7
Pressure drop: ΔP, psi.	4.8367
Mach number at the inlet pipe: M_1	0.3906
Mach number at the outlet pipe: M_2	0.5191
Upstream gas velocity: v_g, ft/sec.	659.332
Gas sonic velocity: v_s, ft/sec.	1686.33
Gas critical pressure: P_c, psia	8.7724
Fluid flow pattern	Subsonic

Table 23.26 Computer Results of Vent Lines to the Flare Stack of Example 23.16.

Line	Stack	AB	BD	DE	DF	BC	CH	CG
Normal size: in.	30	18	12	8	8	12	10	6
Schedule number:	10	20	40	40	40	40	24	40
Internal diameter: d, in.	29,376	17.376	11.958	7,981	7.981	11.958	10.20	6.065
Flow rate: W, lb/hr	350,000	350.000	180.000	60,000	120,000	170,000	100,000	70,000
Compressibility factor: Z	1	1	1	1	1	1	1	1
Gas density: ρ, lb/ft^3	0.122	0.249	0.347	0.263	0.487	0.262	0.250	0.445
Gas viscosity: μ, cP	0.0108	0.0108	0.0118	0.0130	0.0110	0.0099	0.010	0.0098
Ratio of specific heats: k (C_p/C_v)	1.27	1.27	1.27	1.27	1.27	1.27	1.27	1.27
Gas molecular weight: M_w, lb/lb. mole	56.0	56.0	69.5.0	55.0	80.0	46.4	40.0	60.
Gas temperature: °F	186.6	186.6	233.3	340.0	180.0	137.6	150.0	120.0
Actual pipe length: L, ft r = (P_1/P_2 = 1)	250.0	1000.0	200.0	180.0	100.0	115.0	300.0	150.0
MAPB, psia	–	–	–	45.9	45.7	–	44.7	58.7
Equivalent length of pipe: L_{eq}, ft	–	–	–	–	–	–	–	–
Gas Reynolds number: Re	6,961,150	11,768,570	8,049,359	3,649,051	8,625,030	9,061,175	6,297,405	7,431,389
Pipe roughness: ε, ft	0.00015	0.00015	0.00015	0.00015	0.00015	0.00015	0.00015	0.00015
Darcy friction factor: f_D	0.0113	0.0122	0.0131	0.0143	0.0142	0.0131	0.0136	0.0150
Total length of pipe: L, ft	250.0	1000.0	200.0	180.0	100.0	115.0	300.0	150.0
Gas inlet pressure: P_1, psia	15.0632	30,8268	37.0825	40.9915	41.7929	36.1365	40.9063	46.0759

(Continued)

Table 23.26 Computer Results of Vent Lines to the Flare Stack of Example 23.16. (*Continued*)

Line	Stack	AB	BD	DE	DF	BC	CH	CG
Gas outlet pressure: P_2, psia	14.7	15,0632	34.1452	37.8690	37.8689	34.1452	36.1365	36.1365
Pressure drop: ΔP, psi	0.3632	15.7636	2.9373	3.1225	3.9239	1.9913	4.7698	9.9394
Mach number at the inlet pipe: M_1	0.1989	0.2777	0.2330	0.1905	0.2771	0.2566	0.2067	0.2791
Mach number at the outlet pipe: M_2	0.2038	0.5684	0.2531	0.2062	0.3058	0.2716	0.2340	0.3559
Upstream gas velocity: V_g, ft/s	169.973	237.384	185.135	182.740	197.160	231.656	203.006	218.275
Gas sonic velocity: Vg, ft/s	853.838	853.828	793.624	958.319	710.708	901.763	981.254	781.240
Gas critical pressure: P_c, psia	3.3799	9.6546	9.7446	8.8052	13.0602	10.4573	9.5332	14.5016
Fluid flow pattern	Subsonic	Subsonic	Subsonic	Subsonic	Subsonic	Subsonic	Subsonic	Subsonic

straight length of pipe, upstream gas velocity, sonic velocity, and critical pressure. It determines whether fluid flow is sonic or subsonic. Table 23.25 shows the results of the computer program.

Example 23.23: Flare and Relief Blowdown System

Size the flare manifold with relief loads and flow conditions shown in Figure 23.73.

Solution

Table 23.26 summarizes the computer results using the computer program KAFLO. Note that the backpressures are close to but less than the MABPs, showing that the line sizing is acceptable.

A Rapid Solution for Sizing Depressuring Lines [5c]

This is based on the technique developed by Lapple [73]. The methods employ a theoretical critical mass flow based on an ideal nozzle and adiabatic flow conditions and assume a known upstream low velocity source pressure. The mass flux, where $k = C_p/C_v = 1.0$, can be determined by the following:

$$G_{Ci} = 12.6 P_1 \left(\frac{M_w}{ZT_1} \right)^{0.5} \tag{23.173}$$

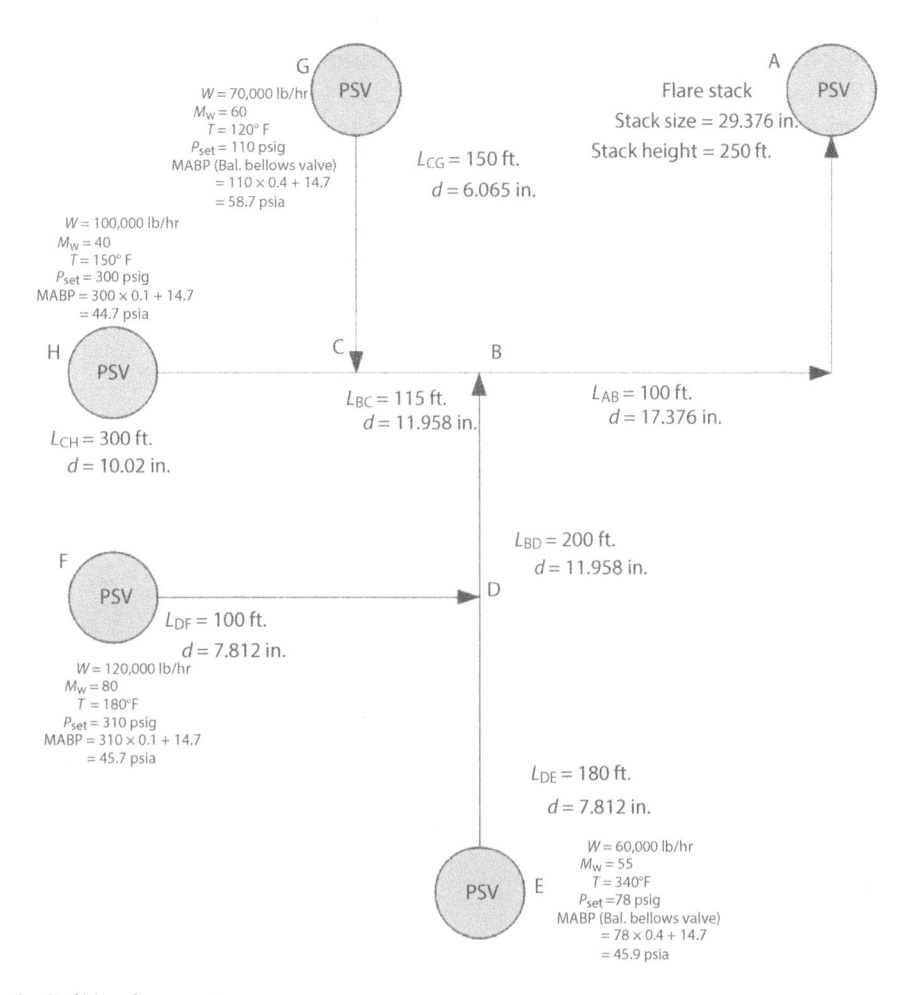

Pressure Relieving Devices and Emergency Relief System Design 209

Figure 23.73 Flare and relief blowdown system.

Flare stack data:
- Stack size = 29.376 in.
- Stack height = 250 ft.

PSV A

PSV G:
- W = 70,000 lb/hr
- M_w = 60
- T = 120° F
- P_{set} = 110 psig
- MABP (Bal. bellows valve) = 110 × 0.4 + 14.7 = 58.7 psia

L_{CG} = 150 ft.
d = 6.065 in.

L_{AB} = 100 ft.
d = 17.376 in.

PSV H:
- W = 100,000 lb/hr
- M_w = 40
- T = 150° F
- P_{set} = 300 psig
- MABP = 300 × 0.1 + 14.7 = 44.7 psia

L_{CH} = 300 ft.
d = 10.02 in.

L_{BC} = 115 ft.
d = 11.958 in.

L_{BD} = 200 ft.
d = 11.958 in.

PSV F:
- W = 120,000 lb/hr
- M_w = 80
- T = 180°F
- P_{set} = 310 psig
- MABP = 310 × 0.1 + 14.7 = 45.7 psia

L_{DF} = 100 ft.
d = 7.812 in.

L_{DE} = 180 ft.
d = 7.812 in.

PSV E:
- W = 60,000 lb/hr
- M_w = 55
- T = 340°F
- P_{set} = 78 psig
- MABP (Bal. bellows valve) = 78 × 0.4 + 14.7 = 45.9 psia

In metric units:

$$G_{Ci} = 6.7 P_1 \left(\frac{M_w}{ZT_1}\right)^{0.5} \qquad (23.174)$$

where

G_{Ci} = critical mass flux, lb/s.ft² (kg/s.m²)
P_1 = pressure at the upstream low velocity source (see Figure 23.74), psia, (kPaa)
M_w = molecular weight of the vapor.
T_1 = upstream temperature, °R = 460 + °F, (K = 273.15 + °C)
Z = compressibility factor.

The compressibility factor Z should be taken at flow conditions and thus will change as the fluid moves down the line with resulting pressure drop. A stepwise calculation may be employed to allow for this variation. An accurate solution using this method is tedious. But sufficiently accurate results can be obtained by performing the calculation over relatively large increments of pipe lengths, using an average compressibility factor over those lengths.

Regardless of which equation is used, actual mass flux (G) is a function of critical mass flux (G_{Ci}), frictional resistance (N), and the ratio of downstream to upstream pressure. Figure 23.74 illustrates these relationships. Lapple [73] has developed similar charts for adiabatic cases with ratios of specific heats of 1.4 and 1.8. In the area below the diagonal line (Figure 23.74), the ratio G/G_{Ci} remains constant, which indicates that sonic flow has been established. The total frictional resistance for use with the chart is expressed by [5c]

$$N = f\frac{L}{D} + \sum K \qquad (23.175)$$

where

N = line resistance factor (dimensionless)
F = Moody friction factor
L = actual length of the line, ft (m)
D = diameter of the line, ft (m)
K = resistance coefficients of fittings (see Chapter 15 in Volume 2).

If a Fanning friction factor is used, N = 4f L/D. These methods assume that there are no enlargements or contractions in the piping and no variation in the Mach number that results from a change in area. Coulter [88] provides a more comprehensive treatment of ideal gas flow through sudden enlargements and contractions. Another method

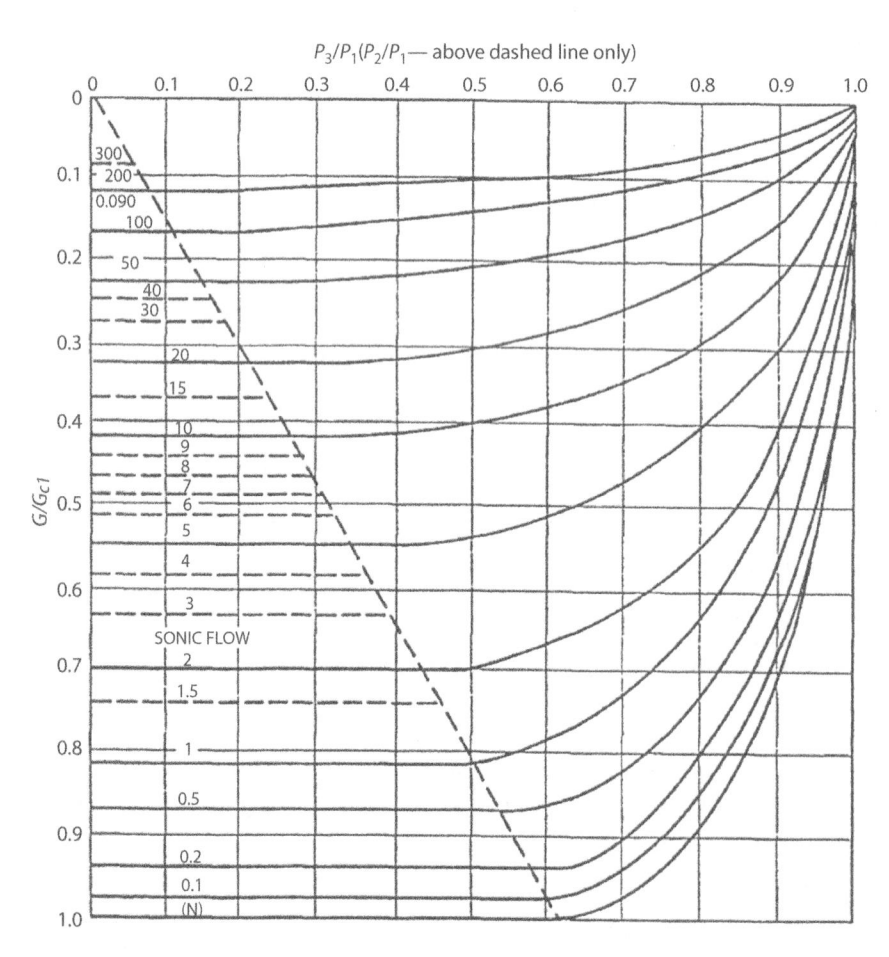

Figure 23.74 Adiabatic flow of k = 1.0. Compressible Fluids Through Pipes at High Pressure. (From API, RP 521, 4th ed., Mar 1997 [5c].)

of calculating pressure drops for ideal gases at high velocities is the use of Fanno lines. Fanno lines are the loci of enthalpy/entropy conditions that result from adiabatic flow with friction in a pipe of constant cross-section. Fanno lines extend into both supersonic and subsonic flow zones. However, for relief disposal systems, only the subsonic is of concern. The use of Fanno lines permits the calculation of pressure drops for ideal gases under adiabatic or isothermal flow conditions, with the total piping resistance as a parameter [89]. Generally, the velocity in gas discharging piping cannot exceed the sonic or critical velocity limit (this limit is shown on Lapple's charts [73] or Fanno lines).

In most disposal systems, the gases being handled are not ideal. For gases, deviations from the ideal are expressed as compressibility factors, which in turn are normally correlated with reduced pressure and reduced temperature. For hydrocarbon gases, the compressibility factor is less than 1.0 if the reduced temperature does not exceed 2.0 and the reduced pressure does not exceed about 6. Since most refinery pressure relief valve disposal system falls within these limits, the compressibility of the gases will usually be less than 1.0. As long as compressibility is less than 1.0, the pressure drop calculated for an ideal gas will be larger than that calculated for the same gas incorporating the compressibility factor [5c]. The following steps are employed in the use of Figure 23.74.

1. Calculate the N (number of velocity heads) from Eq. 23.175
2. Calculate P_3/P_1 or P_2/P_1

where

P_1 = pressure at upstream low velocity source, psia (kPaa).
P_2 = pressure in the pipe at the exit or any point distance L downstream from the source, psia (kPaa).
P_3 = pressure in reservoir into which pipe discharges 14.7 psia (101.35 kPaa) with atmospheric discharge.

3. Calculate G_{Ci} from Eq. 23.173 or Eq. 23.174.
4. From P_3/P_1 or P_2/P_1 and N, read G/G_{Ci}
5. Calculate G in lb/s ft^2 (kg/s m^2)
6. Calculate W in actual flow in lb/s (kg/s).

where

W = G x cross-sectional area of pipe, ft^2 (m^2)

API 520 part 2 refers to ASME B31.1—Power Piping and ASME B31.3—Process Piping—for the design of piping systems to withstand reaction forces from pressure relief devices. ASME B31.1 contains a section on "Dynamic Amplification of Reaction Forces." The internal forces and moments within a piping system are generally larger when the loads in the system are varying with time, as opposed to those produced under a static load. ASME B31.3 contains the term "dynamic load factor (DLF)," which has a value between 1.1 and 2 and can be determined if the engineer knows the valve opening time, although this is not readily available. The forces determined by the appropriate calculations should be multiplied by a DLF to determine the maximum force that the pipework will experience in the initial opening of the relief valve. If engineers are using vendor-supplied figures for the reaction force, then they should check whether a DLF has been used or not [87].

A check of the discharged pipework (Figures 23.75 and 23.76) will determine whether it requires some urgent redesign.

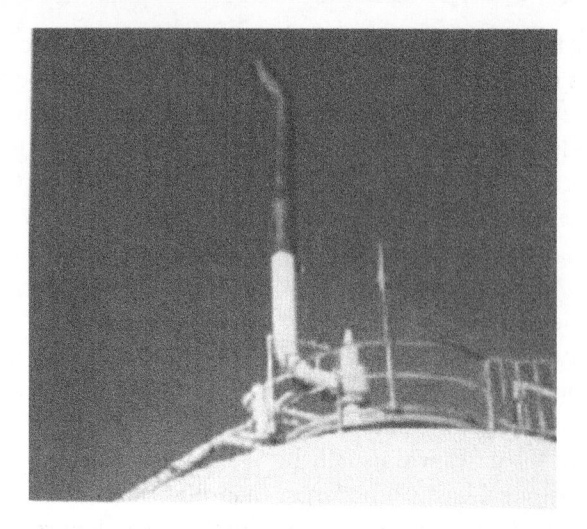

Figure 23.75 Unsupported discharge pipework – reaction forces are likely to cause damage to the relief valve and vessel nozzle.

Figure 23.76 Long unsupported relief discharge lines are likely to fail in a relief vent.

Codes and Standards

When designing pressure relief systems, engineers often apply the common standards such as API RP 521 as it contains the required considerations and information. However, what is missing is that the equipment being protected is also designed to a code that is very likely to have its own rules about excessive under- and overpressure. API RP 521 is based on ASME codes, specifically ASME VIII, but if the equipment isn't designed to this, there may be important elements that would be missing in the design.

An example is the allowable accumulation that is the pressure above the design pressure or MAWP that the equipment is allowed to experience in an emergency situation. ASME VIII will allow 10%, 16%, or 21% depending on the situation and the relief system configuration. European vessel codes limit accumulation to 10%. Furthermore, there is a European cryogenic vessel design code that allows relief devices to be set at the test pressure, rather than the MAWP. Also, ASME I has very different rules to ASME VIII on allowable installation configuration and accumulations. Thus, the key is that the final relief system needs to fulfill the requirements of both the equipment design code and the relief device design code.

Another area in the design is that not all relief valves are designed to a standard specification, e.g., ASME, API 526 or EN 4126 part 1, and so on. These tend to be cheaper, brass, low-lift relief valves for utility systems. The criteria used in API RP 520, such as the maximum built-up and backpressure of 10%, are typical numbers based on valves designed to ASME or API codes. Engineers should check the details of the actual device being installed.

Good engineering practice and experience are essential in meeting the requirements of all codes and standards, and as Special Notes section in API RP 521: "These publications are not intended to obviate the need for applying sound engineering judgement regarding when and where these publications should be utilized."

Discharge Locations

The fluids through the safety relief devices should be made safe as to where they end up, and engineers need to know where the safe location is, as shown on many P & IDs. The considerations as the type of fluid to be discharged are as follows:

- What phase is it—gas, vapor, or liquid?
- What hazardous properties does it have—flammable, noxious, toxic?

Others less obvious considerations may also be required such as noise, light pollution, or odor. Once these have been identified, then the appropriate discharge location can be decided upon.

Figure 23.77 Vent from a hydrocarbon condensate tank with lightning conductor.

Typically, gas or vapor releases are taken to a high point and released to atmosphere via a vent or a flare. If it is a liquid relief, then it needs to be returned to the process or to a collection system. A two-phase relief needs effective separation with the gas and liquid each being collected individually. However, this can be difficult with a tight space and relief systems where this final step has not been fully reviewed. For example, steam relief valves that discharge onto walkways at knee or even face height; explosion panels that exit a building next to the access door; discharge points that have pipework routed above them, which will disrupt the dispersion and may direct hazardous material back down to the ground.

Complying with good engineering practice, an audit of a newly installed relief system is required before it is commissioned to ensure that the requirements match the design. The discharge point is one area that the engineer should review as the P & IDs may not provide information as to where the discharge point will be situated (Figure 23.77).

In general, relief systems should be designed as the last line of defense against potentially catastrophic events, and thus, every effort should be made to ensure that they perform on demand, regardless of the cause of time of failure.

Process Safety Incidents with Relief Valve Failures and Flarestacks

A Case Study on Williams Geismar Olefins Plant, Geismar, Louisiana [95]

This case study reviews the catastrophic equipment rupture, explosion and fire at the Williams Olefins Plant in Geismar, Louisiana, which killed two employees at the facility. The incident occurred during nonroutine operational activities that introduced heat to an exchanger reboiler type, which was offline, creating an overpressure event while the vessel was isolated from its pressure relief device. The introduced heat increased the temperature of the liquid propane mixture confined within the reboiler shell, resulting in a dramatic pressure rise within the vessel due to liquid thermal expansion. The reboiler shell catastrophically ruptured, causing a boiling liquid expanding vapor explosion (BLEVE) and fire.

Process safety management program weaknesses at the Williams Geismar facility during the 12 years leading to the incident caused the reboiler to be unprotected by overpressure. These weaknesses include deficiencies in implementing management of change (MOC), Pre-Startup Safety Review (PSSR), and Process Hazard Analysis (PHA) programs. In addition, the company did not perform a hazard analysis or develop a procedure for the operational activities conducted on the day of the incident. This incident shows the importance of:

- Using the hierarchy of controls when evaluating and selecting safeguards to control process hazards.
- Establishing a strong organizational process safety culture.
- Developing robust process safety management programs.
- Ensuring continual vigilance in implementing process safety management programs to prevent major process safety incidents.

The Williams Geismar Olefins Plant produces ethylene (C_2H_4) and propylene (C_3H_6) for the petrochemical industry. The plant originally produced 600 million pounds of ethylene annually. Over the years, the production capacity increased to 1.35 billion pounds of ethylene and 80 million pounds of propylene per year.

Process Flow of the Olefins

Figure 23.78 shows the process flow diagram of the olefins. At the beginning of the olefins production process, ethane (C_2H_6) and propane (C_3H_8) enter "cracking furnaces" where they are converted to ethylene (C_2H_4) and propylene (C_3H_6) as well as several byproducts, including butadiene (C_4H_6), aromatic compounds (C_nH_{2n-6}, where n = 6, 7, 8), methane (CH_4) and hydrogen (H_2). The furnace effluent gases leave the cracking furnaces and enter heat exchangers

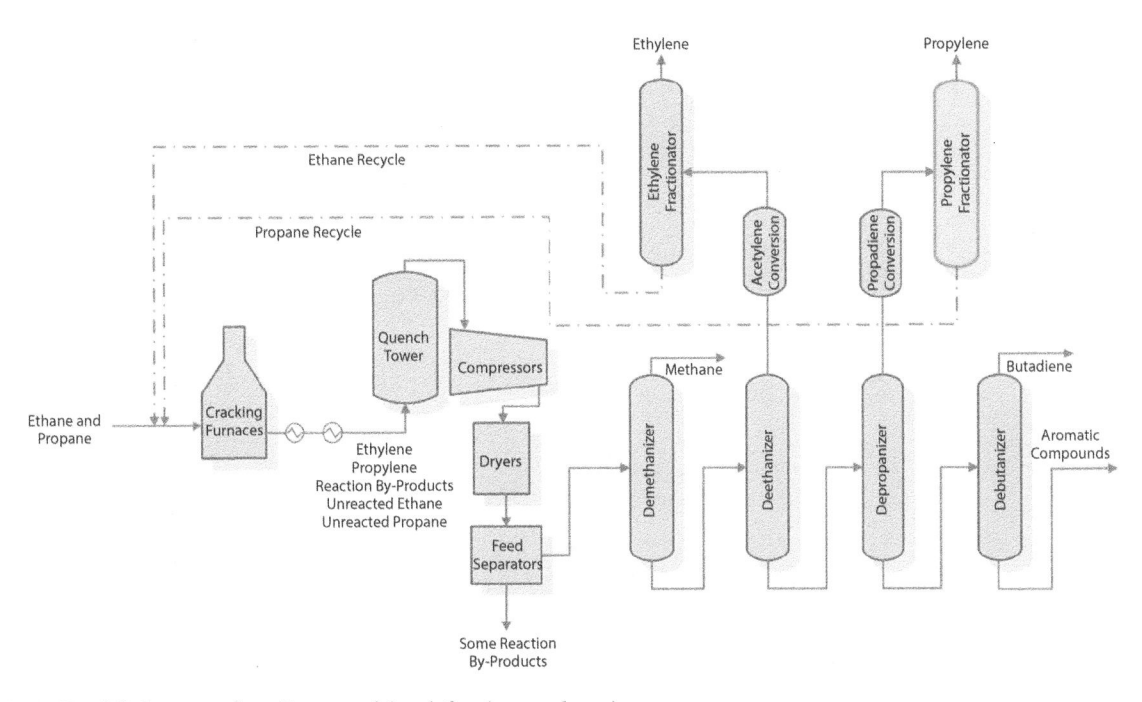

Figure 23.78 Simplified process flow diagram of the olefins (www.csb.gov).

that reduce the temperature of the gases. The furnace effluent gases then enter the quench tower for further cooling by direct contact with quench water, which is sprayed downward from the top of the tower. After additional processing, the cooled gases go to a series of distillation columns, such as the propylene fractionator, which separate the reaction products into individual components. The ethylene, propylene, butadiene and aromatic compound products are then separated and sold to customers. Unreacted ethane and propane are recycled back to the beginning of the process.

The quench water that directly contacts the heated furnace effluent gases is part of a closed – loop water circulation system. As the heated furnace effluent gases are cooled in the quench tower, heat transfers to the quench water. The heated quench water then serves as a heat source in various heat exchangers within the process, heating process streams while also reducing the temperature of the quench water. Finally, a cooling water system further cools the quench water before it circulates back to the quench tower (Figure 23.79).

Because the quench water directly contacts process gases, oily tar products contained in the gas condense into the quench water. The quench water settler removes most of the tar material; however, some oily material remains in the quench water. Over time, some of this material adheres to and builds up on the inside of the process equipment such as heat exchanger tubes, resulting in a decrease in both heat-transfer efficiency and quench water flow rate. The buildup of such material is referred to as fouling (see Volume 4 of the series). When quench water flow through the process periodically decreases due to fouling, William's operational personnel would evaluate the quench water system by analyzing, among other things, flow rates through pumps, and heat exchangers to identify the fouled piece of equipment likely causing the decreasing in quench water flow. William's personnel were performing this type of non-routine operational activity when the incident occurred on June 13, 2013.

The propylene fractionator reboilers A and B are shell and tube heat exchangers, where tube-side hot quench water vaporizes shell-side hydrocarbon process fluid, which is approximately 95% C_3H_8 with the balance composed mostly

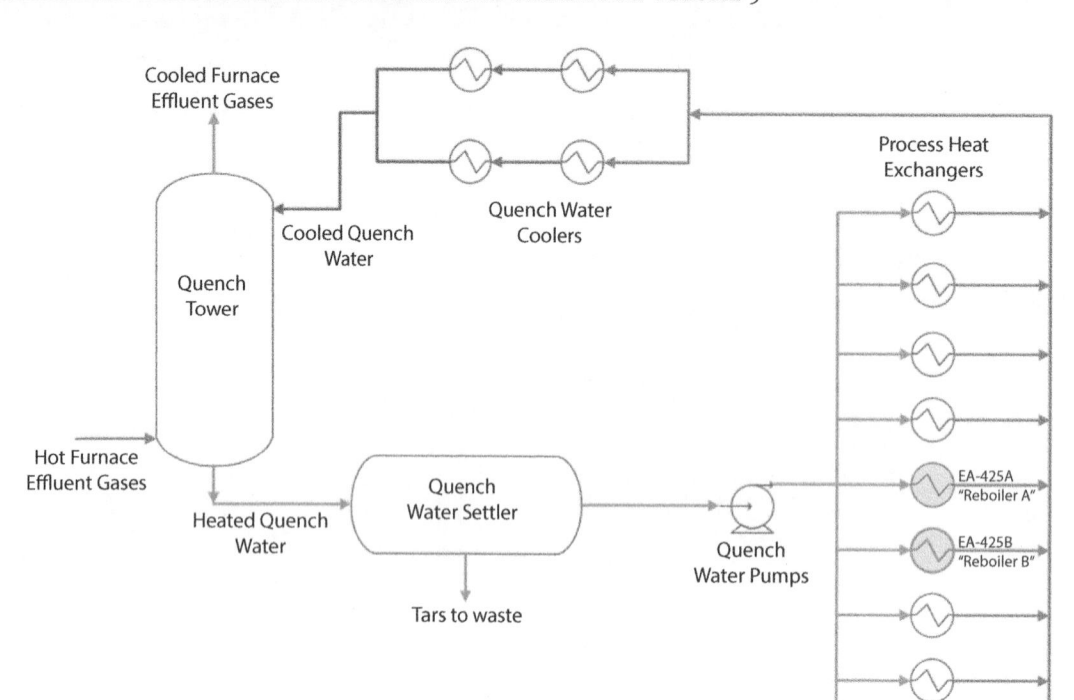

Figure 23.79 Quench water system. The propylene fractionator Reboilers A and B are highlighted in yellow. The reboiler that ruptured, Reboiler B, is indicated with the red outline (Source: www.csb.gov).

of C_3H_6 and C_4H_{10} (Figures 23.80 and 23.81). Quench water enters the propylene fractionator reboilers at approximately 185°F and partially vaporizes the shell-side propane, which enters the reboiler at a temperature of approximately 130°F.

The original propylene fractionator design had both reboilers continuously operating. This process design required periodic propylene fractionator downtime when the reboilers fouled and required cleaning (see Chapter 21, Volume 4 of the series). In 2001, Williams installed valves on the shell-side and tube-side reboiler piping to allow for continuous operation with only one reboiler operating at a time. The other reboiler would be offline as a standby but ready for operation with only one reboiler operating at a time. The other reboiler would be offline but ready for operation, isolated from the process by the new valves. This configuration allowed for cleaning of a fouled reboiler while the propylene fractionator continued to operate. Unforeseen at the time due to flaws in the Williams process safety management program, these valves also introduced a new process hazard. If the new valves were not in the proper position (open or closed) for each phase of operation, the reboiler could be isolated from its protective pressure relief valve located on top of the propylene fractionator (Figure 23.81).

The Incident

On June 13, 2013, during a daily morning meeting with operations and maintenance personnel, the plant manager noted that the quenched water flow through the operating propylene fractionator reboiler (reboiler A) had dropped gradually over the past day (Figure 23.82). The group then analyzed plant data and noticed the entire quench water circulation rate seemed to be impaired. An operation's supervisor informed the group that he would try to determine what caused the drop in flow. After evaluating the quench water system in the field, the supervisor informed several other personnel that fouling within the operating reboiler (reboiler A) could be the problem, and they might need to switch the propylene fractionator reboilers to correct the quench water flow. The supervisor attempted to meet the operations manager to discuss switching the reboilers – a typical chain of communication – so that they could

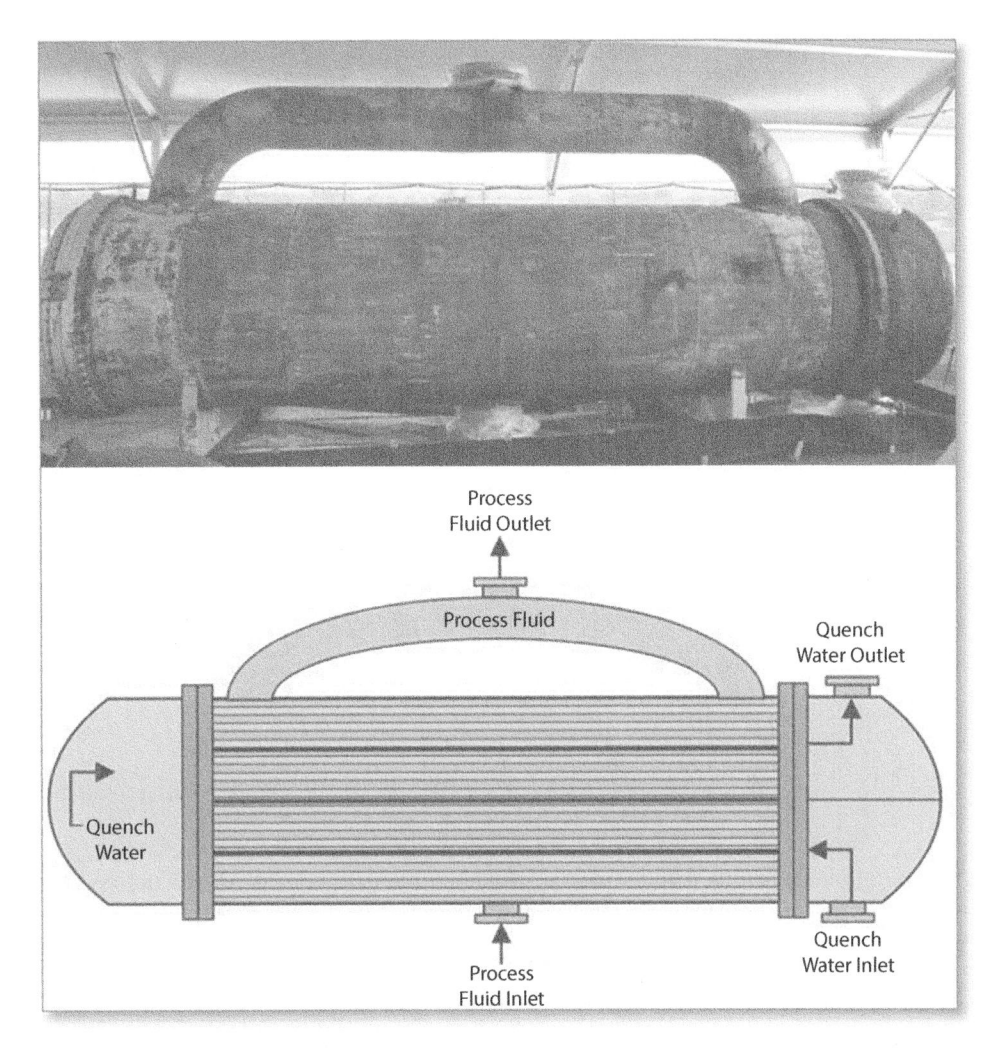

Figure 23.80 Propylene fractionator reboiler (Source: www.csb.gov).

begin getting the necessary maintenance and operations personnel involved who needed to perform the work. The manager was unavailable, but the supervisor decided to return to the field and continued evaluating the quench water system.

The U.S. Chemical Safety and Hazard Investigation Board (CSB) determined that at 8:33 am, the operation's supervisor likely opened the quench water valves on the offline reboiler, reboiler B, as indicated by the rapid increase in quench water flow rate shown in Figure 23.83. Approximately three minutes later, reboiler B exploded (Figure 23.84). Propane and propylene process fluid erupted from the ruptured reboiler and from the propylene fractionator due to failed piping. The process vapor ignited, creating a massive fireball. The force of the explosion launched a portion of the propylene fractionator reboiler piping into a pipe rack approximately 30 ft overhead (Figure 23.85).

An operator working near the propylene fractionator at the time of the explosion died at the scene. The supervisor succumbed to severe burn injuries the next day. The explosion and fire also injured Williams employees and contractors who were working on a Williams facility expansion project. 167 personnel reported injuries, and the fire lasted approximately 3.5 hours, and Williams reported releasing over 30,000 pounds of flammable hydrocarbon during the incident. The plant remained down for 18 months and restarted in January 2015.

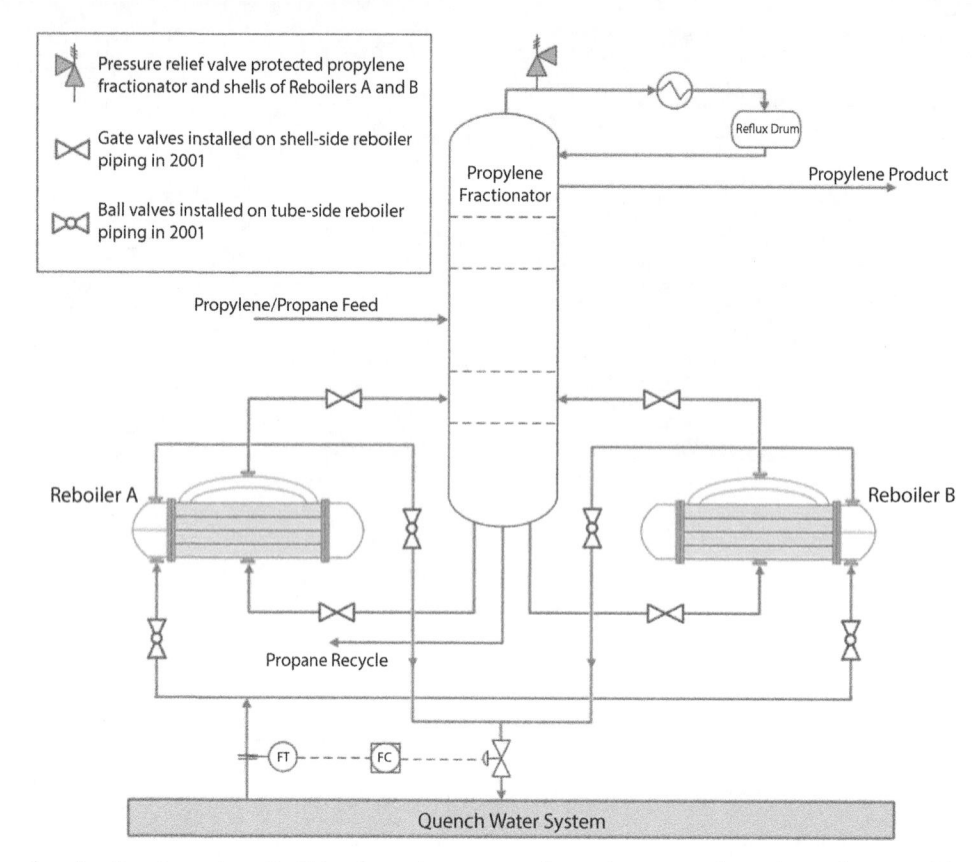

Figure 23.81 Propylene fractionator systematic. This schematic represents the equipment configuration at the time of the incident. The valves (gate valves) isolating the reboilers from the pressure relief valve at the top of the propylene fractionator were not part of the original design, and were installed in 2001. (Source: www.csb.gov).

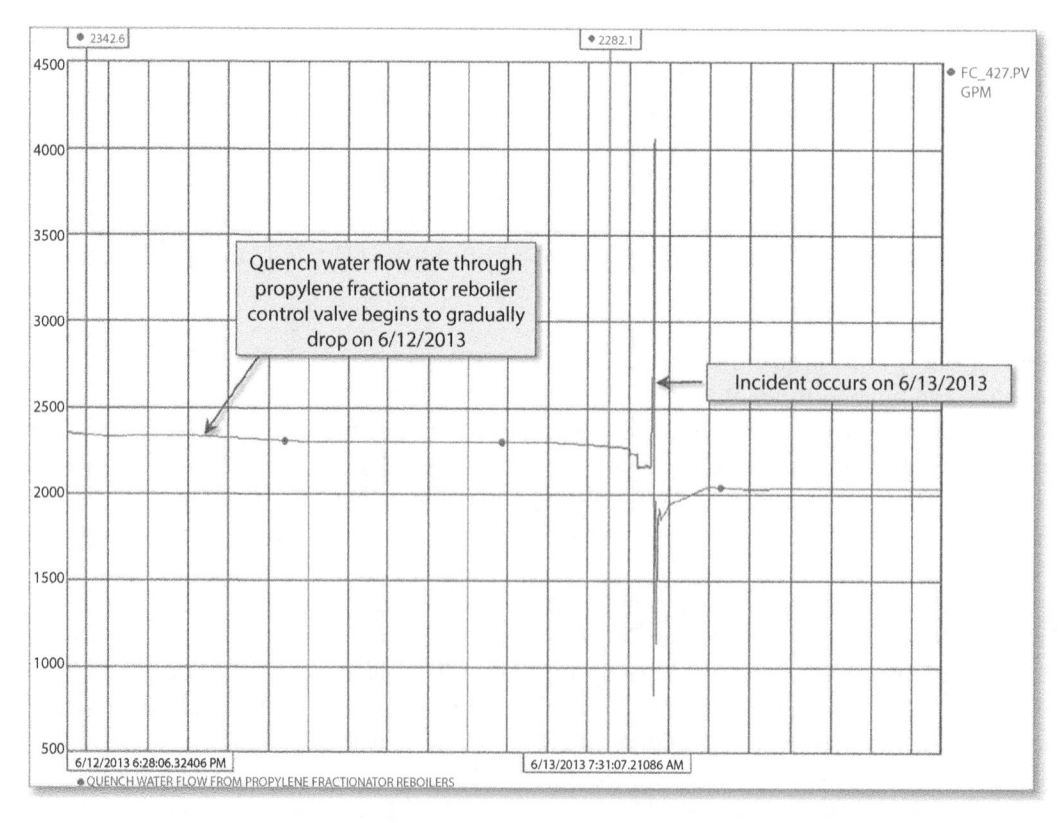

Figure 23.82 Graph of quench water flow rate through propylene fractionator Reboiler A to incident. Williams personnel identified that quench water flow rate had dropped. (Source: www.csb.gov).

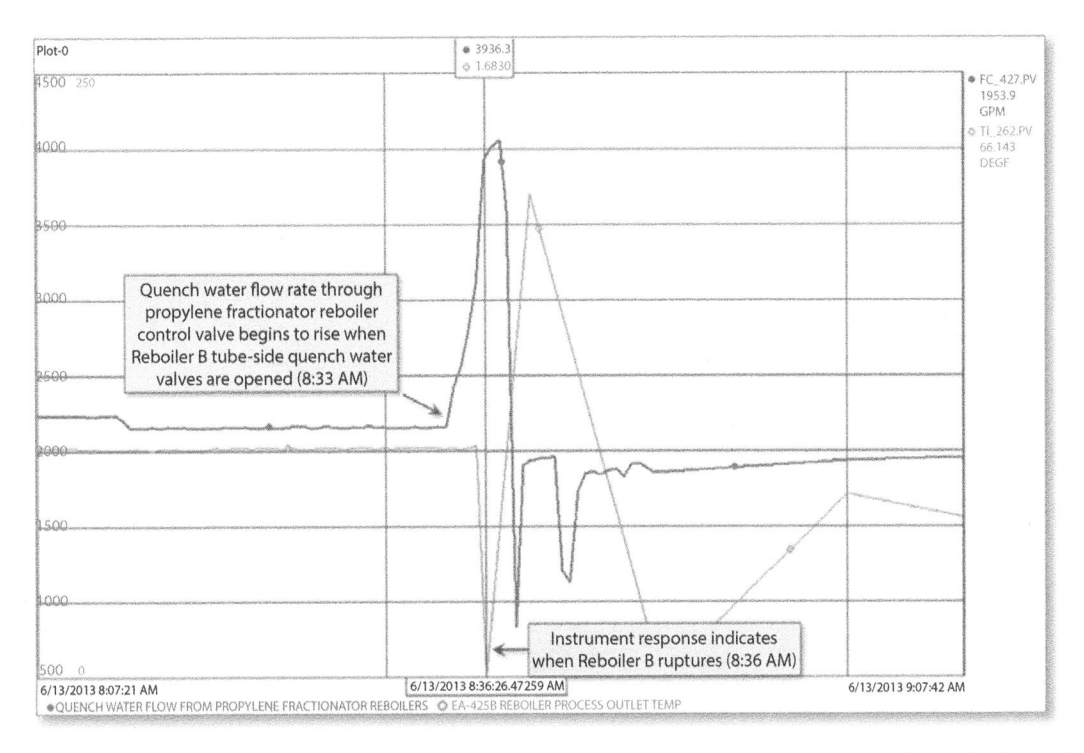

PRESSURE RELIEVING DEVICES AND EMERGENCY RELIEF SYSTEM DESIGN 219

Figure 23.83 Graph of quench water flow rate immediately prior to incident. Quench water flow rate rises when Reboiler B tube side quench water valves are opened. Reboiler B ruptures approximately 3 mins later (Source: www.csb.gov).

Figure 23.84 Post – incident photo of the ruptured Reboiler B. (Source: www.csb.gov).

Technical Analysis

The CSB commissioned metallurgical testing of the ruptured reboiler B and found that the propylene fractionator reboiler B failed, resulting in the formation of a crack, at a high internal pressure estimated to be between 674 and 1,212 psig. The CSB concluded that a pressure of this magnitude was likely the result of liquid thermal expansion in

Figure 23.85 Post-incident photo of the propylene fractionator reboilers and surround area. The reboiler B vapor return piping can be seen overhead in the piperack (red circle). (Source: www.csb.gov.)

the liquid propane filled and blocked-in reboiler B shell. This caused the overpressure of the heat exchanger while it was isolated from its pressure relief device. The initial crack formation quickly progressed to catastrophic vessel failure, which resulting in a boiling liquid expanding vapor explosion (BLEVE)[†].

Williams Geismar performed operation on propylene fractionator reboiler at a time while the other reboiler was on standby. After the operating reboiler fouled, operations staff would put the standby reboiler online. They would then shut down, drain, blind and clean the fouled reboiler. Afterward, they would remove the blinds and pressurize the reboiler with nitrogen, leaving the inlet and outlet block valves isolating the standby, nitrogen – filled reboiler shell from the propylene fractionator process fluid. The reboiler remained on standby, typically for a couple of years until the second, now operating reboiler fouled.

Williams performed maintenance on reboiler B in February 2012. Following the maintenance activity, workers left reboiler B on standby, reported filled with nitrogen and isolated from the process by a single closed block valve on the inlet piping and single closed block valve on the outlet piping. The CSB determined that between the 2012 maintenance activity and the day of the incident – a period of six months, flammable liquid propane accumulated on the shell side of the standby reboiler B (Figure 23.86). The propane could have entered the standby reboiler via a mistakenly opened valve, leaking block valve(s), or another unknown mechanism[††]. Depending on the situation that allowed propane to enter the reboiler, the nitrogen could have compressed and/or been pushed from the reboiler into the process. Williams had not installed instrumentation to detect process fluid within the reboiler. As a result, Williams personnel did not know that the standby reboiler B contained liquid propane[†††].

Post-incident field observations identified that the reboiler B tube-side hot quench water valves were in the open position (Figure 23.87). The shell-side process valves were closed, which isolated the shell of reboiler B from its protective pressure relief valve on the top of the propylene fractionator (Figure 23.81). This valve alignment shows that heat was introduced into a closed system (i.e., the blocked-in reboiler B shell).

When the reboiler B hot quench water valves were opened, the liquid propane within the standby reboiler B shell began to heat up. This caused the liquid propane to increase in volume due to liquid thermal expansion[*], filling any remaining occupiable vapor space within the shell.

When the liquid could no longer expand due to confinement within the blocked-in reboiler B shell, the pressure rapidly increased until the internal pressure exceeded the shell's mechanical pressure limit (Figure 23.88) and the reboiler shell failed[**].

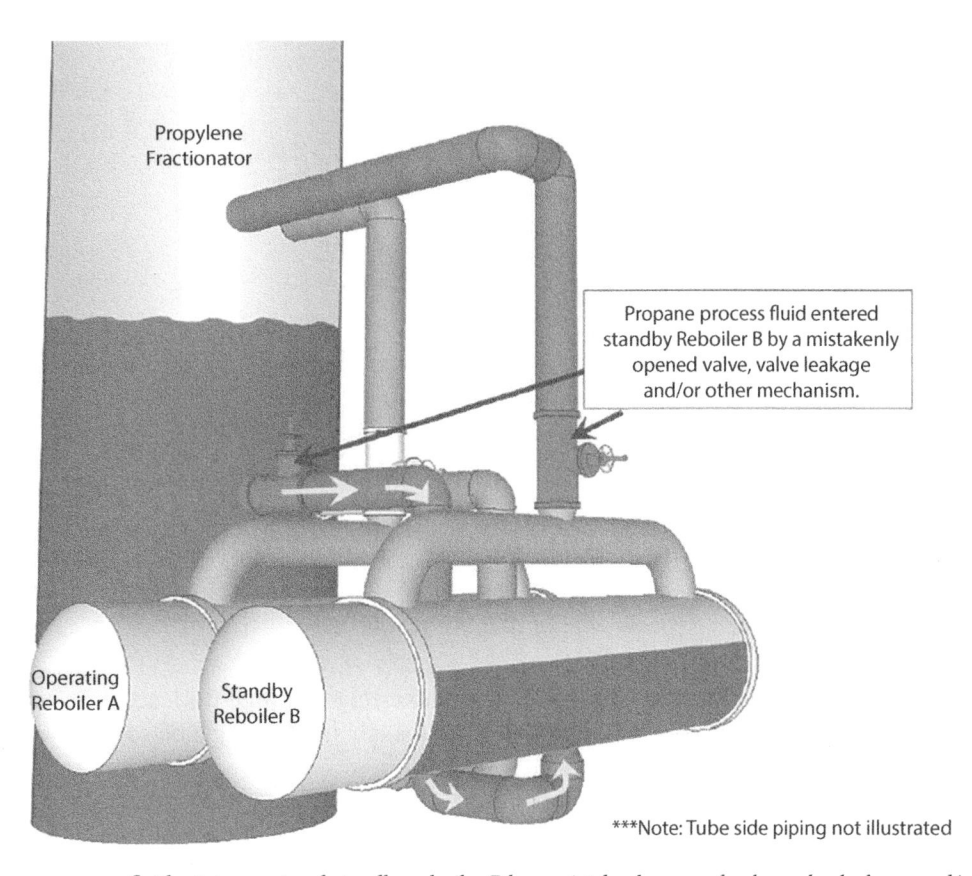

Figure 23.86 Propane process fluid mixture entered standby reboiler B by a mistakenly opened valve, valve leakage, and/or another mechanism. (Source: www.csb.gov.)

Figure 23.87 Post – incident, the reboiler B quench water inlet ball valve was found partially open (left), and the reboiler B quench water outlet ball valve was found fully open (right). When the position indicator is parallel to the pipe, the valve is opened; when the position indicator is perpendicular to the pipe, the valve is closed. (Source: www.csb.gov.)

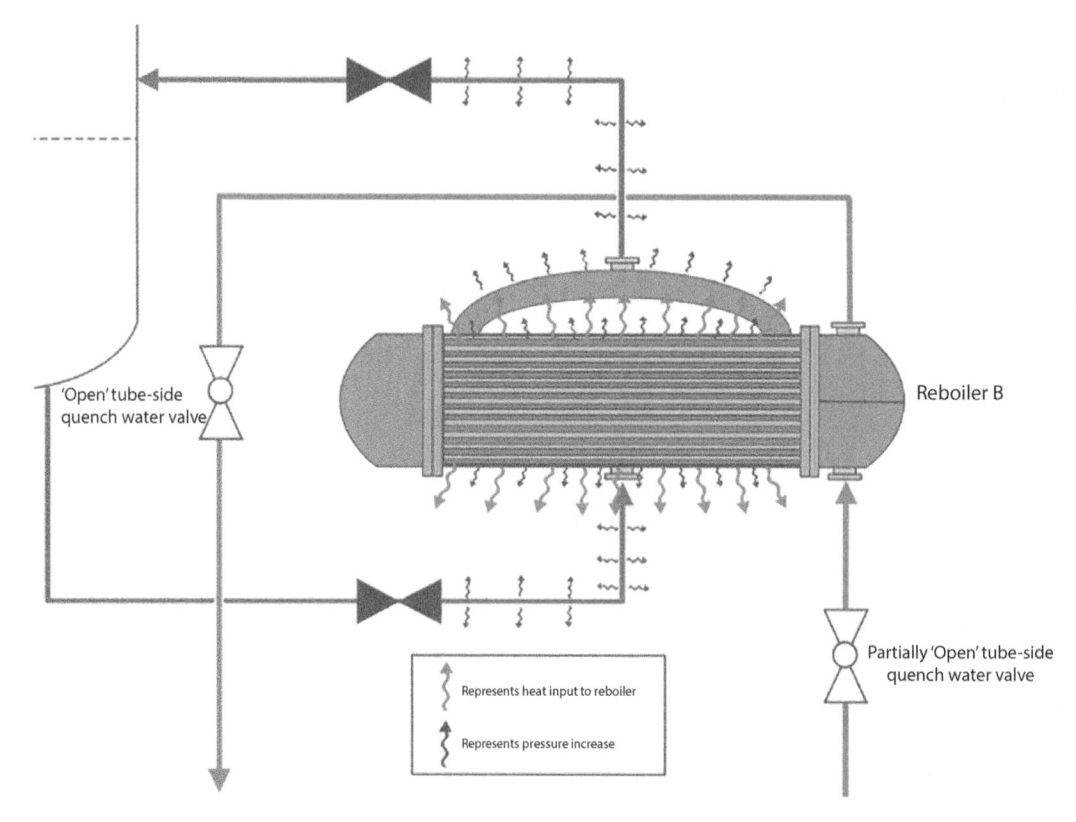

Figure 23.88 Expanding shell-side liquid propane could not sufficiently increase in volume due to the lack of overpressure protection and the closed shell-side process valves. As a result, the shell-side pressure increased until the reboiler shell failed. (Source: www.csb.gov.)

Key Lessons

Closed gate (block) valves leak, and they are susceptible to inadvertent opening. Both scenarios can introduce process fluids to offline equipment. More robust isolation methods, such as inserting a blind, can better protect offline equipment from accumulation of process fluid.

Process safety culture has been a topic of increased focus and applied within the refining, and the chemical process industry. "Safety Culture" is often described as "the way we do things around here," or "how we behave when no one is watching." The chemical process industry has defined process safety culture as "the common set of values, behaviors, and norms at all levels in a facility or in the wider organization that affects process safety."

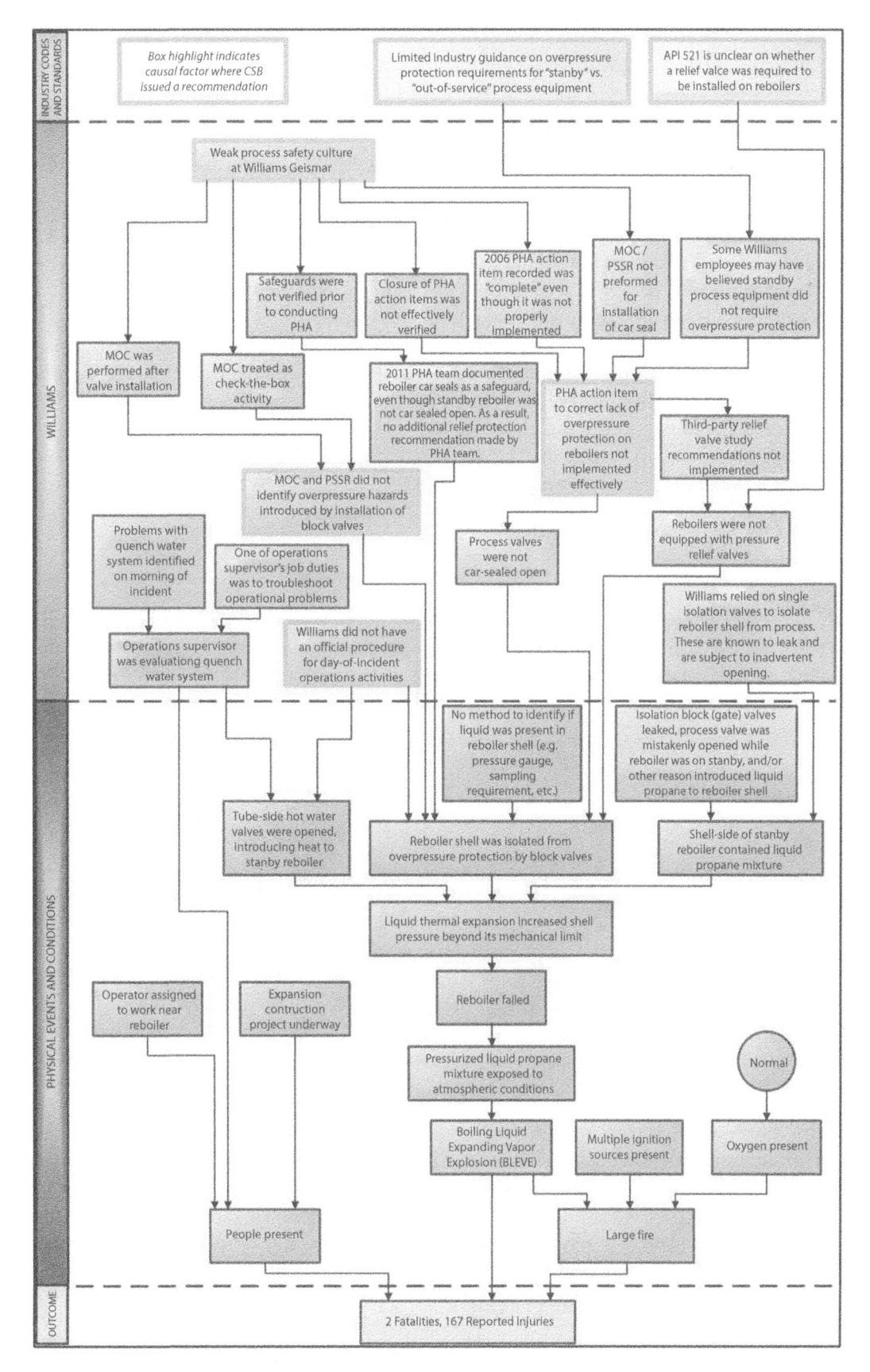

PRESSURE RELIEVING DEVICES AND EMERGENCY RELIEF SYSTEM DESIGN   223

Figure 23.88a shows the casual analysis of the incident

A significant determinant of an organization's process safety culture is the quality of its written safety management programs (e.g., process safety management procedures, including PHA, MOC, PSSR, operating procedures and written corporate policies) and how well individuals within the organization, ranging from the CEO to the field operator, implement those programs.

In this instance, Williams Geismar's process safety management program deficiencies that contributed to the incident are:

1. Williams did not perform adequate Management of Change (MOC) or Pre-Startup Safety Reviews (PSSRs) for two significant process changes involving the propylene fractionator reboilers – the installation of block valves and the addition of car seals. As a result, the company did not evaluate and control all hazards introduced to the process by those changes. Not identifying and controlling the new process over pressurization hazard was casual to the incident.
2. Williams did not adequately implement action items developed during Process Hazard Analyses (PHA) or recommendations from a contracted pressure relief system engineering analysis. Consequently, Williams did not effectively apply overpressure protection by either a pressure relief valve or by administrative controls to the standby reboiler B.
3. Williams did not perform a hazard analysis and develop a procedure prior to the operations activities conducted on the day of the incident.

Figure 23.88a shows the casual analysis of the incident.

Process safety management and its attributes are discussed in Chapter 24.

†BLEVE is the explosive release of expanding vapor and boiling liquid when a container holding a pressure liquefied gas-where the liquefied gas is above its normal atmospheric pressure boiling point temperature at the moment of vessel failure-suddenly fails catastrophically. This explosive release creates an overpressure wave that can propel vessel fragments, damage nearby equipment and buildings, and injure people. If the pressurized liquid is flammable, a fireball or vapor cloud explosion often occurs. BLEVE often result in failed vessel flattened on the ground.

Fireball from propane BLEVE experiment.   Vessel flattened on ground following BLEVE.

††Large gate valves such as the ones installed on the Williams reboilers are known to leak. The American Petroleum Institute (API) specifies allowable leakage rates through closed valves. For 16-in. and 18-in. valves such as the inlet valve and outlet valve on the propylene fractionator reboilers, API specifies an allowable leakage rate of 64 and 72 bubbles of gas per minute, respectively, during leak testing of the valves. (See API Standard 598, 9th ed., Valve Inspection and Testing, September 2009, p. 10.) The reboiler block valves were leak tested following the incident. Their leakage rate was within that allowed by API Standard 598. While valve leakage likely allowed some process fluid to enter reboiler B while it was on standby, a different mechanism could have introduced the bulk of the process fluid to the standby reboiler.

†††Records indicate that Williams filled the reboiler B shell with nitrogen, to a pressure of approximately 50 psig, during a 2012 maintenance activity. Reboiler B did not have a pressure gauge installed on its shell to allow for periodic monitoring. A pressure gauge could have alerted the operations supervisor that the reboiler B shell was at a pressure of at least 124 psig (the equilibrium vapor pressure of the process fluid at ambient temperature). This could have served as an indication that process fluid had entered the reboiler B shell.

*Thermal expansion is the increase in volume of a given mass of a solid, liquid or gas as it is heated to a higher temperature. The liquid propane expanded and pressurized the reboiler faster than the vessel contents could escape through the leaking block valves.

**Equipment or pipelines which are full of liquid under no-flow conditions are subject to hydraulic expansion due to increase in temperature and, therefore, require overpressure protection. Sources of heat that cause this thermal expansion are solar radiation, heat tracing, heating coils, heat transfer from the atmosphere or other equipment. Another cause of overpressure is a heat exchanger blocked-in on the cold side while the flow continues on the hot side [94].

The UC Chemical Safety and Hazard Investigation Board (www.csb.gov) has provided animation of this incident, which is quite informative and educational to readers.

Explosions in Flarestacks

For an explosion to occur, fuel, air (or oxygen) and a source of ignition are required. In a flarestack, the fuel is almost there as the purpose of the flarestack is to burn it, thus as the flare is normally there, all that is required is the air. Air can leak into the flare lines from the equipment that feeds the lines or through leaks in the lines leading to the stack, or can diffuse down the stack if the up flow stops or becomes very low. Another possible cause is development of a vacuum in the flare system so that air is actually sucked into the stack. This may be unlikely, but the following are accounts of two explosions that were caused this way [91].

In the first incident, the pressure control valves on two compressor suction drums should have been set to prevent the pressure falling below 30 psig (2 barg), but were set in error at zero. One of them was probably set slightly below zero. This allowed a slight vacuum to form, and it sucked air down the flarestack. Calculations showed that if the control valve was open for only a few minutes, the stack would be completely filled with air and thus it would fill with a flammable mixture of vapor and air in less time. Additionally, the nitrogen purge, intended to prevent back flow of

Figure 23.89 Base of flarestack.

air down the stack, was only a third of its normal rate. The resulting explosion deformed the base and blew off parts of the tip. They landed 150 ft (45m) away.

The second explosion occurred in a plant in which some equipment operated under vacuum. There were ruptured discs below the relief valves on a section of the plant that was under a slight pressure, but they had failed and the relief valves were leaking. This leak increased the vacuum elsewhere in the plant and pulled air into the stack, despite the presence of a molecular seal and a flame arrestor in the stack. The initial explosion was followed by two others while the plant was shutting down. The stack was ruptured in three places.

The report suggests that there may have been a split in the top hat of the molecular seal and that the flame arrestor should have been nearer the top of the stack. Flame arrestors are more efficient when they are near the end of a pipe, but they are likely to become dirty and produce a pressure drop. They should not be installed in flarestacks such that they provide an uninterrupted path to the atmosphere.

Figure 23.89 shows the results of an explosion in a large flarestack. The stack was supposed to be purged with nitrogen. However, the flow was not measured and had been cut back almost to zero to save nitrogen. Air leaked in through the large bolted joint between unmachined surfaces. The flare had not been lit for some time. Shortly after it was relit, the explosion occurred – the next time some gas was put into the stack. The mixture of gas and air moved up the stack and the pilot flame ignited it.

The following precautions should prevent similar incidents from happening again [92]:

1. Stacks should be welded. They should not contain bolted joints between unmachined surfaces.
2. There should be a continuous flow of gas up every stack to prevent air diffusing down and to sweep away small leaks of air into the stack. The continuous flow of gas does not have to be nitrogen – a waste gas stream is effective. But if gas is not being flared continuously, it is usual to keep nitrogen flowing at a linear velocity of 1.2 to 1.4 in/s (0.03 – 0.06 m/s). The flow of gas should be measured. A higher rate is required if hydrogen or hot condensable gases are being flared. If possible, hydrogen should be discharged through a separate vent stack and not mixed with other gases in a flarestack.
3. The atmosphere inside every stack should be monitored regularly, say daily, for oxygen content. Large stacks should be fitted with oxygen analyzers that alarm at 5% (2% if hydrogen is present). Small stacks should be checked with a portable analyzer.

Another stack explosion occurred nine months later in the same plant, despite the publicity given to the incident just described. To prevent leaks of carbon monoxide (CO) and hydrogen (H_2) from the glands of a number of compressors getting into the atmosphere of the compressor house, they were sucked away by a fan and the discharged through a small vent stack. Air leaked into the duct because there was a poor seal between the duct and the compressor. The mixture of air and gas was ignited by lightning.

The explosion could have been averted if the recommendations made after the first explosion had been followed. If there had been a flow of inert gas into the vent collection system and if the atmosphere inside had been tested regularly for oxygen (O_2). Why were they not followed? Perhaps because it was not obvious that collection system and if the atmosphere inside had been tested regularly for oxygen (O_2). Why were they not followed? Perhaps because it was not obvious that recommendations made after an explosion on a large flarestack applied to a small vent stack.

Vent stacks have been ignited by lightning or in other ways on many occasions. On several occasions, a group of 10 or more stacks have been ignited simultaneously. This is not dangerous provided that the following conditions apply:

1. The gas mixture in the stack is not flammable so that the flame cannot travel down the stack.
2. The flame does not impinge on overhead equipment. (Remember that in a wind, it may bend at an angle of 45°).

3. The flame can be extinguished by isolating the supply of gas or by injecting steam or an increased quantity of nitrogen. The gas passing up the stack will have to contain more than 90% nitrogen to prevent it from forming a flammable mixture with air.

A flarestack and the associated blowdown lines were prepared for maintenance by steaming for 16 hours. The next job was to isolate the system from the plant by turning a figure – 8 plate in the 35 in. (0.9 m) blowdown line. As it was difficult to turn the figure -8 plate while steam escaped from the joint, the steam purge was replaced by a nitrogen purge two hours beforehand.

When the plate had been removed for turning, leaving a gap about 2 in (50 mm) across, there was an explosion. A man was blown off the platform and killed.

The steam flow was 0.55 ton/h, but the nitrogen flow was only 0.4 ton/h, the most that could be made available. As the system cooled, air was drawn in. Some liquid hydrocarbon had been left in a blowdown vessel, and the air and hydrocarbon vapor formed a flammable mixture. According to the report, this moved up the stack and was ignited by the pilot burner, which was still lit. It is possible, however, that it was ignited by the maintenance operations.

As the steam was hot and the nitrogen was cold, much more nitrogen than steam was needed to prevent air from being drawn into the stack. After the explosion, calculations showed that 1.6 tons/h were necessary, four times as much as the amount supplied. After the explosion, the company decided to use only nitrogen in the future, not steam.

Should the staff have foreseen that steam in the system would cool and that the nitrogen flow would be too small to replace it? Probably the method used seemed so simple and obvious that no one stopped to ask if there were any hazards.

Three explosions occurred in a flarestack fitted, near the tip, with a water seal, which was intended to act as a flame arrestor and prevent flames from passing down the stack. The problems started when, as a result of incorrect valve settings, hot air was added to the stack that was burning methane (CH_4). The methane/air mixture was in the explosive range, and as the gas was hot (570°F [300°C]), the flashback speed from the flare (12 m/s) was above the linear speed of the gas (10 m/s in the tip, 5m/s in the stack). An explosion occurred, which probably damaged the water seal, though no one realized this at the time. Steam was automatically injected into the stack and the flow of methane was tripped. This extinguished the flame. When flow was restarted, a second explosion occurred, and as the water seal was damaged, this one traveled right down the stack into the knockout drum at the bottom. Flow was again restarted, and this time the explosion was louder. The operating team then decided to shut down the plant [91]. We should not restart a plant after an explosion (or other hazardous event) until we know why it occurred.

Another explosion occurred like that described above because the nitrogen flow to a stack was too low. It was cut back by an inexperienced operator; there was no low-flow alarm or high-oxygen alarm [93]. The incident was frankly described so that others may learn from it, but nowhere in the report (or editorial comment) is there any indication that the lessons learned were familiar ones, described in published reports decades before.

Relief Valves

Incidents occur because of faults in relief systems such as relief valves, bursting discs, etc. When equipment is damaged because the pressure could not be relieved, someone usually finds afterwards that the relief valve (or other relief devices) had been isolated, wrongly installed or interfered with in some other way. The following incidents are concerned with the peripherals of relief valves rather than the valves themselves [92].

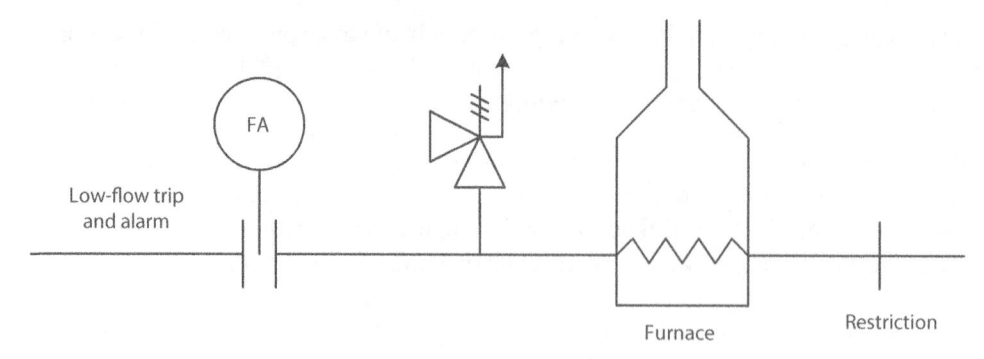

Figure 23.90 When the relief valve lifted, the flow through the furnace was reduced [92].

Location

A furnace was protected by a relief valve on its inlet line as shown in Figure 23.90. A restriction developed after the furnace. The relief valve lifted and took most of the flow. The flow through the furnace tubes fell to such as low level that they overheated and burst. The low – flow trip, which should have isolated the fuel supply to the furnace when the flow fell to a low value could not do so because the flow through it was normal. The relief valve should have been placed after the furnace, or if this was not possible, before the low-flow trip.

Another point on location that is somehow overlooked is that most relief valves are designed to be mounted vertically and should not be mounted horizontally.

Relief Valve Registers

Companies retain a register of relief valves and test them at regular intervals (e.g, every one – two years) and do not allow their sizes to be changed without proper calculation and authorization. However, equipment has been overpressured because the following items were not registered. They had been overlooked because they were not obviously a relief device or part of the relief systems. The following is a list of incidents relating to these issues [92]:

a. A hole or an open vent pipe – the simplest relief device possible. Two men were killed because the size of a vent hole in a vessel was reduced from 6in. to 3in. The tank was used for storing a liquid product that melts at 206°F (97°C). It was therefore heated by steam coil using steam at a gauge pressure of 100 psi (7 bar). At the time of the incident, the tank was almost empty and was being prepared to receive some product. The inlet line was being blown with compressed air to prove that it was clear – the normal procedure before filling the tank. The air was not getting through, and the operator suspected a choke in the pipeline. In fact, the vent on the tank was choked. The gauge air pressure (75 psi or 5 bar) was sufficient to burst the tank (design gauge pressure 5 psi or 0.3 bar). Originally the tank had a 6-in. diameter vent. But at some time, this was blanked off, and a 3-in. diameter dip branch was used instead as the vent. Several other abnormalities are: The vent was not heated; its location made it difficult to inspect. Neither manager, supervisor, nor operators recognized that if the vent choked, the air pressure was sufficient to burst the tank. However, if the 6-in. vent had not been blanked, the incident would not have occurred. Everyone should be aware of the maximum and minimum temperatures and pressures that their equipment can withstand.
b. A restriction orifice plate limiting the flow into a vessel or the heat input into a vessel should be registered if it was taken into account in sizing the vessel's relief valve. Restriction plates are easily removed. A short length of narrow diameter pipe is better.
c. A control valve limiting the flow into a vessel or the heat input into a vessel should be registered if its size was taken into account in sizing the vessel's relief valve. The control valve record sheets or database entries should be marked to show that the trim size should not be changed without checking that the relief valve will still be suitable.
d. Check (nonreturn) valves should be registered and inspected regularly if their failure could cause a relief valve to be undersized. Usually, two check valves of different types in series are used if the check valve forms part of the relief system.

Relief Valve Faults [92]

The following are examples of faults in relief valves, which are not the results of errors in design but of poor maintenance practice:

1. Identification numbers stamped on springs, thus weakening them (Figure 23.91).
2. The sides of springs ground down so that they fit.
3. Corroded springs.
4. A small spring put inside a corroded spring to maintain its strength. Sometimes the second spring was wound the same way as the first spring so that the two interlocked (Figure 23.92).
5. Use of washers to maintain spring strength.
6. Welding of springs to end caps (Figure 23.93).
7. Deliberate bending of the spindle to gag the valve (Figure 23.94).
8. Too many coils allowing little, if any, lift at set pressure (Figure 23.95).

Figure 23.91 Identification marks on body coils could lead to sprint failure.

Figure 23.92 Use of additional inner spring of unknown quality in an attempt to obtain set pressure.

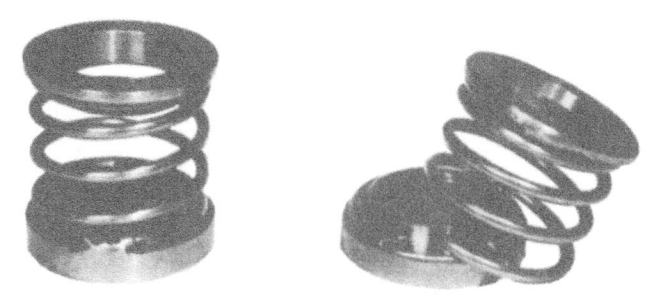

Figure 23.93 End caps welded to sprint. Failure occurred at weld.

Figure 23.94 Deliberate bending of the spindle to gag the valve.

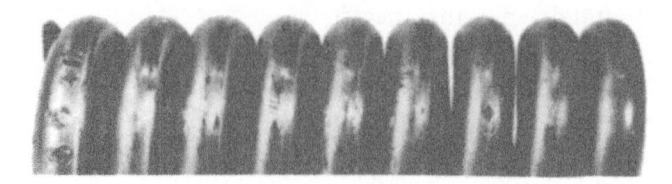

Figure 23.95 Example of too many coils.

Figure 23.96 This relief – valve tailpipe was not adequately supported.

In general, all relief valves should be tested thoroughly and inspected regularly, as we should not assume that the listed items as indicated above could not happen. When a large petroleum company introduced a test program, it was shocked by the results: out of 187 valves sent for testing, 23 could not be tested because they were leaking or because the springs were broken, and 74 failed to open within 10% of the set pressure – that is, more than half of them could not operate as required [96].

Tailpipes [92]

Figure 23.96 shows what happened to the tailpipe of a steam relief valve that was not adequately supported. The tailpipe was not provided with a drain hole (or if one was provided, it was too small), and the tailpipe filled with water. When the relief valve lifted, the water hit the curved top of the tailpipe with great force. Absence of a drain hole in a tailpipe also led to the incident described. On other occasions, drain holes have been fitted in relief-valve tailpipes even though these are discharged into flare system, where the gas then escaped into the plant area. Sometimes relief-valve exit pipes are not adequately supported and have sagged on exposure to fire, restricting the relief-valve discharge.

GLOSSARY

Accumulation: 1. The pressure increase over the maximum allowable working pressure of a vessel during discharge through the pressure relief device, expressed in pressure units or as a percent. Maximum allowable accumulations are established by applicable codes of operating and fire contingencies. **2.** The build-up of unreacted reagent or intermediates, usually associated with reactant added during a semi-batch operation.

Activation energy E_a: The constant E_a in the exponential part of the Arrhenius equation, associated with the minimum energy difference between the reactants and an activated complex (transition state that has a structure intermediate to those of the reactants and the products), or with the minimum collision energy between molecules that is required to enable a reaction to occur.

Adiabatic: A system condition in which no heat is exchanged between the chemical system and its environment.

Adiabatic induction time: Induction period or time to an event (spontaneous ignition, explosion, etc.) under adiabatic conditions, starting at operating conditions.

Adiabatic temperature rise: Maximum increase in temperature that can be achieved. This increase occurs when the substance or reaction mixture decomposes or reacts completely under adiabatic conditions. The adiabatic temperature rise follows from

$$\Delta T_{adia} = x_o (\Delta H_{RX})/C\phi$$

where x_o = Initial mass fraction, ΔH_{RX} = Heat of reaction J/kg, C= Liquid heat capacity J/kg.K, ϕ = Dimensionless thermal inertial factor (Phi- factor).

Atmospheric discharge: The release of vapors and gases from pressure-relieving and depressuring devices to the atmosphere.

Autocatalytic reaction: A reaction, the rate of which is increased by the catalyzing effect of its reaction products.

Autoignition temperature: The autoignition temperature of a substance, whether solid, liquid, or gaseous, is the minimum temperature required to initiate or cause self-sustained combustion (e.g., in air, chlorine, or other oxidant) with no other source of ignition. For example, if a gas is flammable and its autoignition temperature (AIT) is exceeded, there is an explosion.

Backpressure: The pressure that exists at the outlet of a pressure relief device as a result of the pressure in the discharge system. Backpressure can be either constant or variable. Backpressure is the sum of the superimposed and built-up backpressures.

Balanced pressure relief valve: A spring-loaded pressure relief valve that incorporates a means of minimizing the effect of backpressure on the performance characteristics of the pressure relief (see API-RP 520, Part I).

Blowdown: 1. The difference between the set pressure and the closing pressure of a pressure relief valve, expressed as a percentage of the set pressure or in pressure units. **2.** The difference between actual popping pressure of a pressure relief valve and actual re-seating pressure expressed as a percentage of set pressure.

Blowdown pressure: Is the value of decreasing inlet static pressure at which no further discharge is detected at the outlet of a safety relief valve of the resilient disk type after the valve has been subjected to a pressure equal to or above the popping pressure.

Boiling-Liquid-Expanding-Vapor-Explosive (BLEVE): Is the violent rupture of a pressure vessel containing saturated liquid/vapor at a temperature well above its atmospheric boiling point. The sudden decrease in pressure results in explosive vaporization of a fraction of the liquid and a cloud of vapor and mist, with accompanying blast effects.

The resulting flash vaporization of a large fraction of the liquid produces a large cloud. If the vapor is flammable and if an ignition source is present at the time of vessel rupture, the vapor cloud burns in the form of a large rising fireball.

Built-up backpressure: 1. The increase in pressure in the discharge header that develops as a result of flow after the pressure relief device or devices open. **2.** The pressure existing at the outlet of a pressure relief device caused by flow through that particular device into a discharge system.

Burst pressure: The inlet static pressure at which a rupture disc device functions.

Chatter: Is the abnormal rapid reciprocating motion of the movable parts of a pressure relief valve in which the disk contacts the seat.

Closed-bonnet pressure relief valve: A pressure relief valve whose spring is totally encased in a metal housing. This housing protects the spring from corrosive agents in the environment and is a means of collecting leakage around the stem or disk guide. The bonnet may or may not be sealed against pressure leakage from the bonnet to the surrounding atmosphere, depending on the type of cap or lifting-lever assembly employed or the specific handling of bonnet venting.

Closed disposal system: A disposal system capable of containing pressures that are different from atmospheric pressure.

Conventional pressure relief valve: 1. A spring-loaded pressure relief valve whose performance characteristics are directly affected by changes in the backpressure on the valve (see API RP 520 Part I). **2.** A conventional safety relief valve is a closed bonnet pressure relief valve that has the bonnet vented to the discharge side of the valve. The performance characteristics (opening pressure, closing pressure, lift and relieving capacity) are directly affected be changes of the back pressure on the valve.

Combustible: A term used to classify certain liquids that will burn on the basis of flash points. Both the National Fire Protection Association (NFPA) and the Department of Transportation (DOT) define "combustible liquids" as having a flashing point of 100°F (37.8°C) or lower.

Combustible Dusts: Dusts are particularly hazardous; they have a very high surface area-to-volume ratio. When finely divided as powders or dusts, solids burn quite differently from the original material in the bulk. Many combustible dusts produced by industrial processes are explosible when they are suspended as a cloud in air. A spark may be sufficient to ignite them. After ignition, flame spreads rapidly through the dust cloud as successive layers are heated to ignition temperature.

Condensed phase explosion: An explosion that occurs when the fuel is present in the form of a liquid or solid.

Confined explosion: An explosion of a fuel-oxidant mixture inside a closed system (e.g., a vessel or building).

Confined Vapor Cloud Explosion (CVCE): Is a condensed phase explosion occurring in confinement (equipment, building, or/and congested surroundings). Explosions in vessels and pipes and processing or storing reactive chemicals at elevated conditions are examples of CVCE. The excessive build-up of pressure in the confinement leads to this type of explosions leading to high overpressure, shock waves, and heat load (if the chemical is flammable and ignites). The fragments of exploded vessels and other objects hit by blast waves become airborne and act as missiles.

Containment: A physical system in which under all conditions, no reactants or products are exchanged between the system and its environment.

Cubic law: The correlation of the vessel volume with the maximum rate of pressure rise.

$$V^{1/3} (dP/dt)_{max} = \text{constant} = K_{max}$$

Decomposition energy: The maximum amount of energy that can be released upon decomposition. The product of decomposition energy and total mass is an important parameter for determining the effects of a sudden energy release, e.g., in an explosion.

Deflagration: The chemical reaction of a substance in which the reaction front advances into the unreacted substance at less than sonic velocity. Where a blast wave is produced that has the potential to cause damage, the term explosive deflagration is used.

Detonation: A release of energy caused by the extremely rapid chemical reaction of a substance in which the reaction front advances into the unreacted substance at equal to or greater than sonic velocity.

Design Institute for Emergency Relief Systems (DIERS): Institute under the auspices of the American Institute of Chemical Engineers funded to investigate design requirements for vent lines in the case of two-phase venting.

Design Pressure of a Vessel: At least the most severe condition of coincident temperature and gauge pressure expected during operation. It may be used in place of the MAWP in all cases where the MAWP has not been established. The design pressure is the pressure used in the design of a vessel to determine the minimum pressure permissible thickness or other physical characteristics of the different parts of the vessel (see also maximum allowable working pressure).

Disk: Is the pressure containing movable element of a pressure relief valve, which effects closure.

Dow Fire and Explosion Index (F&EI): A method (developed by Dow Chemical Company) for ranking the relative fire and explosion risk associated with a process. Analysts calculate various hazard and explosion indexes using material characteristics and process data.

Dust: Solid mixture with a maximum particle size of 500 μm.

Exotherm: A reaction is called exothermic if energy is released during the reaction.

Explosion: Propagation of a flame in a premixture of combustible gases, suspended dust(s), combustible vapor(s), mist(s), or mixtures of thereof, in a gaseous oxidant such as air, in a closed, or substantially closed vessel.

Explosion rupture disk device: Is a rupture disk device designed for use at high rates of pressure rise.

Fail-safe: Design features that provide for the maintenance of safe operating conditions in the event of a malfunction of control devices or an interruption of an energy source (e.g., direction of failure of a motor operated valve on loss of motive power).

Failure: An unacceptable difference between expected and observed performance.

Fire point: The temperature at which a material continues to burn when the ignition source is removed.

Flammability limits: The range of gas or vapor compositions in air that will burn or explode if a flame or other ignition source is present. *Importance:* The range represents an unsafe gas or vapor mixture with air that may ignite or explode. Generally, the wider the range, the greater the fire potential.

Flammable: A "flammable liquid" is defined as a liquid with a flash point below 100°F (37.8°C). Flammable liquids provide ignitable vapor at room temperatures and must be handled with caution. Flammable liquids are: Class I liquids and subdivided as follows:

Class 1A: Those having flash points below 37°F and having a boiling point below point below 100°F.

Class 1B: Those having flash points below 37°F and having a boiling point at or above 100°F.

Flare: 1. A means of safely disposing of waste gases through the use of combustion. With an elevated flare, the combustion is carried out at the top of a pipe or stack where the burner and igniter are located. A ground flare is similarly equipped except that combustion is carried out at or near ground level. A burn pit differs from a flare in that it is primarily designed to handle liquids. **2.** Flares are used to burn the combustible or toxic gas to produce combustion products, which are neither toxic nor combustible. The diameter of a flare must be suitable to maintain a stable flame and prevent a blowdown (when vapor velocities are greater than 20% of the sonic velocity).

Flash fire: The combustion of a flammable vapor and air mixture in which flame passes through that mixture at less than sonic velocity, such that negligible damaging overpressure is generated.

Flash point: The lowest temperature at which vapors above a liquid will ignite. The temperature at which vapor will burn while in contact with an ignition source, but which will not continue to burn after the ignition source is removed.

Gases: Flammable gases are usually very easily ignited if mixed with air. Flammable gases are often stored under pressure, in some cases as a liquid. Even small leaks of a liquefied flammable gas can form relatively large quantities of gas, which is ready for combustion.

Gassy system: In gassy systems, the pressure is due to a permanent gas that is generated by the reaction.

Hazard: An inherent chemical or physical characteristic that has the potential for causing damage to people, property, or the environment.

Hazard analysis: The identification of undesired events that lead to the materialization of a hazard, the analysis of the mechanisms by which these undesired events could occur and usually the estimation of the consequences.

Hazard and Operability (HAZOP): A systematic qualitative technique to identify process hazards and potential operating problems using a series of guidewords to study process deviations. A HAZOP is used to question every part of the process to discover what deviations from the start of the design can occur and what their causes and consequences may be. This is done systematically by applying suitable guidewords. This is a systematic detailed review technique for both batch and continuous plants, which can be applied to new or existing processes to identify hazards.

Hazardous chemical reactivity: Any chemical reaction with the potential to exhibit rates of increase in temperature and/or pressure too high to be absorbed by the environment surrounding the system. Included are reactive materials and unstable materials.

Hybrid mixture: A suspension of dust in air/vapor. Such mixtures may be flammable below the lower explosive limit of the vapor and can be ignited by low energy sparks.

Hybrid system: Hybrid systems are those in which the total pressure is due to both vapor pressure and permanent gas.

Inherently safe: A system is inherently safe if it remains in a non hazardous situation after the occurrence of non acceptable deviations from normal operating conditions.

Inhibition: A protective method where the reaction can be stopped by addition of another material.

Interlock system: A system that detects out-of-limits or abnormal conditions or improper sequences and either halts further action or starts corrective action.

Isothermal: A system condition in which the temperature remains constant. This implies that temperature increases and decreases that would otherwise occur are compensated by sufficient heat exchange with the environment of the system.

Likelihood: A measuring of the expected frequency that an event occurs. This may be expressed as a frequency (e.g., events per year), a probability of occurrence during a time interval (e.g., annual probability), or a conditional probability (e.g., probability of occurrence, given that a precursor event has occurred).

Lift: The actual travel of the disk away from the closed position when a valve is relieving.

Limiting oxygen concentration (LOC): Minimum concentration of oxygen in a mixture with gas, vapor, or dust that will allow it to burn.

Liquids: A vapor has to be produced at the surface of a liquid before it will burn. Many common liquids give off a flammable concentration of vapor in air without being heated, sometimes at well below room temperature. Gasoline, for example, gives off ignitable vapors above about -40°C, depending on the blend. The vapors are easily ignited by a small spark of flame.

Lower Explosive Limit (LEL): The concentration of a powder finely dispersed in air, below which no mixture likely to explode will be present.

Lower Flammable Limit (LFL): The lowest concentration of a vapor or gas (the lowest percentage of the substance in air) that will produce a flash of fire when an ignition source (heat, arc or flame) is present.

Maximum allowable accumulated pressure: The sum of the maximum allowable working pressure and the maximum allowable accumulation.

Maximum allowable working pressure (MAWP): **1.** The maximum gauge pressure permissible at the top of a completed vessel in its operating position of a designated temperature. The pressure is based on calculations for each element in a vessel using normal thickness, exclusive of additional metal thickness allowed for corrosion and loadings other than pressure. The maximum allowable working pressure is the basis for the pressure setting of the pressure relief devices that protect the vessel. **2.** The maximum allowed pressure at the top of the vessel in its normal operating position at the operating temperature specified for that pressure.

Maximum explosion overpressure, P_{max}: The maximum pressure reached during an explosion in a closed vessel through systematically changing the concentration of dust–air mixture.

Maximum reduced explosion overpressure, $P_{red,\,max}$: The maximum pressure generated by an explosion of a dust–air mixture in a vented or suppressed vessel under systematically varied dust concentrations.

Maximum explosion constant, K_{max}: Dust and test-specific characteristic calculated from the cubic law. It is equivalent to the maximum rate of pressure rise in a 1-m^3 vessel.

Maximum explosion pressure (P_{max}): the maximum expected pressure for an explosion of the optimum concentration of the powder concerned in air, in a closed vessel under atmospheric starting conditions.

Maximum explosion pressure rise $(dP/dt)_{max}$: The maximum pressure rise for an explosion of the optimum concentration of the powder concerned in air, in a closed vessel under atmospheric starting conditions. This explosion property depends on the volume of the vessel.

Minimum Ignition Energy (MIE): Is used to measure the lowest energy at which an electrical discharge is just able to ignite the most sensitive mixture of the material in air.

Minimum ignition temperature (MIT): The lowest temperature of a hot surface, which will cause a dust cloud to ignite and flame to propagate.

Minimum Oxygen Concentration (MOC): the concentration of oxygen in air, below which no mixture likely to explode will be formed with the present of dust/air/inert mixture.

Mitigation: Lessening the risk of an accident event. A sequence of action on the source in a preventive way by reducing the likelihood of occurrence of the event, or in a protective way by reducing the magnitude of the event and for the exposure of local persons or property.

Onset temperature: Is the temperature at which the heat released by a reaction can no longer be completely removed from the reaction vessel, and consequently, results in a detectable temperature increase. The onset temperature depends on detection sensitivity, reaction kinetics, on vessel size and on cooling, flow, and agitation characteristics.

Open-bonnet pressure relief valve: A pressure relief valve whose spring is directly exposed to the atmosphere through the bonnet or yoke. Depending on the design, the spring may be protected from contact with vapors or gases discharged by the valve and will be cooled by the free passage with no containment other than a short tail pipe.

Open disposal system: A disposal system that discharges directly from the relieving device to the atmosphere with no containment other than a short tail pipe.

Operating pressure: The pressure to which the vessel is usually subjected in service. A pressure vessel is normally designed for a MAWP that will provide a suitable margin above the operating pressure in order to prevent any undesirable operation of the relieving device.

Oxidant: Any gaseous material that can react with a fuel (either gas, dust, or mist) to produce combustion. Oxygen in air is the common oxidant.

Overpressure: 1. The pressure increase over the set pressure of the relieving device, expressed in pressure units or as a percent. It is the same as accumulation when the relieving device is set at the MAWP of the vessel, assuming no inlet pipe loss to the relieving device. **2.** A pressure increase over the set pressure of the relief device usually expressed as a percentage of gage set pressure.

Note: When the set pressure of the first, or primary, pressure relief valve to open is less than the vessel's MAWP, the overpressure may be greater than 10% of the valve's set pressure.

Pilot-operated pressure relief valve: A pressure relief valve in which the main valve is combined with and controlled by an auxiliary pressure relief valve.

Phi-factor ϕ: A correction factor that is based on the ratio of the total heat capacity (mass x specific heat) of a vessel and the total heat capacity of the vessel contents.

$$\phi = \frac{\text{Heat capacity of sample} + \text{Heat capacity of vessel}}{\text{Heat capacity of sample}}$$

The ϕ factor enables temperature rises to be corrected for heat lost to the container or vessel. The ϕ factor approaches the value of one for large vessels and for extremely low mass vessels.

Pressure relief valve: A generic term applied to relief valves, safety valves, and safety relief valves. A pressure relief valve is designed to automatically reclose and prevent the flow of fluid.

Pressure relief device: Is designed to open to prevent a rise of internal fluid pressure in excess of a specified value due to exposure to emergency or abnormal conditions. It may also be designed to prevent excessive internal vacuum. It may be a pressure relief valve, a non-reclosing pressure relief device or a vacuum relief valve.

Pressure-relieving system: An arrangement of a pressure-relieving device, piping and a means of disposal intended for the safe relief, conveyance, and disposal of fluids in a vapor, liquid, or gaseous phase. A relieving system may consist of only one pressure relief valve or rupture disk, either with or without discharge pipe on a single vessel or line. A more complex system may involve many pressure-relieving devices manifold into common headers to terminal disposal equipment.

Process safety: A discipline that focuses on the prevention of fires, explosions, and accidental chemical releases at chemical process facilities. Excludes classic worker health and safety issues involving working surfaces, ladders, protective equipment, etc.

Purge gas: A gas that is continuously or intermittently added to a system to render the atmosphere nonignitable. The purge gas may be inert or combustible.

Quenching: Rapid cooling from an elevated temperature, e.g., severe cooling of the reaction system in a short time (almost instantaneously), "freezes" the status of a reaction and prevents further decomposition.

Rated relieving capacity: That portion of the measured relieving capacity permitted by the applicable code or regulation to be used as a basis for the application of a pressure relief device.

Relieving conditions: Relieving conditions pertain to pressure relief device inlet pressure and temperature at a specific overpressure. The relieving pressure is equal to the valve set pressure (or rupture disk burst pressure) plus the overpressure. The temperature of the flowing fluid at relieving conditions may be higher or lower than the operating temperature.

Relieving pressure: Is set pressure plus overpressure.

Relief valve: A spring-loaded pressure relief valve actuated by the static pressure upstream of the valve. The valve opens normally in proportion to the pressure increase over the opening pressure. A relief valve is used primarily with incompressible fluids.

Relieving conditions: Used to indicate the inlet pressure and temperature of a pressure relief device at a specified overpressure. The relieving pressure is equal to the valve set pressure (or rupture disk burst pressure) plus the overpressure. The temperature of the flowing fluid at relieving conditions may be higher or lower than the operating temperature.

Risk: The likelihood of a specified undesired event occurring within a specified period or in specified circumstances.

Risk analysis: A methodical examination of a process plant and procedure that identifies hazards, assesses risks, and proposes measures that will reduce risks to an acceptable level.

Runaway: A thermally unstable reaction system, which shows an accelerating increase of temperature and reaction rate. The runaway can finally result in an explosion.

Rupture disk device: 1. A nonreclosing differential pressure relief device actuated by inlet static pressure and designed to function by bursting the pressure-containing rupture disk. A rupture disk device includes a rupture disk and a rupture disk holder. **2.** A relief device that consists of a thin metal plate or disk designed to burst or fail when a specific pressure differential is imposed on the disk. Once the disk has failed, it must be replaced. It will not stop flowing after the overpressure condition is relieved. These are sometimes used to protect a relief valve from corrosive or otherwise compromising process fluids, in addition to providing full flow relief in a large overpressure event (e.g., failure of a high pressure hydrogen stream into a cooling water exchanger.)

Safety relief valve: Is a pressure relief valve characterized by rapid opening pop action or by opening generally proportional to the increase in pressure over the opening pressure. It may be used for either compressible or incompressible fluids, depending on design, adjustment, or application.

Safety valve: A spring-loaded pressure relief valve actuated by the static pressure upstream of the valve and characterized by rapid opening or pop action. A safety valve is normally used with compressible fluids.

Set pressure: 1. The inlet gauge pressure at which the relief device is set to open (burst) under service conditions. **2.** Set pressure in psig or barg, is the inlet pressure at which the pressure relief valve is adjusted to open under service conditions. In a safety or safety relief valve gas, vapor or steam service, the set pressure is the inlet pressure at which the valve pops under service conditions. In a relief or safety relief valve in liquid service, the set pressure is the inlet pressure at which the valve starts to discharge under service conditions.

Stagnation pressure: The pressure that would be observed if a flowing fluid were brought to rest along an isentropic path.

Stamped burst pressure: The valve of the pressure differential across the rupture disk at a coincident temperature at which the rupture disk is designed to burst. It is derived from destructive tests performed on each rupture disk lot at the time of manufacture. The stamped burst pressure is marked on the rupture disk. Rupture disks that are manufactured at zero manufacturing range will typically be stamped at the specified burst pressure.

Superimposed backpressure: The static pressure that exists at the outlet of a pressure relief device at the time the device is required to operate. It is the result of pressure in the discharge system coming from other sources, and it may be either constant or variable. It may govern whether a conventional or balanced type pressure relief valve should be used in specific applications.

Static activation pressure, P_{stat}: Pressure that activates a rupture disk or an explosion door.

Superimposed backpressure: The static pressure existing at the outlet of a pressure relief device at the time the device is required to operate. It is the result of pressure in the discharge system from other sources.

Temperature of no-return: Temperature of a system at which the rate of heat generation of a reactant or decomposition just exceeds the rate of heat loss and will lead to a runaway reaction or thermal explosion.

Thermally unstable: Chemicals and materials are thermally unstable if they decompose, degrade, or react as a function of temperature and time at or about the temperature of use.

Thermodynamic data: Data associated with the aspects of a reaction that are based on the thermodynamic laws of energy, such as Gibbs' free energy, and the enthalpy (heat) of reaction.

Time to maximum rate (TMR): The time taken for a material to self-heat to the maximum rate of decomposition from a specific temperature.

Unconfined Vapor Cloud Explosion (UCVE): Occurs when sufficient amount of flammable material (gas or liquid having high vapor pressure) gets released and mixes with air to form a flammable cloud such that the average concentration of the material in the cloud is higher than the lower *limit of explosion*. The resulting explosion has a high potential of damage as it occurs in an open space covering large areas. The flame speed may accelerate to high velocities and produce significantly blast overpressure. Vapor cloud explosions in densely packed plant areas (pipe lanes, units, etc.) may show accelerations in flame speeds and intensification of blast.

Upper Explosive Limit (UEL) or Upper Flammable Limit (UFL): The highest concentration of a vapor or gas (the highest percentage of the substance in the oxidant) that will produce a flash or fire when an ignition source (heat, arc, or flame) is present.

Vapor depressuring system: A protective arrangement of valves and piping intended to provide for rapid reduction of pressure in equipment by releasing vapors. The actuation of the system may be automatic or manual. These are often found in hydrotreaters where the hydrogen must be rapidly eliminated to control a runaway temperature.

Vapor specific gravity: The weight of a vapor or gas compared to the weight of an equal volume of air, an expression of the density of the vapor or gas. Materials lighter than air have vapor specific gravity less than 1.0 (examples: acetylene, methane, hydrogen). Materials heavier than air (examples: ethane, propane, butane, hydrogen, sulfide, chlorine, sulfur dioxide) have vapor specific gravity greater than 1.0.

Vapor pressure: The pressure exerted by a vapor above its own liquid. The higher the vapor pressure, the easier it is for a liquid to evaporate and fill the work area with vapors, which can cause health or fire hazards.

Vapor pressure system: A vapor pressure system is one in which the pressure generated by the runaway reaction is solely due to the increasing vapor pressure of the reactants, products, and/or solvents as the temperature rises.

Venting (emergency relief): Emergency flow of vessel contents out of the vessel. The pressure is reduced by venting, thus avoiding a failure of the vessel by overpressurization. The emergency flow can be one-phase or multiphase, each of which results in different flow and pressure characteristics. Multiphase flow, e.g., vapor and or gas/liquid flow, requires substantially larger vent openings than single phase vapor (and /or gas) flow for the same depressurization rate.

Vent area, A: Area of an opening for explosion venting.

Vent stack: The elevated vertical termination of a disposal system that discharges vapors into the atmosphere without combustion or conversion of the relieved fluid.

Acronyms and Abbreviations

AGA	American Gas Association
AIChE	American Institute of Chemical Engineers
AIChE/CCPS	American Institute of Chemical Engineers—Center for Chemical Process Safety
AIChE/DIERS	American Institute of Chemical Engineers—Design Institute for Emergency Relief Systems
AIT	Auto-Ignition Temperature
API	American Petroleum Institute

ARC	Accelerating Rate Calorimeter
ASME	American Society of Mechanical Engineers
ASTM	American Society of Testing Materials
bar-m/sec	Bar-meter per second
BLEVE	Boiling Liquid Expanding Vapor Explosion
CFD	Computational fluid dynamics
CSB	U.S. Chemical Safety and Hazard Investigation Board
CPI	Chemical Process Industry
DIERS	Design Institute for Emergency Relief Systems.
DOE	Department of Energy
EFCE	European Federation of Chemical Engineers
EPA	U.S. Environmental Protection Agency
HAZOP	Hazard and Operability
HAZAN	Hazard Analysis
HMSO	Her Majesty's Stationery Office
HRA	Human Reliability Analysis
HSE	Health and Safety Executive, United Kingdom
IChemE	Institution of Chemical Engineers (U.K.)
ICI	Imperial Chemical Industries
LFL	Lower Flammable Limit
LNG	Liquefied Natural Gas
LPG	Liquefied Petroleum Gas
MSDS	Material safety data sheet
NFPA	National Fire Protection Agency
NIOSH	National Institute for Occupational Safety and Health
OSHA	Occupational Safety and Health Administration
PFD	Process Flow Diagram
PHA	Preliminary Hazard Analysis
P&ID	Piping and Instrumentation Diagram
TNT	Trinitrotoluene
TLV	Threshold Limit Values
UFL	Upper Flammable Limit
VCDM	Vapor Cloud Dispersion Modeling
VCE	Vapor Cloud Explosion
VDI	Verein Deutscher Ingenieure
VSP	Vent Sizing Package

Nomenclature

a	= area, in^2.
a_p	= cross-sectional area of the inlet pipe, ft^2
A	= area, m^2, ft^2, in^2; consistent with equation units
or A	= nozzle throat area, or orifice flow area, effective discharge area (calculations required) or from manufacturer's standard orifice areas, in^2
A_1	= initial vessel relief area, in^2. or m^2
A_2	= second vessel relief area, in^2 or m^2
A_3	= exposed surface area of vessel, ft^2
A_s	= internal surface area of enclosure, ft^2 or m^2
A_v	= vent area, ft^2 or m^2
A_w	= total wetted surface area, ft^2
AIT	= auto-ignition temperature

B	= cubical expansion coefficient per of liquid at expected temperature (see tabulation in text)
BP	= boiling point, °C or °F
B.P.	= burst pressure, either psig or psi abs
Bar	= 14.5 psi = 0.987 atm = 100 k a atm; 14.7 psia = 1.0135 bars
$C1 = c = C$	= gas/vapor flow constant depending on ratio of specific heats C_p/C_v
C_p/C_v	= ratio of specific heats
C_o	= sonic flow discharge orifice constant, varying with Reynolds number
C_h	= specific heat of trapped fluid, Btu/lb/°F
C_2	= subsonic flow constant for gas or vapor, function of $k = C_p/C_v$
c	= orifice coefficients for liquids
d	= diameter, inches (usually of pipe)
d'	= flare tip diameter, ft
$D = d_t$	= flare tip diameter, in.
D_F	= minimum distance from midpoint of flame to the object, ft
dp/dt	= rate of pressure rise, bar/s or psi/s
E	= joint efficiency in cylindrical or spherical shells or ligaments between openings (see ASME Code Par. UW-12 or UG-53)
e	= natural logarithm base, e = 2.718
e_t	= TNT equivalent (explosion)
F	= environment factor
F'_{gs}	= relief valve factor for non-insulated vessels in gas service exposed to open fires
F = Fh	= fraction of heat radiated
F'	= operating environment factor for safety relief of gas only vessels
F	= Flow gas/vapor, cubic feet per minute at 14.7 psia and 60°F
F_u	= The ratio of the ultimate stress of the vessel to the allowable stress of the vessel
F_y	= ratio of the yield stress of the vessel to the allowable stress of the vessel
F_1 or F_2	= relief area for vessels 1 or 2 resp., ft²
F_2	= coefficient of subcritical flow
°F	= temperature, °Fahrenheit
f	= specific relief area, m³/m², or area/unit volume
f_q	= steam quality, dryness fraction
G	= specific gravity of gas (air = 1), or specific gravity of liquid (water = 1) at actual discharge temperature
GPM	= gallons per minute flow
g	= acceleration of gravity, 32.0 ft/s²
H_c	= heat of combustion of gas/vapor, Btu/lb
H	= total heat transfer rate, Btu/h
h_1	= seal, submerged, ft
h_L	= seal pipe clearance,
h	= head of liquid, ft
h_c	= net calorific-heat value, Btu/scf
j	= number of purge cycles (pressurizing and relief)
K	= permissible design level for flare radiation (including solar radiation), Btu/hr/ft²
K_p	= liquid capacity correction factor for overpressures lower than 25% from non-code equations only.
K_b	= vapor or gas flow correction factor for constant backpressures above critical pressure
K_v	= vapor or gas flow factor for variable backpressures. Applies to balanced seal valves only.
K_w	= liquid correction factor for variable backpressures. Applies to balanced seal valves only. Conventional valves require no correction.
K_u	= liquid viscosity correction factor
K_{sh}	= steam superheat correction factor
K_n	= Napier steam correction factor for set pressures between 1500 and 2900 psig.

$K = K_d$	= coefficient of discharge:*
	= 0.975 for air, steam, vapors, and gases
	= 0.724 for ASME Code liquids**
	= 0.64 for non-ASME Code liquids
	= 0.62 for bursting/rupture disk
K_d	= discharge coefficient orifice or nozzle
K_w	= variable or constant backpressure sizing factor, balanced valves, liquids only
k	= ratio of specific heats, C_p/C_v
L	= liquid flow, gallons per minute
L_f	= length of flame, ft
$L_v = L$	= latent heat of vaporization, Btu/lb
L_x	= distance between adjacent vents, m or ft.
LEL	= lower explosive, or lower flammable limit, per cent of mixture of flammable gases only in air
$L_1, L_2,$	= lower flammability limits, vol % for each flammable gas in mixture
L_3	= longest dimension of the enclosure, ft
L/D	= length-to-diameter ratio, dimensionless
M	= molecular weight
m	= meter or percent moisture, or 100 minus steam quality
MAWP	= maximum allowable working pressure of a pressure vessel, psi gauge (or psi absolute if so specifically noted)
MP	= melting point (freezing point), °C or °F
MR	= universal gas constant = 1544 ft-lb_f/lb_m.s^2. Units depend on consistency with other symbols in equation, or manufacturing range for metal bursting/rupture disks.
m_j	= spark energy, milli-joules
m_{TNT}	= mass of TNT, lb
n	= moles of specified components
n_H	= total number moles at pressure or atmospheric condition
n_L	= total number mols at atmospheric pressure or low pressure or vacuum condition
P	= relieving pressure, psia = valve set pressure + permissible overpressure, psig, + 14.7, or any pressure, bar (gauge), or a consistent set of pressure units. Minimum overpressure is 3 psi
$P_1 = P$	= pressure, psia
p''	= maximum header exit pressure into seal, psig
P_b	= stamped bursting pressure, plus overpressure allowance (ASME 10% or 3 psi, whichever is greater) plus atmospheric pressure (14.7), psia
$P_c = P_{crit}$	= critical pressure of a gas system, psia
P_d	= design pressure of vessel or system to prevent deformation due to internal deflagration, psig
P_d	= ASME Code design pressure (or maximum allowable working pressure), psi
P_{do}	= pressure on outlet side of rupture disk, psia
P_e	= exit or backpressure, psia, stamped burst pressure
P_{er}	= perimeter of a cross section, ft or m.
P_H	= initial high pressure, mmHg
P_i	= maximum initial pressure at which the combustible atmosphere exists, psig
P_j'	= initial pressure of system, psia
P_L	= initial low pressure or vacuum, mmHg
P_{max}	= maximum explosion pressure, bar, or other consistent pressure units
P_{uv}	= maximum pressure developed in an unvented vessel, bar (gauge) or psig
P_{op}	= Normal expected or maximum expected operating pressure, psia
$P_i; P_o$	= relieving pressure, psia, or sometimes upstream pressure, psi abs, or initialpressure of system
P_η	= normal operating gas pressure, psia
ΔP_1	= pressure drop at flare tip, inches water

P_t	= pressure of the vapor just inside flare tips (at top), psia
P_1	= upstream relieving pressure, or set pressure at inlet to safety relief device psig (or psia, if consistent)
P_2	= backpressure or downstream at outlet of safety relief device, psig, or psia, depending on usage
p	= rupture pressure for disk, psig or psia
P_o	= overpressure (explosion), lb_f/in^2.
p'	= pressure, psia
ΔP	= pressure differential across safety relief valve, inlet pressure minus back pressure or down stream pressure, psi. Also = set pressure + overpressure, psig-backpressure, psig. At 10% overpressure delta P equals 1.1 P_1 -P_2. Below 30 psig set ΔP equals P_1 + 3 − P_2.
ΔP	= differential pressure across liquid relief rupture disk, usually equals p, psi
ΔP(dusts)	= pressure differential, bar or psi
Q	= total heat absorption from external fire (input) to the wetted surface of the vessel, Btu/h
Q'	= liquid flow, ft^3/s
Q_A	= required flow, ft^3/min at actual flowing temperature and pressure, acfm
Q_f	= heat released by flame, Btu/h
Q_r	= heat release, lower heating valve, Btu/h
Q_r	= required flow, cu ft/min at standard conditions of 14.7 psia and 60°F, scfm
q	= average unit heat absorption, Btu/h/ft^2 of wetted surface
R	= ratio of the maximum deflagration pressure to the maximum initial pressure, as described in NFPA Code-69, Par 5-3.3.1 also R = individual gas constant = MR/M = 1544/M
R'	= adjusted value of R, for NFPA Code-69
Re	= Reynolds number (or sometimes, N_{Re})
R_g	= Universal gas constant = 1544 = MR
R_{gc}	= individual gas constant = MR/M
R_i	= inside radius of vessel, no corrosion allowance added, in.
°R	= temperature, absolute, degrees Rankine
r = rc	= ratio of backpressure to upstream pressure, P_2/P_1, or critical pressure ratio, P_c/P_1
r	= relative humidity, percent
S	= maximum allowable stress in vessel wall, from ASME Code, psi., UCS-23.1-23.5; UHA-23, UHT-23
S' = SpGr	= specific gravity of liquid, referenced to water at the same temperature
S_G = Sg	= SpGr of gas relative to air, equals ratio of mol wt of gas to that of air, or liquid fluid specific gravity relative to water, with water = 1.0 at 60°F
SpGr	= specific gravity of fluid, relative to water = 1.0
SSU	= viscosity Saybolt universal seconds
°S	= degrees of superheat, °F
T	= absolute inlet or gas temperature, degrees Rankin °R = °F + 460, or temperature of relief vapor; °R
T_η	= normal operating gas temperature, °R
T_1	= operating temperature, °C (NFPA Code-59)
T^b	= temperature of service, °R
T_w	= vessel wall temperature, °R
T_1	= gas temperature, °R, at the upstream pressure, determined from

$$T_1 = (P_1/P_\eta)(T_\eta)$$

t	= minimum required thickness of shell of vessel, no corrosion, in.
U_∞	= viscosity at flowing temperature, Saybolt universal seconds
U	= lateral wind velocity, ft/s
U_j	= flare tip velocity, ft/s
UEL	= upper explosive or flammable limit, percent of mixture of flammable gases only in air

V	= velocity, ft/s.
or, V	= vessel volume, m³ or ft³ or required gas capacity in SCFM
or, V	= vapor flow required through valve (sub-critical), -Std ft³/min at 14.7 psia and 60°F
$\bar{V}$	= specific volume of fluid, ft³/lb
V_a	= required air capacity, scfm
V_c	= cubic feet of free air per hour, which is 14.7 psia and 60°F, for wetted area A_w > 2800 sq ft
V'	= venting requirement, cubic feet free air per hour at 14.7 psia and 60°F
V_L	= flow rate at flowing temperature, US gpm, or required liquid capacity in US gpm
v	= shock velocity, ft/s or ft/min (depends on units selected)
V_1	= specific volume of gas or vapor at upstream or relief pressure and temperature conditions, ft³/lb
v_s	= sonic velocity of gas, ft/s
V_1, V_2	= volume percent of each combustible mixture, free from air or inert gas
W	= required vapor capacity in pounds per hour any flow rate in pounds per hour, vapor relief rate to flare stack, lb/h
W_c	= charge weight of explosive, lb
W_e	= effective charge weight, pounds of TNT for estimating surface burst effects in free air
W_s	= required steam capacity flow or rate in lb/h, or other flow rate, lb/h
W_{he}	= hydrocarbon to be flared, lb/h
W_{TNT}	= equivalent charge weight of TNT, lb
W_L	= liquid flow rate, gal per min (gpm)
W_{steam}	= steam injected into flare, lb/h
w	= charge weight of explosives of interest, lb
Y_f	= final oxidant concentration, mol fraction
Y_j	= specified component concentration after "j" purges
y_o	= initial concentration of component (oxidant) under low pressure, mol fraction
Z	= compressibility factor, deviation of actual gas from perfect gas law. Usually = 1.0 at low pressure below 300 psig.
Z, or Z_{TNT}	= scaled distance for explosive blasts, ft/ (lb)$^{1/3}$
Z	= actual distance for explosion damage, ft.

Subscripts

1 = condition 1
2 = condition 2

Greek Symbols

β = beta ratio orifice diameter to pipe diameter (or nozzle inlet diameter)
ε = (epsilon) emissivity value
λ = (lambda) yield factor, (W/W_o) 1/3, with subscript "o" referring to reference value
μ = (mu) absolute viscosity at flowing temperature, cP
π = (pi), 3.1418
ρ = (rho) fluid density, lb/ft³
τ = (tau) fraction heat intensity transmitted

*Where the pressure relief valve is used in series with a rupture disk, a combination capacity factor of 0.8 must be applied to the denominator of the above valve equations. Consult the valve manufacturer (also see specific section this chapter of text). For higher factors based on National Board flow test results conducted with various rupture disk designs/arrangements (see Table 23.14).

**For saturated water see ASME Code, Appendix 11.2.

References

1. American Society of Mechanical Engineers (ASME) Boiler and Pressure Vessel Code, Section VIII, *Pressure Vessels*, published by American Society of Mechanical Engineers, New York, N.Y, 1989.
2. Kirkwood, J. G. and Wood, W. W., Editor, Shock and Detonation Waves, Gordon and Breach, London and New York.
3. Tuve, R. L. *Principles of Fire Protection and Chemistry*, National Fire Protection Association, Inc., 1976.
4. Handbook of Industrial Loss Prevention, Factory Mutual Engineering Corp., 2nd Ed., McGraw-Hill Book Co., 1967.
5. (a) *Sizing, Selection, and Installation of Pressure-Relieving Devices in Refineries, Part I—Sizing and Selection*; API Recommended Practice 520, 5th Ed., July 1990, American Petroleum Institute.
   (b) Ibid. *Part II—Installation*, API Recommended Practice 520, 3rd Ed. Nov. 1988, American Petroleum Institute.
   (c) *Guide for Pressure-Relieving and Depressuring Systems*, API Recommended Practice 521, 3rd Ed. Nov. 1990, American Petroleum Institute.
   (d) *Sizing, Selection, and Installation of Pressure-Relieving Devices in Refineries, Part I—Sizing and Selection*; API Recommended Practice 520, 7th Ed., July 2000, American Petroleum Institute.
6. *Fire Protection Handbook*, 17th Ed., National Fire Protection Association, Quincy, MA. 02269, 1991.
7. American Society of Mechanical Engineers (ASME) Boiler and Pressure Vessel Code, Section VII, Power Boilers, 1974.
8. "Design and Installation of Pressure Relieving Devices", Part I—Design; Part II: Installation, API-RF.520, American Petroleum Institute, New York, 1967.
9. "Guide for Pressure Relief and Depressuring Systems", API-RP-521, American Petroleum Institute, New York, 1990.
10. *Venting of Deflagrations*, NFPA-68, National Fire Protection Association, 1988, Ed., Quincy, Mass 02269.
11. "Terminology of Pressure Relief Devices", American National Standards Institute (ANSI) No. B95, 1 (latest ed.).
12. *Explosions Prevention Systems* (1986) NFPA #69, National Fire Protection Association (1986) Quincy, MA 02259.
13. Fisher, H. G., "An Overview of Emergency Relief System Design Practice", Plant/Operations progress, Vol 10; No. 1, 1991.
14. Papa, D. M., "Clear Up Pressure Relief Sizing Methods", Chem. Eng. Prog., V. 87, No.8, 1991, p. 81.
15. Lewis, B. and Von Elbe, G., *Combustion, Flames and Explosions of Gases*, 2nd Ed., Academic Press, 1961
16. Perry, R H. and Green, Don, *Perry's Chemical Engineers' Handbook*, 6th Ed., 1984.
17. Nazario, F. N., "Rupture Disks: A Primer", Chem. Engr., June 20, 1988, p. 86.
18. Pitman, J. F., *Blast and Fragments from Superpressure Vessel Rupture*, Naval Surface Weapons Center, White Oak Silver Springs, Maryland, Report #NSWC/WOL/TR. 75-87, 1976.
19. *Suppressive Shield Structural Design and Analysis Handbook*, U.S. Army Corps of Engineers, Huntsville, Div. No. HNDM-1110-1-2, 1977.
20. Albaugh, L. R. and Pratt, T. H., "Flash Points of Aqueous Solutions", Newsletter No.6, Hazards Evaluation and Risk Control Services, Hercules, Inc., 1979.
21. *Structure to Resist the Effects of Accidental Explosions*, U.S. Army TM 5-1300, NAVFAC-P-397 (Navy), AFM-88-22 (Air Force).
22. Fisher, H. G., The DIERS Users Group, AICHE, National Meeting, New Orleans, La. March 6, 1988
23. Cousins, E. W. and P. E. Cotton, "Design Closed Vessels to Withstand Internal Explosions", Chem. Eng., No.8, 1951, p.133.
24. Cousins, E. W. and P. E. Cotton, "Protection of Closed Vessels Against Internal Explosions", Presented at. annual meeting National Fire Protection Assoc., May 7-11, 1951, Detroit, Mich.
25. Jacobson, M., *et al.*, "Explosibility of Dusts Used in the Plastics Industry", U.S. Bureau of Mines, RI-5971, 1962.
26. Jacobson, M., *et al.*, "Explosibility of Agricultural Dusts", U.S. Bureau of Mines, RI-5753, 1961.
27. Jacobson, M., *et al.*, "Explosibility of Metal Powders", U.S. Bureau of Mines, RI-6516, 1964.
28. Nagy, J., *et al.*, "Explosibility of Carbonaceous Dusts", U.S. Bureau of Mines, RI-6597, 1965.
29. Stull, D. R., Fundamentals of Fire & Explosion, Monograph Series, No. 10, Vol. 73, The Dow Chemical Co., Published Amer. Inst. Chem. Engrs., 1977.
30. Bartknecht, W., *Explosions-Course Prevention Protection*, Translation of 2nd Ed., Springer-Verlag, 1981.
31. Safety and Relief Valves, Cat. FE-316, Farris Engineering, Palisades Park, NJ.
32. Bravo, F. and Beatty, B. D., "Decide Whether to Use Thermal Relief," Chem. Eng. Prog., V. 89, No. 12, 1993, p. 35.
33. Sylvander, N. E. and D. L. Katz, "Design and Construction of Pressure Relieving Systems", Engineering Research Bu. No. 31, Univ. of Michigan Press, Ann Arbor, Mich., 1948.
34. Ludwig, E. E., " Applied Process Design for Chemical and Petrochemical Plants", Volume 1, 3rd, 1995 Gulf Publishing Company, Houston.
35. Conison, J., "Why a Relief Valve", Inst. & Auto., 28, 1955, 988
36. Bigham, J. E., "Spring-Loaded Relief Valves", Chem. Eng., Feb. 10, 1958, p. 133.

37. Weber, C. G., "How to Protect Your Pressure Vessels", Chem. Eng., Oct. 1955.
38. Cassata, J. R., Dasgupta, S. and Gandhi, S. L., "Modeling of Tower Relief Dynamics", Hydro. Proc., Part I, V. 72, No. 10; Part 2, V. 72, No. 11, 1993.
39. Leung, J. C., "Size Safety Relief Valves for Flashing Liquids", Chem. Eng. Prog. V. 88, No.2, 1992, p. 70.
40. Crane Technical Manual No. 410, Crane Co., Chicago, Il, 1986.
41. Crowl, D. A. and Louvar, J. F., Chemical Process Safety: Fundamentals with Applications, Prentice-Hall, 1990.
42. Teledyne Farris Engineering Catalog 187 C., Teledyne Farris Engineering Co., Palisades Park, NJ.
43. Murphy, T. S. Jr., "Rupture Diaphragms", Chem. and Met., No. 11, 1944, p. 108.
44. Lowenstein, J. G., "Calculate Adequate Disk Size", Chem. Eng., Jan. 13, 1958, p. 157.
45. Zook Graphite Rupture Disks, Bull. 6000-2; Zook Enterprises.
46. Rupture Disk Sizing; Catalog 1-1110, Continental Disc Corporation, 1991.
47. Application, Selection and specification, Cat. 7387-1, 1988, Fike Metal Products Co.
48. "Venting Atmospheric and Low-Pressure Storage Tanks", API Standard 2000, 3rd Ed., January 1982, Reaffirmed December 1987, American Petroleum Institute, Washington, D.C.
49. Coker, A. K., Modeling of Chemical Kinetics and Reactor Design," Gulf Publishing Co., Houston, TX 2001.
50. Barton, J. A., and P. F. Nolan, "Incidents in the Chemical Industry Due to Thermal Runaway Reactions, Hazards, X, Process Safety in Fine and Specialty Chemical Plants", I. ChemE. Symp., Ser. No. 115, pp3-18, 1989.
51. Shabica, A. C., "Evaluating the Hazards in Chemical Processing", Chem. Eng. Prog., Vol. 59, No. 9, pp 57–66, 1963.
52. Mumford, C. J., "PSI Chemical Process Safety", Occupational Health and Safety Training Unit, University of Portsmouth Enterprise Ltd., Version 3, 1993.
53. Wells, G., Major Hazards and Their Managements, IChemE, 1997.
54. Barton, J. A., and R. Rogers, Chemical Reaction Hazards, 2nd., ed., IChemE., 1997.
55. Townsend, D. I., and J. Tou, "Thermal Hazard Evaluation by an Accelerating Rate Calorimeter." Thermochimica Acta., Vol. 27, pp 1-30, 1980.
56. Grolmes, M. A., "Pressure Relief Requirement for Organic Peroxides and Other Related Compounds", Int. Symp. on Runaway Reactions, Pressure Relief Design and Effluent Handling, AIChE, Mar 11–13, pp 219–245, 1998.
57. Gustin J. L., "Choice of Runaway Reactions Scenarios for Vent Sizing Based on Pseudo-adiabatic Calorimeter Technique", Int. Symp. on Runaway Reaction, Pressure-Relief Design, and Effluent Handling, AIChE, pp 11–13, 1998
58. Creed, M. J. and H. K. Fauske, "An Easy Inexpensive Approach to the DIERS Procedure", Chem. Eng., Prog., Vol. No. 3, p. 45, 1990.
59. Burelbach, J. P., "Advanced Reactive System Screening Tool (ARSST)", North American Thermal Analysis Society, 28th Annual Conf. Orlando, FL, pp 1-6, Oct. 2000.
60. Coker, A. K., "Size Relief Valves Sensibly Part 2", Chem. Eng., Prog., pp. 94-102, Nov. 1992.
61. Gibson, N., N. Madison, and R. L. Rogers, "Case Studies in the application of DIERS Venting Methods to Fine Chemical Batch and Semibatch reactors, Hazards from Pressure: Exothermic Reactions, Unstable Substances, Pressure Relief and Accidental Discharge", IChemE, Symp. Ser. No. 102, EFCE Event, No. 359, pp. 157–169, 1987.
62. Ritcher, S. H., and F. Turner, "Properly Program the Sizing of Batch Reactor Relief Systems", Chem. Eng. Prog., pp 46–55, 1996.
63. Fisher, H., et al., "Emergency Relief System Design Using DIERS Technology", AIChE's Design Institute for Emergency Relief Systems, AIChE, New York, 1992.
64. Boyle, W. J., "Sizing Relief Area for Polymerization Reactors", Chem. Eng., Prog., Vol. 63, No. 8, p. 61, 1967.
65. Huff, J. E., CEP Loss Prevention Technical Manual, p. 7.,1993.
66. Fisher, H. G., "An Overview of Emergency Relief System Design Practice", Plant/Operations Prog., Vol. 10, No. 1, pp. 1–12, Jan 1991.
67. Leung, J. C., "Simplified Vent Sizing Equations for Emergency Relief Requirements in Reactors and Storage Vessels", AIChE J., Vol. 32, No. 10, pp 1622–1634, 1986.
68. Fauske, H. K., "Emergency Relief System Design for Runaway Chemical Reactions", Plant / Operation Prog., Vol. 3, No. 3, Oct. 1984.
69. Leung, J. C., and H. K. Fauske, Plant / Operation Prog., Vol. 6, No. 2, pp. 77–83, 1987.
70. Fauske, H. K., "Emergency Relief System Design for Runaway Chemical Reaction", Extension of the DIERS Methodology, Chem. Eng., Res. Dev., Vol. 67, pp. 199–202, 1989.
71. Duxbury, H. A. and A. J., Wilday, "The Design of Reactor Relief Systems," Trans. IChemE, Vol. 68, pp. 24–30, Feb, 1990.
72. Noronha, J. A., et al., "Simplified Chemical Equipment Screening For Emergency Venting Safety Reviews and Based on the DIERS Technology", Proceedings of the Int. Symp. on Runaway Reactions, Cambridge, MA, pp. 660–680, Mar 7–9, 1989.
73. Lapple, C. E., "Isothermal and Adiabatic Flow of Compressible Fluids", Trans. AIChE, Vol. 39, pp. 385–432, 1943.

74. Coker, A. K., "Computer program enhances guidelines for gas-liquid separator designs", Oil Gas J., pp. 55–62, May 10, 1993.
75. Grossel, S. S., "Design and Sizing of Knock-out Drums/Catchtanks for Reactor Emergency Relief Systems", Plant/Operations Prog., Vol. 5, No. 3, pp 129–135, 1986.
76. Arun Datta, Process Engineering and Design Using Visual Basic, 2nd ed., CRC Press, Taylor & Francis Group, 2014.
77. Niemeyer, C. E. and G. N. Livingston, "Choose the Right Flare System Design", Chem. Eng., Prog., Vol. 89, No. 12, p. 39, 1993.
78. Straitz, J. F. III, "Make the Flare Protect the Environment", Hydro. Proc. Oct. 1977.
79. Hayek, J. D and E. E. Ludwig, "How to Design Safe Flare Stacks", Petrol/Chem. Engineer, Jun 1960, p. C-31, and p. C-44, July 1960.
80. Tan, Soen H, "Flare System Design Simplified", Hydroc. Proc. Vol. 46, No. 1, p. 172, 1967.
81. Kent, G. R., "Practical Design of Flare Stacks", Hydroc. Proc., Vol. 43, No. 8, p. 121, 1964.
82. Brzustowski, T. A., and E. C. Sommer Jr., "Predicting Radiant Heating from Flames", Proceedings Div., of Refining, API vs. 53, pp. 865–893, 1973.
83. Straitz, J. F., III and R. J. Altube, "Flares: Design and Operations", Pub. National Air Oil Burner Co. Inc.
84. Schneider, D. F., "How to Calculate Purge Gas Volume", Hydroc. Proc., Vol. 72, No. 11, p. 89, Nov. 1993.
85. Mak, H. Y., "New Method Speeds, Pressure Relief Manifold Design", Oil Gas J., p. 166, Nov. 20, 1978.
86. Coker, A. K., "Program Sizes Compressible Flow for Discharge Piping", Oil & Gas J., pp. 63–67, Dec. 11, 1989.
87. C. Flower, and A. Wills, "Seven Deadly Sins", A guide to avoiding some of the most common mistakes in pressure relief system design, The Chemical Engineer, pp. 24–31, June 2017.
88. Coulter, B. M., Compressible Flow Manual, Fluid Research Publishing Company, Melbourne, FL.
89. Kirkpatrick, D. M., "Simpler Sizing of Gas Piping", Hydroc. Proc., Dec. 1969.
90. Daniel A. Crowl, and Scott A. Tipler, Sizing Pressure -Relief Devices, CEP, pp. 68–76, Oct. 2013.
91. V. M. Desai, Process Safety Progress, Vol. 15, No. 3, p. 166, Fall 1996.
92. Trevor Kletz, What Went Wrong? Case Histories of Process Plant Disasters and How They Could Have Been Avoided, Gulf Professional Publishing, Elsevier, 2009.
93. T. Fishwick, Loss Prevention Bulletin, No. 135, p. 18, June 1997.
94. Center for Chemical Process Safety (CCPS), Guidelines for Engineering Design for Process Safety, 2nd ed., 2012.
95. U.S. Chemical Safety and Hazard Investigation Board, Williams Geismar Olefin Plant, Reboiler Rupture and Fire, Geismar, Louisiana, No. 2013-03-I-LA, October 2016 (www.csb.gov).
96. A. B. Smith., Safety Relief Valves on Pressure Systems, Reprint No. C454/001/93, Institution of Mechanical Engineers, London, 1993.
97. Steven A. Treese, Peter R. Pujado, and David S. J. Jones, Handbook of Petroleum Processing, 2nd., ed., Springer International Publishing, Switzerland, 2015.
98. Lees, F. P., Loss Prevention in the Process Industries, Vol. 1, Butterworth-Heinemann, Ltd.

World Wide Web on Two-Phase Relief Systems

www.csb.gov
www.sciencedirect.com
www.chilworth.com
www.fauske.com
www.helgroup.co.uk
www.chemvillage.org
www.chemeng.ed.ac.uk
www.chemsafety.gov
www.kbintl.com
www.rccostello.com
www.che.ufl.edu
www.osha-slc.gov
www.icheme.org
www.harsnet.de/presEng.htm

24

Process Safety and Energy Management in Petroleum Refinery

24.1 Introduction

Process safety practices and process safety management (PSM) systems have been applied since the various incidents and accidents in the refining and petrochemical industries over many years. PSM is widely credited for reductions in major accident, resulting in an improved process safety performance for the process industry. However, many organizations continue to be challenged by inadequate management system performance, resources, and static process safety results. Organizations in the US and the UK and elsewhere in the world have promoted process safety management frameworks for their industries. In the US, the Center for Chemical Process Safety (CCPS) created risk-based process safety (RBPS) as the framework for the next generation of process safety management [1]. Table 24.1 shows the RBPS elements with the corresponding OSHA PSM/EPA RMP elements, and detailed descriptions of these elements are provided elsewhere [2].

Understanding the concept of risk in a risk-based process safety program is essential and is defined as the possibility of loss or injury or someone or something that creates or suggests a hazard. The CCPS definition has three elements as opposed to two. They are: the hazard (what can go wrong), the magnitude (how bad can it be), and the likelihood (how often can it happen). In the petrochemical, refining and chemical process industries, understanding the risk associated with an activity requires answers to the following:

- What can go wrong? (human injury, environmental damage, or economic loss).
- How bad could it be? (magnitude of the loss or injury).
- How often might it happen? (likelihood of the loss or injury).

Resources such as money and personnel are finite, such that when designing and operating a facility, it is essential to select from a wide range of options in deciding how much technical rigor to incorporate into the process safety management activities with the minimum requirements by complying with regulations. A process with low risk does not require the same amount of details in the application of the process safety elements as one with high risk. An analysis of the failure mechanisms associated with over 100 crude distillation units (CDUs)/vacuum distillation units (VDUs) incidents showed that the largest failure mechanism was associated with human error, followed by corrosion, followed by unforeseen process upsets of varying kinds and as shown in Figure 24.1 [3].

Table 24.1 Comparison of RBPS Elements to OSHA PSM Elements.

CCPS RBPS elements	OSHA PSM/EPA RMP elements
Commit to process safety	
1. Process Safety Culture	
2. Compliance with Standards	Process Safety Information
3. Process Safety Competency	
4. Workforce Involvement	Employee Participation
5. Stakeholder Outreach	Stakeholder Outreach (EPA RMP)
Understand hazard and risk	
6. Process Knowledge Management	Process Safety Information
7. Hazard Identification and Risk Analysis	Process Hazard Analysis
Manage risk	
8. Operating Procedures	Operating Procedures
9. Safe Work Practices	Operating Procedures Hot Work Permits
10. Asset Integrity and Reliability	Mechanical Integrity
11. Contractor Management	Contractors
12. Training and Performance Assurance	Training
13. Management of Change	Management of Change
14. Operational Readiness	Pre-startup Safety Review
15. Conduct of Operations	
16. Emergency Management	Emergency Planning and Response
Learn from experience	
17. Incident Investigation	Incident Investigation
18. Measurement and Metrics	
19. Auditing	Compliance Audits
20. Management Review and Continuous Improvement	

24.2 Process Safety

In petroleum refining and petrochemical industries, companies are required by law to have occupational safety programs, with a focus on personal safety, as the focus of these programs is to prevent harm to personnel from workplace accidents such as fall, cuts, strains, and so on. However, process safety focuses on the prevention of fires, explosions, and accidental chemical releases at chemical process facilities or other facilities dealing with hazardous materials such as refineries and oil and gas (onshore and offshore) production installations and so on.

The BP Texas City refinery explosion in 2005, which killed 15 people and injured over 170, resulted in the formation of an independent commission to examine the process safety culture of the company's operations. The commission

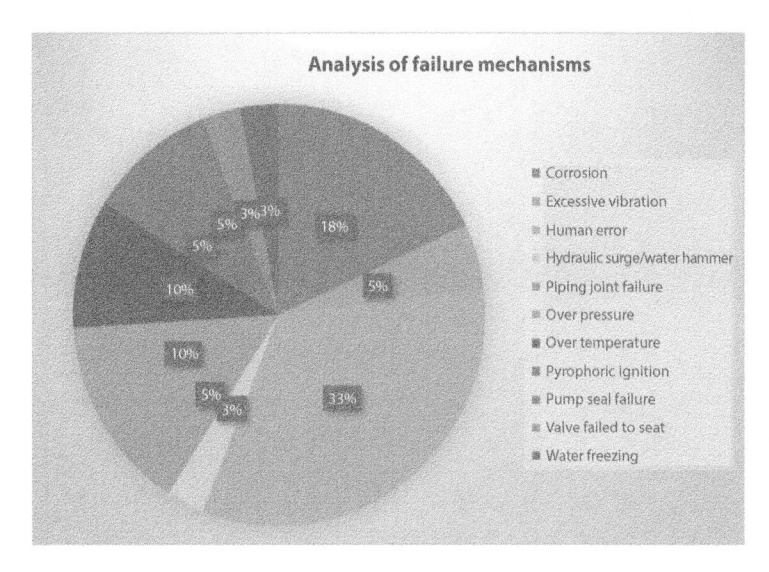

Figure 24.1 Analysis of failure mechanisms [3].

was chaired by James A. Baker and is known as the Baker Panel. The Baker Panel made the following statement on process safety:

"Process safety hazards can give rise to major accidents involving the release of potentially dangerous materials, the release of energy (such as fires and explosions) or both. Process safety incidents can have catastrophic effects and can result in multiple injuries and fatalities, as well as substantial economic property and environmental damage. Process safety refinery incidents can affect workers inside the refinery and members of the public who reside nearby. Process safety in a refinery involves the prevention of leaks, spills, equipment malfunction, over-pressures, excessive temperatures, corrosion, metal fatigue, and other similar conditions. Process safety programs focus on the design and engineering facilities, hazard assessments, management of change (MOC), inspection, testing, and maintenance of equipment, effective alarms, effective process control, personnel training and human factors" [4]. Figure 24.2 shows a photo of the damage to the plant, and Figure 24.3 shows a photo of destroyed trailers west of the blowdown drum (red arrow in the upper left of the figure).

Figure 24.2 Photo of BP Texas City, TX Refinery after the explosion (Source: www.csb.gov).

Figure 24.3 Photo of destroyed trailers west of the blowdown drum (red arrow in upper left of the figure, www.csb.gov).

The quote from the report states that process safety is not limited to the operation of a facility. During the basic research and process research phases, process safety programs cover the operation of pilot facilities. They also cover the selection of the chemistry and unit operations chosen to achieve the design intent of the process. During the design and engineering phase, process safety is involved in choices about the type of unit operations and equipment items to use, the facility layout, etc. Running a facility involves hazard assessments, MOC, inspection, testing, and maintenance of equipment, effective alarms, effective process control, procedures, and training of personnel.

In 1992, the US Occupational Safety and Health Administration (OSHA) issued the Process Safety Management of Highly Hazardous Chemicals (OSHA PSM) regulation, which had its own, although similar, set of process safety management elements for employers and employees in maintaining safety standards. The standard mainly applies to manufacturing industries and those relating to chemicals, transportation equipment, and fabricated metal products. Other sectors include natural gas liquids; electric, gas, and so on. The Environmental Protection Agency (EPA) issued its own version in 1995 under the authority of the Clean-Air Act. This regulation is commonly referred to as RMP, or risk management plan, since the regulation requires the development and submittal of a risk plan based on the regulatory definitions and requirements.

Process safety management system is an analytical tool focused on preventing releases of any substance defined as a highly hazardous chemical including using, storing, manufacturing, handling, or moving such chemicals at the site, or any combination of these activities. PSM clarifies the responsibilities of employers and contractors involved in work that affects or takes place near covered processes to ensure that the safety of both plant and contractor employees is considered. The standard also stipulates written operating procedures; employee training; prestartup safety reviews; evaluation of mechanical integrity of critical equipment; and written procedures of managing change. PSM specifies a permit system for hot work; investigation of incidents involving releases or near misses of covered chemicals; emergency, action plans; compliance audits at least every 3 years; and trade secret protection. PSM is divided into 14 elements as shown in Figure 24.4. These are:

1. Employee involvement
2. Process safety information (PSI)
3. Process hazard analysis (PHA)
4. Operating procedures
5. Training
6. Contractors
7. Pre-startup safety review

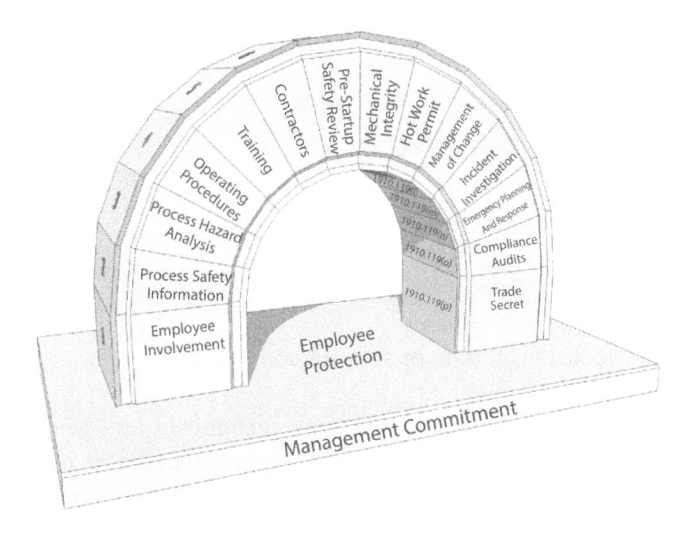

Figure 24.4 Fourteen elements of OSHA's process safety management program (Source: https://en.wikipedia.org/wiki/Process_safety_management).

8. Mechanical integrity
9. Hot work permit
10. Management of change (MOC)
11. Incident investigation
12. Emergency, planning, and response
13. Compliance audits
14. Trade secret

The key provision of PSM is process hazard analysis (PHA), which is a careful review of what could go wrong and what safeguards must be implemented to prevent releases of hazardous chemicals. Employers must identify those processes that pose the greatest risks and evaluate those first.

24.2.1 Process Safety Information

Process safety information (PSI) must include information on the hazards of the highly hazardous chemicals used or produced by the process and information on the technology of the process and on the equipment in the process. Information on the hazards of the highly hazardous chemicals in the process consists of the following:

- Toxicity
- Permissible exposure limits
- Physical data
- Reactivity data
- Corrosivity data
- Thermal and chemical stability data, and hazardous effects of inadvertent mixing of different materials

Information on the technology of the process includes the following:

- A block flow diagram or simplified process flow diagram
- Process chemistry
- Maximum intended inventory
- Safe upper and lower limits for such items as temperatures, pressure, flows, and compositions and
- An evaluation of the consequences of deviations, including those affecting the safety and health of employees

Information on the equipment in the process includes

- Materials of construction
- Piping and instrumentation diagrams (P & IDs)
- Electrical classification
- Relief system design and design basis
- Ventilation system design
- Design codes and standards employed
- Materials and energy balances for processes
- Safety systems (e.g., interlocks, detection, or suppression systems)

The employer shall document that equipment complies with recognized and generally accepted good engineering practices (RARAGEP). For existing equipment designed and constructed in accordance with codes, standards, or practices that are no longer in general use, the employer shall determine and document that the equipment is designed, maintained, inspected, tested, and operated in a safe manner.

The compilation of PSI provides the basis for identifying and understanding the hazards of a process and is necessary in developing the process hazards analysis (PHA) and may be necessary for complying with other provisions of PSM such as management of change (MOC) and incident investigation.

Analysis of these methods is provided in the companion publication OSHA 3133, Process Safety Management Guidelines for Compliance. The process hazard analysis in these methods must ensure the following:

- The hazards of the process.
- The identification of any process incident that had a potential for catastrophic consequences in the work place.
- Engineering and administrative controls applicable to the hazards and their interrelationships, such as appropriate application of detection methodologies to provide early warning of releases. Acceptable detection methods might include process monitoring and control instrumentation with alarms, and detection hardware such as hydrocarbon sensors.
- Consequences of failure of engineering and administrative controls.
- Facility siting.
- Human factors.
- A qualitative evaluation of a range of the possible safety and health effects on employees in the work place if there is a failure of controls.

24.2.2 Conduct of Operations (COO) and Operational Discipline (OD)

Process safety practices and management systems have been employed in the refining and petrochemical industries, and organizations have developed and implemented conduct of operations (COO), and operational discipline (OD) systems. COO involves the ongoing management systems that are developed to encourage performance of all tasks in a consistent, and appropriate manner; OD is the structured execution of the COO and other organizational management systems by personnel throughout the organization. Figure 24.5 shows a process safety pyramid on a triangle, where the minor, serious and catastrophic injuries normally found progressing up to the top of a personal safety triangle have been replaced with appropriate process safety issues. Eliminating these issues at the base of the triangle should result in a reduction in process safety incidents. COO does not focus on basic operations and maintenance elements, such as procedures, training, safe work practices, asset integrity, management of change (MOC), and pre-startup safety review. Rather, it is a management system to aid ensure the effectiveness of these and other PSM systems. COO/OD activities are typically focused on the bottom portion of the triangle with the goal of reducing the number of issues that occur at higher levels of the triangle. Table 24.2 illustrates the indicators of effective COO/OD systems.

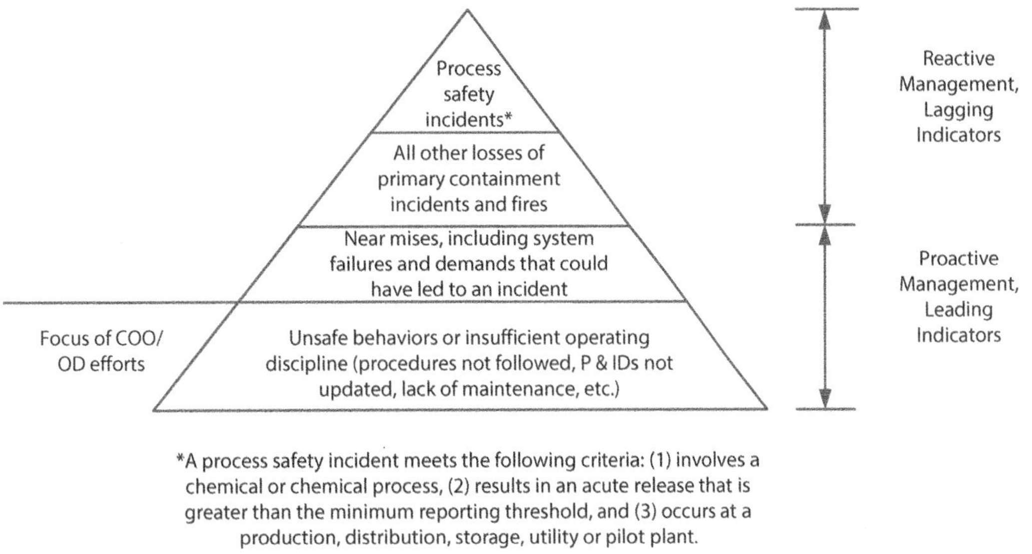
Process safety incidents*
All other losses of primary containment incidents and fires
Near mises, including system failures and demands that could have led to an incident
Unsafe behaviors or insufficient operating discipline (procedures not followed, P & IDs not updated, lack of maintenance, etc.)
Focus of COO/ OD efforts
Reactive Management, Lagging Indicators
Proactive Management, Leading Indicators
*A process safety incident meets the following criteria: (1) involves a chemical or chemical process, (2) results in an acute release that is greater than the minimum reporting threshold, and (3) occurs at a production, distribution, storage, utility or pilot plant.

Figure 24.5 Typical process safety pyramid [39].

The key attributes of COO systems are [41]:

People	**Process**
Clear Authority/Accountability	Process capability
Communications	Safe operating limits
Logs and records	Limiting conditions for operation
Training, Skill maintenance, and Individual competence	**Plant**
Compliance with policies and procedures	Asset ownership/Control of equipment
Safe and productive work environments	Equipment monitoring
Aids to operation – the visible plant	Condition verification
Intolerance of deviations	Management of subtle changes
Task verification	Control of maintenance work
Supervision/Support	Maintaining the capability of safety systems
Assigning qualified workers	Controlling intentional bypasses and impairments.
Access control	
Routines	
Worker fatigue/fitness for duty	

And the key attributes of OD are [41]:

Organizational	Individual
Leadership	Knowledge
Team building and employee involvement	Commitment
Compliance with procedures and standards	Awareness
Housekeeping	Attention to detail

Table 24.2 Indicators of Effective COO/OD System.

Equipment is properly designed and constructed	• Operational, maintenance, safety and environmental considerations are all addressed in the initial design of equipment.
	• Proactive risk analysis results and industry standards are used as inputs to the design process.
	• End users of the equipment (generally operations and maintenance personnel) are involved in the design process.
	• The design process occurs in a controlled manner.
	• The construction occurs in a controlled manner.
Equipment is properly operated	• The proper method for operating equipment has been developed through proactive analysis of the risks and documented in written procedures. Operators are involved in the development of the procedures.
	• Personnel have been trained in normal and abnormal operations, as well as the basis for the procedures and operating limits.
	• Equipment is configured and operated in accordance with procedures.
	• Equipment is returned to service using a controlled process.
	• Changes to operational requirements are appropriately assessed.
Equipment is properly maintained	• Equipment is maintained in accordance with predetermined maintenance strategies developed through a structured assessment process.
	• Personnel are trained to troubleshoot, repair and maintain equipment.
	• Changes to operational conditions are assessed to determine their impact on maintenance requirements.
	• Equipment status is controlled through safe work practices.
	• Equipment failures are analyzed to prevent similar failures.

(Continued)

Table 24.2 Indicators of Effective COO/OD System. (*Continued*)

Management systems are properly executed	▪ Management systems are developed based on the results of proactive analyses and industry best practices.
	▪ Management systems are clearly documented.
	▪ Management systems are executed as written.
	▪ Organizational changes are assessed to determine impacts on existing management systems.
Errors and deviations are consistently addressed	▪ The personnel in the system are always seeking to improve their performance. As a result, there is extensive use of self-checking, peer – checking, audits, incident investigations, management reviews, and metrics to identify and eliminate deviations.
	▪ Personnel are actively seeking discrepancies and resolving issues when identified.
	▪ Personnel take ownership of issues and seek to solve the problem themselves. They involve outside resources to assist them in solving the problems, but retain ownership of the issue.
	▪ Personnel embrace feedback from personnel outside their group as opportunities to improve their systems and processes.

Process Safety Culture: BP Refinery Explosion, Texas City, 2005

On March 23, 2005, an explosion and fires occurred within the Isomerization unit (ISOM) of BP's Texas City Refinery during a startup after a turnaround. Fifteen contractors were killed and over 170 suffered severe injuries. The ISOM unit suffered major damage and adjacent plant and equipment.

The refinery produced about 10 million gallons of gasoline per day (~2.5% of the gasoline sold in the US) for markets in the Southeast, Midwest, and along the East Coast. It also produced jet fuels, diesel fuels, and chemical feed stocks; 29 oil refining units and 4 chemical units cover its 1,200-acre site. The refinery employed approximately 1,800 BP workers and about 800 contractor workers were onsite supporting the turnaround work. The site also had numerous changes in management at both the refinery and corporate levels.

The portable plant buildings where the contractors were located were being used to support an adjacent plant turnaround. They were in an area operated as an uncontrolled area, i.e., a safe area without any Hot Work Permit/Electrical controls imposed.

Detailed Description

The incident occurred while a section of the refinery's ISOM unit was being restarted after a maintenance turnaround that lasted 1 month. The ISOM unit, installed at the refinery to provide higher octane components for unleaded gasoline, consists of four sections: an Ultrafiner desulfurizer, a Penex reactor, a vapor recovery/liquid recycle unit, and a raffinate splitter. Isomerization is a refining process that alters the fundamental arrangement of atoms in the molecule without adding or removing anything from the original material. At the BP Texas City refinery, the ISOM unit converted straight-chain normal pentane and hexane into higher octane-branched isopentane and isohexane for gasoline blending and chemical feedstocks.

During the startup, operations personnel pumped flammable liquid hydrocarbons into a distillation tower for over 3 hours without any liquid being removed, which was contrary to startup procedure instructions. Critical alarms and

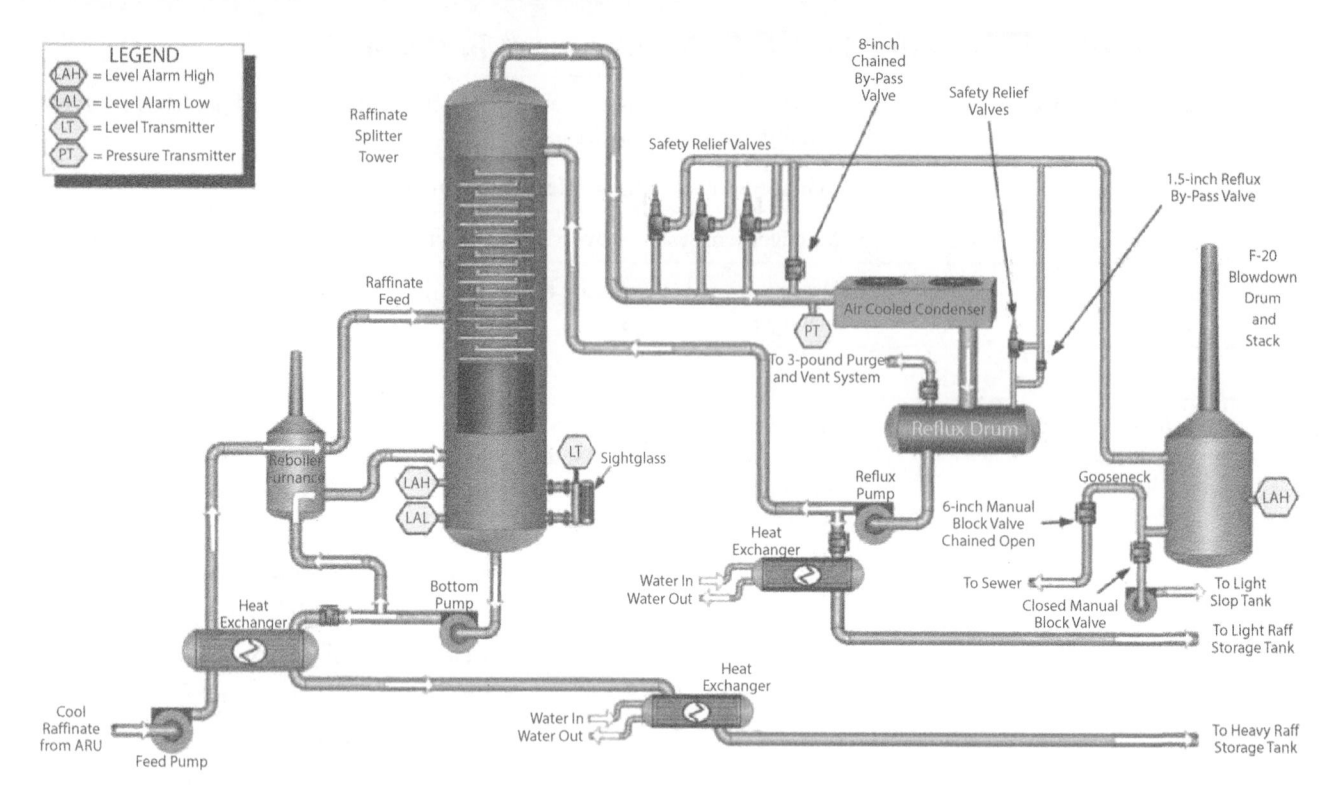

Figure 24.6 Process flow diagram of the Raffinate Column and blowdown drum (Source: www.csb.gov).

control instrumentation provided false indications that failed to alert the operators of the high level in the tower. Consequently, unknown to the operators, the 170-ft (52 m) tall tower was overfilled and liquid overflowed into the overhead pipe at the top of the tower (Figure 24.6).

The overhead pipe ran down the side of the tower to pressure relief valves located 148 ft (45 m) below. As the pipe filled with liquid, the pressure at the bottom rose rapidly from about 21 to 64 psi. The three pressure relief valves opened for 6 minutes, discharging a large quantity of flammable liquid to a blowdown drum with a vent stack open to the atmosphere. The blowdown drum and stack overfilled with flammable liquid, which led to a geyser-like release out the 113 ft (34 m) tall stack. This blowdown system was obsolete and an unsafe design; it was originally installed in the 1950s and had never been connected to a flare system to safely contain liquids and combust flammable vapors released from the process.

The release volatile liquid evaporated as it fell to the ground and formed a flammable vapor cloud. The most likely source of ignition for the vapor cloud was backfire from an idling diesel pickup truck located about 25 ft (7.6 m) from the blowdown drum. The 15 employees killed in the explosion were contractors working in and around temporary trailers that had been previously sited by BP as close as 121 ft (37 m) from the blowdown drum. Figures 24.7–24.9 show the damage to the refinery, blowdown and the trailers, respectively.

Causes

The BP investigation concluded that while many departures to the startup procedure occurred, the key step that was instrumental in leading to the incident was the failure to establish heavy raffinate rundown to tankage, while continuing to feed and heat the tower. By the time the heavy raffinate flow was eventually started, the Splitter bottoms temperature was so high, and the liquid level in the tower rose that this intervention made matters worse by introducing significant additional heat to the feed. The investigation team concluded that the splitter was overfilled and overheated because the Shift Board Operator did not adequately understand the process or the potential consequences of his actions or inactions on the day of the incident.

PROCESS SAFETY AND ENERGY MANAGEMENT IN PETROLEUM REFINERY 259

Figure 24.7 Plant and explosion of BP Texas City refinery (Source: US Chemical Safety board, www.csb.gov).

Figure 24.8 Photo of the blowdown unit of BP Texas City Refinery (Source: US Chemical Safety board, www.csb.gov).

Figure 24.9 Aerial view of the damaged trailers, BP refinery (Source: US Chemical Safety board, www.csb.gov).

Key Lessons

Many Risk Based Process Safety Elements were involved in the BP Texas City explosion and some are listed as follows:

Process Safety Culture

Process safety culture as reported by CSB is the creation of an independent expert to examine BP's corporate safety management systems, safety culture, and oversight of the North American refineries. This became known as the Baker Panel as described earlier. The Panel report focused on safety management systems at BP and resulted in 10 recommendations to the BP Board of Directors.

Selected CSB Findings

- Cost-cutting, failure to invest, and production pressures from BP Group executive managers impaired process safety performance at Texas City.
- The BP Board of Directors did not provide effective oversight of BP's safety culture and major incident prevention programs. The Board did not have a member responsible for assessing and verifying the performance of BP's major incident hazard prevention programs.
- Reliance on the low personal injury rate at Texas City as a safety indicator failed to provide a true picture of process safety performance and the health of the safety culture.
- A "check the box" mentality was prevalent at Texas City, where personnel completed paperwork and checked off on safety policy and procedural requirements even when those requirements had not been met.

Selected Baker Panel Finding

BP has not instilled a common, unifying process safety culture among its US refineries. Each refinery has its own separate and distinct process safety culture. While some refineries are far more effective than others in promoting process safety, significant process safety culture issues exist at all five US refineries, not just Texas City. Although the five refineries do not share a unified process safety culture, each exhibits some similar weaknesses. The Panel found instances of a lack of operating discipline, toleration of serious deviations from safe operating practices, and apparent complacency toward serious process safety risks at each refinery.

Process Knowledge Management

BP acquired the Texas City refinery as part of its merger with Amoco in 1999. Neither Amoco nor BP replaced blowdown drums and atmospheric stacks, even though a series of incidents warned that this equipment was unsafe. In 1992, OSHA cited a similar blowdown drum and stack as unsafe, but the citation was withdrawn as part of a settlement agreement and therefore the drum was not connected to a flare as recommended. Amoco, and later BP, had safety standards requiring that blowdown stacks be replaced with equipment such as a flare when major modifications were made. In 1997, a major modification replaced the ISOM blowdown drum and stack with similar equipment but Amoco did not connect it to a flare. In 2002, BP engineers proposed connecting the ISOM blowdown system to a flare, but a less expensive option was chosen.

Training and Performance Assurance

- The operator training program was inadequate. The central training department staff had been reduced from 28 to 8, and simulators were unavailable for operators to practice handling abnormal situations, including infrequent and high-hazard operations such as startups and unit upsets.
- Supervisors and operators poorly communicated critical information regarding the startup during the shift turnover. BP did not have a shift turnover communications requirement for its operations staff. ISOM operators were likely fatigued from working 12-hour shifts for 29 or more consecutive days.
- A lack of supervisory oversight and technically trained personnel during the startup, and especially hazardous period, was an omission contrary to BP safety guidelines. An extra board operator was not assigned to assist despite a staffing assessment that recommended an additional board operator for all ISOM startups.

Management of Change (MOC)

- BP Texas City did not effectively assess changes involving people, policies, or the organization that could impact process safety. For example, the control room staff was reduced from two people to one, who was overseeing three units.
- Local site MOC rules required that where a portable building was to be placed within 100 m (350 ft) of a process unit, a Facility Siting Analysis had to be carried out. However, this location had already been used many times for these trailers. Not doing an effective MOC put all the people in the portable building at unnecessary risk.

Asset Integrity and Reliability

- The process unit was started despite previously reported malfunction of the tower level indicator, level sight glass, and a pressure control valve.
- Deficiencies in BP's mechanical integrity program resulted in the run to failure process equipment at Texas City.

The COO/OD related issues associated with this incident are:

- An operational check of the independent high-level alarm in the raffinate splitter tower was not performed prior to startup, even though it was required by procedures.
- The operators did not respond to the high-level alarm in the splitter (it was on throughout the incident).
- The level indication available to the operators was useless during most of the startup because they deliberately maintained the level above the indicated range of the level instruments.
- When the day shift supervisor arrived at about 7:15 am, no job safety review or walkthrough of the procedures to be used that day was performed as required by procedures.
- The board operator printed off the wrong startup procedure (although this was not a significant factor because he never referred to it).
- The splitter bottoms were heated at 75°F per hour despite the procedural limit of 50°F per hour.
- The Day Shift Supervisor left the plant during the startup about 3 ½ hours prior to the explosion. No replacement was provided during this period.
- The operational procedures were certified as current, although they did not include changes to relief valve settings made prior to the most recent recertification.
- Outside operators did not report significant deviations of operating parameters (such as rising pressure on the splitter bottoms pumps) to the control room.
- Deficiencies first identified in 2003 and 2004 still existed in training programs for ISOM operators.

24.2.3 Process Hazard Analysis

A Process hazard analysis (PHA) is a rigorous, orderly, and systematic approach for identifying, evaluating, and controlling the hazards of processes involving highly hazardous chemicals. Consequences addressed can include employee safety, environmental impact, public safety, extent of equipment/facility damage, and/or effects on public image. Causes of such situations are identified, and the scenarios are ranked on severity as well as frequency of occurring. The employer must perform an initial process hazard analysis (hazard evaluation) on all processes covered by this standard. The PHA methodology selected must be appropriate to the complexity of the process and must identify, evaluate, and control the hazards involved in the process.

Safeguards currently in place are accounted for and when risk is unmitigated/deemed unacceptable, recommendations for follow up actions are provided. A management review is conducted to determine what changes will be made. The goal of performing a PHA is to detect unprotected situations and trigger the necessary safety improvements to reduce or minimize risk.

Risk Analysis (RA) is to identify hazards within a process. Some industries have regulatory requirements to perform risk assessments. The results of a risk analysis can be used to justify process improvements. Additionally, identifying and mitigating hazardous scenarios through risk analysis is considered a Recognized and Generally Accepted Good Engineering Practice (RAGAGEP), and ultimately makes good practice. Table 24.3 shows select Risk Analysis Types.

A risk analysis for hazardous process should be conducted for each stage of design, operation, and shutdown. PHAs are required to be completed initially and revalidated every 5 years—however, major changes made to a process warrant a total redo to be conducted. Furthermore, when either temporary or permanent modifications are made to a hazardous process, a Management of Change (MOC), PHA should be utilized to evaluate the inherent hazards that can result from the change. Table 24.4 shows the applicability of several PHA techniques during some of the different phases of a process life cycle. The employer must use one or more of the following methods as appropriate to determine and evaluate the hazards of the process being analyzed:

- What-if
- Checklist
- What-if/checklist

Table 24.3 Select Risk Analysis Types and Commonly Used Industry.

Types of risk analysis	Commonly used industry
Probabilistic Risk Analysis (PRA)	Nuclear, Power, Airline Transportation, Urban Planning.
Process Hazard Analysis (PHA)	Required for OSHA PSM [29 CFR 1910.119] or EPA RMP [40 CFR Part 68] covered chemicals; applicable to all hazardous processes.
Dust Hazard Analysis (DHA)	Required for facilities handling combustible particulate solids.

Table 24.4 Applicability of Select PHA Techniques.

Phase of process design or operation	Technique					
	Checklist	What - if	What – if/ checklist	Hazard and Operability Study (HAZOP)	Failure Modes and Effects Analysis (FMEA)	Fault Tree Analysis (FTA)
R & D		√	√			
Design	√	√	√			
Pilot Plant Operation	√	√	√	√	√	√
Detailed Engineering	√	√	√	√	√	√
Construction/ Startup	√	√	√			
Routine Operation	√	√	√	√	√	√
Modification	√	√	√	√	√	√
Incident Investigation		√		√	√	√
Decommissioning	√	√	√			

- Hazard and operability study (HAZOP)
- Failure mode and effects analysis (FMEA)
- Fault tree analysis
- An appropriate equivalent methodology

Table 24.5 shows the comparison of RBPS elements to OSHA PSM elements.

Safe Operating Limits

A way of deciding whether a change requires the use of MOC program is to determine if the proposed change takes the process conditions outside its safe operating limits. This approach generates the following definition:

A change that requires evaluation by the Management of Change program occurs when a critical variable will be taken outside its predefined safe limits.

Table 24.5 Comparison of RBPS Elements to OSHA PSM Elements.

CCPS RBPS Elements	OSHA PSM/EPA RMP Elements
Commit to Process Safety	
1. Process Safety Culture	
2. Compliance with Standards	Process Safety Information
3. Process Safety Competency	
4. Workforce Involvement	Employee Participation
5. Stakeholder Outreach	Stakeholder Outreach (EPA RMP)
Understand Hazard and Risk	
6. Process knowledge Management	Process Safety Information
7. Hazard Identification and Risk Analysis	Process Hazard Analysis
Manage Risk	
8. Operating Procedures	Operating Procedures
9. Safe Work Practices	Operating Procedures Hot Work Permits
10. Asset Integrity and Reliability	Mechanical Integrity
11. Contractor Management	Contractors
12. Training and Performance Assurance	Training
13. Management of Change	Management of Change
14. Operational Readiness	Pre-startup Safety Review
15. Conduct of Operations	
16. Emergency Management	Emergency Planning and Response
Learn from Experience	
17. Incident Investigation	Incident Investigation
18. Measurement and Metrics	
19. Auditing	Compliance Audits
20. Management Review and Continuous Improvement	

Impact on Other Process Safety Elements

The word "change" can be defined in the context of MOC is to determine if the proposed change affects any of the other elements of process safety. For example, if a new operating procedure requires to be written, then the MOC must be followed. Furthermore, if the proposed change requires alteration/update in the P & IDs, then the change must be applied in the MOC.

The proper MOC lies at the heart of any successful risk management or process safety management program. Personnel associated with the design and operation of any industrial facility want to perform well at the job; however, in spite of their best intentions, accidents continue to occur with fatality, or severe injuries, loss of production, and environmental pollution. All of these undesired events are caused by the uncontrolled changes, as someone, somewhere altered the operating conditions outside their safe range without taking the proper precautions (i.e., without implementing the MOC process).

Setting up an MOC system with its accompanying procedures, forms and software is inadequate. The personnel who use the system must understand its intent and the manner in which it is being employed. A facility's MOC program may be sound, but if the personnel involved do not understand its fundamental purpose, then the program will be ineffective. MOC system must not only be a program, but a way of life for all personnel and contractors engaged in the facility.

24.3 General Process Safety Hazards in a Refinery

Refineries include combinations of various unit operations and process units that convert crude oil into light end products, LPG, fuel (gasoline, diesel, kerosene, jet fuel), and heavier products (lube oils, asphalt, coke). The basic operations in a refinery involve separation, breakdown of large molecules (hydrotreaters and hydrocrackers), rearranging molecules (isomerization), and combining molecules (reforming, alkylation) to turn crude oil into components such as propane, gasoline, kerosene, diesel fuel, and so on (Figure 24.10).

Refineries handle large quantities of flammable gases and liquids. Any event that causes loss of containment has the potential to lead to an explosion and fire. Refineries can have flammable gas detectors with water deluge systems to limit the impact of some flammable releases. After the explosion in Texas City, as described earlier, increased attention was placed on the placement and protection of buildings within refineries.

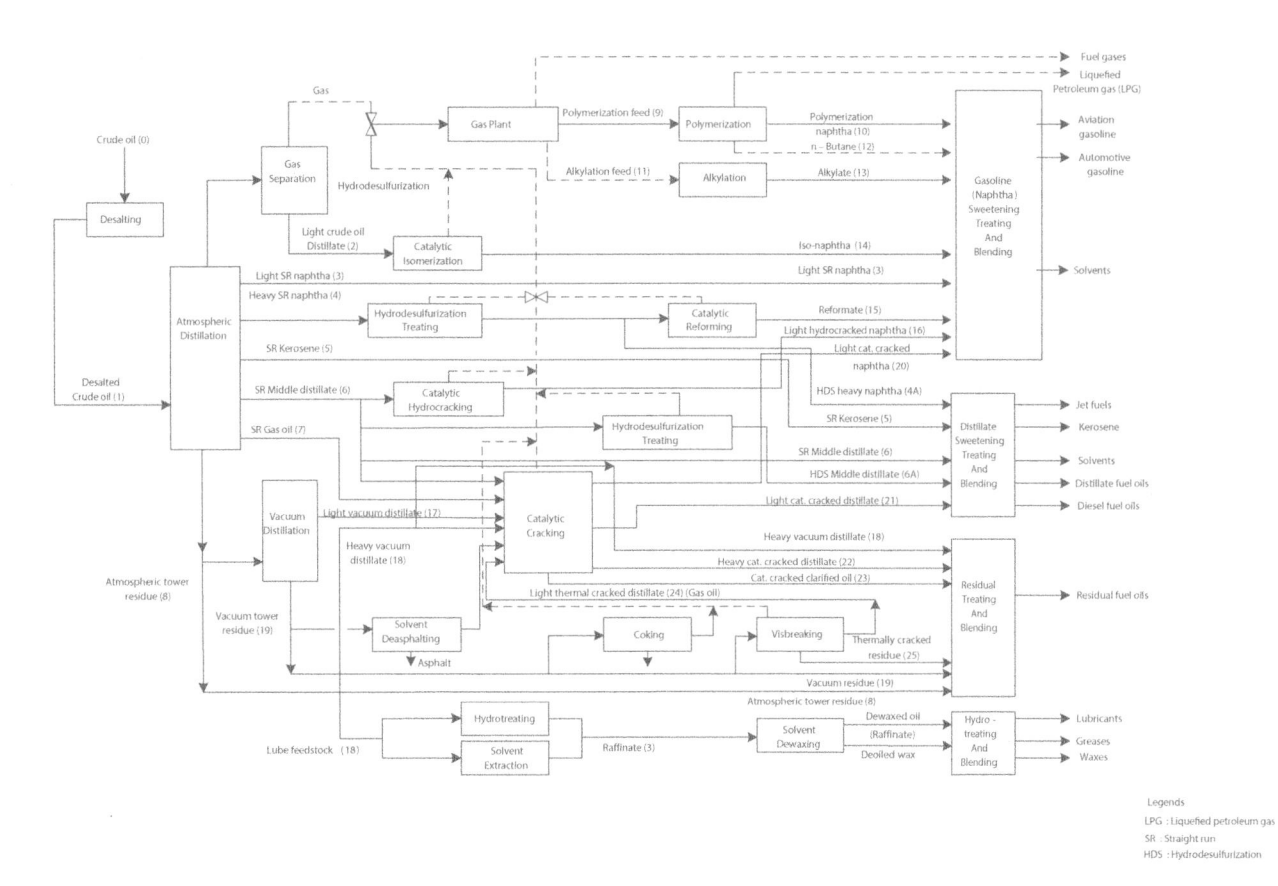

Figure 24.10 Overview of petroleum refinery flow diagram.

Many refineries handle feedstocks that contain various forms of sulfur and produce high levels of hydrogen sulfide (H_2S) as a byproduct in processing sections. H_2S is a highly toxic, dense gas and causes death at concentrations as low as ~400 rpm. Although known for a rotten egg-like odor, at low concentrations, at ~100 ppm people become desensitized to its color, so odor cannot be relied upon to provide adequate warning of exposure, or increasing concentration. In fact, 100 rpm is the Immediately Dangerous to Life and Health (IDLH) level of H_2S. This means that starting at 100 rpm, H_2S can pose an immediate threat to life causing irreversible adverse health effects, or impairing an individual's ability to escape from a dangerous atmosphere. This is because it can cause eye irritation and difficulty breathing at 100 ppm. Also, unlike some other highly toxic materials, such as chlorine (Cl_2), H_2S is an invisible gas. Being a dense gas, it can accumulate in poorly ventilated areas. Therefore, loss of containment of H_2S is a highly hazardous event. Refineries will typically have area gas detectors for H_2S and personnel H_2S monitors that are worn when in the plant. Hydrotreaters remove sulfur and H_2S [2].

A refinery has many process equipment items such as heat exchangers, distillation columns, furnaces, and storage tanks, all containing flammable materials and many containing H_2S, a toxic gas. Operating problems on these equipment items can cause loss of containment, leading to fires and explosion. Furthermore, corrosion of piping, and equipment, due to the impurities in crude, and use of hydrogen in several operations, is a common problem in refineries. Corrosion can be a cause of loss of containment events in any unit in a refinery. Asset Integrity and Reliability are a key PSM element for refineries.

The Material Data Sheet (MSDS) of H_2S provides the following:

- Hazard statements such as extremely flammable gas.
- If venting or leaking gas catches fire, do not extinguish flames. Flammable vapors may spread from leak, creating an explosive re-ignition hazard.
- Contains gas under pressure, and may explode if heated.
- Fatal if inhaled.
- May cause respiratory irritation.
- May form explosive mixture with air and oxidizing agents.

Precautionary statements are:

- Before entering an area, especially a confined area, check the atmosphere with an appropriate device (H_2S sensor).
- Wear protective gloves, protective clothing, eye protection, respiratory protection, and/or face protection.
- Avoid release to the environment.
- Eliminate all ignition sources if safe to do so.
- Use a back flow preventive device in the piping.
- Do not open valve until connected to equipment prepared for use.
- Do not depend on odor to detect the presence of gas.

Desalters

Critical Operating Parameters Impacting Process Safety

Critical operating parameters are included in refining design manuals and individual process unit operating instructions. Some of the critical parameters that plant supervisors and operators require to be aware of are as follows:

Operating personnel should regularly monitor salt, solids, and the water content of the incoming crude mix and desalted crude oil. Other parameters that should be monitored frequently are

- Crude oil salt content.
- Solids and BS & W content.
- Water quality and injection rates.
- Mix valve pressure drop.
- Desalting chemical injection type and flowrates.

- Grid voltage and amperage.
- Desalter temperature and pressure.
- Brine pH and the level of the brine/crude oil interface.
- The quantity of slop oil where this is injected into crude.

The Quality of Aqueous Effluent from Desalters

This must be regularly monitored as sight glasses enable operators to monitor the color of the effluent that are vulnerable to failure if not correctly specified and fitted. It is essential that the anti-corrosion films made of transparent mica are correctly installed on the inside surfaces of these sight glasses.

Desalter Water Supply

Part of the water-supply arrangements to desalters are normally fed from an atmospheric cone roofed tank located on the unit. There have been a number of incidents where the water injection pumps have failed and back flow prevention, non-return valves, and/or low flow trips and have not operated. The consequences are that crude oil may back flow into the desalter water tank and release light ends from the tank vent. There is also a risk of overflow from the tank.

Vibration within Relief Valve (RV) Pipework

Damaging vibration can be set up in relief valve piping where the pressure drop across the inlet piping between the main process inventory and the RV inlet nozzle is greater that 3% of set pressure. The phenomenon is referred to as "machine gunning." Desalters may be prone to this as it is usual to locate the desalter RVs at the CDU main column. This allows the RVs to discharge directly into the column flash zone and avoid the need for large bore piping runs that would be required to carry two-phase flow to the main column if the RVs were to be fitted directly onto the desalter vessel.

Example of Process Safety Incidents and Hazards

A fire occurred in a crude distillation unit (CDU) during start-up after a 4-week maintenance turnaround. Due to excessive leakage from two heat exchangers, it was decided to stop the start-up to rectify these leaks. When the feed rate was reduced, the pressure in the desalter increased inadvertently and the desalter RV opened and relieved into the crude column. There were severe vibrations in the RV and related piping, and a flange leak started at the RV by-pass (see arrow in Figure 24.11). This ignited within seconds, and large flames erupted next to the crude column. The fire was extinguished after about 2 hours. The unit was shut down for a further 4 weeks to complete the necessary repairs.

Hydrotreating [2]

The main purpose of hydrotreating as shown in Figure 24.12 is to remove impurities such as sulfur, nitrogen, oxygen, and metals. Figure 24.12 shows the feed mixed with hydrogen, preheated to 600–800°C (1100–1472°F) and charged at high pressures (up to 69 bar (1,000 psi)) to a catalytic reactor to form hydrogen sulfide (H_2S), ammonia (NH_3), and metal chlorides. The petroleum portion of the feed, containing olefins (C_nH_{2n}) and aromatics (C_nH_{2n-6}), reacts with hydrogen to form saturated compounds. The product stream is depressurized and cooled. Excess hydrogen is recycled, and the rest of the stream is sent to a column to remove butanes from the naphtha product. Figure 24.12 is a process flow diagram of a hydrotreating unit and Figure 24.10 illustrates several hydrotreaters units in a refinery.

24.4 Example of Process Safety Incidents and Hazards

An explosion occurred in the hydrotreater section of the Tesoro Anacortes Refinery in 2010. In this incident, a heat exchanger ruptured, releasing hydrogen and naphtha at 500°C (930°F), which ignited and caused a fire that killed seven people. Figure 24.13 shows an aerial view of the damaged heat exchangers at Tesoro Refinery in Washington, DC.

Figure 24.11 A bypass valve of a CDU unit.

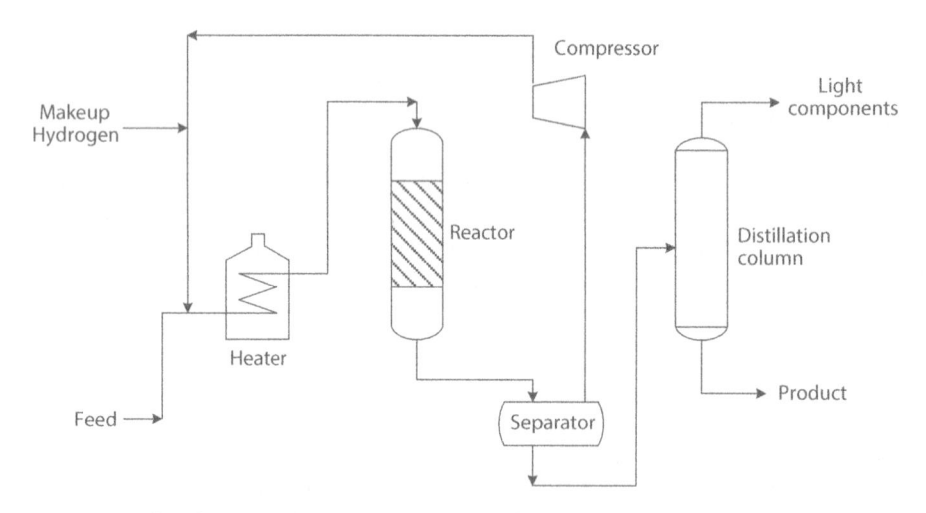

Figure 24.12 Hydrotreater process flow diagram.

The rupture was due to a phenomenon called High-Temperature Hydrogen Attack (HTHA). In HTHA, hydrogen diffuses through the steel walls of equipment at high temperatures and reacts with carbon in the steel, producing methane. This reduces the carbon in the steel, causing pressure inside it. The methane causes fissures to form on the steel, weakening it. The heat exchangers at the Tesoro refinery were carbon steel, which is susceptible to HTHA. HTHA is difficult to identify in its early stages, as the fissures are very small. By the time it can be detected, the equipment already has a higher likelihood of failure. High chromium steel is more resistant to HTHA and is therefore a safer material of construction. Figure 24.14 shows the microscopic cracks in the metal, and Figure 24.15 shows a close-up of the ruptured heat exchanger.

The American Petroleum Institute (API) has a recommended practice regarding HTHA; API RP 941, Steels for Hydrogen Service at Elevated Temperatures, and Pressures in Petroleum Refineries and Petrochemical Plants, 7th edition, 2008. API 941 provides a curve (called Nelson curve) that shows the temperatures and pressures at which HTHA can occur for various metals. The CSB investigation found that the Nelson curve was inaccurate, and API issued an alert, to that effect, in 2011.

Another process hazard is the potential for reverse flow of high-pressure hydrogen from the hydrotreater to the upstream process if forward liquid feed flow is lost (e.g., feed pumps trip off). Check valves and/or chopper valves, which are meant to prevent reverse flow, are used to reduce the risk of this occurring.

Process Safety and Energy Management in Petroleum Refinery 269

Figure 24.13 Aerial view of the damaged heat exchangers at Tesoro Refinery in Washington, DC (Source: www.csb.gov).

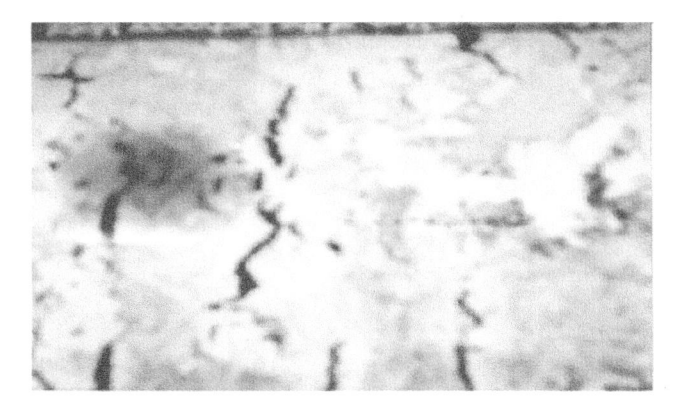

Figure 24.14 Microscopic cracks in the metal (Source: www.csb.gov).

Figure 24.15 Close-up of ruptured heat exchanger (Source: www.csb.gov).

The hydrotreating reaction is exothermic and controlled by maintaining the proper feed rates and temperatures for the composition of the feed. Loss of control can lead to excess heat generation and higher than normal temperatures. The high temperatures can weaken the vessels and potentially lead to loss of containment. Operators can try to control the reaction by adjusting feed rates or preheat temperatures. Quenching system can be installed as a safeguard.

Catalytic Cracking [2]

Catalytic cracking uses a catalyst to break down heavy fractions from the crude distillation unit into lighter ones such as gasoline and kerosene (Figure 24.10). The most common process is fluid catalytic cracking (FCC) and Figure 24.16 shows a process flow diagram of an FCC. The FCC unit is one of the largest physical units in a refinery. The oil flows upwards and a catalyst mixed in a riser at 425–480°C (800–900°F), where the reactions take place. The catalyst is separated from the product in a disengagement chamber and goes to the regenerator where it is regenerated by adding air to burn off coke that has formed on it. Catalyst exit temperatures are 650–815°C (1,200–1,500°F). The regenerated catalyst is then returned to the riser of the reactor. The product flows to a column where product fractions are separated. The slurry oil is recycled back to the reactors.

24.5 Process Safety Hazards

Erosion of the piping by the catalyst can lead to loss of containment. Inspections need to check for leaks due to corrosion. Reverse flow through the reactor slide valves can result in air introduction into the reactor. This can lead to a flammable mixture and ignition of the hot hydrocarbons, resulting in a fire or explosion. Removing spent catalyst is

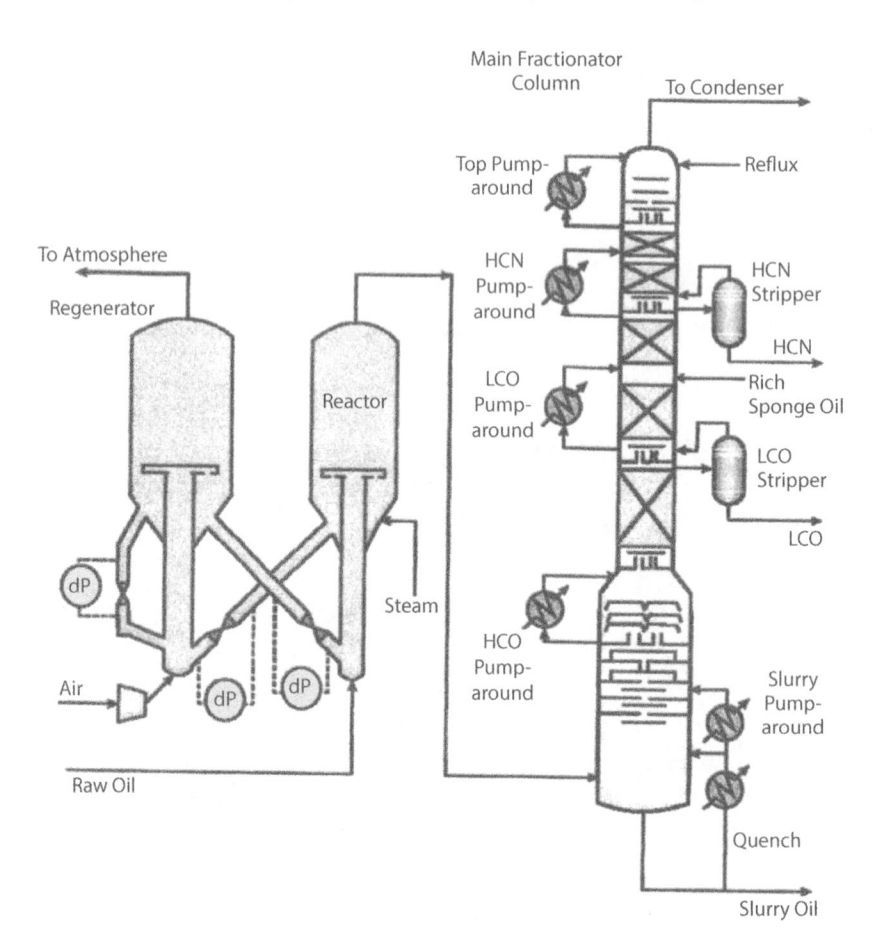

Figure 24.16 Fluid catalytic cracking (FCC) process flow diagram [2].

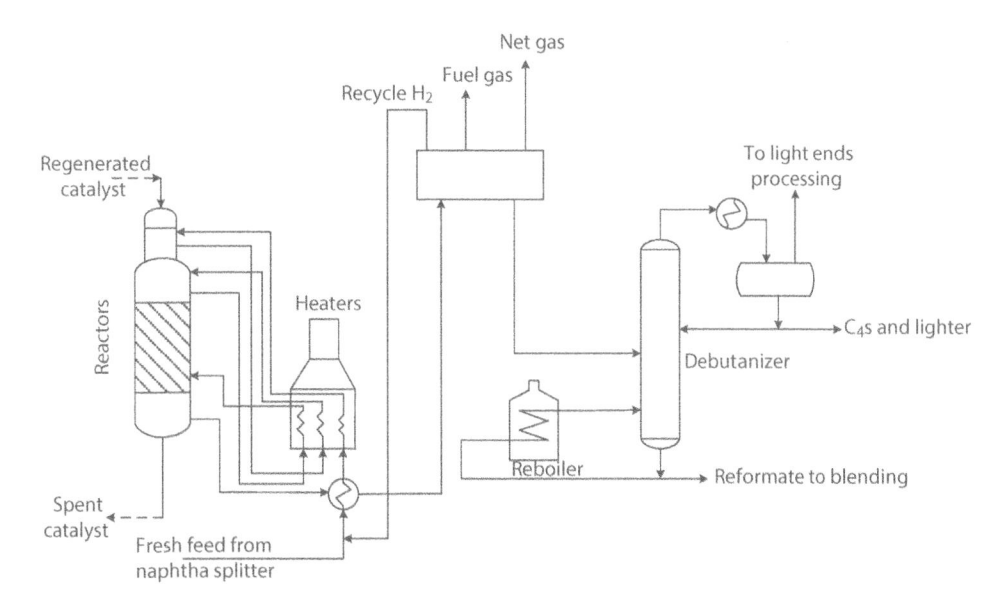

Figure 24.17 Naphtha reformer process flow diagram.

potentially hazardous due to the potential for fires from iron sulfide formation. The choked catalyst must be cooled and wetted before being dumped into containers.

Reforming

Reforming is the process used to convert naphthenes and paraffins to aromatics and isoparaffins, increasing the octane rating. The process also releases hydrogen, which is used in the hydrotreaters. There are two main designs, semi-regeneration and continuous catalyst regeneration (CCR). Figure 24.17 shows a CCR process flow diagram, where the reactor unit is actually a series of reactors.

The reforming section is also subject to HTHA. Furthermore, the hydrogen may combine with chlorine compounds to hydrogen chloride, leading to chloride corrosion. A good inspection program is required to check for leaks due to corrosion. Emissions of carbon monoxide (CO) and hydrogen sulfide (H_2S) can occur during the catalyst regeneration.

Alkylation [2]

In the Alkylation unit, isobutane reacts with propylene or butylene to form the alkylate, which is a mixture of high-octane materials such as isooctane. Sulfuric acid (H_2SO_4) or hydrogen fluoride (HF) is used as the catalyst. Figure 24.18 shows the block diagram of an alkylation unit.

In H_2SO_4 catalyzed alkylation unit, a chiller reduces the petroleum feed to about 4–5°C (40°F) and then the feed is mixed with the acid catalyst in the reactor. The acid is then separated and recycled to the reactor in a settler. A series of fractionators separate propane, butanes, and the alkylate.

Hydrotreating Units

Chloride corrosion is a particular concern for hydrotreating units that consume hydrogen from catalytic reforming units. These units consume chlorine (Cl_2) in the process of regenerating the acid function of the catalyst and thus may be contaminated by the chloride (Cl^-) in the hydrogen produced leading to the contamination of the hydrotreating unit. Austenitic stainless steels are very susceptible to failure due to the chloride attack, and their application should be avoided in regions susceptible to chloride contamination above 60°C.

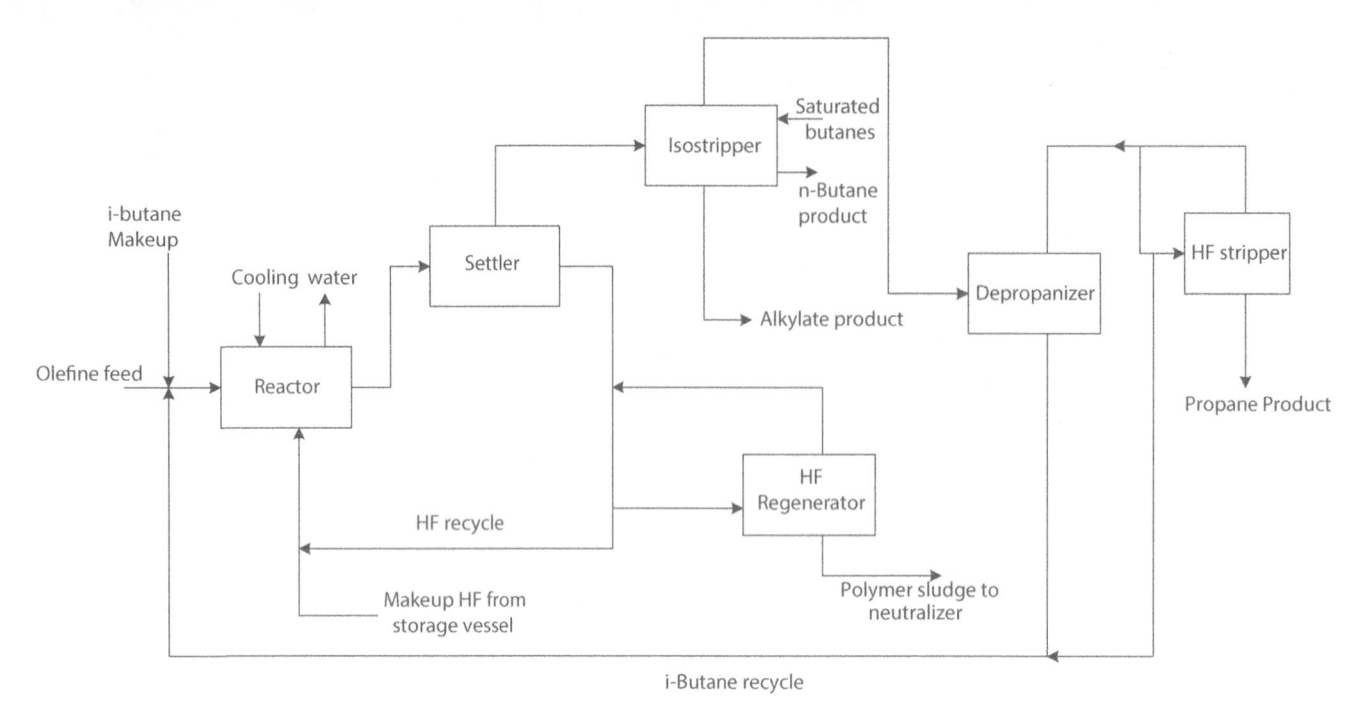

Figure 24.18 HF alkylation process flow diagram.

24.5.1 Examples of Process Safety Incidents and Hazards

The alkylation unit uses large volumes of H_2SO_4 or HF. Both are corrosive and highly hazardous. Loss of containment is a hazardous event. Loss of HF is highly hazardous and can be highly toxic. Absorption through the skin can cause cardiac arrest and inhalation causes damage to the linings of the lungs. HF can form a cloud that can travel outside of a refinery, as happened in the Texas City event as described below. Some units will have automatic systems to detect a release and spray large amounts of water on an HF release to remove or scrub it from the air.

HF release, Texas City, TX, 1987 [2]

A crane lifting a heat exchanger failed, causing it to drop the exchanger, which severed a 4-in. loading line and a 2-in. pressure relief line of an alkylation unit settling drum containing HF plus isobutane. The drum was under pressure and about 39683.2 lb (18,000 kg) of HF and 39462.8 lb (17,900 kg) of isobutylene were released. Concentrations of HF of 50 ppm were noted about three-quarters of a mile from the source of the release, based on damage to vegetation. That level of HF is considered to be the threshold level above which life-threatening effects can be observed. The release was mitigated by transferring as much HF as possible from the settler to railcars and by spraying water on the release. If took 44 hours for the release to stop.

HF release, Corpus Christi, TX, 2009

A control valve failed closed, blocking flow in the process piping. The sudden flow blockage caused violent shaking in the piping, which broke two threaded connections. There was a release of flammable hydrocarbons, which ignited. The fire caused several other failures, including the release of about 42,000 lb (19050 kg) of HF. There was the water mitigation system, which did activate and absorb most of the HF. One worker was critically burned. Citgo reported that 30 lb of HF was not captured by the water mitigation system. Studies have shown that the best these systems can do is 95% removal efficiency. The CSB recommendations state that, at 90% efficiency, the atmospheric release would have been about 4,000 lb (1814 kg). The water supply for the system was nearly used up, and salt water from the adjacent ship channel was used for firefighting. The CSB found that Citgo had never conducted a safety audit of the unit. The API practice regarding alkylation recommends a safety audit every 3 years.

The alkylation reaction is exothermic. Loss of control has the same consequences as in the hydrotreater.

The API has a recommended practice regarding HF Alkylation. API RP 751, Safe Operating of Hydrofluoric Acid Alkylation Units. 3rd ed., June 2007.

HF release at Philadelphia Energy Solutions Refining and Marketing LLC (PES), Philadelphia 2019

On June 21, 2019, a major process loss of containment caused a fire and explosion at the Philadelphia Energy Solutions Refining and Marketing LLC (PES). The PES refinery is an integrated facility of two separate refineries, Girard Point and Point Breeze. The incident occurred in the PES Girard Point refinery hydrofluoric acid (HF[†]) alkylation unit. Figure 24.19 shows the process flow diagram of the PES hydrofluoric acid alkylation unit, and the two equipment failure locations of interest are shown in orange.

During the early hours of Friday, June 21, 2019, the HF alkylation unit was reported operating normally. At 4:00:16 am, there was a sudden loss of containment causing flammable process fluid containing HF to release from the PES alkylation unit, forming a ground – hugging vapor cloud. At 4:02:06 am, the flammable vapor cloud ignited, causing a large fire in the alkylation unit. At 4:02:37 am, the control room operator activated the Rapid Acid Deinventory (RAD) system, which deinventoried bulk HF acid from the V-10 HF settler to the RAD drum. At 4:15 am, during the ongoing fire, an explosion occurred in the alkylation unit. A second explosion in the unit then occurred at 4:19 am.

At 4:22 am, a third and the largest explosion occurred when the V-1 Treater Feed Surge Drum, containing primarily butylene, isobutane and butane violently ruptured (Figures 24.20 and 24.21). A fragment of the vessel weighing approximately 38,000 lb flew across the Schuylkill River, and two other fragments, one weighing about 23,000 lb and the other 15,500 lb, landed in the PES refinery (Figure 24.22), which appeared to be a secondary event caused by the fire. The fire was extinguished the following day on Saturday June 22, at about 8:30 am.

PES estimated that about 676,000 lb. of hydrocarbons was released during the event; an estimated 608,000 lb. were combusted. Low-concentration of HF was also present in some of the process piping and equipment that failed during the incident, causing HF to release to the atmosphere. HF is a highly toxic chemical, as 5,239 lb. of HF were released from piping and equipment during the incident; 1,968 lb. of the released HF was contained by water spray within the unit and was processed in the refinery wastewater treatment plant, and 3,271 lb. of HF released to the atmosphere and was not contained by water spray.

Five workers experienced minor injuries during the incident and response, requiring first aid treatment. On June, 26, 2019, PES shut down the facility and filed for bankruptcy on July 22, 2019.

A ruptured pipe elbow was found in the unit post-incident (Figure 24.23). The rupture of the elbow appeared to be the initiating event causing the process fluid release. The elbow was part of the piping between V-11, the depropanizer accumulator and T-6, the depropanizer distillation column. The elbow, was on the discharge (outlet) piping from a pump (one of two pumps in this system) that was not operating at the time of the incident. At the time of the event, this piping was operating at a pressure of about 380 psig and a temperature of about 100°F. The approximate design composition of process fluid in the piping was:

Material	Weight percent
Propane (C_3H_8)	94.7
Hydrofluoric acid (HF)	2.5
Additional Hydrocarbons	2.8

[†] Hydrofluoric acid (HF) is immediately dangerous to life or health (IDLH) at 30 parts per million (ppm). Upon physical contact with skin, HF penetrates the skin and causes destruction to deep tissue layers and bone. Fatalities have been reported from an HF skin exposure to as little as 2.5% of body surface area. If inhaled, HF can cause severe lung injury and pulmonary edema-fluid in the lungs, which can result in death.

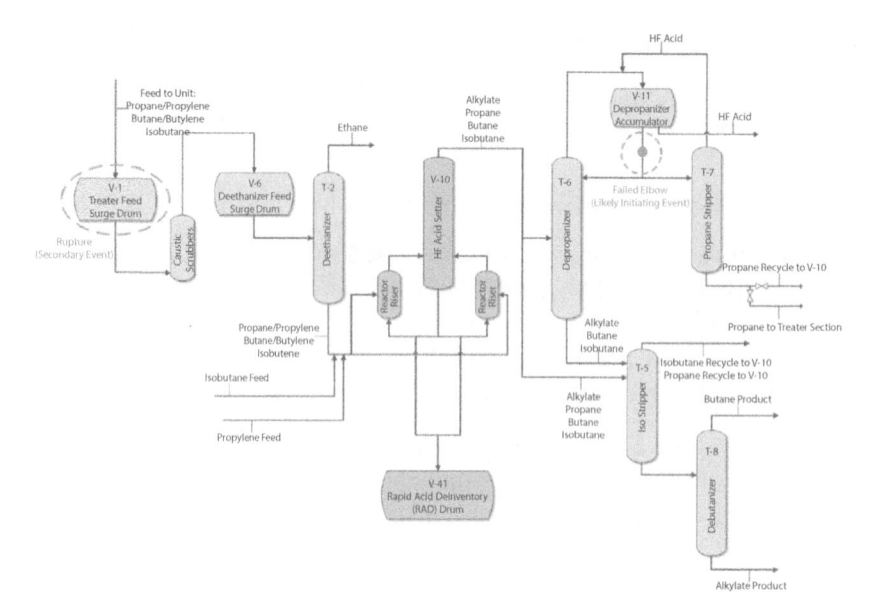

Figure 24.19 Process flow diagram of PES hydrofluoric acid alkylation unit. The equipment shown in red is the hydrofluoric acid alkylation reaction section and the Rapid Acid Deinventory (RAD) drum, to which the hydrofluoric acid can be routed (Source: www.csb.gov).

Figure 24.20 Video still of V-1 explosion (Source: www.csb.gov).

The piping circuit that includes the ruptured elbow was subject to regular ultrasonic thickness measurements at designated monitoring locations (CMLs)[††] as part of the PES inspection program to monitor the rate of piping metal loss due to corrosion. Locations of CMLs and the more recent thickness measurements are shown in Figure 24.24. The default retirement thickness (i.e., the piping thickness by which the piping should be replaced) of this piping in the PES inspection database is 0.18-inch. A CM was not located on the ruptured elbow, so the thickness of this elbow was not monitored. The thinnest measurement taken of the ruptured elbow post-incident was 0.012 inch. This value is less than 7% of the PES default retirement thickness.

[††]A condition monitoring location (CML) is a designated area where periodic thickness examinations are conducted. Each CML represents as many as four inspection locations located circumferentially around the pipe. CMLs are also referred to as thickness monitoring locations (TMLs). CMLs were historically referred to as corrosion (rather than condition) monitoring locations, and that terminology is sometimes still used with the industry.

Figure 24.21 Comparison of incident scene pre- and post-incident (Source: www.csb.gov).

Figure 24.22 Locations and phots of the post-incident V-1 vessel fragments. The fragment 1 photo was taken after the fragment was recovered from the bank of the river and relocated to the PES refinery. The deformation of fragment 1 was likely caused by impact with the river bank (Source: www.csb.gov).

Figure 24.23 Photo of ruptured pipe elbow found post-incident (www.csb.gov).

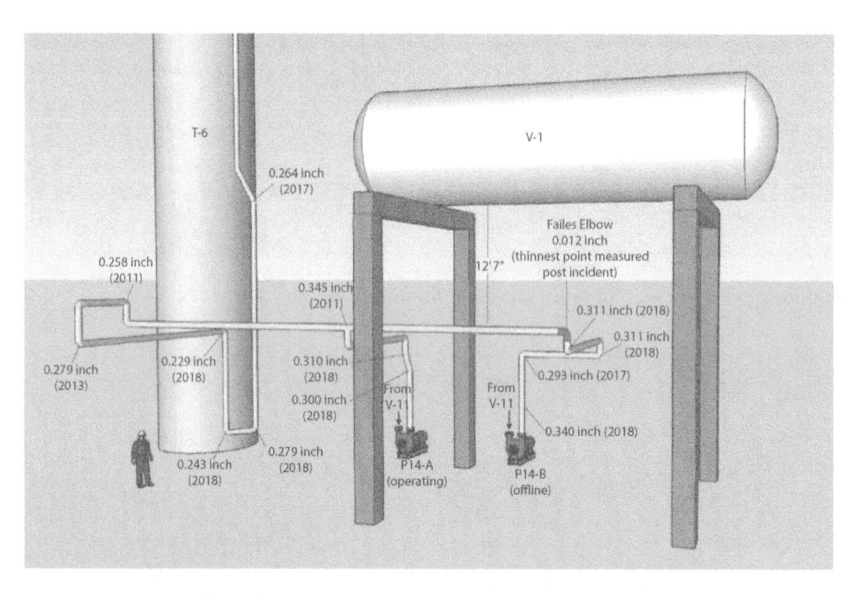

Figure 24.24 Model of the piping circuit containing the ruptured elbow. The most recent thickness measurements at designated Condition Monitoring Locations (CMLs) are shown. The year of each measurement is shown in parentheses. Figure also illustrates that V-1 was directly above the ruptured elbow (Source: www.csb.gov).

Post-Incident Activities

After the incident, PES hired a company that specializes in chemically cleaning alkylation units to develop a process to neutralize the HF contained in the RAD drum by neutralizing it with a base (chemical with a high pH) to produce water and a salt.

The piping circuit containing the ruptured elbow was installed in about 1973, and both the elbow that failed and the adjacent elbow were stamped "WPB", indicating they were constructed to meet the ASTM A 234 WPB material specifications.

The ASTM A 234 Standard Specification for Factory – Made Wrought Carbon Steel and Ferritic Alloy Steel Welding Fittings (1965 version); the applicable version at the time of the pipe installation, required that Grad WPB pipe "permissible raw materials" composition meet the A 106 Grade B chemical composition specification. The 1972 ASTM A 106 standard did not specify nickel (Ni) and copper (Cu) composition requirements. In 1995, ASTM A 234 began specifying nickel and copper composition, as well as compositions of other elements. The WPB composition requirements in 1972 and 1995 are shown below. The chemical composition of the ruptured elbow and the adjacent elbow are also shown. The nickel and copper content of the ruptured elbow exceeds the updated A234 WPB chemical composition requirements.

API RP 751 Safe Operation of Hydrofluoric Acid Alkylation Units states: "HF corrosion has been found to be strongly affected by steel composition and localized corrosion rates can be subtly affected by local chemistry differences. NACE Paper 03651 indicated that the combination of carbon (C) content and residual element (RE) content (Cr, Ni, Cu) could increase non-uniform corrosion by up to five-fold compared to baseline measured corrosion rates".

Coking [2]

The Coker takes the heavy feedstock and thermally cracks them to produce lighter products. The residue is a solid coke. There are two main coking processes: Delayed Coking and Continuous Coking. Figure 24.25 shows the flow diagram for a delayed coking unit. The bottoms from the crude distillation unit are heated to about 842–932°F (450–500°C) and charged to the bottoms of a coke drum. The material sits in the drum for about 24 hours at 40–115 psig (3–8 bar) and thermal cracking to form lighter products continues. The lighter product is drawn off to a fractionator and recovered material sent to other parts of the refinery for processing. When a predetermined level is reached, flow is directed into a second coke drum. Some refineries have several coke drums. Coke drums can be up to 120 ft (37 m) tall and 29 ft (9 m) in diameter. The first drum has to be cooled, the tops and bottoms removed, and then the drum washed with high-pressure fluids to remove the coke.

Equilon Anacortes Refinery Coking Plant Accident, 1998

A storm caused a power interruption during the first hour of filling a drum. The charged line itself became clogged with coke. Operators tired to clear the blockage by steaming out the line and believed they had succeeded. Based on temperature readings, the staff concluded cooling of the drum was done. The top head was removed, and then the bottom head. As the bottom head was removed, hot heavy oil broke through a crust and ignited because it was above its auto-ignition temperature. There was an explosion and a fire. Six people were killed. The staff was misled by the temperature sensors located on the outside of the drum instead of in it. Equilon subsequently installed a remote-controlled cleaning system.

Delayed Cokers have been a source of many serious accidents and were the subject of an OSHA Safety Hazard Information Bulletin (SHIB) on Hazards of Delayed Coker Unit (DCU) Operations.

If the switching valves are not properly aligned, or are leaking through, hot material can be sent to the drum being cleaned, leading to loss of containment and potential fires and explosion. Opening the wrong valve has led to serious incidents per the OSHA (SHIB). Providing interlocks to control valve opening can prevent this from occurring.

In some delayed coking units, the cleaning is done manually. When the coke drum heads are removed, workers can be exposed to geysers of steam, hot water, coke particles, hot tar balls through the top drum, and avalanches of coke from the bottom head. It is difficult to predict when material can be ejected from the drum heads. Operator training to be prepared for the hazards of opening the drum is needed. Shrouds around the drum head or an automated removal system can mitigate the hazard. Some units use remote-controlled cleaning units to avoid exposing operators to potential hazards.

Chemical requirements	A 106 Grade B 1972	A234 WPB 1995-2018 (%)	Ruptured elbow composition (%)	Adjacent elbow composition (%)
Carbon, max %	0.30	0.30	0.14	0.24
Manganese, max %	0.29 to 1.06	0.29 to 1.06	0.80	0.90
Phosphorus, max %	0.048	0.050	≤ 0.005	0.012
Sulfur, max %	0.058	0.058	0.010	0.016
Silicon, min %	0.10	0.10	0.10	0.24
Chromium, max %	No specification	0.40	0.18	0.02
Molybdenum, max %	No specification	0.15	0.06	<0.005
Nickel, max %	No specification	0.40	1.74	≤ 0.01
Copper, max %	No specification	0.40	0.84	0.02
Vanadium, max %	No specification	0.08	< 0.005	< 0.005

With some feeds, foaming can occur causing high drum pressure, level, and plugging of the drum outlets and relief valves. This can result in over pressurization and loss of containment. Antifoams can be added to these feeds to prevent this from happening.

In general, industry experience shows that the frequency of incidents is higher during process transitions such as startups or restarts from temporary idle conditions. Several of the incidents described occurred during a transient operating condition.

Design Considerations

Since engineered safety systems are often offline during transient operations, the role of operators and technical personnel and their knowledge of the process are essential. Written procedures are required for these operating modes and are, in fact, required for processes covered by the OSHA PSM and EPA RMP standards. Emergency or abnormal procedures must include what actions operators should take when process conditions exceed beyond their defined limits.

Risk associated with transient operating states should be identified in the Hazard Identification and Risk Analysis (HIRA). HIRA must include startup and shutdown, loss of utilities, and should define the responses to the identified process upsets. This information can be turned into emergency procedures. The PHA can be used to document the risk in transient operations, and write operating procedures, providing adequate training, and refresher training on the risk when startups and shutdowns occur. Operational readiness reviews should be done before startups, and MOC reviews must be held when unusual situations occur.

24.6 Hazards Relating to Equipment Failure

Process engineers and operating personnel should alert corrosion engineers and plant (metal) inspectors allocated to their process units when safety critical wash water or chemical injection streams do not function as required in plant operating procedures or instructions. An analysis of major accidents' reports on equipment failures shows that the largest failure is process piping and valves followed by pumps, followed by fired heaters, followed by columns/vessels as shown in Figure 24.26.

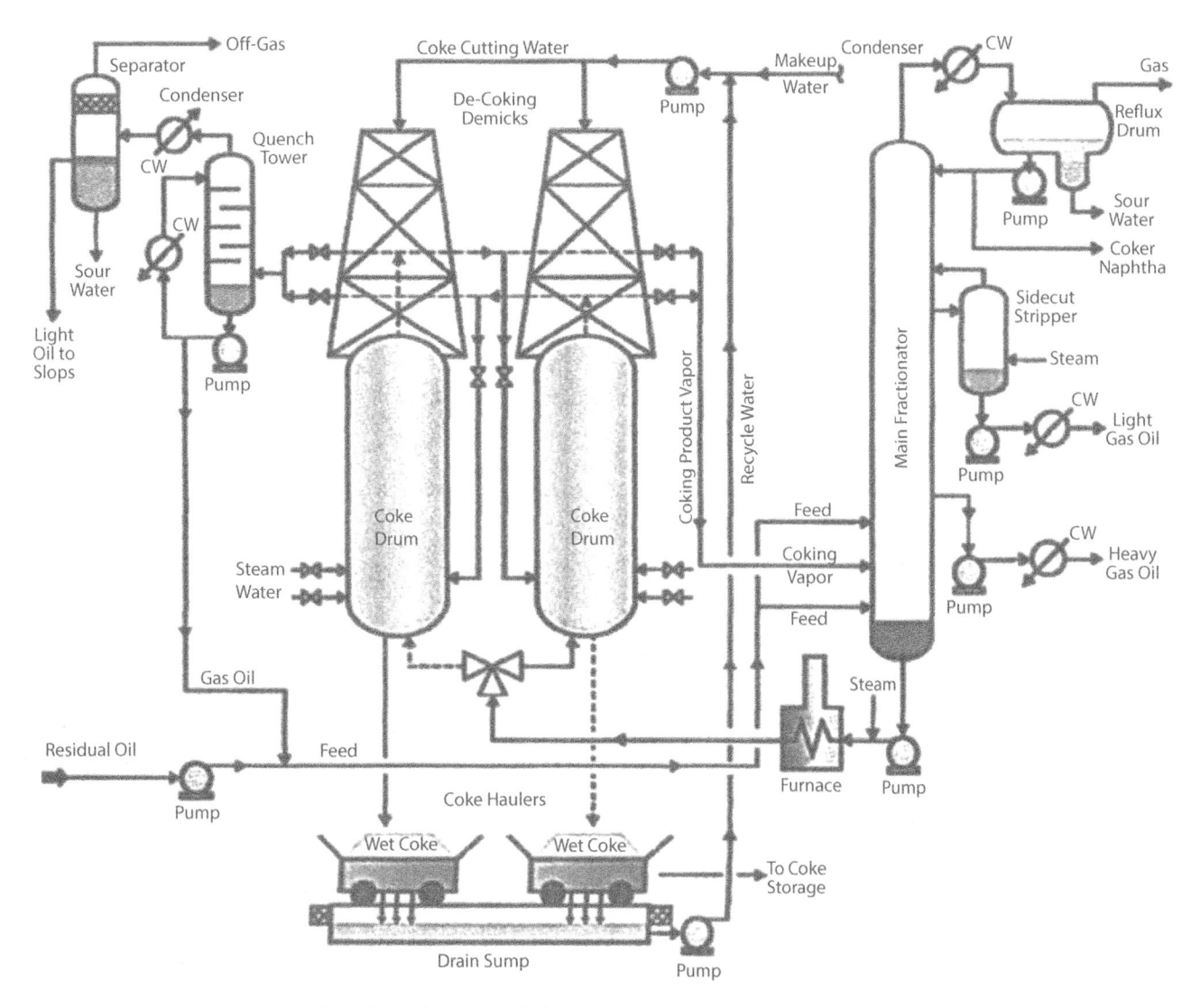

Figure 24.25 Process flow diagram for a delayed coker unit [2].

24.7 Columns and Other Process Pressure Vessels and Piping

Corrosion

Corrosion is the natural process that converts a refined metal into a more chemically stable form such as oxide, hydroxide, or sulfide. It is the gradual destruction or deterioration of materials (metal or non-metal) by chemical and/or electrochemical reaction with their environment. It is a global phenomenon and one of the most serious problems affecting the economic growth of industrialized countries. The global corrosion cost is estimated to be ~US2.2 trillion annually, which is over 3% of the world GDP. For example, the US is losing more than $276 billion on account of corrosion. Corrosion of the equipment occurs because of the environmental condition and fluid in the equipment. Table 24.6 shows the classification of corrosion.

Corrosion may be uniform or localized that is microscopic or macroscopic. The various forms of corrosion that are uniform are: pitting, crevice, galvanic, stress cracking, intergranular, dealloying, or selective corrosion. Others are

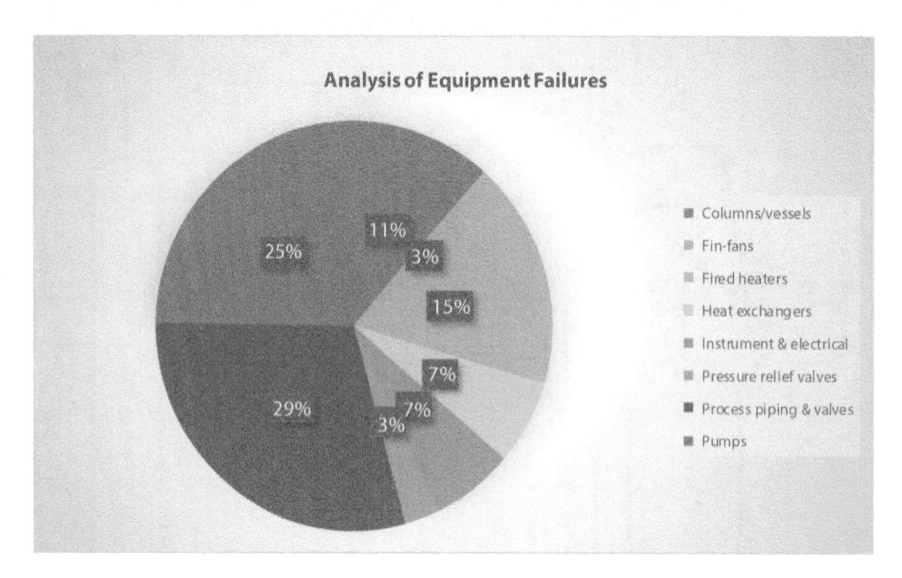

Figure 24.26 Analysis of equipment failure [3].

caustic embrittlement, fatigue, filiform, high temperature, microbiological, erosion, and fretting. Table 24.7 provides a brief description of the various forms of corrosion.

Corrosion Inhibitors

Corrosion of the metallic surfaces can be reduced or controlled by use of inhibitors that form protective film on the surface of the metal. Various types of inhibitors are passivating, organic, and precipitating. Table 24.8 shows inhibitors used in the petroleum refinery and CPIs.

Major areas of concern in the petroleum sector are: Oil and gas exploration and production, petroleum refining, gas and liquid transmission pipelines, gas distribution, and hazardous material transport. Environmental concerns by regulatory authorities are due to release of pollutants in air, soil, or water caused by corrosion leaks resulting in high consequence events. Some of the major corrosives in the petroleum industries are naphthenic acids, sulfur compounds, ammonia, carbon monoxide, carbon dioxide, oxygen, chlorides, cyanides, hydrogen, hydrochloric acid, sulfuric acid, nitric acid, phenols, and various organic chemicals (Table 24.9).

The following is a list of the most corrosion mechanism on CDUs/VDUs with the appropriate API RP 571 reference. Some of the incidents that occurred as a result of undetected corrosion are as follows:

Ammonium Chloride Corrosion (API RP 571 5.1.1.3): Corrosion that forms under ammonium chloride (NH_4Cl) or amine deposits that form on the inside surfaces at the top of a CDU main column, and overheads piping and condensers.

Operators can reduce the likelihood of these deposits occurring through

- Ensuring good desalter operation with acceptable (low ammonia) desalter wash water quality.
- Maintaining wash water to overheads systems (if used) at the required flow rates.
- Ensuring that filming amine addition is maintained at the required levels.
- Be aware of the salt sublimation temperature for the unit. This will depend on the partial pressure of the salt species.

Hydrochloric Acid (HCl) Corrosion (API RP 571 5.1.1.4): This forms in the overheads system of CDU main columns as the first drops of water condense out, and can also be a problem in VDU ejector/condenser sets.

Table 24.6 Classification of Corrosion.

Based on Mechanism of Corrosion	Chemical corrosion: Uniform or non-uniform, Electrochemical corrosion: Galvanic corrosion or bimetallic corrosion, galvanic microcell within a metal, differential aeration cells, uniform corrosion, crevice corrosion, deposit corrosion, water line corrosion.
Based on Nature of Corrosion	Corrosion can be wet or dry. A liquid moisture is necessary for the wet corrosion while dry corrosion usually involves reaction with high temperature gases.
Based on Nature of Corrosion	Atmospheric, marine, underground corrosion, biological corrosion, high temperature corrosion, hydrogen cracking, hot sulfide corrosion, metal salt corrosion, carburizing.
Based on Corrosion Deterioration	General corrosion, localized corrosion, intergranular corrosion.
Based on Mechanical Factors	Stress corrosion cracking fatigue, cavitation.

Operators can reduce the likelihood of HCl attack by

- Maximizing crude oil tank water separation and draining into the tank farm. Many sites have a policy of allowing a minimum settling time of 2 days before allowing the tank to be pumped to a CDU.
- Ensuring good desalter operation.
- Maintaining wash water to overheads systems (if used) at the required flow rates.
- Maintaining combinations of neutralizing and filming amines at the required levels.
- Monitoring the pH of the overheads water that collects in the reflux drum/accumulator boot on a regular basis, typically once per shift, and report any excursions outside a predetermined safe range to plant supervisors.

Sulfidation Corrosion (API RP 571 4.4.2.1): This attacks carbon steel and other alloys as a result of the reactive corrosion of sulfur compounds (e.g., elemental sulfur, H_2S, and mercaptans) in crude oils. It occurs in the hot sections of the unit [generally above 446°F (230°C)]. The higher the reactive sulfur levels in the crude oil feed, the higher the corrosion. The solution is to ensure correct metallurgical selection for the process conditions.

Process operators should ensure full compliance of operating envelopes to avoid sulfidation.

A severe corroded vacuum residue pump suction line suffered major failure after one of the pumps had been isolated following a minor seal fire. Up draught from overhead fin-fans exchanger exacerbated the fire.

A CDU splitter pumparound line failed due to sulfidation corrosion, which had gone undetected despite a piping thickness testing program. The resulting jet fire impinged on another equipment, including overhead fin-fans and was extinguished in 2 ½ hours.

High-temperature sulfide corrosion inside an external CDU overflash line resulted in leakage and fire. Low velocity within this line allowed H_2S at 662°F (350°C) to separate out accelerating corrosion. External overflash piping was subsequently removed.

The effects of failure due to corrosion are clear, but the fate of the products of corrosion such as iron sulfide (FeS) scale can also cause problems downstream with blockage and maintain flow regimes within their safe operating limits.

Opening the bypass around the blocked and infrequently used heavy slop oil recycle flow control valve caused overheating in three heater passes followed by a rapid loss of VDU column vacuum and a transfer line flange fire.

Table 24.7 Types of Corrosion.

Types of corrosion	Description
Uniform corrosion	This is a surface phenomenon and occurs through uniform attack on all surfaces of the metal exposed to acidic, alkaline, humid, or moisture laden environment and is normally characterized by a chemical or electrochemical reaction. Uniform corrosion can be prevented or reduced by proper material of construction, use of inhibitors, surface coating, and cathodic protection
Galvanic corrosion	This type of corrosion occurs when two dissimilar metals are brought into contact in a corrosive or conductive solution resulting in the corrosion of the metal, which is less noble. This type of corrosion can be prevented by selection of combination of metals as close as possible, by insulating two dissimilar metals, applying coating, addition of inhibitors and cathodic protection, etc.
Pitting corrosion	Pitting is a localized form of corrosion and is characterized by surface cavities, which can have different shapes. The process of pitting is slow and results in homogeneity on the metal surface, local loss of passivity, mechanical/chemical rupture of protective oxide film. Pitting is accelerated by more acidic or higher temperature conditions. Prevention can be done by proper selection of material of construction (using metals showing fewer tendencies to pitting) and addition of inhibitors.
Crevice corrosion	This is due to an intensive localized attack in crevices, which exist at lap joints, bolts, rivets, and gaskets. Prevention can be done by proper design and operating procedures.
Stress corrosion cracking (SCC)	SCC is caused by the simultaneous presence of tensile stress, a specific corrosive environment. Chloride SCC and sulfide SCC are two forms of SCC. SCC stressed regions undergo localized attack resulting in hairline cracks. Suitable environment, tensile strength, a sensitive metal, appropriate temperature, and pH are important conditions necessary for stress corrosion cracking. Stress corrosion cracking can be prevented by eliminating the critical environmental species, changing the alloy, applying cathodic protection, adding inhibitors, coating, and shot penning.
Caustic embrittlement	This is a form of corrosion occurring in metals in contact with caustic under certain condition by an alkaline environment. The cracks result from the combined action of tensile stress and corrosion.
Hydrogen damage	Hydrogen blistering, hydrogen embrittlement, decarburization, and hydrogen attack are various forms of hydrogen damage. Hydrogen blistering is caused by the diffusion of atomic or nascent hydrogen in the crystal lattice and collection in fissures or cavity. Hydrogen embrittlement is caused by penetration of atomic or nascent hydrogen through the metal structure, resulting in loss of ductility. Carburization and hydrogen attack occur at a high temperature. Decarburization is the removal of carbon from steel by moist hydrogen and at a high temperature. The hydrogen embrittlement mechanism involves the contamination of steel by molecular hydrogen (H_2) leading to the risk of failure, especially in periods of instability such as startup/shutdown of the process. The procedure of starting the hydrotreating units is carried out at the lowest possible pressure as well as controlling the cooling or heating rate of the reaction system. Figure 24.26a shows the mechanism of the embrittlement of steel by hydrogen.

(Continued)

Table 24.7 Types of Corrosion. (*Continued*)

Types of corrosion	Description
	Figure 24.26a Hydrogen fragilization mechanism [40]. Due to the need of high temperatures, the above mechanism is usually of concern in the load and reaction heating sections.
Corrosion fatigue	This is due to the tendency of metals and alloys to fracture under repeated cyclic stress. Fatigue life, which is the number of cycles needed for failure, is dependent on the stress level. Corrosion fatigue is the cracking of metals resulting from combined action of a corrosive environment and repeated or alternate stress. Corrosion fatigue can be prevented or reduced by eliminating or reducing the stress by use of inhibitors and by coating.
Exfoliation corrosion	Exfoliation is a severe form of intergranular corrosion that raises surface grains from metal by forming corrosion products at grain boundaries under the surface. It occurs mostly in heavily rolled or extruded products where the grains are flattened and elongated in the direction of hot working.
Intergranular corrosion	This is a localized form of attack occurring at grain or adjacent to grain boundaries with little or no attack on grain boundaries themselves resulting in a loss of strength and ductility.
High-temperature corrosion	Selective oxidation of chromium when exposed to low oxygen atmosphere at high temperature. Some of the other form of high temperature corrosion may be oxidation–reduction, sulfidation, carburization, and nitriding. The sulfide corrosion process is common in petroleum refinery facilities and occurs due to the degradation of steel involving the reaction of iron with sulfur compounds contained in the feed streams, usually above 260°C. In the case of regions of the unit without the presence of hydrogen, the application of steels containing 5 – 12% chromium (Cr) is considered robust to ensure sufficient useful life of the process equipment. The higher the Cr content in steel as shown in Figure 24.26b, the greater is the resistance to corrosion by sulfide. Furthermore, the addition of silicon to steels applied to hydrotreating units also contribute to reducing the rate of corrosion by the sulfide.

(*Continued*)

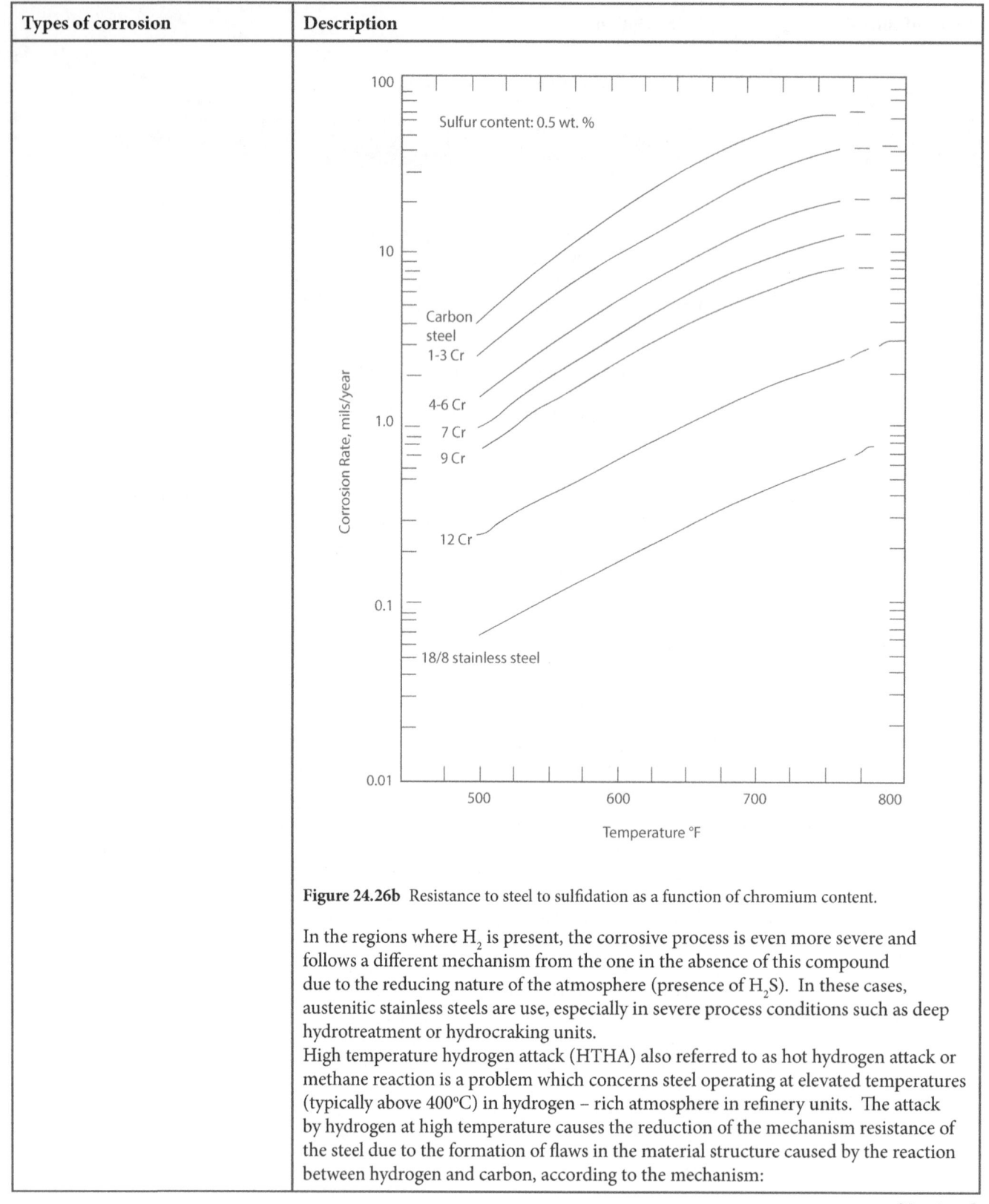

Table 24.7 Types of Corrosion. (*Continued*)

Types of corrosion	Description
	Figure 24.26b Resistance to steel to sulfidation as a function of chromium content. In the regions where H_2 is present, the corrosive process is even more severe and follows a different mechanism from the one in the absence of this compound due to the reducing nature of the atmosphere (presence of H_2S). In these cases, austenitic stainless steels are use, especially in severe process conditions such as deep hydrotreatment or hydrocraking units. High temperature hydrogen attack (HTHA) also referred to as hot hydrogen attack or methane reaction is a problem which concerns steel operating at elevated temperatures (typically above 400°C) in hydrogen – rich atmosphere in refinery units. The attack by hydrogen at high temperature causes the reduction of the mechanism resistance of the steel due to the formation of flaws in the material structure caused by the reaction between hydrogen and carbon, according to the mechanism:

(Continued)

Table 24.7 Types of Corrosion. (*Continued*)

Types of corrosion	Description
	$$8H + C + Fe_3C \Leftrightarrow 2CH_4 + 3Fe$$ The presence of methane in the structure causes the formation of cracks in the steel structure and can result in premature failures. Preventive actions require the selection of adequate alloy steel (Cr, Mo, and V) and operation within the recommended pressure and temperature parameters as shown in Figure 24.26c. **Figure 24.26c** Nelson curve (steel operations limits – temperature vs. H_2 partial pressure., API 571/2011.
	Naphthenic acid corrosion occurs in units when subjected to processing of currents with high acidity. The control parameter used is the total acid number (TAN), which is a measurement of acidity that is determined by the amount of potassium hydroxide (KOH) in mg, which is required to neutralize the acids in 1.0 gm. of oil. This is usually the number of total acidity above 0.30 mg/KOH/g. The TAN value indicates to the crude oil refinery the potential of corrosion problems. Generally, the naphthenic acids in the crude oil that cause the corrosion problems. Higher temperatures at 240°C and above indicate the possibility of corrosion by naphthenic acids. Furthermore, the turbulent flow regimes increase the corrosion rates. Controlling the total acidity of the crude requires adequate measures be applied in the hydroprocessing units. However, the combination of high acidity with reduced levels of sulfur in the crude oil and of the intermediate currents result in greater severity of the corrosive process. In this instance, the low-sulfur content limits the formation of the protective layer of iron sulfide (FeS). Naphthenic corrosion is mitigated by the addition of molybdenum to steel, thus providing greater resistance to attack by naphthenic acids.

(*Continued*)

Table 24.7 Types of Corrosion. (*Continued*)

Types of corrosion	Description
De-alloying or selective on corrosion	Selective removal of one constituent of a metal from the alloy, e.g., dezincification, graphitization.
Stray current corrosion	A form of attack caused by electrical currents.
Filiform of corrosion or Underfilm corrosion	This is a form of corrosion initiated electrolytically due to the presence of moisture, oxygen, and corrosive ions and results in the form of fine trenches under paint, enameled, or lacquered surfaces.
Microbiological corrosion	Deterioration of a metal caused directly or indirectly from the activity of living organisms, which occur as a result of their influence on anodic and cathodic reactions. This type of corrosion is commonly found in water storage tanks, pipelines, and in fuel tanks. This can be prevented or substantially reduced using biocides.
Erosion corrosion	Deterioration or degradation of material surface due to mechanical action, often by impinging liquid, abrasion of slurry, particles suspended in fast flowing liquid or gas, or the attack on metal by contact with high-velocity liquids resulting in pitting. This is characterized in appearance by grooves, gullies, waves, rounded holes, and valleys. Hardness is considered to be a measure of material's erosion resistance. Prevention can be accomplished by reducing velocity, using material with better resistance to erosion, proper design, and coatings. The second-generation duplex alloys, with their relatively highly abrasive solid resistances, provide very good erosion resistance protection.
Cavitation	Cavitation is the formation of growth and collapse of bubbles or cavities in a liquid. This is corrosion of material removed by the formation and collapse of vapor bubbles in a liquid near a metal surface. Various forms of cavitation are traveling, fixed, vortex, or vibratory.
Fretting	It is a form of wear or damage of material occurring in contact between materials under load subjected to vibration and slip. Fretting destroys the dimensional accuracy of closely fitted parts and increases the susceptibility to fatigue failure.

(Source: Fontana [36]. Schweitzer [37]).

Naphthenic Acid Corrosion—NAC (API RP 571 5.1.17): This form of high-temperature corrosion that occurs in CDUs and VDUs when crude oils that contain naphthenic acids are processed. It mostly occurs in the hot sections of CDU/VDUs at above 392°F (200°C) and in areas of high turbulence. Piping systems are particularly vulnerable in areas of high fluid velocity.

Crude oils are normally considered as naphthenic and requiring special material of selection where the total acid number (TAN) is >0.5 mg KOH/gm.

The main mitigation method is through the selection of the process equipment metallurgy. Process engineers and production planners can reduce the impact acid corrosion has on a unit that is not designed to handle it by blending the crude oils to reduce the TAN of the feedstock.

High-temperature NAC inhibitors are available and can be used in conjunction with extensive corrosion monitoring systems to mitigate acid corrosion. The dosage rate must be carefully monitored and controlled in line with the measured corrosion rates.

Wet H_2S Damage (API RP 571 5.1.2.3): This describes a range of damage that can occur to carbon and low alloy steels through blistering or cracking. The basic chemistry is based on the reaction of H_2S with the iron oxides in pipe scale that creates iron sulfide and hydrogen atoms. The hydrogen atoms diffuse into the steel of the pipe or equipment

Table 24.8 Inhibitors in Petroleum Refinery and Chemical Process Industries.

System	Medium	Corrosion inhibitor	Remarks
Oil well operation	Acidic condition, chloride	Mercaptans and glycol-xanthates, oleic and naphthenic acid, derivatives of amine, diamines, zinc metaphosphate, biocides: Biomin—I & II, Scale inhibitors: Scalemin—I & II	
Condensate well	CO_2, organic acid	Cronox film—Plus	
Crude oil pipeline	Crude oil, water, air, salts, sulfur compounds, naphthenic acid.	Water-soluble and oil-soluble inhibitors. Amine and nitrites	Corrosion occurs due to the presence of water and air in the crude oil. CO_2 and H_2S present are highly corrosive.
Natural gas pipeline	Natural gas and condensate CO_2, H_2S	Filming inhibitor	Presence of CO_2, H_2S, and water makes the corrosive environment. Dehydration is an effective means of reducing corrosion.
Oil storage	H_2S, chlorides	Phosphate, nitrites, imidazolines, oleic acid, salts of amines, filming amine	
Atmospheric and vacuum distillation columns	Crude oil, chlorides, H_2S, sulfur compounds, nitrogen compounds, naphthenic acid	Aminoalkyl aryl phosphate	Crude desalting, caustic injection, neutralization of vacuum column overhead vapors and condensed water, water washing in crude column can reduce corrosion.
Fluid catalytic cracking unit	Naphthenic acid, H_2S, chlorides, water/steam, CO_2, O_2.	Diamide, imidazoline, quaternary amine, polysulfide, thiocynate.	Water wash followed by use of inhibitor. Water wash dilutes and scrubs the corrosive matter, H_2S, NH_3, Cl, and cyanide, most FCC corrosion inhibitors or filming. Some of the FCC inhibitors are oil-soluble amide, quaternary amine, oil-soluble imidazoline inhibitor.
Steam and condensate line	Water, CO_2	Oxygen scavenger, inorganic sulfite, hydrazine, carbohydrazine, hydroquinone, and ascorbic acid. Neutralizing type amines: Morpholine, diethylaminoethanol, cyclohexyl amine, ammonia.	Addition of catalyst to the hydrazine mixture ensures completion of the oxygen scavenging and metal passivating reactions. Filming amine reacts with carbonic acid in condensate to form neutral amine salts, thus raising the pH.
Cooling water	Dissolved oxygen, hardness, chloride, sulfate, etc.	Filming type amine: Ethyloxylated soya amine, diadodecylamine, and tridodecylamine. Addition of sodium silicate, chromates, polysulfide, sodium molybdate. Use of biocides (oxidizing and non-oxidizing): oxidizing: Chlorine, ClO_2 (calcium hypochloride), Non-oxidizing: Isothiozoline, dimethyl bisthiocyanate.	Octadecylamine, biocides are used for removal of undesirable formation of biological film.

(Source: Fontana [36]. Schweitzer [37])

Table 24.9 Major Corrosives in Petroleum Industry.

Plant	Corrosive pollutants
Corrosion in drilling process	Oxygen, moisture, CO_2, H_2S, chloride salts, organic acid
Natural gas processing	Moisture, CO_2, H_2S, chloride, amine degradation products, impurities in amine system
Crude oil processing and refining	Naphthenic acid, sulfur compounds, high temperature, chloride, sulfuric acid, HF, caustic soda, MEA, K_2CO_3
Residue upgradation process (Visbreaking, delayed coking, FCC, hydrocracker, thermal cracking processes)	High temperature, naphthenic acid, cyanide, chloride, sulfide, hydrogen sulfide.
Olefin plant (naphtha/gas cracker units)	Inorganic sulfides, mercaptans, CO_2, soluble hydrocarbons, polymerized product, phenolic compounds cyanide, coke, spent caustic, SO_x, NO_x, hydrocarbons, high temperature, and chloride
Aromatic production unit catalytic reforming	Dissolved organics, chlorides, sulfides in naphtha, high temperature, benzene and its homologous, presence of acids, bases, and/or salts in the aromatic hydrocarbons.

well, collecting at a discontinuity or inclusion and then combined to form hydrogen atoms, which become trapped because of their larger size causing blisters or cracking.

Mitigation is through selection of materials of construction, including not using dirty steels. Internal coatings are effective, such as Monel, used at the top of CDU main columns. Post welded heat treatment (PWHT) is effective in reducing wet H_2S damage.

Operators can play an important role in testing aqueous streams for pH and reporting any deviations from predetermined safe limits to plant supervisors.

Caustic Stress Corrosion Cracking (API RP 571 4.5.3): This occurs in carbon and low alloy and 300 series stainless steels exposed to caustic soda (NaOH) and caustic potash (KOH).

Cracking can be prevented in carbon steel piping and vessels by subjecting piping and process vessels to post welded heat treatment (PWHT), which requires heating to 1,150°F (621°C) to relieve stresses created in the fabrication processes.

Particular care is required with non-PWHT carbon steels with steam tracing design and when steaming out.

Caustic stress corrosion cracking leading to a 76 cm (2.5 ft) long rip in the 20″ main crude line occurred at a caustic injection point relocated from downstream to upstream of the pre-heat exchangers. The injection quill was incorrectly positioned.

Crude Oil Distillation Column Unit (CDU): Corrosion in the CDU represents a significant portion of refining operational costs as a result of lost production, inefficient operation, high maintenance and corrosion control chemical costs. Corrosion in the CDU overhead is due to acid attack at the initial water condensation point, where the major corrosives are chlorides and sulfur compounds, and naphthenic acid. Magnesium chloride ($MgCl_2$) that is present in the crude can readily hydrolyze at a temperature of 248°F (120°C) to form hydrogen chloride (HCl). Electrochemical corrosion occurs due to the presence of hydrogen sulfide and hydrogen chloride (H_2S-HCl). Furthermore, the presence of naphthenic in the crude complicates the corrosion problem in the CDU. H_2S-HCl present at the overhead vapors system of the fractionating towers affect the equipment such as the condensers, coolers, reflux drums (accumulators), vapor lines, gasoline reflux pipes and pumps. The naphthenic acid enhances corrosion in the pipes and return bends of the furnace. The attack continues into the flash section of the main fractionating tower, on trays shell and corresponding to the kerosene and gas oil section of the fractionation tower [16].

Caustic solution (NaOH) provides a major corrosion prevention and control in crude oil desalting. Others are neutralization of the crude, vacuum column overhead vapors and condensed water, water washing in crude column overheads and the use of corrosion inhibitors. Controlling corrosion in the overhead condensing of vapor in the crude distillation tower is a major challenge for the refiners. A typical treatment strategy involves neutralizing the acids in condensed water with ammonia (NH_3) and amines ($HOCH_2CH_2NH_2$) while avoiding the formation of corrosive salts via a vapor phase reaction with hydrogen chloride [16].

Fluidized Bed Catalytic Cracking (FCC) Corrosion: Corrosion and fouling significantly reduce the efficiency of the FCC. Steel and copper alloys corrosion, hydrogen blistering of steel in the FCC fractionate and vapor recovery section are shown to contaminate the reactor from the nitrogen and sulfur bearing compounds in the feedstock. Corrosion in the FCC is mainly because of the presence of nitrogen and sulfur compounds that are converted to ammonia and cyanide in case of nitrogen compounds and to hydrogen sulfide in the case of sulfur compounds. Also, some CO_2, chloride and organic acids may be present in the unit. The cracked nitrogen compounds form hydrogen sulfide. The severity of corrosion as to progress to the point of hydrogen blistering depends very much of the temperature and the activity of the catalyst within the reactor and the feedstock contaminants level [43].

Some of the corrosion reactions in the FCC are:

Steel corrosion reactions:

$$Fe + 2HS^- \rightarrow FeS + S^{-2} + 2H^0$$

$$Fe + 6CN^- \rightarrow Fe(CN)_6^{-4} + S^{-2}$$

Copper corrosion reaction:

$$Cu^{+2} + 4NH_4^+ \rightarrow Cu(NH_3)_4^{+2} + 4H^+$$

Effective water wash is a key to any corrosion control program, as this would move the vapor phase corrosive into liquid phase, where they are easily treated. Use of polysulfide such as sodium or ammonium polysulfide and corrosion inhibitors can reduce corrosion in FCC unit. Polysulfide converts cyanides to thiocyanates by the reaction [43]:

$$S_x^{-2}(polysulfide) + CN^- \rightarrow S_{x-1}^{-2} SCN^-$$

Amine Absorber Corrosion: Corrosion and cracking in gas processing using amines solutions are common phenomenon and can results from poor design and operating practices. The corrosivity of material for the vessel such as steel is related to the specific amine chemicals with the operating conditions such as temperature, velocity, and gas loading [44]. Monoethanolamine ($HOCH_2CH_2NH_2$) is used to remove CO_2 and H_2S from the gas. Corrosion in this process can result in unscheduled downtime, production losses, reduced equipment life and possible injury to personnel. Both gas types ratio of H_2S to CO_2 can impact the corrosion rate. Furthermore, corrosion and cracking in gas processing using monoethanolamine for CO_2 and H_2S removal has been a cause of major concern because of the severe corrosivity to carbon alloy steel commonly used for the material of construction. Corrosive condition in gas treating system can result from poor design, operating practices (e.g., too high flow rate or changes in flow direction) and from the contamination of amine solution with sulfide and CO_2. The parameters that affect the carbon steel corrosion in amine system are pH, temperature and velocity. Corrosivity increases with a decrease in pH over the range of 9 – 12. Correspondingly, with an increased temperature and velocity, the corrosion rate is increased, and it is most severe in rich amine solution in the area of high velocity and/or turbulence [44].

Steam and Water Line Corrosion: Corrosion of steam and condensate line provides a major problem in petroleum refining and petrochemical facilities. The major causes of corrosion are the oxygen that results in pitting and a low pH which gives rise to generalized thinning of the piping. Formation of low pH condensate is due to carbon acid which is formed by reaction of CO_2 with condensed steam. CO_2 formation takes place due to the breakdown of bicarbonate and carbonates. Deposit of contaminants and corrosion occur throughout the steam generating system. Boiler tubes with the highest heat transfer are the common location for these deposits from feed water and returning condensate [45].

Cooling water corrosion is also common in petroleum and petrochemical facilities as huge amount of cooling water is used at various stages of operation in these complexes. Cooling water both once through and recirculating is responsible for both corrosion and scaling of piping and the heat exchangers. Crude refining operation in the coastal area causes additional problem of seawater corrosion, pitting, uniform corrosion, and stress corrosion cracking. Important factors that can influence corrosion in cooling water system are dissolved oxygen, temperature, velocity, pH and dissolved solids.

24.8 Inadequate Design and Construction

Corrosion within "dead legs"

Piping "dead legs" are not expected to be found in the original process unit designs. Areas where no flows are expected to occur such as the bypasses around control valves or heat exchangers are designed to be free draining to the main process flow. However, when changes are made to the unit design, dead legs can be formed as sections of redundant piping are isolated rather than being removed.

Piping failure due to corrosion within a "dead leg" section of crossover piping on the suctions of the VDU flash zone reflux pumps resulted in major loss of vacuum. A fire and explosion occurred inside and outside of the column.

Failure through corrosion of a crossover line, normally a "dead leg," on the suction side of the vacuum residue and flash zone pumps released hot hydrocarbon that auto-ignited. The fire was exacerbated by failure of the pump discharge pipework.

Some "dead legs" are formed in piping systems that are only used occasionally. These situations can be overcome by the use of metallurgy or by isolating at the normally operating hot piping and flushing the infrequently used piping with gas oil or similar.

A stagnant line within the CDU/VDU residue systems suffered major failure due to internal sulfide corrosion. This line was only used at startup and shutdown, i.e., for about 2 weeks every 2 years, at which time it operated at 630°F (332°C).

24.9 Inadequate Material of Construction Specification

Many major incidents arise from the wrong material of construction being inadvertently used. Failures through corrosion can occur at seemingly insignificant locations and can take many years, but can be reduced to a matter of weeks when inadequate materials are used. The consequences of failure can lead to major losses.

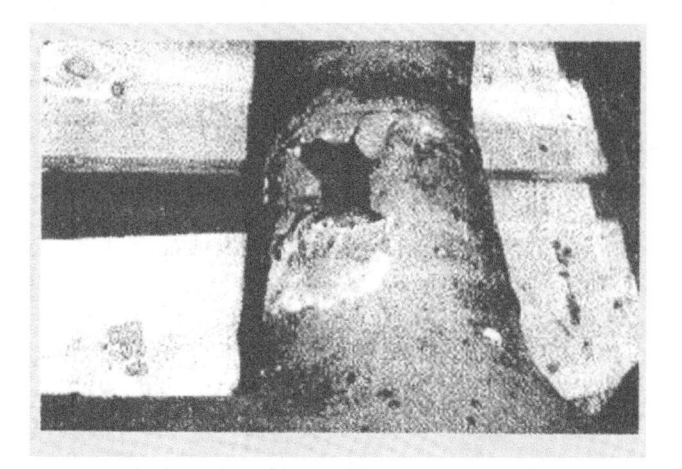

Figure 24.27 Corrosion in a 4" line made of carbon steel on a VDU.

A small section of ½% Mo carbon steel pipe had been installed between the 5% Cr 1/2% Mo CDU transfer line and a thermowell. The piping slowly corroded over 20 years before failure through corrosion occurred.

A leak occurred on a VDU when a 4-in. line that was used to recirculate hot distillate at 700°F (370°C) into the unit feed before the furnace [approximately 480°F (250°C)] ruptured as shown in Figure 24.27. After removal of the insulation, the release auto-ignited, resulting in a serious fire. The 4-in. pipe was made of ordinary carbon steel and installed 10 years earlier.

Failure can also occur at welds of dissimilar materials, particularly if the wrong materials are used.

A flange of the wrong material that had been installed on the VDU residue pump discharge line corroded due to failure at the welded zone. Further investigation showed many other examples of incorrect material usage in the residue piping system.

An error in defining the correct piping material specification break led to carbon steel pipe being used where 5Cr/0.5Mo was the correct specification for a VDU residue recycle line to the charge heater inlet, leading to corrosion and failure.

24.10 Material Failures and Process Safety Prevention Programs

PSM regulations require a facility to maintain mechanical integrity (MI) of the equipment. Facilities should have a well – documented inspection and repair program to ensure the equipment and piping are well maintained to prevent losses of containment and process safety incidents. A primary MI concern is materials damage and prevention. The MI program is usually based on the risks and corrosion rates of the various process services and materials used (e.g., risk – based inspection).

API 510 Pressure Vessel Inspection Code: In – Service Inspection, Rating, Repair and Alteration) defines the U.S. refining industry requirements for mechanical integrity. API 579 (Fitness for Service) provides methods and requirements for evaluation of equipment to continue operating in a facility. API Recommended Practice 571 provides a detailed discussion of the materials damage mechanisms and methods of control for each major type of process unit in a refinery. Many of these can be extrapolated to other types of petrochemical facilities. Table 24.10 provides a summary of the materials damage mechanisms discussed in RP – 571.

Piping Repair Incident at Tosco Avon Refinery, CA, USA

On February 23, 1999, a fire occurred in the crude unit at Tosco Corporation's Avon oil refinery in Martinez, CA, USA. Workers attempted to replace piping attached to a 150 ft tall fractionator (Figure 24.28) tower while the process unit was in operation. During the removal of the piping, naphtha was released onto the hot fractionator and ignited. The flames engulfed five workers located at different heights on the tower causing four fatalities and one sustained serious injury.

A pinhole leak was discovered in the crude unit on the inside of the top elbow of the naphtha piping, near where it was attached to the fractionator at 112 ft (34 m) above ground. Refinery personnel responded immediately, closing four valves in an attempt to isolate the piping, while the unit remained in operation.

Subsequent inspection of the naphtha piping showed that it was extensively thinned and corroded. A decision was made to replace a large section of the naphtha line. Over the 13 days between the discovery of the leak and the fire, workers made numerous unsuccessful attempts to isolate and drain the naphtha piping. The pinhole leak reoccurred three times, and the isolation valves were retightened in unsuccessful efforts to isolate the piping. Nonetheless, refinery supervisors proceeded with scheduling the line replacement while the unit was in operation.

On the day of the incident, the piping contained approximately 90 gallons (340 liters) of naphtha, which was being pressurized from the running process unit through a leaking isolation valve. A work permit authorized maintenance employees to drain and remove the piping. After several unsuccessful attempts to drain the line, a maintenance supervisor directed the workers to make two cuts into the piping using a pneumatic saw. After a second cut, naphtha began to leak. The supervisor directed the workers to open a flange to drain the line. As the line was being drained, naphtha was suddenly released from the open end of the piping that had been cut first. The naphtha ignited, most

Table 24.10 Refinery Materials Damage Mechanism (Reference: API Recommended Practice RP- 571).

Affected unit / Damage Mechanisms	Crude/ vacuum	Delayed coker	Fluid catalytic cracking	FCC light ends recovery	Continuous catalytic reforming (CCR)	Catalytic reforming (cyclic, semi-regent)	Hydroprocessing (treating, cracking)	Alkylation, sulfuric acid	Alkylation, HF	Amine treating	Sulfur recovery	Sour water stripper	Isomerization	Hydrogen plant
Mechanical & Metallurgical Failure														
Graphitization			X											
Softening (Spheroidization)			X											
Temper Embrittlement		X	X		X	X	X							X
Strain Aging														
885°F Embrittlement			X				X							X
Sigma Phase/Chi Embrittlement			X				X							
Brittle Fracture			X				X							
Creep/Stress Rupture		X	X		X	X	X							X
Thermal Fatigue		X	X			X								X
Short-term Overheating - Stress Rupture		X	X				X							X
Steam Blanketing														X
Dissimilar Metal Weld (DMW) Cracking	X								X		X			X
Thermal Shock		X	X											X
Erosion/Erosion-Corrosion	X	X	X		X	X	X		X	X		X		X
Cavitation														
Mechanical Fatigue					X	X								X

(Continued)

Table 24.10 Refinery Materials Damage Mechanism (Reference: API Recommended Practice RP-571). (Continued)

Affected unit	Crude/vacuum	Delayed coker	Fluid catalytic cracking	FCC light ends recovery	Continuous catalytic reforming (CCR)	Catalytic reforming (cyclic, semi-regent)	Hydroprocessing (treating, cracking)	Alkylation, sulfuric acid	Alkylation, HF	Amine treating	Sulfur recovery	Sour water stripper	Isomerization	Hydrogen plant
Vibration – Induced Fatigue														
Refractory Degradation			■			■					■			■
Reheat Cracking			■		■									■
Uniform or Localized Loss of Thickness														
Galvanic Corrosion									■					
Atmospheric Corrosion														
Corrosion Under Insulation (CUI)		■											■	
Cooling Water Corrosion					■									
Boiler Water/ Condensate Corrosion		■	■								■			■
CO_2 Corrosion	■										■			■
Flue Gas Dew Point Corrosion							■							
Microbiologically Induced Corrosion (MIC)		■												
Soil Corrosion														
Caustic Corrosion	■							■						
Dealloying		■												
Graphitic Corrosion														
Amine Corrosion										■				

(Continued)

Table 24.10 Refinery Materials Damage Mechanism (Reference: API Recommended Practice RP-571). *(Continued)*

Affected unit	Crude/vacuum	Delayed coker	Fluid catalytic crackin	FCC light ends recovery	Continuous catalytic reforming (CCR)	Catalytic reforming (cyclic, semi-regent)	Hydroprocessing (treating, cracking)	Alkylation, sulfuric acid	Alkylation, HF	Amine treating	Sulfur recovery	Sour water stripper	Isomerization	Hydrogen plant
Ammonium Bisulfide Corrosion (Alkaline Sour Water)							X			X		X		
Ammonium Chloride Corrosion	X	X	X		X	X	X							
Hydrochloric Acid (HCl) Corrosion	X				X	X	X							X
High Temperature H2/H2S Corrosion						X	X							
Hydrofluoric Acid (HF) Corrosion									X					
Naphthenic Acid Corrosion (NAC)	X	X					X							
Phenol (Carbolic Acid) Corrosion														
Phosphoric Acid Corrosion														
Sour Water Corrosion (Acidic)														
Sulfuric Acid Corrosion								X			X			
High Temperature Corrosion [400°F; 204°C]											X			
Oxidation	X	X	X		X	X					X			X
Sulfidation	X	X	X		X	X	X				X			
Carburization			X		X	X								X
Metal Dusting														X

(Continued)

Table 24.10 Refinery Materials Damage Mechanism (Reference: API Recommended Practice RP-571). (Continued)

Affected unit	Crude/ vacuum	Delayed coker	Fluid catalytic crackin	FCC light ends recovery	Continuous catalytic reforming (CCR)	Catalytic reforming (cyclic, semi-regent)	Hydroprocessing (treating, cracking)	Alkylation, sulfuric acid	Alkylation, HF	Amine treating	Sulfur recovery	Sour water stripper	Isomerization	Hydrogen plant
Fuel Ash Corrosion	■													
Nitriding														
Decarburization			■											
Environment-Assisted Cracking														
Chloride Stress Corrosion Cracking (Cl SCC)	■		■				■					■		■
Corrosion Fatigue														
Caustic Stress Corrosion Cracking (Caustic Embrittlement)	■							■	■				■	
Ammonia Stress Corrosion Cracking		■			■									
Liquid Metal Embrittlement (LME)							■							■
Hydrogen Embrittlement (HE)					■									
Polythionic Acid Stress Corrosion Cracking (PASCC)			■				■							■
Amine Stress Corrosion Cracking										■				

(Continued)

Table 24.10 Refinery Materials Damage Mechanism (Reference: API Recommended Practice RP-571). (Continued)

Affected unit	Crude/ vacuum	Delayed coker	Fluid catalytic crackin	FCC light ends recovery	Continuous catalytic reforming (CCR)	Catalyctic reforming (cyclic, semi-regent)	Hydroprocessing (treating, cracking)	Alkylation, sulfuric acid	Alkylation, HF	Amine treating	Sulfur recovery	Sour water stripper	Isomeriztion	Hydrogen plant
Wet H2S Damage (Blistering, HIC, SOHIC, SCC)	■	■	■	■			■			■	■	■		
Hydrogen Stress Cracking - HF									■					
Carbonate Stress Corrosion Cracking				■										
Other Mechanisms														
High Temperature Hydrogen Attack (HTHA)					■	■	■							■
Titanium Hydriding										■		■		■

Figure 24.28 Photo of the crude distillation tower at Tosco Avon refinery, CA, USA (Source: www.csb.gov).

likely from contacting the nearby hot surfaces of the crude tower, and quickly engulfed the tower structure and personnel. Figure 24.29 shows the arrangement of the tower at the time of the initial leak, Figure 24.30 shows naphtha stripper level control valve manifold removed to ground level, and Figure 24.31 shows the blockage in valve C and close-up of valve stem showing that it was not fully closed.

Lessons Learned from this accident

The opening of plant and equipment requires

Careful work planning with a job hazard analysis

A work permitting system that:

- Assigns authority and accountability for all aspects of the work.
- Provides written confirmation that a work site is prepared and safe before work commences.
- Describes all outstanding potential hazards and required precautionary measures.
- Ensures that plant/equipment is formally handed back to operations in a safe state for recommissioning.
- Ensures higher risk maintenance activities are authorized by a higher level of management staff.
- Controls who does the work and at what time.
- Work permit system must be regularly monitored by management and periodically independently audited to ensure that procedures are being effectively implemented.
- Permit issuing and performing authorities must be certified competent to undertake these duties following formal training.

24.11 Hazard and Operability Studies (HAZOP)

The design and operation of a process plant form an integral part of safety and systematic procedures should be employed to identify hazards and operability, and where necessary, they should be quantified. During the design of a new plant, the hazard identification procedure is repeated at intervals. This is first carried out on the pilot plant before the full-scale version as the design progresses. Potential hazards whose significance can be assessed with the help of experiments are often revealed by this study.

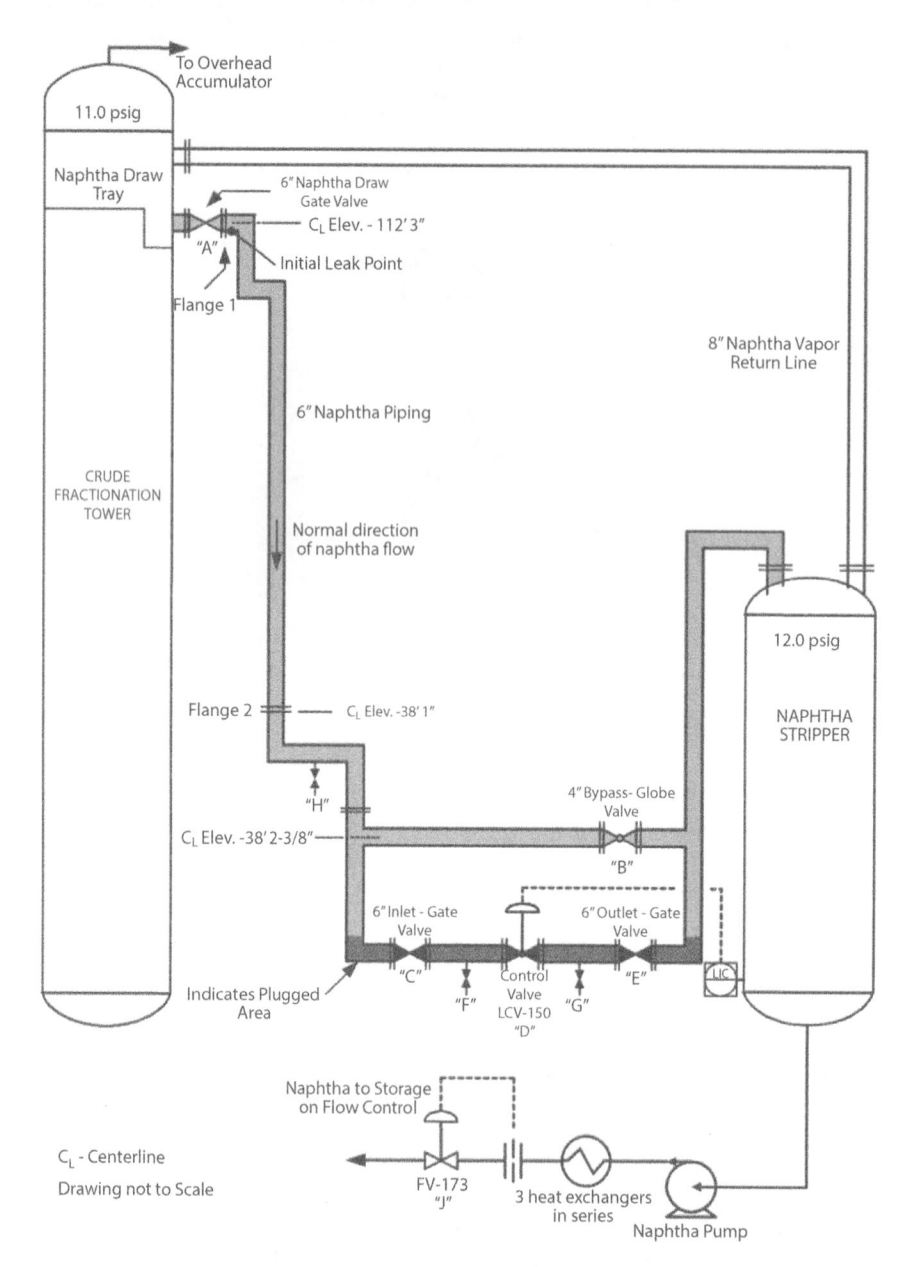

Figure 24.29 Fractionator and naphtha draw arrangement at the time of the initial leak (Source: www.csb.gov).

The HAZOP study identifies a potential hazard. It provides little information on risk and consequences or its seriousness. However, judgment is required, and sometimes the designer may decide that the consequences of the hazard are either trivial or unlikely to be ignored. In certain instances, the solution is obvious and the design is modified. A fault tree analysis (FTA) is useful where the consequences of the hazard are severe, or where its causes are many. The fault tree indicates how various events or combinations of events can give rise to a hazard. It can be employed to identify the most likely causes of the hazard, and thus to show where additional safety precautions will be most effective.

The HAZOP studies have formed an integral part of process safety management (PSM); several books and standards on this aspect have been published since its introduction to the CPI in the 1960s [5–7]. Over the years, practitioners have gained extensive experience with this methodology and in the US, 29 CFR 1910 is mandatory. A HAZOP study is a structured review of the design of a plant and operating procedures. The main aims are the following:

- identify potential maloperations.
- assess their consequences.
- recommend corrective actions.

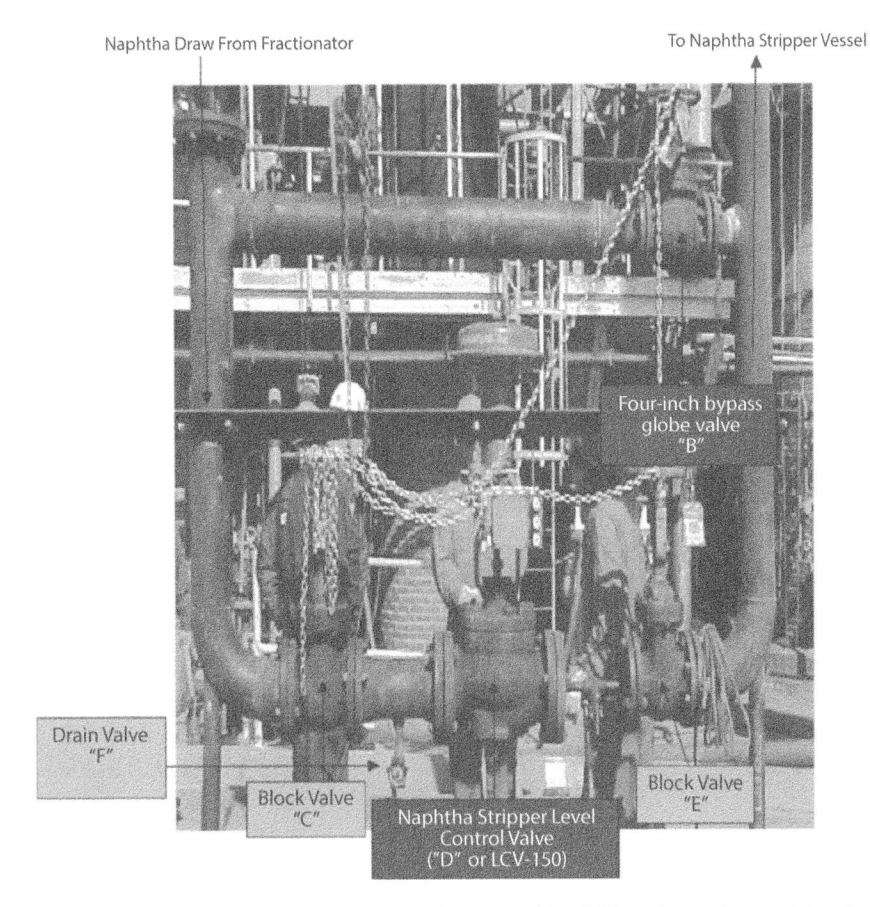

Figure 24.30 Naphtha stripper level control valve manifold removed to ground level. The valve at the top right of manifold is the 4-in. bypass valve.

Blockage found in Valve "C" and close up of valve stem showing that it was not fully closed

Figure 24.31 Blockage in valve C (Source: www.csb.gov).

The recommended actions may eliminate a potential cause or interrupt the consequences. They involve the following:

- pipework or other hardware changes.
- changes in operating conditions.
- more precise operating instructions.
- addition of alarms with prescribed operator responses.
- addition of automatic trip systems.

Table 24.11 lists some common failure modes and design considerations for reactors [2].

Table 24.11 Common Failure Modes, Causes, Consequences, Design Considerations for Reactors [2].

Failure mode	Causes	Consequences	Design consideration
Loss of Cooling	Loss of heat transfer medium from supply Control system failure	Potential runaway reaction	Emergency relief system Dual cooling condenser and reactor jacket Automatic actuation of secondary cooling medium on detection of low coolant flow, or high pressure, or high reactor temperature Automatic stopping of feeds of reactants or catalyst (with semi-batch or continuous reactors).
Loss of agitation	Loss of power Motor failure Agitator blades become loose/fall off	Potential runaway reaction	Emergency relief system Uninterrupted power supply backup to motor Agitator power consumption or rotation indication interlocked to stop the feed of reactants or catalyst or activate emergency cooling
Overcharge of reactant or catalyst	Error in measurement Control system failure	Potential runaway reaction Overflow of reactor	Emergency relief system Dedicated charge tanks sized to hold only the amount of reactant/catalyst needed Quantity of reactant/catalyst added limited by flow totalizer Redundant flow totalizers High level interlock/permissive to limit quantity of reactant/catalyst
Wrong reactant/catalyst	Misidentification Mix-up during product change	Potential runaway reaction	Emergency relief system Dedicated feed tank and reactor train for production of one product Control software preventing charge valve or pump operation until correct material bar code has been scanned.
Step done out of sequence	Poor instructions/training Human error	Potential runaway reaction	Controllers that verify a step has been done before advancing to the next step.

The procedure for a HAZOP study is to apply a number of guide words to various sections of the process design intention. The design intention informs what the process is expected to carry out. Table 24.12 shows these guidewords, and Figure 24.32 summarizes the whole procedure. Some companies have developed their own set of guidewords for particular technologies; however, while clear recommendations can be made as to which guidewords should be considered, it is not possible to provide such firm advice regarding parameters. The selection of parameters is a task each team must address for each system studied. Table 24.13 gives examples of parameters that might be used in the analysis of a process operation. The list is intended to show the depth and breadth of the parameter and guideword search that can be used. Further, the extent of this list emphasizes the need for the HAZOP team to form a clear conceptual model of the step and to use it to decide which parameters should be used in the search for possible deviations. When deviations are being sought, it must be remembered that not every guideword combines with a parameter to give a meaningful deviation. Therefore, it is futile and a waste of time to discuss combinations that do not have a physical meaning. Table 24.13a shows examples of meaningful combinations. In general, HAZOP

Table 24.12 Hazop Record Sheet.

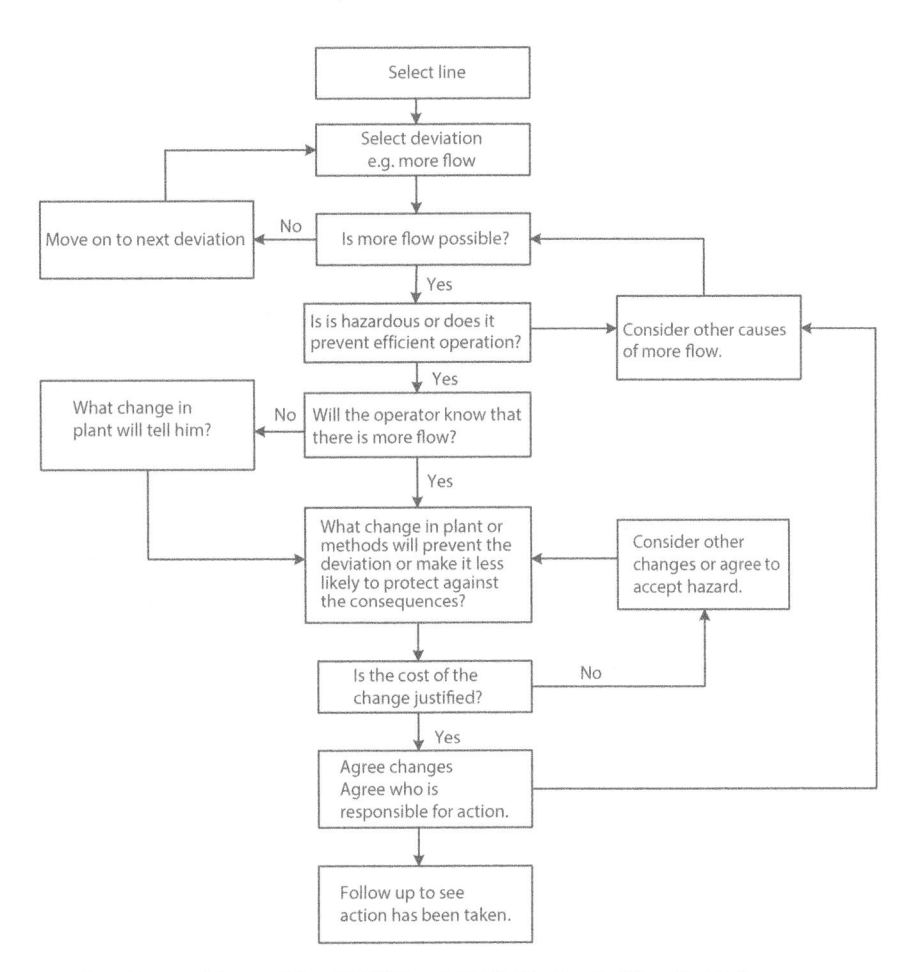

Hazard and operability study report								
Project title:				Sheet of				
Project number:				Date:				
P & ID number:				Chairman:				
Line number:				Study team:				
Guide word	Deviation	Cause	Consequence	Safeguards	Action			
					Number	By	Detail	Reply accepted

Figure 24.32 HAZOP procedure (Source: Trevor Keltz: HAZOP and HAZAN, 4th. ed. IChemE, 1999).

Table 24.13 Examples of Possible Parameters for Process Operations.

• Addition	• Phase
• Composition	• Speed
• Level	• Particle size
• Flow or amount	• Reaction
• Mixing	• Control
• Transfer	• pH
• Temperature	• Sequence
• Viscosity	• Signal
• Measure	• Start/stop
• Stirring	• Operate
• Pressure	• Maintain
• Separation	• Services
• Communication	• Time

(Source [7] By permission of IChemE., All rights reserved).

Table 24.13a Examples of Meaningful Combinations of Parameters and Guide Words.

Parameter	Guidewords that can give a meaningful combinations
Flow	None; more of; less of; reverse; elsewhere; as well as
Temperature	Higher; lower
Pressure	Higher; lower; reverse
Level	Higher; lower; none
Mixing	Less; more; none
Reaction	Higher (rate of); lower (rate of); none; reverse; as well as/other than; part of
Phase	Other; reverse; as well as
Composition	Part of; as well as
Communication	None; part of; more of; less of; other; as well as

(Source [7] By permission of IChemE., All rights reserved).

study is most effective when it is a creative process and the use of checklists for guidewords or parameters can stultify creativity. Nonetheless, checklists can be helpful for an experienced team.

Study Co-ordination

HAZOP is a structured review exercise carried out by a team of between three and six people, one of whom acts as a chairman. Another member of the team acts as a secretary and records the results of the proceeding. A typical team comprises

- Chairman: study leader, experienced in HAZOP.
- Project or design engineer: usually a mechanical engineer, responsible for keeping the costs within budget.
- Process engineer: usually the chemical engineer who drew up the flowsheet.
- Commissioning manager: usually a chemical engineer, who will have to start up and operate the plant.
- Control system design engineer: modern plants contain sophisticated control and trip systems and HAZOPS often result in the addition of yet more instrumentation.
- Research chemist: if new chemistry is involved.
- Independent team leader: an expert in the HAZOP technique, not the plant whose job is to ensure that the team follows the procedure.

The team should have a wide range of knowledge and experience. If a contractor designs the plant, then the HAZOP team should include people from both the contractor and client organizations. On a computer-controlled batch plant, the software engineer should be a member of the HAZOP team, which should include at least one other person who understands the computer logic.

In an existing plant, the team should include several people with experience of the plant. A typical team is as follows:

- Plant manager: responsible for plant operation.
- Process foreman: knows what actually happens rather than what is supposed to happen.
- Plant engineer: responsible for mechanical maintenance such as testing of alarms and trips, as well as installation of new instruments.
- Process investigation manager: responsible for investigating technical problems and for transferring laboratory results to plant-scale operations.
- Independent team leader.

24.11.1 HAZOP Documentation Requirements

The essential document for a HAZOP is the process and instrumentation diagram (P & ID), but other documents are also required at the start of the study [8]:

- Process and instrumentation diagrams (P & IDs).
- Process flow diagrams (PFDs).
- General arrangement drawings.
- Relief/venting philosophy.
- Chemical hazard data.
- Piping specifications.
- Process data sheets.
- Previous safety report.

Other desirable documents include

- Operation and maintenance instructions.
- Safety procedure documents.
- Vendor package information.
- Piping isometrics.

The requirements for each study often vary and the way the information is assembled differs from a project to project. However, sufficient documentation should always be available to indicate clearly the design intent and to provide details of process parameters. Additional information is required for batch processes as discussed later.

24.11.2 The Basic Concept of HAZOP

The concept of HAZOP requires the splitting up of the plant into sections and the systematic application of a series of questions in each section. The study team discovers how deviations from the design intent can occur and can decide the consequences of the deviations from the points of view of hazard and operability.

The following terms are used as a basis for all HAZOP studies:

Design intent—the way in which the plant is intended to operate.

Deviation—any perceived deviations in operation from the design intent.

Cause—the cause of the perceived deviations.

Consequence—the consequences of the perceived deviations.

Safeguards—existing provisions to mitigate the likelihood or consequences of the perceived deviations and to inform operators of their occurrence.

Actions—the recommendations or requests for information made by the study team in order to improve the safety and/or operability of the plant.

Guidewords—simple words used to qualify the intent and hence discover deviations.

Parameters—basic process requirements such as "flow," "temperature," "pressure," and so on.

24.11.3 Division into Sections

In order to proceed in a logical and efficient manner, the P & ID is first divided into sections, as the division into few very large sections can result in important deviations and consequences being missed. The most important point in sectioning is that the guidewords must apply uniformly throughout every part of the section. Factors to be considered when sectioning the plant include

- Purpose/function of the section.
- Material (volume or mass) in the section.
- Material process/state considerations.
- Reasonable isolation/terminal points.
- Consistency of approach.

The general guidelines to be followed can be summarized as follows:

- Define each major process component as a section. Usually anything assigned an equipment number is considered a major component.
- Define one line section between each major component.
- Define additional line sections for each branch off the main process flow.
- Define a process section at each connection to existing equipment.

Use of Guidewords

The questions are constructed using a number of guidewords to ensure a consistent and structured approach. The application of an accepted set of guidewords ensures that every conceivable deviation is considered. The guidewords are normally applied in conjunction in a series of process parameters to arrive at a meaningful deviation. The HAZOP study could be considered as structured or guided brainstorming exercise.

The development of meaningful deviations from the guidewords depends on the nature of the process being studied. Two approaches are possible:

1. Select a guideword—say, NONE—apply it in turn to a number of process parameters, for example, FLOW, TEMPERATURE, PRESSURE—to produce such meaningful deviations as NO FLOW, NO PRESSURE, and so on.
2. Select a process parameter—say, FLOW and—apply the guide words in turn to produce such meaningful deviations as NO FLOW, MORE FLOW and so on.

Both techniques should provide the same results but many practitioners prefer the second approach as it's more logical.

Tables 24.14 and 24.15 list typical guidewords, parameters, deviations, and possible causes.

24.11.4 Conducting a HAZOP Study

In order to make the HAZOP technique work effectively, it is necessary to institute a formal procedure including the following headings:

Table 24.14 Process Parameter-Related Guidewords [8].

NO FLOW	Wrong routing—blockage—incorrect slip plate—incorrectly fitted NRV—burst pipe—large leak—equipment failure—incorrect pressure differential—isolation in error—no material available—vapor lock.
LESS FLOW	Line restriction—filter blockage—defective pumps—fouling—density or viscosity problems—incorrect specification of process fluid.
REVERSE FLOW	Defective NRV—syphon effect—incorrect differential pressure—two-way flow—emergency venting—incorrect routing.
MORE FLOW	Increased pumping capacity—increased suction pressure—reduced delivery head—greater fluid density—exchanger tube leaks—restriction orifice plates deleted—cross connection of systems—control surging—valve(s) failed open.
LESS PRESSURE	Vacuum condition—condensation—gas dissolving in liquid—restricted pump or compressor suction line—undetected leakage—vessel drainage.
MORE PRESSURE	Surge problems—leakage from interconnected hp system—gas breakthrough—isolation procedures for relief valves defective—thermal overpressure—positive displacement pumps—failed open PCVs—uncontrolled reaction.
LESS TEMPERATURE	Ambient conditions—fouled or failed exchanger tubes—fire situation—cooling water failure—defective control—fixed heater control failure—internal fires—reaction control failures.
LESS VISCOSITY	Incorrect material specification or temperature.
MORE VISCOSITY	Incorrect material specification or temperature.
COMPOSITION CHANGE	Leaking isolation valves—leaking exchanger tubes—phase change—incorrect feedstock/specification—inadequate quality control—process control.
MORE THAN	Contamination leaking exchanger tubes or isolation valves—incorrect operation of system—interconnected systems—effect of corrosion—wrong additives—ingress of air—impurities—extra phases.

Table 24.15 General Guidewords [8].

OTHER ACTIVITIES	Start-up and shutdown of plant—testing and inspection—commissioning—decommissioning—demolition—weather—seismic—physical impact—containment loss and consequences—domino effects—purging—washing out—toxicity.
RELIEF	Relief philosophy—type of relief device and reliability—relief valve discharge location—pollution implications.
CONTROL	Control philosophy—location of instruments—response time—set points of alarms and trips—time available for operator intervention—alarm and trip testing—fire protection—electronic trip/control amplifiers—panel arrangement and location—auto/manual facility human error.
SAMPLING	Sampling procedure—time for analysis result—calibration of automatic samplers/reliability—accuracy of representative sample.
CORROSION/EROSION	Cathodic protection arrangements—internal /external corrosion protection—engineering specifications—zinc embrittlement—stress corrosion cracking—fluid velocities—riser splash zones.
SERVICE FAILURE	Instrument air/steam/nitrogen/cooling water—hydraulic power—electric power—electric power—telecommunications—heating and ventilating systems—computers.
MAINTENANCE	Isolation—drainage—purging—cleaning—drying—slip plates—access—rescue plan—training—pressure testing—work permit system—condition monitoring—catalyst change and activation.
STATIC	Earthing arrangements—insulated vessels/equipment—low conductance fluids—splash filling of vessels—insulated strainers and valve components—dust generation—and handling-hoses.
SPARE EQUIPMENT	Installed/non-installed spare equipment—availability—modified specifications—storage of spares.
SAFETY	Fire and gas detection system/alarms—emergency shutdown arrangements—firefighting response time—emergency and major emergency training—contingency plans - TLVs and methods of detection - noise levels - security arrangement—knowledge of hazards of process materials—first aid/medical resources—effluent disposal—hazards created by others (adjacent storage areas/process plant, etc.)—testing of emergency equipment—compliance with local national regulations - environmental considerations.

- Define objectives and scope.
- Prepare for the study.
- Carry out the study.
- Record the results.
- Follow up.

Define Objective and Scope

The objectives and scope should be clearly stated and understood by all concerned, and documented and agreed before the start of the study. The definition should include, but not necessarily be limited to the following:

- Study terminal points, best defined in terms of P & IDs.
- Design status at time of study, defined in terms of P & ID revision status.
- Extent to which effects on and by adjacent plant should be considered.
- Study program including action reply and final reporting dates.
- Links with studies being conducted on adjacent or related plants.

Prepare for the Study

The timing of the study is essential to its success. The timetable depends upon

- Project program dates.
- Availability of documentation.
- Availability of personnel.

The ideal time to carry out the HAZOP is at the process design freeze. Many companies use the completed HAZOP as a signal that the P & IDs are "approved for design." At this stage, the process design should be sufficiently advanced for most of the required information to be available but not sufficiently advanced for alterations to be too costly.

Record the Results

As the study progresses, it is the responsibility of the secretary to record the discussion accurately. The chairman ensures that adequate time is allowed and helps by summarizing the discussion and the agreed actions/recommendations. The chairman ensures that the team is in agreement with the findings recorded. Table 24.16 shows the basic record of work sheet as completed by the team secretary as the study proceeds. A number of formats are used, some omitting the safeguards column. Computerized systems are also available to produce reports interactively. The record sheets are completed as the study proceeds and actions/recommendations soon afterwards.

The HAZOP record sheet forms part of the final plant documentation and should be in a form that can be submitted to the regulatory authorities. In order to keep track of the sections being studied, they are first identified on the master copy of the P & ID and, as each section is completed, are clearly marked as complete. It is possible to see by examining the drawing exactly what has been studied and the order in which the study has been carried out. Thus, any branches, vents, drains, and so on, which may not have been studied, can be clearly identified.

24.11.5 Hazop Case Study [8]

The hydrocarbon is transferred from intermediate storage via the J1 pumps and a 1 km over ground pipeline running adjacent to a public road into a 25 m³ nitrogen blanketed feed/settling tank operating at 20°C and 1 barg. Control

Table 24.16 Applicability of Selecting Process Hazard Analysis Techniques.

Phase of process design or operation	Technique					
	Checklist	What - if	What – if/ checklist	Hazard and Operability Study (HAZOP)	Failure Modes and Effects Analysis (FMEA)	Fault Tree Analysis (FTA)
R & D		√	√			
Design	√	√	√			
Pilot Plant Operation	√	√	√	√	√	√
Detailed Engineering	√	√	√	√	√	√
Construction/Startup	√	√	√			
Routine Operation	√	√	√	√	√	√
Modification	√	√	√	√	√	√
Incident Investigation		√		√	√	√
Decommissioning	√	√	√			

is on liquid level and there is a split range pressure control from the 2 barg site nitrogen supply. Integrated flow measurement is provided at J1 pump common delivery for accountancy purposes, and a manual sample point is provided at the tank inlet for offline analysis. There is a branch, normally closed, to a petrol blending system.

Figure 24.33 shows the P&ID of the hydrocarbon transfer system and Table 24.17 provides an action report that would be expected of a typical HAZOP team. Note that there is no "safeguards" column in this report and a total of 20 actions were generated, which is the typical number for a single plant section.

24.11.6 HAZOP of a Batch Process

A HAZOP study of a batch processing plant involves a review of the process as a series of discrete stages, as the conditions in a single reactor change with time in a given cycle. Variations in the rate of change, as well as changes in the duration of settling, mixing, and reacting, are important deviations. A HAZOP on a batch process includes

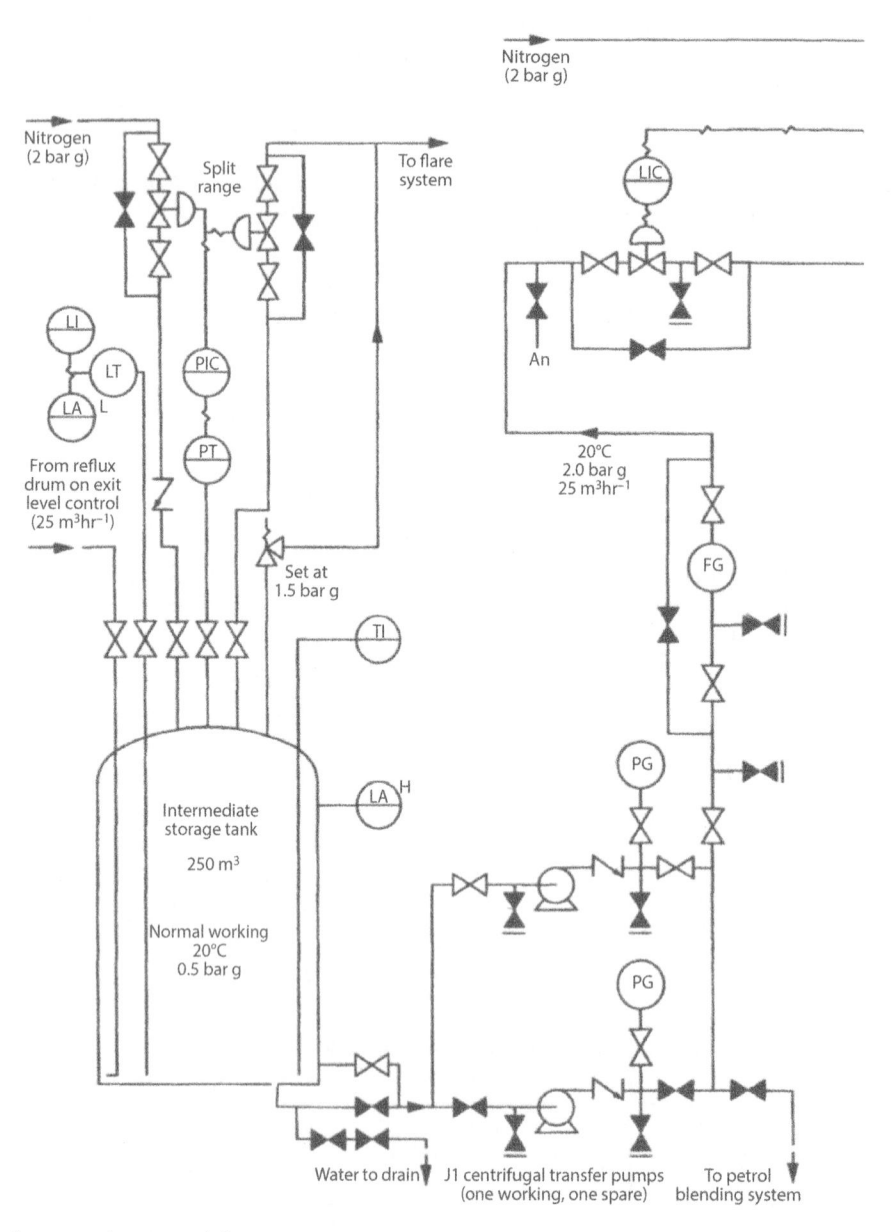

Figure 24.33 Hydrocarbon transfer system [8].

(*Continued*)

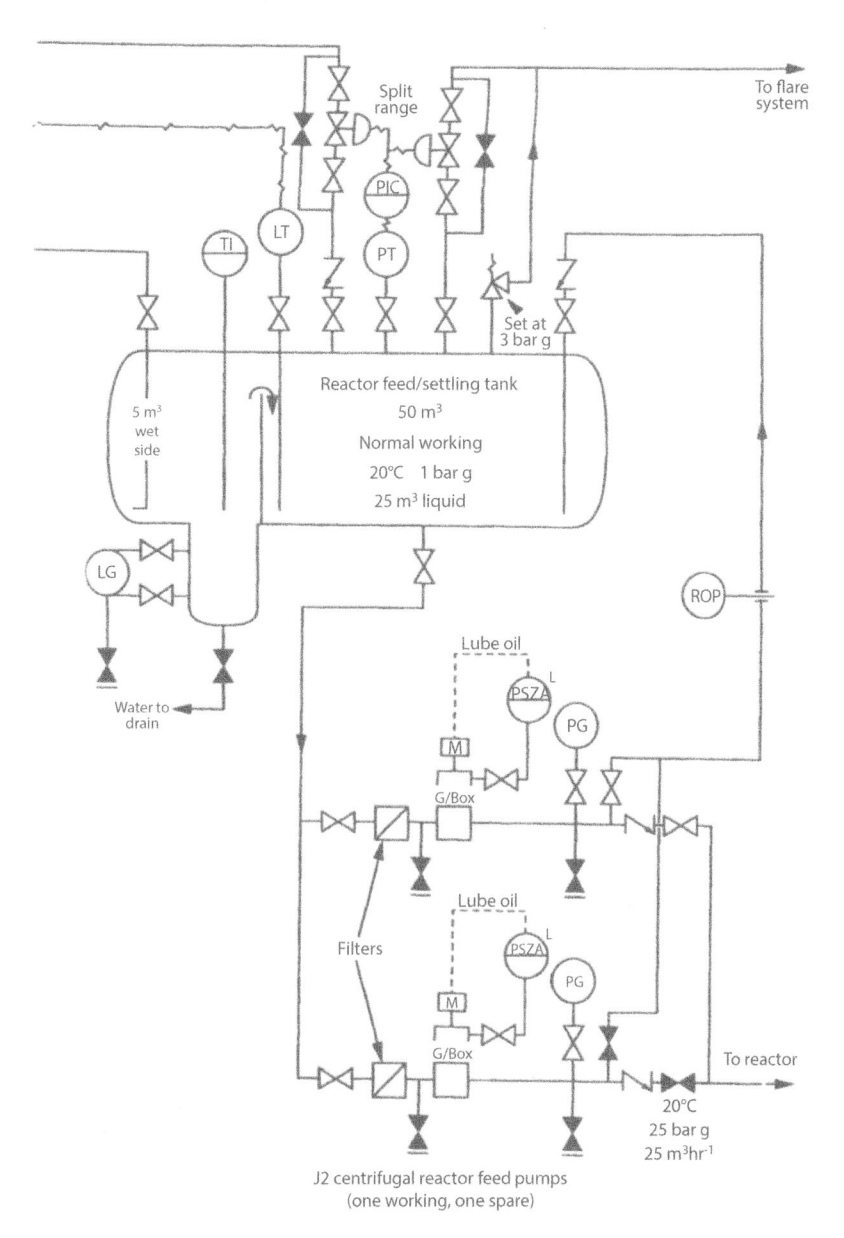

Figure 24.33 (Continued) Hydrocarbon transfer system [8].

the flowrates of service fluids: it is more common to look at the reactants in terms of amounts charged or discharged rather than flowrates.

Sequential operating instructions are applied to identify the design intent through various stages of the process. The HAZOP guidewords can then be applied to each design intent instead of the process line. For example, raise concentration or temperature to a specified value for a given time. The HAZOP would then consider the consequences of not achieving or exceeding the desired concentration or temperature in the specified time. A further example is, if an instruction states that 1 tonne of A has to be charged to a reactor, then the team should consider deviations such as

DON'T CHARGE	A
CHARGE MORE	A
CHARGE LESS	A
CHARGE AS WELL AS	A
CHARGE PART OF	A (if A is a mixture).
CHARGE OTHER THAN	A

Table 24.17 HAZOP Record Sheet.

HAZARD AND OPERABILITY STUDY REPORT

Project title: Hydrocarbon transfer system: Results of line section from intermediate storage to buffer/settling tank.			Sheet 1 of 5
Project number:			Date:
P & ID number:			Chairman:
Line number:			Study team:

Guide word	Deviation	Possible causes	Consequence	Action required
NONE	NO FLOW	1. No hydrocarbon available at intermediate storage	Loss of feed to reaction section and reduced output. Polymer formed in heat exchanger under no flow conditions. Buffer tank level fails.	(a) Ensure good communications with intermediate storage operator. (b) Install low level alarm on settling tank LIC
		2. J1 pump fails (motor fault, loss of drive, impeller corroded, etc.), power failure	As for (1)	Covered by (b) Add pump running indicator lights (also in control room).
		3. Line blockage, LCV fails shut	As for (1) J1 pump overheats Rising level in storage tank?	Covered by (b) (c) Install kickback on J1 pumps
		4. Line fracture	As for (1) Hydrocarbon discharged into area adjacent to public highway	Covered by (b). Consider adding second FQ at buffer tank inlet. (d) Institute regular patrolling and inspection of transfer line.
		5. Valve closure in error	As for (1)	(e) Review operator reliability and provision for J2 pump protection

(*Continued*)

Table 24.17 HAZOP Record Sheet. *(Continued)*

HAZARD AND OPERABILITY STUDY REPORT

Project title: Hydrocarbon transfer system: Results of line section from intermediate storage to buffer/settling tank.				Sheet 2 of 5	
Project number:				Date:	
P & ID number:				Chairman:	
Line number:				Study team:	

Guide word	Deviation	Possible causes	Consequence	Action required
REVERSE	REVERSE FLOW	6. Failure of PIC and higher-than-normal N_2 pressure	N_2 gas breakthrough	(f) Add anti-siphon provisions in buffer tank dip-pipe
		7. Failure or leakage of NRV	N_2 passed to intermediate storage	(g) Check capacity of intermediate storage relief system. Review maintenance program
		8. Backflow through standby pump of standby pump	Reduced delivery to buffer tank	(h) Check operating instructions for isolation
MORE OF	MORE FLOW	9. LCV fails open or LCV bypass open in error or both pumps operating	Settling tank overfills Incomplete separation of water phase in tank leading to reaction problems.	(i) Install high-level alarm on LIC and check sizing of relief opposite liquid overfilling. Consider high-high level trip with auto re-set. (j) Institute locking-off procedure for LCV bypass when not in use (k) Extend J2 pump suction line to 300 mm above tank base

(Continued)

Table 24.17 HAZOP Record Sheet. (*Continued*)

HAZARD AND OPERABILITY STUDY REPORT

Project title: Hydrocarbon transfer system: Results of line section from intermediate storage to buffer/settling tank.

Project number:				Sheet 3 of 5
P & ID number:				Date:
Line number:				Chairman:
				Study team:

Guide word	Deviation	Possible causes	Consequence	Action required
LESS OF	LESS FLOW	10. Leaking flange or valve stub not blanked and leaking	Material loss adjacent to public highway	Covered by (d)
		11. Blocked pump section	Pump damage	(l) Check filter periodically
		12. Impeller wear	Reduced delivery pressure	(m) Review inspection/maintenance program.
		13. Possible vapor locking in hot weather	As for 15.	(n) Provide suitable vent on pump casing
MORE OF	MORE PRESSURE	14. Isolation valve closed in error or LCV close, with J1 pump running	Transfer line subjected to full pump delivery, surge pressure or closed head pressure. Pressure exceeds line specification. Slow closing time of control valve	(o) Covered by (c) except when kickback blocked or isolated. Check line, FQ and flange ratings, and reduce stroking speed of LCV if necessary. Install a PG upstream of LCV and an independent PG on settling tank.
		15. Thermal expansion in an isolated valved section due to fire or strong sunlight	Line fracture or flange leak	(p) Install thermal expansion relief on valved section (relief discharge route to be decided later in study).

(*Continued*)

Table 24.17 HAZOP Record Sheet. (*Continued*)

HAZARD AND OPERABILITY STUDY REPORT				
Project title: Hydrocarbon transfer system: Results of line section from intermediate storage to buffer/settling tank.				Sheet 4 of 5
Project number:				Date:
P & ID number:				Chairman:
Line number:				Study team:
Guide word	Deviation	Possible causes	Consequence	Action required
MORE OF	MORE TEMPERATURE	16. High intermediate storage temperature	Higher pressure in transfer line and settling tank	(q) Check whether there is adequate warning of high temperature at intermediate storage. If not, install a device.
LESS OF	LESS TEMPERATURE	17. Winter conditions	Water sump and drain line freeze-up	(r) Lag water sump down to drain valve, and steam trace drain valve and drain line downstream.
PART OF	HIGH WATER CONTENT IN STREAM	18. High water level in intermediate storage tanks	Water sump fills up more quickly. Increased chance of water phase passing to reaction section.	(s) Arrange for frequent draining off of water from intermediate storage tank. Install high interface level alarm on sump.
	PRESENCE OF MORE VOLATILE COMPONENTS	19. Disturbance on distillation columns upstream of intermediate storage.	Higher system pressure	(t) Check that the design of settling tank and associated pipework, including relief valve sizing, will cope with sudden ingress of more volatile hydrocarbon.

(*Continued*)

Table 24.17 HAZOP Record Sheet. (*Continued*)

HAZARD AND OPERABILITY STUDY REPORT				
Project title: Hydrocarbon transfer system: Results of line section from intermediate storage to buffer/settling tank.				Sheet 5 of 5
Project number:				Date:
P & ID number:				Chairman:
Line number:				Study team:
Guide word	Deviation	Possible causes	Consequence	Action required
OTHER	MAINTENANCE	20. Equipment failure, flange leak, etc.	Line cannot be completely drained or purged.	(u) Install low-point drain and N_2 purge point downstream of LCV. Also N_2 vent on settling tank.

REVERSE CHARGE A (i.e., can flow occur from the reactor to the A container?). This can be the most serious deviation if

A IS ADDED EARLY
A IS ADDED LATE
A IS ADDED TOO QUICKLY
A IS ADDED TOO SLOWLY

Many accidents have occurred because process materials flowed in the wrong direction to that expected. For example, ethylene oxide (C_2H_4O) and ammonia (NH_3) were reacted to make ethanolamine ($HOCH_2CH_2NH_2$). Some ammonia flowed from the reactor in the opposite direction, along the ethylene oxide transfer line into the ethylene oxide tank, past several non-return valves and a positive displacement pump. It got past the pump through the relief valve, which discharged into the pump suction line. The ammonia reacted with 30 m³ of ethylene oxide in the tank, which ruptured violently. The released ethylene oxide vapor exploded causing damage and destruction over a wide area. A HAZOP study might have disclosed the facts that reverse flow could occur [9].

Limitations of HAZOP Studies

While HAZOP can readily identify the majority of operational problems, it is still fallible to what it can miss. The HAZOP-Guide to best practice [6] states, "A HAZOP study is not an infallible method of identifying every possible hazard or operability problem that could arise during the actual operations. Expertise and experience within the team is crucial to the quality and completeness of a study. The accuracy and extend of the information available to the team, the scope of the study and the manner of the study all influence its success". Krishnan [10] has illustrated the limitations of HAZOP studies as exemplified in Table 24.18. The table serves as an illustration to gain insight with regard to accidents and their predictability by HAZOP. Table 24.19 shows pertinent points that should be kept in mind while performing a thorough HAZOP study.

Conclusions

The HAZOP technique has provided a method of safety assurance in the refinery industry and related CPI and is accepted throughout the world as a way of demonstrating that a project has been subjected to a rigorous safety examination. It is agreed that success or failure of such a technique depends upon

- The accuracy of the drawings and other documents used in the study.
- The expertise and experience of the team.
- The ability of the team to visualize deviations, causes, and consequences.
- The ability of the team to assess the seriousness of hazards.
- The skill of the chairman in keeping the study on track.

The skill of the chairman is the most important as the technique will only be successful if the chairman has the required training and experience for the task. However, the contribution by the team in the investigation is essential for a total success of the HAZOP study.

24.12 HAZAN

The Hazard analysis (HAZAN) is a quantitative way of assessing the likelihood of failure. Other names associated with this technique are risk analysis, quantitative risk assessment (QRA), and probability risk assessment (PRA). Kletz [11] expressed the view that HAZAN is a selective technique while HAZOP can be readily applied to new design and major modification. Some limitations of HAZOP are its inability to detect every weakness in design

Table 24.18 Summary of Recent Major Industrial Accidents that are Sourced to Process Safety Failure.

Incident	Causes	HAZOP identification of the cause
Flixborough, UK (1974) Cyclohexane vapor cloud explosion	The cause was determined as the failure of a bypass line connecting two reactors handling high quantities of cyclohexane. The bypass line was installed in a hurry without proper engineering or safety review.	HAZOP would not have been able to predict the actions of the plant staff.
Bhopal, India (1984) Release of poisonous MIC gas (Methyl isocyanate)	The multiple causes that led to this accident included storage of a highly poisonous chemical MIC in a facility that was shut down and idle, failure of the scrubbing and flare system to absorb the toxic vapor, and lack of knowledge on the nature of toxicity of the chemical itself.	A well-done HAZOP would have predicted the hazards due to the non-operability of the relief system. But it would not have predicted the nature of the risk and catastrophe in case of a loss of containment unless all details of the chemical were known. (This raises a question as to how much we know about the long-term and short-term effects of chemicals).
Seveso, Italy (1976) Release of poisonous gas TCDD	A runaway reaction caused the chemical TCDD (2, 3, 7, 8 tetrachlorodibenzo-paradioxin—one of the most potent toxins known to man) to be released through the relief system in a white cloud over the town of Seveso. A heavy rain washed the TCDD into the soil. Lack of knowledge and poor communication with public delayed response from authorities.	HAZOP would have predicted the consequences of runaway reaction and release of the vapor. But the lack of knowledge about the chemical itself and its consequences would not have been predictable by HAZOP.
Three Mile Island nuclear plant (1979) Equipment malfunction and shutdown resulting in partial meltdown of reactor core	In the Three Mile island nuclear plant, shutdown of main feed water pumps caused chain shutdown of steam generators and the reactor. Pressure build-up in the reactor system occurred and the relief valve on system opened to relieve the pressure, but failed to seat back. Result was loss of containment of coolant through the relief valve. But somehow there was no direct way the operators could know that the level of coolant was dangerously low in the reactor and that the reactor was overheating.	In Three Mile Island, there were alarms in the control room but the operator did not know that the relief valve was stuck open and the coolant level was getting low. Alarms in the control room resulted in confused initial actions by the operating staff that actually worsened the situation. Though the reactor core suffered a partial meltdown, worst-case scenarios were avoided. A HAZOP would have predicted lack of critical alarms but would not have predicted the actions by the operating staff. Reports indicate that there were, in fact, too many alarms (nearly 100!) in the control room.
Piper Alpha Offshore platform (1988) Leaked condensate caught fire resulting in massive fire and explosion and loss of the platform	It is believed that the leak came from pipe-work connected to a condensate pump. A safety valve had been removed from the pipe-work for overhaul and maintenance. The pump itself was undergoing maintenance work. When the pipe-work from which the safety valve was removed was pressurized at startup, it is believed that the leak occurred.	HAZOP would have predicted and corrected a lot of items that were found lacking in the platform during the subsequent inquiry including unit spacing and locations, safety provisions, etc., but would not have predicted the causes of the incident and further the reason for escalation of the incident to a catastrophe.

(Source: G. Unni Krishnan [8], Reprinted with permission from Hydrocarbon Processing, By Gulf Publishing Company, copyrighted 2005; All rights reserved).

Table 24.19 Checklist for a Productive HAZOP Study.

1. HAZOP is not an infallible method. It can achieve many items, but users must be aware of study's limitations.
2. An overall Process Safety management (PSM) system should be in place and HAZOP should be a part of the same. For complex systems, always conduct another study augmenting the HAZOP study including Quantitative Risk Analysis (QRA), FMEA, and/or event tree analysis.
3. HAZOP *is not a substitute* for design. The process design should be sound with regard to codes, standards, and good engineering practice. Also, the design stage should not be rushed with the attitude that the HAZOP team will catch any errors or omissions at a later time. This is the biggest judgment lapse an engineering team can make.
4. Composition of the HAZOP team is critical. The best individuals in a positive group dynamics environment are necessary to produce a high-quality HAZOP report.
5. Management must support and act on the HAZOP recommendations. Many recommendations made with much effort and time are forgotten due to lack of follow-up action and no support from management.
6. Low-probability major-consequence hazards such as total power failure, utilities shutdown, location suitability, high loss of containment, natural disasters, etc., should be reviewed separately in a major hazards review.
7. Identify safety critical equipment and control systems, analyze them carefully, and provide redundancy efforts if required.
8. Always consider facility/unit startup and shutdown cases.
9. Operator response in an abnormal or emergency situation should be identified and later tested and verified.
10. Understand double-jeopardy situations correctly. Double-jeopardy definition only includes situations occurring *at the same instant*. Remember: More than one situation can develop over a period of time although not at the same instant as the first event.
11. When a HAZOP study is done on modifications to an existing plant, thoroughly review the history of the facility. Many changes must have occurred. Without understanding the design intentions of each modification, the HAZOP study will not be meaningful. Proper management of change procedures is a must.
12. Adding more hardware to the system is not always the best solution to a problem. When such a situation arises, risk-based assessment would be helpful.
13. Analyze perceived low risks for escalation potential to a major accident.
14. Thoroughly review the "incident database" for similar plant or equipment. Such reviews enable focusing on other potential areas.
15. Remember: A HAZOP cannot predict a deliberate act of sabotage.

(Source: G. Unni Krishnan [8], Reprinted with permission from Hydrocarbon Processing, By Gulf Publishing Company, copyrighted 2005; All rights reserved).

such as weaknesses in plant layout, or miss hazards due to leaks on lines that pass through or close to a unit but carry material that is not used on that unit. However, hazards should generally be avoided by changing the design. Assessing hazard by HAZAN or any other technique should always be the alternative choice.

A small team similar to that used in HAZOP carries out hazard analysis. The five steps in HAZAN are

- Estimate how frequently the incident will occur.
- Estimate the consequences to employees, members of the public.
- Estimate the plant and profits.
- Compare the results of the first two steps with a target or criterion.
- Decide whether it is necessary to act to reduce the incident's frequency or its severity.

24.13 Fault Tree Analysis

Fault tree analysis (FTA) is used to assess the frequency of an incident. A fault tree is a diagram that shows how primary causes produce events, which can contribute to a particular hazard. There are several pathways in which a single primary cause can combine with other primary causes or events. Therefore, a single cause may be found in more than one hazard and may occur at different locations in the fault tree.

The graphical structure of the fault tree enables the primary causes and secondary events are combined to produce the hazards. We can compare the relative contributions of the different events to the probability of the hazardous outcome by employing the probability of occurrence of causes and events on the fault tree.

24.14 Failure Mode and Effect Analysis (FMEA)

Failure Mode and Effect Analysis (FMEA) is an alternative method of hazard identification, and it is less formalized than HAZOP. It involves the consideration of the possible outcomes from all discerned failure modes or deviations within a system. It is often applied at different levels of detail or complexity, which are usually referred to as the hierarchy level.

FMEA systematically identifies the consequences of component failure on that system and determines the significance of each failure mode with regard to the system's performance. The technique is used to study material and equipment failure and can be applied to a wide range of technologies. It is referred to as a "bottom-up" technique where each failure mode within the system is traced forward logically in sequence to the final effect.

The use of FMEA is generally limited by the time and resources available and also by the availability of the necessary data. It is best confined to items that are critical by earlier analyses or by other criteria such as safety cost, and it is often used in conjunction with FTA. In order to ensure the appropriate systems are analyzed, it is advisable to start at the highest possible "hierarchy" or "indenture" level of a system. A broad, probable qualitative, analysis at this level identifies the most important or critical contributors. Once identified, such areas can be analyzed at the next or more detailed hierarchy level, at which stage it should be possible to become more quantitative in the approach.

Methodology of FMEA

The following steps are required [8]:

- Define the system to be evaluated, the functional relationships of the parts of the system, and their performance requirements.
- Establish the level of analysis.
- Identify failure modes, their cause and effects, their relative importance, and their sequence.
- Identify failure detection, rectification, and isolation provisions and methods.
- Identify design and operating provisions against such failures.
- Summarize, recommend corrective actions, and issue report.

Definition of System to be Evaluated

As with all safety, it is first essential to define the extent of the system to be analyzed. Because of the complexity of the technique, FMEA is usually performed in relatively small steps. It can be carried out by analysis with a knowledge of the system concerned and the ways in which its component interacts.

Level of Analysis

The choice of level of analysis can be very difficult. One method is to perform an FMEA based on the functional structure of a system rather than on its physical components. In a functional FMEA, the failure modes are expressed as failure to perform a particular subsystem function: it should consider both primary and secondary functions. The primary function is that for which the subsystem was provided, whereas the secondary function is one that is merely a consequence of the subsystem's presence.

Analysis of Failures

All possible failure modes should be considered. Examples include [8]

- Premature operation.
- Failure to operate when required.
- Intermittent operation.
- Failure to cease operation when required.
- Loss of output or failure during operation.
- Degraded output.

Having identified the various failure modes, the analyst then looks at the likely causes and the effects on both the component concerned and on the system of which it is a part. From this, consideration is given to the relative importance of the effects and the sequence in which they occur. The safeguards against such failures and methods of detecting them are then examined.

It should be possible to identify the most significant failures in terms of their effects on the overall system and to decide whether or not the existing safeguards and detection devices are adequate. In identifying a weak link, it may be necessary to subject that component to more detailed analysis, or perhaps it may be decided to eradicate or reduce the probability of failure by design. Table 24.4 shows the applicability of selecting process hazard analysis techniques.

24.15 The Swiss Cheese Model

The model helps personnel to understand events, failure, and decisions that can cause an incident or near miss occurring. Figure 24.34 depicts layers of protection that all have holes. When a set of unique circumstances occurs, the holes line up and allow an incident to happen. The displayed protection layers in the figure are slices of cheese. The holes in the cheese represent the following potential failures in the protection layers:

- Human errors during design, construction, commissioning, operation, and maintenance.
- Management decisions.
- Single-point equipment failures or malfunctions.
- Knowledge deficiencies.
- Management system inadequacies, such as a failure to perform hazard analyses, failure to recognize and manage changes, or inadequate follow-up on previously experienced incident warning signs.

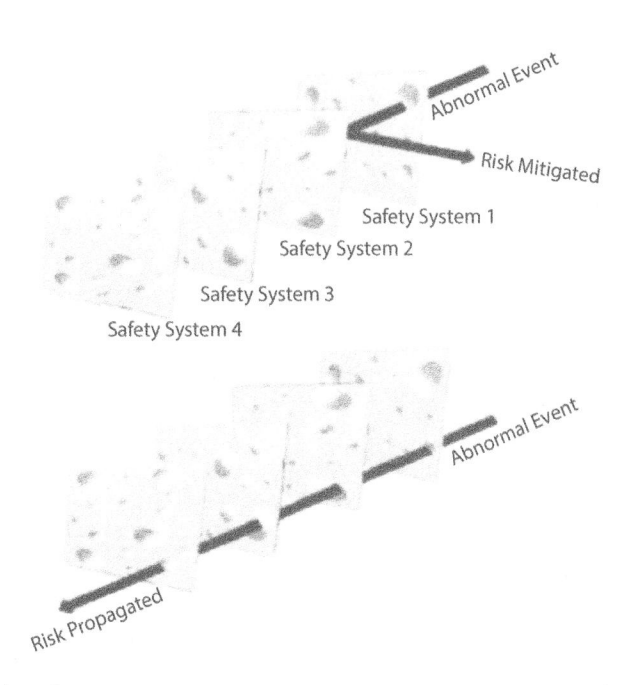

Figure 24.34 Swiss Cheese model of incidents.

Figure 24.34 illustrates incidents that are typically the result of multiple failures to address hazards effectively. A management system may include physical safety devices or planned activities that protect and guard against failure. An effective PSM can reduce the number of holes and the size of the holes in each of the system layers.

24.16 Bowtie Analysis

The form of PHA is widely used to assess the risk of a specific hazard or unit operation rather than to evaluate a whole plant unit. It addresses all threat lines that may lead to loss of containment of a hazard and potential consequences. Furthermore, it identifies relevant barriers and provides recovery measures; it also identifies the so-called escalation factors, which may make the barrier invalid. Bowtie may be used as a management tool to determine critical activities to maintain barrier validity.

A type of qualitative safety review scenarios are identified and depicted on the pre-event side (left side) of a Bowtie diagram, and credible consequences and scenario outcomes are shown on the post-event side (right side) of the diagram, where associated barrier safeguards are included (Figure 24.35).

A barrier is the term used to designate measures to prevent threats from releasing a hazard or measures to limit the consequences arising from the top event. They may be hardware, referred to as critical equipment barriers or human interventions also called critical human barriers. An equipment barrier could be a pressure relief valve, and correspondingly, a human barrier could be following a set of procedures. A combination barrier could be a high-level alarm and the operator responding to the alarm. For a barrier to be considered valid, it must be effective, independent, and auditable. Barriers that prevent threats from releasing the hazards are called controls. They sit between the hazards and the top event, on the left-hand side of the Bowtie (Figure 24.35). Barriers that limit or mitigate the consequences arising from the top event are referred to as recovery measures. They are allocated between the top event and the possible consequences on the right-hand side of the Bowtie.

Validity Rules for Barriers

Valid barriers can fully address the threats or consequences. They must be effective, independent, and auditable. Partial valid interdependent barriers directly address the threat or consequence, but require the assistance and support of another barrier so that the threat or consequence can be fully addressed. When a partial barrier is obtained, an attempt is made to combine it with a measure that would be valid. However, it may be kept separate in order to capture the appropriate health, safety, and environment critical activities, which may be allocated to different department. Figures 24.36 and 24.37 show the controls and recovery measures in the Bowtie analysis, respectively.

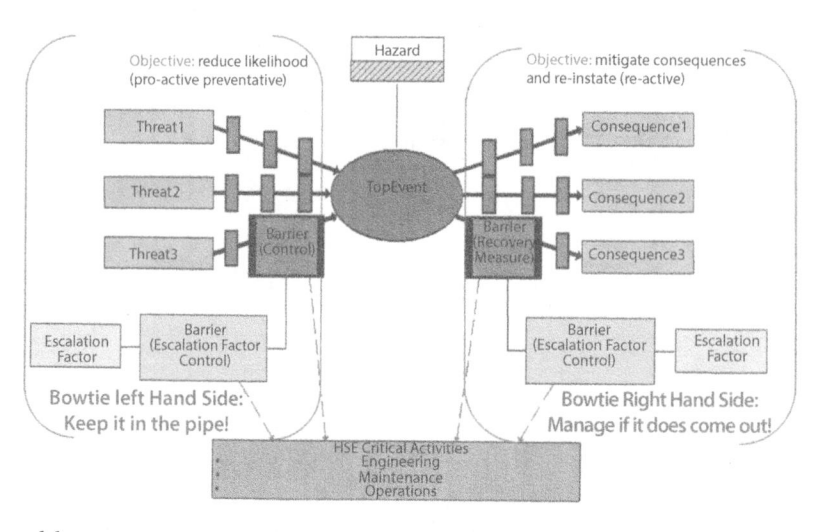

Figure 24.35 The Bowtie model.

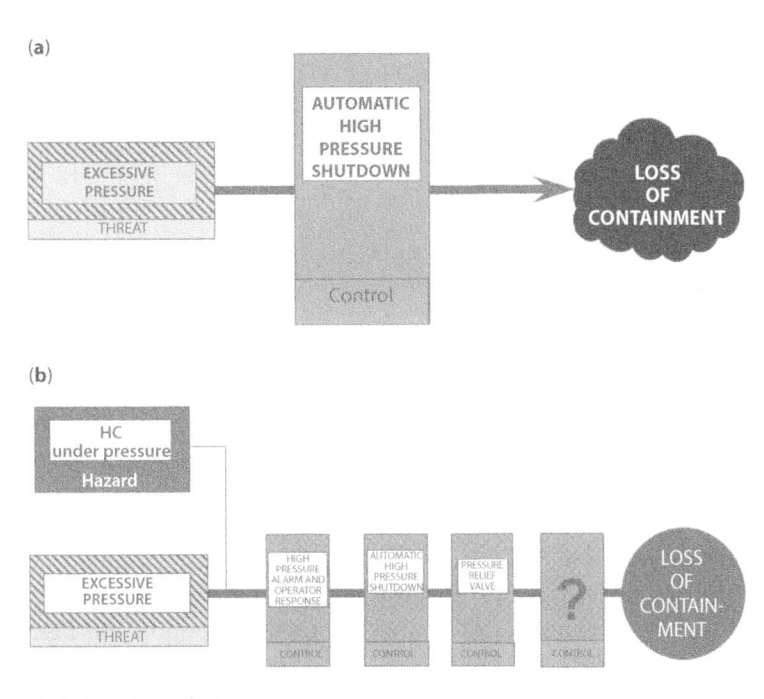

Figure 24.36 (a) and (b) Controls in Bowtie analysis.

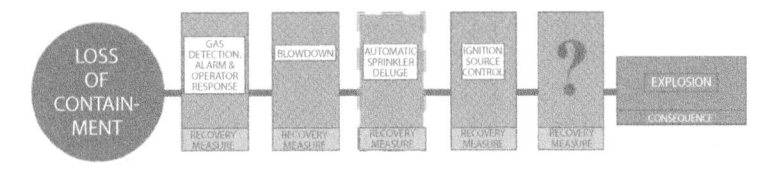

Figure 24.37 Recovery measures in Bowtie analysis.

Recovery measures can vary and be independent on the first release of the hazard and the potential to reduce the risk of escalation or actual full consequence. For example, a bund wall in a tank farm prevents the content of the tanks to flow into area where more damage may occur like a river. A gas detection system detects the first gas release and can initiate an escalation reduction measure like a deluge system, a depressuring system, or any operator intervention.

In general, the bow-tie analysis provides a structured approach in the relation to the measures available and required to keep the process and products contained under the foreseeable circumstances. To contain the feedstocks, processes, and products, the process designer determines the parameters that are required for the basic design of equipment and related systems. The basic design is not particularly a barrier that can be counted as one to be part of the barriers that fulfill the criteria to prevent the release of the hazard. As the basic design is expected to be present, it is inherently required to fulfill the objective to produce oil, manufacture hydrocarbon products, and sell these. Within the overall design, measures are incorporated to fulfill the management structure in order to avoid loss of containment and undesirable consequences (Figure 24.38).

The procedure for conducting a Bowtie Analysis (BTA) is

1. Prepare and organize the study.
2. Select an event to be analyzed.
3. Develop the pre-event side of the diagram.
4. Develop the post-event side of the diagram.
5. Identify any recommendations.
6. Document the results.
7. Resolve recommendations.
8. Follow-up on recommendations.

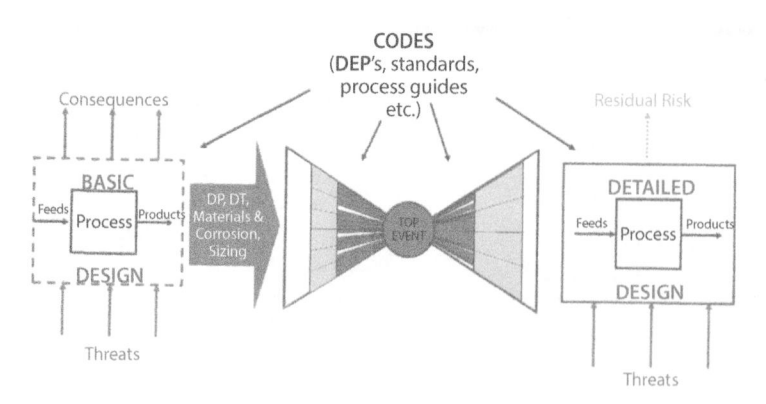

Figure 24.38 Bowtie model overall process.

Example

Cooling water pump trip will cause loss of cooling in a hydrocarbon product rundown cooler to a storage and cause a large product vapor release from the vent tank. The released cloud may be ignited by a passing vehicle. A TZA-HH is provided downstream of the product cooler, which stops the product rundown in case of too high temperature.

- **Hazard**: hydrocarbon product.
- **Cause**: loss of cooling due to pump trip.
- **Consequence**: explosion due to passing vehicle.
- **Barriers**: high-temperature trip in rundown, control of ignition sources in tank area.

Process Safety Isolation Practices in Petroleum Refinery and Chemical Process Industries

Many hazards occur in various facilities in petroleum refinery and chemical process industries, which could be electrical, chemical, thermal or other energy that can cause a serious incident. Generally, the hazards associated with stored energy will always remain, however, the risks associated with the hazards during maintenance can be minimized or mitigated by employing adequate safety procedures and training for an effective lockout/tagout (LOTO) system. The standard for LOTO is in compliance with OSHA 1910.147.

Isolation is a process of making a piece of equipment inoperable, and Lockout uses locks (and probably chains) as part of the isolation process to hold a device in a de-energized state and prevents the re-energization of the equipment unit the isolation is removed. Tags are also applied to inform the operatives of the use of the locks as part of an isolation.

One of the key findings from the Piper Alpha disaster in the North Sea, UK., 1988 in which 167 people died, and was the largest energy industry property damage loss value of US$2,088m based on December 2019 values. During the Piper Alpha incident, it was found that there were shortcomings in the Permit to Work (PtW), isolation, management of change (MOC), and shift handover systems. A large explosion and fire occurred on the offshore platform when a pressure safety valve was removed for routine maintenance. If an effective LOTO had been in place to add an effective layer of protection for the removal of the safety valve, then the incident would not have occurred.

Generally, LOTO is applied to electrical isolations, as if an electrical item of equipment is inadvertently energized whilst maintenance personnel are working on it, then the risk of a fatality is real and therefore, an excellent robust isolation procedure should be applied. However, a defective process isolation fails to a potentially dangerous level and can adversely impact many personnel such as in process safety incidents like Pipe Alpha or the Phillips Petroleum Houston Chemical Complex disaster in 1989 with 23 fatalities and a further 314 injuries, in which isolation practices were leading causes.

A recent incident is the release of H_2S toxic gas at the Aghorn Operating waterflood station in Odessa, Texas in 2019. The Aghorn operating waterflood station is used as part of a process to extract oil from underground reservoirs in West Texas. During extraction, oil comes out of the ground with some water in it. The water is removed from the oil, but it can contain some residual oil and other contaminants such as H_2S, a toxic gas. At the Aghorn waterflood station, pumps in the pump-house are used to pressurize and inject the water back into the oilfield. The injected water adds pressure to the reservoir allowing a larger quantity of oil to be extracted.

The CSB reported that on the night of the incident, the waterflood station's control system activated an oil level alarm on a pump. An Aghorn pumper was notified, and drove to the waterflood station, and attempted to isolate the pump from the process by closing two valves, but failed to isolate the pump from energy sources before performing the work. At some point while he was in the vicinity of the pump, it automatically turned on, and water containing H_2S escaped into the pump house. The pumper was overcome by the toxic gas and was fatally injured. The report stated that the pumper was not wearing his personal H_2S detection device inside the pump house on the night of the incident, and there was no evidence that Aghorn management required the use of these devices. At the time of the incident, Aghorn did not have any written Lockout/Tagout policies or procedures. The pumper did not perform Lockout/Tagout to deenergize the pump before performing work on it. The automatic activation of the pump allowed water containing H_2S to release from the pump. The CSB noted that Aghorn did not comply with OSHA regulation 29 CFR 1910.147 – The Control of Hazardous Energy (Lockout/Tagout) to ensure equipment was isolated from energy sources prior to performing work on it. The CSB also reported that there was inadequate ventilation in the pumphouse leading to fatal concentration of H_2S gas within the building. There was lack of robust safety management program that includes risk identification, assessment, mitigation and monitoring of design procedures, maintenance and training. The CSB recommended to Aghorn Operating Inc. various safety improvements at all waterflood stations where the potential exposure to dangerous levels of toxic H_2S gas exits [39].

Shirley [40] opined the many reasons as to why organizations cannot meet process LOTO requirement:

- Operators know what they need to do as this activity has been done before.
- We just follow the standard operating procedure.
- It is physically not possible to lock or put a chain on a valve.

But an improvement on this position that is prevalent with some frequency is that the energy installations refer to their isolation procedure as LOTO, but the plant does not physically lock the valves with chains and locks. Instead, the site closes the valves and places "do not operate" tags on the isolation points to indicate that the equipment is undergoing maintenance or servicing and cannot be operated until the tag is removed. This approach is correctly referred to as tag out which differs from LOTO as the site cannot claim the full benefit of LOTO without locking the valves. While tagout is a valid method under OSHA regulations for the utility industry, facilities can better safeguard employees by implementing a lockout element.

Where process isolation practices are supported by documentations (e.g., P & IDs), the following summary is offered for the typical approaches taken [40]:

It is not uncommon for the isolation procedure to state that the sketches should be produced when devising the process isolations. However, the sketch is not a controlled document and this approach could cause defective isolations such as a drain line or small instrumentation line that could be missed.

A variation on the above approach is for a simplistic drawing showing the main process lines in a standard operating procedure (SOP). This approach is taken to simplify the process for the operators as they do not refer to complicated drawings as the process and instrumentation diagrams (P & IDs), and if this isolation is a frequent occurrence, then the time is saved as it eliminates the requirement to devise the isolation from the beginning on every occurrence. The approach to simplify the tasks for the operators can be well meaning, but the benefits of simplification may come at the cost of introducing a management of change (MOC) hazard. The site would need to have a robust change management system to ensure that any process changes are captured in the isolation standard operating procedures.

The P & IDs are marked up with the isolation points and when appended to the work permit form and isolation certificate, which is the best practice. In this instance, the P & IDs must be a copy taken from the master "as built" set of P & IDs. As with most management system, there is a cultural element to the implementation of these procedures. Furthermore, there is an underlying process that enforces the successfulness of these procedures as safety barriers.

These documents are critical layer of protection in which the safety of the job should be full assessed and challenged when required. If these layers of protection are not reviewed correctly during the document approval stage, then they will not be correctly reviewed appropriately. An example for this is cross referencing the isolation certificates to the Permit to Work form (a key finding from the Pipe Alpha incident), which is intended to inform all operatives working under the isolations as to all of the work being performed and to be completed before the isolations are removed.

Shirley provided the best practices LOTO program that requires routine training and continuous commitment to safeguarding personnel and plant assets from the unexpected release of hazardous energy [40]:

1. Use a safety lock to hold an energy isolation device in a de-energized state and prevent the re-energization of the machine or equipment until removed. A lock should be used for each trade working under the isolation. The locks can be colored – coded to depict each trade.
2. The keys for the locks should be kept in a locked box under the supervision of the operations supervisor.
3. Tags should be used in the field to denote that a valve is being used for the purpose of isolation. The tags should be marked as "Do Not Operate" and the Permit to Work number/isolation reference number should be recorded on the tags.
4. An isolation certificate should be used which is cross referenced to the Permit to Work forms being used under the process isolation.
5. A copy of the master as built P & IDs should be marked-up with the location of the isolations which should be individually numbered (this number can also be added to the tags in the field.) The marked-up P & IDs should be appended to the isolation certificate and the corresponding Permits to Work. Therefore, it is best practice to devise the isolation from scratch to ensure that the information used is up – to – date and nothing is missed.
6. To further ensure the accuracy of the information used for the preparation of the isolation, it is best practice to line walk the system the P & IDs. It is also best practice to deploy "setting to work" (line walking the system with the issuing authorities and the personnel working under the Permit to Work) prior to commencing the work. This has the added benefit that the final checks are made to the status of the isolations and the personnel working under the isolations are suitably informed of the locations of the isolations, so they can check themselves to look for change at the commencement of work each day as part of their tool box talks. The real important aspect of the setting to work initiative is that it is done as part of the site's procedure.

Figure 24.39 shows a typical isolation tag on a piece of equipment.

Figure 24.39 An isolation tag on a piece of equipment (Source: hazardex, www.hazardexonthenet [40]).

Generally, a testing element is added to the procedure within field equipment isolation. For example, valves do not leak, although they do leak by design. The standards that cover seat leakage for control valves (ANSI/FCI/70-2 1976(R1982)) allow a 4 ml per minute leak rate of a 6" valve. These leakage rates may be insignificant initially, but for an isolation that could be in place for some time before work commences, a significant pressure could build up. This can result in a tragic consequence for the petroleum refining and chemical industries ranging from fires and explosions that lead to fatalities, plant damage and environmental pollutions.

A testing element of LOTO can be a local pressure gauge or opening a drain line and bleeding off the accumulated pressure. Good practice when conducting the required maintenance is to loosen bolts on the cap ends and cracking the flanges open to prevent cap end blowing off under the excessive pressure force that may have built-up [42].

Lock out/tag out has five required components that must be fully compliant with Occupational Safety and Health Administration (OSHA) law. The five components are:

1. Lockout-Tagout Procedures (documentation).
2. Lockout-Tagout Training (for authorized employees and affected employees).
3. Lockout-Tagout Policy (often referred to as a program).
4. Lockout-Tagout Devices and Locks.
5. Lockout-Tagout Auditing – Every 12 months, every procedure must be reviewed as well as a review of authorized employees.

OSHA's standard on the Control of Hazardous Energy (Lockout-Tagout), found in 29 CFR 1910.147 states the steps employers must take to prevent accidents associated with hazardous energy. The standard addresses practices and procedures necessary to disable machinery and prevent the release of potentially hazardous energy while maintenance or servicing activities are carried out.

24.17 Inherently Safer Plant Design

Hazards should be considered and, if possible, eliminated at the design stage or in the process development where necessary. This involves considering alternative processes, reduction or elimination of hazardous chemicals, site selection, or spacing of process units. It is essential to consider inherently safer principles in the design stage, because designers may have various constraints imposed upon them by the time the process is developed. The safety (i.e., safety, health, environmental, and loss prevention) performance of any plant is a function of many factors but can be considered to be primarily dependent on the following aspects [12]:

- The qualities of the people who design, operate, and maintain it.
- The effectiveness of the management and management systems in design, operation, maintenance and incident response.
- The effectiveness of engineered safety systems and control hazards.
- The risk potential of the plant and the process being carried out.

A given level of safety performance can be attained in different ways by allocating different standards to each of these aspects. For example, use of safety systems with good personnel and management systems can result in a high hazard plant being made "tolerably" safe. While a company may place an emphasis on its management systems, another may rely more on engineered safeguards. The exercise becomes a balancing act between the risk potential of the plant, the risk reduction afforded by the various safety control and management efforts, and the cost of these efforts. In reducing or eliminating the hazard potential of the plant by careful selection of the process and good engineering design of the plant, the need for "add-on" safety systems and detailed management controls is reduced. The plant is "inherently safer" because its safety performance is less reliant on "add-on" engineered systems and management controls, which can and do fail. Inherently safer approach has the advantage of providing a means to address safety, health, environmental, and loss prevention issues in a strategic and integrated manner by dealing with the hazards at source, rather than trying to find ways to live with them. Minimizing the inherent hazards of the plant offers savings by reducing the need for expensive safety systems and instrumentation, easing the burden on personnel and procedures, and simplifying on-site and off-site emergency plans. Englund [13] has given details of inherently safer plants of various pieces of equipment in the CPI. He emphasized the importance of user-friendly plants as originally expatiated by Kletz [14]. This strategy is based on a hierarchy of four approaches to process plant safety:

1. **Intensification**. Reducing the hazardous materials.
2. **Substitution**. Substituting the hazardous materials with less hazardous ones. For example, if the hazardous material is an intermediate product, alternate chemical reaction pathways might be used.
3. **Attenuation**. Using the hazardous materials or processes in a way that limits their hazard potential by lowering the temperature/pressure or adding stabilizing additives.
4. **Simplification**. Making the plant and process simpler to design, build, and operate, hence less prone to equipment, control, and human failings.

Table 24.20 The Friendly Plant Concept (A Risk-Based Approach).

Consequence reduction	
Friendly plant Tolerant of People or Equipment Failings	Classical Inherently Safer Plant Less Hazardous: • processes • materials • conditions and lower inventories
Alternative- Unfriendly plant "High Consequence" Plant	Needs Extra: • hardware • control systems • engineered safeguards ….to control hazards, leading to a complex (unfriendly) plant
Frequency reduction	
Friendly plant Less Prone to People or Equipment Failings	Simpler Plant Simpler to: • design • build • operate • maintain
Alternative-Unfriendly plant Prone to Failings	Complex Plant Needs Extra: • hardware • control systems • engineered safeguards ….to prevent or control failings, leading to a complex (unfriendly) plant, and extra: • manpower • training • procedures • management controls ….to make plant operable

(Source: Mansfield, D and K. Cassidy, [9], By permission, IChemE, All rights reserved).

Table 24.21 The Steps to Plant Design.

Decision point	Key questions/decisions	Information used
Initial Specification	What Product What Throughput	Market Research R&D New Product
Process Synthesis Route	How to make the product What route What reactions, materials starting point	R&D Chemists research Known synthesis routes and techniques
Chemical Flowsheet	Flowrates, Conversion Factors, and Basic Unit Operation Selection Temperatures, Pressures solvents and catalyst selection	Process synthesis route Lab and pilot scale trials Knowledge of existing processes
Process Flowsheet	Batch vs Continuous operation Unit operation selection Control/operation philosophy	Information above plus Process engineering design principles and experience
Process Conceptual Design	Equipment selection and sizing Inventory of process Single vs Multiple Trains Utility requirements Overdesign/Flexibility Recycles and Buffer capacities Instrumentation and Control	As above plus equipment suppliers data, raw materials data, Company design procedures and requirements
Process Detailed Design	Location/Siting of plant Preliminary plant layout Materials of construction Detailed specification based on concept design	Process conceptual design and codes/standards and procedures on past projects/designs

(Source [9] By permission of IChemE., All rights reserved).

The goal of these approaches is summarized by the notion of a "friendly plant" as shown in Tables 24.20, which provides a means to visualize the "inherently safer" concept [12]. This concept needs to be integrated into the overall project and addressed together with other constraints and objectives. A framework is required, which enables the basis of the project, process, or plant to be challenged in a systematic way at each of the key steps in the project's life-cycle. This framework considers alternatives and thus allows them to be determined in terms of their inherent safety, feasibility, and cost. Table 24.21 shows the essential steps to be considered when starting from initial specification to process detailed design.

Inherently Safer Plant Design in Reactor Systems

Here, we shall review the practical application to reactor systems.

The following should be considered in applying inherently safer plant involving reactor systems:

1. A good understanding of reaction kinetics is required to establish safe conditions for operation of exothermic reactions.
2. Use continuous reactors if possible. It is usually easier to control continuous reactors than batch reactors. If a batch reaction system is required, minimize the amount of unreacted hazardous materials in the reactor. Figures 24.40 and 24.41 show typical examples.

Methods have been developed for improving batch process productivity in the manufacture of styrene-butadiene latex by the continuous addition of reactants so the reaction takes place as the reactor is being filled. These are not continuous processes even though the reactants are added continuously during most of a batch cycle. The net result is that reactants can be added about as fast as heat can be removed. There is relatively little hazardous material in the reactor at any time because the reactants, which are flammable or combustible, are converted to non-hazardous and non-volatile polymer almost as fast as they are added.

3. If possible, produce and consume hazardous raw materials *in situ*. Some process raw materials are so hazardous to ship and store that it is very desirable to minimize the amount of these materials on hand. Sometimes it is possible to achieve this by using less hazardous chemicals, so there is only a small amount of the hazardous material in the reactor at any time.
4. Liquid phase of solid-phase reactors contains more material than vapor-phase reactors and thus contains more stored energy than vapor-phase reactors.
5. Using high purity raw materials and products can reduce the amount the waste material that must be handled.
6. Consider designing the reactor for the highest pressure that could be expected in case of a runaway reaction to reduce the possibility of release of material to the environment. For example, for a certain process for the manufacture of polystyrene, the composition in the reactor included monomer, polymer, and solvent. The maximum pressure that could be reached by adiabatic polymerization of the mixture in the reactor, beginning at the reaction temperature of 248°F (120°C), was about 300 psig (2068 kPa gage). With this knowledge, it was possible to design polymerization equipment that will withstand this pressure, plus a reasonable safety factor, with considerable confidence that a runaway will not cause a release of material through a pressure relief system or because of an equipment rupture.

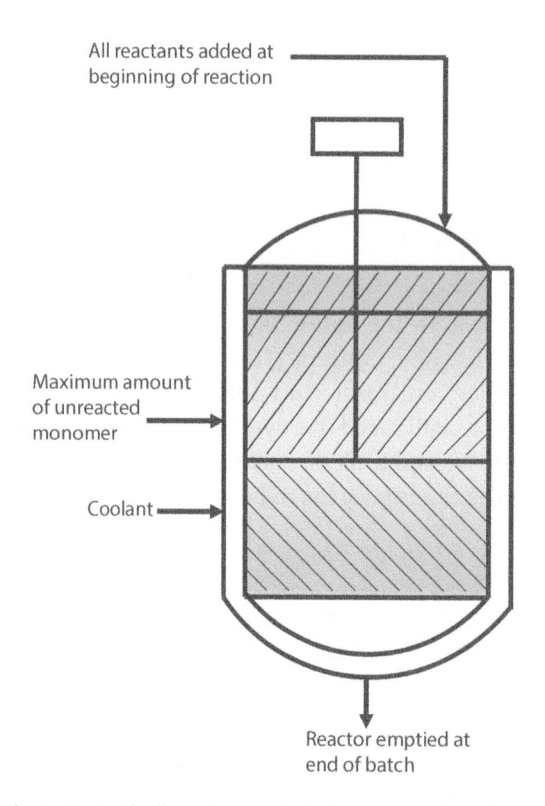

Batch reaction with all reactants added at beginning of reaction.
Consider able amount of flammable and hazardous material in reactor at beginning.

Figure 24.40 Process A (Source: S. M., Englund: Inherently Safer Plants: Practical Applications, Process Safety Progress, vol. 14, No. 1, pp. 63–70, AIChE, 1995).

Figure 24.41 Process B (Source: S. M., Englund: Inherently Safer Plants: Practical Applications, Process Safety Progress, vol. 14, No. 1, pp. 63–70, AIChE, 1995).

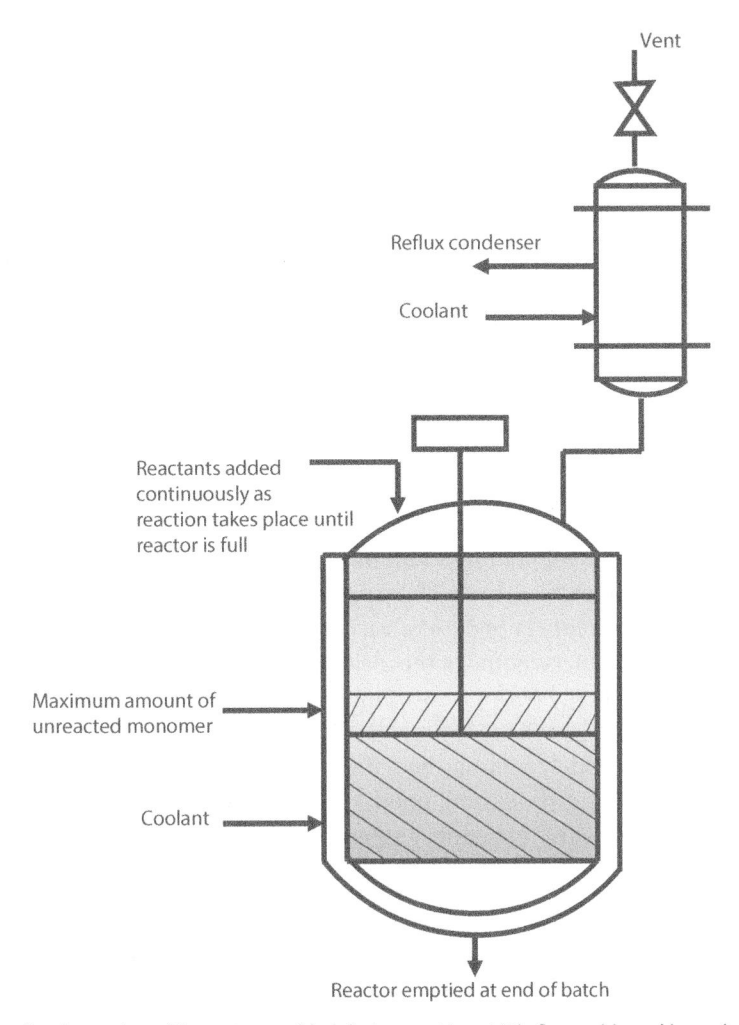

7. Limit the total charge possible to a batch reactor by using a pre-charge or feed tank of limited capacity. Alternatively, limit the rate of addition by selection of a pump with a maximum capacity lower than the safe maximum rate of addition for the process, or by using restriction orifices.
8. The maximum or minimum temperature attainable in a vessel can be limited by properly designed jacket heating systems. If steam heating is used, maximum temperatures can be limited by controlling steam pressure. A steam desuperheater may be needed to avoid excessive temperature of superheated steam from a pressure letdown station.
9. Tubular reactors often offer the greatest potential for inventory reduction. They are usually simple, have no moving parts, and a minimum number of joints and connections that can leak.
10. Mass transfer is often the rate-limiting step in gas–liquid reactions. Novel reactor designs that increase mass transfer can reduce reactor size and may also improve process yields.

Tables 24.13 and 24.13a show examples of parameters for process operations meaningful combinations of parameters and guidewords respectively in a batch process.

The dangerous substances and explosive atmosphere regulations act (DSEAR) is the United Kingdom implementation of the European Directive 1999/92/EC *Minimum requirements for improving the safety and health protection of workers potentially at risk from explosive atmospheres, also known as the ATEX 137 directive*. This directive sets minimum requirements aimed at protecting employees, contractors, visitors, and the public from hazards posed by fire and explosions. The DSEAR is concerned with the storage, handling, and use of dangerous substances, which are defined as substances and preparations classified under the Chemicals Hazards Information and Packaging for Supply (CHIP) regulations. Applying inherent safety principles early in the development of the process can be a useful technique in achieving compliance with DSEAR regulations act. Bell [15] has illustrated how compliance with DSEAR will embed a more inherently safe design basis within the chemical process industries.

Batch reaction with reactants added during reaction. Little flammable and hazardous material present at any time. Reflux (or knockback) condenser used to provide additional heat transfer.

24.18 Energy Management in Petroleum Refinery

Petroleum refineries are one of the major energy-consuming industries that contribute to a major proportion of the production cost. The industry is energy intensive and energy consumption is affected by the refinery configuration, types of feedstock being processed, severity of operation, vacuum system employed, steam and power balance, process integration with petrochemical plants, yield pattern, mandate in product specifications, environmental regulations, and flexibility in operation. Energy in a typical refinery is 60% of variable costs and 1% improvement in energy equals to ~$650,000 year saving for a 100,000 bpsd refinery. Some of the major approaches for energy conservation are energy-efficient design and operation, benchmark energy performance, and continuous energy improvement [16].

- The energy consumption varies widely among the various units and depends on design and the age of the facility, types of crudes being processed, energy management policy, and product profiles of the fuel, and it accounts for over 40% of refinery operating expenses and over 60% of olefin plant operating expenses. Figure 24.42 shows a typical energy consumption of the various units. Energy is consumed in the refinery as follows:
- Indirect fuel for rising steam or generating power.
- Direct fuel in process heaters/boilers
- Hydrocarbon losses during handling/storage of crude oil/products, processing, handling during dispatch, loading and unloading of products, flare losses and leakages, etc.
- Steam and power for drives.
- Circulating cooling water.

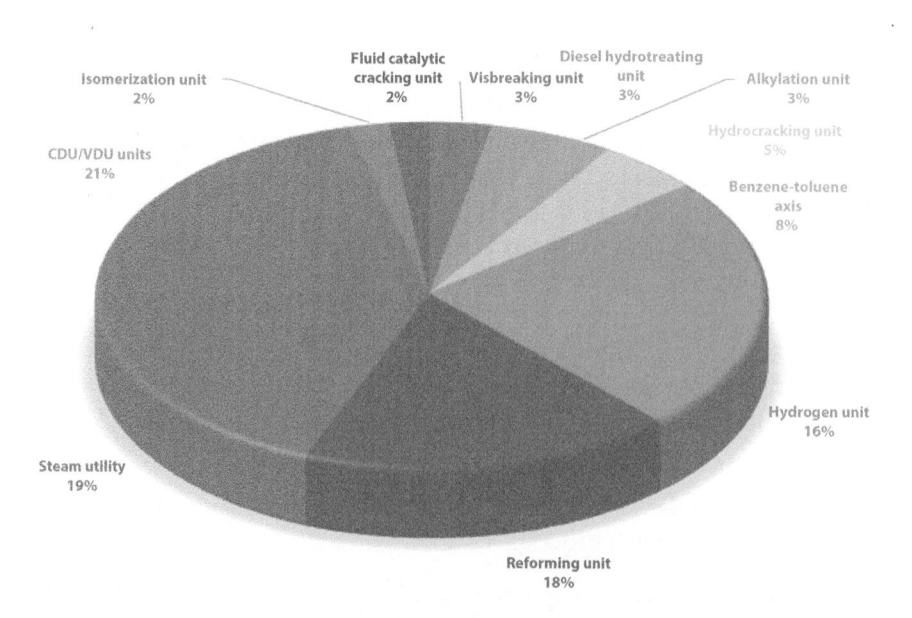

Figure 24.42 Breakup of energy consumption in different sections of refinery [18].

Table 24.22 Energy System that Must be Operated and Maintained.

To use energy	Furnaces, boilers, reboilers, motors, steam, and gas turbines
To recover energy accounting systems	Exchangers, waste heat boilers, economizers, expanders, insulation
To distribute energy	Natural gas, fuel gas, fuel oil, steam, electric power
To remove energy	Cooling towers, recirculating water, once-through water system, air fan/fin coolers, condensers (to water)

Source: Birchfield, G. S, Aspen Tech., 2002.

The total energy consumed in the refinery depends upon the types of crude processed, the processing schemes, plant capacity, equipment and systems, the degree of heat in the systems, mode of product dispatches, the general housekeeping and maintenance practices, the complexity of operation, and the general concern for the energy-efficient operation. Table 24.22 illustrates the energy required to operate refining processes and to generate products of required specifications. Therefore, optimizing refinery energy systems requires an integrated approach as some energy expenditures are independent of process operations and there are various process/energy interactions where yields and energy should be considered simultaneously. Optimizing refinery energy systems requires the following:

- Energy balancing.
- Process analysis.
- Steam power system analysis.
- Equipment level efficiency assessment.
- Rigorous energy economics.
- Efficiency monitoring.
- Analysis of process/energy interactions.
- Use of advance methods/optimization tools.

Ensuring good housekeeping and process safety as discussed earlier are essential and requires little investment in repairing pump seals, leaking joints, proper insulation of hot surface, periodic maintenance of bearings, and so on. The major point for energy conservation through modifications/retrofit on process equipment, namely

- Fired heaters.
- Power/steam drives.
- Steam/power generation systems.
- Heat exchangers.
- Distillation/absorption/extraction/stripping/evaporation system.

Total cost of energy

This includes purchased fuels, steam and power, plant produced, energy consumed, distributing energy in plant, operating energy equipment, boilers, furnaces, exchangers, drivers, and removing waste energy by cooling water/air.

Energy Policy

Energy policy is one of the major steps in energy conservation program that includes

- Replacing energy inefficient equipment with one that is more efficient.
- Adopting energy efficient and environmentally friendly technologies.
- Maximizing the waste energy recovery.
- Promoting the use of renewable sources of energy.

- Creating awareness among the employees.
- Benchmarking performance with the best in the world and endeavoring to be ahead (e.g., Solomon Associates).
- Fostering the culture of participation and innovation, improvement in energy conservation.

Crude Distillation Unit

Crude oil after undergoing the desalting process is sent to via the furnace at a temperature of ~662°F (~350°C) to the crude distillation unit as mists, where physical separation takes place to various fractions such as LPG, naphtha, kerosene, diesel, light gas oil, heavy gas oil, bitumen, and long residue. This unit is energy intensive and the largest energy consumer in the refinery. Therefore, it is essential that ways of improving energy efficiency in this unit are implemented, as its performance is linked to other processes such as the vacuum distillation, visbreaking, thermal cracking, hydrocracking, FCC, hydrotreating, reforming, polymerization, isomerization, alkylation, and so on.

Pinch technology as discussed in Chapter 22 (Volume 3) provides ways of managing the energy in this unit. The efficient operation of the column can result in enhanced yield of the products with reduction in utilities and energy consumption. Many energy savings measures adopted in refineries are [19–21].

- Use of high efficiency/low pressure drop packing for better separation and lower energy consumption.
- Optimizing the reflux ratio for optimum number of trays; lower reflux ratio and low column pressure that result in fewer trays and/or lower reflux ratio for a given separation.
- Use of split tower arrangement.
- Use of divided wall column technology, which offers better separation and energy saving.
- Multiple effect heat cascading for distillation.
- Increased heat recovery from side draw off and pump around.
- Use minimum stripping steam/dry vacuum towers.
- Better control system so as to reduce reboilers and overhead condenser loads.
- Possible reduction in reflux ratio.
- Minimize overflash.
- Installation of absorption heat pump and mechanical vapor compressor using screw compressor.
- Use of structured packing, which results in improved fractionation efficiency, higher capacity, and reduced energy consumption.
- Better instrumentation and control of the distillation column.
- Optimization of insulation thickness.

A distributed distillation allows 10–30% reduction in separation energy, and it enhances opportunities of heat integration by shifting the temperature levels of the heat duty thus yielding different purity products. Reid [22] and Gadalla *et al.* [23] have suggested an optimization framework for the existing distillation system and its heat exchanger network that simultaneously lower energy consumption and capacity at a minimum capital investment.

Heat Exchangers

Heat exchangers form an integral part of equipment that transfer heat from one fluid to another either for heating/cooling/vaporizing or condensing and factors that cause the utilization of this equipment are

- Optimum use of heat exchanger train configurations to increase the crude preheat temperature.
- Closer approach temperature between hot and cold streams in heat recovery.
- Use of welded plate heat exchanger in feed-effluent service.
- Optimum recovery of waste heat from hot product draws off streams.
- Integration of heat exchangers between various processes and utility units to economize on heat inputs, and optimal heat recovery.
- Lower Δp in heat exchangers and simplified piping arrangements.

- Use of antifoulants for control of scaling in heat exchangers.
- Use of high flux tubes for low-temperature boiling.
- Effective utilization of low-level heat.
- Online cleaning system in certain heat exchangers due to fouling.

Steam Traps

A steam trap is a device used to discharge condensates and non-condensable gases with a negligible consumption or loss of live steam. They open, close, or modulate automatically and the three functions of steam traps are [24]

- Discharge condensate as soon as it is formed (unless it is desirable to use the sensible heat of the liquid condensate).
- Have a negligible steam consumption (i.e., being energy efficient).
- Have the capability of discharging air and other non-condensable gases.

An effective maintenance and replacement program can reduce a number of faulty traps to 5–10%, saving about 10% of the energy consumed in a steam distribution system. The most common cause of failure is dirt, which blocks the flow of condensate through the orifice in the steam trap and causes it to fail, close. Accumulated dirt on the seat of a steam trap prevents the valve from closing the orifice resulting in failure and leakage of steam. Dirt also is responsible for corrosion, erosion, and pitting. Improved team traps are being developed with built-in dirt handling capabilities [25].

Optimization of Refinery Steam/Power System

The efficient use of utilities, optimization of steam power balance, and heat exchanger train have been the prime concern in energy conservation areas. The major areas where energy efficiency can be improved are

- Heat integration
- Fired heaters efficiency
- Optimizing steam and power system

Reducing fouling/surface cleaning/surface coating in heat exchanger/furnace

Some of the measures taken for reducing fouling/surface cleaning/surface coating in heat exchanger/furnace are

- Prevent solids from forming.
- Prevent solids from adhering to themselves and to the heat transfer surfaces.
- Remove solids from the surface.
- Efficient desalting.

Pumping System

Some of the measures that are taken during design stage of a project [26] are as follows:

- Ensure that the actual system head requirement of the installed pump and rated heads are as close as possible to avoid mismatch in heads leading to inefficient operation.
- Proper sizing of pumps with respect to flow head as per process demand.
- Incorporate speed control mechanisms like variable speed drives for variable flow requirements instead of throttling of the delivery valves of the pump.
- Incorporate high-capacity pumps instead of parallel operation of pumps where possible.
- Use low friction pipes when required.
- Minimize the number of fittings such as bends and use the shortest routes, when installing pumps.

Electric Drives

Electric drives account for major parts of auxiliary electricity consumption such as in pumps, fans, mixers, and so on. A variable speed or frequency drive serves significantly by allowing soft starts and matching motor torque and speed to the load. This saves energy and accounts for longer life. The following are considered for selecting suitable drives:

- Adequate sizing of the motor.
- Use of high efficiency motor over standard conventional motors.
- Incorporate dual speed motors where applicable.
- Provide proper protection against rain/wind efficiency belt and transmission belts.

Furnace System

In the refinery, a large number of furnaces are used for heating the process fluids. Forced draft burners are increasingly used as they offer important advantages like significantly higher combustion efficiency, reduced particle emission, lower consumption of atomizing steam, better control of flame shape, and an ability to create a more compact furnace. The following measures are taken to enhance the efficiency [20]:

- Monitor CO and excess air to reduce rejected energy and improve efficiency by installing O_2 analyzer.
- Install improved burners, economizer, and air preheaters.
- Preheat boiler feed water with available low-temperature process streams.
- Maximize the use of heat transfer surface by optimizing soot blowing frequency and decoking the tubes.
- Flash blow down to produce low-pressure steam.
- Reduce radiation loss.
- Reduce flue gas temperature.
- Choose right auxiliaries.
- Increase convection section duty/area.
- Generate steam heating boiler feed water.
- Provide additional cleaning facilities.

Possible improvements in furnaces are shown in Table 24.23.

Table 24.23 Possible Improvement in Furnaces.

Equipment system	Purpose	Improvement possible
Burners	To burn oil	To keep excess air as low as possible. Simultaneously promoting flame conditions that result in complete combustion of fuel at lower level of excess air with minimal NO_x and CO emissions.
Air preheater	Recovering energy from stack gases by heating the combustion air.	2.5 % increase in energy improvement for each 55°C drop in stack gases temperature.
Economizer and waste heat boiler	Transfer energy from stack gases to raise steam or heat water	1% increase in boiler efficiency for each 55°C increase in feed water temperature.
Combustion control systems	Regulate flow of fuel and air for desired fuel–air ratio.	0.25% efficiency improvement for each 1% decrease in excess air.
Soot blowers	Remove deposits that cover heat transfer surface.	To keep heat transfer surface clean and available for effective heat transfer both in the convection bank as well as in APH system.

Source [21].

Compressed Air

Compressed air is an essential utility in refinery operation and has a high impact on production processes and overall cost production. Many commercial and industrial compressed air users are improving air system, which is energy efficient, reducing maintenance costs and lowering noise levels with rotary screw compressors and using variable speed. Variable-speed drivers lower compressed air costs and employing advanced VSD technology results in an improved efficiency, flexibility, and noise control. State-of-the-art rotary screw compressors can save users 20–35% on electricity in situations where they have variable loads [24, 27].

Flare System

A gas flare/flare stack is a gas combustion device used in petroleum refineries and chemical plants for burning off flammable gas released by pressure relief valves during unplanned over-pressuring of plant equipment. They are used for the planned combustion of gases over relatively short periods during plant or partial plant startups and shutdowns. At oil and gas extraction sites, gas flares are similarly used for a variety of startup, maintenance, testing, safety, and emergency purposes. In production flaring, it is used to dispose large amounts of unwanted associated petroleum gas.

When industrial plant equipment items are over-pressured, the pressure relief valve is an essential safety device (see Chapter 23) that automatically releases gases and sometimes liquids. Those pressure relief valves are required by industrial design codes and standards as well by law. The released gases and liquids are routed through large piping systems called flared headers to a vertical elevated flare. These gases are burned as they exist the flare stacks. The size and brightness of the resulting flame depend upon the flammable materials flow rate in Btu/h or J/h.

Most industrial plant flares have a vapor–liquid separator (knockout drum) upstream of the flare to remove any large amounts of liquid that may accompany the relieved gases.

Figures 24.43 shows the schematic of a flare system comprising of the following:

- A knockout drum to remove any oil or water from the relieved gases.
- A water seal drum to prevent any flashback of the flame from the top of the flare stack.

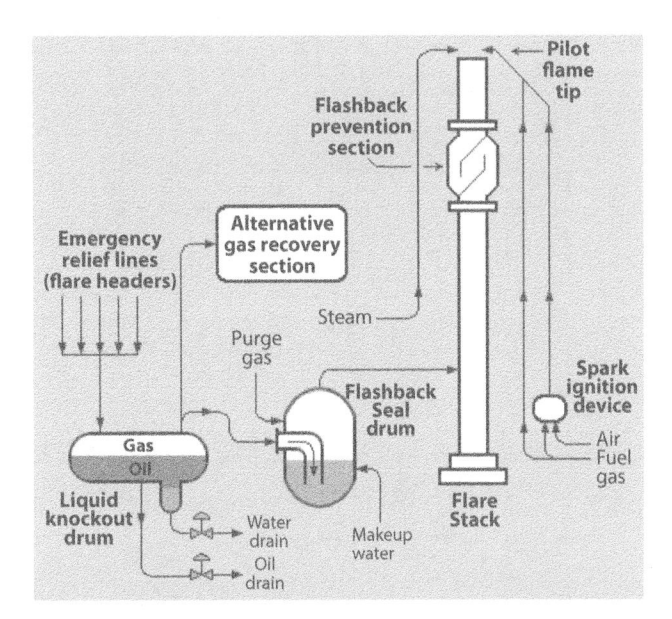

Figure 24.43 Schematic flow diagram of an overall vertical, elevated flare stack system in an industrial plant. (Source: www.en.m.wikipedia.org)

- An alternative gas recovery system for use during partial plant startups and shutdowns as well as other times when required. The recovered gas is routed into the fuel gas system of the overall industrial plant.
- A steam injection system to provide an external momentum force used for efficient mixing of air with the relieved gas, which promotes smokeless burning.
- A pilot flame (with its ignition system) that burns all the time so that it is available to ignite relieved gases when required.
- The flare stack, including a flashback prevention section at the upper part of the stack.

The liquid ring compression systems have proved to be cost-effective and a profitable solution for the recovery of flare gas and other refinery off gases since they act as a scrubber, feeding the process with clean gas. Furthermore, these liquid ring compressors allow plant operators to optimize gas treating capacity and increase profits without new capital equipment. Figures 24.44 and 24.45 show a flash stack at a refinery and of an associated gas from an oil well site, respectively.

24.18.1 Environmental Impact of Flaring

CH_4 estimated global warming potential is 34 times greater than CO_2, and gas flares convert CH_4 to CO_2 before it is released into the atmosphere; they reduce the amount of global warming that would otherwise occur. However, flaring emissions contribute to 270 $MtCO_2$ and reducing flaring emissions is the key to avoid dangerous global warming [28–30].

Improperly operated flares may emit CH_4 and other volatile organic compounds (VOC) as well as SO_2 and other sulfur compounds, which are known to exacerbate asthma and other respiratory problems [31]. Other emissions from these operated flares may include aromatic hydrocarbons (benzene, C_6H_6, toluene C_7H_8, xylene C_8H_{10}) and benzo(a) pyrene, which are known to be carcinogenic. A study found that gas flares contributed to over 40% of the black carbon deposited in the Artic, thus further increasing the rates of snow and the ice melt [32, 33]. Flaring can affect wildlife by attracting birds and insects to the flame, and approximately 7,500 migrating songbirds were attracted to and killed by the flare at the LNG terminal in Saint John, New Brunswick, Canada on September 13, 2013 and similar

Figure 24.44 Flare stack at the Shell Haven refinery in U.K. (Source: www.en.m.wikipedia.org).

Figure 24.45 Flaring of associated gas from an oil well site in Nigeria (Source: www. en.m.wikipedia.org).

incidents have occurred at flares on offshore oil and gas installations [34]. An increasing number of governments and industries are committing to eliminate flaring by 2030 [30].

24.18.2 Environmental Impact of Petroleum Industry

The petroleum industry is essential worldwide as no industrialized country can survive without it. However, in spite of its significant contribution, the industry is among the top highly polluting industries. All upstream and downstream steps from oil exploration to refining and application of petroleum products from automobile to petrochemical production, the hydrocarbon sector produces a large variety of pollutants in the form of liquid, air, solid waste, noise, and thermal. Therefore, sound environmental management to mitigate the pollution has become one of the major issues as described in Chapter 27. Use of heavy crude oil and changing the stringent environmental standards for fuel have further complicated the control strategies requiring enormous investment.

The environmental pollution problem from petroleum industries is rather complex as a wide variety of pollutants are discharged into water stream and emitted into the environment. The environmental pollution from production of crude oil and gas processing also includes the following major areas:

- Oil and gas exploration, drilling, production, pipeline, and marine transportation.
- Storage and handling emissions.
- Process emissions.
- Fugitive emissions.
- Secondary emissions.

Typical sources of emissions in exploration and production are shown in Table 24.24. Offshore activities and transportation of crude oil are also the major sources of marine pollution and there are several statutory and governmental regulations by various countries regarding prevention and pollution (e.g., oil spillage in the coastal zones and at sea). At the planning and design stage of each project, specific provision must be made, which clearly indicates control measures to combat the adverse impact on the marine environment.

Storage tanks, truck, railway, and marine terminals are the sources of VOC emission and oil spillage problems. Storage and handling emissions are dependent upon the construction, and size of the storage tank, the vapor pressure of the stored organic liquid, and the ambient quality of the tank location.

Table 24.24 Typical Sources of Emissions in Exploration and Production and Control Strategies.

Source	Contaminant	Control system
Well-drilling (test, completion, and work over operations)	Hydrocarbons, H_2S, CO_2, SO_2, NO_x, particulate material, fumes	Incineration combustion improvement with special burners
Well-exploration Pumping triphasic separation	Light hydrocarbons Light hydrocarbons	Good maintenance (joints) emissions discharge in gas circuit
Oily water treatment storage tank	Hydrocarbons Hydrocarbons, H_2S	Covering and blanketing. Tank pressure control (floating roof)
Oil and fuel leaks Leading to explosion	Hydrocarbons Hydrocarbons, H_2S	Good house keeping Vapor recovery, discharge to storage tank, incineration
Well-draining Open drain system Cold vent Flares	Hydrocarbons, H_2S Hydrocarbons Hydrocarbons, H_2S Hydrocarbons, H_2S	Closed circuit Hydraulic seals, siphons Good dispersion Flares tip design (improvement of flare nozzles)
Gas turbine exhaust Other combustible units	CO, CO_2, NO_x, SO_2, CO, CO_2, NO_x, SO_2, particulate matter	High stack adjustment, high stack burner adjustment, selection of combustibles, dust removal (system depending on requirements)

(Source: Perspective Plan on Environmental Management, 1985–86 to 1989–90. Safety and Environmental Management Oil and Natural Gas Commission, India).

Process emissions occur from reactors, distillation columns, purification equipment, fire heaters, condensers, reformers, crackers, filters, sulfur recovery units, recovery and control equipment, stacks, vent, and so on.

Fugitive emissions are of two types: low-level leaks from process equipment and episodic fugitive emissions where an equipment failure results in sudden large release [17]. Fugitive emission occurs from pumps, valves, flange, mechanical seal, relief valves, tanks, instrument connections, sample connections, and open-ended lines. Secondary emission occurs from the wastewater treatment unit, cooling tower, boilers, process sewers, and so on. In most cases, fugitive emission for equipment leaks is the largest source typically accounting for 40–60% of total VOC emission [18].

Some of the major sources of pollutants in the petroleum refinery are crude oil and natural-gas processing, crude oil storage, crude processing, desalting and reforming, catalyst regeneration units, hydroprocessing units, lube refining and lube treatment processes, boiler blowdowns, power plants, effluent treatment plant, and so on. The major pollutants are emulsified oils, phenol, cyanides, inorganic salts, naphthenic acids, heavy metals, sulfide, spent catalysts, tars, H_2S, NH_3, NO_x, SO_x, CO, CO_2, hydrocarbons, VOC, particulate matter, fine catalyst dusts, solvents like phenol, etc.

Heavy crudes contain high-sulfur content, which is a major source of H_2S and SO_x emissions. In refineries and petrochemical complexes, SO_x is emitted from a number of process equipment such as heaters, boilers, FCCs, sulfur recovery units, blow down system, vapor recovery and flares, process gas flares, fluid coking, etc. The flue gas leaving the FCC regenerator is a major source of SO_x emission. Hydrotreating units are the largest source of SO_x emission. CO emission takes place from the regenerator where the coke from the catalyst is burned off. In the hydrocarbon industry, process heaters are used at the number of places in the process, which are major source of NO_x emission. NO_x is formed during combustion through thermal fixation of atmospheric nitrogen introduced with the fuel or through the oxidation of nitrogen introduced through the fuel.

The potential sources of liquid pollution may be reactors, equipment water overflows; sample blow downs, distillation column, absorber, product wash, and purification; boiler blowdowns, cooling water; steam-fed vacuum pumps, pumps and compressors, and so on. The potential sources of solid waste from petrochemical industries are the cracking unit, reformer spent catalyst, sludge from process and waste treatment facilities, and catalyst regeneration units.

Sour water generation is also a problem in refinery and petrochemical complexes. The sour water is produced at a number of distinct points within a refinery and petrochemical complex, which is then drained in the sour water collection system. Sour water results in different sections when steam and/or water used for various processes pick up H_2S and NH_3, which if not properly treated affect the waste-water treatment system.

Oily sludge, spent catalyst, and spent caustic are some of the major sources of waste generation in petroleum and petrochemical complexes. Oily sludges are formed through emulsification of oil with water usually in the presence of suspended solids. Various sources of oily sludge are sludge from API separators, slop oil emulsion, heat exchanger bundles cleaning sludge, cooling tower sludge, and tank bottoms. The spent caustic is discharged from gas purification processes, where it is used for scrubbing the dissolved sulfide, H_2S acidic gases, mercaptans, and acid gases. Catalyst deactivation through sintering, poisoning, or build-up of surface deposits and loss due to attrition is the cause of generating the spent catalyst as this is unavoidable. However, the use of catalysts with high selectivity, reactivity, stability, regeneration, and resistance to attrition may result in less generation of the spent catalyst.

24.18.3 Environmental Impact Assessment (EIA)

Environmental impact assessment (EIA) is a framework to the compatibility of projects in terms of locations, suitability of the technology, efficiency in resource utilization, recycling, etc. EIA has been made mandatory in some countries and details of pollutants generated from petroleum refining are given in Table 24.25.

Table 24.25 Water and Air Pollutants Discharge/Emission from Various Processes in Petroleum Refining.

	Source	Pollutants
1.	Crude oil storage	Emulsified oil, oily sludge, VOC
2.	Desalting, atmospheric and vacuum distillation processes	Inorganic salts (NaCl, $MgCl_2$, etc.), sludge, oil and grease, phenols, naphthenic acid, NH_3, VOC, H_2S, SO_x, NO_x, CO
3.	Thermal cracking, visbreaking, coking	Oil and grease, VOCs, H_2S, SO_x, NO_x, CO, particulates.
4.	Fluidized bed catalytic cracking, hydrocracking	Oil and grease, VOCs, H_2S, SO_x, NO_x, CO, spent catalyst, particulates
5.	Catalytic reforming and aromatic separation	Oil and grease, VOCs, aromatics, H_2S, SO_x, NO_x, CO, spent catalyst, particulates and solvents.
6.	Hydrotreating and hydrodesulfurization	Oil and grease, VOCs, H_2S, SO_x, NO_x, CO, spent catalyst.
7.	Lube oil and wax processing	Oil and grease, solvents, clay, bauxite, SO_2
8.	Sweetening and sulfur recovery plant	VOCs, H_2S, SO_x, NO_x, CO, spent catalyst, spent caustic.
9.	Process heaters and flares	VOCs, H_2S, SO_x, NO_x, CO
10.	Power plant cooling tower blowdowns	Oil and grease, VOCs, H_2S, SO_x, NO_x, CO, particulates, chromium, calcium and magnesium salts.
11.	Water treatment plant	Suspended solids, oily sludge, N_2, and phosphorous compounds, chlorides, heavy metals.

Some of the basic steps in EIA involve baseline study, scoping or impact identification, impact measurement, impact analysis, management plant, and green belt design. The environmental issues from various activities of petroleum refinery are as follows:

- Deterioration in air quality due to emission of hydrocarbon, SO_x, NO_x, H_2S, particulate matter, etc.
- Deterioration in water quality due to presence of high BOD, COD, turbidity, oil and grease, dissolved solid, toxic organic compounds, cyanide, heavy metals, carbon, etc.
- Deterioration due to noise pollution.
- Adverse impact on biological and marine life due to discharge of various toxic compounds, such as heavy metal, oil grease, phenolic compounds, etc.
- Impact due to various solid wastes.
- Socioeconomic destruction due to influx of labor force, migration to urban areas, movement of heavy machinery, additional traffic, etc.

Table 24.26 shows the characterization of effluent discharges from the refinery and petrochemical facilities.

Resources that are impacted by location, expansion, and modernization of the petroleum and plants are

- **Physical component**: Metrology, air quality, surface water, hydrology, ground water, topology, geology, soil, and material.
- **Ecological environment**: Fresh water ecology, terrestrial, forest, fauna, sanctuary, natural vegetation, species diversity, fisheries, animals, and so on.
- **Socioeconomic and cultural aspects**: Impact on economic and cultural aspects, economic yield, etc.

24.18.4 Pollution Control Strategies in Petroleum Refinery

At the start of any project in the upstream/downstream in the petroleum refinery, it is essential that adequate planning and scope of work should be provided that clearly indicates control measures to combat the adverse impact on the marine environment. Oil discharges from offshore drilling and production can be controlled by installing equipment and treatment facilities and observance with the required engineering design practices and guidelines. Safety and pollution control on platforms are inter-independent. Advanced standards and practices are required to prevent and control oil spillage from petroleum marine transportation. Furthermore, measures to ensure proper handling of oily wastes should be established for all transportation vessels. Coastal and estuaries are susceptible to serious pollution due to discharge of effluents and wastes from many sources and methods of discharge through

Table 24.26 Typical Characteristics of Combine Effluent from an Integrated Refinery and Petrochemical Complexes.

Parameter	Unit	Refinery	Petrochemical
pH			6.0 – 6.5
Oil and Grease	mg/l	10	-
BOD	mg/l	15	50
COD	mg/l	-	250
TSS	mg/l	20	100
Sulfides	mg/l	0.5	2
Phenols	mg/l	1	5

conduits beyond the lowest tide mark should be established. Water produced at onshore drilling sites should be treated at effluent treatment plants and disposed off by subsurface injection at appropriate depths, or used for pressure maintenance or into abandoned wells. The principal air pollutants from oil and gas processing units are VOC, H_2S, SO_x, NO_x, CO, and particulate matters. Table 24.27 shows typical sources of emissions in exploration of oil and production and control strategies.

Waste management involves identification of the waste generated, efficient collection and handling, optimal reuse and recycling, and effective disposal with no environmental problems. The basic steps for waste reduction are as follows: recognize the waste, determine the cause, plan corrective action, eliminate the cause, and establish controls to prevent its recurrence.

The OSHA, EPA, HSE, and other governmental organizations in other countries have imposed restrictions and regulations on the petroleum refining and CPIs with the view of minimizing emissions and pollutions. These industries will require to reevaluate and revamp the waste management facilities by implementing pollution prevention techniques and improve operating and maintenance procedures. The simple process of removing oil/grease, solids, and COD from wastewater is now more challenging due to new requirements for stricter control of hydrocarbon vapors to atmosphere and elimination of the potential for contaminated water into ground, sewer, and surface water streams.

Environmental strategic planning starts with the scope of work and conceptual design, construction, startup operations, and final shutdown. It is essential to improve environmental regulatory compliance, reduce environmental equipment investment, and facilitate a quick shutdown of the plant. Furthermore, excellent housekeeping procedures require regular visual inspections, spillage prevention program, implementing the pollution/waste training programs, supporting plant wide waste recycling program, and conducting leak detection and repair program as these play a key role in minimizing and controlling the generation of wastewater, and reducing emissions to air, water, sewer, and the soil.

The major loss areas in the petroleum refining and CPIs are measuring and accounting of raw materials and intermediate/finished products, impurities in the feedstocks, effluent streams from process equipment, vents, drains, overflows and leaks, losses due to plant upsets, startup and shutdown, climatic conditions such as high temperature in summer and washing effects in rainy season, and effects of varying throughputs and storage loss.

The wastewater generated from petroleum is quite complex and contains a wide variety of pollutants making the wastewater treatment system further complicated to process. This is compounded with increasing trend of integrating the refinery and petrochemicals for value-added products, which involves reducing the level of BOD, COD, and other specific pollutants.

An efficient wastewater management in any large industries is crucial to controlling costs and satisfying environmental obligations. The basic technique for wastewater management is: minimize generation, segregation, reuse, recycle, treatment, and disposal. Reduction in wastewater generation can be achieved by careful planning and selection of the technology, raw materials, operating conditions, equipment, product substitution, monitoring of water used and wastewater discharge, and taking appropriate corrective measures. Segregation of the wastewater stream and recycling of water in the process can further aid in the reduction of waste. Before treating combined effluents from the

Table 24.27 Typical Sources of VOCs from Process Plants.

Sources	Relative %
Fugitive equipment leaks	50–60
Loading	20–30
Waste water treatment	10–15
Storage tanks	10–15

refinery and petrochemical complex, source treatment of the individual effluent from various plants is provided. To be cost effective, a multilateral approach that characterizes process streams, identifies contaminant sources, quantifies the pollutant level, and so on must be applied. Hydrocarbon losses in the refinery and petrochemical complexes can be substantially reduced by taking preventative measures, which vary from monitoring operating conditions to implementing justified cost-effective capital projects.

The objective of an effective wastewater management plan is to review current and future projects, determine the impacts on wastewater generation, the reuse, and provision of secondary treatment, apply a new technology for further development, and provide a recommendation to minimize wastewater generation and increase its recycle. This also requires examination of solids and sludge generation, disposal and recommendation for reducing and treatment options, and determination of effluent quality criteria to meet current and future generations.

The spent caustic from naphtha cracker, gas crackers, and various sweetening processes are one of the major sources of water pollution in the petroleum and petrochemical complexes and is contaminated with sulfides, mercaptans, and phenols. The spent caustic is considered to be hazardous and cannot be directly managed by biological wastewater treatment and must be treated separately before final biological treatment. Wet air oxidation technology (WAO) is promising and gaining acceptance for treating spent caustic, which detoxifies the spent caustic by oxidizing the sulfide and mercaptans to sulfate and breaks down the toxic naphthenic and cresylic compounds [35].

WAO is a form of hydrothermal treatment. It involves the oxidation of dissolved or suspended components in water using oxygen as the oxidizer. The oxidation reactions occur in the superheated water at a temperature above the normal boiling point of water 212°F (100°C) but below the critical point 705°F (374°C).

Commercial systems of WAO typically use a bubble column reactor, where air is bubbled through a vertical column that is liquid full of hot and pressurized wastewater. Fresh wastewater enters the bottom of the column and oxidized wastewater exits the top. The heat release during the oxidation is used to maintain the operating temperature. WAO is a liquid phase reaction using dissolved oxygen in water to oxidize wastewater contaminants. The dissolved oxygen is typically supplied using pressurized air, but pure oxygen can also be used. The oxidation reaction generally occurs at the moderate temperature of 302–608°F (150–320°C) and pressures from 145 to 3191 lb_f/in^2 (10 to 220 barg). The process converts organic contaminants to carbon dioxide, water, and biodegradable short-chain organic acids. Inorganic constituents such as sulfides and cyanides are converted to non-reactive inorganic compounds. Figure 24.46 shows a diagram of a bubble column reactor, where the introduction of gas takes place at the bottom of the column and causes a turbulent stream to enable an optimum gas exchange. The mixing is carried out by the gas sparging and it requires less than mechanical stirring. The liquid can be in parallel flow or counter-current. Bubble column

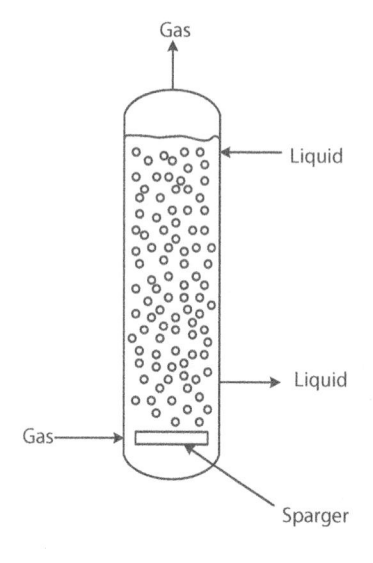

Figure 24.46 A bubble column reactor.

reactors are characterized by a high liquid content and a moderate phase boundary surface. The bubble column is particularly useful in reactions where the gas–liquid reaction is slow in relation to the absorption rate.

After flow equalization, the spent caustic is pumped to charcoal filters for removal of tarry and polymers and is fed to an oxidation system. Process parameters such as low and high temperatures are applied. The low-temperature WAO systems oxidize the sulfides to thiosulfate and sulfate, but high concentrations of thiosulfate are present in the treated effluent. The mid temperature systems fully oxidize the sulfides to sulfate and mercaptans are oxidized to sulfonic acids. For sulfidic spent caustics, this results in a high chemical oxygen demand (COD) destruction (>90%). High-temperature systems are used to oxidize organic compounds that are present in naphthenic and cresylic spend caustic. Table 24.28 shows typical classification of WAO treatment systems.

Stringent restrictions and regulations have required the treatment of wastewater so that it can be recycled or discharged safely. Water use and wastewater monitoring system is the first step in controlling and regulating the discharge of wastewater. Various wastewater treatment technologies are given in Figure 24.47. A typical treatment for petroleum refining may involve primary screening, flow equalization, oil/water separation, biological treatment, clarification, tertiary treatment, and solids handling. Prior to collection for equalization and biological treatment, primary source treatment of the effluents from the individual effluent stream is carried out as per the requirement. Primary and secondary oil separation systems use API oil separators or parallel plate separators while secondary separation may be applied when the oil content is high. This involves coagulating agents like ferrous sulfate and ferric

Table 24.28 Typical Classification of Wet Air Oxidation (WAO) Treatment Systems.

Classification	Temperature °C	Pressure bar	Treatment of compounds
Low	110–150	2–10	Reactive sulfides
Mid	200–220	20–45	Sulfides, Mercaptans
High	240–260	45–100	Naphthenic and Cresylic acids, Sulfides and Mercaptans

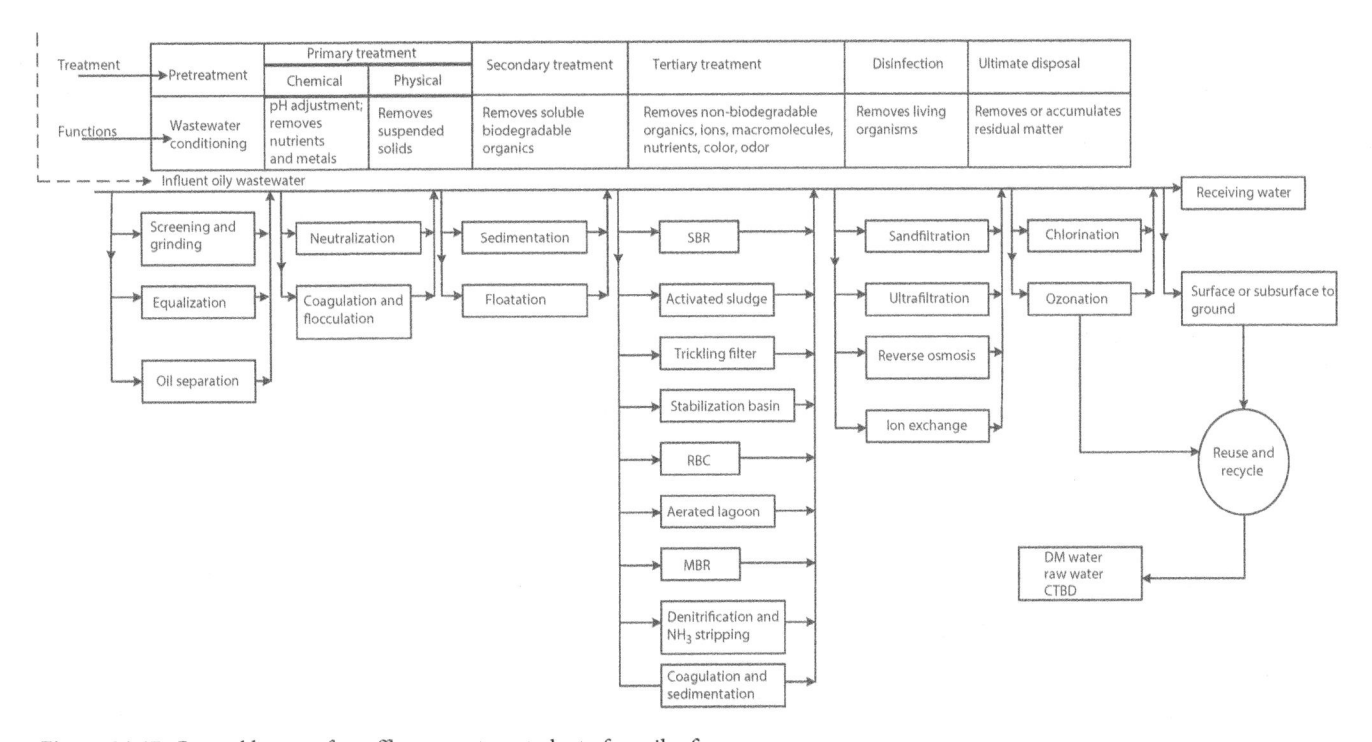

Figure 24.47 General layout of an effluent treatment plant of an oil refinery.

sulfate lime. Poly aluminum chloride has been used to de-emulsify an oil separator effluent. Sulfide precipitation is suited to wastewater containing sulfide and mercaptans. Biological systems including trickling filter, activated sludge process, oxidation ponds, and aerated lagoons are commonly used.

Some of the strategies for minimizing air toxic emission are process chemistry modification, changes in the specific constants responsible for emission, operational modification, and preventive maintenance practice. For reducing the CO emission from FCC recovery regenerator, CO converter can be installed to convert the CO to CO_2.

In petroleum refineries, many plant components are potential sources for emitting VOCs. Monitoring and maintenance use of lower leak equipment collecting emissions and use of leak less technology can control fugitive emission. Some of the VOC control options are recovery vapor balancing, absorption, adsorption, refrigeration and destruction, thermal oxidation, catalytic oxidation, and biofilter. Fugitive emissions can be controlled by elaborate leak detection and repair and equipment modifications. Many plants have implemented leak detection and repair (LDAR) programs where various sources of fugitive emission like leaks from valves, flanges, pump seals, etc. are routinely monitored for leaks and maintained on a regular basis.

Compiling an emission inventory is the first step in reducing or controlling emissions. This also enhances waste minimization, process safety, and attaining the set goals. Catalytic reforming unit is a potential emissions source of a number of compounds and has been identified as a potential source of hazardous air pollutants by the EPA under the Clean Air Act (CAA) Amendments of 1990.

Fired heaters may be the major sources of sulfur emission as they use fuel oil. Fuel oil desulfurization, flue gas desulfurization, and the use of low sulfur may be some of the methods for reducing the sulfur emissions. NO_x can be controlled either by adjusting the combustion process by adjusting the parameters responsible for high NO_x emissions or by flue gas treatment. Low NO_x burner technology for ethylene cracking furnaces has been developed in order to meet the reduction of NOx emission from combustion processes to less than 10 ppm. Additives can be used for feed desulfurization and flue gas desulfurization and are measures for reducing SO_x emissions. Emission from sulfur recovery units, which is a major source of Sox emissions, can be enhanced by improving the sulfur efficiency by integrating Tail gas cleanup unit. IFP Clauspol 99.9 + process can be employed for reducing SO_x emission, and by applying a solvent desaturation loop, a higher sulfur conversion rate is reached.

Biodegradation is another concept being applied to destroy environmental pollutants in different kinds of wastes. For example, bioscrubbing and infiltration are commonly used to remove air pollutants. Bioscrubbing is primarily used for contaminants that are strongly soluble in water. These are then scrubbed from the waste gas in an absorption column and then passed to a separate oxidation reactor using a standard water treatment plant to aerobically degrade the contaminants. The biofiltration process uses a biological microbial film fixed on support media within a single process where the contaminants are both absorbed from the waste gas and converted to benign end products such as water and CO_2. Biofilters have been developed from abating odors to technically sophisticated and control equipment in removing specific chemicals from industrial sources and have achieved 95–99% efficiency in removing VOC.

VOC recovery processes are absorption, adsorption, vapor balance, and refrigeration. VOC destructive processes are: use of flares, incinerators, catalytic oxidation, and biofilters. The choice of the control methods depends on many factors such as the vapors, vapor quantity, number and location of emission points, and viability of appropriate abatement technologies for the vapor under consideration. Wastewater plants in the refinery and CPIs are the sources of VOC emission that include stripping, absorption, chemical oxidation, membrane separation, etc.

Process vents emission can be controlled by various recovery and control devices such as absorbers, chillers, catalytic incinerators, thermal oxidizer, flares, and routing the vents to a boiler or process heaters. Use of floating roof tanks and proper design of storage tank piping to prevent the tank from being over- or under-pressurized and to prevent plugging are some essential issues for controlling storage tank emissions. Controlling emissions during decommissioning and maintenance of various piping require proper decommissioning procedures. An LDAR can be applied for a cost-effective measure to reduce total fugitive emissions, and the LDAR program uses a sensitive gas detection instrument and samples each piping component separately to determine the concentration of hydrocarbon adjacent to a potential leak site. Other methods for fugitive emission control are installation of new packing sets in block and control valves and upgrading of pump seals to multiseal designs.

Storage tanks are essential sources of VOC emissions and they account for 10–15% of total VOC emission from process plants. Three types of tank designs are used for the atmospheric storage of petroleum products: fixed roof, external roof, and internal floating roof. Some of the techniques used for controlling VOC emission from fixed roof tanks are: vapor balancing, vapor recovery and destruction, and installation of internal floating head. Some of the tank emission controls in external floating roof tanks are checking of the condition of existing seals, replacement of vapor mounted primary with liquid mounted primary seal, controlling losses from roof fittings, and installation of vapor recovery [17].

Emission during loading can be controlled by use of submerged loading in place of splash loading, vapor balancing during loading, and installation of vapor recovery system. Emissions from wastewater treating system can be controlled by good housekeeping, by reducing wastewater volume and concentrations, optimizing stripper operation, installation of sewer system suppression, and reducing air/water contact area (covering, nitrogen blanketing, and replacing API separator with covered corrugated plate interceptor) [17].

Solid waste from refineries and CPIs are oily sludge, spent catalyst, coke from the cracking units, effluent treatment sludge, spent carbon and resins, contaminated soil, and tank sediments. Treatment of these wastes includes pretreatment and dewatering, detoxification, incineration, deep well injection, chemical and biological treatment, etc.

24.18.5 Energy Management and CO_2 Emissions in Refinery

Energy costs typically represent 50-60% of non-feedstock refinery operating costs; energy consumptions in a refinery can be classified into 1. thermal energy consumption (e.g., fired heaters, incinerators, boilers and coke burn in fluid catalytic cracking (FCC), 2. electrically energy consumptions: Motors, lighting and power supply. Sound management has direct impact on CO_2 emissions from various processes in the refinery. These are the distilllation columns, catalytic cracking and reforming, hydrodesulfurization, hydrogen production, delayed Coker, alkylation, flares and captive power plants. About 65-70% of the total CO_2 emission from a refinery is contributed by both captive power plant and hydrogen production units respectively. About 10 tons of CO_2 is produced per tons of hydrogen. Crude distillation and FCC are the other prominent emission of CO_2 amongst the process units.

Three major areas where CO_2 reduction opportunity can be realized are:

- Efficiency improvements of various process fired heaters, steam rising and power generation.
- Hydrogen production and coke burn-off from FCC.
- Switching over from refinery fuel to natural gas.

About 15% reduction in total CO_2 emissions from a refinery can be achieved by switching the fuel from regular refinery fuel oil to natural gas. Various hydrogen treatment processes increase the CO_2 emissions from the refinery. Furthermore, hydrogen management involving pinch technology (Chapter 22), and better hydrogen management through optimizing the refinery hydrogen balance and strategies to minimize hydrogen losses, use of more selective catalysts in hydrotreater units, fuel substitution with fuel having higher hydrogen content and lower carbon. This enhances the thermal efficiency and thus maximizes the H_2 production in catalytic reforming unit [16].

24.19 Benchmarking in Refinery

The worldwide refining industry is facing greater competitive pressures in a marketplace increasingly dominated by large new refineries that enjoy both economies of scale and the benefits of modern technology. Larger and newer low-cost refiners can now ship their products economically anywhere in the world as they are no longer hampered by high shipping costs. In order to survive and thrive in this environment, refineries today must strive continually for improved performance in reliability, margin generation, and operating expenses control. Maintaining the status quo and relying on what has worked in the past will be insufficient to remain competitive.

Tables 24.29 Comparison of Actual Energy Consumption for Major Process Units with Benchmark Consumption.

S. no.	Unit	Actual energy consumption		Benchmark energy consumption	
		Btu/bbl	kacl/m³	Btu/bbl	kcal/m³
1.	CDU (stand alone)	74,640–123,900	118,305–196,383	73,600–78,650	116,657–124,661
2.	VDU (stand alone)	86,200–198,400	136,628–314,466	65,330	103,549
3.	C & VDU (combined)	104,900–155,700	166,267–246,786	88,000–109,00	139,481–172,766
4.	Naphtha splitter	102,660–236,740	162,717–375,235	102,150	161,909
5.	FCCU (with coke)	256,675–505,000	406,832–800,430	250,400	396,886
6.	Delayed coker (LR)	370,100–421,140	586,612–667,511	316,710	501,988
7.	Aromatics recovery	654,175	1,036873	505,840	801,761
8.	Hydrocracker unit	433,300	686,784	262,320	415,780
9.	Hydrogen unit	87,387–110,850	138,509–175,698	66,930	106,085
10.	Propane deasphalting	454,380–573,255	720,196–908,614	261,640	414,702

Companies require to understand the factors driving competition an industry progress. All starts with understanding the conditions of their own operations, and benchmarking has become one of the important practices that produce superior performance when adapted and implemented. Benchmarking is recognized as an effective approach towards improving efficiency, productivity, quality, profitability, and others such as dimensions of performance that drives competitiveness. Benchmarking may or may not be achieved because a large number of parameters affect the energy consumption depending on age and capacity of the plant and type of crude oil processed. Target energy consumption may be a more realistic figure. Targeting requires setting an intermediate attainable energy consumption level for process units and steam power plant offsite keeping in view of the present configuration. Feedstock availability, product pattern, and local constraints of each refinery are means of identifying potential energy-saving areas. Table 24.29 shows the comparison of actual energy consumption for major process units with benchmark consumption and energy-saving potential in the refineries [38]. Solomon Associates company provides advisory services for refiners worldwide for reliability, equipment utilization, operating expenses, gross margin, overall performance range, and so on. The company's Comparative Performance Analysis™ (CPA™) methodology normalizes data across all plant sizes, types, and geographies to provide an insight to understand where each refiner stands (www.solomononline.com).

Glossary

Accumulation: The build-up of unreacted reagent or intermediates, usually associated with reactant added during a semi-batch operation.

Activation energy E_a: The constant E_a in the exponential part of the Arrhenius equation, associated with the minimum energy difference between the reactants and an activated complex (transition state that has a structure intermediate to those of the reactants and the products), or with the minimum collision energy between molecules that is required to enable a reaction to occur.

Adiabatic: A system condition in which no heat is exchanged between the chemical system and its environment.

Autocatalytic reaction: A reaction, the rate of which is increased by the catalyzing effect of its reaction products.

Autoignition temperature: The autoignition temperature of a substance, whether solid, liquid, or gaseous, is the minimum temperature required to initiate or cause self-sustained combustion (e.g., in air, chlorine, or other oxidant) with no other source of ignition. For example, if a gas is flammable and its autoignition temperature (AIT) is exceeded, there is an explosion.

Asset integrity: A PSM program element involving work activities that help ensure that equipment is properly designed, installed in accordance with specifications, and remains fit for purpose over its life cycle. Also called asset integrity and reliability.

Atmospheric storage tank: A storage tank designed to operate at any pressure between ambient pressure and 0.5 psig (3.45 kPa gage).

Biochemical oxygen demand (BOD): Is the amount of dissolved oxygen needed (i.e., demanded) by aerobic biological organisms to break down organic material present in a given water sample at certain temperature over a specific time period. The BOD value is most commonly expressed in milligrams of oxygen consumed per liter of sample during 5 days of incubation at 20°C and is often used as a surrogate of the degree of organic pollution of water.

Boiling-liquid-expanding-vapor-explosive (BLEVE): Is the violent rupture of a pressure vessel containing saturated liquid/vapor at a temperature well above its atmospheric boiling point. The sudden decrease in pressure results in explosive vaporization of a fraction of the liquid and a cloud of vapor and mist, with accompanying blast effects. The resulting flash vaporization of a large fraction of the liquid produces a large cloud. If the vapor is flammable and if an ignition source is present at the time of vessel rupture, the vapor cloud burns in the form of a large rising fireball.

Checklist analysis: A hazard evaluation procedure using one or more pre-prepared lists of process safety considerations to prompt team discussions of whether the existing safeguards are adequate.

Chemical oxygen demand (COD): Is an indicative measure of the amount of oxygen that can be consumed by reactions in a measured solution, expressed in mass of oxygen consumed over volume of solution (mg/l). A COD test can be used to easily quantify the amount of organics in water. The most common application of COD is in quantifying the amount of oxidizable pollutants found in surface water (e.g., lakes and rivers) or wastewater. COD is useful in terms of water quality by providing a metric to determine the effect an effluent will have on the receiving body, similar to BOD.

Chemical process industry: The phrase is used to include facilities that manufacture, handle, and use chemicals.

Confined explosion: An explosion of a fuel–oxidant mixture inside a closed system (e.g., a vessel or building).

Confined vapor cloud explosion (CVCE): Is a condensed phase explosion occurring in confinement (equipment, building, or/and congested surroundings). Explosions in vessels and pipes, processing, or storing reactive chemicals at elevated conditions are examples of CVCE. The excessive build-up of pressure in the confinement leads to this type of explosions leading to high overpressure, shock waves, and heat load (if the chemical is flammable and ignites). The fragments of exploded vessels and other objects hit by blast waves become airborne and act as missiles.

Containment: A physical system in which under all conditions, no reactants or products are exchanged between the system and its environment.

Decomposition energy: The maximum amount of energy that can be released upon decomposition. The product of decomposition energy and total mass is an important parameter for determining the effects of a sudden energy release, e.g., in an explosion.

Deflagration: The chemical reaction of a substance in which the reaction front advances into the unreacted substance at less than sonic velocity. Where a blast wave is produced that has the potential to cause damage, the term explosive deflagration is used.

Detonation: A release of energy caused by the extremely rapid chemical reaction of a substance in which the reaction front advances into the unreacted substance at equal to or greater than sonic velocity.

Design Institute for Emergency Relief Systems (DIERS): Institute under the auspices of the American Institute of Chemical Engineers funded to investigate design requirements for vent lines in the case of two-phase venting.

Disk: Is the pressure containing movable element of a pressure relief valve that effects closure.

Dow Fire and Explosion Index (F&EI): A method (developed by Dow Chemical Company) for ranking the relative fire and explosion risk associated with a process. Analysts calculate various hazard and explosion indexes using material characteristics and process data.

Exotherm: A reaction is called exothermic if energy is released during the reaction.

Explosion: Propagation of a flame in a premixture of combustible gases, suspended dust(s), combustible vapor(s), mist(s), or mixtures of thereof, in a gaseous oxidant such as air, in a closed, or substantially closed vessel.

Explosion rupture disk device: Is a rupture disk device designed for use at high rates of pressure rise.

Fail-safe: Design features that provide for the maintenance of safe operating conditions in the event of a malfunction of control devices or an interruption of an energy source (e.g., direction of failure of a motor operated valve on loss of motive power).

Failure mode and effects analysis: A hazard identification technique in which all known failure modes of components or features of a system are considered in turn, and undesired outcomes are used.

Failure: An unacceptable difference between expected and observed performance.

Fire point: The temperature at which a material continues to burn when the ignition source is removed.

Flammability limits: The range of gas or vapor compositions in air that will burn or explode if a flame or other ignition source is present. *Importance:* The range represents an unsafe gas or vapor mixture with air that may ignite or explode. Generally, the wider the range, the greater the fire potential.

Flammable: A "flammable liquid" is defined as a liquid with a flash point below 100°F (37.8°C). Flammable liquids provide ignitable vapor at room temperatures and must be handled with caution. Flammable liquids are: Class I liquids and subdivided as follows:

Class 1A: Those having flash points below 73°F and having a boiling point below point below 100°F.
Class 1B: Those having flash points below 73°F and having a boiling point at or above 100°F.

Flares: Flares are used to burn the combustible or toxic gas to produce combustion products, which are neither toxic nor combustible. The diameter of a flare must be suitable to maintain a stable flame and prevent a blowdown (when vapor velocities are greater than 20% of the sonic velocity).

Flash fire: The combustion of a flammable vapor and air mixture in which flame passes through that mixture at less than sonic velocity, such that negligible damaging overpressure is generated.

Flash point: The lowest temperature at which vapors above a liquid will ignite. The temperature at which vapor will burn while in contact with an ignition source, but which will not continue to burn after the ignition source is removed.

Gases: Flammable gases are usually very easily ignited if mixed with air. Flammable gases are often stored under pressure, in some cases as a liquid. Even small leaks of a liquefied flammable gas can form relatively large quantities of gas, which is ready for combustion.

Hazard: An inherent chemical or physical characteristic that has the potential for causing damage to people, property, or the environment.

Hazard analysis: The identification of undesired events that lead to the materialization of a hazard, the analysis of the mechanisms by which these undesired events could occur, and usually the estimation of the consequences.

Hazard and Operability (HAZOP): A systematic qualitative technique to identify process hazards and potential operating problems using a series of guidewords to study process deviations. A HAZOP is used to question every part of the process to discover what deviations from the start of the design can occur and what their causes and consequences may be. This is done systematically by applying suitable guidewords. This is a systematic detailed review technique for both batch and continuous plants, which can be applied to new or existing processes to identify hazards.

Hazardous chemical reactivity: Any chemical reaction with the potential to exhibit rates of increase in temperature and/or pressure too high to be absorbed by the environment surrounding the system. Included are reactive materials and unstable materials.

Hazard identification: The inventory of material, system, process, and plant characteristics that can produce undesirable consequences through the occurrence of an incident.

Hazard identification and risk analysis (HIRA): A collective term that encompasses all activities involved in identifying hazards and evaluating risk at facilities, throughout their life cycle, to make certain that risks to employees, the public, or the environment are consistently controlled within the organization's risk tolerance.

Hot work: Any operation that uses flames or can produce sparks (e.g., welding).

Incident: An event or series of events, resulting in one or more undesirable consequences, such as harm to people, damage to the environment, or assets/business losses. Such events include fires, explosions, releases of toxic or otherwise harmful substances, and so on.

Inherently safe: A system is inherently safe if it remains in a nonhazardous situation after the occurrence of non-acceptable deviations from normal operating conditions.

Inhibition: A protective method where the reaction can be stopped by addition of another material.

Incident investigation: A systematic approach for determining the causes of an incident and developing recommendations that address the causes to help prevent and mitigate future incidents. See also Root cause analysis and Apparent cause analysis.

Interlock system: A system that detects out-of-limits or abnormal conditions or improper sequences and either halts further action or starts corrective action.

Layer of protection analysis (LOPA): An approach that analyzes one incident scenario (cause–consequence pair) at a time, using predefined values for the initiating event frequency, independent protection layer failure probabilities, and consequence severity, in order to compare a scenario risk estimate to risk criteria for determining where additional risk reduction or more detailed analysis is needed. Scenarios are identified elsewhere, typically using a scenario-based hazard evaluation procedure such as HAZOP study.

Likelihood: A measuring of the expected frequency that an event occurs. This may be expressed as a frequency (e.g., events per year), a probability of occurrence during a time interval (e.g., annual probability), or a conditional probability (e.g., probability of occurrence, given that a precursor event has occurred).

Limiting oxygen concentration (LOC): Minimum concentration of oxygen in a mixture with gas, vapor, or dust that will allow it to burn.

Liquids: A vapor has to be produced at the surface of a liquid before it will burn. Many common liquids give off a flammable concentration of vapor in air without being heated, sometimes at well below room temperature. Gasoline, for example, gives off ignitable vapors above about -40°C, depending on the blend. The vapors are easily ignited by a small spark of flame.

Log out/tag out (LOTO): Is a safety procedure used in industry to ensure that dangerous equipment /machines are properly shut off and not able to be started up again prior to the completion and maintenance or repair work. It requires that hazardous energy sources be "isolated and rendered inoperative" before work is started on the equipment. The isolated power sources are then locked and a tag is placed on the lock identifying the worker who placed it. The worker then holds the key for the lock ensuring that only he/she can remove the lock and start the machine. This prevents accidental startup of a machine while it is in a hazardous state or while a worker is in direct contact with it.

Lower explosive limit (LEL): The concentration of a powder finely dispersed in air, below which no mixture likely to explode will be present.

Lower flammable limit (LFL): The lowest concentration of a vapor or gas (the lowest percentage of the substance in air) that will produce a flash of fire when an ignition source (heat, arc or flame) is present.

Management of change (MOC): A system to identify, review, and approve all modifications to equipment, procedures, raw materials, and processing conditions, other than "replacement in kind," prior to implementation.

Management system: A formally established set of activities designed to produce specific results in a consistent manner on a sustainable basis.

Mitigation: Lessening the risk of an accident event. A sequence of action on the source in a preventive way by reducing the likelihood of occurrence of the event, or in a protective way by reducing the magnitude of the event and for the exposure of local persons or property.

Near-miss: An unplanned sequence of events that could have caused harm or loss if conditions were different or were allowed to progress, but actually did not.

Operating procedures: Written, step-by-step instructions and information necessary to operate equipment, compiled in one document including operating instructions, process descriptions, operating limits, chemical hazards, and safety equipment requirements.

Operational discipline (OD): The performance of all tasks correctly every time; good OD results in performing the task the right way every time. Individuals demonstrate their commitment to process safety through OD. OD refers to the day-to-day activities carried out by all personnel. OD is the execution of the system by individuals within the organization.

Operational readiness: A PSM program element associated with effects to ensure that a process is ready to start-up/restart. This element applies to a variety of restart situations, ranging from restart after a brief maintenance outage to restart of a process has been mothballed for several years.

Organizational change: Any change in position or responsibility within an organization or any change to an organizational policy or procedure that affects process safety.

OSHA Process Safety Management (OSHA PSM): A U.S. regulatory standard that requires use of a 14-element management system to help prevent or mitigate the effects of catastrophic releases of chemicals or energy from processes covered by the regulations 49 CFR 1910.119.

Pressure relief valve (PRV): A pressure relief device that is designed to reclose and prevent the further flow of fluid after normal conditions have been restored.

Pre-startup safety review (PSSR): A systematic and thorough check of a process prior to the introduction of a highly hazardous chemical to a process. The PSSR must confirm the following: construction and equipment are in accordance with design specifications; safety, operating, maintenance, and emergency procedures are in place and are adequate. A process hazard analysis has been performed for new facilities and recommendations and have been resolved or implemented before startup, and modified facilities meet the management of change requirements; training of each employee involved in operating a process has been completed.

Preventive maintenance: Maintenance that seeks to reduce the frequency and severity of unplanned shutdowns by establishing a fixed schedule of routine inspection and repairs.

Probabilistic risk assessment (PRA): A commonly used term in the nuclear industry to describe the quantitative evaluation of risk using probability theory.

Process hazard analysis (PHA): An organized effort to identify and evaluate hazards associated with processes and operations to enable their control. This review normally involves the use of qualitative techniques to identify and assess the significance of hazards. Conclusions and appropriate recommendation are developed. Occasionally, quantitative methods are used to help prioritize risk reduction.

Process knowledge management: A Process Safety Management (PSM) program element that includes work activities to gather, organize, maintain, and provide information to other PSM program elements. Process safety knowledge primarily consists of written documents such as hazard information, process technology information, and equipment-specific information. Process safety knowledge is the product of this PSM element.

Process safety: A discipline that focuses on the prevention of fires, explosions, and accidental chemical releases at chemical process facilities. Excludes classic worker health and safety issues involving working surfaces, ladders, protective equipment, etc.

Process safety culture: The common set of values, behaviors, and norms at all levels in a facility or in the wider organization that affects process safety.

Process safety incident/event: An event that is potentially catastrophic, i.e., an event involving the release/loss of containment of hazardous materials that can result in large-scale health and environmental consequences.

Process safety information (PSI): Physical, chemical, and toxicological information related to the chemicals, process, and equipment. It is used to document the configuration of a process, its characteristics, its limitations, and as data for process hazard analyses.

Process safety management (PSM): A management system that is focused on prevention of, preparedness for, mitigation of, response to, and restoration from catastrophic releases of chemicals or energy from a process associated with a facility.

Process safety management systems: Comprehensive sets of policies, procedures, and practices designed to ensure that barriers to episodic incidents are in place, in use, and effective.

Purge gas: A gas that is continuously or intermittently added to a system to render the atmosphere nonignitable. The purge gas may be inert or combustible.

Quenching: Rapid cooling from an elevated temperature, e.g., severe cooling of the reaction system in a short time (almost instantaneously), "freezes" the status of a reaction and prevents further decomposition.

Reactive chemical: A substance that can pose a chemical reactivity hazard by readily oxidizing in air without an ignition source (spontaneously combustible or peroxide forming), initiating or promoting combustion in other materials (oxidizer), reacting with water, or self-reacting (polymerizing, decomposing, or rearranging). Initiation of the reaction can be spontaneous, by energy input such as thermal or mechanical energy, or by catalytic action increasing the reaction rate.

Recognized and Generally Accepted Good Engineering Practice (RAGAGEP): A term originally used by OSHA, stems from the selection and application of appropriate engineering, operating and maintenance knowledge when designing, operating, and maintaining chemical facilities with the purpose of ensuring safety and preventing process safety incidents.

It involves the application of engineering, operating, or maintenance activities derived from engineering knowledge and industry experience based upon the evaluation and analyses of appropriate internal and external standards, applicable codes, technical reports, guidance, or recommended practices or documents of similar or multiple sources and will vary based upon individual facility processes, materials, service, and other engineering considerations.

Risk Management Program (RMP) Rule: EPA's accidental release prevention rule, which requires covered facilities to prepare, submit, and implement a risk management plan.

Risk-Based Process Safety (RBPS): The Center for Chemical Process Safety's (CCPS) PSM system approach that uses risk-based strategies and implementation tactics that are commensurate with the risk-based need for process safety activities, availability of resources, and existing process safety culture to design, correct, and improve process safety management activities.

Safety Instrumented System (SIS): The instrumentation controls and interlocks provided for safe operation of the process.

Relieving pressure: Is set pressure plus overpressure.

Risk: The likelihood of a specified undesired event occurring within a specified period or in specified circumstances.

Risk analysis: A methodical examination of a process plant and procedure that identifies hazards, assesses risks, and proposes measures that will reduce risks to an acceptable level.

Runaway: A thermally unstable reaction system, which shows an accelerating increase of temperature and reaction rate. The runaway can finally result in an explosion.

Rupture disk device: Is a non-reclosing pressure relief device actuated by inlet static pressure and designed to function by the bursting of a pressure containing disk.

Safety relief valve: Is a pressure relief valve characterized by rapid opening pop action or by opening generally proportional to the increase in pressure over the opening pressure. It may be used for either compressible or incompressible fluids, depending on design, adjustment or application.

Set pressure: The inlet pressure at which the relief device is set to open (burst).

Stagnation pressure: The pressure that would be observed if a flowing fluid were brought to rest along an isentropic path.

Static activation pressure, P_{stat}: Pressure that activates a rupture disk or an explosion door.

Superimposed back pressure: The static pressure existing at the outlet of a pressure relief device at the time the device is required to operate. It is the result of pressure in the discharge system from other sources.

Temperature of no-return: Temperature of a system at which the rate of heat generation of a reactant or decomposition just exceeds the rate of heat loss and will lead to a runaway reaction or thermal explosion.

Thermally unstable: Chemicals and materials are thermally unstable if they decompose, degrade, or react as a function of temperature and time at or about the temperature of use.

Thermodynamic data: Data associated with the aspects of a reaction that are based on the thermodynamic laws of energy, such as Gibbs' free energy and the enthalpy (heat) of reaction.

Time to maximum rate (TMR): The time taken for a material to self-heat to the maximum rate of decomposition from a specific temperature.

Unconfined vapor cloud explosion (UCVE): Occurs when sufficient amount of flammable material (gas or liquid having high vapor pressure) gets released and mixes with air to form a flammable cloud such that the average concentration of the material in the cloud is higher than the lower *limit of explosion*. The resulting explosion has a high potential of damage as it occurs in an open space covering large areas. The flame speed may accelerate to high velocities and produce significantly blast overpressure. Vapor cloud explosions in densely packed plant areas (pipe lanes, units, etc.) may show accelerations in flame speeds and intensification of blast.

Upper explosive limit (UEL) or upper flammable limit (UFL): The highest concentration of a vapor or gas (the highest percentage of the substance in the oxidant) that will produce a flash or fire when an ignition source (heat, arc, or flame) is present.

Vapor specific gravity: The weight of a vapor or gas compared to the weight of an equal volume of air, an expression of the density of the vapor or gas. Materials lighter than air have vapor specific gravity less than 1.0 (examples: acetylene, methane, hydrogen). Materials heavier than air (examples: ethane, propane, butane, hydrogen, sulfide, chlorine, sulfur dioxide) have vapor specific gravity greater than 1.0.

Vapor pressure: The pressure exerted by a vapor above its own liquid. The higher the vapor pressure, the easier it is for a liquid to evaporate and fill the work area with vapors, which can cause health or fire hazards.

Vapor pressure system: A vapor pressure system is one in which the pressure generated by the runaway reaction is solely due to the increasing vapor pressure of the reactants, products, and/or solvents as the temperature rises.

Venting (emergency relief): Emergency flow of vessel contents out of the vessel. The pressure is reduced by venting, thus avoiding a failure of the vessel by overpressurization. The emergency flow can be one-phase or multiphase, each of which results in different flow and pressure characteristics. Multiphase flow, e.g., vapor and or gas/liquid flow, requires substantially larger vent openings than single-phase vapor (and/or gas) flow for the same depressurization rate.

Vent area, A: Area of an opening for explosion venting.

Acronyms and Abbreviations

AGA	American Gas Association
AIChE	American Institute of Chemical Engineers
AIChE/CCPS	American Institute of Chemical Engineers—Center for Chemical Process Safety
AIChE/DIERS	American Institute of Chemical Engineers—Design Institute for Emergency Relief Systems
AIT	Auto-Ignition Temperature
API	American Petroleum Institute
ASME	American Society of Mechanical Engineers
ASTM	American Society of Testing Materials
bar-m/sec	Bar-meter per second
BLEVE	Boiling Liquid Expanding Vapor Explosion
CSB	U.S. Chemical Safety and Hazard Investigation Board
CPI	Chemical Process Industry
DIERS	Design Institute for Emergency Relief Systems
DOE	Department of Energy
EFCE	European Federation of Chemical Engineers
EPA	U.S. Environmental Protection Agency
HAZOP	Hazard and Operability
HAZAN	Hazard Analysis
HRA	Human Reliability Analysis
HSE	Health and Safety Executive, United Kingdom
LFL	Lower Flammable Limit
LNG	Liquefied Natural Gas
LPG	Liquefied Petroleum Gas
MSDS	Material safety data sheet
NFPA	National Fire Protection Agency
NIOSH	National Institute for Occupational Safety and Health
OSHA	Occupational Safety and Health Administration
PE	Process Engineer
PFD	Process Flow Diagram
PHA	Preliminary Hazard Analysis
P&ID	Piping and Instrumentation Diagram
TLV	Threshold Limit Values
UFL	Upper Flammable Limit
VCE	Vapor Cloud Explosion

References

1. U.S. Chemical Safety and Hazard Investigation Board, Interim Investigation Report, Chevron Richmond Refinery Fire, Chevron Richmond Refinery, Richmond, CA, August 6, 2012. (http://www.csb.gov/chevron-refinery-fire).
2. Introduction to Process Safety for Undergraduates and Engineers, CCPS of the American Institute of Chemical Engineers, Wiley, New York, NY, 2016.
3. BP Process Safety Series, Hazard of Oil Refining Distillation Units, IChemE, U.K., 2008.
4. The Report of the BP U.S. Refineries Independent Safety Review Panel, January 2007. (http://www.bp.com/liveassets/bp_internet/globalbp/globalbp_uk_english/SP/STAGING/local_assets/assets/pdfs/Baker_panel_report.pdf).
5. Crawley, F., Preston, M. and B. Tyler, Hazop: Guide to Best Practice, IChemE, U.K. 2000.
6. Guidelines for Hazard Evaluation Procedures, 2nd Ed., AIChE Center for Chemical Process Safety, USA, 1992.
7. Chemical Industries Association (CIA), A Guide to Hazard and Operability Studies, U.K., 1977.
8. Skelton, Bob, Process Safety Analysis—An Introduction, IChemE, U.K., 1997.

9. Barton, J. A., and R. Rogers, Chemical Reaction Hazards, 2nd Ed., IChemE, U.K., 1997.
10. Krishnan, G. U., What HAZOP Studies cannot do? Hydroc. Proc., pp 93 – 95, Oct. 2005.
11. Keltz, T. A., Hazop and Hazan, 4th ed., IChemE, UK., 1999.
12. Mansfield, D., and K. Cassidy, Inherently Safer Approaches to Plant Design, Hazards, XII European Advances in Process Safety, IChemE, pp. 285 – 299, 1994.
13. Englund, S. M., Design and Operate Plants for Inherent Safety, Part 1, Chem. Eng., Prog., pp. 86 – 91, Mar 1991.
14. Keltz, T. A., What You Don't Have, Can't Leak, Chemistry and Industry Jubilee Lecture, "Chemistry and Industry", p. 287, 6 May 1978.
15. Bell, C., Process and Plant Design, The Chemical Engineer, U.K., pp. 38-40, Nov. 2005.
16. Indra Deo Mall, Petroleum Refining Technology, CBS Publisher & Distributors, Pvt Ltd, 2015.
17. Siegell, J. H., Control VOC Emissions, Hydroc. Proc. Vol 76, No. 4, p.119, 1997.
18. Chopra, S. J., Refinery for Future, QIP Short-term Course on Advances in Hydrocarbon Engineering, IIT Roorkee, India, June 23 – July 4, 2003.
19. Krishnan, V., and R. P. Verma, Energy Conservation in Refineries, Hydroc. Technology, p. 47, Aug. 15, 1993.
20. Verma, R. P., Energy Conservation in Refineries, Hydroc. Technology, Special Issue, p. 45, Jan. 1991.
21. EIL, Energy Conservation in Process Industry, 1982.
22. Reid, J. A., Distributed Distillation with Heat Integration, Petroleum Technology Quarterly, p. 85, Autumn 2000.
23. Gadalla, M, |M. Jobson, and R. Smith, Chem. Eng. Progr., p. 44, April 2003.
24. Coker, A. K., Ludwig's Applied Process Design for Chemical and Petrochemical Plants, Vol 3, 4th ed., Elsevier, 2015.
25. Hairston, D., Trapping Steam, Chemical Engineering, p. 23, January 2003.
26. Harindranath, N, and M. Kamath, Integration of Energy Efficiency Concepts at the Design Stage in Process Industries. Chemical Industry Digest, p. 76, July 2007.
27. Perry, W., Variable Speed Drives Lower Compressed Air Cost., Hydroc. Proc., p. 52, July 2002.
28. Jain, Atul, L, et al., Radiative forcings and global warming potentials of 39 greenhouse gases, Journ. of Geophysical Research: Atmospheres, 105, (D16): 20773 – 20790, August 27, 2000.
29. Natural gas—Gas Flaring and gas venting—Eniscuola, Eniscuola Energy and Environment, Retrieved, 23, June, 2018.
30. Flaring emissions—Tracking Fuel Supply—Analysis, IEA, Retrieved, 2020.
31. Frequent, Routine Flaring May Cause Excessive, Uncontrolled Sulfur Dioxide Releases, Enforcement Alert, Washington, D. C., EPA, October 2000.
32. Stohl, A., Kilmont, Z., Eckhardt, S., Kupianen, K. Chevckenko, V. P., Kopeikin, V. M., Novigatsky, A. N., Black carbon in the Artic: The underestimated role of gas flaring and residential combustion emissions, Atmos. Chem. Phy., 13 (17): 8833-8855, 2013.
33. Michael Stanley, Gas flaring: An industry practice faces increasing global attention, World Bank, Retrieved 2020-01-20.
34. Seabirds at Risk around Offshore Oil Platforms in the North-west Atlantic, Marine Pollution Bulletin, Vol. 42, No. 12, pp. 1,285 – 1,290, 2001.
35. Carlos, T. M. S., Manage Refinery Spent Caustic Efficiency, Hydroc. Proc., p. 89, Feb. 2002.
36. Fontana, M. G., Corrosion Engineering, 3rd ed., McGraw Hill Book Co., 1987.
37. Schweitzer, P. A., Encyclopedia of Corrosion Technology, Marcel Dekker, 1998.
38. Sil, K., Dey, G. K. and C. S.S, Naryana, Benchmarking in Refinery Area. Workshop on Energy Management and Conservation, Lovraj Kumar Memorial Trust, New Delhi, Nov. 11-12, 2003.
39. Hydrogen Sulfide Release at Aghorn Operating Waterflood Station, CSB report May, 2021, https://www.csb.gov/cbs-release-final-aghorn-investigation-report
40. Jason Shirley, Process isolation practices: An industry wide view, hazardex-the journal for hazardous area environments, pp.24-29, www.hazardexonthenet.net, December 2020.
41. Conduct of Operations and Operational Discipline For Improving Process Safety in Industry, AIChE Center for Chemical Process Safety (CCPS), CCPS-Wiley, New York, NY, 2011.
42. Zhang, W., Evaluation of Susceptibility to Hydrogen Embrittlement – A Rising step load testing method., Material Sciences and Applications, 7, 389-395, 2016.
43. Walker, H. B., Reduce FCC Corrosion, Hydrocarbon Processing, p. 81, Jan 1984.
44. Kane, R. D., and M. S. Cayard., Select Materials for High Temperature. Chemical Engineering Progress, p. 83, March 1995.
45. Huchler, L. A., Select the Best Boiler Water Chemical Treatment Program. Chemical Engineering Progress, p. 45, Aug., 1998.

25

Product Blending

25.0 Introduction

One of the critical economic issues for the refinery is selecting the optimal combination of components for the products. Refining does not provide commercially viable products directly, but semi-finished products, which must be blended to meet the correct customers' specifications. Blending is an important operation in the refinery as it is a physical process in which accurately weighed quantities of two or more components are mixed thoroughly to form a homogeneous phase, which can be either similar or dissimilar in nature.

Most of the products obtained from the distillation/fractionation columns are blended with fractions obtained from other units to help in keeping the wastage minimum and invariably increasing the quantity of the products. Almost all products from gas to lube oil are not only blended of fractions but additives as well. All such blends are formulated to have the required properties conforming to the correct specifications.

The products of blending are gasoline, jet fuels, heating oils, and diesel fuels. The objective of product blending is to allocate the available blending components so that demands and specifications are met at the least cost and to give products, which maximize overall profit.

Gasoline blending is much more complicated than a simple mixing of the components. A typical refinery may have as many as 8 to 15 hydrocarbon streams to consider as blend stocks. These may range from butane, the most volatile component to a heavy naphtha and include several gasoline naphthas from crude distillation, catalytic cracking and thermal processing units in addition to alkylate, polymer, and reformate. Modern gasoline may be blended to meet simultaneously 10 to 15 different quality specifications, such as vapor pressure, initial, intermediate and final boiling points, sulfur content, color, stability, aromatic content, olefin content, octane measurements for several different portions of the blend, and other local governmental or market restrictions. Since each of the individual components contributes uniquely in each of these quality areas and each bears a different cost of manufacture, the proper allocation of each component into its optimal disposition is of major economic importance.

Gasoline blending is a refinery operation that blends different component streams into various grades of gasoline. Typical grades include 83 octane (blended with an oxygenated fuel such as ethanol), regular 87 octane, and premium 92 octane. The Reid vapor pressure (RVP) is set depending on the average temperature of the location the gasoline will be used (cold temperatures require higher RVP than warmer climates). These two specifications are the most significant, and they are documented with each blend, to minimize the potential of octane giveaways.

If the octane specification is 87, then each 0.1 octane over this target value incurs further costs to the refiner. For example, in the U.S., this cost calculates to approximately $1,000,000 per 0.1 octane giveaway per 100,000 bpd crude capacity. The RVP is slightly different, as refiners aim to blend as much low value normal-butane (component RVP of

52.0 psi) into the final blend without going over the specification. For example, the cost of n-butane is $7 per barrel that can be sold as gasoline at $25 per barrel just by blending. The $18 per barrel profit is significant to the refiner, making RVP economics important.

Distillate fuel blending has other specifications that must be ascertained. Distillate blending includes jet fuels, diesel fuels, kerosene and No. 1 and No. 2 fuel oils. Diesel fuels properties that are measured include cetane number (analogous to octane number for the gasoline engine); flash points (relates to fire hazard in storage); low temperature properties; including cloud point; and pour point and sulfur content (see Glossary of Technical Terminology in this text). In blending some products such as residual fuel oils or asphalt, viscosity is one of the specifications that must be met.

The objective of product blending is to assign all available blend components to satisfy the product demand and specifications to minimize cost and overall profit. Almost all refinery products are blended for the optimal use of the intermediate product streams for the most efficient and profitable conversion of petroleum to marketable products. For example, typical gasolines may consist of straight-run naphtha from distillation, crackate from fluidized catalytic cracking (FCC) unit, reformate, alkylate, isomerate, and polymerate from other processing units in proportion to make the desired grades of gasolines and specifications. Furthermore, intermediate streams can be blended into different finished products; for example, naphtha can be blended into gasoline or jet fuel streams depending on the demand. Previously, the blending was performed in batch operations. However, with on-line equipment, computerization, and improved techniques, blending is readily performed with greater efficiency and accuracy. Keeping inventories of the blending stocks along with cost and physical data have increased the flexibility and profits from on-line blending through optimization programs. In most cases, the components blend nonlinearly for a given property (for example, vapor pressure, octane number, cetane number, viscosity, pour point, and so on), and correlations and programming techniques are required for reliable predictions of the specified properties in the blend.

Blends of petroleum-based gasoline with 10% ethanol (referred to as E10) account for more than 95% of the fuel consumed in motor vehicles with gasoline engines in the U.S. Ethanol blended fuels are a pathway to compliance with elements of the federal renewable fuel standard (FRS). The total volume of ethanol blended into motor fuels used in the U.S. has increased since 2010 although at a declining rate of growth; meanwhile, the use of ethanol-free gasoline (E0) by fuel consumers has declined.

U.S. Energy Information Administration (EIA) tracks fuel components through data it collects from refiners, importers, large blending terminals, and ethanol producers. U.S. refiners produce large volumes of blendstocks for oxygenate blending (BOB) that are referred to as RBOB or CBOB depending on whether they are formulated to be blended with ethanol to make reformulated or conventional gasoline, respectively. An increase in ethanol use in the motor vehicle fleet is to adopt fuel blends containing a higher volume such as E15 and E85. However, not all gasoline-powered vehicles can use these fuels. The U.S. Environmental Protection Agency (EPA) issued a partial waiver that allows the use of E15 in model year 2001 on newer vehicles. Fuels marketed as E85, which contains between 51% and 83% ethanol by volume can only be used in flex fuel vehicles, which make up about 7% (16.3 million) of the current on-road fleet of light-duty vehicles. Sales of E15 and E85 remain very limited because of a variety of economic, environmental and distribution system challenges. EPA estimated use of 320 million gallons of E15 and 200 million gallons of E85 in 2016, and these combined would represent only 0.4% of the total 142 billion gallons of fuel use by vehicles and other equipment with gasoline-burning engines [13].

There are also volumes of gasoline containing no ethanol often referred to as E0. EIA surveys indicated U.S. E0 refinery and blender production of 894,000 bpd of E0 imports in 2015. U.S. Census Bureau data showed exports of 476,000 bpd of finished motor gasoline in 2015, and an average of 6000 bpd of E0 was withdrawn from stocks held at refineries and terminals or was in pipelines during 2015. Domestic disposition of E0 was calculated to 494,000 bpd (7.6 billion gallons). However, actual use of E0 in vehicles, boats and other equipment with gasoline-burning engines was below that level because some vehicles of E0 that entered domestic market might have been blended with ethanol at smaller terminals that were not reviewed by EIA or blended at the point of retail sale.

EIA data on ethanol supply and disposition could be used together with the E0 domestic disposition data to develop an estimate of E0 use in vehicles and other equipment with gasoline-burning engines. EIA developed fuel ethanol

Figure 25.1 A view of petroleum refinery unit with storage blending tanks (Source: www.briannica.com).

Figure 25.1a Refinery tank farm blending and shipping facilities.

Figure 25.1b Refinery inline blending.

balances, which accounted for production, imports, exports, stock change and blending, and nearly all fuel ethanol was blended with a BOB to produce E10. However, not all ethanol use was captured in EIA's blending data, and that ethanol supply within the U.S. exceeded identified uses by 17,000 bpd (254 million gallons). Figure 25.1 shows a view of petroleum facility with storage blending tanks; Figures 25.1a and 25.1b show photos of refinery tank farm blending and shipping facilities and refinery inline blending facility, respectively.

25.1 Blending Processes

There are two ways of blending; by batch and continuous. Batch blending starts with mixing known amounts of components in a tank mixer using an agitator and other accessories as pressure gauge, liquid level indicators, and so on. Agitation can also involve air where toxic materials like lead, biocides are blended. Mixing in tanks is also accompanied by heating or cooling coils.

Most refiners employ computer-controlled on-line blending for blending gasoline and distillates. All the components to be blended are pumped simultaneously into a common header at rates specified as per the formulations as shown in Figure 25.2. The rate of flow is controlled by a valve operated by a pneumatic or electric relay system. The signals received correspond to the flow rates, and these can accurately modulate the flow rates by adjusting the valve. The long pipeline through which all these proportioned components travel across acts as a mixer to produce the blend. Additives can also be injected into the system. Figures 25.2a and 25.2b show a schematic diagram of a blending process and representation of the online blending system for diesel product.

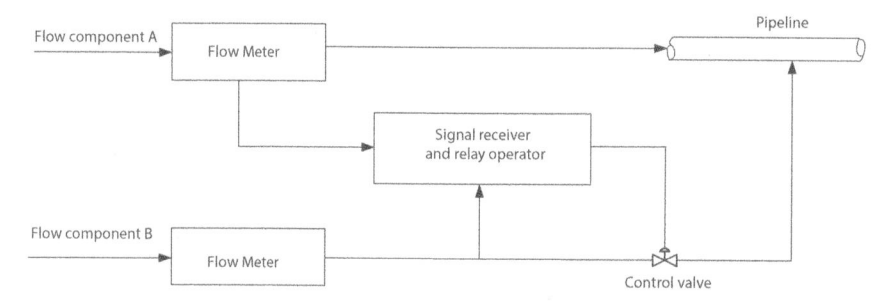

Figure 25.2 Schematic diagram for blending components.

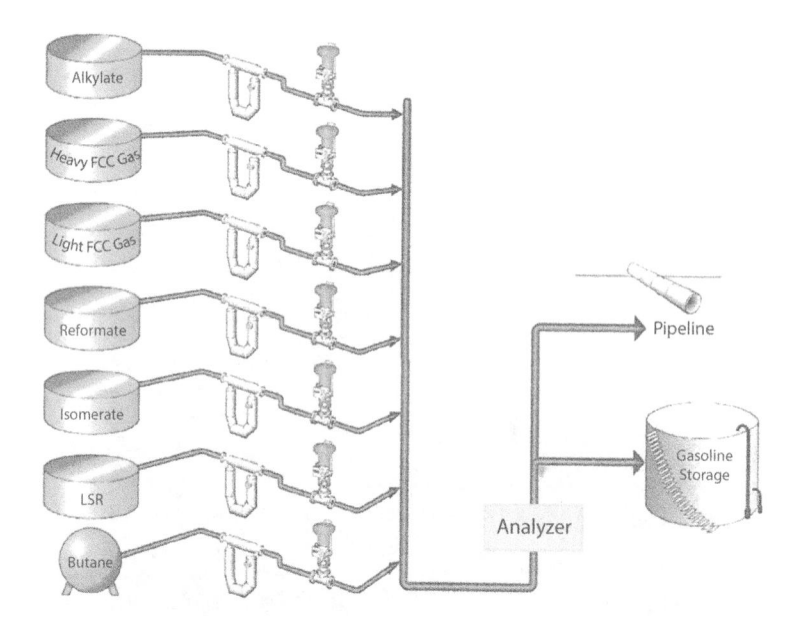

Figure 25.2a Schematic diagram of a blending process.

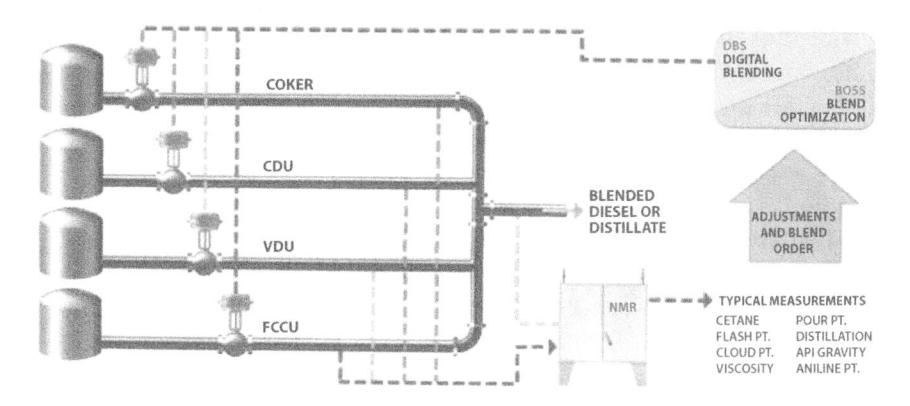

Figure 25.2b Schematic representation of the online blending systems for diesel product.

Inventories of blending stocks together with cost and physical property data are maintained in a database. Many of the properties of the blending components are non-linear, such as the octane number, therefore, estimating final blend properties from the components can be quite complex. When a certain volume of given quality product is specified, linear programming models (LP) are employed, which is a mathematical technique that permits the rapid selection of an optimal solution from a multiplicity of feasible alternative solutions. The LP can be used to optimize the blending operations to select the blending components to produce the required volume of the specific product at the lowest cost. However, non-linear programming is preferred as enough data are available to define the equations because the components blend non-linearly, and values are functions of the quantities of the components and their properties.

Ensuring that the blended streams meet the desired specifications, stream analyzers, such as boiling point, specific gravity, Reid vapor pressure and research and motor octane numbers are installed to provide feedback control of blending streams and additives. The blending components involve an iterative process (i.e., trial-and-error) to achieve all critical specifications most economically, as the large number of variables leads to several similar solutions that give the approximate equivalent total overall cost or profit, and is easily handled by a computer.

Each component is characterized by its specific properties and cost of manufacture, and each gasoline grade requirement is similarly defined by quality requirements and relative market value. The linear programming solution specifies the unique disposition of each component to achieve maximum operating profit. The next stage is to measure carefully the rate of addition of each component in the blend and collect it in storage tanks for final inspection before delivering it for sale. The problem is not fully resolved until the product is actually delivered into customer's tanks; frequently, last-minute changes in shipping schedules or production qualities require the reblending of the finished gasoline of the substitution of a high-quality (and therefore costlier) grade for one or more immediate demand even though it may generate less income for the refinery.

25.1.1 Gasoline Blending

Different gasolines like alkylates, reformates, polymerate, crackate, straight runs, and so on are blended along with various additives to boost the performance value of gasoline. Such additives include octane enhancers, metal deactivators, anti-oxidants, anti-knock agents, gum and rust inhibitors, detergents, and so on are added during and/or after blending to provide specific properties not inherent in hydrocarbons. However, the blend should be to specifications and the two essential properties on which blends are critically constituted are vapor pressure and octane number. The vapor pressure of a mixture can be estimated by Raoult's law, but scant information on the molecular composition of a blend does not permit it; and laborious experimentation for evaluating molecular composition is unwise.

25.2 Ternary Diagram of Crude Oils

25.2.1 Elemental Analysis and Ternary Classification of Crude Oils

Crude oils found in various parts of the world can vary considerably in characteristics as these can readily be observed in the differences in specific gravity that exists between the crudes. For example, Tapis crude (Malaysia) has an API gravity of 46, while Boscan (Venezuela) has a gravity of 9.2. This difference is because although each crude oil contains basically the same hydrocarbon compounds the proportion of these hydrocarbons varies considerably from one crude to another. Some crudes are relatively rich in paraffins, and this is reflected by the waxy nature of the crude (most Middle East crudes fall into this group), while others contain more cycloparaffins (i.e., naphthenes) and aromatics (such as Nigerian and some American West Coast crudes).

Although it is theoretically possible to produce any type of refined product from any crude, it is not economically feasible; for instance, better yield of reformer stock for aromatic production is obtained from Nigerian crude than from Kuwait, while considerable, more residuum for fuel oil is obtained from Kuwait than Nigeria. To satisfy the demand for these two products, refineries often blend two such crudes, changing the proportion of the blend to satisfy the need. Furthermore, if product demands are seasonal as in the case of many gas oils, import of selected crudes is scheduled to optimize the production of such cuts for the season [3].

Despite a wide variety of crude oil found in different parts of the earth, the elemental composition of most crude changes in narrow ranges, as shown in Table 25.1.

With such narrow ranges of change in elemental contents, elemental composition does not have much utility for classification of crude oil. Instead, variations in hydrocarbon composition (paraffins, naphthenes, and aromatics) are used to classify crude oils, using a ternary diagram, shown in Figure 25.3. Each apex of the triangle represents 100% weight of the corresponding compounds, and 0% of these hydrocarbons on the side of the triangle across from the apex. For example, the side at the bottom of the triangle (across from the apex of 100% aromatics) represents binary mixtures of paraffins and naphthenes.

Table 25.3 shows the six classes of crude oil that are defined using a ternary diagram. These classes are shown as areas on the ternary diagram for paraffins, and it is generally accepted that Class 1 (rich in paraffins) represents the most desirable type of crude oil because refining these crudes would readily lead to high yields of light and middle distillates that constitute the fuels such as gasoline, diesel fuel, and jet fuel, which are in high demand. Extensive refining would be required to produce high yields of distillate fuels from aromatic crudes (For example, Classes 4–6). Class 1 crudes tend to have high °API and low sulfur contents and tend to be more expensive than the other types of crude oils.

Table 25.1 Elemental Composition of Crude Oil.

Element	% wt
C	84–86
S	11–14
H	0–6
N	0–1
O	0–2

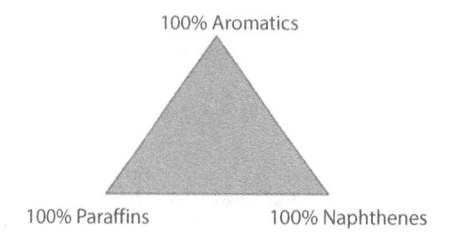

Figure 25.3 A ternary diagram for classification of crude oils.

25.2.2 Reading a Ternary Diagram

A ternary diagram is a triangle, with each of the three apexes representing a composition, such as Aromatics, paraffins and naphthenes. For the moment they are labeled A, B, and C.

Illustrating how the triangular diagram works is as follows:

The drawing to the left has only the skeleton of the triangle present as we concentrate on point A. Point A is at the top of the heavy vertical red line (arrow). Along this, line is indicated percent of A. A point plotted at the top of the vertical line nearest A indicates 100% A. A horizontal bar at the bottom of the line (farthest from A) represents 0% of A. Any other percentage can be indicated by a line appropriately located along the line between 0% and 100%, as shown by the numbers off to the right (Figure 25.3a).

The horizontal lines that represent various percentages of A can be of any length since they run parallel to the base line and remain the same distance from the bottom and top of the triangle. The lines are projected out to the right of the red arrow line just as far as where the imaginary side of the triangle will be, and their percentage abundances written along the right side of the triangle. By doing this, the right side of the triangle becomes the scale for percent abundance of A. To be complete; the horizontal lines also extend to the left until they contact the left side of the imaginary triangle, but no percent abundances are written there. In the final ternary diagram, the red vertical arrow is removed.

Point B is at the lower left apex of the triangle. We construct a percent abundance scale for B by rotating the heavy red scale line 120° counter clock wise so that it runs from the right side of the triangle to the lower left corner. The right side of the triangle now becomes the base line for the percent scale for B, and a series of red lines has been drawn parallel to the triangle's right side to mark off the percentages. These lines are projected out to the left and bottom sides of the triangle, and the percent scale for B laid out along the left side (Figure 25.3b).

Point C is at the lower right apex of the triangle. We construct the percent abundance scale for C by rotating the heavy red scale line another 120° so that it runs from the left side of the triangle to the lower right corner, and the percent scale lines and percent abundance numbers rotate with it. The sum result is the ternary diagram to the right with all the scales present. Note that the heavy red lines are not included in this final triangle. Also observe that the ternary diagram is read counter clockwise.

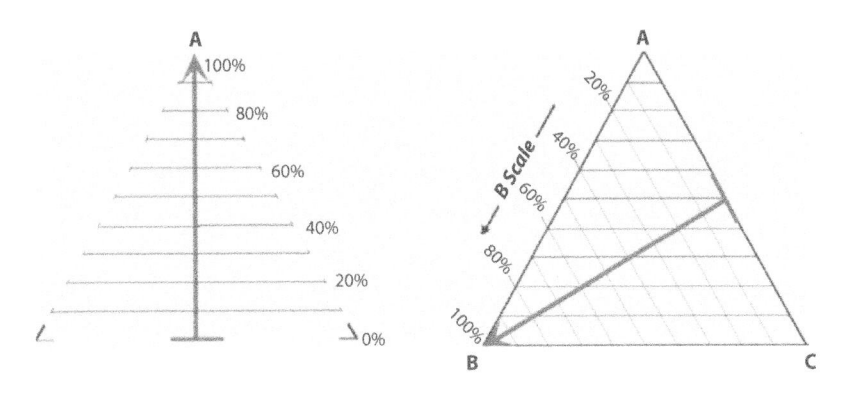

Figure 25.3a A ternary diagram for classification of crude oils.

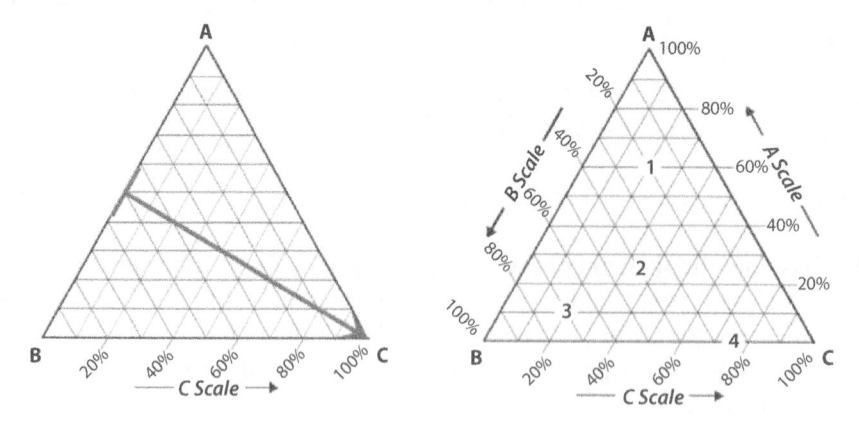

364   PETROLEUM REFINING DESIGN AND APPLICATIONS HANDBOOK VOLUME 5

Figure 25.3b A ternary diagram for classification of crude oils.

1.	60% A	20% B	20% C	= 100%
2.	25% A	40% B	35% C	= 100%
3.	10% A	70% B	20% C	= 100%
4.	0.0% A	25% B	75% C	= 100%

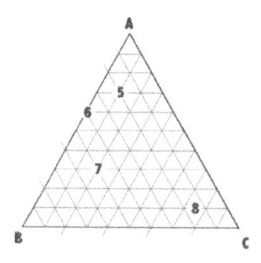

5.	?% wt A	?% wt B	?% wt C	= 100%
6.	?% wt A	?% wt B	?% wt C	= 100%
7.	?% wt A	?% wt B	?% wt C	= 100%
8.	?% wt A	?% wt B	?% wt C	= 100%

Solution

5.	70% wt A	20% wt B	10% wt C	= 100%
6.	60% wt A	40% wt B	0% wt C	= 100%
7.	30% wt A	50% wt B	20% wt C	= 100%
8.	10% wt A	15% wt B	75% wt C	= 100%

Example 25.1

A refinery has access to two different crude oil stocks A and B with the following compositions:

	Naphthenes, %wt	Aromatics, %wt
Crude A	10	60
Crude B	60	10

a) What type of crude oils are A and B according to the ternary classification are based on this composition?

b) What type of crude would be obtained if A and B are blended in a proportion of A/B = 2/3?

Solution

	Naphthenes, %wt	Aromatics, %wt	Paraffins, %wt
Crude A	10	60	30
Crude B	60	10	30

a) What type of crude oils are A and B according to the ternary classification are based on this composition?

Solution

A: Aromatics: 60%wt, Naphthenes: 10%wt

From Table 25.2,

Aromatic-Intermediate crude (Aromatics > 50%, Paraffins > 10%)

B: Aromatics: 10%, Naphthenes: 60%

Naphthenic crude (P + N > 50%, N > P, N > 40%)

b) What type of crude would be obtained if A and B are blended in a proportion of A/B = 2/3?

In the blend with A/B = 2/3: A: 40%, B = 60%

Binary Blend C:

Naphthenes = (0.4)(10) + (0.6)(60) = 40%

Aromatics = (0.4)(60) + (0.6)(10) = 30%, Paraffins = 30%

Binary Blend C: Border-line between Paraffinic–Naphthenic (A < 50%, P < 40, N < 40%)

and Naphthenic ((P + N > 50%, N > P, N > 40%) crude oils.

Table 25.2 Ternary Classification of Crude Oils.

1. **Paraffinic Crudes** Paraffins + naphthenes > 50% Paraffins > Naphthenes Paraffins >40%	2. **Naphthenic Crudes** Paraffins + naphthenes > 50% Naphthenes > Paraffins Naphthenes > 40%
3. **Paraffinic—Naphthenes Crudes** Aromatics < 50% Paraffins < 40% Naphthenes < 40%	4. **Aromatic—Naphthenic Crudes** Aromatics > 50% Naphthenes > 25% Paraffins < 10%
5. **Aromatic—Intermediate Crudes** Aromatics > 50% Paraffins > 10%	6. **Aromatic—Asphaltic Crudes** Naphthenes < 25% Paraffins < 10%

We can use the ternary diagram in solving the problem, and as described below.

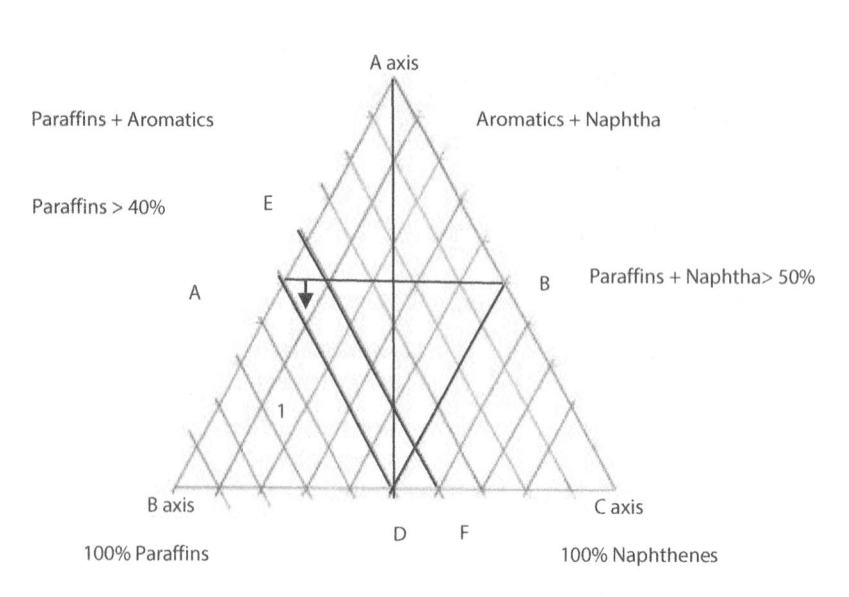

```
                        Aromatics, 100%
                              C
                           A axis
       Paraffins + Aromatics    Aromatics + Naphtha

       Paraffins > 40%    E
                                        Paraffins + Naphtha > 50%
                      A           B
                         1
       B axis                           C axis
       100% Paraffins  D    F    100% Naphthenes

                    Paraffins > Naphthenes

                    Paraffins + Naphthenes
```

Paraffin crudes
Paraffins + Naphthenes > 50%
Paraffins > 40%
Paraffins > Naphthenes
Below line AB a. Contents of paraffins + naphthenes > 50% b. Left of CD, paraffins > naphthenes c. Line EF, paraffins > 40%

PRODUCT BLENDING 367

Aromatics, 100%
C
A axis

Naphthenes > Paraffins

Paraffins + Naphthenes > 50%

A
B

2

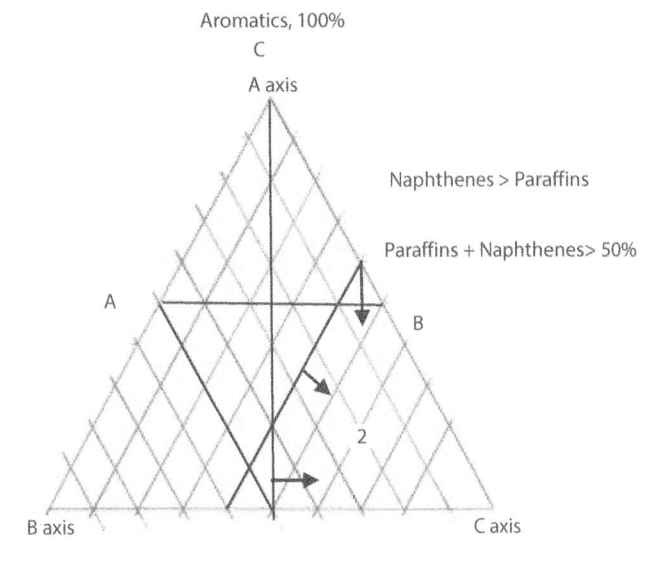

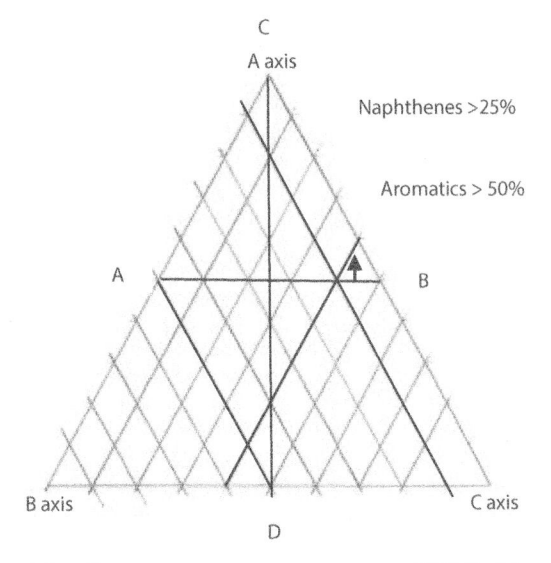

B axis
C axis
100% Paraffins
100% Naphthenes

Naphthenes > Paraffins

Naphthene crudes
Paraffins + Naphthenes > 50%
Naphthenes > 40%
Naphthenes > Paraffins

Aromatics, 100%
C
A axis

Naphthenes > 25%

Aromatics > 50%

A
B

B axis
C axis
D
100% Paraffins
100% Naphthenes

Aromatic + Naphthenic crudes
Aromatics > 50%
Paraffins < 10%
Naphthenes > 25%

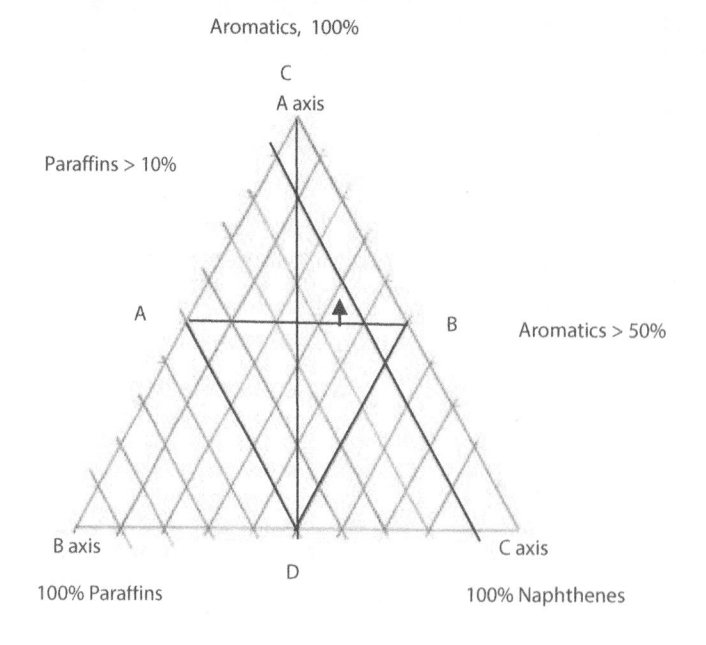

Aromatics, 100%
C
A axis

Paraffins > 10%

A ─────── B   Aromatics > 50%

B axis   C axis
       D
100% Paraffins   100% Naphthenes

Aromatic—Intermediate crudes
Aromatics > 50%
Paraffins >10%

Example 25.2

The compositions of three crude oils available to a refinery are as follows:

	Aromatics, % wt	Naphthenes, % wt	Paraffins, % wt
Crude A	?	20	60
Crude B	50	?	30
Crude C	?	20	10

The refiners would like to maintain a weight ratio of 1/1, Crude B/Crude C in a ternary blend of the oils A, B, and C. What would be the minimum concentration of Crude A (% wt) in a ternary blend that could be classified as paraffinic oil?

Solution

	Aromatics, % wt	Naphthenes, % wt	Paraffins, % wt
Crude A	20	20	60
Crude B	50	20	30
Crude C	70	20	10

Set A + B + C = 100 g and B = C

Paraffin balance:

$$0.60A + 0.3(100 - A)/2 + 0.1(100 - A)/2 > 40$$

$$1.2A + 30 - 0.3A + 10 - 0.1A > 80$$

$$.8A > 40$$

A > 50, therefore, A must be greater than 50% to maintain a paraffinic crude blend.

Product qualities are estimated through correlations that depend on the quantities and the properties of the blended components. Mixing rules together with correlations are employed to estimate the blend properties such as specific gravity, RVP, viscosity, flash point, pour point, cloud point, and aniline point. The octane number for gasoline is correlated with corrections based on aromatic and olefin content. The desired property of the blend is determined by

$$P_{Blend} = \frac{\sum_{i=1}^{n} q_i P_i}{\sum_{i=1}^{n} q_i} \qquad (25.1)$$

where

P_i is the value of the property of component i and q_i is the mass, volume or molar flow rate of component i contributing to the total amount of the finished product. For example, q_i can be volume fraction x_{vi}, therefore, the denominator in Eq. 25.1 = 1. Eq. 25.1 assumes that the given property is additive (or linear). Additive properties include specific gravity, boiling point and sulfur content. However, properties such as Reid vapor pressure (RVP), viscosity, flash temperature, pour point, aniline point, and cloud point are not additive. Table 25.3 lists typical properties of pure components and petroleum cuts that can be blended for gasoline to meet market specification.

25.3 Reid Vapor Pressure Blending

RVP is the vapor pressure at 100°F of a product determined in a volume of air four times the liquid volume. RVP is not an additive property, thus RVP blending indices are used. A commonly used RVP index is based on an empirical method developed by Chevron Oil Trading Company and is defined by

$$BI_{RVP_i} = RVP_i^{1.25} \qquad (25.2)$$

where

BI_{RVP_i} is the RVP blending index for component i and RVP_i is the RVP of component i in psi.

Using the index, the RVP of a blend is:

$$BI_{RVP_i,Blend} = \sum_{i=1}^{n} x_{vi} BI_{RVP_i} \qquad (25.3)$$

re x_{vi} is the volume fraction of component i.

Example 25.4

Calculate the RVP of a blend of light straight run gasoline (LSR) and heavy straight run gasoline (HSR), reformate and FCC gasoline using Table 25.3.

The quantities and RVP values of the blending components are:

Component	Quantity (bpd)	RVP (psi)
LSR gasoline	5000	11.1
HSR gasoline	4000	1.0
Reformate 94 RON	6000	2.8
FCC gasoline	7000	13.9

Solution

The RVP of the blend is calculated through the following steps:

- The volume fraction of each component x_{vi} is calculated and listed in Table 25.4.
- The RVP index of each component is calculated. For example, the calculation of LSR gasoline is $BI_{RVP_i} = 11.1^{1.25} = 20.26$
- The volume fraction is multiplied by the index for each component as shown in Table 25.4.

From the summation, $BI_{RVP_i, Blend} = 14.305$, then

$$RVP_{Blend} = (14.305)^{1/1.25} = 8.4 \text{ psi}$$

Therefore, the RVP of the blended product is 8.4 psi.

Additives such as propane (C_3H_8), i-butane (iC_4H_{10}), and n-butane (nC_4H_{10}) can be added to the gasoline blend to adjust the RVP requirements. The Excel spreadsheet program (Example 25.4.xlsx) shows the calculations of Example 25.4.

Example 25.5

Determine the amount of n-butane required to produce a gasoline blend with RVP = 10 psi from the components listed in Table 25.3. The RVP of n-butane is 52 psi.

Solution

Assume V_{Butane} is the volume flow rate (bpd) of n-butane required to be added to the blend. RVP indices are calculated for all components including n-butane and listed in Table 25.5.

The volume is first multiplied by the index and then divided by the total volume. The summation gives the RVP index for the blend.

Table 25.3 Typical Properties for Gasoline Blending Components [3].

No.	Component	RVP (psi)	MON	RON	API
1.	iC_4	71.0	92.0	93.0	
2.	nC_4	52.0	92.0	93.0	
3.	iC_5	19.4	90.8	93.2	
4.	nC_5	14.7	72.4	71.5	
5.	iC_6	6.4	78.4	79.2	
6.	LSR gasoline	11.1	61.6	66.4	78.6
7.	LSR gasoline isomerized once-through	13.5	81.1	83.0	80.4
8.	HSR gasoline	1.0	58.7	62.3	48.2
9.	Light hydrocracker gasoline C_5–C_6	12.9	82.4	82.8	79.0
10.	Hydrocrackate, C_5–C_6	15.5	85.5	89.2	86.4
11.	Hydrocrackate, C_6–190°F	3.9	73.7	75.5	85.0
12.	Hydrocrackate, 190–250°F	1.7	75.6	79.0	55.5
13.	Heavy hydrocracker gasoline	1.1	67.3	67.6	49.0
14.	Coker gasoline	3.6	60.2	67.2	57.2
15.	Light thermal gasoline	9.9	73.2	80.3	74.0
16.	C_6^+ Light thermal gasoline	1.1	68.1	76.8	55.1
17.	FCC Light gasoline, 200–300°F	1.4	77.1	92.1	49.5
18.	Hydrog. light FCC gasoline, C_5^+	13.9	80.9	83.2	51.5
19.	Hydrog. C_5—200°F FCC gasoline	14.1	81.7	91.2	58.1
20.	Hydrog. light FCC gasoline, C_6^+	5.0	74.0	86.3	49.3
21.	Hydrog. C_5^+ FCC gasoline	13.1	80.7	91.0	54.8
22.	Hydrog. 300–400°F FCC gasoline	0.5	81.3	90.2	48.5
23.	Reformate 94 RON	2.8	84.4	94.0	45.8
24.	Reformate 98 RON	2.2	86.5	98.0	43.1
25.	Reformate, 100 RON	3.2	88.2	100.0	
26.	Aromatic concentrate	1.1	94.0	107.0	
27.	Alkylate, $C_3^=$	5.7	87.3	90.8	
28.	Alkylate, $C_4^=$	4.6	95.9	97.3	70.3
29.	Alkylate, $C_3^=$, $C_4^=$	5.0	93.0	94.5	
30.	Alkylate $C_5^=$	1.0	88.8	89.7	
31.	Polymer	8.7	84.0	96.9	59.5

Table 25.4 Summary of RVP Calculations.

Component	Quantity (bpd)	x_{vi}	BI_{RVPi}	$x_{vi} \times BI_{RVPi}$
LSR gasoline	5000	0.227	20.26	4.599
HSR gasoline	4000	0.182	1.0	0.182
Reformate 94 RON	6000	0.273	3.62	0.989
FCC gasoline	7000	0.318	26.84	8.535
Total	22,000	1.000		14.305

$$BI_{RVP,Blend} = 10^{1.25} = \frac{314{,}912 + 139.64\,V_{Butane}}{22{,}000 + V_{Butane}}$$

$$17.78\,(22{,}000 + V_{Butane}) = 314{,}912 + 139.64\,V_{Butane}$$

Solving the above equation gives V_{Butane} = 626.3 bpd. This is the amount of n-butane required to adjust the RVP of the blend to 10 psi.

The total amount of gasoline at 10 psi RVP = 22,000 + 626.3 = 22,626 bpd.

The Excel spreadsheet program (Example 25.5.xlsx) shows the calculations of Example 25.5.

Table 25.3 shows blending property data for many refinery streams, and theoretical method for blending to the desired Reid vapor pressure requires the average molecular weight of each of the streams be known. There are several ways of estimating the average molecular weight of a refinery stream from the characterization factor, boiling point, and so on; however, a more convenient way is to use the empirical method developed by Chevron Research Co. Vapor pressure blending indices (VPBIs) have been compiled as a function of the RVP of the blending streams and shown in Table 25.6. The Reid vapor pressure of the blend is closely approximated by the sum of all the products of the volume fraction (v) times the VPBI for each component. In equation form:

$$RVP_{blend} = \sum v_i (VPBI)_i \qquad (25.4)$$

Table 25.5 Summary of RVP Calculations.

Component	Quantity, V_i (bpd)	BI_{RVPi}	$V_i \times BI_{RVPi}$
LSR gasoline	5000	20.26	101,300
HSR gasoline	4000	1.0	4000
Reformate 94 RON	6000	3.62	21,732
FCC gasoline	7000	26.84	187,880
n-Butane	V_{Butane}	139.64	139.64 V_{Butane}
Total	22,000 + V_{Butane}		314,912 + 139.64 V_{Butane}

Table 25.6 Reid Vapor Blending Index Numbers for Gasoline and Turbine Fuels [4].

Vapor pressure, psi	0.0	0.1	0.2	0.3	0.4	0.5	0.6	0.7	0.8	0.9
0	0.00	0.05	0.13	0.22	0.31	0.42	0.52	0.64	0.75	0.87
1	1.00	1.12	1.25	1.38	1.52	1.66	1.79	1.94	2.08	2.23
2	2.37	2.52	2.67	2.83	2.98	3.14	3.30	3.46	3.62	3.78
3	3.94	4.11	4.28	44.4	4.61	4.78	4.95	5.13	5.30	5.48
4	5.65	5.83	6.01	6.19	6.37	6.55	6.73	6.92	7.10	7.29
5	7.47	7.66	7.85	8.04	8.23	8.42	8.61	8.80	9.00	9.19
6	9.39	9.58	9.78	9.98	10.2	10.4	10.6	10.8	11.0	11.2
7	11.4	11.6	11.8	12.0	12.2	12.4	12.6	12.8	13.0	13.2
8	13.4	13.7	13.9	14.1	14.3	14.5	14.7	14.9	15.2	15.4
9	15.6	15.8	16.0	16.2	16.4	16.7	16.9	17.1	17.3	17.6
10	17.8	18.0	18.2	18.4	18.7	18.9	19.1	19.4	19.6	19.8
11	20.0	20.3	20.5	20.7	20.9	21.2	21.4	21.6	21.9	22.1
12	22.3	22.6	22.8	23.0	23.3	23.5	23.7	24.0	24.2	24.4
13	24.7	24.9	25.2	25.4	25.6	25.9	26.1	26.4	26.6	26.8
14	27.1	27.3	27.6	27.8	28.0	28.3	28.5	28.8	29.0	29.3
15	29.5	29.8	30.0	30.2	30.5	30.8	31.0	31.2	31.5	31.8
16	32.0	32.2	32.5	32.8	33.0	33.2	33.5	33.8	34.0	34.3
17	34.5	34.8	35.0	35.3	35.5	35.8	36.0	36.3	36.6	36.8
18	37.1	37.3	37.6	37.8	38.1	38.4	38.6	38.9	39.1	39.4
19	39.7	39.9	40.2	40.4	40.7	41.0	41.2	41.5	41.8	42.0
20	42.3	42.6	42.8	43.1	43.4	43.6	43.9	44.2	44.4	44.7
21	45.0	45.2	45.5	45.8	46.0	46.3	46.6	46.8	47.1	47.4
22	47.6	47.9	48.2	48.4	48.7	49.0	49.3	49.5	49.8	50.1
23	50.4	50.6	50.9	51.2	51.5	51.7	52.0	52.3	52.6	52.8
24	53.1	53.4	53.7	54.0	54.2	54.5	54.8	55.1	55.3	55.6
25	55.9	56.2	56.5	56.7	57.0	57.3	57.5	57.9	58.1	58.4

(*Continued*)

Table 25.6 Reid Vapor Blending Index Numbers for Gasoline and Turbine Fuels [4]. (*Continued*)

Vapor pressure, psi	0.0	0.1	0.2	0.3	0.4	0.5	0.6	0.7	0.8	0.9
26	58.7	59.0	59.3	59.6	59.8	60.1	60.4	60.7	61.0	61.3
27	61.5	61.8	62.11	62.4	62.7	63.0	63.3	63.5	63.8	64.1
28	64.4	64.7	65.0	65.3	65.6	65.8	66.1	66.4	66.7	67.0
29	67.3	67.6	67.9	68.2	68.4	68.8	69.0	69.3	69.6	69.9
30	70.2									
40	101									
(nC4) 51.6	138									
(iC4) 72.2	210									
(C3) 190.0	705									

Example: Calculate the vapour-pressure of a gasoline blend as follows

Equation: $VPBI = VP^{1.25}$

Component	Volume Fraction	Vapor Pressure psi	Vapor Pressure Blending Index No.	Volume Friction x VPBI
n-butane	0.050	51.6	138	6.90
Light Straight Run	0.450	6.75	10.9	4.90
Heavy Refined	0.500	1.00	1.00	0.50
Total	1.000	7.45	12.3	12.3

From the brochure, "31.0°API Iranian Heavy Crude Oil," by arrangement with Chevron Research Company. Copyright © 1971 by Chevron Oil Trading Company.

In the case where the volume of a component (n-butane) to be blended for a given RVP is required:

$$A(VPBI)_a + B(VPBI)_b + \ldots\ldots\ldots + W(VPBI)_w = (Y + W)(VPBI)_m \qquad (25.5)$$

where

- A = bbl of component a and so on
- W = bbl of n-butane (w)
- Y = A + B + C ….. (all components except n-butane)
- $(VPBI)_m$ = VPBI corresponding to the desired RVP of the mixture
- w = subscript indicating n-butane (n-C_4H_{10})

Example 25.6

Determine the amount of n-butane (n-C_4H_{10}) to be blended with the components listed at 12psi RVP using Table 25.6.

Table 25.7 Vapor Pressure vs. RVP Index of Gasolines [2].

Vapor pressure, kPa	RVP index									
	0	1	2	3	4	5	6	7	8	9
0	0.00	0.09	0.21	0.35	0.51	0.67	0.84	1.02	1.20	1.40
10	1.59	1.79	2.00	2.21	2.42	2.64	2.86	3.09	3.32	3.55
20	3.79	4.02	4.26	4.51	4.75	5.00	5.25	5.51	5.77	6.02
30	6.28	6.55	6.81	7.08	7.35	7.62	7.89	8.17	8.44	8.72
40	9.00	9.29	9.57	9.86	10.14	10.43	10.72	11.01	11.31	11.60
50	11.90	12.20	12.50	12.80	13.10	13.41	13.71	14.02	14.33	14.64
60	14.95	15.26	15.57	15.89	16.20	16.52	16.84	17.16	17.48	17.80
70	18.12	18.45	18.77	19.10	19.43	19.76	20.09	20.42	20.75	21.08
80	21.42	21.75	22.09	22.42	22.76	23.10	23.44	23.78	24.12	24.47
90	24.81	25.16	25.50	25.85	26.20	26.55	26.90	27.25	27.60	27.95
100	28.30	28.66	29.01	29.37	29.73	30.08	30.44	30.80	31.16	31.52
110	31.89	32.25	32.61	32.98	33.34	33.71	34.07	34.44	34.81	35.18
120	35.55	35.92	36.29	36.66	37.04	37.41	37.78	38.16	38.54	38.91
130	39.29	39.67	40.05	40.43	40.81	41.19	41.57	41.95	42.34	42.72
140	43.10	43.49	43.87	44.26	44.65	45.04	45.43	45.81	46.20	46.59
150	46.99	47.38	47.77	48.16	48.56	48.95	49.35	49.74	50.14	50.54
160	50.93	51.33	51.73	52.13	52.53	53.93	53.33	53.73	54.14	54.54
170	54.94	55.35	55.75	56.16	56.56	56.97	57.38	57.79	58.19	58.60
170	54.94	55.35	55.75	56.16	56.56	56.97	57.38	57.79	58.19	58.60
190	63.14	63.55	63.97	64.39	64.80	65.22	65.64	66.06	66.48	66.90
200	67.32	67.74	68.16	68.58	69.01	69.43	69.85	70.28	70.70	71.13

RVP index of lpg gases for gasoline blending.

Component	Vapor pressure, kPa	RVP index*
PROPANE	1310	705.42
i-BUTANE	497.8	210.46
n-BUTANE	355.8	138.31

*RVP index $(VP/6.8947)^{1.25}$

Component	bpcd	RVP, psi.	VPBI	vol. x VPBI
n-Butane	V	52	139.64	139.64V
LSR gasoline	5,000	11.1	20.26	101,300
HSR gasoline	4,000	1.0	1.0	4,000
Reformate -94	6,000	2.8	3.62	21,720
FCC gasoline	7,000	14.1	27.32	191,240
Total	22,000 + V			318,260 + 139.64V

For 12 psi RVP, (VPBI)m = 22.3 (Table 25.6)

22.3 (22,000 + V) = 318,260 + 139.64V

490600 + 22.3V = 318,260 + 139.64V

117.34V = 172340

V = 1469 bpcd of n-butane required.

Total amount of gasoline at 10 psi RVP = 22,000 + 1469 = 23,469 bpcd.

The Excel spreadsheet program (Example 25.6.xlsx) shows the calculations of Example 25.6.

25.3.1 Reid Vapor Pressure Blending for Gasolines and Naphthas

Gasolines of different Reid vapor pressures (RVPs) do not blend linearly. For accurately estimating the RVP of the blends, RVP blend indices are used as presented in Table 25.7.

Example 25.7

Determine the RVP of a blend of n-butane, alkylate, and cat reformate with the following properties:

Component	Volume fraction, x_{vi}	Vapor pressure, kPa	VP blend index (VPBI)	$x_{vi} \times$ VPBI
n-Butane	0.02	355.8	138	2.76
Alkylate	0.45	29.7	6.19	2.79
Reformate	0.53	35.2	7.66	4.06
Blend	1.00	42.1		9.61

Solution

Given the RVP of the blend components, the vapor pressure blend index for the individual components is read from the RVP vs. RVP indices (Table 25.7). The RVP index for the blend is next estimated by linear blending the component RVP indices. Therefore, a blend index of 9.6 corresponds to an RVP of 42.1 kPa for this blend.

25.4 Flash Point Blending

The flash point is the lowest temperature at which vapor arises from oil and ignites. It indicates the maximum temperature at which a fuel can be stored without a serious hazard. If the flash point of a petroleum product does not meet the required specification, it can be adjusted by blending it with other fractions. Flash point is not an additive property and thus requires a blending index, which is linear on a volume basis. The flash point of a blend is determined by:

$$BI_{FP_i,Blend} = \sum_{i=1}^{n} x_{vi} BI_{FP_i} \qquad (25.6)$$

where x_{vi} is the volume fraction of component i, and BI_{FP_i} is the flash point index of component i that can be determined from the following correlation [4]

$$BI_{FP_i} = FP_i^{1/x} \qquad (25.7)$$

FP_i is the flash point temperature of component i, in K, and the best value of x is −0.06.

Another relation to estimate the flash point blending index is based on the flash point experimental data and expressed by [5]

$$BI_{FP_i} = 51708 \times \exp\left[\frac{(\ln(FP_i) - 2.6287)^2}{(-0.91725)}\right] \qquad (25.8)$$

FP_i is the flash point temperature of component i, in °F. The flash point blending index is blended based on wt% of components.

The flash point of the blend is:

$$FP_{Blend} = \exp\left[2.6287 + \left\{\ln\left(\frac{BI_{FP,Blend}}{51708}\right)(-0.91725)\right\}^{1/2}\right] \qquad (25.9)$$

Table 25.8 Quantities and Flash Points of the Blending Components.

Component	bpd	SpGr	Flash point, °F
A	2500	0.80	120
B	3750	0.85	100
C	5000	0.90	150

Example 25.8

Calculate the blend flash point of the components as shown in the Table 25.8. If the resulted temperature is lower than 130°F, component D is added to the blend to increase the flash point. where its flash point is 220°F. Then, calculate the amount in bpd of component D (SpGr = 0.95) to be added to adjust the flash point.

Solution

The flash point blending indices are from Eq. 25.8 are calculated as follows:

For component A

$$BI_{FP_A} = 51708 \times \exp\left[\frac{(\ln(120) - 2.6287)^2}{(-0.91725)}\right]$$
$$= 321.36$$

For component B

$$BI_{FP_B} = 51708 \times \exp\left[\frac{(\ln(100) - 2.6287)^2}{(-0.91725)}\right]$$
$$= 731.07$$

For component C

$$BI_{FP_C} = 51708 \times \exp\left[\frac{(\ln(150) - 2.6287)^2}{(-0.91725)}\right]$$
$$= 106.47$$

The mass flow rates are calculated for each component based on the given specific gravity and bpd as shown in Table 25.9.

For component A, w_A is:

$$w_A = \left(2500\frac{bbl}{day}\right)\left(\frac{5.6\,ft^3}{1\,bbl}\right)\left(0.8 \times 62.4\frac{lb}{ft^3}\right)\left(\frac{day}{24h}\right)$$
$$= 29120\ lb/h$$

Table 25.9 Summary of the Above Calculations.

Component	x_{wi}	BI_{FPi}	$x_{wi} \times BI_{FPi}$
A	0.206	321.36	66.20
B	0.329	731.07	240.52
C	0.465	106.47	49.51
Total			356.23

For component B, w_B is:

$$w_B = \left(3750\frac{bbl}{day}\right)\left(\frac{5.6\,ft^3}{1\,bbl}\right)\left(0.85\times 62.4\frac{lb}{ft^3}\right)\left(\frac{day}{24h}\right)$$

$$= 46410.0\ lb/h$$

For component C, w_C is:

$$w_C = \left(5000\frac{bbl}{day}\right)\left(\frac{5.6\,ft^3}{1\,bbl}\right)\left(0.90\times 62.4\frac{lb}{ft^3}\right)\left(\frac{day}{24h}\right)$$

$$= 65520.0\ lb/h$$

The total mass flow rate of components A, B, and C = 141,050 lb/h

The weight fractions x_{wi} for components A, B, and C are:

$$x_{wA} = \frac{29120}{141050} = 0.206$$

$$x_{wB} = \frac{46410.0}{141050} = 0.329$$

$$x_{wC} = \frac{65520.0}{141050} = 0.465$$

$$BI_{FP,Blend} = \sum_{i=1}^{n}(x_{wi})(BI_{FPi}) = 356.23$$

$$FP_{Blend} = \exp\left[2.6287 + \left\{\ln\left(\frac{356.23}{51708}\right)(-0.91725)\right\}^{1/2}\right]$$

$$= 117.39°F$$

The calculated flash point is less than 130°F required, therefore, component D is added to increase the flash point.

The flash point index for component D at the flash point of 220°F is:

$$BI_{FP_D} = 51708 \times \exp\left[\frac{(\ln(220)-2.6287)^2}{(-0.91725)}\right]$$

$$= 12.41$$

The blend flash point index at 130°F is:

$$BI_{FP_i} = 51{,}708 \times \exp\left[\frac{(\ln(130)-2.6287)^2}{(-0.91725)}\right]$$
$$= 218.93$$

If w_D is the mass flow rate of component D, then from the mass balance:

$$(w_A)(BI_{FPA}) + (w_B)(BI_{FPB}) + (w_C)(BI_{FPC}) + (w_D)(BI_{FPD})$$
$$= (w_A + w_B + w_C + w_D)(BI_{FP@130°F})_{mixture} \tag{25.10}$$

That is:

$(321.36)(29{,}120) + (731.07)(46{,}410) + (106.47)(65{,}520) + (12.41)(w_D) = (141{,}050 + w_D)(218.93)$

$w_D = 93{,}940.\ \text{lb/h}$

The volumetric flow rate of component D is:

$$93940\,\frac{\text{lb}}{\text{h}} \times 24\,\frac{\text{h}}{\text{day}} \times \frac{1\,\text{bbl}}{5.6\,\text{ft}^3} \times \frac{\text{ft}^3}{0.95 \times 62.4\,\text{lb}} = 6791.5\ \text{bbl/day}$$

25.5 Alternative Methods for Determining the Blend Flash Point

Example 25.9

Kerosene (2000 bpsd) with a flash point of 120°F is to be blended with 8000 bpsd of fuel oil with a flash point of 250°F. What will be the flash point of the final blend?

Solution

Using Table 25.10, the flash index

Component	Volume	Fraction (A)	Flash (°F)	Flash point blending index (B)	Factor (A × B)
Kerosene	2000	0.2	120	311	62.200
Fuel oil	8000	0.8	250	5.56	4.448
Total	10,000	1.0	-	-	66.648

Table 25.10 Flash Point Blending Index Numbers [4].

Flash point, °F	0	1	2	3	4	5	6	7	8	9
0	168,000	157,000	147,000	137,000	128,000	120,000	112,000	105,000	98,600	92,400
10	86,600	81,200	76,100	71,400	67,000	62,900	59,000	55,400	52,100	49,000
20	46,000	43,300	40,700	38,300	36,100	34,000	32,000	30,100	28,400	26,800
30	25,200	23,800	22,400	21,200	20,000	18,900	17,800	16,800	15,900	15,000
40	14,200	13,500	12,700	12,000	11,400	10,800	10,200	9,680	9,170	8,690
50	8,240	7,810	7,410	7,030	6,670	6,330	6,010	5,700	5,420	5,150
60	4,890	4,650	4,420	4,200	4,000	3,800	3,620	3,441	3,280	3,120
70	2,970	2,830	2,700	2,570	2,450	2,330	2,230	2,120	2,020	1,930
80	1,840	1,760	1,680	1,600	1,530	1,460	1,400	1,340	1,280	1,220
90	1,170	1,120	1,070	1,020	978	935	896	857	821	**786**
100	**753**	722	692	662	635	609	584	560	537	515
110	495	475	456	438	420	404	388	372	358	344
120	331	318	305	294	283	272	261	252	242	233
130	224	216	350	200	193	186	179	172	166	160
140	154	149	144	138	134	129	124	120	116	112
150	108	104	101	97.1	93.8	90.6	87.5	84.6	81.7	79.0
160	76.3	73.8	71.4	69.0	66.7	64.5	62.4	60.4	58.4	56.5
170	54.7	52.9	51.3	49.6	48.0	46.5	45.1	43.6	42.3	40.9
180	39.7	38.4	37.3	36.1	35.0	33.9	32.9	31.9	30.9	30.0
190	29.1	28.2	27.4	26.6	25.8	25.0	24.3	23.6	22.9	22.2
200	21.6	20.9	20.3	19.7	19.2	18.6	18.1	17.6	17.1	16.6
210	16.1	15.7	15.2	14.8	14.4	14.0	13.6	13.3	12.9	12.5
220	12.2	11.9	11.6	11.2	10.9	10.6	10.4	10.1	9.82	9.56
230	9.31	9.07	8.83	8.60	8.37	8.16	7.95	7.74	7.55	7.85
240	7.16	6.98	6.80	6.63	6.47	6.30	6.15	5.99	5.84	5.70
250	5.56	5.42	5.29	5.16	5.03	4.91	4.79	4.68	4.86	4.45
260	4.35	4.24	4.14	4.04	3.95	3.86	3.76	3.68	3.59	3.51
270	3.43	3.35	3.27	3.19	3.12	3.05	2.98	2.91	2.85	2.78
280	2.72	2.66	2.60	2.54	2.48	2.43	2.37	2.32	2.27	2.22
290	2.17	2.12	2.08	2.03	1.99	1.95	1.90	1.86	1.82	1.79

(Continued)

Table 25.10 Flash Point Blending Index Numbers [4]. (*Continued*)

Flash point, °F	0	10	20	30	40	50	60	70	80	90
300	1.75	1.41	1.15	0.943	0.777	0.643	0.535	0.448	0.376	0.317
400	0.269	0.229	0.196	0.168	0.145	0.125	0.108	0.094	0.082	0.072
500	0.063	0.056	0.049	0.044	0.039	0.035	0.031	0.028	0.025	0.022

May be used to blend flash temperature, determined in any apparatus but, preferably, not to blend closed cup with open cup determinations.

Using Eq. 25.9, the flash point of the blend is

$$FP_{Blend} = \exp\left[2.6287 + \left\{\ln\left(\frac{66.648}{51,708}\right)(-0.91725)\right\}^{1/2}\right]$$

$$= 164°F$$

Example

Component	Volume frac.	Flash point, °F	Blending index	Volume × blending index
A	0.3	100	753	225.9
B	0.1	90	1,170	117
C	0.6	130	224	134.4
Total	1.0	111	477	477

Example 25.10

Determine the flash point of a blend containing 40 vol % component A with a flash point of 100°F, 10% component B with a flash point of 90°F and 50% component C with a flash point of 130°F.

Solution

From the flash point blending table (Table 25.11), the blend indices for the three components that blend linearly with the volumes are as follows:

Component	Vol %	x_{vi}	Flash point, °F	Flash index	x_{vi} × Flash index
A	40	0.4	100	754.2	301.68
B	10	0.1	90	1168.6	116.86
C	50	0.5	130	224.6	112.3
Blend	100	1.0	108.3		530.84

FLASH POINT,

Table 25.11 Flash Point (Abel) vs. Flash Blending Index [2].

Flash point, °F	0	1	2	3	4	5	6	7	8	9
80	1845.4	1761.4	1681.6	1605.7	1533.5	1464.9	1399.6	1337.5	1278.4	1222.2
90	1168.6	1117.6	1069.0	1022.8	978.7	936.6	896.6	858.4	822.0	787.3
100	754.2	722.6	692.4	663.7	636.2	610.0	584.9	561.0	538.1	516.3
110	495.4	475.5	456.4	438.2	420.8	404.1	388.2	372.9	358.3	344.3
120	331.0	318.1	305.9	294.1	282.9	272.1	261.8	251.9	242.4	233.3
130	224.6	216.2	208.2	200.5	193.1	186.0	179.2	172.7	166.5	160.4
140	154.7	149.1	143.8	138.7	133.8	129.0	124.5	120.1	115.9	111.9
150	108.0	104.3	100.7	97.3	93.9	90.7	87.7	84.7	81.8	79.1
160	76.5	73.9	71.5	69.1	66.8	64.6	62.5	60.5	58.5	56.6
170	54.8	53.0	51.3	49.7	48.1	46.6	45.1	43.7	42.3	41.0
180	39.7	38.5	37.3	36.2	35.1	34.0	32.9	31.9	31.0	30.0
190	29.1	28.3	27.4	26.6	25.8	25.1	24.3	23.6	22.9	22.2
200	21.6	21.0	20.4	19.8	19.2	18.7	18.1	17.6	17.1	16.6
210	16.2	15.7	15.3	14.9	14.4	14.0	13.7	13.3	12.9	12.6
220	12.2	11.9	11.6	11.3	11.0	10.7	10.4	10.1	9.8	9.6
230	9.3	9.1	8.8	8.6	8.4	8.2	8.0	7.8	7.6	7.4
240	7.2	7.0	6.8	6.6	6.5	6.3	6.2	6.0	5.9	5.7
250	5.6	5.4	5.3	5.2	5.0	4.9	4.8	4.7	4.6	4.5
260	4.4	4.3	4.1	4.1	4.0	3.9	3.8	3.7	3.6	3.5
270	3.4	3.4	3.3	3.2	3.1	3.1	3.0	2.9	2.9	2.8
280	2.7	2.7	2.6	2.5	2.5	2.4	2.4	2.3	2.3	2.2
290	2.2	2.1	2.1	2.0	2.0	2.0	1.9	1.9	1.8	1.8
300	1.8	1.7	1.7	1.6	1.6	1.6	1.5	1.5	1.5	1.4

Notes:

Flash index = $10^{(-6.1188 + 4345.2/(\text{FLASH POINT} + 383))}$

Flash point = $4345.2/(\text{LOG (Flash index)} + 6.1188) - 383.0$.

Where

Flash point (Abel) is in °F.

Table 25.12 Flash Point vs. Flash Index (for 154 and 144 Indices) [2].

Flash point, °C	0.0	0.5	1.0	1.5	2.0	2.5	3.0	3.5	4.0	4.5
10	784.80	762.07	740.07	718.78	698.17	678.23	658.92	640.22	622.12	604.59
15	587.61	571.16	555.23	539.80	524.85	510.36	496.31	482.70	469.51	456.72
20	444.33	432.31	420.65	409.35	398.38	387.75	377.43	367.42	357.71	348.29
25	339.15	330.27	321.66	313.30	305.18	297.30	289.65	282.22	275.00	267.99
30	261.18	254.57	248.14	241.90	235.83	229.94	224.21	218.64	213.23	207.97
35	202.85	197.88	193.04	188.34	183.77	179.32	174.99	170.88	166.69	162.70
40	158.83	155.05	151.38	147.81	144.33	140.95	137.65	134.44	131.32	128.28
45	125.31	122.43	119.62	116.88	114.22	111.62	109.09	106.63	104.22	101.88
50	99.66	97.38	95.21	93.10	91.04	89.03	87.08	85.17	83.31	81.49
55	79.72	77.99	76.31	74.67	73.07	71.50	69.98	68.49	67.04	65.62
60	64.24	62.89	61.57	60.28	59.03	57.80	56.60	55.43	54.59	53.18
65	52.09	51.03	49.99	48.98	47.98	47.02	46.07	45.15	44.24	43.36
70	42.50	41.66	40.83	40.03	39.24	38.47	37.72	36.99	36.27	35.57
75	34.88	34.21	33.55	32.91	32.28	31.66	31.06	30.47	29.90	29.34
80	28.79	28.25	27.72	27.20	26.70	26.20	25.72	25.25	24.78	24.33
85	23.88	23.45	23.02	22.61	22.20	21.80	21.41	21.07	20.65	20.28
90	19.92	19.57	19.22	18.88	18.55	18.22	17.91	17.59	17.29	16.99
95	16.69	16.41	16.12	15.85	15.58	15.31	15.05	14.79	14.54	14.30
100	14.06	13.82	13.59	13.36	13.14	12.92	12.71	12.50	12.29	12.09
105	11.89	11.70	11.51	11.32	11.14	10.96	10.78	10.61	10.44	10.27
110	10.10	9.94	9.78	9.63	9.48	9.33	9.18	9.04	8.90	8.76
115	8.62	8.49	8.36	8.23	8.10	7.98	7.85	7.73	7.61	7.50
120	7.38	7.27	7.16	7.06	6.95	6.85	6.74	6.64	6.54	6.45
125	6.35	6.26	6.17	6.07	5.99	5.90	5.81	5.73	5.64	5.56
130	5.48	5.40	5.33	5.25	5.17	5.10	5.03	4.96	4.89	4.82
135	4.75	4.68	4.62	4.55	4.49	4.43	4.37	4.30	4.25	4.19
140	4.13	4.07	4.02	3.96	3.91	3.85	3.80	3.75	3.70	3.65
145	3.60	3.55	3.51	3.46	3.41	3.37	3.32	338	3.24	3.19
150	3.15	3.11	3.07	3.03	2.99	2.95	2.91	2.88	2.84	2.80

(Continued)

Table 25.12 Flash Point vs. Flash Index (for 154 and 144 Indices) [2]. (*Continued*)

Flash point, °C	0.0	0.5	1.0	1.5	2.0	2.5	3.0	3.5	4.0	4.5
155	2.77	2.73	2.70	2.66	2.63	2.59	2.56	2.53	2.50	2.47
160	2.44	2.41	2.38	2.35	2.32	2.29	2.26	2.23	2.20	2.18
165	2.15	2.12	2.10	2.07	2.05	2.02	2.00	1.97	1.95	1.93
170	1.90	1.88	1.86	1.84	1.82	1.79	1.77	1.75	1.73	1.71
175	1.69	1.67	1.65	1.63	1.61	1.60	1.58	1.56	1.54	1.52
180	1.51	1.49	1.47	1.45	1.44	1.42	1.41	1.39	1.37	1.36
185	1.34	1.33	1.31	1.30	1.28	1.27	1.26	1.24	1.23	1.22
190	1.20	1.19	1.18	1.16	1.15	1.14	1.13	1.11	1.10	1.09
195	1.08	1.07	1.06	1.04	1.03	1.02	1.01	1.00	0.99	0.98
200	0.97	0.96	0.95	0.94	0.93	0.92	0.91	0.90	0.89	0.88
205	0.87	0.86	0.86	0.85	0.84	0.83	0.82	0.81	0.80	0.80
210	0.79	0.78	0.77	0.76	0.76	0.75	0.74	0.73	0.73	0.72
215	0.71	0.71	0.70	0.69	0.69	0.68	0.67	067	0.66	0.65
220	0.65	0.64	0.63	0.63	0.62	0.62	0.61	0.60	0.60	0.59
225	0.59	0.58	0.58	0.57	0.57	0.56	0.55	0.55	0.54	0.54
230	0.53	0.53	0.52	0.52	0.52	0.51	0.51	0.50	0.50	0.49
235	0.49	0.48	0.48	0.47	0.47	0.47	0.46	0.46	0.45	0.45
240	0.45	0.44	0.44	0.43	0.43	0.43	0.42	0.42	0.41	0.41

Flash index = $10^{((2050.86/F+273.16)-4.348)}$
F = Flash Point, °C

a. The flash index of the blend is calculated at 530.8, which corresponds to a flash point of 108°F.

b. The flash index is first determined from Table 25.12. Two empirical indices are worked out, the 154 index and the 144 index, respectively. The 154 index is a criterion for meeting the 154°F flash point and 144 index is criterion for meeting the 144°F flash point.

If the value of the 154 index is positive for any component or blend, it will meet the 154°F flash criterion; that is the flash will be equal to or higher than 154°F. Similarly, if the 144 flash index is positive, it will meet the 144°F flash criterion. If the 144 index is negative, the corresponding flash point will be lower than 144°F.

$$144 \text{ index} = (0.6502 - 0.01107 \times \text{FI}) \times \text{MB} \qquad (25.11)$$

$$154 \text{ index} = (0.4240 - 0.0098 \times \text{FI}) \times \text{MB} \qquad (25.12)$$

where

FI = the flash index (Table 25.12) and MB is moles/bbl.

This estimation requires data on molecular weight of the fraction. For routinely blended stocks, the values of the 144 and 154 indices are prepared and these can be used to determine whether the given blend will meet the flash index. Each index blends linearly with volume and has zero as a reference point.

Example 25.11

Determine whether the following fuel oil blend will meet the 154°F and 144°F flash criteria:

Stream	Vol. %	Specific gravity	API	MB	Flash, °C	Flash index	144°F index	154°F index
Vacuum Resid.	63.53	1.0185	7.43	0.361	250.0	0.40	0.2332	0.1518
FCC Cutter	25.20	0.9348	19.87	0.772	89.0	20.6	0.3258	0.1829
Kerosene	3.12	0.7901	47.59	1.687	41.0	20.6	0.7120	0.3996
LT. Diesel	8.15	0.8428	36.39	1.317	90.0	20.0	0.5647	0.3192
Blend	100.0						1.8357	1.0535

As the 144 and 154 indices are positive for this blend, the blend meets both 144 and 154 flash specifications.

Flash Index = $10^{((2050.86/(F + 273.16) - 4.348))}$

F = Flash Point, °C

25.6 Pour Point Blending

The pour point is the lowest temperature at which oil can be stored and still capable of flowing or pouring, when it is cooled without stirring under standard cooling conditions. Pour point is not an additive property and pour point blending indices are used that blend linearly on a volume basis.

Pour point is an essential property for diesel and fuel oil blends. Pour point blending is also non-linear and pour point blending indices are developed to enable reliable calculation of the pour points of the blends. A blending margin of 10PI (pour index) is allowed between the guaranteed specification and the refinery blending. For example, to guarantee a pour specification of −6°C (21.2°F), pour index 336.3), the blending target would be 326.3PI. In terms of the pour point, this corresponds to a blending margin of 1°F.

The pour point of a blend is determined by:

$$BI_{PP,Blend} = \sum_{i=1}^{n} x_{vi} BI_{PPi} \tag{25.13}$$

where x_{vi} is the volume fraction of component i, and BI_{PPi} is the pour point index of component i that can be determined from the following correlation [5]

$$BI_{PPi} = 3,262,000 \times \left(\frac{PP_i}{1000}\right)^{12.5} \tag{25.14}$$

where PP_i is the pour point of component i, in °R = (°F + 460)

The pour point of the product, PP_{Blend} is then evaluated using the reverse form of Eq. 25.14. Another relation to estimate the pour point blending index is [3]:

$$BI_{PPi} = PP_i^{1/x} \tag{25.15}$$

PP_i is the pour point temperature of component i, in K and the best value of x is 0.08.

Example 25.12

What is the pour point of the following blend from Table 25.13?

Solution

The given pour point temperatures are converted to °F and °R, respectively, and afterward, the pour point indices are calculated for each component as shown in Tables 25.14 and 25.15, respectively.

The blending index of the mixture pour point is:

$$BI_{PP,Blend} = \sum_{i=1}^{n} x_{vi} BI_{PPi} = 1680.82$$

Table 25.13 Quantities and Pour Points of the Blending Components.

Component	bpd	Pour point (°C)
Catalytic cracked gas oil	2000	−15
Straight run gas oil	3000	−3
Light vacuum gas oil	5000	42
Heavy vacuum gas oil	1000	45

Table 25.14 Calculations of Components at °F and °R.

Component	bpd	Pour point (°C)	Pour point (°F)	Pour point (°R)
Catalytic cracked gas oil	2000	−15	5	465
Straight run gas oil	3000	−3	26.6	486.6
Light vacuum gas oil	5000	42	107.6	567.6
Heavy vacuum gas oil	1000	45	113	573

Table 25.15 Summary of Calculations.

Component	Pour point (°R)	BI_{ppi}	bpd	x_{vi}	$x_{vi} \times BI_{ppi}$
Catalytic cracked gas oil	465	227.3	2000	0.1818	41.33
Straight run gas oil	486.6	401.	3000	0.2727	109.35
Light vacuum gas oil	567.6	2747.98	5000	0.4545	1248.96
Heavy vacuum gas oil	573	3093.26	1000	0.0909	281.18
Total			11000.0	1.000	1680.82

The blend pour point is calculated by rearranging Eq. 25.14 as:

$$PP_{Blend} = \left(\frac{BI_{pp,Blend}}{3262000}\right)^{1/12.5} \times 1000$$

$$= \left(\frac{1680.82}{3262000}\right)^{1/12.5} \times 1000 = 545.7°R \ (85.7°F)$$

$$PP_{Blend} = 85.7°F \ (29.8°C)$$

The Excel spreadsheet (Example 25.12.xlsx) shows the calculations of Example 25.12.

A blending margin of 10PI (pour index) is allowed between the guaranteed specification and the refinery blending. For example, to guarantee a pour specification of –6°C (21.2°F, pour index 336.3), the blending target would be 326.3PI. In terms of the pour point, this corresponds to a blending margin of 1°F. Tables 25.16 and 25.17 show the blending indices used to estimate the pour point of distillate petroleum products.

Determining pour point for a blend of two or more products is rather difficult. In this instance, blending indices are used as shown in Figure 25.4.

Example 25.13 [2]

Determine the amount of kerosene that must be blended into diesel with a 43°F pour point to lower the pour point to 21°F. The properties of kerosene and diesel stream are as follows:

	Kerosene	Diesel
Specific gravity	0.7891	0.8410
Pour point	–50°F	43°F

Solution

Determining the pour point of the blend; first, determine the pour indices from the pour point blend table (Table 25.16), corresponding to the pour points of kerosene and diesel, then the target pour point and blend linearly are as follows:

Table 25.16 Pour Point of Distillate Blends [2].

Pour point, °R	Index	Pour point, °R	Index	Pour point, °R	Index
360	8.99	394	27.18	428	78.16
361	9.31	395	28.67	429	80.48
362	9.63	396	29.59	430	82.85
363	9.97	397	30.54	431	85.30
364	10.32	398	31.51	432	87.80
365	10.68	399	32.52	433	90.38
366	11.05	400	33.55	434	93.02
367	11.44	401	34.62	435	95.74
368	11.83	402	35.71	436	98.52
369	12.24	403	36.84	437	101.39
370	12.66	404	38.00	438	104.32
371	13.10	405	39.19	439	107.34
372	13.54	406	40.41	440	110.44
373	14.01	407	41.68	441	113.62
374	14.48	408	42.97	442	116.88
375	14.97	409	44.31	443	120.23
376	15.48	410	45.68	444	123.67
377	16.00	411	47.10	445	127.19
378	16.54	412	48.55	446	130.81
379	17.10	413	50.04	447	134.53
380	17.67	414	51.58	448	138.34
381	18.26	415	53.16	449	142.25
382	18.87	416	54.78	450	146.26
383	19.50	417	56.45	451	150.37
384	20.14	418	58.17	452	154.59
385	20.81	419	59.93	453	158.92
386	21.49	420	61.74	454	163.37
387	22.20	421	63.60	455	167.92

(Continued)

Table 25.16 Pour Point of Distillate Blends [2]. (*Continued*)

Pour point, °R	Index	Pour point, °R	Index	Pour point, °R	Index
388	22.93	422	65.52	456	172.59
389	23.68	423	67.49	457	177.38
390	24.45	424	69.51	458	182.30
391	25.24	425	71,59	459	187,34
392	26.06	426	73.72		
393	26.91	427	75.91		
460	192.50	496	493.72	532	1185.38
461	197.80	497	506.31	533	1213.53
462	203,23	498	519.19	534	1242.30
463	208.80	499	532.38	535	1271.70
464	21450	500	545.87	536	1301.73
465	220.36	501	559.67	537	1332,42
466	226.35	502	573,80	538	1363.77
467	232.50	503	588,25	539	1395.79
468	238.80	504	603,04	540	1428.51
469	245.26	505	618.16	541	1461.93
470	251.88	506	633.64	542	1496.07
471	258.66	507	649.47	543	1530.95
472	265.61	508	665.67	544	1566.56
473	272.73	509	682,24	545	1602.94
474	280.02	510	699.18	546	1640.10
475	287.50	511	716.51	547	1678.04
476	295.15	512	734.24	548	1716.80
477	303.00	513	752.37	549	1756.37
478	311.04	514	770.90	550	1796.78
479	319.27	515	78986	551	1838.05
480	327.70	516	809.25	552	1880.18
481	336.34	517	829.07	553	1923.21

(*Continued*)

Table 25.16 Pour Point of Distillate Blends [2]. (*Continued*)

Pour point, °R	Index	Pour point, °R	Index	Pour point, °R	Index
482	345.18	518	849.34	554	1967.13
483	354.24	519	870,07	555	2011.98
484	363.52	520	891.26	556	2057.77
485	373.02	521	912.92	557	2104.51
486	382.75	522	935.07	558	2152.23
487	392.71	523	957.71	559	2200,95
488	402.91	524	980.85	560	2250.67
489	413.35	525	100451		
490	424.05	526	1028,69		
491	434.99	527	1053.40		
492	446.20	528	1078.66		
493	457.67	529	110448		
494	469.41	530	1130.86		
495	481.42	531	115783		

Notes:
Also applicable to freeze points and fluidity blending is on a volume basis.
Pour point blend index = 3262000* × (Pour point, °R/1000)$^{1.25}$
Pour point, °R = 1000 × (Index/316000)$^{0.08}$
Pour point, °F = Pour point (°R) − 460
*Correlation of hu and burns.

Blend component	Pour point,°F	Pour point,°R	Blend index	Vol %
Diesel	43	503	588.25	53.6
Kerosene	−50	410	45.68	46.4
Blend	21	481	336.34	100.0

Therefore, to lower the pour point to 21°F, 46.4% kerosene by volume must be blended.

Example 25.14 [3]

A 2000 BPSD of a gas oil having a pour point of −5°F and ASTM distillation 50% point of 500°F is to be blended with 4000 BPSD of a waxy distillate having a pour point of 30°F and a 50% ASTM distillation of 700°F. The pour point of the final blend is predicted as follows:

Table 25.17 Pour Point Blending Indices for Distillate Stocks [1].

ASTM 50% temp	300	350	375	400	425	450	475	500	525	550	575	600	625	650	675	700
Pour point																
70	133	131	129	128	127	125	123	120	118	115	113	110	108	105	103	100
65	114	111	109	107	105	103	101	98	96	94	91	88	85	82	79	76
60	99	94	92	90	87	85	82	80	77	74	72	69	67	64	62	60
55	88	79	77	75	73	71	68	66	63	61	58	56	53	50	48	46
50	72	68	66	63	61	59	56	54	52	49	47	44	42	39	37	35
45	60	56	54	52	50	48	46	44	42	40	38	35	33	31	29	27
40	52	48	46	44	42	40	38	36	34	32	30	28	26	24	22	21
35	44	41	39	37	35	33	32	30	28	26	24	23	21	19	18	16
30	37	34	32	31	29	27	26	24	23	21	19	18	16	15	14	13
25	32	29	27	26	24	23	21	20	18	17	15	14	13	12	11	10
20	27	24	23	21	20	19	17	16	15	14	12	11	10	9.1	8.3	7.5
15	23	20	19	18	17	16	14	13	12	11	10	9.0	8.1	7.2	6.4	5.8 V
10	20	17	16	15	14	13	12	11	9.8	8.8	8.0	7.1	6.3	5.6	5.0	4.5
5	17	15	14	13	12	11	9.7	8.8	7.9	7.1	6.3	5.6	5.0	4.4	3.8	3.5

(Continued)

Table 25.17 Pour Point Blending Indices for Distillate Stocks [1]. (Continued)

ASTM 50% temp	300	350	375	400	425	450	475	500	525	550	575	600	625	650	675	700
Pour point																
0	14	12	11	10	9.6	8.7	7.9	7.1	6.3	5.6	5.0	4.4	3.8	3.4	3.0	2.7
-5	12	10	9.5	8.7	8.0	7.2	6.5	5.8	5.1	4.5	3.9	3.4	3.0	2.7	2.4	2.1
-10	10	8.8	8.0	7.3	6.6	5.9	5.3	4.7	4.1	3.6	3.2	2.8	2.5	2.2	1.9	1.6
-15	8.8	7.4	6.8	6.1	5.5	4.9	4.4	3.9	3.4	3.0	2.6	2.2	1.9	1.7	1.4	1.2
-20	7.5	6.3	5.7	5.1	4.6	4.1	3.6	3.2	2.8	2.4	2.1	1.8	1.5	1.3	1.1	0.94
-25	6.4	5.3	4.7	4.2	3.7	3.3	2.9	2.5	2.2	1.9	1.7	1.4	1.2	1.0	0.90	0.72
-30	5.5	4.5	4.0	3.6	3.2	2.8	2.4	2.1	1.8	1.5	1.3	1.1	0.96	0.80	0.67	0.56
-35	4.6	3.7	3.3	2.9	2.6	2.3	2.0	1.7	1.4	1.2	1.0	0.90	0.75	0.62	0.51	0.43
-40	4.0	3.2	2.8	2.5	2.2	1.9	1.6	1.4	1.2	1.0	0.86	0.73	0.62	0.51	0.41	0.33
-45	3.3	2.7	2.4	2.1	1.8	1.5	1.3	1.1	0.98	0.82	0.68	0.58	0.48	0.38	0.31	0.25
-50	2.8	2.3	2.0	1.7	1.5	1.3	1.1	0.93	0.78	0.66	0.56	0.47	0.38	0.31	0.25	0.20
-55	2.5	1.9	1.7	1.4	1.2	1.1	0.90	0.77	0.65	0.55	0.46	0.37	0.30	0.24	0.19	0.15
-60	2.1	1.6	1.4	1.2	1.0	0.87	0.74	0.62	0.52	0.43	0.36	0.30	0.24	0.19	0.14	0.10
-65	1.8	1.4	1.2	1.0	0.85	3.72	0.60	0.50	0.41	0.34	0.28	0.23	0.18	0.14	0.10	0.07
-70	1.5	1.1	0.99	0.84	0.71	0.60	0.50	0.42	0.36	0.30	0.25	0.20	0.15	0.11	0.08	0.05

From Gary & Handwerk

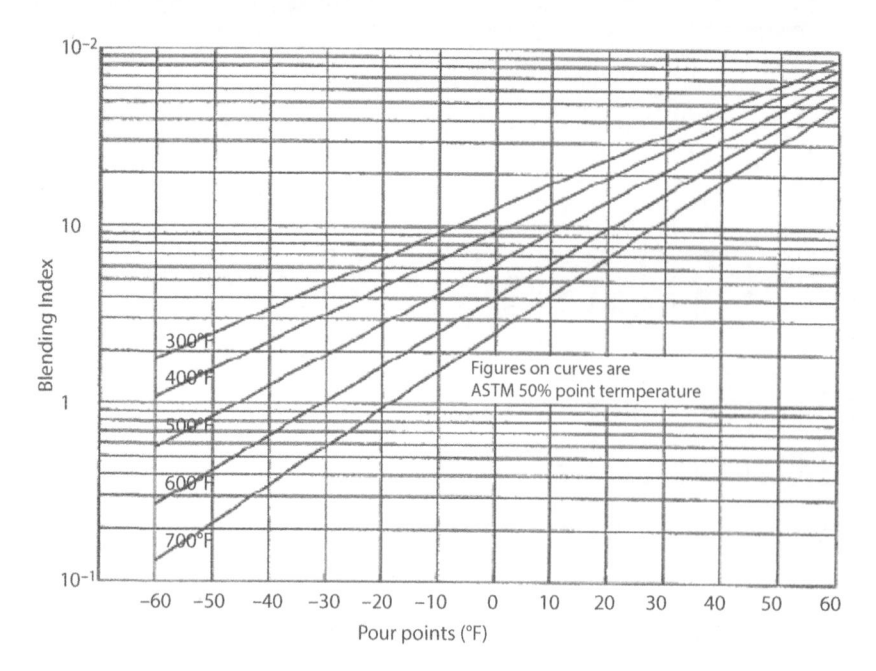

Figure 25.4 Pour point blending chart [3].

	Composition		ASTM 50%		Pour point		
Component	BPSD	Fraction	50%	Factor	Pour pt.	Index	Factor
		(A)	(B)	(A) x (B)		(C)	(A x C)
Gas oil	2000	0.33	500	165	-5	5.8	1.9
Waxy dist.	4000	0.67	700	469	+30	12.7	8.5
Total	6000	1.00	-	634	-	-	10.4

From Figure 25.4, the pour point corresponds to an index of 10.4 and ASTM 50% of 634°F is 22°F. This is the pour point of the blend.

Table 25.17 shows the pour point indices for some distillate fuels, where the blending indices are tabulated as a function of ASTM 50 % temperature, °F (first horizontal listing) and Pour Point, °F (first vertical listing). For example, pour point blending index for a distillate that has an ASTM 50% temperature of 500°F and a pour point of 40°F would be 36. This index can be used in calculating the pour point of blends using this distillate as a component.

Example 25.15

Using blending indices to calculate the pour point of a blend, consider blending a straight-run gas oil (50% ASTM, T = 470°F and pour point = −6°F) and a hydrotreated heavy gas oil (50% ASTM, T = 620°F and pour point = 40°F). What would be the pour point of a binary blend that consists of 68.7% volume of straight-run gas oil and 31.3% vol. of hydrotreated heavy gas oil?

The procedure to calculate the pour point of the blend is summarized as follows:

1. Read the pour point blending indices for the two distillates in Table 25.17 (PPI column). Straight-run gas oil PPBI = 6.3 (using double interpolation), and hydrotreated heavy gas oil PPBI = 26 (Table 25.18).

Table 25.18 Pour Point Blending (non-linear).

Component	Pour point	Volume fraction	50% ASTM, T, °F	Pour point index (PPI)*	Pour factor (PPI × volume fraction)	Blend (50% ASTM × volume fraction)
Straight-run gas oil	−6°F	0.687	470	6.3	4.33	323
Hydrotreated heavy gas oil	40°F	0.313	620	26	8.14	194
Blend					12.47	517

*From the blending chart (Table 25.17) Read chart to get: Pour point of blend = 15°F
ASTM 50% T for blend = 517°F

2. Multiply PPBI for individual distillates with their respective volume fraction to calculate the Pour factor. For example, the Pour factor for straight-run gas oil = 0.687 × 6.3 = 4.33. The Pour factor for hydrotreated heavy gas oil = 0.313 × 26 = 8.14. Add the pour factors for each component, to calculate the blending index for the blend = 12.47.
3. Calculate blend 50% ASTM temperature (linearly additive) of the blend by multiplying the volume fraction with ASTM 50% T of each component and adding them together (0.687 × 470 + 0.313 × 620 = 517°F).
4. Using the blend ASTM 50 % just calculated (517°F) and the blending index from Table 25.17 is 15°F.

Note: If it is assumed linear addition of the pour points, then the calculated blending pour point is: 0.687 (−6) + 0.313 (40) = 8.4°F. This gives an underestimation of the pour point. Thinking that a diesel fuel has a point 8.4°F, one may try to start a diesel truck on a 12°F, but to no avail. Not knowing that the fuel tank has a gel, and not a liquid that can be easily pumped to the combustion cylinder.

25.7 Cloud Point Blending

Cloud point is the lowest temperature at which oil becomes cloudy and the first particles of wax crystals are observed as the oil is cooled gradually under standard conditions. Cloud point is not an additive property and as such the cloud point blending indices are used. This blend linearity on a volume basis by:

$$BI_{CP,Blend} = \sum_{i=1}^{n} x_{vi} BI_{CPi} \tag{25.16}$$

where x_{vi} is the volume fraction of component i and BICPi is the cloud point blending index of component i that can be determined from [3, 5]:

$$BI_{CP,i} = CP_i^{1/x} \tag{25.17}$$

CPi is the cloud point temperature of component i, in K and the value of x = 0.05

Example 25.16

Calculate the cloud point of the following blend

Component	bpd	Cloud point (°F)
A	1000	−5
B	2000	5
C	3000	10

Solution

The given cloud point temperatures are converted to K; then the cloud point indices are calculated for each component as listed in Table 25.19.

The blend cloud point is from Eqs. 25.16 and 25.17:

$$BI_{CP,Blend} = \sum_{i=1}^{n} xv_i \, BI_{CPi} = 1.8317 \times 10^{48}$$

$$CP_{Blend} = (BI_{CP,Blend})^{0.05} = (1.8317 \times 10^{48})^{0.05} = 258.9 K = -14.25°C = 6.35°F$$

The Excel spreadsheet Example 25.16.xlsx shows the calculations of Example 25.16.

25.8 Aniline Point Blending

The aniline point indicates the degree of aromaticity of a petroleum fraction. It is the minimum temperature at which equal volumes of the aniline, and the oil are completely miscible. Aniline point is not an additive property and therefore, aniline point blending indices are used based on blend linearity on a volume basis. The aniline point of a blend is determined by:

$$BI_{AP,Blend} = \sum_{i=1}^{n} x_{vi} \, BI_{APi} \qquad (25.18)$$

where x_{vi} is the volume fraction of component i and BI_{APi} is the aniline point blending index of component i that can be determined from [6]:

Table 25.19 Summary of Calculations.

Component	x_{vi}	Cloud point (°F)	Cloud point (K)	$BI_{cpi} \times 10^{48}$	$x_{vi} \times BI_{cpi} \times 10^{48}$
A	0.167	−5	252.59	1.118	0.1867
B	0.333	5	258.15	1.728	0.575
C	0.500	10	260.93	2.140	1.070
Total	1.000				1.8317

$$BI_{AP,i} = 1.124[\exp(0.00657\, AP_i)] \tag{25.19}$$

AP_i is the aniline point of component i, °C

Example 25.17

What is the aniline point of the following blend?

Component	bpd	Aniline point (°C)
Light diesel	4000	71.0
Kerosene	3000	60.7
Light cycle gas oil	3000	36.8

Solution

The volume fraction and aniline point indices are calculated and shown in Table 25.20, and Table 25.21 shows the summary of the calculations.

The aniline blend point is from Eqs. 25.18 and 25.19:

$$AP_{Blend} = \frac{\ln\left[\dfrac{\sum_{i=1}^{n} x_{vi}\, BI_{APi}}{1.124}\right]}{0.00657}$$

$$= \ln(1.6486/1.124)/0.00657 = 58.3°C$$

25.8.1 Alternative Aniline Point Blending

The aniline point of a gas oil is indicative of the aromatic content of the gas oil. The aromatic hydrocarbon exhibits the lowest and paraffins the highest values. Aniline point (AP) blending is not linear; therefore, the blending indices are used. The following function converts aniline point to the aniline index as follows:

$$\text{ANLIND} = 1.25 \times AP + 0.0025 \times (AP)^2 \tag{25.20}$$

Table 25.20 Summary of Calculations of Example 25.17.

Component	x_{vi}	BI_{APi}
Light diesel	0.4	1.792
Kerosene	0.3	1.675
Light cycle gas oil	0.3	1.431
Total	1.0	

Table 25.21 Summary of Calculations of Example 25.17.

Component	xv_i	Aniline point (°C)	BI_{Api}	$x_{vi} \times BI_{Api}$
Light diesel	0.4	71	1.792	0.7168
Kerosene	0.3	60.7	1.675	0.5025
Light cycle gas oil	0.3	36.8	1.431	0.4293
Total	1.0			1.6486

where ANLIND is the aniline index and AP is the aniline point, °F

The aniline index can be converted back into the aniline point by the following:

$$AP = 200 \times \left[\left(1.5625 + \frac{ANLIND}{100} \right)^{0.5} - 1.25 \right] \quad (25.21)$$

The diesel index is determined by:

$$\text{Diesel Index} = \frac{(\text{API Gravity} \times \text{Aniline Point})}{100} \quad (25.22)$$

Table 25.22 shows the Aniline point blending index numbers.

Example 25.18

Determine the aniline point and diesel index of the following gas oils blend.

Blend components	Vol %	Specific gravity	Aniline point, °F
Light diesel	0.500	0.844	159.8
Kerosene	0.200	0.787	141.3
Light cycle gas oil	0.300	0.852	98.3
Blend	1.000	0.835	138.57

Table 25.22 Aniline Point Blending Index Numbers.

Aniline point, °F	0	-1	-2	-3	-4	-5	-6	-7	-8	-9
-10	20.0	17.4	14.9	12.6	10.3	8.10	6.06	4.17	2.46	1.00
0	49.1	46.0	42.8	39.8	36.8	33.8	30.9	28.1	25.3	22.6

Aniline point, °F	0	-1	-2	-3	-4	-5	-6	-7	-8	-9
0	49.1	52.4	55.6	58.9	62.3	65.7	69.1	72.6	76.1	79.6
10	83.2	86.8	90.5	94.2	97.9	102	105	109	113	117
20	121	125	129	133	137	141	145	149	153	157
30	162	166	170	174	179	183	187	192	196	200
40	205	209	214	218	223	227	232	237	241	246
50	250	255	260	264	269	274	279	283	288	293
60	298	303	308	312	317	322	327	332	337	342
70	347	352	357	362	367	372	377	382	388	393
80	398	403	408	414	419	424	429	435	440	445
90	451	456	461	467	472	477	483	488	494	491
100	505	510	516	521	527	532	538	543	549	554
110	560	566	571	577	582	588	594	599	605	611
120	617	622	628	634	640	645	651	657	663	669
130	674	680	686	692	698	704	710	716	722	727
140	733	739	745	751	757	763	769	775	781	788
150	794	800	806	812	818	824	830	836	842	849
160	855	861	867	873	880	886	892	898	904	911
170	917	923	930	936	942	948	955	961	967	974
180	980	986	993	999	1,006	1,012	1,019	1,025	1,031	1,038
190	1,044	1,050	1,057	1,064	1,070	1,077	1,083	1,090	1,096	1,103
200	1,110	1,116	1,122	1,129	1,136	1,142	1,149	1,156	1,162	1,169
210	1,176	1,182	1,189	1,196	1,202	1,209	1,216	1,222	1,229	1,236
220	1,242	1,249	1,256	1,262	1,269	1,276	1,283	1,290	1,297	1,303
230	1,310	1,317	1,324	1,331	1,337	1,344	1,351	1,358	1,365	1,372
240	1,379	1,386	1,392	1,400	1,406	1,413	1,420	1,427	1,434	1,441

(Continued)

Table 25.22 Aniline Point Blending Index Numbers. (*Continued*)

Mixed aniline point, °F	0	1	2	3	4	5	6	7	8	9
0	-736	-730	-723	-716	-709	-703	-696	-689	-682	-675
10	-668	-660	-653	-646	-639	-631	-623	-616	-608	-600
20	-593	-584	-577	-569	-561	-552	-544	-536	-528	-519
30	-511	-503	-494	-486	-477	-468	-460	-451	-442	-433
40	-425	-416	-407	-398	-389	-380	-371	-361	-352	-343
50	-334	-324	-315	-306	-296	-287	-277	-267	-258	-248
60	-239	-229	-219	-210	-200	-190	-180	-170	-160	-150
70	-140	-130	-120	-110	-100	-89.6	-79.4	-69.2	-58.9	-48.6
80	-38.3	-27.9	-17.5	-7.06	3.39	13.9	24.4	35.0	45.5	56.1
90	-66.8	77.4	22.1	98.8	110	120	131	142	153	164
100	175	186	197	208	219	230	241	252	263	274
110	285	297	308	319	330	342	353	364	376	387
120	399	410	422	433	445	456	468	479	491	503
130	514	526	538	550	561	573	585	597	609	620
140	632	644	656	668	680	692	704	716	728	741

Example	Component	Volume frac.	Aniline point, °F	Index	Volume frac. × index
	A	0.8	70	387	278
	B	0.2	40 (Mixed)	−425	−85
	Total	1.0	37 (or 102 Mixed)	193	193

Solution

The aniline point of the blend is determined by estimating the aniline blend index of each component using Eq. 25.20. Then blending the components volumetrically, the diesel index is calculated as a function of aniline point and API gravity using Eq. 25.22. Table 25.23 shows the summary of the calculations.

The blend index is:

$$\text{Diesel Index} = \frac{(\text{API Gravity} \times \text{Aniline Point})}{100}$$
$$= (37.96 \times 138.57)/100$$
$$= 52.60$$

Table 25.23 Summary of the Calculations.

Blend components	Vol %	Specific gravity	Aniline point, °F	AP index	Vol*API
Light diesel	0.500	0.844	159.8	263.59	131.795
Kerosene	0.200	0.787	141.3	226.54	45.307
Light cycle gas oil	0.300	0.852	98.3	147.03	44.109
Blend	1.000	0.835	138.57		221.215

25.9 Smoke Point Blending

The smoke point is the maximum flame height in millimeter at which the oil burns without smoking when tested at standard specified conditions. The smoke point of a blend is determined by [7]:

$$SP_{Blend} = -255.26 + 2.04\, AP_{Blend} - 240.8 \ln(SpGr_{Blend}) + 7727\left(SpGr_{Blend}/AP_{Blend}\right) \quad (25.23)$$

where SP_{Blend} is the blend smoke point in mm, AP_{Blend} and $SpGr_{Blend}$ are the aniline point and the specific gravity of the blend respectively. The aniline point for the blend is calculated in an earlier section. Specific gravity is an additive property and can be blended linearly on a volume basis. The specific gravity of a blend is estimated using the mixing rule:

$$SpGr_{Blend} = \sum_{i=1}^{n} x_{vi}\, SpGr_i \quad (25.24)$$

where x_{vi} is the volume fraction of component i, and SpGri is the specific gravity of component i. API is not an additive property and it does not blend linearly. Therefore, API is converted to specific gravity, which can be blended linearly.

Example 25.19

What is the smoke point of the following blend?

Component	bpd	Aniline point (°C)	SpGr
Kerosene A	4000	50	0.75
Kerosene B	3000	60	0.8
Kerosene C	3000	55	0.85

Solution

The volume fractions, the specific gravity of the blend and aniline point indices are calculated as follows from Eqs. 25.23, 25.24, respectively. Tables 25.24 and 25.25 show the summary of the calculations of Example 25.19.

$$SpGr_{Blend} = \sum_{i=1}^{n} x_{vi}\, SpGr_i = 0.795$$

Table 25.24 Summary of Calculations.

Component	xv_i	Aniline point (°C)	$SpGr_i$	$x_{vi} \times SpGr_i$
Kerosene A	0.4	50	0.75	0.300
Kerosene B	0.3	60	0.8	0.240
Kerosene C	0.3	55	0.85	0.255
Total	1.0			0.795

Table 25.25 Summary of Calculations.

Component	x_{vi}	Aniline point (°C)	BI_{Api}	$x_{vi} \times BI_{Api}$
Kerosene A	0.4	50	1.561	0.6244
Kerosene B	0.3	60	1.667	0.5001
Kerosene C	0.3	55	1.613	0.4839
Total	1.0			1.6084

$$BI_{AP,i} = 1.124[\exp(0.00657\, AP_i)] \tag{25.19}$$

where

AP_i is the aniline point of component i, °C

$$AP_{Blend} = \frac{\ln\left[\dfrac{\sum_{i=1}^{n} x_{vi}\, BI_{APi}}{1.124}\right]}{0.00657}$$

$$= \ln(1.6084/1.124)/0.00657 = 54.54°C$$

The smoke point blend is from Eq. 25.23

$$SP_{Blend} = -255.26 + 2.04(54.54) - 240.8 \ln(0.795) + 7727(0.795/54.54) = 23.88 \text{ mm.}$$

25.9.1 Smoke Point of Kerosenes

The smoke point (S) of a kerosene can be estimated by the following correlation with the aromatic content of the stream.

$$S = \left(\frac{53.7}{\text{arom}^{0.5}}\right) + 0.03401 \times API^{1.5} + 1.0806 \tag{25.25}$$

where

S = smoke point, mm
arom = aromatics content, LV%
API = API gravity of the cut.

25.10 Viscosity Blending

Viscosity is not an additive property; therefore, viscosity blending indices are used to determine the viscosity of the blended products. Several correlations and tables are available for determining the viscosity indices. The viscosity index of the blended product is determined by:

$$BI_{vis,Blend} = \sum_{i=1}^{n} x_{vi} BI_{visi} \qquad (25.26)$$

where x_{vi} is the volume fraction of component i, and BI_{visi} is the viscosity index of component i that can be determined by [8]:

$$BI_{visi} = \frac{\log_{10} v_i}{3 + \log_{10} v_i} \qquad (25.27)$$

where v_i is the viscosity of component in cSt.

The viscosity of the blended product is then calculated by rearranging Eq. 25.27:

$$\log_{10} v_i = \frac{3 BI_{visi}}{[1 - BI_{visi}]}$$

and

$$v_{Blend} = 10^{\left[3 BI_{vis,Blend} / \left(1 - BI_{vis,Blend}\right)\right]} \qquad (25.28)$$

Example 25.20

Estimate the viscosity of a blend of 2000 barrels of Fraction-1 with a viscosity of 75 cSt, 3000 barrels of Fraction-2 with a 100 cSt and 4000 barrels of Fraction-3 with 200 cSt all at 130°F.

Solution

The viscosity index is calculated from Eq. 25.27 and the blend viscosity is calculated from the blend index as shown in Table 25.26.

The blend viscosity is calculated from the blend index (Eq. 25.28) as:

$$v_{Blend} = 10^{[3 \times 0.41137/(1-0.41137)]}$$

$$= 124.91 \text{ cSt } (125 \text{ cSt})$$

For example,

a light naphtha stream with a blending octane number = 75

a reformate with a blending octane number = 86

Table 25.26 Summary of Calculations.

Component	Quantity, bpd	xv_i	vi (cSt)	BI_{visi}	$x_{vi} \times BI_{Api}$
Fraction-1	2000	0.222	75	0.385	0.08547
Fraction-2	3000	0.333	100	0.400	0.13320
Fraction-3	4000	0.444	200	0.434	0.19270
Total	9000	0.999			0.41137

and the true octane number of gasoline is, ON = 83. What would be the volume fraction of reformate (x) in the blend?

$$ON = x(86) + (1 - x)(75) = 83$$

$$x = 0.73$$

Therefore, 73% by volume of reformate is required in the blend. Additive concentration may be calculated the same way and ON for multicomponent blend can be calculated the same way for research and motor octane numbers.

25.11 Regular Gasoline

The total crude throughput to the refinery is calculated as follows [2]:

Capacity required = 3 million tons/year

In barrels per calendar day (BPCD), this is:

$$\frac{3{,}000{,}000 \times 2205}{365} = 18.1 \text{ million lb per day}$$

Density of crude, lb/gal = 7.22

Then $BPCD = \dfrac{(18.1) \times (10^6)}{(7.22)(42)} = 59{,}689 \text{ BPCD}$

The stream yields are summarized from the yield calculations as follows [2]:

	% vol. on crude	BPCD
Gas	2.8	1671
Light naphtha	6.2	3701
Heavy naphtha	15.8	9431
Kerosene	3.0	1791
Light gas oil	18.0	10744
Heavy gas oil	8.2	4894
Residue	46.0	27457
Total	100.0	59689

From these base yields, the blending recipes and unit capacities can be calculated.

25.12 Product Blending

The yield of reformate is given by the licensor as 83.2% on feed. Consider that the regular gasoline will be made up of all the light naphtha make and with sufficient reformate to meet 80 octane number clear and will then be blended with about 3.0 ml/IG of TEL to make the specification octane. The light naphtha octane number is 75 RES clear.

Let x be the volume of reformate in the blend. Then:

$$x(86) + (1-x)(75) = 80$$

$$x = 0.454 \text{ vol.}$$

Total light naphtha = 3701 BPCD = 54.6 % of the gasoline stream.

Total regular gasoline = $\dfrac{3701}{0.546}$ = 6778 BPCD

Reformate to regular gasoline = 3077 BPCD

25.12.1 Premium Gasoline

Total reformate make = 0.832 (9431) = 7847 BPCD

Reformate to regular gasoline = 3077 BPCD

Reformate to premium gasoline = 4770 BPCD

This reformate when blended with 1.5 ml/IG TEL will meet 93 ON (RES)

25.13 Viscosity Prediction From the Crude Assay

Determining the viscosity of a blend of two or more components, a blending index must be used. Figure 25.5 shows a graph of these indices as given in Maxwell's Data Book on Hydrocarbons. Using the blending index and having split the fraction into components, the viscosity of the fraction can be predicted as shown in the following example.

Component	Volume %	Mid-BPT (°F)	Viscosity (cSt) at 100°F	Blending index	Viscosity factor
	(A)			(B)	(A × B)
1	13.0	420 (210°C)	1.49	63.5	825.5
2	16.5	460 (238°C)	2.0	58.0	957.0
3	21.0	490 (254°C)	2.4	55.0	1155.0
4	18.0	520 (271°C)	2.9	52.5	945.0
5	18.5	550 (288°=C)	3.7	49.0	906.5
6	13.0	591 (311°C)	4.8	46.0	598.0
Total	100.0	-	-		5387.0

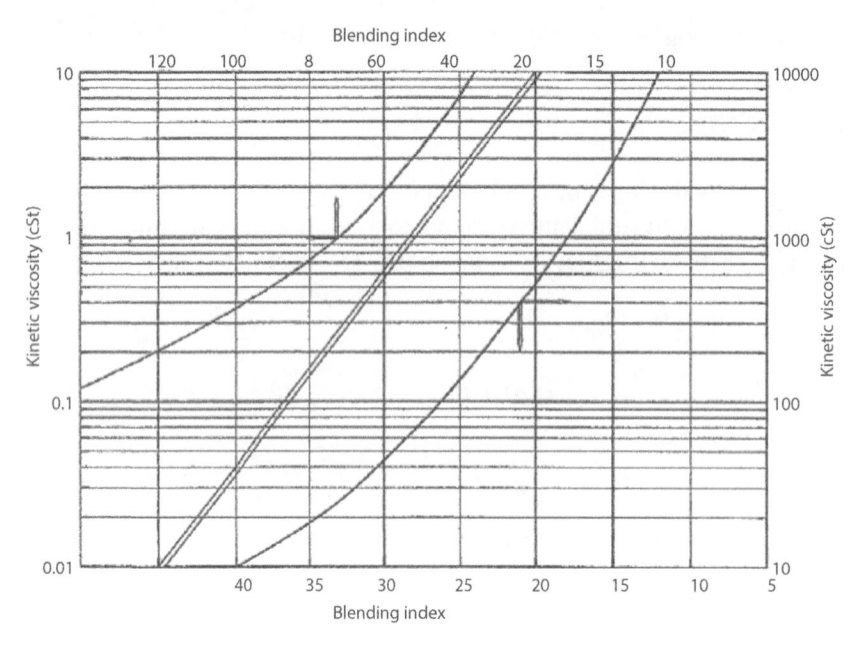

Figure 25.5 Viscosity blending chart (source: Maxwell, Data Book on Hydrocarbons).

Table 25.27a Octane Numbers of Some Oxygenates (alcohols and ethers) [9].

Compound	RON	MON
Methanol	125-135	100-105
Ethanol	120-130	98-103
Methyl-tertiary-butyl-ether (MTBE)	113-117	95-101
Ethyl-tertiary-butyl-ether (ETBE)	118-122	100-102
Tertiary-butyl alcohol (TBA)	105-110	95–100
Tertiary-amyl – methyl ether (TAME)	110-114	96-100

Table 25.27b Blending Values of Octane Improvers (boosters/additives).

Compound	Formula	MW	API	TB (°F)	RVP (psi)	Flash point (°F)	RON	MON
Methanol	CH_4O	32	46.2	148.5	40	53.6	135	105
Ethanol	C_2H_6O	46.1	46.1	173	11	53.6	132	106
TBA	$C_4H_{10}O$	74.1	47.4	180.4	6	39.2	106	89
MTBE	$C_5H_{12}O$	88.1	58.0	131.4	9	−18.4	118	101
ETBE	$C_6H_{14}O$	102.2	56.7	159.8	4	−2.2	118	102
TAME	$C_6H_{14}O$	102.2	53.7	185	1.5	12.2	111	98
TEL	$C_8H_{20}Pb$	323.4	3.143	239	0	199.4	10,000	13,000

Overall viscosity index $= \dfrac{5387.0}{100} = 53.87$

From Figure 25.5, viscosity = 2.65 cSt (actual plant result was 2.7 cSt).

25.14 Gasoline Octane Number Blending

The octane number is a characteristic of spark engine fuels such as gasoline. Octane number is a measure of the fuel's tendency to knock in a test engine compared to other fuels. The posted octane number (PON) is commercially used for gasoline (known as the road octane number) is the average of its research octane number (RON) and motor octane number (MON). The difference between RON and MON is referred to as fuel sensitivity (S).

There are several additives such as oxygenated alcohols and ethers that can enhance a gasoline octane number. Tables 25.27a and 25.27b show a list of oxygenates with their octane numbers and blending values of octane additives/improvers, respectively. Table 25.28 lists RONs for selected hydrocarbons.

Table 25.28 Research Octane Number of Hydrocarbons [10].

Paraffins	RON	Naphthenes	RON
n-Butane	94	Cyclopentane	>100
i-Butane	>100	Cyclohexane	83
n-Pentane	61.8	Methylcyclopentane	91.3
2-Methyl – 1- butane	92.3	Methylcyclohexane	74.8
n – Hexane	24.8	1,3-Dimethylcyclopentane	80.6
2 – Methyl – 1 - pentane	73.4	1,1,3-Trimethylcyclopentane	87.7
2,2- Dimethyl – 1 – butane	91.8	Ethylcyclohexane	45.6
n- Heptane	0	Isobutylcyclohexane	33.7
2 – Methylhexane	52	Aromatics	
2,3 - Dimethylpentane	91.1	Benzene	-
2,2,3 – Trimethylbutane	>100	Toluene	>100
n-Octane	<0	o-Xylene	-
3,3-Dimethylhexane	75.5	m-Xylene	>100
2,2,4 – Trimethylpentane	100	p-Xylene	>100
n-Nonane	<0	Ethylbenzene	>100
2,2,3,3 Tetramethylpentane	>100	n-Propylbenzene	>100
n-decane	<0	Isopropylbenzene	>100
Olefins		1-Methyl-3-ethylbenzene	>100
1-Hexene	76.4	n-butylbenzene	>100
1-Heptene	54.5	1-Methyl-3-isopropylbenzene	-
2-Methyl-2-hexene	90.4	1,2,3,4-Tetramethylbenzene	>100
2,3-Dimethyl-1-pentene	99.3		

Octane numbers are blended on a volumetric basis using the blending octane numbers of the components. True octane numbers do no blend linearly, thus it is necessary to use blending octane numbers in calculating the octane number of the blend. Blending octane numbers can be estimated from empirical correlations that have been developed; blending octane numbers when added on a volumetric average basis, will give the true octane of the blend as obtained from standard test using CFR test engines. The following equation in determining the octane number of the blend is:

True Octane Number of a blend

$$ON_{Blend} = \sum_{i=1}^{n} x_{vi} ON_i \qquad (25.29)$$

where x_{vi} is the volume fraction of component i and ON_i is the octane number of component i.

Many alternative methods have been proposed for estimating the octane number of gasoline blends, since the mixing rule requires corrections. One such correction method that uses the octane number index [2]. The following octane index correlations depend on the octane number range as follows:

For $11 \leq ON \leq 76$

$$BI_{ONi} = 36.01 + 38.33(ON/100) - 99.8(ON/100)^2 + 341.3(ON/100)^3 \\ - 507.02(ON/100)^4 + 268.64(ON/100)^5 \qquad (25.30a)$$

For $76 \leq ON \leq 103$

$$BI_{ONi} = -299.5 + 1272(ON/100) - 1552.9(ON/100)^2 + 651(ON/100)^3 \qquad (25.30b)$$

For $103 \leq ON \leq 116$

$$BI_{ONi} = 2206.3 - 4313.64(ON/100) + 2178.57(ON/100)^2 \qquad (25.30c)$$

The octane number index for a blend can be determined by:

$$BI_{ON,Blend} = \sum_{i=1}^{n} x_{vi} BI_{ONi} \qquad (25.31)$$

where x_{vi} is the volume fraction of component i, and BI_{ONi} is the octane number index of component i that can be determined from Eqs. 25.30a–c.

Example 25.21

For a blend of alkylate $C_4^=$, coker gasoline and FCC gasoline. Calculate the RON by

    a. Linear mixing of octane numbers.
    b. Linear mixing of octane number indices.

Table 25.29 Quantities and RON of the Blending Components.

Component	Quantity, bpd	RON
Alkylate $C_4^=$	6000	97.3
Coker gasoline	4000	67.2
FCC gasoline	5000	83.2

Properties and quantities of the components are shown in Table 25.29.

Solution

Using Eq. 25.29 for the mixing rule, the RON for the blend is shown in Table 25.30:

a. Using Eq. 25.29, the linear mixing octane number = 84.59
b. Using the index method

For $76 \leq 97.3 \leq 103$, Using Eq. 25.30b

$$BI_{RON,Alkylate} = -299.5 + 1272(0.973) - 1552.9(0.973)^2 + 651(0.973)^3$$

$$= 67.66$$

For $11 \leq 67.2 \leq 76$, Using Eq. 25.30a

$$BI_{RON,Coker} = 36.01 + 38.33(67.2/100) - 99.8(677.2/100)^2 + 341.3(67.2/100)^3$$
$$- 507.02(67.2/100)^4 + 268.64(67.2/100)^5$$
$$= 53.65$$

For $76 \leq 83.2 \leq 103$, Using Eq. 25.30b

$$BI_{RON,FCC} = -299.5 + 1272(0.832) - 1552.9(0.832)^2 + 651(0.832)^3$$

$$= 58.78$$

Table 25.30 Summary of Calculations.

Component	bpd	x_{vi}	RON	$x_{vi} \times$ RON
Alkylate $C_4^=$	6000	0.400	97.3	38.920
Coker gasoline	4000	0.267	67.2	17.942
FCC gasoline	5000	0.333	83.2	27.730
Total	15000	1.000		84.592

Table 25.31 Summary of Calculations.

Component	Bpd	x_{vi}	BI_{ON}	$x_{vi} \times BI_{ONi}$
Alkylate $C_4^=$	6000	0.400	67.66	28.064
Coker gasoline	4000	0.267	53.65	14.325
FCC gasoline	5000	0.333	58.78	19.574
Total	15000	1.000		60.92

The octane number index for a blend can be determined by Eq. 25.29 as shown in Table 25.31.

$$BI_{RON,Blend} = 0.4(67.66) + 0.267(53.65) + 0.33(58.78) = 60.98$$

Example 25.22

It is required to produce gasoline with a RON of 98 from the blending components in Table 25.31 in Example 25.21. Calculate the amount of the oxygenate MTBE with a RON of 115 to be added in order to adjust the blend RON of 98.

Solution

Using the index method of Eq. 25.30B, the octane number index for the gasoline blending is

$$BI_{RON,Blend} = -299.5 + 1272(0.98) - 1552.9(0.98)^2 + 651(0.98)^3$$

$$= 68.37$$

The octane number index for the MTBE at RON = 115 using Eq. 25.30c is:

$$BI_{RON,MTBE} = 2206.3 - 4313.64(1.15) + 2178.57(1.15)^2$$

$$= 126.77$$

Table 25.32 shows the summary of the calculations.

Table 25.32 Summary of the Calculations.

Component	bpd	RON	BI_{ON}
Alkylate $C_4^=$	6000	97.3	67.66
Coker gasoline	4000	67.2	53.65
FCC gasoline	5000	83.2	58.78
Gasoline blending	15000	98.0	68.37
MTBE	V_{MTBE}	115.0	126.77

The RON blending indices of alkylate, coker gasoline, and FCC gasoline are calculated in Example 25.21. The blending equation is:

$$BI_{Blend,gasoline}(15{,}000 + V_{MTBE}) = (6000 \times BI_{RON,C_4^=}) + (4000 \times BI_{RON,Coker}) + (5000 \times BI_{RON,FCC})$$
$$+ (V_{MTBE} \times BI_{RON,MTBE}) \qquad (25.32)$$

$$(15{,}000 + V_{MTBE})(68.37) = (6000)(67.66) + (4000)(53.6) + (5000)(58.78) + (126.77\, V_{MTBE})$$

$$1025550 + 68.37 V_{MTBE} = 405960 + 214400 + 293900 + 126.77 V_{MTBE}$$

$$58.4 V_{MTBE} = 111290$$

$$V_{MTBE} = 1905.65 \text{ bpd } (1906 \text{ bpd})$$

This is the amount required to adjust the blend RON of 98.

Depending on the availability of olefin and aromatic contents in the blended components, the octane number of the blend can be calculated using the linear mixing rule method with a correction factor [11].

$$RON_{Blend} = \overline{R} + 0.03324(\overline{RS} - \overline{R} \times \overline{S}) + 0.00085(\overline{O^2} - \overline{O}^2) \qquad (25.33)$$

$$MON_{Blend} = \overline{M} + 0.04285(\overline{MS} - \overline{M} \times \overline{S}) + 0.00066(\overline{O^2} - \overline{O}^2) - 0.00632\left(\frac{\overline{A^2} - \overline{A}^2}{100}\right)^2 \qquad (25.34)$$

where the terms represent volumetric average values of given properties of components as follows:

R = Research octane number (RON)
M = Motor octane number (MON)
S = Sensitivity (RON−MON)
RS = RON × Sensitivity
MS = MON × Sensitivity
O = Volume percent olefins
A = Volume percent aromatics

Example 25.23

Twu and Coon [12] reported RON of 91.8 for a blend with the composition given in Table 25.33.

a. Calculate the RON and MON for the blend using Eqs. 25.33 and 25.34. Then compare with the reported experimental value.
b. Calculate the RON of the blend using the blending index method of Eq. 25.33 and compare with the reported experimental value.

Table 25.33 Composition, RON, MON, Olefin, and Aromatic Contents of the Blending Components [12].

Component	x_{vi}	RON	MON	Olefin content (%)	Aromatic content (%)
Reformate	0.41	97.8	87	1.4	63.1
Thermally cracked	0.04	70.4	65.1	32.5	9.8
Catalytically cracked	0.32	92.6	80.2	53.3	23.9
Polymerized	0.09	96.8	82.3	100	0
LSR	0.14	58	58	1.3	3.8

Solution

All the terms in Eqs. 25.33 and 25.34 are calculated and listed in Tables 25.34–25.36, respectively.

a. From Eq. 25.33, Tables 25.34–25.36

$$\text{RON}_{\text{Blend}} = 89.38 + 0.03324(941.74 - 89.38 \times 9.92) + 0.00085(1851.65 - 28.116^2)$$

$$= 92.113$$

From Eq. 25.34, Tables 25.32–25.34

$$\text{MON}_{\text{Blend}} = 79.464 + 0.04285(824.67 - 79.464 \times 9.92) + 0.00066(1851.65 - 28.116^2) - 0.00632\left(\frac{1821.11 - 34.443^2}{100}\right)^2$$

$$= 81.469$$

Table 25.34 Summary of Calculations.

Component	x_{vi}	S (R–M)	RS	MS	$\overline{R}$ ($x_{iv} \times R$)	$\overline{M}$ ($x_{iv} \times M$)
Reformate	0.41	10.8	1056.24	939.6	40.1	35.67
Thermally cracked	0.04	5.3	373.12	345.03	2.82	2.604
Catalytically cracked	0.32	12.4	1148.24	994.48	29.63	25.66
Polymerized	0.09	14.5	1403.6	1193.35	8.71	7.41
LSR	0.14	0		0	8.12	8.12
Total					89.38	79.464

Table 25.35 Summary of Calculations.

Component	x_{vi}	$\overline{S}$ $(\overline{R}-\overline{M})$	$\overline{RS}$ $(x_{iv} \times RS)$	$\overline{MS}$ $(x_{iv} \times MS)$
Reformate	0.41	4.43	1056.24	939.6
Thermally cracked	0.04	0.220	373.12	345.03
Catalytically cracked	0.32	3.97	1148.24	994.48
Polymerized	0.09	1.3	1403.6	1193.35
LSR	0.14	0		0
Total		9.92	941.74	824.67

Table 25.36 Summary of Calculations.

Component	x_{vi}	% A	$\overline{A}$ $(x_{iv} \times A)$	% O	$\overline{O}$ $(x_{iv} \times O)$	$\overline{A^2}$ $(x_{iv} \times (\%A)^2)$	$\overline{O^2}$ $(x_{iv} \times (\%O)^2)$
Reformate	0.41	63.1	25.871	1.4	0.574	1632.46	0.0784
Thermally cracked	0.04	9.8	0.392	32.5	1.3	3.842	42.25
Catalytically cracked	0.32	23.9	7.648	53.3	17.06	182.79	909.08
Polymerized	0.09	0	0.000	100.0	9.00	0.00	900.00
LSR	0.14	3.8	0.532	1.3	0.182	2.02	0.24
Total			34.443		28.116	1821.11	1851.65

The above calculations show that the RON without correction $(\overline{R})$ is 89.38, and with olefin and aromatic corrections, it is 92.11. The laboratory tested RON for the blend was 91.8 [12]. Therefore, the correction using the olefin and aromatic contents adjusts the calculations to be closer to the experimental value of 0.36% difference.

b. The blending index method can be used to calculate the RON of the above blend as follows:

The blending index is 63.6.

Component	x_{vi}	BI_{RON}	$x_{vi} \times BI_{RON}$
Reformate	0.41	68.16	27.946
Thermally cracked	0.04	54.53	2.1812
Catalytically cracked	0.32	63.71	20.214
Polymerized	0.09	67.17	6.045
LSR	0.14	51.52	7.213
Total			63.599

25.15 Other Blending Correlations

Cetane Index

The cetane index (CI) is estimated from the API gravity and 50% ASTM distillation temperature in °F of the diesel, by the following correlation

$$CI = -420.34 + 0.016 \times (API)^2 + 0.192 \times (API) \times \log(M) \\ + 65.01(\log M)^2 - 0.0001809 \times M^2 \qquad (25.35)$$

or

$$CI = -454.74 - 164.416 \times D + 774.74 \times D^2 - 0.554 \times B \\ + 97.803 - (\log B)^2 \qquad (25.36)$$

where

API = API gravity
M = mid-boiling temperature, °F
D = density at 15°C
B = mid-boiling temperature, °C

Diesel Index

The diesel index (DI) is correlated with the aniline point of diesel by

$$DI = \text{aniline point} \times \text{API gravity}/100 \qquad (25.37)$$

The diesel index can be approximately related to cetane index of diesel by:

$$DI = (CI - 21.3)/0.566 \qquad (25.38)$$

U.S. Bureau of Mines Correlation Index (BMCI)

$$BMCI = 87{,}552/(VABP) + 473.7 \times (SpGr) - 456.8$$

where VABP is the volume average boiling point, °R (°F + 460), and SpGr is the specific gravity. The BMCI has been correlated with many characteristics of a feed, such as crackability, in-steam cracking, and the paraffinic nature of a petroleum fraction. Paraffinic compounds have low BMCIs and aromatics high BMCIs.

Aromaticity Factor

The aromaticity factor is related to the boiling point, specific gravity, aniline point, and sulfur as follows:

$$AF = 0.2514 + 0.00065 \times VABP + 0.0086 \times S + 0.00605 \times AP + 0.00257 \times \left(\frac{AP}{SpGr}\right) \qquad (25.39)$$

where

AF = aromaticity factor
VABP = volume average boiling point, °F
S = sulfur, wt%
AP = aniline point, °F
SpGr = specific gravity

25.16 Fluidity of Residual Fuel Oils

The low-temperature flow properties of a waxy fuel oil [19] depend on its handling and storage conditions. These properties may not be truly assessed by the pour point. The pour point of residual fuel oils is mainly influenced by the previous thermal history of the oil. A waxy structure built on cooling of oil can normally be broken by the application of relatively little pressure. The usefulness of pour point test in relation to residual fuel oils is questionable and the tendency to regard pour point as the limiting temperature at which a fuel oil will flow can be erroneous.

Furthermore, the pour point test does not indicate what happens when an oil has a considerable head of pressure behind it, such as when gravitating from a storage tank or being pumped from a pipeline. Failure to flow at the pour point is normally attributed to a separation of wax from the fuel. This can also be due to viscosity in the case of very viscous fuel oils.

25.16.1 Fluidity Test

Specifying the handling of fuel oils is essential and because of the limitation of the pour point test, other tests have been devised [19]. However, most of these test methods are time-consuming and a method that is relatively quick and easy is based the ASTM specification D-1659-65. This method covers the determination of fluidity of a residual fuel oil at a specified temperature in an "as received" condition. The procedure is as follows:

A sample of fuel oil in its as received condition is cooled at the specified temperature for 30 min. in a standard U-tube and tested for movement under prescribed pressure conditions.

The sample is considered fluid at the temperature of the test, if it will flow 2 mm under a maximum pressure of 152 mmHg. the U-tube for the test is 12.5 mm dia.

This method may be applied for operational situations where it is necessary to ascertain the fluidity of a residual oil under prescribed conditions in an as-received state.

The conditions of the method simulate those of a pumping situation, where oil is expected to flow through a 12 mm pipe under slight pressure at specified temperature.

Fluidity, like the pour point of specification D-97 is used to define cold flow properties. It differs from D-97 in that

1. It is restricted to fuel oils.
2. A prescribed pressure is applied to the sample. This represents an attempt to overcome the technical limitation of pour points method, where gravity-induced flow is the criterion.

The pumpability test (ASTM specification D-3245) represents another method for predicting field performance in cold-flow conditions. The pumpability test, however, has limitations and may not be suitable for very waxy fuel oils, which solidify so rapidly in the chilling bath that no reading can be obtained under conditions of the test. It is also time-consuming and therefore, not suitable for routine control testing.

Table 25.37 Summary of the Calculations of Example 25.25.

Grade	vol %	Fluidity temperature, °R	Fluidity index	Blend
964	40.0	492	446.20	178.48
965	60.0	519	870.07	522.04
Blend				700.52

25.16.2 Fluidity Blending

Fluidity, like pour points do not blend linearly. Blending indices are therefore used for blending the fluidity of two fuel oils. The blending indices used are the same as those for distillate fuels (Table 25.16). The fluidity index is determined by:

$$\text{Fluidity index} = 3162000 \times \left[\frac{(\text{Fluidity, °R})}{1000}\right]^{12.5} \quad (25.40)$$

$$\text{Fluidity(°R)} = 1000 \times \left[\frac{(\text{Fluidity index})}{3162000}\right]^{0.08} \quad (25.41)$$

Example 25.24

Determine the fluidity of a blend of two fuel oil grades. The blend contains 40 vol% grade 964 and 60 vol% grade 965. As per fluidity test, grade 964 is fluid at 32°F and grade 965 is fluid at 59°F.

Solution

The fluidity of the blend is determined by first determining the fluidity index of the two grades and blending linearly. The fluidity index can be read directly from Table 25.16 or determined by Eqs. 25.40 and 25.41, respectively. Table 25.37 shows the summary of the calculations

The corresponding fluidity of the blend is 510°R (50.1°F)

25.17 Conversion of Kinematic Viscosity to Saybolt Universal

Viscosity or Saybolt Furol Viscosity

Saybolt universal viscosity in the efflux time in seconds for a 60-cm³ sample to flow through a standard orifice in the bottom of a tube. The orifice and tube geometry are specified in the ASTM standards. Saybolt Furol viscosity is determined in the same manner as Saybolt universal viscosity using a larger orifice size.

25.17.1 Conversion to Saybolt Universal Viscosity

Kinematic viscosity in centistokes (mm²/s) can be converted to Saybolt universal viscosity in Saybolt universal seconds (SUS) units at the same temperature by API databook [20] procedure or using ASTM conversion tables [21]:

$$SUS = 4.6324 \times cSt + \frac{[1 + 0.03264 \times cSt]}{[3930.2 + 262.7 \times cSt + 23.97 \times cSt^2 + 1.646 \times cSt^3] \times 10^{-5}} \quad (25.42)$$

where

SUS = Saybolt universal viscosity in Saybolt universal seconds units at 37.8°C
cSt = kinematic viscosity, centistokes (mm²/s)

SUS viscosity at 37.8°C can be converted to SUS viscosity at another temperature by following relationship:

$$SUS_t = [1 + 0.000110 \times (t - 37.8)] \times SUS \qquad (25.43)$$

where

SUS = Saybolt universal viscosity in Saybolt universal seconds units at 37.8°C
SUS_t = SUS viscosity at the required temperature
t = temperature at which SUS viscosity is required, °C

Example 25.25

An oil has a kinematic viscosity of 60 centistokes at 37.8°C. Calculate the corresponding Saybolt universal viscosity at 37.8°C and 98.9°C.

Solution

Using Eq. 25.42:

$$SUS = 4.6324 \times 60 + \frac{[1 + 0.03264 \times 60]}{[3930.2 + 262.7 \times 60 + 23.97 \times 60^2 + 1.646 \times 60^3] \times 10^{-5}}$$

$$= 277.94 \text{ s.}$$

Using Eq. 25.43, the viscosity at 98.9°C is:

$$SUS_t = [1 + 0.000110 \times (98.9 - 37.8)] \times 277.94$$

$$= 279.81 \text{ s}$$

25.17.2 Conversion to Saybolt Furol Viscosity

Kinematic viscosity in centistokes (mm²/s) at 50 and 98.9°C can be converted to Saybolt Furol viscosity in Saybolt Furol seconds (SFS) units at the same temperature by API databook [22]

$$SFS_{50} = 0.4717 \times VIS_{50} + \frac{13.924}{[(VIS_{50})^2 - 72.59 \times VIS_{50} + 6.816]} \qquad (25.44)$$

$$SFS_{98.9} = 0.4792 \times VIS_{98.9} + \frac{5.610}{[(VIS_{98.9})^2 + 2.130]} \qquad (25.45)$$

where

SFS_{50} = Saybolt Furol viscosity at 50°C, Saybolt Furol s
VIS_{50} = kinematic viscosity at 50°C, centistokes (mm²/s)

$SFS_{98.9}$ = Saybolt Furol viscosity at 98.9°C, Saybolt Furol s
$VIS_{98.9}$ = kinematic viscosity at 98.9°C, centistokes (mm²/s)

Example 25.26

1. Estimate the Saybolt furol viscosity of an oil at 50°C, if the kinematic viscosity at 50°C is 3000 centistokes.
2. Estimate the Saybolt Furol viscosity of an oil at 98.9°C, if the kinematic viscosity at 98.9°C is 120 centistokes.

Solution

1. Using Eq. 25.44,

$$SFS_{50} = 0.4717 \times 3000 + \frac{13.924}{[(3000)^2 - 72.59 \times 3000 + 6.816]}$$

$$= 1415.1 \text{ sec}$$

2. Using Eq. 25.45,

$$SFS_{98.9} = 0.4792 \times 120 + \frac{5.610}{[(120)^2 + 2.130]}$$

$$= 57.50 \text{ sec}$$

25.17.3 Refractive Index of Petroleum Fractions

The refractive index of a petroleum fraction can be calculated from its mean average boiling point, molecular weight, and relative density using the API databook procedure [23]

The method may be used to predict refractive index for a petroleum fraction with normal boiling point up to 1100K.

$$n = \left(\frac{1+2\times I}{1-I}\right)^{0.5} \tag{25.46}$$

$$I = 3.587 \times 10^{-3} \times T_b^{1.0848} \times \left(\frac{M}{d}\right)^{-0.4439} \tag{25.47}$$

where

n = refractive index at 20°C
I = characterization factor of Huang [23] at 20°C
T_b = mean average boiling point, K
M = molecular weight of petroleum fraction
d = liquid density at 20°C and 101.3 kPa, in kg/dm³

Example 25.27

Calculate the refractive index of a petroleum fraction with a liquid density of 0.7893 kg/dm³, molecular weight of 163.47, and mean average boiling point of 471.2K.

Solution

Using Eqs. 25.47 and 25.46, respectively:

$$I = 3.587 \times 10^{-3} \times (471.2)^{1.0848} \times \left(\frac{163.47}{0.7893}\right)^{-0.4439}$$

$$= 0.267$$

$$n = \left(\frac{1 + 2 \times 0.267}{1 - 0.267}\right)^{0.5}$$

$$= 1.447$$

25.18 Determination of Molecular-Type Composition

An estimate of fractional composition of paraffins, naphthenes, and aromatics contained in light and heavy petroleum fractions can be obtained if the data on viscosity, relative density, and refractive index of the desired fraction are available. The algorithm is based on a procedure of API databook [24]

$$X_p = a + b\,(R_i) + c\,(VG) \tag{25.48}$$

$$X_n = d + e\,(R_i) + f\,(VG) \tag{25.49}$$

$$X_a = g + h\,(R_i) + I\,(VG) \tag{25.50}$$

where

X_p	= mole fraction of paraffins
X_n	= mole fraction of naphthenes
X_a	= mole fraction of aromatics
R_i	= refractivity intercept as given by Eq. 25.51
a, b, c,	= constants varying with molecular weight range
d, e, f, g,	= constants varying with molecular weight range
h, i	= constants varying with molecular weight range
	= constants varying with molecular weight range
	= constants varying with molecular weight range
VG	= viscosity gravity function VGC as given by Eqs. 25.52 and 25.53 for light fractions.

$$R_i = n - d/2 \tag{25.51}$$

Table 25.38 Constants of Light and Heavy Fractions.

Constant	Light fraction (MW = 80–200)	Heavy fraction (MW = 200–500)
a	−23.940	−9.000
b	24.210	12.530
c	−1.092	−4.228
d	41.140	18.660
e	−39.430	−19.900
f	0.627	2.973
g	−16.200	8.660
h	15.220	7.370
i	0.465	1.255

where

n = refractive index at 20°C and 101.3 kPa and d is the liquid density at 20°C and 101.3 kPa in kg/dm³

Table 25.38 shows the constants in Eqs. 25.48–25.50.

For heavy fractions (molecular weight 200–500), the viscosity gravity constant is determined by:

$$VGC = \frac{10 \times d - 1.0752 \times \log(V_{311} - 38)}{10 - \log(V_{311} - 38)} \qquad (25.52)$$

or

$$VGC = \frac{d - 0.24 - 0.022 \times \log(V_{372} - 35.5)}{0.755} \qquad (25.53)$$

where d is the relative density at 15°C and 101.3 kPa and V is the Saybolt universal viscosity at 311 or 372K in Saybolt universal seconds.

For light fractions (molecular weight 80–200), the viscosity gravity constant is:

$$VGF = -1.816 + 3.484 \times d - 0.1156 \times \ln(v_{311}) \qquad (25.54)$$

or

$$VGF = -1.948 + 3.535 \times d - 0.1613 \times \ln(v_{372}) \qquad (25.55)$$

where v is the kinematic viscosity at 311 and 372K, in mm2/s.

The viscosity gravity (VG) constant is a useful function for approximate characterization of viscous fractions of petroleum. It is relatively insensitive to molecular weight and related to the composition of the fraction. Values of VG near 0.8 indicate samples of paraffinic character, while those close to 1.0 indicate an aromatic character. The VG should not be applied to residual oils or asphaltic materials.

Example 25.28

Calculate the molecular type distribution of a petroleum fraction of relative density 0.9433, refractive index 1.5231, normal boiling point 748K, and a viscosity of 695 Saybolt universal seconds at 311K.

Solution

First determine the molecular weight of the fraction as a function of the mean average boiling point and relative density at 15°C by API databook procedure [25]

$$MW = 2.1905 \times 10^2 \times \exp(0.003924 \times T) \times \exp(-3.07 \times d) \times T^{0.118} \times d^{1.88} \qquad (25.56)$$

where MW is the molecular weight, d is the relative density at 15C and T is the mean average boiling point in K.

$$MW = 2.1905 \times 10^2 \times \exp(0.003924 \times 748) \times \exp(-3.07 \times 0.9433) \times 748^{0.118} \times 0.9433^{1.88}$$

$$= 445.69$$

As the molecular weight is greater than 200, the fraction is termed heavy. Viscosity gravity constant VGF is determined by Eq. 25.52.

$$VGC = \frac{10 \times 0.9433 - 1.0752 \times \log(695 - 38)}{[10 - \log(695 - 38)]}$$

$$= 0.8916$$

The refractive index, R_i from Eq. 25.51:

$$R_i = 1.5231 - 0.9433/2$$

$$= 1.0515$$

Using Table 25.38, constants for heavy fractions and Eqs. 25.48–25.50, the mole fraction of paraffin, naphthenes, and aromatics are:

$$X_p = -9.00 + 12.53 \times (1.0515) - 4.228 (0.8916)$$

$$= 0.4056$$

$$X_n = 18.66 - 19.9 (1.0515) + 2.973 (0.8916)$$

$$= 0.3859$$

$$X_a = -8.66 + 7.37\,(1.0515) + 1.225\,(0.8916)$$

$$= 0.1818$$

Example 25.29

Calculate the molecular type distribution of a petroleum fraction of relative density 0.8055, refractive index 1.4550, normal boiling point 476K, and a viscosity of 1.291 mm²/s at 311K.

Solution

First determine the molecular weight of the fraction as a function of the mean average boiling point and relative density at 15°C by API databook procedure [25]

$$MW = 2.1905 \times 10^2 \times \exp(0.003924 \times 476) \times \exp(-3.07 \times 0.8055) \times 476^{0.118} \times 0.8055^{1.88}$$

$$= 164.86$$

As the molecular weight is less than 200, the fraction is termed light. Viscosity gravity constant VGF is determined by Eq. 25.54 or 25.55.

$$VGF = -1.816 + 3.484 \times 0.8055 - 0.1156 \times \ln(1.291)$$

$$= 0.9608$$

The refractive index is:

$$R_i = 1.4550 - 0.8055/2$$

$$= 1.0523$$

Using Table 25.38, constants for light fractions, and Eqs. 25.48–25.50, the mole fraction of paraffin, naphthenes, and aromatics are:

$$X_p = -23.94 + 24.21 \times (1.0522) - 1.092\,(0.9608)$$

$$= 0.4846$$

$$X_n = 41.14 - 39.43\,(1.0522) + 0.627\,(0.9608)$$

$$= 0.2542$$

$$X_a = -16.2 + 15.22\,(1.0522) + 0.465\,(0.9608)$$

$$= 0.2613$$

25.19 Determination of Viscosity From Viscosity/Temperature Data at Two Points

Kinematic viscosity/temperature charts [24] are a convenient way to determine the viscosity of a petroleum oil or liquid hydrocarbon at any temperature within a limited range, provided viscosities at two temperatures are known. The procedure is to plot two known kinematic viscosity/temperature points on the chart and draw a straight line through them. A point on this line within the range defined shows the kinematic viscosity at the corresponding desired temperature and vice versa.

The kinematic viscosity of a petroleum fraction can also be determined by a linear function of temperature by the following equations. These equations agree closely with the chart scales. They are necessary when calculations involve viscosities smaller than 2cSt.

$$\log \log Z = A - B \log T \tag{25.57}$$

$$Z = v + 0.7 + \exp(-1.47 - 1.84 \times v - 0.51 \times v^2) \tag{25.58}$$

$$v = (Z - 0.7) - \exp[-0.7487 - 3.295 \times (Z - 0.7)$$
$$+ 0.6119 \times (Z - 0.7)^2 - 0.3193 \times (Z - 0.7)^3] \tag{25.59}$$

where

v = kinematic viscosity, cSt (mm²/s)
T = temperature, K or °R
A and B = constants

Constants A and B for a petroleum fraction can be calculated when the temperature and corresponding viscosity are available for two points and are expressed by:

$$\log \log Z_1 = A - B \log T_1 \tag{25.60}$$

$$\log \log Z_2 = A - B \log T_2 \tag{25.61}$$

where

$$Z_1 = v_1 + 0.7 + \exp\left(-1.47 - 1.84 v_1 - 0.51 v_1^2\right) \tag{25.62}$$

$$Z_2 = v_2 + 0.7 + \exp\left(-1.47 - 1.84 v_2 - 0.51 v_2^2\right) \tag{25.63}$$

v_1 and v_2 are the kinematic viscosities at temperatures t_1 and t_2 in K or °R, respectively.

Let

$$k_1 = \exp\left(-1.47 - 1.84 v_1 - 0.51 v_1^2\right) \tag{25.64}$$

$$k_2 = \exp\left(-1.47 - 1.84 v_2 - 0.51 v_2^2\right) \tag{25.65}$$

$$y_1 = \log Z_1 = \log(v_1 + 0.7 + k_1) \tag{25.66}$$

$$y_2 = \log Z_2 = \log(v_2 + 0.7 + k_2) \tag{25.67}$$

Therefore, Eqs. 25.60 and 25.61 can be rewritten as follows:

$$\log y_1 = A - B \log T_1 \tag{25.68}$$

$$\log y_2 = A - B \log T_2 \tag{25.69}$$

Substituting Eq. 25.68 into Eq. 25.69 gives:

$$B = \frac{\log\left(\dfrac{Z_1}{Z_2}\right)}{\log\left(\dfrac{T_2}{T_1}\right)} \tag{25.70}$$

$$A = \log(Z_1) + B \log(T_1) \tag{25.71}$$

with the values of A and B known, viscosity at any other temperature in K or °R can be determined from Eqs. 25.57–25.59.

Example 25.30

A kerosene stream of crude has a kinematic viscosity of 1.12 cSt at 323K and a viscosity of 0.7cSt at 371.9 K. Determine the kinematic viscosity of the stream at 311K.

$$v_1 = 1.12 \text{ cSt}$$

$$v_2 = 0.70 \text{ cSt}$$

Solution

Calculate k_1 and k_2 from Eqs. 25.64 and 25.65 as follows:

$$k_1 = \exp(-1.47 - 1.84(1.12) - 0.51(1.12)^2)$$

$$= 0.0154$$

$$k_2 = \exp(-1.47 - 1.84(0.70) - 0.51(0.70)^2)$$

$$= 0.0494$$

Calculate Z_1 and Z_2 from Eqs. 25.66 and 25.67

$$\log Z_1 = \log(1.12 + 0.7 + 0.0154)$$

$$= 0.2637$$

$$Z_1 = 10^{0.2637} = 1.8354$$

$$\log Z_2 = \log(0.7 + 0.7 + 0.0494)$$

$$= 0.1612$$

$$Z_2 = 10^{0.1612} = 1.4494$$

The constants A and B for the petroleum fraction from Eqs. 25.70 and 25.71:

$$B = \frac{\log\left(\dfrac{1.8354}{1.4494}\right)}{\log\left(\dfrac{371.9}{323}\right)}$$

$$= 1.6749$$

$$A = \log(1.8354) + 1.6749 \log(323)$$

$$= 4.4664$$

The viscosity at 311K from Eq. 25.68

$$\log Z_{311} = A - B \log T_{311}$$

$$= 4.4664 - 1.6749 \log(311)$$

$$\log Z_{311} = 0.2912$$

$$Z_{311} = 10^{0.2912} = 1.9556$$

The corresponding viscosity at 311K from Eq. 25.59:

$$v_{311} = (1.9556 - 0.7) - \exp[-0.7487 - 3.295 \times (1.9556 - 0.7)]$$

$$+ 0.6119 \times (1.9556 - 0.7)^2 - 0.3193 \times (1.9556 - 0.7)^3]$$

$$= 1.2451 \text{ cSt}$$

25.20 Linear Programming (LP) for Blending

Linear programming is an optimization modeling technique in which a linear function is maximized or minimized when subjected to various constraints. This technique has been useful for guiding quantitative decisions in business planning in industrial engineering, refineries and chemical plants. Petroleum refining involving gasoline blending bears a different cost of manufacture; the proper allocation for each component into its optimal disposition is of major economic importance. In order to address this problem, most refiners employ LP that permits the rapid selection of an optimal solution from a multiplicity of feasible alternative solutions, as each component is characterized by its gasoline blending in petroleum refining. A few major areas where this technique is employed are [17]:

1. blending gasolines.
2. refinery models.
3. allocation of transportation facilities for shipping products from refineries to terminals.
4. integrated operations—simultaneously consider the complete refinery models (item 2) along with the transportation models (item 3).
5. allocation of transportation facilities for shipping products from the crude-oil field to refineries
6. integrated operations, which encompass items 1–5.

The LP models in the areas listed are viewed as top-management planning tools, which can be used to balance various aspects of a company's operations. However, progress has been made in using LP models for shorter and longer-range problems. Figure 25.6 shows a schematic that indicates the material flow from the crude field to refinery to market. The arrows indicate the transportation links.

Consider the cash-flow model within the framework of the Figure 25.6. If we keep an account and ask, at strategic check-points in the fluid flow model, whether cash flows into or out of it, we must know for each operating phase, whether the cash is generated and flows into the account and how much or vice-versa. This constitutes the cash-flow model. The contention is that it will be feasible to construct a mathematical model that will observe all fluid flow requirements (or commitments) and at the same time, will keep track of all cash flows into and out the account. In a

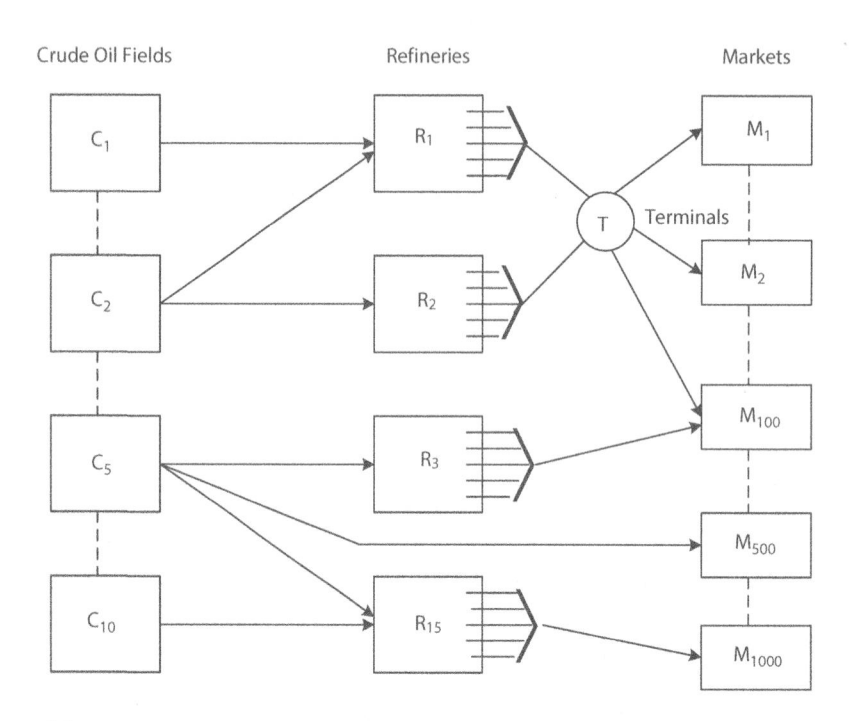

Figure 25.6 Fluid – flow model.

refinery, it will spell out in detail what crudes should go to what refinery, the throughput and product yields at each refinery, the complete allocation of product distribution all the way to the consumer and a summation of essential operating budgets broken down for each important facility [17].

It is essential that the proposed model as shown in Figure 25.6 can simulate and optimize with reasonable accuracy the fluid flow and cash flow of the corporation. Such a model could be used to determine the most economical means of selecting the type and source of the crude that is best suited for each refinery. Furthermore, it would help decide what products to make and how much of them to make for each market resulting in a low-cost and way to transport the product to the market.

LP is an optimizing model that could be used to good advantage despite the highly nonlinear characteristics of the fluid flow-cash flow model. These nonlinearities can be resolved within the framework of a linear program model by adding more constraints in order to make the model piece-wise linear. This results in the requirement of a very large computer to solve such linear program models.

The structure of the linear program model follows very closely the schematic of Figure 25.6. For instance, each refinery is represented by a sublinear program model of 100 or more constraints. Furthermore, sublinear models are used to represent the movement of crude oil from the fields to the refineries, and transportation of finished products from the refineries to the markets.

The general linear programming problem is to find a vector $(x_1, x_2, \ldots x_n)$, where n is the independent variable, which minimizes or maximizes the linear form or the objective function, P as:

$$P = c_1 x_1 + c_2 x_2 + c_3 x_3 + \ldots \ldots c_j x_j + \ldots \ldots + c_n x_n \tag{25.72}$$

subject to the linear constraints

$$x_j \geq 0, j = 1, 2, \ldots \ldots, n$$

and

$$a_{11} x_1 + a_{12} x_2 + \ldots \ldots a_{ij} x_j + \ldots \ldots + a_{1n} x_n \leq b_i$$

$$a_{i1} x_1 + a_{i2} x_2 + \ldots \ldots a_{ij} x_j + \ldots \ldots + a_{in} x_n \leq b_i \tag{25.73}$$

$$a_{m1} x_1 + a_{m2} x_2 + \ldots \ldots a_{mj} x_j + \ldots \ldots + a_{mn} x_n \leq b_m$$

where a_{ij}, b_i, and c_i are given constants and m<n.

One way of looking at the model, is that the structure of the fluid-flow portion is represented by the following constraints

$$A \times X = b \tag{25.74}$$

subject to

$$X \geq 0 \tag{25.75}$$

On the other hand, the cash-flow aspect of the model is represented by minimizing or maximizing C × X, where C = $(c_1, c_2, \ldots c_m)$ is a row vector, X = $(x_1, x_2, \ldots x_m)$ is a column vector, A = (a_{ij}) is a matrix and B = $(b_1, b_2, \ldots b_m)$ is a column vector.

The objective function (P) is usually the octane number of the blend that needs to be maximized. In some instances, the objective function can be the minimized cost or setting the RVP to a certain value. The primary constraints variables (x_1 to x_n) are the volume of each cut of a blend which must be greater than zero. Other constraints can be the capacity of the tank in which the summation of xi,s will not exceed. Additional constraints can be any of the blend properties that we discussed in the chapter. Therefore, a_{i1} to a_{mn} are the properties or their indices for components 1 to n and b_i to b_m are the targeted blending properties of their indices.

In most blending cases, two or more properties are required in order to be adjusted by the addition of some modifiers to the blend. For example, oxygenates are added to enhance gasoline octane number and n-butane is added to adjust the RVP. Determining the quantities of the additives becomes more difficult in non-linear properties cases. Determining the amounts of these blends is based on solving two or more equations with two or more unknowns depending on the property equations.

LP Software

LP software includes two kinds of programs. The first is a solver program, which takes data specifying an LP and a mixed integer linear program (MILP) as input, solves it and returns the results. Solver software may contain one or more algorithms (simplex and interior point LP solvers and branch-and-bound methods for MILPs, which call an LP solver many times). Some LP solvers also include facilities for solving some types of nonlinear problems, usually quadratic programming problems (quadratic objective function, linear constraints) or separable nonlinear problems in which the objective or some constraint functions are a sum of nonlinear functions, each of a single variable.

A feature of LP programs is the inclusion of modeling systems, which provide an environment for formulating, solving, reporting on, analyzing and managing LP and MILP models. Modeling systems are designed as a language for formulating optimization models, and most are capable of formulating and solving both linear and nonlinear problems. Algebraic modeling systems represent optimization problems using algebraic notation and a powerful indexing capability. This allows a set of similar constraints to be represented by a single modeling statement, regardless of the number of constraints in the set.

Another type of widely used modeling system is the spreadsheet solver. Microsoft Excel contains a module called the Excel Solver, which allows the user to enter the decision variables, constraints and objective of an optimization problems into the cells of a spreadsheet and then invoke an LP, MILP, or NLP solver.

The Excel Solver

Microsoft Excel, incorporates an NLP solver that operates on the values and formulas of a spreadsheet model. Versions 4.0 and later include an LP solver and mixed-integer programming (MIP) capability for both linear and nonlinear problems. The user specifies a set of cell addresses to be independently adjusted (the decision variables), a set of formula cells whose values are to be constrained (the constraints), and a formula cell designated as the optimization objective. The solver uses the spreadsheet interpreter to evaluate the constraint and objective functions, and approximates derivatives using finite differences. The NLP solution engine for the Excel Solver is the generalized reduced gradient (GRG2) algorithm, first developed in the late 1960s by Jean Abadie and has since been refined by several researchers [35].

Other spreadsheets contain similar solvers. LP software vendors are:

Company name	Solver name	Web addresses/e-mail address
CPLEX Division of ILOG	CPLEX	www.cplex.com
IBM	Optimization Software Library (OSL)	www.research.ibm.com/osl/
LINDO Systems Inc.	LINDO	www.lindo.com
Dash Associates	XPRESS-MP	www.dashopt.com
Sunset Software Technology	AXA	Sunsetw@ix.netcom.com
Advanced Mathematical Software	LAMPS	infor@amsoft.demon.co.uk

Refinery works on very small Gross Refinery Margin (GRM) and off specification product blends result in considerable overhead cost because product blending is the last operation in the refinery process. Refineries operate with high volume and limited GRM and thus a small improvement in GRM can have a significant impact on the profit. Optimization literature on refinery shows that studies have been carried out at three major directions [27]:

1. Improving process models of the non-linear blending and other refinery processes.
2. Modeling the uncertainty in the refinery blending process using stochastic programming framework.
3. Different possibilities of integrating Model Predictive Control (MPC), product blending optimization and scheduling problems.

A typical refinery optimization problem can be split into sub-problems using spatial decomposition scheme proposed by Jia and Ierapetritou [28]. These sub-problems are solved as independent optimization problems, and the combined results are considered as approximate solution to the overall refinery optimization problem. There are several commercial tools for solving stand-alone refinery optimization problems, such as Aspen Blend and Aspen PIMS-MBO™ from Aspen Tech. These are software products for online and offline blending optimization problems.

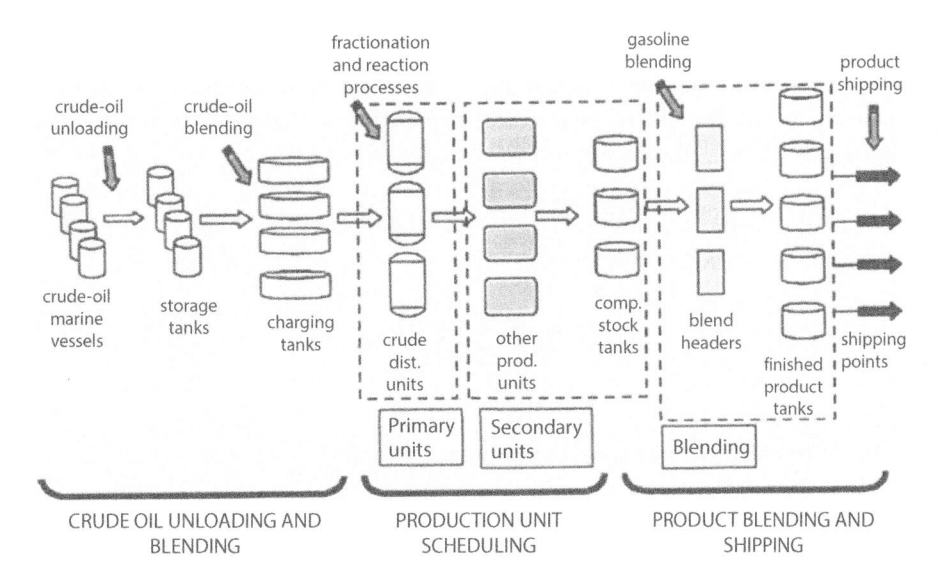

Figure 25.7 Schematic division of refinery wide operation [29].

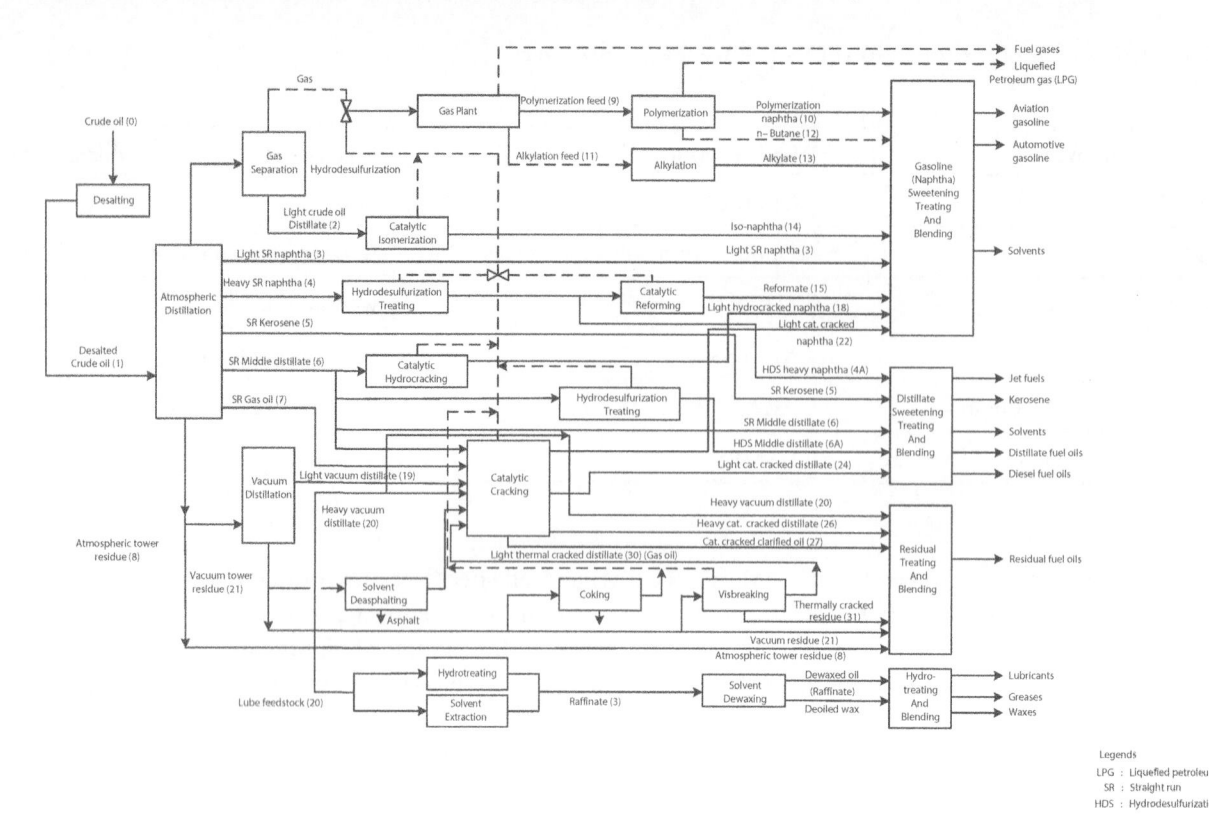

Figure 25.8 Overview of petroleum refining and blending processes.

EBCTM and OpenBPCTM and BlendTM are Honeywell products for online and offline blending optimization; Profit ControllerTM from Honeywell and ROMeoTM from Invensys are online optimization solutions for primary and secondary process units. Figure 25.7 shows a schematic decomposition of the refinery optimization problem and Figure 25.8 shows an overview of petroleum refining and blending processes.

25.20.1 Mathematical Formulation

The problem formulation includes multiple blenders and storage tanks as shown in Figure 25.7 [27]. Properties of the final products are modeled using blend laws which relate the final product property to quantities and properties of feed components. Blend product properties are either linearly or non-linearly blended based on volume or mass of the feed components. For example, RON of the blend is modeled by linear blend law while RVP of the blend is modeled by non-linear blend law. Blend law used to estimate product property based on volume/mass fraction or volume/mass flow and property of feed components are as follows:

1. Linearly blended by component fractions in tanks: the property of product blend is a linear function of component volume/mass fractions and properties.

$$PE_j = \sum_i X_i BV_{i,j} \tag{25.76}$$

where X_i is the volume/mass fraction of component i in the product tank.

2. Linearly blended by component flows in blend headers:

The instantaneous property of product blend is a linear function of component volume/mass flow rates and properties

$$PE_j = \frac{\sum_i F_i BV_{i,j}}{\sum_i F_i} \tag{25.77}$$

where F_i is the volume/mass flow rate of component i to the blend header.

Blend optimization is subjected to different constraints types, which are operational (equipment limits on component flow), inventory (volume limits on feed components) and quality (analyzer limits and tank property specification). The following are the constraints for blending optimization:

Average property constraint: When blending into a destination tank, the product property in tank cannot violate the product specifications. These limits are transformed as average property constraints.

Instantaneous property constraint: When an analyzer is used to detect product property at blend header, analyzer has limits imposed on its measurements. The property detected by the analyzer should not cross the specified analyzer limits. These limits are considered as instantaneous property constraints.

Average composition constraint: It is a constraint on concentration of each component in destination tank when destination tank property integration is nonlinear.

Equipment constraint: It is a constraint on component flows based on hydraulic constraints of the equipment.

Cost constraint: It is a constraint on the cost function value. In the case of "Minimum Giveaway" optimization mode, the value of cost function should not exceed "minimum cost calculated in 'Cost' mode.

Material balance constraint: It is an overall material balance constraint for each blender used in blending operation.

Component balance constraint: It is the component balance constraint for each feed component used in the blending operation.

Rundown constraint: It is the material balance constraint on rundown stream used in the blending problem.

Blend volume constraint: It is the constraint on total blend quantity specification for the blending operation.

Component volume availability constraint: It is the constraint on the component volume available for blending operation.

The following are decision variables for blending optimization:

1. Feed component volume fraction (R_i)
2. Feed component volumetric flow (F_i)
3. Product property variable (W_j)
4. Total flow to blender (TF_b)
5. Segregation flow–flow from rundown stream to segregation tank (FS_k)
6. Blend volume-predicted volume in destination tank over which the product is expected to be on-specification.
7. Feed component property ($BV_{i,j}$)
8. Rundown total flow (FR_k)

Blending optimization problem is formulated as a multi-objective optimization. Its primary objective is to control the blending process so that the products are on-specification. Once the products are on-specification, the secondary objective is to minimize cost or giveaway or distance of decision variable from target value (minimum distance). The following combination of objectives are:

1. Control–Cost.
2. Control –Giveaway
3. Control–Minimum Distance
4. Control–Cost–Giveaway

1. Control Objective

Objective of control mode is to maintain product property in specified range (make it on—specification). Objective function when blending into destination tank is:

$$\text{MinF} = \sum_b \left(\sum_j (PE_j - W_j)^2 P\_Cost_j OR_j \right)_b \tag{25.78}$$

Decision variables—all variables mentioned in the notation.

Constraints—all constraints mentioned in the notation, except the cost constraint.

2. Cost Objectives

The objective of this mode is to minimize the cost of feed components. Objective function is:

$$\text{MinF} = \sum_b \left(\sum_i C\_Cost_i F_i \right)_b \tag{25.79}$$

3. Giveaway Objective

Objective of this model is to minimize quality giveaway. In refinery terms, giveaway is giving away a product of better quality than specifications. Minimizing the giveaway is necessary to reduce the money lost because of giving product of better quality than required by product specifications. Objective function is:

$$\text{MinF} = \sum_b \left(\sum_j (PE_j - P\_Lim_j)^2 P\_Cost_j \right)_b \tag{25.80}$$

4. Minimum Distance Objective

Objective of the mode is to minimize the distance between current value of the variable and its target value. The objective function for rundown flow is:

$$\text{MinF} = \sum_b \left(\sum_k (FS_k - FS\_Tgt_k)^2 Rundown\_Cost_k \right)_b \tag{25.81}$$

Decision variables for all objectives except the control objective are all variables mentioned in the notation, except the total flow and product property variables.

25.20.2 Problem Solution

One of the main requirements is to solve the proposed problem in real time. However, if this is impossible, then the intermediate iteration result after a fixed time interval should be feasible and better than the previous iteration results. This requirement ensures that a feasible solution is provided for cases where the solver cannot converge in available time for a real time optimization. The proposed problem is solved using MINOS solver because it is a feasible region search robust nonlinear programming (NLP) solver and meets all the solution requirements [28].

NLP is an optimization problem that seeks to minimize (or maximize) a nonlinear objective function subject to linear or nonlinear constraints. Problems formulated as NLP are more accurate compared to their LP counterparts since most chemical processes are nonlinear in nature. Refinery models that account for nonlinear relationship of process variables are more reliable and represent the refinery systems more closely. The general representation of NLP problems is as shown in Eq. 25.82.

$$\text{Minimize/Maximize } f(x) \dots\dots\dots\dots\dots x = [x_1, x_2, \dots\dots\dots\dots x_n]^T$$

$$\text{Subject to } h_i(x) = b_i \dots\dots\dots\dots i = 1, 2, \dots\dots\dots\dots m \quad (25.82)$$

$$g_j(x) \leq c_j \dots\dots\dots\dots j = 1, 2, \dots\dots\dots\dots r$$

In this formulation, bilinear and trilinear terms, and basic mathematical functions can be found. In Eq. 25.82, at least the objective function $f(x)$, the equality constraint $h_i(x)$ or the inequality constraint $g_j(x)$ must be nonlinear.

The challenge in modeling using NLP formulation is that of achieving a reasonable convergence. This is largely because many real-valued functions are non-convex. Convexity of a feasible region can only be guaranteed if constraints are all linear. Furthermore, it is difficult knowing whether an objective function or inequality constraints are convex or not. However, convexity test can be carried out to satisfy conditions of optimality referred to as Kuhn-Tucker conditions (also called KKT conditions). Most algorithms embedded in commercial solvers terminate when these conditions are satisfied within some tolerance. For problems with a few numbers of variables, KKT solutions can sometimes be found analytically and the one with the best objective function value is chosen.

Unlike LP, NLP is reported in a significant number of publications in refinery problems involving blending relation and pooling. Moro *et al.* [30] developed a non-linear optimization model for the entire refinery topology with all the process units considered and non-linearity due to blending included. Pinto *et al.* [31] extended this work and by Neiro and Pinto [32] for multiperiod and multi-scenario cases involving non-linear models. They considered non-linearity in the development of the scheduling model for refinery production with product blending. Furthermore, Hamisu [33] considered nonlinearity in the development of a scheduling model for refinery production with product blending.

The objective function is formulated as augmented Lagrangian form by introducing Lagrangian multipliers and penalty parameters. The sequence of iterations is performed, and each one requiring a solution of linearly constrained sub-problems. These sub-problems contain the original linear constraints and bounds on variables as well as linearized form of nonlinear constraints.

$$\min_x F(x) - \lambda_k^T (f - \bar{f}) + \frac{1}{2}\rho(f - \bar{f})^T (f - \bar{f}) \quad (25.83)$$

Subject,

$$\begin{aligned} \bar{f} &= b_1 \\ A_2 x &= b_2 \\ l &\leq (x) \leq u \end{aligned} \qquad (25.84)$$

The values of Lagrangian multipliers are calculated by the solver. Penalty parameters are specified to the solver. Reduced-gradient algorithm is used to minimize the objective function.

Notation

Parameters and variables

PE_j	estimate of property j
W_j	property variable j
P_Cost_j	cost of property j
OR_j	off-spec ratio for property j
F_i	feed flow rate of component i.
BV_{ij}	property j of feed component i
C_Cost_i	cost of component i
P_Lim_j	property j limit at which giveaway is minimized
FS_k	segregation flow for rundown k
TF_b	total flow of blender b
$*\_Tgt$	target value of variable
$F(x)$	nonlinear objective function
f	nonlinear constraint function
$\bar{f}$	linearized form of nonlinear constraint function f
x	variables in nonlinear objective/constraint function
λ	vector of estimated Lagrangian multipliers
ρ	penalty parameter
A_2	coefficients of nonlinear variables x in linear constraint
b_1, b_2	equality to the constraints
l, u	lower and upper bounds on constraints

Indices

b	blenders
i	intermediate components
j	properties or qualities
k	rundown streams

Example 25.31

Maximize: $f(x) = 2x1 + x2$

Subject to: $x_1 + x_2 \leq 5$
$x_1 - x_2 \geq 0$
$6x_1 + 2x_2 \leq 21$
$x_1, x_2 \geq 0$ and integer

Solution

The Excel spreadsheet (Example 25.31.xlsx) with Solver is used to maximize the objective function to give the maximum $y = 7.75$, $x_1 = 2.75$, $x_2 = 2.25$. Figures 25.9, 25.10, and 25.11 show a spreadsheet screenshot showing a solve parameter constraints, results, and answer report, respectively, of Example 25.31.

Example 25.32

Minimize: $f(x) = x_1 + 4x_2 + 2x_3 + 3x_4$
Subject: $-x_1 + 3x_2 - x_3 + 2x_4 \geq 2$
$x_1 + 3x_2 + x_3 + x_4 \geq 3$
$x_1, x_2 \geq 0$ and integer
$x_3, x_4 \geq 0$

Solution

The Excel spreadsheet (Example 25.32.xlsx) with Solver is used to maximize the objective function to give the minimum $y = 3.83$, $x_1 = 0.5$, $x_2 = 0.833$, $x_3 = 0$, $x_4 = 0$. Figures 25.12, 25.13, and 25.14 show spreadsheet screenshots and the Solver with parameter constraints, results and answer report, respectively, of Example 25.32.

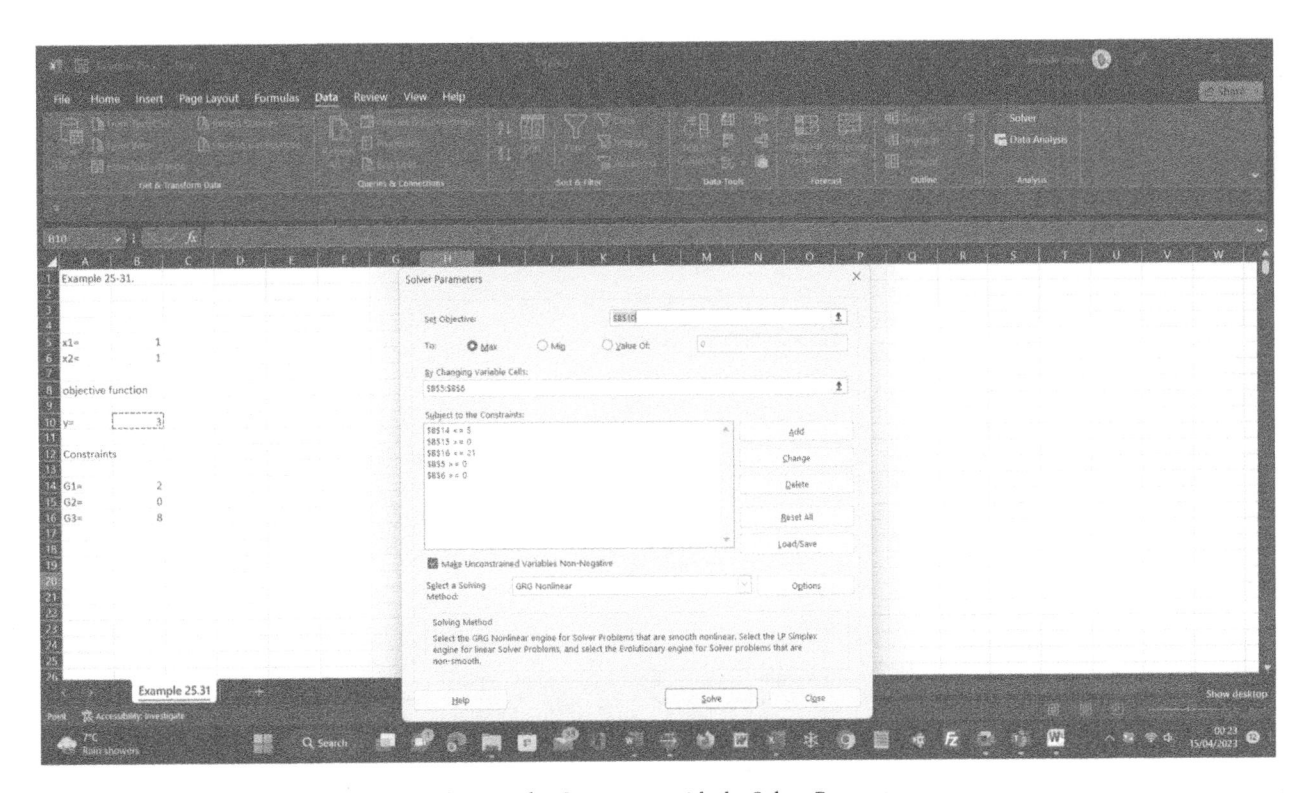

Figure 25.9 Screenshot of the Example 25.31 showing the Constraints with the Solver Parameters.

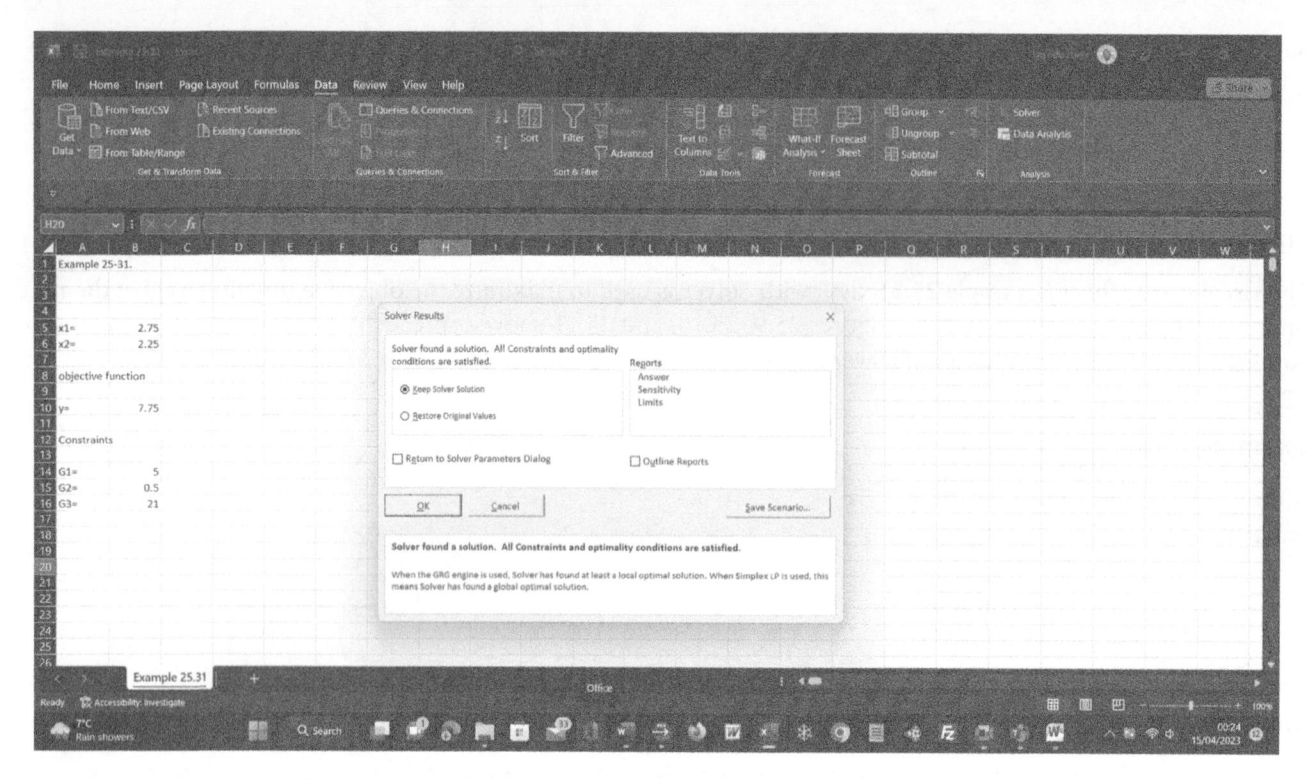

Figure 25.10 Screenshot of the Example 25.31 showing the Solver results.

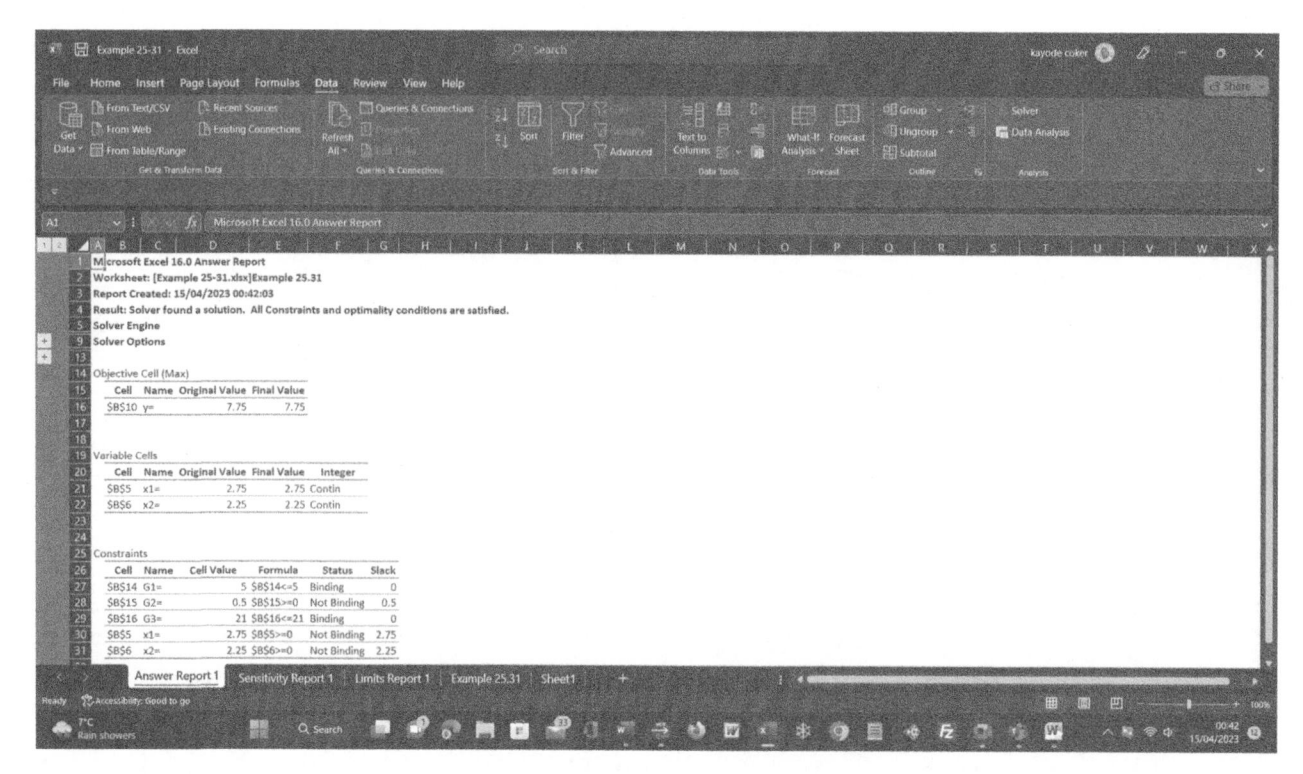

Figure 25.11 Screenshot of Answer Report of Example 25.31 from the Solver results.

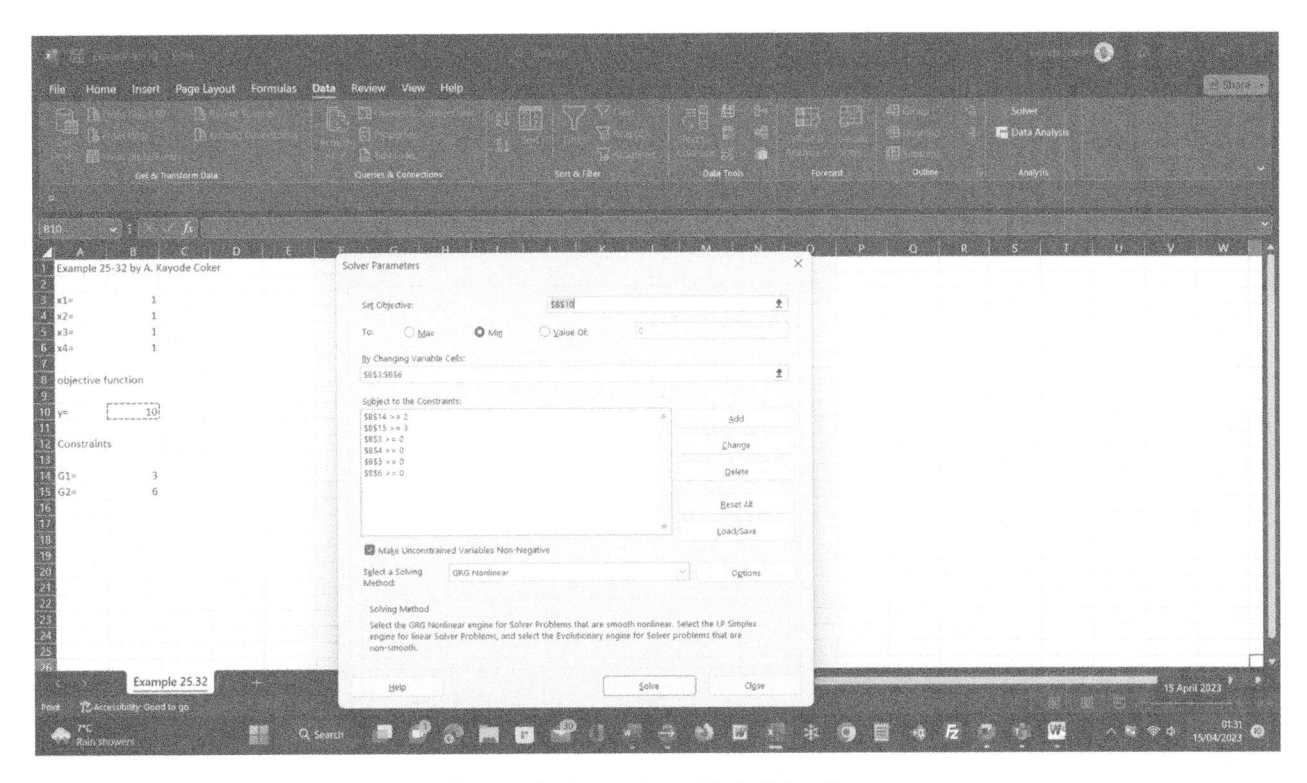

Figure 25.12 Screenshot of the Example 25.32 showing the Constraints with the Solver Parameters.

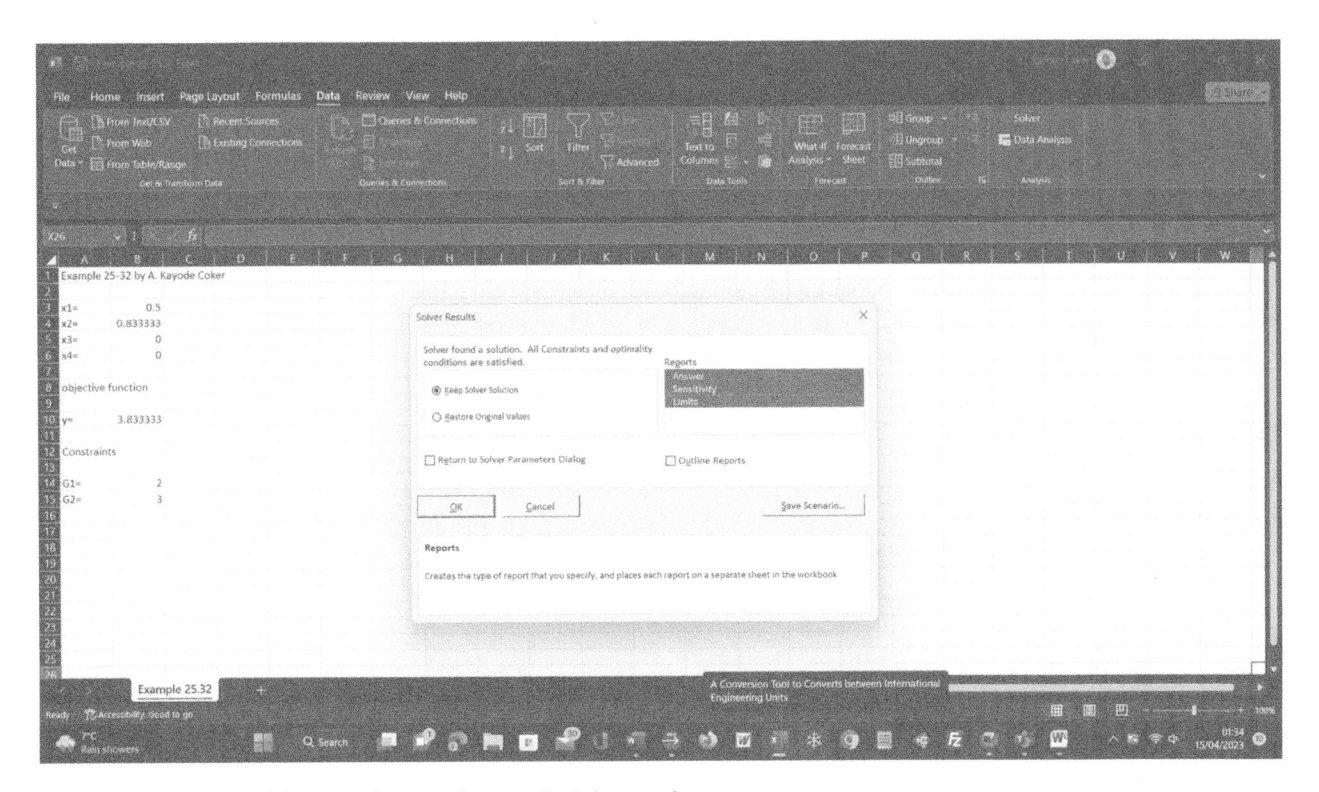

Figure 25.13 Screenshot of the Example 25.32 showing the Solver results.

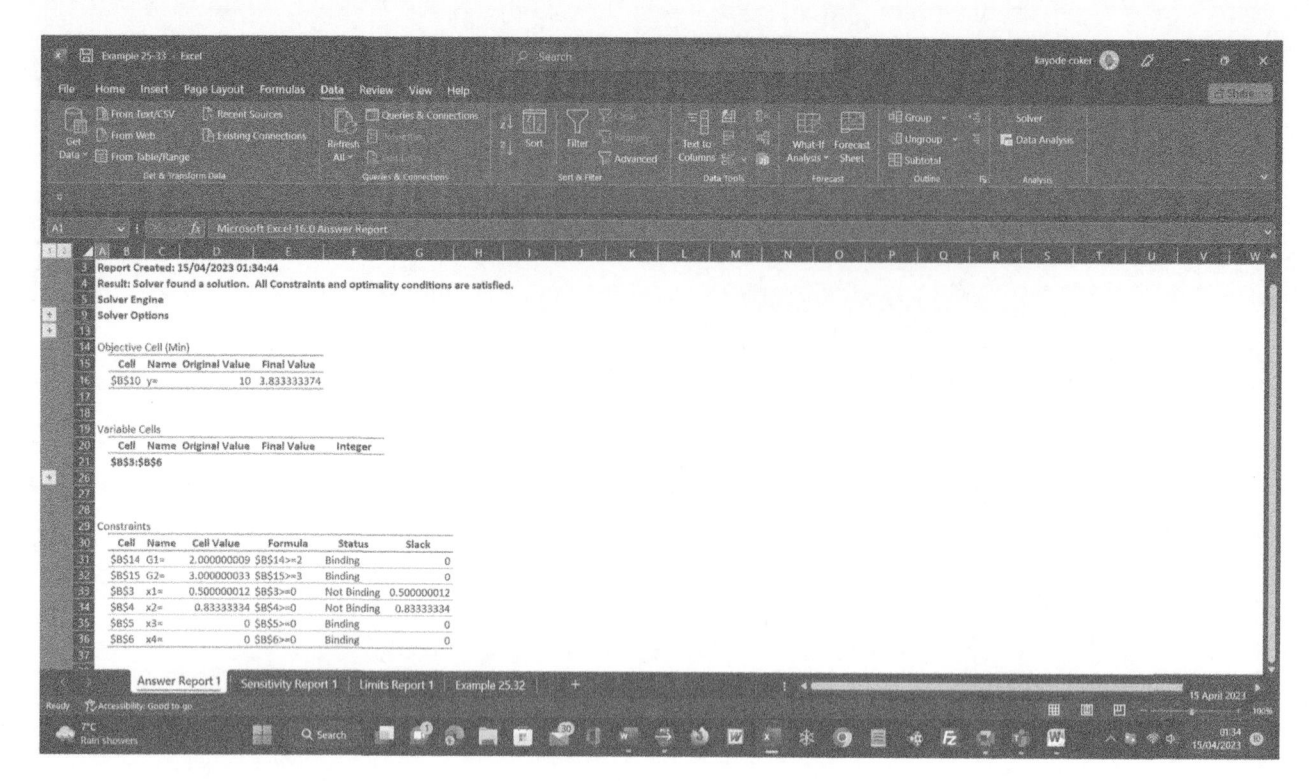

Figure 25.14 Screenshot of Answer Report of Example 25.32 from the Solver results.

Example 25.33

Maximize: $f(x) = x_1 + x_2 + x_3$
Subject to: $x_1 + 2x_2 + 2x_3 + 2x_4 + 3x_5 \leq 18$
$2x_1 + x_2 + 2x_3 + 3x_4 + 2x_5 \leq 15$
$x_1 - 6x_4 \leq 0$
$x_2 - 8x_5 \leq 0$
all $x_j \geq 0$, integer

Solution

The Excel spreadsheet (Example 25.33.xlsx) with Solver is used to maximize the objective function to give the maximum $y = 8.927$, $x_1 = 3.073$, $x_2 = 5.854$, $x_3 = 0$, $x_4 = 0.512$, $x_5 = 0.732$. Figures 25.15, 25.16, and 25.17 show spreadsheet screenshots and the Solver with parameter constraints, results, and answer report, respectively, of Example 25.33.

Example 25.34

A blend is performed of 3000 barrels of LSR gasoline and 2000 barrels of hydrocracker gasoline. Determine the amount of FCC gasoline and n-butane that must be added to the blend to produce a gasoline of 12 psi RVP and 0.525 cSt viscosity. The properties of all the components are shown in Table 25.39.

Solution

Calculate the RVP index and viscosity index using Eqs. 25.2 and 25.27, and the results are shown in Table 25.40.

We consider RVP blend as the objective function:

$$12^{1.25}(5000 + X + Y) = (3000)(12.1^{1.25}) + (2000)(15.5^{1.25}) + (X)(4.8^{1.25}) + (Y)(51.6^{1.25}) \quad (25.85)$$

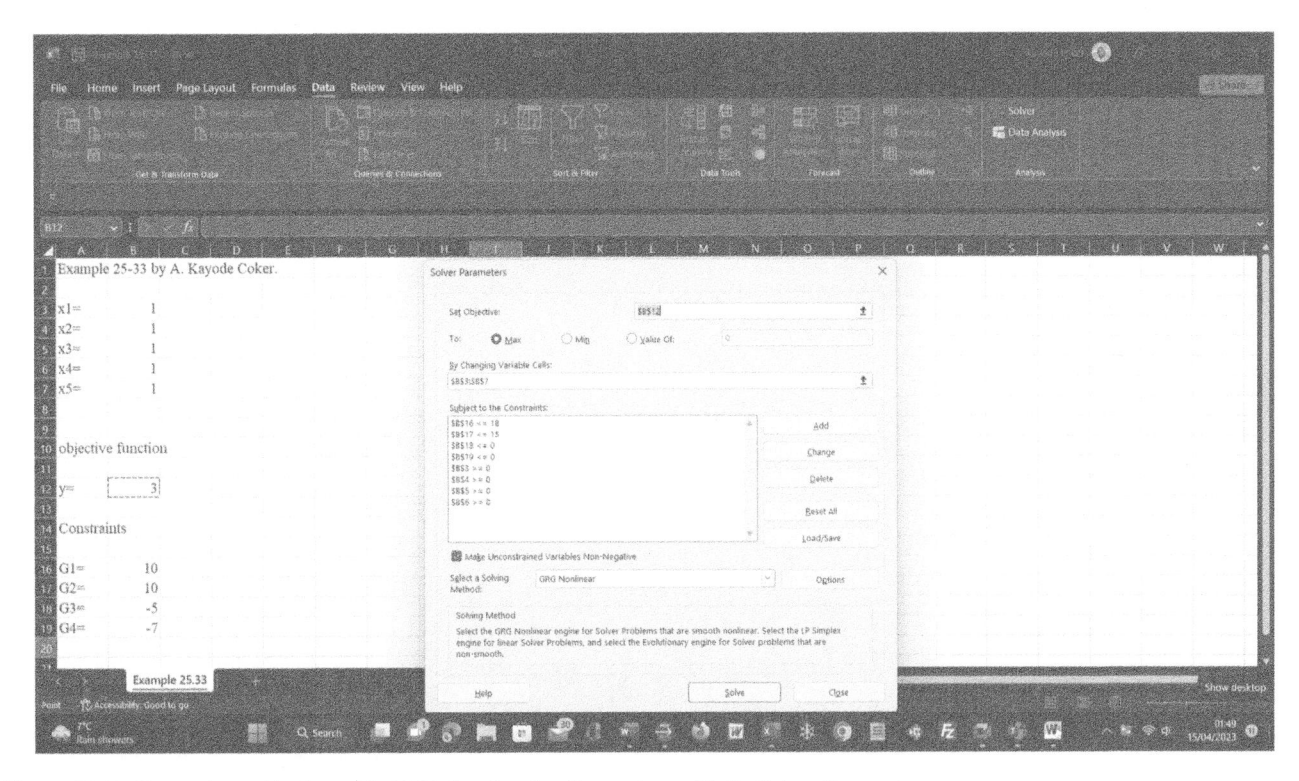

Figure 25.15 Screenshot of the Example 25.33 showing the Constraints with the Solver Parameters.

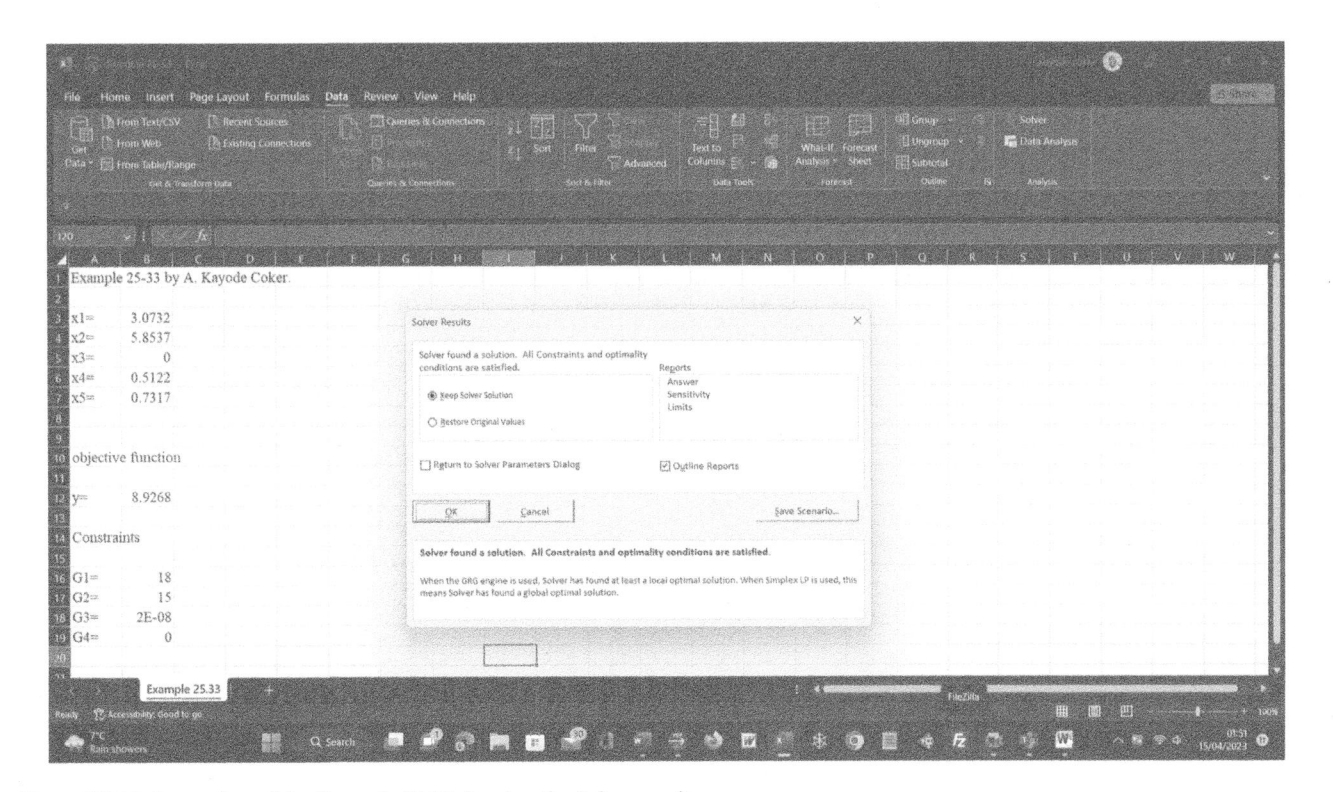

Figure 25.16 Screenshot of the Example 25.33 showing the Solver results.

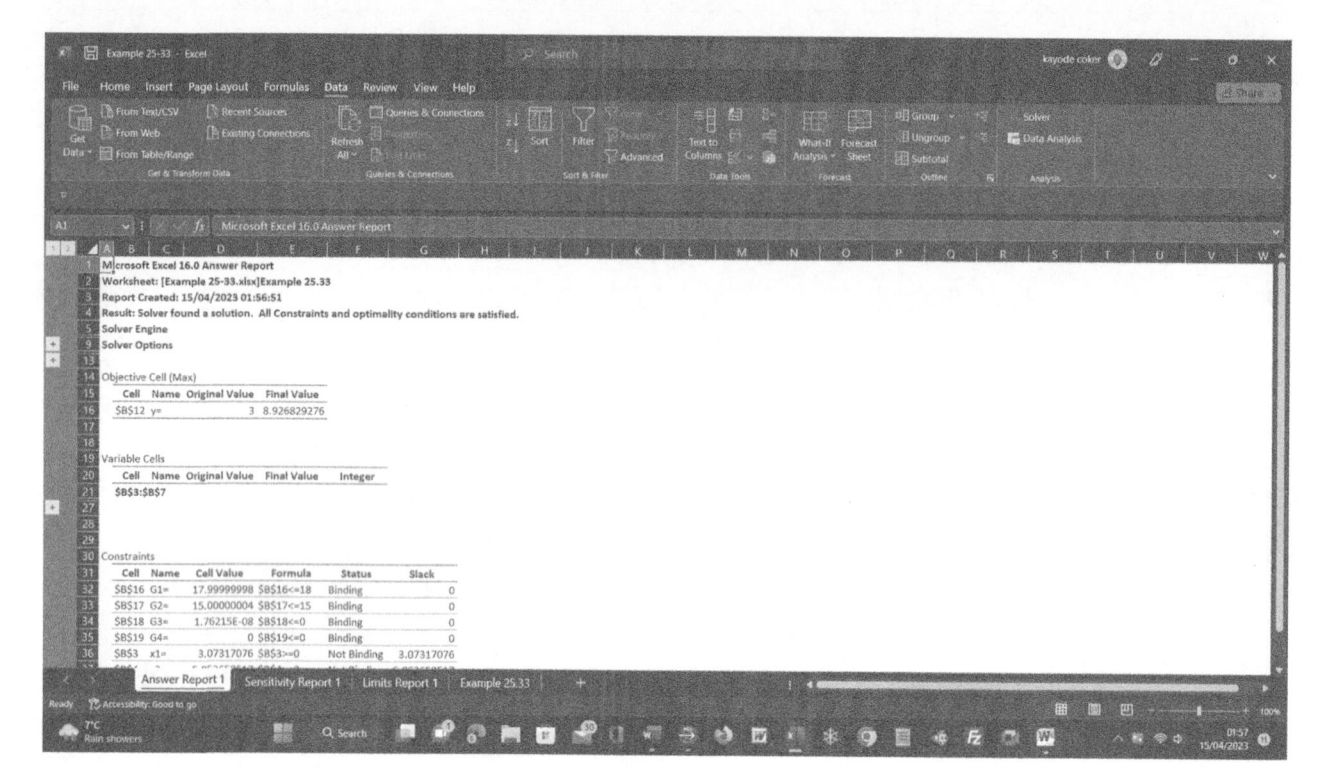

Figure 25.17 Screenshot of Answer Report of Example 25.33 from the Solver results.

Table 25.39 Component Amounts and Properties of the Blend.

Component	Barrels	RVP, psi	Viscosity, cSt
LSR gasoline	3000	12.1	0.2886
Hydrocracker gasoline	2000	15.5	0.4140
FCC gasoline	X	4.8	0.8607
n-Butane	Y	51.6	0.3500

Table 25.40 Results of Calculations of RVP and Viscosity Indices.

Component	Barrels	RVP, psi	Viscosity, cSt	BI_{RVP}	BI_{vis}
LSR gasoline	3000	12.1	0.2886	22.56	−0.21936
Hydrocracker gasoline	2000	15.5	0.4140	30.76	−0.14635
FCC gasoline	X	4.8	0.8607	7.11	−0.0222
n-Butane	Y	51.6	0.3500	138.3	−0.1792

Rearranging Eq. 25.85 gives:

$$\text{RVP Blend} = ((5000 + X + Y) * 12.1^{1.25}) - (3000 * 12.1^{1.25}) - (2000 * 15.5^{1.25})$$

$$-(X*4.8^{1.25}) - (Y*51.6^{1.25}) \quad (25.86)$$

Viscosity index of the blend from Eq. 25.27 is:

$$BI_{visi} = \frac{\log_{10}(0.525)}{3 + \log_{10}(0.525)} = -0.1028$$

We shall consider the blend viscosity at the first constraint equation given by:

$$(5000 + X + Y)(-0.1028) = (3000)(-0.21936) + (2000)(-0.14635)$$

$$+ (X)(-0.0222) + (Y)(-0.1792) \quad (25.87)$$

Rearranging Eq. 25.87 gives:

$$\text{Vis. Blend} = 0.1028 * (5000 + X + Y) + (3000 * (-0.21936)) + (2000 * (-0.14635))$$

$$+ (X * (-0.0222)) + (Y * (-0.1792)) \quad (25.88)$$

$$\text{Other constraints are } X > 0 \text{ and } Y > 0 \quad (25.89)$$

Using the Solver in the Excel spreadsheet with the object function and constraints gives:

X = 6025.899 barrels

Y = 460.1502 barrels

Solving Eqs. 25.86 and 25.88 simultaneously and algebraically as follows:

$$22.33 (5000 + X + Y) = 67680 + 61520 + 7.11X + 138.3Y$$

Or

$$111673 + 22.33X + 22.33Y - 67680 - 61520 - 7.11X - 138.3Y = 0$$

$$f(Z) = 15.22X - 115.97Y - 17527 = 0 \quad (25.86)$$

Expressing Eq. 25.88 gives:

$$-0.1028(5000 + X + Y) + (3000)(-0.21936) + 2000(-0.14635)$$

$$+ X(-0.0222) + Y(-0.1792) = 0$$

$$-514 - 0.1028X - 0.1028Y + 658.08 + 292.7 + 0.0222X + 0.1792Y = 0$$

$$-0.0806X + 0.0764Y + 436.78 = 0$$

or

$$X = \frac{436.78 + 0.0764Y}{0.0806} \qquad (25.90)$$

Substituting Eq. 25.90 into Eq. 25.86 gives

$$15.22\left(\frac{436.78 + 0.0764Y}{0.0806}\right) - 115.97Y - 17527 = 0$$

$$6647.79 + 1.1628Y - 9.347Y - 1412.676 = 0$$

$$8.184Y = 5253.114$$

$$Y = 639.86 \text{ (640 barrels)}$$

Substituting Y = 639.68 into Eq. 25.90 gives

$$X = \frac{436.78 + 0.0764(639.68)}{0.0806}$$
$$= 6025.45 \text{ barrels (6025 barrels)}$$

The Excel spreadsheet (Example 25.34.xlsx) shows the results of Example 25.34 and Figures 25.18, 25.19 and 25.20 show spreadsheet screenshots and the Solver with parameter constraints and solution, respectively.

Example 25.35

A blend is performed for LSR gasoline, hydrocracker gasoline, FCC gasoline and n-butane. Calculate the amount of each component that must be added to produce gasoline blend of maximum octane number and 12 psi RVP. The properties of each cut are shown in Table 25.41. The olefin contents in each cut is assumed to be zero. The maximum blend storage capacity is 15,000 barrels. Recalculate if the available amount of FCC gasoline is 6000 barrels.

Solution

The objective function is maximizing octane number. Since the octane contents are zero, Eq. 25.33

$$\text{RON}_{\text{Blend}} = \bar{R} + 0.03324\left(\overline{RS} - \bar{R} \times \bar{S}\right) \qquad (25.91)$$

PRODUCT BLENDING 443

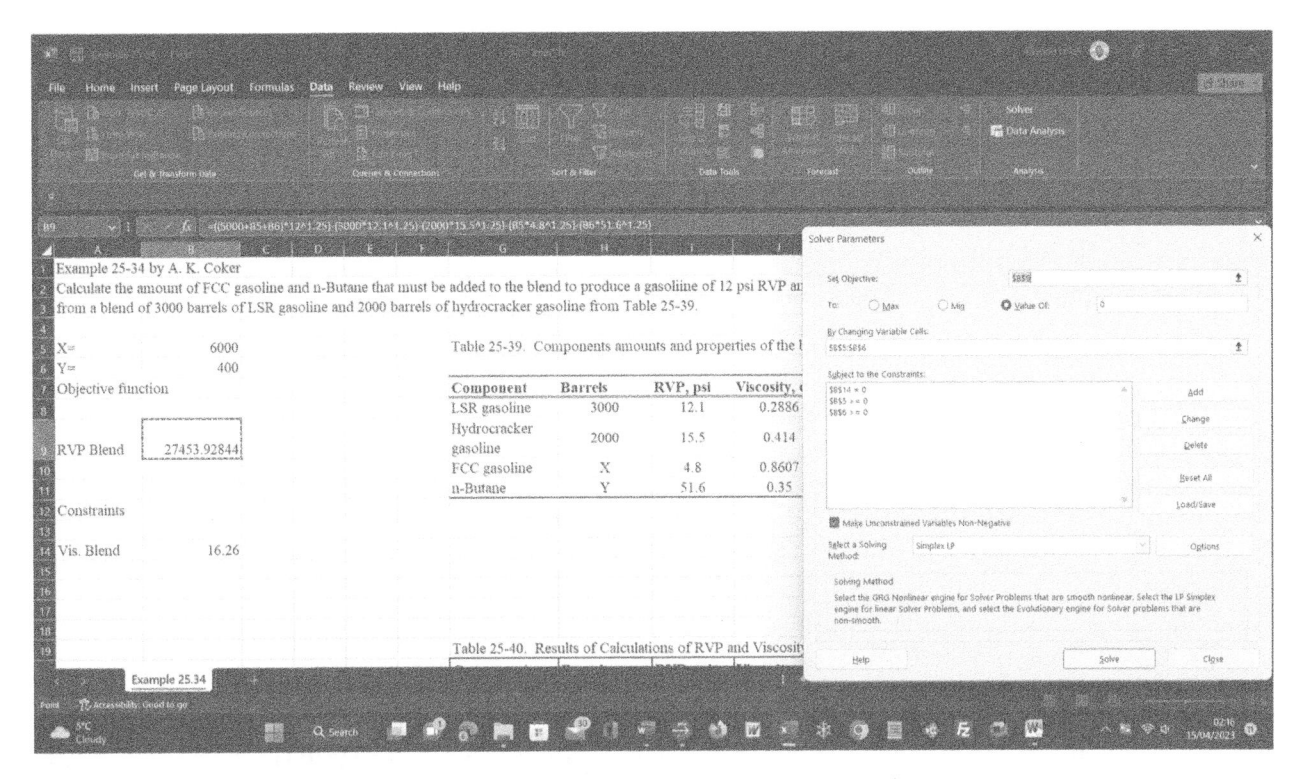

Figure 25.18 Screenshot of the Example 25.34 showing the Constraints with the Solver Parameters.

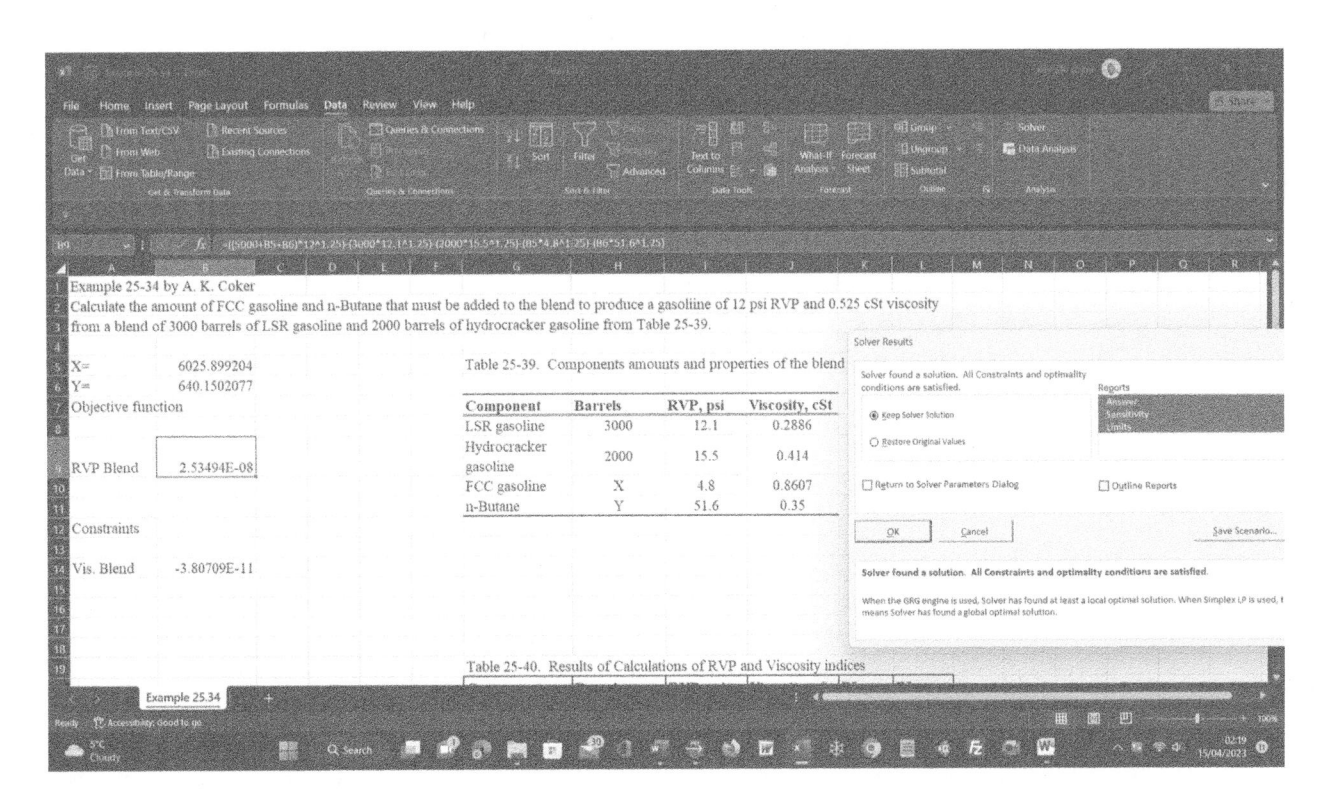

Figure 25.19 Screenshot of the Example 25.34 showing the results.

Table 25.41 Components Properties for the Blend.

Component	Barrels	RVP, psi	RON	MON
LSR gasoline	X_1	12.1	91	81
Hydrocracker gasoline	X_2	15.5	89.5	85.5
FCC gasoline	X_3	4.8	97	96
n-Butane	X_4	51.6	93	92
Blend	X_T	12	Max	

where

$$\bar{R} = (91)\frac{X_1}{X_T} + (89.5)\frac{X_2}{X_T} + (97)\frac{X_3}{X_T} + (93)\frac{X_4}{X_T} \tag{25.92}$$

$$\bar{S} = (91-81)\frac{X_1}{X_T} + (89.5-85.5)\frac{X_2}{X_T} + (97-96)\frac{X_3}{X_T} + (93-92)\frac{X_4}{X_T} \tag{25.93}$$

$$\overline{RS} = (91-81)91\frac{X_1}{X_T} + (89.5-85.5)89.5\frac{X_2}{X_T} + (97-96)\frac{X_3}{X_T} + (93-92)\frac{X_4}{X_T} \tag{25.94}$$

The following are four constraints:

$$X_1 \geq 0, X_2 \geq 0, X_3 \geq 0 \text{ and } X_4 \geq 0 \tag{25.95}$$

An additional constraint is:

$$X_1 + X_2 + X_3 + X_4 = 15{,}000 \text{ barrels} \tag{25.96}$$

The constraint equation for the RVP of the blend can be obtained by calculating the RVP index from Eq. 25.2

$$12.1^{1.25} X_1 + 15.5^{1.25} X_2 + 4.8^{1.25} X_3 + 51.6^{1.25} X_4 - 12.0^{1.25} X_T = 0 \tag{25.97}$$

or

$$22.56 X_1 + 30.755 X_2 + 7.105 X_3 + 138.3 X_4 - 22.33 X_T = 0 \tag{25.98}$$

Using the solver in Excel spreadsheet (Example 25.35.xlsx) to maximize RON, the objective function of Eq. 25.91 with the above constraints Eqs. 25.95, 25.96, and 25.97, the results are:

RON = 96.53, and X_1 = 0, X_2 = 0, X_3 = 13258.69 and X_4 = 1741.31 barrels.

Figures 25.21–25.23 show snapshots of the Excel spreadsheet showing the Solver constraints and solution, respectively.

PRODUCT BLENDING   445

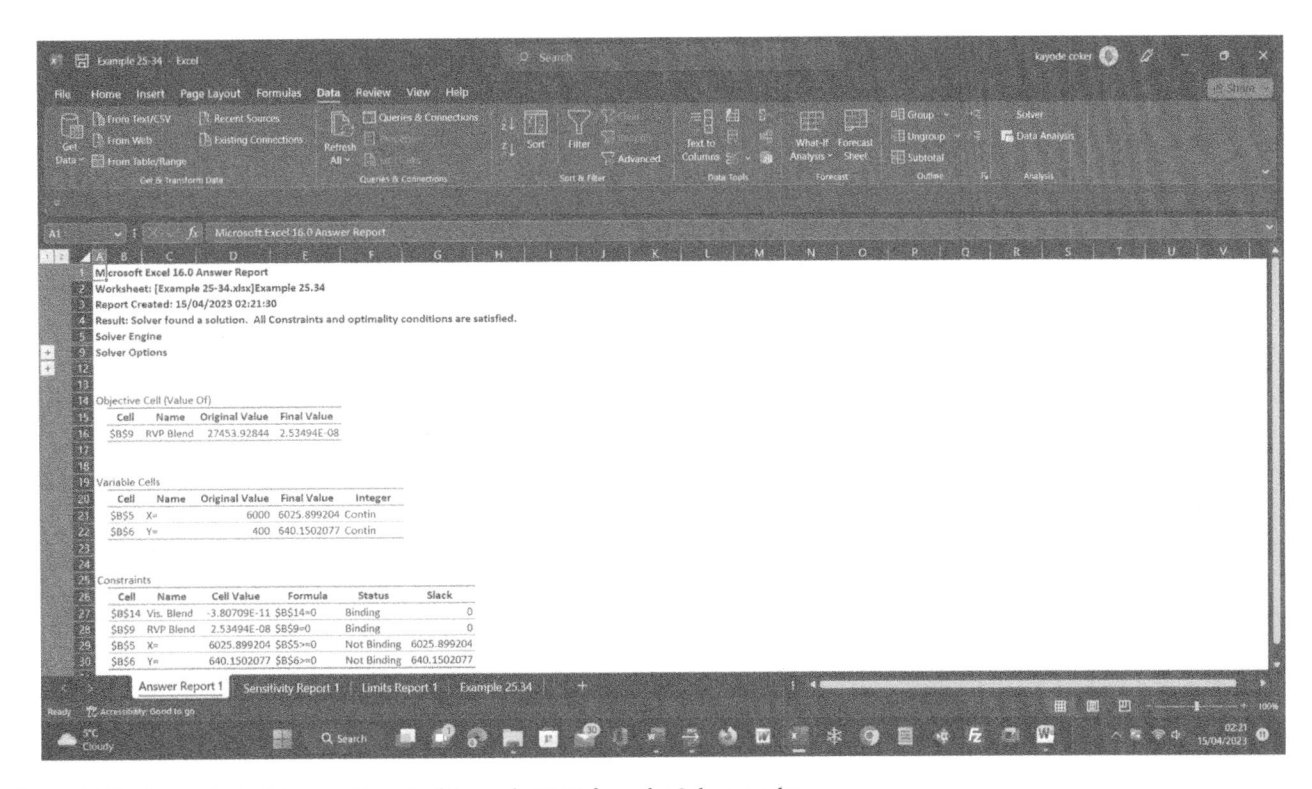

Figure 25.20 Screenshot of Answer Report of Example 25.34 from the Solver results.

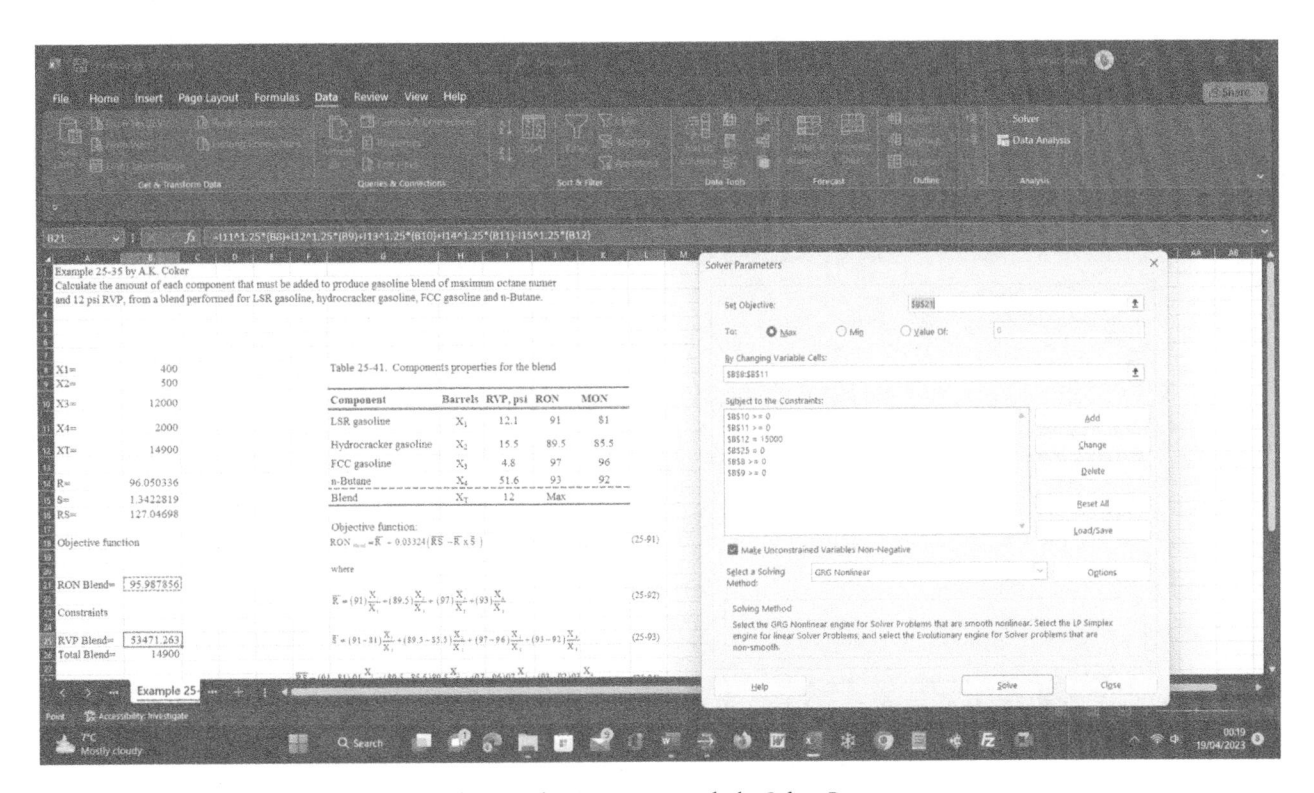

Figure 25.21 Screenshot of the Example 25.35 showing the Constraints with the Solver Parameters.

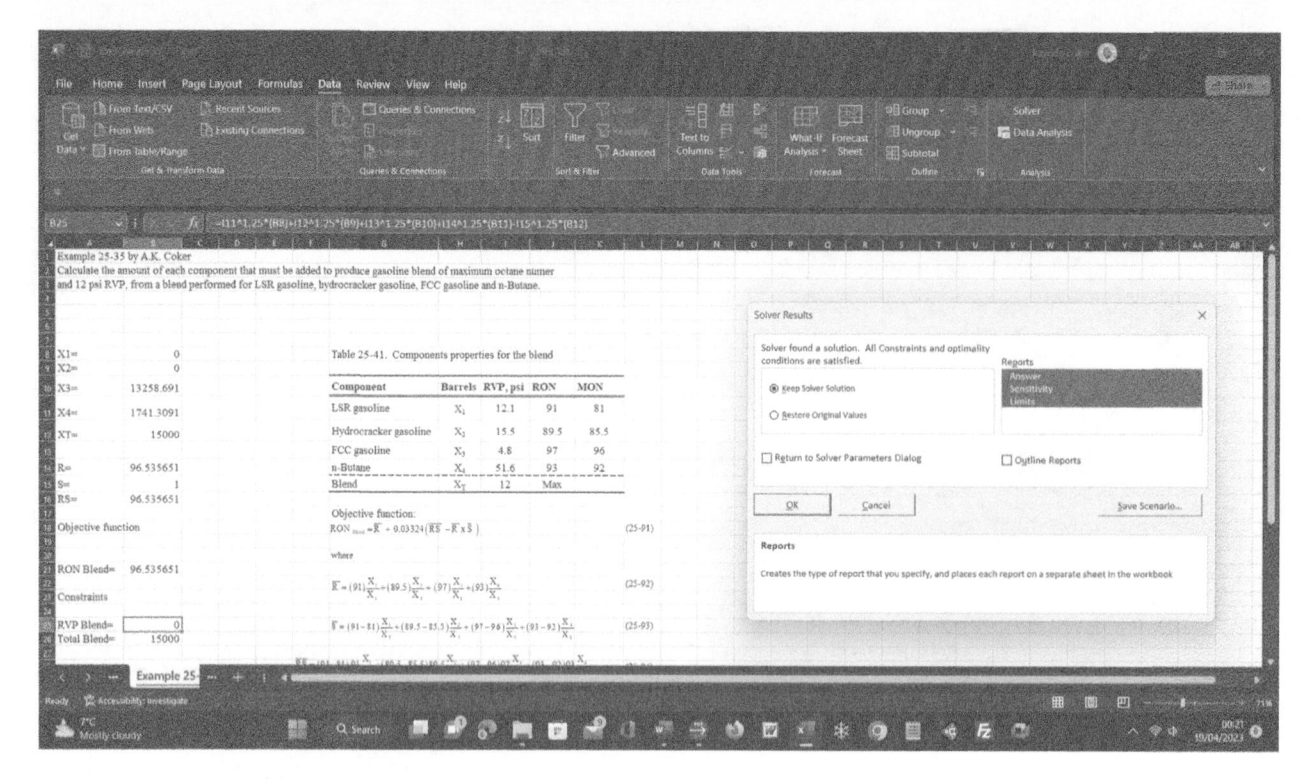

Figure 25.22 Screenshot of the Example 25.35 showing the Solver results.

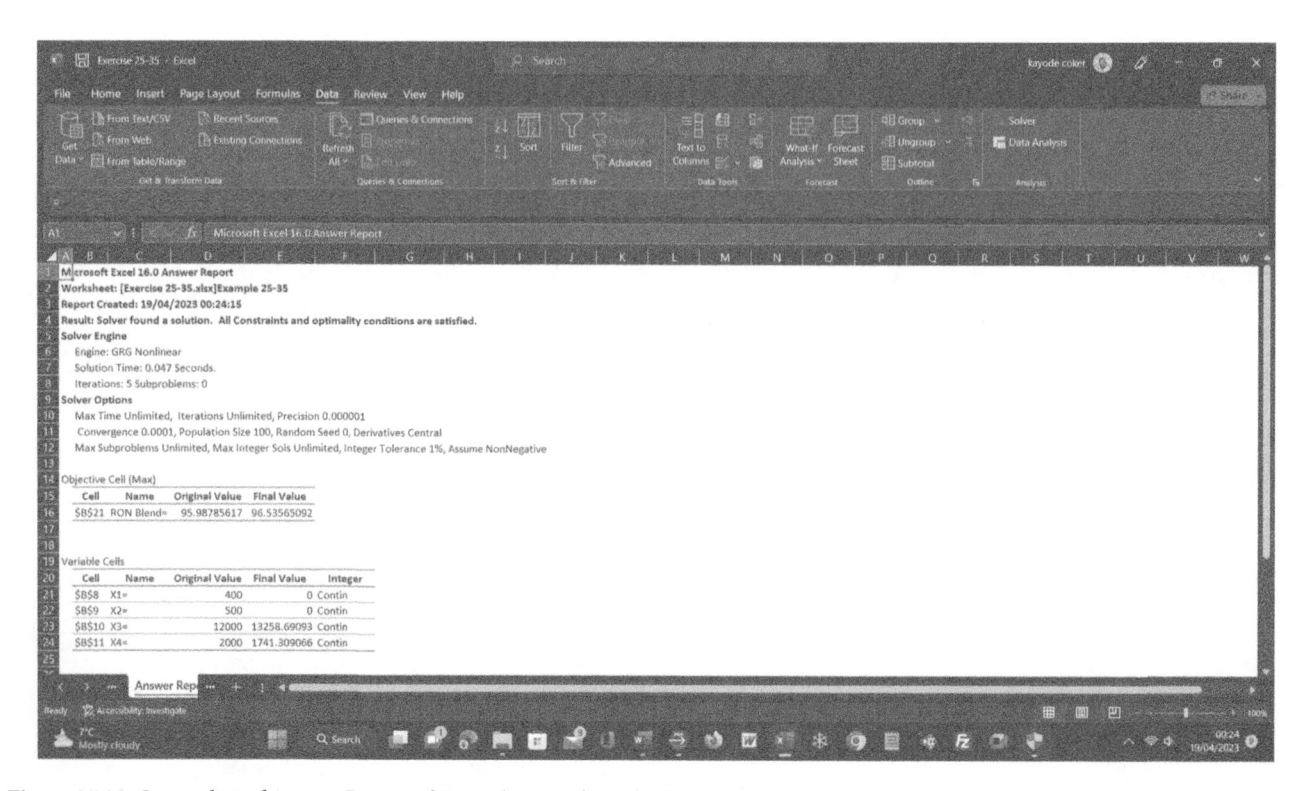

Figure 25.23 Screenshot of Answer Report of Example 25.35 from the Solver results.

PRODUCT BLENDING 447

If the constraint for FCC gasoline (6000 barrels) is added, the results for the maximum blend storage are:

$$\text{RON} = 93.50 \text{ and } X_1 = 0, X_2 = 8228.5, X_3 = 6000 \text{ and } X_4 = 771.5 \text{ barrels.}$$

Figures 25.24–25.27 show the Excel spreadsheet calculations of the results with FCC=6000 barrels.

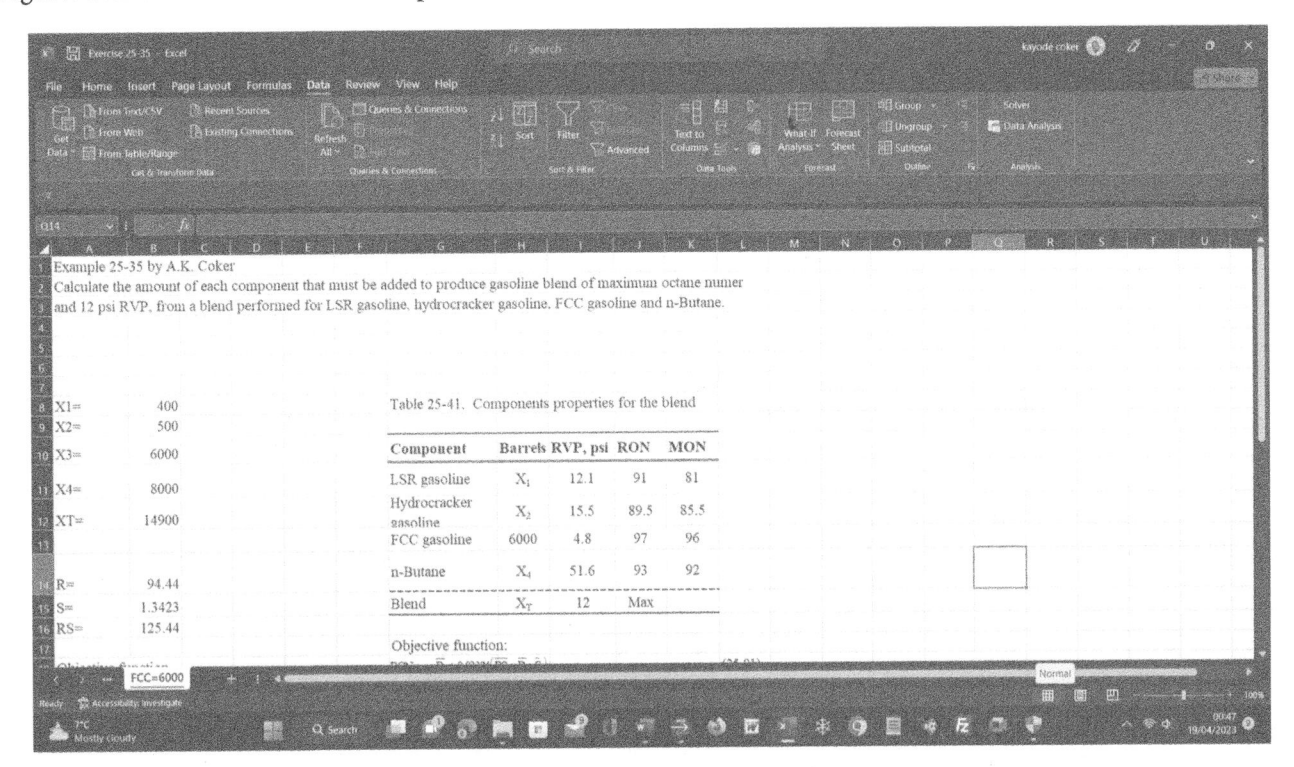

Figure 25.24 Screenshot of Answer Report of Example 25.35 with FCC = 6000 barrels.

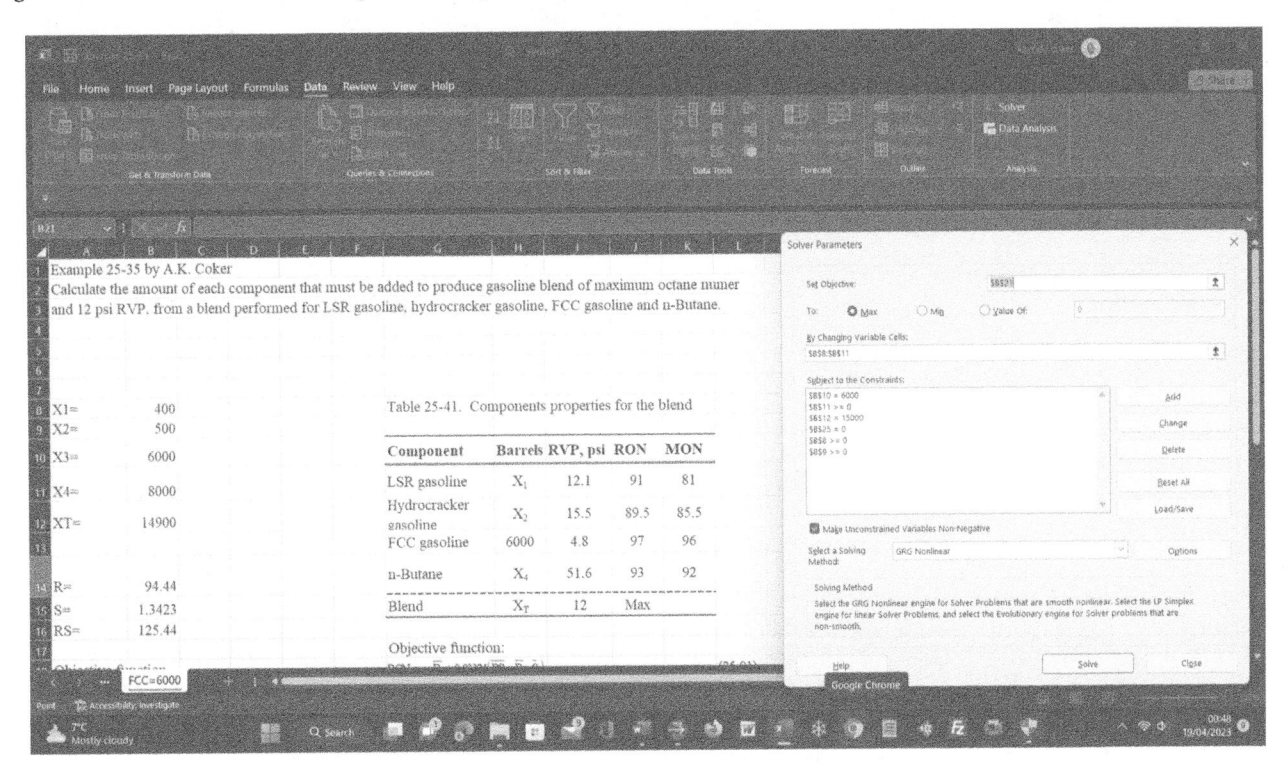

Figure 25.25 Screenshot of Example 25.35 with FCC = 6000 barrels with the constraints with the Solver Parameters.

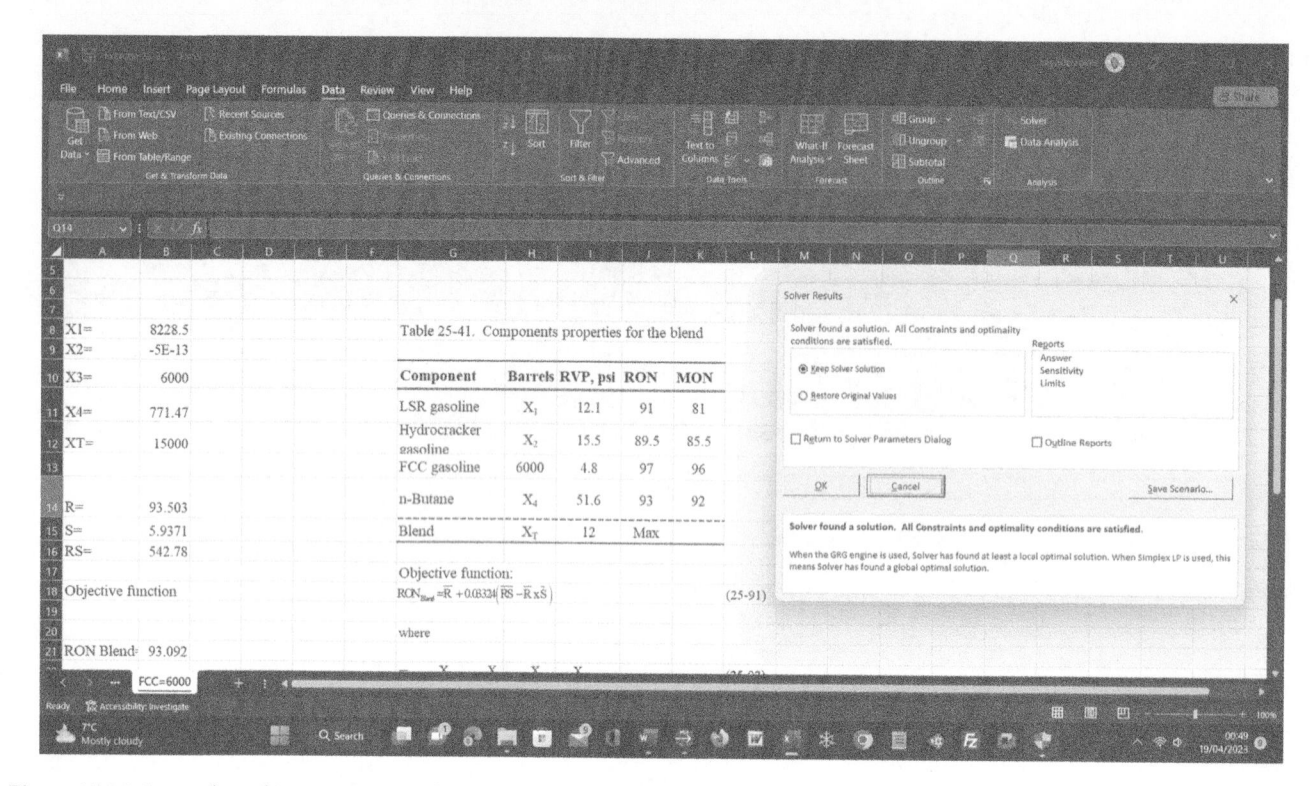

Figure 25.26 Screenshot of Answer Report of Example 25.35 from the Solver results with FCC = 6000 barrels.

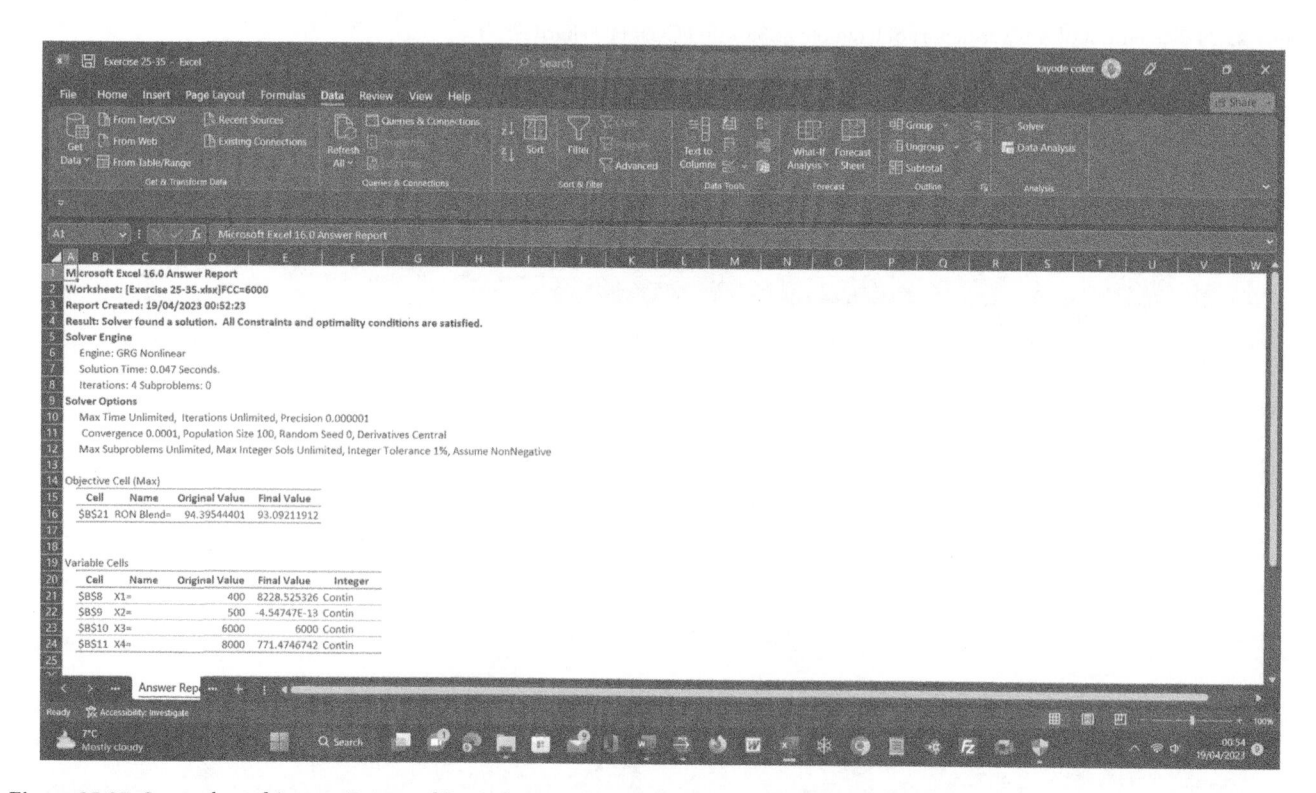

Figure 25.27 Screenshot of Answer Report of Example 25.35 from the Solver results with FCC = 6000 barrels.

Example 25.36

Calculate the amount of each blending stock that would produce a 400,00 bbls gasoline product with the following specifications, °API = 70, ON = 95, RVP = 9 psig max. Available blends are as follows:

	Component	Volume bbl	°API	ON	RVP psi	SpGr	RVPI = RVP$^{1.25}$
Tank 1	Reformate	500,000	70	94	10	0.7022	17.7828
Tank 2	Isomerate	400,000	69	92	9	0.7057	15.5885
Tank 3	Alkylate	600,000	72	96	8	0.6953	13.4543
	Desired Blend	400,000	70	96	9	0.7022	15.5885

$$°API = \frac{141.5}{SpGr} - 131.5$$

or

$$SpGr = \frac{141.5}{(°API + 131.5)}$$

Solution

Let

N1 = bbls of tank 1 Reformate

N2 = bbls of tank 2 Isomerate

N3 = bbls of tank 3 Alkylate

Objective function: N1 + N2 + N3 = 400,000

Constraints

N1 ≤ 500,000

N2 ≤ 400,000

N3 ≤ 600,000

N1, N2, N3 ≥ 0 (non-negative)

for API gravity 0.7022 X1 + 0.7057 X2 + 0.6953X3 ≤ 0.7022
for ON   94 X1 + 92 X2 + 96X3 ≥ 95
for RVP   17.7828 X1 + 15.5885X2 + 13.4543 ≤ 15.5885

where

X1 = N1/(N1 + N2 + N3)

$$X2 = N2/(N1 + N2 + N3)$$

$$X3 = N3/(N1 + N2 + N3)$$

The Excel spreadsheet Example 25.36.xlsx shows the calculation using Solver to determine the amount of each blending stock that would produce 400,000 bbls gasoline. The Solver solution indicates that

$$N1 = 133,333.4$$

$$N2 = 33,332.83$$

$$N3 = 233,333.8$$

Blend properties:

API = 71.1

ON = 95.0

RVP = 8.76 psig.

Figures 25.28–25.31 show the screenshots of the Solver with the Constraint Parameters and the Answer report of Example 35.36.

A Case Study

An oil refinery produces a stream of hydrocarbon base fuel from each of four different processing units [34]. The processing units have capacities of C_1, C_2, C_3, and C_4 barrels per day, respectively. A portion of each base fuel is fed to a central blending station where the base fuels are mixed into three grades of gasoline. The remainder of each base

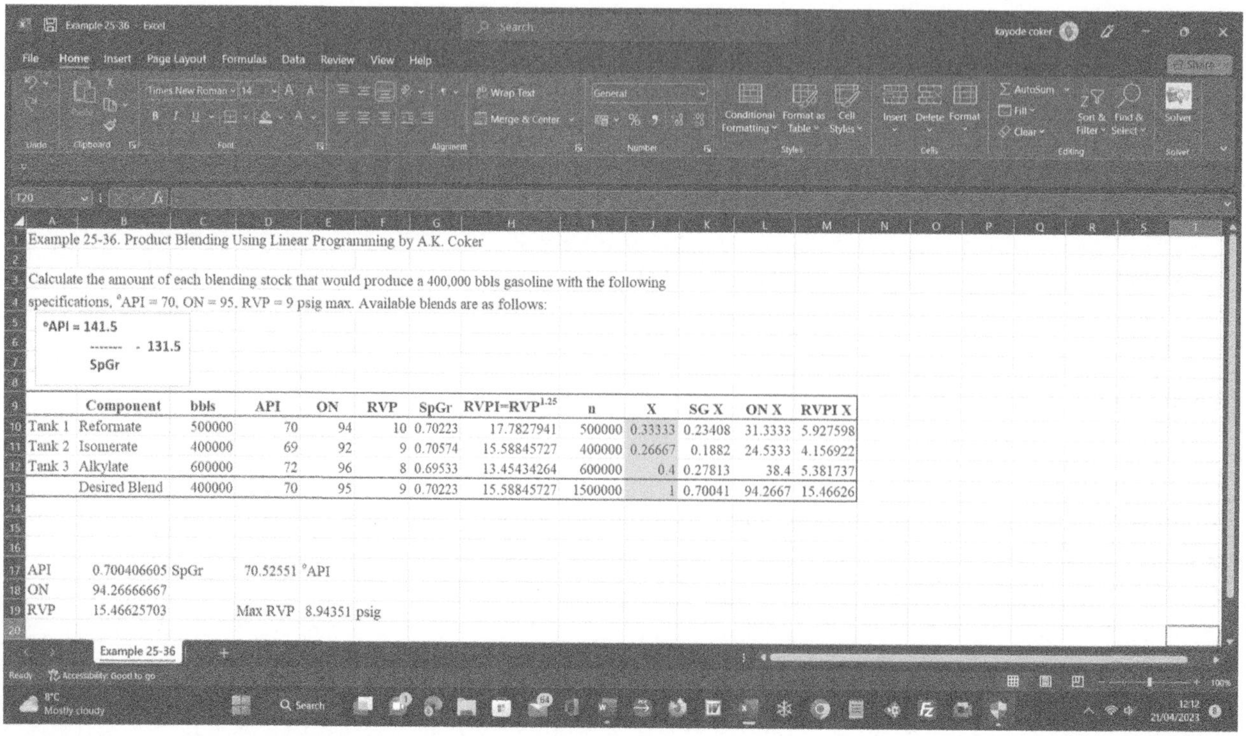

Figure 25.28 Screenshot of Example 25.36.

PRODUCT BLENDING 451

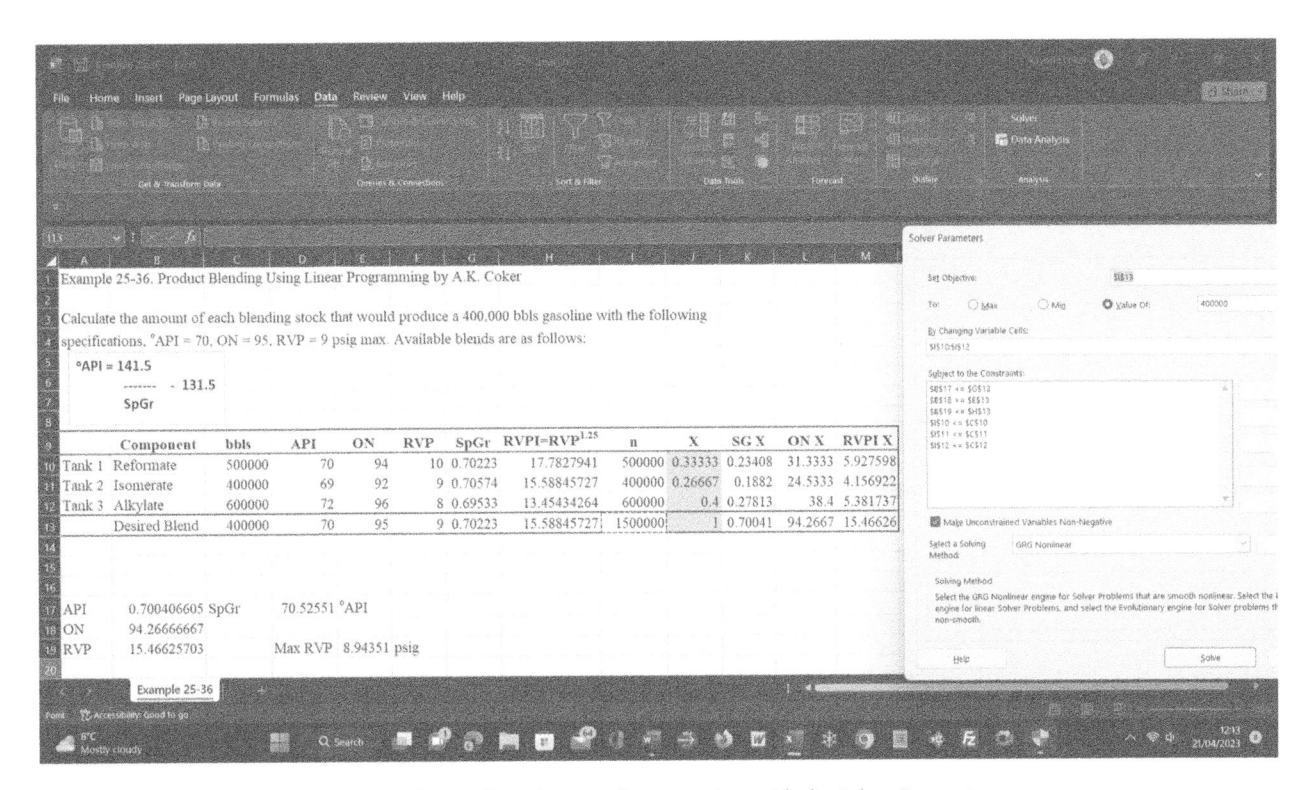

Figure 25.29 Screenshot of Example 25.36 from Solver showing the constraints with the Solver Parameters.

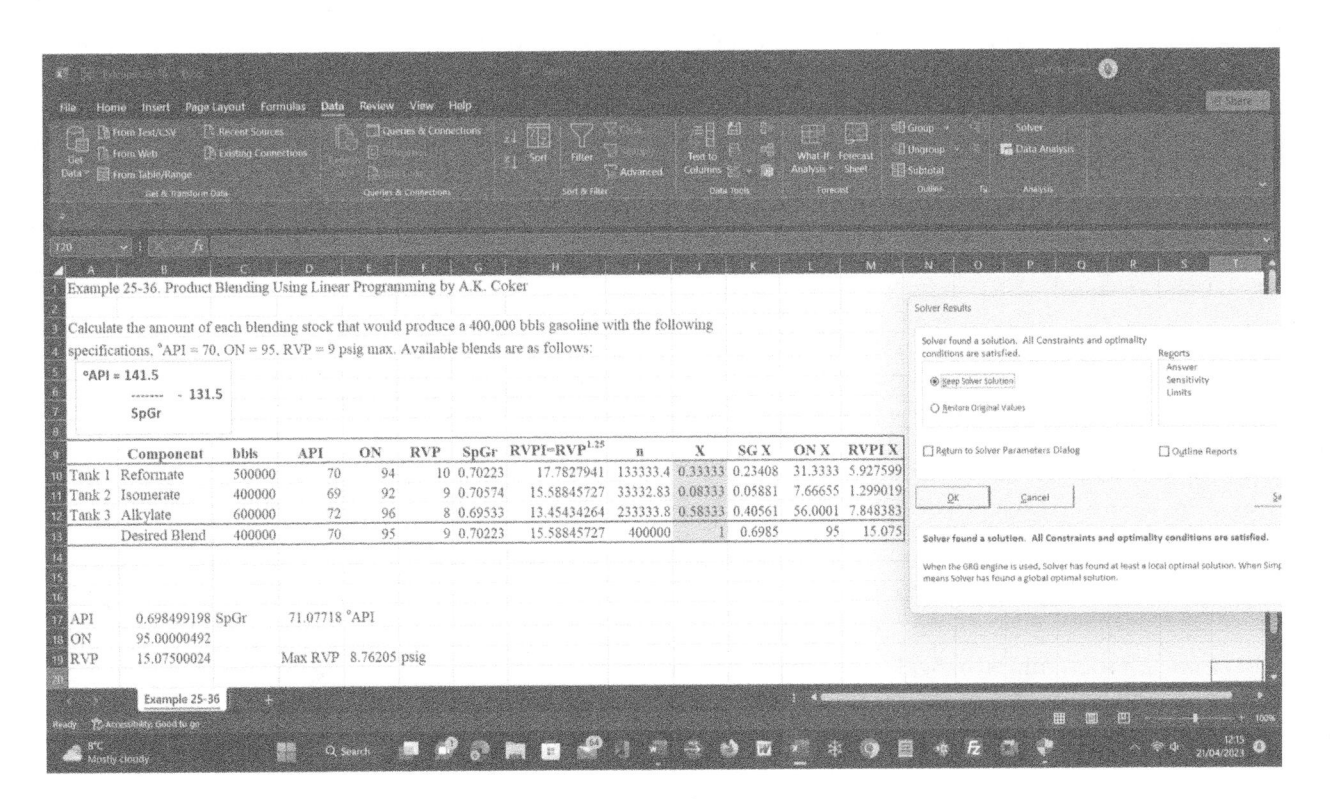

Figure 25.30 Screenshot of Example 25.36 from Solver showing the results.

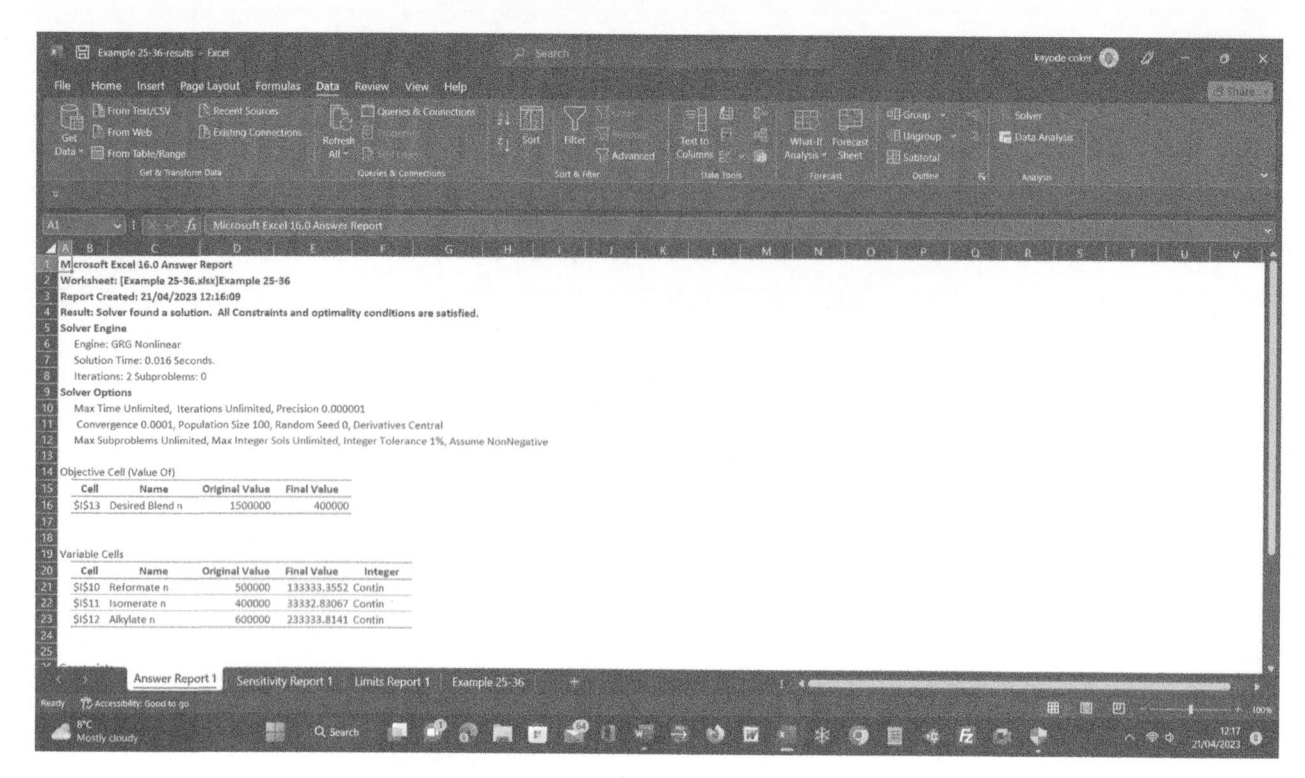

Figure 25.31 Screenshot of Answer Report of Example 25.36.

fuel is sold "as is" at the refinery. Storage facilities are available at the blending station for temporary storage of the blended gasolines, if required.

Let N_1, N_2, N_3, and N_4 be the octane ratings of the respective base fuels, and let S_1, S_2, S_3, and S_4 be the profit derived from their direct sale. Let O_1, O_2, and O_3 be the octane numbers of the three grades of gasoline that must be blended on any given day to satisfy customer demands D_1, D_2, and D_3. Let R_1, R_2, and R_3 be the minimum octane requirements of the three grades of blended gasoline and let P_1, P_2, and P_3 be the profit per gallon that is realized from the sale of each grade of blended gasoline. Finally, let xij be the quantity of the ith base fuel used to blend the jth gasoline.

The octane number of each gasoline can be expressed as the weighted average of the octane numbers of the constituent base fuels. The weighting factors are the fractions of the base fuels in each gasoline. Thus,

$$O_1 = \frac{x_{11}N_1 + x_{21}N_2 + x_{31}N_3 + x_{41}N_4}{x_{11} + x_{21} + x_{31} + x_{41}}$$

$$O_2 = \frac{x_{12}N_1 + x_{22}N_2 + x_{32}N_3 + x_{42}N_4}{x_{12} + x_{22} + x_{32} + x_{42}}$$

$$O_3 = \frac{x_{13}N_1 + x_{23}N_2 + x_{33}N_3 + x_{43}N_4}{x_{13} + x_{23} + x_{33} + x_{43}}$$

Develop a linear programming model to determine how much of each base fuel should be blended into gasoline and how much should be sold "as is" in order to maximize profit. Solve the model using the following numerical data:

$C_1 = 13{,}000$ bbl/day $N_1 = 82$ octane $S_1 = \$0.90$/bbl

$C_2 = 7{,}000$ $N_2 = 95$ $S_2 = 1.05$

$C_3 = 25{,}000$ $N_3 = 102$ $S_3 = 1.25$

$C_4 = 15{,}000$ $N_4 = 107$ $S_4 = 1.60$

$D_1 = 13{,}000$ bbl/day $R_1 = 87$ octane $P_1 = 3.5$ ¢/gal

$D_2 = 25{,}000$ $R_2 = 89$ $P_2 = 4.5$

$D_3 = 18{,}000$ $R_3 = 93$ $P_3 = 6.0$

Note: 42 gal = 1 bbl

Solution

Maximize $y = P_1(x_{11} + x_{21} + x_{31} + x_{41}) + P_2(x_{12} + x_{22} + x_{32} + x_{42})$
$+ P_3(x_{13} + x_{23} + x_{33} + x_{43}) + (S_1/42)(42C_1 - x_{11} - x_{12} - x_{13})$
$+ (S_2/42)(42C_2 - x_{21} - x_{22} - x_{23}) + (S_3/42)(42C_3 - x_{31} - x_{32} - x_{33})$
$+ (S_4/42)(42C_4 - x_{41} - x_{42} - x_{43})$

Subject to $x_{11} + x_{21} + x_{31} + x_{41} \geq 42D_1$
$x_{12} + x_{22} + x_{32} + x_{42} \geq 42D_2$
$x_{13} + x_{23} + x_{33} + x_{43} \geq 42D_3$
$x_{11}N_1 + x_{21}N_2 + x_{31}N_3 + x_{41}N_4 \geq R_1(x_{11} + x_{21} + x_{31} + x_{41})$
$x_{12}N_1 + x_{22}N_2 + x_{32}N_3 + x_{42}N_4 \geq R_2(x_{12} + x_{22} + x_{32} + x_{42})$
$x_{13}N_1 + x_{23}N_2 + x_{33}N_3 + x_{43}N_4 \geq R_3(x_{13} + x_{23} + x_{33} + x_{43})$

where

x_{11} = gallons of base fuel 1 used in gasoline 1
x_{21} = gallons of base fuel 2 used in gasoline 1
x_{31} = gallons of base fuel 3 used in gasoline 1
x_{41} = gallons of base fuel 4 used in gasoline 1
x_{12} = gallons of base fuel 1 used in gasoline 2
x_{22} = gallons of base fuel 2 used in gasoline 2
x_{32} = gallons of base fuel 3 used in gasoline 2
x_{42} = gallons of base fuel 4 used in gasoline 2
x_{13} = gallons of base fuel 1 used in gasoline 3
x_{23} = gallons of base fuel 2 used in gasoline 3
x_{33} = gallons of base fuel 3 used in gasoline 3
x_{43} = gallons of base fuel 4 used in gasoline 3

The model assumes full capacity production.

The Excel spreadsheet (Optimization-Case-Study-akc.xlsx) using the Solver: What-if—analysis tool that finds the optimal value of a target cell by changing values in cells used to calculate the target cell.

The results of the optimization shown in the worksheet Case-study-1a are:

Solution: y_{max} = 111,836,36 at x_{11} = 436,800, x_{41} = 109,200, x_{12} = 756,000, x_{42} = 294,000

x_{13} = 423,360, x_{43} = 332,640, all other x_{ij} = 0.

All x_{ij} expressed in terms of gallons per day.

How would the profit and the optimum policy be affected if the selling prices of the gasolines were changed to 3 ¢/bbl, 4.5 ¢/bbl (as before) and 5.5 ¢/bbl, respectively? What would be the consequences of decreasing D2 by 7000 bbl/day while simultaneously increasing D3 by 2000 bbl/day?

The results of the optimization shown in the worksheet Case-study-1b are:

Solution: y_{max} = 10,805,636 at x_{11} = 436,800, x_{41} = 109,200, x_{12} = 756,000, x_{42} = 294,000

x_{13} = 423,360, x_{43} = 332,640, all other x_{ij} = 0.

Table 25.42 LINGO Input and Output for the Case Study [37].

```
Input
! Given;
P1=3.5;
P2=4.5;
P3=6;
S1=0.9;
S2=1.05;
S3=1.25;
S4=1.60;
C1=13000;
C2=7000;
C3=25000;
C4=15000;
N1=82;
N2=95;
N3=102;
N4=107;
D1=13000;
D2=25000;
D3=18000;
R1=87;
R2=89;
R3=93;
! Maximize;
max=P1*(X11 + X21 + X31 + X41) + P2*(X12 + X22 + X32 + X42) + P3*(X13 + X23 + X33 + X43) + (S1/42) *(42*C1 - X11
    - X12 - X13)+ (S2/42) *(42*C2 - X21 - X22 - X23) + (S2/42) *(42*C3 - X31 - X32 -X33)+ (S4/42) *(42*C4 - X41 - X42
    - X43);
! Subject to;
X11 + X21 + X31 + X41 = 42*D1;
X12 + X22 + X32 + X42 = 42*D2;
X13 + X23 + X33 + X43 = 42*D3;
X11*N1 + X21*N2 + X31*N3 + X41*N4 >= R1 *(X11 + X21 + X31 + X41);
X12*N1 + X22*N2 + X32*N3 + X42*N4 >= R2 *(X12 + X22 + X32 + X42);
X13*N1 + X23*N2 + X33*N3 + X43*N4 >= R3* (X13 + X23 + X33 + X43);
```

All x_{ij} expressed in terms of gallons per day.

What would be the consequences of decreasing D2 by 7000 bbl/day while simultaneously increasing D3 by 2000 bbl/day?

The results of the optimization shown in the worksheet Case-study-1c are:

Solution: y_{max} = 3,788,692 at x_{11} = 436,800, x_{41} = 109,200, x_{12} = 211,680, x_{42} = 82,320

x_{13} = 47,040, x_{43} = 36,960, all other x_{ij} = 0.

All x_{ij} expressed in terms of gallons per day

Alsuhaibani [37] employed a linear optimization programming method (LINGO) for the case study. Table 25.42 shows the code and results and are in good agreement with the results obtained from the Excel spreadsheet.

25.21 Environmental Concern of Gasoline Blending

There are growing environmental issues of air pollution of diesel vehicles and other blending fuel resulting in fatalities and impairment of people's health worldwide. Ethanol is widely used as an oxygenate that enhances the fuel combustion and reduces exhaust emissions. However, the addition of ethanol in the U.S. gasoline as a means of enhancing renewable fuel content results in an increase in CO_2 emissions. Ethanol is also a cost-effective high-octane fuel component enabling fuel refiners to substitute lower-cost, lower-octane hydrocarbons in the blend stock while maintaining the octane ratings of the finished fuel. A gasoline blend stock specifically formulated to be blended with ethanol is referred to as a blend stock for oxygenated blending (BOB).

The US Environmental Protection Agency (EPA) has conducted air toxic modeling for vehicles fuels and concluded that an increase in ethanol volume in fuels causes an increase in toxic emissions both in the running and starting of the engine. Although, neat ethanol is a non-toxic additive to gasoline, while gasoline, particularly the aromatic portion of each gallon is toxic, carcinogenic and mutagenic.

Gasoline was originally oxygenated with lead, despite evidence of lead's toxicity stretching back more than a century; it remained in the gasoline supply until the Clean-Air Act Amendments of 1990, which mandated its removal by 1995. When lead was finally removed from gasoline, the oil industry insisted on a petroleum product as a replacement such as methyl tertiary butyl ether (MTBE). MTBE was eventually replaced due to groundwater and soil contamination concerns by the BTEX complex (benzene, toluene, ethylbenzene and xylene). BTEX, particularly benzene was also known to be toxic when it was added to gasoline. Today, millions live near roadways where the toxic BTEX complex is burned by gasoline vehicles. Children and fetal development are especially sensitive to vehicle pollution, but no one is immune to its effects.

Table 25.43 shows the parameters of selected gasoline for global markets, where the range of the research octane number (RON) as suggested by the World Fuels Charter (WFC) for gasoline is 91–98. The range covers global gasoline octanes except for China V gasoline of the minimum 81 RON. Most of the gasoline limits in the table were established to reduce tailpipe emissions from automotive vehicles that are harmful to the atmosphere, humans and animals resulting in drastic global climate change.

A catalytic converter is a technology that is installed in automotive exhaust systems with the goal of minimizing tailpipe emissions. Unreacted hydrocarbons or particulates, volatile organic compound (VOC), carbon monoxide (CO), nitrogen oxides (NO_x) and sulfur oxides are emitted from automotive vehicles. These converters are emission control devices that reduce toxic gases and pollutants in the exhaust gas from an internal-combustion engine into less toxic pollutants by catalyzing a redox reaction (i.e., an oxidation and a reduction reaction), installed to convert harmful gases and hydrocarbons in the exhaust gas. The two-way converters remove carbon monoxide (CO) and unburned hydrocarbons (C_xH_x) by combining oxygen (O_2) with CO and unburned C_xH_x to produce carbon dioxide (CO_2) and water (H_2O).

Table 25.43 Parameters for Selected Gasoline for Global Markets [15].

Fuel parameter		China V	Bharat IV (1)	Euro IV	Euro V	EPA RFG (2)	EPA conv (3)	WFC (4)
RON, min	(5)	81/92/95	91.0	91–95	91–95	NA	NA	91/95/98
MON, min	(6)	NS	81.0	81–85	81–85	NS	NS	82.5/85/88
Sulfur	wppm	50		10	10	10		10
AKI	(7)	84/87/91	NS	NS	NS	87/87/91	87/87/91	NS
Aromatics (8)	vol. %	40.0	35.0	35.0	35.0	20.7/19.5	27.7/24.7	35.0
Olefin (9)	vol. %	25.0	21.0	18.0	18.0	11.9/11.2	12.0/11.6	10.0
Benzene (10)	vol. %	1.0	1.0	1.0	1.0	0.66/0.66	1.21/1.15	1.0
Lead	mg/l	5.0	5.0	5.0	5.0			NS
Density	kg/m^3	NS	720–775	NS	720–775	NS	NS	715–770
RVP (11)	kPa	(12)	60.0	60–70	60–70	47.6/82.0	57.2/83.6	(13)

Notes:
1. Bharat IV is the set of specification for gasoline in India.
2. United States EPA reformulated gasoline.
3. United States conventional gasoline.
4. World Fuels Charter category.
5. RON is research octane number.
6. MON is motor octane number.
7. AKI is antiknocking index.
8. The aromatic concentration varies seasonally for US EPA RFG and EPA Conv. The first number is for the summer months and the second number is for the winter season.
9. 11. Olefin and benzene concentrations also vary seasonally and the RVP, Reid vapor pressure, limit varies seasonally in the U.S.
12. The limits of the Reid vapor pressures for the summer and winter months for the China V gasolines are 40–65 and 45–95 kPa, respectively.
13. For WFC, the RVP specifications vary with the temperature of the atmosphere. For temperatures that are greater than 15°C, RVP is 45–60.

For temperatures that are less than 15 C and greater than 5°C, RVP is 65–80.
For temperatures that are greater than –15°C and less than –5°C, RVP is 75–90.
For temperatures less than –15°C, RVP is 85–105.

The two-way catalytic converter has two simultaneous reactions, namely:

1. **Oxidation of carbon monoxide to carbon dioxide**

$$2\ CO + O_2 \rightarrow 2\ CO_2$$

2. **Oxidation of hydrocarbons (unburnt and partially burned fuel) to carbon dioxide and water**

$$C_xH_{2x+2} + [(3x+1)/2]\ O_2 \rightarrow x\ CO_2 + (x+1)\ H_2O \text{ (a combustion reaction)}$$

This type of catalytic converter is widely used on diesel engines to reduce hydrocarbon and carbon monoxide emissions. They were also used on gasoline engines in North America market automobiles. Because of their inability to control NOx, they were superseded by the three-way converters.

In 1981, two-way catalytic converters were rendered obsolete by three-way catalytic converters that reduce oxides of nitrogen (NOx). Three-way catalytic converters (TWC) have the additional advantage of controlling the emission of NOx (nitric oxide, NO and nitrogen dioxide, NO2), which are precursors to acid rain and smog.

In the three-way converters, NO_x, CO and C_xH_x gases enter the converters and a redux action involving reduction and oxidation reactions, respectively, with platinum plus rhodium catalysts in the former reaction where NO_x is reduced to N_2 and platinum plus palladium catalysts in the latter where CO and unburned C_xH_x are oxidized to CO_2 and H_2O. However, two-way converters are still used for lean-burn engines. This is because three-way converters require either rich or stoichiometric combustion to successfully reduce NO_x.

25.21.1 Operation of Catalytic Converter

Before catalytic converters were developed, waste gases made by a vehicle engine blew straight down the exhaust tailpipe and into the atmosphere. the catalytic converter sits between the engine and the tailpipe (Figure 25.32), but it does not work like a simple filter as it changes the chemical composition of the exhaust gases by rearranging the atoms from, they are made by the following:

1. Molecules of polluting gases are pumped from the engine past the honeycomb catalyst, made from platinum, palladium, and rhodium.
2. The catalyst splits up the molecules into their atoms.
3. The atoms then recombine into molecules of relatively harmless substances such as carbon dioxide, nitrogen, and water, which blow out safely through the exhaust.

Figures 25.33 and 25.34 show the schematic of the catalytic converter and the stages of the operation as stated above. Figure 25.35 shows a cutaway section of a catalytic converter, and Figure 25.36 shows a section of the material of zirconium dioxide ceramic with alumina washcoat and the precious metals (Pt, Pd, and Rh) in the outer washcoat. When a section of the material is being heated simulating a combustion process, the platinum and rhodium react with oxygen (O_2) and propane (C_3H_8) causing it the burn as without a flame as shown in Figure 25.37.

The reactions taking place in the converters are as follows:

1. **Reduction of nitrogen oxides (NO_x) to nitrogen (N_2)**

   - $2\ CO + 2\ NO \rightarrow 2\ CO_2 + N_2$
   - hydrocarbon + NO $\rightarrow CO_2 + H_2O + N_2$
   - $2\ H_2 + 2\ NO \rightarrow 2\ H_2O + N_2$

Figure 25.32 A three-way catalytic converter on a gasoline powered vehicle (Source: www.en.m.wikipedia.org).

Figure 25.33 Schematic of a catalytic converter (Source: www.explainthatstuff.com).

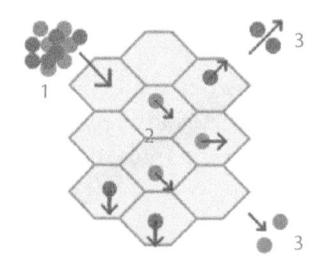

Figure 25.34 Schematic of the operation of the process (Source: www.explainthatstuff.com).

Figure 25.35 Cutaway section of two-chamber catalytic converters.

2. **Oxidation of carbon monoxide to carbon dioxide**

    - $2\,CO + O_2 \rightarrow 2\,CO_2$

3. **Oxidation of unburnt hydrocarbons (C_xH_x) to carbon dioxide and water, in addition to the above NO reaction**

    - hydrocarbon + $O_2 \rightarrow H_2O + CO_2$

Unwanted reactions can occur in the three-way catalytic converters, such as the formation of odoriferous hydrogen sulfide (H_2S) and ammonia (NH_3). Formation of each can be limited by modifications to the wash coat and precious

Figure 25.36 Honeycomb material of zirconium dioxide ceramic with alumina washcoat, and the precious metals (Pt, Pd, Rh) are in the outer washcoat.

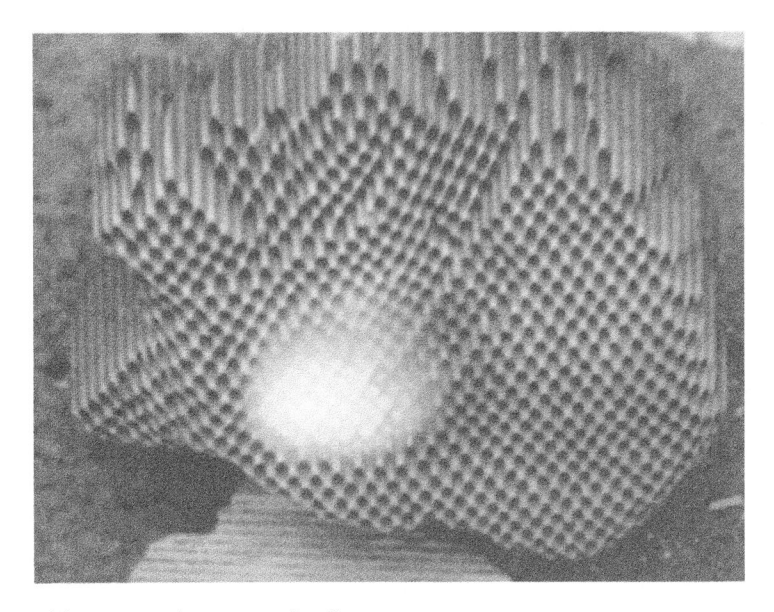

Figure 25.37 A heated section of the ceramic showing a red-yellow spot.

metals used. It is difficult to eliminate these byproducts entirely. For example, when control of H_2S emissions is desired, nickel or manganese is added to the wash coat. Both substances act to block the absorption of sulfur by the wash coat. H_2S forms when the wash coat has absorbed sulfur during a low-temperature part of the operating cycle, which is then released during the high-temperature part of the cycle, and the sulfur combines with the hydrocarbons.

These three reactions occur most efficiently when the catalytic converter receives exhaust from an engine running slightly above the stoichiometric point. For gasoline combustion, this ratio is between 14.6 and 14.8 parts air to one-part fuel by weight. The ratio for autogas (LPG), natural gas and ethanol fuels can be significantly different for each, notably oxygenated or alcohol-based fuels with E85 requiring approximately 34% more fuels to reach the stoichiometric point, require modified fuel system tuning and components when using those fuels. Generally, engines fitted with three-way catalytic converters are equipped with a computerized closed-loop feedback fuel-injection system using one or more oxygen sensors though early in the deployment of three-way converters, carburetors equipped with feedback mixture control were used.

Regulations of fuel quality vary world-wide; gasoline and diesel fuels are highly regulated, and compressed natural gas and LPG are being reviewed. In Asia and Africa, the regulations are often relaxed; in some places, sulfur content of the fuel can reach 20,000 parts per million (2%). Any sulfur in the fuel is oxidized to SO_2 or SO_3 in the combustion chamber. If the sulfur passes over a catalyst, it may be further oxidized in the catalysts, i.e., SO_2 to SO_3. SO_x are precursors to H_2SO_4, a major component of acid rain. While it is possible to add substances such as vanadium to the catalyst washcoat to combat SO_x formation, such as addition will reduce the effectiveness of the catalyst. The most effective solution is to further refine the fuel at the refinery to produce ultra-low-sulfur-diesel (ULSD). Regulations in Europe, Japan, and North America tightly restrict the amount of sulfur permitted in motor fuels. However, the direct financial expense of producing such clean fuel may make it impractical for use in developing countries. As a result, cities in these countries with high levels of vehicles suffer from acid rain, which damages stones and woodwork of buildings, poisons humans, animals, and ecosystems at a high financial cost.

25.21.2 Effectiveness of Catalytic Converters

Catalysts make a big difference to emissions, with three-way converters giving a significant extra benefit over two-way converters. Figure 25.38 shows the pollutants in grams per km of 80,000 km, using data for light-duty gasoline fuel vehicles from US EPA (1990) [36]. Catalytic converters are mainly designed to reduce immediate, local air pollution and dirty air when driving and Figure 25.38 shows that they are effective. However, they do not eliminate emissions. They operate at high temperatures over 300°C (572°F), when the engine has warmed up. The early types took about 10–15 min to warm up, and were ineffective for the first few kilometers/miles of a journey. Modern types take 2–3 min to warm up, but significant emissions can still occur during this period.

Another issue is whether they increase greenhouse-gas emissions, since CO_2 is emitted from the tailpipe of the exhaust, which is the major cause of global warming and climate change. There is the notion that catalytic converters make climate change worse because they turn carbon monoxide (CO) to carbon dioxide (CO_2). The CO that vehicles produce during combustion by itself would eventually be converted to CO_2 in the atmosphere; however, the converter improves the air quality in the streets by reducing CO that the vehicles emit.

Figure 25.39 shows the analysis of the metals of the catalytic converter using a spectrophotometer and Figures 25.40–25.46 show arrangements of catalytic converters and a simulation flow inside a catalytic converter respectively.

As in the case with catalytic processes (see Vol. 1), catalyst life in these catalytic converters is shortened due to the deposition and poisoning effects of sulfur. As the concentration of sulfur contaminant increases on the catalytic converter catalysts, the activity and efficiency of the catalyst are reduced resulting in its losses. When the catalyst in

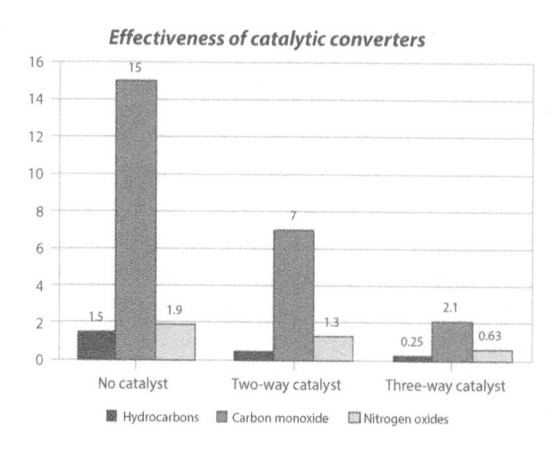

Figure 25.38 Effectiveness of catalytic converters (Source: www.explainthatstuff.com).

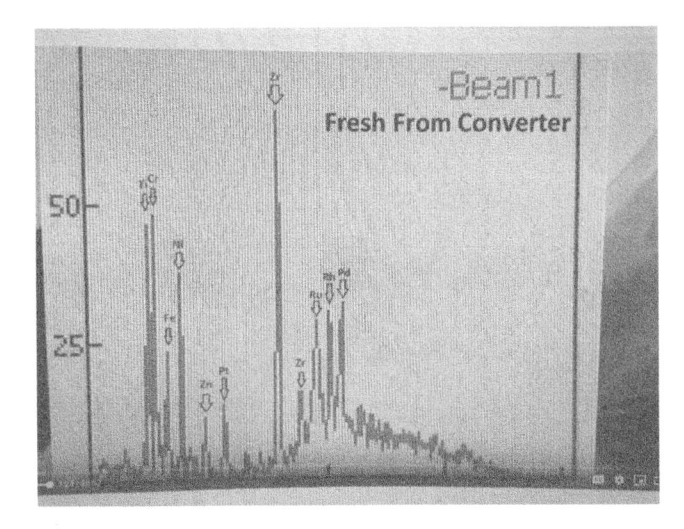

Figure 25.39 Analysis of the metals of the catalytic converter using a spectrophotometer.

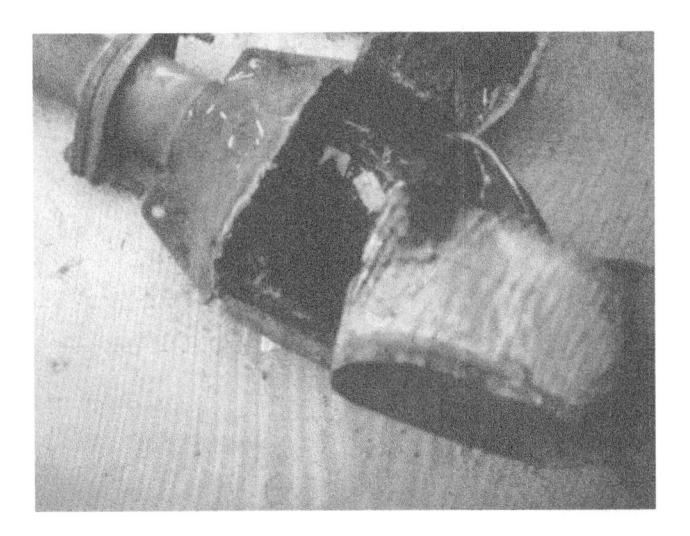

Figure 25.40 A cutaway section of the catalytic converter.

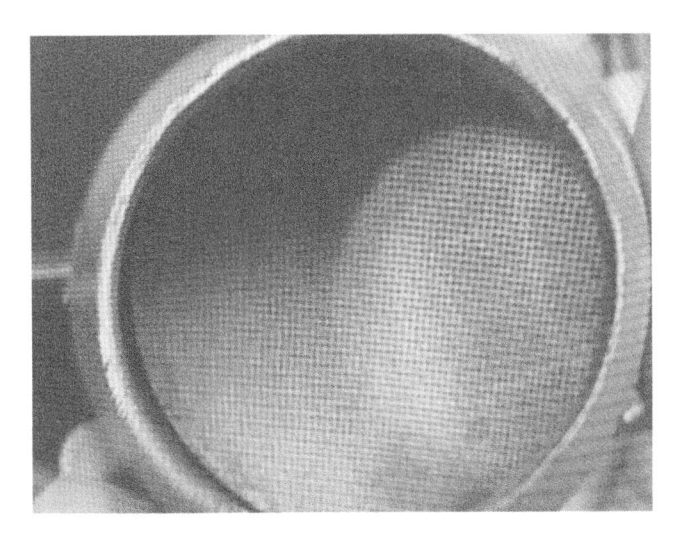

Figure 25.41 Metal casing with spiral pattern.

Figure 25.42 Metal casings with ceramic substrate of catalytic converters.

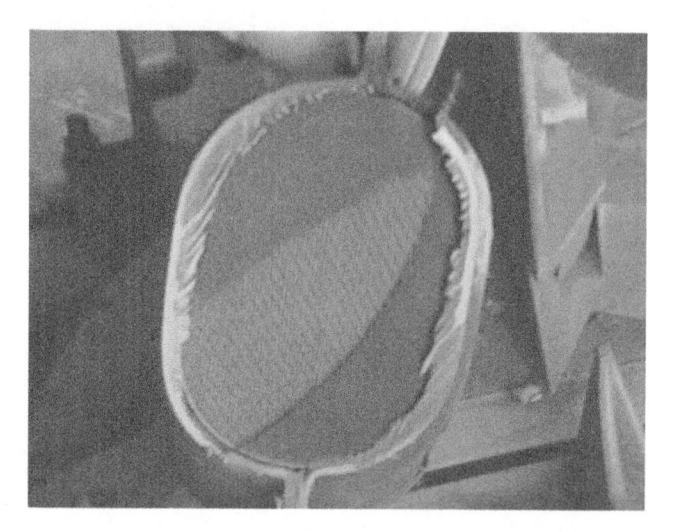

Figure 25.43 Substrate removed from the catalytic converter casing.

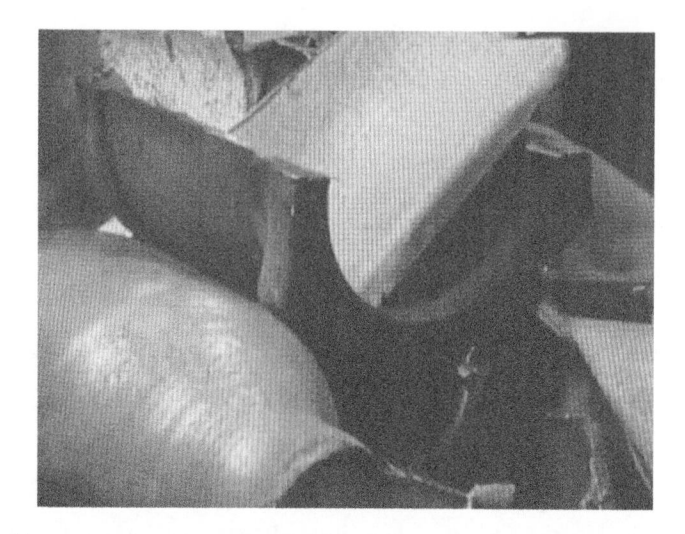

Figure 25.44 Cutaway sections of the scrapped metal casings from the substrates of the catalytic converters.

Figure 25.45 Substrates from the metal casings of the catalytic converters.

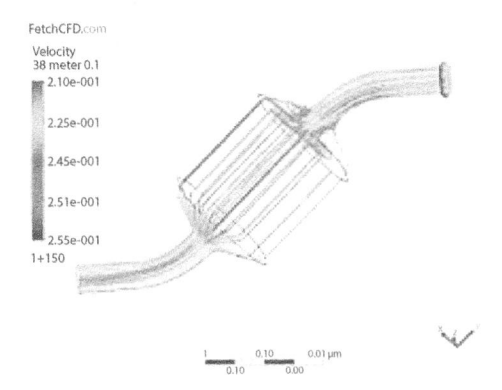

Figure 25.46 Simulation of flow inside a catalytic converter (Source: www.en.m.wikipedia.org).

the converter is spent, untreated tailpipe contaminants would then be continuously released into the environment, and since the rate of catalyst deactivation is dependent on the concentration of sulfur oxides in the exhaust gases, minimizing its concentration via lowering the sulfur concentrations in gasoline greatly extends the effective life of catalysts in the converters [15, 16].

Oyekan [15] suggested the need in gasoline to pursue a similar biodiesel fuel blend strategy by replacing higher percentages of the fossil-fuel contribution in gasoline with blends from renewable resources as biofuels. These are fuels produced from plants, crops such as sugar beet, rape seed oil or reprocessed vegetable oils or fuels made from gasified biomass; fuels made from renewable biological sources include ethanol, methanol, and biodiesel, thereby reducing the greenhouse gas (GHG) emissions. The added oxygenates enhance the combustion of gasoline fuel and minimize tailpipe emissions of carbon monoxide and particulate matters. Some countries add oxygenates for octane enhancement contributions in gasoline and alcohols such as methanol, ethanol, isopropyl alcohol, normal butanol, tertiary butanol, and ethers. The common ethers are methyl tertiary butyl ether (MTBE), tertiary amyl butyl ether (TAME), and ethyl tertiary butyl ether (ETBE). However, as explained earlier MTBE has caused deleterious effects to groundwater and soil contamination.

Environmental and Energy Study Institute (EESI)argues the removal of BTEX from gasoline and identifies between two methods of fuel blending and their use or misuse in the EPA's research on gasoline emissions. Many studies have examined the effect of ethanol content in gasoline on exhaust emissions. These studies can be classified into two types:

a. studies in which ethanol is simply added "splash blend" to a fixed-composition hydrocarbon blendstock.
b. studies in which after adding ethanol, the fuel is "match-blended" for one or more properties by varying the BOB composition to maintain certain characteristics of the finished fuel, such as octane ratings, vapor pressure or parameters of the distillation profile from ASTM D86.

For typical full-boiling range gasolines, the distillation profile from an ASTM D86 test is a smooth, roughly linear relationship of temperature vs. percent fuel distilled. Thus, the use of three discrete points on the distillation curve T10, T50, and T90, defined as the 10%v, 50%v, and 90%v distilled temperatures together with the Reid vapor pressure (RVP), and vapor-liquid ratio temperature has been sufficient to characterize the volatility and related behavior of gasoline in engines. These parameters have also been used as "match" criteria for blending fuels in studies intended to evaluate the effects of fuel composition on vehicle emissions.

The difference between "splash blending" and "match blending" of gasoline makes or breaks ethanol from a toxicity standpoint. Splash blending is how most gasoline in the U.S. is formulated and is carried out by adding 10% by volume ethanol to finished gasoline. A recent study has incorporated match blending of T50 and T90 to evaluate the effect of fuel ethanol content on vehicle emission is the EPAct/V2/E-89 conducted by the U.S. EPA in partnership with the Department of Energy and the Coordinating Research Council (CRC). The study was to generate updated fuel effects models of the on-road gasoline vehicle fleet, as required by the Energy Policy Act of 2005 (EPAct). The aim is to illustrate the substantial difference in the distillation characteristics of typical gasoline without ethanol compared to ethanol-gasoline blends, and to illustrate potential issues with match-blending practices for ethanol–gasoline blends used in testing of fuel effects on emissions, particularly the use of T50 and T90 as match–blend criteria. However, studies have shown that the use of splash blending reduces air toxicity and pollution overall since the addition of a cleaner-burning compound lowers the volume of toxics in the gasoline [14]. The degradation of emissions which can result is primarily due to the added hydrocarbons but has often been incorrectly attributed to the ethanol. Studies to determine the effects of ethanol at up to 30% v with an E10 blendstock should generally require only minimal changes in composition to meet ASTM D4814.

Match blending is when the blendstock (gasoline) composition is modified for each ethanol-gasoline blend to match one or more fuel properties, such as a specific boiling point, generally by adding more aromatics to gasoline (the BTEX complex). The effects on emissions depend on which fuel properties are matched and what modifications are made, making the trends difficult to interpret.

However, the problem with this blending type is it raises the total volume of aromatics, the most toxic portion of each gallon of gasoline, because it uses a dirtier gasoline blendstock. If emissions tests were carried out with match blended fuels, one would expect the toxics to be higher than if the same tests were carried out is splash blended fuels, and this conclusion has been supported by Anderson et al. [14].

References

1. Jones, D. S. J., Elements of Petroleum Processing, John Wiley & Sons Ltd., 1995.
2. Surinder Parkash, Refining Processes Handbook, Gulf Professional Publishing, Elsevier, 2003.
3. Riazi, M. R., Characterization and Properties of Petroleum Fractions, ASTM. International, Philadelphia, 2005.
4. James H. Gary, Glenn E. Handwerk, and Mark J. Kaiser, Petroleum Refining Technology and Economics, 5th ed., CRC Press, Taylor & Francis Group, 2007.
5. Hu, J., and A. Burn, Index predicts cloud, pour and flash points of distillates fuel blends. Oil & Gas Journ., 68 (45), 66, 1970.
6. Baird, C. T., Guide Oil Yields and Product Properties, Ch. De Lahaute-Belotte 6, Cud Thomas Baird IV, 1222, Veznaz, Geneva, Switzerland, 1981.

7. Jenkis, G. I. and R. P., Walsh, Quick measure of jet fuel properties, Hydrocarbon Proc., 47, 5, 1968.
8. Baird, C. T., Guide to Petroleum Product Blending, HPI Consultants, Inc., Austin, Chevron Oil Trading Company., 31.0°API Iranian Heavy Crude Oil, 1989.
9. Guibet, J. Characteristics of petroleum products for energy use. In "Crude Oil Petroleum Products Process Flowsheets", Petroleum Refining (Wauquier, J. Ed.) Chapter 5, TECHNIP, France, 1995.
10. Antos, G. J., et al., Catalytic Naphtha Reforming, Marcel Dekker, New York, USA, 1995.
11. Healy, W. C., Maasen, C. W., and R. T. Peterson, Predicting Octane Numbers of Multicomponent Blends, Report Number RT-70, Ethyl Corporation, Detroit, 1959.
12. Twu, C., and J. Coon., A Generalized Interaction Method for the Prediction of Octane Number of Gasoline Blends, Simulation Science Co., Inc., Brea, CA, pp. 1–18, 1998.
13. U.S. Energy Information Administration, Almost all U.S. gasoline is blended with 10% ethanol, https://www.eia.gov/today-inenergy/detail.php?ie=26092, May 4, 2016.
14. James E. Anderson, Wallington, T., Stein, R. A. and William M. Studzinski, Issues with T50 and T90 as Match Criteria for Ethanol-Gasoline Blends, https://www.sae.org/publications/technical-papers/content/2104-01-9080.
15. Oyekan, Soni, O., Catalytic Naphtha Reforming Process, CRC Press, Taylor & Francis Group, 2019.
16. Private communications with Dr. Soni Oyekan, 2020.
17. Aronofsky, J. S., Linear Programming—A Problem Solving Tool for Petroleum Industry Management., Journ. of Petroleum Technology, pp729–736, SPE 315, July 1962.
18. ASTM specification D-97.
19. ASTM specification D-3245.
20. API databook procedure 11A1.1.
21. ASTM specification D-2161.
22. API databook procedure11A1.4
23. Huang, P, K., Characterization and Thermodynamic Correlations for Undefined Hydrocarbon Mixtures, Ph.D. Thesis, Department of Chemical Engineering, Pennsylvania State University, 1977.
24. API databook procedure 2B4.1. See also M. R. Raizi and T. E. Daubert. Prediction of Composition of Petroleum Fractions, Ind. Eng. and Chem. Proc. Des. and Dev., 16, p. 289, 1980.
25. API databook, procedure 2B2.1
26. ASTM specification D-341
27. Purohit, Amit, and Tukaram Suryawanshi, Integrated Product Blending Optimization for Oil Refinery Operations, 10th IFAC International Symp. On Dynamic and Control of Process Systems.,Mumbai, India., pp 343, Dec. 18–20, 2013.
28. Jia, Z., and M. Ieraptetritou., Mixed-integer linear programming model for gasoline blending and distribution scheduling. Ind. Eng. Chem. Res. 42, 825–835, 2003.
29. Mendez, C. A., Grossman, I. E., Harjunkoski, I, and P. Kabor., A simultaneous optimization approach for offline blending and scheduling of oil-refinery operations. Computers and Chemical Engineering, 30, 614–634, 2006.
30. Moro, L., Zanin, A. and J. Pinto, A planning model for refinery diesel production. Computers & Chemical Engineering, 22, pp. 1039–1042, 1998. Available at: http://www.sciencedirect.com/science/article/pii/S0098135498002099 [Accessed September 6, 2013].
31. Pinto, J.M., Joly, M. & Moro, L.F.L., Planning and scheduling models for refinery operations. Computers & Chemical Engineering, 24(9–10), pp.2259–2276. 2000. Available at: http://linkinghub.elsevier.com/retrieve/pii/S0098135400005718.
32. Neiro, S. and J. Pinto, J. Multiperiod optimization for production planning of petroleum refineries. Chem. Eng. Comm., 192(1), pp.62–88, 2005. Available at: http://www.tandfonline.com/doi/abs/10.1080/00986440590473155 [Accessed September 6, 2013].
33. Hamisu, A. A., Petroleum Refinery Scheduling with Consideration for Uncertainty. Ph.D. Thesis, Cranfield University, 2015.
34. Gottfried, Byron S., Spreadsheet Tools for Engineers, Excel 5.0 Version, 1996.
35. Abadie, J., and J. Carpentier, "Generalization of the Wolfe Reduced Gradient Method to the Case of Nonlinear Constraints", In Optimization, R. Fletcher, ed., Academic Press, New York, pp 37–47, 1969.
36. Faiz, et al., Air Pollution from Motor Vehicles: Standards and Technologies for Controlling Emissions, World Bank, 1996.
37. Abdulrahman, S. Alsuhaibani, Texas A & M University. Private communications.

Bibliography

Shixun Jiang, Optimisation of Diesel and Gasoline Blending Operations, Ph.D. Thesis, Centre for Process Integration, School of Chemical Engineering and Analytical Science, The University of Manchester, 2016.

26

Cost Estimation and Economic Evaluation

26.1 Introduction

Recently, mergers and downsizing in the various petroleum refining and chemical process industries (CPIs) have required management to make effective decisions regarding investments in strategic assets, sometimes with limited engineering participation. Due to limited budgets, potential projects are scrutinized stringently before funds are allocated to proceed with them; management requires reasonably accurate cost estimates at each stage of the funding. The choice as to which estimation method to use at each stage depends on the information available at the time of preparation, its desired accuracy and the end use of the estimate.

Many industries use some form of classification system to identify the various types of estimates that may be prepared during the life cycle of a project, and also to indicate the overall maturity and quality of the estimates produced. However, the process industries in general and individual companies and organizations, have not been consistent and tend not to have a firm understanding of the terminology used during classification. The Association for Advancement of Cost Engineering International (AACE) recently developed recommended practices for cost estimate classification for the process industries [1]. Table 26.1 shows a list of the recommended practices, which are intended to be the principal AACE's certification products and services.

The AACE 18R-97 is the recommended practice cost estimate classification for the process industries. It is a reference document that provides extensions and additional detail for applying the principles of estimate classification specifically to project estimates for engineering, procurement, and construction (EPC) work. It also describes and differentiates between various types of project estimate by identifying five classes of estimates as shown in Table 26.2. Class 1 estimate is associated with the highest level of project definition or maturity; a Class 5 estimate is assigned with the lowest level. Five characteristics are used to distinguish one class of estimate from another, namely: degree of project definition, end use of the estimate, estimating methodology, estimating accuracy, and effort to prepare the estimate. However, the degree of project definition is the primary characteristic used to identify an estimate class [2].

The CPI depends on process flow diagrams (PFDs), and piping and instrumentation diagrams (P&IDs) as the primary scope defining documents. These documents form the principal deliverables in determining the level of project definition, the maturity of the information used to perform the estimate, and subsequently the estimate class. An estimate input checklist, including items such as PFDs, process and utility equipment lists, instrumentation and control system diagrams, is incorporated into AACE 18R-97, which identifies the engineering deliverables used to prepare a project estimate. Further information on ACCE 18R-97 can be obtained at aacei.org.

Table 26.1 AACE Recommended Practices for Cost Estimate Classification.

Documents	Recommended practices
Cost Engineering Terminology	10S – 90
Required Skills and Knowledge of a Cost Engineer	11R – 88
Model Master's Degree Program with Emphasis in Cost Engineering	12R – 89
Recommended Method for Determining Building Area	13S - 90
Roles and Duties of a Planning and Scheduling Engineer	14R – 90
Profitability Methods	15R – 81
Conducting Technical and Economic Evaluations in the Process and Utility Industries	16R – 90
Cost Estimation Classification System	17R – 97
Cost Estimation Classification System – As Applied in Engineering, Procurement, and Construction for the Process Industries	18R – 97
Estimate Preparation Costs in the Process industries	19R – 97
Project Code of Accounts	20R – 98
Project Code of Accounts – As Applied in Engineering, Procurement, and Construction for the Process Industries	21R – 98
Direct Labor Productivity Measurement as Applied on Construction and Major Maintenance Projects	22R – 01
Planning & Scheduling – Identification of Activities	23R – 02
Planning & Scheduling – Developing Activity Logic	24R – 03
Estimating Lost Labor Productivity in Construction Claims	25R – 03

26.2 Refinery Operating Cost

Petroleum refining is a capital-intensive enterprise as a grass root refinery of average complexity processing 100 million barrel of crude per day may cost a billion dollars to construct. Therefore, for a refinery to be economically viable, its operating cost must be minimized. There are cases where refineries are forming an integral part of larger petrochemical plants, where the products such as the aromatics from the reforming unit are fed directly to the CPI plants involving further processing to yield products such as ethyl benzene, styrene, ethylene glycol, plastics, and so on. Many operating cost elements such as depreciation, insurance, and personnel are used; the cost remains constant with refinery throughput and the operating cost per barrel of crude being processed is reduced and thereby enhancing the profitability of the facility. The refinery operating cost can be classified as follows:

Personnel cost. This includes salaries and wages of regular employees, employee benefits, contract maintenance labor, and other services.

Maintenance cost. This includes maintenance materials, contract maintenance labor and equipment rental.

Table 26.2 Cost Estimate Classification Matrix.

Estimate class	Project definition (% of complete definition)	Purpose of estimate	Estimating method	Accuracy range (variation in low and high ranges)	Preparation effort (index relative to project cost)
Class 5	0 – 2	Screening	Capacity-factored, parametric models	L: -20 to –50% H: 30 to 100%	1
Class 4	1 – 15	Feasibility	Equipment-factored, parametric models	L: -15 to –30% H: 20 to 50%	2 – 4
Class 3	10- 40	Budget authorization or cost control	Semi-detailed unit cost estimation with assembly-level line items	L: -10 to –20% H: 10 to 30%	3 – 10
Class 2	30 – 70	Control of bid or tender	Detailed unit-cost estimation with forced, detailed take off	L: -5 to –15% H: 5 to 20%	4 – 20
Class 1	50 - 100	Check estimate, bid or tender	Semi-detailed unit cost estimation with detailed take off.	L: -3 to –10% H: 3 to 15%	5 - 100

The state of process technology and the availability of cost data strongly affect the accuracy range of an estimate. Plus or minus high (H) and low (L) values represent the variation in actual costs versus estimated costs, after applying contingency factors. The "preparation effort" uses an index to describe the cost required to prepare an estimate, relative to that for preparing a Class 5 estimate. For example, if it costs 0.005% of the project cost to develop a Class 5 estimate, then a Class 1 estimate could require as much as 100 times that, or 0.5% of the total project cost.

(Source: Dysert, L., By permission of Chemical Engineering, Oct. 2001, pp 70-81).

Insurance. This is required for the fixed assets of the refinery and its hydrocarbon inventory.

Depreciation. This must be assessed on refinery assets, plant machinery, storage tanks, marine terminal and so on.

General and administrative costs. These include all office and other administrative expenses.

Chemicals and additives. These are the compounds used in processing petroleum and final blending, such as antioxidants, antistatic additives and anti-icing agents, pour points depressants, anticorrosion agents, dyes, water treatment chemicals and so on.

Catalysts. Proprietary catalysts used in various process units.

Royalties. These are paid either in a lump sum or running royalty purchased for know-how.

Purchased utilities. These include electric power, steam, water, etc.

Purchased refinery fuel. This includes natural gas for use as finery fuel and feedstock for hydrogen production.

26.2.1 Theoretical Sales Realization Valuation Method

The theoretical sales realization valuation method (TSRV) is when the total expense is allocated to the participants in the ratio of TSRV of its product. Example 26.1 illustrates how this is used.

Example 26.1

The total operating expenses of a marine terminal of a refinery during a month were $1.3 million. This amount is to be allocated to participants using the TSRV method. The product shipments during the month from the terminal were as follows:

Product	AOC's shipments, bbl	BOC's shipments, bbl
Naphtha	817, 149	511,711
Gasoline	412,477	78,417
Kerosene	632,858	101,675
Diesel	1,900, 245	460,552
Fuel oil	1,706,555	376,461
Asphalt	29,221	50,832
Total	5,498,505	1,579,648

The first step is to calculate the value of the product shipped by both the participants, and this is done by multiplying the shipment volumes by the unit cost of the product as follows:

Product	MOP price, $/bbl	AOC's shipments, $ millions	BOC's shipments, $ millions
Naphtha	18.681	15.265	6.373
Gasoline	25.761	10.626	1.347
Kerosene	27.252	17.247	1.847
Diesel	23.234	44.149	7.133
Fuel oil	13.422	22.905	3.369
Asphalt	15.000	0.438	0.508
Total		110.631	20.577

The total value of the product shipped over marine terminal = $131.208 million
Value of the product shipped by participant AOC = $110.631 million
Value of the product shipped by participant BOC = $20.577 million
Participant AOC's product share = $110.631/$131.208 =84.0%
Participant BOC's product share = $20.577/$131.208 =16.0%
Total operating expenses of the marine terminal = $1.31 million
Participants AOC's share of operating cost (84%) = $1.092 million
Participants BOC's share of operating cost (16.0%) = $0.121 million

26.2.2 Cost Allocation for Actual Usage

The following cost items are allocated to the participants as per their actual usage:

1.	The cost of the chemicals and additives, such as antiknock compounds, pour point depressants, and antistatic dissipaters. It is possible to accurately estimate the quantity of antiknock compound, pour point depressants, and other additives used in the final blending of their products from shipment and quality data records.
2.	All operating expenses involved in receiving crude oil and other feedstocks in each operating period are segregated and allocated to the participants on the basis of that received by each in the period. e.g., if a participant brings a crude or another feedstock for processing in its share of refining capacity, all expenses related to receiving the crude is allocated to that participant. If a crude is brought in the pipeline for processing by both participants, the pipeline related expense is allocated to the participants in the ratio of the crude received.
3.	All operating expenses involved in the manufacture and shipping of solid products, such as asphalt and sulfur, in each operating period are segregated and allocated to the participants on the basis of their respective shares of shipment of such products.

26.3 Capital Cost Estimation

Capital cost estimation is an essential part of investment appraisal. Many types of capital cost estimate are made, ranging from initial order-of-magnitude estimates to detailed estimates which require the collection of accurate technical data. The American Association of Cost Engineers (AACE) defines five types of cost estimates as follows:

1. **Order of magnitude estimate (ratio estimate)** is based on similar previous cost data, where an approximate forecast of fixed investment can be determined without flowsheet, layout or equipment analysis by applying overall ratios (to account for differences in scale of production), and applying appropriate escalation factors to update from previous installations. The probable accuracy of estimate is within ± 30%.
2. **Study estimate (factored estimate)** is based on knowledge of major equipment items. This is applicable when a scheme has been developed to the stage of preliminary flowsheets, with the duty rating of principal items of equipment specified, and a geographical location for the construction of the facility known. A cost estimate can be prepared based on estimates for each main plant item or group of items. The factored estimate accuracy is within ± 30%.
3. **Preliminary estimate (sanction, budget authorization estimate, or scope estimate)** is applied after a study estimate has been accepted. The further engineering work, which has been authorized aims to obtain and present information such as preliminary material and energy balances, P&IDs, equipment lists and material specifications, duty rating and sizing of all process equipment, instrumentation and control devices. A cost estimate which may be achieved has a probable error of less than ± 20%. For several items such as insulation, electrical services, instrumentation and pipework, estimates are derived by applying factors to the estimated costs of the main equipment items.
4. **Definitive estimate (project control estimate)** is based on almost complete data, but before completion of drawings and specifications. Initially derived from project funds allocation, which is based upon the preliminary stage estimate and the authorization to proceed, estimates can be refined as design work proceeds and decisions are made. A definitive estimate can be determined shortly before complete drawings and specifications have been discussed with vendors. The estimate is aimed to achieve a probable error in the limited range of ± 10%.
5. **Detailed estimate (tender or contractor's final cost estimate)**, requires completed engineering drawings, specifications and site surveys. Process and mechanical datasheets are prepared and vendors'

Table 26.3 Probable Accuracy Relative to the Cost of the Estimate.

Recommended nomenclature	Probable range of accuracy (%)	Cost as % of project expenditure (%)
Detailed estimate	±2 to ±5	5 to 10
Definitive estimate	±5 to 15	1 to 3
Preliminary estimate	±10 to 25	0.4 to 0.8
Study estimate	±20 to ±30	0.1 to 0.2
Order of magnitude estimate	±30 to ±50	0 to 0.1

(Source: Gerrard, A.M. [3]).

quotations for selected equipment items are obtained such that competitive prices, which are compatible with quality and delivery, are available for the estimate. The probable accuracy of estimate is within ± 5%. Each type of estimate and their probable range of accuracy are summarized in Table 26.3.

Process plant designs commence with preliminary designs based on approximate technical data, calculations, and cost data and proceed to final designs that require detailed and accurate data, calculations and quotations. Cost estimates of a proposed plant are continuously carried out during the development of a process from laboratory to construction. The total capital cost C_{TC}, of a project consists of the fixed capital cost, C_{FC}, plus the working capital, C_{WC}, plus the cost of land and any other non-depreciable assets, C_L. This is given by

$$C_{TC} = C_{FC} + C_{WC} + C_L \tag{26.1}$$

The fixed capital cost, C_{FC}, is the capital required to provide all the depreciable facilities; C_{FC} may be divided into two classes known as the battery limits and auxiliary facilities. The boundary of battery limits includes all manufacturing and processing equipment. The auxiliary facilities are the storage areas, administration offices, utilities and other essential and non-essential supporting facilities.

Generally, as the size of project under consideration increases, the cost of preparing an estimate tends to decrease as a percentage of project total cost. Table 26.3 illustrates a summary of the expected accuracy ranges and approximate costs of the preparation of estimates for the various classifications. The cost of preparing an estimate exceeding the bands shown in Table 26.3 depends on uncertainty of basic data, inadequate management associated with the project, development into new areas of technology, frequent changes of scope of work, and so forth. The cost of preparing cost estimates is dependent not only on the nature and scale of the project, but also on the experience and overhead structure of the organization preparing the estimate [3].

26.4 Equipment Cost Estimations by Capacity Ratio Exponents

The process engineer often plays a key role in the preparation of cost estimates and in their development. From a first draft flowsheet and a preliminary plot plan, a preliminary cost estimate can be prepared by the "factoring" or equivalent method. This basically accumulates the individual costs of each item of major equipment and then multiplies by an experience factor to produce one or all of (i) total plant cost installed with or without overhead costs (ii) piping installed (iii) equipment installed. For accuracy, these factors must be developed from actual plant costs, and are often peculiar to a specific construction type or engineering approach to the project. The factor of 2.5 to 6.0 usually covers most petroleum refining and petrochemical processing plants. This factor times the costs of major equipment (pumps, compressors, tanks, columns, exchangers) but not instruments will give total plant costs. The plant will include usual control buildings, structure, foundations, overhead charges, construction fees, engineering costs, and so on. A value of 4.0 may usually be realistic [4].

The process designer must be aware of costs as reflected in the (i) selection of a basic process route (ii), the equipment to be used in the process, and (iii) the details incorporated into the equipment. The designer must not arbitrarily select equipment, specify details or set pressure or temperature levels for design without recognizing the relative effect on the specific cost of an item as well as associated equipment such as relieving devices, instruments, and so on [4].

With more comprehensive and better information regarding the process and layout plans, estimating engineers can prepare detailed estimates, which are often quite accurate, usually ±10% for the best. It is the duty of the process designer to supply the best information in order to contribute to better or improved estimates.

Estimating equipment costs is a specialty field in itself. The estimator must therefore have access to continuously updated basic reference costs and to graphical costs relations, which are a function of capacity of this equipment. Page's *Estimator's Manual of Equipment and Installation Costs* [5] is a helpful reference. Since the equipment is only a portion of the total cost of a plant, or an addition to a chemical project, installation costs, which reflect the labor portion of the total cost, must also be determined. Useful and comprehensive data for such needs are presented for equipment [5], general construction [6], heating, air-conditioning, ventilating, plumbing [7], piping [8], electrical-related areas [9] and all disciplines [10] in the cited references.

Even an inexperienced estimator should be able to establish an approximation of the costs, provided he/she adequately visualizes the work functions and steps involved. From the same type of work reference, an experienced estimator can develop a realistic cost, usually expressed with certain contingencies to allow for unknown factors and changing conditions. The professional estimators will normally develop cost charts and tables peculiar to the nature of their responsibilities and requirements of their employers.

It is often necessary to calculate the cost of a piece of equipment when there are no available cost data for the particular size of capacity. If the cost of a piece of equipment or plant size or capacity Q_1 is C_1. The cost C_2 of a similar piece of equipment or plant size or capacity Q_2 can be calculated from the equation

Table 26.4 Typical Exponents for Equipment Cost vs Capacity.

Equipment	Exponent (m)
Reciprocating compressor	0.75
Turbo blowers compressor	0.5
Electric motors	0.8
Evaporators	0.5
Heat exchangers	0.65 to 0.95
Piping	0.7 to 0.9
Pumps	0.7 to 0.9
Rectangular tanks	0.5
Spherical tanks	0.7
Towers, constant diameter	0.7
Towers, constant height	1.0

(Source: Institution of Chemical Engineers [12]).

$$C_2 = C_1 \left(\frac{Q_2}{Q_1}\right)^m \tag{26.2}$$

where

C_1 = Cost of plant or section of plant of original capacity "1"
C_2 = Cost of plant or section of plant of new capacity "2"
Q_1 = Capacity of plant or section of original requirements
Q_2 = Capacity of plant or section of new requirements
m = cost exponent (or capacity factor)

the value m depends on the type of equipment or plant. It is generally taken as 0.6, the well-known six-tenths rule [11]. This value can be used to get a rough estimate of the capital cost, if there are insufficient data to calculate the index for the particular size of equipment required. Tables 26.4 and 26.5 list values of m for various types of equipment and products respectively. The value of m typically lies between 0.5 and 0.85, depending on the type of plant,

Table 26.5 Capacity Factors (m) for Process Units.

Product	Factor
Acrolynitrile	0.60
Butadiene	0.68
Chlorine	0.45
Ethanol	0.73
Ethylene oxide	0.78
Hydrochloric acid	0.68
Hydrogen peroxide	0.75
Methanol	0.60
Nitric acid	0.60
Phenol	0.75
Polymerization	0.58
Polypropylene	0.70
Polyvinyl chloride	0.60
Sulfuric acid	0.65
Styrene	0.60
Thermal cracking	0.70
Urea	0.70
Vinyl acetate	0.65
Vinyl chloride	0.80

(Source: Guthrie, K.M. [13]).

and must be assessed carefully for its applicability to each estimating situation. The m used in the capacity factor equation is actually the slope of the log-curve that has been drawn to reflect the change in the cost of plant as it is made larger or smaller. These curves are typically drawn from the data points of the known costs of completed plants. With an exponent less than 1, scales of economy are achieved such that as plant capacity increases by a percentage, say, by 20%, the costs to build the larger plant increases by less than 20%. With an exponent of 0.6, doubling the capacity of a plant invariably increases costs by approximately 50%, and tripling the capacity of a plant increases costs by approximately 10% [2].

Dysert [2] reported that as plant capacities increase, the exponent tends to increase, and as the plant capacity increases to the limits of existing technology, the exponent approaches a value of 1. At this point, it becomes theoretically as economical to build two plants of a similar size rather than one large plant on the same specific site. Table 26.6

Table 26.6 Percentage Error (in %) if Capacity Factor (m) of 0.7 is Used for Estimate.

Actual exponent	Capacity–increase multiplier (Q_2/Q_1)							
	1.5	2	2.5	3	3.5	4	4.5	5
0.20	23	41	58	73	88	100	113	124
0.25	20	36	51	64	75	87	97	106
0.30	18	32	44	55	64	74	83	91
0.35	16	28	38	47	55	63	70	76
0.40	13	23	32	39	46	52	57	63
0.45	11	18	26	32	36	41	46	50
0.50	9	15	20	25	28	32	35	38
0.55	6	11	15	18	21	23	25	28
0.60	4	7	10	12	13	15	16	18
0.65	2	3	5	6	6	7	8	8
0.70	0	0	0	0	0	0	0	0
0.75	-2	-4	-5	-5	-6	-7	-7	-8
0.80	-4	-7	-9	-10	-12	-13	-14	-15
0.85	-6	-10	-13	-15	-17	-19	-20	-21
0.90	-8	-13	-17	-20	-22	-24	-26	-28
0.95	-10	-16	-21	-24	-27	-29	-31	-33
1.00	-11	-19	-24	-28	-31	-34	-36	-38
1.05	-13	-22	-28	-32	-36	-39	-41	-43
1.10	-15	-24	-31	-36	-40	-43	-45	-47
1.15	-16	-27	-34	-39	-43	-46	-49	-52
1.20	-18	-30	-37	-42	-47	-50	-53	-55

(Source: Dysert, L., By permission of Chemical Engineering, Oct. 2001, pp 70-81).

shows the percent error that may occur if an assumed capacity factor of 0.7 is used, and the actual value is different. For example, if the new plant is triple the size of an existing plant, and the actual capacity factor is 0.85 instead of assumed 0.7, the cost of the new plant is underestimated by only 15%. Similarly, for the same threefold scale-up in plant size, if the capacity factor should be 0.6 instead of the assumed 0.7, then the plant cost is overestimated by only 12%. Dysert [2] generated these data from equation (26.2); the capacity-increase multiplier is Q_2/Q_1 and in the base m is 0.7. The error occurs as m varies from 0.7.

Cost indices should be used to derive the cost data to a desired year. A cost index is a value for a given point in time showing the cost at that time relative to a certain base time. If the cost at some time in the past is known, the equipment cost at the present time can be determined from the equation. This is applicable for any given year of installation but does not correct for the differences in cost from year to year. This is conveniently done as described in the next section.

Experience has indicated that this six-tenths rule is reasonably accurate for capacity scale-up of individual items of equipment. Thus, if the cost of one size of a piece of equipment is known, an estimating figure for one twice as large can be found by multiplying by $(2)^{0.6}$.

The most difficult feature of this method is that for each type of plant or plant product, as well as for each type of equipment, there is a break point where the 0.6 no longer correlates the change in capacity. For small equipment or plants in reasonable pilot or semi-works size, the slope of the cost curve increases and the cost ratio is greater than 0.6, sometimes 0.75, 0.8 or 0.9. From several cost values for respective capacities a log-log plot of capacity versus cost will indicate the proper exponent by the slope of the resultant curve. Extrapolation beyond eight - or tenfold may usually introduce an unacceptable error [4].

26.5 Yearly Cost Indices

The three most-used cost indices for the chemical, petrochemical, petroleum refining industry for relating the cost level of a given year or month to a reference point are the following:

1. **Chemical Engineering Plant Cost Index** [10]. Probably the most commonly used cost-adjusting index printed/updated monthly is in *Chemical Engineering* magazine and has established continuity over many years. Its breakdown component costs apply to plants and plant equipment/systems. This can be accessed from the website: www.chem.com
2. **Marshall and Swift Equipment Cost Index** [14]. Commonly used for process industry equipment and index numbers presented by industries in *Chemical Engineering* magazine on a monthly basis.
3. **Nelson Farrar Cost Index** [15]. This is generally suited to petroleum refining plants and is referenced to them. It is updated and published regularly in *The Oil & Gas Journal*. It can be accessed from the website: www.ogj.com

These indices are used to update costs when values at some earlier date are known. The new costs are of estimating accuracy and should be verified whenever possible, as should the results of using the 0.6 power for correlating cost and capacity. The new cost can be obtained by

$$EC_2 = EC_1 \left(\frac{I_2}{I_1}\right) \tag{26.3}$$

where

I_2 = index value for year represented by 2, (usually current)

COST ESTIMATION AND ECONOMIC EVALUATION   477

I_1 = index value for earlier year represented by 1
EC_2 = equipment estimated cost for year represented by 2
EC_1 = equipment purchased cost (when available) for year represented by 1

Cost indices are used to give a general estimate, but no index can account for all the factors. Many different types of cost index are published in journals such as the *Chemical Engineering plant cost index, Engineering News-Records Construction index, the Oil & Gas Journal and Process Engineering.* Table 26.7 shows the international plant cost indices from March 1992 to February 1993 and Table 26.8 gives the annual plant cost indices for the years 1963 to 2000. Figure 26.1 shows historic trends and values of cost indices pertinent to the chemical process industry. The Chemical Engineering Plant Cost Index (CEPCI) can be used to account for changes that result from inflation, and is accessed from the website: www.che.org.

Example 26.2

A plant to produce 100,000 bbl/d of ethanol is to be completed in Houston in 2004. A similar plant with a capacity of 150,000 bbl/d and a final cost of $50 million, was completed in 2002. Determine the cost of the 100,000 bbl/d plant.

Solution

Using the capacity factor algorithm in Equation (26.2), with m = 0.73 gives:

$$C_2 = C_1 \left(\frac{Q_2}{Q_1}\right)^m$$

$$C_2 = \$50\,M \left(\frac{100,000}{150,000}\right)^{0.73}$$

$$= \$37.2M$$

Table 26.7 International Plant Cost Indexes Mar 1992-Feb 1993 Base Date Jan 1990 = 100.

Country	Mar	Apr	May	Jun	Jul	Aug	Sep	Oct	Nov	Dec	Jan	Feb
Australia	107.7	107.9	107.9	108.2	108.4	108.6	108.7	108.8	108.8	109.0	109.3	109.4
Belgium	108.6	109.2	109.3	109.2	110.4	110.08^R	110.8	111.4	111.2	111.2	111.2	111.2
Canada	107.0	107.0	106.9	106.3	105.7	106.3	106.9	107.5	107.5	108.7	108.8	108.8
Denmark	107.0	108.4	108.5	109.3	110.4	108.0	108.5	108.9	108.5	110.4	110.7	-
France	106.5	106.9	107.0	107.1	107.5	107.5	108.1	108.3	108.5	108.6	109.2	109.5
Germany	113.8	114.0	114.0	114.9	115.0	115.0	115.4	115.4	115.4^F	116.2	116.3	116.4
Italy	109.5	109.6	109.6	109.9	110.1	110.0	110.2	110.5	110.5	110.7	-	-
Japan	103.0	102.8	102.8	110.4	102.7	101.3	103.1	103.3	103.4	102.5	100.3	103.0
Netherlands	106.8	108.0	108.1	108.1	108.8	108.8	108.8	108.7	108.7	108.7	109.5	109.5
New Zealand	103.9	104.4	104.4	104.4	105.0	105.0	105.0	105.3	105.4	105.4	-	-
Spain	111.1	113.2	113.2	113.3	114.4	114.4	114.4	114.7	114.8	114.8	-	-
Sweden	106.2	107.9	108.8	109.2	109.6	107.9	108.4	109.0	109.5	111.3	-	-
USA	101.7	101.1	101.6	101.4	100.4	100.8	100.7	100.9	100.5	100.7	100.7	100.7
UK	117.2	115.0	115.1	117.0	116.8	117.1	118.2	116.2	114.4	114.6	114.9	115.8

R = revised value, F = forecast value.
(Source: Process Engineering, June 1993).

Table 26.8 Annual Plant Cost Indices.

Year	Composite CE index	Equipment	Construction labor	Buildings	Engineering and supervision
1963	102.4	100.5	107.2	102.1	103.4
1964	103.3	101.2	108.5	103.3	104.2
1965	104.2	102.1	109.7	104.5	104.8
1966	107.2	105.3	112.4	107.9	106.8
1967	109.7	107.7	115.8	110.3	108.0
1968	113.7	109.9	121.0	115.7	108.6
1969	119.0	116.6	128.3	122.5	109.9
1970	125.7	123.8	137.3	127.2	110.6
1971	132.3	130.4	146.2	135.5	111.4
1972	137.2	135.4	152.2	142.0	111.9
1973	144.1	141.8	157.9	150.9	122.8
1974	165.4	171.2	163.3	165.8	134.4
1975	182.4	194.7	168.6	177.0	141.8
1976	192.1	205.8	174.2	187.3	150.8
1977	204.1	220.9	178.2	199.1	162.1
1978	218.8	240.3	185.9	213.7	161.9
1979	238.7	264.7	194.9	228.4	185.9
1980	261.2	292.6	204.3	238.3	214.0
1981	297.0	323.9	242.4	274.9	268.5
1982	314.0	336.2	263.9	290.1	304.9
1983	317.0	336.0	267.6	295.6	323.3
1984	322.7	344.0	264.5	300.3	336.3
1985	325.3	347.2	265.3	304.4	338.9
1986	318.4	336.3	263.0	303.9	341.2
1987	323.8	343.9	262.6	309.1	346.0
1988	342.5	372.7	265.6	319.2	343.3
1989	355.4	391.0	270.4	327.6	344.8
1990	357.6	392.2	271.4	329.5	355.9
1991	361.3	396.9	274.8	332.9	354.5
1992	358.2	392.2	273.0	334.6	354.1
1993	359.2	391.3	270.9	341.6	352.3

(Continued)

Table 26.8 Annual Plant Cost Indices. (*Continued*)

Year	Composite CE index	Equipment	Construction labor	Buildings	Engineering and supervision
1994	368.1	406.9	272.9	353.8	351.1
1995	381.1	427.3	274.3	362.4	347.6
1996	381.7	427.4	277.5	356.1	344.2
1997	386.5	433.2	281.9	371.4	342.5
1998	389.5	436.0	287.4	374.2	341.2
1999	390.6	435.5	292.5	380.2	339.9
2000	394.1	438.0	299.2	385.6	340.6

(Source: Vatavuk, W.M., By permission of Chemical Engineering, Jan. 2002, pp 62-70 [16].

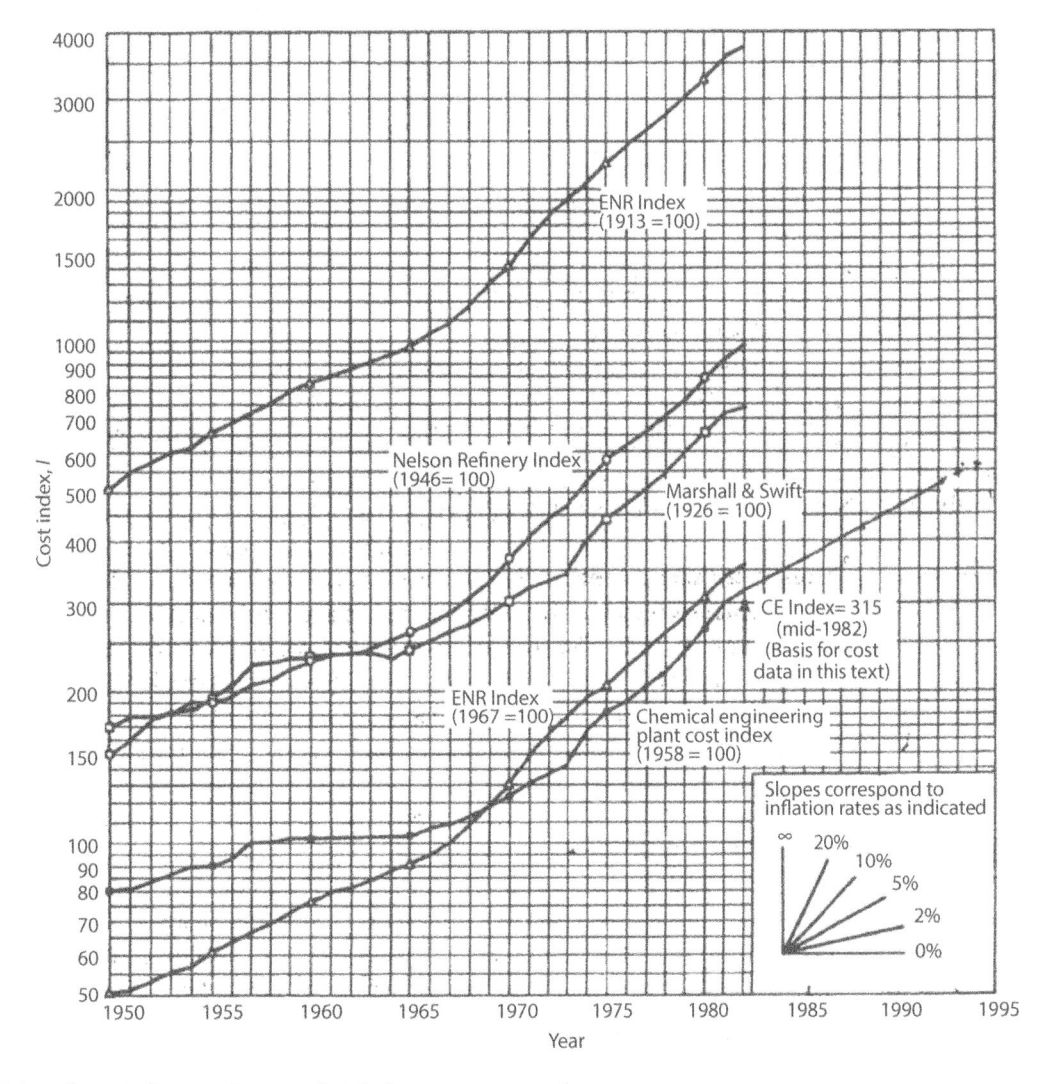

Figure 26.1 Selected cost indices pertinent to chemical process construction.

Example 26.3

A 100,000 bbl/d ethanol plant is to be completed in Houston in 2004. A similar plant in Indonesia with a capacity of 150,000 bbl/d and a final cost of $50 million, was completed in 2002. Estimate the cost of the 100,000 bbl/d plant in Houston.

Solution

A better estimate of the cost of the plant in Houston is obtained by adjustments for differences in scope, location and time. The plant in Indonesia includes piling, tankage, and owner's costs that are included in the plant to be built in Houston. However, construction costs in Houston are expected to be 1.25 times those in Indonesia. Escalation will be included as a 1.06 multiplier from 2002 to 2004. Additional costs for the Houston plant (not included in the cost estimate of the Indonesian plant) will probably involve stringent pollution control measures.

The revised estimate is as follows:

Plant in Indonesia = $50M	Escalate to 2004 $50M x 1.06 = $53M
Deduct $10M for piling, tankage and owner costs = $40M	Capacity factor estimate = $$53M \left(\frac{100,000}{150,000} \right)^{0.73} = \$39.4M$
Indonesia to Houston adjustment $40M x 1.25 = $50M	Add $5M for pollution requirements = $44.4M

26.6 Factored Cost Estimate

The purchased cost of an item of equipment, free on board (FOB) is quoted by a supplier, and may be multiplied by a factor of 1.1 to give the approximate delivered cost. The factorial methods for estimating the total installed cost of a process plant are based on a combination of materials, labor and overhead cost components. The fixed capital cost C_{FC} of a plant based on design can be estimated using the Lang factor method [17] given by the equation

$$C_{FC} = f_L \sum C_{EQ} \qquad (26.4)$$

or

Total plant cost (TPC) = Total equipment cost (TEC)

$$\text{x Equipment factor} \qquad (26.5)$$

where

f_L = 3.10 for solids processing
f_L = 3.63 for mixed solids-fluid processing
f_L = 4.74 for fluid processing

$\sum C_{EQ}$ is the sum of the delivered costs of all the major items of process equipment

The major advantage of the Lang method is that it benefits from having cost of equipment available, which is reflected in increased accuracy.

26.7 Detailed Factorial Cost Estimates

To increase accuracy, the cost factors that are compounded into the Lang factor are considered individually. Direct-cost items due to which cost is incurred in the construction of a plant, in addition to equipment items are the following:

- Equipment erection, including foundations and minor structural work.
- Electrical-related, power and lighting, including stand-by provisions.
- Process buildings, including control rooms, and structures.
- Piping, including insulation and painting.
- Instruments, local and control room.
- Storages, raw materials and finished products.
- Ancillary buildings, offices, laboratory buildings, workshops, gatehouses.
- Utilities (services), provision of plant for steam, cooling water, air, firefighting services, inert gas, effluent treatment (if not in plant costs), e.g. lagoons, holding pits, process water supplies.
- Site, site preparation, landscaping, internal roads and fencing.

Table 26.9 shows typical factors for the components of the capital cost, and these can be used to make an approximate estimate of it using equipment cost data published in the literature. In addition to the direct cost of the purchase and installation of equipment, the capital cost of a project will include the indirect costs list in Table 26.9, which can be estimated as a function of direct costs [18].

Other methods for estimating capital investment consider the fixed-capital investment required as a separate unit. These are known as the functional-unit estimates, the process step scoring method, and the modular estimate.

The functional unit may be characterized as a unit operation, unit process, or separation method, which involves energy transfer, moving parts or a high level of internals. The unit includes all process streams together with side or recycle streams. Bridgwater [19] proposed seven functional units, namely: compressor, reactor, absorber, solvent extractor, solvent recovery column, main distillation column, furnace and waste heat boiler. Taylor [20] developed the step counting method, based on a system in which a complexity score accounting for factors such as throughput, corrosion problems and reaction time is estimated for each process step. The modular estimate considers individual modules in the total system with each module consisting of a group of similar items. For the modular estimate, all heat exchangers are classified in one module, all furnaces in another, all vertical process vessels in another, and so on. The total cost estimate considers fewer than six general groupings. These are chemical processing, solids handling, site development, industrial buildings, offsite facilities, and project indirects [21]. Table 26.10 gives a more detailed explanation and definition of a functional unit.

The principle of the step count functional unit estimating methods is that the average cost of a functional unit in a process is a function of various process parameters:

$$\text{Capital cost per functional unit} = f(Q, T, P, M) \tag{26.6}$$

where

Q = capacity or throughput
T = temperature
P = pressure
M = material of construction

A number of authors have published correlations based on a step counting approach: Zevnik and Buchanan [22], Taylor [20], Timms [23], Wilson [24], Bridgwater [25] and so on. These and other correlations are reviewed and compared by Gerrard [3]. Correlations by some of these authors are as follows:

Table 26.9 Typical Factors for Estimation of Fixed Capital Cost of a Project.

Item	Process type		
	Liquids	Liquids-solids	Solids
1. Major equipment, total purchase cost	PCE	PCE	PCE
f_1 Equipment erection	0.40	0.45	0.50
f_2 Piping	0.70	0.45	0.20
f_3 Instrumentation	0.20	0.15	0.10
f_4 Electrical	0.10	0.10	0.10
f_5 Building, process	0.15	0.10	0.05
*f_6 Utilities	0.50	0.45	0.25
*f_7 Storages	0.15	0.20	0.25
*f_8 Site development	0.05	0.05	0.05
*f_9 Ancillary buildings	0.15	0.20	0.30
2. Total physical plant cost (PPC) PPC = PCE $(1+f_1+.......+f_9)$=PCE x	3.40	3.15	2.80
f_{10} Design and Engineering	0.30	0.25	0.20
f_{11} Contractor's fee	0.05	0.05	0.05
f_{12} Contingency	0.10	0.10	0.10
Fixed capital = PPC $(1 + f_{10} + f_{11} + f_{12})$ = PPC$_x$	1.45	1.40	1.35

(Source: Sinnott. R.K. [18]).
*Omitted for minor extensions or additions to existing sites.

Table 26.10 Explanation and Definition of a Functional Unit.

- A functional unit is a significant step in a process and includes all equipment and ancillaries necessary for operation of that unit. Thus, the sum of the costs of all functional units in a process gives the total capital cost.
- Generally, a functional unit may be characterized as a unit operation, unit process, or separation method that has energy transfer, moving parts and /or a high level of "internals".
- Pumping and heat exchanger are ignored as they are considered as part of a functional unit unless substantial special loads such as refrigeration are involved.
- Storage 'in process', is ignored, unless mechanical handling is involved – that is, for solids – as the cost of storage is relatively low and tends to be a constant function of the process. Large storages of raw materials, intermediates or products are usually treated separately from 'the process' in the estimate.
- Multi-stream operation is taken as one unit.
- Simple 'mechanical' separation where there are no moving parts is ignored – that is cyclone, gravity settler – as the cost is usually relatively insignificant.

(Source: Gerrard, A.M. [3]).

Zevnik and Buchanan's Method

$$CF = 2 \times 10^{(F_t + F_p + F_m)} \tag{26.7}$$

where

CF = complexity factors
F_t = temperature factor, read from a graph of F_t versus maximum (or minimum) process temperature
F_p = pressure factor, read from a graph of F_p versus maximum (or minimum) process pressure
F_m = materials of construction factor read from a table of factors from 0 for mild steel and wood to 0.4 for precious metals.

Gerrard has updated the Zevnik and Buchanan's method to 2000 with an adjusted *Engineering News Record* (ENR) construction cost index.

For plant capacity above 10 million pounds per year (4536 tonnes/year) and temperature and pressure above ambient:

$$C = 7470 \, N \, Q^{0.6} \, 10^{\left[(0.1 \log P_{max}) + \left(1.8 \times 10^{-4} (T_{max} - 300)\right) + (F_m)\right]} \tag{26.8}$$

where

C = estimated capital cost, million £ in 2000
Q = plant capacity, tonnes per year
N = number of functional units
P_{max} = maximum process pressure, atm
T_{max} = maximum process temperature, K
F_m = materials of construction factor
0 for mild steel and wood
0.1 for aluminium, brass, lower grade stainless steel
0.2 for monel, nickel, higher grade stainless steel
0.3 for hastelloy
0.4 for precious metals

For plant capacities below 10 million pounds per year (4536 tonnes/year), the equation is:

$$C = 17{,}280 \, N \, Q^{0.5} \, 10^{\left[(0.1 \log P_{max}) + \left(1.8 \times 10^{-4} (T_{max} - 300)\right) + (F_m)\right]} \tag{26.9}$$

For sub-ambient temperatures, the temperature factor becomes:

$$(0.57 - (1.9 \times 10^{-2} \, T_{min})) \text{ in place of } (1.8 \times 10^{-4} (T_{max} - 300))$$

For sub-ambient pressures, the pressure factor becomes:

$$0.1 \log\left(\frac{1}{P_{min}}\right) \text{ in place of } (0.1 \log(P_{max}))$$

Zevnik and Buchanan's definition of a functional unit refers to all equipment necessary to carry out a single significant process function.

Timm's Method

This involves a simpler approach for gas phase processes only, including both organic and inorganic chemical products. The following equations have been updated from the original work by applying an adjusted US plant cost index and converting at $2/£ to account for exchange rate and location effects.

$$C = 10560 \, NQ^{0.615} \qquad (26.10)$$

with materials of construction, temperature and pressure effects.

$$C = 4910 \, NQ^{0.639} \, F_m \, (T_{max})^{0.066} \, (P_{max})^{-0.016} \qquad (26.11)$$

where

- C = capital cost in UK £$_{2000}$, battery limits
- N = number of functional units
- Q = plant capacity, tonnes/year
- F_m = materials of construction factor
  - 1.0 for carbon steel
  - 1.15 for low grade stainless steel
  - 1.2 for medium grade stainless steel
  - 1.3 for high grade stainless steel
- T_{max} = maximum process temperature, K
- P_{max} = maximum process pressure, bar

Bridgwater's Method

For processes with predominantly liquid and/or solid handling phases, the correlation equation is:

$$C = 1930 \, N \left(\frac{Q}{s} \right)^{0.675} \qquad (26.12)$$

where

- C = capital cost £$_{2000}$, battery limits
- N = number of functional units
- Q = plant capacity, tonne/year, above 60,000
- s = reactor "conversion" = $\dfrac{\text{mass of desired product}}{\text{mass reactor input}}$

Plant capacities below 60,000 tonnes/year

$$C = 169{,}560 \, N \left(\frac{Q}{s} \right)^{0.30} \qquad (26.13)$$

Gerrard [3] developed a generalized approach based upon the principle that the capital cost is a function of a number of steps and basic process parameters, particularly capacity or throughput, can be applied to any special situation to derive a model for that industry or group of processes.

The correlations are for refuse sorting and separating processes, and for non-biological effluent treatment.

(a) Refuse sorting and separating processes:

$$C = 3250 \, N \, (686 + Q) \tag{26.14}$$

where

C = capital cost, £$_{2000}$, battery limits basis
N = number of functional units
Q = plant capacity in tonnes refuse feed per day

(b) Non-biological effluent treatment

$$C = 1900 \, N_e \, Q^{0.453} \tag{26.15}$$

where

C = capital cost, £$_{2000}$, battery limits basis, fully automated plant, Buildings are not included
Q = design throughput, imperial gal/h
N_e = number of effluent treatment steps.

Effluent treatment steps are:

- Acid/alkali neutralization.
- Chrome reduction, aqueous (if gaseous sulfur dioxide is used, add a further half step).
- Cyanide oxidations to cyanate, aqueous (if gaseous chlorine used add a further half step).
- Demulsification.
- Filter press.
- Ion exchange.
- Lime reagent preparation.
- Settlement.
- Water recycle system.

In SI units, the correlation is:

$$C = 1750 \, N_e \, Q^{0.453} \tag{26.16}$$

where

Q = throughput, m³/h

The capital cost includes all equipment, materials, labor, civils installation, commissioning and cubicle for the control panel. Reagent warehousing would cost about 20% more, and complete enclosure up to 100% more.

Equipment factored estimates (EFEs, Class4) in Table 26.2 are typically prepared during the feasibility stages of a project, when engineering is approximately 1-15% complete. They are used to determine whether there is sufficient reason for funding the project. This estimate is used to justify the funding required to complete additional engineering and design for a Class 3 or budget estimate. The first steps when preparing an EFE are to estimate the cost for each item of process equipment, to examine the equipment list carefully for completeness, and to further compare it against the PFDs and P&IDs. However, the equipment list is often in a preliminary stage when an EFE is prepared and even when the major equipment is identified, it may be necessary to assume a cost percent for auxiliary equipment that remains to be identified.

Equipment is often sized at 100% of normal operating duty; however, by the time the purchase orders have been issued, some percentages of over-sizing would have been added to the design specifications. The purchase cost of the

Table 26.11 Example of Equipment Factored Estimation.

Item description	Equipment cost ($)	Equipment factor	Total, ($)	Derived multiplier
Columns	650,000	2.1	1,365,000	
Vertical vessels	540,000	3.2	1,728,000	
Horizontal vessels	110,000	2.4	264,000	
Shell-and-tube heat exchangers	630,000	2.5	1,575,000	
Plate heat exchangers	110,000	2.0	220,000	
Pumps, motors	765,000	3.4	2,601,000	
Raw equipment cost (TEC)	2,805,000			
Direct field cost (DFC) =	2,805,000 x 2.8		7,754,000	2.8
Direct field labor (DFL) cost =	DFC x 25%		1,938,000	
Indirect field costs (IFC): Temporary construction facilities; construction services, Supplies and consumables; field staff and subsistence expenses; payroll, benefits, insurance; construction equipment and tools				
IFC =	DFL x 115%		2,229,000	
Total field costs (TFC) =	DFC + IFC		9,982,000	3.6
Home office costs (HOC): Project management, controls and estimating criteria, procurement, construction management, engineering and design, and home-office expenses				
HOC =	DFC x 30%		2,326,000	
Subtotal project cost=	TFC + HOC		12,308,000	4.4
Other project costs (OTC), including project commissioning costs				
Commissioning =	DFC x 30%		233,000	
Contingency =	(TFC + HOC) x 15%		1,846,000	
Total OTC			2,079,000	
Total Installed project cost (TIPC) =			14,387,000	5.1

Note: The multiplier is the ratio of DFC, TIPC and other costs to the raw total equipment cost of $2,805,000.
The cost of each type of equipment was multiplied by a factor to derive the installed DFC for that unit. For instance, the total cost of all vertical vessels ($540,000) was multiplied by an equipment factor of 3.2 to obtain an installed DFC of $1,728,000. The total installed cost (TIC) for this project is $14, 387,000.

(Source: Dysert, L., By permission of Chemical Engineering, Oct. 2001, pp 70-81).

Table 26.12 Heat Exchanger Discipline Equipment Factors.

	Factor	Cost $
Equipment cost	1.0	10,000
Installation labor	0.05	500
Concrete	0.11	1,100
Structural steel	0.11	1,100
Piping	1.18	11,800
Electrical parts	0.05	500
Instrumentation	0.24	2,400
Painting	0.01	100
Insulation	0.11	1,100
Total DFC	2.86	$28,600

This illustrates equipment factors for a Type 316 stainless steel heat exchanger with a surface area of 2,400 ft². The purchase cost of $10,000 is multiplied by each factor to generate the DFC for that discipline

(Source: Dysert, L., By permission of Chemical Engineering, Oct. 2001, pp 70-81).

equipment is obtained from purchase orders, published equipment cost data, and vendor quotations. It is essential to accurately determine the equipment costs as the material cost of equipment often represents 20-40% of the total project costs for process plants. Once the equipment cost is determined, the appropriate equipment factors may be generated and applied, by applying the necessary adjustments for equipment size, metallurgy and operating conditions. Tables 26.11 shows examples of equipment factored estimates, and Table 26.12 shows heat exchanger equipment factors respectively. Table 26.13 summarizes the various factors affecting the capital cost of chemical plants.

26.8 Bare Module Cost for Equipment

The bare module equipment cost represents the sum of direct and indirect costs, and are shown in Table 26.13. The conditions specified for the base case are [26]

1. Unit fabricated from most common material, usually carbon steel (SS).
2. Unit operated at near ambient temperature.

The equation used to calculate the bare module cost for each piece of equipment is

$$C_{BM} = C_p F_{BM} \qquad (26.17)$$

where

C_{BM} = bare module equipment cost – consist of both direct and indirect costs for each unit

F_{BM} = bare module cost factor – multiplication factor to account for the items in Table 26.14, plus the specific materials of construction and operating pressure
C_p = purchased cost for base conditions – cost of equipment made of the most common material, usually carbon steel and operating at near ambient pressures.

The entries in Table 26.14 can be explained as follows:

Column 1: Lists the factors given in Table 26.13.
Column 2: Provides equations used to evaluate each of the costs. These equations introduce multiplication cost factors, α_i. Each cost item, other than the purchased equipment cost, introduces a separate factor.
Column 3: For each factor, the cost is related to the purchased cost C_p by an equation of the form

$$C_{XX} = C_p f(\alpha_i) \tag{26.18}$$

The function, $f(\alpha_i)$, for each factor is given in Column 3. Using Table 26.13, Equations (26.17) and (26.18), the bare module factor can be expressed by

$$F_{BM} = [1 + \alpha_L + \alpha_{FIT} + \alpha_O \alpha_L + \alpha_E][1 + \alpha_M] \tag{26.19}$$

26.9 Summary of the Factorial Method

A quick and approximate estimate of the investment which would be required for a project can be achieved from the following procedures:

1. Prepare material and energy balances, draw up preliminary flowsheets, size major equipment items and select materials of construction.
2. Estimate the purchase cost of the major equipment items (using the general literature).
3. Calculate the total physical plant cost (PPC), using the factors given in Table 26.9.

$$PPC = PCE(1 + f_1 + \ldots\ldots\ldots + f_9) \tag{26.20}$$

4. Calculate the indirect costs from the direct costs using the factors given in Table 26.9.
5. The direct plus indirect costs give the total fixed capital.
6. Estimate the working capital as a percentage of the fixed capital (10 - 20%).
7. Add the fixed and working capital to obtain the total investment required.

A summary of the costs to consider in the evaluation of the total capital cost of a chemical plant is shown in Table 26.13.

26.10 Computer Cost Estimating

There are two main methods of using a computer in cost estimates. The first involves calculating the cost given a correlation, a curve fit, or a more rigorous model. The second is to use it to analyze a set of historical costs, with the aim of showing important cost trends.

Many simulation software packages with mass/energy balance simulators include costing routines. Some of these packages are developed in-house or alternatively from many vendors of cost estimators. The AACE provides an Excel spreadsheet on cost estimates and website links to vendors who provide computer packages on cost estimation. Most spreadsheets have statistical packages and numerical optimizers to analyze records. The power law or exponential equation is expressed by

$$C = kS^n \tag{26.21}$$

where n represents index of that type of equipment.

This can be extended to a multivariable form as:

$$C = kS_1^{n1} S_2^{n2} S_3^{n3} \ldots \tag{26.22}$$

Table 26.13 Factors Affecting the Capital Cost of Chemical Plants.

Factors associated with the installation of equipment	Symbol	Comments
(1) Direct Project Expenses (a) Equipment f.o.b. cost	C_P	Purchased cost of equipment at manufacturer's site. (FOB = free on board).
(b) Materials required for installation	C_M	Includes all piping, insulation and fire proofing, foundations and structural supports, instrumentation and electrical, and painting associated with the equipment.
(c) Labor to install equipment and material	C_L	Includes all labor associated with installing the equipment and materials mentioned in (a) and (b).
(2) Indirect Project Expenses (a) Freight, insurance, and taxes	C_{FIT}	Includes all transportation costs for shipping equipment and materials to the plant site; all insurance on the items shipped; and any purchases taxes that may be applicable.
(b) Contractor's overhead	C_O	Includes all fringe benefits such as vacation, sick leave retirement benefits, etc.; labor burden such as social security and unemployment insurance, etc.; and salaries and overhead for supervisory personnel.
(c) Contractor engineering expenses	C_E	Includes salaries and overhead for the engineering, drafting, and project management personnel on the project.
(3) Contingency and Fee (a) Contingency	C_{Cont}	A factor included to cover unforeseen circumstances. These may include loss of time due to storms, and strikes, small changes in the design, and unpredicted price increases.
(b) Contactor's fee	C_{Fee}	This fee varies depending on the type of plant and variety of other factors.

(Continued)

Table 26.13 Factors Affecting the Capital Cost of Chemical Plants. (*Continued*)

Factors associated with the installation of equipment	Symbol	Comments
(4) Auxiliary Facilities (a) Site development	C_{site}	Includes the purchase of land; grading and excavation of the site; installation and hook-up of electrical, water, and sewer systems; and construction of all internal roads, walk ways, and parking lots.
(b) Auxiliary Buildings	C_{Aux}	Includes administration offices, maintenance shop and control rooms, warehouses and service buildings (e.g., cafeteria, dressing rooms, and medical facility).
(c) Offsites and Utilities	C_{Off}	Includes raw material and final product storage, raw material and final product loading and unloading facilities, all equipment necessary to supply required process utilities (e.g., cooling water, steam generation, fuel distribution systems, etc.), central environmental control facilities (e.g., waste water treatment, incinerators, flares, etc), and fire protection systems.

(Sources: Turton, R., et al. [27]).

Table 26.14 Equations for Evaluating Direct, Indirect, Contingency, and Fee Costs.

Factor	Basic equation	Multiplying factor to be used with purchased cost, C_P
1. Direct (a) Equipment (b) Materials (c) Labor	$C_P = C_P$ $C_M = \alpha_M C_P$ $C_L = \alpha_L (C_P + C_M)$	1.0 $(1 + \alpha_M) \alpha_L$ $(1 + \alpha_M)(1 + \alpha_L)$
Total Direct	$C_{DE} = C_P + C_M + C_L$	$(1 + \alpha_M)(1.0 + \alpha_L)$
2. Indirect (a) Freight, etc (b) Overhead (c) Engineering	$C_{FIT} = \alpha_{FIT}(C_P + C_M)$ $C_O = \alpha_O C_L$ $C_E = \alpha_E (C_P + C_M)$	$(1.0 + \alpha_M) \alpha_{FIT}$ $(1.0 + \alpha_M) \alpha_L \alpha_O$ $(1.0 + \alpha_M) \alpha_E$
Total Indirect	$C_{IDE} = C_{FIT} + C_O + C_E$	$(1.0 + \alpha_M)(\alpha_{FIT} + \alpha_L \alpha_O + \alpha_E)$
Bare Module	$C_{BM} = C_{IDE} + C_{DE}$	$(1.0 + \alpha_M)(1.0 + \alpha_L + \alpha_{FIT} + \alpha_L \alpha_O + \alpha_E)$
3. Contingency & Fee (a) Cont. (b) Fee	$C_{Cont} = \alpha_{Cont} C_{BM}$ $C_{Fee} = \alpha_{Fee} C_{BM}$	$(1.0 + \alpha_M)(1.0 + \alpha_L + \alpha_{FIT} + \alpha_L \alpha_O + \alpha_E) \alpha_{Cont}$ $(1.0 + \alpha_M)(1.0 + \alpha_L + \alpha_{FIT} + \alpha_L \alpha_O + \alpha_E) \alpha_{FEE}$
Total Module	$C_{TM} = C_{BM} + C_{Cont} + C_{Fee}$	$(1.0 + \alpha_M)(1.0 + \alpha_L + \alpha_{FIT} + \alpha_L \alpha_O + \alpha_E)(1.0 + \alpha_{Cont} + \alpha_{Fee})$

(Source: Turton, R, et al. [27]).

where S_1, S_2, and so on are independent measures of size, for example, height and diameter of a vessel. This equation can easily be linearized, and an extension of these forms is to add a further constant, k', to represent the fixed cost element.

$$C = k' + kS^n \tag{26.23}$$

and

$$C = k' + kS_1^{n1} S_2^{n2} S_3^{n3} \ldots\ldots\ldots \tag{26.24}$$

Recently, Petley and Edwards [27] have shown the use of fuzzy matching to be useful in costing complete chemical plant, and Gerrard and Brass [28] have used this approach together with neural networks and rational polynomials to provide equipment cost predictions.

26.11 Project Evaluation

Introduction

Project evaluation enables the technical and economic feasibility of a chemical process to be assessed using preliminary process design and economic evaluations. Once a process flowsheet is available, these evaluations can be classified into a number of steps: material balance calculations, equipment sizing, equipment cost determination, utilities requirements, investment cost estimation, sales volume forecasting, manufacturing cost estimation, and finally profitability and sensitivity analysis.

The results of an economic evaluation are reviewed together with other relevant aspects, for example, competition and likely product life, in arriving at project investment decisions. The decisions are essential in order to plan and allocate the available resources for long-term use. The objectives of an economic appraisal are the following:

- To ensure that the expected future benefits justify the expenditure of resources.
- To choose the best project from among alternatives to achieve future benefits.
- To utilize all available resources.

These aim to ensure so far as is practicable that the right project is undertaken and is likely to attain the desired profitability.

In petroleum refining and CPIs, an investment project may arise from any of a range of activities. It may be a minor modification to an existing plant, a major plant expansion or revamping, a completely new plant (on an existing or greenfield site) or the development of an entirely new process or product. Economic assessment for a major or a new plant may be carried out with increasing degrees of accuracy at different stages as it progresses. This may be based on research and development (R & D), through various stages of the project (e.g., pilot plants, material evaluation, preliminary plant design), and leads to the decision whether or not to proceed with investment in a full-scale plant for the process. Table 26.15 lists examples of the use of engineering economics. The approach used for economic evaluation depends on the quality of the information, which depends on the stage in the project at which it is being undertaken.

Investment decisions are often based on several criteria such as: annual return on investment (ROI), payback period (PBP), net present value (NPV), the average rate of return (ARR), present value ratio (PVR) or the internal rate of return (IRR). Discounted cash flow rate on return (DCFRR) is another popular means of evaluating the economic viability of a proposed project. Hortwiz [29] recommended the DCFRR as the best means to determine the ROI since it accounts for the time value of money (the cost of money), which states that the value of a sum of money is time dependent. For example, given the choice of receiving say $100 now or $100 in a year's time, it is preferred to receive the money now since it could be invested (at no risk) to obtain interest until it is required. Therefore, the $100 received now is worth more than the $100 received in a year's time by the amount of interest it could earn in the intervening year. Conversely, $100 received in a year's time is worth less than $100 received now, and $100 received in 2 year's time is less than this, and so on. The time value of money is the result of the availability of investment

Table 26.15 Use of Chemical Engineering Economics.

1. Production or plant technical services
(i) Plant equipment continuously needs repair, scheduled or unscheduled replacement, or modernization. The engineer responsible should know approximately what the comparative performance, costs and payout periods are, even if there is a plant engineering group specializing in that type of analysis, or later a firm price quotation will be obtained.
(ii) Plant changes, e.g., initiated by the increasing costs of energy and to meet environmental requirements, necessitates that many energy saving, pollution, and hazardous waste control possibilities must be considered. As a basis for recommendations the responsible engineer should personally conduct design and cost estimates, pay-back, and economic calculations on the alternatives before making even preliminary recommendations.
(iii) Competitors' processing methods, as well as R&D, sales, or management suggested changes must be continuously examined. Managers or other groups may be responsible, but the engineer can help by making preliminary cost estimates and economic analyses of the changes as a guide to his/her own thinking and assessment of the group's position.
(iv) All engineers should have some feel for their company's business, products, and economics. This requires a basic understanding of company annual reports and of general industry economic news reinforced by regular reading.
2. Research and Development
(i) During the analytical phase of creative thinking many novel ideas require a rapid cost estimation and economic analysis to provide a clearer idea of their merit. Managers or others may be assigned to do this work, but the chances of novel proposals being accepted increases if some economic screening is performed by the originator.
(ii) Many obstacles, or alternative directions that may be taken to attempt to solve the problems, arise during R & D. Brief cost estimates and economic analyses often help in determining which are the optimum.
(iii) Following success during an early or intermediate stage of an R&D program, new funding requests are usually required to continue the study. These requests benefit from support by preliminary economic analyses. In the final stages of a project the engineer may be part of a team assigned to provide a more definitive preliminary economic projection and analysis.
(iv) Credibility with production, sales, or management personnel is enhanced by demonstrating a reasonable knowledge of the costs and economics of the projects under study, and general industry economics.
3. Sales
(i) A general knowledge of company costs, profits, and competition are essential for more effective salesmanship.
(ii) Salesmen may recommend new products, improvements, or pricing ideas to their management. A cost and economic estimate for these ideas should be helpful in the proposal report.
(iii) Salesmen may perform market surveys. A general economic knowledge of the industries and companies surveyed may be essential, and is always useful.
(iv) Salesmen, may progress into management, where economic knowledge is essential.
4. Engineering
(i) As a result of the specialization within most engineering companies and engineering departments, cost estimating and economics may not be required directly in many engineering company or departmental jobs. Other jobs will however deal exclusively with cost estimating and economic analysis, and all benefit from a good, fluent knowledge of the basic economic procedures. Engineering departments or companies usually have very well-developed in-house methods and data that must be used, but the basics are still applicable.

(Continued)

Cost Estimation and Economic Evaluation 493

Table 26.15 Use of Chemical Engineering Economics. (*Continued*)

5. General
(i) All chemical engineers are assumed to know the rudiments of cost estimating, economic evaluation, and the economics of their industry. A high percentage will find this knowledge useful or necessary throughout their careers.
(ii) Progression into management in associated areas will necessitate further understanding of microeconomics.
(iii) All work situations are competitive. One means of enhancing advancement potential is to demonstrate to superiors knowledge and an interest in management, business, and economics. Associated with this is the demonstration of ability, an interest in accepting responsibility, and ability to communicate. Promotion may go to those perceived as able to "manage", as those with MBA (Master of Business Administration) degrees, or equivalent capabilities. A confidence in the ability to acquire managerial skills as required, and a knowledge of economics, should render our chemical engineer an equal or preferred candidate for advancement.

(Source: After Garret [39]).

possibilities, which generates returns on any investment made. This time value of money does not account for inflation, which is a separate factor. The IRR as an investment criterion introduces the possibility that given cash flows may result in more than one IRR. Cannaday *et al.* [30] developed a method for determining the relevance of an IRR. They inferred that an IRR is relevant, if its derivative with respect to each of the cash flows is positive.

Powell [31] reviewed the basics of various discounted cash flow techniques for project evaluation. Discounting is a method that accounts for the time value of money to provide either for the capital, or to convert the cash flows to a common point in time so that they are summed. Ward [32] proposed a new concept known as the net return rate (NRR) that provides a better indication of a project's profitability. Techniques and criteria for economic evaluation of projects are widely available in the literature and texts [33–38]. A summary of the conventional decision criteria is given here.

26.12 Cash Flows

Return on Investment (ROI)

In engineering economic evaluation, rate of return on investment (ROI) is the percentage ratio of average yearly profit (net cash flow), over the productive life of the project, divided by the total initial investment. This is calculated after income tax has been deducted from the gross or pre-tax income. The remaining net income may be used either for paying dividends for re-investment or spent for other means. ROI is defined by

$$\text{ROI} = \frac{\text{Annual return}}{\text{Investment}} \times 100 \qquad (26.25)$$

The annual return may be the gross income, net pre-tax income, net after-tax income, cash flow or profit. These may be calculated for one particular year or as an average over the project life. Investment may be the original total investment, depreciated book-value investment, lifetime average investment fixed capital investment or equity investment. The investment includes working capital and sometimes capitalized expenses for example interest on capital during construction.

The proper evaluation of costs as they affect the selection of processes and equipment is not included in this book. Every process engineer should however be cognizant of the relationships. There are several methods to evaluate return on invested money; the nomograph of Figure 26.2 represents one. It is a useful guide [39] to estimate the order

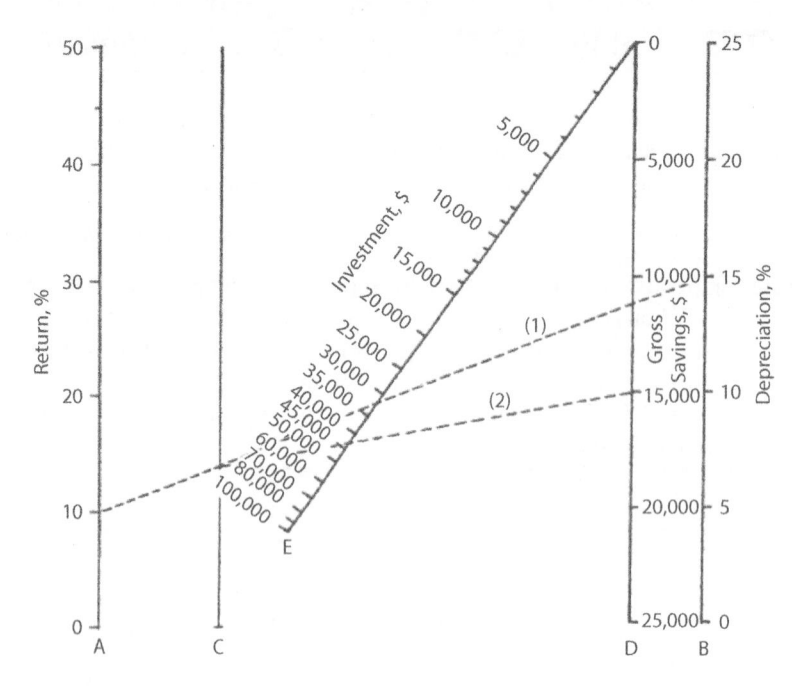

Figure 26.2 Annual savings, return, and depreciation fix justifiable investment. (By permission from G. A., Larson, Power, Sept. 1995.)

of magnitude of a return on expenditure to gain savings in labor and/or material costs. The nomograph is used to determine the investment justified by gross annual savings, assuming a percent return, a percent annual depreciation charge, and a 50% Federal tax on net savings.

$$\text{Return} = \frac{(\text{Gross savings} - \text{Depreciation} \times \text{Investment})(1 - \text{Federal Tax})}{\text{Investment}} \quad (26.26)$$

Example 26.4: Justifiable Investment for Annual Savings [39]

Find the justifiable investment for a gross annual savings of $15,000 when a return of 10% and a depreciation rate of 15 percent are specified.

1. From Figure 26.2, connect scales A and B.
2. From the intersection with the C scale, connect a line to the D scale.
3. At the intersection of line (2) with the inclined investment scale, E, read that a $43,000 investment is justified to save $15,000 gross per year.

Accounting Coordination

All new plants, and changes to existing facilities and plants must be coordinated with a cost accounting system. Often costs associated with the building, services, utilities, and site development must be separated from each other. Each company has reason and need for various arrangements in order to present proper information for tax purposes and depreciation. The project engineer is usually responsible for this phase of coordination through the engineering groups, but the process engineer may need to present proper breakdown details; these then serve to coordinate the cost breakdowns. Figure 26.3 is an example of such an accounting diagram.

Payback Period (PBP)

Payback period (PBP) is widely used when long-term cash flows, that is, over a period of years are difficult to forecast, since no information is required beyond the breakeven point. It may be used for preliminary evaluation or as a

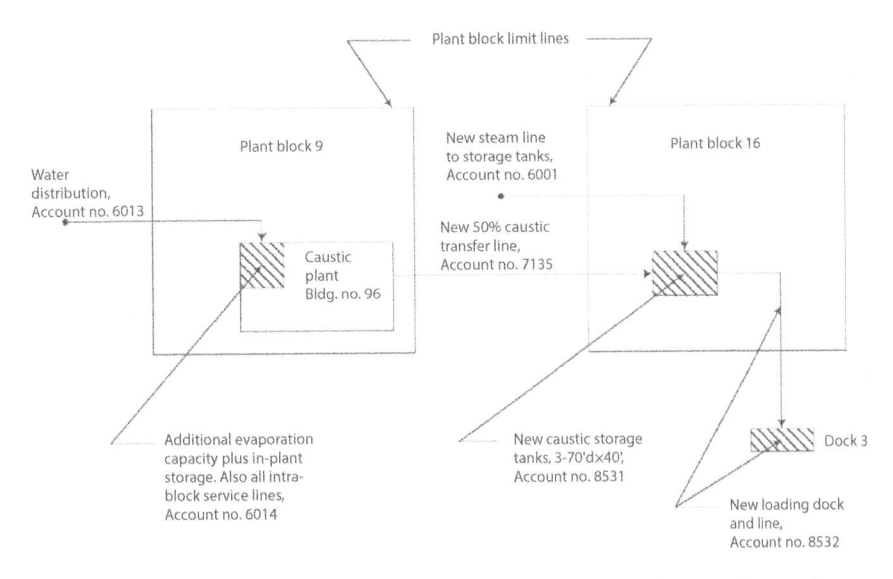

Figure 26.3 Account diagram for accumulation of project costs. Cost estimates must be made to confirm to the same scope.

project-screening device for high-risk projects in times of financial uncertainty. PBP period is usually measured as the time from the start of production to recovery of the capital investment. The PBP period is the time taken for the cumulative net cash flow from start-up of the plant to equal the depreciable fixed capital investment ($C_{FC} - S$). It is the value of t that satisfies the equation

$$\sum_{t=0}^{t=(PBP)} C_{CF} = (C_{FC} - S) \qquad (26.27)$$

where

C_{CF} = net annual cash flow.
C_{FC} = fixed capital cost.
S   = salvage value.

Figure 26.4 shows the cumulative cash flow diagram for a project. The PBP is the time that elapses from the start of the project A, to the breakeven point E, where the rising part of the curve passes the zero cash position line. The PBP thus measures the time required for the cumulative project investment and other expenditure to be balanced by the cumulative income.

Example 26.5

Consider the following cash flow:

Year	0	1	2	3	4
Cash flow	−$8,000	$3,000	$4,000	$5000	$5000

Here, the cumulative cash flow at the end of the second year is $3,000 + $4,000 = $7,000, which is less than the initial investment, but the cumulative cash flow at the end of the third year is $3,000 + $4,000 + $5,000 = $12,000, which is more than the initial investment. The payback period is thus between 2 and 3 years, assuming that the cash flow of year 3 is received uniformly throughout the year. The PBP is calculated as follows:

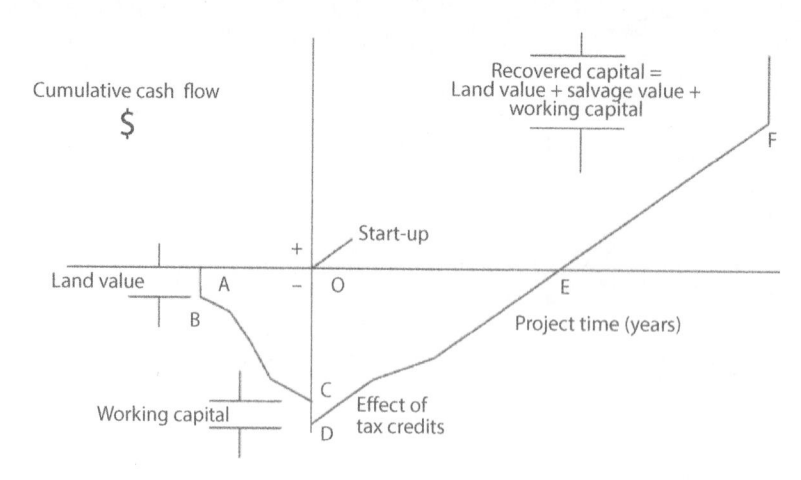

Figure 26.4 Cumulative cash flow diagram.

1	Initial investment	$8,000
2	Cash flow recovered to end of second year	$7,000
3	Amount still to be recovered (line 1 – line 2)	$1,000
4	Amount recovered in third year	$5,000
5	Line (3) divided by line (4) ($1,000/$5,000)	0.2 year
6	Payback period (2 years + number of years in line 5)	2.2 years

In this example, 0.2 is the fraction of year number 3 that it will take to recover $1,000. Adding this fraction to the 2 years during which $7,000 is recovered yields a PBP of 2 + 0.2 = 2.2 years.

The PBP method of evaluating investments has a number of flaws, and is inferior to other methods. A major disadvantage is that after the PBP, all the cash flows are completely ignored. It also ignores the timing of the cash flows within the PBP.

Payback period and ROI select particular features of the project cumulative cash flow and ignore others. They take no account of the pattern of cash flow during a project. Other techniques such as net present value (NPV) and discounted cash flow rate of return (DCFRR) are more comprehensive because they take account of the changing pattern of project net cash flow with time. They also take account of the "time value" of money.

Present Worth (or Present Value)

In an economic evaluation of a project, it is often necessary to evaluate the present value of funds that will be received at some definite time in the future. The present value (PV) of a future amount can be considered, as the present principal at a given rate and compounded to give the actual amount received at a future date. The relationship between the indicated future amount and the present value is determined by a discount factor. Discounting evaluates each year's flow on an equal basis. It does this by means of the discount, or present value factor, which is the reciprocal of the compound interest factor $(1 + i)^n$, where

i = interest rate
n = the year in which the interest is compounded

$$\text{The discount factor} = \frac{1}{(1+i)^n} \quad (26.28)$$

If C_n represents the amount available after n interest periods, p is the initial principal and the discreet compound interest rate is i, then PV can be expressed as:

$$PV = p = \frac{C_n}{(1+i)^n} \quad (26.29)$$

Net Present Value (NPV)

The net present value (NPV) of a project is the cumulative sum of the discounted cash flows including the investment. The NPV corresponds to the total discounted net return, above and beyond the cost of capital and the recovery of the investment. The NPV represents a discounted return or profit, but is not a measure of the profitability. Each cash flow is evaluated by computing its present value. This is done by taking a cash flow of year n and multiplying it by the discount factor for the nth year.

$$\text{Present value of } p_n = C_n \left[\frac{1}{(1+i)^n} \right] \quad (26.30)$$

The present value p at year 0 of a cash flow C_t in year t at an annual discount rate of i is

$$p = \frac{C_t}{(1+i)^t} \quad (26.31)$$

For a complete project, the earlier cash flows are usually negative and the later ones are usually positive. The NPV is the sum of the individual present values of the yearly cash flows. This is expressed as:

$$NPV = C_0 + \frac{C_1}{(1+i)} + \frac{C_2}{(1+i)^2} + \ldots\ldots\ldots + \frac{C_n}{(1+i)^n} \quad (26.32)$$

Eq. (26.32) can be expressed as:

$$NPV = \sum p = \sum_{i=0}^{i=n} \frac{C_t}{(1+i)^n} \quad (26.33)$$

where

NPV = net present value
C_0 = initial investment
C_n = cash flow
n = year n
i = interest rate of return (ROI/100)

The life of the project n years must be specified together with the estimated cash flows in each year up to n.

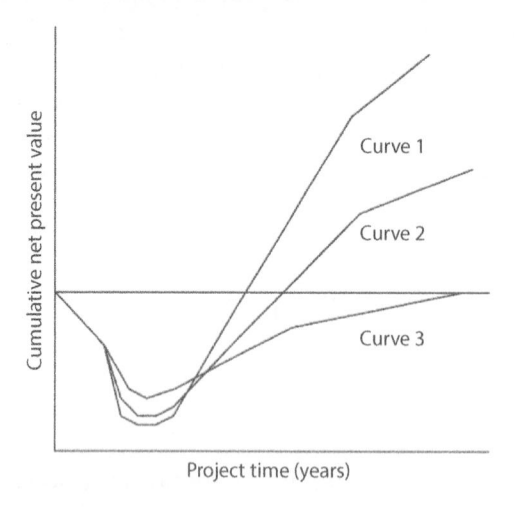

Figure 26.5 Cumulative net present value (NPV) of a project.

If a project includes a series of identical yearly cash flows, such as the net inflow for several years, their present values can be obtained in one calculation instead of obtaining these individually. For yearly cash flows C from year m to year n, then the sum of their present values is:

$$p = \frac{C}{i(1+i)^{m-1}} \left[1 - \frac{1}{(1+)^{n-m+1}} \right] \qquad (26.34)$$

Figure 26.5 shows the cumulative NPV stages in a project, and Table 26.16 shows the discount factors for computing the NPV.

Assuming that the investment is made in year 0 ($C_0 = I$), and the cash flows over the project life are constant, then Eq. (26.32) simplifies to give

$$NPV = \left[C \sum_{n=1}^{n} \left(\frac{1}{1+i} \right)^n \right] - I \qquad (26.35)$$

The term $\frac{1}{(1+i)}$ is a geometric progression whose sum can be expressed as the single term

$$\sum_{n=1}^{n} \frac{1}{(1+i)^n} = \frac{(1+i)^n - 1}{i(1+i)^n} \qquad (26.36)$$

where Eq. (26.36) becomes the present value of an annuity. Eq. (26.32) can thus be expressed as:

$$NPV = C \left[\frac{(1+i)^n - 1}{i(1+i)^n} \right] - I \qquad (26.37)$$

The IRR is that value of i that makes NPV equal to 0. Therefore, if NPV is set to 0, the IRR that makes the future cash flows equal to the investment (the "breakeven" point) can be estimated. Eq. (26.37) then becomes:

Table 26.16 Discount Factors (This Table Shows the Present Value of Unity Discounted for Different Numbers of Years and at Different Rates of Discount).

Rate of discount/year	1%	2%	3%	4%	5%	6%	7%	8%	9%	10%
0	1.0000	1.0000	1.0000	1.0000	1.0000	1.0000	1.0000	1.0000	1.0000	1.0000
1	0.9901	0.9804	0.9709	0.9615	0.9524	0.9434	0.9346	0.9259	0.9174	0.9091
2	0.9803	0.9612	0.9426	0.9246	0.9070	0.8900	0.8734	0.8573	0.8417	0.8264
3	0.9706	0.9423	0.9151	0.8890	0.8638	0.8396	0.6163	0.7938	0.7722	0.7513
4	0.9610	0.9238	0.8885	0.8548	0.8227	0.7921	0.7629	0.7350	0.7084	0.6830
5	0.9515	0.9057	0.8626	0.8219	0.7835	0.7473	0.7130	0.6806	0.6499	0.6209
6	0.9420	0.8880	0.8375	0.7903	0.7462	0.7050	0.6663	0.6302	0.5963	0.5645
7	0.9327	0.8706	0.8131	0.7599	0.7107	0.6651	0.6227	0.5835	0.5470	0.5132
8	0.9235	0.8535	0.7894	0.7307	0.6768	0.6274	0.5820	0.5403	0.5019	0.4665
9	0.9143	0.8368	0.7664	0.7026	0.6446	0.5919	0.5439	0.5002	0.4604	0.4241
10	0.9053	0.8203	0.7441	0.6756	0.6139	0.5584	0.5083	0.4632	0.4224	0.3855
11	0.8963	0.8043	0.7224	0.6496	0.5847	0.5268	0.4751	0.4289	0.3875	0.3505
12	0.8874	0.7885	0.7014	0.6246	0.5568	0.4970	0.4440	0.3971	0.3555	0.3186
13	0.8787	0.7730	0.6810	0.6006	0.5303	0.4688	0.4150	0.3677	0.3262	0.2897
14	0.8700	0.7579	0.6611	0.5775	0.5051	0.4423	0.3878	0.3405	0.2992	0.2633
15	0.8613	0.7430	0.6419	0.5553	0.4810	0.4173	0.3624	0.3152	0.2745	0.2394
16	0.8528	0.7284	0.6232	0.5339	0.4581	0.3936	0.3387	0.2919	0.2519	0.2176
17	0.8444	0.7142	0.6050	0.5134	0.4363	0.3714	0.3166	0.2703	0.2311	0.1978
18	0.8360	0.7002	0.5874	0.4936	0.4155	0.3503	0.2959	0.2502	0.2120	0.1799
19	0.8277	0.6864	0.5703	0.4746	0.3957	0.3305	0.2765	0.2317	0.1945	0.1635
20	0.8195	0.6730	0.5537	0.4564	0.3769	0.3118	0.2584	0.2145	0.1784	0.1486
	11%	12%	13%	14%	15%	16%	17%	18%	19%	20%
0	1.0000	1.0000	1.0000	1.0000	1.0000	1.0000	1.0000	1.0000	1.0000	1.0000
1	0.9009	0.8929	0.8850	0.8772	0.8696	0.8621	0.8547	0.8475	0.8403	0.8333
2	0.8116	0.7972	0.7831	0.7695	0.7561	0.7432	0.7305	0.7182	0.7062	0.6944
3	0.7312	0.7118	0.6931	0.6750	0.6575	0.6407	0.6244	0.6086	0.5934	0.5787
4	0.6587	0.6355	0.6133	0.5921	0.5718	0.5523	0.5337	0.5158	0.4987	0.4823
5	0.5935	0.5674	0.5428	0.5194	0.4972	0.4761	0.4561	0.4371	0.4190	0.4019

(Continued)

Table 26.16 Discount Factors (This Table Shows the Present Value of Unity Discounted for Different Numbers of Years and at Different Rates of Discount). (*Continued*)

	11%	12%	13%	14%	15%	16%	17%	18%	19%	20%
6	0.5346	0.5066	0.4803	0.4556	0.4323	0.4104	0.3898	0.3704	0.3521	0.3349
7	0.4817	0.4523	0.4251	0.3996	0.3759	0.3538	0.3332	0.3139	0.2959	0.2791
8	0.4339	0.4039	0.3762	0.3506	0.3269	0.3050	0.2848	0.2660	0.2487	0.2326
9	0.3909	0.3606	0.3329	0.3075	0.2843	0.2630	0.2434	0.2255	0.2090	0.1938
10	0.3522	0.3220	0.2946	0.2697	0.2472	0.2267	0.2080	0.1911	0.1756	0.1615
11	0.3173	0.2875	0.2607	0.2366	0.2149	0.1954	0.1778	0.1619	0.1476	0.1346
12	0.2858	0.2567	0.2307	0.2076	0.1869	0.1685	0.1520	0.1372	0.1240	0.1122
13	0.2575	0.2292	0.2042	0.1821	0.1625	0.1452	0.1299	0.1163	0.1042	0.0935
14	0.2320	0.2046	0.1807	0.1597	0.1413	0.1252	0.1110	0.0985	0.0876	0.0779
15	0.2090	0.1827	0.1599	0.1401	0.1229	0.1079	0.0949	0.0835	0.0736	0.0649
16	0.1883	0.1631	0.1415	0.1229	0.1069	0.0930	0.0811	0.0708	0.0618	0.0541
17	0.1696	0.1456	0.1252	0.1078	0.0929	0.0802	0.0693	0.0600	0.0520	0.0451
18	0.1528	0.1300	0.1108	0.0946	0.0808	0.0691	0.0592	0.0508	0.0437	0.0376
19	0.1377	0.1161	0.0981	0.0829	0.0703	0.0596	0.0506	0.0431	0.0367	0.0313
20	0.1240	0.1037	0.0868	0.0728	0.0611	0.0514	0.0433	0.0365	0.0308	0.0261
	21%	22%	23%	24%	25%	26%	27%	28%	29%	30%
0	1.0000	1.0000	1.0000	1.0000	1.0000	1.0000	1.0000	1.0000	1.0000	1.0000
1	0.8264	0.8197	0.8130	0.8065	0.8000	0.7937	0.7874	0.7813	0.7752	0.7692
2	0.6830	0.6719	0.6610	0.6504	0.6400	0.6299	0.6200	0.6104	0.6009	0.5917
3	0.5645	0.5507	0.5374	0.5245	0.5120	0.4999	0.4882	0.4768	0.4658	0.4552
4	0.4665	0.4514	0.4369	0.4230	0.4096	0.3968	0.3844	0.3725	0.3611	0.3501
5	0.3855	0.3700	0.3552	0.3411	0.3277	0.3149	0.3027	0.2910	0.2799	0.2693
6	0.3186	0.3033	0.2888	0.2751	0.2621	0.2499	0.2383	0.2274	0.2170	0.2072
7	0.2633	0.2486	0.2348	0.2218	0.2097	0.1983	0.1877	0.1776	0.1682	0.1594
8	0.2176	0.2038	0.1909	0.1789	0.1678	0.1574	0.1478	0.1388	0.1304	0.1226
9	0.1799	0.1670	0.1552	0.1443	0.1342	0.1249	0.1164	0.1084	0.1011	0.0943
10	0.1486	0.1369	0.1262	0.1164	0.1074	0.0992	0.0916	0.0847	0.0784	0.0725
11	0.1228	0.1122	0.1026	0.0938	0.0859	0.0787	0.0721	0.0662	0.0607	0.0558
12	0.1015	0.0920	0.0834	0.0757	0.0687	0.0625	0.0568	0.0517	0.0471	0.0429

(*Continued*)

Table 26.16 Discount Factors (This Table Shows the Present Value of Unity Discounted for Different Numbers of Years and at Different Rates of Discount). (*Continued*)

	21%	22%	23%	24%	25%	26%	27%	28%	29%	30%
13	0.0839	0.0754	0.0678	0.0610	0.0550	0.0496	0.0447	0.0404	0.0365	0.0330
14	0.0693	0.0618	0.0551	0.0492	0.0440	0.0393	0.0352	0.0316	0.0283	0.0254
15	0.0573	0.0507	0.0448	0.0397	0.0352	0.0312	0.0277	0.0247	0.0219	0.0195
16	0.0474	0.0415	0.0364	0.0320	0.0281	0.0248	0.0218	0.0193	0.0170	0.0150
17	0.0391	0.0340	0.0296	0.0258	0.0225	0.0197	0.0172	0.0150	0.0132	0.0116
18	0.0323	0.0279	0.0241	0.0208	0.0180	0.0156	0.0135	0.1180	0.0102	0.0089
19	0.0267	0.0229	0.0196	0.0168	0.0144	0.0124	0.0107	0.0092	0.0079	0.0068
20	0.0221	0.0187	0.0159	0.0135	0.0115	0.0098	0.0084	0.0072	0.0061	0.0053
	31%	32%	33%	34%	35%	36%	37%	38%	39%	40%
0	1.0000	1.0000	1.0000	1.0000	1.0000	1.0000	1.0000	1.0000	1.0000	1.0000
1	0.7634	0.7576	0.7519	0.7463	0.7407	0.7353	0.7300	0.7246	0.7194	0.7143
2	0.5827	0.5739	0.5653	0.5569	0.5487	0.5407	0.5328	0.5251	0.5176	0.5102
3	0.4448	0.4348	0.4251	0.4156	0.4064	0.3975	0.3889	0.3805	0.3724	0.3644
4	0.3396	0.3294	0.3196	0.3102	0.3011	0.2923	0.2839	0.2757	0.2679	0.2603
5	0.2592	0.2495	0.2403	0.2315	0.2230	0.2149	0.2072	0.1998	0.1927	0.1859
6	0.1979	0.1890	0.1807	0.1727	0.1652	0.1580	0.1512	0.1448	0.1386	0.1328
7	0.1510	0.1432	0.1358	0.1289	0.1224	0.1162	0.1104	0.1049	0.0997	0.0949
8	0.1153	0.1085	0.1021	0.0962	0.0906	0.0864	0.0806	0.0760	0.0718	0.0678
9	0.0880	0.0822	0.0768	0.0718	0.0671	0.0628	0.0588	0.0551	0.0516	0.0484
10	0.0672	0.0623	0.0577	0.0536	0.0497	0.0462	0.0429	0.0399	0.0371	0.0346
11	0.0513	0.0472	0.0434	0.0400	0.0368	0.0340	0.0313	0.0289	0.0267	0.0247
12	0.0392	0.0357	0.0326	0.0298	0.0273	0.0250	0.0229	0.0210	0.0192	0.0176
13	0.0299	0.0271	0.0245	0.0223	0.2020	0.0184	0.0167	0.0152	0.0138	0.0126
14	0.0228	0.2050	0.0185	0.0166	0.0150	0.0135	0.0122	0.0110	0.0099	0.0090
15	0.0174	0.0155	0.0139	0.0124	0.0111	0.0099	0.0089	0.0080	0.0072	0.0064
16	0.0133	0.0118	0.0104	0.0093	0.0082	0.0073	0.0065	0.0058	0.0051	0.0046
17	0.0101	0.0089	0.0078	0.0069	0.0061	0.0054	0.0047	0.0042	0.0037	0.0033
18	0.0077	0.0068	0.0059	0.0052	0.0045	0.0039	0.0035	0.0030	0.0027	0.0023
19	0.0059	0.0051	0.0044	0.0038	0.0033	0.0029	0.0025	0.0022	0.0019	0.0017
20	0.0045	0.0039	0.0033	0.0029	0.0025	0.0021	0.0018	0.0016	0.0014	0.0012

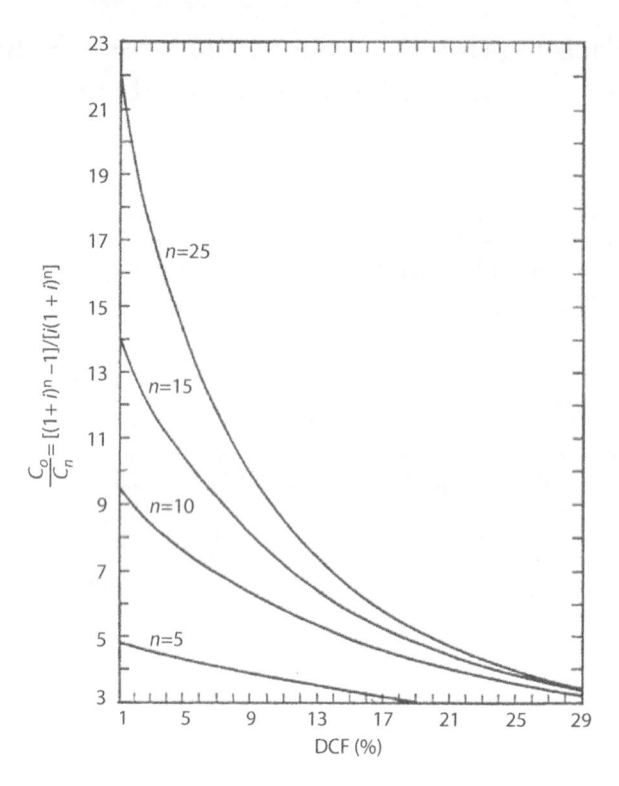

Figure 26.6 Graphical solution for simplified DCF calculation (constant cash flow, compounded annually, no salvage working capital return.) (Source: Horwitz [29]).

$$\frac{I}{C} = \frac{(1+i)^n - 1}{i(1+i)^n} \qquad (26.38)$$

where

I = investment
C = cash flow for each year
i = rate of return (IRR/100)
n = years of project life

Eq. (26.38) is defined as the annuity with I as the present value and C as the equal payments over n years at interest i. The I/C term is referred to as the PBP. If I and C are evaluated and the life of the project is known, i can be computed by trial and error. The value of i can be determined from Figure 26.6.

The NPV measures the direct incentive to invest in a given proposal as a bonus or premium over the amount an investor could otherwise earn by investing the same money in a safe alternative, which would yield a return calculated at the rate i. The resulting NPV from a project's cash flows is a measure of the cash profit that the project will produce after recovering the initial investment and meeting all costs, including the cost of capital. The more positive NPV is, the more attractive the proposition. If NPV is 0, the viability of the project is marginal; if it is negative, the proposal is unattractive.

The Profitability Index (PI)

The profitability index (PI) is defined as the present value of an investment's cash flow divided by the initial outlay (investment). If the present value of the cash flow is PV, then the PI is:

$$PI = \frac{PV}{I} \quad (26.39)$$

where

I = the initial outlay (investment)
PV = net present value + initial outlay (i.e., NPV + I)

Example 26.6

Suppose the cost of capital is 15%, and the cash flows on a project are as follows:

Year							
	0	1	2	3	4	NPV (at 10%)	IRR
Cash flow	–$600	$250	$250	$250	$250	$192.47	24.1%

The project's NPV is $192.47, and the PV is $792.47 = (192.47 + 600). The PI is:

$$PI = \frac{PV}{I} = \frac{\$792.47}{\$600} = 1.32$$

This can be interpreted as a 32% return on a $600 investment. Alternatively, it can also be interpreted that the investment will return the initial amount plus the NPV that is equal to 32% of the initial investment.

The PI method has an advantage over the IRR method in that the latter uses the IRR itself to adjust for the time value of the money whereas the former uses the cost of capital. However, PI is still inferior to the NPV as a criterion for evaluating investment, because like the IRR, it does not directly take into account the difference in investment scale. Table 26.17 summarizes the discussion of the NPV, IRR, and PI techniques for project evaluation. The Excel spreadsheet Example 26.6.xls shows the economic calculations and Figures 26.7a and 26.7b respectively show the snapshots of the Excel spreadsheet calculations.

Discounted Cash Flow Rate of Return (DCFRR)

The discounted cash flow rate of return (DCFRR) is known by other terms, for example, the profitability index, the true rate of return, the investor's rate of return, and the internal rate of return. It is defined as the discount rate i that makes the NPV of a project equal to zero. This can be expressed mathematically by

$$NPV = \sum p = \sum_{t=0}^{t=n} \frac{C_t}{(1+i)^t} = 0 \quad (26.40)$$

Relationship between PBP and DCFRR

For the case of a single lump-sum capital expenditure C_{FC} that generates a constant annual cash flow, A_{CF} in each subsequent year, the PBP is expressed by Perry [40]:

$$PBP = \frac{C_{FC}}{A_{CF}} \quad (26.41)$$

Table 26.17 Methods of Evaluating Investment Proposal.

	NPV	IRR	PI	PBP
Accept-reject decision for independent projects	Accept if NPV > 0 Reject if NPV < 0	Accept if IRR > k Reject if IRR < k	Accept if PI > 1 Reject if PI < 1	Accept if PBP < managerial limit. Reject if PBP > managerial limit
Decision when choosing from mutually exclusive proposals	Choose proposal with high NPV provided it is greater than zero.	Choose proposal with highest IRR provided it is higher than the cost of capital, k.	Choose proposal with highest PI provided it is greater than 1.	Choose proposal with shortest PBP provided it is less than managerial limit.
Advantages	1. Is a direct measure of a project's contribution to stockholders' wealth. 2. Uses the opportunity cost of capital. 3. Assumes that interim cash flows are reinvested at a rate of return equal to the cost of capital. 4. Accept-reject decisions, and ranking of investments, are consistent with maximization of stockholders' wealth.	1. Accept-reject decisions or independent projects are consistent with maximization of stockholders' wealth. 2. Measured in percentage, which is appealing to many people.	1. Accept-reject decisions are the same as that of NPV and IRR- therefore they are consistent with maximization of stockholders' wealth. 2. The assumption about reinvestment rate is as made by the NPV. 3. Expresses the NPV in relation to the initial investment study.	1. Simple and easy to use. 2. If used sensibly, may provide a first screen that eliminates further (costly) analysis of high-risk projects.

(Continued)

Table 26.17 Methods of Evaluating Investment Proposal. (*Continued*)

	NPV	IRR	PI	PBP
Disadvantages	1. The only real disadvantage occurs when projects with unequal lifetimes are compared. This disadvantage is not unique to NPV.	1. Not a direct measure of the effect of a project on stockholders' wealth.	1. Not a direct measure of the effect of a project on stockholders' wealth.	1. Does not take the time value of money into consideration.
	2. Although not really a disadvantage, the NPV is not expressed in percent and does not reveal what resources are needed to generate a given NPV.	2. Makes the assumption that cash flows can be reinvested to yield the IRR.	2. Ranking is not consistent with maximization of stockholders' wealth.	2. Ignores cash flows beyond the payback period.
		3. Ranking is inconsistent with maximization of stockholders' wealth.		3. Does not measure the effect of a project on stockholders' wealth.
		4. Some projects may have more than one IRR and others may have no IRR at all.		4. Ranking is generally inconsistent with maximization of stockholders' wealth.

NPV = Net Present Value
IRR = Internal Rate of Return
PI = Profitability Index
PBP = Pay back Period
(Source: Ben-Horim. M, Essential of Corporate Finance, Allyn, and Bacon, Inc. 1987).

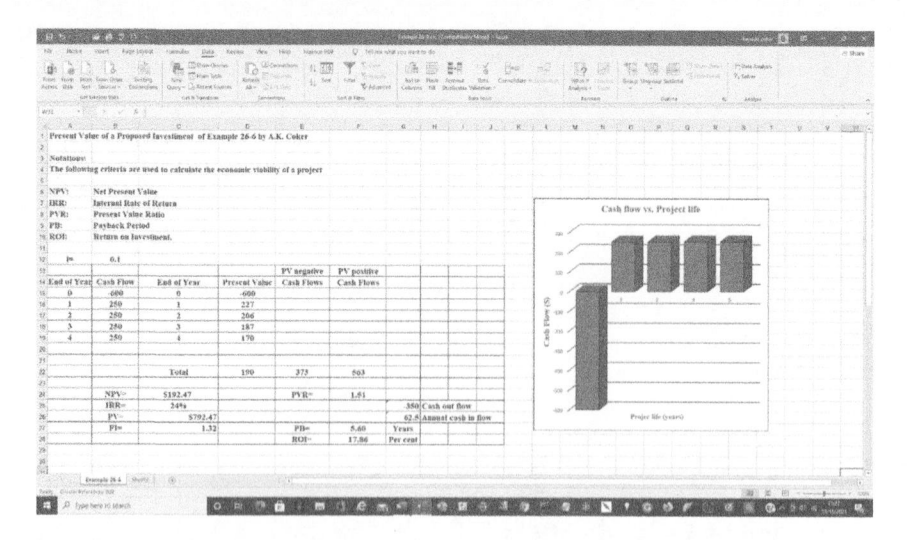

Figure 26.7a Screenshot of Excel worksheet for Example 26.6, with an embedded bar graph of the cash flow components.

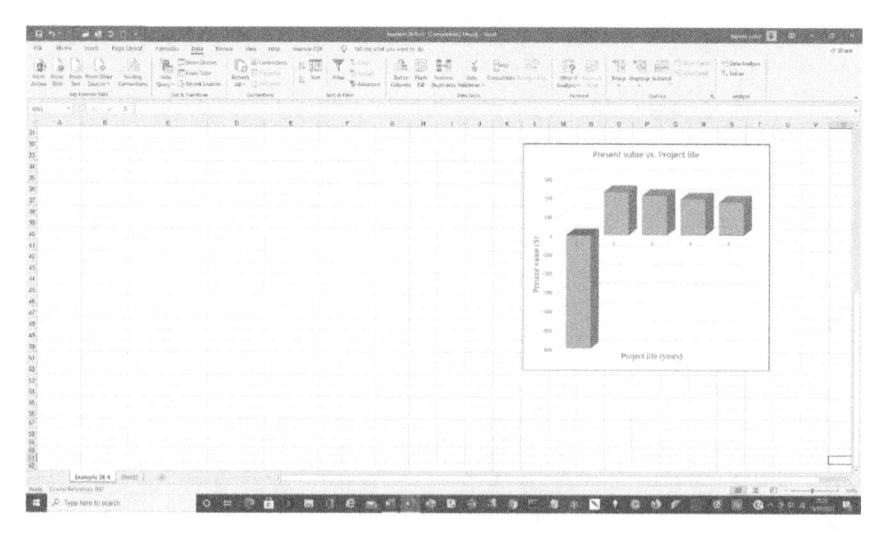

Figure 26.7b Screenshot of Excel worksheet for Example 26.6, with an embedded bar graph of the cash flow components.

If the scrap value of the capital outlay may be taken as zero, then in the limiting case when n approaches infinity (i.e. $n \rightarrow \infty$), the maximum DCFRR is defined as:

$$(\text{DCFRR})_{\max} = \frac{1}{\text{PBP}} \qquad (26.42)$$

This means, for example, that if the PBP is 5 years, the maximum possible DCFRR that can be reached is 20 %. The corresponding (DCFRR) for a PBP of 10 years is 10 %.

The main advantage of DCFRR over NPV is that it is independent of the zero or base year that is chosen. In contrast, the value of NPV varies according to the zero-year chosen. In calculating the NPV, the cost of capital has to be explicitly included as i in the discounting calculations. In computing the DCFRR, the cost of capital is not included. Instead, the value calculated for i is compared with the cost of capital to see whether the project is profitable. The DCFRR for a project is the rate of ROI. It measures the efficiency of the capital and determines the earning power of the project investment. Therefore, a DCFRR of 20% implies that 20 % per year will be earned on the investment, in addition to which the project generates sufficient money to repay the original investment plus any interest payable on borrowed capital and all taxes and expenses.

Example 26.7

A manufacturing company is considering bringing out a new product. The company plans to invest $12 million at the beginning of year 1 (i.e., at the end of year 0) and another $7 million 1 year later. Marketing studies suggest that this investment will generate an irregular series of revenues from year 3 through 12. The cash flow components are summarized below:

End of year	Revenue ($)	End of year	Revenue ($)
0	−12,000,000	7	5,000,000
1	−7,000,000	8	6,000,000
2	0	9	5,500,000
3	1,000,000	10	3,500,000
4	2,000,000	11	2,500,000
5	3,000,000	12	2,000,000
6	4,000,000	13	1,000,000

Calculate the following:

1. The present value of the proposed investment, assuming the company normally receives a return of 8% per year, compounded annually.
2. The IRR (i.e., the interest rate at which the present value of the cash flow is zero).

Solution

The cash flow was entered into an Excel worksheet (Example 26.7.xls) to determine the present value of this proposed investment, assuming the company normally receives a return of 8 % per year, compounded annually. The initial investments are shown as negative cash flow components at the end of years 0 and 1 respectively. Figures 26.8a and 26.8b show the snapshots of the Excel worksheet containing the cash flow components, where columns A and B

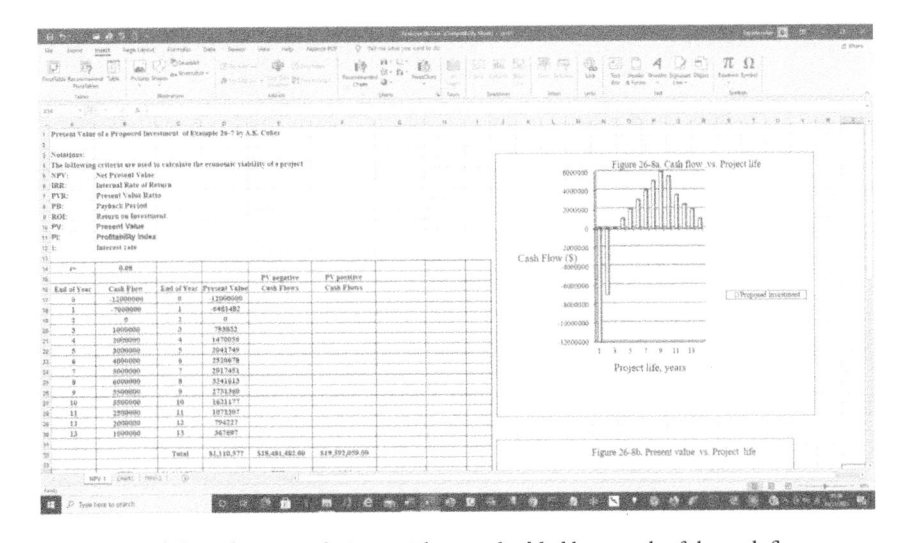

Figure 26.8a Screenshot of Excel worksheet for Example 26.7, with an embedded bar graph of the cash flow components.

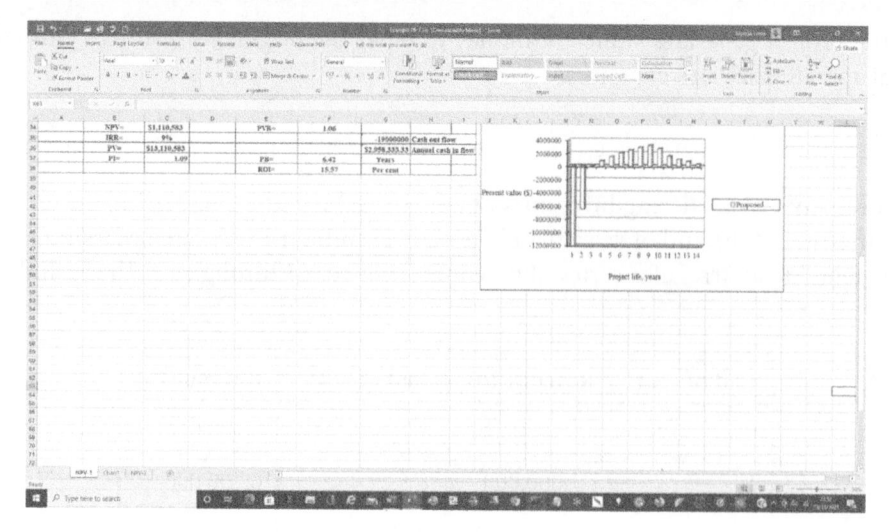

Figure 26.8b A snapshot of the Excel spreadsheet program of Example 26.7.

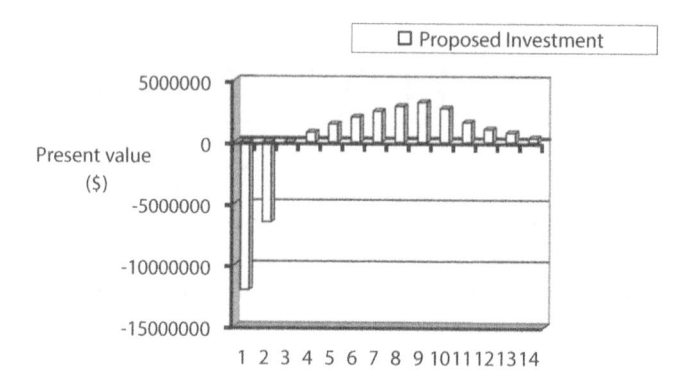

Figure 26.8c Present value vs. project life.

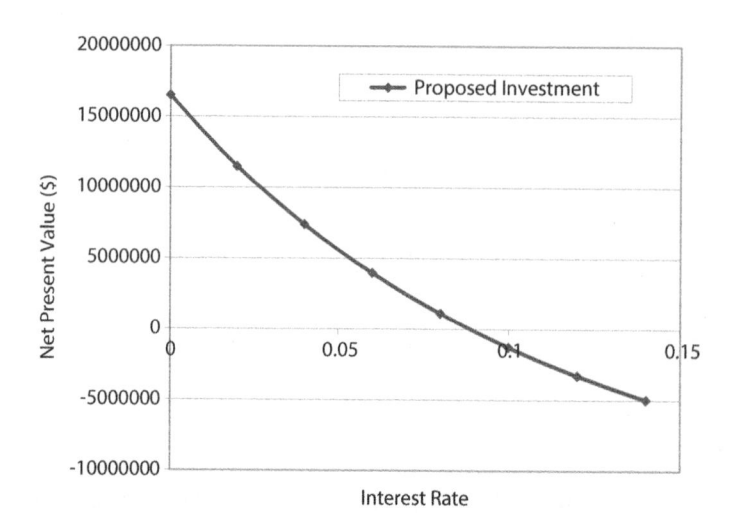

Figure 26.8d Net present value vs. interest rate.

show the project life and the cash flow respectively. A bar graph of the cash flow components (column D) is also shown. Figure 26.8c shows a plot of the present value (PV) against project life, and Figure 26.8d illustrates the NPV at varying interest rate. The plot indicates that the NPV decreases with increasing interest rate, eventually crossing the abscissa and becoming negative. The IRR is the crossover point; that is, the value of the interest rate at which the NPV is zero. Figure 26.8d shows an IRR of 9%.

Cell D32 is the NPV, which contains the sum of the present values of the cash flow components. The NPV is $1.11 x 10^6 indicating that the present value of the future cash inflows exceeds the present value of the initial investments. Additionally, cell E32 in the spreadsheet is the PVR and is calculated to equal 1.06, which is greater than 1.0. Hence, this new product represents an attractive investment opportunity. Further calculations show that the PBP is 6.4 years and the ROI is 15.6%.

Example 26.8

The following cash flows describe two competing investment opportunities.

End of year	Revenue proposal A ($)	Revenue proposal B ($)
0	−500,000	−500,000
1	150,000	50,000
2	150,000	70,000
3	150,000	100,000
4	150,000	200,000
5	150,000	400,000

Determine which investment is more attractive, based on

1. Present value using an interest rate of 5%, compounded annually.
2. The IRR, NPV and PVR, PV and PI.

Solution

The Excel worksheet (Example 26.8.xls) calculates the NPV and IRR for investment plans A and B. Figure 26.9a shows the cash flows against the project life of investment plans A and B, and Figure 26.9b shows the present value against the project life of the two investment plans A and B, respectively. The NPV at 5% interest rate shows that proposed plan B ($175,446) is higher than that of proposed plan A ($149,421). However, the calculated IRR shows that proposed plan A (15%) is higher than that of proposed plan B (13%). Further calculation on the PVR shows that plan B (PVR = 1.35) is higher than plan A (PVR = 1.3). The present value of B (PV = $675,446) is higher than plan A (PV = $649,422) and the profitability index of B (PI = 1.35) is also higher than plan A (PI = 1.30), which further indicates that proposed investment plan B is more attractive. Figure 26.9c shows plots of NPV as a function of interest rate for plans A and B respectively. Figures 26.9d–f show snapshots of the Excel spreadsheet calculations respectively. These profitability criteria as illustrated in Table 26.18 indicate that proposed plan B is the more attractive investment.

Net Return Rate (NRR): The net return rate (NRR) is analogous to the rate of return and is the net average discounted "return" on the investment over and above the cost of capital. This is defined by

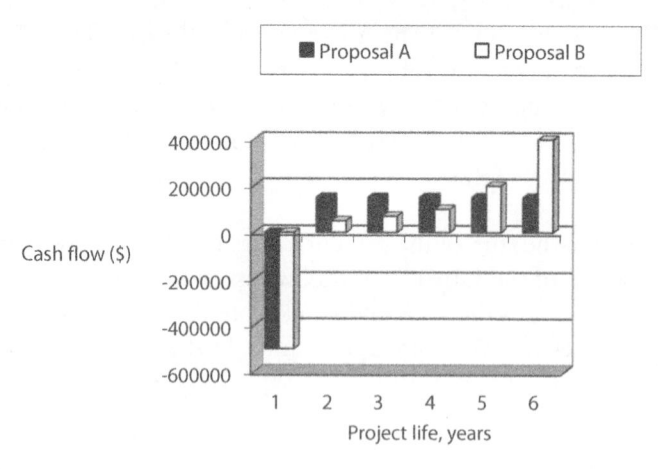

Figure 26.9a Cash flow vs project life.

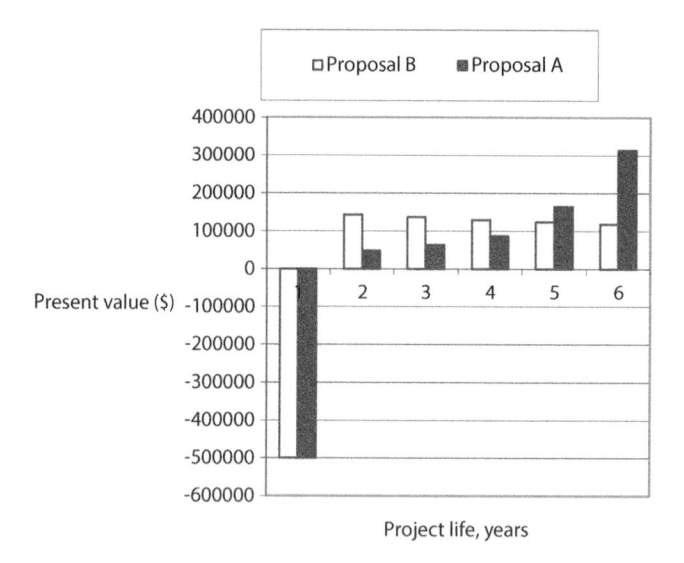

Figure 26.9b Present value vs. project life.

$$NRR = \frac{NPV}{\left(\begin{array}{c}\text{Discounted}\\\text{investment}\end{array}\right)\left(\begin{array}{c}\text{Project}\\\text{life}\end{array}\right)} \times 100 \qquad (26.43)$$

where the investment is discounted to the same point as the NPV. Holland *et al.* [35] introduced the NPV/Investment ratio as a normalized measure of the total discounted return over the life of the investment. Ward [32] showed that the NPV can be divided by the number of cash flow increments (venture lifetime) such that the NRR corresponds to the average discounted net ROI. The cost of capital is already accounted for by the discount rate in the NPV computation, and therefore the NRR is the true return rate.

26.12.1 Incremental Criteria

Incremental Net Return Rate: In an incremental-investment analysis, the various options are first ranked by size and the venture with the smallest investment is selected as the base case. Then, the incremental net present value (INPV)

COST ESTIMATION AND ECONOMIC EVALUATION 511

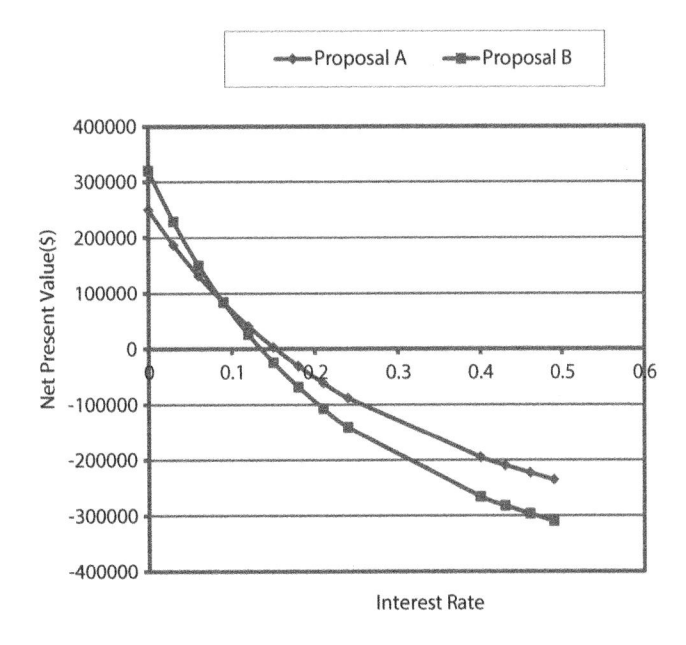

Figure 26.9c Net present value vs. interest rate.

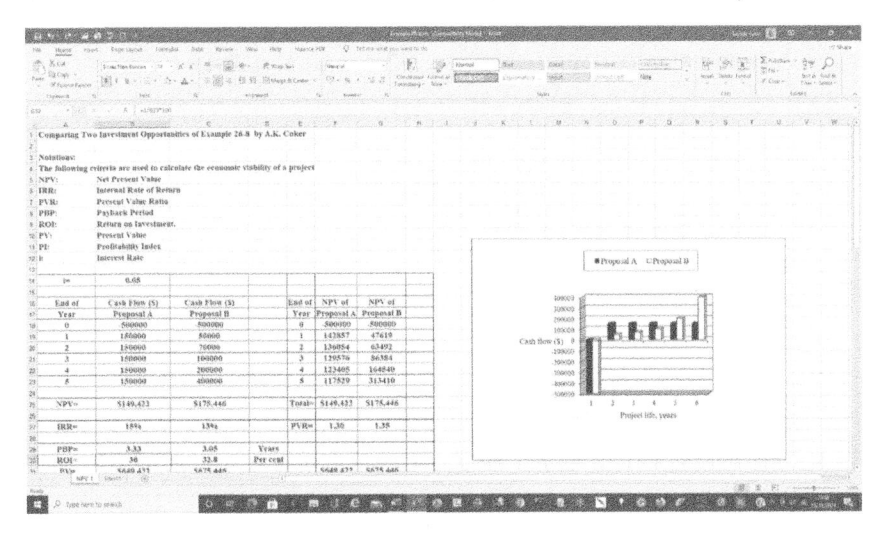

Figure 26.9d A snapshot of the Excel spreadsheet calculations of Example 26.8.

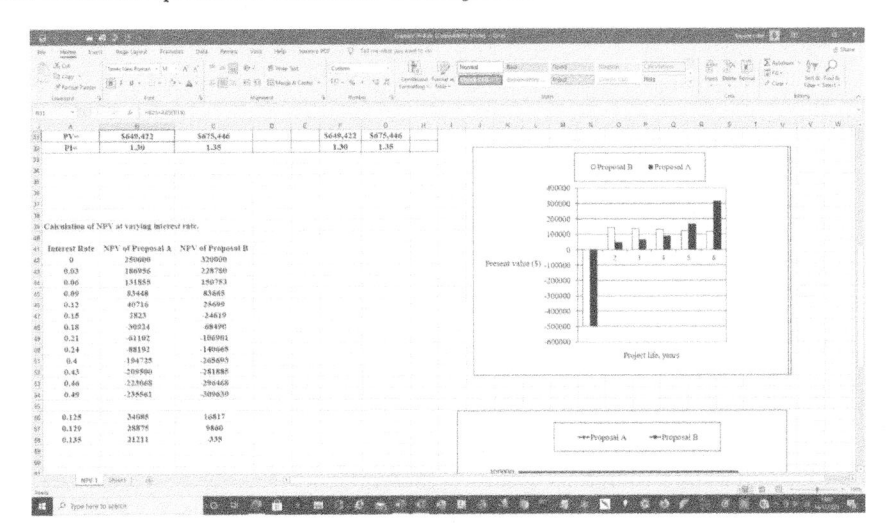

Figure 26.9e A snapshot of the Excel spreadsheet calculations of Example 26.8.

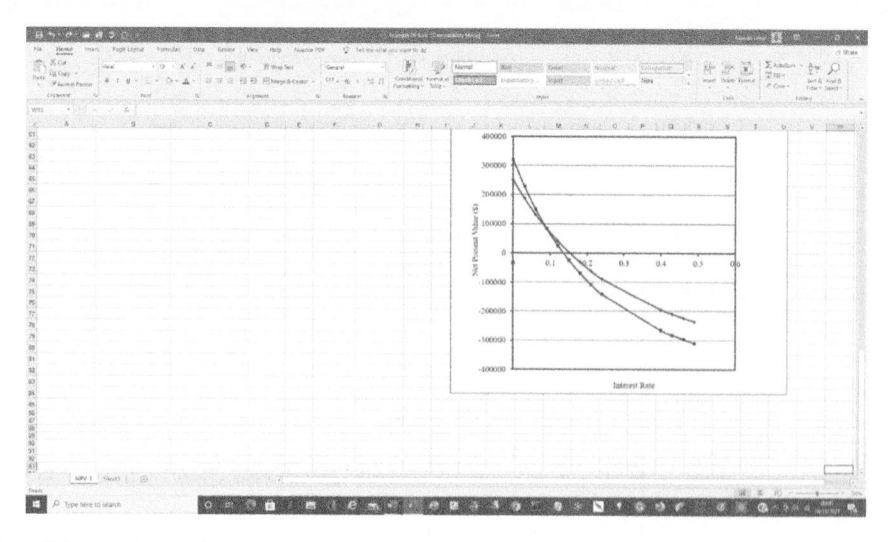

Figure 26.9f A snapshot of the Excel spreadsheet calculations of Example 26.8.

Table 26.18 Profitability Criteria of Proposal A and B.

Profitability criteria	Proposal A	Proposal B
Net Present Value (NPV), $	1,149,222	1,75,446
Present Value Ratio (PVR)	1.3	1.35
Internal Rate of Return (IRR), %	15	13
Return on Investment (ROI), %	30	32.8
Payback Period, years	3.3	3.05

and incremental net return rate (INRR) can be calculated. The INPV is the NPV of the incremental cash flows, and is defined as

$$\text{INRR} = \frac{\text{INPV}}{\left(\text{Incremental investment}\right)\left(\text{Project life}\right)} \times 100 \qquad (26.44)$$

The INRR is recommended as the primary criterion of profitability, and is appropriate when the alternatives being considered have different investment levels.

Depreciation

Estimation of depreciation charges may be based on

1. The cost of operation.
2. A tax allowance.
3. A means of building up a fund to finance plant replacement.

or,

4. A measure of falling value.

The annual depreciation charge can be calculated using "straight line" depreciation, and is expressed as:

$$D = \frac{C_{FC} - S}{n} \qquad (26.45)$$

where

D = annual depreciation
C_{FC} = initial fixed capital cost
n = number of years of projected life
S = salvage value

Assuming that C_{FC} is the initial fixed capital investment, and S is the projected salvage value at the end of n years of projected life, the depreciated rate d_j for any particular year j is

$$D_j = (C_{FC} - S)d_j \qquad (26.46)$$

where

D_j = Annual depreciation charge.

With the straight-line calculation procedure, where d_j is constant, combining Eq. (26.45) and Eq. (26.46) gives

$$d = \frac{1}{n} \qquad (26.47)$$

The various component of a plant such as equipment, buildings, and improvements are characterized by projected lifetimes. During this period, each item depreciates from its initial investment cost C_{FC} to a salvage value S over the period of n years of its projected lifetime. At the end of any particular year k, the depreciated value or book value V_k is

$$V_k = C_{FC} - \sum_{1}^{k} D_j \qquad (26.48)$$

where D_j is the annual depreciation charge for year j.

Substituting Eq. (26.46) into Eq. (26.48) gives

$$V_k = C_{FC} - \sum_{1}^{k} (C_{FC} - S)d_j$$

$$= C_{FC} - (C_{FC} - S)\sum_{1}^{k} d_j \qquad (26.49)$$

For the straight-line depreciation procedure,

$$\sum_{1}^{k} d_j = \sum_{1}^{k} d = kd = \frac{1}{n} \qquad (26.50)$$

and

$$V_k = C_{FC} - \frac{k}{n}(C_{FC} - S) \qquad (26.51)$$

Double Declining Balance (DDB) Depreciation: In practice, equipment and complete plants depreciate and lose value more rapidly in the early stages of life. The depreciation based on the declining book value balance can be expressed as:

$$D_j = d_j \cdot V_{j-1} \qquad (26.52)$$

The rate of depreciation d_j is the same for each year j; however, the depreciation charges decrease each year since the book value decreases each year. For a declining balance method, the depreciation rate of decline is up to, but no more than twice the straight-line rate. This is given by

$$d_j = d = \frac{2}{n} \qquad (26.53)$$

Capitalized Cost: The capitalized cost C_K of a piece of equipment of a fixed capital cost C_{FC} having a finite life of n years and an annual interest rate i is defined by

$$(C_K - C_{FC})(1+i)^n = C_K - S \qquad (26.54)$$

where

S = salvage or scrap value

C_K is in excess of C_{FC} by an amount which, when compounded at an annual interest rate i for n years will have a future worth of C_K less the salvage or scrap value S. If the renewal cost of the equipment and the interest rate are constants at $(C_{FC} - S)$ and i, then C_K is the amount of the capital required to replace the equipment in perpetuity.

Re-arranging Eq. (26.54) gives

$$C_K = \left[C_{FC} - \frac{S}{(1+i)^n} \right] \left[\frac{(1+i)^n}{(1+i)^n - 1} \right] \qquad (26.55)$$

or

$$C_K = (C_{FC} - S \cdot f_d) f_K \qquad (26.56)$$

where

f_d = discount factor
f_K = the capitalized cost factor

$$= \frac{(1+i)^n}{(1+i)^n - 1}$$

The **Average Rate of Return (ARR):** The average rate of return (ARR) method averages out the cash flow over the life of the project. This is defined by

$$ARR = \frac{\text{The average cash flow}}{\text{Original Investment}} \times 100 \qquad (26.57)$$

The higher the percentage value of the average rate of return (ARR), the better the profitability of the project.

Present Value Ratio (Present Worth Ratio): This commonly used profitability index (PI) in conjunction with the NPV method shows how closely a project has met the criterion of economic performance. This index is known as the present value ratio (PVR) or present worth ratio (PWR), and is defined as

$$PVR = \frac{\text{present value of all positive cash flows}}{\text{present value of all negative cash flows}} \qquad (26.58)$$

The discounted cumulative cash position, which is commonly referred to as the NPV or net present worth (NPW) of the project is defined as:

NPV = Cumulative discounted cash position at the end of the project.

NPV of a project is greatly influenced by the level of fixed capital investment and a better criterion for comparison of projects with different investment levels may be the PVR.

The PVR gives an indication of how much the project makes relative to the investment. A ratio of unity shows that the income just matches the expected income from capital invested for a given interest rate. A ratio of less than unity indicates that the income does not come up to the minimum expectations. A ratio of more than unity means that the project exceeds the minimum expectations.

26.12.2 Profitability

A project is profitable if its net earnings are greater than the cost of capital. In addition, the larger the additional earnings, the more profitable the venture, and therefore the greater the justification for putting the capital at risk. A profitability estimate attempts to quantify the risk taken. The methods used to assess profitability are as follows:

1. Return on investment (ROI).
2. Payback period (PBP).
3. Net present value (NPV).

4. Discounted cash flow rate of return (DCFRR).
5. Net return rate (NRR).
6. Equivalent maximum investment period (EMIP).
7. Interest recovery period (IRP).
8. Rate of return on depreciated investment.
9. Rate of return on average investment.
10. Capitalized cost.
11. Average rate of return (ARR).
12. Present value ratio (PVR).

Abrams [41] has listed other methods for assessing project profitability.

Example 26.9

Consider a plant costing $1,000,000 to build that produces a product A. For the same capital outlay, a different plant can be erected to produce an alternative product B. Conditions are such that each plant will only be in operation for 8 years and then both will be scrapped. The cash flows in each of the 8years obtained by selling A and B are shown in Table 26.19. Determine which plant is more profitable.

Solution

The Excel worksheet (Example 26.9.xls) and Fortran computer programs (Project1.for and Project2.for) from the companion website calculate the various profitability criteria for investment plans A and B at 10% discount rate. Table 26.20 shows the profitability criteria of products A and B.

Figures 26.10a and 26.10b show plots of the cash flow and NPV against the project life, respectively, and Figure 26.10c illustrates plots of the NPV at varying interest rate for products A and B. The IRR values at which NPV = 0 using **Goal Seek** from the Data menu of **What If analysis** are 28.3% and 64.9% of products A and B respectively. The results in Table 26.20 show that product B is a better investment choice than product A. Figures 26.10d–26.10f show the snapshots of the Excel spreadsheet calculations of Example 26.19.

Example 26.10

A project requires the initial capital expenditure of $1,000, 000 on plant and $100,000 on industrial buildings, all incurred at the same time. The project is expected to have constant annual operating costs of $150,000 and to

Table 26.19 Annual Cash Flows and Cumulative Cash Flows.

End of year	Product A cash flows ($)	Product A cumulative cash flows ($)	Product B cash flows ($)	Product B cumulative cash flows ($)
0	-1,000,000	-1,000,000	-1,000,000	-1,000,000
1	100,000	-900,000	800,000	-200,000
2	200,000	-700,000	700,000	500,000
3	300,000	-400,000	600,000	1,100,000
4	400,000	0	500,000	1,600,000
5	500,000	500,000	400,000	2,000,000
6	600,000	1,110,000	300,000	2,300,000
7	700,000	1,800,000	200,000	2,500,000
8	800,000	2,600,000	100,000	2,600,000

Table 26.20 Profitability Criteria of Products A and B.

Profitability criteria	Product A	Product B
Net Present Value (NPV), $	1,136,360	1,665,074
Present Value Ratio (PVR)	2.14	2.67
Net Return Rate (NRR), %	14.20	20.81
Average Rate of Return (ARR), %	45.0	45.0
Internal Rate of Return (IRR), %	28.0	65.0
Discounted Cash Flow Rate of Return (DCFRR), %	28.26	64.86
Payback Period, years	3.3	3.1
Return on Investment, %	30	32.8

generate an annual income of $500,000. The project is expected to have a 5-year life. Calculate the after-tax project cash flows, NPV, IRR, and PVR if:

The plant will have no final scrap value.
The final plant scrap value is $200,000.
The final plant scrap value is $400,000.

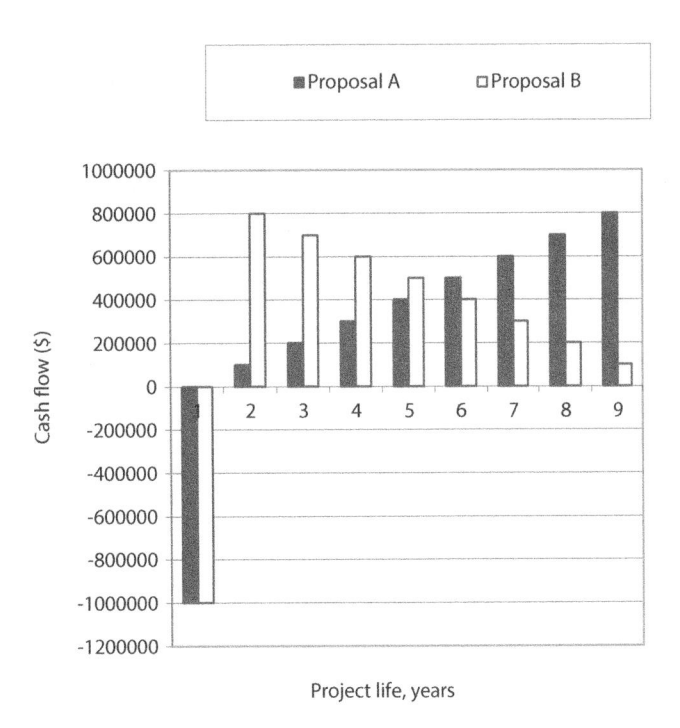

Figure 26.10a Cash flow vs. project life.

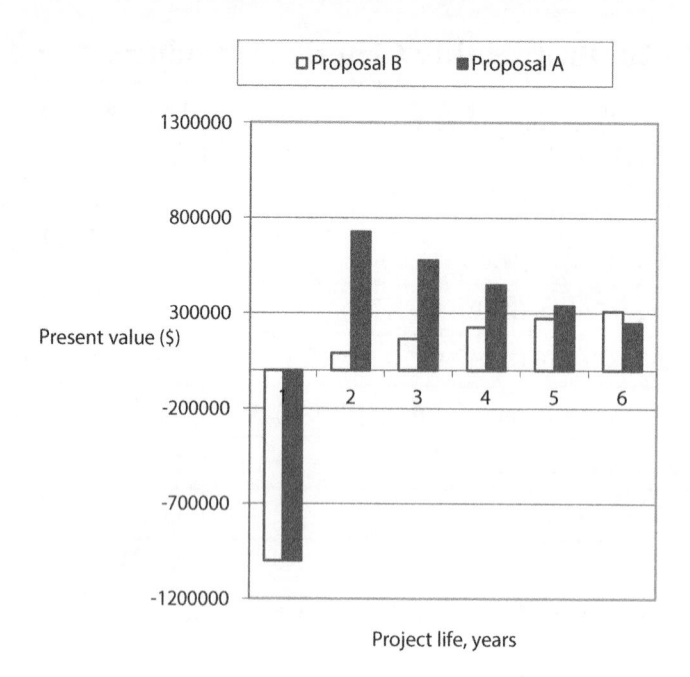

Figure 26.10b Present value vs. project life.

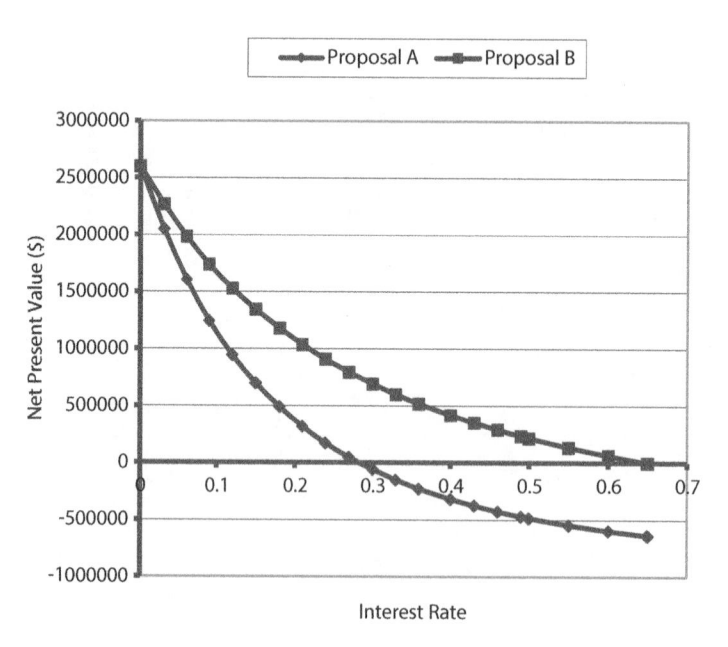

Figure 26.10c Net present value vs. interest rate.

COST ESTIMATION AND ECONOMIC EVALUATION 519

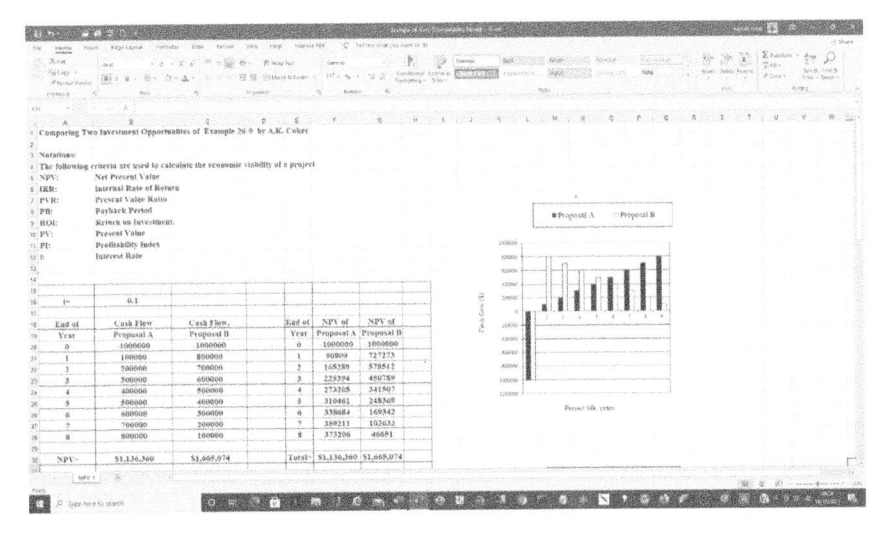

Figure 26.10d A snapshot of the Excel spreadsheet calculations of Example 26.9.

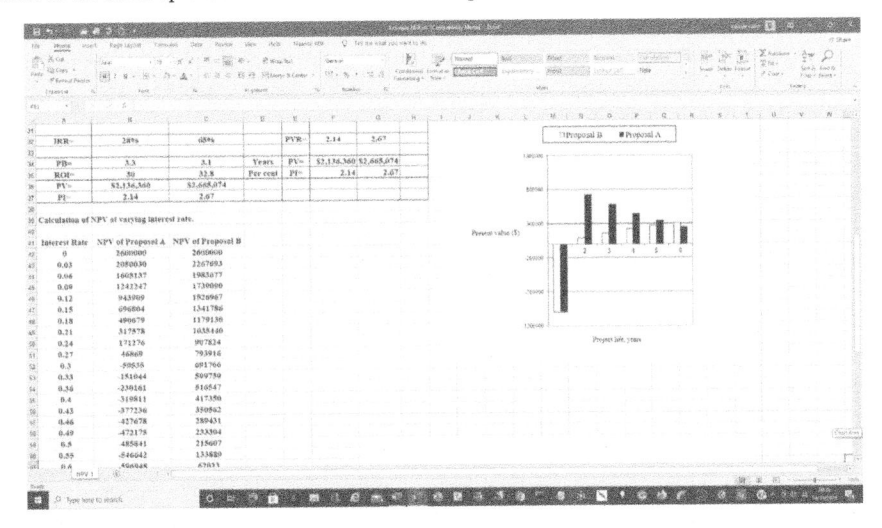

Figure 26.10e A snapshot of the Excel spreadsheet calculations of Example 26.9.

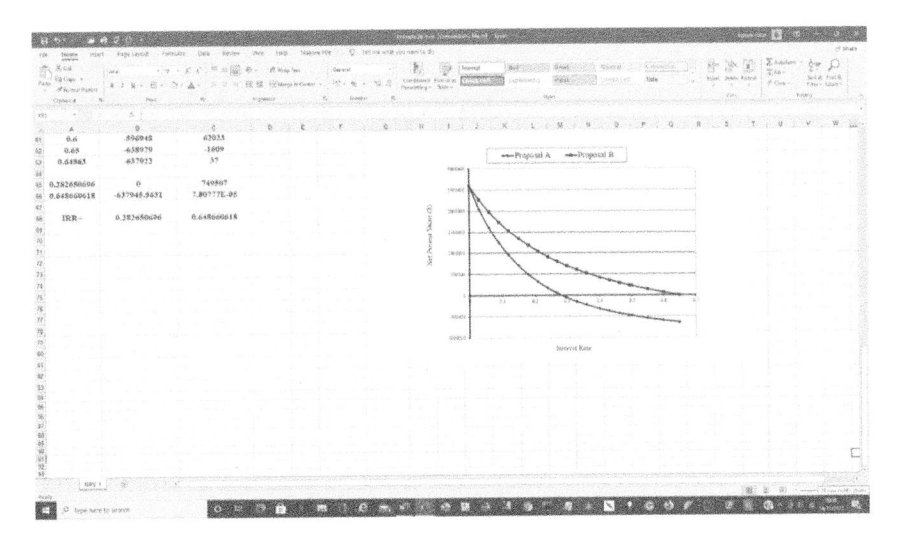

Figure 26.10f A snapshot of the Excel spreadsheet calculations of Example 26.9.

Table 26.21 Profitability Analysis of After-Tax Cash Flows.

Scrap value ($)	IRR (%)	NPV ($)	PVR	DCFRR
0	11	32,446	1.03	0.0104
200,000	14	113,166	1.1	0.1376
400,000	16	193,886	1.18	0.1608

Year no.	Plant allowance qualifying expenditure ($)	Allowance ($)		Industrial allowance qualifying expenditure ($)	Building allowance ($)	Total capital allowances ($)
1	1000000	250000	(25%)	100000	4000 (4%)	254000
2	750000	187500		100000	4000 (4%)	191500
3	562500	140625		100000	4000 (4%)	144625
4	421875	105469		100000	4000 (4%)	109469
5	316406	316406	balancing	100000	4000 (4%)	320406
Total	allowance	1000000			20000	1020000

Solution

The Excel worksheet (Example 26.10.xls) provides cells for the initial capital plant investment, expenditure buildings, operating cost, income, before-tax operating profit, capital allowances, taxable profit, tax at 35%, profit after tax and after-tax cash flow. The spreadsheet calculates the various profitability criteria before and after tax. Table 26.21 shows the profitability criteria of after-tax cash flows with the plant having no scrap value, a scrap of $200,000, and $400,000 respectively. Figures 26.11a–f show plots of cumulative cash flow vs. project life, NPV vs. interest rate, cash flow vs. project life for the plant having no scrap value, and with scrap values of $200,000 and $400,000 respectively.

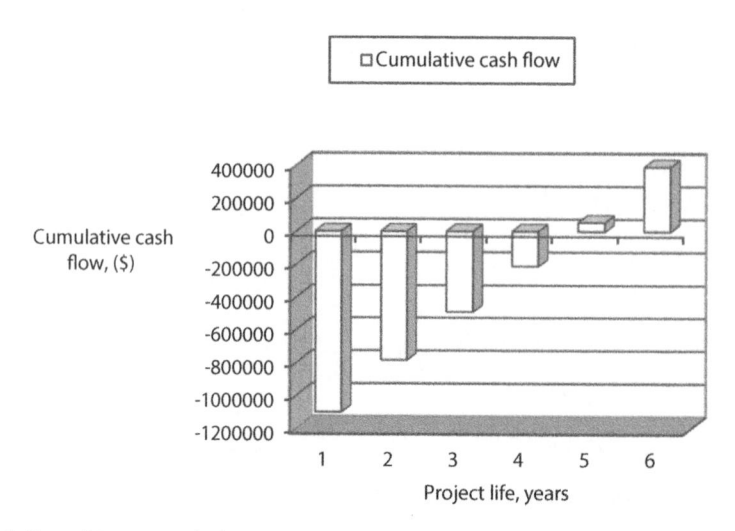

Figure 26.11a Cumulative cash flows (No scrap value).

Economic Analysis

Computer programs (project1.for and project2.for) have been developed to estimate the NPV, PVR, NRR, ARR, PBP, and DCFRR. These analyses are performed for a given cash flow over the operating life of a project. These programs can be incorporated as subroutines into larger programs, if required. In addition, a detailed computer program has been developed to review an economic project using Kirkpatrick's [42] input data: These data are defined as follows:

Annual Revenue, $: The money received (sales minus the cost of sales) for 1-year production from the plant. This is assumed as being constant for life of the project.

Annual Operating Cost, $: The cost of e.g., raw materials, labor, utilities, administration, insurance, and royalties, but does not include debt service payments.

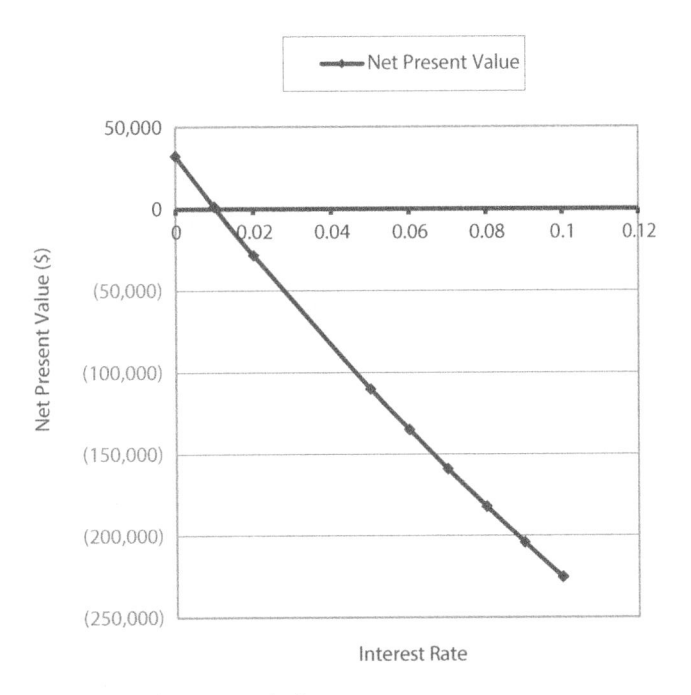

Figure 26.11b Net present value vs. interest rate (No scrap value).

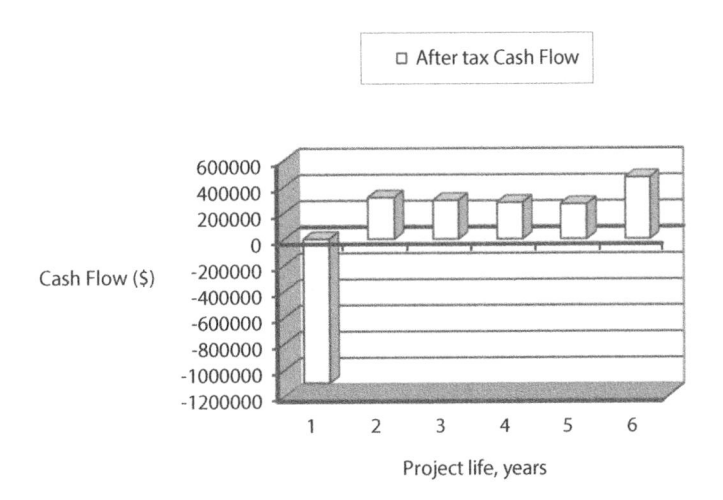

Figure 26.11c After – tax cash flow (Scrap value = $200,000).

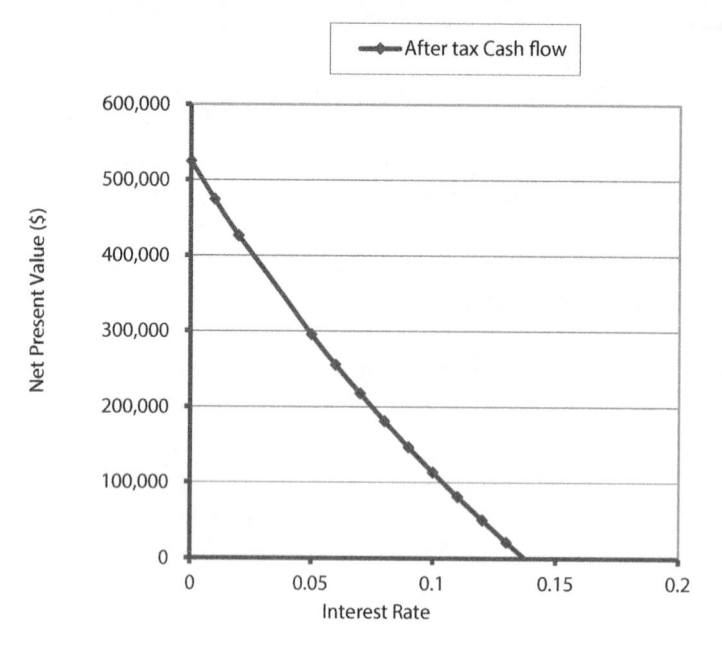

Figure 26.11d Net present value vs. interest rate (Scrap value = $200,000).

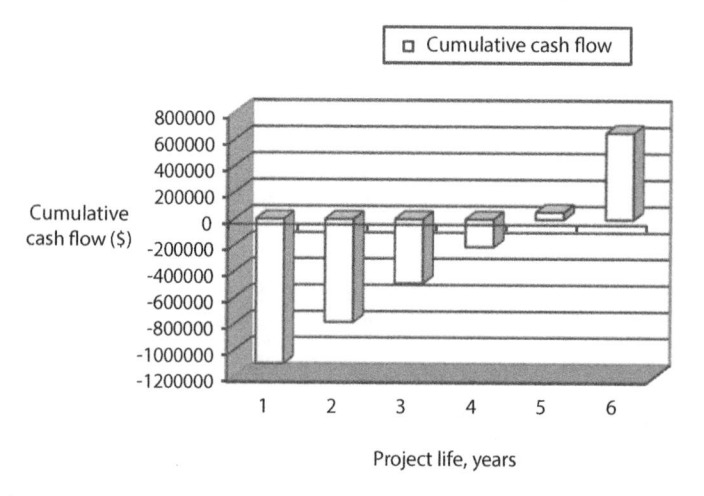

Figure 26.11e Cumulative cash flow (Scrap value = $400,000).

Depreciating Base, $: The capitalized cost of the facility, less non-depreciable items, e.g., land and inventory. No salvage is subtracted since DDB depreciation is used.

Project Life, years: The length of time for which the facility is to be operated. It is also the term of the loan and the depreciation time.

Initial Loan, $: The capitalized cost minus owner equity.

Payments/Year: The number of payments on the loan per year: 1-annually, 2-biannually, 4-quarterly, and 12-monthly.

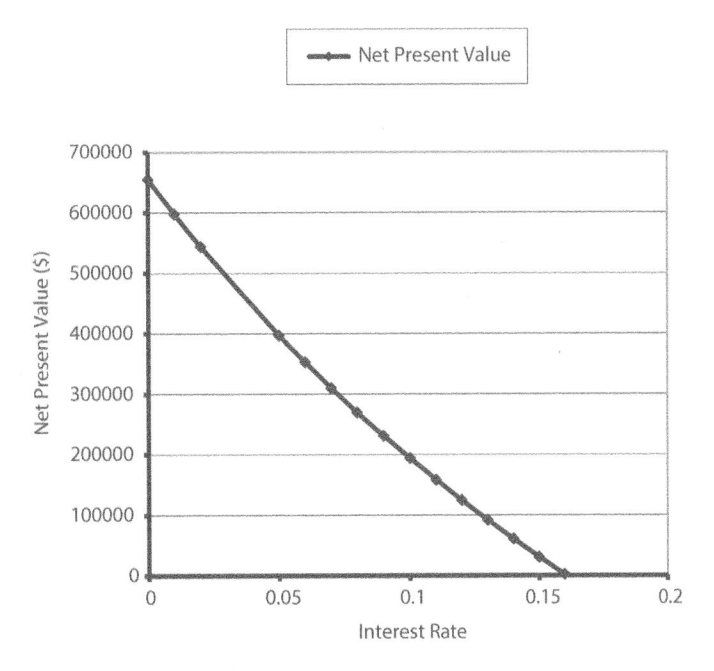

Figure 26.11f Net present value vs. interest rate (Scrap value = $400,000).

Periodic Interest Rate: The annual interest rate divided by the payments per year. For a 10% annual interest rate and monthly repayments, this is 0.10/12.

Investment Tax Credit, $: The percentage of the initial investment allowed as a tax credit in the year the investment is made.

Tax Rate: The percentage tax that must be paid on the project's pre-tax income. This rate is assumed to remain constant during the life of the project.

Debt Service/Period, $: The amount of each loan payment, i.e.:

$$\frac{(\text{Loan})\left[\dfrac{\text{Annual interest}}{\text{Payments per year}}\right]}{1-\left[1+\dfrac{\text{Annual interest}}{\text{Payments per year}}\right]^{-\left(\text{years}\times\frac{\text{Payments}}{\text{year}}\right)}}$$

The Salvage Value (or Scrap Value): The projected salvage value of equipment at the end of the project lifetime is that portion of the fixed capital that cannot be depreciated.

Land Value: Land is not considered depreciable, on the assumption that it can be used indefinitely for succeeding projects on a specific site, or it can be sold.

Working Capital: The capital invested in various necessary inventoried times, which are recoverable.

The calculations for every year of the project life are as follows:

1. Calculate the depreciation as double declining (that is, twice the depreciation based divided by project life) and subtract it from the depreciation base.
2. Calculate the yearly interest and principal of the debt service payments.
3. Subtract from the revenue, the operating cost, annual depreciation, and interest.
4. Subtract from positive pre-tax income, until no negative pre-tax income remains (if the pre-tax income is negative or has been negative in a prior year).
5. Calculate the income tax due at the given rate and apply the investment tax credit against the income tax due until the credit is exhausted (if after step 4, the pre-tax income is still positive).
6. Deduct the income tax remaining after step 5 from the pre-tax income, leaving the after-tax income.
7. Add the after-tax income to the depreciation for the year, yielding the cash flow.
8. Determine the loan balance at the end of the year.
9. Determine the amount of unused depreciation after the last year of the project's life. Since double declining depreciation does not totally exhaust the depreciation account, the unused depreciation should be added to the cash flow of the last year as salvage value recovered at the termination of the project under consideration.
10. Calculate the NPV, for a known discount rate.
11. Determine the PVR.
12. Calculate the NRR.
13. Calculate the ARR.
14. Estimate the PBP.

For a given discount rate, calculate the discount factor during the operating life of the project. Multiply the discount factor by the cash flow and obtain the NPV and NRR. Valle-Riestra [36] has listed alternative methods of calculating cash flow, as shown in Table 26.22.

Example 26.11

Calculate the yearly return of investment on the following financial data as shown in Table 26.23.

Solution

The computer program (project3.for) calculates the following: the yearly cash flows, cumulative cash flows, PV, NPV, PVR, ARR, NRR, and the PBP. Table 26.24 gives both the input data and the computer output for the cash flows during 10 years of the project life, and at a discount rate of 5%. The calculated salvage value of the investment is $4,294,967. The cash flows generated by the computer program (project3.for) using the financial data in Table 26.23 are then used to determine the DCFRR using the computer program (project2.for). Table 26.25 lists the input data and the computer output for DCFRR of the investment as 13.42%. The Excel worksheet (Example 26.11.xls) calculates NPV, IRR, PVR, PB and ROI using the cash flows generated from the fortran program (project3.for). Figures 26.12a–c respectively show the various plots of the economic evaluations. Figures 26.12d–f show the snapshots of the Excel spreadsheet calculations of Example 26.11.

Cash flows into a project can be time-dependent, where the cash flows occur in a continuous process rather than on a one-time basis. Table 26.22 shows different methods of calculating cash flow from a project, and Figure 26.13 illustrates a continuous cash flow diagram. Continuous cash flows into the project are from sales revenues, and cash flows out are for out-of-pocket expenses. The difference between the incoming and outgoing cash flow is the net cash flow generated by the project. The combined net cash flow generated by all the company's projects is reduced by the income tax payment to yield the net continuous cash flow. Expressing the definition of profit with the continuous cash flow results in the following:

$$\text{Profit before taxes} = (\text{continuous cash flow before taxes}) - (\text{depreciation}) \tag{26.59}$$

$$\text{Profit after taxes} = (\text{continuous cash flow after taxes}) - (\text{depreciation}) \tag{26.60}$$

Table 26.22 Methods of Computing Cash Flow from a Project.

Method 1	Method 2
1. Cost of Manufacture (COM) includes depreciation	1. Cost of Manufacture (COM) includes depreciation
2. General expenses	2. Depreciation
3. Total operating expense (3) = (1) + (2)	3. COM less depreciation (3) + (1) -(2)
4. Total sales	4. General expenses
5. Profit before taxes	5. Out of pocket expenditure
6. Income taxes (6) ≈ 1/2 (5)	6. Total sales
7. Profit after tax (7) = (5) − (6) ≈ 1/2 (5)	7. Cash flow before taxes (7) = (6) -(5)
8. Depreciation	8. Profit before taxes (8) = (7) -(2)
9. Continuous cash flow from project (9) = (7) + (8)	9. Income tax (9) ≈ 1/2 (8)
	10. Profit after tax (10) = (8) − (9) ≈ 1/2 (8)
	11. Continuous cash flow from project (11) = (10) + (2)

Methods 1 and 2 in Table 2.21 give identical results.
(Source: Valle-Riesta [37]).

Table 26.23 Financial Data.

1	Annual revenue, $	9,000,000
2	Annual operating cost, $	1,200,000
3	Depreciated base, $	40,000,000
4	Project life, years	10
5	Initial loan, $	30,000,000
6	Number of payments per year	12
7	Annual interest, %	10
8	Investment credit, %	10
9	Tax rate, %	50
10	Rate of return, %	5

Table 26.24 Input Data and Computer Results of Example 26.11.

```
DATA3.DAT
-------------------------------------------------------
9000000
1200000
40000000
10
30000000
12
10
10
50
5
```

```
                           Computer Results

                    ECONOMIC EVALUATION OF A PROJECT:
********************************************************************************
        1           ANNUAL REVENUE $:                      9000000.00
        2           ANNUAL OPERATING COST $:               1200000.00
        3           DEPRECIATION BASE $:                  40000000.00
        4           PROJECT LIFE, YEARS:                       10.
        5           INITIAL LOAN $:                       30000000.00
        6           NUMBER OF PAYMENTS PER YEAR:               12.
        7           ANNUAL INTEREST, %:                        10.
        8           INVESTMENT CREDIT, %:                      10.
        9           TAX RATE, %:                               50.
       10           DISCOUNT RATE %:                            5.00
********************************************************************************



                                 YEAR 1
********************************************************************************
        REVENUE $:                                        9000000.00
        OPERATING COST $:                                 1200000.00
        DEPRECIATION $:                                   8000000.00
        INTEREST $:                                       2917172.00
        PRE-TAX INCOME $:                                -3117172.00
        TAX AT 50.%, $:                                         0.00
        AFTER TAX INCOME $:                              -3117172.00
        CASH FLOW $:                                      4882828.00

********************************************************************************



                                 YEAR 2
********************************************************************************
        REVENUE $:                                        9000000.00
        OPERATING COST $:                                 1200000.00
        DEPRECIATION $:                                   6400000.00
        INTEREST $:                                       2724474.40
        PRE-TAX INCOME $:                                -1324474.40
        TAX AT 50.%, $:                                         0.00
        AFTER TAX INCOME $:                              -1324474.40
        CASH FLOW $:                                      5075526.00

********************************************************************************
```

(continued)

Table 26.24 Input Data and Computer Results of Example 26.11. (*Continued*)

```
                              YEAR 3
****************************************************************************
     REVENUE $:                                           9000000.00
     OPERATING COST $:                                    1200000.00
     DEPRECIATION $:                                      5120000.00
     INTEREST $:                                          2511598.20
     PRE-TAX INCOME $:                                     168401.74
     TAX AT 50.%, $:                                            0.00
     AFTER TAX INCOME $:                                   168401.74
     CASH FLOW $:                                         5288402.00
****************************************************************************

                              YEAR 4
****************************************************************************
     REVENUE $:                                           9000000.00
     OPERATING COST $:                                    1200000.00
     DEPRECIATION $:                                      4096000.00
     INTEREST $:                                          2276432.20
     PRE-TAX INCOME $:                                    1427568.00
     TAX AT 50.%, $:                                            0.00
     AFTER TAX INCOME $:                                  1427568.00
     CASH FLOW $:                                         5523568.00
****************************************************************************

                              YEAR 5
****************************************************************************
     REVENUE $:                                           9000000.00
     OPERATING COST $:                                    1200000.00
     DEPRECIATION $:                                      3276800.00
     INTEREST $:                                          2016641.00
     PRE-TAX INCOME $:                                    2506559.20
     TAX AT 50.%, $:                                            0.00
     AFTER TAX INCOME $:                                  2506559.20
     CASH FLOW $:                                         5783359.00
****************************************************************************

                              YEAR 6
****************************************************************************
     REVENUE $:                                           9000000.00
     OPERATING COST $:                                    1200000.00
     DEPRECIATION $:                                      2621440.00
     INTEREST $:                                          1729646.00
     PRE-TAX INCOME $:                                    3109796.00
     TAX AT 50.%, $:                                            0.00
     AFTER TAX INCOME $:                                  3109796.00
     CASH FLOW $:                                         5731236.00
****************************************************************************
                                                          (continued)
```

Table 26.24 Input Data and Computer Results of Example 26.11. (*Continued*)

```
                                    YEAR 7
***************************************************************************
        REVENUE $:                                          9000000.00
        OPERATING COST $:                                   1200000.00
        DEPRECIATION $:                                     2097152.00
        INTEREST $:                                         1412599.00
        PRE-TAX INCOME $:                                   4290249.00
        TAX AT 50.%, $:                                           0.00
        AFTER TAX INCOME $:                                 4290249.00
        CASH FLOW $:                                        6387401.00
***************************************************************************


                                    YEAR 8
***************************************************************************
        REVENUE $:                                          9000000.00
        OPERATING COST $:                                   1200000.00
        DEPRECIATION $:                                     1677722.00
        INTEREST $:                                         1062353.00
        PRE-TAX INCOME $:                                   5059926.00
        TAX AT 50.%, $:                                     2229985.20
        AFTER TAX INCOME $:                                 2829940.20
        CASH FLOW $:                                        4507662.00
***************************************************************************


                                    YEAR 9
***************************************************************************
        REVENUE $:                                          9000000.00
        OPERATING COST $:                                   1200000.00
        DEPRECIATION $:                                     1342177.00
        INTEREST $:                                          675431.50
        PRE-TAX INCOME $:                                   5782391.00
        TAX AT 50.%, $:                                     2891195.40
        AFTER TAX INCOME $:                                 2891195.40
        CASH FLOW $:                                        4233373.00
***************************************************************************


                                    YEAR 10
***************************************************************************
        REVENUE $:                                          9000000.00
        OPERATING COST $:                                   1200000.00
        DEPRECIATION $:                                     1073742.00
        INTEREST $:                                          247994.34
        PRE-TAX INCOME $:                                   6478264.00
        TAX AT 50.%, $:                                     3239132.00
        AFTER TAX INCOME $:                                 3239132.00
        CASH FLOW $:                                        4312874.00
***************************************************************************
        SALVAGE VALUE $:                                    4294967.00
```

(*continued*)

Table 26.24 Input Data and Computer Results of Example 26.11. (*Continued*)

```
                        ECONOMIC PROJECT EVALUATION
                     NET PRESENT VALUE AT A GIVEN DISCOUNT RATE
********************************************************************************
   11 YEARLY CASH FLOWS INCLUDING YEAR 0

    5.00 PERCENTAGE ANNUAL DISCOUNT RATE
--------------------------------------------------------------------------------
                             YEARLY CASH FLOWS
--------------------------------------------------------------------------------

  YEAR        CASH FLOW       CUMULATIVE CASH FLOW    DISCOUNT FACTOR    PRESENT VALUE
--------------------------------------------------------------------------------

   0        -30000000.0          -30000000.0              1.0000         -30000000.0
   1          4882828.0          -25117172.0              0.9524           4650313.0
   2          5075526.0          -20041650.0              0.9070           4603652.0
   3          5288402.0          -14753244.0              0.8638           4568321.0
   4          5523568.0           -9229676.0              0.8227           4544254.0
   5          5783359.0           -3446317.0              0.7835           4531414.0
   6          5731236.0            2284919.0              0.7462           4276738.0
   7          6387401.0            8672320.0              0.7107           4539409.0
   8          4507662.0           13179982.0              0.6768           3050964.2
   9          4233373.0           17413360.0              0.6446           2728871.2
  10          4312874.0           21726230.0              0.6139           2647732.2
  11          8589934.0           30316163.0              0.5847           5022359.0
--------------------------------------------------------------------------------

   THE NET PRESENT VALUE ($):        15164030.
   PRESENT VALUE RATIO:                  1.505
   THE NET RETURN RATE:                  4.60 %
   THE AVERAGE RATE OF RETURN:          18.28 %
   THE PAYBACK PERIOD IS BETWEEN:  5 AND 6 YEARS
```

Table 26.25 Input Data and Computer Results for the Discounted Cash Flow Rate of Return (DCFRR) of Example 26.11.

```
   DATA3.DAT
--------------------------------------------------------------------------------
   12
   -30000000
   4882828
   5075526
   5288402
   5523568
   5783359
   6387401
   4507662
   4233373
   4312874
   8589934
                    DISCOUNT CASH FLOW RATE OF RETURN CALCULATION
********************************************************************************
                        12 YEARLY CASH FLOWS INCLUDING YEAR 0
--------------------------------------------------------------------------------
   YEAR                                                              CASH FLOW
********************************************************************************

    0                                                              -30000000.0
    1                                                                4882828.0
    2                                                                5075526.0
    3                                                                5288402.0
    4                                                                5523568.0
    5                                                                5783359.0
    6                                                                5731236.0
    7                                                                6387401.0
    8                                                                4507662.0
    9                                                                4233373.0
   10                                                                4312874.0
   11                                                                8589934.0
--------------------------------------------------------------------------------

   The Discount Cash Flow Rate of Return (%): 13.424
```

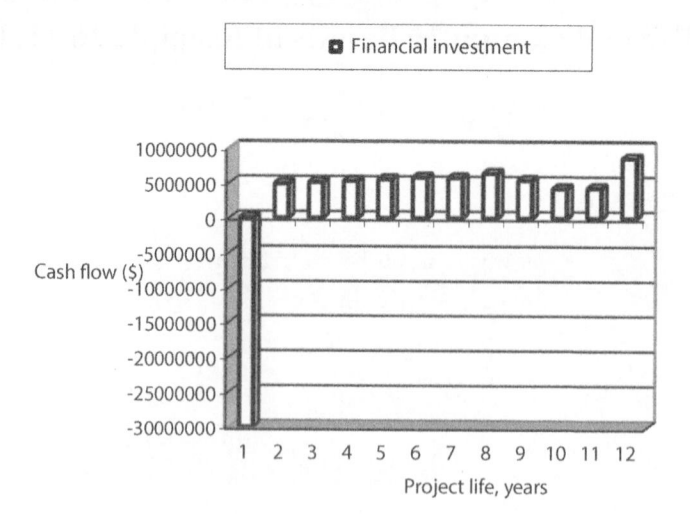

Figure 26.12a Cash flow vs. project life.

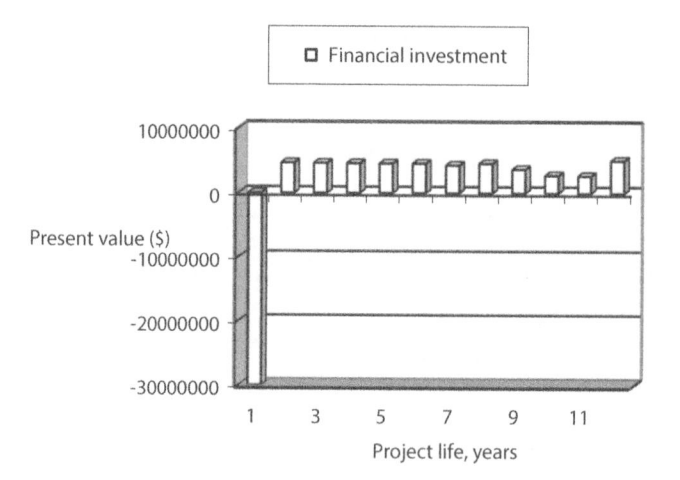

Figure 26.12b Present value vs. project life.

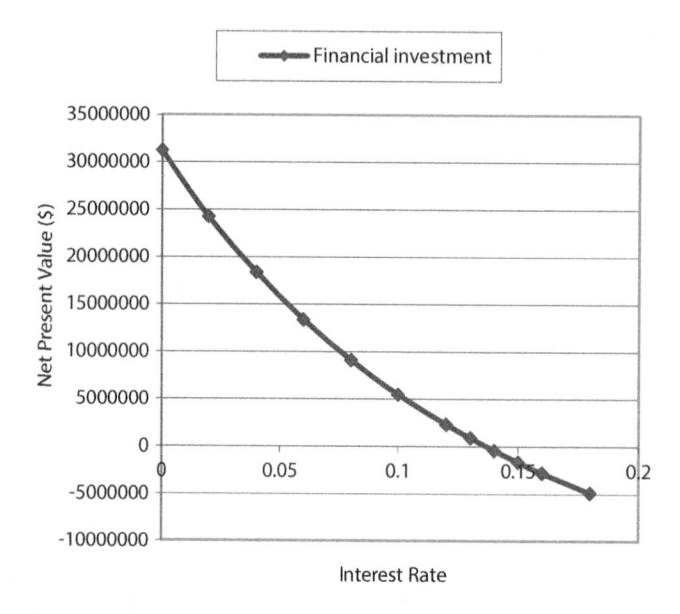

Figure 26.12c Net present value vs. interest rate.

Cost Estimation and Economic Evaluation 531

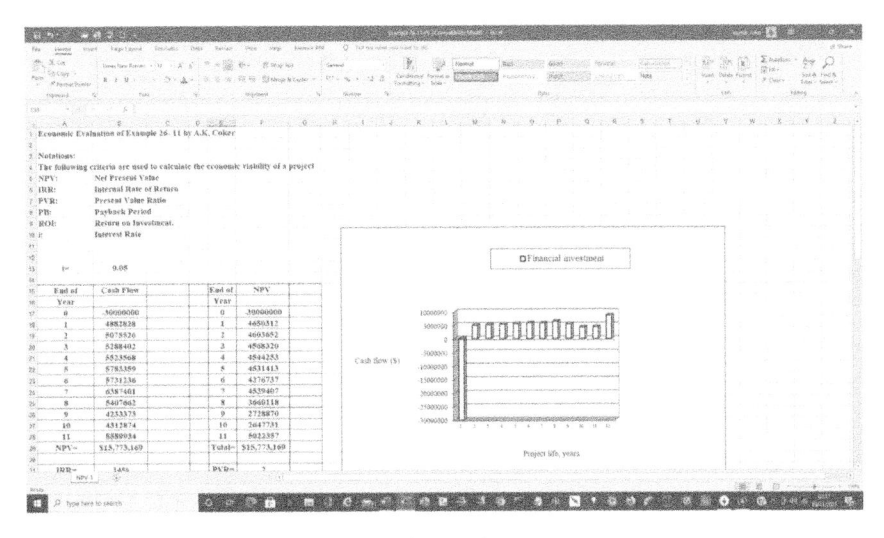

Figure 26.12d A snapshot of the Excel spreadsheet calculations of Example 26.11.

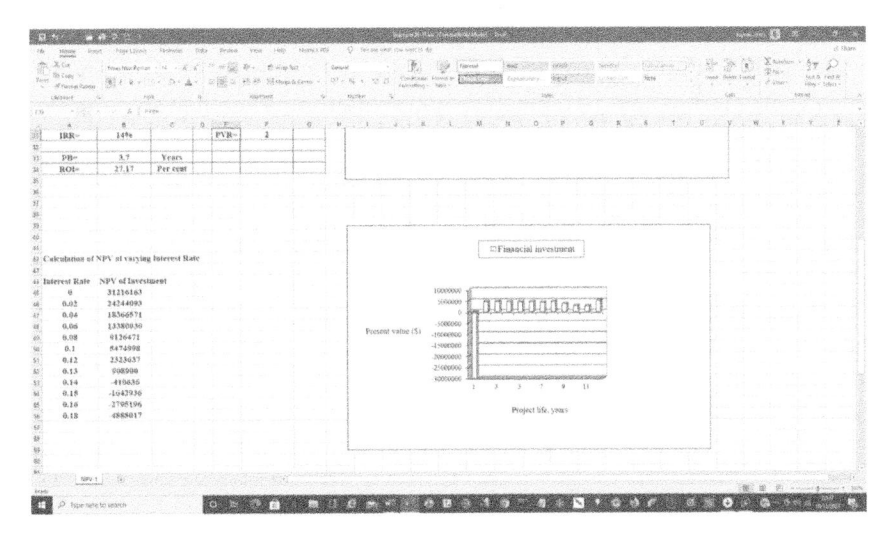

Figure 26.12e A snapshot of the Excel spreadsheet calculations of Example 26.11.

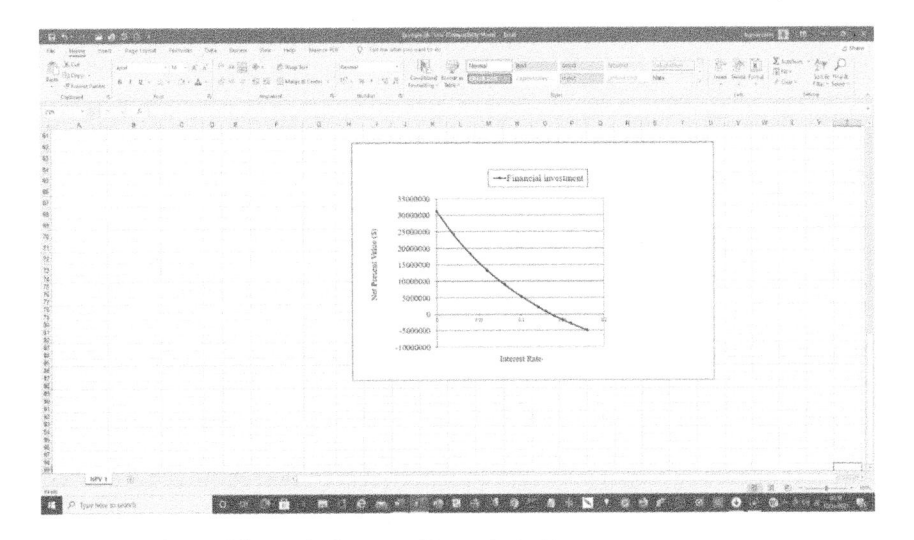

Figure 26.12f A snapshot of the Excel spreadsheet calculations of Example 26.11.

It is important to remember what has sometimes been overlooked in unsuccessful ventures

$$\text{Cash flow} \neq \text{Profit} \qquad (26.61)$$

Allen [43] has provided a systematic procedure for assessing investment proposals for new plant and equipment, exploiting new technology and replacing uneconomic, inefficient, and obsolete plants, process, and equipment.

26.12.3 Inflation

A decrease in the average purchasing value of currency is referred to as inflation. If a given product costs $100 last year and now costs $200, then the product has suffered a 20% rate of inflation. That is, the purchasing power of the currency (i.e., of the $120) has consequently fallen by a factor of ($120 - $100)/$120 or 16.7%. An inflation rate, e.g., 15% means that the average cost of goods and services will increase 15% in one year. The result is that commencing construction one year early will reduce the amount of money expended by 15%. Since the 1980s, inflation has been considered in most economic project evaluations.

When inflation is used in economic evaluations, all items except interest on a loan and depreciation are considered to increase in value at the same rate as inflation. Generally, interest is set at the time a loan is negotiated and does not change with inflation. In addition, depreciation depends on the method (e.g., straight line or DDB) used, and the capital charges incurred before start-up are not affected by the inflation rate after start-up. While determining the profitability of a project (e.g., NPV), the interest rate is assumed to be greater than the inflation rate. Money may be lost on the project while the NPV indicates the opposite, if the inflation rate is greater than the interest rate. There are cases where the interest rate is set at the expected inflation rate plus a real expected interest rate. The real expected interest rate is the interest rate that is used to calculate the NPV when there is no inflation.

Alternatively, the present value is calculated using the inflation rate as the interest rate, the NPV is then determined using the real expected interest rate.

Effect of Inflation on NPV: The effect of inflation on NPV for a proposed project can be expressed by

$$NPV = \sum_{t=0}^{t=n} \frac{C_t}{(1+i)^t (1+i_i)^t} \qquad (26.62)$$

where

C_t = cash flow.
i = interest rate.
i_i = fractional rate of inflation.
n = number of years of project life.

Eq. (26.62) allows all the net annual cash flows to be corrected to their purchasing power in year zero, Eq. (26.62) becomes identical with Eq. (26.40).

For many years, companies and countries have lived with the problem of inflation, or the falling value of money, while other factors (e.g., labor costs) tend to rise each year. Failure to account for these trends in predicting cash flows can lead to serious and erroneous results; therefore, giving misleading profitability estimates.

26.12.4 Sensitivity Analysis

When an economic evaluation has been carried out using single-value forecasts and estimates for everything that contributes to the yearly cash flows, the information obtained from it for project decision-making purposes can be extended by carrying out an economic evaluation sensitivity analysis. This explores the relative effects on the economic viability of a project of possible changes in the forecast data that contribute to the project cash flows. It focuses on the areas that are most critical in terms of any uncertainty, and it indicates where confidence in forecasts is most vital. Sensitivity analysis also enables the economic effects of changes in a project to be reviewed; for example, changes in fixed and variable costs resulting from the use of different equipment types, different phasing of investment, delays in plant start-up, and the effect of possible different market growth patterns. It can also be used to explore the effects on the economic viability of a project of uncertainty in different areas, but does not attempt to quantify the uncertainty in an area [37, 43]. In general, it is worthwhile to make tables or plot curves that show the effect of variations in costs and prices on profitability. Its purpose is to determine to which factors the profitability of a project is most sensitive. Sensitivity analysis should always be carried out to observe the effect of departures from expected values.

Sensitivity analysis consists of finding the effects on project NPV, PVR and IRR of changes in the input factors selected, taken one at a time. Once the Excel worksheets (Examples 26.8.xls and 26.9.xls) have been set up with the series of data tables, cash flow tables, and NPV, IRR, PVR calculations for the project, the entry in any cell in the input tables can be changed and the consequent effects will automatically be recalculated and will ripple through the following tables to give the corresponding new project NPV, PVR and IRR values. Sensitivity analysis thus consists of systematically organizing a sequence of changes in the input data and displaying the resulting changes to the project NPV, IRR, and PVR.

26.13 Refining Economics

The overall economics or viability of a refinery depends on the interaction of three key elements, namely: the choice of crude oil used (i.e., crude slates), the complexity of the refining equipment (refinery configuration) and the desired type and quality of products produced (product slate). Environmental considerations and utilization rates also influence refinery economics. Crude slates and refinery configurations take into account the type of products that are required in the market. The quality and specifications of the final products are essential as the environmental requirements become more stringent.

The use of more expensive crude oil (e.g., lighter, sweeter) requires less refinery upgrading, but the supplies of light, sweet crude oil are decreasing and the differential between heavier and more sour crudes is increasing. Using cheaper heavier crude oil means more investment in upgrading processes. Thus, costs and payback periods for refinery processing units must be weighed against anticipated crude oil costs and the projected differential between light and heavy crude oil prices.

Crude Slates

Different types of crude oil yield a different mix of products depending on the crude oil's natural qualities. Crude oil types are typically differentiated by their density (measured as API gravity) and their sulfur content. Crude oil with a low API gravity is considered a heavy crude oil and typically has a higher sulfur content and a larger yield of lower valued products. Therefore, the lower the API of a crude oil, the lower the value it has to a refiner as it will either require more processing or yield a higher percentage of lower-valued byproducts such as heavy fuel oil, which usually sells for less than crude oil.

Crude oil with a high sulfur content is referred to as a sour crude while sweet crude has a low sulfur content. Sulfur is an undesirable characteristic of petroleum products, particularly in transportation fuels. It can hinder the efficient operation of some emission control technologies, and when burned in a combustion engine, is released into the atmosphere where it can form sulfur dioxide (SO_2). With increasingly restrictive sulfur limits on transportation fuels, sweet crude oil sells at a premium. Sour crude oil requires more severe processing to remove the sulfur. Refiners are generally willing to pay more for light, low sulfur crude oil.

Refinery Configuration

A refiner's choice of crude oil is influenced by the type of processing units at the refinery. Refineries fall into three broad categories. The simplest is a topping plant, which consists only of a distillation unit, and possibly a reformer to provide the octane rating. Yields from this plant would most closely reflect the natural yields from the crude processed. Typically, only condensates or light sweet crude would be processed at this type of facility unless markets for heavy fuel oil (HFO) are readily and economically viable. For example, asphalt plants are topping refineries that run heavy crude oil because they are only viable for producing asphalt.

The next level of refining is a cracking refinery. Here, the gas oil portion from the crude distillation unit (a stream heavier than diesel fuel but light than heavy fuel oil) is cracked/broken further into gasoline and distillate components using catalysts, high temperature and/or pressure.

The last level of refining is the coking. This refinery processes residual fuel, the heaviest material from the crude unit and thermally cracks it into lighter product in a coker or a hydrocracker. The addition of a fluid catalytic cracking unit (FCCU) or a hydro cracker significantly increases the yield of higher-valued products like gasoline and diesel oil from a barrel of crude, allowing a refinery to process cheaper, heavier crude while producing an equivalent or greater volume of high-valued products.

Hydrotreating is a process used to remove sulfur from finished products (see volume 1 of these volume sets). As the requirement to produce ultra-low sulfur products increases, additional hydrotreating capability is being added to refineries. Refineries that currently have large hydrotreating capability have the ability to process crude oil with a higher sulfur content.

Product Slates

Refinery configuration is also influenced by the product demand in each region. Refineries produce a wide range of products including propane (C_3H_8), butane (C_4H_{10}), petrochemical feedstock, gasolines (naphtha, specialties, aviation gasoline, motor gasoline), distillates (jet fuels, diesel, stove oil, kerosene, furnace oil), heavy fuel oil, lubricating oils, waxes, asphalt and still gas. Gasoline accounts for about 40% of demand with distillate fuels representing about 1/3 of product sales and heavy fuel oil accounting for only 8% of sales.

The relationship between gasoline and distillate sales can create challenges for refiners. A refinery has a limited range of flexibility in setting the gasoline to distillate production ratio. Beyond a certain point, distillate production can only be increased by also increasing gasoline production. Europe is a major gasoline exporter followed by the U.S.

Refinery Utilization

Another critical component of refining economics is the utilization rate, or how efficiently the refining complex is operating. The oil price shocks in the 70s led to improvements in the efficiency of vehicles and to fuel switching from oil to natural gas and electricity. This reduced the demand for petroleum products and resulted in a substantial surplus of refining capacity. The spare capacity resulted in increased competition among refiners, which further curtailed refining margins. Less efficient, smaller refiners were closed in favor of new larger facilities. Weak economic conditions in the early 1980s put additional pressure on the industry to rationalize their operations, resulting in a significant number of refinery closures.

In recent years, growth in the demand for petroleum products has led to an improvement in capacity utilization, increasing operating efficiency and reducing costs per unit of output. As a result, refinery utilization rates have been above 90% in some countries, and a utilization rate of about 95% is considered optimum as it allows for normal shut downs required for maintenance and seasonal adjustments.

Refinery capacity is based on the designed size of the crude distillation units (often referred to as nameplate capacity). Through upgrades or de-bottlenecking procedures, refineries can process more crudes than the nameplate size of the distillation unit would indicate. In such cases, a refinery is able to achieve a utilization rate > 100% for short periods of time.

Environmental Initiatives

Refiners also make investment decisions because of environmental impact of the processing the crudes and voluntary actions or legislative and regulatory requirements by agencies, governments and so on. In recent years, these establishments have directed considerable effort toward reducing the environmental impact of burning fossil fuels. Many of these initiatives are towards reducing emissions due to greenhouse gas (GHG) and the increase in global warming and at providing cleaner fuels. Petroleum refining is a very complicated and capital-intensive industry and new regulations require industry to make additional investment to meet the more stringent standards.

Because of stricter environmental regulations, refiners have to add units and processes to ensure that the gas emissions are within the regulations. This applies to water discharges into rivers, and the marine environment, solid waste in the form of spent catalysts, sludge or coke, which has to be disposed of without causing harm to the environment.

In the oil refining business, refiners are exposed to significant risk due to the volatile market segments of crude oil and refined products, which are influenced by global, regional and local supply and demand changes. It is therefore a challenge for the refiners to rely on operational efficiency for their competitive edge through constant innovation, upgrading and optimization to maximize the difference between cost and price.

26.13.1 Refinery Margin Definitions

Determining the profitability for a specific refinery is very difficult since the data on operational and environmental compliance costs are generally unavailable. A rough measure is obtained by calculating the cost of crude oil feedstock (though to achieve this with precision would require knowledge of the crude blends used in a specific refinery) and comparing that cost with the market value of the suite of products that are produced at the refinery. However, this still requires more information than might be publicly available for a typical refinery, and is subject to market conditions for the various products that are produced.

The profitability of a refinery is determined by calculating the refining margin. Two types of margins are defined. The gross margin is the difference between the composite value of the refined products at the refinery gate and the cost of the crude oil delivered to the refinery.

$$\text{Gross margin} = \sum_{i}^{N} (\text{price of product i} \times \text{yield of product i}) - \text{crude price} \qquad (26.63)$$

where N represents all products produced. The price of product i is the spot price per barrel of product i. The yield of product i is the volume per cent of product i per barrel of crude oil feed to the refinery. The crude price is the price paid by the refiner for a barrel of delivered crude oil.

The net refining margin is the gross margin less the variable refining costs, that is, chemicals, catalysts, fuel and working capital charges:

$$\text{Net margin} = \text{gross margin} - \text{variable cost} \tag{26.64}$$

For a refinery to be profitable, the net margin has to be greater than the total fixed costs per barrel of crude oil processed. The difference between the net margin and the fixed cost per barrel of crude oil is called the cash margin and represents the profit the refinery makes per barrel of processed crude oil.

$$\text{Cash margin} = \text{net margin} - \text{fixed cost} \tag{26.65}$$

In addition, petroleum analysts use the spread as an indicator of refining profitability. A product spread is the difference between the products price and the price of crude oil. A useful and simplified measure of refinery profitability is the "crack spread". The crack spread is the difference in the sales price of the refined product (gasoline and fuel oil distillates) and the price of crude oil. An average refinery would follow what is known as the 3-2-1 crack spread, which assumes that three barrels of crude can be refined to produce two barrels of gasoline and one barrel of distillates. It is calculated by:

$$\text{3-2-1 Crack spread (\$ bb-1)} = (2 \times \text{gasoline price} + 1 \times \text{distillate price} - 3 \times \text{crude oil price})/3 \tag{26.66}$$

The higher the crack spread, the more money the refinery will make, so it will be utilizing as much capacity it has available. Inversely, at some lower crack spread prices, it actually may be in the refinery's best interest due to costs of the plant to scale back the amount of capacity utilized.

Calculating the 3-2-1 crack spread typically uses published prices for crude oil, gasoline and distillates. These prices are typically taken from the New York Mercantile Exchange (NYMEX). The NYMEX has traded contracts for crude oil and gasoline but no contract for diesel fuel (the most produced of the distillate fuel oils). In calculating the 3-2-1 crack spread, prices for heating oil futures are typically used instead.

Example 26.12

Calculate the 3-2-1 crack spread of the following:

Oil price:	$84.54/barrel
Gasoline price:	$2.57/gallon
Heating oil price:	$2.79/gallon

Note: 42 gallons = 1 barrel

Solution

Using Eq. 26.66, the 3-2-1 crack spread is:
(2 barrels x 42 gal/barrel x $2.57/gal of gas) + (1 barrel x 42 gal/barrel x $2.79/gal of heating oil)
− (3 barrels x $84.54/ barrel of oil)
= $79.44 profit/3 barrels of oil

The crack spread is: $26.48/barrel of oil.

The crack spread is not a perfect measure of refinery profitability. It measures whether the refinery will make money at the margin, i.e., whether an additional barrel of crude oil purchased upstream will yield sufficient revenues from saleable products downstream. Existing refineries must consider their refining costs in addition to the cost of crude oil. These costs include labor (though it is generally a small part of refinery operations); chemical catalysts; utilities and any short-term financial costs such as borrowing money to maintain refinery operations. These variable costs of refining may amount to perhaps $20 per barrel (depending on conditions in utility and financial markets). In the example above, the true margin on refining would be $6.58 per barrel of crude oil, which is lower than the simple crack spread.

The crack spread tends to be sensitive to the slate of products produced from the refinery. In the example, we used gasoline and distillate fuel oil (heating oil) because those are two typically high-valued products and U.S. refineries are generally engineered to maximize production of gasoline and fuel oil.

The crack spread is also sensitive to the selection of the oil price used. Here, the NYMEX future price for crude oil is based on the West Texas Intermediate blend, which is a fairly light crude oil. Many U.S. refineries are engineered to accept heavier crude oils as feedstocks. If there are systematic differences in the prices of heavy crude oils, versus West Texas Intermediate, then the crack spread calculation may not be sensible for a particular refinery.

Example 26.13

Calculate the gross and net margin for an average refinery using the following data:

	Product yield %	Price per barrel (US $)
Gasoline	52.5	72
Distillates	29.0	73
Other products	19.0	48.5
Crude price		58
Variable refining costs		6.9

Using Eq. 26.63, the gross margin is:
Gross margin = (72 x 0.525 + 73 x 0.29 + 48.5 x 0.19) − 58
= $10.2
Net margin from Eq. 26.64 is:
Net margin = 10.2 − 6.9 = $3.3

26.13.2 Refinery Complexity

The following classification can be used to define the complexity of a refinery:

Simple refinery	It has atmospheric crude distillation, a catalytic reformer to produce high octane gasoline, and middle distillate hydrotreating units.
Complex refinery	It has in addition to the units of a simple refinery, conversion units such as hydrocrackers and fluid catalytic cracking units.
Ultra-complex refinery	The refinery has all of the units above in addition to deep conversion units, which convert atmospheric or vacuum residue into light products.

The complexity of a refinery can be assessed by calculating the complexity factor. Each unit has a coefficient of complexity (CC_i) defined as the ratio of the capital cost of the unit per ton of feedstock to the capital cost of the crude distillation unit (CDU) per ton of feedstock. The complexity factor (Cf_i) of the whole refinery is then calculated from the coefficients of complexity for the units in the refinery as follows:

$$CF = \sum_i^{F_i} \frac{F_i}{F_{CDU}} CC_i \qquad (26.67)$$

where F_i and F_{CDU} are the feed rate to unit i and CDU respectively.

Example 26.14

Calculate the complexity factor for a refinery which has the following unit data.

Unit	Complexity coefficient (CC_i)	Percent capacity relative to CDU capacity (F_i/F_{CDU})
Vacuum distillation unit	1.5	40
Fluid catalytic cracking	7	25
Hydrocracking	7	15
Hydrotreating	5	50
Reforming	3.4	30
Alkylation	6	5

Solution

From Eq. 26.67

$$CF = 0.40 \times 1.5 + 0.25 \times 7 + 0.15 \times 7 + 0.50 \times 5 + 0.3 \times 3.4 + 0.05 \times 6 = 7.8$$

The following terms are part of the refinery margins:

Hydrocarbon Margin (HCM)	The difference between the aggregate value of the products and the aggregate value of the crudes and feedstocks required for the refining. It is not the true measure of the refinery's profit, but it can only give a good approximation of the refinery's profitability for a particular configuration. It is used to approximate the profitability if a change in refinery configuration or operating modes and conditions are being considered. It is used to calculate the "Gross Refinery Margin" or GRM.
Gross Refinery Margin (GRM)	Is the refinery's margin after deducting all variable cost from the Hydrocarbon Margin. It is the preferred parameter for indicating the refining profitability as it includes all variable cost incurred during the refining process.
Variable Cost	This varies with throughput of the refinery such as imported feedstock, chemicals, catalyst, maintenance, utilities and purchased energy (e.g., natural gas and electricity).
Net Refinery Margin	
Fixed cost	This does not vary with refinery throughput. The fixed costs in the short term are manpower, maintenance, insurance, administration and depreciation.

Stream that can be readily valued are namely:

- Ready products, crude oil and imported feedstocks.
- Blending components
- Intermediate feedstocks.

Ready products, crude oil and other feedstocks are normally valued at market price. If we consider the following:

26.13.3 Supply and Demand Balance

Refinery bulk products such as gasoline, diesel, and hydrodesulfurization fuel oil (HSFO) have large markets. However, changes on the production of these products will not affect the overall supply/demand balance.

Other products such as aviation fuel, some gasoline choice grades and hydrocarbon solvents have limited market demand, therefore attention on the economic effects if a decision to increase production of such products is being considered, confirm the availability of outlets at an early stage.

Product Quality

Products need to be within specified ranges to avoid "quality give away".

Normally, any extra does not command a premium.

Standard Density

Most products are sold by volume, and a correction based on a standard density must be carried out before valuation if the quoted price is in $/t.

Density correction is significant in the optimization of distillation units, i.e., in optimizing the cut point between naphtha and middle distillate pool, the standard density correction between the two products is around 12%.

Blending Components

Blending components are valued at its blending value in the poor starting from the value of a final product, the value of the pool is corrected for all constraining properties, then subsequently corrected for its standard density if the product is sold by volume.

Constraining Properties

The constraints may differ for different grades, season, refinery configuration, crude diet and product specification.

Quality Premiums/Discounts

These are typical quality premium/discount values per degree of increase/decrease of a particular property per ton of final product.

Intermediate Feedstocks: The marginal value of an intermediate can be calculated by its marginal impact on the refinery mass balance assuming there is a change in the volume of such stream.

There are software programs that are available to calculate the marginal economics and are based on comparison of the full refinery yield expense. But these programs are not sufficiently detailed to capture condition effects as a result of the proposal that may lead to erroneous results. However, designers can check the accuracy of such programs to account for the following:

- Product quality changes as a function of cut point.
- Yield pattern as a function of throughput.
- Yield and product quality as a function of feed quality.

A Case Study [44]

Application of Cost Estimation Techniques and Economic Evaluations

An example illustrating the methods to estimate capital and operating costs and ROI is included using a simplified refinery facility.

Problem Statement

For the following of a simplified refinery, calculate

1. The products available for sale
2. Investment
3. Operating costs

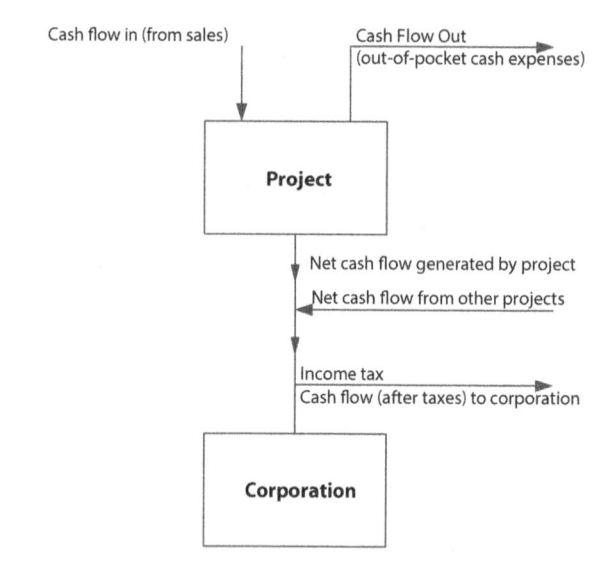

Figure 26.13 A continuous cash flow diagram. (Source: Valle-Riestra [36]).

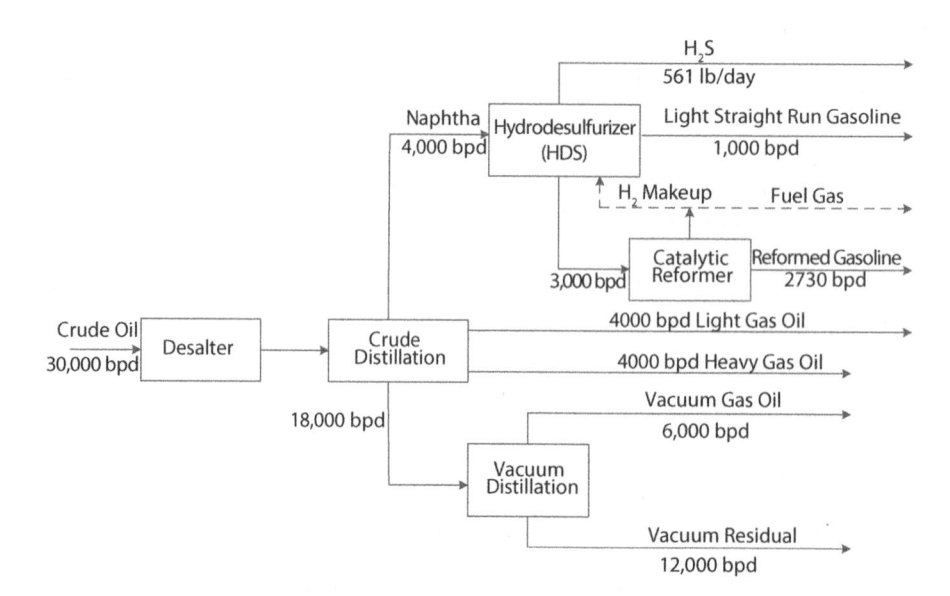

Figure 26.14 Block flow diagram for the case study example.

4. Simple rate of ROI
5. True rate of ROI

Using the block flow diagram (Figure 26.14). The following data are

1.	Crude charge rate: 30,000 bpsd
2.	Crude oil sulfur content: 1.0 wt%
3.	Full – range naphtha in crude: 4,000 bpsd 240°MBP 56°API gravity 11.8 K_w
4.	Light gas oil in crude: 4,000 bpsd
5.	Heavy gas oil in crude: 4,000 bpsd
6.	Vacuum gas oil in crude: 6,000 bpsd
7.	Vacuum residual in crude: 12,000 bpsd
8.	On – stream factor: 93.15%
9.	Cost of makeup water: $0.32/1000 gal
10.	Cost of power: $0.12 kWh
11.	LHV of heavy gas oil: 5.5 MMBtu/bbl.
12.	Replacement cost of desulfurizer catalyst is $1.60/lb
13.	Replacement cost of reformer catalyst is $8/lb
14.	Insurance annual cost is 0.5% of plant investment
15.	Local taxes annual cost is 1.0% of plant investment
16.	Maintenance annual cost is 5.5% of plant investment
17.	Miscellaneous supplies annual cost is 0.15% of plant investment
18.	Average annual salary plus payroll burden for plant staff and operators is $72,000
19.	Value of crude oil and products at refinery is: 　　　　　　　　$/bbl 　　Crude　　　　　　　60.00 　　Gasoline　　　　　　83.00 　　Light gas oil　　　　79.00 　　Heavy gas oil　　　　75.00 　　Vacuum gas oil　　　73.00 　　Vacuum residual　　 55.00
20.	Depreciation allowance: 15 year, straight -line

21.	Corporate income tax: 50% of taxable income
22.	Location: St. Louis 2007
23.	Construction period: 2007
24.	Escalation rate (applicable to construction costs only) is 3% per year

Process Description

The crude oil is to be desalted and fractionated to produce full-range naphtha, light gas oil, heavy gas oil, and "atmospheric bottom". The latter cut is fed to a vacuum unit for fractionation into vacuum gas oil and vacuum residuals. The full – range naphtha is to be hydrodesulfurized. After desulfurization, the light straight – run (LSR) portion (i.e., material boiling below 180°F (82.2°C) of the full-range naphtha is separated for blending into the gasoline product. The balance of the naphtha is fed to a catalytic reformer, which is operated to produce a reformate having a research octane number (clear) of 93. The reformate plus the LSR is mixed to make the final gasoline product. Propane and lighter hydrocarbons, including the hydrogen, which are produced in the catalytic reformer are consumed as fuel. The necessary hydrogen makeup for the hydrodesulfurizer is taken from these gases before they are burned as fuel. The balance of the fuel requirement is derived from light gas oil. The hydrogen sulfide produced in the hydrodesulfurizer is a relatively small amount and is burned in an incinerator. No other product treating is required. It can be assumed that sufficient tankage of approximately 12 days' storage of all products is required. The total storage requirement will thus be approximately 360,000 barrels. Figure 26.14 shows the block diagram of the process.

Catalytic Reformer

Calculate the properties of feed to the reformer, given total naphtha stream properties as follows:

Mid. Boiling point	240°F
API gravity	56°API
K_w	11.8

The material boiling below 180°F (LSR) is not feed to the reformer. After desulfurizing the total naphtha, the LSR is fractionated out. The problem is to estimate the volume and weight of the LSR, assuming a distillation curve is not available. The LSR is then deducted from the total naphtha to find the net reformer feed. This is done as shown in the following steps:

1. Assume the butane (C_4H_{10}) and lighter hydrocarbons in the naphtha are negligible. Therefore, the lightest material would be isopentane (iC_5H_{12}) with a boiling point of 82°F (27.8°C). Hence, the mid. boiling point of the LSR would be approximately (82 + 180) (0.5) = 131°F (55°C).
2. Assume that the LSR has the same K_w as the total naphtha (i.e., 11.8).
3. From the general charts [Ref. 44] relating K_w, mean average boiling point and gravity, find the gravity of the LSR. This is 76.5°API.
4. The gravity of the naphtha fraction boiling above 180°F (82.2°C) is next determined by a similar procedure. The mid. boiling point for the total naphtha is given above as 240°F (115.6°C), and the initial boiling point was estimated in Step 1 as 82°F (27.8°C). Therefore, the naphtha end point can be estimated as shown.

$$240 + (240 - 82) = 398°F (203.3°C)$$

Now, the approximate mid. boiling point of the reformer feed is estimated as

$$(180 + 398)(0.5) = 289°F \ (142.8°C)$$

Using a K_w of 11.8, the reformer feed gravity is found as in Step 3 to be 52.5°API.

5. With the above estimates of gravity of both the LSR and reformer feed, it is now possible to estimate the relative amounts of each cut that will exist in the total naphtha stream. This is done by weight and volume balances as shown below.

V_{LSR}	= gal. LSR
V_{RF}	= gal. reformer bed
V_N	= gal. total naphtha
W_{LSR}	= lb LSR
W_{RF}	= lb reformer bed
W_N	= lb total naphtha
lb/gal LSR	= 5.93 (67.5°API)
lb/gal RF	= 6.41 (52.5°API)
lb/gal. V_N	= 6.29 (56°API)

Volume balance, hourly basis:
$V_N = (4000)(42/24) = 7000$ gal/h
$V_{LSR} + V_{RF} = 7000$ gal/h

Weight balance, hourly basis:
$V_N (6.29) = (7000)(6.29) = 44{,}030$ lb/h
$V_{LSR}(5.93) = V_{RF}(6.41) = 44{,}030$ lb/h

Solving the two equations for V_{LSR} and V_{RF} gives:
$V_{LSR} = 1750$ gal/h = 1000 bpd
$V_{RF} = 2520$ gal/h = 3000 bpd

6. The above information can be tabulated as

Stream	°API	lb/gal	gal/h	lb/h	bpd
LSR	67.5	5.93	1750	10,378	1000
Reformer feed	52.5	6.41	5250	33,652	3000
Total naphtha			7000	44,030	4000

7. It should be emphasized that the above methods for approximating the naphtha split into LSR and reformer feed are satisfactory for preliminary cost and yield computations before final design calculations.

8. The reformer feed properties can now be used with the yield curves [Ref. 44]. The following yields are based on production of 93 RON reformate. From the yield curves:

vol % C_5^+	86.0
vol % C4,s	5.0 (iC_4H_{10}/nC_4H_{10} = 41.5/58.5)
wt % CH_4 and C_2H_6	1.1
wt % C_3H_8	1.92
wt % H_2	1.75

With the above data, the following table is

Component	gal/h	lb/gal	lb/h	bpd	Mscf/day
H_2			589		2682
CH_4 and C_2H_6			370		145[a]
C_3H_8	153	4.23	646		133
iC_4H_{10}	109	4.69	511	62	
nC_4H_{10}	154	4.86	748	88	
C_5^+	4515	6.82	30,788[b]	2580	
Total			33,652	2730	2960
Feed	5250	6.29	33,652	3000	

[a] Assume lb CH_4/lb C_2H_6 = 0.5 (i.e., CH_4 and C_2H_6 = 23.3 mol. wt.)
[b] lb/h C_5^+ obtained by difference from total feed less other products.

Naphtha Desulfurizer

Assume crude oil = 1.0% sulfur. Then from the curve for miscellaneous crudes, naphtha contains 0.05% sulfur (240°F MBP). Calculate the amount of sulfur produce. Assume K_w for desulfurized feed is 11.8. This is combined with 240°F MBP, gives a naphtha API gravity of 56°API (see reformer calculations).

56 °API	= 6.29 lb/gal
wt % S in naphtha	= S_N = (4000) (42) (6.29) (0.0005) = 528 lb/day
Maximum H_2S formed	= 32 (528) = 561 lb/day
Theoretical H_2 required	= 561 − 528
	= 33 lb/day
	= 16.5 mol/day
	= 6.26 Mscf/day

The makeup H_2 required is about 100 to 150 scf/bbl or (4000) (0.15) = 600 Mscf/day

Summary of Investment and Utilities Costs

	bpsd	$ (x10³) 2005 Gulf Coast	Cooling water (gpm)	lb/h stm	kW	Fuel (MMBtu/h)
Desalter	30,000	2,800	63		20	
Crude unit	30,000	50,000	3,125	12,5000	1125	63
Vacuum unit	18,000	22,000	1,875	7,500	225	23
Naphtha desulfurizer	4,000	10,000	833	1,000	333	17
Reformer	3,000	17,000	833	3,750	375	38
Initial catalyst (desulfurizer)		Included				
Initial catalyst (reformer)		960				
Subtotal		102,760	6,729	24,750	2078	141
Cooling water system, 7,740 gpm[a]		1,071			194	
Steam system, 30900 lb/h[b]		3,214				37
Subtotal		107,045	6,729	24,750	2272	178
Storage[c]	12 days[d]	23,400				
Subtotal		130,045				
Offsites[c]	(30%)	39,134				
Subtotal		169,579				
Location factor	1.5					
Spec cost factor	1.04[e]					
Contingency	1.15[e]					
Escalation	(1.03)[f]					
Total		313,047				

[a] Add 15% excess capacity to calculated cooling water circulation.
[b] Add 25% excess capacity to calculated steam supply.
[c] Individual values for utilities in the storage and offsite categories are accounted for by notes a and b.
[d] 360,000 barrels at an average cost of $50.00/bbl.
[e] These factors are compounded.
[f] This is the projected cost at the location in St. Louis in 2007. No paid-up royalties are included.

Hydrogen from catalytic reformer is 2682 Mscf/day, which is more than adequate.

Calculation of Direct Annual Operating Costs

After completing the investment and yield calculations, the annual operating costs of the refinery can be determined as considered into three major categories:

1. Costs that vary as a function of plant throughput and on-stream time. These include water makeup to the boilers and cooling tower, electric power, fuel, running royalties, and catalyst consumption
2. Costs that are a function of the plant investment. These include insurance, local taxes, maintenance (both material and labor), and miscellaneous supplies.
3. Costs that are determined by the size and complexity of the refinery. These include operating, clerical, technical, and supervisory personnel.

The following sections illustrate the development of these costs.

On-Stream Time

Refineries generally has an on-stream (full capacity) factor of ~ 92 to 96%. For this example, a factor of 93.15% (340 days per year) is used.

Water Makeup

1. To cooling tower (30°F Δt)
   1 % evaporation for 10°F Δt
   1/2 % windage loss
   1 % blowdown to control solids concentration
   Cooling tower, makeup = (3) (1 %) + ½ % + 1% = 4 ½%
   Makeup = 0.045 × 7740 gpm = 348 gpm
2. To boiler
   Average boiler blowdown to control solids concentration can be assumed to be 5%.
   Boiler makeup = (0.05) (30,900) = 1545 lb/h = 3.1 gpm
   Total makeup water = 351 gpm
   Average cost to provide makeup water is approximately $0.32/1000 gal.
   Therefore, annual water makeup cost is (351) (1440) (340) (0.32) (10^{-3}) = $25,068

Power

Industrial power costs range from $0.10/kWh (in locations where there is hydroelectric power) to $0.24/kWh. For this example, use $0.24/kWh

Power cost = (2212) (24) (340) (0.12) = $2,040,653 per year

Fuel

Here, no separate charge will be made for fuel, because it is assumed that the refinery will use some of the heavy gas oil products for fuel. The amount of gas oil consumed must be calculated, so that this quantity can be deducted from the products available for sale.

From this summary tabulation of utilities, we require 178 MMBtu/h for full-load operation. This fuel is supplied by combustion of reformer off-gas supplemented with heavy gas oil. Some of the reformer off-gas is consumed in the hydrodesulfurizer, and this quantity (hydrogen portion only) must be deducted from available fuel. A fuel balance is made to determine the amount of heavy gas oil consumed as fuel.

Step. 1. From the reformer calculations the available fuel gas is

Component	Total lb/h	HDS[a] usage lb/h	Available for fuel lb/h
H_2	589	132	457
C_1	123		123
C_2	246		246
C_3	646		646
Total	1604	132	1472

[a]From desulfurizer calculations, hydrogen makeup was 600 Mscfd.

In petroleum work, "standard conditions" are 60°F and 14.7 psig. At these conditions, 1 lb mole = 379.5 scf. Therefore, hydrogen consumed in the HDS unit is

$$\frac{60,000}{(24)(379.5)}(2) = 132 \text{ lb/h}$$

Step 2. The calculated lower heating value (LHV) of available fuel gas is

Component	Total lb/h	LHV[a] Btu/lb	LHV MMBtu/h
H_2	457	51,600	23.6
C_1	123	21,500	2.6
C_2	246	20,420	5.0
C_3	646	19,930	12.9
Total	1472		44.1

[a]From Note 11 and Note 12 or other convenient source.

Step. 3. Heavy gas oil required for fuel. Assume 5.5 MMBtu/bbl LHV, then:

$$\frac{178 - 44.1}{5.5}(24) = 584 \text{ bpsd}$$

Step. 4. Heavy gas oil remaining for sale:

$$4000 - 584 = 3416 \text{ bpsd}$$

Royalties

The reformer is a proprietary process, and therefore, royalties must be paid. On a running basis, these range from $0.08 to $0.15 per barrel of feed. For this example, use a value of $0.10.

$$\text{Annual cost} = (0.10)(3000)(340) = \$102,000$$

Catalyst Consumption

Catalyst consumption costs are as follows:

Desulfurizer	0.002 lb/bbl; $1.60/lb
Annual cost	= (4000) (340) (0.002) (1) = $4,320
Reformer	0.004 lb/bbl; $7 per lb
Annual cost	= (3000) (340) (0.004) (7) = $28,560
Total catalyst cost	$4,320 + $28,560 = $32,880/yr.

Insurance

This cost usually is 0.5% of the plant investment per year
($313,047,000) (0.005) = $1,565,235

Local Taxes

Local taxes account for 1% of the plant investment per year
($313,047,000) (0.01) = $3,130,470/yr.

Maintenance

This cost varies between 3 and 8% of plant investment per year. For this example, use an average value of 5.5% (includes material and labor)
($313,047,000) (0.055) = $17,217,585/yr.

Miscellaneous Supplies

This item includes miscellaneous chemicals used for corrosion control, drinking water, office supplies, and so on. An average value of 0.15% of the plant investment per year
($313,047,000) (0.0015) = $469,705/yr.

Plant Staff and Operators

The number of staff personnel and operators depends on plant complexity and location. For this example, the following staff could be considered typical of a modern refinery.

	Number per shift	Total payroll
Refinery manager		1
Operations manager		1
Maintenance manager		1
Engineers		3
Operators	4	18
Lab personnel		2
Technicians		2
Clerical personnel		4
Total		32

Assume the average annual salary plus payroll burden is $72,000 per person. Then the total annual cost for staff and operators is:

($72,000) (32) = $2,304,000/yr.

Note: Maintenance personnel are not listed above because this cost was included with the maintenance item. Also note that it takes about 4 ½ workers on the payroll for each shift job to cover vacations, holidays, illness and fishing time.

Calculations of Income before Income Tax

Sales are summarized in the following table:

Product	bpd	Mbpy	$/bbl	$/yr x 10^3
Gasoline				
LSR	1,000			
Reformate	2,730			
Total	3,730	1,268	83.00	105,244
Light gas oil	4,000	1,360	79.00	107,440
Heavy gas oil	3,416	1,161	75.00	87,075
Vacuum gas oil	6,000	2,040	73.00	148,920
Vacuum residual	12,000	4,080	55.00	224,400
Total				673,079
Crude cost	30,000	10,200	60.00	612,000
Direct operating cost				29,577
Income before tax				31,502

Summary of Direct Annual Operating Costs

	$/yr x 10^3
Makeup water	25
Power	2,041
Fuel[a]	-
Royalties	102
Catalyst	33
Insurance	1,565
Local taxes	3,130
Maintenance	17,218
Miscellaneous supplies	470

Plant staff and operators	2,304
Subtotal	26,888
Contingency (10%)	2,689
Total[b]	29,579

[a]Fuel quantity is deducted from available heavy gas oil for sale.
[b]Additional items such as corporate overhead, research and development, and sales expense are omitted.

Calculation of ROI

Investment = $313,047,000

	$/yr × 10^3
Income before tax	31,502
Less depreciation allowance[a]	20,870
Taxable income	10,632
Income tax at 50%	5,316
Income after tax	5,316
Plus depreciation allowance	20,870
Cash flow	26,186
Return on investment (% per year)	8.00
Payout period (years)	11.95

[a]15 year, straight-line.

Note: True rate of return (discounted cash flow) basis: 20-year life, no salvage value. Interest on capital during construction period and average feedstock and product inventories are not considered in the above product. These items would result in an increase in investment and a decrease in the rate of return.

The computer program Project1.for was developed and used to determine:

1. The net present value (NPV)
2. Present value ratio (PVR)
3. The net return rate (NRR)
4. The payback period (PBP)

Table 26.26 shows the Input Data and Results of the program.

Table 26.26 Discounted Cash Flow Rate of Return Calculation.

15 yearly cash flows including year 0	
Year	Cash flow
0	-313047000.0
1	26186000.0
2	26186000.0
3	26186000.0
4	26186000.0
5	26186000.0
6	26186000.0
7	26186000.0
8	26186000.0
9	26186000.0
10	26186000.0
11	26186000.0
12	26186000.0
13	26186000.0
14	26186000.0

The discount cash flow rate of return (%): 2.178.

INPUT DATA
14
8
-313,047,000.0
26,186,000.0
26,186,000.0
26,186,000.0
26,186,000.0
26,186,000.0
26,186,000.0
26,186,000.0
26,186,000.0
26,186,000.0
26,186,000.0
26,186,000.0
26,186,000.0
26,186,000.0
26,186,000.0

NET PRESENT VALUE CALCULATION				
15 YEARLY CASH FLOWS INCLUDING YEAR 0				
8.00 PERCENTAGE ANNUAL DISCOUNT RATE				
Year	Cash flow ($)	Cumulative cash flow ($)	Discount factor	Present value ($)
0	-313,047,000.00	-313,047,000.00	1.0000	-313,047,000.00
1	26,186,000.00	-286,861,000.00	0.9259	24,246,295.89
2	26,186,000.00	-260,675,000.00	0.8573	22,450,273.45
3	26,186,000.00	-234,489,000.00	0.7938	20,787,289.54
4	26,186,000.00	-208,303,000.00	0.7350	19,247,489.22
5	26,186,000.00	-182,117,000.00	0.6806	17,821,749.56
6	26,186,000.00	-155,931,000.00	0.6302	16,501,618.81
7	26,186,000.00	-129,745,000.00	0.5835	15,279,275.75
8	26,186,000.00	-103,559,000.00	0.5403	14,147,476.68
9	26,186,000.00	-77,373,000.00	0.5002	13,099,516.37
10	26,186,000.00	-51,187,000.00	0.4632	12,129,180.47
11	26,186,000.00	-25,001,000.00	0.4289	11,230,722.86
12	26,186,000.00	1,185,000.00	0.3971	10,398,817.29
13	26,186,000.00	27,371,000.00	0.3677	9,628,533.95
14	26,186,000.00	53,557,000.00	0.3405	8,915,309.04

The net present value ($):   -97163451
Present value ratio: 0.69
The net return rate: 2.22 %
The average rate of return: 8.36%
The payback period is between: 11 AND 12 YEARS.

The computer program Project2. for was used to determine the DCFRR for the 15-year cash flow at $313,047,000.0 initial expenditure and a yearly cash flow of $26,186,000.0 for the 30,000 bpd refinery. The result from Table 26.27 shows the DCFRR of 2.2%. The Excel spreadsheet (**Case-Study.xlsx**) is used to calculate the IRR, by using the Goal Seek from What If analysis from the Data menu. This is carried out by setting the cell D61 = 0 (i.e., NPV=0) and changing the cell E61. This gives a value of 0.02179, indicating that the IRR = 2% and the NPV = -$97,163410 which are the same results from the computer program, indicating that the project is not economically viable and should be rejected. Figures 26.15a– 26.15c respectively show the various plots of the economic evaluations. Figures 26.15d-e show the snapshots of the Excel spreadsheet calculations of the case study.

26.14 Global Effects on Refining Economy

With the implementation of the Paris Agreement (L'accord de Paris [45]), agreement within the United Nations Framework Convention on Climate Change (UNFCCC), dealing with greenhouse -gas- (GHG) emissions mitigation, resulting in global warming and the need to limit carbon dioxide (CO_2) to which petroleum – based fuels are

Table 26.27 Discounted Cash Flow Rate of Return Calculation.

15 yearly cash flows including year 0	
Year	Cash flow
0	-313,047,000.0
1	26,186,000.0
2	26,186,000.0
3	26,186,000.0
4	26,186,000.0
5	26,186,000.0
6	26,186,000.0
7	26,186,000.0
8	26,186,000.0
9	26,186,000.0
10	26,186,000.0
11	26,186,000.0
12	26,186,000.0
13	26,186,000.0
14	26,186,000.0

The discount cash flow rate of return (%): 2.178.

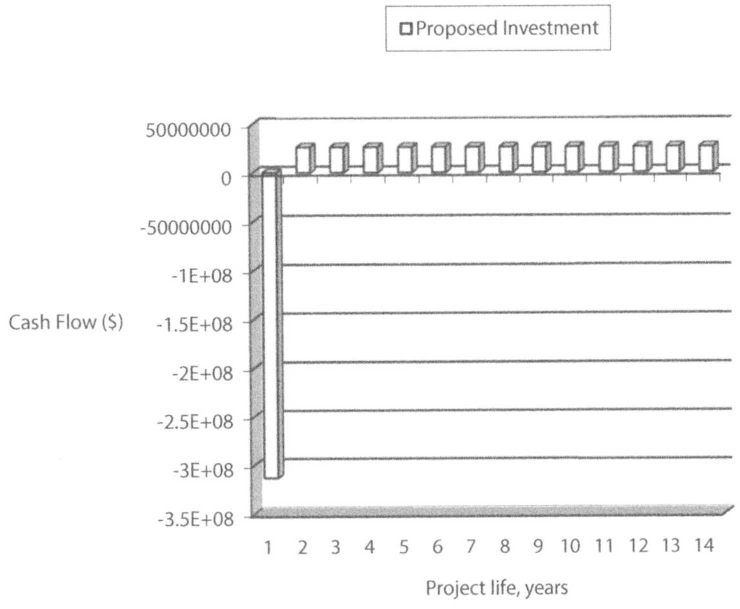

Figure 26.15a Cash flow vs. project life.

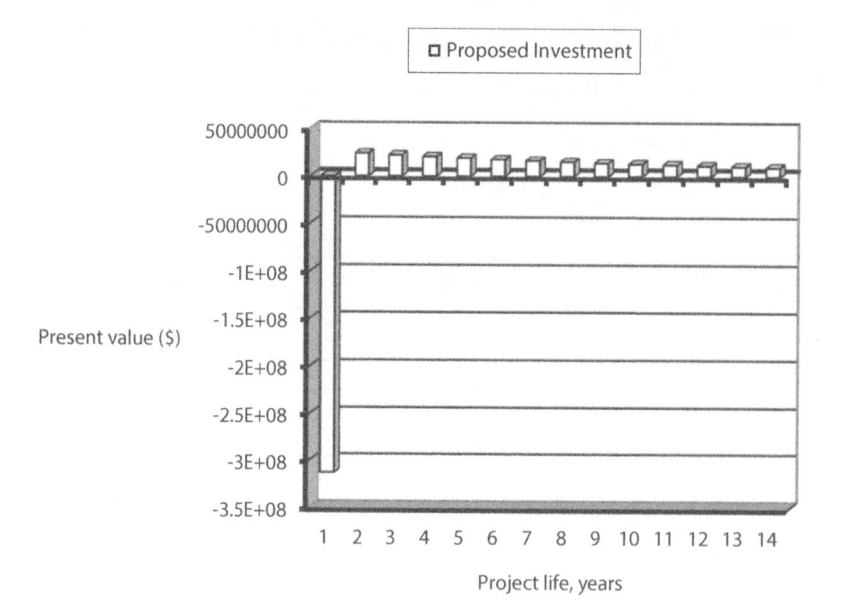

Figure 26.15b Present value vs. project life.

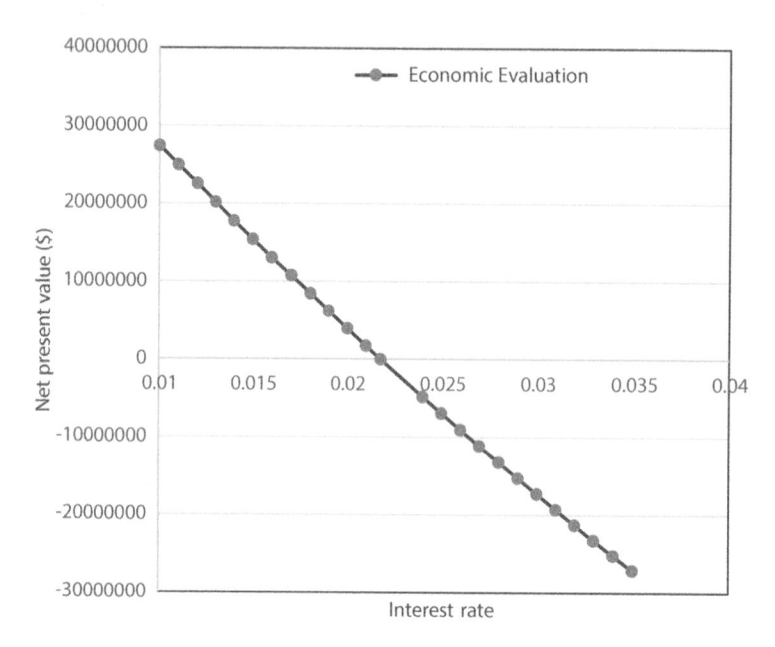

Figure 26.15c Present value vs. interest rate.

major contributors. Refineries worldwide would need to reduce the emission and reduce products such as diesel and the enhancement of aromatics and ethenes for petrochemical products. The COVID-19 pandemic has slowed the world economy since 2019 as the demand for aviation fuel has reduced, which has led to contraction in major industrialized countries. The future outlook of the refining industry depends on the demand for transportation fuel as alternative sources of energy are presently being explored. As the nations of the world are moving towards environmentally friendly transportation fuels (e.g., H_2, ethanol, biofuels, electricity, etc.), new transportation fuel specifications are being put into effect.

Biomass is a renewable resource that produces a range of liquid fuels that exhibit a wide range of physical and chemical properties. Biomass is clean for it has negligible content of sulfur (S), nitrogen (N_2) and ash-forming

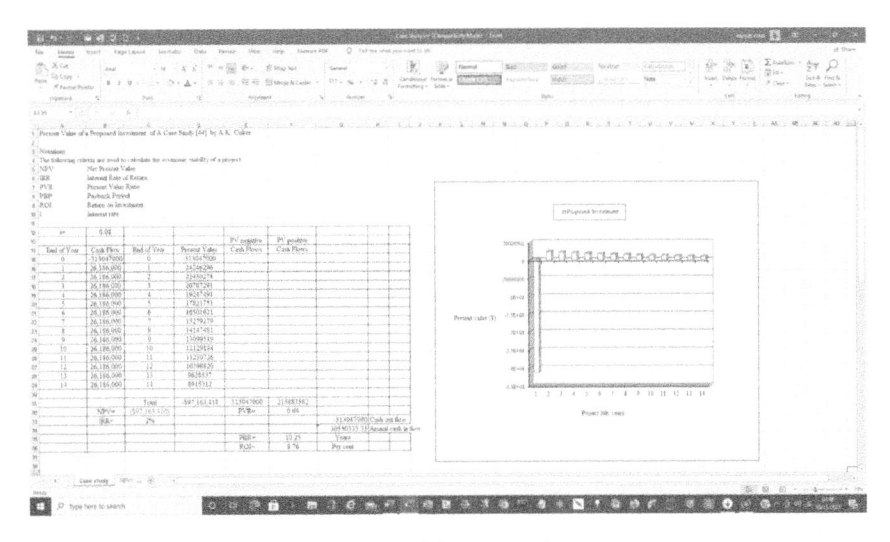

Figures 26.15d A snapshot of the Excel spreadsheet calculations of the case study.

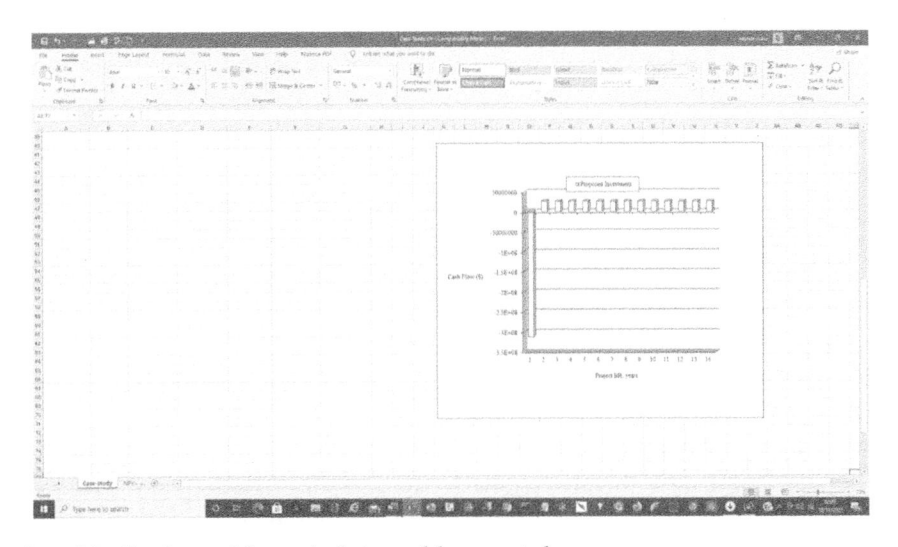

Figures 26.15e A snapshot of the Excel spreadsheet calculations of the case study.

constituents, which give lower emissions of sulfur dioxide (SO_2), nitrogen oxides (NO_x) and soot than convectional fossil fuels. The main biomass resources are: forest and mill residues, agricultural crops and wastes, wood and wood wastes, animal wastes, livestock operation residues, aquatic plants, fast-growing trees and plants, and municipal and industrial wastes. When biomass materials are heated in the absence of oxygen, they can produce a liquid (bio-oil or bio-crude) as well as some gas and solid fuels. This liquid can be processed into low-carbon fuels using existing petroleum refining technologies; this project provides the following.

- Lowers the cost of producing advanced biofuels by using existing infrastructure.
- Enables repurposing of existing petroleum refineries to produce low carbon intensity fuels.
- Reduces the carbon footprint of aviation and long-distance travel
- Enables the use of non-food biomass to produce transport fuels.

The global biofuel production increased 10 billion liters in 2018 to reach a record 154 billion liters, doubling the growth of 2017, and the output is forecast to increase 25% to 2024, an upwards revision from 2018 owing to better market prospect in Brazil, the U.S. and China (Figure 26.16) [48].

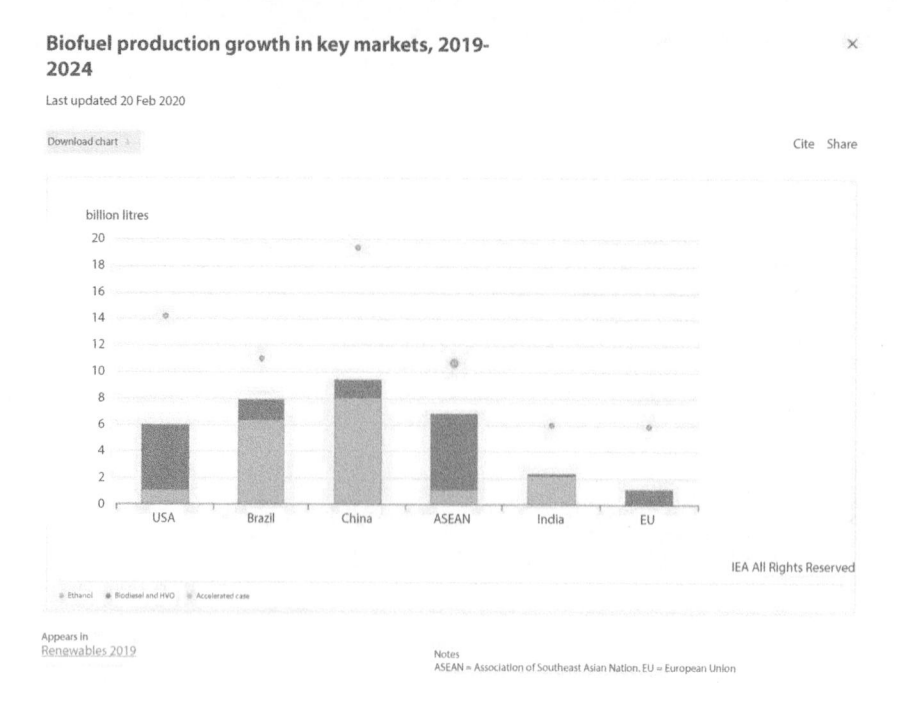

Figure 26.16 Biofuel growth in key markets, 2019 – 2024 (Source: www.iea.org.).

Alcohols are oxygenated fuels produced from biomass and practically any of the organic molecules of the alcohol family can be used as a fuel. The alcohols that can be used for motor fuels are methanol (CH_3OH), ethanol (C_2H_5OH), propanol (C_3H_7OH), butanol (C_4H_9OH). However, only methanol and ethanol fuels are technically and economically suitable for internal combustion engines.

Currently, the production of ethanol (i.e., bio-ethanol) by fermentation of corn-derived carbohydrates is the main technology used to produce liquid fuels from biomass resources. Furthermore, amongst different biofuels, suitable for application in transportation, bioethanol and biodiesel are the most feasible at present. The key advantage of bioethanol and biodiesel is that they can be mixed with convectional petrol and diesel respectively, which allows using the same handling and distribution infrastructure. The advantage of bioethanol and biodiesel is that when they are mixed at low concentrations ($\leq$ 10% bioethanol in petrol and $\leq$ 20% biodiesel in diesel), no engine modifications are required.

Ethanol can be blended with gasoline to create E85, a blend of 85% ethanol and 15% gasoline. E85 and blends with even higher concentration have been used in Brazil. More widespread practice has been to add up to 20% to gasoline line (E20-fuel or gasohol) to avoid engine changes. However, E100-fueled vehicles have difficulty starting in cold weather, but this is not a problem for E85 vehicles because of the presence of gasoline.

In comparison to gasoline, ethanol contains 35% oxygen by weight; gasoline contains none. The presence of oxygen promotes more complete combustion, which results in fewer tailpipe emissions. Compared to the combustion of gasoline, the combustion of ethanol substantially reduces the emission of carbon monoxide (CO), volatile organic compounds, particulate matter and greenhouses gases. However, a unit of ethanol contains about 32% less energy than a unit of gasoline. One of the best qualities of ethanol is its octane rating [47]. As the biorefinery matures, well-integrated sites will be able to profitably produce a diversity of products from renewable feedstocks. This product slate will encompass clean transportation fuels – ethanol, gasoline, diesel and jet fuel, as well as a complete range of chemical products and precursors to be used to produce renewable plastics, fibers, and rubbers. All of these factors will impact the economy of petroleum refining in the coming years and for the foreseeable future.

There are several fluctuations in oil prices due to the COVID-19 pandemic since 2019, as the demand for crude oil has declined. In 2021, the Brent crude oil spot price averaged at $72/barrel as the global production is 210,000 barrel

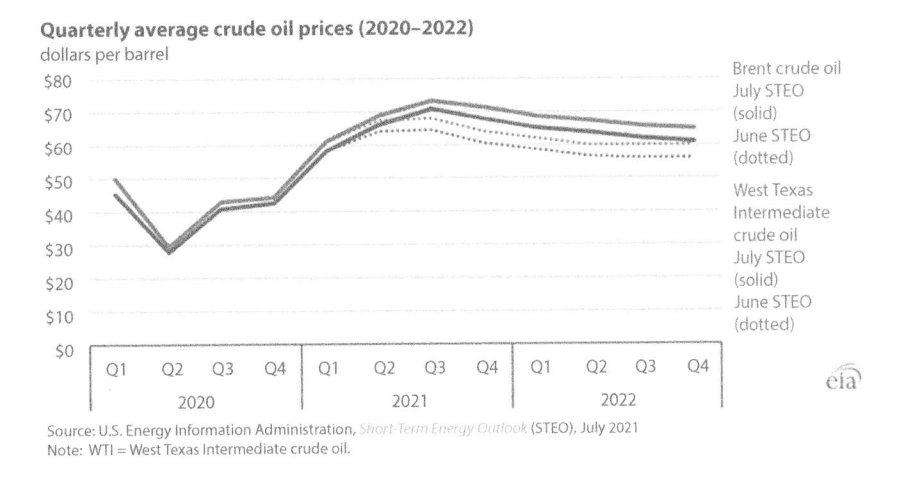

Figure 26.17 Quarterly average crude oil prices (2020-2022) (Source: www.eia.gov.).

per day [46], and some refineries are incorporating integrating their facilities with petrochemical plants to enhance their margins. The U.S. Energy Information Administration (EIA) [46] expects that global oil production, largely from OPEC + members (OPEC and partner nonmember countries), will increase by more than global oil consumption. The EIA forecasts that rising production will reduce the persistent global oil inventory draws that have occurred for much of the past 2020, and keeps prices similar to current levels, averaging $72/b during the second half of 2021 (Figure 26.17). However, in 2022, the continued growth in production from OPEC + and accelerating growth in U.S. tight oil production, along with other supply growths will outpace growth in global oil consumption and contribute to declining oil prices, which will invariably affect the profit margins of these refineries.

While historically gasoline was the largest driver of overall refining margins, distillate is playing a stronger role, as the annual average gasoline and distillate crack spreads have been increasing since the 1990s. The gasoline margin has been higher than the distillate crack, but latterly, the distillate crack spread exceeded gasoline. The distillate is becoming a much stronger contributor to margins than in the past. The strong growth in distillate demand worldwide and in particular, in Europe is a major driver behind this change.

For the refiners with bottom upgrading, the margins have become very attractive. The current light-heavy crude differential provides a very strong incentive for added bottom processing conversion capacity.

26.14.1 Carbon Tax

The impact of rising greenhouse gas levels on climate change is one of the biggest issues facing the world. A carbon tax (pollution tax) is a tax levied on the carbon emissions required to produce goods and services. Unlike command-and-control regulations that limit or prohibit emissions by each individual polluter, a carbon tax aims to allow market forces to determine the most efficient way to reduce pollution. It is an indirect tax – i.e., a tax on transaction as opposed to direct tax, which taxes income. Carbon taxes are price instruments since they set a price rather than an emission limit. In addition to creating incentives for energy conservation, a carbon tax puts renewable energy such as wind, solar and geothermal on a more competitive footing.

Carbon taxes are intended to make visible the hidden social costs of carbon emissions, which are otherwise felt only in indirect ways like more severe weather events. They are designed to reduce carbon dioxide (CO_2) emissions by increasing prices. This invariably decreases demand for such goods and services and incentivizes efforts to make them less carbon-intensive. A carbon tax covers only CO_2 emissions; however, they can also cover other greenhouse gases such as methane (CH_4) or nitrous oxide (N_2O) by calculating their global warming potential relative to CO_2. Research has shown that carbon taxes effectively reduce emissions and economists argue that carbon taxes are the most efficient (lowest cost) way to curb climate change as countries worldwide are committed to achieving net zero emissions by 2050.

Chemical/process engineers could incorporate carbon tax into the cashflow and profitability analyses, which enable them to carry out any major decision. A carbon tax would allow the ingenuity of the engineering profession to flourish and thus find the best solution through the innovative power of the crowd.

26.15 Economic Terminologies on Sustainability

Carbon footprint

The concept and name of the carbon footprint derived from the ecological footprint concept was developed by Rees and Wackernagel [49] in the 1990s. While carbon footprints are usually reported in tons of emissions (CO_2-equivalent) per year, ecological footprints are usually reported in comparison to what the planet can renew. This assesses the number of "earths" that would be required if everyone on the planet consumed resources at the same level as the person calculating their ecological footprint. The carbon footprint is one part of the ecological footprint. Carbon footprints are more focused than ecological footprints since they measure merely the emissions of gases that cause climate change into the atmosphere.

A carbon footprint is the total greenhouse gas (GHG) emissions caused by an individual, event, organization, service, place or product, expressed as carbon dioxide equivalent. Greenhouse gases, including the carbon-containing gases, carbon dioxide (CO_2) and methane (CH_4) can be emitted through the burning of fossil fuel, land clearance and the production and consumption of food, manufactured goods, materials, wood, roads, buildings, transportation and other services.

A carbon footprint is measure of the total amount of carbon dioxide (CO_2) and methane (CH_4) emissions of a defined population, system or activity, considering all relevant sources, sinks and storage within the spatial and temporal boundary of the population, system or activity of interest. It is calculated as carbon dioxide equivalent using the relevant 100-year global warming potential (GWP 100) [50].

Global Warming Potential (GWP)

The Global warming potential (GWP) of a greenhouse gas is its ability to trap extra heat in the atmosphere over time relative to carbon dioxide (CO_2). This is most often calculated over 100 years, and is known as the 100-year GWP. It is the heat absorbed by any greenhouse gas in the atmosphere, as a multiple of the heat that would be absorbed by the same mass of carbon dioxide. GWP is 1 for CO_2. For other gases it depends on the gas and the time frame. The GWP depends on the following factors [51]:

- The absorption of infrared radiation by a given gas.
- The spectral location of its absorbing wavelengths.
- The atmospheric lifetime of the gas.

A high GWP correlates with a large infrared absorption and a long atmospheric lifetime. The dependence of GWP on the wavelength of absorption is more complicated. Even if a gas absorbs radiation efficiently at a certain wavelength, this may not affect its GWP much if the atmosphere already absorbs most radiation at that wavelength. A gas has the most effect if it absorbs in a "window" of wavelengths where the atmosphere is fairly transparent. The dependence of GWP as a function of wavelength has been found empirically and published as a graph.

Because the GWP of a greenhouse gas depends directly on its infrared spectrum, the use of infrared spectroscopy to study greenhouse gases is centrally important in the effort to understand the impact of human activities on global climate change.

The influence of a factor that can cause climate change, such as a greenhouse gas, is often evaluated in terms of its radiative forcing. Radiative forcing is a measure of how the energy balance of the Earth-atmosphere system is influenced when factors that affect climate are altered. The word radiative arises because these factors change the balance between incoming solar radiation and outgoing infrared radiation within the Earth's atmosphere. This radiative balance controls the Earth's surface temperature. The term forcing is used to indicate that Earth's radiative balance is being pushed away from its normal state. Radiative forcing is usually quantified as the 'rate of energy change per unit area of the globe as measured at the top of the atmosphere', and is expressed in units of 'Watts per square meter. When radiative forcing from a factor or group of factors is evaluated as positive, the energy of the Earth-atmosphere system will ultimately increase, leading to a warming of the system.

As radiative forcing provides a simplified means of comparing the various factors that are believed to influence the climate system to one another, global warming potentials (GWPs) are one type of simplified index based upon radiative properties that can be used to estimate the potential future impacts of emissions of different gases upon the climate system in a relative sense. GWP is based on a number of factors, including the radiative efficiency (infrared-absorbing ability) of each gas relative to that of carbon dioxide, as well as the decay rate of each gas (the amount removed from the atmosphere over a given number of years) relative to that of carbon dioxide

The radiative forcing capacity (RF) is the amount of energy per unit area, per unit time, absorbed by the greenhouse gas that would otherwise be lost to space. It can be expressed by the formula

$$RF = \sum_{i=1}^{100} abs_i \bullet F_i / (1-d) \qquad (26.68)$$

Where the subscript i represents an interval of 10 inverse centimeters. abs represents the integrated infrared absorbance of the sample in that interval, and F_i represents the RF for the interval.

The Intergovernmental Panel on Climate Change (IPCC) provides the generally accepted values for GWP, which changed slightly between 1996 and 2001. An exact definition of how GWP is calculated is to be found in the IPCC's 2001 Third Assessment Report. The GWP is defined as the ratio of the time-integrated radiative forcing from the instantaneous release of 1 kg of a trace substance relative to that of 1 kg of a reference gas:

$$GWP(x) = \frac{\int_0^{TH} a_x \bullet [x(t)] dt}{\int_0^{TH} a_r \bullet [r(t)] dt} \qquad (26.69)$$

where TH is the time horizon over which the calculation is considered; a_x is the radiative efficiency due to a unit increase in atmospheric abundance of the substance (i.e., $Wm^{-2} kg^{-1}$) and [x(t)] is the time-dependent decay in abundance of the substance following an instantaneous release of it at time t=0. The denominator contains the corresponding quantities for the reference gas (i.e., CO_2). The radiative efficiencies a_x and a_r are not necessarily constant over time. While the absorption of infrared radiation by many greenhouse gases varies linearly with their abundance, a few important ones display non-linear behavior for current and likely future abundances (e.g., CO_2, CH_4, and N_2O). For those gases, the relative radiative forcing will depend upon abundance and hence upon the future scenario adopted.

Since all GWP calculations are a comparison to CO_2 which is non-linear, all GWP values are affected. Assuming otherwise as is done above will lead to lower GWPs for other gases than a more detailed approach would. Clarifying this, while increasing CO_2 has less and less effect on radiative absorption as ppm concentrations rise, more powerful greenhouse gases like methane and nitrous oxide have different thermal absorption frequencies to CO_2 that are not filled up (saturated) as much as CO_2, so rising ppms of these gases are far more significant.

An Improved Method of Using GWPs

Cain [52] provides an improved technique of using GWP. Her approach builds on previous work [53] which equates a "pulse" emission of CO_2 with an increase in the emission rate of methane. GWP is typically defined to compare pulses of emissions with each other. A pulse is when a specified mass of gas is released into the atmosphere instantaneously. Over the coming years, the CO_2 remains in the atmosphere, as it is a long-lived gas, and so leads to a permanent increase in the CO_2 concentration. A change in the methane emission rate also leads to higher concentrations of methane in the atmosphere, assuming the sinks remain constant, as the source is larger. This usage is referred to as "GWP*", as it still uses GWP_{100}, but instead of comparing two pulses it effectively spreads the methane emission out evenly over the 100-year time-horizon.

By using GWP*, emissions of methane expressed as CO_{2e} relate much more closely to temperature response. This can be seen in the figure below. The top panels use GWP_{100}, while the lower panels use GWP*. The left panels show annual emissions of CO_{2e} (upper left) and CO_{2eq*} (lower left) (Figure 26.18). In the right panels, temperature is shown in the dashed lines alongside the cumulative CO_{2eq}/CO_{2eq*} emissions in the solid lines.

When using GWP_{100}, the temperature response to methane (CH_4) does not track the cumulative methane emissions (blue) (Figure 26.18). The agreement is much better when using GWP* (panel d, bottom-right). Due to the correspondence between cumulative CO_{2e*} emissions and temperature response, the temperature peaks when CO_{2eq*} emissions reach net-zero (marked by arrows). This is not the case for the usual CO_{2eq} emissions, where continued CO_{2eq} emissions lead to a slight cooling in the final decades of the century. Any CO_{2eq} emission that leads to cooling is not actually equivalent to CO_2 on a temperature basis.

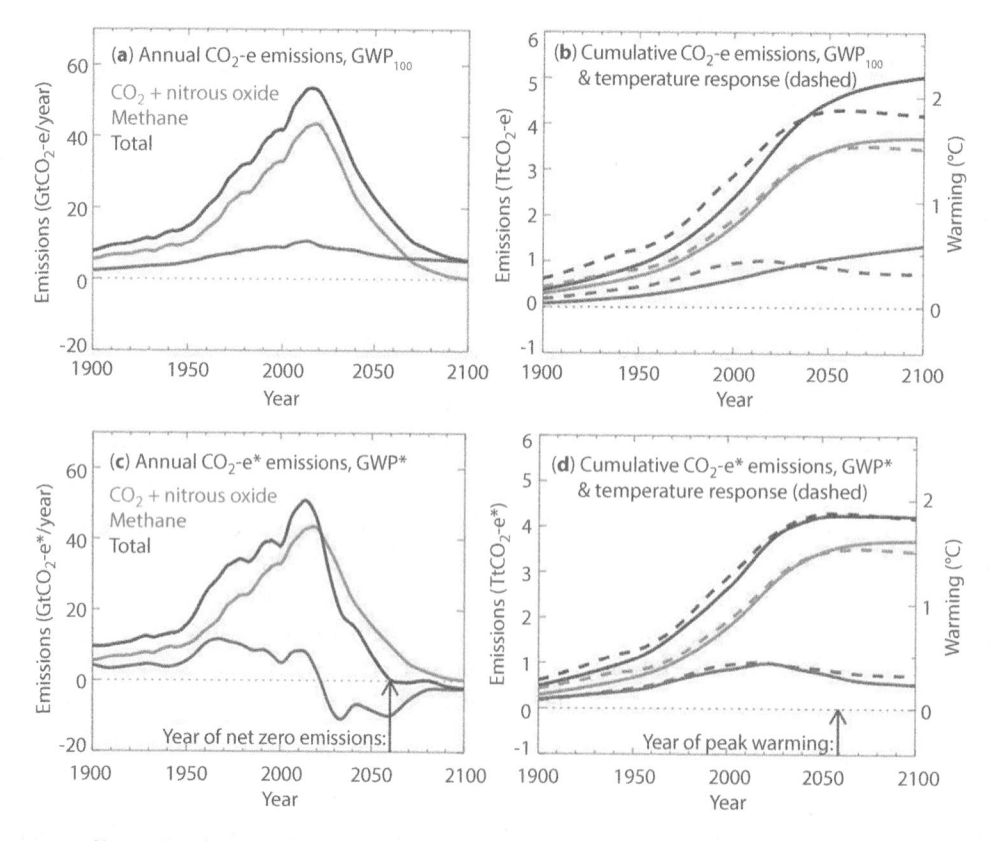

Figure 26.18 Emissions of long – lived CO_2 and N_2O (red), short-lived CH_4 (blue), and their sum (black) for the main ambitious mitigation scenario in the last IPCC report (RCP 2.6) (Source: Briefing paper, "Climate metrics under ambitious mitigation" [52]).

The product of GWP and the mass emission rate of a greenhouse gas results in the equivalent emission of carbon dioxide, the benchmark compound, that would result in the same radiative forcing. Performing this calculation, a direct connection is made between mass emission from a process of any greenhouse gas and global warming impact. The global warming index for the entire product system is the sum of the emission-weighted GWPs for each chemical.

The product of GWP and the mass emission rate of a greenhouse gas results in the equivalent emission of carbon dioxide, the benchmark compound, that would result in the same radiative forcing. Performing this calculation, a direct connection is made between mass emission from a process of any greenhouse gas and global warming impact. The global warming index for the entire product system is the sum of the emission-weighted GWPs for each chemical.

$$I_{GW} \sum_i (GWP_i \times m_i) \tag{26.70}$$

where m_i is the mass emission rate of chemical I from the entire product system (kg/h). This step will provide the equivalent process emissions of greenhouse chemicals in the form of the benchmark compound, CO_2.

The global warming index from Equation 26.70 accounts for direct effects of the chemical, but most chemicals of interest are so short-lived in the atmosphere (because the action of hydroxyl radicals in the troposphere) that they disappear (become converted to CO_2) long before any significant direct effect can be felt. Organic chemicals will have an indirect global warming effect because of the CO_2 released upon oxidation within the atmosphere and other compartments of the environment. Shonnard and Hiew [60] defined an expression to account for this indirect effect for organic compounds with atmospheric reaction residence times of less than half a year as

$$GWP(\text{Indirect}) = N_c \frac{MW_{CO_2}}{MW_i} \tag{26.71}$$

where N_c is the number of carbon atoms in the chemical I and the molecular weights (MW) convert from a molar to a mass basis for GWP, as originally defined. Organic chemicals whose origins are in renewable biomass (plant materials) will have no net global warming impact because the CO_2 released upon environmental oxidation of these compounds will replace CO_2 removed from the atmosphere during photosynthesis of the biomass.

Example 26.15: Global warming index or air emissions of 1,1,1-trichloroethane from a production process [61]

1,1,1-Tricholoroethane (1,1,1-TCA) is used as an industrial solvent for metal cleaning, as a reaction intermediate, and for other important uses (World Health Organization (WHO), 2000). Sources for air emissions include distillation, condenser, vents, storage tanks, handling and transfer operations, fugitive sources, and secondary emissions from wastewater treatment. This example will estimate the global warming impact of the air emissions from this process. Include direct impacts to the environment (from 1,1,1-TCA) and indirect impacts from energy usage (CO_2 and N_2O release) in your analysis. The following data show the major chemicals that impact global warming when emitted from the process.

Determine the global warming index for the process and the percentage contribution for each chemical.

Data: Air Emissions (Based on a 15,500 kg 1,1,1-TCA/h Process)

Chemical	m, kg/h	GWP_i
TCA	10	100
CO_2	7,760	1
N_2O	0.14	298

(Source: U.S EPA, 1979-1991: Allen and Rosselot, 1997; Boustead, 1993).

Solution

Using equation 26.70, the process global warming index is

IGW = (10 kg/h) (100) + (7760 kg/h) (1) + (0.14 kg/h) (298)
    = 1000 + 7760 + 41.7
    = 8801.7 kg/h

The percent of the process IGW for each chemical is

1,1,1 - TCA : (1000/8801.7) x 100 = 11.4%
CO_2: (7760/8801.7) x 100 = 88.1%
N_2O: (43.4/8801.7) x 100 = 0.5%

Discussion: In this exercise, the majority of the global warming impact from the production of 1,1,1-TCA is from the energy requirement of the process and not from the emission of the chemical with the highest global warming potential. This analysis assumes that a fossil fuel was used to satisfy the energy requirements of the process. If renewable resources were used (biomass-based fuels), the impact of CO2 on global warming would be significantly reduced. Finally, the majority of the global warming impact of 1,1,1-TCA could very well be felt during the use stage of its life cycle if the compound evaporates when it is used. This possibility was not included in this exercise.

Note: $CO_{2e} = CO_{2eq} = CO_{2-e}$ is calculated from GWP. It can be measured in weight or concentration. For any amount of any gas, it is the amount of CO_2 which would warm the earth as much as that amount of that gas. Thus, it provides a common scale for measuring the climate effects of different gases. It is calculated as GWP times amount of the other gas. For example, if a gas has GWP of 100, two tonnes of the gas have CO_2 of 200 tonnes, and 1 part per million of the gas in the atmosphere has CO_{2e} of 100 parts per million. CO_2 is the reference. It has a GWP of 1 regardless of the time period used. CO_{2eq} emissions cause increase in atmosphere concentration of CO_2 that will last thousands of years. Estimates of GWP values over 20, 100 and 500 years are periodically compiled and revised in reports from the Intergovernmental Panel on Climate Change.

The table below shows the 2007 estimates used for international comparisons of GWP of greenhouse gases.

Greenhouse gas	Chemical formula	100-year Global warming potentials (2007 estimates, for 2013-2020 comparisons)
Carbon dioxide	CO_2	1
Methane	CH_4	25
Nitrous oxide	N_2O	298
Hydrofluorocarbons (HFCs)		
HFC-23	CHF_3	14800
Difluoromethane (HFC-32)	CH_2F_2	675
Fluoromethane (HFC-41)	CH_3F	92
HFC-43-10mee	$CF_3CHFCHFCF_2CF_3$	1640
Pentafluoroethane (HFC-125)	C_2HF_5	3500
HFC-134	$C_2H_2F_4$ (CHF_2CHF_2)	1100

Greenhouse gas	Chemical formula	100-year Global warming potentials (2007 estimates, for 2013-2020 comparisons)
1,1,1,2-Tetrafluoroethane (HFC-134a)	$C_2H_2F_4$ (CH_2FCF_3)	1430
HFC-143	$C_2H_3F_3$ (CHF_2CH_2F)	353
1,1,1-Trifluoroethane (HFC-143a)	$C_2H_3F_3$ (CF_3CH_3)	4470
HFC-152	CH_2FCH_2F	53
HFC-152a	$C_2H_4F_2$ (CH_3CHF_2)	124
HFC-161	CH_3CH_2F	12
1,1,1,2,3,3,3-Heptafluoropropane (HFC-227ea)	C_3HF_7	3220
HFC-236cb	$CH_2FCF_2CF_3$	1340
HFC-236ea	CHF_2CHFCF_3	1370
HFC-236fa	$C_3H_2F_6$	9810
HFC-245ca	$C_3H_3F_5$	693
HFC-245fa	$CHF_2CH_2CF_3$	1030
HFC-365mfc	$CH_3CF_2CH_2CF_3$	794
Perfluorocarbons		
Carbon tetrafluoride – PFC-14	CF_4	7390
Hexafluoroethane – PFC-116	C_2F_6	12200
Octafluoropropane – PFC-218	C_3F_8	8830
Perfluorobutane – PFC-3-1-10	C_4F_{10}	8860
Octafluorocyclobutane – PFC-318	$c\text{-}C_4F_8$	10300
Perfluouropentane – PFC-4-1-12	C_5F_{12}	9160
Perfluorohexane – PFC-5-1-14	C_6F_{14}	9300
Perfluorodecalin – PFC-9-1-18b	$C_{10}F_{18}$	7500
Perfluorocyclopropane	$c\text{-}C_3F_6$	17340
Sulphur hexafluoride (SF_6)		
Sulphur hexafluoride	SF_6	22800
Nitrogen trifluoride (NF_3)		
Nitrogen trifluoride	NF_3	17200
Fluorinated ethers		
HFE-125	CHF_2OCF_3	14900
Bis(difluoromethyl) ether (HFE-134)	CHF_2OCHF_2	6320
HFE-143a	CH_3OCF_3	756

Greenhouse gas	Chemical formula	100-year Global warming potentials (2007 estimates, for 2013-2020 comparisons)
HCFE-235da2	$CHF_2OCHClCF_3$	350
HFE-245cb2	$CH_3OCF_2CF_3$	708
HFE-245fa2	$CHF_2OCH_2CF_3$	659
HFE-254cb2	$CH_3OCF_2CHF_2$	359
HFE-347mcc3	$CH_3OCF_2CF_2CF_3$	575
HFE-347pcf2	$CHF_2CF_2OCH_2CF_3$	580
HFE-356pcc3	$CH_3OCF_2CF_2CHF_2$	110
HFE-449sl (HFE-7100)	C_4F9OCH_3	297
HFE-569sf2 (HFE-7200)	$C_4F9OC_2H_5$	59
HFE-43-10pccc124 (H-Galden 1040x)	$CHF_2OCF_2OC_2F_4OCHF_2$	1870
HFE-236ca12 (HG-10)	$CHF_2OCF_2OCHF_2$	2800
HFE-338pcc13 (HG-01)	$CHF_2OCF_2CF_2OCHF_2$	1500
	$(CF_3)_2CFOCH_3$	343
	$CF_3CF_2CH_2OH$	42
	$(CF_3)_2CHOH$	195
HFE-227ea	$CF_3CHFOCF_3$	1540
HFE-236ea2	$CHF_2OCHFCF_3$	989
HFE-236fa	$CF_3CH_2OCF_3$	487
HFE-245fa1	$CHF_2CH_2OCF_3$	286
HFE-263fb2	$CF_3CH_2OCH_3$	11
HFE-329mcc2	$CHF_2CF_2OCF_2CF_3$	919
HFE-338mcf2	$CF_3CH_2OCF_2CF_3$	552
HFE-347mcf2	$CHF_2CH_2OCF_2CF_3$	374
HFE-356mec3	$CH_3OCF_2CHFCF_3$	101
HFE-356pcf2	$CHF_2CH_2OCF_2CHF_2$	265
HFE-356pcf3	$CHF_2OCH_2CF_2CHF_2$	502
HFE-365mcfʼll t3	$CF_3CF_2CH_2OCH_3$	11
HFE-374pc2	$CHF_2CF_2OCH_2CH_3$	557
	$-(CF_2)_4CH(OH)-$	73
	$(CF_3)_2CHOCHF_2$	380
	$(CF_3)_2CHOCH_3$	27

Greenhouse gas	Chemical formula	100-year Global warming potentials (2007 estimates, for 2013-2020 comparisons)
Perfluoropolyethers		
PFPMIE	$CF_3OCF(CF_3)CF_2OCF_2OCF_3$	10300
Trifluoromethyl sulphur pentafluoride (SF_5CF_3)		
Trifluoromethyl sulphur pentafluoride	SF_5CF_3	17

Carbon Dioxide Equivalent

U.S. Environmental Protection Agency (EPA) defines carbon dioxide equivalent as the number of metric tons of CO_2 emissions with the same global warming potential as one metric tone of another greenhouse gas, and is calculated using Equation A-1 in 40 CFR Part 98 [54].

Carbon dioxide equivalent (CO_{2e} or CO_{2eq} or $CO_{2\text{-}e}$) is calculated from GWP. It can be measured in weight or concentration. For any amount of any gas, it is the amount of CO_2 which would warm the earth as much as that amount of the gas. Thus, it provides a common scale for measure the climate effects of different gases. It is calculated as GWP times the amount of other gas. As weight, CO_{2e} is the weight of CO_2 which would warm the earth as much as the amount of the gas. Thus, it provides a common scale for measuring the climate effects of different gases. It is calculated as GWP times the amount of the other gas. As weight, CO_2 is the weight of CO_2 which would warm the earth as much as a particular weight of some other gas. It is calculated as GWP times the weight of the other gas [55]. For example, if a gas has GWP of 100, two tonnes of the gas will have CO_{2e} of 200 tonnes, and 9 tonnes of the gas has CO_{2e} of 900 tonnes.

As concentration, CO_{2e} is the concentration of CO_2 which would warm the earth as much as a particular concentration of some other gas or of all gases and aerosols in the atmosphere. It is calculated as GWP times concentration of the other gas(es). For example, CO_{2e} of 500 parts per million would reflect a mix of atmospheric gases which warm the earth as much as 500 parts per million of CO_2 would warm it. CO_2 calculations depend on the time-scale chosen, typically 20 years or 100 years, since gases decay in the atmosphere or are absorbed naturally at different rates.

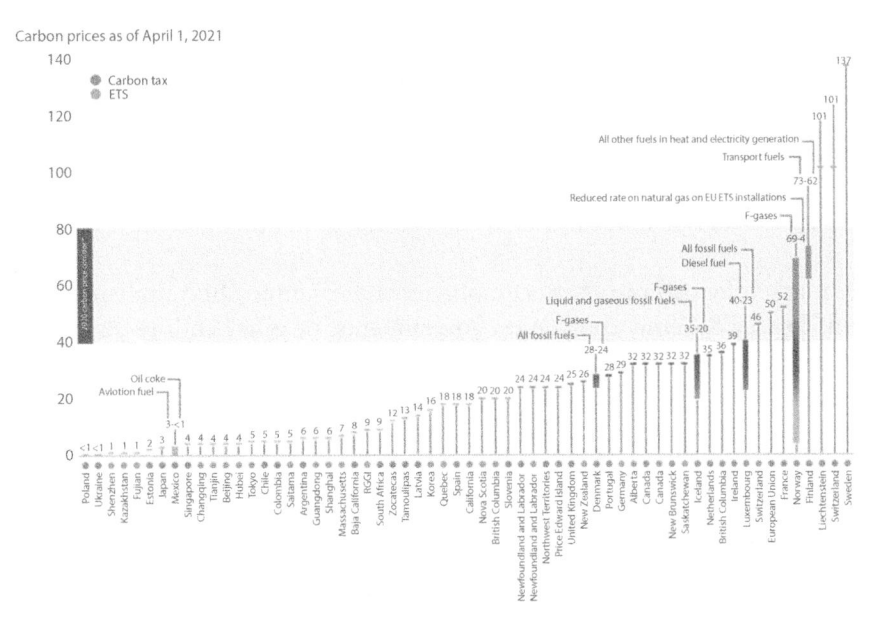

Figure 26.19 Carbon prices as of April 1, 2021 [59].

The following units are commonly used:

- By the UN Climate change panel (IPCC): billion metric tonnes = n x 10^9 tonnes of CO_2 equivalent ($GtCO_{2eq}$).
- In industry: million metric tonnes of CO_2 equivalents (MMTCDE) and $MMTCO_{2eq}$.
- For vehicles: grams of CO_2 equivalent per mile (gCO_{2e}/mile) or per kilometer (gCO_{2e}/km).

For example, GWP for methane (CH_4) over 20 years at 86 and nitrous oxide (NO_x) at 289; the emissions of 1 million tonnes of CH_4 and NO_x are equivalent to emissions of 86 or 289 million tonnes of CO_2 respectively [55].

Carbon Credit

Carbon credit is defined as "a certificate showing that a government or company has paid to have a certain amount of carbon dioxide removed from the environment". It is a term used for any tradable certificate or permit representing the right to emit one tonne of CO_2 or the equivalent amount of a different greenhouse gas (tCO_{2e}).

Carbon credits and carbon markets are a component of national and international attempts to mitigate the growth in concentrations of greenhouse gases (GHGs). One carbon credit is equal to one tonne of CO_2, or in some markets, CO_2 equivalent gases. Carbon trading is an application of an emissions trading approach. Greenhouse gas emissions are capped and then markets are used to allocate the emissions among the group of regulated sources.

The goal is to allow market mechanisms to drive industrial and commercial processes in the direction of low emissions or less carbon intensive approaches than those used when there is no cost to emitting CO_2 and other GHGs into the atmosphere. Since GHG mitigation projects generate credits, this approach can be used to finance carbon reduction schemes between trading partners and around the world.

The concept of carbon credits came into existence as a result of increasing awareness of the need for controlling emissions. The Intergovernmental Panel on Climate Change (IPCC) observed that policies that provide a real or implicit price of carbon could create incentives for producers and consumers to significantly invest in low GHG product, technologies and processes. Such policies could include economic instruments, government funding and regulation, while noting that a trade permit system is one of the policy instruments that has been shown to be environmentally effective in the industrial sector, as long as there are reasonable levels of predictability over the initial allocation mechanism and long-term price. The mechanism was formalized in the Kyoto Protocol, an international agreement between more than 170 countries and the market mechanisms were agreed through the subsequent Marrakesh Accords.

Carbon Offset

A carbon offset is a reduction in emissions of CO_2 or other greenhouse gases made in order to compensate for emissions made elsewhere. Offsets are measured in tonnes of CO_2 equivalent (CO_{2e}). For example, one tonne of carbon offset represents the reduction of one tonne of CO_2 or its equivalent in other greenhouses gases.

There are two types of markets for carbon offsets, compliance and voluntary. In compliance market like the European Union (EU) Emission Trading Scheme, companies, governments, or other entities buy carbon offsets in order to comply with mandatory and legally binding caps on the total amount of carbon dioxide they are allowed to emit per year. Failure to comply with these mandatory caps within compliance markets results in fines or legal penalty. According to the World Bank State and Trends 2020 Report, carbon pricing initiatives are in place or are scheduled for implementation globally. These include both emission trading schemes like cap-and-trade systems as well as carbon taxes. While these initiatives represent markets for carbon, however, not all incorporate provisions for carbon offsets, instead a greater emphasis is being placed on achieving emissions reductions within the operations of regulated entities. Compliance markets for carbon offsets comprise both international carbon markets developed through the Kyoto Protocol and Paris Agreement, and domestic carbon pricing initiatives that incorporate carbon offset mechanisms.

Many entities exist within the voluntary carbon market. For example, carbon offset vendors offer direct purchase of carbon offsets, often also offering other services such as designating a carbon offset project to support or measure a purchaser's carbon footprint. In 2016, about US$191.3 million of carbon offsets were purchased in the voluntary market, representing about 63.4 million metric tons of CO_{2e}. In 2018 and 2019, the voluntary carbon market transacted 98 and 104 million metric tonnes of CO_{2e} respectively [56, 57].

Carbon Price

Carbon pricing is a method for nations to reduce global warming. The cost is applied to greenhouse gas emissions in order to encourage polluters to reduce the combustion of coal, oil and gas, the main source of global warming and the driver for climate change. The method is widely agreed and considered efficient. Carbon pricing seeks to address the economic problem that emissions of CO_2 and other GHGs are a negative external, which are a detrimental product that is not charge by any market. A carbon price usually takes the form of a carbon tax or carbon emission trading, a requirement to purchase allowances to emit.

21.7% of global GHG emissions are covered by carbon pricing in 2021; a major increase due to the introduction of Chinese national carbon trading scheme. Regions with carbon pricing include most European countries, and Canada. However, top emitters like India, Russia, the Gulf states and many US states have not yet introduced carbon pricing. In 2020, carbon pricing generated $53 billion revenue [58]. A price level of $135-5.500 in 2030 and $245-13.0 per tonne of CO_2 in 2050 would be required to drive carbon emissions to stay below 1.5°C limit.

About one-third of the systems stays below $10/tCO_2$, the majority is below $40. Once exception is the steep incline in the EU-ETS reaching $60 in September 2021. Sweden and Switzerland are the only countries with more than $100/tCO_2$. Figure 26.19 shows the carbon prices as of April 2021 [59].

Nomenclature

ARR	= The average rate of return.
C	= Cash flow for each year.
C_{CF}	= Net annual cash flow.
C_{FC}	= Fixed capital cost.
C_K	= Capitalized cost.
C_L	= Land cost.
C_0	= Initial investment cost.
C_n	= Cash flow.
C_{TC}	= Total capital cost.
C_{WC}	= Working capital cost.
C_1	= Capital cost of the existing plant.
C_2	= Capital cost of the designed plant.
D	= Annual depreciation.
DCFRR	= Discounted cash flow rate of return.
D_j	= Annual depreciation charge.
d_j	= Depreciation rate.
EMIP	= Equivalent maximum investment period.
f_d	= Discount factor.
f_K	= Capitalized cost factor.
I	= Investment cost.
IRP	= Interest recovery period.
IRR	= Internal rate of return.
i	= Interest rate of return (ROI/100).
m	= Exponential power for cost capacity relationships, project life.

NPV   = Net present value.
NRR   = Net return rate.
n   = Years of project life.
P   = Initial principal.
PBP   = Payback period.
PV   = Present value.
PVR   = Present value ratio.
PWR   = Present worth ratio.
Q_1 = Capacity of the existing plant.
Q_2 = Capacity of the designed plant.
ROI   = Return of investment.
S   = Salvage value.

References

1. SSVR 18-R97, "Recommended Practice for Cost Estimate Classification – As Applied in Engineering, Procurement, and Construction for the Process Industries", AACE International, Morgantown, W. Va., 1997.
2. Dysert, L., "Sharpen Your Capital Cost Estimation Skills", Chemical Engineering, p. 70, Oct. 2001.
3. Gerrard, A.M., Guide to Capital Cost Estimating, 4th ed., IChemE, 2000.
4. Ludwig, E. E., Applied Process Design for Chemical and Petrochemical Plants, volume 1, 3rd edition, Gulf Publishing Company, Houston, 1995.
5. Page, J. S., Estimator's Manual of Equipment and Installation Cost, Gulf Publishing Co. Houston, Texas, 1963.
6. Page, J. S., Estimator's General Construction Man-Hour Manual, 2nd ed., Gulf Publishing Co., Houston, Texas, 1977.
7. Page, J. S., Estimator's Man-Hour manual on Heating, Air Conditioning, Ventilation and Plumbing, 2nd ed., Gulf Publishing Co., Houston, Texas, 1977.
8. Page, J. S., and J. G. Nation, Estimator's Piping Man-Hour Manual, Gulf Publishing Co., Houston, Texas, 1976.
9. Page, J. S., and J. G. Nation, Estimator's Electrical Man-Hour Manual, Gulf Publishing Co., Houston, Texas, 1959.
10. "Chemical Engineering Plant Cost Index" and "Marshall and Swift (M&S) Equipment Cost Index" appear regularly in Chemical Engineering.
11. Chilton, C.H., "Six Tenths Factor", Applies to Complete Plant Costs, Chem. Eng., reprinted in Cost Engineering in the Process Industries Estimation, Mc Graw-Hill Book, Co., 1960.
12. Institution of Chemical Engineers, "A New Guide to Capital Cost Estimating", IChemE, 1977.
13. Guthrie, K. M., "Capital and Operating Costs for 54 Chemical Processes", Chem. Eng., June 1970.
14. "Marshall & Swift Equipment Cost Index," Chem. Engr. Mc-Graw Hill Publishing Co., Published regularly in specific issues.
15. "Nelson Refinery Construction Index", Oil & Gas Journal. Published on specific schedules during the year.
16. Vatavuk, W. M., "Updating the CE Plant Cost Index," Chem. Eng., p. 62, January 2002.
17. Lang, H. J., "Simplified Approach to Preliminary Cost Estimates," Chem. Eng., 55, p 112, June, 1948.
18. Sinnott, R. K., Coulson & Richardson's Chemical Engineering, Vol. 6, revised 2nd ed., Butterworth-Heinemann, 1998.
19. Bridgwater, A. V., "The Functional Unit Approach to Rapid Cost Estimation," ACCE Bull., Vol. 18, No. 5, p 153, 1976.
20. Taylor, J. H., "The Process Step Scoring Method for Making Quick Capital Estimates," Eng. & Process Econ., Vol. 2, pp 259-267, 1977.
21. Dodge, W. J., et al., The Module Estimating Technique as an Aid in Developing Plant Capital Costs, Trans, AACE, 1962.
22. Zevnik, F. C., and Buchanan, R. L., "Generalized Correlation for Process Investment," Chem. Eng., Prog., Vol. 59, No. 2, pp 70-77, 1963.
23. Timms, S. R. M., M. Phil, Thesis (Aston Univ.), 1980.
24. Wilson, G. T., "Capital Investment for Chemical Plant," Brit Chem. Eng., Vol. 16 No. 10, pp 931-934, 1971.
25. Bridgwater, A. V., and C. J. Mumford, Waste Recycling and Pollution Control Handbook, Chapter 20, George Godwin, UK, 1979.
26. Turton, R., Bailie, R. C., Whiting, W. B., and J. A. Shaeiwitz, Analysis, Synthesis, and Design of Chemical Processes. Prentice Hall International Series, New Jersey, 1998.
27. Petley, G. J., and D. W. Edwards, "Further Developments in Chemical Plant Costing Using Fuzzy Matching," Comput. Chem. Eng., Vol. 19 (supplement), S675-S680. (Part of ESCAPE 5 conference, 11-14 June 1995).
28. Gerrard, A. M, and J. Brass, "Preliminary Cost Modeling for Process Vessels," Proceedings of the 9th International Conference on Flexible Automation and Intelligent Manufacturing, Tilburg, Netherlands (Begell House, New York), 1999.

29. Horwitz, B. A., "The Mathematics of Discounted Cash Flow Analysis," Chem. Eng., pp 169-174, May 1980.
30. Cannaday, R. E., Colwell, P. F., and H. Paley, "Relevant and Irrelevant Internal Rates of Return," The Eng. Econ., Vol. 32, No. 1, pp 17-33, 1986.
31. Powell, T. E., "A Review of Recent Developments in Project Evaluation," Chem. Eng., pp 187-194, Nov. 1985.
32. Ward, T. J., "Estimate Profitability Using Net Return Rate," Chem. Eng., pp 151-155, March 1989.
33. Klumpar, I. V., "Project Evaluation By Computer," Chem. Eng., pp 76-84, June 29, 1970.
34. Linsley, J., "Return on Investment Discounted and Undiscounted," Chem. Eng., pp 201-204, May, 1979.
35. Holland, F. A., Watson, F. A., and J. K. Wilkinson, Introduction to Process Economics, 2nd ed., John Wiley & Sons Ltd., New York, 1976.
36. Valle-Riestra, J. F., Project Evaluation in the Chemical Process Industries, McGraw-Hill Book Company, New York, 1983.
37. Allen, D. H., A Guide to the Economic Evaluation of Projects, 2nd. ed., IChemE England 1980.
38. Garrett, D. E., Chemical Engineering Economics, Van Nostrand Reinhold, New York, 1989.
39. Larson, G. A., "Annual Savings, Return and Depreciation Fix Justifiable Investment", Power, Sept., pp 103, 1955.
40. Perry, R. H., and D. W. Green, Eds., Perry's Chemical Engineers' Handbook, CD –ROM, Mc. Graw-Hill Co., 1999.
41. Abrams, H. J., The Chemical Engineer, No. 241 1970.
42. Kirkpatrick, D. M., "Calculator Program Speeds up Project Financial Analysis," Chem., Eng., pp 103-107, August, 1979.
43. Allen, D.H., Economic Evaluation of Projects- A Guide, IChemE, 3rd ed., 1991.
44. Gary, James H, Handwerk, Glenn, E., and Mark J. Kaiser, Petroleum Refining Technology and Economics, 5th ed., CRC Press, Taylor & Francis Group, 2007.
45. Paris climate accord or Paris climate agreement, https://treaties.un.org/pages/ViewDetails.aspx?
46. U.S. Energy Information Administration, Brent crude oil price forecast to average $72 per barrel in the second half of 2021., Jul 15, 2021.
47. James G. Speight, An Introduction to Petroleum Technology, Economics and Politics, Scrivener-Wiley, 2011.
48. International Energy Agency., Biofuel production using petroleum refining technologies, www.iea.gov.
49. Rees, W. and Mathis Wacknergel, https://en.wikipedia.org/wiki/Carbon_footprint.
50. Wright, L, Kemp, S., and I. Williams, "Carbon footprinting: Towards a universally accepted definition", Carbon Management. 2 (1): 61-72. doi:10.4155/CMT.10.39. S2CID 154004878
51. Glossary: Global Warming Potential (GWP), U.S. Energy Information Administration. Retrieved 2011-04-26., https://en.wikipedia.org/wiki/Global_warming_potential.
52. M. Cain, A new way to assess global warming potential of short-lived pollutants (carbonbrief.org/guest-post-a-new-way-to-assess-global-warming-potential-of-short-lived-pollutants).
53. Myles, R. Allen, et al., New use of global warming potentials to compare cumulative and short-lived climate pollutants, Nature Climate Change, Letters, 2nd May 2016, DOI:10.1038/NCLIMATE2998.
54. https://www.3.epa.gov/carbon-footprint-calculator/tool/definitions/co2e.html
55. https://en.wikipedia.org/wiki/Global_warming_potential
56. Hamrick, Kelley; Gallant, Melissa, Unlocking Potential: State of the Voluntary Carbon Market, Forest Trends, Ecosystem Marketplace, p. 3, May 2017, Retrieved 2019-01-29.
57. "Demand for Voluntary Carbon Offsets Holds Strong as Corporates Stick With Climate Commitments". Ecosystem Marketplace. Retrieved 2020-12-30.
58. "State and Trends of Carbon Pricing, 2021. The World Bank, 2021. Doi:10.1596/978-1-4648-1728-1.
59. https://en.wikipedia.org/wiki/Carbon_price.

Bibliography

Ranade, S.M., S. C., Sherck, and D. H. Jones, "Know Marginal Utility Costs", Hydrocarbon Processing, V. 68, No. 9, p. 81, 1989.
Moshe, Ben-Horim., "Essentials of Corporate Finance", Allyn and Bacon, Inc, 1987.
Slack, J. B., "Steam Balance: A New Exact Method", Hydrocarbon Processing, Mar., p. 154, 1969.
Ranade, S.M., S.C. Shreck, and D. H. Jones, "Know Marginal Utility Costs", Hydrocarbon Processing, V. 68, No. 9, p. 81, 1989.
Gottfried, B.S., Spreadsheet Tools for Engineers Excel 2000 Version, Mc. Graw-Hill, 2000.
Taylor, J. H., "The Process Step Scoring Method for Making Quick Capital Estimates," Eng. and Process Econ. Vol 2, p. 259, 1977.
Peters, M.S., and K. D. Timmerhaus, Plant Design and Economics For Chemical Engineers, 3rd. ed., McGraw-Hill Int. Book Company, N.Y. 1981.

27

Sustainability in Engineering, Petroleum Refining and Alternative Fuels

27.0 Introduction

The Paris Agreement (L'accord de Paris [1]) is an agreement within the United Nations Framework Convention on Climate Change (UNFCCC), dealing with greenhouse gas (GHG) emission mitigation, adaptation and finance, which was signed in 2016. The agreement was negotiated by representatives of 196 state parties and adopted by consensus on 12 December 2015; by February 2020, all UNFCCC members have signed the agreement. 189 have become a party to it, except for Iran and Turkey. The agreement's long-term temperature goal is to keep the increase in global average temperature to well below 2°C above pre-industrial levels, and to pursue efforts to limit the increase to 1.5°C, recognizing that this would substantially reduce the risks and impacts of climate change. This recent COP26, the International Climate Change conference in Glasgow, UK under UNFCCC reaffirmed the commitments by countries in article 2 of the Convention is to stabilize greenhouse gas concentrations in the atmosphere at a level that would prevent dangerous anthropogenic interference with the climate system by committing to further cuts in carbon emissions by 2030, that all countries must commit to reach net zero emissions as soon as possible, developed countries must honor their commitments, including meeting the 2020 US100bn dollar a year goal for climate finance and seeking to agree a package which takes forward the Paris Agreement, which entered into force on 4 November 2016 and 191 parties have ratified the agreement.

There have been a number of developments since 2005, including the publication of the Intergovernmental Panel on Climate Change (IPCC) Special Report on Global Warming of 1.5°C above pre-industrial levels and related greenhouse gas emissions pathways in 2018. The IPCC noted that limiting warming to 1.5°C is possible but would require rapid and far-reaching transitions across multiple sectors.

This should be done by reducing emissions as soon as possible in order to achieve a balance between anthropogenic emissions by sources and removals by sinks of greenhouse gases in the second half of the 21st century. It also aimed to increase the ability of parties to adapt to the adverse impacts of climate change and make finance flows consistent with a pathway toward low greenhouse-gas emissions and climate-resilient development [2]. Figure 27.1 shows the global carbon dioxide emissions of countries and Figure 27.2 shows the world of annual CO_2 emissions from the burning of fossil fuels for energy and cement productions. Cumulative emissions include land-use change, and are measured between the years 1950 and 2000. Under the Paris Agreement, each country must determine, plan, and regularly report on the contribution that it undertakes to mitigate global warming, although no mechanism is

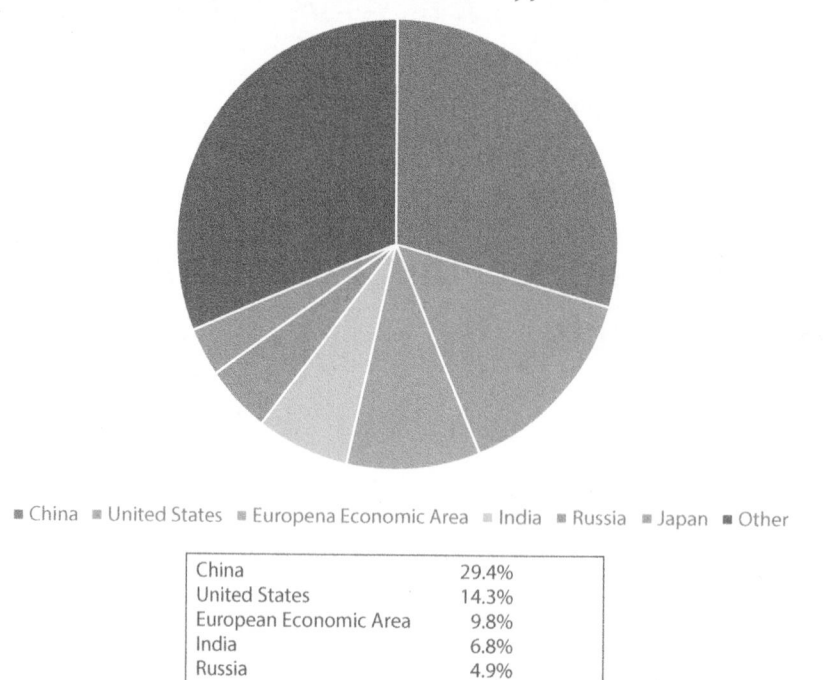

Global carbon dioxide emissions by jurisdiction

China	29.4%
United States	14.3%
European Economic Area	9.8%
India	6.8%
Russia	4.9%
Japan	3.5%
Other	31.3%

Figure 27.1 Global carbon dioxide emissions by jurisdiction (Source: https://en.wikipedia.org/wiki/Paris_Agreement).

Figure 27.2 World map of CO_2 emissions from the burning of fossil fuels for energy and cement productions (Source: Our World Data).

imposed to set a specific emissions target by a set date; however, each target should go beyond previously set targets [3–5].

The strategy involved energy and climate policy, including the so-called 20/20/20 targets, namely the reduction of CO_2 emissions by 20%, the increase of renewable energy's market share to 20% and a 20% increase in energy-efficiency [6]. Countries are further encouraged to reach global peaking of GHG emissions as soon as possible, and the agreement has been described as an incentive for and driver of fossil-fuel divestment [7, 8].

Greenhouse gases are those that absorb and emit infrared radiation in the wavelength range emitted by earth. CO_2 (0.04%), nitrous oxide (N_2O), methane (CH_4), ozone (O_3) are trace gases that account for almost one-tenth of 1% earth's atmosphere and have appreciable greenhouse effect.

The most abundant greenhouse gases in the earth's atmosphere are:

- Water vapor (H_2O)
- Carbon dioxide (CO_2)
- Methane (CH_4)
- Nitrous oxide (N_2O)
- Ozone (O_3)
- Chlorofluorocarbons (CFCs)
- Hydrofluorocarbons (includes HCFCs and HFCs)

Atmospheric concentrations are determined by the balance between sources (emissions of the gas from human activities and natural systems) and sinks (the removal of the gas from the atmosphere by conversion to a different chemical compound or absorption by bodies of water). The proportion of an emission remaining in the atmosphere after a specified time is the airborne fraction (AF). The annual AF is the ratio of the atmospheric increase in a given

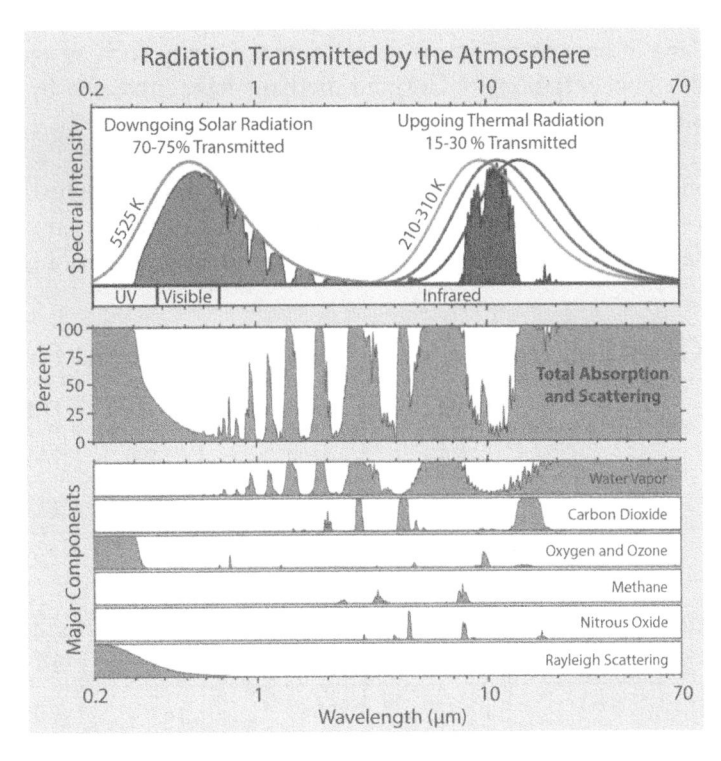

Figure 27.3 Radiation transmitted by the atmosphere (Source: www.wikepedia.com).

year to that year's total emissions. As of 2006, the annual AF for CO_2 was 0.45 and this has increased at a rate of $0.25 \pm 0.21\%$ per year over the period of 1959–2006. Figure 27.3 shows the atmospheric absorption and scattering at different wavelengths of electromagnetic waves. The largest absorption band of carbon dioxide is not far from the maximum in the thermal emission from the ground, and it partly closes the window of transparency of water, and thus its major effect.

A greenhouse gas (GHG) is a gas that absorbs and emits radiant energy within the thermal infrared range. Greenhouse gases cause the greenhouse effects on the planets. The greenhouse is a warming effect caused by certain gases that retain heat from sunlight. Without such gases, the average surface temperature of the earth would be below freezing. As a result of this effect, the earth is ~33°C warmer than it would be without it. Without it, the average temperature of the earth's surface would be below 0°C, and human beings would find it uninhabitable. The presence of greenhouse gases in the atmosphere has reached an alarming situation during the last one and a half century, although some of the greenhouse gases are natural such as water vapor, which causes ~ 36–70% of the greenhouse effect. CO_2 is the next important that causes 9–26%, followed by CH_4, about 4–9% and O_3 about 3–7%. Table 27.1 shows the yearly emissions of the gases left into the atmosphere by different fossil carbon [9].

The total CO_2 emissions are contributed by different attributes, namely:

Forest and soil	100×10^6 tons
Human activity	30 giga tons
Geological	3000–3,200 giga tons
Oceans	1400–2,000 giga tons.

Brown coal emits three times as much CO_2 as natural gas; black coal emits twice as much CO_2 per unit of electric energy. Coal produces 1.7 to 2.2 times CO_2 by wt/wt, gas produces 2.7–3.0 and oil at about 3–4 times wt/wt. CO_2 is the most ubiquitous and interesting such that ~3 billion tons per year are being vented. Other GHGs are N_2O, NO_2, N_2O_3 and sulfur are increasing in concentration owing to human activity such as agriculture, power houses and automobiles. The atmospheric concentrations of CO_2 and methane have increased by 31% and 149% respectively above pre-industrial levels since 1750.

Fossil-fuel burning is responsible for 3/4 of the increase in CO_2 from human activity over the past 20 years. Other parts are due to elimination of mature forests and other biological activities, which cannot be eliminated. Fossil fuels coal, oil and natural-gas release carbon dioxide during production and consumption. Fossil fuels are contributing

Table 27.1 Gases Left into the Atmosphere by Different Fossil Carbon.

Pollutant	Hard coal	Brown coal	Fuel oil	Other	Oil
CO_2 (g/GJ)	94600	10100	77400	74100	56100
SO_2 (g/GJ)	765	1361	1350	228	0.68
NOx (g/GJ)	292	183	195	129	93.3
CO (g/GJ)	89.1	89.1	15.7	15.7	14.5
Non-methane organic (g/GJ)	4.92	7.78	3.70	3.24	1.58
Particulate g/GJ	1203	3254	16	1.91	0.1
Flue gas volume total (m³/GJ)	360	444	279	276	272

EPA Home (Global Emissions Source: IPCC (2007); based on global emissions.

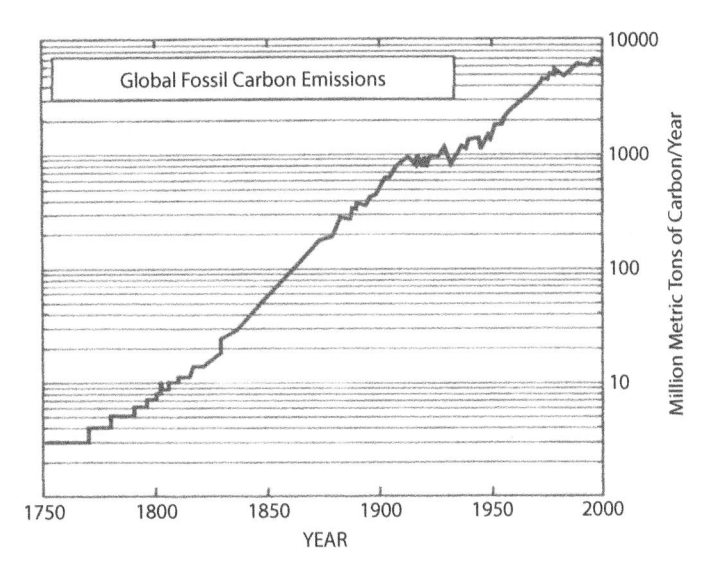

Figure 27.4 Fossil carbon emission in metric tonnes vs. year.

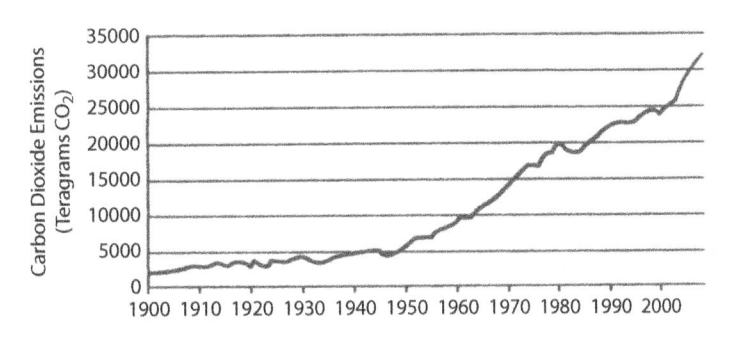

Figure 27.5 Global carbon dioxide (CO_2) emissions from fossil fuels (Source: Boden, T.A., Marland G., and R. J. Andres, Global Regional and National Fossil-Fuel CO_2 Emissions, Carbon Dioxide Information Analysis Center, National Laboratory, U.S. Department of Energy, Tenn., U.S.A., Oak Ridge).

to a rising quality of life in many parts of the world, and based on current projections of population and economic growth, the world's demand for energy will increase substantially over the next 30 years. The majority of that energy will be provided by fossil fuels, even as lower carbon alternatives continue to emerge. United Nation Environment Programme (UNEP) stated that emission levels driven by burning fossil fuels need to reduce by 14% by 2020 for the world to reach a pathway that could keep the global temperature rise below 2°C compared with pre-industrial levels.

Table 27.2 Gases Contribution to Trapping Radiation.

Gas and its presence in the atmosphere (vol)		% greenhouse effect that would be absent if all gases were removed from Earth's atmosphere	% in atmosphere due to human activity
Water vapor -	1–3%	36%	0%
Clouds		14%	
Carbon dioxide	0.04%	12%	26%
Ozone	3%	Not known	
Methane	2 ppm	?	60%

Table 27.3 Mass of CO_2 Emitted per Quantity of Energy for Various Fuels.

Fuel name	CO_2 emitted (lb/10^6 Btu)	CO_2 emitted (g/MJ)	CO_2 emitted (g/kWh)
Natural gas	117	50.30	181.08
Liquefied petroleum gas	139	59.76	215.14
Propane	139	59.76	215.14
Aviation gasoline	153	65.78	236.81
Automobile gasoline	156	67.07	241.45
Kerosene	159	68.36	246.10
Fuel oil	161	69.22	249.19
Tires/tire derived fuel	189	81.26	292.54
Wood and wood waste	195	83.83	301.79
Coal (bituminous)	205	88.13	317.27
Coal (sub-bituminous)	213	91.57	329.65
Coal (lignite)	215	92.43	332.75
Petroleum coke	225	96.73	348.23
Coal (anthracite)	227	97.59	351.32

Note: One liter of gasoline, when used as a fuel produces 2.32 kg (~1300 liters or 1.3m^3) of CO_2, a greenhouse gas. 1US gallon produces 19.4 lb (1291.5 gallons or 172.65 ft^3).

Figure 27.4 shows the rise in global fossil carbon emissions over 250 years and Figure 27.5 shows the rise of CO_2 emissions over the period of 100 years respectively. The current 100 billion tons of carbon (gigatons of carbon only) are emitted annually as CO_2 (37.5 billion tons of anthropogenic CO_2) into the atmosphere, while the rest is absorbed by the oceans and the land to about equal proportions. The greenhouse gases that are present in the atmosphere account for human activity or from nature, and their contribution to approximate greenhouse effect is shown in Table 27.2 and Table 27.3 illustrates the relative CO_2 emission from various fuels.

27.1 Impacts on the Overall Greenhouse Effect

Schmidt *et al.* [10] analyzed the impact of individual components of the atmosphere to the greenhouse effect. They estimated that water vapor accounts for ~50% of the earth's greenhouse effect, with clouds contributing 25%, CO_2, contributing 20% and the minor greenhouse gases and aerosols accounting for the remaining 5%. The contribution of each gas to the greenhouse effect is determined by the characteristics of that gas, its abundance, and any direct effects it may cause. For example, the direct radiative effect of a mass of methane is ~84 times stronger than the same mass of CO_2 over a 20-year time frame, but it is present in much smaller concentrations so that its total direct radiative effect has been smaller, in part due to its shorter atmospheric lifetime in the absence of additional carbon sequestration. In addition to its direct radiative impact, CH_4 has a large indirect radiative effect because it contributes to O_3 formation. Shindell *et al.* [11] argue that the contribution to climate change from CH_4 is at least double the previous estimates as a result of this effect. When ranked by their direct contribution to the greenhouse effect, the most important are:

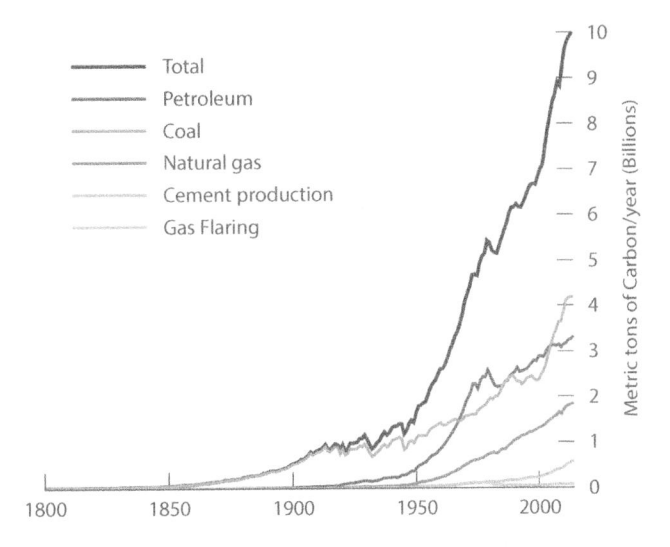

Figure 27.6 Modern global CO_2 emissions from the burning of fossil fuels (Source: https://en.wikipedia.org/wiki/Greenhouse_gas#cite_note-21).

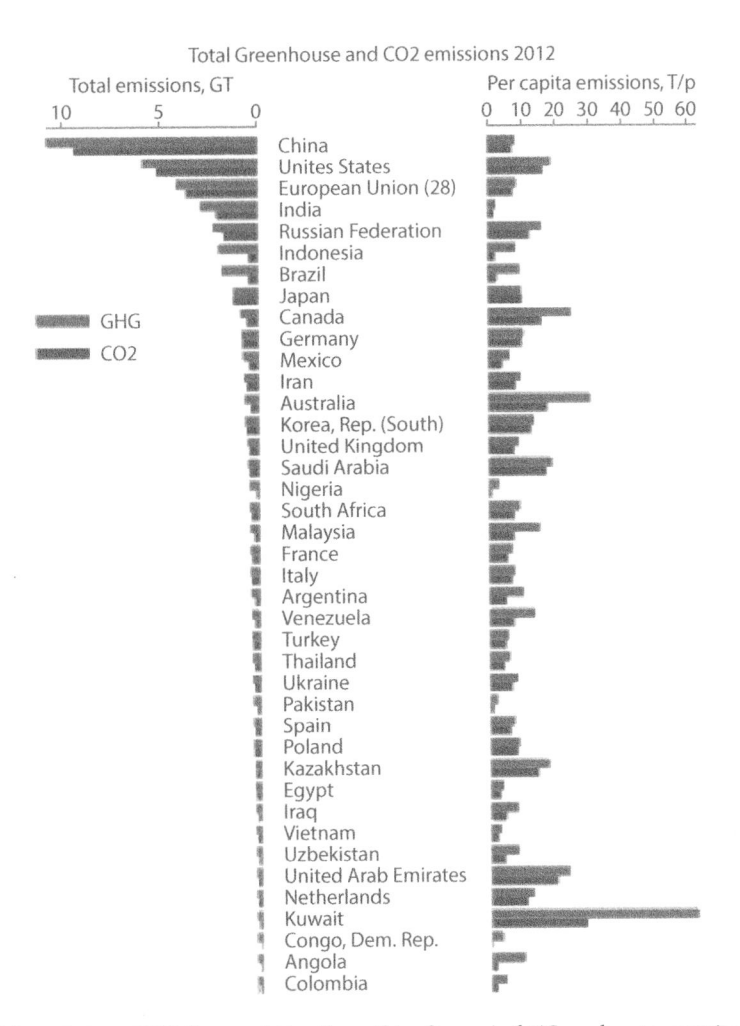

Figure 27.7 Total GHG and CO_2 emissions 2012 (Source: https://en.wikipedia.org/wiki/Greenhouse_gas#cite_note-21).
Note: The top 40 countries emitting all greenhouse gases, showing both that derived from all sources including land clearance and forestry and also the CO_2 component excluding those sources.

Compound	Formula	Concentration in atmosphere, ppm	Contribution %
Water vapor and clouds	H_2O	10–50,000*	36–72
Carbon dioxide	CO_2	~400	9–26
Methane	CH_4	~1.8	4–9
Ozone	O_3	2–8**	3–7

Notes:
* Water vapor varies locally.
** The concentration in stratosphere. About 90% of the ozone earth's atmosphere is contained in the stratosphere.

Figure 27.6 shows the modern global CO_2 emissions from the burning of fossil fuels and Figure 27.7 illustrates the top 40 countries emitting all greenhouse gases.

27.2 Carbon Capture and Storage in Refineries

Carbon capture and storage (CCS) is an essential element of various measures needed to reduce greenhouse gas (GHG) emissions, without which the cost of reaching targets will increase by 40%, which is more costly than for any other low-carbon technology. CCS is a key technology to reduce CO_2 emissions across various sectors of the economy while providing other societal benefits, namely: energy security and access, air pollution reduction, grid stability and jobs preservations and creation. These form a part of a broad range of measures, which are aimed at reducing CO_2 emissions.

Recently, CCS projects in Europe were targeted at reducing GHG emission from the power sector, where the largest emissions' points are found. However, the last few years have witnessed significant changes in the power sector, including an enhanced renewable energy, a rapid phase-out of coal-fired power plants and the emergence of nuclear power for medium-term reform of the energy system.

CO_2 capture is a process that involves the separation of CO_2 from gas streams. These gas streams could include but are not limited to combustion flue gases, process off gases (i.e., by-product gases from blast furnaces and basic oxygen furnaces, tail gases from steam methane reforming and various refinery processes and so on) or natural gas (i.e., from natural-gas processing). For many decades, CO_2 capture processes have been used in several industrial applications at a scale close to those required in CCS applications.

In general, CO_2 capture processes can be classified according to their gas separation principle, namely: chemical absorption, physical absorption, adsorption, calcium and reversible chemical loops, membranes, and cryogenic separation [12]. The chemical absorption process uses the reversible chemical reaction of CO_2, with an aqueous solvent usually an amine or ammonia. CO_2 is separated by passing the flue gas through a scrubbing system. The absorbed CO_2 is stripped from the solution in a desorber, and a pure stream of CO_2 is sent for compression while the regenerated solvent is sent back to the absorber.

On a global level, the total CO_2 emitted by refineries in 2015 was estimated to be around 970 Mt per year. The total processed crude oil was ~82 Mb/d (total capacity ~ 97.5 Mb/d), which results in a CO_2 intensity of around 200 kg CO_2/t crude oil. This intensity varies for each refinery and depends on the complexity, energy efficiency and ratio of feedstocks other than crude oil [13]. The global refining sector contributes ~4% of the total anthropogenic CO_2 emissions. CCS has been recognized as one of the technologies that could be deployed to achieve a deep reduction of greenhouse-gas emissions. In the EU countries, the verified emissions from 79 mainstream refineries were 138 Mt CO_2/y in 2015 (~14% of the world emissions from refineries). Recently, the EU refining industry has reduced its

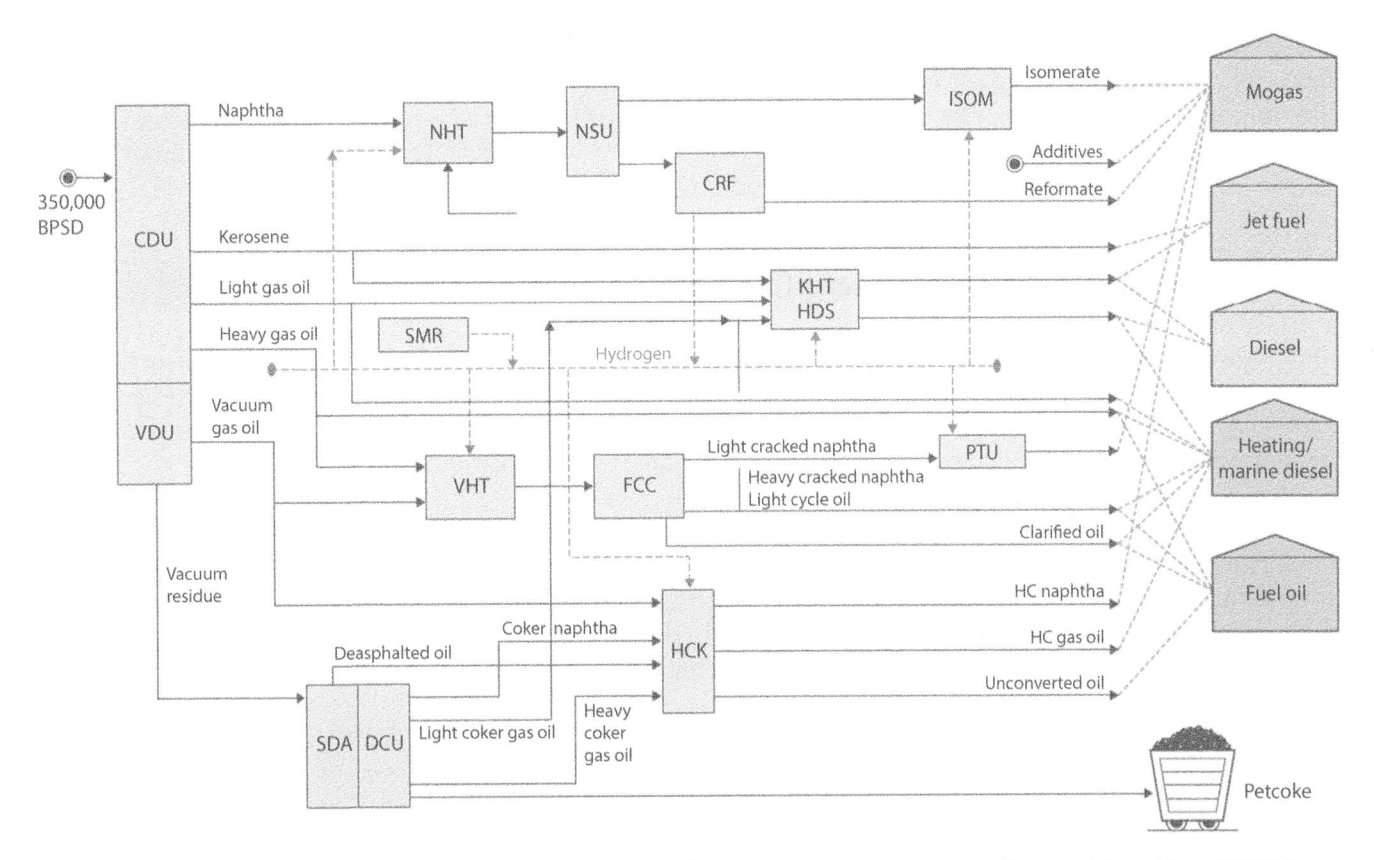

Figure 27.8 Simplified flow diagram of a typical complex refinery with more than 10 emission sources. (Source: Adapted from SINTEF (2017). ReCAP Project—Evaluating the Cost of Retrofitting CO_2 Capture in an Integrated Oil Refinery: https://www.sintef.no/recap).

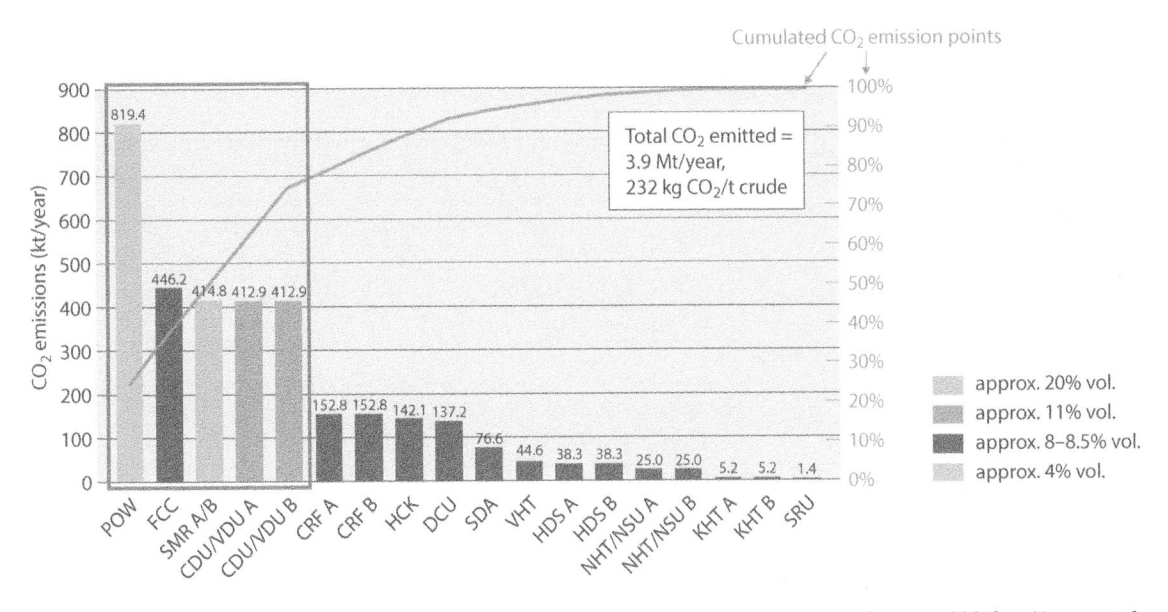

Figure 27.9 The main CO_2 emission sources for a typical complex refinery with a normal capacity of 350,000 bbl/day. (Source: Adapted from SINTEF (2017). ReCAP Project --Evaluating the Cost of Retrofitting CO_2 Capture in an Integrated Oil Refinery: https://www.sintef.no/recap).

environmental footprint by continually increasing its energy integration and investments in efficiency. Furthermore, the use of cogeneration and advanced catalyst technology has allowed for further energy reduction gains, resulting in an improved energy efficiency by ~1% per annum since 1990 [14]. Figure 27.8 presents a simplified flow diagram of a typical complex refinery with more than 10 emission sources, and Figure 27.9 shows that five of these sources represent 75% of the total CO_2 emitted. The CO_2 concentration varies between 5–20% vol.

27.3 Sustainability in the Refinery Industries

The growing populations and affluence around the globe have put an increasing pressure on air, water, arable land and raw materials toward the latter half of the 20th century and the beginning of the 21st century. Concerns over the ability of natural resources and environmental systems present the requirements of global populations now and in the future as a part of an emerging awareness to the concept of sustainability.

The 2005 World Summit on Social Development identified development goals as economic development, social development and environmental protection as illustrated in Figure 27.10 [15]. This view has been expressed using three overlapping ellipses indicating that the three pillars of sustainability are not mutually exclusive and can be mutually reinforcing as illustrated in Figure 27.11 [16]. The three pillars have served as a common ground for numerous sustainability standards and certification systems.

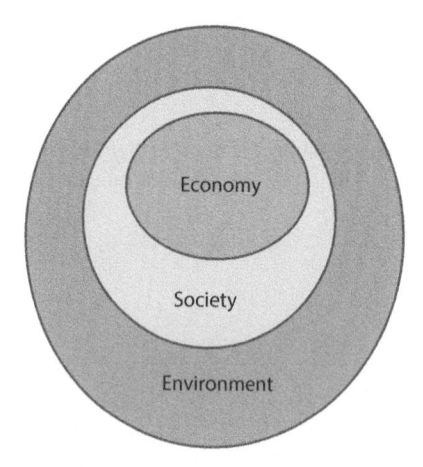

Figure 27.10 The relationship between the "three pillars of sustainability" in which both economy and society are constrained by environmental limits [15].

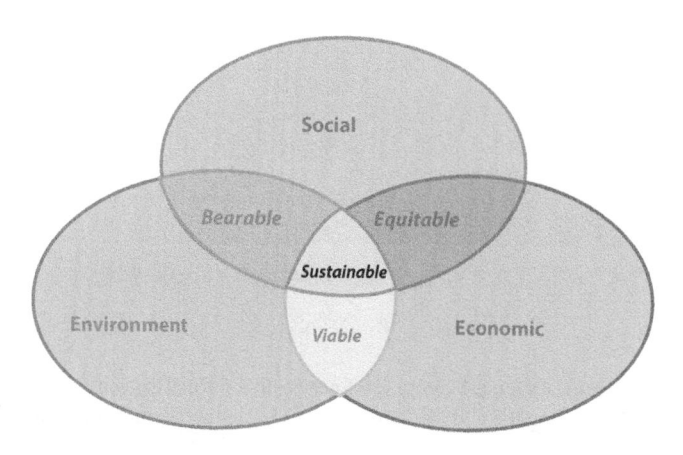

Figure 27.11 Venn diagram of sustainability development at the confluence of three constituent parts [16].

A computer simulation software has been implemented in a refinery that uses digital twin – enabled virtual sensors where the model calculates NO_x, CO_x and other emissions at many points across the process units than would be possible with physical sensors. The information is then presented to operators, technicians and plant management as an easy-to-interpret dashboard. The digital twin-based solution enables it to ensure compliance with air quality regulations, and target and tracks its sustainability progress. From an economic view point, since the operators have a high level of confidence in the model, they can run the plant closer to limits, which has resulted in improvement in yields. The solution of the digital twin-based virtual sensors has resulted in improved worker's job satisfaction, since the transparency it provides allows employees know they are working in a sustainability – conscious organization that is addressing the global climate change threat and improving air quality.

The term sustainability has many different meanings to people. To sustain is defined by some as to "support without collapse". The Bruntland Commission was assigned to create a global agenda for change by the General Assembly of the United Nations in 1984, and they defined sustainability very broadly: "Humanity has the ability to make development sustainable- to ensure that it meets the needs of the present without compromising the ability of future generations to meet their own needs [17]. However, the search on the definition of sustainability will provide many variations on this basic concept.

Sustainability involves simultaneous progress in four principal areas:

- Human
- Economic
- Technological
- Environmental

It requires conservation of resources while minimizing depletion of nonrenewable resources and using sustainable practices for managing renewable resources. There can be no product development or economic activity without the presence of available resources, except for solar or nuclear energy as the resources are finite. Furthermore, efficient designs applying processing integration, energy management, pinch analysis or inherently safer design can invariably conserve resources while also reducing impacts caused by material extraction and related activities. However, the depletion of nonrenewable resources and overuse of otherwise renewable resources could limit their availability for future generations.

In engineering, incorporating sustainability into products, processes, technology systems and services generally requires integrating environmental, economic and social factors in the evaluation of designs. Applying the concepts of sustainability into the quantitative design tools and performance metrics that can be applied in engineering design

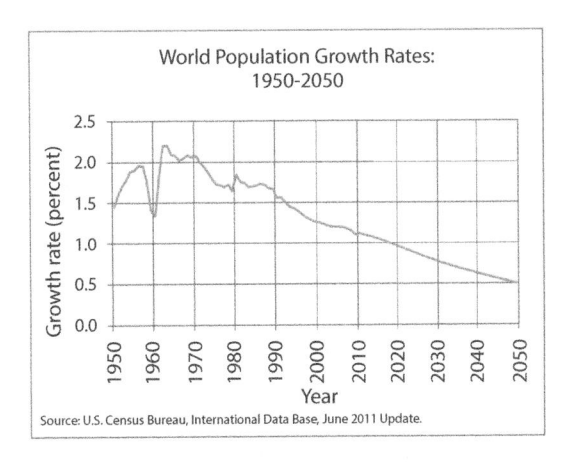

Figure 27.12 Population growth rate vs. year. (Source: U.S. Census Bureau, International Data Base).

is often a challenge. Quantitative tools that are available to engineers seeking to design for sustainability are ever changing; however, they focus on natural resources conversion such as fossil fuels (e.g., coal, crude oil and natural gas) and emission reduction such as GHG (e.g., CO_2, H_2S, N_2O, SO_2, and so on). Figure 27.12 shows the world population growth rate, 1950-2050, as estimated in 2011 by the U.S. Census Bureau, International database. Although, as the rate of growth decreases, population continues to rise. In 2050 still growing by over 45 million per annum.

27.4 Sustainability in Engineering Design Principles

From the introduction of sustainability development concepts in Brundtland Report (WCED, 1987), there have been many attempts to incorporate sustainability principles into engineering design. Allen and Shonnard [18] have described the various principles; however, the Sandestin, sustainable engineering principles and the nine principles of Green Engineering are illustrative of the multiple sets of principles available as shown in Table 27.4. Table 27.5 shows the twelve principles of green engineering. The descriptions of these principles are as follows [18]:

Principle 1: Engineering process and products holistically, use system analysis and integrate environmental impact assessment tools. These concepts resonate in a number of Green and Sustainable Engineering principles and are addressed at length in various textbooks, including Green Engineering: Environmentally Conscious Design of Chemical Processes by Allen and Shonnard. The principle points out the importance of systematic evaluation and reduction of human health and environmental impacts of designs, products, technologies, processes, and systems. The use of system-based techniques such as heat and mass integration techniques (e.g., pinch analysis, see chapter 22)

Table 27.4 Sandestin Sustainable Engineering Principles.

Green Engineering transforms existing engineering disciplines and practices to those that promote sustainability. Green Engineering incorporates development and implementation of technologically and economically viable products, processes and systems that promote human welfare while protecting human health and elevating the protection of the biosphere as a criterion in engineering solutions. To fully implement green engineering solutions, engineers use the following principles:	
Principle 1:	Engineer process and products holistically, use system analysis and integrate environmental impact assessment tools.
Principle 2:	Conserve and improve natural ecosystems while protecting human health and well-being.
Principle 3:	Use life cycle thinking in all engineering activities.
Principle 4:	Ensure that all materials and energy inputs and outputs are as inherently safe and benign as possible.
Principle 5:	Minimize depletion of natural resources.
Principle 6:	Strive to prevent waste.
Principle 7:	Develop and apply engineering solutions, while being cognizant of local geography, aspirations, and cultures.
Principle 8:	Create engineering solutions beyond current or dominant technologies; improve, innovate and invent technologies to achieve sustainability.
Principle 9:	Actively engage communities and stakeholders in development of engineering solutions.
There is a duty to inform society of the practice of Green Engineering.	

Table 27.5 Principles of Green Engineering.

Principle 1:	Designers need to strive to ensure that all materials and energy inputs and outputs are as inherently non-hazardous as possible.
Principle 2:	It is better to prevent waste than to treat or clean up waste after it is formed.
Principle 3:	Separation and purification operations should be designed to minimize energy consumption and materials use.
Principle 4:	Products, processes and systems should be designed to maximize mass, energy, space and time efficiency.
Principle 5:	Products, processes and systems should be "output pulled" rather than "input pushed" through the use of energy and materials.
Principle 6:	Embedded entropy and complexity must be viewed as an investment when making design choices on recycle, reuse, or beneficial disposition.
Principle 7:	Targeted durability, not immortality, should be a design goal.
Principle 8:	Design for unnecessary capacity or capability (e.g., "one size fits all") solutions should be considered a design flaw.
Principle 9:	Material diversity in multicomponent products should be minimized to promote disassembly and value retention.
Principle 10:	Design or products, processes, and systems must include integration and interconnectivity with available energy and materials flows.
Principle 11:	Products, processes and systems should be designed for performance in a commercial "afterlife"
Principle 12:	Material and energy inputs should be renewable rather than depleting.

is essential to minimize human health and environmental impacts of designs through material and energy optimization. The principle also conveys the importance of not shifting risk (e.g., reducing releases to one environmental medium may increase the risk to another medium and/or increase the likelihood of worker exposures and jeopardize worker safety. For example, a consequence of shifting the risk is the use of methyl-tert-butyl-ether (MTBE) as a gasoline additive. MTBE was added to gasoline to reduce emissions of carbon monoxide from automobile tailpipes, thereby protecting human health. Its greater mobility in soil and water environments, however, meant that spills of MTBE could more readily migrate to and disperse in water supplies than spills of gasoline.

Principle 2: Conserve and improve natural ecosystems while protecting human health and well-being: This principle expresses the importance of understanding environmental processes for engineers involved in design of chemicals, automobiles, buildings, and other manufactured goods. There are many examples where a lack of understanding caused severe environmental harm and raised the level of health risk to humans and other forms of life. E.g., Chlorofluorocarbons (CFCs) were thought to be ideal refrigerants. They replaced dangerous refrigerant fluids like ammonia and made storage of food and building climate control far safer. Their benefits to human health and well-being were clear; however, once the role of CFCs in stratospheric ozone destruction chemistry was worked out, it became clear that there were hazards to human health associated with CFS use. Not every engineer needs to be an expert in environmental processes and health effects, but designers should be aware of the potential harm that can be caused and work with multidisciplinary experts to achieve more sustainable solutions.

Principle 3: Use life-cycle thinking in all engineering activities: This principle complements Principle 1. Every engineered product is created, functions over a useful life, and is eventually disposed of to the environment. Life-cycle thinking can help avoid a narrow outlook on environmental, social and economic concerns and help make informed decisions. Life-cycle thinking that gets incorporated into design will help identify design alternatives that minimize environmental impacts at the various life stages. This same kind of thinking can also consider economic and societal aspects. The importance of life-cycle thinking can be illustrated through the greenhouse gas emissions associated with biofuels.

Principle 4: Ensure that all material and energy inputs and outputs are as inherently safe and benign as possible: This principle complements Principle 3. These characteristics of materials and chemicals must be applied to all stages of a product's life, from extraction to use and disposal. The following are some questions that must be asked in each stage of the life cycle: Are the materials toxic? Are there inherently benign (in terms of toxicity) materials that can be used as substitutes? Will exposure during manufacturing be a health problem to workers? Does the product pose minimal impact during recycle and disposal? Will an unintentional release of material quickly degrade in the environment? Properties of materials relevant to safety, beyond toxicity, must also be considered, such as flammability, explosivity, and corrosivity.

Principle 5: Minimize depletion of natural resources: As the world's population continues to grow and become more affluent, natural resources will be used at ever greater rates, and the importance of this principle is raised. Efficient use of non-renewable energy resources is of primary interest, and development of renewable alternatives for energy and materials should be given a high priority in engineering design.

Principle 6: Strive to prevent waste. Waste not only represents material that takes up space in landfills but more importantly represents a loss of efficiency in a production system that includes many input materials and energy sources. When waste is avoided through design, the environmental impacts associated with the input materials and the energy that went into producing the discarded product are also avoided. The following are some questions that must be asked during design: Are there ways (e.g., procedures, engineering) to improve the yield or efficiency of raw materials? How can the releases or wastes be recycled and reused? Can the product be reused after its normal commercial life, hence minimizing the raw materials needed to manufacture new products?

Principle 7: Develop and apply engineering solutions, while being cognizant of local geography, aspirations, and cultures. Engineering designs are directed toward meeting individual human and societal needs, and in order to better achieve this goal; awareness of the societal context of the design is crucial. An engineering design in one society, such as rapid public transportation systems, may not meet the aspirations and needs in another society, even though the design achieves environmental objectives. The main point is to move each society, through engineering design, toward more sustainable utilization of resources in a way that achieves that society's or individual's aspirations.

Principle 8: Create engineering solutions beyond current or dominant technologies; improve, innovate, and invent (technologies) to achieve sustainability. Sustainability can be a powerful motivation for change in engineering designs, technologies, processes, and products. This principle emphasizes the importance of being innovative (i.e., "outside-the-box" thinking) in the development of new technologies. The knowledge gained through considering the many dimensions of sustainability should be reflected in how engineering designs accomplish societal objectives.

Principle 9: Actively engage communities and stakeholders in development of engineering solutions. There are many examples of stakeholder and community engagement in the development of engineered solutions in a wide range of activities, including city planning, infrastructure development, and production of manufactured goods. One illustrative example is in the mining industry where *Seven Questions to Sustainability* from the Mining Minerals Sustainable Development North America project (IISD, 2002) has been adopted by key members from project inception to mine closure. During this engagement, the communities surrounding the proposed mine development express their wishes with regard to managing the economic development and any concerns over local environmental consequences.

A second set of engineering design principles, the 12 Principles of Green Engineering, from Anastas and Zimmerman [19] is as follows:

Principle 1: Designers need to strive to ensure that all material and energy inputs and outputs are as inherently non-hazardous as possible. This principle recognizes that significant costs and hazards results from the selection of sources of materials and energy. Additional control systems are required to capture and destroy hazardous materials during production, use and disposal, all of which add to the cost of the design. If inputs to the system are inherently less hazardous, the risks of failure will be reduced and the amount of resources expended on control, monitoring, and containment will be less.

Principle 2: It is better to prevent waste than to treat or clean up waste after it is formed. The creation of waste in engineered systems adds to the complexity, effort, and expense of the design. This is especially true for hazardous wastes, which require extraordinary measures for their control, monitoring, transport, and disposal. To reduce waste generation, the design must strive to incorporate as much of the input materials as possible into final products. This strategy can be applied at many scales, for example, at the molecular level in the design of chemical reactions, at larger scales such as in machining of parts, and further in the assembly of discrete parts. Any waste that is generated should be considered as a raw material to be used again in the current product system or as an input to a separate product system.

Principle 3: Separation and purification operations should be designed to minimize energy consumption and materials use. In the chemical-and mineral processing industries, large – scale separation processes are among the largest energy-consuming units and generate a significant proportion of emissions and wastes. Even in industry sectors where the mass of the products produced is not large, separation processes can be significant. In electronics manufacturing, the generation of ultra-pure waste and the creation of ultra-clean work environments require separation processes with significant costs and energy demands. Design for efficient separation is very important in these industries, and several approaches can be investigated during design.

Gains in energy efficiency can be attempted through heat integration by considering all streams needing to gain or lose energy in the process and also outside the process if in close proximity to other facilities (e.g., pinch analysis, see chapter 22). Similarly, pollution can be prevented by considering mass integration, taking waste streams from one process and using them as raw materials for another. In certain instances, products can be induced to self-separate by adjusting conditions to take advantage of chemical and physical properties of the chemicals. As another example, in mechanical systems, reversible fasteners can be used to encourage the disassembly of manufactured parts at the end of life.

Principle 4: Products, processes, and systems should be designed to maximize mass, energy, space, and time efficiency. Energy and mass efficiency were dealt with in Principle 3, so this discussion will focus on space and time efficiency. Space and time are interrelated in many engineered systems, but most obviously in the transport of raw material and products. Reducing the distance between points of use for materials can save time and reduce pollution. Close proximity can facilitate exchanges of waste heat and materials in highly integrated production systems. Colocation of manufacturing and recycle facilities can also lead to efficiency gains in many production systems. Industrial parks near residential areas can lead to sharing of excess heat with communities. However, these proximity opportunities that take advantage of space and time factors also must consider safety concerns due to potential exposure to emissions and industrial accidents. The excess time that a product sits in inventory can run up against storage stability limits and could lead to excess waste generation.

Principle 5: Products, processes, and systems should be "output pulled" rather than "input pushed" through the use of energy and materials. It is well known that some chemical reactions can be "pulled" to completion by removing certain co-products from the reaction mixture. This chemical phenomenon, which is termed Le Chatelier's principle, can be applied to engineering design across scales of production. "Just-in-time" manufacturing is an example of this principle where only the necessary units are produced in the necessary quantities at the necessary time by bringing production rates exactly in line with demand. Planning manufacturing systems for final output eliminates the wastes associated with over-production, waiting time, processing, inventory, and resource inputs.

Principle 6: Embedded entropy and complexity must be viewed as an investment when making design choices on recycle, reuse, or beneficial disposition. Entropy and complexity are related concepts when considering engineered systems. Products having a high degree of order and structure in a product, the greater is the amount of energy invested to create such structure and complexity. When considering end-of-life options for products, the degree of complexity and structure should point the way to proper reuse, recycle and remanufacturing options. Highly complex parts should be reused, if at all possible, in order to avoid the investment required to create a replacement part from virgin (newly extracted from nature) resources.

Principle 7: Targeted durability, not immortality, should be a design goal. Many products last much longer than the expected commercial life. There can be multiple impacts from this extended durability. For example, buildings with inefficient energy systems, designed when energy was relatively inexpensive, may become inoperable in times of expensive energy. At a different scale, the challenge in the design of molecules and of manufactured parts is to create products that are durable yet do not persist indefinitely in the environment. Durability means that the products last for the intended commercial life and are thereafter readily reconfigured or degraded at the end of life into harmless substances that assimilate easily into natural cycles.

Principle 8: Design for unnecessary capacity or capability (e.g., "one size fits all") solutions should be considered a design flaw. Most products are overdesigned to cover a wide application range and settings. Automobiles must be designed to function not only in warm temperatures but also in extremes of cold. However, there are instances where design for "one size fits all" does not make the most sense and is potentially wasteful. For example, a jacket could be designed for the coldest possible climate, but this garment would not be much use for a stroll along the beach on a breezy evening in Florida in the winter. Likewise, lighting of a large classroom, office space, or a room at home with a single light switch would not make as much sense if only one person were in the room at a given time. In such a case, district lighting would save on energy when only one or a few occupants are present in a large space.

Principle 9: Material diversity in multicomponent products should be minimized to promote disassembly and value retention. This design principle has elements in common with Principle 6. Increasing material diversity in products has the effect of making recycling more difficult and expensive because the number of recycling options and their complexity increase as material diversity increases. Different kinds of materials and the use of additives have a strong influence on recycling methods and costs.

Principle 10: Design of products, processes, and systems must include integration and interconnectivity with available energy and materials flows. This design principle states that products, processes, and entire engineered systems should be designed to use the existing infrastructure of energy and material flows. Integration with existing infrastructure can occur at the scale of a unit operation, production line, manufacturing facility, or industrial park. Taking advantage of existing energy and material flows will minimize the need to generate energy and/or acquire and process raw materials. Applications of this principle include the recovery and use of heat from exothermic chemical reactions, the cogeneration of heat and power, and the recovery of electrical energy by regenerative braking in hybrid vehicles.

Principle 11: Products, processes and systems should be designed for performance in a commercial "afterlife." Designing components for a second, third, or even longer lifer is an important strategy in product design. When components are recovered and reused in next-generation products, the environmental impacts of raw material acquisition from virgin resources and conversion are eliminated, and the overall life-cycle environmental impacts are reduced. This strategy is especially important for products that become obsolete prior to component failure, such as cell phones and other electronics devices. Important examples of this principle also include the recovery and recycle of spent copy toner cartridges, the renovation of industrial buildings for housing, and reuse of beverage containers as practiced in Germany, where bottles are more substantially in their construction to allow for collection, washing, sterilization, refilling and relabelling.

Principle 12: Material and energy inputs should be renewable rather than depleting. The use of non-renewable raw materials from nature in the design of engineered systems moves the Earth system incrementally toward depletion of finite resources and is therefore, unsustainable by definition. All renewable resources derive their usefulness and energy from the sun, and as a result these system inputs can be sustainable, if used at a level consistent with their rate of removal, for the foreseeable future. Biomass is an important form of renewable resources in that it can serve as not only an energy source but also as feedstock for design of material products. One important form of a biomass product is liquid transportation fuels. Biofuels are renewable on relatively short timescales, and the cycling of biofuel carbon between the atmosphere, biomass/biofuel, and back to the atmosphere again is readily integrated into natural cycles in a way that might not cause accumulation of CO_2 in the atmosphere. An offshoot of this principle is the importance of design for products and systems that integrate well with natural cycles of elements across the life cycle, from raw material acquisition to end-of-life processes. A detailed description in the use of biomass/biofuel as a renewable resource is provided in the text.

These engineering design principles, in addition to the others mentioned in the beginning of this section, establish a framework for designing more sustainable products and processes. At first, changes in engineering design are likely to be improvements to inherently unsustainable products and system, but over time, it is hoped that these design principles will move industry and consumers toward inherently sustainable products and production systems. However, there will be tensions in applying these principles. What if making a process inherently safer requires more energy? What if minimizing water use requires more energy?

The Sandestin Green Engineering Principles were developed based on a list of principles collated from reviews of sustainability or green related principles and statements including the Hannover Principles, CERES [20]. However, the goal of sustainable engineering design is to create products that meet the needs of today in an equitable fashion while maintaining healthy ecosystems and without compromising the ability of future generations to meet their resource needs.

27.5 Alternative Fuels (Biofuels)

The cost of crude oil has reduced to $25.0 per barrel due to the world events caused by the Coronavirus and its devastation to the world economy in 2020, with countries worldwide having to suspend production due to the health of their people. However, with time, the cost of the crude will eventually increase from its present state to a profitable one for both producers and refiners. There had been an interest in recent years for the production of biofuels in the form of gasohol and biodiesel for improving the quality of the environment as well as to provide environmentally friendly fuels. Renewable fuels are expanding worldwide due to increasing petroleum processing regulations and commitment to reduce greenhouse gases.

A biofuel is a fuel that is produced through processes from biomass rather than a fuel produced by the very slow geological processes involved in the formation of fossil fuels, such as oil. Since biomass can be used as a fuel directly (e.g., wood logs), the terms biomass and biofuel can be misconstrued. Biomass denotes the biological raw material the fuel is made of, or some form of thermally/chemically altered solid end product, like torrefied pellets or briquettes. Biofuel is reserved for liquid or gaseous fuels used for transportation. The U.S. Energy Information Administration (EIA) states that if the biomass used in the production of biofuel can regrow quickly, the fuel is generally considered to be a form of renewable energy.

Biofuels are produced from plants (i.e., energy crops), or from agricultural, commercial, domestic and/or industrial wastes (i.e., if the waste has a biological origin). Renewable biofuels generally involve contemporary carbon fixation, such as those that occur in plants or microalgae through the process of photosynthesis. There are suggestions that biofuel can be carbon-neutral because all biomass crops sequester carbon to a certain extent. This is because all crops move CO_2 from above-ground circulation to below-ground storage in the roots and the surrounding soil. McCalmont *et al.* [21] found below-ground accumulation ranging from 0.42 to 3.8 tonnes per hectare per year of soils below Miscanthus x giganteus energy crops with a mean accumulation rate of 1.84 tonne (0.74 tonnes per acre per year) or 20% of total harvested carbon per year.

However, the proposal that biofuel is carbon-neutral has been superseded by the more subtle proposal that for a particular biofuel project to be carbon-neutral, the total carbon sequestered by the energy crop's root system must compensate for all the above-ground emissions (related to this particular biofuel project). This includes any emissions caused by direct or indirect land-use change. Many first-generation biofuel projects are not carbon neutral given these demands. Some have had higher total greenhouse-gas emissions than some fossil-based alternatives [22, 23]. Some are carbon neutral or even negative, though especially perennial crops. The amount of carbon sequestered, and the amount of GHG emitted will determine if the total GHG life-cycle cost of a biofuel project is positive, neutral or negative. A carbon negative life cycle is possible if the total below ground accumulation more than compensates for the total life-cycle GHG emissions above ground. To achieve carbon neutrality, yields should be high and emissions should be low.

Biofuels or biodiesel is a renewable fuel for diesel engines made from soybean oil, natural oils and yellow greases. Biodiesel is mono-alkyl esters of long chain fatty acids derived from vegetable oils and animal fats, designated B100, and must meet the specifications of ASTM D 6751 like petroleum diesel. Bio-fuels have passed three generations; the first generation of fuels referred to as biofuels that are made from sugar, starch, vegetable oil, or animal fats using conventional technology [24]. This generation of fuels includes biodiesel, methanol, ethanol, butanol, alcohols, biogas and solid biofuels. The first-generation biofuels are costly as not only that of raising food costs by forcing limited land availability for food crops and so on. In the US, the main food crop is corn with the diversion for alcohol. The food cost is increased by more than 100%, which was reflective elsewhere.

The second-generation fuels are made from lignocellulosic biomass using the advanced technical process. This generation uses biomass to liquid technology and the ones undergoing development are biohydrogen, bio-DME, biomethanol and high-temperature upgrading diesel, Fischer-Tropsch (FT) diesel, and mixed alcohols (mixture of mostly ethanol, propanol and butanol with some pentanol, hexanol, heptanol, and octanol). In the second generation, biofuels are based on gasification of biomass waste. Cellulosic biomass can be separated from lignin and subjected to enzymatic conversion. Alternatively, the biomass can be gasified to carbon monoxide (CO) and hydrogen (H_2), which can then react to form a wide variety of hydrocarbons through FT process (biomass to liquids).

Biofuels value chain is a multicomponent industry as numerous organic feeds can be converted into various fuels and blending streams with established processing technologies. Figure 27.13 illustrates a typical biofuel value chain. Starting from the feedstock, it progresses to various processes via biotransformation or thermochemical to the principal biofuel products. Biomass gasification is receiving increasing importance as the most potential feedstock. Biomass transformation offers numerous processing routes for various feedstocks into desired transport fuels and blending components. The system has flexibility from varying feedstocks and processing technologies. Presently, the US and Brazil are the two largest biofuel producers in the world.

Torrefaction of biomass, e.g., wood or grain is a mild form of pyrolysis at temperatures typically between 392 and 608°F (200 and 320°C). Torrefaction changes biomass properties to provide a better fuel quality for combustion and gasification applications. Torrefaction produces a relatively dry product, which reduces or eliminates its potential for organic decomposition. Torrefaction biomass can be used as an energy carrier or as a feedstock used in the production of bio-based fuels and chemicals [96].

There is renewed interest in the use of ethanol gasoline blend, petroleum and biodiesel blend and due to the availability of alcohol from sugar producing countries. Although the US has been using ethanol as an oxygenate and for blending in their gasoline, the use of ethanol has received momentum in recent years. Furthermore, new biofuels are emerging in the market, especially biodiesel which includes hydrogenated vegetable oil and biomass-to-liquid diesel produced from cellulosic biomass [25–27]. The driving force for biofuel development is the rising greenhouse-gas emissions from conventional fuels and the need for reduction in dependence on imported crude oil.

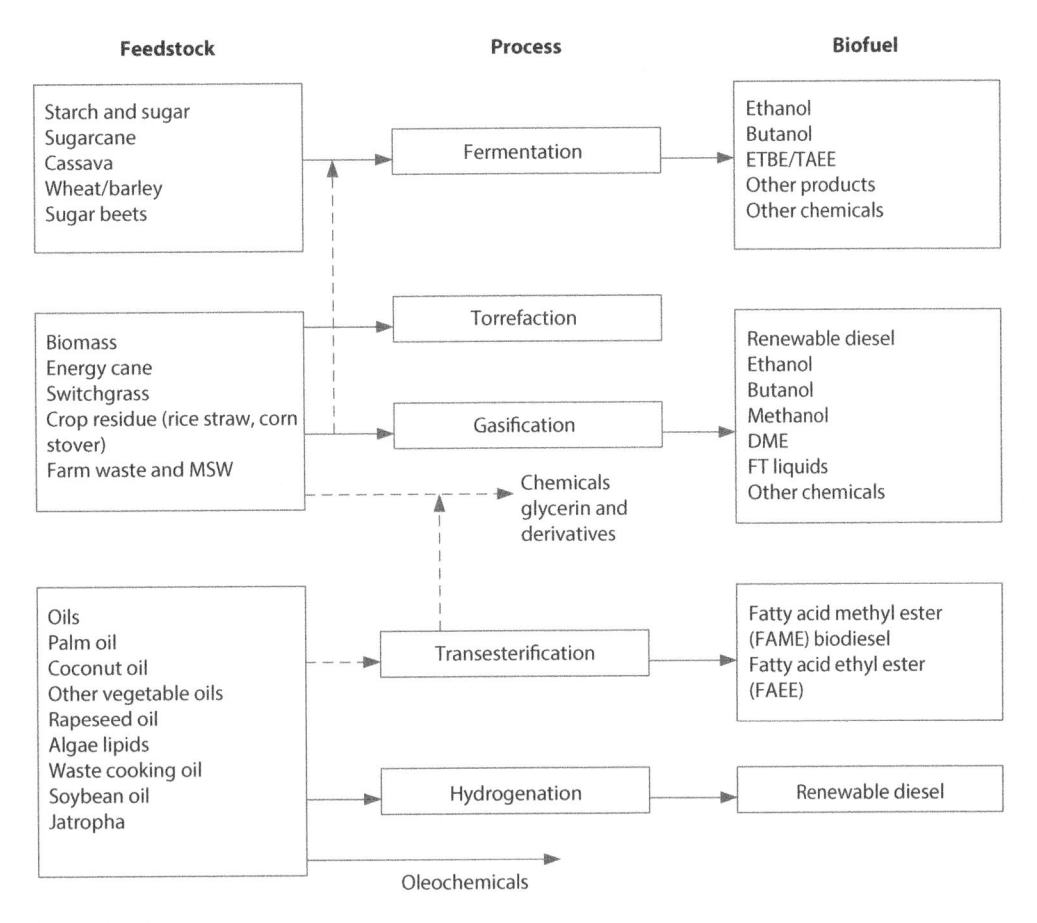

Feedstock	Process	Biofuel
Starch and sugar, Sugarcane, Cassava, Wheat/barley, Sugar beets	Fermentation	Ethanol, Butanol, ETBE/TAEE, Other products, Other chemicals
Biomass, Energy cane, Switchgrass, Crop residue (rice straw, corn stover), Farm waste and MSW	Torrefaction, Gasification	Renewable diesel, Ethanol, Butanol, Methanol, DME, FT liquids, Other chemicals
	Chemicals glycerin and derivatives	
Oils, Palm oil, Coconut oil, Other vegetable oils, Rapeseed oil, Algae lipids, Waste cooking oil, Soybean oil, Jatropha	Transesterification, Hydrogenation	Fatty acid methyl ester (FAME) biodiesel, Fatty acid ethyl ester (FAEE), Renewable diesel
	Oleochemicals	

Figure 27.13 Biofuels value chain.

27.6 Process Intensification (PI) in Biodiesel

A wide range of process intensification techniques can be used in bioprocessing, ranging from high gravity fields, electric fields and ultrasound to membrane processes and some reactors. For example, enzymatic micro-reactors could be employed as sensors for energy production, chemical synthesis and environmental clean-up. The inherent advantage of enzymes in reactors is that they operate at ambient temperatures and pressures. This results in less complex engineering than that necessary for most chemical micro reactors.

An oscillatory flow reactor (OFR) is a type of continuous flow reactor consisting of tubes containing equally spaced orifice plate baffles that impose an oscillatory motion upon the net flow of the process fluid. This creates flow patterns leading to efficient heat and mass transfer while maintaining an overall plug flow regime [88]. The use of an OFR device allows for a given reaction volume, longer residence times as the mixing is not dependent on the net flow and the reactor length-diameter ratio can be reduced, as this forms an important factor for the commercialization of the process at industrial scale. This configuration achieves easier process control, and reduction in capital and pumping costs. The reactor provided ~ 99% oil conversion with negligible amounts of di – or triglycerides in less than 30 mins. at 50°C, showing the efficiency of the reaction system. OFR allows reactor design with a shorter length to diameter ratio. Other reactor types that provide continuous production of biodiesel include static mixers, micro-channel reactors, cavitational reactor, rotating spinning tube reactors and microwave reactors.

A.P. Harvey [88] has employed oscillatory baffle reactors to intensify the production of biodiesel fuels using rapeseed oil as the feedstock, as an attraction of a renewable energy source, thus reducing CO_2 emissions and pollution. The range of PI projects in this area includes a portable plant, solid catalysts (which allow a reduced number of process steps compared to liquid catalysts), the development of a reactive extraction process direct from the oilseed, examination of cold flow properties and the production of biodiesel from algae.

Biodiesel is referred to as methyl or ethyl esters produced by trans-esterifying the triglycerides that constitute most natural oils and fats. The molecules produced are typically of chain length 16-22, and are suitable for use in modern diesel engines. The raw oils and fats cannot be used in modern diesel engine as their viscosity is too high. But biodiesel can be used directly as a replacement for diesel as its viscosity is a factor of at least 6 lower than the feed vegetable oil.

The conversion of triglycerides to methyl esters are as follows:

$$\text{Triglyceride} + \text{Methanol} \xleftrightarrow{OH^-} \text{Diglyceride} + \text{Methyl Ester}$$

$$\text{Diglyceride} + \text{Methanol} \xleftrightarrow{OH^-} \text{Monoglyceride} + \text{Methyl Ester}$$

$$\text{Monoglyceride} + \text{Methanol} \xleftrightarrow{OH^-} \text{Glycerol} + \text{Methyl Ester}$$

$$\text{Overall: Triglyceride} + 3\text{Methanol} \xleftrightarrow{OH^-} \text{Diglyceride} + 3\text{Methyl Ester}$$

Biodiesel is the product of renewable sources as it is CO_2 neutral, thus does not add to global warming. Its use rather than crude oil-derived diesel is not only good for the environment in terms of its effect on CO_2, but also directly on-air quality, as biodiesel is a cleaner fuel than fossil fuel derived diesel, thereby generating lower emissions of hydrocarbons, particulates and carbon monoxides. Its lubricity ensures that the wear on engines is reduced.

The oscillatory baffle reactor/oscillatory flow reactor (OBR/OFR) types are novel applications where a long residence time batch process is converted to a continuous one. For a biodiesel, Harvey indicates that a conversion could be carried out in 10 minutes, compared to 1 – 6 hours in continuous industrial processes. Figure 27.14 shows the flow diagram of the process and Figure 27.15 shows components of the OFR. The ultimate aim is to make the plant portable and could be used worldwide to farmers to produce their own fuel locally.

A continuous oscillatory baffled reactor (COBR) is a specially designed reactor to achieve plug flow under laminar flow conditions. The technology incorporates annular baffles to a tubular reactor framework to create eddies when liquid is pushed up through the tube. Also, when liquid is on a downstroke through the tube, eddies are created on the other side of the baffles. Eddy generation on both sides of the baffles creates very effective mixing while still maintaining plug flow. A higher yield of the product can be obtained with greater control and reduced waste. Each baffled cell acts as a continuous flow stirred tank (CSTR) and because a secondary pump is creating a net laminar flow, much longer residence times can be achieved relative to turbulent flow systems. By changing variable parameters such as baffle spacing or thickness, COBRs can operate with much better mixing control, as it has been found that a spacing of 1.5 times tube diameter size is the most effective mixing condition, and vortex deformation increases with increase in baffle thickness greater than 3 mm.

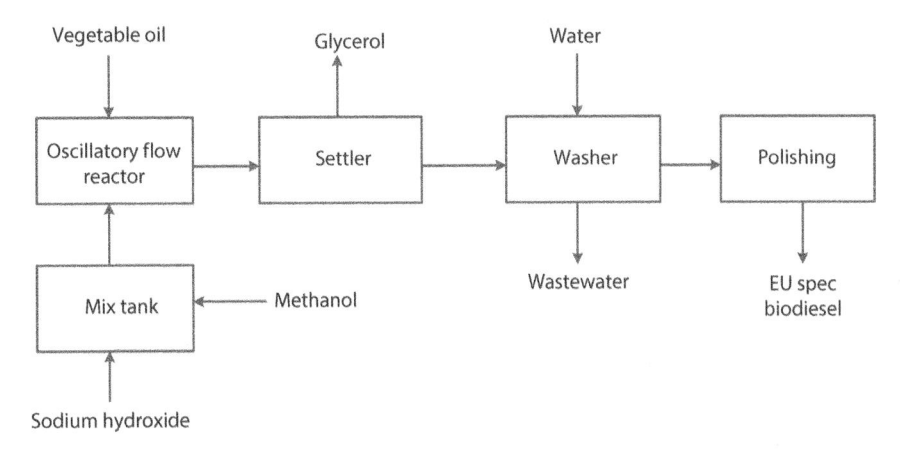

Figure 27.14 The flowsheet of intensified biodiesel plant [88].

Figure 27.15 Harvey, A.P. beside an oscillatory flow reactor for the biodiesel plant [88].

The low shear rate and enhanced mass transfer provided by the COBR ensure that it is an ideal reactor for various biological processes. It has been shown that COBR provides an evenly distributed five-fold reduction in shear rate to conventional tubular reactors, which is especially essential for biological process given that high shear rate can damage microorganisms.

For the case of mass transfer, COBR fluid mechanics allow for an increase in oxygen gas residence time. Furthermore, the vortexes created in the COBRs cause a gas bubble break-up and thus result in an increase in surface area for gas transfer. COBRs are especially advantageous for aerobic biological processes and a very promising aspect with its ability to scale-up while still retaining the advantages in shear rate and mass transfer.

The prospects for COBR applications in fields like bioprocessing are very promising, improvements are being implemented for global use. There is additional complexity in the COBR design relative to other bioreactors, which can introduce complications in operation. Furthermore, for bioprocessing it is possible that fouling of baffles and internal surfaces become prevalent. Figure 27.16 shows a standard design of a continuous oscillatory baffled reactor including a pump and equally spaced baffles.

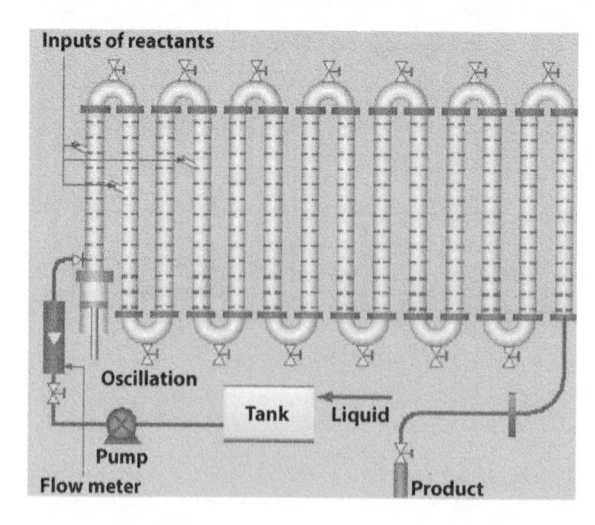

Figure 27.16 Standard design of a continuous oscillatory baffle reactor (COBR) including pump and equally spaced baffles. (Source: https://en.wikipedia.org/wiki/Oscillatory_baffled_reactor).

27.7 Biofuel from Green Diesel

Algae can be used to produce "green diesel" referred to as renewable diesel, hydrotreating vegetable oil or hydrogen-derived renewable diesel through a hydrotreating refinery process that breaks molecules down into shorter hydrocarbon chains used in diesel engines. It has the same chemical properties as petroleum-based diesel meaning that it does not require new engines, pipelines or infrastructure to distribute and use. However, it has yet to be produced at a cost that is competitive with petroleum. While hydrotreating is the most common pathway to produce fuel-like hydrocarbons via decarboxylation*/decarbonylation**, there is an alternative process that offers a number of important advantages over hydrotreating, where the studies of Crocker *et al.* [28] and Lercher *et al.* [29] are highly noteworthy. Oil refining study for catalytic conversion of renewable fuels by decarboxylation is also being conducted. As the oxygen is present in the crude oil at rather low levels (~0.5%), deoxygenation in petroleum refining is not noticeable as no catalysts are specifically formulated for oxygenates hydrotreating. Therefore, one of the critical technical challenges to make the hydrodeoxygenation of the algae oil process economically feasible is related to the research and development of an effective catalyst [30, 31].

Furthermore, green diesel is produced through hydrocracking biological oil feedstocks such as vegetable oils and animal fats—the cracking of larger molecules into smaller hydrocarbon chains used in diesel engines. It is also referred to as renewable diesel or hydrotreated vegetable oil. Green diesel has the same chemical properties as petroleum-based diesel and is developed by Nestle Oil, Valero, Dynamic Fuels and Honeywell UOP.

Analysis

ASTM D6751-07 is the standard specification for bio-diesel fuel blend stock with middle distillate fuels in the US. Blend is 0–40%, and is determined by using FT-IR spectroscopy. Methanol traces cannot influence vapor pressure but can influence flash point (10% can reduce FP from 120–60°C).

Processing of Biodiesel

Biodiesel is renewable, eco-friendly, clean burning fuel that is produced from virgin or used vegetable oils (both edible and non-edible) and animal fats. Countries worldwide are using biodiesel blended with petroleum diesel in different compositions, and the national standard for this is shown in Table 27.6. Vegetable oil-based diesel can offer better integration within crude oil refineries for fuel blending. Biodiesel consists of the monoalkyl esters formed by

---

*decarboxylation is a chemical reaction that removes a carboxyl group and releases CO_2. Usually, decarboxylation refers to a reaction of carboxylic acids, removing a carbon atom from a carbon chain.

**decarbonylation is a type of organic reaction that involves the loss of CO. It is often an undesirable reaction, since it represents a degradation. It describes a substitution process, where a CO ligand is replaced by another ligand.

Table 27.6 National Standards for Biodiesel [22].

	Austria	Czech Republic	France	Germany	Italy	Sweden	USA
Standard/specification	ONC1191 6507	CSN 65 Official	Journal 51605	DIN V 10635	UNI 155436	SS D6751	ASTM
Date	July 1997	Sept. 1998	Sept. 1997	Sept. 1997	April 1997	Nov. 1996	2001
Application	Fame	Rme	Vome	Fame	Vome	Vome	Femae
Density @ 15°C, gm/cm^3	0.85 – 0.89	0.87 – 0.89	0.87 – 0.90	0.875 – 0.90	0.86 – 0.90	0.87- 0.90	-
Visc. @ 40°C, mm^2/s	3.6 – 5.0	3.5 – 5.0	3.5 – 5.0	3.5 – 5.0	3.5 – 5.0	3.5 – 5.0	1.9 – 6.0
Distillation (95%, °C)	-	-	<360	-	<360	-	360
Flash point, °C	>100	>110	>100	>110	>100	>100	>130
CFPP (Cold filter plug pt, (°C)	0/ - 15	-5	-	0/-10/-20	-	-5	-
Pour point, °C	-	-	<-10	-	<0/<-15	-	-
Sulfur (% wt)	<0.02	<0.02	-	<0.01	<0.01	<0.01	<0.05
Water (mg/kg)	-	<500	<200	<300	<700	<300	500
Cetane No.	>49	>48	>49	>49	-	>48	>47
Neutral No. mg KOH/g	<0.8	<0.5	<0.5	<0.5	<0.5	<0.6	-
Methanol (% wt)	<0.2	-	<0.1	<0.3	<0.2	<0.2	-
Ester content (% wt)	-	-	>96.5	-	>98	>98	-

a catalyzed reaction of triglycerides in the oil or fat with a simple monohydric alcohol [32]. Ray [33] provides an option for renewable diesel as:

Feedstock:	Jatropha, rapeseed, soybean, tallow, algae oils, palm oils
Technologies:	Biodiesel (FAME), ecofining™ process, stand-alone hydroprocessing/isomerization, co-processed hydroprocessing.

Fats and oils are triglycerols and are separated at room temperature. In triglycerols of vegetable origin, fatty acids esterified onto position 2 are significantly different from those esterified onto positions 1 and 3, which exhibit little overall difference in substitution pattern, whereas in products of animal origin, random substitution predominates. Animal and plant fats and oils are composed of triglycerides, which are esters containing three fatty acids, alcohol and glycerol. In the trans esterification process, the alcohol is first de-protonated with a base to make it a stronger nucleophile. Commonly, methanol or ethanol is used for esterification. The slow reaction is between triglyceride and the alcohol. Heat is supplied as the catalyst (acid and/or base) that is used to enhance the rate of reaction. Common catalysts in transesterification* are sodium hydroxide (NaOH), potassium hydroxide (KOH) and sodium methoxide (CH_3ONa).

Almost all biodiesel is produced from virgin vegetable oils using the base-catalyzed technique. It is the most economical process for treating virgin vegetable oils requiring only low temperature and pressure and producing over 98% conversion yield (provided the starting oil is low in moisture and free fatty acids). However, biodiesel produced from other sources or by other methods may require acid catalysis; the process is, however, much slower. Since the base catalyzed transesterification process is the predominant method for commercial production, the process is described as follows:

Triglycerides [1] are reacted with an alcohol such as ethanol [2] to give ethyl esters of fatty acids [3] and glycerol [4]. The reaction sequence is

R^1, R^2, R^3 are radicals

or

| Vegetable oil (Triglyceride) | Alcohol | Catalyst | Glycerol | Alkyl ester (Biodiesel) |

CH_2COOR'''
$|$
CH_2COOR'' + $3ROH$ $\underset{}{\overset{NaOH}{\rightleftharpoons}}$ CH_2OH $RCOOR'''$
$|$ $|$
CH_2COOR' $CHOH$ + $RCOOR''$
                                                                                        $|$
                                                                                        CH_2OH $RCOOR'$

Sodium hydroxide (NaOH) is the catalyst ~ 3-5% in methanol. The catalyst is added to 60 – 70 liters vegetable oil in a tank reactor. The oil is kept at 45–65°C and the mixture is esterified as shown in Figure 27.17. Table 27.7 shows the properties of biodiesel and green diesel.

An alternative process consists of catalyzed-free method; the one-stage transesterification with super critical methanol (4). The reaction is carried out under supercritical conditions, i.e., at temperatures higher than the critical temperature of methanol. The raw materials are methanol; and triglycerides with some amount of free fatty acids, thereby allowing the production of biodiesel from cheap feedstocks without the aid of alkaline or acid catalysts. This eliminates the need for neutralization steps downstream of the reactor. Minimizing the heat of consumption and pumping power that is usually high requires a one-reactor configuration of all supercritical processes, two medium pressure successive reactors with intermediate glycerol removal and a heat recovery scheme composed of heat exchangers and adiabatic flash drums. Glycerol is retained in adsorption beds, desorbed in a swing step and recycled to the first reactor. Design parameters are obtained from experimental data and estimation. An advantage of this process is that no process water effluents are produced, and the whole system is essentially "dry" and only small water is produced. Glycerol purification is simplified by the absence of a catalyst and low water content. At high

---

*Transesterification or alcoholysis is defined as the process in which nonedible oil is allowed to chemically react with an alcohol. A process of exchanging the organic functional group R" of an ester with the organic group R' of an alcohol. These reactions are often catalyzed by the addition of an acid or base catalyst. The reaction can also be accompanied with the help of other enzymes, particularly lipases. In this reaction, methanol (CH_3OH) and ethanol (C_2H_5OH) are the most commonly used alcohols because of their low cost and availability.

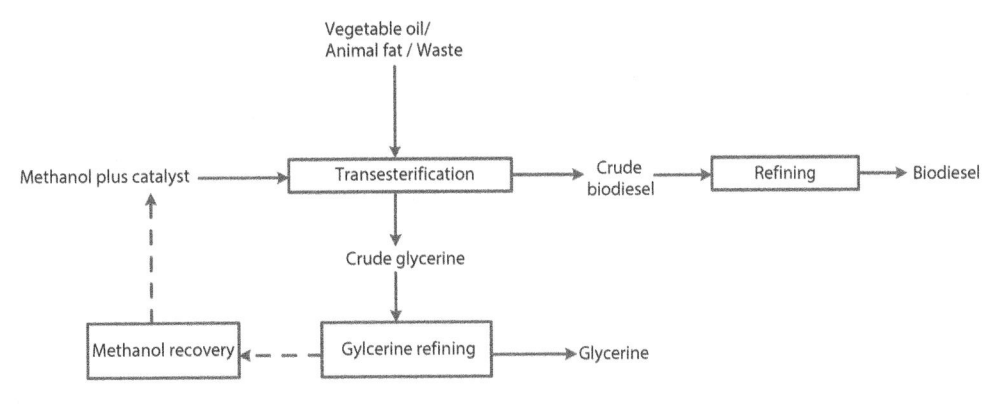

Figure 27.17 Schematic production of biodiesel.

Table 27.7 Comparison of Biodiesel and Green Diesel Properties.

	Mineral ultra-low sulfur diesel	Biodiesel (FAME)	Green diesel
% O_2	0	11	0
Density g/ml	0.84	0.883	0.78
Sulfur content	<10	<10 ppm	<10 ppm
Heating value (lower) MJ/kg	43	38	44
Cloud point, °C	-5	-5	-5 to -30
Distillation (10 – 90% pt.)	200 – 350	340 – 355	265 – 320
Cetane	40	50	80 – 90
Stability	Good	Marginal	Good

temperatures (T_c = 235°C) and high pressures 35–40 mpa, the oil and methanol are in a single phase, and reaction occurs spontaneously and rapidly. The free fatty acids are converted to methyl esters instead of soap, as a wide variety of feedstocks can be processed and the catalyst removal can be eliminated.

Complications can occur when the triglycerides contain free fatty acids, which form soap and water when they react stoichiometrically with sodium or potassium hydroxide. Yellow greases contain between 5–15% free fatty acids, while brown greases contain in excess of 15% free fatty acids. Depending on the level of free fatty acids present, an acid-catalyzed esterification pre-treatment step may be required. Heterogeneous catalysts are needed that can handle the simultaneous conversion of triglycerides as well as free fatty acids to fatty acid methyl ester (FAME).

Fatty acid methyl esters (FAME) are a type of fatty acid ester that are derived by transesterification of fats with methanol. The molecules in biodiesel are primarily FAME that are obtained from vegetable oils by transesterification. FAME is typically produced by an alkali – catalyzed reaction between fats and methanol in the presence of a base such as sodium hydroxide (NaOH), potassium hydroxide (KOH) and sodium methoxide (CH_3ONa). FAMEs have an advantage in their use in biodiesel instead of free fatty acids is that it nullifies any corrosion that free fatty acids would cause to the metals of engines, production facilities and so on. This is because, free fatty acids are mildly acidic and with time can cause cumulative corrosion unlike the esters. FAMEs also have about 12 – 15 units higher cetane number than free fatty acids. The chemical reaction is shown below.

$$\text{H}_2\text{C}-\text{O}-\overset{\text{O}}{\underset{}{\text{C}}}-\text{R} \quad \text{HC}-\text{O}-\overset{\text{O}}{\underset{}{\text{C}}}-\text{R} \quad \text{H}_2\text{C}-\text{O}-\overset{}{\underset{\text{O}}{\text{C}}}-\text{R} \quad + \quad 3\,\text{HO}-\text{CH}_3 \quad \xrightarrow{\text{Cat.}} \quad \text{H}_2\text{C}-\text{O}-\text{H} \quad \text{HC}-\text{O}-\text{H} \quad \text{H}_2\text{C}-\text{O}-\text{H} \quad + \quad 3 \quad \text{R}-\overset{}{\underset{\text{O}}{\text{C}}}-\text{O}-\text{CH}_3$$

Renewable diesel is defined as diesel that meets all specifications of typical petroleum diesel but is produced by hydrotreating non-petroleum materials, such as vegetable oils, animal fats or biomass. Renewable diesel is completely interchangeable with petroleum diesel and is completely compatible on a 100% basis in existing diesel engines.

It is different from biodiesel, since biodiesel is produced from many of the same animal and vegetable oils as renewable diesel, but by a different process called trans-esterification. The process adds oxygen to these oils, and when blended with petroleum diesel improves its emissions characteristics. Furthermore, it has certain properties that make it inconvenient to store and limits its content in diesel fuels to 5-20% due to its incompatibility with existing diesel engines. Because of this limitation, its production has been limited in the U.S.

Renewable diesel produced from vegetable oils (e.g., used cooking oil, distillers corn oil or tallow) is generally more chemically homogeneous than petroleum diesel. Further, renewable diesel has a higher cetane number than petroleum diesel. Cetane number is a measure of how efficiently a diesel engine can generate power with that fuel. As a result, more energy is derived from less fuel, reducing emissions per unit amount of energy. Additionally, renewable diesel has essentially zero sulfur and other impurities found in petroleum diesel.

Renewable diesel is low carbon because the feedstocks used to make the diesel, such as distillers corn oil, tallow and used cooking oil are byproducts from other processes. Thus, renewable diesel produced from these feedstocks has a low-carbon intensity (CI). Carbon intensity (CI) is a measure of lifecycle emissions from extraction or growth, refinement, distribution, storage and combustion, and is referred to as grams of carbon dioxide (CO_2) equivalent per megajoule (MJ) of energy. Renewable diesel made from the above by-product feedstocks can be in the range of 22 gCO_2/MJ-25 CO_2/MJ depending on the specific by-product feedstock [83, 84]. By comparison, petroleum diesel has a CI of 102, while renewable diesel produced from soybean oil has a CI of 53, as The CO_2 emitted from growing soybeans must be included in the CI calculation.

A renewable diesel unit addition could provide required strength for the refinery that has suffered in volatile market as many refineries are investigating spending their CAPEX budgets on renewable diesel and related projects. These projects are in the form of a grassroot renewable facility, the revamp of an existing unit or adjusting an existing plant to co-process both petroleum and renewable fuels. Preston *et al.* [85] have described the essential involvement in a renewable diesel project and its feature.

27.7.1 Specifications of Biodiesel

Biodiesel oil specifications are mostly in agreement with petroleum diesel, as shown in Table 27.8. B100 intended for blending into diesel fuel that is expected to give satisfactory vehicle performance at fuel temperatures at or below 10°F (-12°C) shall comply with a cold soak filterability limit of 200 s max. Biodiesel can be blended and used in many different concentrations, including B100 (pure biodiesel), B20 (20% biodiesel, 80% petroleum diesel), B5 (5% biodiesel and 95% petroleum diesel), and B2 (2% biodiesel, 98% petroleum diesel). B20 is a common biodiesel blend in the US.

Table 27.8 Requirements for Biodiesel (B100) Blend Stock ASTM D6751.

Property		Limits	Units
Calcium and magnesium combined	EN14538	5 max.	ppm
Kinematic viscosity, 40 oD	D93	93.0 min.	°C
Sulfur	D445	1.9 – 6.0	Mm²/s
0.05 max (S500)	% mass		
Copper strip corrosion	D130	0.020 max.	
Cetane number	D613	47 min.	
Cloud point	D2500	Report to customer	°C
Phosphorus content	D4951	0.001 max.	% mass
Oxidation stability	EN14112	3 min.	hours
Cold Soak stability	Annex A1	360 max.	seconds
Alcohol control	**One of the following**	**must be met:**	
(1) Methanol content	EN14110	0.2 max.	vol %
(2) Flash point	D93	130 min.	°C

The following advantages and disadvantages are:

Advantages

- Produced from non-petroleum, renewable resources.
- Less air pollutants (other than nitrogen oxides (NO_x)).
- Biodegradable, non-toxic, and safer to handle.
- Can be used in most diesel engines.

Disadvantages

- Lower fuel economy and power (10% lower for B100, 2% for B20).
- Currently more expensive.
- Use of blends above B5 is not yet approved by many auto manufacturers.
- B100 mostly not suitable for use in low temperatures.

27.7.2 Bioethanol

Alcohol has been globally accepted as an alternative to gasoline and alcohol blend gasoline is being used in many countries in varying proportion ranging from 5–20% as ethanol (C_2H_5OH). In Brazil, C_2H_5OH is sold pure as E-100 fuel, and accounts for about 40% of the fuel consumed by vehicles. Brazil and the US account for ~70% of the world's ethanol production. C_2H_5OH, though, is produced from food-based feedstocks like molasses and corn. However, as the demand for ethanol increases, continuous efforts are being made all over the world to utilize alternative feedstock like biomass, which is sustainable and an inexpensive feedstock. Figure 27.18 shows biomass transformation routes to biofuel and Figure 27.19 shows the manufacture of biofuel.

Ethanol has been used in gasoline for the last two decades, and ethanol fuel has a gasoline gallon equivalency (GGE) value of 1.5 US gallons. This means that 1.5 gallons of ethanol produce the energy of one gallon of gasoline. E10 is a

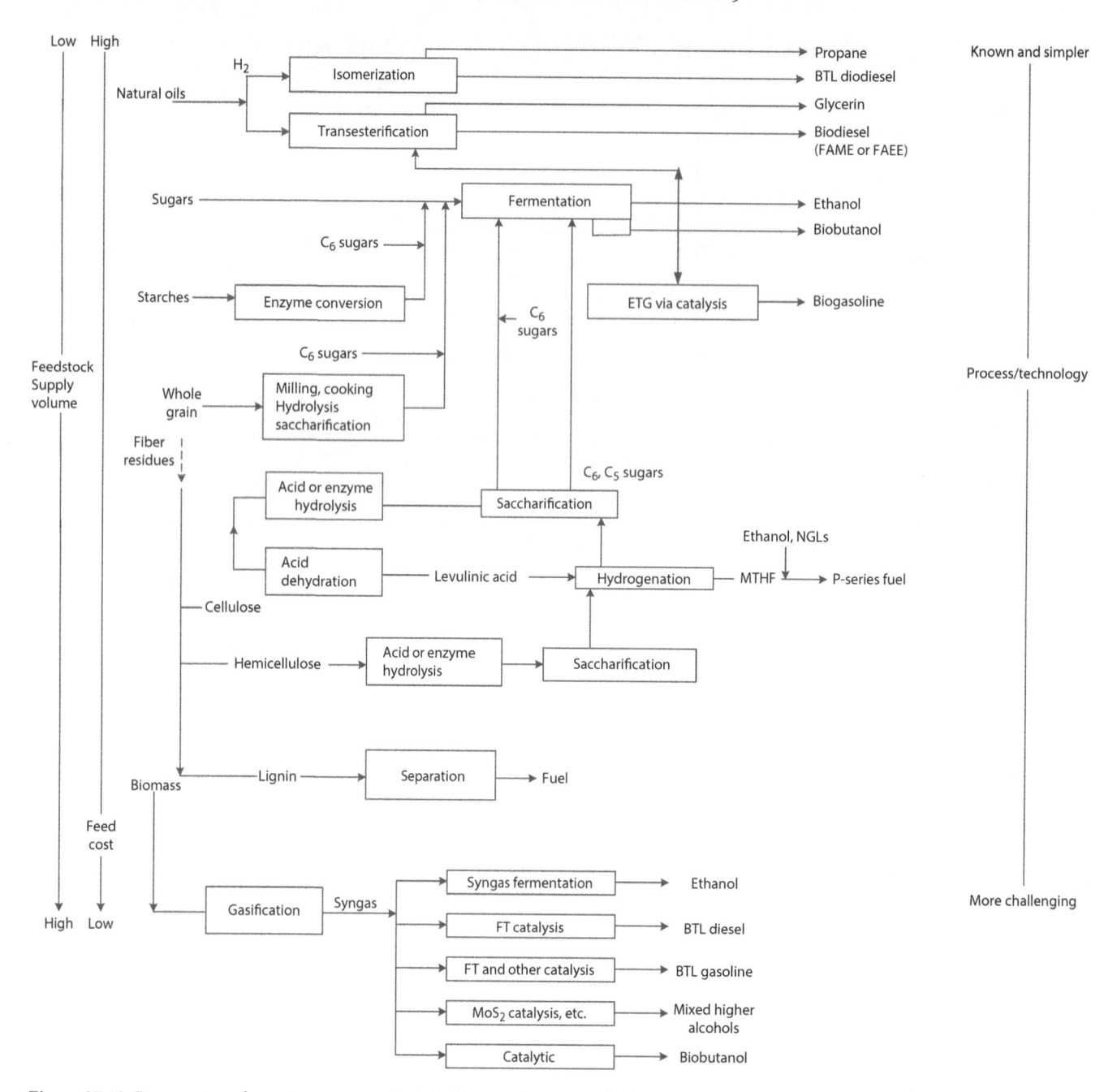

Figure 27.18 Biomass transformation routes to biofuel. (Source: Casone; Biofuels: What is Beyond Ethanol and Biodiesel, Hydroc. Processing, p. 95 – 109, 2005).

low – level blend composed of 10% ethanol and 90% gasoline. Environmental Protection Agency (EPA) has legalized its use in any gasoline - powered vehicle E15 is a low-level blend composed of 15% ethanol and 85% gasoline. E85 is a high-level gasoline blend containing 51–83% ethanol, depending on geography or season and qualifies as an alternative fuel under the EPA act. E85 can be used in flexible fuel vehicles; however, it cannot be legally used in conventional gasoline-powered vehicles. Ethanol can influence vapor pressure of gasoline, and thus the blend must be checked for vapor pressure. It is tested using ASTMD4806-06c like regular gasoline for stability, antioxidants, metal deactivators and biocides (not different from petroleum fuels). The use of E10 was spurred by the Clean-Air Act Amendments of 1990 that mandated the sale of oxygenated fuels in areas with unhealthy levels of carbon monoxide.

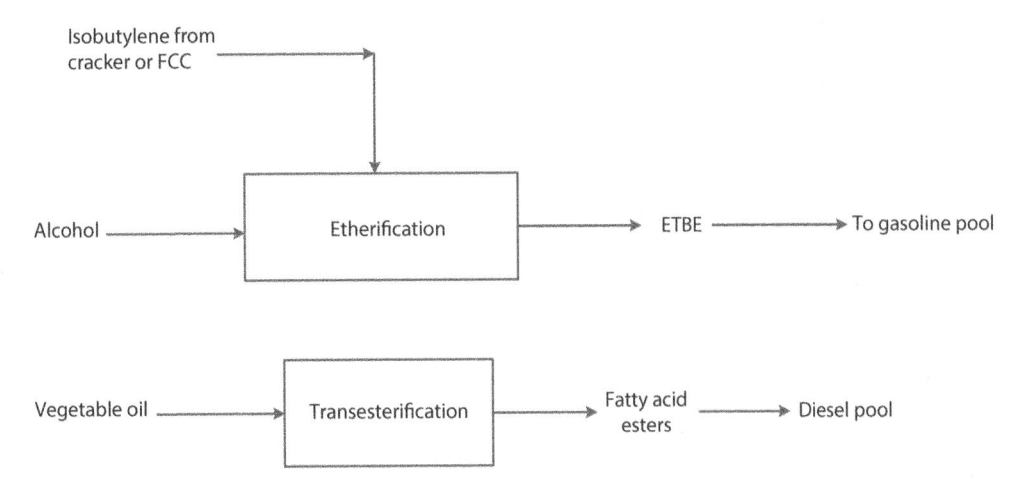

Figure 27.19 Schematic production of biofuels.

The estimated total global production of ethanol is ~11 billion gallons per annum. International Fuel Prices (IFP) estimates that the present production of ethanol can meet ~2.7% of total global fuel demand. Being an oxygen carrier, ethanol helps fuel burn more fully and substantially reduces emissions of CO and CO_2 resulting in a 78% reduction of carbon emission as compared to fossil fuels as coal or crude oil. Ethanol blended fuels are environmentally friendly and produce 13% less greenhouse gases than fossil-fuel [34].

Although ethanol is produced from conventional raw materials from sugarcane, sugar beet, corn, wheat and so on, sugar cane is the largest source for ethanol production. Another route that has been studied is biomass that is available in large quantities. The types of feedstocks for ethanol production are [35]:

Sugars	Molasses, sugarcane, beet sweet sorghum and fruits
Starches	Corn, wheat, rice, potatoes, cassava, sweet potatoes, and so on
Lignocellulosic	Straw, bagasse, other agricultural residues, wood, energy crops
Algae	Ethanol production

Molasses is the major route for the manufacture of alcohol, and Figure 27.20 shows the process flow diagram comprising of the following:

Prefermenter	Growing of yeast
Fermentation	Molasses handling, fermenter feeding and fermentation of molasses
Yeast treatment	Treatment of yeast cell generated during fermentation and recycling system
Fermentation	Fermentation of molasses after addition of yeast and adjusting pH and alcohol recovery, recovery of alcohol from fermentation section, washing of fermenter alcohol
Distillation	Distillation of fermented wash, alcohol separation and separation of spent wash

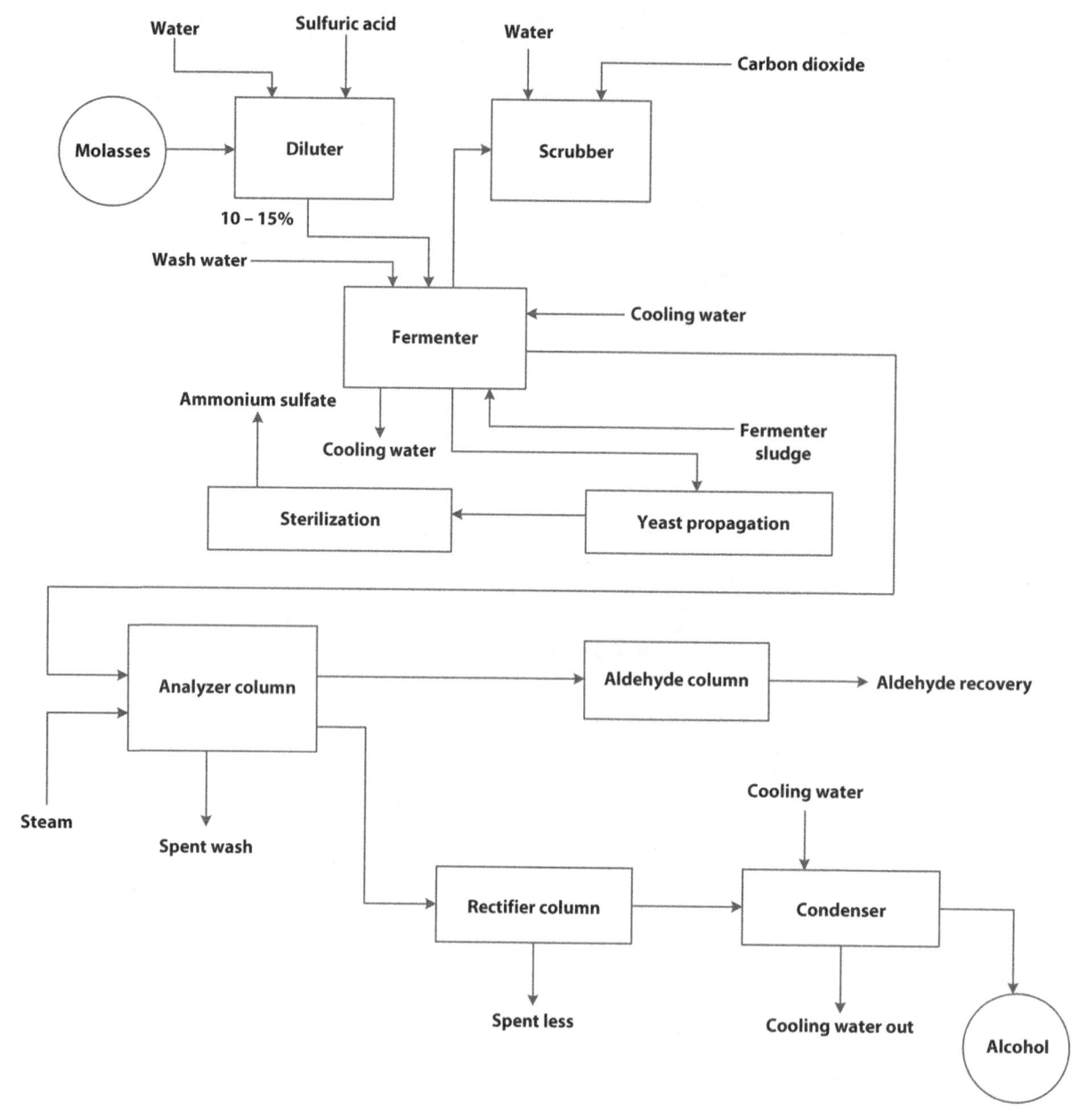

Figure 27.20 Process of manufacturing of alcohol from molasses.

The cost of ethanol production from this feedstock can be reduced by enhancing agricultural methods to increase sugarcane production and deploying energy-efficient ethanol dehydration methods such as the pressure swing adsorption and membrane separation.

Lignocellulosic biomasses refer to organic material such as wood chips, bagasse, straw, corn stalks, grass, and so on, and are increasingly employed as feedstocks for bioethanol. They are composed of cellulose (40–60%), hemicellulose (20–40%) and lignin (10–25%). Table 27.9 shows the potential for ethanol from cellulosic matter. The nature and availability of lignocellulosic feedstocks in different parts of the world depend on the climate, agricultural practices, environmental factors and technological developments. However, the obstacles acquired in efficient conversion of biomass to ethanol are [24]:

- Pretreatment.
- Sachatification of cellulose and hemicellulose matrix.
- Simultaneous fermentation and hexose and pentose sugars.

Table 27.9 Potential for Ethanol from Cellulosic Matter.

Feedstock	Gallons/day
Bagasee	112
Cornstover	113
Rice straw	110
Forest thinning	82
Hardwood dust	101
Mixed paper	116

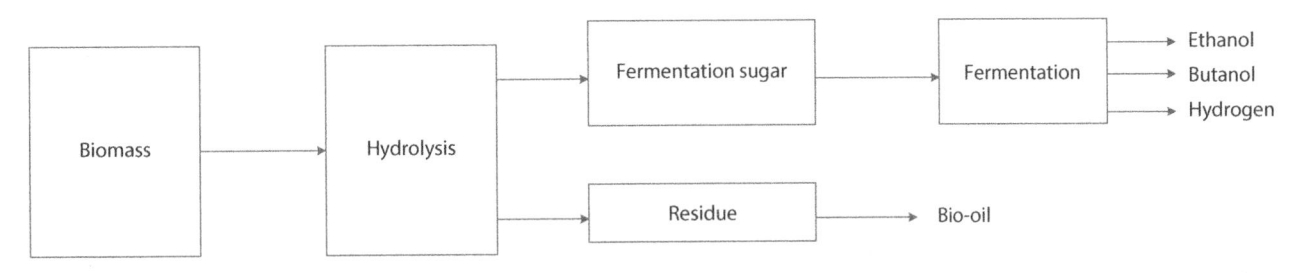

Figure 27.21 Alcohol from biomass.

The most commonly used pretreatment methods are steam explosion and dilute acid prehydrolysis, which are then followed by enzymatic hydrolysis cellulose in the lignocellulosic materials to fermentable reducing sugars, fermentation of sugar into ethanol and downstream processing of ethanol.

Figure 27.21 shows a typical flow diagram for the manufacture of alcohol from biomass.

The advantages of lignocellulosic material are that it is renewable, inexpensive, and locally and domestically available.

27.7.3 Biodiesel Production

Application

The refining hydrocarbon technology – Biodiesel process is optimized to produce biodiesel from palm oil, rape-seed oil, vegetable and animal products that contain fatty acids with even number of carbon atoms (12 – 22). The lack of sulfur in the biodiesel enables complying with many international fuel specifications.

The biodiesel is comparable to petroleum – based diesel. Triglycerides are reacted with methanol (CH_3OH), ethanol (C_2H_5OH) or higher alcohols to yield biodiesel within the acceptable boiling range. Methanol is commonly used for the biodiesel production since it is the most cost-effective of alcohols, and it can provide better economics for the biodiesel producers. Biodiesel is produced by reacting vegetable oils and animal fats (triglycerides) with CH_3OH in the presence of highly alkaline heterogeneous catalyst at moderate pressure and temperature. Pretreatment may be required if the vegetable oil has a high free-fatty acids content to optimize methyl esters yield. If free fatty acids are present in the feed, the first step is esterification of the free fatty acid with CH_3OH. However, if the free-fatty acids concentrations are low, then this step is not required.

The triglycerides and CH_3OH are converted by transesterification reaction to yield methyl esters of the oils and fats, and glycerine is produced as a by-product. The glycerine is separated from the methyl esters (biodiesel) by phase separation via gravity settling. The methyl esters and glycerine are purified to meet the product specifications.

Process

Figure 27.21a shows a simplified process flow diagram where the feed, vegetable oil or animal fats are pumped from storage and mixed with CH_3OH in the required molar ratio (vegetable oil/CH_3OH) at moderate operating pressure. The feed is heated to the reaction temperature and is sent to esterification reactor. Free-fatty acids are pre-treated if the concentration exceeds 3% of the feed. The reactor contains an acid catalyst, which for this reaction can remove 99.9% of free fatty acids from the vegetable oils. *(Note: the pretreatment is only required when the feed contains free fatty acids; otherwise, this step is neglected).*

The effluent from the first reactor (if free fatty acids are present) or the heated feed is sent to the transesterification reactor, where 3 moles of CH_3OH react with the triglycerides to produce 3 moles of methyl ester oil (biodiesel) and one mole of glycerine. The transesterification reactor uses a highly alkaline heterogeneous catalyst and provides essentially 100% conversion. The transesterification reactor effluent is sent to the gravity-separator settler. The biodiesel product is taken from the top of the separator and is washed with water. The washed biodiesel product is taken from the top of the drum. Water washing removes excess CH_3OH from the reaction products, which is recovered by normal distillation. The pure CH_3OH is recycled back to the reactor. The bottoms from the separator/settler are sent to the purification unit to remove impurities and residual CH_3OH is then recycled back to the reactor. Pure glycerine product is sent to storage.

Figure 27.21b is an alternate flow scheme where a spare transesterification reactor is added to remove glycerine from the reactor to sustain the reaction rates. Once the reaction rates are reduced, the reactor is switched off and washed with hot solvent to remove residual glycerine and biodiesel. This extra reactor provides higher reaction rates and onstream capability while enhancing yield and productivity. Glycerine purity can exceed 99.8% after distillation.

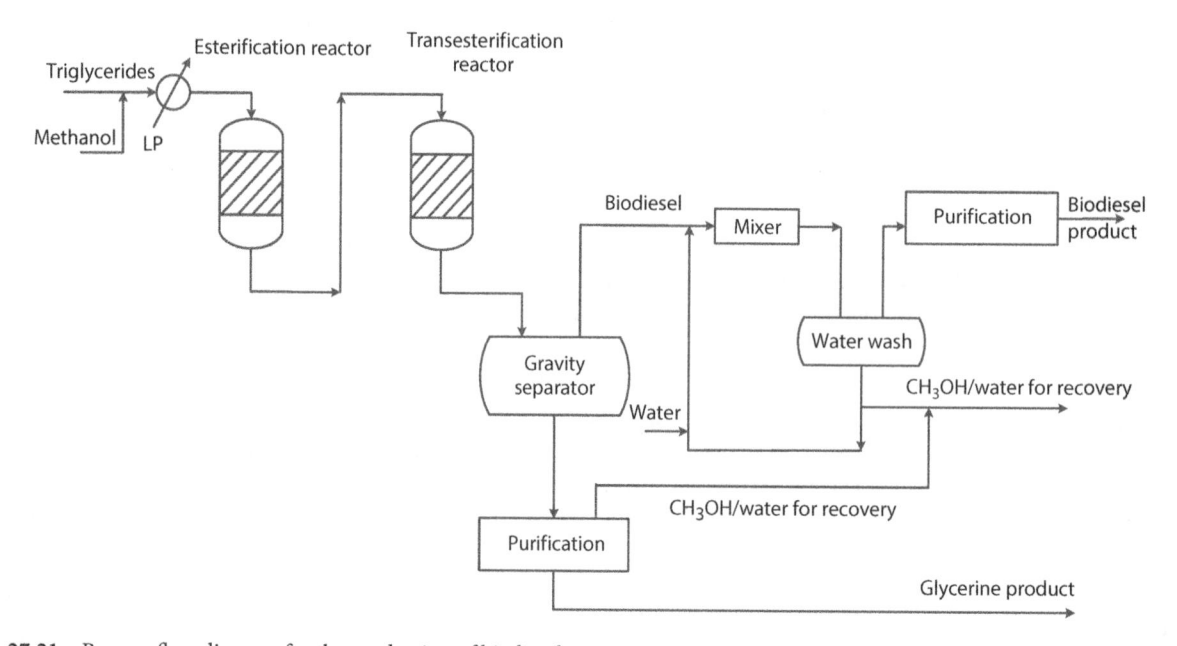

Figure 27.21a Process flow diagram for the production of biodiesel.

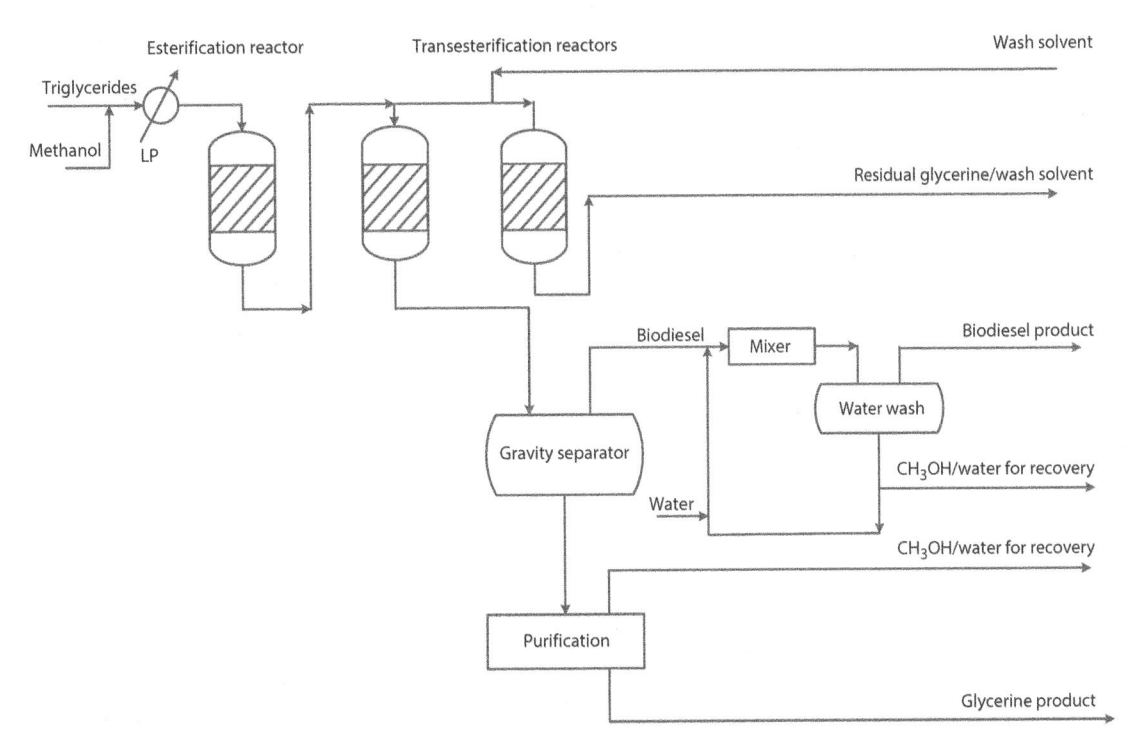

Figure 27.21b An alternative process flow diagram for the production of biodiesel.

Reaction Chemistry

$$\text{Triglycerides} + 3CH_3OH \rightarrow \text{Methylester of the oil (biodiesel)} + \text{Glyerol}$$

Table 27.9a1 shows the comparison between diesel from refining of petroleum crude oil and biodiesel from the above process.

Economics

The normal utilities for continuous biodiesel unit based on heterogeneous catalyst for tonnes/h of biodiesel capacity excluding the glycerine purification utilities are shown below.

CAPEX ISBL plant: USD/ton Biodiesel	235-265
Steam, lb/h	368
Water, cooling gpm	64
Power, kWh	9

Some of the advantages of biodiesel are: low toxicity, derived from renewable sources, biodegradable, reduced engine due to lubrication, reduced emission level of particulates, carbon oxides (CO, CO_2), sulfur oxides (SO_x) and under some conditions, nitrogen oxides (NO_x). The cost of biodiesel as compared to petroleum-based diesel is presently a major barrier to its commercialization as it is 60-90% of the feedstock. Table 27.9a2 shows the properties of biodiesel and green diesel, and Table 27.9a3 provides a comparison between synthetic diesel and Fatty acid methyl

Table 27.9a1 Comparison of the Diesel with Biodiesel Properties.

Fuel property	Diesel	Biodiesel
Fuel standard	ASTM D 975	ASTM P S 121
Fuel composition	C10-C21 HC	C12-C22 FAME
Lower heating value, Btu/gal	131	117
Kinematic viscosity at 40°C	1.3 – 4.1	1.9-6
Specific gravity at 60°F	0.85	0.88
Water, wppm	161	500
Carbon	87	77
Hydrogen	13	12
Oxygen	0	11
Sulfur, wppm	15-500	0
Boiling point, °F	380-650	370-340
Flash point, °F	140-175	210-140

Table 27.9a2 Comparison of Biodiesel with Green Diesel Properties.

	Mineral ultra-low sulfur diesel	Biodiesel, (Fatty acid methyl ester (FAME))	Green diesel
% O_2	0	11	0
Density, g/ml	0.84	0.883	0.78
Sulfur content	< 10 ppm	< 10 ppm	< 10 ppm
Heating value (lower) MJ/kg	43	38	44
Cloud point, °C	-5	-5	-5 to – 30
Distillation (10-90% pt)	200–350	340–355	265 – 320
Cetane	40	50	80 – 90
Stability	Good	Marginal	Good

(Source: A Ray, 2009 [94].)

ester (FAME) manufacturing. The properties for FAME originating from different raw materials are shown in Table 27.9a4 and Table 27.9a5 compares the synthetic diesel product with other diesel fuels.

27.7.4 An Alternative Process of Manufacturing Biodiesel

The methanol esterification of vegetable oil can also be achieved using a homogeneous catalyst operated in either a batch or continuous mode. Alkali catalyzed systems use sodium hydroxide (NaOH), potassium hydroxide (KOH) or

Table 27.9a3 Comparison Between Synthetic Diesel with FAME Manufacturing.

Feature	BTL - diesel	Biodiesel (FAME)
Raw material origin	Flexible to use several vegetable oils and animal fats.	Fixed to specific vegetable oil
Additional feeds	Hydrogen	Methanol
Product quality	Excellent blending properties	Limitations with blending, stability issues, cold property issues
By-product	Handled in refinery	Glycerine
NO_x aspects	Reduces NO_x emissions	Increases NO_x emissions
CO_2 balance (kg) CO_2/kg of fuel	0.5 – 1.5	1.4 – 2
Logistics	Can use refinery logistics; no restrictions for blending	Requires investment for separate logistics
Investment considerations	Large units utilizing integration with oil refinery	Small independent units

(Source: Koskinen *et al.*, February 2006, Hydrocarbon Processing [95].)

Table 27.9a4 Properties of FAME Originating from Different Raw Materials.

Raw material	Melting point (°C)	Cetane number
Rapeseed oil, soybean oil	–10 – 0	55–58
Sunflower oil	–12	52
Olive oil	–6	60
Cotton seed oil	–5	55
Corn oil	–10	53
Coconut oil	–9	70
Palm oil	14	65

(Source: Koskinen *et al.* February 2006, Hydrocarbon Processing [95].)

sodium methoxide (i.e., sodium methylate and sodium methanolate) while acid catalyzed system uses HCl, H_2SO_4 or sulfonic acid. The process steps are as follows:

- Pretreatment of vegetable oil.
- Catalyst synthesis and mixing of vegetable oil with methanol in separate reactors.
- Transesterification.
- Methanol recovery.
- Glycerine separation washing and refining of biodiesel.

Figure 27.21c shows the block diagram of the transesterification process.

Table 27.9a5 Synthetic Diesel Product in Comparison with Other Diesel Fuels.

	Synthetic diesel	GTL diesel*	FAME	EN590/2005 diesel fuel, summer grade
Density at 15°C (kg/m³)	775 – 785	770 – 785	~ 885	~ 835
Viscosity at 40°C (mm²/s)	2.9 – 3.5	3.2 – 4.5	~ 4.5	~ 3.5
Cetane number	84 – 99**	73 – 81	~ 51	~ 53
10% distillation (°C)	260-270	~ 260	~340	~ 200
90% distillation (°C)	295 – 300	325 – 330	~355	~350
Cloud point, °C	-5 – 30	+ 5 – 25	0 – 5	~ - 5
Heating value (lower) MJ/kg	~ 44	~ 43	~ 38	~ 43
Heating value (MJ/L)	~ 34	~ 34	~ 34	~ 36
Polyaromatics content (% wt)	0	0	0	~ 4
Oxygen content (% wt)	0	0	~ 11	0
Sulfur content (mg/kg)	< 10	< 10	< 10	< 10

* gas to liquid (GTL)
** Blending cetane number
(Source: Koskinen *et al.* February 2006, Hydrocarbon Processing [95].)

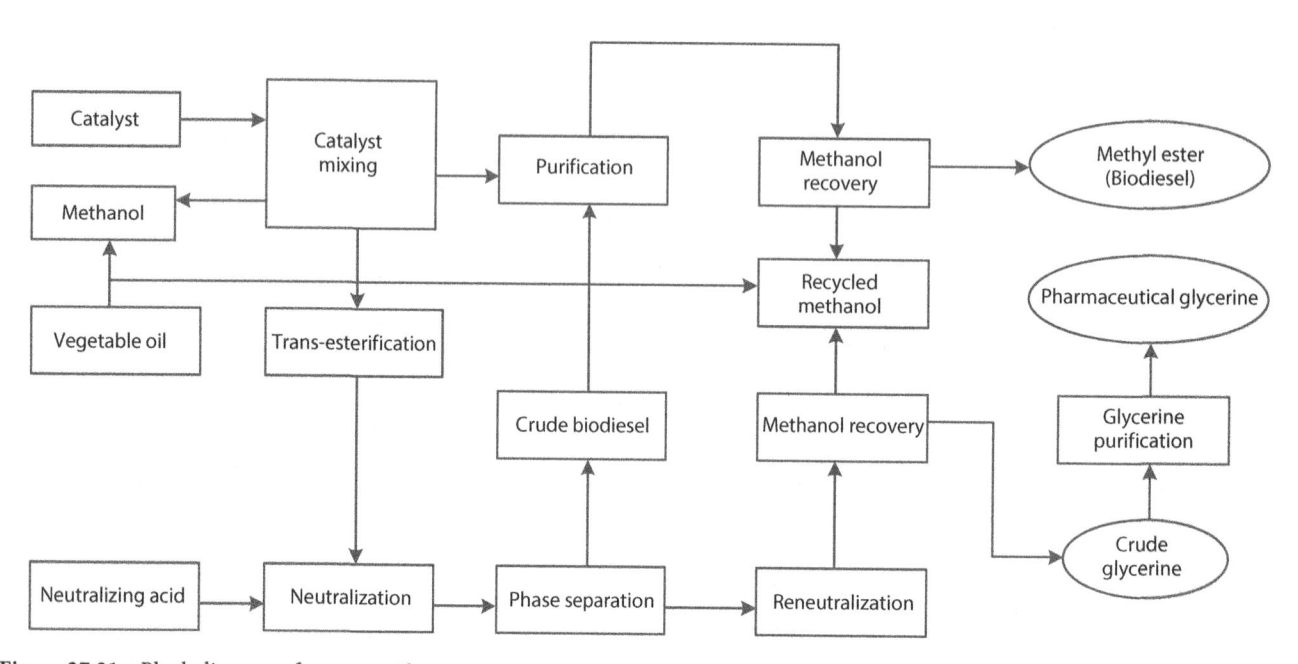

Figure 27.21c Block diagram of transesterification process.

Reaction Chemistry

$$\begin{array}{c} CH_2COOR''' \\ | \\ CH_2COOR'' \\ | \\ CH_2COOR' \end{array} + 3ROH \underset{}{\overset{NaOH}{\rightleftharpoons}} \begin{array}{c} CH_2OH \\ | \\ CHOH \\ | \\ CH_2OH \end{array} + \begin{array}{c} RCOOR''' \\ RCOOR'' \\ RCOOR' \end{array}$$

Vegetable oil (Triglyceride)   Alcohol   Catalyst   Glycerol   Alkyl ester (Biodiesel)

27.7.5 Biofuel from Algae

Algae fuel/algae biofuel or algal oil is an alternative liquid fossil fuel that uses algae as its source of energy-rich oils. These fuels are an alternative to commonly known biofuel sources such as corn and sugarcane. When these are made from seaweed (macroalgae), it is referred to as seaweed fuel or seaweed oil. Several companies and government agencies are enduring to reduce capital and operating costs by making algae fuel production commercially viable. Like fossil fuel, algae fuel releases CO_2 when burnt, but unlike fossil fuel, algae fuel and other biofuels only release CO_2 removed from the atmosphere via photosynthesis as the algae or plant grew. The energy crisis and the world food crisis have ignited interests in algaculture (farming algae) for making biodiesel and other biofuels using land unsuitable for agriculture. The advantages of this method are that they can be grown with minimal impact on fresh water resources and can be produced using saline and wastewater, and have a high flash point. Furthermore, they are biodegradable and relatively harmless to the environment if spilled. They cost more per unit mass than second-generation biofuel crops due to high capital and operating costs, but are claimed to yield between 10 and 100 times more fuel per unit area.

The oil-rich algae can be extracted from the system and processed into biofuels, with the dried product further reprocessed to ethanol. The production of algae to harvest oil for biofuels has not been on a commercial scale, but feasibility studies have been carried out to provide an estimate of the yield. In addition to its high yield, algae culture, unlike crop-based biofuels does not require a decrease in food production since it requires neither farmland nor fresh water. A new method of production uses freshwater alga Chlorella vulgaris, grown using flue gas from a gas-fired power station as the carbon source. Cultivation using a two-stage method involves the cells initially grown to a high concentration of biomass under nitrogen-required conditions, where the cells accumulate triglycerides (i.e., an ester derived from glycerol and three-fatty acids). Cultivation in typical raceways and air-lift tubular bioreactors was investigated, as well as different methods of downstream processing. The results showed that 40 tons per hectare per annum can be achieved from microalgae such as C. vulgaris [9].

Algae can be cultivated and harvested in support of a wide range of biofuel products. Additionally, algae biofuels systems hold promise to enable rapid production of high quality, high throughput biofuels systems in support of carbon emission reduction targets and in support of clean fuel production.

Algae can be converted into various types of fuels, depending on the technique and the part of the cells used. The lipid or oily part of the algae biomass can be extracted and converted into biodiesel through a process similar to that used for any other vegetable oil, or converted in a refinery into "drop-in" replacements for petroleum-based fuels. Alternatively, after the lipid extraction, the carbohydrate content of algae can be fermented into bioethanol or butanol fuel.

Microalgae are capable of producing more than 30 times the amount of oil (per annum per unit area of land) when compared to oil seed crops. They are more efficient converters of solar energy than any known plant because they grow in suspensions where they have unlimited access to water and more efficient access to CO_2 and dissolved nutrients. Microalgae are the fastest growing photosynthesizing organisms and can complete an entire growing cycle every few days. Figure 27.22 shows a photograph of green algae.

Figure 27.22 Photograph of green algae from Ernst Haeckel's Kunstformen der Natur, 1904. (Source: https://en.wikipedia.org/wiki/Plant#Algae).

27.7.6 Economic Viability of Algae

There is a demand for sustainable biofuel production; however, the use of a particular biofuel depends on its cost efficiency. Research is now focusing on cutting the cost of algal biofuel production to compete with conventional petroleum [36, 37]. The production of several products from algae has been mentioned as the most essential factor for making algae production economically viable. Other factors are the improvement of solar energy to biomass conversion efficiency and oil extraction production from the algae. A formula was derived in estimating the cost of algal oil for it to be a viable substitute to petroleum diesel as:

$$C_{(algal\ oil)} = 25.9 \times 10^{-3}\ C_{(petroleum)} \tag{27.1}$$

where

$C_{(algal\ oil)}$ is the price of microalgal oil in \$/gal. and $C_{(petroleum)}$ is the price of crude oil in \$/bbl. Equation 27.1 assumes that the algal oil has roughly 80% of the caloric energy value of crude petroleum [38].

With current available technology, it is estimated that the cost of producing microalgal biomass is \$2.95/kg of photobioreactors and \$3.80/kg for open-ponds. These estimates assume that CO_2 is available at no-cost [36]. If the annual biomass production capacity is increased to 10,000 tonnes, the cost of production per kg reduces to ~\$0.47 and \$0.6 respectively. Assuming that the biomass contains 30% oil by weight, the cost of biomass for providing a liter of oil would be ~\$1.40 (\$5.30/gal) and \$1.81 (\$6.85/gal) for photobioreactors and raceways respectively. Oil recovered from the lower-cost biomass produced in photobioreactors is estimated to cost \$2.80/l, assuming the recovery process contributes 50% to the cost of the final recovered oil. However, if the existing algae projects can achieve biodiesel production price targets of less than \$1.0/gal, the US may realize its goal of replacing up to 20% of its transport fuels by 2020 by using environmentally and economically sustainable fuels from algae production [39].

Algae biodiesel is still a fairly new technology; although research began over 30 years ago, it was delayed due to a lack of funding and a relatively low petroleum cost. It was in the early 2000s when the gas peaked that it eventually had a revitalization in research for alternative-fuel sources [40]. While the technology exists to harvest and convert algae into a usable source of biodiesel, it still has not been implemented on a large scale to support the current energy needs. Further research would be required to make the production from corn and grain, although Solazyme [41] and Sapphire Energy [42] began small-scale commercial sales in 2012 and 2013 respectively, but most efforts had been abandoned or changed to other applications by 2017 [43]. It is expected that, due to economies of scale and mechanization, the price of seaweed fuel production costs can still be reduced by up to 100% [44].

27.8 Fast Pyrolysis

The word "pyrolysis" is used for the degradation of a substance upon exposure to high temperatures. Exclusion of air of oxygen is essential in case ignition and combustion of the primary products is considered undesirable. Slow pyrolysis of wood at temperatures up to 400°C is a process that has been exploited for thousands of years. Charcoal is a smokeless fuel that is still used widely for cooking and heating purposes. It has been used industrially as well, viz. in blast furnaces for the making of iron (Fe). Before the eighties of the last century, 'fast' pyrolysis research was mostly dealing with the thermal decomposition of organic compounds, polymer, salts, etc. Radlein and Quignard provide a review on biomass fast pyrolysis [87].

The production of pyrolysis oil is an efficient method to utilize biomass as a renewable energy resource for energy, chemicals and/or materials. Raw biomass is mostly collected from distant areas in relatively small quantities; however, the biomass is often difficult to handle due to unfavourable properties as:

- A fluffy structure and low volumetric density
- The presence of contaminants such as sand and ash
- A high moisture contents
- A complex chemical structure, built up mainly form cellulose, hemicellulose and lignin.

In fast pyrolysis a uniform liquid is produced with improved properties*. It enables the distributed production of pyrolysis liquids at a relatively small scale and in remote areas, for centralized utilization elsewhere at a much larger scale and any desirable time. In this way, the biomass usage is fully decoupled from its production. The improved properties enable a more efficient transport and processing if compared to the original biomass.

27.8.1 Fast Pyrolysis Principle

Pyrolysis is a process in which organic materials are heated in the absence of air/oxygen. Under these conditions the organic material decomposes, forming vapors, permanent gases and charcoal. The vapors can be condensed to form the main product: pyrolysis liquid. In order to maximize the liquid production, the biomass heating as well as the vapor condensing needs to be done quickly. Hence the name **fast** pyrolysis. Alternatively, the biomass conversion can be directed at producing charcoal. In this case heating is less rapid and the process is called slow pyrolysis or carbonization. The latter is usually carried out at temperatures below 400°C.

Fast pyrolysis is employed to convert the biomass to a maximum quantity of liquid of around 60 to 70 wt.% of the feedstock. A more uniform, stable and cleaner-burning product is obtained, that could serve as an intermediate

---

*Definition of Fast Pyrolysis Liquid by IEA Task 34: Liquid condensate recovered by thermal treatment of lignocellulosic biomass at short hot vapor residence time (typically less than ~ 5seconds) typically at between 450 – 600°C at near atmospheric pressure or below, in the absence of oxygen, using small (typically less than 5mm) dry (typically less than 10% water) biomass particles. A number of engineered systems has been used to effect high heat transfer into the biomass particle and quick quenching of the vapor product, usually after removal of solid by-product "char" to recover a single-phase liquid product. Bio-oil is a complex mixture of, for the most part, oxygenated hydrocarbon fragments derived from the biopolymer structures. It typically contains 15-30% water. Common organic components include acetic acid (CH_3COOH), methanol (CH_3OH), aldehydes (RCHO) and ketones (RCOR´), cyclopentenones, furans, alkyl-phenols, alkyl-methoxy-phenols, anhydrosugars, and oligomeric-sugars and water – insoluble lignin derived compounds. Nitrogen and sulfur containing compounds are also sometimes found depending on the biomass source.

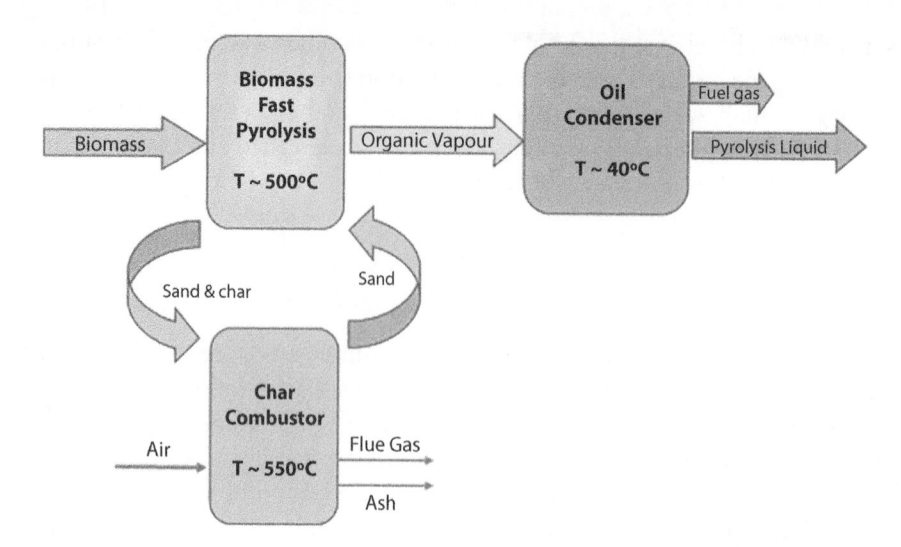

Figure 27.23 Flow diagram of fast pyrolysis of biomass to fuel gas and pyrolysis liquid. (Source: Welcome to PyroWiki - PyroWiki (pyroknown.eu).

energy carrier and feedstock for subsequent processing. The essential conditions of fast pyrolysis for the production of pyrolysis liquids are:

- a very fast heating of relatively small biomass particles (order of seconds)
- controlling the pyrolysis reactor temperature at a level around 500°C
- a short vapor residence time to avoid further cracking to permanent gases
- rapid cooling of all the vapors to form the desired pyrolysis liquid.

By-products in the form of char and non-condensable gases are produced as well. In an industrial process, these two by-products (both 10 to 20 wt.%) would be used primarily as a fuel for the generation of the required process heat (including feedstock drying). The char is also proposed to be applied as a ('biochar soil improver) or as a substitute for metallurgic coke in the steel industry. Alternatively, for specific purposes (and reasons), it can be recombined with the fast pyrolysis oil to form a char-oil slurry (for example the BioLiq process, formerly Dynamotive's Bio-oil Plus). The gaseous by-product essentially is a mixture of CO and CO_2. Apart from possible flue gas emissions resulting from the char combustion, there are no waste streams. The biomass ash will be largely concentrated in the char by-product. It is separated when the char is combusted in the process, viz. to generate the heat for drying and heating of the biomass feedstock. Ash separation enables recycling of the minerals as a natural fertilizer to the soil on which the biomass was grown originally. Figure 27.23 shows the flow diagram of fast pyrolysis of biomass to fuel gas and pyrolysis liquid.

27.8.2 Fast Pyrolysis Technologies

Various fast pyrolysis technologies have been developed over the last decades on basis of different reactor types. The essential characteristics of a fast pyrolysis reactor for maximal oil production are the very rapid heating of the biomass, an operating temperature around 500°C, and a rapid quenching of the produced vapors. In small dedicated laboratory reactors high oil yields can be obtained when very small particles are used while imposing high heat (and mass) transfer rates, and low vapor residence times. In real scale installation however, the consequences of heat transfer limitations and excessive vapor residence times may be significant.

Indeed, high interparticle heat transfer rates are required (heat transfer coefficients of more than 500 W/m²K). Moreover, intra-particle biomass heat transfer limitations should be avoided, which would require particles sized to less than 3 mm for the effective heat penetration depth. Vapor phase residence times should be kept within a few seconds in order to maintain the oil yield. In case a pyrolysis plant is meant to produce a liquid fuel for combustion or gasification, the process could be designed in a way that maximizes the energy conversion to the liquid product. However, when the bio-oil product is meant to derive bio-fuels or chemicals from it, other factors than just the vapor residence time should be considered as well. The composition of the oil obviously depends on the type of feedstock used, but can be steered further by process conditions, equipment dimensions and the application of catalysis.

Several fast pyrolysis technologies have been developed (e.g., laboratory, pilot, demo or commercial scale); One way to distinguish the different technology is on the basis of reactor types or mixing employed:

- Ablative
- Fluidized bed
- Circulating fluidized bed
- Auger reactor
- Rotating cone
- Vacuum
- New developments (e.g., microwave pyrolysis and hydropyrolysis)

27.8.3 Minerals of Biomass

Depending on the type of biomass, the ash content roughly varies from 0.2 (softwood) up to 10 wt.% (herbaceous materials). Salts containing sodium, potassium, magnesium and calcium are the major minerals found in biomass ash. An advantage of fast pyrolysis is the separation of inorganic compounds from the main liquid product. Fast pyrolysis oils are essentially mineral free; typically, more than 95% of the minerals end up in the char. However, the presence of minerals in the biomass does adversely affect the performance of the pyrolysis processing. During pyrolysis, the alkali and alkaline earth metals act as catalysts, which affect the resulting bio-oil composition.

27.8.4 Applications of Fast Pyrolysis Liquid

Heat and Power

Assuming a typical composition for a pyrolysis liquid (including 25wt% of water) the combustion can be written as:

$$C_{10}H_{15}O_8 \text{ (l)} + 9.75\, O_2 \text{ (g)} \rightarrow 10\, CO_2 \text{ (g)} + 7.5\, H_2O \text{ (g)} \quad (27.2)$$

$$(\Delta H_R) = 16.5\, \frac{MJ}{kg} \text{ of pyrolysis liquid}$$

For the stoichiometric combustion of 1 kg of pyrolysis liquid about 6 Nm³ of air is needed, and 1.7 kg of CO_2 is produced.

Biomass derived fast pyrolysis oil is a renewable fuel that can be used for the production of heat, steam and power. After relatively small modifications, good quality pyrolysis oil can be burnt quite well in traditional boilers and furnaces, and even in turbines. For small units, it usually requires a redesign of the burner and its operation mode, next to the application of corrosion resistant materials for all the equipment that is contacted with the pyrolysis oil. Boiler combustion [87] is well developed after various small and larger scale testing during the past 15 to 20 years, amongst others by VTT and Oilon in Finland, Canmet Energy in Canada, Fortum in Finland, and Stork with BTG-BTL in The Netherlands.

More challenging is the combustion of fast pyrolysis oil in internal combustion engines. Problems identified are related to various pyrolysis oil properties such as: particulates (causing erosion), acidity (corrosion), poor lubrication (injection needle friction), thermal instability (deposits), ignition (delay), viscosity (atomization), combustion (emissions) and water content (reduced caloric value). Various methods, either related to modifying the pyrolysis (blending with diesel, esterification, hydro-deoxygenation, etc.) or the engine, have been proposed to solve or avoid the problems.

The properties of fast pyrolysis liquids are very different from those of crude oil derived transportation fuels. Modification/upgrading is required to improve the compatibility with such fuels. As shown in Figure 27.24, various approaches are possible ranging from simple physical treatment to severe thermochemical chemical treatment. Physical treatment and/or mild chemical treatment of pyrolysis liquid is meant to improve specific properties like acidity, viscosity and stability. It should simplify the application of the pyrolysis liquid in stationary engines and turbines. Examples of such treatments are blending, emulsifying and esterification. It is important to note that they never enable automotive applications.

To produce real transportation fuels (diesel, kerosene, gasoline) from fast pyrolysis oil, a severe thermochemical treatment of the fast pyrolysis is needed (upgrading). It includes a full deoxygenation of the pyrolysis oil in a catalytic hydrotreatment process. Because full deoxygenation requires quite some hydrogen, an alternative approach is to just partially de-oxygenate the pyrolysis oil and finish the conversion to transportation fuel in an existing crude oil refinery unit.

Research [86] studies are being carried out to investigate the various aspects to the upgrading processes, where a two-step process is applied. The first occurs at 200-250 °C to stabilize the reactive components in the oil, and the second is at 350-400 °C to deoxygenate the stabilized product in a further hydrotreating step. The presence of a proper catalyst in both steps is essential. Presumably carbohydrate chemistry is involved in the low temperature step, while lignin chemistry prevails at higher temperatures. The cellulose-derived fraction of the pyrolysis oil needs to be transformed first, preferably to alcohols, in a 'mild hydrogenation' step. Then, dehydration and hydrogenation can take place at more severe conditions. The high temperature step is similar to petrochemical hydrotreatment. Full deoxygenation should be carried out while using active and selective catalysts to suppress charring, promote hydrogenation and reduce methanation. It may eventually result in carbon yields of above 50 wt.% (see Figure 27.25). As an alternative to full deoxygenation of the pyrolysis oil in the catalytic hydrotreatment process, mixtures of transportation fuels can also be obtained by co-feeding partially upgraded fast pyrolysis oil to an FCC refinery unit.

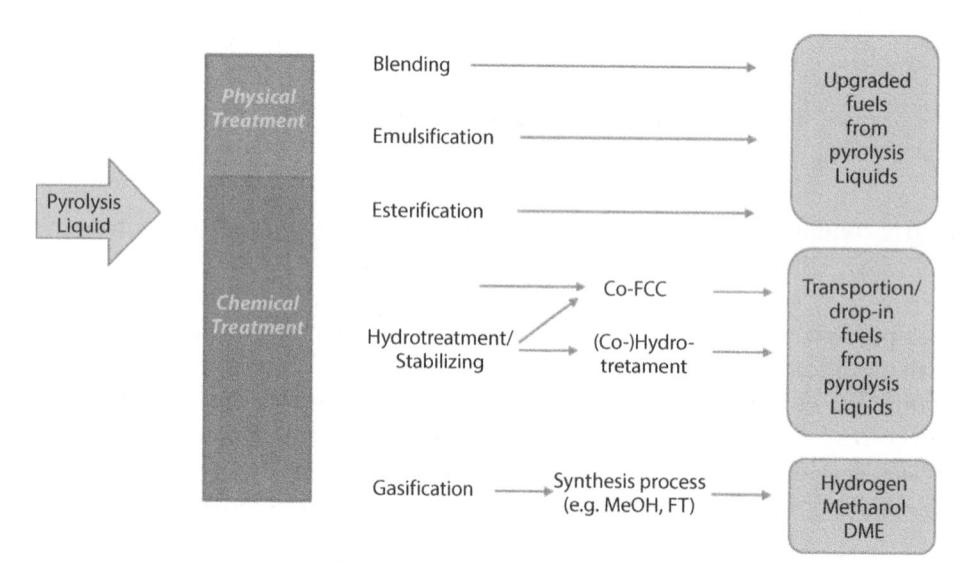

Figure 27.24 Pyrolysis liquid with physical and chemical treatments. (Source: www.Welcome to PyroWiki - PyroWiki (pyroknown.eu)).

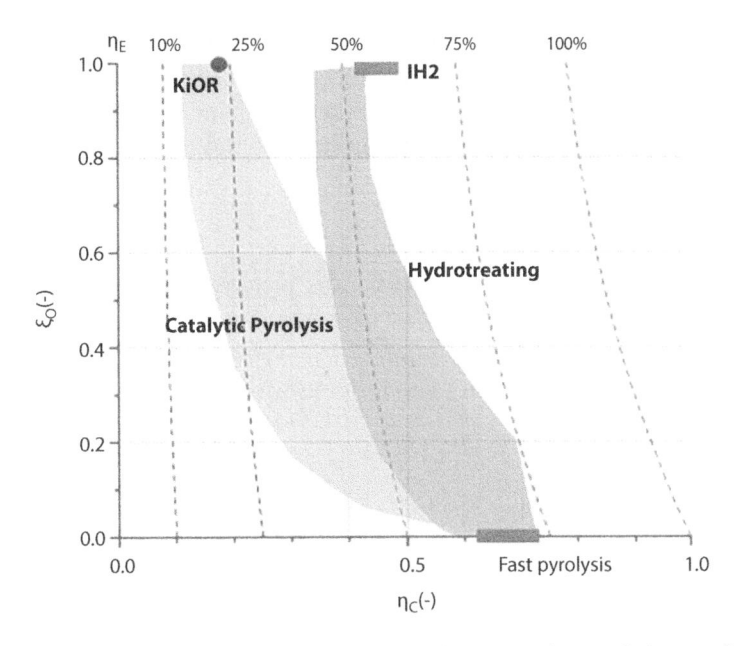

Figure 27.25 For various biofuel production routes the final deoxygenation degree correlates with the overall carbon yield. (Source: Welcome to PyroWiki - PyroWiki (pyroknown.eu)).

To produce real transportation fuels (diesel, kerosene, gasoline) from fast pyrolysis oil, a severe thermochemical treatment of the fast pyrolysis is needed (upgrading). It includes a full deoxygenation of the pyrolysis oil in a catalytic hydrotreatment process. Because full deoxygenation requires quite some hydrogen, an alternative approach is to just partially de-oxygenate the pyrolysis oil and finish the conversion to transportation fuel in an existing refinery unit.

27.8.5 Chemicals and Materials

Pyrolysis oil is a mixture of cracked components originating from the pyrolysis of the three main building blocks of biomass; cellulose, hemicellulose and lignin. Pyrolysis offers a good pretreatment to facilitate the fractionation of biomass. After pyrolysis the (ash free) oil can easily be fractionated into three product streams namely; pyrolytic lignin (from lignin), pyrolytic sugars (from cellulose) and a watery phase containing smaller organic components e.g., acetic acid (mainly from hemicellulose). Pyrolysis oil is considered to be a good renewable resource for renewable chemicals and materials. The three fractions obtained from pyrolysis oil are:

- Pyrolytic lignin (~ 35% of the original carbon)
- Sugar syrup (~ 45% of the original carbon)
- Aqueous phase (~20% of the original carbon).

After the fractionation pyrolysis process, the fractions can be further processed to produce chemicals or green products. A highly potential application for the pyrolytic lignin can be a source for renewable substituent for fossil phenol in phenol/formaldehyde resins and derivatives. These types of resins are widely used in wood products like particle boards, plywood, and so on. Another application of the pyrolytic lignin is in replacing fossil bitumen in various bitumen-based materials, e.g., in asphalt and roofing materials. It can also be used in the production of green phenolic derivatives as a possible raw material for various coatings, composites and preservatives.

27.8.6 Bio-Fuels-Fast Pyrolysis Bio-Oil (FPBO) from Biomass Residues

A technology using lignocellulosic biomass residues as feedstocks can be converted into a dark-brown bioliquid known as fast pyrolysis bio-oil (FPBO), a homogeneous energy carrier. The byproducts of the fast pyrolysis process are heat (i.e., steam) and power (electricity). The fast pyrolysis process consists of a thermochemical decomposition of biomass through rapid heating, at a temperature of 450 – 600°C in the absence of oxygen.

The fast pyrolysis technology converts up to 70% of the dry basis biomass feedstock into bio-oil and the remaining parts into char and gas. The heat produced by the combustion of pyrolysis char and non-condensable gases is recovered as high-pressure steam and can be utilized in a steam turbine system for electric power generation and feedstock drying. Excess steam can be sold to nearby industrial facilities or district heating grids.

Feedstocks

A large number of different lignocellulosic feedstocks can be processed in the fast pyrolysis plants. These feedstocks range from wood, sunflower husk, bagasse, tobacco, energy crops, straw, olive stone residues and many more, to establish their pyrolysis oil yield and quality (Figure 27.26). Typically, woody biomass gives the highest yields. Before entering the reactor, the particles in the biomass residues will be reduced to a size below 3 mm to allow rapid

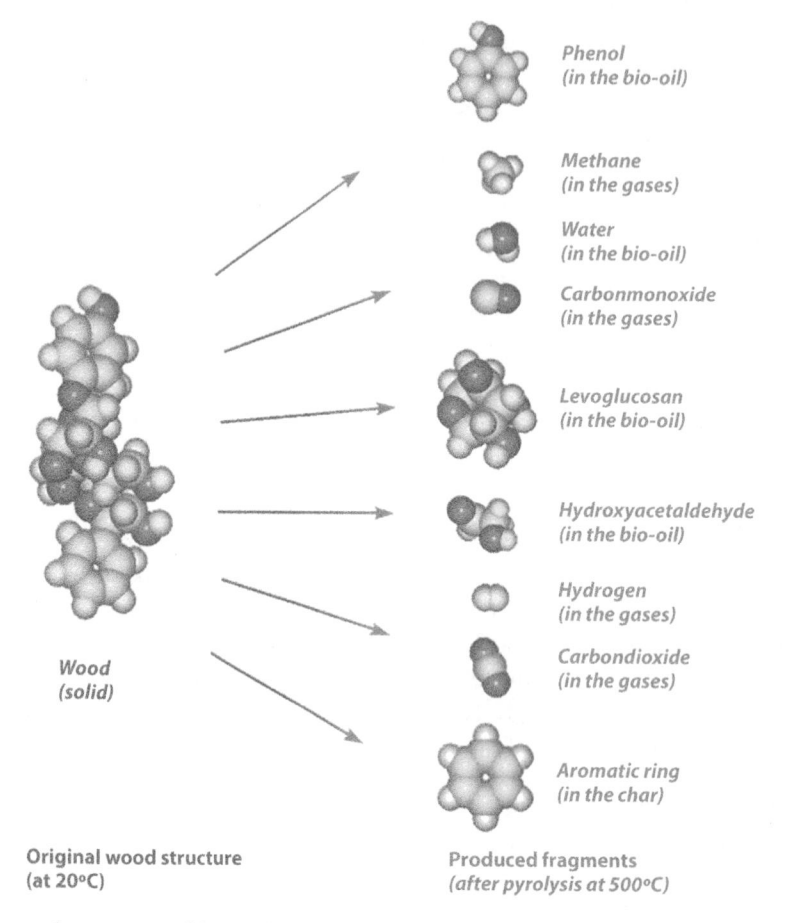

Figure 27.26 Wood structure to the structure of the products.

Figure 27.27 Photographs of feedstocks for FPBO technology.

conversion. At the same time, its moisture content will be brought below 3 wt.%, to avoid too much water in the pyrolysis oil. Figure 26.27 shows the photograph of feedstocks for FPBO technology.

27.8.7 Properties of Pyrolysis Oil

The technology produces an FPBO (fast pyrolysis bio-oil) that contains a low ash and solid concentration. The energy density of FPBO is 5 to 20 times higher than the original biomass residue. The heating value (LHV) of pyrolysis oil is 16–23 MJ/l, compared to 37 MJ/l for fossil fuels. The density of the liquid is about 1170 kg/m^3, which is denser than fuel oil and significantly denser than the original biomass residues.

Pyrolysis oil has a low pH-value of around 2.5–3. Due to the large amount of oxygenated components, the oil has a polar nature and does not mix with hydrocarbons. The degradation products from the original biomass include organic acids (like formic and acetic acid), giving the oil its low pH value. Water is an integral part of the single-phase chemical solution. The bio-oil has water contents of typically 15–30 wt%. Phase separation occurs when the water content is higher than about 30%.

Pyrolysis liquid is a dark-brown, free flowing, oil-like substance with a pungent odor and several distinctive properties as shown in Table 27.10. Pyrolysis liquid is a derived ligno-cellulosic biomass and contains 20 to 30 wt % water and fragments of the cellulose, hemicelluloses, lignin, and extractives. The organics in pyrolysis liquid include hundreds of different compounds (e.g., ketones, aldehydes, sugars, acids, phenols, aromatics, extractives, and so on). The density of pyrolysis liquid is about 1.2 kg/l and its lower heating value is 15 – 17 MJ/kg. The corresponding crude oil values are 0.85 kg/l and 42 MJ/kg respectively. Pyrolysis liquid contains 40% of the energy of crude oil on weight basis, and roughly 60% on a volumetric basis. It's acidic with a pH of 2.5 – 3, due to the presence of organic acids. The analysis of the liquid is shown in Table 27.11.

Table 27.10 Physical Properties of Pyrolysis Liquid.

Property	Unit	Value
C-Carbon	wt %	46
H-Hydrogen	wt %	7
N-Nitrogen	wt %	<0.01
O (Balance)	wt %	47
Water content	wt %	25
Ash content	wt %	0.02
Solids content	wt %	0.04
Density	kg/m^3	117
LHV	MJ/kg	16
LHV	MJ/liter	19
pH		2.8
Kinematic viscosity (40°C)	cSt	30

Table 27.11 Analysis Methods for Different Properties of Fast Pyrolysis Liquid [87].

Property	Unit	Typical value	Method ASTM	EN ISO	DIN
Water content	wt %	15 – 30	E203		
pH		2 – 3	E70		
C - Carbon	wt % (dry basis)	50 – 60	D5291		
H- Hydrogen	wt % (dry basis)	6–8	D5291		
N- Nitrogen	wt % (dry basis)	< 0.5	D5291		
S Sulfur	wt % (dry basis)	<0.05	D5453	20846	
O (Balance)	wt % (dry basis)	47	-	-	-
HHV – Higher heating value	MJ/kg	14 – 19	D240		51900
LHV – Lower heating value	MJ/kg	13 – 18	D240		51900
TAN - Total Acid Number	mg KOH/g	70 – 100	D664		
Density at 15°C	kg/m³	1,100 – 1,300	D4052	12185	
Kinematic viscosity (40°C)	cSt	15 – 40	D445	3104	
Pour point	°C	-9 – (-36)	D97	3016	
Flash point	°C	40–110	D93B	2719	
Sustained combustion	-	does not sustain		9038	
MCR/CCR Carbon residue	wt %	15–25	D4530-D189		
Solids	wt %	<1	D7579		
Ash	wt %	<0.3		6245	
Na, K, Ca, Mg	wt % (dry basis)	<0.06		16476	

Fast pyrolysis technology design offers several advantages and these are as follows:

Main advantages

Absence of inert carrier gas	Instead of carrier gas, mechanical mixing is used for the rapid heating of biomass particles. As a result, gas volume flows are smaller, resulting in smaller downstream equipment.
High feedstock flexibility: suitable for a wide range of biomass residues	The technology can handle biomass residues with low ash melting temperatures, as temperatures in the heat carrier cycle can be controlled to a large extent.
	Due to the patented cyclone design and low amounts of solids in the oil (down to 0.01% wt). The high sand-to-biomass ratio, the produced pyrolysis oil is stable and has a very low solids content.

Lower CAPEX and OPEX	As explained, the RCR principle inherently results in more compact equipment. In combination with a standardized but flexible design, the CAPEX is lower. Optimizing the design with efficient and integrated heat recovery results in lower OPEX.
Shorter construction time on location	With the modular building approach, a plant can be erected on site within two weeks.
High energy efficiency: up to 85-90% (biomass in – oil/heat/electricity out)	The excess heat of the process produces enough steam to bring the moisture content of biomass residues (up to wet basis 55%) down to the required level, in combination with power generation. In general, more electricity can be produced than required for the entire plant.

27.9 Acid Gas Removal

Gases from various operations in a refinery that processes sour crudes contain hydrogen sulfide (H_2S) and occasionally carbonyl sulfide (COS). Some hydrogen sulfide in refinery gases is formed as a result of conversion of sulfur compounds in processes such as hydrotreating, cracking and coking. The earlier conventional method of treatment was to burn the hydrogen sulfide along with other light gases as refinery fuel, because its removal from the gases and conversion to elemental sulfur was not economical. However, air pollution regulations require that most of the hydrogen sulfide be removed from refinery fuel gas and converted to elemental sulfur (see Chapter 12, Volume 1).

Amine gas treating referred to as amine scrubbing, gas sweetening and acid gas removal, is a group of processes that use aqueous solutions of various alkylamines to remove H_2S and carbon dioxide (CO_2) from gases. It is a common unit process used in petroleum refineries, and is also used in petrochemical plants, natural-gas processing plants and other CPIs. Processes within oil refineries or chemical processing plants that remove H_2S are known as "sweetening" processes because the odor of the processed products is improved by the absence of H_2S. An alternative to the use of amines is membrane technology, but it is less attractive due to the relatively high capital and operating costs.

In addition to H_2S, many crudes contain some dissolved CO_2 that through distillation finds its way into the refinery fuel gas. These components as hydrogen sulfide and carbon dioxide are generally termed acid gases. They are removed simultaneously from the fuel gas by a number of different processes, some of which are the following [12].

Chemical Solvent Processes

1. Monoethanolamine (MEA)
2. Diethanolamine (DEA)
3. Methyl-diethanolamine (MDEA)
4. Diglycolamine (DGA)
5. Hot potassium carbonate (K_2CO_3)

Physical Solvent Processes

1. Selexol
2. Propylene
3. Sulfinol
4. Rectisol

The most commonly used amines are the alkanolamines DEA, MEA, and MDEA. These amines are used to remove sour gases and liquid hydrocarbons such as liquefied petroleum gas (LPG).

27.9.1 Process Description of Amine Gas Treating

Gas containing H_2S or both H_2S and CO_2 are referred to as sour gases or acid gases in the refining or hydrocarbon processing industries. The chemistry involved in the amine treating of such gases varies with the particular amine being used.

Chemical Reactions

The overall chemical reactions applicable for H_2S and CO_2 reacting with primary and secondary amines are:

For hydrogen sulfide H_2S removal:

$$RNH_2 + H_2S \rightleftharpoons RNH_3^+ + HS^- \quad \text{Fast} \tag{27.3}$$

$$RNH_2 + HS^- \rightleftharpoons RNH_3^+ + S^{--} \quad \text{Fast} \tag{27.4}$$

$$H_2S_{(gas)} \rightleftharpoons H_2S_{(Solution)} \tag{27.5}$$

$$H_2S_{(Solution)} \rightleftharpoons H^+ + HS^- \tag{27.6}$$

$$H_2O \rightleftharpoons H^+ + OH^- \tag{27.7}$$

$$pp_{(H_2S)} = H_{(H_2S)}[H_2S] \tag{27.8}$$

For carbon dioxide (CO_2) removal

$$2RNH_2 + CO_2 \rightleftharpoons RNH_3^+ + RNHCOO^- \quad \text{Fast} \tag{27.9}$$

$$RNH_2 + CO_2 + H_2O \rightleftharpoons RNH_3^+ + HCO_3^- \quad \text{Slow} \tag{27.10}$$

$$RNH_2 + HCO_3^- \rightleftharpoons RNH_3^+ + CO_3^{--} \quad \text{Slow} \tag{27.11}$$

$$H_2O + CO_2 \rightleftharpoons 2H^+ + CO_3^{--} \tag{27.12}$$

$$H_2O \rightleftharpoons H^+ + OH^- \tag{27.13}$$

$$pp_{(CO_2)} = H_{(CO_2)}[CO_2] \tag{27.14}$$

The above reactions proceed to the right at high pressure and/or low temperatures and to the left at high temperatures and/or low pressures. Secondary amines undergo similar reactions as primary amines. Tertiary amines can only react with CO_2 via the acid/base reaction in Equation 27.3. The CO_2 reaction is slowed by the time required to dissolve CO_2 and its conversion to bicarbonate. This may be the reason that tertiary amines exhibit a greater selectivity of H_2S in the presence of CO_2. Table 27.12 lists approximate guidelines for a number of alkanolamine (amine) processes.

Table 27.12 Approximate Guidelines for Amine Processes [45].

	MEA	DEA	DGA	Sulfinol	MDEA
Acid gas pickup, scf/gal @ 100°F [2]	3.1–4.3	6.7–7.5	4.7–7.3	4–17	3–7.5
Acid gas pickup, mols/mol amine, normal range [3]	0.33–0.40	0.20–0.80	0.25–0.38	NA	0.20–0.80
Lean solution residual acid gas, mol/mol amine, normal range [4]	0.12 ±	0.01 ±	0.06 ±	NA	0.005–0.01
Rich solution acid gas loading, mol/mol amine, normal range [3]	0.45–0.52	0.21–0.81	0.35–0.44	NA	0.20–0.81
Solution concentration wt% normal range	15–25	30–40	50–60	3 comps., varies	40–50
Approximate reboiler heat duty, Btu/gal lean solution [5]	1,000–1,200	840–1,000	1,100–1,300	350–750	800–900
Steam heated reboiler tube bundle, approx. average heat flux, Q/A = Btu/h-ft² [6]	9,000–10,000	6,300–7,400	9,000–10,000	9,000–10,000	6,300–7,400
Direct fired reboiler fire tube, average heat flux, Q/A = Btu/h-ft² [6]	8,000–10,000	6,300–7,400	8,000–10,000	8,000–10,000	6,300–7,400
Reclaimer, steam bundle or fire tube, average heat flux, Q/A = Btu/h-ft² [6]	6–9	NA [7]	6–8	NA	NA [7]
Reboiler temperature, normal operating range, °F [8]	225–260	230–260	250–270	230–280	230–270
Heats of reaction; [10] approximate Btu/lb H_2S	610	720	674	NA	690
Btu/lb CO_2	660	945	850	NA	790
NA – not applicable or not available					

Notes:
1. These data alone should not be used for specific design purpose. Many design factors must be considered for actual plant design.
2. Dependent upon acid gas partial pressures and solution concentrations.
3. Dependent upon acid gas partial pressures and corrosiveness of solution. Might be only 60% or less of value shown for corrosive systems.
4. Varies with stripper overhead reflux ratio. Low residual acid gas contents require more stripper trays and/or higher reflux ratios yielding larger reboiler duties.
5. Varies with stripper overhead reflux ratios, rich solution feed temperature to stripper and reboiler temperature.
6. Maximum point heat flux can reach, 20,000–25,000 But/h-ft² at highest flame temperature at the inlet of a direct fired fire tube. The most satisfactory design of firetube heating elements employs a zone by zone calculation base on thermal efficiency desired and limiting the maximum tube wall temperature as required by the solution to prevent thermal degradation. The average heat flux, Q/A, is a result of these calculations.
7. Reclaimers are not used in DEA and MDEA systems.
8. Reboiler temperatures are dependent on solution conc. flare/vent line back pressure and/or residual CO_2 content required. It is good practice to operate the reboiler at as low a temperature as possible.
9. According to Total.
10. B.I. Crynes and R.N. Maddox, Oil Gas J., p 65–67, Dec. 15 (1969). The heats of reaction vary with acid gas loading and solution concentration. The values shown are average.

Figure 27.28 shows a typical amine gas treating process, which includes an absorber unit and a regenerator unit with ancillary equipment. In the absorber, the downflowing amine solution absorbs H_2S and CO_2 from the up flowing sour gas to produce a sweetened gas stream (i.e., a gas free of H_2S and CO_2) as a product and an amine solution rich in the absorbed acid gases. The resultant "rich" amine is then routed into the regenerator (a stripper with a reboiler) to produce regenerated or "lean" amine that is recycled for reused in the absorber. The stripped overhead gas from the regenerator is concentrated H_2S and CO_2.

Alternative stripper configurations include matrix, internal exchange, flashing feed and multipressure with split feed using seven model solvents that approximate the thermodynamic and rate properties of 7m (30 wt%) MEA, K_2CO_3 promoted by PZ and hindered amines. Their results showed that solvents with high heat of absorption (MEA, MEA/PZ) favor operation at the normal pressure. Many of these configurations offer more energy efficiency for specific solvents or operating conditions. Vacuum operation favors solvent with low heats of absorption, while operation at the normal pressure favors solvents with high heats of absorption. Solvents with high heats of absorption require less energy for stripping from temperature swing at fixed capacity. The stripper recovers 40% of CO_2 at a higher pressure and does not have inefficiencies associated with multipressure stripper. Energy and costs are reduced since the reboiler duty cycle is slightly less than the normal pressure stripper. An internal exchange stripper has a smaller ratio of water vapor to CO_2 in the overhead's stream, and thus fewer streams are required. The multipressure configuration with split feed reduces the flow into the bottom section, which also reduces the equivalent work. Flashing feed

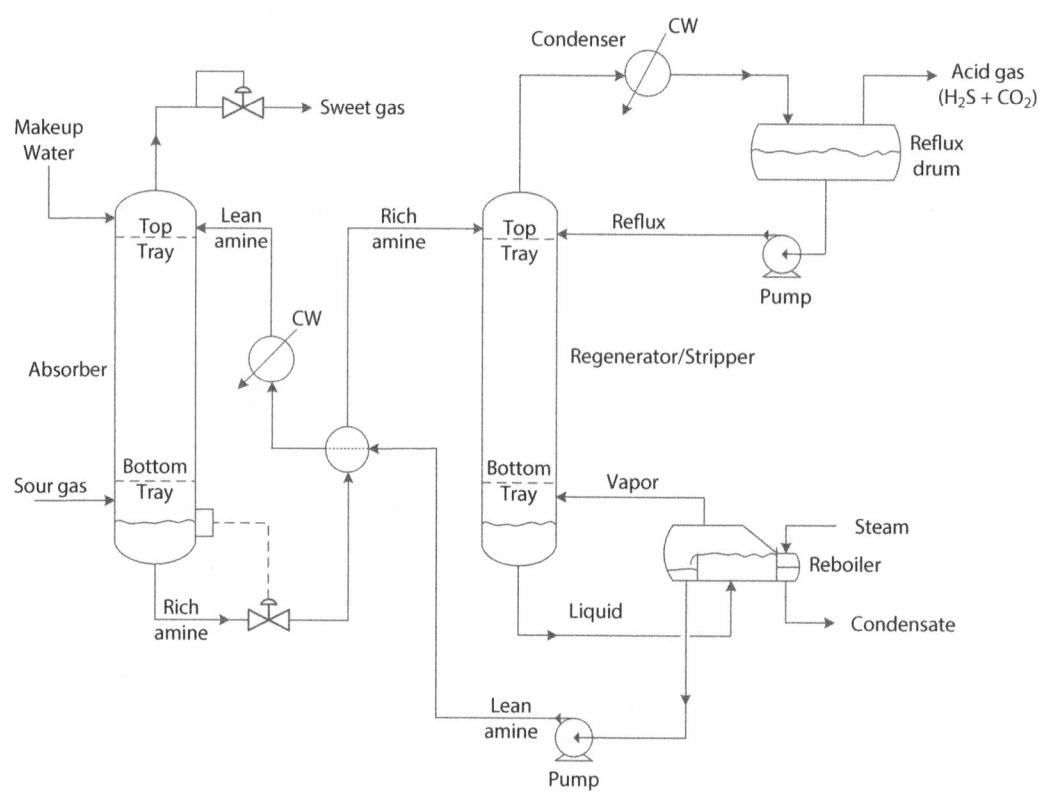

Typical operating range

Absorber: 35 to 50°C and 5 to 205 atm. absolute pressure
Regenerator: 115 to 126°C and 1.4 to 1.7 atm. absolute pressure at tower bottom

Figure 27.28 Process flow diagram of a typical amine treating process used in petroleum refineries, natural gas processing plants and other industrial facilities.

requires less heat input because it uses the latent heat of water vapor to help strip some of the CO_2 in the rich stream entering the stripper at the bottom of the column. The multipressure configuration is more attractive for solvents with higher heats of absorption [45].

The amine concentration in the absorbent aqueous solution is an important parameter in the design and operation of an amine gas treating process. Depending on which one of the following four amines the unit was designed to use and what gases it was designed to remove; there are some typical amine concentrations, expressed as weight percent of pure amine in the aqueous solution [46].

• Monoethanolamine	About 20% for removing H_2S and CO_2 and about 32% for removing CO_2
• Diethanolamine	About 20–25% for removing H_2S and CO_2
• Methyldiethanolamine	About 30–55% for removing H_2S and CO_2
• Digylcolamine	About 50% for removing H_2S and CO_2

The choice of amine concentration in the circulating aqueous solution depends upon a number of factors, and these include whether the amine unit is treating raw natural gas or petroleum refinery by-product gases that contain relatively low concentrations of both H_2S and CO_2 or whether the unit is treating gases with a high percentage of CO_2, such as the off gas from the steam reforming process used in ammonia production or the flue gases from power plants.

Both H_2S and CO_2 are acid gases and are corrosive to carbon steel. In an amine treating unit, CO_2 is the stronger acid, and H_2S forms a film of iron sulfide on the surface of the steel that acts to protect the steel. In treating gases with a high percentage of CO_2, corrosion inhibitors are used that permit the use of higher concentrations of amine in the circulating solution.

Another factor involved in choosing an amine concentration is the relative solubility of H_2S and CO_2 in the selected amine. The choice of the type of amine will affect the required circulation rate of amine solution, the energy consumption for the regeneration and the ability to selectively remove either H_2S or CO_2.

In oil refineries, the stripped gas is mostly H_2S, much of which comes from a sulfur-removing process called hydrodesulfurization. This rich H_2S-rich stripped gas stream is then routed into a Claus process to convert it into element sulfur [12]. The vast majority of the 64×10^6 metric tons of sulfur produced worldwide was the by-product of sulfur from refineries and other hydrocarbon processing plants [47].

Amines Used [48]

Table 27.13 Illustrates the Physical Properties of Gas Treating Chemicals.

| Monoethanolamine | Gas sweetening with monoethanolamine (MEA) is used where there are low contactor pressures and/or stringent acid gas specifications. MEA removes both H_2S and CO_2 from gas streams. H_2S concentrations well, below 4.0 ppm can be achieved. CO_2 concentrations as low as 100 ppm can be obtained at low to moderate pressures. COS and CS_2 are removed by MEA, but the reactions are irreversible unless a reclaimer is used. Even with a reclaimer, complete reversal of the reactions may not be achieved. The result is solution loss and build-up of degradation products in the system. Total acid gas pickup is traditionally limited to 0.3 – 0.35 moles of acid gas/mole of MEA, and solution concentration is usually limited to 10 – 20wt %. Inhibitors can be used to allow much higher solution strengths and acid gas loadings. Because MEA has the highest vapor pressure of the amines used for gas treating solution, losses through vaporization from the contactor and stripper can be high. This problem can be minimized by using a water wash. |

(Continued)

Table 27.13 Illustrates the Physical Properties of Gas Treating Chemicals. (*Continued*)

Diethanolamine	This process employs an aqueous solution of diethanolamine (DEA). DEA will not treat to pipeline quality gas specifications at as low a pressure as will MEA.
	Among the processes, using DEA is the SNPA – DEA process developed by Societe Nationale des Petroles d'Aquitaine (Total) to treat the very sour gas which was discovered in Lacq France in the 1950s. The original patents covered very high acid gas loading of 0.9 to 1.3 moles per mole of amine. This process is used for high pressure, high acid gas content streams having relatively high ratio of H_2S/CO_2. The original process has been progressively improved, and Total through Prosernat is now proposing high DEA solution concentrations up to 40 wt % with the high acid gas loading together with corrosion control by appropriate design and operating procedures.
	Maximum attainable loading is limited by the equilibrium solubility of H_2S and CO_2 at the absorber bottoms conditions. Below are equilibrium solubility values for 40 wt% DEA solutions at 190°F.
	<table><tr><td>H_2S partial pressure, psia</td><td>45</td><td>145</td><td>220</td></tr><tr><td>Mole H_2S/mole amine</td><td>0.66</td><td>0.80</td><td>0.97</td></tr><tr><td>CO_2 partial pressure, psia.</td><td>45</td><td>90</td><td>145</td></tr><tr><td>Mole CO_2/mole amine</td><td>0.49</td><td>0.55</td><td>0.60</td></tr></table>
	Although mole/mole loadings as high as 0.8 – 0.9 have been reported, most conventional DEA plants still operate at significantly lower loadings.
	The process flow scheme for conventional DEA plants resembles the MEA process. The advantages and disadvantages of DEA as compared to MEA are:
	• The mole/mole loadings typically used with DEA (0.35 – 0.82 mole/mole) are much higher than those normally used (0.3 – 0.4) for MEA. • Because DEA does not form a significant amount of non-regenerable degradation products, a reclaimer is not required. Also, DEA cannot be reclaimed at reboiler temperature as MEA can. • DEA is a secondary amine and is chemically weaker than MEA, and less heat is required to strip the amine solution. • DEA forms a regenerable compound with COS and CS_2 and can be used for the partial removal of COS and CS_2 without significant solution losses.

(*Continued*)

Table 27.13 Illustrates the Physical Properties of Gas Treating Chemicals. (*Continued*)

Diglycolamine®	This process uses Diglycolamine® brand [2-(2-aminoethoxy)] ethanol in an aqueous solution. DGA® is a primary amine capable of removing not only H_2S and CO_2, but also COS and mercaptans from gas and liquid streams. Because of this, DGA® has been used in both natural and refinery gas applications. DGA® has been used to treat natural gas to 4.0 ppmv at pressures as low as 125 psig. DGA® has a greater affinity for the absorption of aromatics, olefins, and heavy hydrocarbons than the MEA and DEA systems. Therefore, adequate carbon filtration should be included in the design of a DGA® treating unit. The process flow for the DGA® treating process is similar to that of the MEA treating process. The three major differences are: • Higher acid gas pick-up per gallon of amine can be obtained by using 50 – 60% solution strength rather than 15 – 20% for MEA (more moles of amine per volume of solution). • The required treating circulation rate is lower. This is a direct function of higher amine concentration. • Reduced reboiler steam consumption. Typical concentrations of DGA® range from 50 - 60% DGA® by weight while in some cases as high as 70 wt% has been used. DGA® has an advantage for plants operating in cold climates where freezing of the solution could occur. The freezing point for 50% DGA® solution is – 30°F. Because of the high amine degradation rate DGA® systems require reclaiming to remove the degradation product. DGA® reacts with both CO_2 and COS to form N, bis (hydroxyethoxyethyl) urea, generally referred to as BHEEU. DEA is recovered by reversing the BHEEU reaction in the reclaimer.
Methyldiethanolamine	Methyldiethanolamine (MDEA) is a tertiary amine, which can be used to selectively remove H_2S to pipeline specifications at moderate to high pressure. If increased concentration of CO_2 in the residue gas does cause a problem with contract specifications or downstream processing, further treatment will be required. The H_2S/CO_2 ratio in the acid gas can be 10 – 15 times as great as the H_2S/CO_2 ratio in the sour gas. Some of the benefits of selective removal of H_2S include: • Reduced solution flow rates resulting from a reduction in the amount of acid gas removed. • Smaller amine regeneration unit. • Higher H_2S concentrations in the acid gas resulting in reduced problems in sulfur recovery. CO_2 hydrolyzes much slower than H_2S. This makes it possible for a significant selectiveness of tertiary amines for H_2S. This fact is used by several companies who provide process designs using MDEA for selective removal of H_2S from gases containing both H_2S and CO_2. A feature of MDEA is that it can be partially regenerated in a simple flash. As a consequence, the removal of bulk H_2S and CO_2 may be achieved with a modest heat input for regeneration. However, as MDEA solutions react, only slowly with CO_2, activators must be added to the MDEA solution to enhance CO_2 absorption and the solvent is then called activated MDEA.

(*Continued*)

Table 27.13 Illustrates the Physical Properties of Gas Treating Chemicals. (*Continued*)

Triethanolamine	Triethanolamine (TEA) is a tertiary amine and has exhibited selectivity for H_2S over CO_2 at low pressures. TEA was the first amine commercially used for gas sweetening. It was replaced by MEA and DEA because of its inability to remove H_2S and CO_2 to low outlet specifications. TEA has potential for the bulk removal of CO_2 from gas streams. It has been used in many ammonia plants for CO_2 removal.
Diisopropanolamine	Diisopropanolamine (DIPA) is a secondary amine which exhibits, though not as great as tertiary amines, selectivity for H_2S. This selectivity is attributed to the steric hinderance of the chemical.
Formulated Solvents and Mixed Amines	Formulated Solvents is the name given to a new family of amine-based solvents. Their popularity is primarily due to equipment size reduction and energy savings over most of the other amines. All the advantages of MDEA are valid for the Formulated Solvents, usually to a greater degree. Some formulations are capable of slipping larger portions of inlet CO_2 (than MDEA) to the outlet gas and at the same time removing H_2S to less than 4 ppmv. For example, under conditions of low absorber pressure, and high CO_2/H_2S ratios, such as Claus tail gas clean-up units, certain solvent formulations can slip upwards to 90% of the incoming CO_2 to the incinerator. While at the other extreme, certain formulations remove CO_2 to a level suitable for cryogenic plant feed. Formulations are also available for CO_2 removal in ammonia plants. Finally, there are solvent formulations which produce H_2S to 4 ppmv pipeline specifications, while reducing high inlet CO_2 concentrations to 2% for delivery to a pipeline. This case is sometimes referred to as bulk CO_2 removal. This need for a wide performance spectrum has led formulated solvent suppliers to develop a large stable of different MDEA – based solvent formulations. Most formulated solvents are enhancements to MDEA. Thus, they are referred to as MDEA-based solvents or formulations. Benefits claimed by suppliers are: *For New Plants* - Reduced corrosion - Reduced circulation rate - Lower energy requirements. - Smaller equipment due to reduced circulation rates. *For Existing Plants* - Increase in capacity, i.e., gas through put or higher inlet acid gas composition. - Reduced corrosion. - Lower energy requirements and reduced circulation rates. Formulated Solvents are proprietary to the specific supplier offering the product. Companies offering these products and /or processes include INEOS, Hunstman Corporation, Dow Chemical Company, UOP, BASF, Shell Global Solutions and TotalFinalElf via Prosernat.
Sterically Hindered Amines	Other amines have been used to treat sour gas. One specialty amine has been constructed by a process defined as steric hindrance. The actual structure of the amine has been formed to accommodate a specific process requirement. This type of amine and the associated technology is different than Formulated Solvents, which create the desired formulations by blending different components with a standard amine such as MDEA. An example of this technology is FLEXSORB® solvents, marketed by ExxonMobil Research and Engineering Company.

27.9.2 Equilibrium Data for Amine–Sour Gas Systems

A distinctive characteristic of amine treating systems is the interactive effects of one acid gas constituent with amine upon the equilibrium partial pressures of the other constituent. The most commonly encountered sour gas constituents are H_2S and CO_2. The capacity of a given amine for either one of the acid gas constituents alone is much greater than when the two occur together. Jones et al. [49] have presented data to confirm the interactive effect of H_2S and CO_2 in monoethanolamine solutions. Lee et al. [50] have presented similar data for diethanolamine solutions. Dingman et al. [51] have presented data for H_2S and CO_2 in equilibrium with commercially used concentrations of DGA®. These data provide the basis for predicting the equilibrium concentrations of MEA, DEA and DGA® solutions when in contact with sour gas containing both H_2S and CO_2.

Jou, Otto and Mather [52] investigated the solubility of H_2S and CO_2 in MDEA and further investigated the solubility of mixtures of H_2S and CO_2 in MDEA [53] respectively. The first paper presents solubilities of H_2S in MDEA and solubility of CO_2 in MDEA, whereas the second report presents the results of varying mixtures of H_2S and CO_2 and their solubility in MDEA.

Kent and Eisenberg [54] proposed a reaction equilibrium model to correlate/predict the vapor-liquid equilibrium between H_2S and CO_2 and primary or secondary ethanolamines. They tested the model extensively against the data for MEA and DEA with good results. This model allows for interpolation/extrapolation of equilibrium data to compositions and temperatures where no measurements have been made.

27.9.3 Emerging Technologies [48]

Alkanolamine Processes	A compact alkanolamine process (CAP) based on the ProPure co-current contacting technology is being developed by ProPure Purification, a subsidiary of Statoil. It is a continuous selective trace H_2S removal process. It relies on the rapid and selective tertiary alkanolamine chemistry with H_2S used in a novel co-current short residence time contactor, further enhancing the H_2S selectivity. The process is ideally suited for offshore applications where low weight and foot print is a premium. It competes well with conventional scavenging processes as the H_2S containing solvent is regenerated and thus scavenger chemical consumption is avoided.
Hybrid Solvent Processes	A modern version of the Amisol process has been developed by IFP together with Total called Hybrisol [55]. This process combines the physical solvency attributes of methanol with the chemical reactivity of alkanolamines (DEA or MDEA). It is integrated upstream with the Ifpex-1 dehydration technique and when coupled with a downstream cold process achieves simultaneous dehydration, complete acid gas and mercaptan removal together with sweet NGL extraction. It also produces dry acid gas at some pressure.
Highly Sour Gas Pretreatment Processes	In the Sprex process developed by IFP together with Total [55], a substantial part of the acid gases, notably the H_2S from the inlet gas is pre-extracted in a cyclone or column by cold reflux as a bottoms pumpable liquid at line pressure and ambient temperature. The extracted liquid also contains essentially all the water of saturation of feed gas so no further dehydration nor inhibition is required for the downstream cold condensing process. The liquid also contains some limited amount of hydrocarbons.
	An Acid Gas Fractionating process developed by Kvaerner Process Systems [56] utilizes refrigeration and a downstream fractionation column to liquefy the acid gas components as a bottoms product for disposal. The process claims high extraction levels of H_2S and CO_2 with limited hydrocarbon losses with use of the fractionation column.

Claus Reaction Processes	An alternative to liquid redox processes has recently been developed [57, 58]. CrystaSulf, marketed by CrystaTech, Inc., is a technology that utilizes a non-aqueous solution to absorb H_2S from the process stream, and reacts it with dissolved SO_2 to form dissolved elemental sulfur in the absorber via the modified Claus reaction. The proprietary non-aqueous solution does not absorb CO_2, and therefore is not affected by high partial pressures of CO_2.
	After contact with the process stream, the solution is flashed to lower pressure. The flash gas may be vented to fuel gas, or to a flare, or may be compressed and returned to the inlet, if desired. The solution keeps elemental sulfur dissolved until the temperature is reduced from the absorber temperature of 150–160°F to the crystallizer temperature of 90–120°F. There, elemental sulfur precipitates from the solution. Since the elemental sulfur is kept dissolved in the absorber and other pressurized portions of the plant, the problem of sulfur plugging in high-pressure equipment is thus avoided, and the process can be applied to high-pressure gas streams.
Molecular Sieves	Engelhard has developed the Molecular Gate™ technology using titanium silicate materials whose pore size can be adjusted to +/- 0.1 Angstrom. These materials can be used to preferentially adsorb molecules on a size exclusion basis. For example, a material with 3.7 Angstroms pore diameter will adsorb nitrogen with a molecular diameter of 3.6 Angstroms but pass methane with a molecular diameter of around 3.8 Angstroms
	The Molecular Gate™ system utilizes a pressure swing adsorption (PSA) process. It can also be designed for the simultaneous removal of CO_2 and nitrogen from natural gases containing both impurities [59].
Gas/Liquid Membranes	Kavaerner Process Systems and its joint industry partners are in the final stages of commercializing a novel membrane contactor technology based on highly resistant inert PTFE materials from WL Gore and Associates [60].
	The contactor can be applied in a range of gas processing applications with amines including both low pressure removal of CO_2 from exhaust gases and high-pressure natural gas sweetening. Benefits include substantial reduction of volume and weight contactor and reduced energy requirements.
Membranes for CO_2 Removal from Liquid Ethane	Studies conducted in a pilot plant utilizing a full-scale membrane element have verified that membranes can be used to remove CO_2 from a liquid ethane stream. The membrane module consists of an asymmetric hollow fiber made of cellulose-acetate. Tests have shown acceptable ethane losses, while providing a secondary benefit of removing some methane from the ethane product [61].

Batch Scavenging Processes	The Gas Research Institute (GRI) has sponsored an evaluation program for H_2S scavenging technologies [62]. Among the novel/emerging scavengers tested are the following: • Sulfa-Scrub HSW-700L from Baker-Petrolite. • GasTreat 136 from Champion Technologies. • GasTreat 155 from Champion Techologies. • Swan MSS-58 from Swan Industries Inc. • Sulfarid 8411 from Edmunds and Associates Inc. • Quaker Enviro-Tek from Quaker Chemicals Corp. • DM – 5927 from BetzDearborn. • Nalco/Exxon EC-5492A from Nalco/Exxon Energy Chemical LP. • Dynea and Statoil have patented and developed an H_2S scavenger called Dyno HR2707 which is reported to be very efficient at low temperatures and very low levels of H_2S [63].
Controlled – Freeze Zone (CFZ) Process	This ExxonMobil process removes CO_2, H_2S and other impurities from natural gas. The unique aspect of the process is that it induces CO_2 freezing in an open area of a cryogenic fractionation column [64]. The Ryan Holmes process, described adds a hydrocarbon stream to suppress CO_2 freezing in the distillation column. The solid CO_2 melts in the stripping section of the CFZ column, and flows out as a liquid along with the other contaminants. The liquid stream can then be easily pumped for downhole disposal. Methane (CH_4), nitrogen (N_2) and helium (He) come out overhead.

Reactions of MEA with CO_2 produce two chemical reactions namely: the carbamate and bicarbonate salts. In carbon capture and storage, the bicarbonate is needed as it is heat sensitive and it is easy to reverse the process and break down the reaction into the starting amine and carbon dioxide, while the carbamate salt is rather stable and easy to break up. However, more energy is required for the process and the disadvantage of MEA solvent is the formation of the carbamate which impacts the efficiency of the process and thus requires more energy input.

Chemistry

27.9.4 Advanced Amine Based Solvents

The advanced based solvents reduce the amount of carbamate and generate more of the bicarbonate.

Amine blends that are activated by concentrated piperazine are used extensively in commercial CO_2 removal for carbon capture and storage (CCS) because piperazine allows for protection from significant thermal and oxidative degradation at typical coal flue gas conditions. The thermal degradation rates for methyl diethanolamine (MDEA) and piperazine (PZ) are negligible and PZ, unlike other metals protect MDEA from oxidative degradation. This increased stability of the MDEA/PZ solvent blend over MDEA and other amine solvents provides for greater capacity for and requires less work to capture a given amount of CO_2.

Piperazine's solubility is low, and it is used in relatively small amounts to supplement another amine solvent. It is low in concentration, and the CO_2 absorption rate, heat of absorption and solvent capacity are increased through the addition of piperazine to amine gas treating solvents, the most common is MDEA due to its unmatched high rate and efficiency.

Chemistry

Piperazine chemical formula $(C_4H_{10}N_2)$ is formed as a co-product in the ammoniation of 1,2 dichloroethane or ethanolamine. These are the only routes to the chemical that is used commercially. Piperazine is separated from the product stream, which contains ethylenediamine, diethylenetriamine, and other related linear and cyclic chemicals of this type.

The amine groups on piperazine react readily with carbon dioxide to produce PZ carbamate at a low loading (mol CO_2/equiv. PZ) range and PZ dicarbamate at an operating range of 0.31-0.41 mol CO_2/equiv. PZ, enhancing the rate of overall CO_2 absorbed under operating conditions. Due to these reactions, there is limited free piperazine present in the solvent, resulting in its low volatility and rates of precipitation as PZ-$6H_2O$.

(a) At low loading
(b) At operating range of concentrated piperazine process.

Piperazine (PZ) reacts with carbon dioxide to produce PZ carbamate and PZ bicarbamate at low loading and operating range respectively.

Disadvantages of Amine Solvents

- Cost of capture is high ($50 – 90/tonne CO_2 captured), about 50% on the cost of electricity.
- Certain amines and their degradation products are hazardous requiring high expensive emission control.

- Corrosive, especially when adding CO_2 thus high-grade material of construction such as high grade stainless, which can be rather expensive.
- Difficult to recycle to dispose of, and end of life.

To avoid the carbamate and bicarbonate reactions with the amines, use of carbonate solvents are considered. A typical carbonate is potassium carbonate (K_2CO_3) as used in the oil and gas sector as in UOP Benfield process. It is cheap, readily available and non-toxic. Furthermore, degradation issues are largely avoided as K_2CO_3 is degraded to carbon dioxide (CO_2) and water (H_2O). Furthermore, it reacts very slowly and thus tricks are required for working at low pressure P_{CO_2}.

27.10 Alkaline Salt Process (Hot Carbonate)

The basic process was developed by the US Bureau of Mines and employs an aqueous solution of potassium carbonate (K_2CO_3). The contactor and stripper both operate at temperatures in the range of 110–116°C. The process is not suitable for gas streams containing only H_2S. If H_2S is to be removed to pipeline specification or there are low CO_2 outlet specifications, special designs or a two-stage system may have to be used. The hot potassium carbonate process employs an aqueous solution of potassium carbonate (K_2CO_3) to remove both CO_2 and H_2S. It also removes some carbonyl sulfide (COS) and carbon disulfide (CS_2) through hydrolysis. The following reactions are reversible based on the partial pressures of the acid gases occur in this process.

The overall reactions for CO_2 and H_2S with potassium carbonate can be represented by:

$$K_2CO_3 + CO_2 + H_2O \rightleftharpoons 2KHCO_3 \tag{27.15}$$

$$K_2CO_3 + H_2S \rightleftharpoons KHS + KHCO_3 \tag{27.16}$$

It has been demonstrated that the process works best near the temperature of reversibility of the reactions. However, a high partial pressure of CO_2 is required to keep $KHCO_3$ in solution and H_2S will not react if the CO_2 pressure is not high. For this purpose, this process cannot achieve a low concentration of acid gases in the exit stream. Special designs or a two-stage system are therefore required for removing H_2S to pipeline specification or to reduce CO_2 to low levels.

The hot carbonate process as shown in Figure 27.29 is referred to as the "hot" process because both the absorber and stripper operate at elevated temperatures in the range of 230°F – 240°F (110 °C– 116°C). In this process, the sour gas enters at the bottom of the absorber and flows countercurrently to the carbonate liquid stream. The sweet gas exits at the top of the absorber. The rich carbonate solution exits from the bottom of the absorber and is flashed in the stripper, where acid gases are driven off. The lean carbonate solution is pumped back to the absorber. The lean solution may or may not be cooled slightly before entering the absorber. The main disadvantage of this solution is linked to the need to replace the water that saturates the treated gas at high temperatures. To reduce this phenomenon while simultaneously improving the unit's energy balance a common practice is to install a heat exchanger upstream of the absorber, between the sour gas and the sweet gas, with the recovery of condensed water from the latter.

In this process, the entire system is operated at high temperature to increase K_2CO_3 solubility. Thus, a dead spot where the solution is likely to cool and precipitate should be avoided. If solids do precipitate the system may suffer from plugging, erosion or foaming. Potassium carbonate (K_2CO_3) causes general stress corrosion of the unit. Therefore, all carbon steel must be stress-relieved to limit corrosion. Various corrosion inhibitors are available to decrease corrosion. However, the solvents react with some corrosion inhibitors and cause erosion of the unit. The hot potassium carbonate process is particularly useful for removing large quantities of CO_2. The main advantages of carbonate solutions for CO_2 removal are the high chemical solubility of CO_2 in the carbonate/bicarbonate system and low solvent costs. The major difficulty is a relatively slow reaction in the liquid phase, causing low mass transfer rates and therefore requiring a large contact surface.

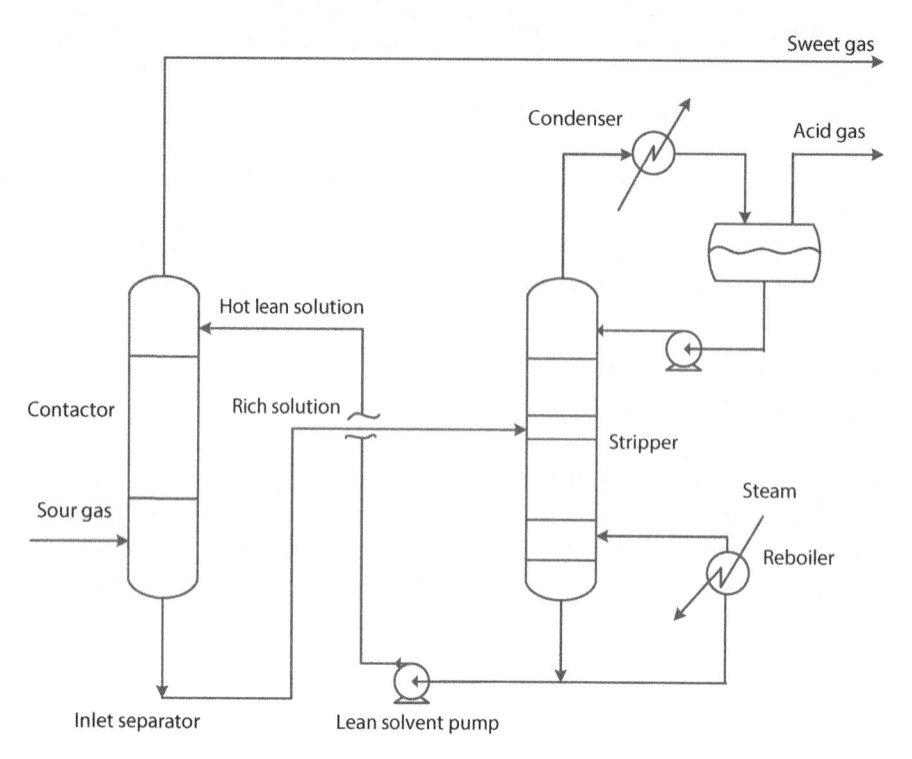

Figure 27.29 Alkaline salt: A single stage process.

Process improvements have been obtained through the use of catalytic additions and promoters to the solution, such as DEA, arsenic trioxide, selenous acid and tellurous acid. In applications for the removal of hydrogen sulfide (H_2S), tripotassium phosphate (K_3PO_4) may be used. These activators increase the performance of the hot potassium carbonate system by increasing the reaction rates both in the absorber and stripper. In general, these processes also decrease corrosion in the system.

Split Flow Process of Potassium Carbonate Process

In this process scheme (Figure 27.30) the lean solution stream is split. The hot solution is fed to the middle of the contact for bulk removal. The remainder is cooled to improve equilibrium and is fed to the top of the contactor for trim acid gas removal.

Two Stage Process

Here, the contactor is like the split flow process. In addition, the stripper is in two sections (Figure 27.31). A major portion of the solution is removed at the mid-point of the stripper and pumped to the lower section of the contactor. The remainder is further stripped with steam and then cooled prior to entering the top of the contactor.

Numerous improvements have been achieved to the potassium carbonate process resulting in significant reduction in capital (CAPEX) and operating costs (OPEX). At the same time, lower acid gas concentration in the treated gas can now be achieved. The most popular of the carbonate processes are:

Benfield Process: The Benfield process is licensed by UOP. Several activators are used to enhance the performance of the potassium carbonate solution.

Hi-Pure Process: The Hi-Pure process is a combination conventional Benfield K_2CO_3 process and alkanolamine process. The gas stream is first contacted with K_2CO_3 followed by contacting with an amine. The process can achieve outlet CO_2 concentrations as low as 30 ppmv and H_2S concentration of 1 ppmv.

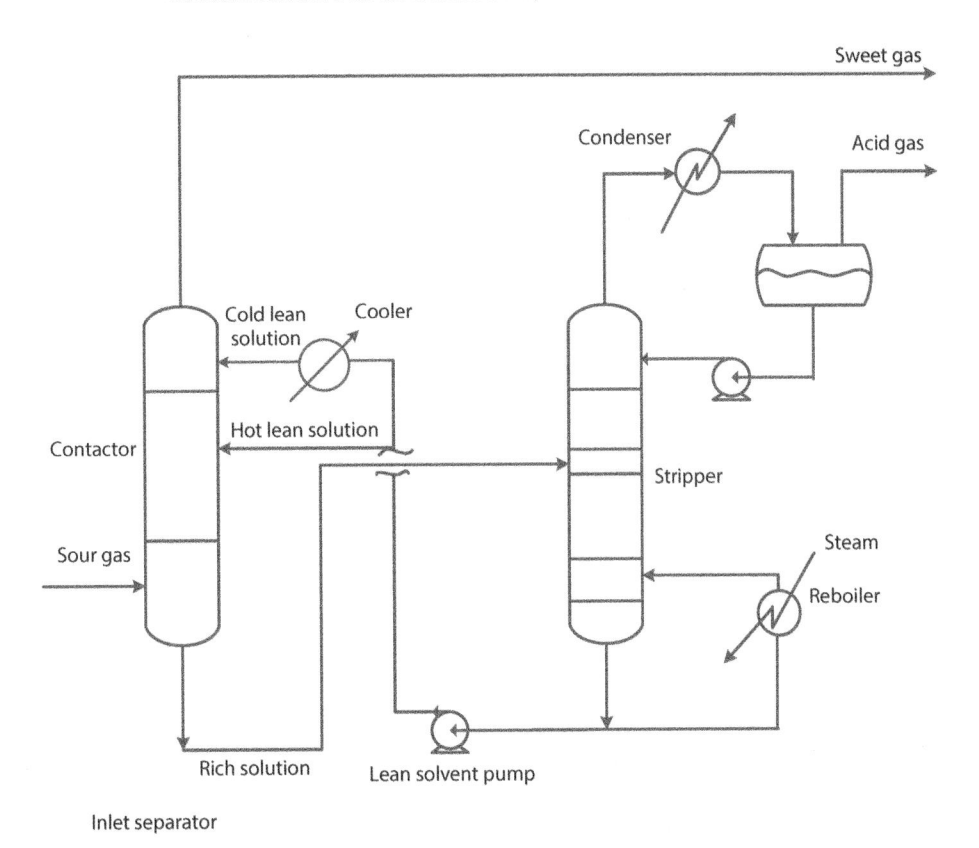

Figure 27.30 Alkaline salt: A split flow process.

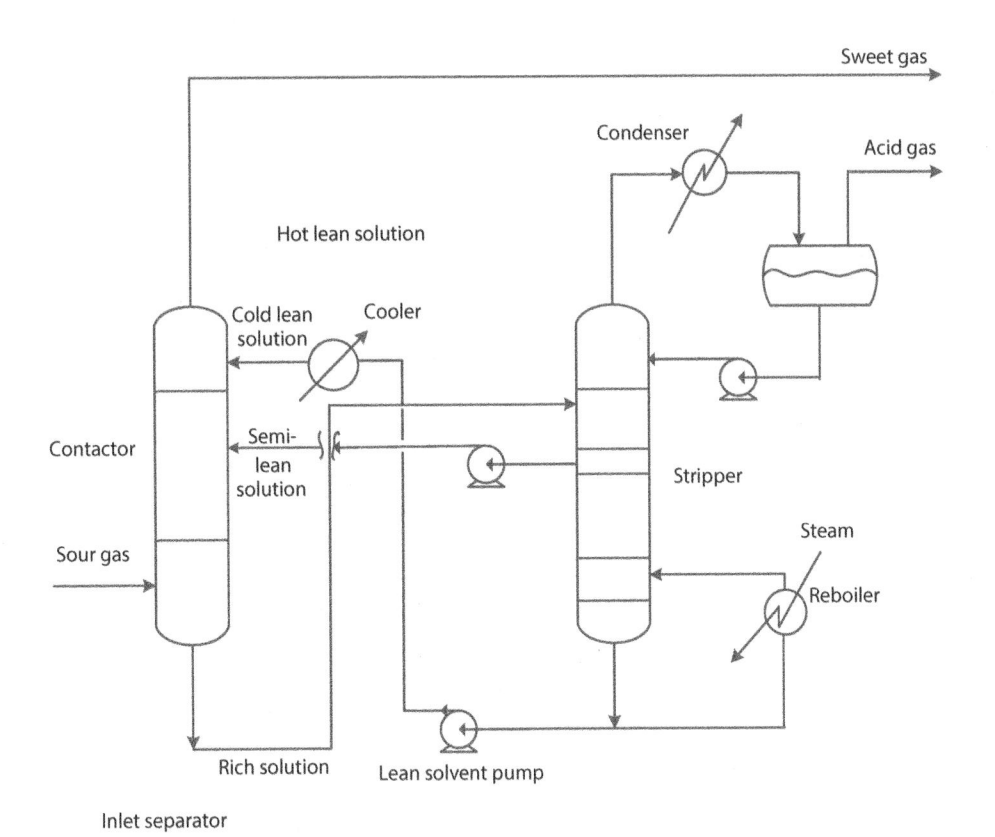

Figure 27.31 Alkaline salt: A two-stage process.

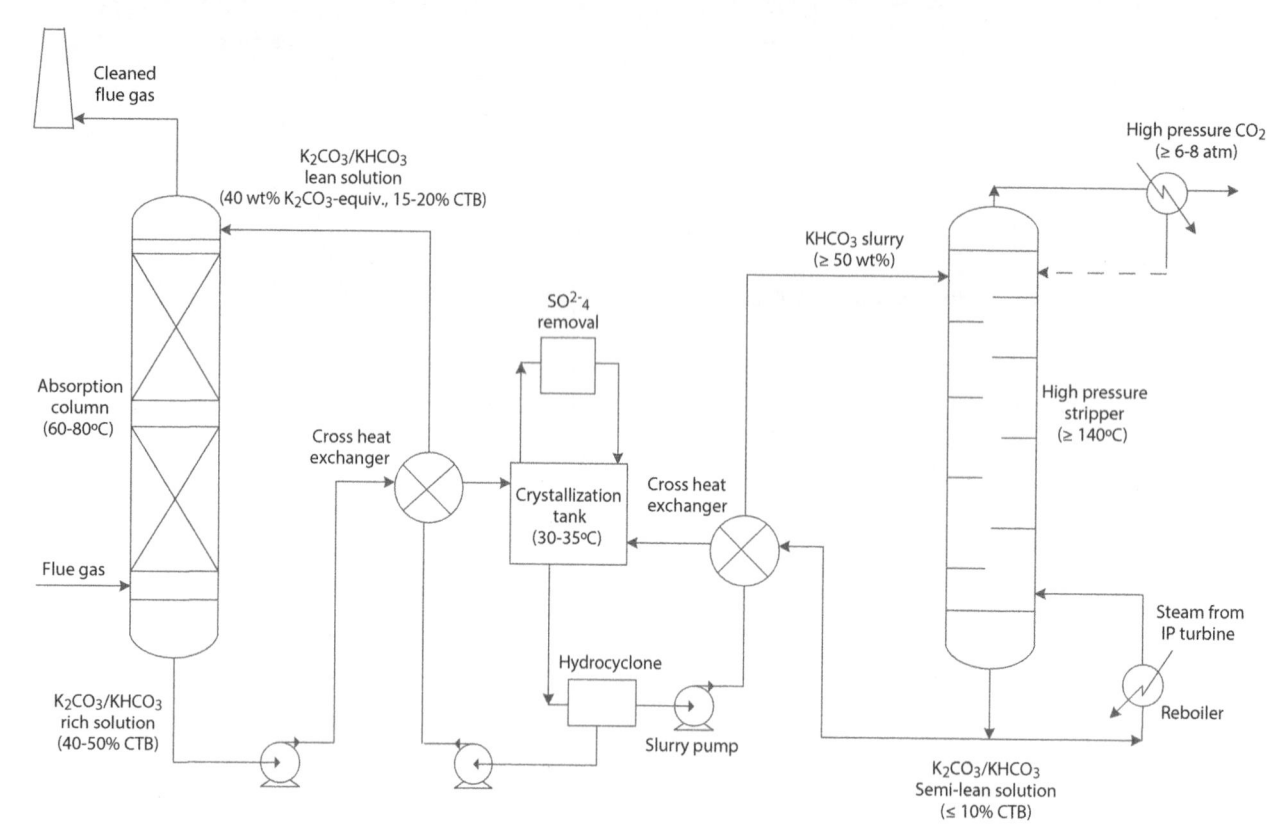

Figure 27.32 Flow diagram of the carbonates process.

Catacarb Process: The Catacarb process is licensed by Eickmeyer and Associates. Activators, corrosion inhibitors, potassium salts, and water are contained in the solution. This process is mostly used in the ammonia industry.

Figure 27.32 shows the flow diagram of the carbonates solution process.

27.11 Ionic Liquids

Ionic liquids are salts that are liquid at or near to ambient temperature (< 100°C). CO_2 is readily absorbed into the liquid as it has a very low vapor pressure. The use of ionic liquids in carbon capture is a potential application of ionic liquids as absorbents for use in carbon capture and sequestration. They are polar, non-volatile materials that have been considered for many applications. MEA has been used in industrial scales in post combustion carbon capture as well as in other CO_2 separations such as "sweetening" of natural gas. However, amines are corrosive, degrade over time and require large industrial facilities. Ionic liquids on the other hand have low vapor pressures. This property results from their strong Coulombic attractive force. Vapor pressure remains low through the substance's thermal decomposition point (> 300°C). This low vapor pressure simplifies their use and make them "green" alternatives. Additionally, it reduces the risk of contamination of the CO_2 gas stream and of leakage into the environment.

The solubility of CO_2 in ionic liquids is governed primarily by the anion and less by the cation. The hexafluorophosphate (PF_6^-) and tetrafluoroborate (BF_4^-) anions have been shown to be especially amenable to CO_2 capture.

Disadvantages

In carbon capture, an effective absorbent is one which demonstrates a high selectivity, this implies that CO_2 will preferentially dissolve in the absorbent compared to other gaseous components. In post-combustion carbon capture, the most salient separation is CO_2 from N_2, whereas in pre-combustion separation CO_2 is primarily separated from H_2. Other components and impurities may be present in the flue gas such as hydrocarbons, SO_2, or H_2S. Before selecting the appropriate solvent to use for carbon capture, it is essential to ensure that at the given process conditions and flue gas composition, CO_2 maintains a much higher solubility in the solvent than the other species in the flue gas and thus a high selectivity.

The selectivity of CO_2 in ionic liquids has been widely studied and generally, polar molecules with an electric quadrupole moment are highly soluble in liquid ionic substances. At high process temperatures, the solubility of CO_2 decreases while the solubility of other species such as CH_4 and H_2 may increase with increasing temperature, thus reducing the effectiveness of the solvent. However, the solubility of N_2 in ionic liquids is relatively low and does not increase with increasing temperature, so the ionic liquids in post-combustion carbon capture may be appropriate due to the consistently high CO_2/N_2 selectivity. The presence of common flue gas impurities such as H_2S severely inhibits CO_2 solubility in ionic liquids and great caution is required when choosing an appropriate solvent for a particular flue gas.

Viscosity

Ionic liquids are highly viscosity compared with that of commercial solvents. Ionic liquids which employ chemisorption depend on a chemical reaction between the solute and solvent for CO_2 separation. The rate of this reaction is dependent on the diffusivity of CO_2 in the solvent and is inversely proportional to viscosity. The self-diffusivity of CO_2 in ionic liquids are generally to the order of 10^{-10} m^2/s, approximately an order of magnitude less than similarly performing commercial solvents used on CO_2 capture. The viscosity of an ionic liquid can vary significantly depending on the type of anion and cation, the alkyl chain length and the amount of water or other impurities in the solvent.

Tunability

Ionic liquids exhibit selectivity towards one or more of the phases of a mixture. 1-Butyl-3-methlyimidazolium hexafluorophosphate (BMIM-PF_6) is a room-temperature ionic liquid that was used as a viable substitute for volatile organic solvents in liquid-liquid separations. Others, such as hexafluorophosphate (PF_6^-) and tetrafluoroborate (BF_4^-) containing ionic liquids have been studied for their CO_2 absorption properties as well as 1-ethyl-3-methylimidazolium (EMIM) and unconventional cations like trihexyl (tetradecyl) phosphonium (P_{66614}). Selection of different anion and cation combinations in ionic liquids affects their selectivity and physical properties. Additionally, the organic cations in ionic liquids can be "tuned" by changing chain lengths or by substituting radicals. Ionic liquids can be mixed with other ionic liquids, water, or amines to achieve different properties in terms of absorption capacity and heat of absorption. This tunability has led some to refer ionic liquids as "designer solvents". 1-butyl-3-propylamineimidazolium tetrafluoroborate was specifically developed for CO_2 capture as it is designed to employ chemisorption to absorb CO_2 and maintain efficiency under repeated absorption/regeneration cycles. Other ionic liquids have been studied and simulated for potential use of as CO_2 absorbents.

Currently, CO_2 capture uses mostly amine-based absorption technologies, which are energy intensive and solvent intensive. Volatile organic compounds alone in chemical processes represent a multibillion-dollar industry. Therefore, ionic liquids present an attractive alternative should their deficiencies be resolved.

During the capture process, the anion and cation play a crucial role in the dissolution of CO_2. Spectroscopic results suggest as favorable interaction between the anion and CO_2, wherein CO_2 molecules preferentially attach to the anion. Furthermore, intermolecular forces, such as hydrogen bonds, van der Waals bonds, and electrostatic attraction, contributes to the solubility of CO_2 in ionic liquids. This makes ionic liquids promising candidates for CO_2 capture because the solubility of CO_2 can be modeled accurately by the regular solubility theory (RST), which reduces operational costs in developing more sophisticated model to monitor the capture process.

The structure of an ionic liquid 1-butyl-3-propylamineimidazolium tetrafluoroborate is:

1 Butyl-3-propylamineimidazolium tetrafluoroborate is a task-specific ionic liquid for use in CO_2 separation.

A Case Study of Acid Gas Sweetening with DEA (Schlumberger and Honeywell UniSim® Design Suite R470 Technology)

A typical acid gas treating facility is simulated using Honeywell UniSim® Design Suite R470. A water-saturated natural gas is fed to an amine contactor. Diethanolamine (DEA) at a strength of 28 wt% in water is used as the absorbing medium. The contactor consists of 20 real stages. The rich amine is flashed from the contactor pressure of 6900 kPa to 620 kPa before it enters the rich/lean amine exchanger, where it is heated to the regenerator feed temperature of 95°C. The regenerator also consists of 20 real stages. Acid gas is rejected from the regenerator at 50°C, while the lean amine is produced at approximately 110°C. The lean amine is cooled and recycled to the contactor. Figure 27.33 shows the snapshot of the process flow diagram generated by Honeywell UniSim Design R470 software and Figure 27.34 highlights the tables of the main equipment items showing the process parameters such as molar flow rates, pressures and temperatures.

Learning Objectives:

- Simulate Amine towers in UniSim Design.

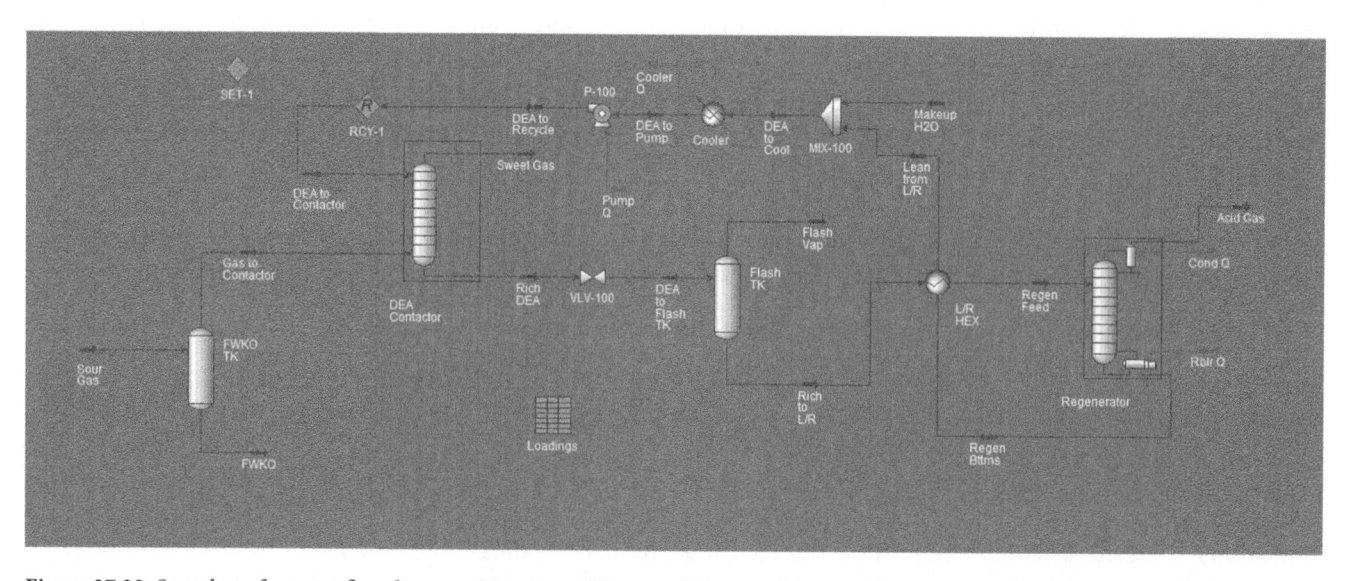

Figure 27.33 Snapshot of process flow diagram. (Courtesy of Honeywell Process Solution, UniSim Design® R470, Honeywell® and UniSim® are registered trademarks of Honeywell International Inc.)

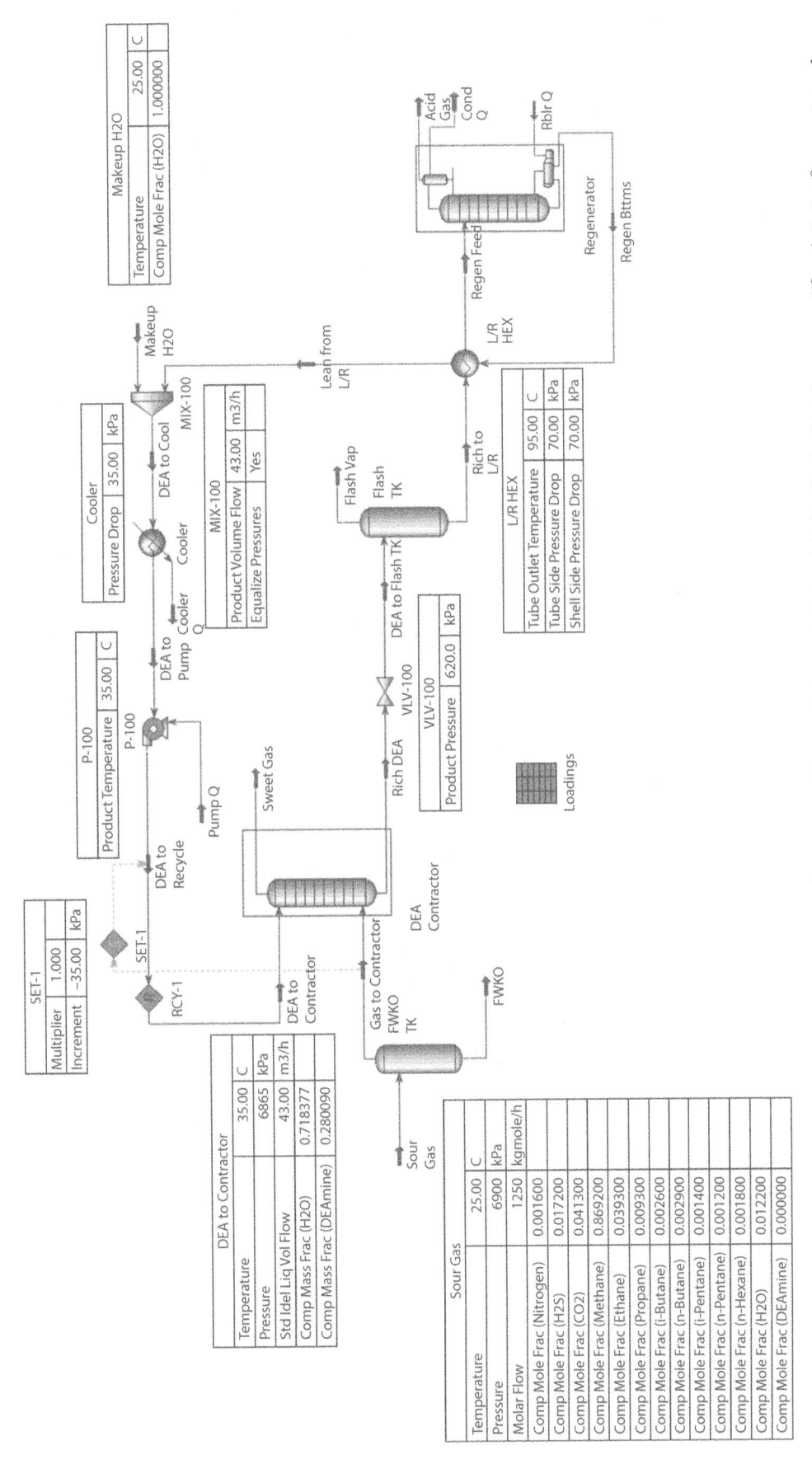

Figure 27.34 Process flow diagram with tables of the equipment items. (Courtesy of Honeywell Process Solution, UniSim Design® R470, Honeywell® and UniSim® are registered trademarks of Honeywell International Inc.)

- Supply tray dimensions to calculate component efficiencies for Amine towers.
- Use the Set operations.
- Use the Spreadsheet.

Building the Simulation

Defining the Simulation Basis

Use the DBRAmine property package with the following components:

Nitrogen	n-Butane
H_2S	i-Pentane
CO_2	n-Pentane
Methane	n-Hexane
Ethane	H_2O
Propane	DEAmine
i-Butane	

1. Set Up a New Case.
2. **Use the DBRAmine** propery package with the components above. Use the **Kent-Eisenberg** model.

Amines Property Package

The Amines package contains the thermodynamic models developed by D.B. Robinson & Associates for their proprietary amine plant simulator, AMSIM. Their equilibrium acid gas solubility and kinetic parameters for aqueous alkanolamine solutions in contact with H_2S and CO_2, have been incorporated into this property package. The Amines property package has been fitted to extensive experimental data gathered from a combination of D. B. Robinson's in-house data and numerous technical references.

The Amines package incorporated a specialized stage efficiency model to permit the simulation of columns on a real tray basis. The stage efficiency model calculates H_2S and CO_2 component stage efficiencies based on the tray dimensions and the calculated internal tower conditions for both absorbers and strippers. Figures 27.35 – 27.54 show the various steps in the simulation exercise of the case study of Acid Gas Sweetening with DEA.

Column Overview

Figures 27.35 and 27.36 show the flow diagrams of the contactor and regenerator respectively with specifications of the temperatures and pressures.

Contactor

Adding the Basics

Adding the feed streams

3. Add a new stream for the inlet gas with the following values:

TS-1		
Current Number of Stages	20	
Top Stage Temperature	35.22	C
Bottom Stage Temperature	56.78	C
Top Stage Pressure	6850	kPa
Bottom Stage Pressure	6900	kPa

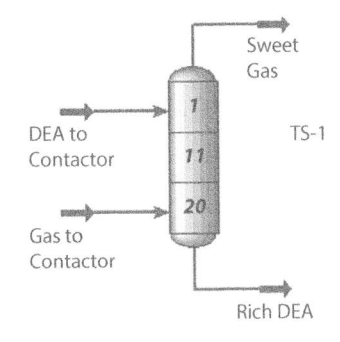

Figure 27.35 Contactor showing the inlet and outlet streams.

Main TS		
Current Number of Stages	18	
Top Stage Temperature	101.6	C
Bottom Stage Temperature	124.4	C
Top Stage Pressure	205.0	kPa
Bottom Stage Pressure	220.0	kPa

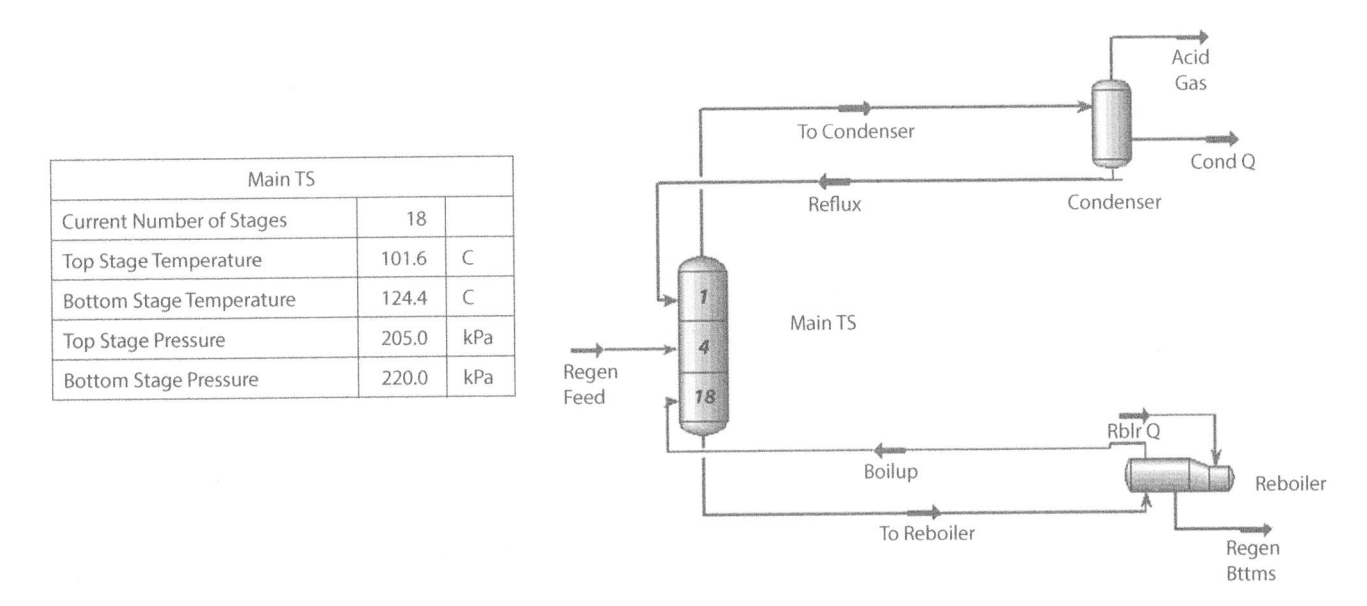

Figure 27.36 Regenerator showing the inlet and outlet streams.

In this cell...	Enter...
Name	Sour Gas
Temperature	25°C (75°F)
Pressure	6900 kPa (1001 psia)
Molar Flow	1250 kgmole/h (25 MMSCFD)
Component	**Mole Fraction**
Nitrogen	0.0016
H2S	0.0172
CO2	0.0413
Methane	0.8692
Ethane	0.0393
Propane	0.0093
i-Butane	0.0026

n-Butane	0.0029
i-Pentane	0.0014
n-Pentane	0.0012
n-Hexane	0.0018
H2O	0.0122
DEAmine	0.0

4. Add a second stream for the lean amine feed to the amine contactor with the following values:

In this cell...	Enter...
Name	DEA to Contactor
Temperature	35°C (95°F)
Pressure	6850 kPa (994 psia)
Std Ideal Liq Vol Flow	43 m3/h (190 USGPM)
Component	**Mass Fraction**
H2O	0.72
DEAmine	0.28

Note: Ensure that the DEAmine and H_2O compositions are entered on a weight (mass) basis.

The values for the stream DEA to Contactor will be updated once the recycle operation is installed and has been calculated.

Physical Unit Operations

This section shows the separator and the contactor vessels. The sour gas is fed to the separator and the outlet streams are the gas to the contactor and FWKO liquid. In the contactor, DEA solution is used to strip the H_2S and CO_2 from the gas. The overhead outlet is the sweet gas, and the bottoms liquid outlet is the rich DEA solution.

Separator Operation

Any free water carried with the gas is first removed in a Separator operation, FWKO TK.

5. Add a Separator and provide the following information.

In this cell...	Enter...
Connections	
Name	FWKO TK
Inlet	Sour Gas
Vapour outlet	Gas to Contactor
Liquid Outlet	FWKO

The flow rate of water in FWKO is 14.26 kg mole/h.

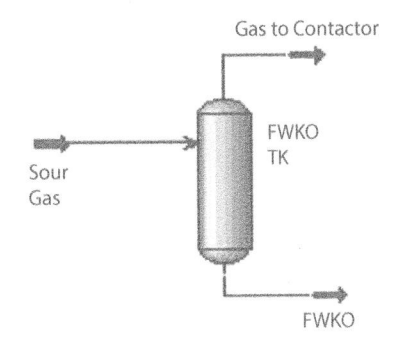

Sustainability in Engineering, Petroleum Refining and Alternative Fuels

Figure 27.37 FWKO separator with inlet and outlet streams.

Figure 27.37 shows the flow diagram of FWKO units.

Contactor Operation

The amine contactor is simulated using an Absorber operation in UniSim Design.

6. Add an Absorber column operation with the following specifications:

In this cell...	Enter...
Connections	
Name	DEA Contactor
No. of Stages	20
Top Stage Inlet	DEA to Contactor
Bottom Stage Inlet	Gas to Contactor
Ovhd Vapour Outlet	Sweet Gas

Note: The Input Experts are toggled on and off from the Simulation tab of the Preferences view.

In this cell...	Enter...
Bottoms Liquid Outlet	Rich DEA
Pressures	
Top	6850 kPa (994 psia)
Bottom	6900 kPa (1001 psia)
Estimates	
Top Temperature	40°C (100°F)
Bottom Temperature	70°C (160°F)

The Amines property package requires that real trays be used in the contactor and regenerator operations. To model this in UniSim Design, component specific efficiencies are required for H_2S and CO_2 on a tray-by-tray basis. These proprietary efficiency calculations are provided in the column as part of the Amines package. Tray dimensions must be supplied to enable this feature. Tray dimensions enable component specific efficiencies to be calculated by

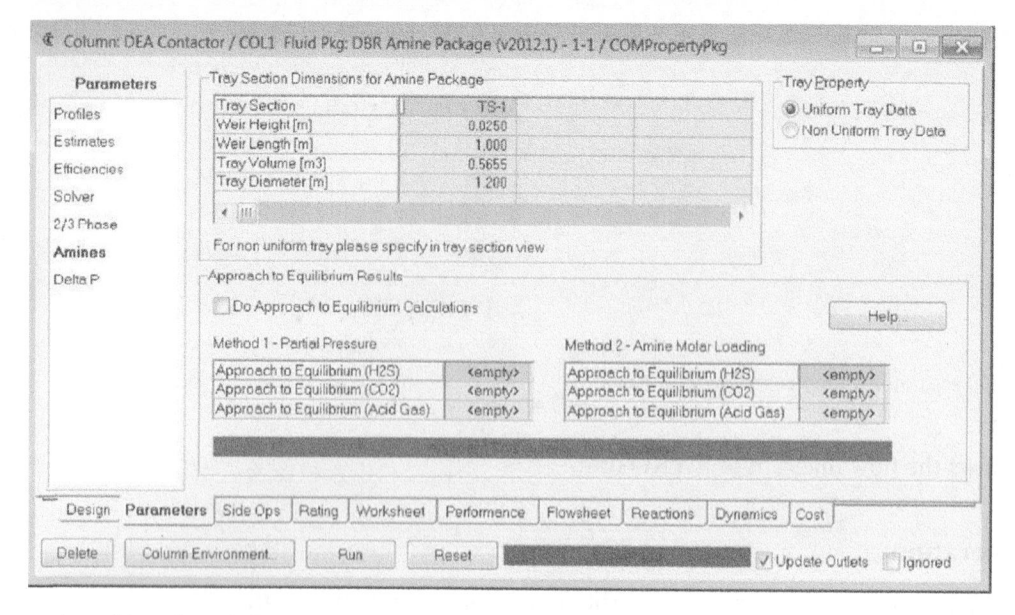

Figure 27.38 Screenshot of the column DEA contactor showing the parameters.

estimating the height of liquid on the tray and the residence time of vapor in the liquid. To supply the dimensions for Amines calculation, switch to the **Parameters** tab, **Amines** page. Figure 27.38 shows the screenshot of the Column DEA Contactor with Parameters tab.

7. Enter the following Tray Section dimension:

In this cell...	Enter...
Amines	
Weir height	0.025 m (0.082 ft)
Weir length	1.0 m (3.3 ft)
Tray Diameter	1.2 m (4.0 ft)

Figure 27.39 The screenshot of column DEA showing the efficiencies of the parameters tab.

8. **Run** the Column.
9. Once the Column has converged, move to the **Efficiencies** page on the **Parameters** tab.
10. Select the **Component** radio button in the **Efficiency Type** group to view the component efficiencies.

Figure 27.39 shows the screenshot of the Column DEA with efficiencies of the Parameters tab.

11. Switch to the Worksheet tab to view the concentrations of H_2S and CO_2 in the product streams from the column.

The concentrations of H_2S and CO_2 in the Sweet Gas are 0.000 and 0.001708 respectively from Honeywell UniSim Design R470 software **Case-Study-akc.usc**

Valve Operation

Rich DEA from the Contactor is directed to a Valve, VLV-100, where the pressure is reduced to 620 kPa (90 psia), which is close to the Regenerator operating pressure.

12. Add a Valve with the following values:

In this cell...	Enter...
Connections	
Inlet	Rich DEA
Outlet	DEA to Flash TK

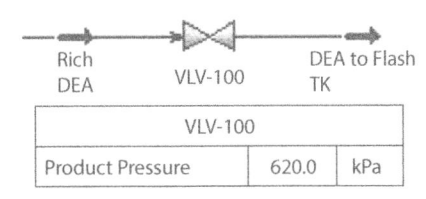

Figure 27.40 Valve VLV-100 with inlet Rich DEA stream and outlet DEA stream to Flash tank.

In this cell...	Enter...
Worksheet	
Pressure, DEA to Flash TK	620 kPa (90 psia)

Separator Operation

Gases which are flashed off from Rich DEA are removed using the rich amine flash tank, Flash TK, which is installed as a Separator operation.

13. Add a Separator with the information shown below:

In this cell...	Enter...
Connections	
Name	Flash TK
Inlets	DEA to Flash TK
Vapour Outlet	Flash Vap
Liquid Outlet	Rich to L/R

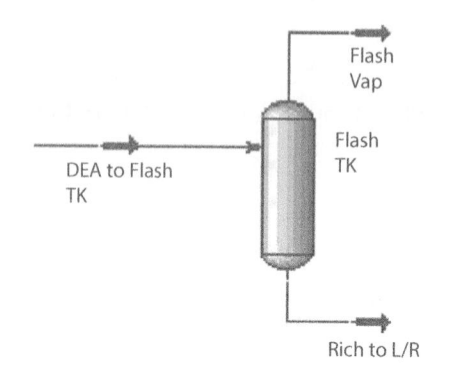

Figure 27.41 Flash tank TK with inlet stream DEA and outlets Flash vapor stream and liquid Rich to L/R stream.

Heat Exchanger Operation

Regen Feed is heated to 95°C (200°F) in the lean/rich exchanger, L/R HEX, prior to entering the Regenerator, where heat is applied to break the amine–acid gas bonds, thereby permitting the DEA to be recycled to the contactor.

14. Add a Heat Exchanger with the following values:

In this cell...	Enter...
Connections	
Name	L/R HEX
Tube Side Inlet	Rich to L/R
Tube Side Outlet	Regen Feed
Shell Side Inlet	Regen Bttms
Shell Side Outlet	Lean from L/R
Parameters	
Tubeside Delta P	70 kPa (10 psi)
Shellside Delta P	70 kPa (10 psi)
Heat Exchanger Model	Exchanger Design (Weighted)
Worksheet	
Regen Feed, Temperature	95°C (203°F)

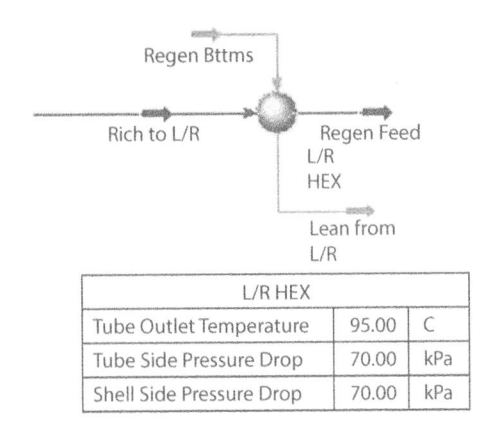

L/R HEX		
Tube Outlet Temperature	95.00	C
Tube Side Pressure Drop	70.00	kPa
Shell Side Pressure Drop	70.00	kPa

Figure 27.42 A heat exchanger with inlet and outlet streams respectively.

Regenerator Operation

The Amine Regenerator is modeled as a distillation column. There are 20 real stages – 18 stages in the Tray Section plus a Reboiler and a Condenser. The component efficiencies for this tower are assumed to be constant at 0.8 for H_2S and 0.15 for CO_2. The efficiencies of the Condenser and Reboiler must remain at 1.0, so only stages 1-18 should have efficiencies entered for them. A Damping factor of 0.4 will provide a faster, more stable convergence. The Damping factor controls the step size used in the outer loop when updating the thermodynamic models in the inner loop.

Note: Damping factors will have no effect on problems where the heat and spec. error does not converge. Certain columns require the use of a damping factor. Amine Regenerators, TEG Strippers and Sour Water Strippers use damping factors in the 0.25–0.5 range.

15. Add a Distillation column with the following information:

In this cell...	Enter...
Design\Connections	
Name	Regenator
No. of Stages	18
Inlet Streams/Stage	Regen Feed/4
Condenser Type	Full Reflux
Ovhd Vapour Outlet	Acid Gas
Bottoms Liquid Outlet	Regen Bttms
Reboiler Energy Stream	Rblr Q
Condenser Energy Stream	Cond Q
Parameters\Solver	
Fixed Damping Factor	0.40
Solving Method	Legacy Inside-Out
Parameters\Profiles	

Condenser Pressure	190 kPa (27.5 psia)
Condenser Delta P	15 kPa (2.5 psi)
Reboiler Pressure	220 kPa (31.5 psia)
Top Stage Temperature	100°C (210°F)
Reboiler Temperature	125°C (260°F)
Parameters\Efficiencies	
CO2	0.15
H2S	0.8

Note: You will need to click the **Reset H$_2$S, CO$_2$** button before you can enter the specified H$_2$S and CO$_2$ efficiencies.

In this cell...	Enter...
Design\Monitor	
Overhead Vap Rate (Estimate)	75 kgmole/h (1.5 MMSCFD)
Reflux Ratio (Estimate)	1.5
Design\Monitor	
Column Temperature, Condenser	50°C (120°F)
Column Duty, Reboiler	1.3e7 kJ/h (1.2e7 Btu/hr)

Note: Remember to activate the Condenser Temperature and Reboiler Duty specifications once they are added. An alternative specification that could be used is the Component Recovery for DEA, because all of the DEA (100%) should be recovered in the bottom product stream.

16. **Run** the Column.

Mixer Operation

Water make-up is necessary, since water is lost in the Absorber and Regenerator overhead streams. A Mixer operation combines the lean amine from the Regenerator with a water makeup. These streams mix at the same pressure.

17. Add a new stream.

In this cell...	Enter...
Connections	
Name	Makeup H2O
Temperature	25°C (77°F)
Component	**Mole Fraction**
H2O	1.0

Note: The Mixer will adjust the water flow rate to achieve the circulation rate. An adjust operation could have been used but is not necessary.

In this cell...	Enter...
Connections	
Name	Makeup H2O
Temperature	25°C (77°F)
Component	**Mole Fraction**
H2O	1.0

18. Add a Mixer with the following information:

In this cell...	Enter...
Connections	
Inlets	Makeup H2O Lean from L/R
Outlet	DEA to Cool
Parameters	
Pressure Assignment	Equalize All
Worksheet	
Std Ideal Liq Vol Flow, DEA to Cool	43 m3/h (190 USGPM)

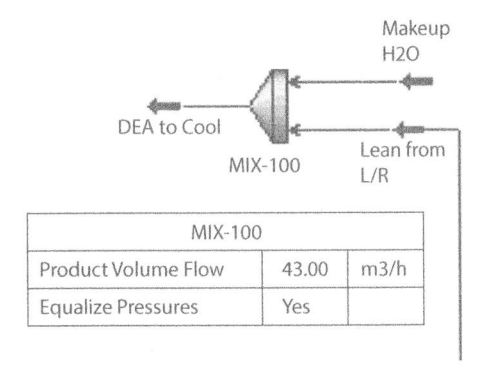

MIX-100		
Product Volume Flow	43.00	m3/h
Equalize Pressures	Yes	

Figure 27.43 A Mixer showing two inlet, makeup H_2O and Lean amine from L/R streams and an outlet DEA stream.

The flow rate of the Makeup H_2O is 0.7065 kg mole/h, and after closing the recycle loop this becomes 5.454 kg mole/h from Honeywell UniSim Design R470 software **Case-Study-akc.usc**.

Cooler Operation

19. Add a Cooler with the values given below:

In this cell...	Enter...
Connections	
Name	Cooler
Feed Stream	DEA to Cool
Product Stream	DEA to Pump
Energy Stream	Cooler Q
Parameters	
Pressure Drop	35 kPa (5 psi)

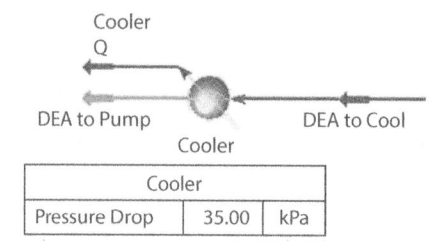

Figure 27.44 A cooler with inlet and outlet streams, with a pressure drop = 35 kPa.

Pump Operation

20. Add a Pump with the following information:

In this cell...	Enters...
Connections	
Inlet	DEA to Pump
Outlet	DEA to Recycle
Energy	Pump Q
Worksheet	
Temperature, DEA to Recycle	35°C (95°F)

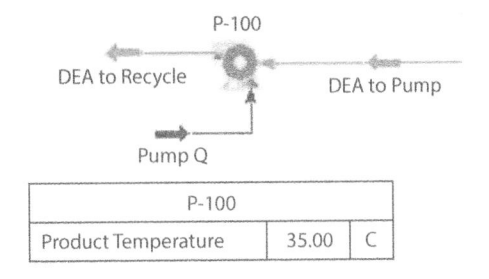

Figure 27.45 Pump P-100 with DEA solution at 35°C.

Adding Logical Unit Operations

Set Operation

The Set is a steady-state operation used to set the value of a specific Process Variable (PV) in relation to another PV. The relationship is between the same PV in two like objects; for instance, the temperature of two streams, or the UA of two exchangers.

21. Double-click on the **Set** icon. Complete the **Connections** tab as shown in the following figure.

Figure 27.46 shows the screenshot of the Target Variable and Source Variable in the Connections tab.

22. Go to **Parameters** tab. Set the Multiplier to **1**, and the Offset to **-35 kPa** (-5 psi) as shown below:

Figure 27.47 shows the screenshot of the Parameters in the Parameters tab.

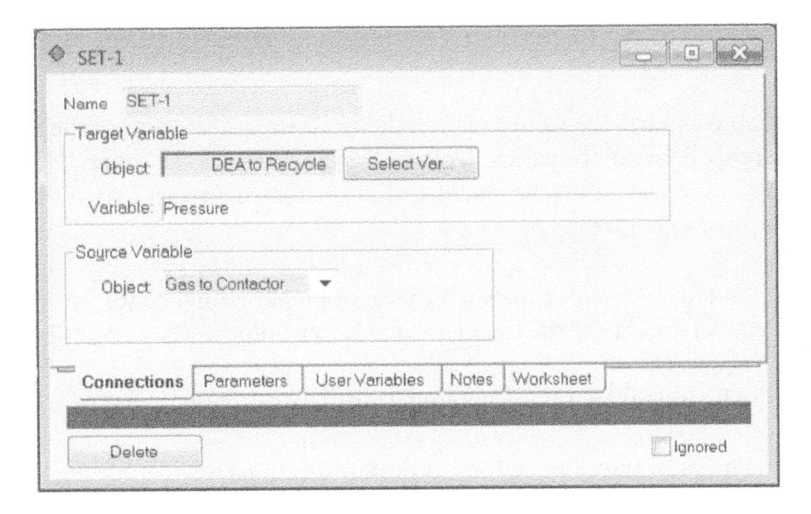

Figure 27.46 A window showing the Connections tab with Source and Target Variables.

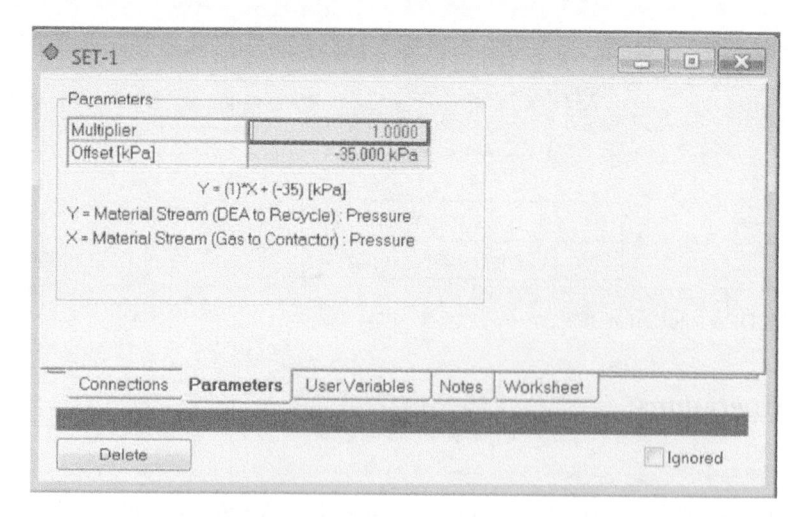

[Figure showing SET-1 window with Parameters tab:
Multiplier: 1.0000
Offset [kPa]: -35.000 kPa
Y = (1)*X + (-35) [kPa]
Y = Material Stream (DEA to Recycle) : Pressure
X = Material Stream (Gas to Contactor) : Pressure]

Figure 27.47 A simulation window showing the Parameters tab.

[Recycle icon]

Figure 27.48 A recycle icon in the simulation program from the pallets.

Recycle Operation

The Recycle installs a theoretical block in the process stream. The feed into the block is termed the calculated recycle stream, and the product is the assumed recycle stream. The following steps take place during the convergence process:

- UniSim Design uses the conditions of the assumed stream and solves the flowsheet up to the calculated stream.
- UniSim Design then compares the values of the calculated stream to those of the assumed stream.
- Based on the difference between the values, UniSim Design modifies the values in the assumed stream.
- The calculated process repeats until the values in the calculated stream match those in the assumed stream within specified tolerances.

In this case, the lean amine (DEA to Contactor) stream which was originally estimated will be replaced with the new calculated lean amine (DEA to Recycle) stream and the Contactor and Regenerator will be run until the recycle loop converges.

23. Double-click on the **Recycle icon**. On the **Connections** tab, select the connections from the

Figure 27.49 shows the screenshot of the Connections in the Connections tab.

24. Switch to the **Parameters** tab. Ensure the tab is as shown in the figure below:

Figure 27.50 shows the screenshot of the Recycle RCY-1 Parameters in the Parameters tab.

Note: The smaller the Sensitivity the tighter the tolerance.

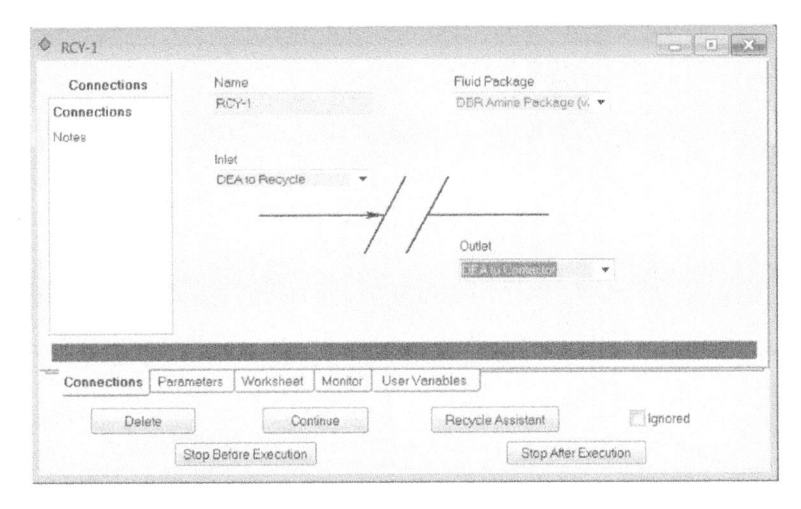

Figure 27.49 A simulation window showing the Connections tab between the inlet and outlet streams.

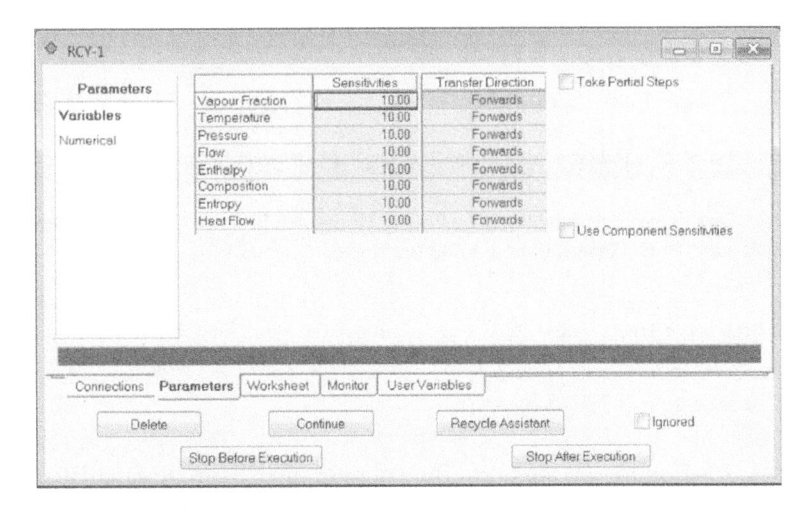

Figure 27.50 A simulation window showing the Parameters tab.

Save your case

Analyzing the Results

The incoming sour gas contained 4.1% CO_2 and 1.7% H_2S. For the inlet gas flow rate of 1250 kg mole/h (25 MMSCFD), a circulating solution of approximately 28 wt% DEA was used to remove H_2S and CO_2. The conventional pipeline gas specification is no more than 2.0 vol % CO_2 and 4 ppm (volume) H_2S.

The concentration of H_2S and CO_2 in the Sweet Gas are: 0.00 and 0.001708 respectively.

CO_2 vol% in the Sweet Gas.

Note: For gas streams vol% does not refer to liquid vol%. For gas streams volume fraction = mole fraction, and thus to calculate vol%, multiply mole fraction by 100.

CO_2 (mole fraction) = 0.001708

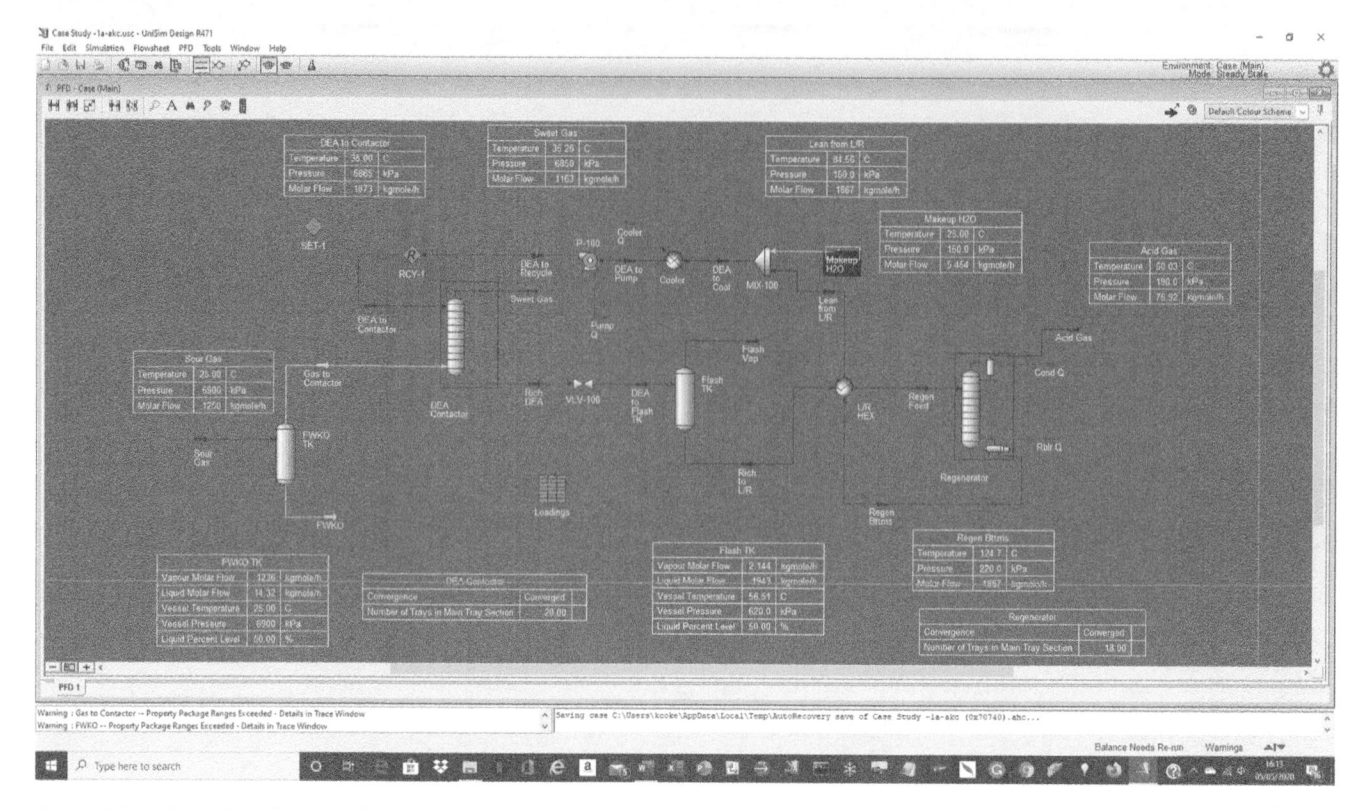

Figure 27.51 Snapshot of process flow diagram with tables. (Courtesy of Honeywell Process Solution, UniSim Design® R470, Honeywell® and UniSim® are registered trademarks of Honeywell International Inc.)

CO_2 (vol%) = 0.001708 × 100 = 0.1708

H_2S vol ppm level in the Sweet Gas = 0.2391 vol ppm

The volume fraction = mole fraction for gas streams. The vol ppm of H_2S can be calculated by either:

1. Multiplying the mole fraction of H_2S by 1e6 using a spreadsheet, or
2. Changing the basis to mole flow which will display more significant figures, then calculate: (H_2S mole flow/Total mole flow) * 1e6

The pipeline gas specifications have been met since:

CO_2: 0.1708 vol% < 2 vol%

H_2S: 0.2391 ppm (volume) < 4 ppm (volume)

The simulation software (**Case-Study-akc.usc**) provides the acid-gas sweetening simulation with DEA and Figure 27.51 shows the snapshot of the process flow diagram using Honeywell UniSim Design® suite R470.

27.12 Advanced Modeling

Concentrations of acid gas components in an amine stream are typically expressed in terms of loadings of amine. Loadings are defined as moles of the particular acid gas divided by moles of the circulating amine. The Spreadsheet

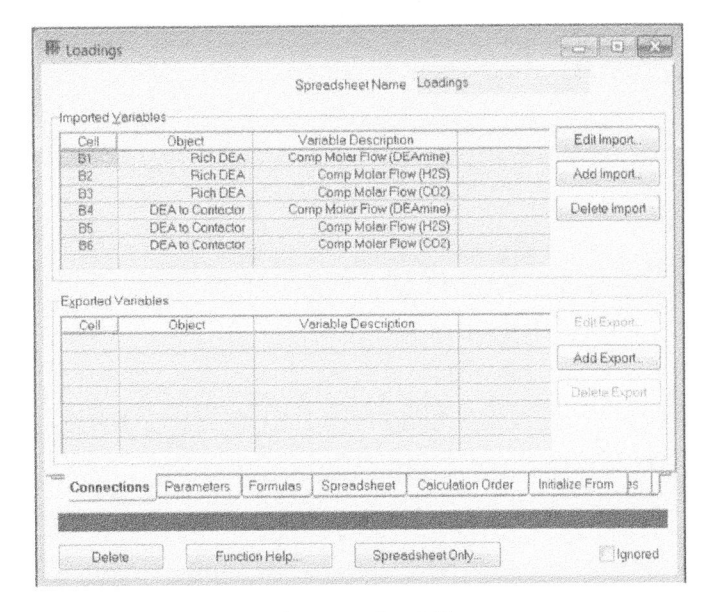

Figure 27.52 A simulation window with the Connections and imported variables.

in UniSim Design is well-suited for this calculation. Not only can the loadings be directly calculated and displayed, but they can be incorporated into the simulation to provide a "control point" for optimizing the amine simulation.

25. Add a spreadsheet using the following variables for the loading calculations:

Figure 27.52 shows the screenshot of the Loadings Window in the Connection tab.

26. On the Spreadsheet page, enter the formulae for the loading calculations.

In this cell...	Enter...
Spreadsheet	
D2	=b2/b1
D3	=b3/b1
D5	=b5/b4
D6	=b6/b4

The acid gas loadings can be compared to values recommended by D. B. Robinson.

Maximum acid gas loadings (moles acid gas/moles of amines)		
	CO2	H2S
MEA, DEA	0.50	0.35
DEA	0.45	0.30
TEA, MDEA	0.30	0.20

27. Enter appropriate text labels for the imported variables/formulae. The spreadsheet should appear as shown.

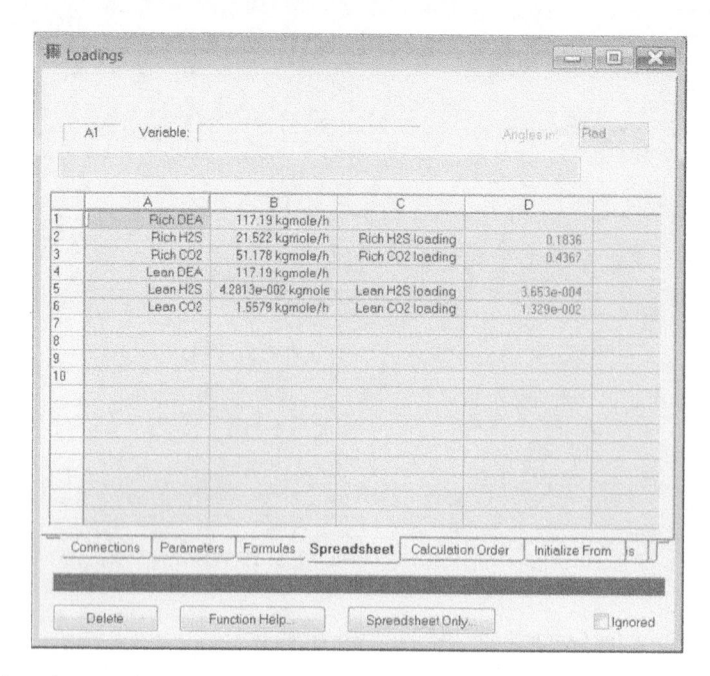

Figure 27.53 A simulation window showing the Spreadsheet tab.

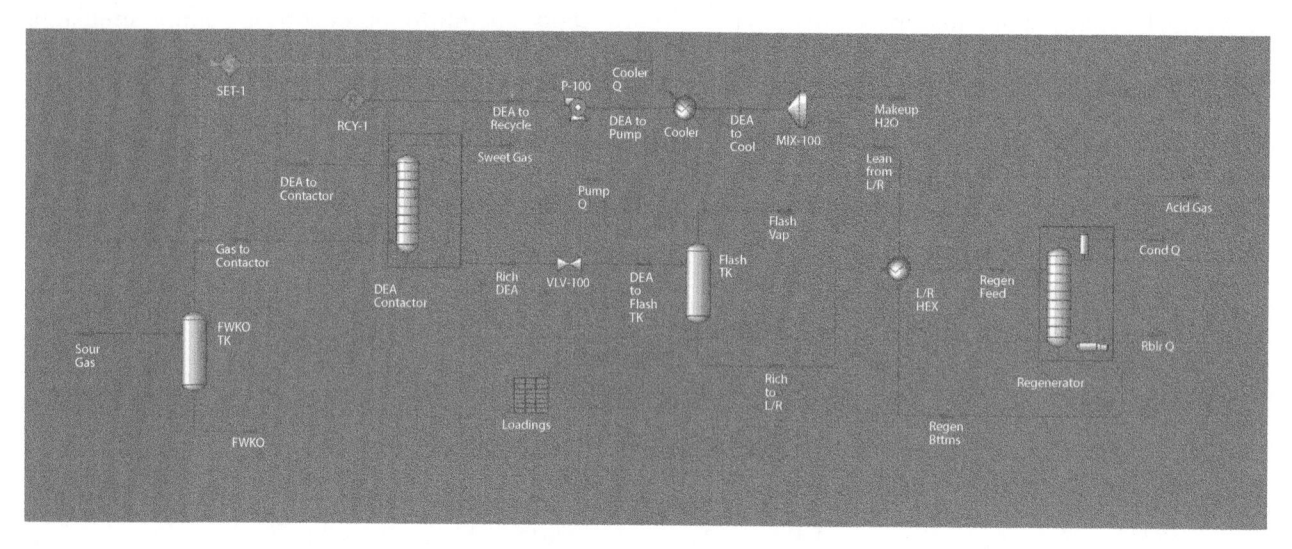

Figure 27.54 Snapshot of process flow diagram. (Courtesy of Honeywell Process Solution, UniSim Design® R470, Honeywell® and UniSim® are registered trademarks of Honeywell International Inc.)

Figure 27.53 shows the screenshot of the Loadings Window in the Spreadsheet tab.

Figure 27.54 shows the screenshot of the process flow diagram.

27.13 Carbon Capture and Storage (CCS)

Amines are used to remove CO_2 in various areas ranging from natural-gas production to the food and beverage industry. There are multiple classifications of amines, each of which has different characteristics relevant to CO_2 capture. For example, MEA reacts strongly with acid gases like CO_2 and has a fast reaction time and an ability to remove high percentages of CO_2, even at the low CO_2 concentrations. Typically, MEA can capture 85–90% of the CO_2 from the flue gas from a coal-fired plant, which is one of the most effective solvents to capture CO_2 [65].

Challenges of carbon capture using amine solvents include the following:

- Low pressure gas increases difficulty in transferring CO_2 from the gas into amine.
- Oxygen content of the gas can cause amine degradation and acid formation.
- CO_2 degradation of primary (and secondary) amines.
- High energy consumption.
- Very large facilities.
- Finding suitable location for the removed CO_2.

The partial pressure is the driving force to transfer CO_2 into the liquid phase. Under the low pressure, this transfer is difficult to achieve without increasing the reboiler's heat duty, which will result in a higher-cost.

Primary and secondary amines, e.g., MEA and DEA will react with CO_2 and form degradations products O_2 from the inlet gas will cause degradation. The degraded amine is no longer able to capture CO_2, which decreases the overall carbon capture efficiency [66].

Currently, the varieties of amine mixtures are synthesized and tested to achieve a more desirable set of overall properties for use in CO_2 capture systems. A major challenge is on lowering the energy required for solvent regeneration, which has an impact on process costs. However, tradeoffs are considered such as the energy required for regeneration being related to the driving forces for achieving high capture capacities. Therefore, reducing the regeneration energy can lower the driving force and thus increase the amount of solvent and size of the absorber needed to capture a given amount of CO_2, thus increasing the capital cost [65].

The current legislation in various developed countries suggests reducing the amount of CO_2 that industry will be allowed to emit as the world's energy demand is expected to increase by nearly 50% by 2030. Fossil fuels such as Natural Gas are likely to be a significant portion of the world's energy mix for many years to come. Shell Catalysts and Technology has developed a patented CO_2 capture technology that utilizes a regenerable amine that offers cutting-edge performance including low energy consumption, fast kinetics and extremely low volatility. The proprietary amine technology captures the CO_2 from the flue gas and releases it as a pure stream, which is then delivered to the client for sale into the EOR and commodity markets or for eventual sequestration. The CANSLOV® technology has the potential to help oil and gas, power and other industries:

- Lower carbon intensity and meet stringent greenhouse gas abatement regulations by removing CO_2 from their exhaust streams.
- Lower SO_2 and NO_2 emissions.
- Sell or reuse pure CO_2 as a marketable by-product.
- Avoid landfill and legacy environmental issues.

Shell CANSOLV technology has shown capabilities for an expansive list of industry including:

- Process heaters, furnaces and boilers
- Sulfur recovery units (SRU)
- Fertilizer plants
- Chemical plants
- Steel and cement
- Metallurgical plants

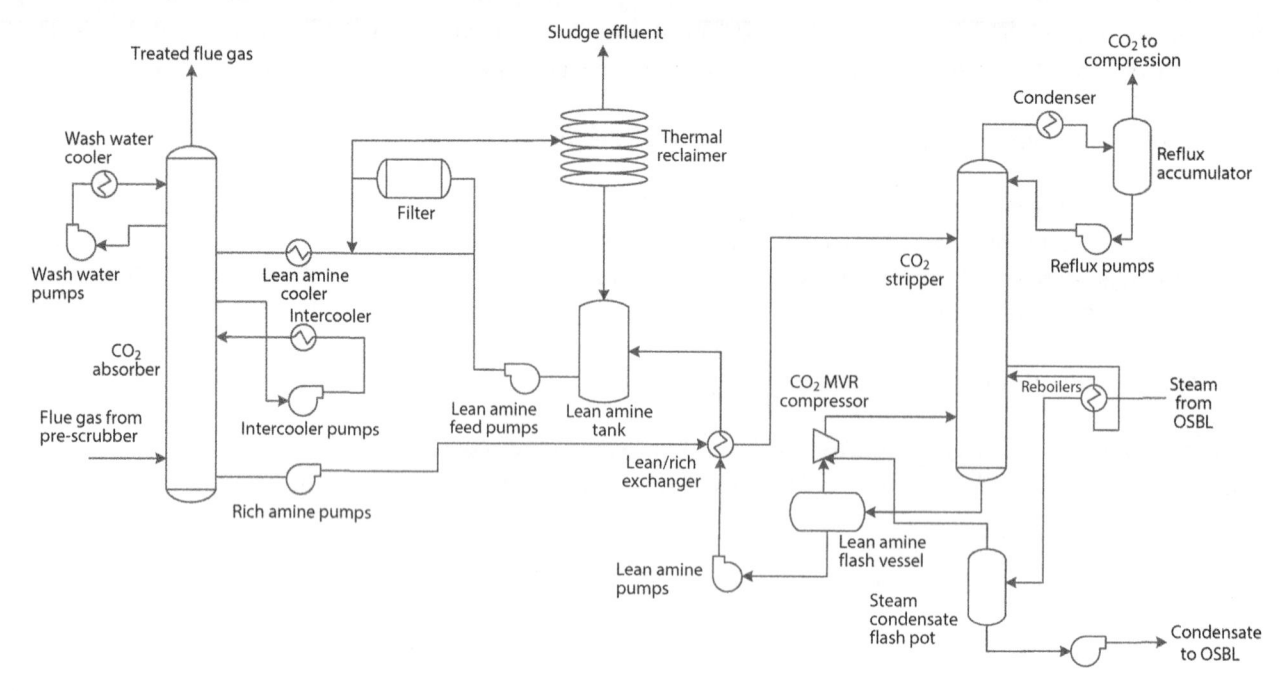

Figure 27.55 CANSOLV® CO_2 capture system flow diagram.

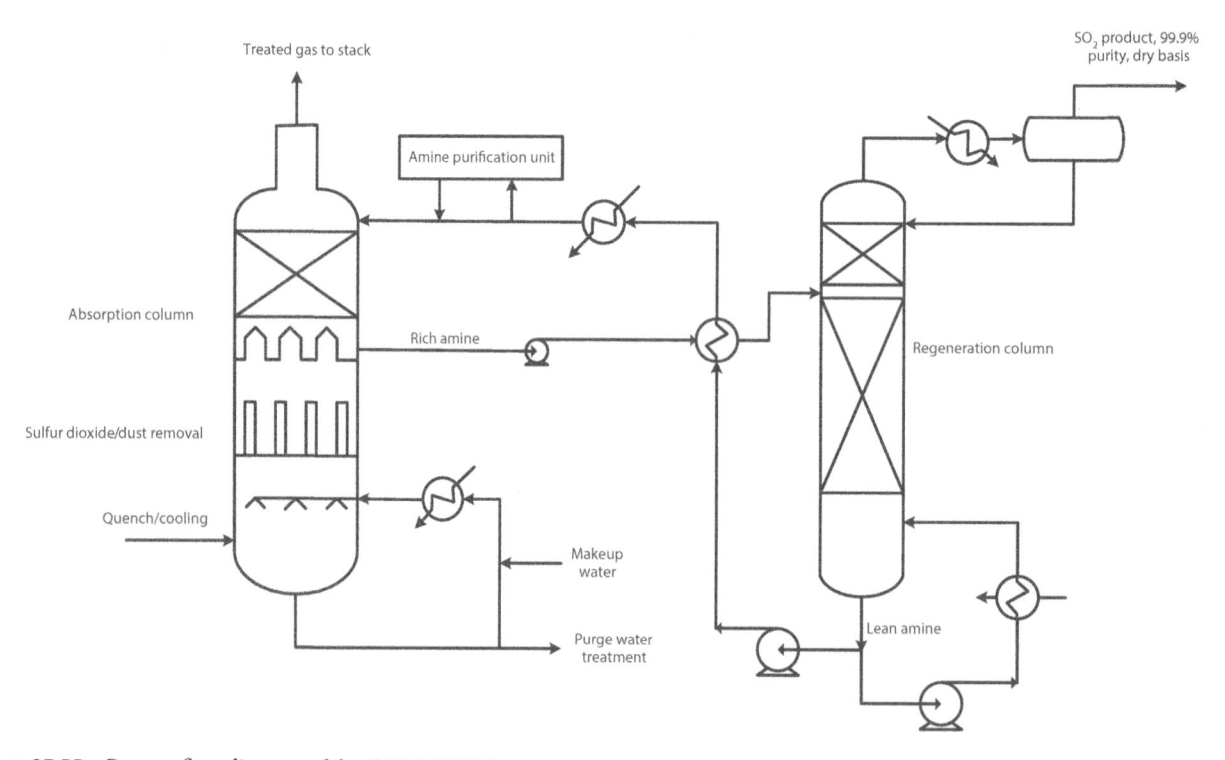

Figure 27.55a Process flow diagram of the CANSOLV SO_2 capture system.

Both sulfur dioxide (SO_2) scrubbing applications and carbon dioxide (CO_2) capture applications are flexible solutions that can be applied to achieve significant containment removal. For example, coal-fired power generation is recognized as being the largest contributor to anthropogenic CO_2 emissions. Using a Shell CANSOLV CO_2 capture solution to decarbonize emissions can enable the extension of an existing asset or implementation of a new one for coal-fired power generators.

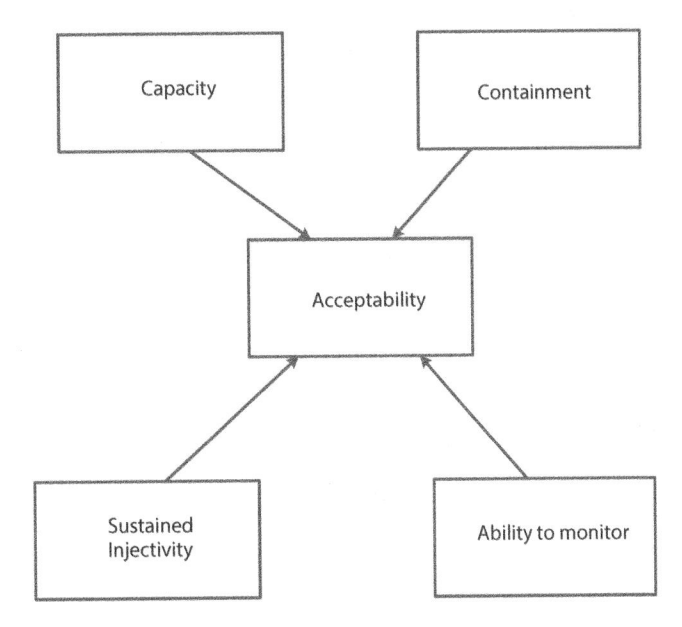

Figure 27.56 Risk management of a CCS project.

The CANSOLV SO_2 scrubbing system control emissions can capture additional value from the SO_2 emitted in various flue gas streams, such as those generated by fluidized catalytic cracking units (FCC), process heaters and boilers, power generation plants, sulfur plants, and spent acid regeneration units. The SO_2 can then be recycled to the sulfur recovery unit (see Chapter 12, Volume 1 of these series) to produce marketable sulfur, converted to saleable sulfuric acid, or liquefied for sale on the market.

The pure CO_2 and SO_2 captured avoid landfill and legacy environmental issues and can be sold or reused as marketable by-products. The CANSOLV technology is highly adaptable, reusable, sustainable, profitable and can be added to an existing plant or incorporated in a new installation across a wide range of industries. Figure 27.55 shows a process flow diagram of CANSOLV CO_2 capture system and Figure 23.55a shows the process flow diagram of SO_2 capture system.

27.14 Risk Management

The risk management requires that Shell and other carbon capture sequestration (CCS) energy companies need to demonstrate:

- The storage site can take the required volume of CO_2 at the required rates "on demand".
- Guarantee containment of the CO_2.
- Ongoing monitoring to prove that all the CO_2 is being contained.
- Confidence of all key stakeholders

Shell utilizes a risk-based approach that allows continuous risk education throughout all stages of a CCS project (see Figure 27.56).

27.15 The Institution of Chemical Engineers (IChemE, U.K.) Position on Climate Change

The Institution of Chemical Engineers (IChemE) is taking a leading role in tackling climate change by committing to work with stakeholders, from governments to communities around the world to deliver a fair, safe and sustainable future in which people worldwide can all thrive. In consultation with members, IChemE's position is based upon nine key principles, including the endorsement of the UN, Sustainable Development Goals (SDGs) and the acknowledgement that to support the aims of the Paris Agreement; net emissions of CO_2 must be reduced to zero and action must start now.

We are now in the amidst of a climate emergency, as human activity is causing the climate to change with significant adverse consequences. IChemE accepts the veracity of the science, and its conclusions published by the Intergovernmental Panel on Climate Change (IPCC). To avoid irreparable social, economic and environmental damage as illustrated earlier in the chapter, it is essential that we accelerate efforts to decarbonize our economic systems and stabilize the levels of greenhouse gases in the earth's atmosphere, if people are to have a chance of limiting the global average temperature rise to 1.5°C, beyond which extreme and irreversible consequences are more likely. Action needs to be global and fair, recognizing the relative differences between regions, both in terms of historic contributions to emissions and vulnerability to the consequences of a warming planet.

In this instance, chemical/process engineers are uniquely placed to take action across industrial sectors to arrest and reverse the damage humans have caused to the life-support systems of our single, shared planet and to contribute to improving food, security, energy and water availability and human health and well-being.

Technologists, biochemical, process and chemical engineers are equipped to imagine, design and implement:

- Means of combating the causes of climate change through reducing the anthropogenic[†] emissions of greenhouse gases.
- Means to mitigate against the effects of climate change through adaption and developing resilient and robust processes.
- Means of halting or reversing the effects of climate change by further developing carbon capture and storage processes, both technological and nature based.

IChemE aims to take a leading role in tackling climate change and to commit to the principles listed below and to work collaboratively as members, through education, research and sustainable engineering practice in contributing globally to the transition to a net zero carbon world by 2050. The following IChemE's principles are:

27.15.1 Net Zero Carbon Emissions

IChemE fully supports the aims of the Paris Agreement to pursue efforts to limit the global temperature increase to 1.5°C relative to pre-industrial levels. Achieving this climate goal will require net emissions of carbon dioxide and other greenhouse gases to be reduced to zero.

Emissions Reduction must Start NOW

IChemE agrees that serious action to combat climate change is urgent and must start immediately and accelerate. IChemE will work with associated industries and governments to achieve the rate of change needed to remain below 1.5°C. The IPCC articulates this as reducing global anthropogenic greenhouse gas (GHG[‡]) emissions by at least 7.6% year on year to 2030 (as an interim target) or reducing total emissions by at least 50% each decade henceforth to 2050.

27.15.2 Guided by UN Sustainable Development Goals

Climate change, its mitigation and adaptation to its impacts does not exist in isolation. The institution endorses the UN Sustainable Development Goals (SDGs) to address climate change, end all forms of poverty and inequality while making sure that no one is left behind. This means that the actions of chemical engineers should minimize adverse impact and not shift impact elsewhere-either geographically, socially, economically or environmentally.

[†] *Anthropogenic refers to impact caused by humans or their activities. This may be direct or indirect.*
[‡] *A greenhouse gas (GHG) is any gas in the atmosphere, which absorbs and re-emits heat, thereby keeping the planet's atmosphere warmer than it otherwise would be. Carbon dioxide (CO_2) is the most common GHG produced by human activity. Carbon dioxide equivalent (CO_2e) is a measure used to compare the emissions from various greenhouse gases based upon their global warming potential (GWP).*

Systems Thinking

To achieve the desired outcomes, a global system thinking approach is essential. Full and robust assessment of life cycles, their emissions and any other potential adverse impacts, together with the drive to a circular economy, is essential practice and must be encoded in industry standards for planning, design, construction, operation and decommissioning.

Global Mechanisms

IChemE endorses the view that governments must take responsibility for the total emissions of greenhouse gases from their economies and must work to meet the goal of net zero by 2050 by introducing and implementing appropriate policies on taxation, carbon pricing and other policy tools.

Best Available Techniques

IChemE believes that we should make use of best available techniques to mitigate and adapt to the effects of climate change. Technologies must be chosen to ensure that they do not entrench the status quo but adapt to changing circumstances. Solutions must be designed to demonstrate the greatest positive outcomes for the environment and society and thus the economy and take into account longevity and operability in a changed environment over the life of the project (e.g., differing rainfall, temperature profiles).

Innovation

IChemE supports the development of new technologies and processes to deliver the transition to net zero emissions by 2050 at the pace required. Innovation§ will be needed and IChemE encourages research and development work to find the new best solutions to deploy.

27.15.3 Training and Application of Skills

The transition to a net zero carbon economy will bring opportunities and challenges. IChemE will work with members and the industries they work in to support the education, training and application of skills of the current and future workforce.

Education

IChemE will continuously work to ensure that the fundamental principles of sustainability, social responsibility and ethics are embedded in the education and training of chemical engineers. This will be mandatory in accredited education and through continuous professional development.

IChemE as a global professional membership organization through its Royal Charter commits to bring community benefit through chemical engineering and safeguard the public interest in matters of safety, health and the environment. The organization recognizes that urgent action is required to address the challenges of climate change and enhances opportunities between members and their employers in establishing pathways to net zero carbon emissions¶ and therefore assisting businesses to adapt to changing environment.

Engineering in general and chemical engineering, in particular have a pivotal role to play in responding to and addressing the threats posed by climate change. The impact that a process has on the environment is established during design and delivered during operation. The impact of chemical engineers on all these decisions is profound, as chemical engineers hold positions of influence in many of the industries and sectors that are the biggest contributors to greenhouse-gas emissions and also have the unique knowledge and expertise to address the challenges represented by climate change. The challenge of climate change cannot be successfully addressed without the meaningful commitment and engagement of the chemical engineering profession [68].

---

§ *An innovation is a new idea, product, process, service or way of doing something that either improves performance, adds value or achieves a desired purpose in a new or better way.*

¶ *Net zero carbon emissions include scope 1 and 2 emissions. Scope 3 emissions are not included in this definition.*

27.16 Oil & Gas and Petrochemical Companies with Zero Carbon Emissions Targets by 2050

Major oil and gas and petrochemical companies are making progress to a transition to zero-carbon emissions by 2050 [69].

BP proposes to invest trillions of dollars in replumbing and rewiring the world's energy system for a net zero carbon emissions. It will involve addressing all the carbon obtain out of the ground as well as the greenhouse gases emitted from operations, resulting in absolute reductions if this applies to every barrel of oil, and gas produced thereby eliminating the emissions' problem.

BP has ten-point plans for net-zero emissions by 2050, namely: net zero across BP's operations on an absolute basis by 2050 or sooner, net zero on carbon in its oil and gas production, 50% cut in the carbon intensity of products, installation of methane measurement at major oil and gas processing sites by 2030, reduction of methane intensity of operations by 50% and increase the proportion of investment into non-oil and gas businesses over time. The remaining five aims for a net zero emissions are more active advocacy for policies that support net zero, including carbon pricing, further incentivise workforce to deliver aims and mobilize them to advocate for net zero, net new expectations for relationships with trade associations, aim to be recognized as a leader for transparency of reporting, including supporting the recommendations of the TCFD and the launch of a new team to help countries, cities and large companies decarbonize.

Shell aims to reduce the net carbon footprint by 65% by 2050 that involves an ambition to a net zero on all the emissions from the manufacture of all its products (scope one and two) by 2050, accelerating Shell's net carbon footprint ambition to be in step with society's aim to limit the average temperature rise to 1.5°C in line with the goals of the Paris Agreement on climate change. This requires reducing the net carbon footprint of the energy products Shell sells to its customers by around 65% by 2050 (increased by 50%), and by around 30% by 2035 (increased from around 20%); and a pivot towards serving businesses and sectors that by 2050 are also net – zero emissions.

Total proposes net – zero carbon emissions by 2050 together with society for its global business across its production and energy products used by its customers. Their major steps are: (i) net zero across Total's worldwide operations by 2050, or sooner (scope 1 + 2); (ii) net zero across all its production and energy products used by its customers in Europe by 2050 or sooner (scope 1 + 2 + 3); and (iii) 60%, or more reduction in the average carbon intensity of energy products used worldwide by Total customers by 2050 (less than 27.5 gCO_2/MJ) – with intermediate steps of 15% by 2030 and 35% by 2040 (scope 1 + 2 + 3).

PKN ORLEN is the principal oil company in Central Europe to declare its aspiration to achieve zero-carbon emissions by 2050. Its primary aim by 2030 is the reduction of CO_2 emissions from its current refining and petrochemical assets by 20% and emissions from power generation by 33% CO_2/MWh. Their emission neutrality strategy is based on four pillars: energy efficiency in production; zero-and low-emission power generation, alternative fuels and green financing. The company proposes to invest more than $6. 46bn in projects that will facilitate mitigation of its environmental impacts and opening up to new business models. The zero-carbon emission strategy is based on the business pillars with extensive experience and a strong market position.

ExxonMobil plans to reduce the intensity of operated upstream greenhouse gas emissions by 15 to 20% by 2025, compared to 2016 levels. This will be supported by a 40 – 50% decrease in methane intensity, and a 35 – 45% decrease in flaring intensity (see chapter 23 of this volume series) across its global operations. The emission reduction plans, which cover scopes 1 and 2 emissions from operated assets, are projected to be consistent with the goals of the Paris Agreement. The company also plans to align with the World Bank's initiative to eliminate routine flaring by 2030.

ExxonMobil business planning process and plants with new treating technology to help customers lower sulphur, reduce the carbon footprint and thus working towards industry – leading greenhouse gas performance with an ambition to achieving net zero emissions by 2050. The company shall continue to advocate for policies that promote cost-effective, market – based solutions to address the risks of climate change.

Other measures include (i) continued investments in lower-emission technologies, such as carbon capture, manufacturing efficiencies, and advanced biofuels; (ii) increased cogeneration capacity at manufacturing facilities; (iii) continued support for sound policies that put a price on carbon and (iv) continued accounting for environmental performance as part of executive compensation.

ExxonMobil plans to provide scope 3 emissions on an annual basis with caution that reporting of these indirect emissions does not ultimately incentivize reduction of the actual emitters. Thus, meaningful reduction in global GHG emissions will require changes in society's energy choices coupled with the development and deployment of affordable lower-emission technologies. Since 2000, the company has invested more than $10 billion in R&D and deploying lower-emissions technologies, including nearly $3 billion at cogeneration facilities that more efficiently produce electricity and reduce related emissions.

The company plans to achieve 15% decrease in methane emissions and a 25% reduction in flaring by 2020 as detailed emissions performance is reported in annual publications, including the Energy and Carbon Summary. ExxonMobil has supported the Paris Agreement from its inception and the Oil and Gas Climate Initiative's announcement to reduce methane and carbon intensity for upstream operations. It has deployed new technologies throughout its operations to reduce flaring and methane emissions, while working to test new technologies to detect and measure fugitive emissions. The company supports the regulation of methane from new and existing sources and issued a methane regulatory framework for governments to consider as they draft new policies.

Dow announced aggressive new commitments to address both climate change and plastic waste on its path toward becoming the most innovative, customer-centric, inclusive and sustainable materials' science company in the world. Its 2019 Sustainable Report outlines progress and results aligned to 2050 sustainable goals.

The company plans to reduce its net annual carbon emissions by 2030 by 5MM tonnes, or 15% from its 2020 baseline. Additionally, Dow intends to be carbon neutral by 2050 in alignment with the Paris Agreement as its committed to implementing and advancing technologies to manufacture products using fewer resources (see Principles of Green Engineering) that help customers reduce their carbon footprints. Furthermore, by 2030, Dow will ensure to 'stop the waste' by enabling 1MM tonnes of plastic to be collected, reused or recycled through its direct actions and partnerships. It is investing and collaborating in key technologies and infrastructure to increase global recycling significantly.

By 2035, Dow will help 'close the loop' by having 100% of its products sold into packaging applications, be reusable, or recyclable. Furthermore, it is committed to redesigning and offering reusable, or recyclable solutions for packaging applications

Haldor-Topsoe plays an important role with its technologies to reduce carbon emissions. Its aims are to deliver technologies that reduce, or even eliminate carbon emissions from fuels and chemicals. The company has for decades perfected chemistry for a better world; In the 80s, acid rain was one of the most pressing environmental problems and today's it is under control. Its world-leading technologies have played a significant role to remove sulfur from fuels and emissions. The company's biggest aim is the reduction in carbon emissions, and it is employing the technologies to overcome this challenge.

Hydroflex™ Technology

Legislations in various countries aimed at reducing emissions is affecting refiners and refinery facilities worldwide. A way of meeting these demands is to add the ability to process renewable feedstocks to the refinery. From tallow to tall oil, HydroFlex™ technology provides the full feedstock flexibility. Any renewable feedstock can be transformed into drop-in, ultra-low-sulfur gasoline, jet fuel or diesel with the aid of this technology. The result is consistently high-grade, clean fuel. The renewable feedstock does not replace fossil fuels but runs alongside it. Thus, there is greater diversity; capacity and the opportunity for growth to the refinery performance.

Renewable feedstocks offer several advantages over first generation fatty acid methyl esters (FAME) or conventional non-renewable feedstocks as shown in the table below. Figure 27.57 shows the simplified processing scheme of Hydroflex™ technology from renewable feedstocks.

	Conventional diesel	1st generation diesel FAME	Renewable diesel
Cetane	40-55	50 – 60	70 – 80
Cold flow properties	++	+	+++
Blend in	-	Max. 7%	0 – 100%
Market value	++	+	++++

Air Products announced a new sustainability goal to reduce its carbon dioxide (CO_2) emission's intensity (kg CO_2/MM Btu) by one – third by 2030 from a 2015 baseline. The key steps toward the goal include carbon capture projects; low – carbon and carbon-free projects, operational excellence and increased use of renewable energy, incorporating Air Products technologies. Air Products personnel are collaborating and innovating solutions to provide significant

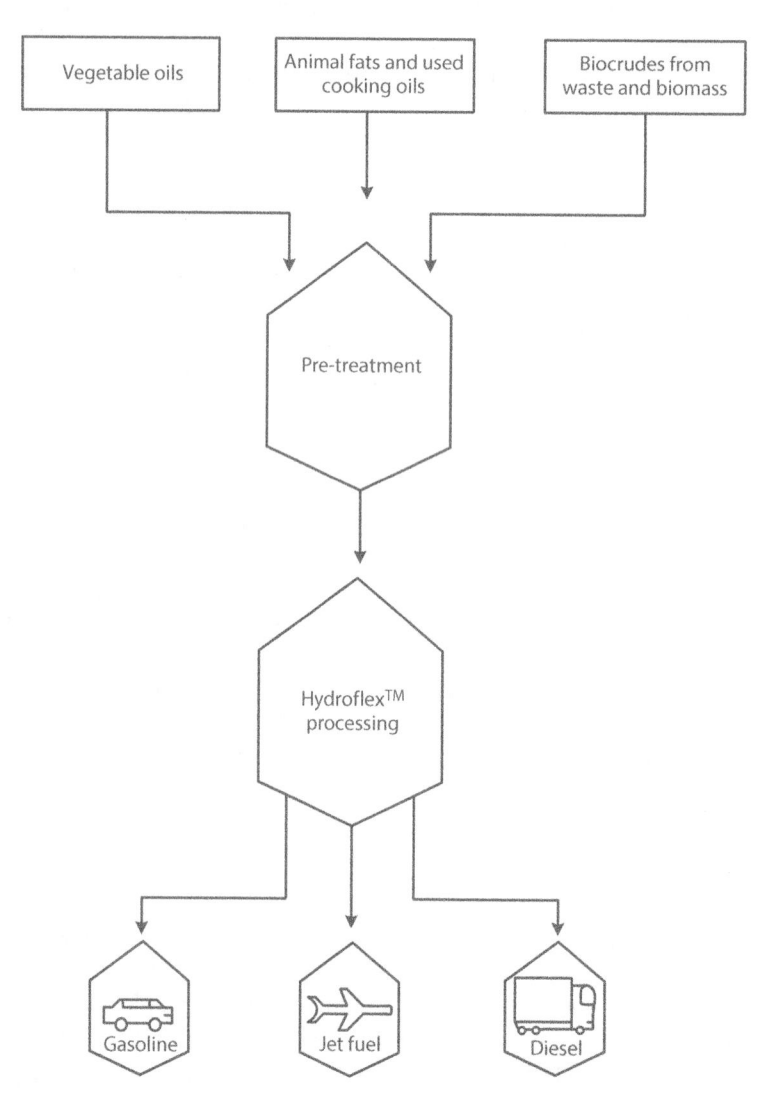

Figure 27.57 Flow scheme of Hydroflex™ processing.

energy and environmental solutions to various challenges, and thereby ensuring that sustainability goals beyond the emissions' reduction targets are achieved.

Evonik and Siemens Energy Partnership

Evonik and Siemens Energy commissioned a pilot plant sponsored by the German Federal Ministry of Education and Research (BMBF) that uses carbon dioxide and water to produce chemicals. The necessary energy is supplied by electricity from renewable sources. The pilot plant is located in Marl, in the northern Ruhr area and its innovative technology of artificial photosynthesis should contribute to the success of the energy revolution. This is an essential part of the Rheticus I and II research projects, which are sponsored by BMBF with a total of $7.39mn. This is part of setting up climate friendly production processes in the chemical industry and at the same time manufacture new innovative products.

Climate protection is not possible without chemistry and Evonik supplies and develops solutions for the energy turnaround. Furthermore, the research projects such as Rheticus are a motivation and innovation driver for a sustainable society. However, Harald Schwager [72], the deputy chairman of the executive board of Evonik cautioned against the speed when phasing out fossil fuels as the security of supply and reliability in political decisions set the framework in which new things are created.

Siemens Energy uses innovative technologies to enable new, more sustainable solutions. Using hydrogen and carbon monoxide electrolysis, Siemens Energy in partnership with Evonik is able to build a bridge from green electricity to sustainable material applications.

Researchers in the Rheticus facility took nature as a model for the idea of artificial photosynthesis. Just as plants use solar energy to produce sugar from carbon dioxide (CO_2), and water in several steps, artificial photosynthesis uses renewable energies to produce valuable chemicals from CO_2 and water through electrolysis with the help of bacteria. This type of artificial photosynthesis can serve as an energy store and thus help to close the carbon cycle and reduce CO_2 pollution in the atmosphere.

The process consists of CO electrolyser developed by Siemens Energy, a water electrolyser and the bioreactor with Evonik's technology. In the electrolysers, CO_2 and water are converted into carbon monoxide (CO) and hydrogen (H_2) with electricity in a first step. This synthesis gas is used by special microorganisms to produce specialty chemicals, initially for research purposes. These are starting materials for special plastics, or food supplements. Plans are to optimize the process by interacting the composition of the synthesis gas, electrolysis and fermentation, and a unit for processing the liquid from the bioreactor to produce the pure chemicals.

Successful completion of the Rheticus project phase between Evonik and Siemens Energy is for a unique platform technology that can produce energy-rich and valuable substances such as specialty chemicals, or artificial fuels from CO_2 in a modular and flexible manner.

27.17 Offshore Petroleum Regulator for Environment and Decommissioning (OPRED), UK

Offshore Petroleum Regulator for Environment and Decommissioning (OPRED) is the environmental and decommissioning regulator for offshore and carbon capture and storage facilities. OPRED will increase its focus on the further reduction of greenhouse gas emissions from all offshore oil and gas operations and will put in place a regulatory framework to support emerging decarbonization technologies. This will include OPRED utilizing its existing data collection system to track progress on emissions reduction.

The UK government is working with OPRED to ensure that their roles, powers and priorities reflect the government's policy for delivering net zero emissions, without imposing significant additional regulatory burdens. The UK is establishing Emissions Trading Scheme (ETS), a market-based measure which provides continuity for businesses. A cap is set on the greenhouse gases that businesses can emit (via the total number of allowances in circulation), which will decrease over time. Businesses then buy and sell emissions allowances through government auctions or secondary markets. The UK ETS will initially apply to energy-intensive industries, electricity generation and aviation.

The mechanism of carbon pricing supports businesses to decarbonize at the least cost. Businesses that can abate cheaply will endeavor to lower their carbon footprint, and thus unable others to purchase additional allowances to cover their emissions. Knowing that the ceiling on emissions will lower transparently over time enables business to plan and invest to decarbonize, while at the same time protecting the competitiveness of businesses and minimizing the risk of carbon leakage.

In order to meet net zero, the oil and gas sector will require to reduce its emissions from offshore production and operations to .05 $MtCO_2e$ by 2050, from 19$MtCO_2e$ today. Methane will be a special focus, given its potency as a greenhouse gas. The industry has signed up to the OGUK led ambition to achieve net zero emissions across all its upstream activities as set out in the OGUK's Roadmap 2035 [73]. The UK government wants the sector to reduce emissions from direct operations, known as Scope 1 and 2 emissions, and to address embodied emissions from the consumption of their products or from supply chain activities, referred to as scope 3 emissions. These scopes are defined as:

> The Greenhouse Gas Protocol Corporate Standard classifies a company's GHG emissions into three scopes:
>
> **Scope 1:** Emissions directly from owned or controlled sources
>
> **Scope 2.** Indirect emissions from the generation of purchased energy
>
> **Scope 3.** Indirect emissions (not included in scope 2) that occur in the value chain of the reporting company, including both upstream and downstream emissions.

The principal objective Oil and Gas Association (OGA), the upstream licensing regulator is to maximize the economic recovery of petroleum from the UK Continental Shelf, a statutory goal established in the wake of the oil price crash in 2014. The OGA has supported the production of oil and gas in the most cost-effective way, but its focus has been evolving with the imposing of net zero carbon emissions target by the UK government. OGA is investigating on possible ways in reducing its own carbon footprint or risk losing its social license to operate. The changes proposed in the OGA's consultation have the potential to make a significant contribution to achieving the UK government's goals. An example is reducing flaring and venting, the resulting greenhouse gas emissions, through its consents, field development process and project stewardship role. This practice is aimed to end before 2030. Furthermore, OGA will tackle regulatory and policy barriers to the use of clean electricity, such as offshore wind, to power offshore oil and gas facilities, as opposed to the current practice of using diesel or gas generators on platforms.

The industry has the skills, technology and capital to unlock innovative solutions which could be instrumental in helping to deliver net zero emissions successfully for the UK economy. It can also play a critical role in the deployment of CCUS, hydrogen production and renewable electricity generation, particularly offshore wind. The OGA aims to maximize economic recovery and will take wide ranging action to implement its revised strategy. This includes benchmarking greenhouse gas emissions to drive performance and creating a new asset stewardship expectation for net zero. Further updates of its guidance and its economic assessments will support regulatory decisions, to include full carbon costs. This approach will allow the OGA to take a much greater role in driving the sector's contribution to the clean energy transition. This will enhance the OGA's role as an environmentally responsible organization, and with a view of delivering net zero target by 2050.

Wood Plc, UK

Wood Plc, UK has taken substantial step to transform its business from a traditional oilfield services provider into broader engineering and consultancy work operating across the energy sector. In 2014, 96% of Wood's revenue was derived from oil and gas work that includes 65% from upstream activity compared with 2020 where upstream activity now accounts for 1/3 of its total revenue.

Wood offers a blend of consulting, projects and operations solutions including a fast – growing renewables business. It has been involved in solar projects, increasing global wind capacity as well as CCUS studies and has an increased presence in the hydrogen market. The company also commits to the oil and gas sector through aiding partners achieve their own energy transition goals.

Tata Chemicals Europe (TCE)

TCE is constructing the U.K. first industrial scale of carbon capture and utilizations (CCU) with the support of a £4.2 million grant from the Energy Innovation Programme at their site in Northwich for the manufacture of high purity sodium bicarbonate. The plant is scheduled for commission in 2021 and is capable of capturing up to 40,000 tCO_2e per year, and will reduce carbon emissions at the plant by 11%. TEC exports 60% of its sodium bicarbonate ($NaHCO_3$) production in the UK to over 60 countries across the globe. The CCU project will be a springboard for TCE to expand its export markets.

Hengli Petrochemical (Dalian) Co. Ltd. (HPDC)

HPDC has commissioned seven new pressure swing adsorption (PSA) units from Honeywell UOP LLC to supply high purity hydrogen at its 20-million tonnes/year (tpy) crude-to-paraxylene integrated refining and petrochemical complex in Hengli Petrochemical Industrial Park (HPIP) at Changxing Island Harbor Industrial Zone in Dalian, Liaoning Province, China.

HPDC will use the new Honeywell UOP Polybed PSA units to produce about 1.4 million m^3/h high purity hydrogen in its downstream hydrotreating operations to help produce diesel, gasoline and jet fuel as well as to create feedstock for petrochemical products producing approximately 1.4 million Nm^3/h (normal cubic metres per hour) of hydrogen. The PSA skid-mounted units use proprietary UOP adsorbents to remove impurities at high pressure from hydrogen-containing process streams, allowing hydrogen to be recovered and upgraded to more than 99.9% purity to meet refining needs. In addition to recovering and purifying hydrogen from steam reformers and refinery off-gases, the Polybed PSA system can be used to produce hydrogen from other sources such as ethylene off gas, methanol off-gas and partial oxidation synthesis gas.

Since its introduction in 1966, UOP has improved Polybed PSA technology with new generations of adsorbents, enhanced cycle configurations, modified process and equipment designs and more reliable control systems and equipment. Today, Honeywell UOP has installed more than 1,100 Polybed PSA units in more than 70 countries, and as a result, Polybed PSA is a proven technology with dozens of large-scale unit references globally.

HPDC also uses Honeywell UOP to provide its proprietary Callidus advanced flares and low – nitrogen oxide (NO_x) burner technology to help reduce NO_x emissions at the Dalian integrated complex.

Saudi Aramco

Many blue chip and energy leading companies are exploring the potential of green hydrogen solutions in their transition and decarbonization strategy. Saudi Aramco has a wide range of research and innovation efforts in the areas of sustainability and hydrogen, as the company is making strides to produce hydrogen from hydrocarbons economically, as well as capturing and utilizing carbon dioxide in a cost-competitive manner. Companies are in partnership with Saudi Aramco for a $5bn production facility in NEOM, Saudi Arabia, powered by renewable energy for the production and export of green hydrogen to global markets. Green hydrogen is a clean-burning fuel that eliminates emissions by using renewable energy to electrolyze water, separating the hydrogen atom within it from its molecular twin oxygen. Currently, not much green hydrogen is being produced, as it accounts for less than one percent of annual hydrogen production. According to International Renewable Energy Agency (IRENA), the world will require 19 exajoules of green hydrogen in the global energy system in 2050, between 133.8 and 158.3 million tonnes per annum. During 2000 – 2020, over 250 MW of green hydrogen projects have been implemented. While the number of green hydrogen electrolyzer projects nearly tripled in 2020 up to 8.2 GW, an increase in those can be over 1, 200% in the next five years. Presently, investment in green hydrogen is relatively small compared to other sustainable and renewable sources.

The demand for hydrogen technologies is rising given their potential to accelerate the transition to more sustainable forms of energy while still supporting current energy models with all their regional variations. Hydrogen is a zero-emissions source of fuels for vehicles and trains. It can be used as a feedstock gas for petroleum and petrchochemical industries and steel. Additionally, it is a source of heat and power for buildings, and can buffer energy generated from renewable sources.

Hydrogen offers compelling benefits as it supports a gradual transition towards lower-carbon sources of energy; it can be generated from natural gas and other non-renewable by-products. Furthermore, it can be used as an energy carrier, as a medium to store energy from renewable and other sources. It can be generated at scale with a zero-carbon footprint by using renewable energy such as solar or wind power, for instance to split water (electrolysis).

Processing

Depending on the target application, the hydrogen produced will require further processing and typical steps include the removal of impurities, the separation of carbon dioxide (CO_2), compression and/or cryogenic liquefaction.

Steam reforming is the main method used to produce hydrogen on an industrial scale. The feedstocks such as natural gas, LPG or naphtha are combined with steam to produce synthesis gas with the aid of heterogeneous catalyst. This mixture of carbon monoxide and hydrogen is then further processed. Since fossil fuels are used in this production method, the end product is called grey hydrogen**. Grey hydrogen can also be produced through the partial oxidation of refinery residues. This residue material is heated to a very high temperature with oxygen and steam to produce a raw synthesis gas. If the carbon dioxide (CO_2) contained in this gas is removed in a downstream carbon capture process, the resulting hydrogen is called blue. Green hydrogen (H_2) is obtained either by steam reforming, if bio-based feedstock is available, or by splitting water by electrolysis. The electricity needed for this electrolysis process is generated exclusively from renewable sources.

Steam reforming intially produces synthesis gas – a mixture of hydrogen and carbon monoxide and carbon dioxide. Cryogenic processes (e.g., condensation or methane scrubbing) to separate these two gases post CO_2 removal. Pressure swing adsoption (PSA)[††] plants are used to obtain H_2 from hydrogen – rich synthesis gases or refinery and petrochemical gases. An alternative hybrid process requires combining membrane and PSA technologies for new found levels of flexibility and efficiency in the production of H_2. PSA can also be used to remove or recover carbon

---

**Grey hydrogen is obtained from fossil fuels. Steam reformers are used to convert natural gas by means of steam addition and a catalyst. The resulting hydrogen-rich synthesis gas is then further processed. During the grey H_2 generation process, the carbon dioxide (CO_2) by-product escapes into the atmosphere. By contrast, with blue hydrogen, the CO_2 is captured in a downstream process. And green hydrogen is produced exclusively from renewable sources. Although, it has not been possible to produce the green hydrogen in sufficient quantities.

††PSA technology works by physically binding gas molecules to an adsorbing material. The binding force between the molecules and the absorbent depends on a number of factors, including the type of gas. This binding process creates a separation effect. Unlike carbon monoxide (CO), carbon dioxide (CO_2) and nitrogen (N_2), highly volatile components with low polarity like hydrogen have a negligible binding force. So, while the impurities adhere to the adsorbent, hydrogen is able to flow right through.

dioxide from proces gas streams at synthesis gas plants. CO_2 can also be recovered from the flue gas of hydrogen plants post combustion capture.

RECTISOL® wash technology is a physical acid gas removal process using an organic solvent (typically methanol) for segregated removing sulfur and CO_2 from the synthesis gas at subzero temperatures. The use of low-energy coil – wound heat exchangers makes this a particularly economical gas purification method. The captured CO_2 can be used for enhanced oil recovery (EOR) or fed into a purification or liquefaction plant to enable further uses.

Synthesis plants for the production of ammonia (NH_3) or methanol (CH_3OH), converting the produced hydrogen and nitrogen, respectively to a syngas stream. These products may be called green ammonia and green methanol, if green hydrogen is utilized as a feedstock. Cryogenic plants are used to liquefy hydrogen so it can be transported and stored efficienctly. They cool the volatile gas down to -253°C to create liquefied hydrogen (LH_2), as this process increases the density of the gas.

Commercial scale RECTISOL® wash units are operated world wide for the purification of H_2, NH_3, CH_3OH syngas and the production of pure CO and oxogases. Due to the physical nature of the process high pressure and high sour gas concentrations are particularly favorable. The technology is frequently used to purify shifted, partially shifted or unshifted gas downstream residue oil, coal or lignite gasification. Due to the low operation temperature, RECTISOL® is also favorable for cryogenic downstream processes like liquid nitrogen wash, cryogenic recovery of CO and oxogas. Figure 27.58 shows the process flow diagram of a 1-stage Rectisol consisting of four sub units, absorption, H_2S enrichment, regeneration and miscellaneous.

To reduce the carbon emissions of an existing gray H_2 asset, CO_2 can be captured from three locations:

- Shifted syngas.
- PSA tail gas.
- Flue gas.

The cost of CO_2 capture depends on the pressure and concentration of the CO_2 in the source steam plus the product specifications for the H_2 and CO_2. The most cost effective location to remove CO_2 is from the pre-combustion streams. The CO_2 can be removed by a variety of means, including solvent-based absorption, PSA or cyogenic fractionation as

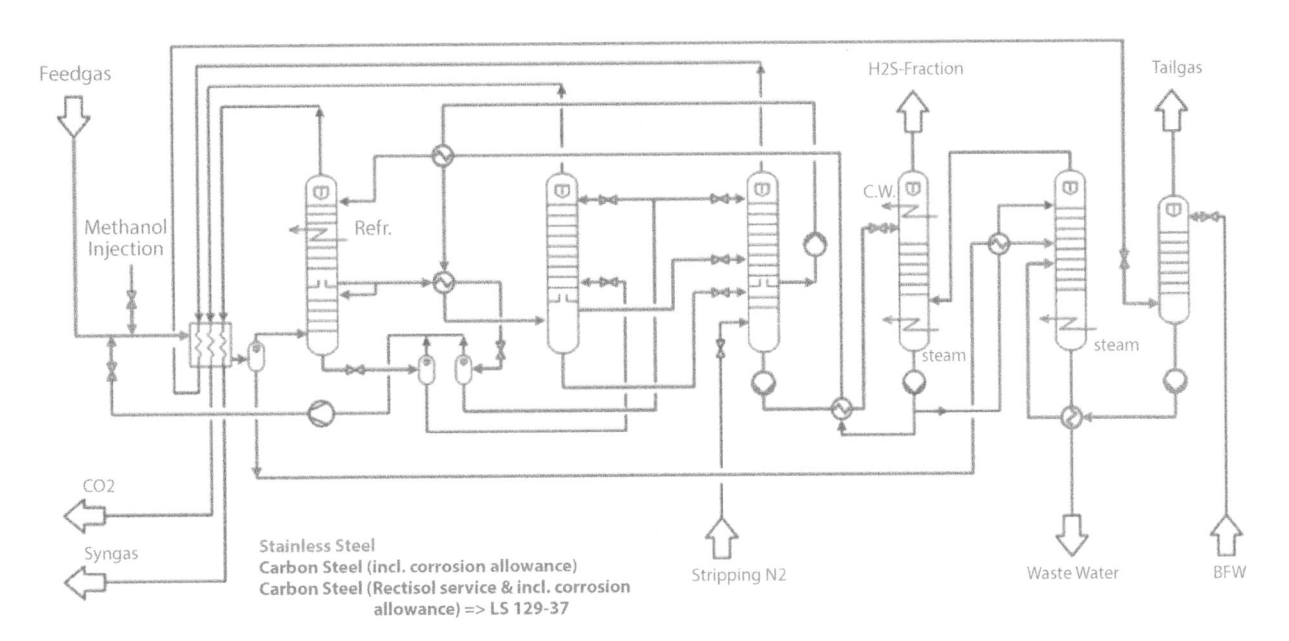

Figure 27.58 Linde's RECTISOL® wash process (Source: Linde AG Linde Engineering Division).

discussed later in the chapter. The option that provides the lowest overall cost of CO_2 captured is cryogenic fractionation, which also achieves additional high-purity H_2 yield as illustrated in Figure 27.59. In this option, the H_2, PSA tail gas is compressed, dried, condensed and fractionated, resulting in high-purity liquid CO_2 stream. Combining separation and liquefaction in a single unit operation saves utilities when a liquid product is required [75].

An alternative option for CO_2 capture from the PSA tail gas is a CO_2 PSA unit. A CO_2 PSA unit can be installed on the shifted syngas or the H_2 PSA tail gas, although the latter is preferred primarily due to a simpler revamp and ease of operation in the event that the CO_2 capture unit is bypassed as shown in Figure 27.60. The CO_2 PSA is the lowest CAPEX and OPEX carbon-capture option and can remove 99% of the CO_2 in the pre-combustion stream, but the extracted product is low pressure and low purity, requiring drying and liquefaction, or contaminant polishing via catalytic oxidation, followed by drying and multiple stages of compression to be transported.

The third option for CO_2 recovery is amine-based solvent capture of the shifted syngas. This established technology can achieve 99% CO_2 removal from the shifted stream as shown in Figure 27.61. However, this option requires the use of steam for solvent regeneration. The carbon emissions associated with steam generation erode the net benefit. Furthermore, by removing the CO_2 upstream of the H_2 PSA, the overall H_2 recovery will be eroded. This deficit could be mitigated with PSA adsorbent reload and cycle modification; however, such changes would make it very difficult to continue operation if the CO_2 removal unit were bypassed.

Finally, the low-pressure CO_2 product requires drying and multistage compression or liquefaction to be ready for transport. For end users that want gas-phase CO_2, and are long on steam, an amine unit is a reliable, proven choice for CO_2 recovery although at a higher cost of capture than tail gas recovery. Table 27.12 shows approximate guidelines for amine processes, and Table 27.13 shows the physical properties of these processes respectively.

In all options, the composition of the fuel gas recycled to the Steam methane reformer (SMR) furnace is significantly altered. As a result, burner revamp is required. CO_2 removal from the fuel gas requires advanced burner technology for stability and to achieve low NO_x emissions [75].

Further reduction in emissions involves eliminating CO_2 from the furnace flue gas. Two options exist to reduce flue gas emissions:

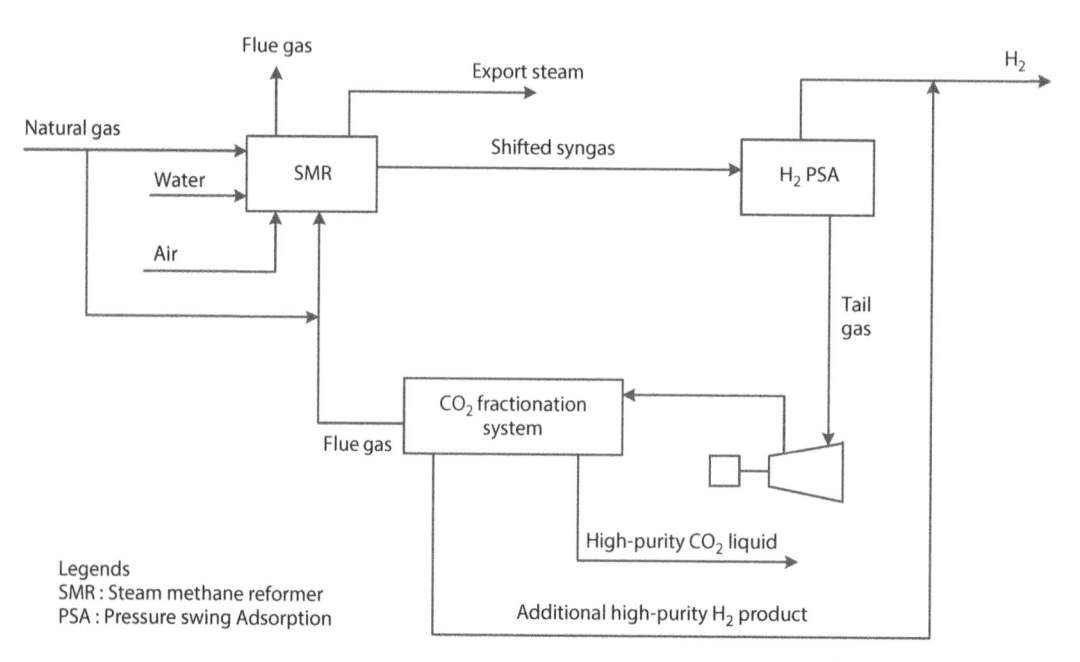

Figure 27.59 Steam methane reformer (SMR) retrofit CO_2 capture option 1 [75].

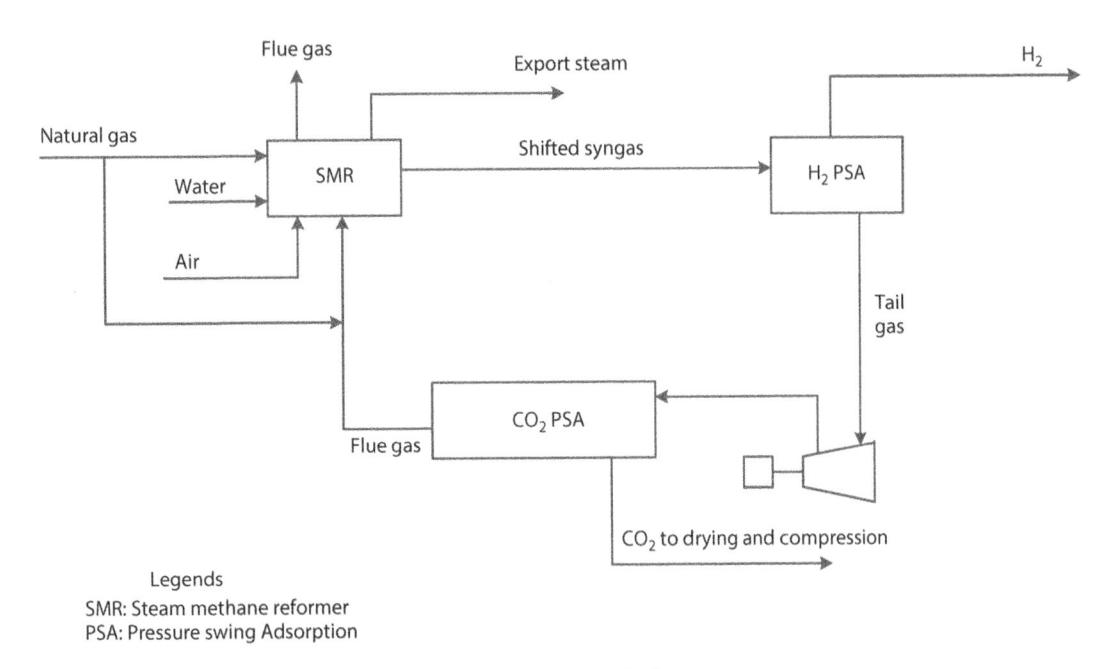

Figure 27.60 Steam methane reformer (SMR) retrofit CO_2 capture option 2 [75].

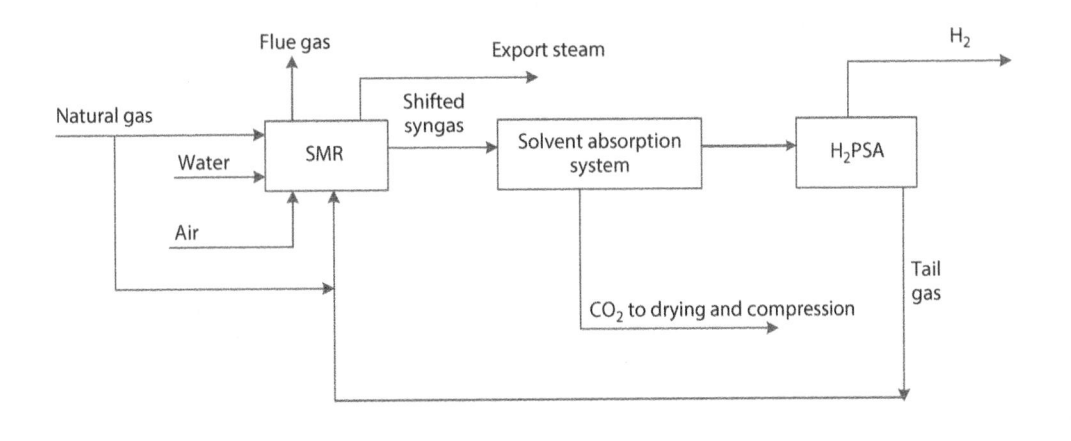

Figure 27.61 SMR retrofit CO_2 capture option 3 [75].

- Solvent-based post – combustion CO_2 absorption.
- SMR revamp options to minimize furnace fireing and use of an H_2-rich fuel.

The flue gas stream is the most expensive stream to scrub CO_2 due to the low pressure and low concentration. Current best-in-class solvent technology for flue gas capture results in costs 2-4 times more per metric of CO_2 captured than pre-combustion capture due to the low CO_2 partial pressure and solvent degradation. To reduce the cost of SMR flue gas CO_2 capture, advanced solvents with high solvent stability, improved mass transfer properties and low heat of regeneration are required.

Flue gas emssions can be further reduced through steam methane reformer revamp options that minimize furnace firing and use of an H_2-rich fuel. Various options result in > 90% CO_2 capture without requiring expensive post-combustion capture. These options include operating the reformer at high methane conversion in a pre-reformer, a primary reformer and optionally secondary gas heated reformer, and optionally, a secondary gas heated reformer in series or in parallel; eliminating excess steam export; using a structured catalyst insert; adding low-temperature water-gas shift to minimize CO content in the shifted syngas; removing the pre-combustion CO_2 from the syngas or the tail gas, and diverting a slipstream of H_2-rich fuel to the furnace [76]. This type of optimized SMR design that concentrates CO_2 for precombustion capture is expected to play an essential role for new assets.

27.18 Gas Heated Reformer (GHR)

Gas heated reformer is a pre-reformer referred to as a convective reformer. It is in contrast from the traditional steam methane reformer (SMR) due to the range of temperature and the method of heat transfer it operates on. Where the

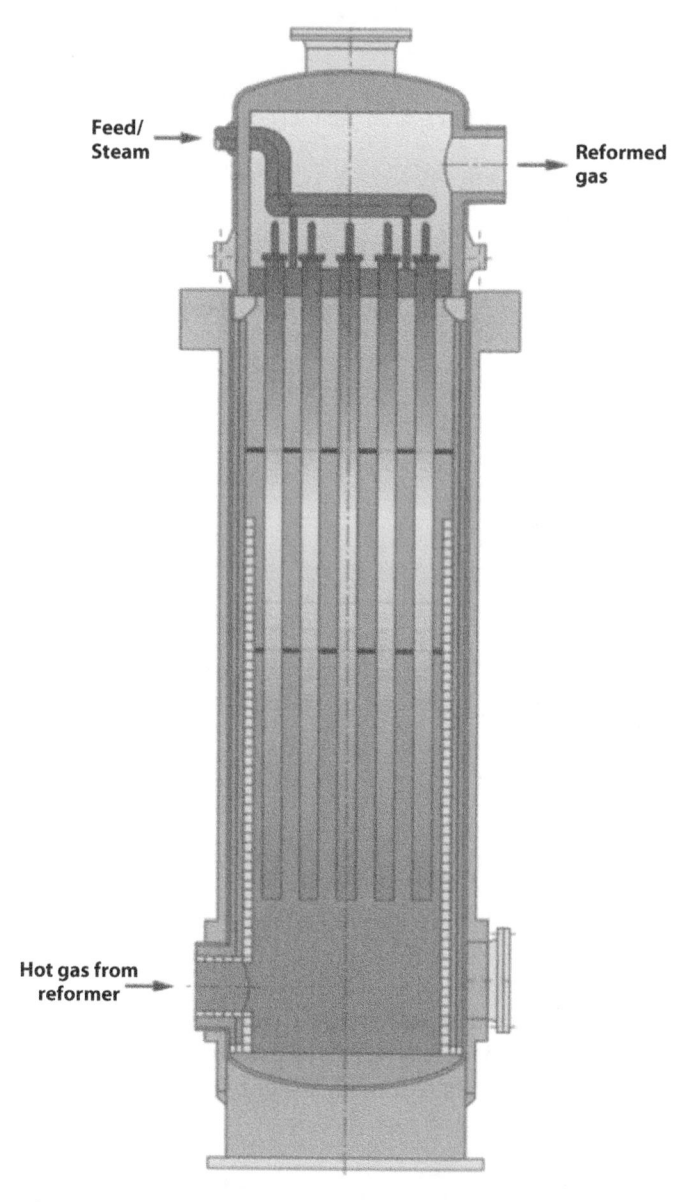

Figure 27.62a Gas heated reformer. (Source: TyssenKrupp Industrial Solution [78]).

SMR reactors are heated by external combustion of natural gas in a system of reactor tubes and burners, the GHR operates as a heat exchanger, absorbing energy by convective heat transfer with another gas [78]. Figure 27.62a shows the basic principles of the GHR. In an integrated system, the GHR is used as both a pre-reformer and a heat exchanger cooling the syngas prior to the water gas shift reactors.

Figure 27.62b show a steam methane reformer reactor (SMR), where the reactor consists of several reactor tubes filled with reforming catalysts and kept in a furnace that provides the necessary heat for the reaction. The reaction is a catalytic reaction supported by nickel based catalyst, which are cost efficient and have adequate activity. When more activity is required, a more noble catalyst can be used. Nobel catalysts provide higher activities and faster reactions but are expensive.

The reaction is normally carried out at pressures > 20 bar, steam to carbon (S/C) ratio of 3:4 on a molar basis and temperatures between 500 – 900°C. A higher S/C ratio is partially to reduce the risk of carbon desposition on the catalyst surface. Higher conversion is thermodynamically favored by low pressures; high S/C ratio and high temperatures. From an energy efficiency and economic view point, low S/C ratio is preferred and modern SMR plants have been designed to withstand higher temperatures. The upper temperature limit is due to material limitations. SMR operates at an energy efficiency up to 80-85% and generally produces more hydrogen per carbon than both partial oxidation (POX) and autothermal reforming reactor (ATR). H_2/CO ratio is typically between 3.5 and 5.5 in the reformed products.

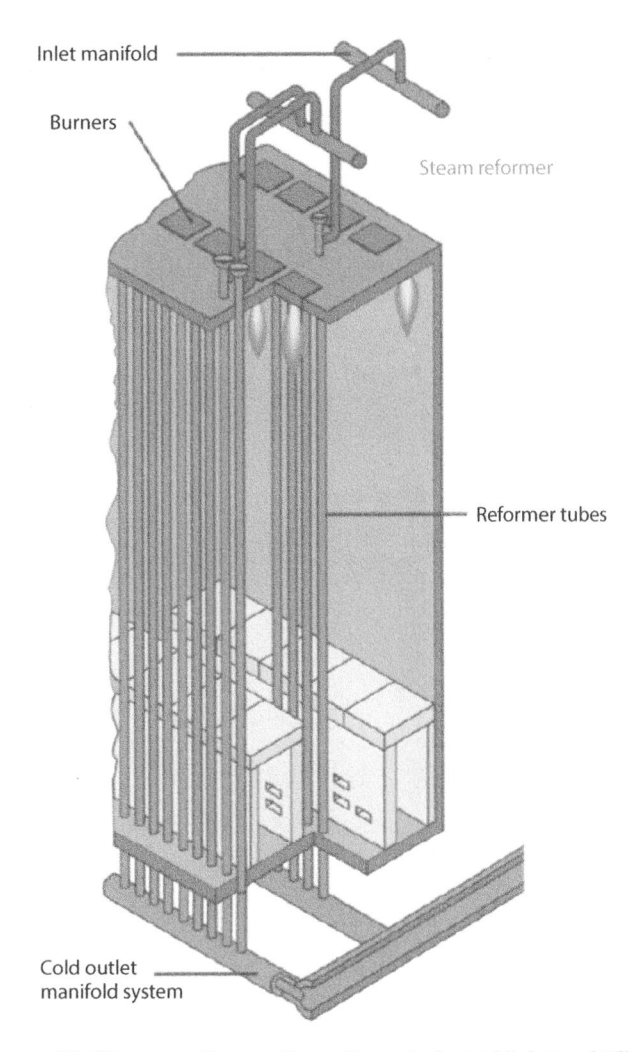

Figure 27.62b Steam methane reformer (SMR) reactor. (Source: TyssenKrupp Industrial Solution [78]).

Partial Oxidation (POX) is a method of producing hydrogen from natural gas, where the reaction is exothermic as compared to SMR reaction. By burning the natural gas with a limited oxygen supply, the products are H_2 and CO as shown by the following equation.

$$C_mH_n + \frac{m}{2}O_2 \rightleftharpoons mCO + \frac{n}{2}H_2 \qquad (27.17)$$

A POX reactor has a hydrocarbon and an oxygen input. Most large scale systems include air separation plant in order to supply clean oxygen. This increases the purity of the output as well as reducing the size of the reactor. The energy efficiency of POX is ~ 70-80% with the reactors usually operating at temperature between 2102 – 2732°F (1150-1500°C) [77].

Figure 27.62c shows autothermal reforming reactor (ATR). ATR is the combination of POX and SMR in one reactor. Natural gas (NG) is partially oxidized in a combustion zone, while steam is injected in a SMR zone. Thus, both the POX and the SRM reactions are active simultaneously. This requires pure oxygen as an input as well as a catalyst bed in the steam reforming section of the reactor. The core benefit of this system is that the heat generated by the POX reaction is consumed by the endothermic SMR reaction, which enables a closed system, and is insulated from external heat supply. Additionally, since the oxidation occurs within the reaction chamber, flue gas is not produced, resulting in no local emissions. Table 27.14 shows that retrofitting an SMR with cyrogenic fractionation on tail gas can provide solid payback in the U.S and EU.

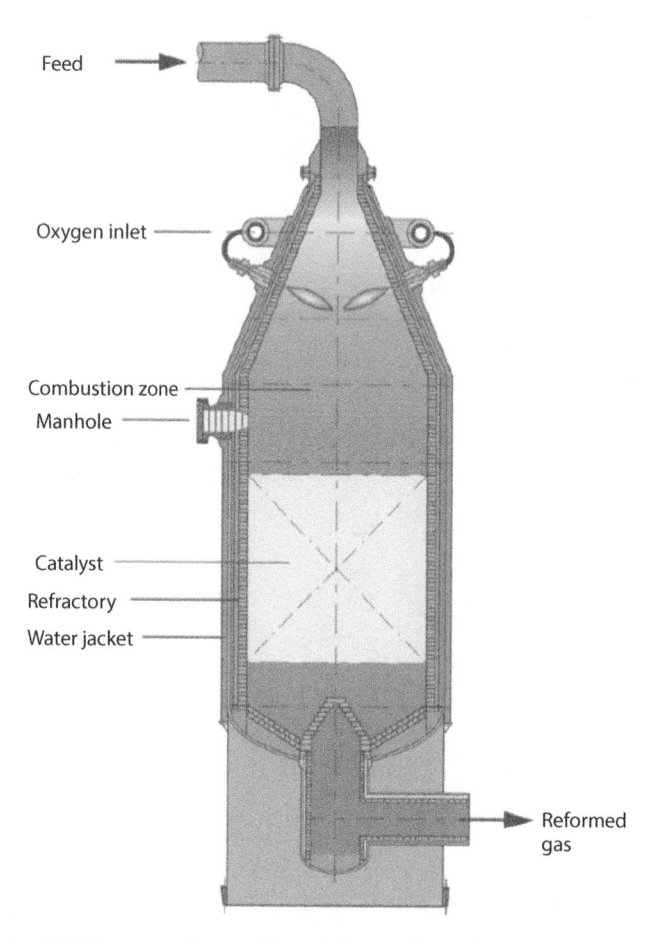

Figure 27.62c Autothermal reforming (ATR) reactor. (Source: TyssenKrupp Industrial Solution [78]).

Table 27.14 Gray H_2 Steam Methane Reactor (SMR) Process Stream Pressure and CO_2 Concentrations.

Pre-combustion		Post-combustion	
	Shifted syngas	PSA tail gas	Flue gas
CO_2 content, mol%	12 – 18%	50 – 60%	15 – 22%
Pressure, barg	20 – 30	0.3 – 0.5	0.1

Blue H_2 producers aiming for carbon intensities approaching that of green H_2 will need to choose among SMR technology that is optimized to minimize radiant firing while using H_2-rich fuel, autothermal reforming (ATR) and partial oxidation (POX). The latter two technologies offer a similar advantage to the optimized SMR design of enabling more than 90% CO_2 capture without costly, post-combustion capture, but they achieve this by eliminating

Table 27.15 Comparison of CO_2 Capture Options for Blue H_2.

	Cryogenic fractionation on tail gas	CO_2 PSA on tail gas	Amine on shifted syngas	Amine on flue gas
% CO_2 recovery from steam	> 99%	> 99%	> 99%	90 - 99%
CO_2 phase	Liquid	Gas	Gas	Gas
Ultra-high-purity CO_2	Yes	No	No	No
Steam required	No	No	Yes	Yes
Burner revamp	Yes	Yes	Yes	No
H_2 yield	+10%	No change	-1%	No change
CAPEX/OPEX	Medium	Low	Medium	High
Cost of CO_2 captured, $/t	20 - 40	35 - 50	45 - 60	70 - 100

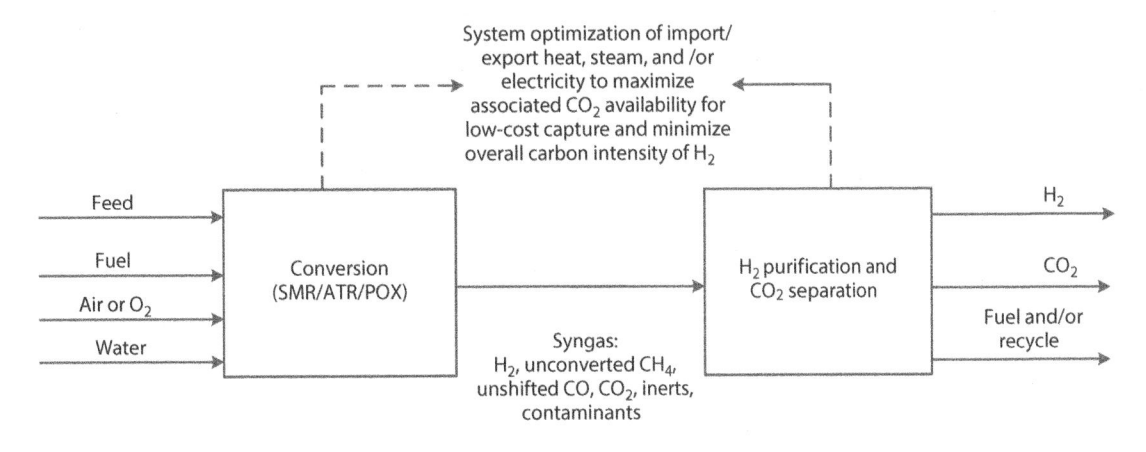

Legends:
SMR: Steam methane reformer
ATR: Autothermal reforming
POX: Partial oxidation

Figure 27.63 New blue H_2 unit landscape [75].

Table 27.16 Example Financials for Viable SMR Retrofit Projects.

Cost of CO_2 captured, $/t	U.S.	EU
Carbon capture plant costs		
Utility cost	-21	-45
Fixed cost (maintenance and overhead)	-11	-11
Annualized capital cost	-18	-18
Product values		
Value of 10% additonal H_2 recovery	28	38
Total		
Net cost of carbon captured	-22	-37
CO_2 transport and storage	-10	-10
CO_2 price	50[1]	48[2]
Net value	18	1

Basis: Negative values are costs and positive values are revenues on $/t CO_2 basis
U.S.: $3/GJ (LHV) natural gas price: $1.35/kg H_2 value
EU: $6.6/GJ (LHV) natural gas price: $1.8/kg H_2 value
[1]Tax credit in U.S. under IRC Section 45Q for carbon captured in permanent geological storage
[2]EU allowance unit trading at €40/t CO_2 in Q1 2021.

the furnace and its associated flue gas at the expense of requiring pure O_2 as a reagent and reduced production of H_2 per mole of methane processed. Table 25.15 shows the comparison of CO_2 capture options for blue H_2.

In these options, > 90% CO_2 capture can be achieved with a single capture step on a pre-combustion stream. The CO_2 capture technologies covered in the SMR retrofit analysis are applicable for new installations, irrespective of reformer selection. The most appropriate pre-combustion technology for carbon capture and H_2 purification will depend greatly on the required phase, purity, pressure, storage and means of transport for both the CO_2 and H_2 (Figure 27.63) [75]. Table 25.16 provides examples of financial for steam reforming methane retrofit projects.

27.19 Pressure Swing Adsorption (PSA)

Pressure swing adsorption (PSA) is one of the important gas separation processes. In PSA, the adsorbent is regenerated by reducing the partial pressure of the adsorbed component. This involves lowering the total pressure or using a purge gas reduction in partial pressure. PSA is useful for bulk separation, applied to a concentrated feed stream, as well as for purifications.

When PSA is used to purify hydrogen (H_2), the syngas is sent through an adsorption column at high pressures letting through H_2 while adsorbing carbon dioxide (CO_2) and other impurities. The pressure inside the column is then lowered near atmospheric pressure desorbing the impurities from the adsorption material. Several columns operate simultaneously making the hydrogen purification process semi-continuous. Columns with multiple adsorbents are normally used when purifying hydrogen, and typical adsorbents are silica gel, alumina (Al_2O_3), activated carbon and zeolite. Operating temperature in PSA units is typically ambient, receiving the feed syngas at a pressure between 290 – 870 psi (20-60 bar); the off-gas exits the unit with pressures between 14.5 – 29 psi (1-2 bar). The PSA unit produces H_2 with a purity ~ 99.9% and with a hydrogen recovery between 60 – 95%. However, the hydrogen recovery decreases with an increase in the hydrogen purity demand. Where methane (CH_4), carbon dioxide (CO_2), and carbon monoxide (CO) are easily adsorbed, oxygen (O_2), argon (Ar) and nitrogen (N_2) are more difficult to adsorb and may reduce the purity of H_2 produced. The principle of PSA is shown in Figure 27.64.

Some of the applications of PSA processes in the separation of gas mixture are:

- Air drying
- Hydrogen purification
- Production of H_2, CO, CO_2 and syngas from steam reformer off-gas.
- Air fractionation to produce O_2 and N_2 – rich streams.
- Acid gas removal: Separation of CO_2 from CH_4
- Flue gas desulfurization – removal of SO_2 and removal of CO_2 from flue gases.
- Bulk separation of normal paraffins.
- Hydrogen recovery from coke oven gas.
- Solvent vapor recovery by PSA.

Various essential characteristics of the adsorbents are pore volume, pore distribution, surface area. Other characteristics include molasses number, iodine number, bulk density, compressive strength and abrasion strength. Figure 27.65 shows UOP sorbex simulated moving bed (SMB), which has achieved commercial success, and some of the commercially available UOP sorbex processes are based on adsorption using molecular sieves as given in Table 27.17. Adsorber design consideration criteria are provided in Table 27.18.

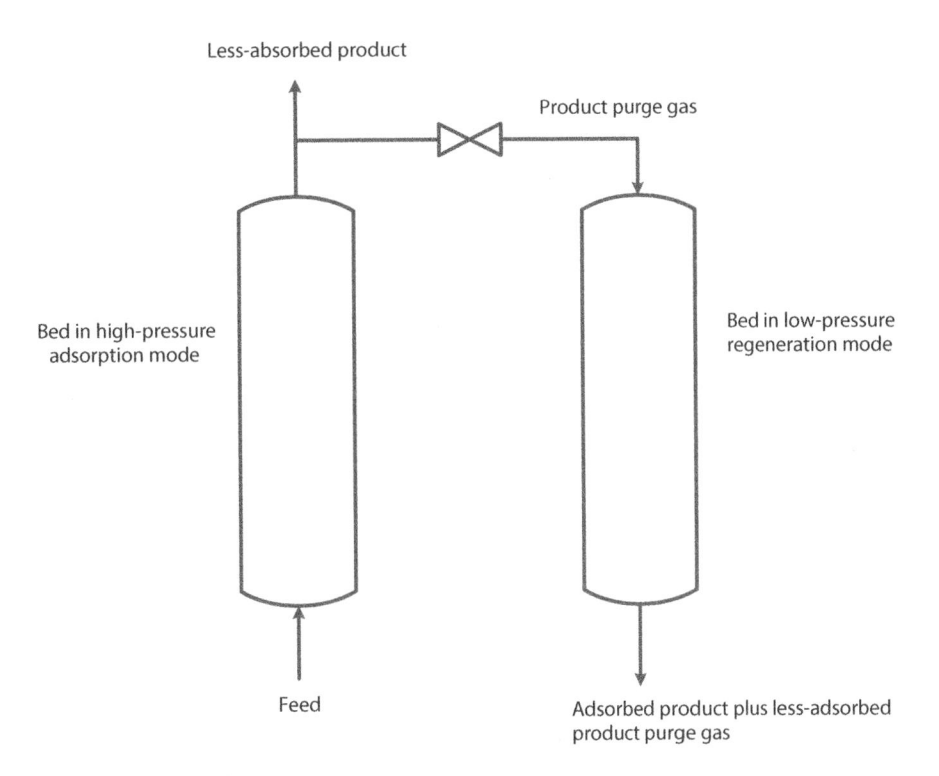

Figure 27.64 Principle of pressure swing adsorption.

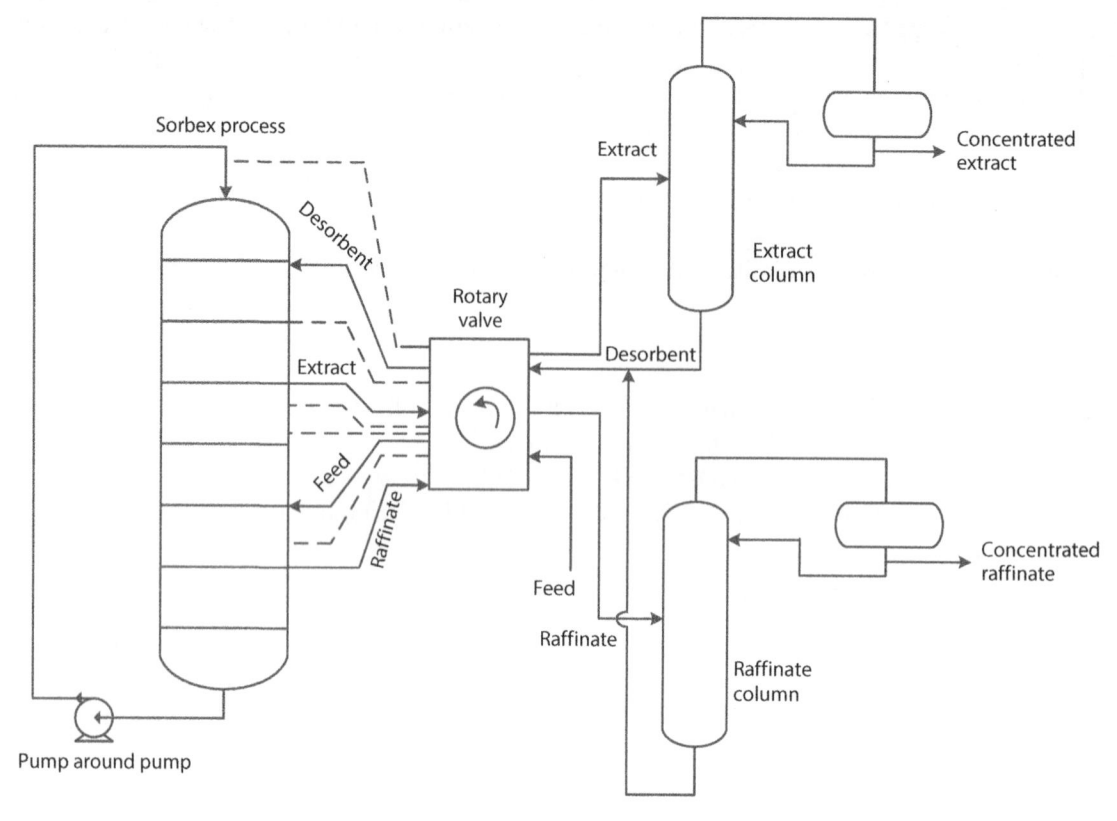

Figure 27.65 UOP sorbex simulated moving bed process.

Table 27.17 Commercial Adsorption Process.

Sorbex process	Application
Parex	Separation of para-xylene from mixed C_8 aromatic isomers
MX sorbex	Meta-xylene from mixed C_8 aromatics isomers
Molex	Linear paraffins from branched and cyclic hydrocarbons
Olex	Olefins from paraffins
Cresex	Para-cresol or meta-cresol isomers
Cymex	Para-cymerne or meta-cymene from cymene isomers
Sarex	Fructose from mixed sugar
UOP ISOSIV processor	Separation of normal praffins from hydrocarbon mixture
Kerosene ISOIV process	For separation of straight chain normal paraffins from the kerosene range (C_{10}-C_{18}) used for detergent industry.

Table 27.18 Adsorber Design Considerations.

Parameters	
Basic adsorbent properties	
Isothermal data	Uptake, release measurements; hysteresis observed, pretreatment conditions, aging upon multiple cycles, multi-components effects
Mass transfer behavior	Interface character; intraparticle diffusion, film diffusion, dispersion
Particle characteristics	Porosity, pore size distribution, specific surface area, density, particle size distribution, particle shape, abrasion resistance, crush, strength, composition, stability, hydrophobicity
Application considerations	
Operating conditions	Flow rate, feed and product concentrations, pressure, temperature, density recovery, cycle time, contaminants
Regeneration	Thermal: Steam, hot fluid, kiln chemical: Acid, base, solvent pressure shift regenerant, adsorbate recovery or disposal
Energy requirement	
Adsorbent life	Attrition, swelling; aging, fouling
Equipment, flow sheet	
Contactor type	Fixed, axial, radial flow pulsed, fluidized bed
Geometry	Number of beds, bed dimensions, flow distribution, dead volumes
Column internals	Bed support, ballast, flow distribution, insulation
Miscellaneous	Instrumentation, materials of construction, safety, maintenance, operation, start-up, shut down

27.20 Distribution and Storage

A challenge in the hard-to-decarbonize sectors are industrial, residential and heavy transportation. Reducing emissions in these sectors requires either large-scale and distributed deployement of carbon capture and storage (CCS) or switching to renewable or clean-burning fuels. Hydrogen is a promising fuel because it generates no greenhouse gas emissions at the point of use. Demand for H_2 is expected to increase up to 10-fold, from 75 MMtpd as it replaces natural gas, diesel and jet fuel [91].

For H_2 to be a critical factor for the energy transition in the coming decades, its production must be decarbonized. Today, H_2 is produced mainly from fossil fuels, resuting in 800 MMtpy of CO_2 emissions – 2% of total global CO_2 emissions. This traditional production scheme is referred to as "gray" H_2 and has lifecycle greenhouse gas emissions of 9 – 11 kg CO_2 equivalent (CO_{2eq})/kg H_2, depending on the method of production and transportation distance [92].

Several methods exist to produce H_2 (see Figure 27.66) with low carbon intensity, including the addition of CCS for "blue" H_2 (1.2 – 1.5 CO_{2eq}/kg H_2 at 90% - 98% CCS rates), use of renewable feedstocks such as biogas or biofuels

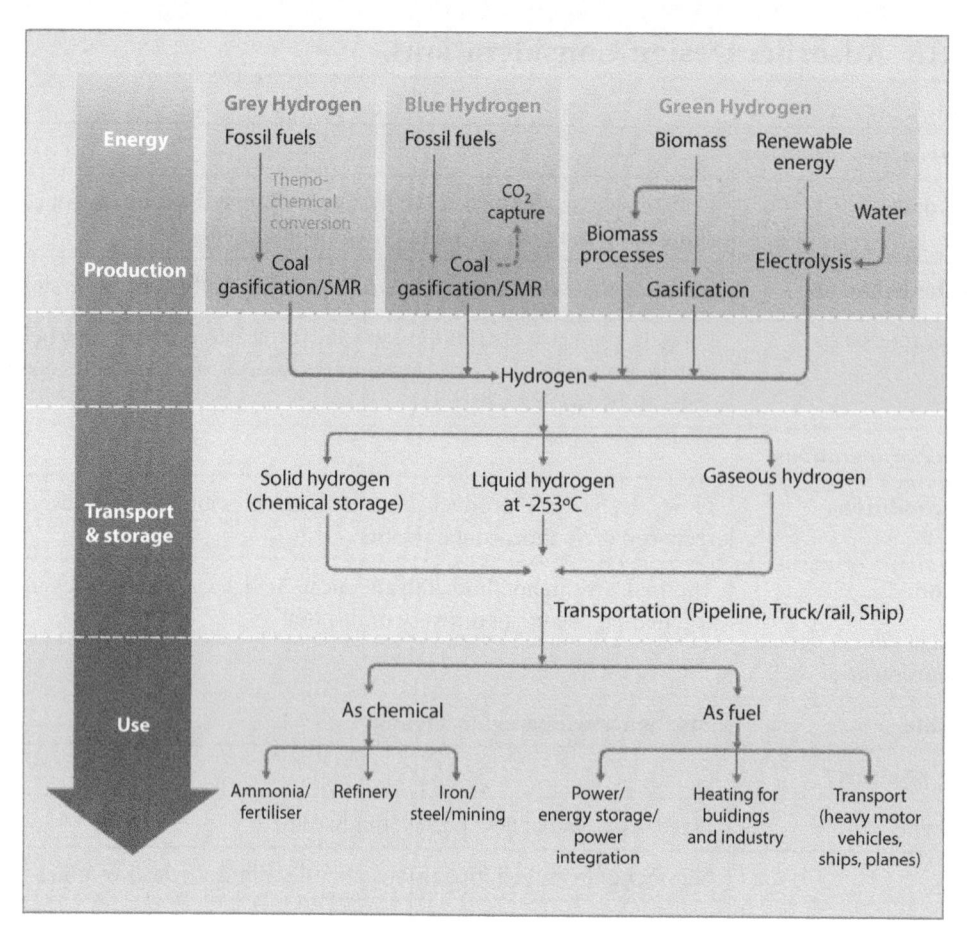

Figure 27.66 Complete life cycle of hydrogen (Source: Kalpana Gupta *et al.* [71]).

(1 – 3.3 CO_{2eq}/kg H_2), or water electrolysis using 100% renewable electricity for "green" H_2 (0.3 – 1 kg CO_{2eq}/kg H_2). Even 100% renewable electricity does not have zero life cycle greenhouse gas emissions because energy is required to produce wind turbines and solar panels. Although green H_2 production remains expensive at 4 – 6 times the cost of steam methane reforming with CCS, however blue H_2 can address the urgent challenge of decarbonization as it is commercially proven and an economic alternative to CO_2 emitting processes. The H_2 production and carbon capture technologies that enable blue H_2 are commercially sound and economical at CO_2 prices that are available in Europe and North America. Decarbonizing H_2 production of existing refining and chemical feedstock assets can be expedited with carbon capture revamps [93].

All three of these low-carbon intensity H_2 production path ways can achieve very low lifecycle emissions compared to gray H_2. The lowest lifecycle emissions are for 100% wind-powered electrolysis of water, coming in at 0.3 kg CO_{2eq}/kg H_2. Green H_2 is at its infancy with the world's largest operating plant having an electrolyzer capacity of 20MW. To generate the same amount of H_2 as a typical refinery steam methane reformer (100 ktpy or 125 MMsft³/d) with renewable-powered electrolysis of water approximately 50 times that electrolyzer capacity would be required (1 GW), along with an equal amount of installed renewable power.

Once the hydrogen has been processed, it requires to be transported to the point of use. Equipment and technologies to efficiently transport both gaseous and liquid hydrogen to its destinations or to storage until needed must be readily available.

A pipeline network is a viable option if multiple customers on an industrial site require H_2. Several production facilities can feed H_2 into the network and in turn can supply the gas to multiple customers at different locations.

Cryogenic tanks can be provided to store the H_2, if it is not going to be used directly. The liquid hydrogen (LH_2)[‡‡] is efficiently stored in vacuum-insulated tanks, which can be installed either vertically or horizontally, with capacities ranging from 3000 – 10,000 liters.

For bulk storage of gaseous hydrogen, underground salt caverns are an option. The gas has to be purified and compressed before it can be injected into the cavern. Hydrogen filled cavities can act as a backup for a pipeline network. Another transportation option for hydrogen is to deliver it to the point of use in converted form, i.e., as ammonia (NH_3) or methanol (CH_3OH).

Applications

Hydrogen is essential across a wide range of industrial processes, such as chemicals, refining, metalworking and glass, or as a future proof source of fuel for a more sustainable future.

H_2 is the key to a more environmentally friendly mobility ecosystem. Hydrogen FuelTech provides high-performance refueling concepts and technologies, as over 160 H_2 fueling stations world-wide are equipped with this technology. Ionic technology is used to compress gaseous H_2 to up to 100 MPa and the cryopump efficiently supplies hydrogen in liquid form ready for refueling. The potential of hydrogn extends far beyond the mobility ecosystem as it is also invaluable across a wide range of industrial applications.

Hydrogen is an alternative transition to a more sustainable energy economy. It can be used as a zero-emission fuel for vehicles, trains, ships and also as a feedstock for various industries including refining, chemical and steelmaking. Additionally, hydrogen can provide a source of energy and heat for buildings and can store energy produced from renewable sources.

Green hydrogen solutions have the potential to revolutionize the global energy system and lead to decarbonization of heavy industry and transport within the next decade, especially in hard to abate sectors. However, there is urgent requirement for continuous funding of innovation with respect to green hydrogen.

The global economy and countries worldwide are dependent on the flows of fossil (e.g., oil, coal) fuels, and natural gas to keep their economies growing. As countries move to decarbonize and adopt renewable energy, many are encountering problems in achieving this aim because of fundamental limitations in wind and solar resources. For these countries to fully decarbonize without encountering financial hardships, they must develop innovative renewable energy carriers (e.g., pipelines, ships, electricity and so on) and build new zero-carbon energy supply chains. Grejtak *et al.* [70] examined these renewable energy carriers, the countries and companies developing them. They evaluated the lifetime costs of 15 different renewable energy carriers ranging from conventional carriers like electricity, hydrogen, synthetic methane, and ammonia, to more advanced energy carrier concepts like liquid organic hydrogen carriers (LOHCs), vanadium and aluminum.

David Mackay of Oxford University illustrated the challenges of meeting energy demand solely through domestic renewable energy. He analyzed population density and per capita energy demand, finding that the power used per unit area in some regions exceeded wind and solar production (Figure 27.67). Furthermore, countries representing $9 trillion of global GDP would face difficulties in meeting energy demand with domestic renewable production alone requiring the import of future energy carriers [70].

[‡‡] *A wide range of industries including metalworking, medical technology, electronics and food processing use liquefied gases. These gases are always supplied to customers in liquid form so that they can be stored on site for later use. Cryogenic solutions for the liquefaction of H_2 and suitable tanks can be delivered for storage and reliably. The inner tanks and piping systems are manufactured from stainless steel to guarantee a high purity grade H_2. This is important for the foodstuff and electronic industries. The outer shell has a special coating that guarantees excellent insulation from ambient effects.*

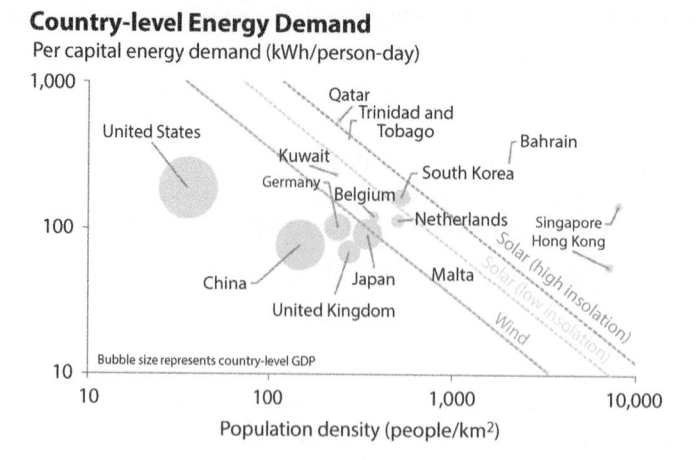

Figure 27.67 Country – level Energy demand vs. Population density [69].

Electricity is the sole means of transmitting renewable energy and approximately there are 5 million kilometers of high-voltage power lines around the world, with roughly 200,000 km added each year. The majority of this transmission line growth is to support renewable energy capacity additions, with China as the leader in new transmission line building to support its wind and solar aspirations.

Grejtak *et al.* [70] predicted the first tipping point for deploying renewable energy import infrastructure will be in 2030, when imported electricity via new high voltage direct current (HVDC) power lines becomes cheaper than low-carbon natural gas turbines. Furthermore, the next tipping point will occur in 2040, when imported liquid hydrogen becomes cheaper than low-carbon steam methane reformation, as this gives companies just 10 years to develop the partnerships and pilot projects required to demonstrate such a transformative energy paradigm. Major companies such as Shell, Mitsui & Co., Equinor, Kawasaki Heavy Industries are developing their own decarbonized energy trade routes in Europe, Japan and Southeast Asia for a $500 bn worth of energy imports.

27.21 Steam Methane Reforming (SMR) for Fuel Cells

Several leading companies are exploring the potential of green hydrogen solutions in their energy transition and decarbonization strategy. For example, Saudi Aramco has a wide range of research, innovation and development programs in the areas of hydrogen and sustainability and the company is making efforts to producing hydrogen from hydrocarbons economically, as well as capturing and utilizing carbon dioxide in a cost competitive manner. Furthermore, Air Products, ACWA Power and NEOM signed an agreement for a $5bn production facility, Saudi Arabia powered by renewable energy for the production and export of green hydrogen to global markets.

Gupta *et al.* [71] showed that hydrogen is very promising as a decarbonizing fuel with high potential for transportation and power generation. However, they expressed that many challenges must be overcome before establishing itself as the cleanest available fuel. Production of low carbon, cost effective hydrogen is the foremost challenge. Figure 27.66 shows the complete life cycle of hydrogen.

The most reliable and efficient process for hydrogen production is steam methane reforming (SMR) of fossil fuels. This process can be divided into the following steps:

- Feed pretreatment
- Steam reforming
- Shift process
- Synthesis gas cooling
- Purification

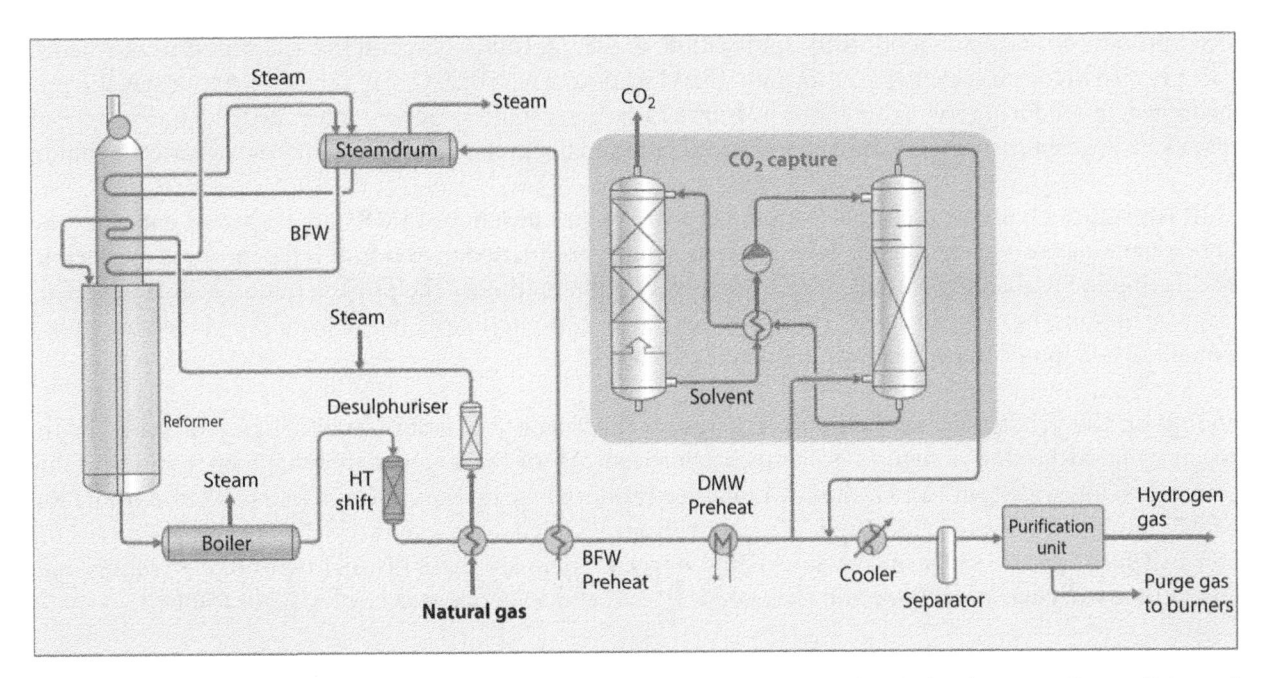

Figure 27.68 Process flow diagram of steam methane reforming (SMR) process with carbon dioxide (CO_2) capture. (Source: Kalpana Gupta et al. [71]).

Figure 27.68 shows the process flow diagram of SMR process with carbon dioxide (CO_2) capture. The primary reaction of reforming is strongly endothermic (see Chapter 11, volume 1 of the volume series). The heat needed to drive the reaction forward is usually supplied by burning natural gas and thus producing CO_2. Carbon monoxide (CO) in the output stream from the primary reaction is usually converted to CO_2 via the water gas shift reaction to increase hydrogen production.

The reactions taking place are as follows:

$$CH_4 + H_2O + \text{heat} \rightleftharpoons CO + 3H_2 \quad (\Delta H_R = +\text{ve}) \tag{27.18}$$

$$CO + H_2O + \rightleftharpoons CO_2 + H_2 + \text{heat} \quad (\Delta H_R = -\text{ve}) \tag{27.19}$$

The reformed gas is cooled and routed to a shift reactor to maximize the hydrogen content. The produced syngas is further cooled and process condensate is separated out. The reformed gas has an approximate composition of the following:

Composition	Mol %
H_2	74
CH_4	7
CO	1
CO_2	18
Total	100

The exact proportions depending on freed composition, operating conditions, and the selected process scheme. The gases are purified in the pressure swing adsorption (PSA) section to remove CO, CO_2, and CH_4 impurities and produce grey hydrogen. In producing low carbon "blue hydrogen", a carbon capture process is integrated into the base scheme. A solvent based CO_2 capture process is shown in Figure 27.68, and the capture CO_2 can be used in a variety of industries.

The SMR is an efficient, widely used and economical process; the efficiency of SMR and its specific energy consumption are best among current commercially available hydrogen production methods. It is the most reliable technology and has the highest availability among all hydrogen production methods. The process can be easily integrated with CO_2 capture options, has very low NO_x emissions, and there is no liquid discharge from the processing unit. SMR has a small carbon footprint per tonne of hydrogen produced.

Depending on the production method and feed source, hydrogen is classified as blue, grey or green hydrogen. Hydrogen from coal, oil, and natural gas is grey hydrogen. All of these non-renewable sources and production methods, when integrated with a CO_2 capture unit, then produce blue hydrogen. Hydrogen produced from biomass, wind solar and hydro-powered electrolysis is green hydrogen. Green hydrogen is typically produced via electrolysis of water in which waster is split into hydrogen and oxygen. There are three distinct types of electrolysis, namely: alkaline, proton exchange membrane and electrolysis (PEM) and solid oxide electrolysis cells (SOECs).

Gupta *et al.* [71] compared SMR with other emerging technologies as fuel cell grade hydrogen, fuel cell electric vehicle (FCEV), battery electric vehicle (BEV), biomass gasification or biogas reforming, and from electrolysis. They showed that SMR with carbon capture (blue hydrogen) is the most suitable option in the near term while biomass gasification and electrolysis are mid-term solutions. Other pathways such as fermentation and solar-hydrolysis are projected as long-term solutions.

Furthermore, the SMR process used in refineries and other applications require modification in its purification section to produce fuel cell grade hydrogen. The purification step has a solution based on PSA with or without a catalytic process for producing fuel cell grade hydrogen. A small, modular SMR base hydrogen unit can produce fuel cell grade hydrogen at 350 barg pressure at a cost <$2.8/kg with an investment of $9000-15000/t/y H_2.

A small hydrogen unit of 100-500 Nm^3/h can fuel 10 – 50 city buses or 50-250 passenger cars. A fuel cell powered bus has an operating cost of $0.2 – 0.3/km, and reduces CO_2 emissions compared with fossil fuelling by 60 – 80%. Blue hydrogen should help to create the infrastructure required for the use of fuel cells in the transportation sector.

27.22 New Technologies of Carbon Capture Storage

Many establishments are being created worldwide with a view of combating and mitigating CO_2 emissions in the petroleum and petrochemical industries in agreement with the Paris Net zero of carbon footprint. One such establishment is The National Carbon Capture Center (NCCC) sponsored by the U.S. Department of Energy (DOE), operated and managed by Southern Company. The center evaluates carbon capture processes form third-party developers by focusing on the early-stage development of the most promising cost-effective technologies for future commercial deployment.

The center has hosted many national and international technology developers by fostering the commercialization of new materials and processes for power generation that can meet future environmental standards while limiting the increased cost of electricity. These developers have conducted various pilot plant studies using their technologies and experience to refine and, in many cases, scale-up their technologies, and data generated at the site have proven to be reliable and accurate. With over sixty technologies, the NCCC has participated on the projected cost of carbon capture by one-third as shown in Figure 27.69.

The NCCC is focused on carbon capture technologies for natural gas and coal power plants. However, a significant amount of progress was achieved on gasification and pre-combustion carbon capture technologies. Table 27.19 shows the various research studies carried out since its inception in 2009 [74].

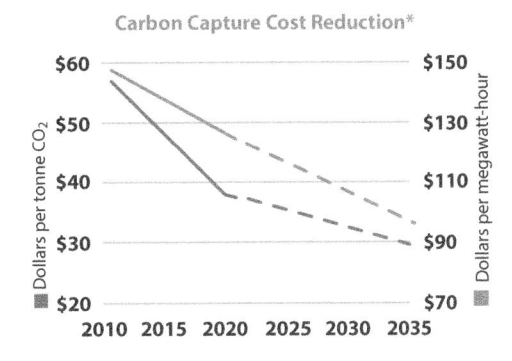

Figure 27.69 Carbon Capture Cost Reduction* (Source: National Energy Technology Laboratory, 2015, Cost and Performance Baseline for Fossil Energy Plants (Vol. 1, Rev. 3)/2018 CURC-EPRI Advanced Fossil Energy Technology Roadmap).
* Supercritical pulverized coal 2011 dollars.

27.23 Carbon Clean Process Design (CC)

As shown in Table 27.19, refinery point sources emit a variety of CO_2 containing gas streams, where gas quality is a major concern. This is because of high concentrations of potentially corrosive, toxic and flammable components that this gas emits. The sector faces increasing pressure to lower carbon emissions as net zero initiatives are widely adopted worldwide. CCS technology provides one of the best ways refineries can decarbonize, and can be paired with hydrogen production to provide new opportunities and diversify value chains. Conventional solvents for carbon capture are amine based with the most common being monoethanolamine (MEA). The CDRMax™ process uses an improved solvent-based carbon capture technology that employs formulation of amines and salts referred to as Amine-Promoted Buffer Salts (APBS). This innovation has been developed into two widely – used commercial solvents, namely: APBS-CDRMax® developed to extract CO_2 from flue gas in large scale industrial plants and APBS-CARBex® designed for biogas/RNG upgrading. These unique solvents can be used as either a drop-in substitute into existing systems such as biogas separation or used with an integrated carbon capture system.

Table 27.19 Research/Investigations on Carbon Capture Storage at the National Carbon Capture Center.

Solvent – Based Carbon Capture Projects	While solvent – based carbon capture, particularly amine scrubbing has a high degree of technical applications, the increase in cost of electricity (COE) using the industry – standard solvent monoethanolamine (MEA) could exceed 70%. A critical area of research for solvent – based carbon capture is the identification of advanced solvents with high capacity for carbon dioxide (CO_2) loading and lower regeneration energy requirements than MEA. Next generation solvents must also be low-cost, non-corrosive, fast reacting and degradation-resistant. Process improvements are required since operating solvent systems with conventional scrubbing, even with advanced solvents could increase COE by 40% or more. Therefore, optimized equipment as well as hybrid systems involving membranes and enzymes are being investigated.
Sorbent-Based Carbon Capture Projects	Unlike most solvent, CO_2 solid sorbents contain no water and therefore offer much lower heating and regeneration energy requirements. To advance sorbents as a viable carbon capture solution, research and development is being carried out to demonstrate sorbent's low cost, thermal and chemical stability, resistance to attrition, low heat capacity, high CO_2 loading capacity, and high selectivity for CO_2. Optimization of process equipment designs is also required to suit the characteristics of each type of sorbent.

(Continued)

Table 27.19 Research/Investigations on Carbon Capture Storage at the National Carbon Capture Center. (*Continued*)

Membrane – Based Carbon Capture Projects	Gas separation membranes offer several notable advantages for carbon capture applications: simple, modular designs; no need for steam or chemicals; and unit operation as opposed to complex processes. Membranes are being researched as a step – change improvement for carbon capture. The scopes of work in this area are to develop membranes with low cost and durability, enhanced permeability and selectivity, thermal stability, and tolerance to flue gas contaminants.
Solvent Testing in the Pilot Solvent Test Unit	The National Carbon Capture Center's Pilot Solvent Test Unit (PSTU) consists of the necessary equipment to test the adsorption and regeneration characteristics of carbon capture solvents and has the flexibility to operate at varying conditions and process configurations. The PSTU conducted several tests with MEA solvent, demonstrating mass and energy balances of nearly 100%, the testing provided a baseline regeneration energy of 1,500 – 1,720 Btu/lb of CO_2.
Babcock & Wilcox OptiCap™ Solvent	B & W's OptiCap solvent demonstrated regeneration energy values in the range of 1,100 – 1,125 Btu/lb CO_2 at a CO_2 removal efficiency of 90%, with further energy savings expected through B & W's optimized heat exchanger design. B & W considers the solvent ready for commercial demonstration, noting that OptiCap performance solvent compares favorably with other commercially ready solvents in areas of regeneration energy, corrosivity and solvent degradation.
Hitachi H3-1 Solvent	The H3-1 solvent showed that compared to MEA, a 37% lower solvent flow rate is needed to achieve 90% CO_2 capture, and the regeneration energy is at least 34% lower. Further studies are carried out with a view for scale-up to commercial operation.
Cansolv DC-201 and DC-103 Solvents	Cansolv performed four tests involving long – term and parametric tests with cool and hot climate conditions. The testing demonstrated a 40% reduction in energy requirements and a 50% reduction in the required liquid-to-gas ratio over MEA. Cansolv's process using the DC-103 solvent is being demonstrated commercially at the SaskPower Boundary Dam Power Station.
Chiyoda T-3 Solvent	Testing of the Chiyoda T-3 solvent for 1,500 hours of operation showed that in comparison to MEA, the optimum liquid-to-gas ratio was about 50% lower, regeneration energy for 90% capture at 1,110 Btu/lb was around 30% lower and corrosivity was significantly lower. Further investigation is being carried on the solvent.
Carbon Clean Solutions (CCSL) Solvent	CCSL estimated that 90% CO_2 removal could be achieved with 1,160 Btu/lb CO_2. CCSL's solvent was further tested at Technology Centre Mongstad (TCM) in Norway and is currently being employed for carbon capture from three commercial facilities: an alkali chemicals and fertilizers plant with a coal-fired boiler in India, a coal – fired combined heat and power plant in Eastern Europe and a chemical plant with a coal – fired boiler in India. New technologies by CCSL in CO_2 capture are described later in the book.
DoE Carbon Capture Simulation Initiative (CCSI)	Testing was conducted with MEA solvent for two campaigns that supported DOE's CCSI and CCSI*, the second phase of the project. The goal of the CCSI is to significantly reduce the time required to develop and scale up new technologies in the energy sector. The testing resulted in a CCSI Toolset that can be used to gain more data and information during carbon capture test campaigns. The CCSI project continues to engage with technology developers of the computational tool.

(*Continued*)

Table 27.19 Research/Investigations on Carbon Capture Storage at the National Carbon Capture Center. (*Continued*)

GE Global Adsorber and GAP-1 Aminosilicone Solvent	GE Global tested its continuous stirred tank reactor (CSTR) and non-aqueous GAP-1 solvent with the PSTU. The CSTR, a one-stage separation unit that is smaller and less costly than conventional columns, achieved 95% capture. Based on solvent operation with the PSTU regenerator, GE showed that the GAP-1 solvent could achieve a 20 – 30% improvement in COE over MEA. GE received DOE phase 1 funding to evaluate a demonstration – scale – 10-megawatt (MW) test at TCM, but the development is suspended.
ION Engineering Solvent	During a 1,100-hour campaign, ION's advanced solvent consistently demonstrated the potential to substantially reduce capital and operating costs. Results indicated at least 30% reduction in regeneration energy requirements, 35% higher CO_2 solvent-carrying capacities and significantly less solvent degradation in comparison with MEA. Larger-scale testing was conducted at TCM, and ION is seeking further pursuing opportunities with development.
AECOM/University of Texas at Austin (UT-Austin) Advanced Flash Stripper (AFS) with Piperazine Solvent	The AFS is an energy-efficient alternative to conventional strippers that is projected to achieve optimal performance with piperazine solvent, which features fast kinetics, high capacity, low volatility and degradation resistance. The AFS skid was integrated with the PSTU to bypass the standard regenerator. The AFS demonstrated more than 40% energy reduction over the PSTU regenerator while operating with piperazine solvent. UT-Austin will conduct further testing of the AFS with piperazine solvent using simulated natural gas flue gas in a separate project sponsored by the CO_2 Capture Project.
Technology Development Solvent Units and Processes	
Aker Clean Carbon Mobile Test Unit	Aker Clean Carbon demonstrated its pilot – scale solvent system during more than 2,500 hours of operation with carbon capture. Test results reported included a significant reduction in emissions of solvent components when using Aker's low-emissions, anti-mist design and reduced energy consumption (about 20% lower specific reboiler duty) using Aker's CCAmine solvent compared to MEA solvent at 90% CO_2 capture. Aker has advanced the technology through larger-scale testing at TCM.
Trimeric/UT-Austin NOx Reduction	Trimeric and UT-Austin performed testing of a chemical process for removal of nitrogen dioxide (NO_2) from amine-based solvent systems to prevent nitrosamine accumulation in the solvent system and thus minimize solvent oxidation and degradation. The process involved the use of a low – cost additive, thiosulfate, in an existing sulfur dioxide (SO_2) pre-scrubber, combining NO_2 and sulfur removal into one step and achieving 90% NO_2 removal. Trimeric is deploying the integrated scrubbing chemistry at a commercial site due to a high result at the small pilot-scale project.
Codexis Enzymes	Codexis performed tests of a bench-scale system using carbonic anhydrase enzymes to accelerate the rate of carbon capture for low-energy solvents that have desirably low heats of reaction such as methyl diethanolamine (MDEA). The testing confirmed the stability of the enzyme in the presence of trace contaminants and demonstrated robust system operation, with CO_2 capture averaging around 65% at a rate of capture more than twenty-five times greater than for MDEA alone.

(*Continued*)

Table 27.19 Research/Investigations on Carbon Capture Storage at the National Carbon Capture Center. (*Continued*)

Akermin Enzymes	Akermin successfully confirmed proof of concept of its pilot-scale process featuring immobilized carbonic anhydrase enzymes using low-cost, non-volatile potassium carbonate solvent. Data showed that the enzyme operation resulted in a seven-fold increase in flue gas flow rate while maintaining 90% CO_2 capture in the same column and a six-fold increase in the mass transfer coefficient. Akermin was able to scale-up the process for further studies. However, with the scaled – up process, lack of catalyst circulation prevented steady-state operation. Akermin was unable to support further development despite promising results from the first testing.
Linde-BASF Solvent Process	Linde and BASF's solvent-based technology incorporated BASF's advanced amine solvent OASE® Blue and novel process along with Linde's process and engineering innovations, such as a gravity-flow interstage cooler and unique reboiler design. Operating the solvent process at a 1.5-MW pilot-scale for more than 4,000 hours, results showed a regeneration energy as low as 1,140 Btu/lb. CO_2 with at least 90% CO_2 capture. Testing also proved out the design of the unique equipment features incorporated in the pilot plant. The technology was selected by DOE for Phase 1 funding of a project to capture 90% of the CO_2 from an existing coal-fired plant at a U.S. site, and Linde has submitted a Phase 2 proposal for implementation of the project.
University of Edinburgh Solvent Analyzer	The University of Edinburgh tested a real-time solvent analysis device, developed to allow for rapid process responses to changes in process variables such as load demand in a commercial carbon capture system. The device proved capable of determining solvent concentration and loading within a 10-second response window, a significant improvement over current state-of-the-art technologies for solvent analysis that can take up to 30 minutes to provide data. The University has designed a commercial version that is commercially available.
Carbon Capture Scientific Gas Pressurized Stripping (GPS)	Carbon Capture Scientific completed testing of the bench-scale GPS process using on amine-based proprietary blended solvent. The GPS system integrates carbon capture and compression into one step. The energy consumption of the GPS process was much lower than that of the MEA baseline case, with the total heat requirement for the GPS process at about 1,310 Btu/lb of CO_2 captured. Carbon Capture Scientific has identified ways to optimize the GPS process to achieve further energy cost reductions, and the group is continuing the commercialization process.
Gas Technology Institute (GTI) Membrane Contactor	GTI is developing a hollow-fiber gas-liquid membrane contactor to replace conventional packed-bed columns in solvent systems to improve CO_2 absorption efficiency. GTI resolved all technical challenges and performed parametric and long-term testing. The system demonstrated 90% CO_2 capture with CO_2 purity greater than 97%. Further testing is planned at the NCCC.
Solvent Testing in the Slipstream Solvent Test Unit	The NCCC's slipstream Solvent Test Unit (SSTU) is conceptually and functionally similar to the PSTU, at about one-tenth the scale, making it well-suited for testing solvents that are in early stages of development. Commissioning of the SSTU with MEA solvent was conducted to provide baseline performance values against which to compare developers' solvents, and results showed a regeneration energy value of 2,200 Btu/lb CO_2 at optimum conditions.

(*Continued*)

Table 27.19 Research/Investigations on Carbon Capture Storage at the National Carbon Capture Center. (*Continued*)

Research Triangle Institute International (RTI) Non-Aqueous Solvent	RTI's solvent is an alternative to conventional aqueous solvents that lowers regeneration energy by eliminating the energy penalty associated with vaporizing water and by increasing the regeneration pressure at lower temperatures. The project, funded by DOE, is the result of collaboration of RTI and Norway's SINTEF organization. Though the SSTU was not optimized for the RTI solvent, around 75% CO_2 capture was accomplished after suitable pressure and temperature combinations were experimentally identified.
Solvent Emissions Studies	Amine emissions from carbon capture systems in the form of aerosols leaving the absorber are a common challenge with commercial solvent processes where sulfur trioxide (SO_3) is present in the flue gas. The formation of aerosols has been found to correlate with the concentration of SO_3 present in the flue gas. Solvent losses from this effect are significant and must be addressed if a commercial system is manufactured.
Cansolv Aerosol Mitigation Equipment	Cansolv Technologies' Thermal Swing Adsorber and Brownian Demister Unit were installed on the SSTU. The processes were designed to mitigate amine losses in carbon capture process by reducing the formation of amine-containing aerosols. Initial testing was conducted while the SSTU operated with MEA.
Effect of Activated Carbon Baghouse	Using an electrical low-pressure impactor (ELP+) and a phase doppler interferometer, aerosol measurements were taken in 2015 as baseline data to compare against results obtained in 2016, when a new activated carbon injection baghouse was brought online at Alabama Power's Plant Gaston Unit 5, upstream of the National Carbon Capture Center. In addition to a drastic shift to smaller particle sizes, the total particle count for aerosols dropped by three to five orders of magnitude compared to operation without the baghouse. Amine measurements showed that MEA emissions from the SSTU were between 100 and 110 parts per million (ppm) prior to the installation of the baghouse, dropping to between 5 and 10 ppm after the baghouse startup.
SO_3 Generator	Continuing with amine aerosol research, an SO_3 generator was installed to inject SO_3 directly into the flue gas stream. SO_3 injection testing was performed in conjunction with emissions monitoring during the AECOM/UT-Austin AFS campaign. Through this testing, effective strategies for solvent emissions reductions were identified, and it was shown that piperazine solvent emissions at the wash water outlet can be maintained below 1 ppm for inlet flue gas containing 2 ppm SO_3. Further testing is planned for further solvent campaigns.
Gas Separation Membranes	
Membrane Technology and Research (MTR) Polaris CO_2 Membranes	Because of the large volumes of low-pressure flue gas generated by power plants, creating an affordable pressure ratio to drive membrane separation is a challenge. MTR is developing a two-step membrane, with the first step operating at vacuum and at a low stage cut, and the second step incorporating sweep gas to provide a final CO_2 capture rate of 90%. After successfully operating a bench-scale unit at the NCCC, MTR employed the lessons learned to construct and test a pilot-scale version. Further development includes operation of the large-scale unit at a Babcock & Wilcox pilot coal fired boiler for the first operation with CO_2 recycle to a boiler by a membrane process, larger-scale operation at TCM and participation in a DOE Phase 1 project for demonstration at a commercial NRG Energy coal-fired power plant.

(*Continued*)

Table 27.19 Research/Investigations on Carbon Capture Storage at the National Carbon Capture Center. (*Continued*)

Ohio State University (OSU) Membranes	OSU has been testing a novel prototype membrane with a thin selective amine – contacting layer over a nanoporous polymer support, designed to be easily manufactured in a continuous process while achieving high CO_2 permeance* and selectivity. After conducting membrane testing in 2015, OSU returned in 2018 and demonstrated improved stability and long-term performance even with several interruptions due to power plant unit shutdowns. The membrane achieved the targeted CO_2 permeance and CO_2 selectivity, and consistent results with duplicate modules proved the reproducibility of the design and viability of the fabrication process. OSU will continue the membrane development program with sponsorship from American Electric Power.
National Energy Technology Laboratory (NETL) Post-Combustion Membrane Skid	NETL has been operating its post-combustion membrane skid, which accommodates flat-sheet and hollow-fiber membranes. NETL aims to assess new materials designed to increase CO_2 permeability while maintaining selectivity, thereby reducing the size and cost of membrane capture systems. Testing at the center has encompassed multiple campaigns with materials such as standard polydimethylsiloxane membranes for baseline evaluations, mixed – matrix designs and NETL- developed polymer materials. Further testing is still planned at NCCC.
Air Liquide Cold Membrane	Air Liquide is evaluating a cold membrane process that combines high-permeance* membrane materials with high CO_2 selectivity at sub-zero temperatures to efficiently separate CO_2 from flue gas. Current testing is focused on development and scaleup of the novel PI-2 membrane material featuring significantly higher CO_2 flux than commercially available material. The PI-2 module achieved ten times the normalized CO_2 permeance* of the commercial module. This improvement dramatically decreases the module count for a commercial power plant and results in significant capital cost reductions. Testing is continuing at the site.
Carbon Capture Sorbents	
SRI International (SRI) Sorbent	The SRI bench-scale sorbent process using carbon microbead sorbents offers several advantages, including low heat requirements, high CO_2 adsorption capacity and excellent selectivity. During operation at the NCCC, performance indicators were lower than expected based on previous testing of SRI's smaller unit at the Univesity of Toledo, with CO_2 capture efficiency at 70% and the CO_2 outlet concentration at 93%. Design modifications and process optimization are expected to improve performance and increase the capture rate to the targeted value of 90%.
DoE Sorbents	Researchers from NETL's Research and Innovation Center operated a bench-scale sorbent unit at the NCCC to evaluate accumulation of trace elements and sorbent degradation with silica supported amine sorbents. The unit operated in circulating and batch modes, with post-test thermogravimetric analysis of sorbent samples showing no permanent loss of carbon capture capacity. NETL continued testing of sorbent to improve material characteristics, although the group's current carbon capture work is focused on membrane material development.

(*Continued*)

Table 27.19 Research/Investigations on Carbon Capture Storage at the National Carbon Capture Center. (*Continued*)

TDA Research Alkalized Alumina Sorbent	TDA is developing a carbon capture process using dry, alkalized alumina sorbent, featuring low cost, low heat of adsorption and capability of near isothermal, low-pressure operation to achieve lower regeneration energy than solvent-based processes. Formal testing of the small pilot-scale sorbent system is planned for 2019.
	The NCCC has evaluated gasification and pre-combustion carbon capture processes to support the development of the next-generation power generation. This is part of accelerating the commercialization of advanced technologies to enable fossil fuel-based power plants to achieve near zero emissions. More than 50,000 hours of technology testing was achieved utilizing syngas generated from the Center's Transport Gasifier to evaluate low-carbon energy options and processes to improve the environmental and reliability aspects of gasification. Scale-ups and process intensification have been implemented, and test priorities have now focused more on post-combustion carbon capture for natural gas and coal power plants. Gasification technology testing is focused on the following: • Bio-mass co-gasification to reduce the carbon footprint of gasification – based power systems. • Sensors and instrumentation for assisting in process control systems. • Syngas utilization, involving projects to convert coal-derived syngas into liquid fuels, and hydrogen. • Syngas conditioning, including processes for gas cleaning and facilitating downstream carbon capture. Pre-combustion carbon capture development addressed key challenges, namely: improving capacity, efficiency, robustness of materials, gas selectivity, footprint and costs – involved in the major processes common to post-combustion carbon capture, which include solvents, membrane and sorbents.
Biomass Co-Gasification	The NCCC conducted two gasifiers runs with biomass testing to support the DOE goal of development of gasification technologies for the conversion of biomass into clean, sustainable energy and other products. With the gasifier feed consisting of 20 weight percent (wt%) raw, pelletized wood biomass and 80 wt% Powder River Basin (PRB) coal and Mississippi lignite, carbon conversions ranged from 97-99.9% with stable operation of the entire gasifier train.
Connecticut Center for Advanced Technology (CCAT) Biomass Operation	Another run with biomass was conducted while operating the gasifier in oxygen – blown mode on behalf of CCAT, which received funding through the Department of Defense (DOD) as part of the department's initiative to develop domestic, renewable feedstock for liquid fuel production. Using both raw and torrefied (heat treated) biomass at concentrations ranging from 10 – 30 wt% of the total feed rate, carbon conversion ranged from 97.6 – 98.7%. These data enabled CCAT team to produce a study that concludes that blending various grades of coal with biomass presents a credible approach for reducing CO_2 emissions and producing liquid fuel, although further cost reductions for associated unit operations are required.

(*Continued*)

Table 27.19 Research/Investigations on Carbon Capture Storage at the National Carbon Capture Center. (*Continued*)

Sensors and Instrumentation	
Stanford University Tunable Diode Laser (TDL)	Stanford University tested the TDL with particulate free syngas for real-time, *in situ* monitoring of water, CO_2, carbon monoxide (CO), methane (CH_4) and gas temperature in a gasification process. Operation was conducted throughout two gasification runs, from startup to shutdown. The TDL data showed excellent agreement with the existing analyzers and *in-situ* measurements, and the rapid time resolution of the TDL allowed it to capture variations in syngas conditions that were previously not evident. The success of the TDL sensor campaigns at the NCCC showed that laser absorption sensing is not only possible in engineering – scale gasifiers, but could be an important new diagnostic tool with the potential for new control strategies in future gasifier utilization and development.
NETL Mass Spectrometer	NETL's gas chromatography inductively coupled plasma mass spectrometer (GC-ICPMS) was used for the first time in the field to measure mercury concentrations directly during NCCC testing of the Johnson Matthey mercury sorbent. Although the measured mercury (Hg) levels from the GC-ICPMS were higher than those determined by Environmental Protection Agency Method 29, the results demonstrated the potential to gather real-time data and facilitate control over key parameters.
Emerson Rosemount Sapphire Thermowells	Multiple tests were conducted with Emerson's sapphire thermowells in gasifier service. The thermowells operated reliably in the highly erosive and corrosive environment. Temperature readings with the sapphire thermowell units tracked well with reference thermocouples, but were slightly lower, with the differences at about 3 and 5 % relative to the reference thermocouples.
Syngas Utilization	
Southern Research Fischer-Tropsch (F-T) Catalyst and High – Temperature Reformer	The Southern Research F-T catalyst process for converting syngas into liquid fuel eliminates the conventional product upgrading and refining steps and enhances the ability to coal-to-liquids and coal/biomass –to –liquids processes to compete with petroleum-based processes. Following initial testing at the NCCC, Southern Research incorporated a 2 to 1 scale up of the F-T reactor and began parallel tests of a high-temperature steam reformer. The reformer, operating upstream of the F-T catalyst in a commercial process for syngas pre-conditioning, showed 90% methane conversion and greater than 95% conversion of tars, with the exit gas containing the desired 2 – to -1 hydrogen to CO ratio (i.e. H/CO). The F-T catalyst exceeded project goals, achieving 75% selectivity for jet fuel-range hydrocarbons, and performing equally well with 100% coal feed in the facility's gasifier and with 20% biomass co-feed. A techno-economic analysis indicated significant capital cost and product cost reductions with the technology, but competition with petroleum-based liquids will require improvements in other major coal-and coal/biomass-to-liquids plant unit operations.
OSU Syngas Chemical Looping (SCL)	OSU is developing a high-pressure syngas-to-hydrogen chemical looping process for effectively converting carbon-based fuels to electricity, hydrogen and/or liquid fuels while simultaneously capturing all the carbon emissions. Employing counter-current moving beds and iron-based composite oxygen carriers under reduction-oxidation conditions, SCL was successfully demonstrated at bench-and sub pilot scale. The testing demonstrated greater than 98% conversion of syngas components (CO, CH_4 and H_2) in the reducer reactor as well as hydrogen production from the oxidizer reactor during steam injection. The steam injection experiment confirmed that the SCL system is capable of converting syngas and steam into pure hydrogen. OSU is progressing with the technology.

(*Continued*)

Table 27.19 Research/Investigations on Carbon Capture Storage at the National Carbon Capture Center. (*Continued*)

NETL Solid Oxide Fuel Cell (SOFC)	Since SOFCs produce electricity through an electrochemical reaction as opposed to combustion, they are more efficient and environmentally benign than conventional power generation processes. Operation of NETL's SOFC multi-cell array at the National Carbon Capture Center demonstrated over 450 hours of continuous operation, with over 4,500 cell-hours of data collected and more than 1 kW of power produced. Performance showed remarkable robustness of SOFC materials to trace material exposure as well as acceptable power density given the modest heating value of the supplied syngas. The test represented the longest duration continuous SOFC test conducted using direct coal syngas as fuel. NETL's SOFC program maintains a diverse portfolio of cell development projects focused on improving electrochemical performance and cell power density, reducing long-term degradation, developing more robust cells and reducing costs.
Syngas Conditioning	
Water – Gas Shift (WGS) Catalysts	Studies to optimize WGS catalyst operation showed that CO conversions adequate to facilitate high carbon capture rates can be achieved at lower steam-to CO molar ratios than those traditionally recommended. Evaluation of the impact of these test results for a commercial 500-MW gasification-based power plant showed the acceptable reduction in steam-to – CO molar ratio (from 2.6 to 1.6) corresponds to a substantial 40 MW increase in net electrical output.
Johnson Matthey Mercury Sorbent	In collaboration with NETL, Johnson Matthey has been developing a palladium-based sorbent to remove mercury and other trace contaminants, such as arsenic and selenium, at high temperature in coal gasification processes. Compared to low-temperature capture by activated carbon, high-temperature capture of these trace elements retains the high thermal efficiency of coal gasification power generation. With more than 6,000 hours of operation, the sorbent demonstrated between 96 and 100% removal of mercury, arsenic and selenium from syngas at 500°F and pressure of 150 – 200 psig. The results also show that the sorbent is not only regenerable, but it remains as effective at capturing metals after regeneration.
TDA Research Ammonia Sorbent	TDA Research conducted initial testing of a sorbent – based gas clean-up technology to remove ammonia and hydrogen cyanide as well as mercury and other trace metals from syngas at high temperatures in a single process step. The sorbent operates in a regenerable manner to remove ammonia and mercury, while irreversibly absorbing hydrogen cyanide and other contaminants. TDA completed 50 hours of testing and demonstrated greater than 99% ammonia removal. TDA was awarded a DOE small Business Innovation Research Phase IIB option to further develop the sorbent.
MHI Carbonyl Sulfide (COS) Hydrolysis Catalysts	Multiple tests were conducted with MHI's COS hydrolysis catalyst to facilitate sulfur removal from syngas. Testing of the catalyst – which features a honeycomb configuration, high performance at around 570°F and high durability in the presence of halogens – was performed for a total of 4,000 hours, demonstrated long-term stability. The catalyst achieved 80% conversion. MHI is making further progress with the development of the catalyst.

(*Continued*)

Table 27.19 Research/Investigations on Carbon Capture Storage at the National Carbon Capture Center. (*Continued*)

Carbon Capture Solvents	
Batch Reactor Solvent Testing	Establishing a database of solvent characteristics at NCCC, researchers evaluated well-known chemical and physical solvents such as ammonia (NH_3) (used to define solvent testing and sampling techniques), potassium carbonate (K_2CO_3), potassium prolinate and diemethyl ether of polyethylene glycol (DEPG). In addition to CO_2 absorption characteristics, solvents were evaluated for co-absorption of hydrogen sulfide (H_2S), regeneration characteristics and performance with water addition. Much of the solvent testing supported DoE studies at the University of Pittsburgh.
University of Alabama Solvents	The NCCC conducted a series of tests in support of the University of Alabama's research on alkylimidazoles, a group of low-volatility, low-viscosity liquids as physical solvents for economical CO_2 separation. Results showed the CO_2 capacity was in the range of other physical solvents tested previously, and regeneration of the solvent renewed its CO_2 capacity but not that of H_2S. The investigation in solvent research is continuing.
Air Products Pressure – Swing Adsorption (PSA)	Air Products designed an alternative to acid gas removal with physical solvents that consists of two process blocks: PSA that separates CO_2 and H_2S from the desired products, and a tailgas disposition block that separates the sulfur containing compounds and purifies the CO_2 to a sequestration-grade product. During testing, the PSA system separated more than 95% of the CO_2 and 99.7% of the H_2S in the feed gas, and post-test evaluations indicated the adsorbent material maintained its capacity. Air Products is continuing with the development of the PSA process as a carbon capture solution applicable to a wide range of gasification technologies, including power plants and hydrogen fuel production.
SRI International Solvent and Bechtel Pressure Swing Claus Processes	SRI's ammonium carbonate-ammonium bicarbonate (AC-ABC) solvent process features high capacity for CO_2 and H_2S, high regeneration pressure (which lessens downstream compression requirements) and thermal stability of the regenerated solvent solution. Operation of SRI's pilot-scale system demonstrated greater than 99% capture of CO_2 and H_2S, and regeneration of high-purity CO_2 and H_2S at pressure. The downstream Claus system showed a conversion of more than 99.5% of H_2S gas to elemental sulfur, with H_2S in the product gas at ppm levels. SRI is pursuing ways to continue development of the AC-ABC solvent process.
Gas Separation Membranes	
MTR Proteus™ Hydrogen and Polaris™ CO_2 Membranes	Both CO_2 and hydrogen – selective membranes offer potentially high-efficiency gas separation in various gasification applications. A dual-membrane design concept developed by MTR, which combines high-temperature hydrogen – and – CO_2 – selective membranes, may offer significant cost and energy savings over conventional acid gas removal. Development of MTR's membranes led to several scale-ups of both membrane types and the demonstration of a pilot-scale membrane – assisted CO_2 liquefaction process. The membranes are being operated in a biowaste-to-ethanol process and in a gas-to-liquids process.
Media & Process Technology (MPT) Carbon Molecular Sieve (CMS) and Palladium – Based Hydrogen Membranes	MPT successfully scaled up its CMS from single tubes to a full-scale 86 tube bundle. MPT also incorporated WGS functionality into the CMS bundle to produce a catalytic membrane reactor, providing separation of hydrogen simultaneously with its formation. Test results validated the membrane's high stability in the presence of aggressive gas-phase contaminants such as sulfur-and – nitrogen – based species, allowing it to operate with untreated syngas. Treatment of the desulfurized permeate stream with MTR's palladium membrane achieved 99% hydrogen purity.

(*Continued*)

Table 27.19 Research/Investigations on Carbon Capture Storage at the National Carbon Capture Center. (*Continued*)

SRI Polybenzimidazole (PBI) Hydrogen Membranes	SRI conducted the first syngas testing of a hydrogen membrane fabricate with spun hollow fibers of the temperature-and – chemical-resistant polymer PBI. Testing confirmed that greater that 90% recovery of CO_2 is possible at temperatures above 375°F.
Worcester Polytechnic Institute Palladium-Based Hydrogen Membranes	WPI's membrane testing led to a seven-fold-scale-up, and with only 1% deviation in thickness among the tubes, the consistent fabrication of the membranes demonstrated the replicability of the technology. Syngas operation demonstrated material robustness and steady permeance* values, with hydrogen product purity as high as 99.9%.
Eltron Hydrogen Transport Membrane (HTM)	Eltron's HTM uses a multi-layer metal alloy tube for hydrogen separation at high pressures typical of gasification environments. Testing showed the membrane to be capable of producing 99.9% pure hydrogen, and the module was in good conditions following testing.
Carbon Capture Sorbents	
TDA Research Sorbents	TDA Research's testing of a solid CO_2 sorbent consistently demonstrated the capability to remove more than 90% CO_2. TDA also tested a combined WGS/CO_2 sorbent system with an innovative heat management component. When parameters were adjusted to achieve 90% CO conversion in the WGS stage, the overall carbon capture rate was greater than 95%. TDA scaled up testing from bench-to-small pilot-scale with a 0.1 MW CO_2 sorbent (without WGS) process, again demonstrating high CO_2 capture and stable operation. After tests at the center, TDA began testing the 0.1-MW system at China's Sinopec facility.

*Permeance is the degree to which a material admits a flow of matter or energy.

In materials science, "permeance" can refer to the ability of a material to allow a substance (such as a gas or liquid) to pass through it. For example, the permeance of a membrane or film might describe how easily oxygen can diffuse through it. Permeance in this sense is often measured in units of moles per second per square meter per pascal (mol/s/m²/Pa).

In electrical engineering, "permeance" (also called "magnetic permeance" or "magnetic reluctance") is a measure of the ease with which a magnetic circuit allows magnetic flux to flow through it. It is the reciprocal of magnetic reluctance, which is analogous to electrical resistance. Permeance is measured in units of henries per meter (H/m) or webers per ampere-turn (Wb/At).

For example, In electromagnetism, pemeance is the inverse of reluctance. Magnetic permeance P is defined as the reciprocal of magnetic reluctance, R.

i.e. $P = 1/R$

The solvent can be paired with the CDRMax® process, or used as a drop replacement for alternative solvents such as methyl diethanolamine (MDEA) for low pressure gas separations. CDRMax produces CO_2 with a purity of 95-99% and offers a $40/tonne cost of capture when used with CDRMax process. APBS-CDRMax® has high solvent stability, low corrosivity, low regeneration energy requirements, and holds up well in oxygenated environments. As a drop-in replacement, APBS-CDRMax® can reduce energy costs, and operational cost can be reduced 20-40% compared to conventional options. Furthermore, APBS-CDRMax® is less reactive to oxygen and twenty times less corrosion and ten times less degradation compared to conventional solvents, which helps lower waste solvent disposal costs to a negligible amount. With APBS-CDRMax®, it is possible to construct with less expensive materials such as carbon steel rather than a lower grade of stainless steel.

APBS-CDRMax® provides higher performance efficiency with less foaming leading to a 50% reduction in ongoing chemical requirement and waste disposal costs. This reduces amine carryover and the need for anti-foaming

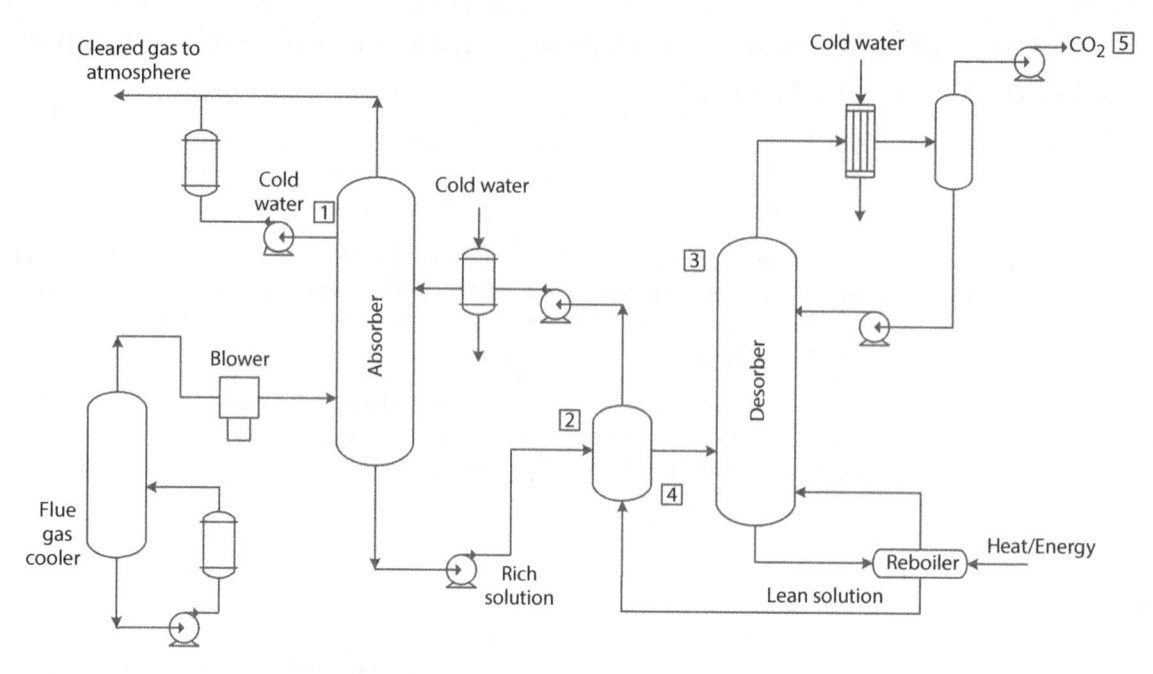

1. **Absorption.** In the absorber, the solvent extracts CO_2 from the feed gas.
2. **Heating:** CO_2-rich solvent is partially heated in the heat exchanger using hot lean solvent from the desorber.
3. **Separation:** CO_2-rich solvent is further heated within the desorber, where CO_2 molecules are released from the solvent.
4. **Regeneration:** CO_2-lean solvent passes back through the heat exchanger to the absorber for reuse.
5. **Storage:** Once isolated, CO_2 can be safely stored or converted into new products for resale in the circular carbon economy.

Figure 27.70 Flow diagram of the CDRMax™ process (Source: Carbon Clean, www.carbonclean.com).

additives. It reduces solvent emissions to parts per billion (ppb) levels, exceeding environmental regulatory requirements and making approval easier. It has five times longer solvent life and 86% less solvent makeup. It can be used as a drop-in replacement of conventional solvents as MEA, MDEA or DGA and reduce thermal energy usage by 20% or more. The unit operation process of absorption is provided in chapter 20 of Volume 3 of these series. Figure 27.70 shows the CDRMax® process flow diagram.

Advantages

- CDRMax™ process achieves 90%+ capture rates and delivers industrial quality CO_2 for re-use or sequestration.
- It offers the lowest corrosion rates, highest health safety executive (HSE) standards.
- It provides uptime rate of 98%+, and can operate for more than 200,000 hours on commercial plants.
- It offers cost-effective modular technology to achieve $30/tonne cost of CO_2 capture.
- It is possible to construct with less expensive materials such as carbon steel rather than a lower grade of stainless steel.

APBS-CARBex® is a proprietary solvent that reduces the cost of upgrading biogas compared to alternative technologies such as membranes or pressure swing adsorption (PSAs). APBS-CARBex® uses thermal energy to remove CO_2 concentrations of up to 50% volume from biogas and landfill gas streams at low pressure. Over 99% of the raw methane is recovered to produce pipeline-quality natural gas (RNG) or compressed natural gas (CNG). It offers improved operating performance compared to alternative solvents such as MEA, MDEA and DGA for CO_2 removal systems operating at low pressures.

Advantages

- APBS-CARBex® requires less compression, resulting in savings of energy usage and cost.
- It requires around 15% less thermal energy compared to methods such as membrane, water scrubbing, and PSA.
- It reduces electricity usage by 70 – 80%.
- It results in a higher methane quality of greater than 99% volume with less than 1% of raw methane lost.
- It provides higher performance efficiency with less foaming leading to 50% reduction in on going chemical requirement and waste disposal costs.
- It resists thermal and chemical degradation, lasting five times longer than conventional solvents.
- It can be used as a drop-in replacement to upgrade performance without any equipment changes or capital costs.

27.23.1 Cyclone Carbon Clean Technology

A new technology referred to as Cyclone carbon clean (CycloneCC) has been developed by Carbon Clean Solutions Ltd. (CSSL), in partnerships with Newcastle and Sheffield Universities, the industry and the U.K. government, and in the US, by a Department of Energy sponsored project known as ROTA-CAP, led by Gas Technology Institute (GTI) in Chicago, tested carbon clean's next-generation solvents and the rotating packed bed (RPB) process on a mobile development skid. It will be the smallest industrial capture technology by overcoming a key barrier to widespread carbon capture utilization and storage and industrial decarbonization. The prefabricated, modular solution is expected to make carbon capture simple, affordable and scalable.

GTI's objective of the project is to develop and validate a transformational carbon capture technology—ROTA-CAP. This will be achieved by the design, construction, testing, and simulation modelling of novel rotating packed bed (RPB) absorbers and regenerators in an integrated, process-intensified carbon capture system using advanced solvents at bench-scale. The performance of the integrated hardware and solvent will be assessed under a range of operating conditions with simulated flue gases and GTI's natural gas burner flue gas to optimize the process, ahead of long-term testing with coal-fired flue gas at the National Carbon Capture Center (NCCC).

The rotating packed bed (RPB) is a novel and compact process intensification (PI) operating unit that utilizes centrifugal acceleration to intensify mass transfer. Its low investment cost leads to growing adoption in gas purification, distillation and nano-material preparation applications. The energy of high – gravity fields generated by centrifugal operation considerably increases the molecular diffusion between gas-liquid counter current flows and consequently improves the heat and mass transport. RPB rotates at hundreds of rpm as compared to centrifuges that rotate in thousands. If one thinks of RPB as being like pumps, which are ideally are about the reliability of pump seals. The ones in the chemical industries have 5 plus seal life and the food industry 15 plus years seal life indicating that RPB is a reliable technology. Further research involves reducing the size of the absorber in the RPB and on the regenerator column where there is opportunity to integrate the heat regenerator with the heat transfer system of the carbon capture plant thereby saving energy.

In the RPB, as the packing is rotated, the liquid and the gas move past each other much more rapidly and thus provide a good mass transfer between the two-phases. Reactions in the liquid phase require more material from the gas phase, and if there is a high rate of mass transfer, the CO_2 can move more easily from the gas phase to the liquid phase, which is ideally suited to carbon capture, which can handle high gas-liquid volume. Furthermore, better packing materials in the RPB result in low pressure drop (Δp), higher surface area and therefore higher rate of mass transfer.

CCSL completed a research and development program with Newcastle University and the University of Hull in the U.K., where a bench-scale prototype RPB absorber (RPBA) was evaluated in CO_2 capture applications using CCSL's solvent (APBS 2) and the industry standard solvent, monoethanolamine (MEA) at various concentrations. The results show close to 50% smaller height of transfer unit (HTU) for APBS 2 solvent compared to 30% MEA solvent for the same absorption rate.

Figure 27.71 shows the schematic of a classical packed column and a single – stage RPB for counter-current gas liquid contacting, and Figure 27.72 shows photograph of rotating packed bed machine capable of handling 3000 m³/h of gas during an absorption process involving a tail gas cleaning of sulfuric acid plant respectively. Figures 27.73a-b show a typical (RPB) zigzag bed HiGee and its packing; Figures 27.74a-d show RPB used in GTI's bench scale and pilot-plant scale studies of carbon capture. Figure 27.75 shows the height of a transfer unit (HTU) for MEA and APBS solvents tested in a prototype RPB absorber, and Table 27.20 shows the project risks and mitigation strategies in RPB.

GTI and Carbon Clean Solutions Ltd. (CCSL) are to develop a compact carbon capture system that uses a RPB contactor absorber and rotating bed contactor regenerator. The two sections of the system need to be connected together and operated continuously to validate and optimize the equipment as a cost and energy effective carbon capture system. Ultimately, the system is indifferent to the solvent used. Combining ROTA-CAP with CCSL's proprietary solvent formulation solvent will further improve capture efficiency. Together with CCSL, GTI will develop both the

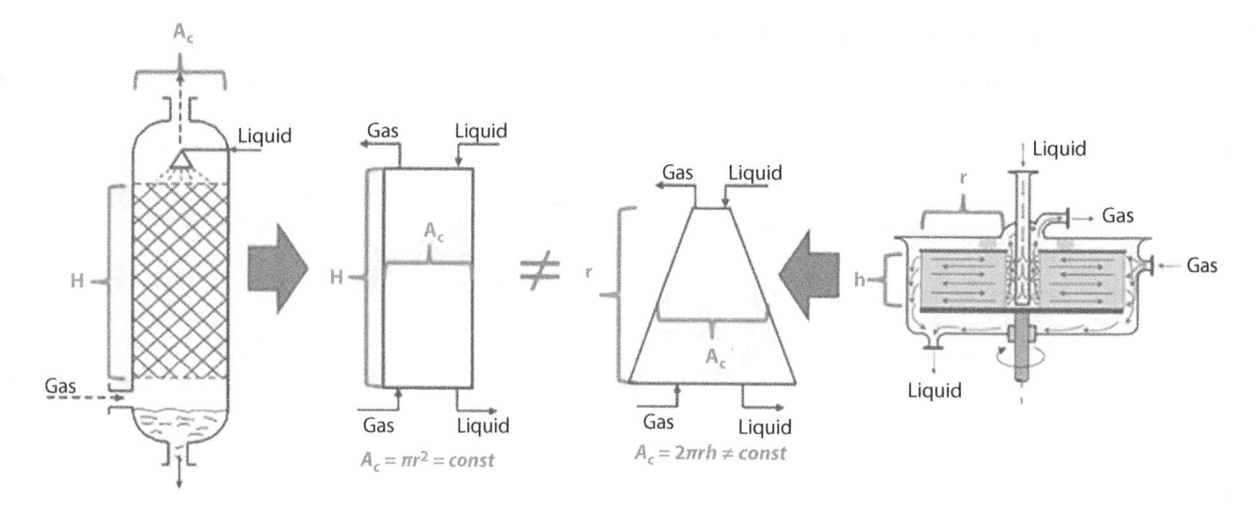

Figure 27.71 Schematic of a classical packed column and a single – stage RPB for counter-current gas liquid contacting (Source: Kolja Neumann *et al.*, A guide on the industrial application of rotating packed beds, Chemical Engineering Research and Design, https://doi.org/10.1016/j.cherd.2018.04.024).

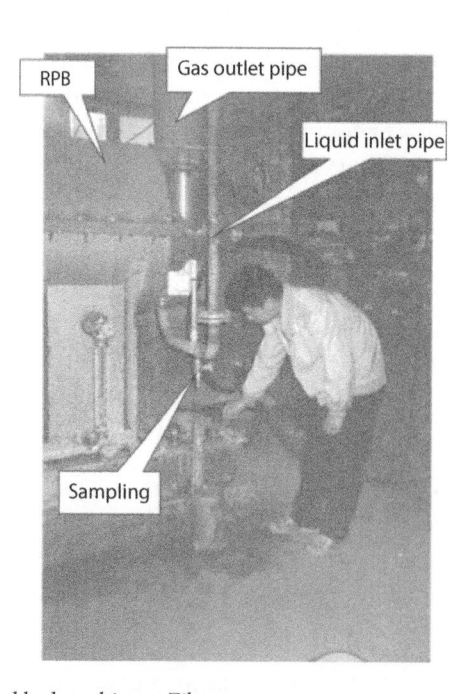

Figure 27.72 A photograph of rotating packed bed machine at Zibo.

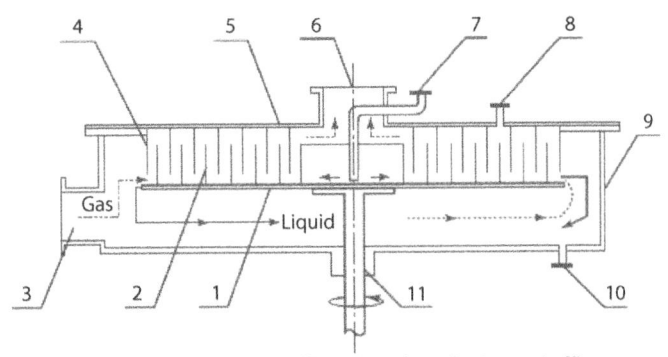

1. Rotational disc 2. Rotational baffle 3. Gas inlet 4. Stationary baffle
5. Stationary disc 6. Gas outlet 7. Liquid inlet 8. Intermediate feed
9. Rotor casing 10. Liquid outlet 11. Rotating shaft

Figure 27.73a The rotating zigzag bed HiGee variant of Wang *et al.* [87].

Figure 27.73b The packing typically used in the HiGee machines.

absorber and regenerator through bench scale with testing at GTI and perform reliability and long-term testing at NCCC.

Many industrial facilities have limited spaces and the biggest barriers to widespread CCUS have been the size and cost of existing technology. Cyclone carbon clean technology is to break down these barriers as the world's smallest industrial carbon capture solution. Cyclone carbon clean is expected to make carbon capture simple, affordable and scalable, thus bringing it within reach of a huge number of industrial emitters such as those with small to mid-size emission point sources.

Cyclone carbon clean (CycloneCC) has a footprint that will be ten times smaller than conventional carbon capture and it could easily be deployable in a short time (< 8 weeks). The solution is expected to reduce CAPEX and OPEX by up to 50%, eventually driving the cost of carbon capture to $30/tonne average, a cost that is well below the current EU carbon price and thus makes the economic case for carbon capture, utilization and storage (CCUS) undeniable.

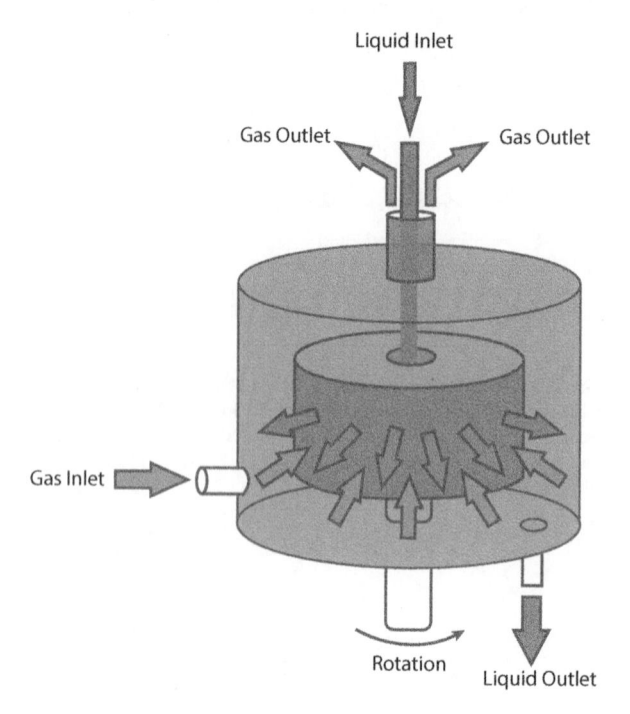

Figure 27.74a Rotating packed bed reactor (Source: GTI-ROTA – CAP: An Intensified Carbon Capture System Using Rotating Packed Beds).

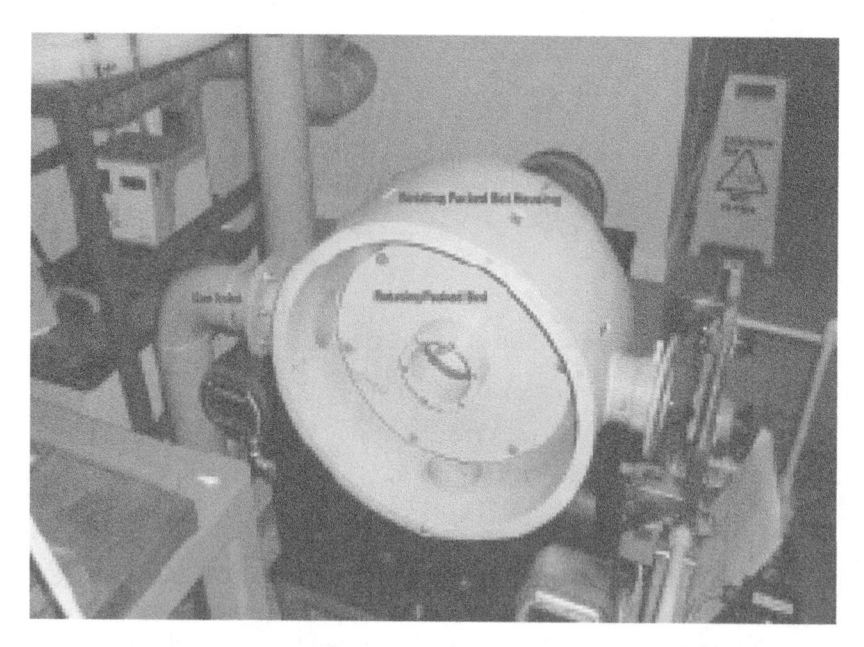

Figure 27.74b Laboratory scale rotating packed bed absorber (Source: GTI-ROTA – CAP: An Intensified Carbon Capture System Using Rotating Packed Beds).

The International Energy Agency (IEA) (CCUS in Clean Energy Transitions) stated that CCUS capacity deployment must be 50% higher than currently predicted trajectories if the world is to reach net zero by 2050. CycloneCC is expected to make this achievable and play a significant role in helping industrial companies realize their net zero ambitions.

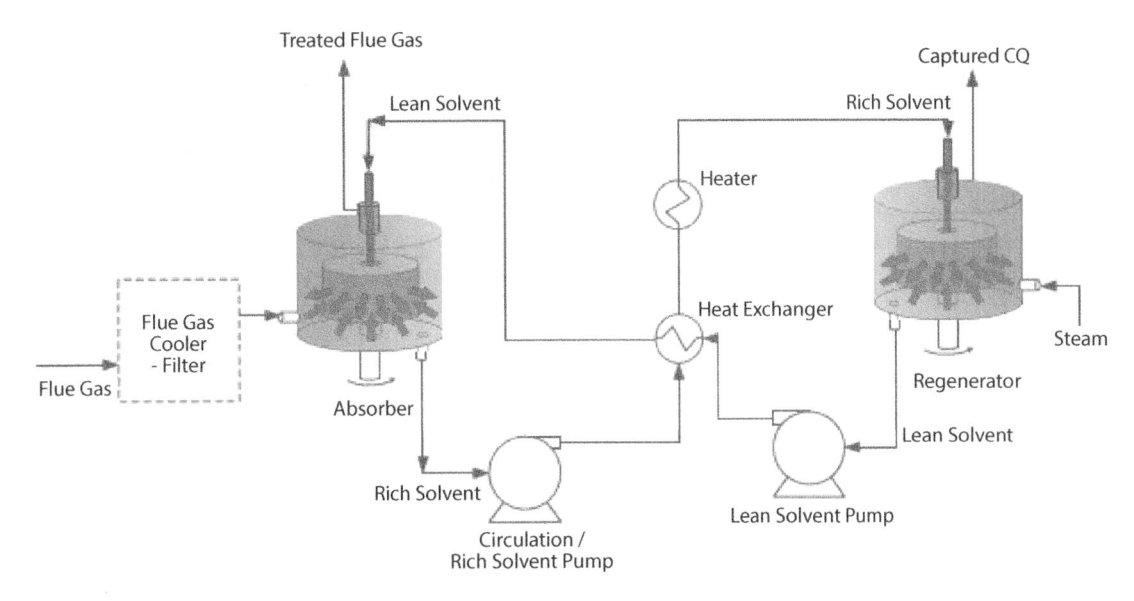

Figure 27.74c ROTA-CAP system process flow diagram (Source: GTI-ROTA – CAP: An Intensified Carbon Capture System Using Rotating Packed Beds).

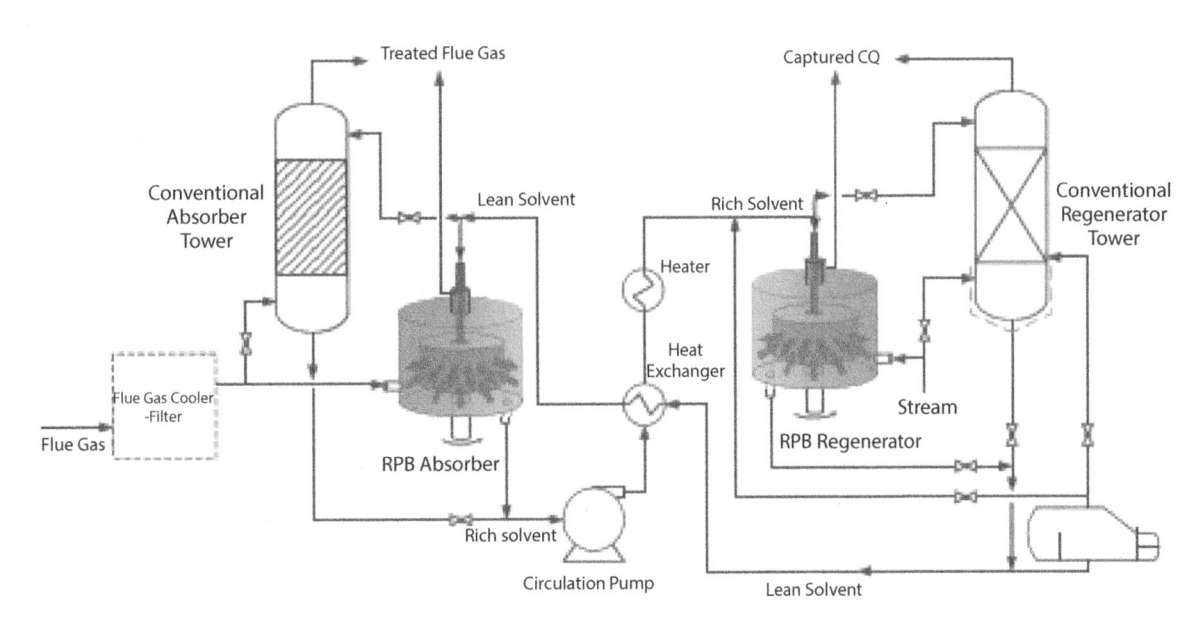

Figure 27.74d Integrated Bench-Scale ROTA-CAP Test skid with conventional tower sections (Source: GTI-ROTA – CAP: An Intensified Carbon Capture System Using Rotating Packed Beds).

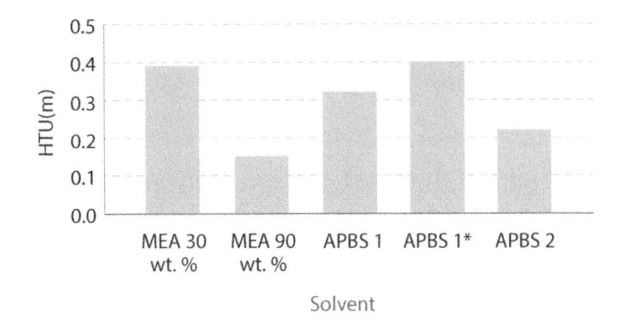

Figure 27.75 Height of a Transfer Unit (HTU) for MEA and APBS Solvents Tested in a prototype RPB absorber (Source: GTI-ROTA – CAP: An Intensified Carbon Capture System Using Rotating Packed Beds).

Table 27.20 Project Risks and Mitigation Strategies (Source: ROTA-CAP: An Intensified Carbon Capture System Using Rotating Packed Beds).

Description of risk	Probability	Impact	Risk management mitigation and response strategies
Technical Risks:			
Scale up of rotating packed bed reactor is too problematic	Low	Moderate	• Previous experience in design of lab scale equipment and commercial equipment manufacturers available for consultation. • GTI's experience on evaluation of high-efficiency gas-liquid contactors for natural gas processing including RPB reactors.
Energy use of RPB reactors is too high	Low	Moderate	• Reactor design will balance the size of reactor and energy use to achieve economic scale-up.
Flue gas contaminants degrade solvent or solvent aerosols form on RPB reactor exit	Moderate	Low	• Solvent analysis to monitor degradation. • Liquid carryover measurement will be done at the exit of the RPB reactor.
Not high enough capture efficiency	Low	Moderate	• Previous work with CCSL solvent APBS matched MEA performance. • Solvent concentration can be adjusted to achieve the desired efficiency.
Safety Risks:			
Rotating Equipment	Low	High	• CCSL has previous experience with designing rotating packed bed equipment. All necessary mechanical engineering calculations will be verified by a commercial equipment manufacturer.
Chemical	Low	High	• HAZOP reviews will be conducted prior to skid fabrication and again prior to initiation of the testing to identify and mitigate and safety risks associated with handling methane and potential reaction products.

27.23.2 CycloneCC Technology

CycloneCC with rotating packed bed (RPB) is a process intensification technology that improves the absorption of CO_2 into the solvent. RPB contains disk packing material that rotates about its axis, thus generates a centrifugal force within the packing and invariably enhances the CO_2 absorption process. The solvent is introduced at the center into the RPB at its center (Figure 27.76), and sprayed onto the packing by a liquid distributor. When the solvent contacts the packing, the centrifugal force appointed by the solvent from the rotational motion forces the solvent to travel radially toward the outer end of the packing where it is drained down to the sump before being pumped to the next stage of the process.

The flue gases which are introduced to the RPB at the outer end of the packing and exit at the inner edge where the solvent enters therefore creates a gas and liquid contact in a counter current mode (Figure 27.77). The flue gases are absorbed by the solvent and the CO_2 present selectively reacts with the components of the solvent thereby temporarily locking the CO_2 within the solvent.

CycloneCC uses carbon clean's advance proprietary amine-promoted buffer salt solvent (APBS-CDRMAx® and rotating packed beds (RPBs). When used together, they ensure CycloneCC is far more efficient than conventional carbon capture methods, reducing costs while matching the performance. Using RPBs instead of a conventional large chimney stack, improves the absorption of CO_2 into the solvent, and has a ten times smaller footprint.

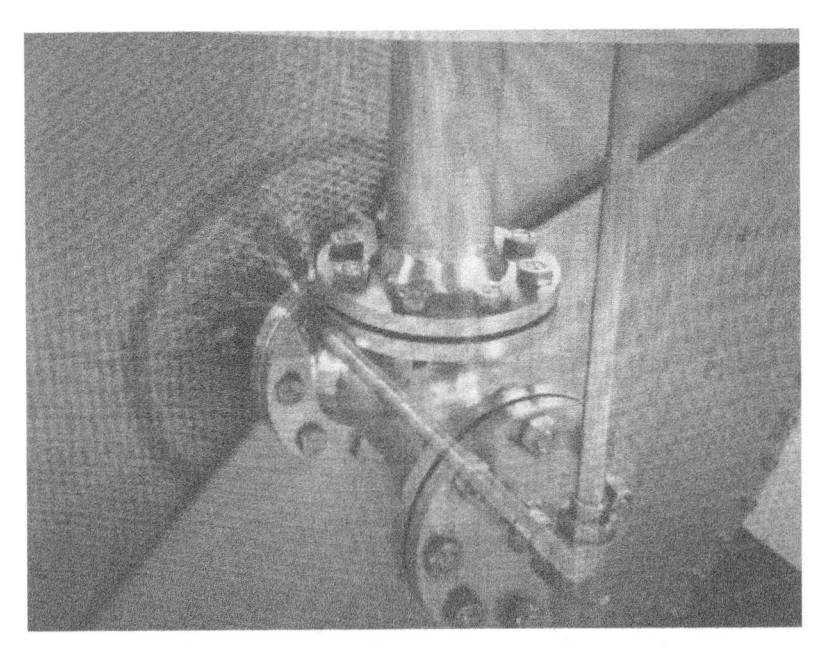

Figure 27.76 CycloneCC showing the solvent introduced at the center through a liquid distributor onto the packed bed (Source: Carbon Clean Solutions Ltd.)

Figure 27.77 CycloneCC showing its operation with flue gas introduced on the left side and solvent at its center and the exit absorbed CO_2 gas (colored red) (Source: Carbon Clean Solutions Ltd.).

Figure 27.78 CycloneCC (Source: Carbon Clean Solution Ltd.).

Figure 27.79 Large scale of CycloneCC (Source: Carbon Clean Solution Ltd.).

CycloneCC is currently being commercialized at 10tpd and 100 tpd with selected partners and Carbon Clean company projects a final product roll out by 2nd quarter of 2022 and market roll out in 2023. Figures 27.78 and 27.79 show CycloneCC in small-scale and large-scale sizes respectively.

27.24 Electrochemically Mediated Amine Regeneration (EMAR)

Another current research investigation in carbon capture is the development of Electrochemically Mediated Amine Regeneration (EMAR) for the capture of CO_2 from flue gases. The absorption step in the EMAR process is the same as in the widely used amine process as discussed earlier, but the desorption step is accomplished electrochemically as opposed to a thermal swing used in the amine process.

EMAR was recently developed to avoid the use of thermal means to release CO_2 captured from post combustion flue gas in the benchmark amine process. A mixture of ethylenediamine (EDA) and aminoethylethanolamine (AEEA) was investigated in order to address concerns related to the high vapor pressure of the former as the primary amine used in EMAR. Furthermore, the properties of the mixed amine systems, including the absorption rates, electrolyte pH and conductivity, and CO_2 capacity, were evaluated in comparison with those of solely EDA. The mixed amine system had similar properties to that of EDA, indicating no significant changes would be necessary for the future implementation of the EMAR process with mixed amines as opposed to that with just EDA. The electrochemical performance of the mixed amines in terms of the cell voltage, gas desorption rate, electron utilization, and energetics was also investigated.

Electrochemically mediated amine regeneration (EMAR) was recently developed to avoid the use of thermal means to release CO_2 captured from post combustion flue gas in the benchmark amine process. To address concerns related to the high vapor pressure of ethylenediamine (EDA) as the primary amine used in EMAR, a mixture of EDA and aminoethylethanolamine (AEEA) was investigated. The properties of the mixed amine systems, including the absorption rates, electrolyte pH and conductivity, and CO_2 capacity, were evaluated in comparison with those of solely EDA. The mixed amine system had similar properties to that of EDA, indicating no significant changes would be necessary for the future implementation of the EMAR process with mixed amines as opposed to that with just EDA. The electrochemical performance of the mixed amines in terms of the cell voltage, gas desorption rate, electron utilization, and energetics was also investigated.

A 50/50 mixture of EDA and AEEA displayed the lowest energetics: ~10% lower than that of 100% EDA. With this mixture, a continuous EMAR process, in which the absorption column was connected to the electrochemical cell as the desorption stage, was tested over 100 h. The cell voltage was very stable and there was a steady gas output close to theoretical values. The desorbed gas was further analyzed and found to be 100% CO_2, confirming no evaporation of the amine. The mixed absorbent composition was also characterized using titration and nuclear magnetic resonance (NMR) spectroscopy, and the results showed no amine degradation. These findings that demonstrate a stable, low vapor pressure absorbent with improved energetics are promising and could be a guideline for the future development of EMAR for CO_2 capture from flue gas and other sources.

Studies have been carried in laboratories where operations have run for over 200 h, spanning 130 absorption/desorption cycles. In this process, an understanding of the thermodynamics of the EMAR process has realized to provide engineering estimates of key features (energetics and sizing) of the technology. The EMAR system does not rely on steam integration, thus showing a plug – and play unit that can be readily deployed for a range of applications. The EMAR system can be scaled-down for distributed, modular and small-scale operations (e.g., stainless steel, mini-mills, etc.). Results from the technology indicate that the EMAR has the potential to be another viable option for post-combustion CO_2 capture [89].

Mechanism

The scheme consists of three chemical or electrochemical transitions, which take place in an absorber, an anode and cathode chambers. The CO_2 is absorbed by MEA lean solvent from the flue gas in an absorber similar to the traditional thermal scrubbing process. After the absorption process, the produced CO_2 rich solvent is then sent to the anode chamber of the EMAR system where a copper electrode is oxidized under electrical polarization to release cupric ions into the solution. These released cupric ions bind to MEA molecules and displace the CO_2 leading to the CO_2 desorption from the solution. The gas/liquid mixture is then separated into a pure CO_2 stream and the copper saturated solution is sent to the cathode where the cupric ions are plated out onto another copper electrode. Then the regenerated MEA solution can be used for the next absorption cycle. The anode and cathode can be easily changed by adjusting the direction of applied voltage, and then the copper recycling can be easily achieved.

Equation 27.20 shows the forward and reverse reaction of the MEA scrubbing process. The process works as a temperature swing cycle where the absorption of CO_2 takes place at a low temperature (~ 40°C) and desorption at a higher temperature (~ 120°C). The amount of thermal energy required for CO_2 capture is about 3.5 – 4.0 GJ/t CO_2, which will consume 30% of the output of a traditional power plant, greatly affecting the economics of the capture process. This process has been studied at a pilot scale, but it cannot be implemented for low concentration (< 15vol %) [90].

$$2HO(CH_2)_2 NH_2 + CO_2 \rightleftharpoons HO(CH_2)_2 NH_3^+ + HO(CH_2)_2 NHCO_2^- + \text{Heat} \qquad (27.20)$$

The EMAR is introduced as an emerging technology to reduce the energy consumption of the amine regeneration process. The process combines the high removal efficiency of amine scrubbing with the high efficiency of an electrochemical separation system, and it does not require the steam source. The electrically driven separation process shows the potential to reduce the difficulty of retrofitting CO_2 capture units to existing fossil fuel fire power plants. The design of such a system requires careful consideration as it involves both heterogeneous electrochemical activation/deactivation of sorbents and homogeneous complex of the activated absorbents with CO_2 molecules. Figure 27.80 shows the cyclic reaction scheme of an EMAR process using MEA as an absorbent and copper as ligand.

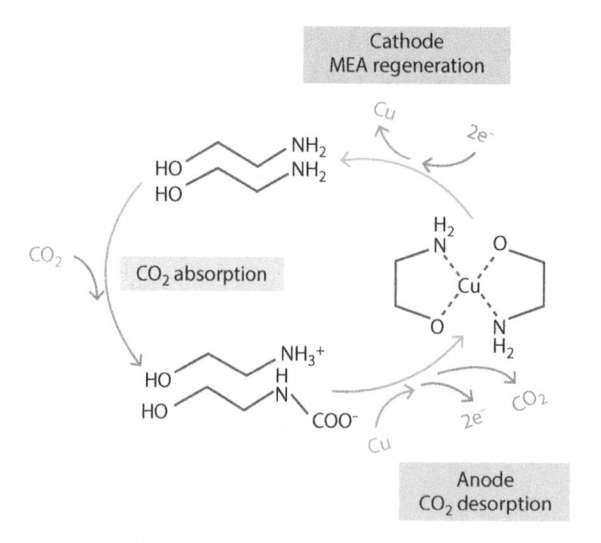

Figure 27.80 The mechanism of chemical reaction in the EMAR cycle based on MEA [90].

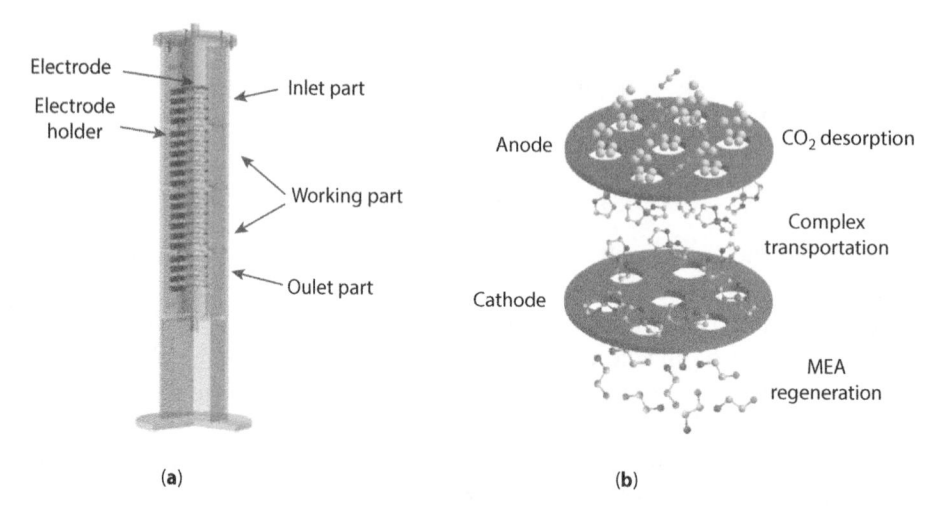

Figure 27.81 (a) (MTFECs) and (b) Structure of a pair of electrodes [90].

Table 27.21 Complexes of Cu(II)-MEA and their Stability Constants [90].

Formula	Structural formula	Stability constant (ionic strength)
MEAH$^+$	HO-CH$_2$CH$_2$-NH$_3^+$	9.51 (0 M), 9.52 (0.1 M), 9.62 (0.5 M), 9.79 (1 M)
[Cu(MEA)]$^{2+}$	[HO-CH$_2$CH$_2$-NH$_2$···Cu]$^{2+}$	4.47 (0 M), 4.5 (0.1 M), 4.6 (0.5 M), 4.63 (1 M)
[Cu(MEA)$_2$]$^{2+}$	[HO-CH$_2$CH$_2$-NH$_2$···Cu···NH$_2$-CH$_2$CH$_2$-OH]$^{2+}$	8.32 (0 M), 8.55 (0.1), 8.25 (0.5 M), 8.4 (1 M)
[Cu(MEA)$_3$]$^{2+}$	tris(ethanolamine)Cu complex	10.5 (0 M), 10.8 (0.5 M)
[Cu(MEA)$_4$]$^{2+}$	tetrakis(ethanolamine)Cu complex	11.1 (0 M), 11.5 (0.5 M)
[Cu(MEA)$_{2\text{-H}}$]$^{+}$	chelate structure	1.43 (0 M), 1.5 (1 M)
[Cu(MEA)$_{2\text{-2H}}$(OH)]^0	bis-chelate hydroxo structure	15.42 (0 M)
[Cu(MEA)$_{2\text{-2H}}$(OH)$_2$]^0	bis-chelate dihydroxo structure	19.58 (0 M), 19.9 (0.5)

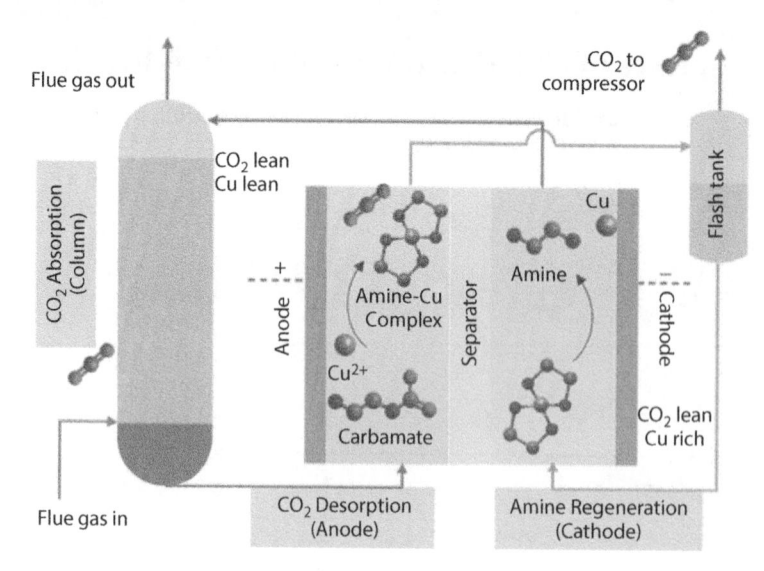

Figure 27.82 The schematic of Electrochemically Mediated Amine Regeneration (EMAR) [89]. (Source: Mio Wang, Howard J. Herzog, and Alan Hatton, Ind. Eng. Chem. Res. 59, 15, 7087-7096, 2020).

The EMAR cycle for the MEA scrubbing process is based on three reactions that occur in absorber, anode chamber and the cathode chamber respectively.

$$\text{Absorption: } 4\text{MEA (aq)} + 2\text{CO}_2 \text{ (g)} \rightleftharpoons 2[\text{MEA}_2\text{CO}_2](\text{aq}) \tag{27.21}$$

$$\text{Anode: } 2[\text{MEA}_2\text{CO}_2](\text{aq}) + \text{Cu(s)} \rightleftharpoons [\text{CuMEA}_4]^{2+} \text{(aq)} + 2\text{CO}_2 \text{ (g)} + 2e^- \tag{27.22}$$

$$\text{Cathode: } [\text{CuMEA}_4]^{2+} \text{(aq)} + 2e^- \rightleftharpoons \text{Cu (s)} + 4\text{MEA} \tag{27.23}$$

Figure 27.81a shows a home-designed bench-scale modular flowing through electrolysis cells (MFTECs) and Figure 27.81b shows the structure of a pair of electrodes respectively. There is one -NH$_2$ and one -OH group in an MEA molecule, and both N and O atoms can be donors to coordinate with Cu(II). However, the O atom is the weaker electron donor, and -OH dissociates only at high pH, therefore the N atom is the main donor to coordinate with Cu(II). At different pH and MEA concentrations, the central metal cation Cu(II) coordinates with different amounts of MEA ligands (including MEACOO$^-$ and MEAH$^+$), forming monovalent cationic complexes, divalent cationic complexes and neutral complexes as shown in Table 27.21. Figure 27.82 shows the schematics of the EMAR process.

27.25 Refinery of the Future

The global pandemic that started in November 2019 has impacted the world's refineries with a reduction in utilization while OPEC + supply restraint narrowed crude price differentials. This dual crisis highlighted some facts that now signal the direction for downstream assets as the energy transition progresses. The oil market is expected to be back at pre-crisis highs by the 4th quarter of 2022, while a resurgent Brent price and record upstream profitability are good indicators of the success OPEC + has had with its strategy to rebalance the market. The recent International Climate Change Conference COP26 in Glasgow, UK, November 2021, reinforced the commitment by countries that signed to 2015 Paris Climate Change agreement, and G7 countries achieving net zero by 2050 at the latest, and providing a vital step forward towards reducing global emissions. This chapter has provided detailed description of novel technologies from research institutes, universities, and companies in carbon capture and storage with the means of achieving net zero by 2050 in the refining sector and other chemical process industries.

The refining margins are above the depths of 2021 and still very low. The recovery has been uneven for refining; gasoline demand has recovered with China and the US leading the way; whereas jet fuel demand globally is still well below pre crisis level. Refiners have been forced to blend jet fuel with diesel, increasing its supply, which has pushed the margin down. Structural overcapacity is a factor that reduces global refining margins (see Figure 27.83) [79].

Integration between crude oil refining processes and petrochemicals is currently being implemented in various companies worldwide. The main focus is to promote and seize the opportunities that exist between both downstream sectors to generate value to the whole crude oil production chain. Petrochemical demand growth is more robust than transport fuels; the commercial viability of many downstream assets is demonstrably threatened in the market downturn, and integrated refinery-petrochemical sites can outperform their fuels. Table 27.22 shows the main characteristics of the refining and petrochemical industry and the synergies between them.

The energy transition and electrification of transport such as electric vehicles (EV) are important, as the society now recognizes that climate change is a major problem. This resulted in new policies (i.e., the Paris Agreement and COP26 in Glasgow, as well as countries, on the national, states and city levels); investment in technology and innovation (such as EV and green hydrogen); and energy companies are shifting strategic direction to embrace decarbonization.

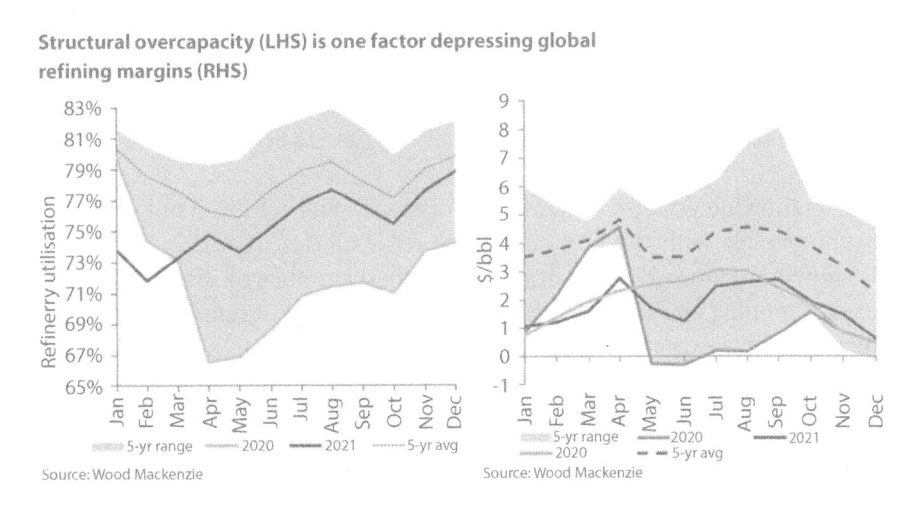

Figure 27.83 Structural overcapacity (LHS) as a factor depressing refining margins (RHS). (Source: Wood Mackenzie REM-Chemicals) [79].

Table 27.22 Refining and Petrochemical Industry Characteristics [79].

Refining industry	Petrochemical industry
Large Feedstock Flexibility	Raw Material from Naphtha/NGL
High Capacities	High Operations Margins
Self Sufficient in Power/Steam	High Electricity Consumption
High Hydrogen Consumption	High Availability of Hydrogen
Streams with low added Value (Unsaturated Gases & C_2)	Streams with Low Added Value (Heavy Aromatics, Pyrolysis Gasoline, $C_{4's}$)
Strict Regulations (Benzene in Gasoline, etc)	Strict Specifications (Hard Separation Processes)
Transportation Fuels Demand in Declining at Global level	High Demand Products

The energy transition and electrification of transport will further slow the pace of global gasoline demand growth whereas the versatility and durability of petrochemicals ensure sustained demand growth, particularly in developing countries of Asia. This will be affected by an increase in recycling being driven by the global war on plastic waste.

The petrochemicals industry already accounts for 13 million b/d of global oil demand and is expected to be the main driver of demand growth for the next two decades. The 83 % of plastic production that is not from recycled feedstock is made from naphtha or natural-gas liquids. If plastic recycling reaches 67% under a sustainable future scenario, it would take 1.5 million b/d a year from global oil demand by 2040 (Figure 27.84) [80].

The shift away from fuels to petrochemical integration poses the following key questions [81]:

- Where are these sites and how different are they to fuels refineries?
- How much does petrochemicals contribute to overall site value?
- What is their competitive position as standalone refiner and as an integrated site?
- What are the key regional trends?
- What is the emerging trend for new integrated sites?

As input to developing sustainable strategies, refiners would need to find solutions to these questions. The refining industry became established in North America and has some of the world's most attractive refineries when assessed on a Net Cash Margin (NCM) basis (Figure 27.85). The emergence of China during this century as a global economic superpower and global refining center has shifted the dynamic from refining crude oil for fuels to integrated petrochemical refineries where petrochemicals contribute more value than fuels. Unlike North America, in China, the integrated petrochemical refinery sites are now dominant (Figure 27.86).

During 2019, second generation integrated sites were commissioned and achieved full operations in 2020, and China is achieving earnings comparable with the best sites in North America. Asia's significance in integrated refiner-petrochemical sites is shown in Figure 27.87, as it has the highest number of integrated sites globally. The Middle East has invested in export refining and recently become more interested in refinery-petrochemical integration as a means of adding value to its crude exports as the supply of gas-based feedstocks is diminishing.

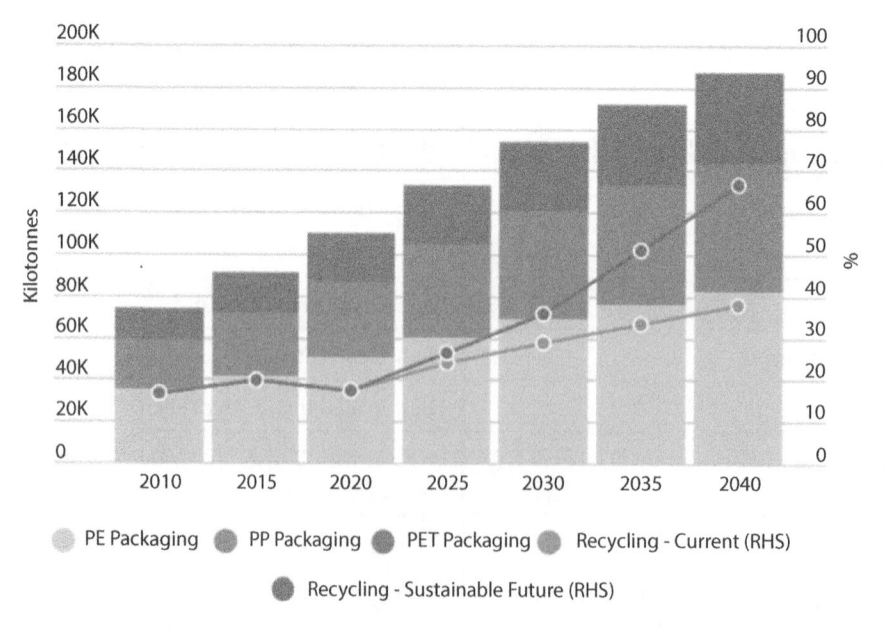

Figure 27.84 Doubling the rate of plastic recycling would by 2040 reduce oil demand by 1.5 million/day [80].

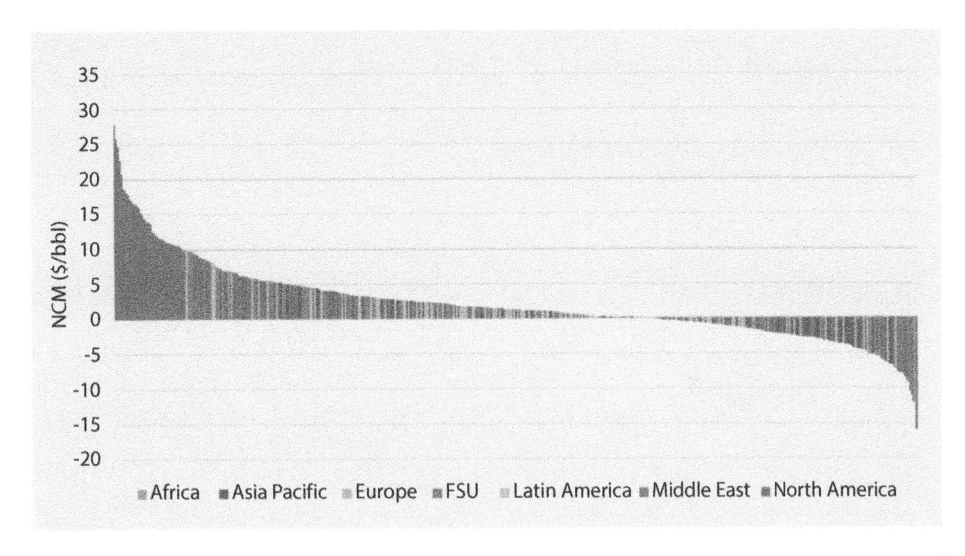

Figure 27.85 Global refinery only Net Cash Margin, 2019, US $/bbl. (Source: Wood Mackenzie REM – Chemicals) [81].

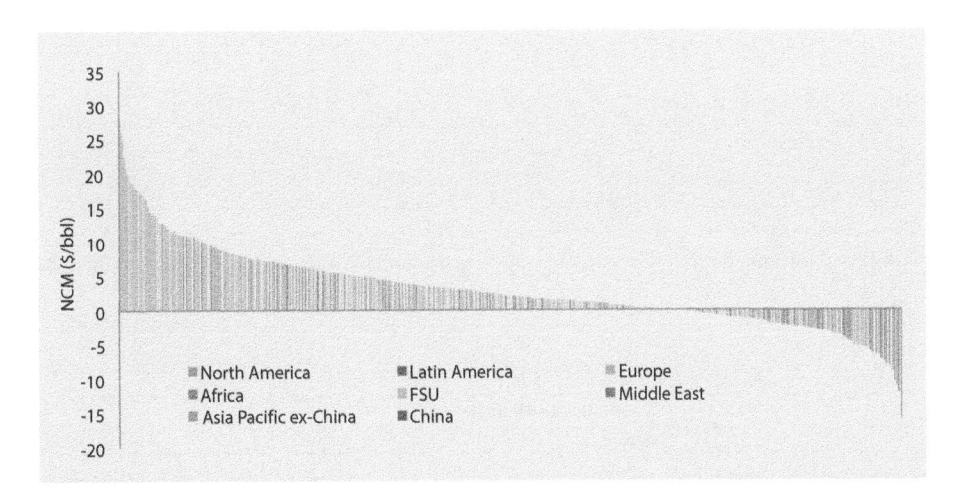

Figure 27.86 Global integrated refinery-petrochemical site Net Cash Margin profile, 2019, US$/bbl. (Source: Wood Mackenzie REM – Chemicals) [81].

Apart from being typically larger, integrated sites differ from fuels refineries through the co-location of petrochemical production facilities, so directly converting the feedstocks into end products of polymers and aromatics. This results in these integrated sites having high severity reformers for aromatics production along with the blending of returns streams from petrochemical production. Many sites also include production of paraxylene derivates, however the bulk of the value contribution is derived from the supply of the commodity petrochemical. Figure 27.88 shows the flow diagram of an integrated site. Each integrated site is unique, as petrochemicals add value to fuels refining and the trend from the European industry is shown in Figure 27.89. It is important to understand the role of the different petrochemicals within the overall site economics, as the contributions will vary over time due to the respective petrochemical commodity cycles.

The petrochemical industry has been growing at a high rate with production of polymers, plastics, glycols, methanol, fertilizers, etc. as compared with the transportation fuels, coupled with the environmental concerns of the GHG emissions from crude oil production. Since the technological bases of the refining and petrochemical industries are similar, leading to possibilities of synergies capable of reducing operational costs and add value to derivatives produced in the refineries. Figure 27.90 presents a block diagram that shows some integration possibilities between refining and petrochemical processes [81].

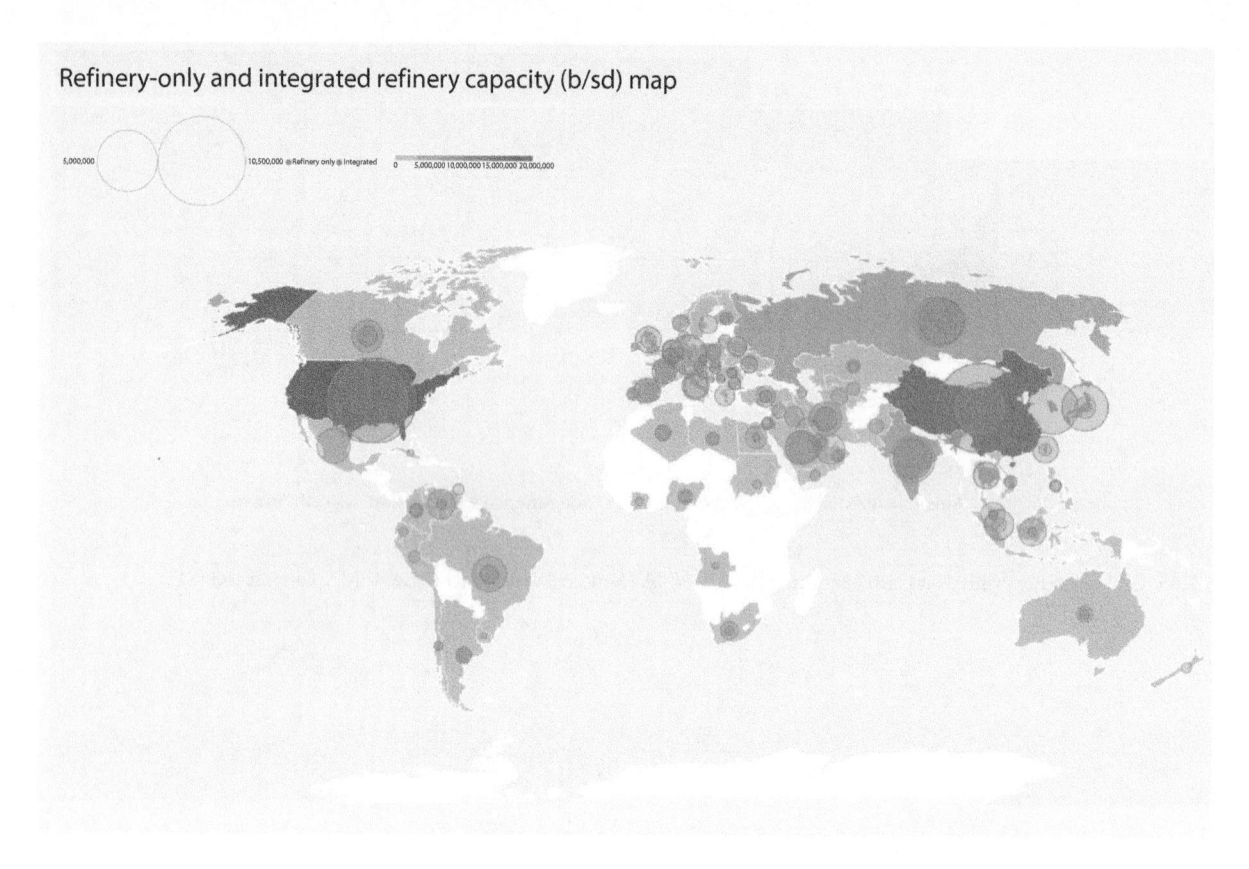

Figure 27.87 Refinery – only and integrated refinery capacity (b/sd) map. (Source: Wood Mackenzie REM – Chemicals) [81].

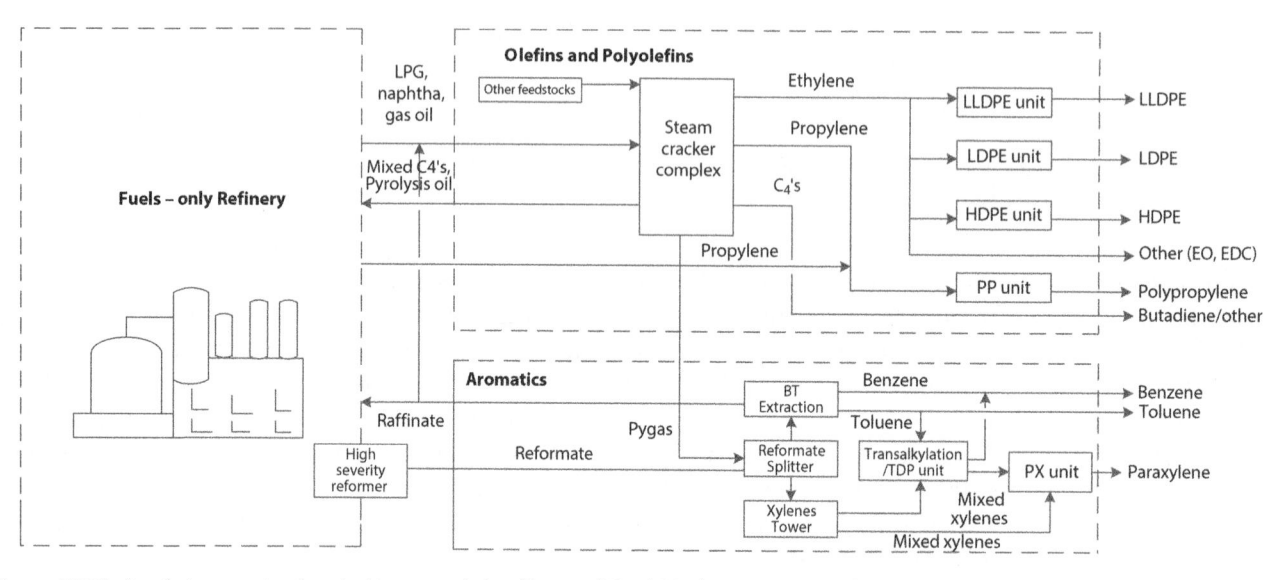

Figure 27.88 A refining – petrochemical integrated site. (Source: Wood Mackenzie REM – Chemicals) [81].

Process streams considered with low added value to refiners like fuel gas (C_2H_6) are attractive raw materials to the petrochemical industry, as well as streams considered residual to petrochemical industries (C_4H_{10}, pyrolysis, gasoline, and heavy aromatics). These can be applied to refiners to produce high-quality transportation fuels, which can help the refining industry attain environmental and quality regulations to derivatives.

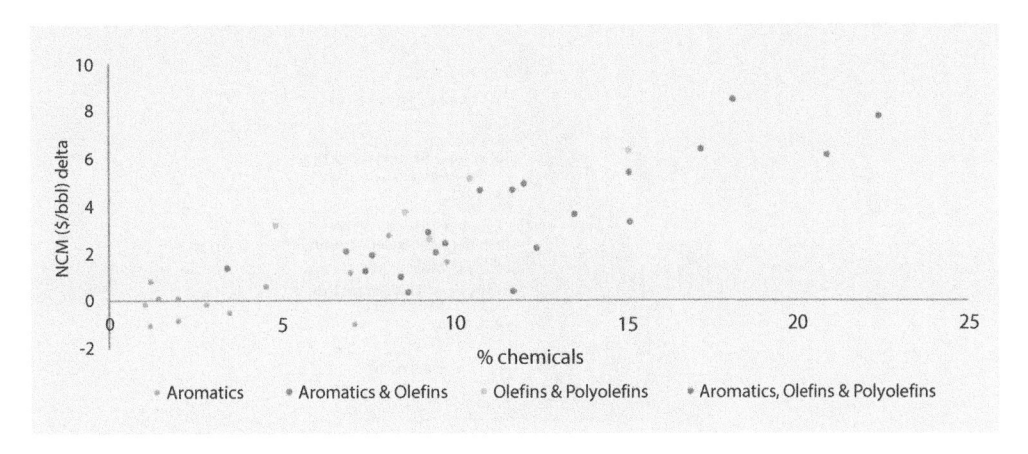

Figure 27.89 European NCM uplift from petrochemicals against petrochemical yield, 2019. (Source: Wood Mackenzie REM – Chemicals) [81].

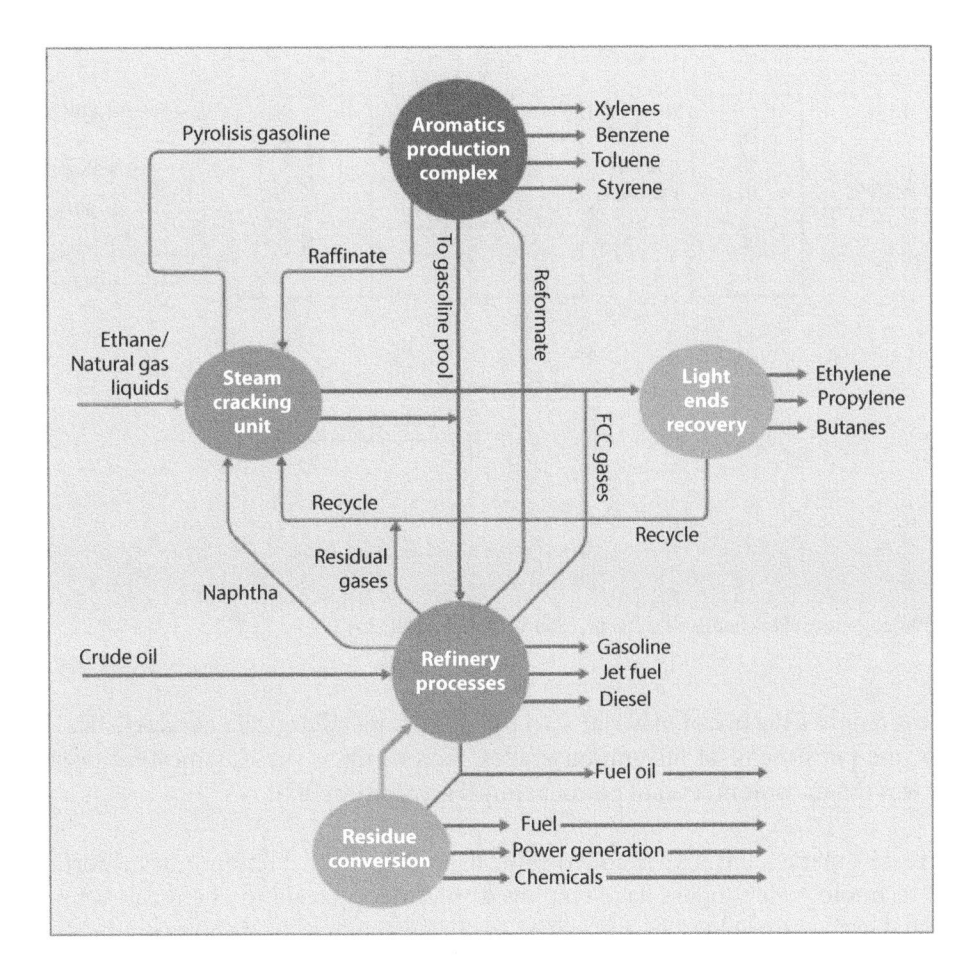

Figure 27.90 Synergies between Refining and Petrochemical Processes [81].

The integration potential and the synergy between these processes rely on the refining scheme adopted by the refinery and the consumer market, process units as fluid catalytic cracking unit (FCCU), and catalytic reforming unit can be optimized to produce petrochemical intermediates to the detriment of streams that will be incorporated to fuels pool. In the FCCU, installation of these units dedicated to produce petrochemical intermediates, which aim to reduce to the minimum the generation of streams to produce transportation fuels. However, the capital investment

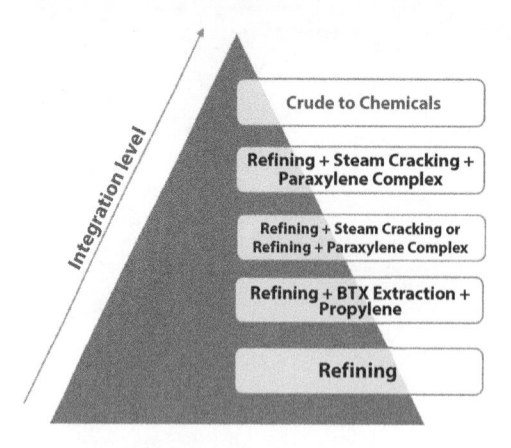

Figure 27.91 Petrochemical Integration Levels. (Source: HIS Markit, 2018).

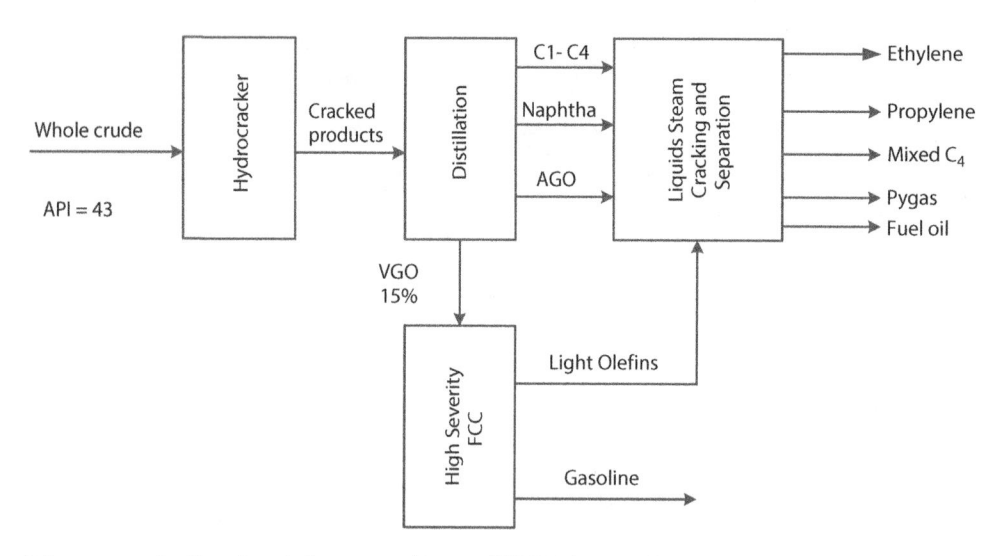

Figure 27.92 Saudi Aramco crude oil to chemicals concept. (Source: IHS Markit, 2017).

is high as the process requires the use of material with the noblest metallurgical characteristics. Figure 27.91 shows the classification of the petrochemical integration grades. According to the classification proposed, the crude to chemicals refinery was the maximum level of petrochemical integration [82].

Because of the increased market and higher added value as well as the reduction in transportation fuels demand, some refiners and technology developers have employed to develop crude to chemicals refining. Saudi Aramco Company has invested in this technology, based on the direct conversion of crude oil to petrochemical intermediates as shown in Figure 27.92.

27.26 The Crude Oil to Chemical Strategy (COC)

The process is based on the quality of the crude oil and deep conversion technology as high severity or petrochemical FCCU and deep hydrocracking technology. The processed crude oil is light with low residual carbon that is a common characteristic of Middle East crude oils. The process involves deep catalytic conversion that aims to reach maximum conversion to light olefins. In this configuration, the petrochemical FCC units have an important role to ensure the high added value to the processed crude oil. An example of FCCU technology developed to maximize the production of petrochemical intermediates is the RxPRO™ process by UOP company. This process combines a petrochemical FCC and a separation process, optimized to produce raw materials to the petrochemical process plants as shown in Figure 27.93. Other technologies are the HS-FCC™ process by Axens company and INDMAX™ process licensed by Lummus company. Figure 27.94 shows the process flow diagram of HS-FCC™ technology. It is essential that both technologies presented in Figures 27.93 and 27.94 are based on petrochemical FCC units that require special design due to the severe operating conditions. The reaction temperature reaches 1112°F (600°C) and higher catalyst circulation rate raises the production of the gases, which require a scaling up of the gas separation unit. The higher thermal demand makes it advantageous to operate the catalyst regenerator in total combustion mode that requires the installation of a catalyst cooler system.

The installation of petrochemical catalytic cracking units requires a deep economic study that accounts for the high capital investment (CAPEX) and high operating costs (OPEX). However, there is the forecast growth of 4.0% per year of petrochemical intermediates in the market until 2025. This indicates an attractive CAPEX that aims to raise the market share in the petrochemical sector with a favorable competitive edge to the refiner for maximizing the intermediates. Figure 27.95 shows a block diagram that demonstrates how the petrochemical FCC unit (the INDMAX™) technology by Lummus company can maximize the yield of petrochemicals in the refining hardware.

In crude oil refining with conventional FCC units involving the higher temperature and catalyst circulation rates, catalyst additives such as zeolitic material ZSM-5 is applied to raise the yield of olefins to 9.0% as compared with the original catalyst. However, this raises the operational costs, but is economically attractive considering the

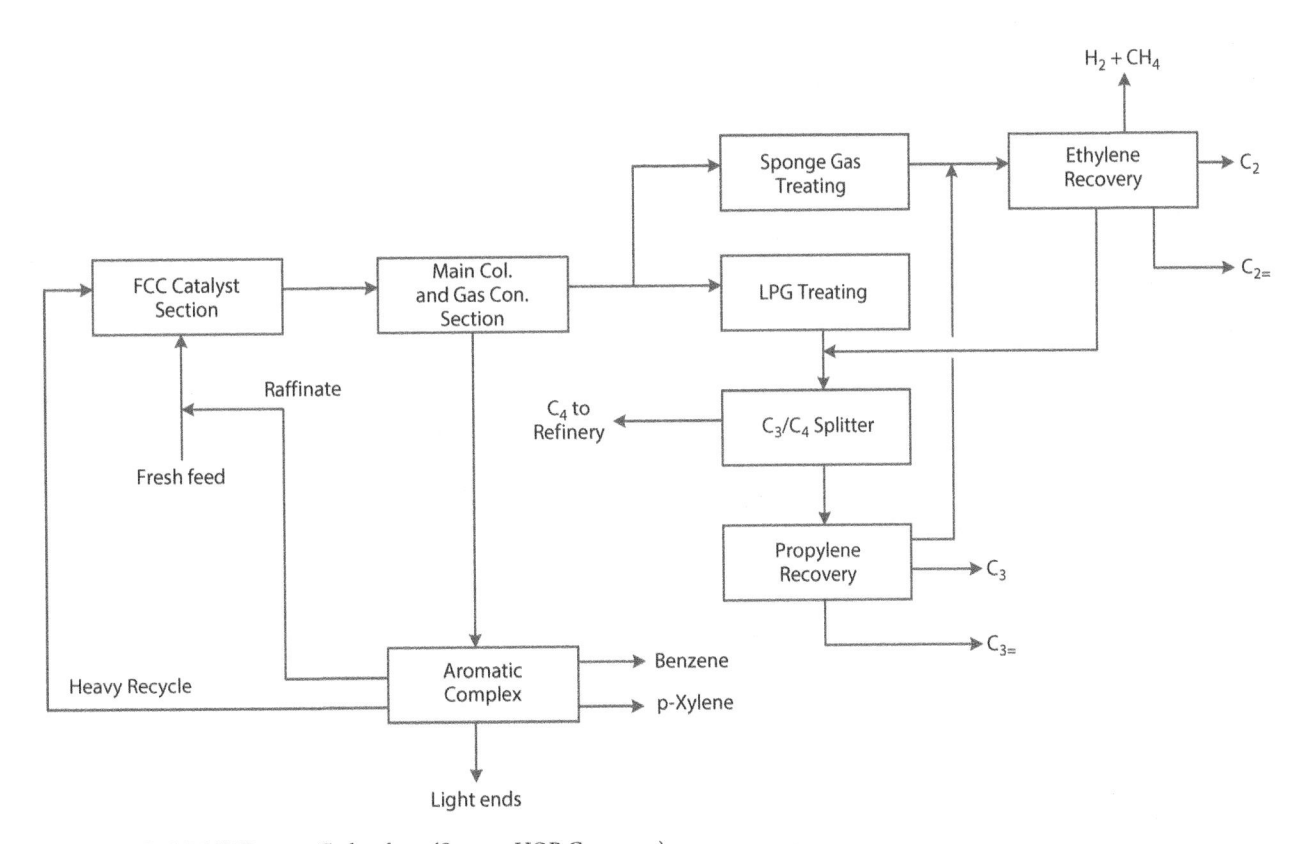

Figure 27.93 RxPRO™ Process Technology (Source: UOP Company).

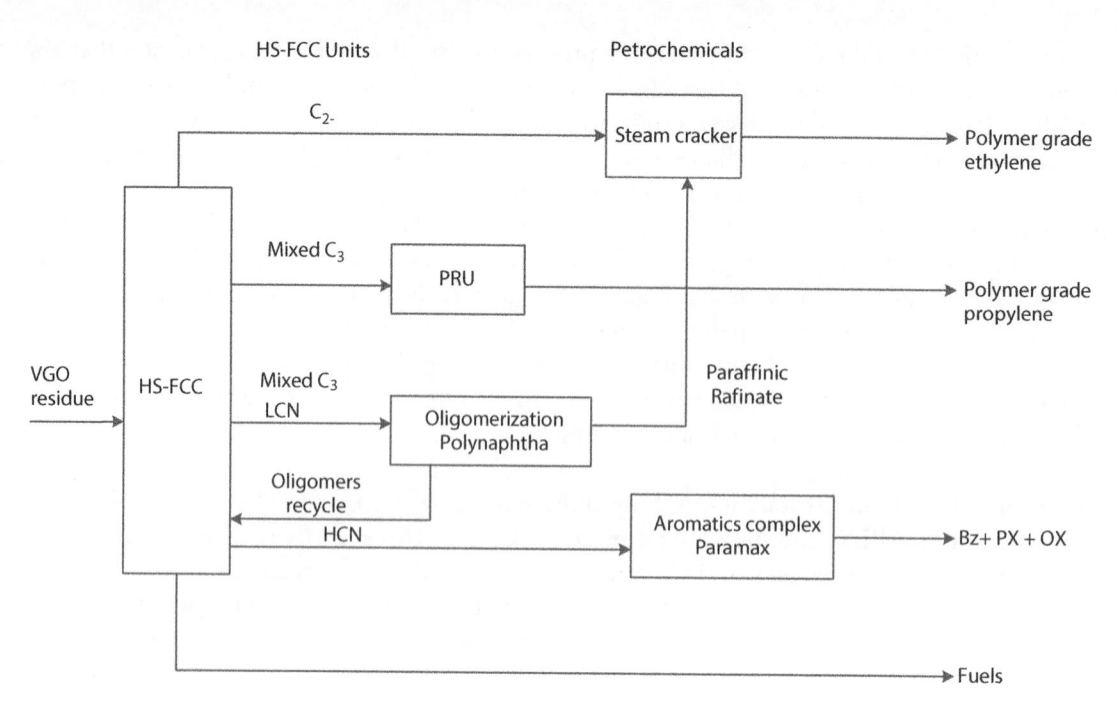

Figure 27.94 HS-FCC™ Process Technology (Source: Axens Company).

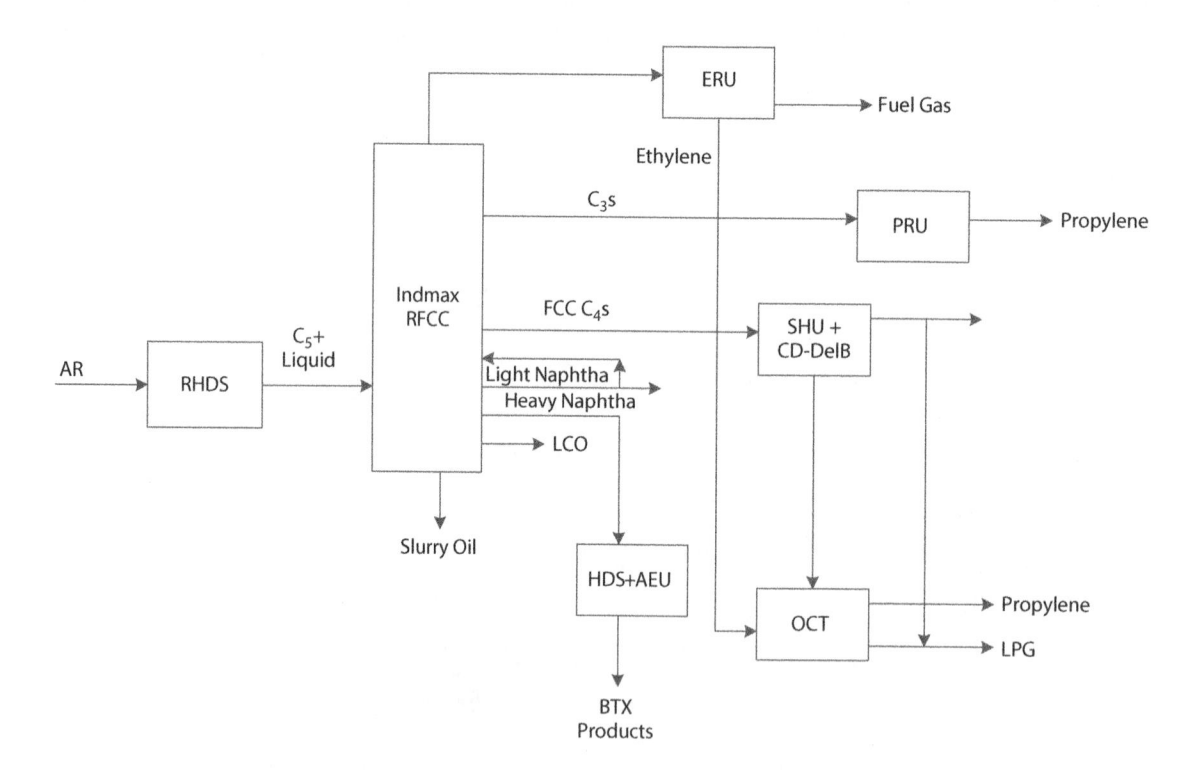

AR:	Atmospheric Residue
RHDS:	Residue Hydro Desulfurization
ERU:	Ethylene Recovery Unit
PRU:	Propylene Recovery Unit
OCT:	Olefin Conversion Technology
HDS+AEU	Hydrodesulfurization, Aromatics Extraction Unit
SHU+DC-DelB	Selective Hydrogenation and CD Deisobutylene Unit

Figure 27.95 Olefins Maximization in the Refining Hardware with INDMAX™ FCC Technology by Chevron Lummus Global Company (Source: SANIN, A.K., 2017).

petrochemical market forecast. Furthermore, the catalyst cooler system enhances the process unit profitability through the total conversion and selectivity to noblest products as propene (C_3H_6) and naphtha against gases and coke production. The catalyst cooler is essential when the unit is designed to operate under total combustion mode due to the exothermic reaction and high release rate generated. This is represented below:

$$C + \frac{1}{2}O_2 \rightarrow CO \text{ (Partial combustion)}, \Delta H = -27 \text{ kcal/mol} \tag{27.24}$$

$$C + O_2 \rightarrow CO_2 \text{ (Total combustion)}, \Delta H = -94 \text{ kcal/mol} \tag{27.25}$$

In this case, the temperature of the regeneration vessel can reach values close to 1400°F (760°C) resulting to higher risks of catalyst damage, which is minimized through a catalyst cooler installation. The option of total combustion mode needs the refinery thermal balance; once in this state will not require the possibility to produce steam in the CO boiler. Furthermore, the higher temperature in the regenerator requires materials with noblest metallurgy, which significantly increases the installation cost of these units and thus prohibitive to some refiners.

Another refining technology to crude-oil to chemicals is the hydrocracking units. Despite the high performance, the fixed bed hydrocracking technologies are not economically effective to treat crude oils directly due to the possibility of short operating life cycle. Technologies that employ ebullated bed reactors and continuum catalyst replacement allow higher campaign period and higher conversion rates. Among these are the H-Oil™ and Hyvahl™ developed by Axens company; the LC-Fining process by Chevron-Lummus, and the Hycon™ by Shell Global Solutions. The reactors operate at temperatures above 450°C and pressure of 250 bar. Figure 27.96 shows a typical process flow diagram of a LC-Fining™ unit, developed by Chevron -Lummus company while the H-Oil™ process by Axens Company is shown in Figure 27.97. The catalysts used in hydrocracking processes can be amorphous (alumina and silica-alumina), and crystalline (zeolites) and have bifunctional characteristic, once the cracking reactions (in the acid sites) and hydrogenation (in the metal sites) occur simultaneously.

An improvement in ebullated bed technologies is the slurry phase reactors where the conversion is > 95%. The technology is the HDH™ process (hydrocracking-distillation-hydrotreatment) developed by PDVSA – Intevep, VEBA-Combicracking Process (VCC)™ commercialized by KBR company; the EST™ process (Eni Slurry Technology) developed by Italian state oil company ENI, and the Uniflex™ technology developed by UOP Company. Figure 27.98 shows the process flow diagram of the VCC™ technology by KBR company.

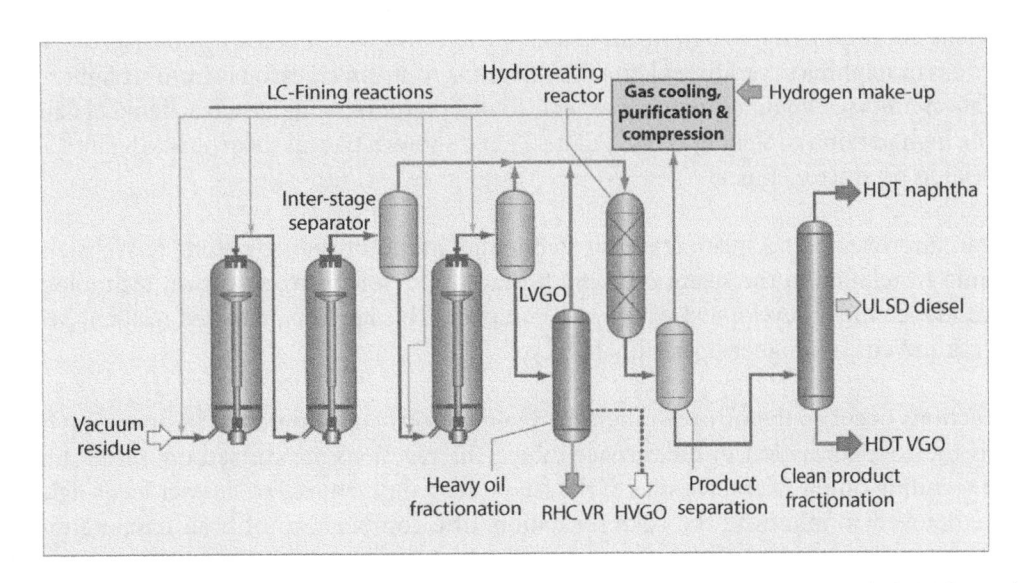

Figure 27.96 Process flow diagram for LC-Fining™ Technology by CLG Company (Source: MUKHERJEE & GILLIS, 2018)

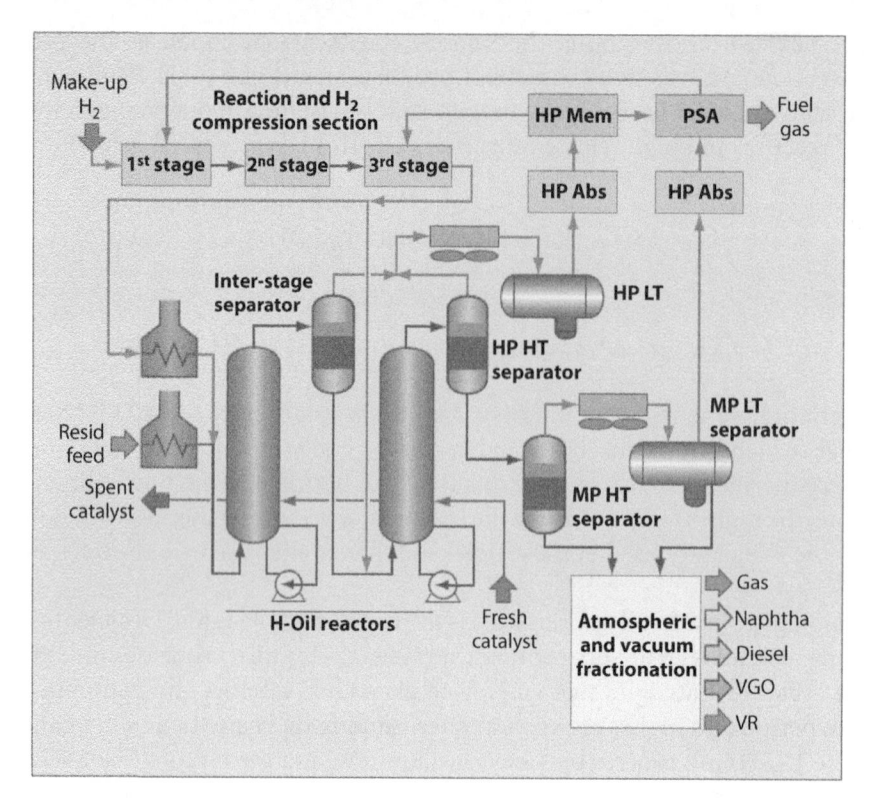

Figure 27.97 Process flow diagram for H-Oil™ Process by Axens Company (Source: FRECON *et al.*, 2019).

In the slurry phase hydrocracking units, the catalysts are injected with the feedstock activated *in situ*. The reactions are carried out in the slurry reactors thus minimizing the reactivation issue and ensuring higher conversion and operating life cycle. Figure 27.99 illustrates the process flow diagram of the Uniflex™ slurry hydrocracking technology by UOP company.

Other commercial technologies to slurry hydrocracking process are the LC-Slurry™ technology by Chevron Lummus company and the Microcat-RC™ by Exxon Mobil company.

The steam cracking process has a fundamental role in the petrochemical industry, as part of light olefins; light ethylene and propylene are produced through steam cracking route. The steam cracking consists of a thermal cracking process that uses gas or naphtha to produce olefins. The naphtha to steam cracking is from straight run naphtha from the crude distillation unit (see Volume 1 of these series). To meet requirements as petrochemical naphtha, the stream needs to present a high paraffin content (> 66%). Figure 27.100 shows a typical steam cracking unit applying naphtha as the raw material to produce olefins.

There has been an improvement in steam cracking technology in recent years by Stone & Webster, Lummus, KBR, Linde and Technip in relation to the steam cracking furnaces. One of the most known technologies is the SRT™ process (short residence time) developed by Lummus company. This applies a reduced residence time to minimize the coking process and ensures higher operational lifecycle.

The cracking reactions occur in the furnace tubes; the main concern and limitation to operating lifecycle of steam cracking units is the coke formation in the furnace tubes. The reactions are carried out under high temperatures (500 – 700°C) depending on the characteristics of the feed inlet temperature. For heavier feeds light gas oil, a lower temperature is employed to minimize the coke formation. The combination of high temperatures and low residence time is the main characteristic of the steam cracking process. Figure 27.101 presents the concept of crude to

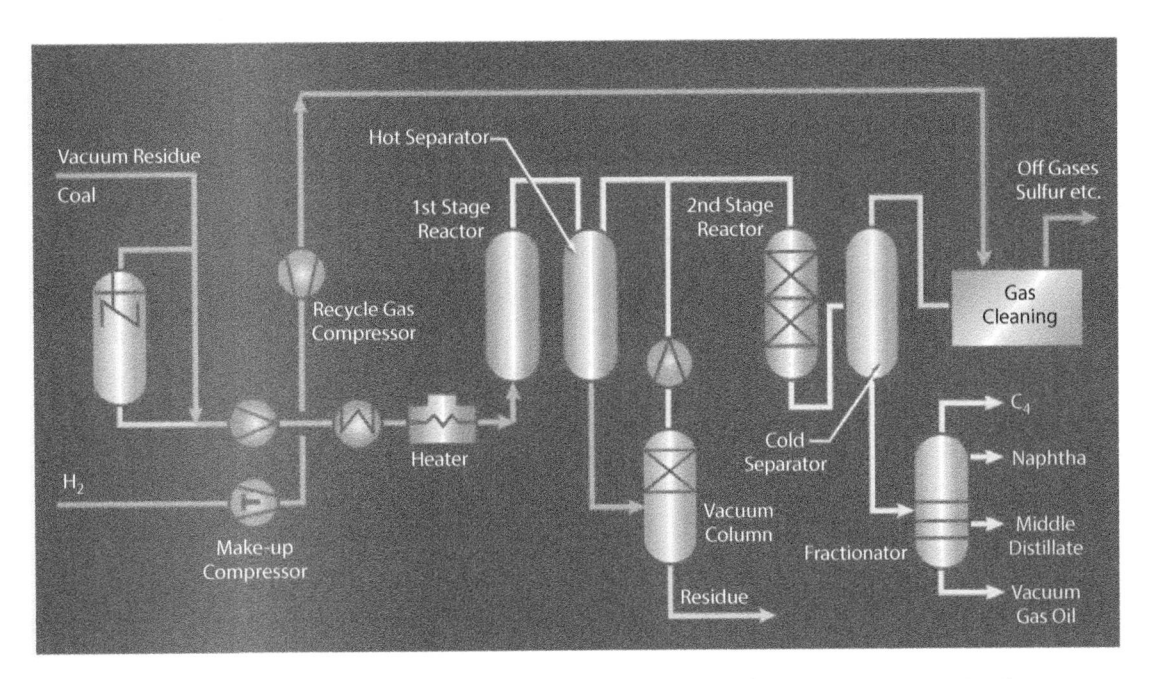

Figure 27.98 Process flow diagram of VCC™ slurry hydrocracking by KBR. company (Source: KBR Company, 2019).

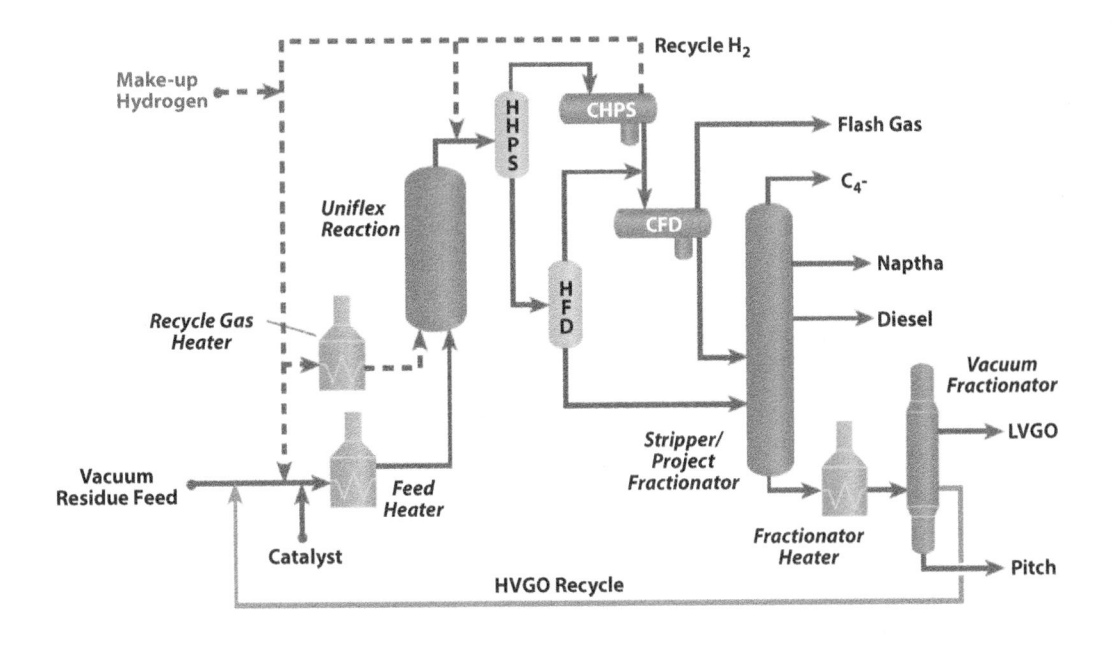

Figure 27.99 Process flow diagram for Uniflex™ slurry phase hydrocracking technology by UOP company. (Source: UOP Company, 2019).

chemicals refining scheme by Chevron-Lummus company, and Figure 27.102 shows a highly integrated refining configuration capable to convert crude oil to petrochemicals developed by UOP company.

Figure 27.103 shows the production focus change to the maximum adding value to the crude oil through the production of high added value petrochemical intermediate or chemicals to general purpose leading to a minimum production of fuels. Saudi Aramco has invested in crude to oil (COC) technology and aims to achieve more integrated refineries and petrochemical plants, and raises competitiveness in the downstream market.

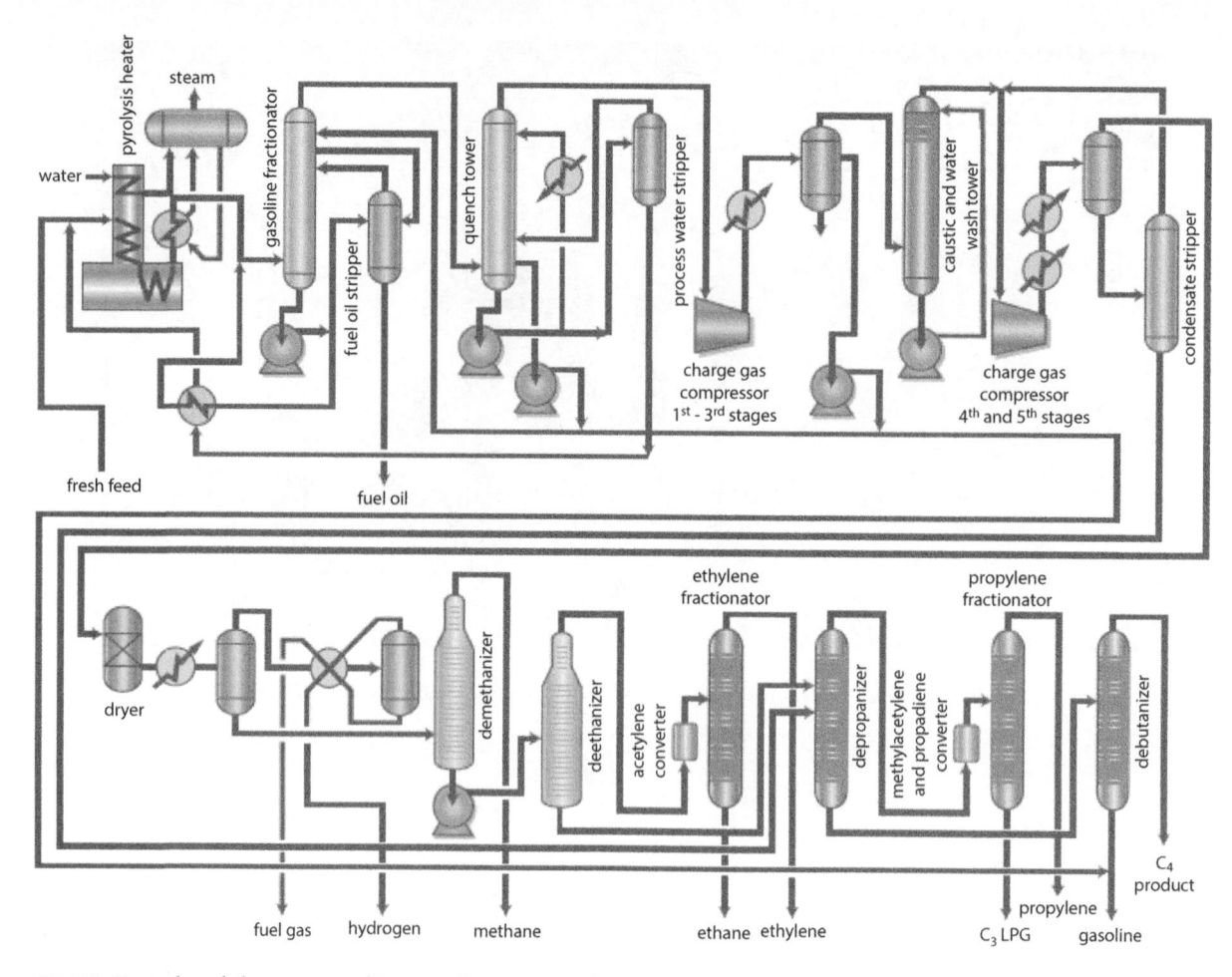

Figure 27.100 Typical naphtha steam cracking unit (Source: Encyclopedia of Hydrocarbons, 2006).

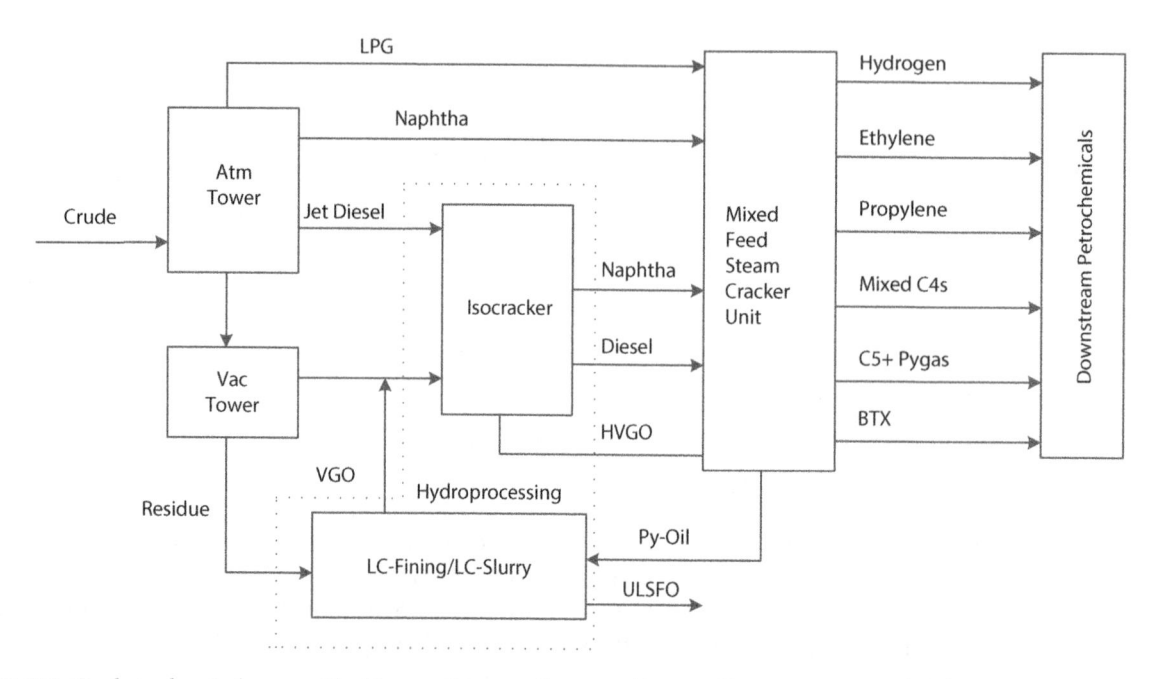

Figure 27.101 Crude to chemicals concept by Chevron Lummus Company (Source: Chevron Lummus Global Company, 2019).

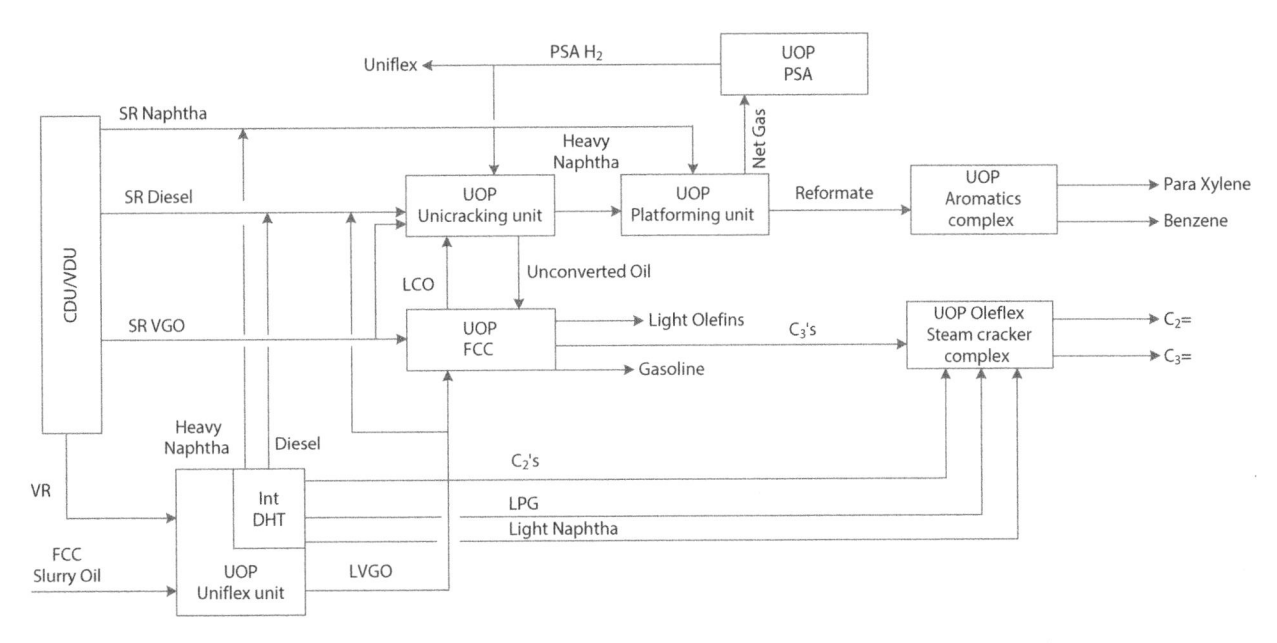

Figure 27.102 Integrated refining configuration based in crude to chemicals (Source: Concept by UOP Company).

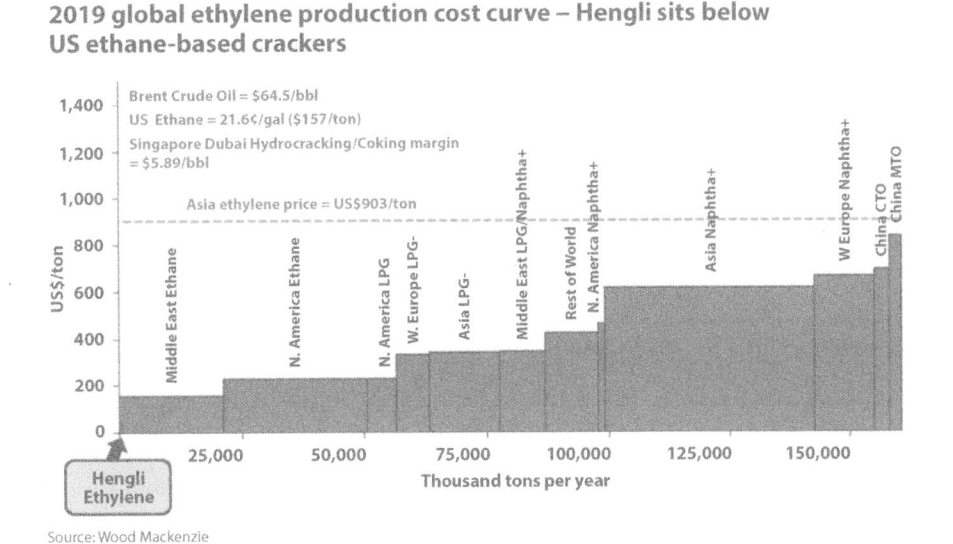

Figure 27.103 Ethylene production cost comparison (Source: Wood Mackenzie REM-Chemicals, 2019).

The major licensors such as Axens, UOP, Lummus, Shell, ExxonMobil, etc., have applied resources to develop technologies that are capable to allow a closer integration in the downstream sector, which enables refiners to maximize the added value from the processed crude oil, where the refining margins are low. Table 27.23 shows the data from The Catalyst Group Company (TCGR) representing some capital investments in crude to chemical projects. Some of these investments were postponed due to the world economic crisis by the COVID-19 pandemic, but these data show the trend in the market. A typical concern of the crude to chemicals enterprises is the operation costs in comparison with the traditional routes. Figure 27.103 shows a comparative study of the operation costs of Hengli crude to chemicals enterprise in relation with traditional ethylene production routes. It is essential to consider that the cost composition comprises of various factors, and the scenario can be different due to the local business situation. Figure 27.104 presents a comparison between the petrochemicals yield of traditional refineries, a benchmark integrated refinery and Hengli crude to chemical complex. From Figure 27.104, it is possible to note that the higher added value

Table 27.23 Announced Oil – to – Chemicals Investments 2019, $billion.

Zhejang Petroleum and Chemical	Zhousha, China	$26	Greenfield	2019 (Phase 1)
Hengl Petrochemical	Changxin Island, China	$11	Greenfield	2019
Shenghong Petrochemical	Lianyungang, China	$11.84	Greenfield	2019
Ningbo Zhongjin Petrochemical (subs Rongsheng Petrochemical	Ningbo, China	$5 (est)	Revamp	2018
Saudi Aramco/NORINCO/Panjun Sincen (Huajin Aramco Petrochemical)	Liaoning Province, China	$10+	Greenfield	2024
SABIC/Fuhaichuang Petrochemical	Zhangzhou, China	NA	Greenfield	NA
SINOPEC/SABIC (Tianjin Petrochemical)	Tianjin, China	$45	Revamp	Operating, pre-2017
PetroChina	Dalian, China	combined	Revamp	Operating, pre-2017
PetroChina	Yunnan, China	(est)	Revamp	Operating, pre-2017
CNOOC	Huizhou, China		Revamp	Operating, pre-2017
SINOPEC	Lianyungang, China	$2.80	Greenfield	NA
SINOPEC	Caofeidian, China	$4.2	Greenfield	NA
SINOPEC	Culie, China	$4.26	Greenfield	2020
	Total China	$120.1		
Other Asia				
Hengyi Group	Pulau Muara Besar, Brunei	$20	Greenfield	2020
Saudi Aramco/ADNOC/India Consortium	Raigad, India	$44	Greenfield	2025
Petronas/Saudi Aramco (RAPID)	Pengerang, Malaysia	$2.7	Greenfield	2019
ExxonMobil (Singapore Chemical Plant)	Jurong Island, Singapore	<$1	Revamp	2023
Peretamina/Rosneft	Tuban, East Java, Indonesia	$15	Greenfield	2025
	Total other Asia	$82.7		
Middle East				
ADNOC	Al Ruwais, UAE	$45	Revamp	2025
Saudi Aramco/SABIC	Yanbu, Saudi Arabia	$30	Greenfield	2025
Saudi Aramco/Total	Jubail, Saudi Arabia	$5	Greenfield	2025
KNPC/KIPIC (Al-Zour Refinery)	Al Ahmadi, Kuwait	$13	Greenfield	2024

(Continued)

Table 27.23 Announced Oil – to – Chemicals Investments 2019, $billion. (*Continued*)

Oman Oil Company/Kuwait Petroleum International (Duqm Refinery)	Oman	$15	Greenfield	2019
	Total Middle East	$108		
Europe				
MOL Group	Hungary, Croatia	$4.5	Revamp	2030
	Total Europe	$4.5		
Total Greenfield $215	Total revamps $100		Total global $315	

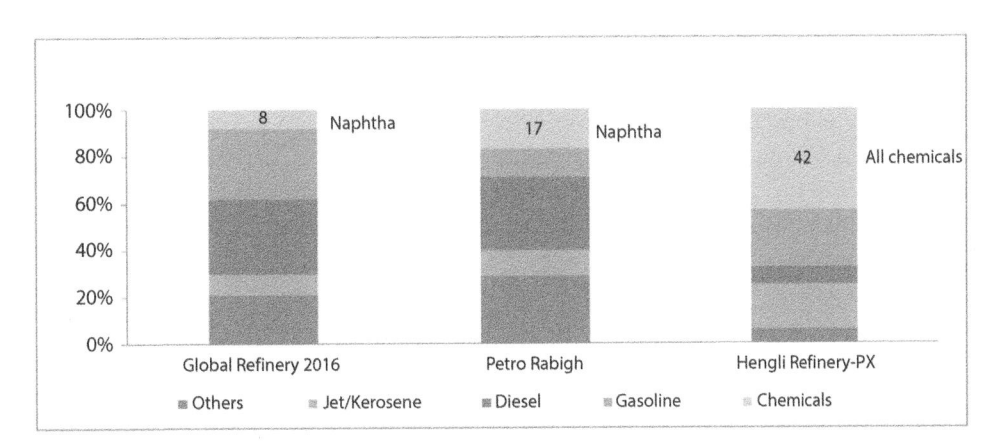

Figure 27.104 Petrochemicals yield comparison (Source: IHS Markit, 2018).

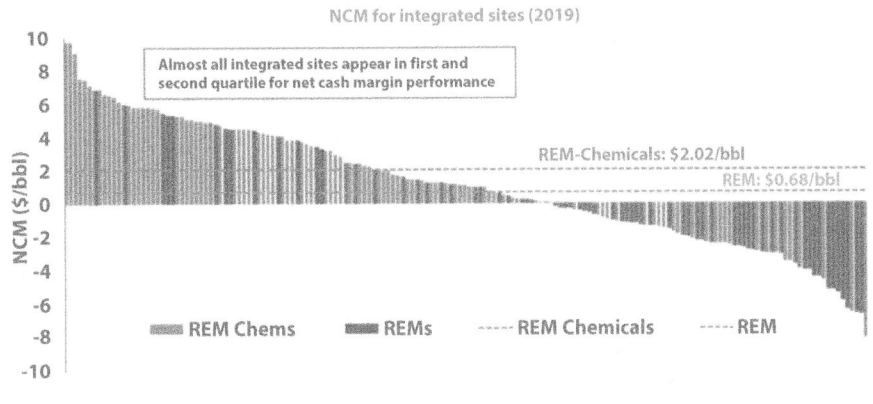

Figure 27.105 Average margins of integrated refining sites (Source: Wood Mackenzie REM-Chemicals, 2021).

reached in the crude to chemicals refineries when compared with highly integrated refineries. Figure 27.105 presents a comparison between the petrochemicals yield of traditional refineries, a benchmark integrated refinery and Hengli crude to chemicals complex. Figure 27.105 illustrates the highly integrated refiners can add from US $0.68/bbl to $2.02/bbl for 169 refineries; the Asian market represents the major concentration of integrated refining plants (close to 64% of the global investments in crude to chemicals).

27.27 Available Crude to Chemicals Processing Routes

There are three viable routes that are being considered to capital investments of crude to chemicals refining complexes. Figure 27.106 shows the concepts of these routes. The conventional routes consider the processing of crude oil in a crude distillation unit (CDU), producing petrochemical intermediates such as naphtha which is then supplied to a petrochemical facility like a steam cracking unit. The ExxonMobil route is based on the direct feed of selected crude oils, normally light and low contaminant crudes, to petrochemical facilities, while the Chines enterprise Hengli, Zhejiang, Shenghong Henyi project considers the feed of mixed crude oil slate to a crude to paraxylene (PA) complex in order to ensure the domestic market that presents high demand by light aromatics (benzene, toluene, xylene (BTX)). A conventional highly integrated refining hardware is adequate to achieve 15 – 20% of petrochemicals yield, while a crude to chemicals refinery can reach up to 40% as shown in Figure 27.107.

The Aramco/Sabic concept is based on a high complexity refining hardware to convert selected light crude oil to maximize the yield of petrochemical intermediates mainly light olefins (e.g., C_2H_4, C_3H_6). There are advantages in integrating the refining with petrochemicals, however it is essential to define a transition strategy where economic sustainability is achieved where the current status (transportation fuels) needs to be invested to build the future thus maximizing the petrochemicals.

In petrochemical FCC units, the reaction temperature reaches 600°C and higher catalyst circulation rate raises the production of the gases, which requires a scale-up of gas separation section. The higher thermal demand requires the catalyst regenerator in total combustion mode that results in the installation of a catalyst cooler system. Figure 27.107 shows the results of a comparative study showing the yields obtained by conventional FCC units, optimized to olefins (FCC to olefins) and the HS-FCC™ designed to maximize the production of petrochemical intermediates.

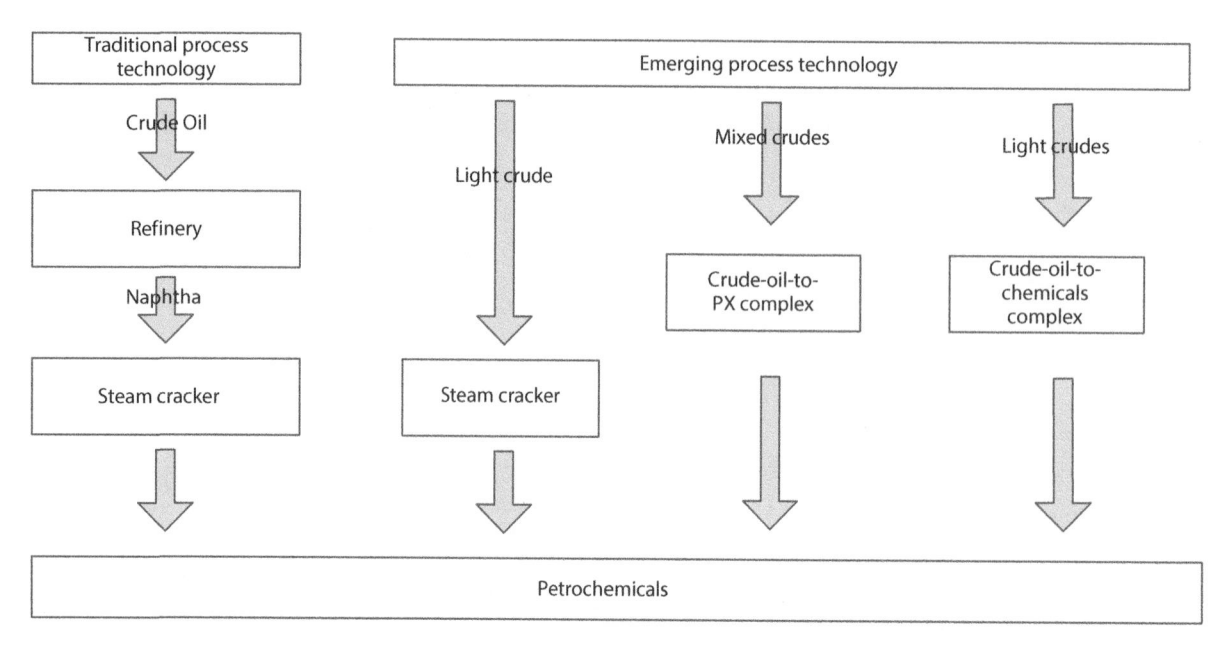

Figure 27.106 Crude to chemicals concepts (Deloitte, 2019).

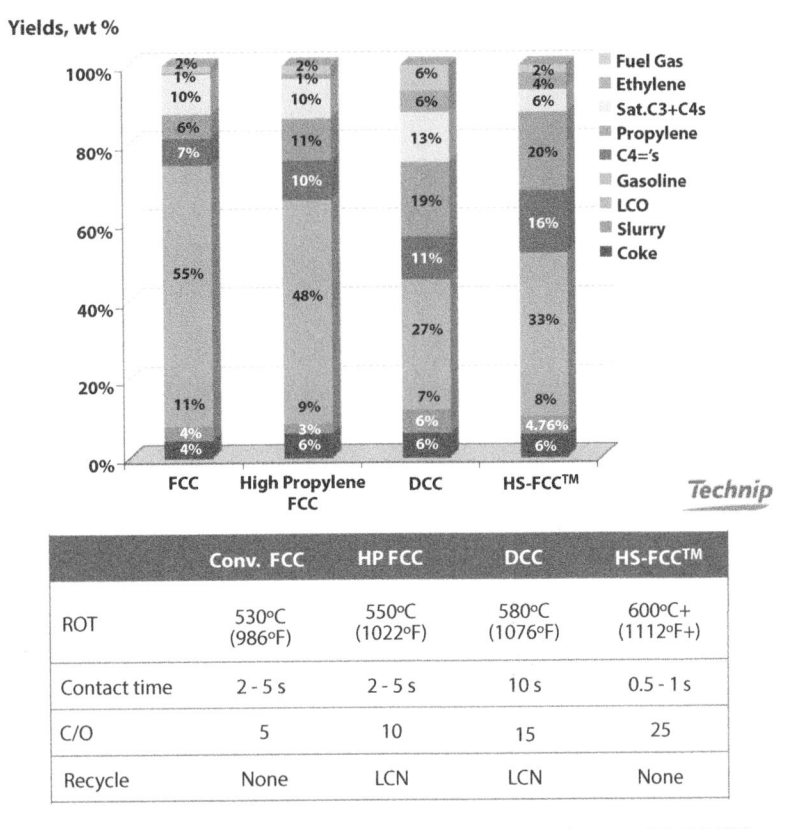

	Conv. FCC	HP FCC	DCC	HS-FCC™
ROT	530°C (986°F)	550°C (1022°F)	580°C (1076°F)	600°C+ (1112°F+)
Contact time	2 - 5 s	2 - 5 s	10 s	0.5 - 1 s
C/O	5	10	15	25
Recycle	None	LCN	LCN	None

Figure 27.107 Comparative study between conventional FCCs and Petrochemical FCC (Source: HS-FCC™).

A higher reaction temperature (TRX) and a catalyst/oil ratio with contact time 0.5 – 1sec. yields five times higher when compared with the conventional process units and the petrochemical FCC (HS-FCC™), leading to a growth of the light olefins yield (Ethylene + Propylene + C_4='s) from 14 to 40%.

27.28 Chemical Looping

A novel technique that uses a solid carrier material to transport a compound (i.e. O_2 or CO_2) from one part of a chemical process or fossil fueled power station to another in a cyclical manner. It is a process that involves employing a solid material known as oxygen carrier to transfer oxygen from air to a fuel source, such as coal or natural gas in a series of chemical reactions. The oxygen carrier is a typically a metal oxide material iron oxide (Fe_2O_3) or copper oxide (CuO), which can be easily reduced to release oxygen and then oxidized again by the fuel source to release heat for energy production. The process of chemical looping involves circulating the oxygen carrier between two separate reaction vessels, one where it is oxidized by the fuel source to release heat and another where it is re-oxidiized by air to release oxygen. This approach enables for the fuel to be efficiently burned while minimizing the formation of harmful pollutants such as nitrogen oxides (NOx) and carbon dioxide (CO_2). The completed cycle is where the reduced oxygen carrier is regenerated back to its higher oxidation state, as only a few transition metals exhibit these properties such as iron (Fe), copper (Cu) and nickel (Ni). Furthermore, the temperatures at which the reactions occur must allow a sensible cycle to be set up.

Chemical looping is typical performed in a fluidized bed reactor so that the carrier particles must be dense enough to withstand this and not agglomerate readily. The materials must not be harmful to the environment, or toxic, scarce or expensive. For example, Ni which is very promising has been found to be carcinogenic and it is thus developed outside a laboratory [xx]. Many different processes can employ similar looping cycles as shown in Figures 27.108a and 27.108b. Such processes take place at high temperatures 500-1000°C, so they are frequently referred to as high temperature looping cycles. Figure 27.108b shows a hydrogen production process, where a fuel is used to reduce iron oxide (Fe_2O_3) down to iron (Fe), which is then oxidized back to Fe_3O_4 using steam, before the loop is finally completed by regeneration to Fe_2O_3.

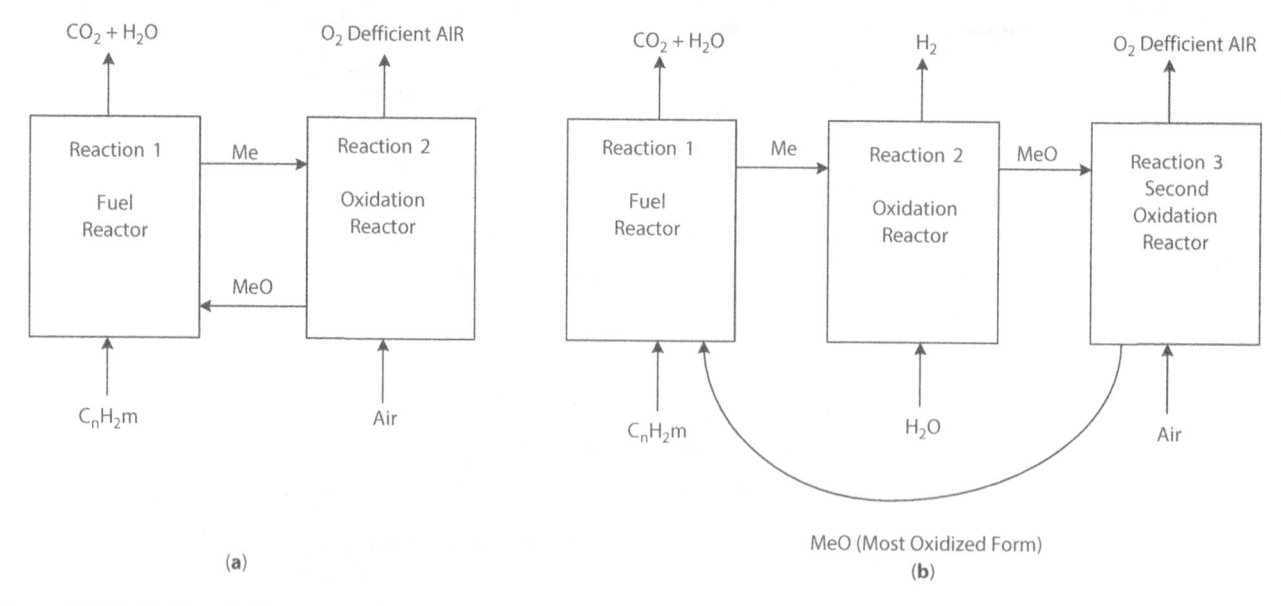

Figure 27.108 (a) Chemical looping combustion. (b) Chemical looping hydrogen production [97].

Chemical looping has been studied as a potential alternative to traditional combustion methods for power generation and industrial processes, as it has the potential to reduce greenhouse gas emissions and increase fuel efficiency. However, the technology is still in the experimental stage and has yet to be widely adopted on a commercial scale.

27.29 Conclusions

Sustainability within the energy, petroleum refinery and petrochemical process industries is a multi-faceted concept that encompasses more than economic, environment, safety and governance. The overall business sustainability through operational excellence and margin optimization remains the essential objectives. Additionally, due to the various challenges each of these industries encounters, each has a different objective and focus for its sustainability initiatives [67].

Petroleum and chemical process industries have put in place plans for sustainability; however, the initiatives at energy companies focus largely on the transition to a lower carbon future via increased used of wind, solar, hydro, geothermal, and other renewable-energy sources; while chemical companies tend to focus their sustainability initiatives on developing and producing more sustainable, circular products characterized by low–carbon footprint and extensive use of renewable and/or recycled raw materials that can support reusable and/or recyclable end products for customers. For both industry segments, enhancing operational safety plays a significant role for many industries to operate successfully, and for both segments, digital transformation is an important source for business and environmental, social and governance (ESG) sustainability.

IChemE commits to provide policy advice to governments based on chemical engineering experience and expertise consistent with its commitment to net zero carbon emissions and the UN SDGs. It will proactively engage with research facilities, industry, government reviews, consultations and policy debates in a manner consistent with its commitment to net zero carbon emissions. The organization will offer training courses on-line and personally that will help educate, re-skill and promote key carbon reduction and adaption technologies. IChemE proposes to develop plans for achieving net zero carbon emissions from its direct operations globally by 2025 and to publish greenhouse-gas emission's data and progress against this target each year that includes considerations of efficiencies, reductions and offsets. It shall establish practical investment criteria that would enable the Institution's funds to be invested in alignment with climate-change goals.

IChemE commits to update its Code of Conduct to include an obligation on all professional members to act in accordance with the principles of sustainability including the UN SDGs, prevent avoidable adverse impact on the environment and society, act to mitigate greenhouse-gas emissions and adapt to a changing climate, and protect and where possible improve, the quality of built and natural environments. The Institution provides various training courses to engineers, and mandates a Continuous Development programme (CDP) to provide the knowledge and skills to support members in transition to a net zero carbon economy and in climate-change adaptation. Additionally, design guidelines, tools and project evaluation techniques are provided to assist practicing engineers to apply sustainable design principles. Finally, accreditation of university degrees enhances the requirements for the treatment of sustainability, energy efficiency, resource efficiency, climate change, environment and biodiversity, thereby preparing graduating chemical engineers for their role in the transition to a zero-carbon world and a changed environment.

As the decade progresses, new blue assets in the form of steam methane reformer (SMR), partial oxidation (POX) and autothermal reforming (ATR) will be built to realize the potential of H_2 that addresses decarbonize sectors that are problematic. Escalating the carbon price coupled with emerging technological advances will further drive the investment. Furthermore, the technology of choice for the syngas separation will vary depending on the end uses of the H_2 and CO_2. A thoughtful pairing of carbon capture and H_2 purification technology can achieve economic differentiation in delivering a significant step in the CO_2 countdown to net zero [75].

The interface between refining and petrochemical processes raises the availability of raw materials to petrochemical plants and makes the supply of energy to these processes more reliable, which ensures better refining margin to refiners due to added value of petrochemical intermediates when compared with transportation fuels. Furthermore, the development of crude to chemicals technologies reinforces the requirements for close integration of refining and petrochemical assets by the brownfield refineries that aim to face the new market focusing in petrochemicals as opposed to transportation fuels.

There is a competitive advantage from the Middle East refiners with ready access to light crude oils, which can be easily applied in crude to chemical refineries. Crude oil to chemicals refineries is dependent on deep conversion processes that requires high capital expenditure. This imposes additional strain on the refiners thus reinforcing the necessity to investigate the possibility for close integration with petrochemical sector with the aim in achieving competitiveness.

Chemical looping technology is very promising as the major advantage is that it uses fluidized beds, which are versatile and can be scaled from the 1MW(t) to the 500 MW(e) level [97]. The technique should be able to efficiently combust solid fuels, including biomass, with a potential route for negative emission (i.e., biomass energy with combustion capture and storage, BECCS), and it is compatible with hydrogen production. The future of chemical looping will be in integration with iron (Fe) and/or hydrogen production, chemicals and energy storage.

Glossary

Absorption	A separation process involving the transfer of a substance from a gaseous phase to a liquid phase through the phase boundary.
Acid Gases	Impurities in a gas stream usually consisting of CO_2, H_2S, COS, RSH and SO_2. Most common in natural gas are CO_2, H_2S and COS.
Acid Gas Loading	The amount of acid gas, on a molar or volumetric basis, that will be picked up by a solvent.
Adsorption	The process by which gaseous components adhere to solids because of their molecular attraction to the solid surface.
Alkanolamines	An organic nitrogen bearing compound related to ammonia having at least one, if not two or three of its hydrogen atoms substituted with at least one, if not two or three linear or branch alkanol groups where only one or two could also be substituted with a linear or branched alkyl group (i.e., methyldiethanolamine MDEA). The number of hydrogen atoms substituted by alkanol or alkyl groups at the amino site determines whether the alkanolamine is primary, secondary or tertiary.

Bioenergy	Refers to heat or electricity produced using biomass or gaseous and liquid fuels with a biological origin such as biomethane produced from biomass.
Biomass	Any material of biological origin used as a feedstock for products (e.g., wood in construction to make chemicals and materials, like bio-based plastics), or as a fuel for bioenergy (heat, electricity and gaseous fuels such as biomethane and hydrogen) or biofuels (transport fuels).
Capture and Storage (BECCS)	during which carbon is captured and stored. If carefully managed, using sustainable biomass, BECCS can generate 'negative emissions' because while providing energy it also captures and stores the atmospheric carbon monoxide (CO) that is absorbed by plants as they grow.
Carbon intensity	The amount of CO_2 emitted when generating a unit of electricity, measured in gram of CO_2 per kWh of electricity produced.
Carbon capture readiness	Is a requirement imposed on thermal plants (such as coal and gas plants) to enable future capturing and storing of carbon following a plant upgrade. Such plants currently emit CO_2 directly into the atmosphere.
Carbon Capture Utilization and Storage (CCUS)	The process of capturing carbon dioxide from industrial processes, power generation, certain hydrogen production methods and greenhouse gas removal technologies such as bioenergy with carbon capture and storage and direct air capture. The captured CO_2 is then either used, for example in chemical processes, or stored permanently in disused oil and gas fields or naturally occurring geological storage sites. CO_2 can be used for enhanced recovery of crude oil, i.e. to extract more oil from existing well, production of petrochemicals, e.g., methanol, urea, polycarbonate and so on.
Carbon Leakage	The situation that may occur, if, for reasons of costs related to climate pricing policies, businesses were to transfer production or reallocate future investments to other countries with laxer emission constraints or carbon pricing. This could lead to an increase in total global carbon emissions.
Carbon Price	A cost applied to carbon pollution to encourage polluters to reduce the amount of greenhouse gases they emit into the atmosphere.
Clean Electricity	Types of electricity generating technologies that emit little or no fossil fuel derived greenhouse gas from generation.
Competitive tendering	A process inviting eligible organizations to compete to carry out work or supply services, with the winner decided by who can offer the best price and quality.
Contracts for Difference Scheme (CfD)	The main support mechanism for large scale low-carbon electricity generation projects. Successful projects are awarded a long-term contract which secures a price to which they will either be topped up if electricity prices are low, or pay back to if electricity prices are high.
Contract for Difference allocation round	The competitive allocation process of CfD contracts. Participants bid the strike price they require to build their project with the cheapest ones winning on a pay-as-clear system.
Decarbonization	A process of reducing the amount of carbon dioxide released into the atmosphere.
Degradation Products	Impurities in a treating solution that are formed from both reversible and irreversible side reactions.
Digitalization	Is the integration of digital technologies into a process, organization, or system. For example, smart meters which automatically send meter readings to energy suppliers, meaning more accurate bills for customers.

Direct Air Carbon Capture and Storage	Use of engineered processes to capture carbon dioxide (CO_2) directly from the atmosphere, for storage or use.
Dispatch signals	A pricing mechanism designed to encourage power stations to produce and send electricity to the grid when it is needed.
Distribution networks	Regional networks that transport gas or electricity into homes and businesses and import electricity from small-scale generation.
Downstream Oil and Gas	The industries and processes in which oil and gas are converted into finished products.
Electricity Capacity	The amount of electrical power a generator can produce when it is running at maximum output.
Electricity Generation	The total electrical energy created over a period of time.
Electricity System	A system consisting of generators, interconnectors, transmission and distribution networks, and storage that deliver electricity to the final consumer (businesses, industry, public sector and homes.) As well as the markets and control infrastructure such as smart and digital technologies, that play a key role in making sure the system balances supply and demand.
Electrification	Switching from using fossil fuels such as gas or petroleum, to using electricity. For example, switching from a petrol car to an electric car.
Emissions Trading System (ETS)	A method of putting a price on emissions. A cap is set on the total amount of certain greenhouse gases that can be emitted by participants. The cap is reduced over time so that total emissions fall. Within the cap, companies receive or buy emissions allowances, which they can trade with one another as needed.
Energy codes	The detailed technical and commercial rules of the energy system.
Energy data	Historical, current and future information covering things such as how, where and when energy is generated, transported used and stored.
Energy efficiency	When something performs better using the same amount of energy, or delivers the same performance for less. The principles of energy efficiency can be applied to many things: buildings, products, appliances, manufacturing processes, and so on.
Energy Performance Certificate	Energy Performance Certificates (EPCs) are required in the UK to provide a prospective owner or tenant with information on the energy performance of a building and recommendations for improvement. EPCs use an A-G rating scale based on the modeled energy bill costs of running the building.
Engineering standards	The specifications to which the energy system is designed and operated.
Flaring	The controlled burning of unwanted or excess natural gases.
Fossil fuels	Oil (and fuels derived from oil), coal and natural gas.
Flue gas	Is the gas exiting to the atmosphere via a flue, which is a pipe or channel for conveying exhaust gases from a furnace, boiler or steam generator. It often refers to the combustion exhaust gas produced at power plants or refinery facilities. Its composition depends on what is being burned, but it usually consists mostly nitrogen (typically more than two-thirds) derived from the combustion of air, carbon dioxide (CO_2), and water vapor as well as excess oxygen (O_2) (also derived from the combustion of air). It further contains a small percentage of a number of pollutants such as particulate matter (e.g., soot), carbon monoxide (CO), nitrogen oxides (NOx) and sulfur dioxide (SO_2).

Gas quality standards	Are rules to ensure that the gases we use, including natural gas, biogas and hydrogen, meet all of the specifications needed to be safe and effective as an energy source.
Gas system	A system consisting of gas producers, refineries, interconnectors, transmission and distribution networks that delivers gas from its original sources to the final consumer (business, industry, public sector and homes). As well as the physical infrastructure, markets play a key role in making sure the system balances supply and demand.
Greenhouse Gas Emissions	Addition to the atmosphere of gases that are a cause of global warming, including carbon dioxide, methane and others.
Greenhouse Gas Removal Technologies (or negative emissions)	Methods that actively remove greenhouse gases from the atmosphere, ranging from engineered to nature – based solutions.
Heat network	A heat network, sometimes called district heating, is a system of insulated pipes that takes heat or cooling generated from a central source and distributes it to a number of domestic and non-domestic buildings.
Heat pump	A device that extracts heat from the air, ground or water and concentrates it to a higher temperature and delivers it elsewhere, for example to a central heating system. It can replace traditional fossil fuel heating, such as a gas or oil boiler. Heat pump systems are designed to extract a greater amount of heat energy from the surrounding environment than the energy they consume in doing so, therefore, they can act as a more efficient source of heat than a conventional electric heater, producing two to three times (or more for every efficient systems) as much heat outputs as they consume in electricity input.
Hybrid interconnector projects	Projects that combine electricity generation with the ability to feed electricity to two (or more) different markets. For example, an offshore wind project that has multiple connections and is able to provide electricity to both the U.K. market and to European markets.
Hydrogen for heat	The combustion of hydrogen produces no long-lived greenhouse gas emissions at point of use, making it a possible low-carbon replacement for natural gas as a fuel source for heating homes and other buildings.
Clean hydrogen	Hydrogen that is produced with significantly lower greenhouse gas emissions compared to current methods of production – methods include reacting methane with steam to form hydrogen and then capturing the carbon dioxide by product (steam methane reformation with CCUS) or using renewable electricity to split water into hydrogen and oxygen (electrolysis).
Low – carbon electricity generating technologies	Types of electricity generating technologies that emit little or no carbon, which include renewables, nuclear CCUS.
Negative Emission	Achieved by removing greenhouse gases from the atmosphere, for example, through direct air capture or bio-energy production with carbon capture.
Net zero	Refers to a point at which the amount of greenhouse gas being put into the atmosphere by human activity in the UK equals the amount of greenhouse gas that is being taken out of the atmosphere.
Nuclear fusion	Is the process that powers the sun: the fusing of hydrogen atoms into helium, which releases large amount of energy. Scientists are developing technology to use this process to provide fusion energy, which could be clean, safe and inexhaustible with no long – lived radioactive waste.

Permeance	Permeance is the degree to which a material admits a flow of matter or energy. In materials science, "permeance" can refer to the ability of a material to allow a substance (such as a gas or liquid) to pass through it.
	For example, the permeance of a membrane or film might describe how easily oxygen can diffuse through it. Permeance in this sense is often measured in units of moles per second per square meter per pascal (mol/s/m²/Pa).
	In electrical engineering, "permeance" (also called "magnetic permeance" or "magnetic reluctance") is a measure of the ease with which a magnetic circuit allows magnetic flux to flow through it. It is the reciprocal of magnetic reluctance, which is analogous to electrical resistance.
	Permeance is measured in units of henries per meter (H/m) or webers per ampere-turn (Wb/At).
	For example, In electromagnetism, pemeance is the inverse of reluctance. Magnetic permeance P is defined as the reciprocal of magnetic reluctance, R.
	i.e. $P = 1/R$
Physical Solvent	A liquid capable of absorbing selected gas compounds by solubility alone without associated chemical reactions.
ppmv	A volume concentration of a species in a bulk fluid measured in parts per million.
Process Intensification	Any chemical engineering development that leads to a substantially smaller, cleaner, safer, sustainable and more energy efficient technology. Process Intensification (PI) can be employed in inherently safer plant grouped into four major strategies: • Minimize-use small quantities of hazardous materials, reduce the size of equipment operating under hazardous conditions such as high temperature or pressure • Substitute-use less hazardous materials, chemistry, and processes • Moderate-reduce hazards by dilution, refrigeration, process alternatives which operate at less hazardous conditions • Simplify-eliminate unnecessary complexity, design "user friendly" plants.
	PI can allow one to moderate conditions to minimize risk of explosions and to simplify processes by having fewer unit operations and less complex plant.
Refineries	Industrial facilities which convert crude oil and gas into specific products such as jet fuel or diesel.
Renewable Energy	Energy that is collected from resources which are naturally replaced in human timescales such as sunlight, wind, rain, tides and waves.
R & D	Research and development: thinking up innovative and new ideas and applying them.
Residence Time	The time period for which a fluid will be contained within a specified volume.
Selective Treating	Preferential removal of one acid gas component, leaving at least some of the other acid gas components in the treated stream.

Small Modular Reactors (SMRs)	SMRs are usually based on proven water-cooled reactors similar to current Nuclear Power station reactors, but on a smaller scale. They use nuclear fission to generate low-carbon electricity. SMRs are called modular reactors as their components can be manufactured in factories using innovative techniques and then transported to site to be assembled.
Smart charging	Connecting an electric vehicle to the electricity grid using a charging device which includes a data connection. This allows electric vehicles that are plugged in using smart chargers to be charged when it is the most efficient, in terms of cost for the consumer and/or from the point of view of balancing supply and demand across the electricity system.
Smart meters	The next generation of gas and electricity meters, which use a secure smart data network to automatically and wirelessly send meter reading to energy suppliers, enable remote topping up of balances for pre-payment customers and near real time energy consumption and expenditure to be visible to domestic energy consumers via an in – Home Display. Smart meters also enable innovations such as time of use tariffs which will help support delivery of our net zero objectives.
Sour Gas	Gas containing undesirable quantities of hydrogen sulfide (H_2S), mercaptans (RSH) and/or carbon dioxide (CO_2).
Sweet Gas	Gas which has no more than the maximum sulfur content defined by: (1) the specifications of the sales gas from a plant, (2) the definition by a legal body.
Syngas	Syngas or synthesis gas, is a fuel gas mixture consisting primarily of hydrogen (H_2), carbon monoxide (CO), and very often some carbon dioxide (CO_2). The name comes from its use as intermediates in creating synthetic natural gas (SNG) and for producing ammonia (NH_3) or methanol (CH_3OH). Syngas is usually a product of coal gasification and the main application is electricity generation. Syngas is combustible and can be used as a fuel of internal combustion engines.
System cost	The annualized costs of building and operating the energy system, including generation, transmission and distribution, balancing and carbon costs.
System operators	Manage the whole energy system and keep it in balance so that gas and electricity are available when needed.
Transesterification	Transesterification or alcoholysis is defined as the process in which nonedible oil is allowed to chemically react with an alcohol. A process of exchanging the organic functional group R" of an ester with the organic group R' of an alcohol. These reactions are often catalyzed by the addition of an acid or base catalyst. The reaction can also be accompanied with the help of other enzymes, particularly lipases. In this reaction, methanol (CH_3OH) and ethanol (C_2H_5OH) are the most commonly used alcohols because of their low cost and availability.
Transmission networks	National networks that transport gas and electricity long distances across Great Britain; the motorways of our energy network.
Threshold Limit Value	The amount of a contaminant to which a person can have repeated exposure for an eight-hour day without adverse effects.
Unabated (gas) generation	Electricity generation where carbon from burning natural gas is not captured and stored.
Wholesale costs	The amount of energy companies pays to buy gas and electricity.

References

1. Paris climate accord or Paris climate agreement, https://treaties.un.org/pages/ViewDetails.aspx?
2. Paris Agreement. United Nations Treaty Collection, https://treaties.un.org/pages/ViewDetails.aspx?, 8 July 2016
3. Article 3, Paris Agreement, 2015.
4. Paris climate accord marks shift toward low – carbon economy, Globe and Mail. Toronto, Canada, 14 December 2015, https://www.theglobeandmail.com/news/world/optimism-in-paris-as-final-draft-of-global-climate-deal-tabled-article27739122/), 14 December 2015.
5. Mark Kinver, COP21: What does the Paris climate agreement mean for me? https://www.bbc.com/news/science-environment-35092127, 14 December 2015.
6. European 20-20-20 Targets, RECS International, http://www.recs.org/glossary/european-20-20-20-targets, 13 April 2019.
7. Vidal, John, and Adam Vaughan, Paris climate agreement may signal end of fossil fuel era, 13 December 2015, https://www.theguardian.com/environment/2015/dec/13/paris-climate-agreement-signal-end-of-fossil-fuel-era.
8. New Paris climate agreement ratifications reaffirm necessity to divest and break free from fossil fuels, 350.org, https://350.org/press-release/paris-climate-agreement-ratifications-divest, 21 September 2016,.
9. B. K. Bhaskara Rao, Modern Petroleum Refining Processes, 6th ed., Oxford & IBH Publishing Co., Pvt. Ltd., 2018.
10. Schmidt, G. A., Ruedy, R., Miller, R. L., and A. A. Lacis, The attribution of the present – day total greenhouse effect, J. Geophys. Res., 115 (D20), pp. D20106, 2010.
11. Shindell, Drew T., An emission – based view of climate forcing by methane and tropospheric ozone, Geophysical Research Letters. 32 (4): L04803, Bibcode: 2005GeoRL. 32.4803S.
12. Coker, A. K., Petroleum Refining & Design Applications Handbook, Vol. 1, Scrivener-Wiley, 2018.
13. Ensys Energy. Supplemental Marine Fuel Availability Study. MARPOL Annex VI Global Sulphur Cap Supply – Demand Assessment, Final Report, https://globalmaritimehub.com/wp-content/uploads/attach_786.pdf, 15 July 2016.
14. Concawe Review, The importance of carbon capture and storage technology in European refineries, Vol. 27, No. 1, July 2018.
15. Scott Cato, M., Green Economics. London: Earthscan, pp 36-37, ISBN 978-1-84407-571-3, 2009.
16. Adams, W. M., The Future of Sustainability: Re-thinking Environment and Development in the Twenty-First Century. Report of the IUCN Renowned Thinkers Meeting, 29-31, January 2006.
17. United Nations, Report of the World Commission on Environment and Development, General Assembly Resolution 42/187, Dec., 11, 1987 (retrieved Oct. 31, 2007).
18. Allen, David T., and David R. Shonnard, Sustainable Engineering Concepts, Design, and Case Studies, Pearson Education, Inc., 2012.
19. Anastas, Paul T., and J. B. Zimmerman., "Design Through the 12 Principles of Green Engineering.", 2003. Environmental Science & Technology 37 (5): 94A – 101A.
20. CERES (Coalition for Environmentally Responsible Economics) Principles, www.lgc.org/ahwahnee/principles.html.
21. Mc.Calmont, J. P., Hastings, A., Mc.Namara, N. P., Richter, G. M., Robson, P., Donnison, I. S., and J. Clifton – Brown, Environmental costs and benefits of growing Miscanthus for bioenergy in the U.K., GCB Bioenergy, 9, pp 493, (2017), https://doi.org/10.1111/gcbb.12294.
22. Whitaker, J., Field, J. L., Bernacchi, C. J., Cerri, C. E., Ceulemans, R., Davies, C. A., DeLucia, E. H., Donnison, I., S., McCalmont, J., P., Paustian, K., Rowe, R. L., Smith, P., Thornley, P., and N. P. McNamara., Consensus, uncertainties and challenges for perennial bioenergy crops and land use., GCB Bioenergy, 10: 150-164, 2018., https://doi.org/10.1111/gcbb.12488.
23. Milner, S., Holland, R. A., Lovett, A., Sunnenberg, G., Hastings, A., Smith, P., Wang, S., and G. Taylor, Potential impacts on ecosystem services of land use transitions to second-generation bioenergy crops in GB. GCB Bioenergy, 8:317-333, 2016, https://doi.org/10.1111/gcbb.12263.
24. Indra Deo Mall, Petroleum Refining Technology, CBS Publishers & Distributors, Pvt. Ltd., 2015.
25. Casone, R., Biofuels: What is Beyond Ethanol and Biodiesel? Hydrocarbon Processing, p. 95, September 2005.
26. DePlan, A., Integrating Biofuels in Refinery Optimization Models. Hydrocarbon Processing, p. 337, September 2008.
27. Pramanik, T. and S. Tripathi, Biodiesel: Clean Fuel of the Future. Hydrocarbon Processing, p. 50, Feb. 2005.
28. Crocker, Mark, *et al.*, CO_2 Recycling Using Microalgae for the Production of Fuels, Applied Petrochemical Research, 4:41-53, 21 March 2015. https://doi.org/10.1007%2Fs13203-014-0052-3.
29. Lercher, Johannes A, Bruch, Thomas; Zhao Chen, Catalytic deoxygenation of microalgae oil to green hydrocarbon – Green Chemistry (RSC Publishing), Green Chemistry, 15 (7): 1720 – 1739, 21 June 2013, https://doi.org/10.1039%2FC3GC40558C.

30. Zhou, Lin., Evaluation of Presulfided NiMon/y-Al_2O_3 for hydrodeoxygenation of Microalgae Oil to Produce Green Diesel, Energy & Fuels, 29: 262-272, 2015, https://doi.org/10.1021%2Fef502258q.
31. Zhou, Lin., Hydrodeoxygenation of microalgae oil to green diesel over Pt, Rh and presulfided NiMo catalysts., Catalysis Science & Technology, 6 (5): 1442 – 1454, (2016).
32. Gerpen, J. V., Biodiesel Processing and Production. Fuel Processing Technology, Vol. 86, p. 1097, 2005.
33. Ray, A., Latest Technology Developments in Biofuels. Lovraj Kumar Memorial Lecture Annual Workshop, 2009. Managing Carbon Footprints in the Process Industry, New Delhi, Nov. 26 – 27, 2009.
34. Sehagal, J. M., Emergence of Ethanol as Global Alternative to Gasoline, Chemical Weekly, p. 193, September 26, 2006.
35. Singh, M. P., Tuli, D. K., Malhotra, R. K., and A. Kumar, Ethanol from Lignolcellulosic Biomass: Prospects and Challenges. Journal of the Petrotech Society, p. 39, June 2008.
36. Chisti, Y, Biodiesel from microalage, Biotechnology Advances, 25 (3): 294 – 306, 2007.
37. Stephens, E., Ross, I. L, Mussgnug, J. H., Wagner, L. D., Borowitzka, M. A., Posten, C., Kruse, O., and B. Hankamer, Future prospects of microalgal biofuel production systems, Trends in Plant Science, 15 (10): 554 – 564, October 2010.
38. Organization of Petroleum Exporting Countries (OPEC): Basket Prices, http://www.opec.org/opec_web/en/data_graphs/40.htm?, (accessed 01/29, 2013).
39. Ghasemi, Y, Rasoul-Amini, S., Naseri, A. T., Montazeri-Najafabady, N., Mobasher, M. A., and F. Dabbagh, Microalgae biofuel potentials – Review, Applied Biochemistry and Microbiology, 48 (2):126 – 144, 2012.
40. Pienkos, P. T., Darzins, A., The promise and challenges of microalgal – derived biofuels, Biofuels, Bioproducts and Biorefining, 3 (4): 431 – 440, 2009. https://doi.org/10.1002%2Fbbb.159.
41. Voegele, Erin, Propel, Solazyme make algae biofuel available to the public, Biomass Magazine, 15 November 2012.
42. Herndon, Andrew, Tesoro is first customer for Sapphire's algae – derived crude oil., https://www.bloomberg.com/new/2103-03-20/tesoro-is-first-customer-for -sapphire-s-algae-derived-crude-oil.html.
43. Wesoff, Eric., Hard lessons from the Great Algae Biofuel., https://www.greentechmedia.com/articles/read/lessonss-from-the-great-algae-biofuel-bubble, 19 April 2017.
44. Could seaweed be the fuel of the future? https://www.euronews.com/2020/02/17/could-seaweed-be-the-fuel-of-the-future.
45. Oyenekan, Babatunde, and Gary T. Rochelle., Alternative Stripper Configurations for CO_2 Capture by Aqueous Amines, AIChE Journ. 53 (12): 3144-154, 2007, https://doi.org/10.1002%2Faic.11316.
46. Arthur Kohl, Richard Nielson, Gas Purification, 5th ed., Gulf Publishing, 1997.
47. Sulfur production report. http://minerals.usgs.gov/minerals/pubs/commodity/sulfur/sulfumcs06.pdf
48. Engineering Data Book, Gas Processing Suppliers Association, Vol. 2, 12th ed., Section 21 Hydrocarbon Treating, Tulsa, OK, 2004.
49. Jones, J. H., Froning, J. R., and E. E. Claytor, Jr., Chemical Engineering Data, p. 85, 4 January 1959.
50. Lee, J. I., Otto, F. D., and A. E. Mather, Gas Processing / Canada, p. 26, March – April 1973.
51. Dingman, J. C., Jackson, J. L., Moore, T. F., and J. A. Branson, Equilibrium Data for the H_2S – CO_2 Diglycolamine Agent – Water System, 62nd Gas Processors Association Annual Meetings, San Francisco, March 14 – 16, 1983.
52. Jou, F. – Y., Otto, F. D., and A. E. Mather, Solubility of H_2S and CO_2 in MDEA Solutions, AIChE annual meeting, 1981.
53. Jou, F. – Y., Otto, F. D., and A. E. Mather, Solubility of mixtures of H_2S and CO_2 in MDEA Solutions, AIChE annual meeting, 1986.
54. Kent, R. L. and Eisenberg, Hydrocarbon Processing, p. 87, February 1976.
55. Lallemand, F., and A. Minkkinen, Highly Sour Gas Processing in an Ever – Greener World, presented at the 80th Annual, GPA convention in San Antonio, Texas, March 12–14, 2001.
56. Sterner, Anthony, J., Acid Gas Fractionation, Laurance Reid Gas Conditioning Conference 2001, p. 17, February 25 – 28, 2001.
57. Darymple, D. A., and G. Srinivas, G., CrytalSulf Liquid Redox and TDA Gas Phase H2S Conversion Technologies for Sour Gas Treating, GPA 78th Annual Convention, Nashville, Tennessee, March 1-3, 1999.
58. McIntush, K. E., et al., Status of the First Commercial Applications of the CrystaSulf, 81st GPA Annual Convention, Dallas, Texas, March 2002.
59. Mitariten, M., Dolan, W., and A. Malio, Innovative Molecular Gate Systems for Nitrogen Rejection Carbon Dioxide Removal and NGL Recovery, 81st GPA Annual Convention, Dallas, Texas, March 2002.
60. Stuksrud, D. B., and H. Dannstrom, Membrane Gas – Liquid Contactor for Natural Gas Sweetening, GPA Europe Spring Meeting, Aberdeen, Scotland, May 17-19, 2000.
61. Gall, G. H., and E. S. Sanders, Removal of Carbon Dioxide form Liquid Ethane Using Membrane Technology, 81st GPA Annual Convention, Dallas, Texas, March 2002.
62. Fisher, K. S., Killion, S. J., and J. E. Lundeen, GRI Filed Testing of Direct-Injection H2S Scavenging, presented at the 78th Annual GPA convention in Nashville, Tennessee, March 1 – 3, 1999.

63. Knudsen, B. and A. F. Mo, Use H_2S Scavenger for Onshore Applications, GPA Europe annual conference, Rome, Italy, September 25–27, 2002.
64. Haut, R. D., Denton, R. D., and E. R. Thomas, Development and Application of the Controlled – Freeze – Zone – Process, SPE Production Engineering, p. 265, August 1989.
65. Folger, P., Carbon Capture: a Technology Assessment, Congressional Research Service Report for Congress, 5: 26-44, 2009.
66. Wu, Ying, John J. Carroll, Carbon Dioxide Sequestration and Related Technologies, John Wiley & Sons, pp. 128–131, July 2011.
67. Peter Reynolds, The Sustainability Future for Energy and Chemicals, ARC Strategies, September 2020. ARC Advisory Group. Three Allied Drive, Dedham, MA 02026 USA, www.arcweb.com.
68. IChemE position on Climate Change, IChemE Advancing Chemical Engineering Worldwide, November 2020.
69. Martin Menachery, The net-zero ambition, Refining & Petrochemicals, 6 October 2020.
70. Grejtak, Tim, van Berkel, Arij, and Yuan-Sheng Yu, 'Evolution of Energy Networks: Decarbonizing the Global Energy Trade', Lux Research, 01 April 2020, www.luxresearchinc.com
71. Kalpana Gupta, Ishita Aggarwal and Maruthi Ethakota, digital refining, PTQ Q4, pp 89–96, 2020, www.digitalrefining.com/article/1002563.
72. Martin Menachery, Evonik, Siemens Energy put pilot plant using CO2 and hydrogen as raw materials for sustainable chemicals into operations. Artificial photosynthesis closes carbon dioxide cycle. Refining & Petrochemicals, 23 September 2020.
73. The Energy White Paper – Powering our Net Zero Future, Presented to Parliament by the Secretary of State for Business, Energy and Industrial Strategy by Command of Her Majesty, Queen's Printer and Controller of HMSO, December 2020.
74. National Carbon Capture Center, 10 Years of Technology Development – Advancing Fossil Energy Technology Solutions, Wilsonville, Alabama. December 2018_5849.
75. E. Carter, and A. Hickman, Ready – now blue Hydrogen leads the way to decarbonization, H2-Tech.com, Q2, 2021.
76. Jakobsen, D. and V. Atland, Concepts for large scale hydrogen production, Master's thesis, Norwegian University of Science and Technology, 2016, Online: https://ntnuopen.ntnu.no/ntnu-xmlui/handle/11250/2402554.
77. K. Liu, C. Song and V. Subramani, Hydrogen and Syngas Production and Purification Technologies, Wiley, 2010.
78. ThyssenKrup Industrial Solutions, https://www.thyssenkrupp-industrial-solutions.com
79. Simon Flowers, Oil refining's for big challenges, Profitability, rationalisation, decarbonisation and EVs., The Edge, A Verisk Business, Wood Mackenzie, 02, July, 2021.
80. Simon Flowers, Stemming the tide of plastic waste – And how increased recycling will cut oil demand., The Edge, A Verisk Business, Wood Mackenzie, 18 September, 2020.
81. Alan Gelder, Petrochemical Integration defines long term downstream winners and losers, February, 2021, Wood Mackenzie, www.woodmac.com
82. Marcio Wagner da Silva, How the crude to chemicals refineries can change the downstream industry, August 2021.
83. U.S. Department of Energy Alternative Fuels Data Center, "Renewable hydrocarbon biofuels," online https://afdc.energy.gov/fuels/emerging_hydrocarbon.html
84. S&P Global Platts, "Evolve or die: U.S. refiners grasp renewables lifeline to stay viable," December 2020, online: https://www.spglobal.com/platts/en/market-insights/latest-new/oil/110420-evolve-or-die-us-refiners-graps-renewables-lifeline-to-stay-viable.
85. Preston, W. E., McColl, Y., and D. Schnittker, Renewable diesel: The latest buzzword in the downstream sector, Hydrocarbon Processing, pp 47, April 2021.
86. Alan H. Zacher, Mariefel V. Olarte, Daniel M. Santosa, Douglas C. Elliott and Susanne B. Jones, A review and perspective of recent bio-oil hydrotreating research, Green Chem., 16, 491-5152014.
87. Wang, G.Q., Xu, Z.C. Yu, Y.L. and Ji, J.B., Performance of rotating zigzag bed- new HiGee, Chemical Engineering and Processing, doi:10.106/j.cep. 2007.11.001, 2007.
88. Harvey, A. P., Biodiesel process intensification projects at Newcastle University. Proc. PIN Meeting, 16 November, www.pinetwork.org, 2006.
89. Mio Wang, Howard J. Herzog and Alan Hatton, Ind., Eng., Chem. Res. 59, 15, 7087-7096, 2020.
90. Xiaomei Wu, Huifeng Fan, Maimoona Sharif, Yunsong Yu, Keming Wei, Zaoxiao Zhang, and Guangxin Liu., Electrochemically-mediated amine regeneration of CO_2 capture: From electrochemical mechanism to bench-scale visualization study, Applied Energy 302 (2021) 117554, https://doi.org/10.1016/j.apenergy.2021.117554.
91. E. Carter and A. Hickman, Ready-now blue hydrogen leads the way to decarbonization, H2T Special Focus: Pathways for Sustainable Hydrogen, H_2 TECH, Q2 2021, H2-Tech.com

92. Hydrogen Council, Hydrogen decarbonization pathways: A life-cycle assessment, 2021, Online: https://hydrogencouncil.com/wp-content/uploads/2021/01/Hydrogen-Council-Report_Decarbonization-Pathways_Part-1-Lifecycle-Assessment.pdf
93. International Energy Agency, The future of hydrogen, 2019, Online: https://www.iea.org/reports/the-future-of-hydrogen.
94. Ray, A., Latest Technology Developments in Biofuels. Lovraj Kuman Memorial Lecture Annual Workshop, 2009. Managing Carbon Footprints in The Process Industry, New Delhi, Nov. 26-27, 2009.
95. Koskinen, M, Sourander, M, and M. Nurminen, Apply a Comprehensive Approach to Biofuels, Hydrocarbon Processing, p. 81, February 2006.
96. Koukoulas, A.A., Torrefaction: A Pathway Towards Fungible Biomass feedstocks, Advanced Bioeconomy Feedstocks conference, 2016.
97. Anthony, B., and P. Fennel, A Virtuous Circle: Chemical Looping has Solid Potential for Optimising Processes, The Chemical Engineer, p 18- 20, April 2023.

Bibliography

1. Karine Ballerat-Busserolles, Ying Wu and John J. Carroll, Cutting -Edge Technology for Carbon Capture Utilization and Storage, Scrivener-Wiley, 2018.
2. Hydrogen Council, Study Task Force, "Hydrogen scaling up: A sustainable pathway for the global energy transition," 2017, Online: https://hydrogencouncil.com/wp-content/uploads/2017/11/Hydrogen-scaling-up-Hydrogen-Council.pdf
3. Hydrogen Council, "Hydrogen decarbonization pathways: A life-cycle assessment," 2021, Online https://hydrogencouncil.com/wp-content/uploads/2021/01/Hydrogen-Council-Report_Decarbonization-Pathways_Part1-Lifecycle-Assessment.pdf
4. International Energy Agency, "The future of hydrogen," 2019, Online: https://www.iea.org/reports/the-future-of-hydrogen.
5. International Renewable Energy Agency (IRENA), "Hydrogen from renewable power: Technology outlook for the energy transition," September 2018., Online: https://www.irena.org/-/media/Files/IRENA/Agency/Publication/2018/Sep/IRENA_Hydrogen_from_renewable_power_2018.pdf
6. International Energy Agency, "Hydrogen," 2021, Online: https://www.iea.org/fuels-and-technologies/hydrogen
7. IEA Greenhouse Gas R & D Programme (IEAGHG), "Reference data and supporting literature reviews for SMR based hydrogen production with CCS", IEAGHG Technical Review 2017-TR3, 2017, Online: https//ieaghg.org/publications/technical-reports/reports-list/10-technical   reviews/778-2017-tr3-reference-data-supporting-literature-reviews-for-smr-based-hydrogen-production-with-ccs.

Appendix D

D-1 Process Flow Diagrams Using VISIO 2002 Software

Figure D-1. Process Flow diagram (Feed & Fuel Desulfurization Section).
Figure D-2. Typical process flow diagram for the production of Methyl Tertiary Butyl Ether (MTBE).
Figure D-3. Piping & Instrumentation diagram for Ammonia plant CO_2 removal.
Figure D-4. Piping & Instrumentation diagram (Ammonia synthesis and refrigeration unit).

D-2 Process Data Sheets

1. Air cooled heat exchanger process data sheet
2. Centrifugal pump schedule: driver
3. Centrifugal pump schedule: pump
4. Centrifugal pump summary
5. Column schedule
6. Construction Commissioning Start-up Checklist
7. Deaerator process data sheet: Deaerator water storage tank
8. Deaerator process data sheet: Deaerator head
9. Drum process data sheet
10. Effluent schedule
11. Equilibrium flash calculation
12. Fabricated equipment schedule
13. Fan/Compressor process duty specification
14. Fractionator calculation summary
15. General services and utilities checklist
16. Hazardous chemical and conditions schedule
17. Heat and mass balances
18. Heat exchanger rating sheet
19. Hydrocarbon dew point calculation
20. Line list schedule
21. Line schedule
22. Line schedule sheet
23. Line summary table
24. Mass balance
25. Mechanical equipment schedule
26. Pipe line list
27. Pipe list
28. Piping process conditions summary
29. Plate heat exchanger data sheet
30. Calculation of pressure drop in fixed catalyst beds
31. Process engineering job analysis summary
32. Pump calculation sheet
33. Pump schedule
34. Relief device philosophy sheet
35. Tank and vessel agitator data sheet
36. Tank process data sheet
37. Tank schedule
38. Tie – in – schedule
39. Tower process data sheet
40. Tray loading summary
41. Trip schedule
42. Utility summary sheet
43. Vessel and tank schedule
44. Vessel and tank summary: driver
45. Vessel schedule
46. Water analysis sheet

734 APPENDIX D

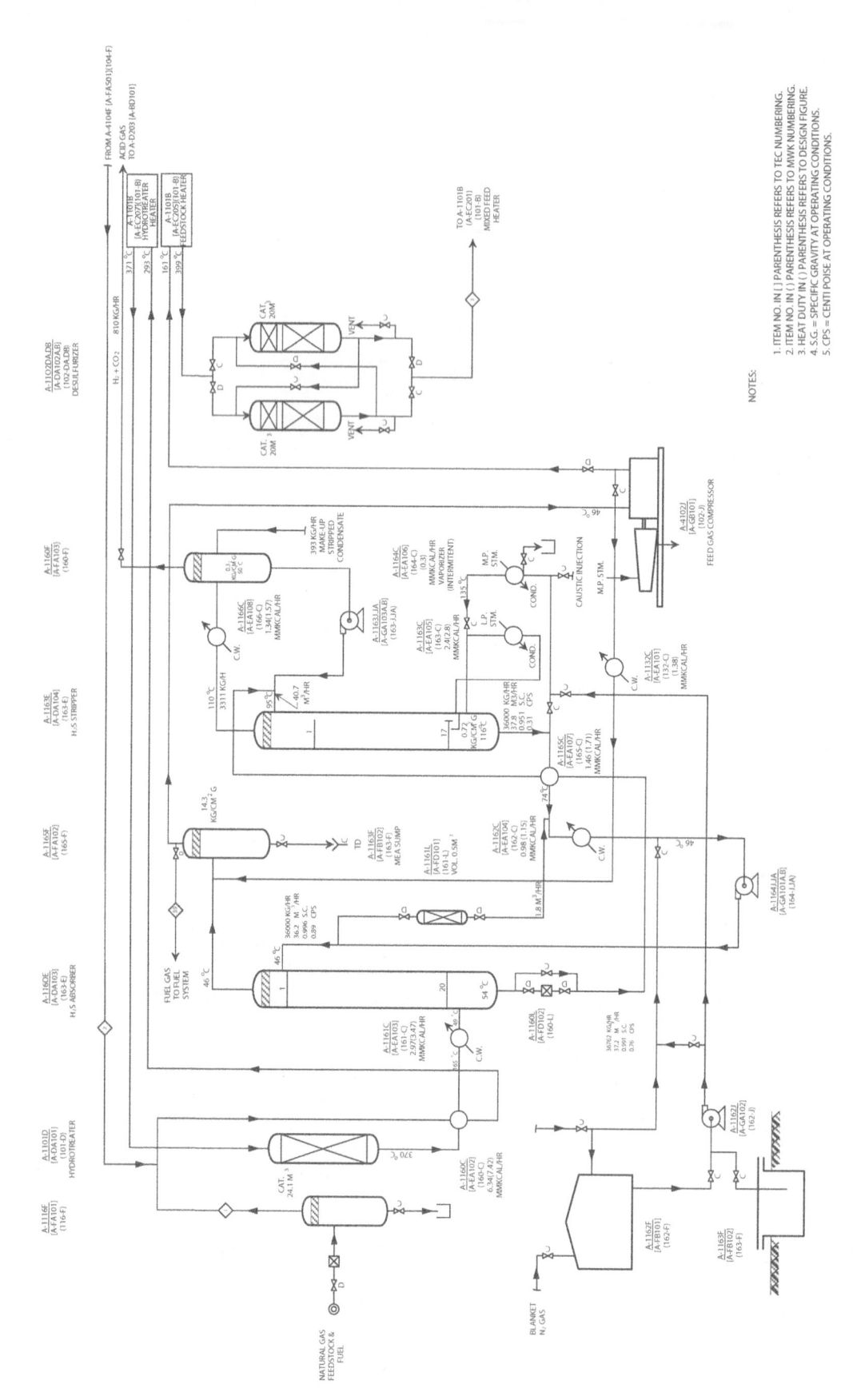

Figure D-1. Process Flow diagram (Feed & Fuel Desulfurization Section).

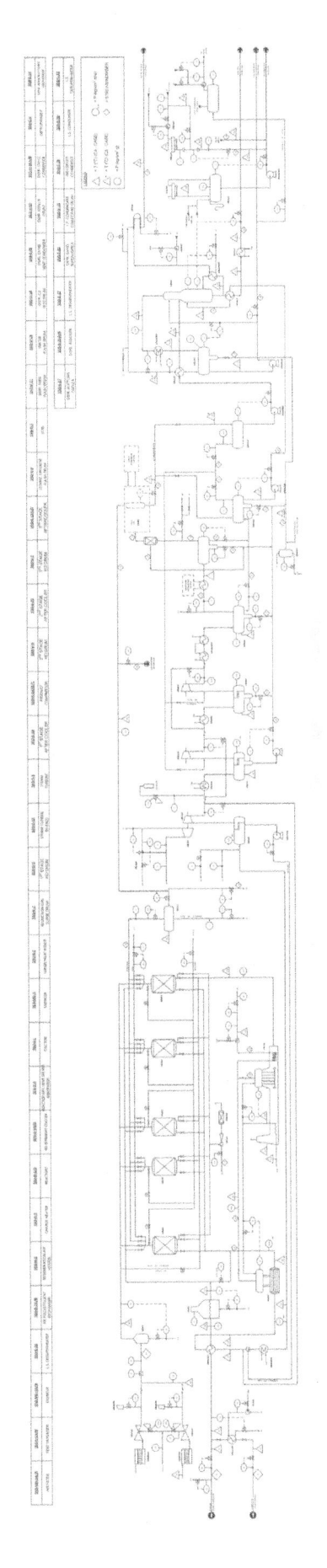

Figure D-2. Typical process flow diagram for the production of Methyl Tertiary Butyl Ether (MTBE).

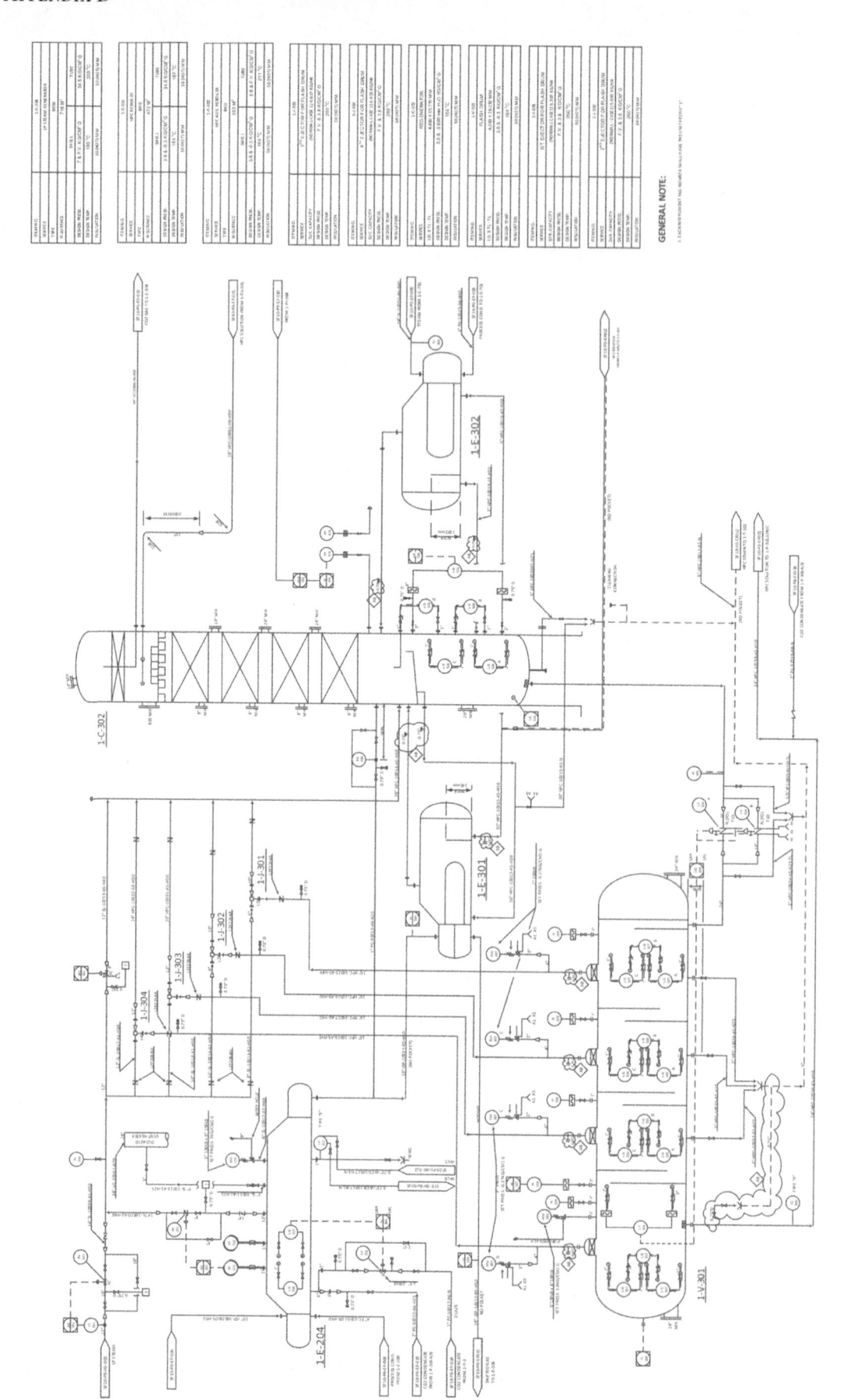

Figure D-3. Piping & Instrumentation diagram for Ammonia plant CO_2 removal.

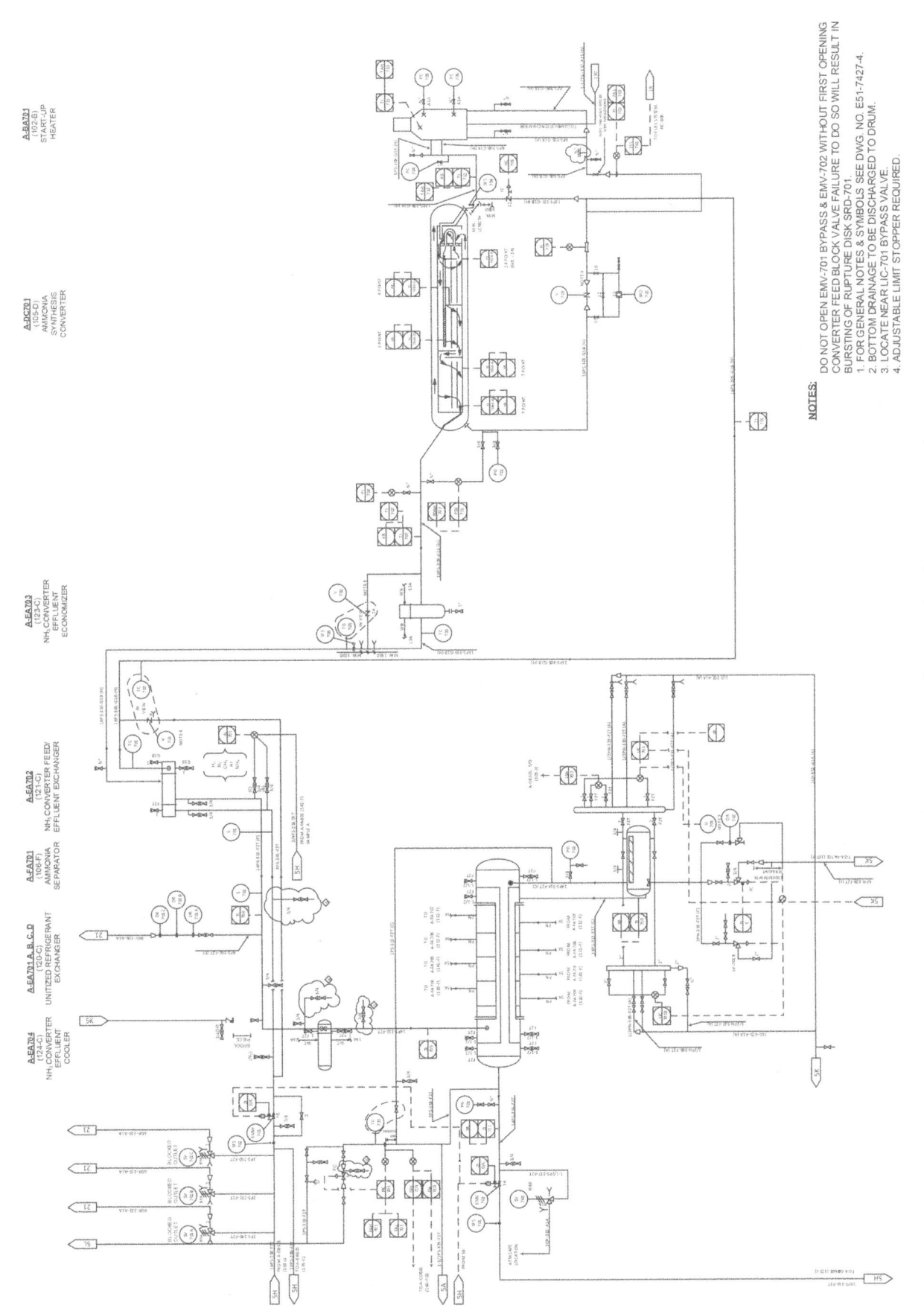

Figure D-4. Piping & Instrumentation flow diagram (Ammonia synthesis and refrigeration unit).

AIR COOLED HEAT EXCHANGER PROCESS DATA SHEET

Document No.
Sheet of   Rev.
Job
Item Name
Item No.(s)
No. Working   Total No. Off

Unless otherwise stated, fluid properties are for mean fluid temperature. For lines 30-36 this is mean from dewpoint to outlet temperatures.

#	TUBE SIDE DATA	UNITS			MATERIALS	
1	TUBE SIDE DATA	UNITS			MATERIALS	
2	Fluid Circulated				Tubes	
3	Total Fluid Entering (Normal)	kg/h	lb/h		Fins	
4	Flow Margin	%	%		Header Boxes	
5	Temperature: In/Out	°C	°F		Tubesheets	
6	Max. Pressure Drop at Line 4	bar	psi		Stress Relief Yes/as Codes	
7	Inlet Pressure: Operat./Design	barg	psig		Radiography Yes/as Codes	
8	Normal Heat Load	kW	Btu/h		Sour Service Yes/No	
9	Heat Load Margin	%	%			
10	Design Temperature	°C	°F			
11	Corrosion Allowance	mm	in.			
12	Fouling Resistance	$m^2 °C/W$	$°Fft^2h/Btu$			
13	Line N.B.: In/Out	mm.	in.			
14	GAS (AND VAPOR)				PROCESS CONTROL REQUIREMENTS	
15	Flow of Vapor & Gas at Inlet	kg/h	lb/h		Local / Remote Set	
16	Molecular Weight: In/Out				Hand / Automatic	
17	Thermal Conductivity: In/Out	W/m °C	Btu/h.ft.°F		Adjustable Pitch Fans	
18	Specific Heat	kJ/kg °C	Btu/lb.°F		Variable Speed Motor	
19	Compressibility Factor				Louvres	
20	Viscosity In / Out	cP	lb/ft.h		Air Recirculation	
21	LIQUID IN					
22	Total Flow of Liquid at Inlet	kg/h	lb/h			
23	Thermal Conductivity: In/Out	W/m °C	Btu/h.ft.°F			
24	Specific Heat	kJ/kg °C	Btu/lb.°F			
25	Density	kg/m^3	lb/ft^3			
26	Viscosity	cP	lb/ft.h			
27	CONDENSATION					
28	Fluid Condensed	kg/h	lb/h			
29	Molecular Weight					
30	Thermal Conductivity	W/m °C	Btu/h.ft.°F			
31	Specific Heat	kJ/kg °C	Btu/lb.°F			
32	Density	kg/m^3	lb/ft^3			
33	Viscosity	cP	lb/ft.h			
34	Latent Heat	kJ/kg	Btu/lb			
35	%Condensate Forming Film	%	%			
36	Temps. for Phase Change: In/Out	°C	°F			
37	COOLING CURVE DATA					
38	%Heat Load					
39	Condensing Fluid Temp. °C					
40						
41	AIR SIDE DATA					
42	Design Inlet Temp. Max/Min	°C	°F			
43	No. of Fans Assumed					
44	Estimated Power per Fan	kW				
45	SITE DATA					
46	Altitude					
47	Plot Size Limitations	Length:			Width:	
48	Noise Limits					
49	Atmospheric Contamination					
50						
51	NOTES					
52						
53						
54						
55						

	1	Date	2	Date	3	Date	4	Date	5	Date
Description										
Made/Revised by										
Checked by										
Approved Process										
Approved by										

CENTRIFUGAL PUMP SCHEDULE							Sheet	of		Issue No.:		Date	1		Date	2		Date	3		Date	
				DRIVER						Description												
Job No.:										Made/Revised by												
Job Name:										Checked by												
Document No.										Approved- Process												
Project No.										Approved												
Client																						
Location											STEAM CONDITIONS											
Plant				ELECTRICAL	NEMA						PSIG	TEMP. °F	MAKE		MODEL		STEAM RATE		WEIGHT		P.O. No.	
ITEM No.	TYPE	H.P.	RPM	ROTATION CW-CCW	CHARACTERISTICS	FRAME																

Centrifugal pump schedule

Centrifugal pump schedule

Centrifugal pump summary

COLUMN SCHEDULE

		Units	Top	Bottom	Top	Bottom	Top	Bottom	Top	Bottom
1	Item Number.									
2	Item Name									
3	Number Required									
4										
5	Shell Diameter (I.D. /O.D)	mm								
6	Shell Length (T.L. – T.L.)	mm								
7	Overall Shell Length - (T.L. - T.L.)	mm								
8	Base Elevation	mm								
9	Trays - Type									
10	– Number									
11	– Spacing	mm								
12	– Liquid Flow									
13	Packing - Type									
14	– Number of Beds									
15	– Bed Height / Volume	mm/m³								
16	– Density	kg/m³								
17	Other Internals									
18										
19										
20										
21	Operating - Pressure									
22	– Temperature									
23	Design - Pressure	bar g								
24	– Temperature	°C								
25	Vacuum Design									
26	Material: Shell									
27	Liner									
28	Internals									
29										
30	Shell Corrosion Allowance	mm								
31	Sour Service									
32	Stress Relieved for Process Reasons									
33	Insulation									
34										
35										
36	Notes									

Sheet of

Issue No.:	1		2		3	
Description						
Made/Revised by	Date		Date		Date	
Checked by						
Approved- Process						
Approved						

Overview Construction - Precommissioining - Commisssioining - Start-up

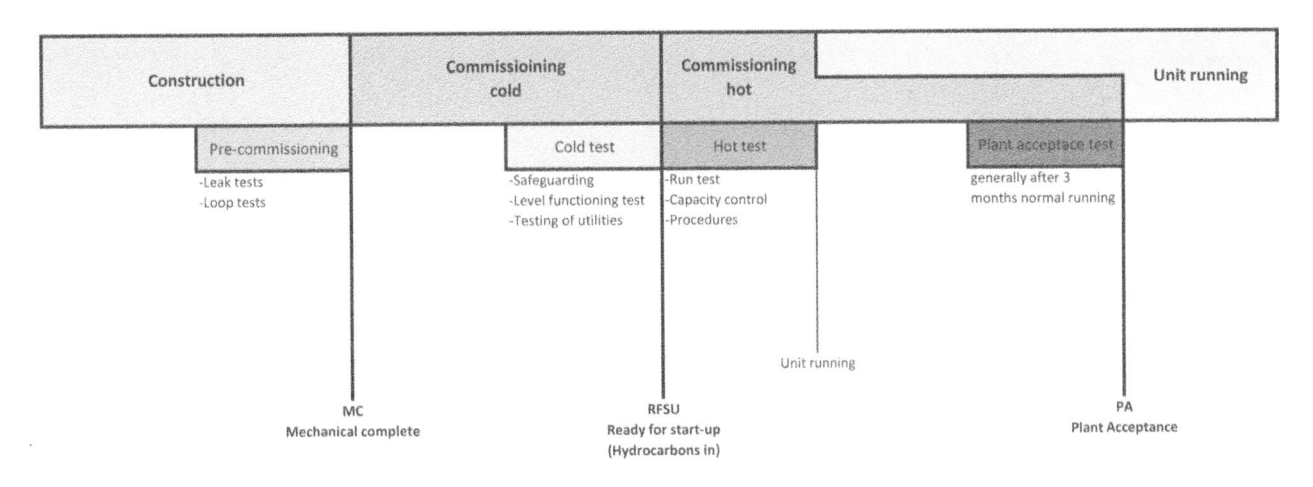

Construction	Commissioining cold	Commissioning hot		Unit running
Pre-commissioning	Cold test	Hot test	Plant acceptace test	
-Leak tests -Loop tests	-Safeguarding -Level functioning test -Testing of utilities	-Run test -Capacity control -Procedures	generally after 3 months normal running	

Unit running

- MC — Mechanical complete
- RFSU — Ready for start-up (Hydrocarbons in)
- PA — Plant Acceptance

References

Gen : Installation of Rotating equipment (amendments/supplements to API RP 686)
Reference to installation of rotating equipment
* foundation
* grouting
* flange alignment
* shaft alignment
* rotation check
* oil flushing
* strainers
Reference to commissioning checklist for rotating equipment in general

Gen : Pumps - Type selection and procurement procedure
Reference to selection of pumps including
* data check
* scope vendor
* scope contractor

Gen : Compressors - Selection, testing and installation
Reference to selection of compressors including
* data check
* scope vendor
* scope contractor
* strainers
* air inlet filters

Gen : Centrifugal pumps (amendments/supplements to ISO 13709:2003)
Reference to design of pumps including
* vibration limits
* baseplate design
* lubrication
* nameplate check
* rotation check

Gen : Reciprocating positive displacement pumps and metering pumps (amendments/supplements to API 674 and API 675)
Reference to design of pumps including
* pump dynamics
* baseplate design
* lubrication
* nameplate check
* piston
* crackcase
* dampener

Gen : Axial, centrifugal, and expander compressors (amendments/supplements to API Std 617)
Reference to design of reciprocating compressors including
* vibration limits
* baseplate design
* lubrication
* nameplate check
* seals
* rotation check

Construction Commissioning Start-up Checklist

Gen : Reciprocating compressors (amendments/supplements to API 618) *Reference to design of reciprocating compressors including* * dynamics * baseplate design * lubrication * nameplate check * valves * piston * crackcase * pulsation dampeners
Gen : Rotary-type positive displacement compressors (amendments/supplements to API 619) *Reference to design of positive displacement compressors including* * dynamics * baseplate design * lubrication * nameplate check
Gen : Centrifugal fans (amendments/supplements to ISO 13705, Annex E) *Reference to design of centrifugal fans* * dynamics * baseplate design * lubrication * nameplate check
#NAME?
Gen : Field inspection prior to commissioning of mechanical equipment *Reference to MC checklist for mechanical equipment, for each piece of equipment specific* * rotating equipment * centrifugal compressors * reciprocating compressors * rotary positive displacements pumps * blowers/fans * centrifugal pumps * reciprocating pumps * rotary positive displacement pumps * mixers
Gen : Preservation of new and old equipment standing idle *Reference to preservation of rotating equipment* * preservation oils * cleaniness

APPENDIX D 745

Centrifugal pumps - Construction and Pre-commissioning checklist

General information	
Requisition number	
Plant	
Unit	

Equipment information	Tag, manufacturer, type
Pump	
Driver	
Coupling	
Gearbox	
Mechanical seal	
Seal liquid system	
Lubrication system	

Construction - General checks	Remarks	Signed	Signed	Signed	Date
1 Check foundation	Roughen, oil and dust free				
2 Check equipment assembly on foundation	25-50 mm between concrete and base plate, pre-align				
3 Check leveling of base plate					
4 Check installation of foundation bolts					
5 Check equipment for transport damage					
6 Check unit completeness					
7 Check free movement of rotor assembly	E-motor de-energised				

Construction - Checks after grouting	Remarks	Signed	Signed	Signed	Date
8 Check proper grouting is applied	Grout type, sealed anchor bolts, humidity, temperature				
9 Check whether foundations bolts are tightened					
10 Check correct bolts and gaskets are used					
11 Check correct installation coupling					
12 Check coupling guard and mounting					
13 Check cooling water system					
14 Check lubrication system					
15 Check bearings and bearing house are clean					
16 Check oil rings free movement					
17 Check oiler for location and level					
18 Check packing rings in stuffing box					
19 Check mechanical seal					
20 Check seal locks are removed					
21 Check seal liquid system	see installation notes manufacturer				
22 Check equipment for soft feet					
23 Shaft alignment	only metal shims are allowed				
24 Check piping and support attached to pump	Flange deviation				
25 Vents, drain correctly installed					
26 Check earth bosses					
27 Tracing installed correctly					
28 Insulation installed correctly					
29 Check documentation	shaft alignment, foundation measurements, nameplate				
30 P&ID / PEFS check					

Construction - Presommissioning	Remarks	Signed	Signed	Signed	Date
31 Check direction of rotation	uncoupled or avoid mechanical seal runs dry (OH5)				
32 Check N2 pre-charge					
33 Check barrier oil fill					
34 Check lubricant and lubricant levels					

Preservation	Remarks	Signed	Signed	Signed	Date
35 Check equipment for internal cleanliness					
36 Check proper preservation	nozzles blinded off, isolated during flushing				

Other items	Remarks
Punch item list attached	

	Name	Department	Initial	Signed	Date
Engineering contractor					
Rotating Equipment engineer					
Supervisor unit/plant					

Reference should be made to the latest version of the following publications: DEP 31.29.00.10-Gen, DEP 31.29.02.11-Gen, DEP 31.29.02.30-Gen, DEP 31.38.01.11-Gen, DEP 61.10.08.11-Gen, DEP 70.10.70.11-Gen

Centrifugal pumps - Commissioning and Start-up checklist

General information	
Requisition number	
Plant	
Unit	

Appendix D

Equipment information	Tag, manufacturer, type
Pump	
Driver	
Coupling	
Gearbox	
Mechanical seal	
Seal liquid system	
Lubrication system	

Commissioning - Checks before initial run	Remarks	Signed	Signed	Signed	Date
1 Check start-up strainers					
2 Check trip and alarm settings					
3 Check instrumentation and safety devices					

Start-up - Initial run	Remarks	Signed	Signed	Signed	Date
4 Check vibration levels					
5 Check bearing temperatures					
6 Check power consumption					
7 Check mechanical performance					
8 Check seal liquid system temperatures					
9 Check unit for leakages					
10 Remove start-up strainers					
11 Make hot check coupling alignment					

	Name	Department	Initial	Signed	Date
Engineering contractor					
Rotating Equipment engineer					
Supervisor unit/plant					

Reference should be made to the latest version of the following publications: DEP 31.29.00.10-Gen, DEP 31.29.02.11-Gen, DEP 31.29.02.30-Gen, DEP 31.38.01.11-Gen, DEP 61.10.08.11-Gen, DEP 70.10.70.11-Gen

References

DEP 31.29.00.10-Gen : Installation of Rotating equipment (amendments/supplements to API RP 686)
Reference to installation of rotating equipment
* foundation
* grouting
* flange alignment
* shaft alignment
* rotation check
* oil flushing
* strainers
Reference to commissioning checklist for rotating equipment in general

DEP 31.29.02.11-Gen : Pumps - Type selection and procurement procedure
Reference to selection of pumps including
* data check
* scope vendor
* scope contractor

DEP 31.29.02.30-Gen : Centrifugal pumps (amendments/supplements to ISO 13709:2003)
Reference to design of pumps including
* vibration limits
* baseplate design
* lubrication
* nameplate check
* rotation check

DEP 31.38.01.11-Gen : Piping - general requirements
Reference to piping requirements
* inlet piping design
* discharge piping design

DEP 61.10.08.11-Gen : Field inspection prior to commissioning of mechanical equipment
Reference to MC checklist for mechanical equipment, for each piece of equipment specific
* rotating equipment
* centrifugal compressors
* reciprocating compressors
* rotary positive displacements pumps
* blowers/fans
* centrifugal pumps
* reciprocating pumps
* rotary positive displacement pumps
* mixers

DEP 70.10.70.11-Gen : Preservation of new and old equipment standing idle
Reference to preservation of rotating equipment
* preservation oils
* cleaniness

Reciprocating pumps - Construction and Pre-commissioning checklist

General information	
Requisition number	
Plant	
Unit	

Equipment information	Tag, manufacturer, type
Pump	
Driver	
Coupling	
Gearbox	
Mechanical seal	
Seal liquid system	
Lubrication system	

	Construction - General checks	Remarks	Signed	Signed	Signed	Date
1	Check foundation	Roughen, oil and dust free				
2	Check equipment assembly on foundation	25-50 mm between concrete and base plate, pre-align				
3	Check leveling of base plate					
4	Check installation of foundation bolts					
5	Check equipment for transport damage					
6	Check unit completeness					
7	Check free movement of piston/ motor rotor	E-motor de-energised				

	Construction - Checks after grouting	Remarks	Signed	Signed	Signed	Date
8	Check proper grouting is applied	Grout type, sealed anchor bolts, humidity, temperature				
9	Check whether foundations bolts are tightened					
10	Check correct bolts and gaskets are used					
11	Check crack case for cleanliness					
12	Check suction and discharge valve assembly					
13	Check tension of V-belts and pulley alignment					
14	Check valve gear					
15	Check pulsation dampeners					
16	Check correct installation coupling					
17	Check coupling guard and mounting					
18	Check cooling water system					
19	Check lubrication system					
20	Check bearings and bearing house are clean					
21	Check stuffing box materials					
22	Check seal					
23	Check seal liquid system	see installation notes manufacturer				
24	Check equipment for soft feet					
25	Shaft alignment	only metal shims are allowed				
26	Check piping and support attached to pump	Flange deviation				
27	Vents, drain correctly installed					
28	Check earth bosses					
29	Tracing installed correctly					
30	Insulation installed correctly					
31	Check documentation	shaft alignment, foundation measurements, nameplate				
32	P&ID / PEFS check					

	Construction - Presommissioning	Remarks	Signed	Signed	Signed	Date
33	Check direction of rotation	uncoupled or avoid seal/bearing runs dry (OH5)				
34	Check N2 pre-charge					
35	Check barrier oil fill					
36	Check lubricant and lubricant levels					

	Preservation	Remarks	Signed	Signed	Signed	Date
37	Check equipment for internal cleanliness					
38	Check proper preservation	nozzles blinded off, isolated during flushing				

Other items	Remarks
Punch item list attached	

	Name	Department	Initial	Signed	Date
Engineering contractor					
Rotating Equipment engineer					
Supervisor unit/plant					

Reference should be made to the latest version of the following publications: DEP 31.29.00.10-Gen, DEP 31.29.02.11-Gen, DEP 31.29.12.30-Gen, DEP 31.38.01.11-Gen, DEP61.10.08.11-Gen, DEP 70.10.70.11-Gen

Reciprocating pumps - Commissioning and Start-up checklist

General information	
Requisition number	
Plant	
Unit	

Appendix D

Equipment information	Tag, manufacturer, type
Pump	
Driver	
Coupling	
Gearbox	
Mechanical seal	
Seal liquid system	
Lubrication system	

Commissioning - Checks before initial run	Remarks	Signed	Signed	Signed	Date
1 Check start-up strainers					
2 Check trip and alarm settings					
3 Check instrumentation and safety devices					

Start-up - Initial run	Remarks	Signed	Signed	Signed	Date
4 Check vibration levels					
5 Check bearing temperatures					
6 Check power consumption					
7 Check mechanical performance					
8 Check proper functioning pulsation dampeners					
9 Check seal or stuffing box temperatures					
10 Check unit for leakages					
11 Remove start-up strainers					
12 Make hot check coupling alignment					

	Name	Department	Initial	Signed	Date
Engineering contractor					
Rotating Equipment engineer					
Supervisor unit/plant					

Reference should be made to the latest version of the following publications: DEP 31.29.00.10-Gen, DEP 31.29.02.11-Gen, DEP 31.29.12.30-Gen, DEP 31.38.01.11-Gen, DEP61.10.08.11-Gen, DEP 70.10.70.11-Gen

References

DEP 31.29.00.10-Gen : Installation of Rotating equipment (amendments/supplements to API RP 686)
Reference to installation of rotating equipment
* foundation
* grouting
* flange alignment
* shaft alignment
* rotation check
* oil flushing
* strainers
Reference to commissioning checklist for rotating equipment in general

DEP 31.29.02.11-Gen : Pumps - Type selection and procurement procedure
Reference to selection of pumps including
* data check
* scope vendor
* scope contractor

DEP 31.29.12.30-Gen : Reciprocating positive displacement pumps and metering pumps (amendments/supplements to API 674 and API 675)
Reference to design of pumps including
* pump dynamics
* baseplate design
* lubrication
* nameplate check
* piston
* crackcase
* dampener

DEP 31.38.01.11-Gen : Piping - general requirements
Reference to piping requirements
* inlet piping design
* discharge piping design

DEP 61.10.08.11-Gen : Field inspection prior to commissioning of mechanical equipment
Reference to MC checklist for mechanical equipment, for each piece of equipment specific
* rotating equipment
* centrifugal compressors
* reciprocating compressors
* rotary positive displacements pumps
* blowers/fans
* centrifugal pumps
* reciprocating pumps
* rotary positive displacement pumps
* mixers

DEP 70.10.70.11-Gen : Preservation of new and old equipment standing idle
Reference to preservation of rotating equipment
* preservation oils
* cleaniness

Rotary positive displacement pumps - Construction and Pre-commissioning checklist

General information	
Requisition number	
Plant	
Unit	

Equipment information	Tag, manufacturer, type
Pump	
Driver	
Coupling	
Gearbox	
Mechanical seal	
Seal liquid system	
Lubrication system	

Construction - General checks	Remarks	Signed	Signed	Signed	Date
1 Check foundation	Roughen, oil and dust free				
2 Check equipment assembly on foundation	25-50 mm between concrete and base plate, pre-align				
3 Check leveling of base plate					
4 Check installation of foundation bolts					
5 Check equipment for transport damage					
6 Check unit completeness					
7 Check free movement of rotor assembly	E-motor de-energised				

Construction - Checks after grouting	Remarks	Signed	Signed	Signed	Date
8 Check proper grouting is applied	Grout type, sealed anchor bolts, humidity, temperature				
9 Check whether foundations bolts are tightened					
10 Check correct bolts and gaskets are used					
11 Check correct installation coupling					
12 Check coupling guard and mounting					
13 Check cooling water system					
14 Check lubrication system					
15 Check bearings and bearing house are clean					
16 Check oil rings free movement					
17 Check oiler for location and level					
18 Check packing rings in stuffing box					
19 Check mechanical seal					
20 Check cleanliness internals					
21 Check seal liquid system	see installation notes manufacturer				
22 Check equipment for soft feet					
23 Shaft alignment	only metal shims are allowed				
24 Check piping and support attached to pump	Flange deviation				
25 Vents, drain correctly installed					
26 Check earth bosses					
27 Tracing installed correctly					
28 Insulation installed correctly					
29 Check documentation	shaft alignment, foundation measurements, nameplate				
30 P&ID / PEFS check					

Construction - Presommissioning	Remarks	Signed	Signed	Signed	Date
31 Check direction of rotation	uncoupled or avoid dry run				
32 Check N2 pre-charge					
33 Check barrier oil fill					
34 Check lubricant and lubricant levels					

Preservation	Remarks	Signed	Signed	Signed	Date
35 Check equipment for internal cleanliness					
36 Check proper preservation	nozzles blinded off, isolated during flushing				

Other items	Remarks
Punch item list attached	

	Name	Department	Initial	Signed	Date
Engineering contractor					
Rotating Equipment engineer					
Supervisor unit/plant					

Rotary positive displacement pumps - Commissioning and Start-up checklist

General information	
Requisition number	
Plant	
Unit	

Equipment information	Tag, manufacturer, type
Pump	
Driver	
Coupling	
Gearbox	
Mechanical seal	
Seal liquid system	
Lubrication system	

Commissioning - Checks before initial run	Remarks	Signed	Signed	Signed	Date
1 Check start-up strainers					
2 Check trip and alarm settings					
3 Check instrumentation and safety devices					

Start-up - Initial run	Remarks	Signed	Signed	Signed	Date
4 Check vibration levels					
5 Check bearing temperatures					
6 Check power consumption					
7 Check mechanical performance					
8 Check seal liquid system temperatures					
9 Check liquid has sufficient lubricating properties					
10 Check unit for leakages					
11 Remove start-up strainers					
12 Make hot check coupling alignment					

	Name	Department	Initial	Signed	Date
Engineering contractor					
Rotating Equipment engineer					
Supervisor unit/plant					

References

Gen : Installation of Rotating equipment (amendments/supplements to API RP 686)
Reference to installation of rotating equipment
* foundation
* grouting
* flange alignment
* shaft alignment
* rotation check
* oil flushing
* strainers
Reference to commissioning checklist for rotating equipment in general
Gen : Pumps - Type selection and procurement procedure
Reference to selection of pumps including
* data check
* scope vendor
* scope contractor
Gen : Piping - general requirements
Reference to piping requirements
* inlet piping design
* discharge piping design
Gen : Field inspection prior to commissioning of mechanical equipment
Reference to MC checklist for mechanical equipment, for each piece of equipment specific
* rotating equipment
* centrifugal compressors
* reciprocating compressors
* rotary positive displacements pumps
* blowers/fans
* centrifugal pumps
* reciprocating pumps
* rotary positive displacement pumps
* mixers
Gen : Preservation of new and old equipment standing idle
Reference to preservation of rotating equipment
* preservation oils
* cleaniness

Centrifugal Compressors - Construction and Pre-commissioning checklist

General information	
Requisition number	
Plant	
Unit	

Equipment information	Tag, manufacturer, type
Pump	
Driver	
Coupling	
Gearbox	
Mechanical seal	
Seal liquid system	
Lubrication system	

Construction - General checks	Remarks	Signed	Signed	Signed	Date
1 Check foundation	Roughen, oil and dust free				
2 Check equipment assembly on foundation					
3 Check leveling of base plate					
4 Check installation of foundation bolts					
5 Check equipment for transport damage					
6 Check unit completeness					
7 Check free movement of rotor assembly	E-motor de-energised				

Construction - Checks after grouting	Remarks	Signed	Signed	Signed	Date
8 Check proper grouting is applied	Grout type, sealed anchor bolts, humidity, temperature				
9 Check whether foundations bolts are tightened					
10 Check correct bolts and gaskets are used					
11 Check correct installation coupling					
12 Check coupling guard and mounting					
13 Check cooling water system					
14 Check lubrication system					
15 Check cleanliness lube oil tank					
16 Check bearings and bearing house are clean					
17 Check oil rings free movement					
18 Check functioning sour oil traps					
19 Check casing can expand					
20 Check mechanical seal					
21 Check seal locks are removed					
22 Check seal liquid system	see installation notes manufacturer				
23 Check proper functioning barring gear					
24 Check axial displacement rotor					
25 Check equipment for soft feet					
26 Shaft alignment	only metal shims are allowed				
27 Check piping and support attached to pump	Flange deviation				
28 Vents, drain correctly installed					
29 Check earth bosses					
30 Tracing installed correctly					
31 Insulation installed correctly					
32 Check documentation	shaft alignment, foundation measurements, nameplate				
33 P&ID / PEFS check					

Construction - Presommissioning	Remarks	Signed	Signed	Signed	Date
34 Check direction of rotation					
35 Check N2 pre-charge					
36 Check barrier oil fill					
37 Check lubricant and lubricant levels					

Preservation	Remarks	Signed	Signed	Signed	Date
38 Check equipment for internal cleanliness					
39 Check proper preservation	nozzles blinded off, isolated during flushing				

Other items	Remarks
Punch item list attached	

	Name	Department	Initial	Signed	Date
Engineering contractor					
Rotating Equipment engineer					
Supervisor unit/plant					

Centrifugal Compressors - Commissioning and Start-up checklist

General information	
Requisition number	
Plant	
Unit	

Equipment information	Tag, manufacturer, type
Pump	
Driver	
Coupling	
Gearbox	
Mechanical seal	
Seal liquid system	
Lubrication system	

Commissioning - Checks before initial run	Remarks	Signed	Signed	Signed	Date
1 Check start-up strainers					
2 Check trip and alarm settings					
3 Check instrumentation and safety devices					
4 Check cut-in oil systems					

Start-up - Initial run	Remarks	Signed	Signed	Signed	Date
5 Check vibration levels					
6 Check bearing temperatures					
7 Check power consumption					
8 Check mechanical performance					
9 Check seal liquid system temperatures					
10 Check unit for leakages					
11 Remove start-up strainers					
12 Make hot check coupling alignment					

	Name	Department	Initial	Signed	Date
Engineering contractor					
Rotating Equipment engineer					
Supervisor unit/plant					

Gen : Installation of Rotating equipment (amendments/supplements to API RP 686)
Reference to installation of rotating equipment
* foundation
* grouting
* flange alignment
* shaft alignment
* rotation check
* oil flushing
* strainers
Reference to commissioning checklist for rotating equipment in general
Gen : Compressors - Selection, testing and installation
Reference to selection of compressors including
* data check
* scope vendor
* scope contractor
* strainers
* air inlet filters
Gen : Axial, centrifugal, and expander compressors (amendments/supplements to API Std 617)
Reference to design of reciprocating compressors including
* vibration limits
* baseplate design
* lubrication
* nameplate check
* seals
* rotation check
Gen : Piping - general requirements
Reference to piping requirements
* inlet piping design
* discharge piping design
Gen : Field inspection prior to commissioning of mechanical equipment
Reference to MC checklist for mechanical equipment, for each piece of equipment specific
* rotating equipment
* centrifugal compressors
* reciprocating compressors
* rotary positive displacements pumps
* blowers/fans
* centrifugal pumps
* reciprocating pumps
* rotary positive displacement pumps
* mixers
Gen : Preservation of new and old equipment standing idle
Reference to preservation of rotating equipment
* preservation oils
* cleaniness
Gen : Preservation of new and old equipment standing idle
Reference to preservation of rotating equipment
* preservation oils
* cleaniness

Reciprocating Compressors - Construction and Pre-commissioning checklist

General information	
Requisition number	
Plant	
Unit	

Equipment information	Tag, manufacturer, type
Pump	
Driver	
Coupling	
Gearbox	
Mechanical seal	
Seal liquid system	
Lubrication system	

Construction - General checks		Remarks	Signed	Signed	Signed	Date
1	Check foundation	Roughen, oil and dust free				
2	Check equipment assembly on foundation	25-50 mm between concrete and base plate, pre-align				
3	Check leveling of base plate					
4	Check installation of foundation bolts					
5	Check equipment for transport damage					
6	Check unit completeness					
7	Check alignment crosshead guides					

Construction - Checks after grouting		Remarks	Signed	Signed	Signed	Date
8	Check proper grouting is applied	Grout type, sealed anchor bolts, humidity, temperature				
9	Check whether foundations bolts are tightened					
10	Check correct bolts and gaskets are used					
11	Check crackshaft deflection					
12	Check suction and discharge valve assembly					
13	Check piston travel clearance					
14	Check pulsation dampeners					
15	Check funtioning clearance pockets					
16	Check tension and alignment of V-belt					
17	Check support flywheel					
18	Check cilinder mounting and alignment					
19	Check correct installation coupling					
20	Check coupling guard and mounting					
21	Check cooling water system					
22	Check lubrication system					
23	Inspect lube oil tank					
24	Inspect lubricator					
25	Check seal					
26	Check seal purch					
27	Check crackcase and crosshead cleanliness					
28	Check proper functioning barring gear					
29	Check equipment for soft feet					
30	Shaft alignment	only metal shims are allowed				
31	Check piping and support attached to pump	Flange deviation				
32	Vents, drain correctly installed					
33	Check earth bosses					
34	Tracing installed correctly					
35	Insulation installed correctly					
36	Check documentation	shaft alignment, foundation measurements, nameplate				
37	P&ID / PEFS check					

Construction - Presommissioning		Remarks	Signed	Signed	Signed	Date
38	Check direction of rotation	(jacking oil pump)				
39	Check N2 pre-charge					
40	Check barrier oil fill					
41	Check lubricant and lubricant levels					

Preservation		Remarks	Signed	Signed	Signed	Date
42	Check equipment for internal cleanliness					
43	Check proper preservation	nozzles blinded off, isolated during flushing				

Other items	Remarks
Punch item list attached	

	Name	Department	Initial	Signed	Date
Engineering contractor					
Rotating Equipment engineer					
Supervisor unit/plant					

Reciprocating Compressors - Commissioning and Start-up checklist

General information	
Requisition number	
Plant	
Unit	

Equipment information	Tag, manufacturer, type
Pump	
Driver	
Coupling	
Gearbox	
Mechanical seal	
Seal liquid system	
Lubrication system	

Commissioning - Checks before initial run	Remarks	Signed	Signed	Signed	Date
1 Check start-up strainers					
2 Check trip and alarm settings					
3 Check instrumentation and safety devices					

Start-up - Initial run	Remarks	Signed	Signed	Signed	Date
4 Check vibration levels					
5 Check bearing temperatures					
6 Check power consumption					
7 Check mechanical performance					
8 Check seal or stuffing box temperatures					
9 Check loading/unloading devices					
10 Check unit for leakages					
11 Remove start-up strainers					
12 Make hot check coupling alignment					

	Name	Department	Initial	Signed	Date
Engineering contractor					
Rotating Equipment engineer					
Supervisor unit/plant					

References

Gen : Installation of Rotating equipment (amendments/supplements to API RP 686)
Reference to installation of rotating equipment
* foundation
* grouting
* flange alignment
* shaft alignment
* rotation check
* oil flushing
* strainers
Reference to commissioning checklist for rotating equipment in general
Gen : Compressors - Selection, testing and installation
Reference to selection of compressors including
* data check
* scope vendor
* scope contractor
* strainers
* air inlet filters
Gen : Reciprocating compressors (amendments/supplements to API 618)
Reference to design of reciprocating compressors including
* dynamics
* baseplate design
* lubrication
* nameplate check
* valves
* piston
* crackcase
* pulsation dampeners
Gen : Piping - general requirements
Reference to piping requirements
* inlet piping design
* discharge piping design
Gen : Field inspection prior to commissioning of mechanical equipment
Reference to MC checklist for mechanical equipment, for each piece of equipment specific
* rotating equipment
* centrifugal compressors
* reciprocating compressors
* rotary positive displacements pumps
* blowers/fans
* centrifugal pumps
* reciprocating pumps
* rotary positive displacement pumps
* mixers
Gen : Preservation of new and old equipment standing idle
Reference to preservation of rotating equipment
* preservation oils
* cleaniness

Screw Compressors - Construction and Pre-commissioning checklist

General information	
Requisition number	
Plant	
Unit	

Equipment information	Tag, manufacturer, type
Pump	
Driver	
Coupling	
Gearbox	
Mechanical seal	
Seal liquid system	
Lubrication system	

	Construction - General checks	Remarks	Signed	Signed	Signed	Date
1	Check foundation	Roughen, oil and dust free				
2	Check equipment assembly on foundation	25-50 mm between concrete and base plate, pre-align				
3	Check leveling of base plate					
4	Check installation of foundation bolts					
5	Check equipment for transport damage					
6	Check unit completeness					

	Construction - Checks after grouting	Remarks	Signed	Signed	Signed	Date
7	Check proper grouting is applied	Grout type, sealed anchor bolts, humidity, temperature				
8	Check whether foundations bolts are tightened					
9	Check correct bolts and gaskets are used					
10	Check correct installation coupling					
11	Check coupling guard and mounting					
12	Check internals for cleanliness					
13	Check suction line for cleanliness					
14	Check tention and alignment V-belt					
15	Check cooling water system					
16	Check lubrication system					
17	Check bearings and bearing house are clean					
18	Check seal					
19	Check seal liquid system	see installation notes manufacturer				
20	Check equipment for soft feet					
21	Shaft alignment	only metal shims are allowed				
22	Check piping and support attached to pump	Flange deviation				
23	Vents, drain correctly installed					
24	Check earth bosses					
25	Tracing installed correctly					
26	Insulation installed correctly					
27	Check documentation	shaft alignment, foundation measurements, nameplate				
28	P&ID / PEFS check					

	Construction - Presommissioning	Remarks	Signed	Signed	Signed	Date
29	Check direction of rotation	uncoupled or avoid mechanical seal runs dry				
30	Check N2 pre-charge					
31	Check barrier oil fill					
32	Check lubricant and lubricant levels					

	Preservation	Remarks	Signed	Signed	Signed	Date
33	Check equipment for internal cleanliness					
34	Check proper preservation	nozzles blinded off, isolated during flushing				

Other items	Remarks
Punch item list attached	

	Name	Department	Initial	Signed	Date
Engineering contractor					
Rotating Equipment engineer					
Supervisor unit/plant					

Screw Compressors - Commissioning and Start-up checklist

General information	
Requisition number	
Plant	
Unit	

Equipment information	Tag, manufacturer, type
Pump	
Driver	
Coupling	
Gearbox	
Mechanical seal	
Seal liquid system	
Lubrication system	

Commissioning - Checks before initial run	Remarks	Signed	Signed	Signed	Date
1 Check start-up strainers					
2 Check trip and alarm settings					
3 Check instrumentation and safety devices					

Start-up - Initial run	Remarks	Signed	Signed	Signed	Date
4 Check vibration levels					
5 Check bearing temperatures					
6 Check power consumption					
7 Check mechanical performance					
8 Check seal liquid system temperatures					
9 Check unit for leakages					
10 Remove start-up strainers					
11 Make hot check coupling alignment					

	Name	Department	Initial	Signed	Date
Engineering contractor					
Rotating Equipment engineer					
Supervisor unit/plant					

References

Gen : Installation of Rotating equipment (amendments/supplements to API RP 686)
Reference to installation of rotating equipment
* foundation
* grouting
* flange alignment
* shaft alignment
* rotation check
* oil flushing
* strainers
Reference to commissioning checklist for rotating equipment in general

Gen : Compressors - Selection, testing and installation
Reference to selection of compressors including
* data check
* scope vendor
* scope contractor
* strainers
* air inlet filters

Gen : Rotary-type positive displacement compressors (amendments/supplements to API 619)
Reference to design of positive displacement compressors including
* dynamics
* baseplate design
* lubrication
* nameplate check

Gen : Piping - general requirements
Reference to piping requirements
* inlet piping design
* discharge piping design

#NAME?

Gen : Preservation of new and old equipment standing idle
Reference to preservation of rotating equipment
* preservation oils
* cleaniness

Blowers-Fans - Construction and Pre-commissioning checklist

General information	
Requisition number	
Plant	
Unit	

Equipment information	Tag, manufacturer, type
Pump	
Driver	
Coupling	
Gearbox	
Mechanical seal	
Seal liquid system	
Lubrication system	

	Construction - General checks	Remarks	Signed	Signed	Signed	Date
1	Check foundation	Roughen, oil and dust free				
2	Check equipment assembly on foundation	25-50 mm between concrete and base plate, pre-align				
3	Check leveling of base plate					
4	Check installation of foundation bolts					
5	Check equipment for transport damage					
6	Check unit completeness					
7	Check free movement of rotor assembly	E-motor de-energised				

	Construction - Checks after grouting	Remarks	Signed	Signed	Signed	Date
8	Check proper grouting is applied	Grout type, sealed anchor bolts, humidity, temperature				
9	Check whether foundations bolts are tightened					
10	Check correct bolts and gaskets are used					
11	Check radial and axial clearance impeller-volute					
12	Check correct installation coupling					
13	Check coupling guard and mounting					
14	Check cooling water system					
15	Check lubrication system					
16	Check bearings and bearing house are clean					
17	Check mechanical seal					
18	Check seal locks are removed					
19	Check seal liquid system	see installation notes manufacturer				
20	Check equipment for soft feet					
21	Shaft alignment	only metal shims are allowed				
22	Check piping and support attached to pump	Flange deviation				
23	Vents, drain correctly installed					
24	Check earth bosses					
25	Tracing installed correctly					
26	Insulation installed correctly					
27	Check documentation	shaft alignment, foundation measurements, nameplate				
28	P&ID / PEFS check					

	Construction - Presommissioning	Remarks	Signed	Signed	Signed	Date
29	Check direction of rotation	uncoupled or avoid mechanical seal runs dry				
30	Check N2 pre-charge					
31	Check barrier oil fill					
32	Check lubricant and lubricant levels					

	Preservation	Remarks	Signed	Signed	Signed	Date
33	Check equipment for internal cleanliness					
34	Check proper preservation	nozzles blinded off, isolated during flushing				

Other items	Remarks
Punch item list attached	

	Name	Department	Initial	Signed	Date
Engineering contractor					
Rotating Equipment engineer					
Supervisor unit/plant					

Blowers-Fans - Commissioning and Start-up checklist

General information	
Requisition number	
Plant	
Unit	

Equipment information	Tag, manufacturer, type
Pump	
Driver	
Coupling	
Gearbox	
Mechanical seal	
Seal liquid system	
Lubrication system	

Commissioning - Checks before initial run	Remarks	Signed	Signed	Signed	Date
1 Check start-up strainers					
2 Check trip and alarm settings					
3 Check instrumentation and safety devices					

Start-up - Initial run	Remarks	Signed	Signed	Signed	Date
4 Check vibration levels					
5 Check bearing temperatures					
6 Check power consumption					
7 Check mechanical performance					
8 Check seal liquid system temperatures					
9 Check unit for leakages					
10 Remove start-up strainers					

	Name	Department	Initial	Signed	Date
Engineering contractor					
Rotating Equipment engineer					
Supervisor unit/plant					

References
Installation of Rotating equipment (amendments/supplements to API RP 686)
Reference to installation of rotating equipment
* foundation
* grouting
* flange alignment
* shaft alignment
* rotation check
* oil flushing
* strainers
Reference to commissioning checklist for rotating equipment in general
Centrifugal fans (amendments/supplements to ISO 13705, Annex E)
Reference to design of centrifugal fans
* dynamics
* baseplate design
* lubrication
* nameplate check
Gen : Piping - general requirements
Reference to piping requirements
* inlet piping design
* discharge piping design
Gen : Field inspection prior to commissioning of mechanical equipment
Reference to MC checklist for mechanical equipment, for each piece of equipment specific
* rotating equipment
* centrifugal compressors
* reciprocating compressors
* rotary positive displacements pumps
* blowers/fans
* centrifugal pumps
* reciprocating pumps
* rotary positive displacement pumps
* mixers
Preservation of new and old equipment standing idle
Reference to preservation of rotating equipment
* preservation oils
* cleanliness

Mixers - Construction and Pre-commissioning checklist

General information	
Requisition number	
Plant	
Unit	

Equipment information	Tag, manufacturer, type
Pump	
Driver	
Coupling	
Gearbox	
Mechanical seal	
Seal liquid system	
Lubrication system	

Construction - General checks	Remarks	Signed	Signed	Signed	Date
1 Check installation of bolts					
2 Check equipment for transport damage					
3 Check unit completeness					
4 Check whether mounting flange is correctly attached					
5 Check free movement of rotor assembly	E-motor de-energised				

Construction - Checks after grouting	Remarks	Signed	Signed	Signed	Date
10 Check correct bolts and gaskets are used					
11 Check correct installation coupling					
12 Check coupling guard and mounting					
14 Check lubrication system					
19 Check seal					
21 Check seal liquid system	see installation notes manufacturer				
23 Shaft alignment	only metal shims are allowed				
24 Check piping and support attached to pump	Flange deviation				
25 Vents, drain correctly installed					
26 Check earth bosses					
27 Tracing installed correctly					
28 Insulation installed correctly					
29 Check documentation	shaft alignment, foundation measurements, nameplate				
30 P&ID / PEFS check					

Construction - Presommissioning	Remarks	Signed	Signed	Signed	Date
31 Check direction of rotation	uncoupled or avoid mechanical seal runs dry (OH5)				
32 Check N2 pre-charge					
33 Check barrier oil fill					
34 Check lubricant and lubricant levels					

Preservation	Remarks	Signed	Signed	Signed	Date
35 Check equipment for internal cleanliness					
36 Check proper preservation	nozzles blinded off, isolated during flushing				

Other items	Remarks
Punch item list attached	

	Name	Department	Initial	Signed	Date
Engineering contractor					
Rotating Equipment engineer					
Supervisor unit/plant					

Mixers - Commissioning and Start-up checklist

General information	
Requisition number	
Plant	
Unit	

Equipment information	Tag, manufacturer, type
Pump	
Driver	
Coupling	
Gearbox	
Mechanical seal	
Seal liquid system	
Lubrication system	

Commissioning - Checks before initial run	Remarks	Signed	Signed	Signed	Date
1 Check start-up strainers					
2 Check trip and alarm settings					
3 Check instrumentation and safety devices					

Start-up - Initial run	Remarks	Signed	Signed	Signed	Date
4 Check vibration levels					
5 Check bearing temperatures					
6 Check power consumption					
7 Check mechanical performance					
8 Check unit for leakages					

	Name	Department	Initial	Signed	Date
Engineering contractor					
Rotating Equipment engineer					
Supervisor unit/plant					

References
Gen : Installation of Rotating equipment (amendments/supplements to API RP 686)
Reference to installation of rotating equipment
* foundation
* grouting
* flange alignment
* shaft alignment
* rotation check
* oil flushing
* strainers
Reference to commissioning checklist for rotating equipment in general
Gen : Piping - general requirements
Reference to piping requirements
* inlet piping design
* discharge piping design
Gen : Field inspection prior to commissioning of mechanical equipment
Reference to MC checklist for mechanical equipment, for each piece of equipment specific
* rotating equipment
* centrifugal compressors
* reciprocating compressors
* rotary positive displacements pumps
* blowers/fans
* centrifugal pumps
* reciprocating pumps
* rotary positive displacement pumps
* mixers
Gen : Preservation of new and old equipment standing idle
Reference to preservation of rotating equipment
* preservation oils
* cleaniness

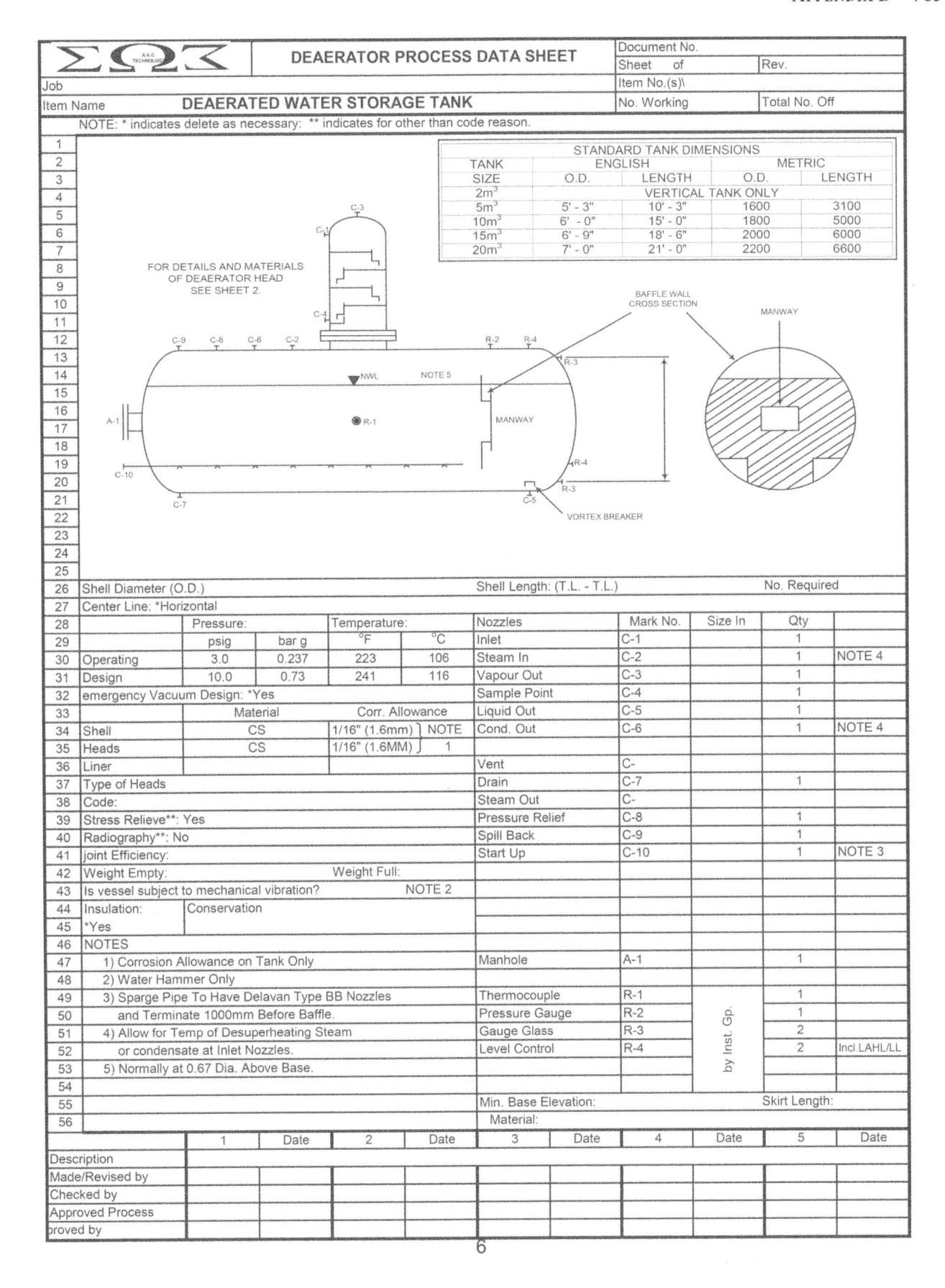

	DEAERATOR PROCESS DATA SHEET	Document No.		
ΣΩΣ A.K.C. TECHNOLOGY		Sheet of		Rev.
Job		Item No.(s)\		
Item Name	DEAERATED WATER STORAGE TANK	No. Working		Total No. Off

NOTE: * indicates delete as necessary; ** indicates for other than code reason.

FOR DETAILS AND MATERIALS OF DEAERATOR HEAD SEE SHEET 2.

	STANDARD TANK DIMENSIONS			
TANK	ENGLISH		METRIC	
SIZE	O.D.	LENGTH	O.D.	LENGTH
2m³	VERTICAL TANK ONLY			
5m³	5' - 3"	10' - 3"	1600	3100
10m³	6' - 0"	15' - 0"	1800	5000
15m³	6' - 9"	18' - 6"	2000	6000
20m³	7' - 0"	21' - 0"	2200	6600

Shell Diameter (O.D.)				Shell Length: (T.L. - T.L.)			No. Required	
Center Line: *Horizontal								
	Pressure:		Temperature:		Nozzles	Mark No.	Size In	Qty
	psig	bar g	°F	°C	Inlet	C-1		1
Operating	3.0	0.237	223	106	Steam In	C-2		1	NOTE 4
Design	10.0	0.73	241	116	Vapour Out	C-3		1
emergency Vacuum Design: *Yes					Sample Point	C-4		1
	Material		Corr. Allowance		Liquid Out	C-5		1
Shell	CS		1/16" (1.6mm) ⎫ NOTE		Cond. Out	C-6		1	NOTE 4
Heads	CS		1/16" (1.6MM) ⎭ 1		Vent	C-		
Liner					Drain	C-7		1
Type of Heads					Steam Out	C-		
Code:					Pressure Relief	C-8		1
Stress Relieve**: Yes					Spill Back	C-9		1
Radiography**: No					Start Up	C-10		1	NOTE 3
joint Efficiency:								
Weight Empty:			Weight Full:					
Is vessel subject to mechanical vibration?			NOTE 2					
Insulation:	Conservation							
*Yes								
NOTES								
1) Corrosion Allowance on Tank Only					Manhole	A-1		1
2) Water Hammer Only								
3) Sparge Pipe To Have Delavan Type BB Nozzles					Thermocouple	R-1		1
and Terminate 1000mm Before Baffle.					Pressure Gauge	R-2	by Inst. Gp.	1
4) Allow for Temp of Desuperheating Steam					Gauge Glass	R-3		2
or condensate at Inlet Nozzles.					Level Control	R-4		2	Incl.LAHL/LL
5) Normally at 0.67 Dia. Above Base.								
					Min. Base Elevation:		Skirt Length:	
					Material:			

	1	Date	2	Date	3	Date	4	Date	5	Date
Description										
Made/Revised by										
Checked by										
Approved Process										
proved by										

DEAERATOR PROCESS DATA SHEET

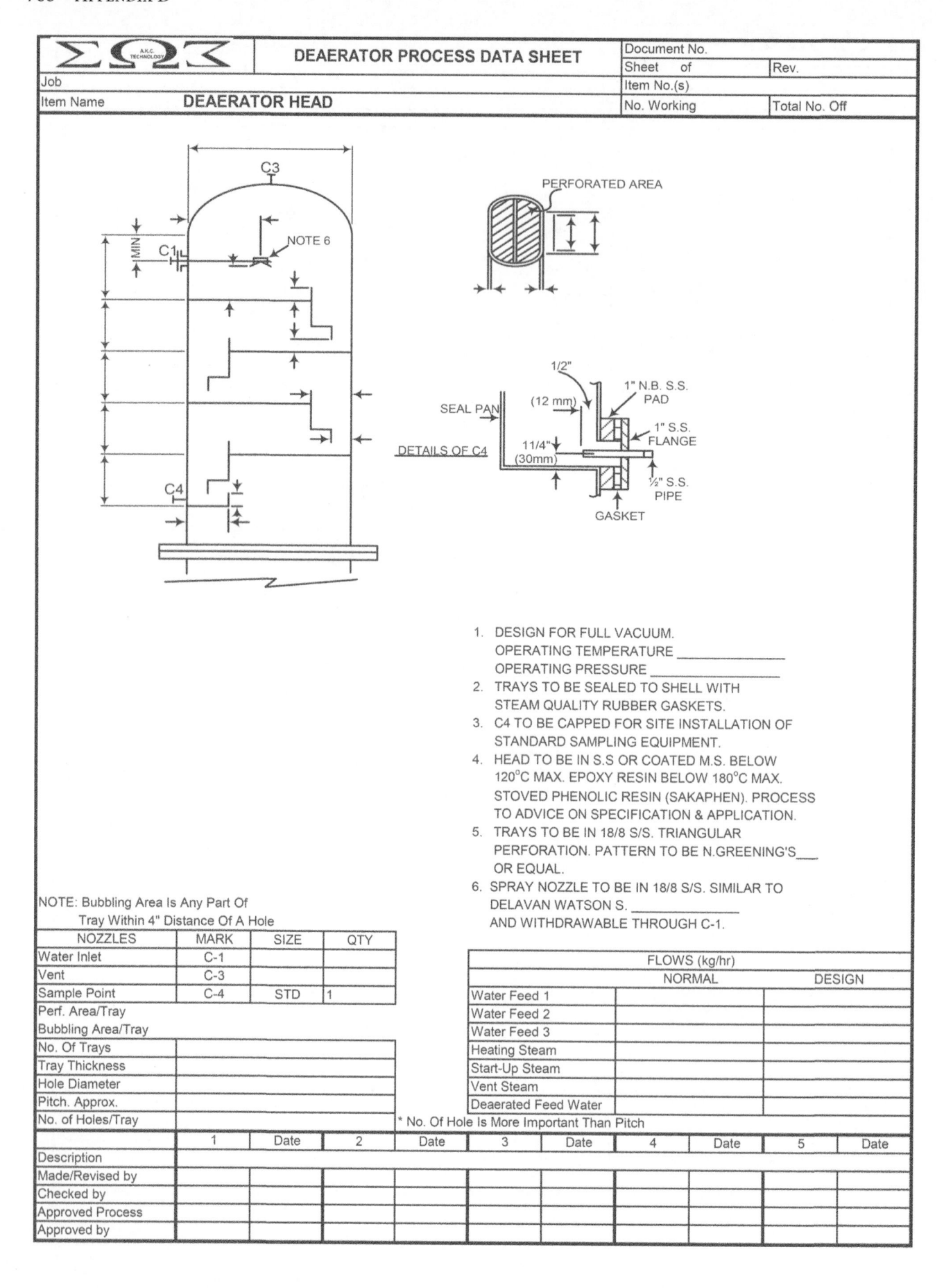

Item Name: DEAERATOR HEAD

			Document No.	
			Sheet of	Rev.
Job			Item No.(s)	
			No. Working	Total No. Off

PERFORATED AREA

SEAL PAN
DETAILS OF C4
1/2" (12 mm)
1 1/4" (30mm)
1" N.B. S.S. PAD
1" S.S. FLANGE
1/2" S.S. PIPE
GASKET

1. DESIGN FOR FULL VACUUM.
   OPERATING TEMPERATURE _____
   OPERATING PRESSURE _____
2. TRAYS TO BE SEALED TO SHELL WITH STEAM QUALITY RUBBER GASKETS.
3. C4 TO BE CAPPED FOR SITE INSTALLATION OF STANDARD SAMPLING EQUIPMENT.
4. HEAD TO BE IN S.S OR COATED M.S. BELOW 120°C MAX. EPOXY RESIN BELOW 180°C MAX. STOVED PHENOLIC RESIN (SAKAPHEN). PROCESS TO ADVICE ON SPECIFICATION & APPLICATION.
5. TRAYS TO BE IN 18/8 S/S. TRIANGULAR PERFORATION. PATTERN TO BE N.GREENING'S ___ OR EQUAL.
6. SPRAY NOZZLE TO BE IN 18/8 S/S. SIMILAR TO DELAVAN WATSON S. _____ AND WITHDRAWABLE THROUGH C-1.

NOTE: Bubbling Area Is Any Part Of Tray Within 4" Distance Of A Hole

NOZZLES	MARK	SIZE	QTY
Water Inlet	C-1		
Vent	C-3		
Sample Point	C-4	STD	1
Perf. Area/Tray			
Bubbling Area/Tray			
No. Of Trays			
Tray Thickness			
Hole Diameter			
Pitch. Approx.			
No. of Holes/Tray			

FLOWS (kg/hr)		
	NORMAL	DESIGN
Water Feed 1		
Water Feed 2		
Water Feed 3		
Heating Steam		
Start-Up Steam		
Vent Steam		
Deaerated Feed Water		

* No. Of Hole Is More Important Than Pitch

	1	Date	2	Date	3	Date	4	Date	5	Date
Description										
Made/Revised by										
Checked by										
Approved Process										
Approved by										

	ΣΩΣ A.K.C. TECHNOLOGY		**DRUM PROCESS DATA SHEET**		Doc. No:		
					Item No.		
		Job:			Sheet of		

Distribution: Item Name:

NOTE: * indicates delete as necessary;   ** indicates for other than Code reason

26	Shell Diameter (O.D. / I.D.):				Shell Length:		No. Required:	
27	Center+C71 Line: *Horizontal / Vertical							
28		Pressure: g	Temperature: °F / °C		Nozzles	Mark No.	Size	Number
29	Item Number				Inlet	C-		
30	Operating							
31	Design				Vapor Out	C-		
32	Emergency Vacuum Design: * Yes / No							
33		Material	Corr. Allowance		Liquid Out:	C-		
34	Shell							
35	Heads							
36	Liner				Thermocouple	R-	by Inst. Op.	
37	Type of Heads				Pressure Gauge	R-		
38	Code:				Gauge Glass	R-		
39	Stress Relieve**: Radiography**:				Level Control	R-		
40	Joint Efficiency:				Safety Valve	R-		
41	Density of Contents:		at	*°F / °C				
42	Weight Empty:		Weight Full:					
43	Is vessel subject to mechanical vibration? *Yes / No				Vent	C-		
44	insulation: Type: *Frost and Personnel Protection / Cold				Drain	C-		
45	*Yes / No / Anticondensation / Heat Conservation				Steam Out	C-		
46	REMARKS:							
47					Manhole	A-		
48								
49								
50								
51								
52								
53					min. Base Elev'n:			
54					Material:			

Issue No.		1	Date	2	Date	3	Date	4	Date
Made/Revised by									
Checked by									
Approved - Process									
Approved by									

EFFLUENT SCHEDULE

STREAM NUMBER	EFFLUENT SOURCE	FLUID	TOTAL FLOWRATE tonne/h.	MAIN CONTAMINANTS	CONTAMINANT QUANTITY kg/h.	CONTAMINANT CONCENTRATION	STREAM TEMP. °C
1							
2							
3							
4							
5							
6							
7							
8							
9							
10							
11							
12							
13							
14							
15							
16							
17							
18							
19							
20							
21							
22							
23							
24							
25							
26							
27							
28							
29							
30							
31							
32							
33							
34							
35							
36							
37							
38							
39							
40							
41							

Job No.:
Job Name:
Document No.
Project No.
Client
Location
Plant

Sheet of
Issue No.:
Description
Made/Revised by
Checked by
Approved- Process
Approved

1 Date
2 Date

42 Notes:
43
44
45
46
47
48

	PROCESS DATA SHEET	Job No.
ΣΩΚ A.K.C. TECHNOLOGY		Item No.
	Job:	Sheet of

EQUILIBRIUM FLASH CALCULATION (Form 1 of 2)

Temperature: _____ °F
Pressure: _____ psia
Pg: _____ psia

Assumed $\frac{V}{L}$ =

	m	K	$K\frac{V}{L}+1$	$\frac{m}{K\frac{V}{L}+1}$	$K\frac{V}{L}+1$	$\frac{m}{K\frac{V}{L}+1}$	$K\frac{V}{L}+1$	$\frac{m}{K\frac{V}{L}+1}$	$K\frac{V}{L}+1$	$\frac{m}{K\frac{V}{L}+1}$	$K\frac{V}{L}+1$	$\frac{m}{K\frac{V}{L}+1}$
H_2												
CO												
CH_4												
CO_2												
C_2H_6												
C_3H_8												
$i\text{-}C_4H_{10}$												
$n\text{-}C_4H_{10}$												
C_2H_4												
C_3H_6												
C_6H_6												

$\sum \frac{m}{K\frac{V}{L}+1} =$

Calculated $\frac{V}{L} =$

$$\text{Calculated} \frac{V}{L} = \frac{1}{\sum \frac{m}{K\frac{V}{L}+1}} - 1$$

Issue No.	1	Date	2	Date	3	Date	4	Date	5	Date
Made/Revised by										
Checked by										
Approved - Process										
Approved by										

ΣΩΣ A.K.C. TECHNOLOGY	**PROCESS DATA SHEET**	Job No.
		Item No.
	Job:	Sheet of

EQUILIBRIUM FLASH CALCULATION (Form 2 of 2)

Temperature: _____ °F
Pressure: _____ psia
Pg: _____ psia

	MOLES			MOL. FRACTION	
	Feed m	Liquid L_x	Gas V_y	Liquid x	Gas y
H_2					
CO					
CH_4					
CO_2					
C_2H_6					
C_3H_8					
$i\text{-}C_4H_{10}$					
$n\text{-}C_4H_{10}$					
C_2H_4					
C_3H_6					
C_6H_6					

$$L_x = \frac{m}{K\frac{V}{L}+1} \qquad\qquad V_y = m - L_x$$

Issue No.	1	Date	2	Date	3	Date	4	Date	5	Date
Made/Revised by										
Checked by										
Approved - Process										
Approved by										

FABRICATED EQUIPMENT SCHEDULE

Job No.:			Sheet of		Issue No.:		1 Date	2 Date	3 Date
Job Name:					Description				
Document No.					Made/Revised by				
Project No.					Checked by				
Client					Approved-Process				
Location					Approved				
Plant									

ITEM No.	No. OFF	DESCRIPTION	DIAMETER (m)	LENGTH (m)	SURFACE (m²/SHELL)	WEIGHT tonne/SHELL	INTERNALS/ H.E. TYPE (See Note 1)	FLUID T = Tube S = Shell	MATERIAL	DESIGN PRESSURE (kg/cm², g)	CONDITIONS TEMPERATURE (°C)	REMARKS
								T				
								S				
								T				
								S				

Notes: 1. H.E. Type Coding - A.C. = Air Cooler; K = Kettle; U.B. = U-Tube Bundle; F.H. = Floating Head; F.T.S. = Fixed Tube Sheet

FAN/COMPRESSOR PROCESS DUTY SPECIFICATION

Document No.
Sheet of   Rev.
Job   Item No.(s)\
Item Name   No. Working   Total No. Off

OPERATING CONDITIONS PER UNIT

#									
1	UNITS								
2	Operating Case								
3	Mass Flowrate	kg/h	lb/h						
4	Standard Volumetric Flowrate	nm^3/h	std ft^3/h						
5	Volume at Suction	m^3/h	ft^3/h						
6	INLET CONDITIONS								
7	Pressure	bara	psia						
8	Temperature	°C	°F						
9	Molecular Weight								
10	Cp/Cv								
11	Compressibility								
12	DISCHARGE CONDITIONS								
13	Pressure	bara	psia						
14	Temperature	°C	°F						
15	Cp/Cv								
16	Compressibility								
17	GAS ANALYSIS								
18		Mol %	Mol %						
19									
20									
21									
22									
23									
24									
25									
26									
27									
28									
29									
30									
31	Entrained Liquids/Solids								
32	PERFORMANCE								
33	Compression Ratio								
34	Estimated Efficiency	%	%						
35	Estimated Absorbed Power	kW	hp						
36	Recommended Driver Power	kW	hp						
37	MECHANICAL ARRANGEMENT								
38	Compressor/Fan Type								
39	Number of Stages			SKETCH					
40	Gas to be kept Oil Free	Yes / No							
41									
42	Driver Type								
43									
44	Casing Design Pressure								
45	Design Pressure								
46									
47	Material - Casing								
48	- Impeller								
49									
50	Sour Service	Yes / No							
51	NOTES								
52	1. Cooling medium available:-								
53									
54									
55									

	1	Date	2	Date	3	Date	4	Date	5	Date
Description										
Made/Revised by										
Checked by										
Approved Process										
Approved by										

FRACTIONATOR CALCULATION SUMMARY

Job No: _____
Item No: _____
Sheet ____ of ____

Job: _____

Distribution

Item Name: _____

#										
1	Operate at		psig		Units: Metric / British (delete one)					
2	Fractionation:					Reflux Ratio & No. of Plates:				
3	Ovhd Prod:	C <		Reflux Ratio:			Actual:		; Use	
4	Bottoms:	C <		Rectifying Plates:			Theoretical:		; Use	
5				Stripping Plates:			Theoretical:		; Use	

#		Feed				Net Bottoms				
7-8	Component	Mols	Mol %		Mols	Mol %	K ($Ksca\,°C$ / $psia\,°F$)	KX	Pat $°C$/$°F$	Vapor Pressure
9	1									
10	2									
11	3									
12	4									
13	5									
14	6									
15	7									
16	8									
17	9									
18	10									

#		Net Overheads					Ext. Rflx.	Gross Overheads		
21-22	Component	Mols	Mol %	K ($Ksca\,°C$ / $psia\,°F$)	KX or $\frac{Y}{K}$		Mols	Mols	Mol %	K ($Ksca\,°C$ / $psia\,°F$) ; $\frac{Y}{K}$
23	1									
24	2									
25	3									
26	4									
27	5									
28	6									
29	7									
30	8									
31	9									
32	10									

34 Thermal Conditions of Feed: _____ Reboiler Vapor Quantity: _____

#	Stream	°API	Sp.Gr. at 15°C at 60°F	Temp. °C/°F	Sp.Gr. at T°	Sp. Vol. At T° cu. m/kg cu. Ft/lb	Mols/hr	Mol. Wt.	kg/hr lb/hr
44	Feed								
45	Net Bottoms								
46	Net Overheads								
47	Gross Overheads								
48	Gross Bottoms								
49	Reboiler Vapor								
50									
51									
52									

53 REMARKS

	Issue No.	1	Date	2	Date	3	Date
57	Made/Revised by						
58	Checked by						
59	Approved - Process						
60	Approved by						

GENERAL SERVICES AND UTILITIES CHECK LIST

Job No.:				Sheet of	PREPARED BY:
Job Name:					
Document No.					DATE:
Project No.					
Client					
Location					
Plant					

SERVICE	QUANTITY SUSTAINED	QUANTITY PEAK	UNIT	OPERATING PRESSURE & TEMPERATURE	REMARKS
ELECTRICAL 13.8 KV			*KVA		
ELECTRICAL 2300V			*KVA		
ELECTRICAL 440V			*KVA		
MISC. LIGHTING, ST., ETC.] (110 V)			*KVA		
WELL WATER (CITY)			GPM		
SEA WATER			GPM		
RIVER WATER - PROCESS			GPM		
RIVER WATER - FIRE PROT.			GPM		
FUEL GAS			SCFM		
ODORIZED GAS			SCFM		
COMPRESSED AIR			SCFM		
STEAM			LB/HR		
(400 OR 475) PSIG			LB/HR		
235 PSIG			LB/HR		
150 PSIG			LB/HR		
30 PSIG			LB/HR		
CONDENSATE RETURN			LB/HR		
CONDENSATE USAGE			LB/HR		
PROCESS TRANSFER & RAW MATERIALS					
CAUSTIC (_%)			LB/HR		
BRINE (_%)			LB/HR		
LPG			LB/HR		
SITE DEV (AREA FILL)					
RAILROADS					
BLOCK ACCESS ROADS					
SANITARY SEWERS			GPM		
WASTE WATER			GPM		

*ASSUME 1 HORSE POWER = 1KVA

General services and utilities checklist.

HAZARDOUS CHEMICAL & CONDITIONS SCHEDULE

Job No:
Job Name:
Document No.
Project No.
Client
Location
Plant

Sheet ___ of ___
Issue No.:
Description
Made/Revised by
Checked by
Approved- Process
Approved

Date ___ Date ___ Date ___

1	2	3	4	5	6	7	8	9	10	11	12	13
Chemical or Hazard	Chemical Symbol	Phase	Location in Plant	Toxicity (Note 3)	Flash Point °C — Closed Cup	Flash Point °C — Open Cup	Fire Hazard — Explosive Limits % by Volume in Air — Upper	Fire Hazard — Explosive Limits % by Volume in Air — Lower	Autoignition Temp. °C	Suitable Extinguishing Agents	Protective Equipment	Remarks

Notes:
1. This Schedule is given without any Legal responsibility on the part of AKC Technology.
2. Column 3: Vapor-V, Liquid-L, Solid-S
3. Toxicity: Denotes Threshold Limit (T.L) under which it is believed nearly all workers may be exposed day after day without adverse effect. The figures relate to average concentration for a normal working day.

4. Column 11 A - Water E - Foam
   B - Powered Talc F - See Remarks
   C - CO_2
   D - Dry Chemical

5. Column 12 B.A. - Breathing Apparatus G.V. - Gloves
   F.C. - Full Clothing X - See Remarks
   F.W. - Footware
   G.O. - Goggles

Hazardous chemical and conditions schedule.

HEAT & MASS BALANCES

Job No.:		Sheet of	Issue No.:	1	Date	2	Date	3	Date
Job Name:			Made/Revised by						
			Checked by						
			Approved- Process						
			Approved by						

* delete unit as necessary

	Stream ↓	Units →		°API	K	M.W.	FLOWING Sp.Gr.	°F* °C	barg* psig	B.P.S.D. at 60°F	FLOWING GPM	Mol. %	Mol/h	kg/h* lb/h	kcal/kg* Btu/lb	106 kcal/h* 106 Btu/h
1																
2																
3																
4																
5																
6																
7																
8																
9																
10																
11																
12																
13																
14																
15																
16																
17																
18																
19																
20																
21																
22																
23																
24																
25																
26																
27																
28																
29																
30																
31																
32																
33																
34																
35																
36																
37																
38																
39																
40																
41																
42																
43																
44																
45																
46																
47																

HEAT EXCHANGER RATING SHEET

Document No.
Sheet of **Rev.**
Items No. (s)

PART 1 - TO BE FILLED IN COMPLETELY

- Arrangement: Horiz = 0; Vert = 1
- Units: S.I. 0; Engineering = 1

#		Units	S.I.	Engineering	HOT FLUID	COLD FLUID
1	Exchanger Name:	*				
2	Job: *			Location: *		
3	Type of Unit: F.T.S.=0; FLTG.HEAD=1; U-TUBE=2; FOR KETTLES ADD 3					
4	Hot Side: Either=0; Shell=1; Tubes=2					
6	Fluid Circulated		-	-		
7	Total Fluid Entering (Normal)		kg/hr	lb/hr		
8	Flow Margin		%	%		
9	Temperature In/Out		°C	°F		
10	Max Pressure Drop at line 8		bar	psi		
11	Inlet Pressure Operating/Design		barA	psia		
12	Normal Heat Load		kW	Btu/hr		
13	Heat Load Margin		%	%		
14	Design Temperature		°C	°F		
15	Corrosion Allowance		mm	ins		
16	Fouling Resistance		°Cm²/W	°F hr ft²/Btu		
17	Line N.B. In/Out		mm	ins		
18	Part 2					
19	GAS (AD VAPOR) IN		-	-	░░░░	
20	Flow of Vapor and Gas at Inlet		kg/hr	lb/hr		
21	Code Number		-	-		
22	Molecular Weight		-	-		
23	Thermal Conductivity		W/m°C	Btu/hr ft°F		
24	Specific Heat		kJ/kg°C	Btu/lb°F		
25	Compressibility Factor		-	-		
26	Viscosity		cP	lb/ft hr		
27	LIQUID IN		-	-	░░░░	
28	Total Flow of Liquid at Inlet		kg/hr	lb/hr		
29	Code Number		-	-		
30	Thermal Conductivity		W/m°C	Btu/hr ft °F		
31	Specific Heat		kJ/kg°C	Btu/lb °F		
32	Density		kg/m³	lb/ft³		
33	Viscosity		cP	lb/ft hr		
34	Viscosity at Ave. Temp. other side		cP	lb/ft hr		
35	CONDENSATION AND VAPORIZATION		-	-	░░░░	░░░░
36	Fluid Condensed or Vaporized		kg/hr	lb/hr		
37	Code Number		-	-		
38	Molecular Weight		-	-		
39	Thermal Conductivity		W/m°C	Btu/hr ft °F		
40	Specific Heat		kJ/kg°C	Btu/lb °F		
41	Density (Hot Fluid)		kg/m³	lb/ft³		░░░░
42	Compressibility Factor (Cold Fluid)		-	-		
43	Viscosity		cP	lb/ft hr		
44	Latent Heat		kJ/kg	Btu/lb		
45	Surface Tension		N/m	lb/ft	░░░░	░░░░
46	Expansion Coefficient		1/°C	1/°F		
47	% Condensate Forming Film		%	%		░░░░
48	Thermal Conductivity		W/m°C	Btu/hr ft °F		
49	Density		kg/m³	lb/ft³		
50	Viscosity		cP	lb/ft hr		
51	Temps for Phase Change Start/Finish		°C	°F		
52	LMTD weighting Factor 'F'		-	-		

#	PART 3											
54	Number of Points in Table Below											
55	% Heat Load	%	%	1	2	3	4	5	6	7	8	9	10
56	Hot Fluid Temperature	°C	°F										
57	Cold Fluid Temperature	°C	°F										

PART 4 - CODE NUMBERS TO BE FILLED IN COMPLETELY

#	Material for	Description	Code No.	O.D.	Thickness	Length	Passes	Number	Contents Lethal?
60	Tubes								Yes/No
61	Shell							░░░	Yes/No
62	Floating Head					Tube Pitch = mm/in.	o	Δ	□
63	Channel								
64	Tube Sheets								

NOTES

	Issue No.	1 Date	2 Date	3 Date	4 Date	5 Date
	Made/Revised by					
	Checked by					
	Approved- Process					
	Approved by					

APPENDIX D

HYDROCARBON DEWPOINT CALCULATION

Job No:
Item No.
Job:
Page of

Heaviest Compound: PROPANE

Pressure: _____ psia
Convergence Pressure: _____ psia

	y	Tb	LIGHT COMPONENT		HEAVY COMPONENT		ASSUMED TEMPERATURE									
							°F		°F		°F		°F		°F	
			FeL	TbFeL	FeH	TbFeH	K	y/K	K	y/K	K	y/K	K	y/K		x
H_2		67	1.000	67.000												
N_2		130	0.0229	2.972	0.105	13.61										
CO		172	0.0046	0.797	0.180	31.00										
CH_4		201	0.0019	0.383	0.244	49.02										
C_2H_4		305			0.548	167.0										
CO_2		330			0.638	210.6										
C_2H_6		332			0.646	214.3										
C_2H_2		340			0.676	229.9										
C_3H_6		406			0.954	387.3										
C_3H_8		416			1.000	416.0										
	1.000															1.000
					Σ y/K											
					x for H_2											
					Σ (yFe$_L$/K)											
					Σ (yFe$_H$/K)											
					E.B.PL											
					E.B.PH											
					P_g											

Estimate of Dewpoint: _____ °F
at Estimate Pg of: _____ psia

Issue No.	1	Date	2	Date	3	Date
Made/Revised by						
Checked by						
Approved - Process						
Approved by						

LINE LIST SCHEDULE

Job No.								Issue No.				Sheet of			
Job Name:								Description							
Document No.								Made/Revised by							
Project No.								Checked by							
Client								Approved - Process							
Location								Approved							
Plant															

1	2	2A	3	4	5		6		7	8		9		10			11						12	13	14	14a	15		16	
ISS	LINE		NOM. PIPING	FLUID		PHASE	LINE ROUTING		PLANT	OPERATING		DESIGN		INSULATION			TEST	FIELD TEST					FIELD TEST LIMITED BY	CHEMICAL CLEANING	REMARKS	STRESS CATEGORY	ELD DRG. No.		ISO. DRG. No.	
COLUMN	No.		PIPE SIZE	SPEC.	DENSITY		FROM	TO	CONDITION	PRESS.	TEMP.	PRESS.	TEMP.	CLASS	NOM. THICK		CLASS	PRESS.	CLASS	PRESS.	CLASS	PRESS.								
DATA CHANGES																														

UNITS: PRESSURE bar g p.s.i g kg/cm² g TEMPERATURE °C °F NOMINAL PIPE SIZE Inches mm Insulation Thickness inches mm.

Column

1	Enter Issue Number in small box. Indicate in outside box number(s) of column(s) containing revision
2A	Blocking this column indicates that line is Process Critical: I.e. requires routing feedback to Pressure.
5	Indicate V for vapors: L for liquids: L/V for mixed phases or make special reference to other conditions (e.g. slurries, etc.)
7	NO = Normal Operation, defined as the coincident pressure and temperature conditions prevailing for the majority of the plant life (insulation to be related to the temperature stated here) MO = Maximum operation, defined as the maximum sustained conditions (material selection is usually related to the temperature stated here). SV = Relief Valve Set Pressure and coincident temperature (minimum basis for "Design P and T") PSO = Pump Shut-Off head (at reduced or no flow) where no pressure relief is provided XT ( ) = Transient conditions which can occur for the total period of time shown in paranthesis, in hours, during an operational year (abbreviated to T_{10} or T_{50}, as appropriate. If Job Piping Code incorporates ANSI B31.3 RT = Relief Valve Blowing conditions, taken as allowing maximum pressure / stress relaxation permitted by Job Piping Code.
8	Coincident pressures and temperatures corresponding to the process cases quoted under Column 7 (the highest temperature) quoted under Column 8 is used for calculating thermal stresses.
9	The pressure and temperature upon which the mechanical design of the line will be based (used for pipewall thickness, flange, gasket and bolting design).

10	Insulation class and thickness to contract requirements:
	Classs 0 No insulation
	        1 Heat Conservation
	        2 Special Heat Conservation (e.g. controlled heat loss for maintaining temperatures).
	        3 Personnel Protection Insulation
	        4 Personnel Protection Guards only (I.e. Insulation not permissible).
	        5 Cold Conservation
	        6 Special Cold Conservation (e.g. anti-condensation only or controlled heat gain).
	        7 Frost Protection
	        8 Acoustic Lining
	        9 Acoustic Lagging
	      11 To indicate Jacketed Line
	      12 To indicate Steam Traced Line
	      14 To indicate Electrically Traced Line
	Example: Class 2/12 denotes special heat conservation insulation for a steam traced line.

11	Field testing to Contract Specification No:

13	Cleaning of Pipework (in addition to standard pre-commissioning clean-out). For co-ordination
	Type 1 Manual (wire brush or equivalent)
	        2 Pigging (including wire brush pig)
	        3 Sandblast or Shotblast
	        4 Pickling - Shop
	        5 Pickling - Site
	        6 Chemical Cleaning (by circulation)
	        7 Steam Blowing
	        8 By Equipment Vendor

14A	Enter the Stress Category as advised by Piping Streee Engineer.
15	For latest issue of listed drawing numbers see Drawing Number Register.
16	For latest issue of listed drawing numbers see Drawing Number Register.

Line Schedule.

Line schedule sheet.

Line summary table

APPENDIX D

MASS BALANCE

Job No.:		Sheet of		Issue No.:			1 Date		2 Date		3 Date	
Job Name:				Description								
Document No.				Made/Revised by								
Project No.				Checked by								
Client				Approved- Process								
Location				Approved								
Plant												

STREAM NUMBER													
STREAM NAME													
PHASE													
TEMPERATURE (°C)													
COMPONENT	MOL. WT.	kg. mol/hr	Mol. %	kg. mol/hr	Mol. %	kg. mol/hr	Mol. %	kg. mol/hr	Mol. %	kg. mol/hr	Mol. %	kg. mol/hr	Mol. %
1 Water													
2 Hydrogen													
3 Nitrogen													
4 Carbon monoxide													
5 Carbon dioxide													
6 Methane													
7 Acetylene													
8 Ethylene													
9 Ethane													
10 Propyne													
11 Propadiene													
12 Propylene													
13 Propane													
14 Isobutane													
15 Isobutene													
16 n-Butane													
17 1-Butene													
18 1,3-Butadiene													
19 1-Pentene													
20 neo-Pentane													
21 Isopentane													
22 n-Pentane													
23 Toluene													
24 m-Xylene													
25 Methanol													
26 MTBE													
27 TBA													
28 DIISO													
29 TOTAL (kg mol/hr)													
30													
31													
32													
33													
34													
35													
36													
37													
38													
39													
40													
41													
42 Notes:													
43													
44													
45													
46													
47													
48													

MECHANICAL EQUIPMENT SCHEDULE

Job No.:		Sheet of	Issue No.:		1	Date	2	Date	3	Date
Job Name:			Description							
Document No.			Made/Revised by							
Project No.			Checked by							
Client			Approved- Process							
Location			Approved							
Plant										

ITEM No.	No. OFF	SERVICE DESCRIPTION	FLUID	S.G/M.W	CAPACITY (m^3/h) (Turbine - Te/hr)	PRESSURE (Kg/cm^2 g) IN	PRESSURE OUT	INLET TEMP. (°C)	NPSHA (m)	TYPE (see Note 1)	MATERIALS	ABSOLUTE POWER (kW)	SPEED (RPM)/ REMARKS

Notes: 1. Rotary Equipment Type Coding: - A = Axial; C = Centrifugal; M = Metering; R = Reciprocating; S = Screw
2. All drives electric motor unless specified otherwise.

Appendix D

Pipe line list

Job No.:		PIPE LINE LIST TITLE:			Sheet of	Issue No.:	1	Date	2	Date	3	Date
Job Name:						Description						
Document No.						Made/Revised by						
Project No.						Checked by						
Client						Approved-Process						
Location						Approved						
Plant												

LINE NUMBER	FROM	TO	TEMPERATURE (°F) OPERATING / DESIGN	PRESSURE (PSIG) OPERATING / DESIGN	VELOCITY FT/SEC.	PRESSURE/100 FT PD/100 FT	FLOW$^{(1)}$	P & ID NUMBER

NOTES: 1. FLOW UNITS : G= gpm; C= acfm; P = lbs/hr.

PIPE LIST

Job No.:				Issue No.:				
Job Name:				Description	1	2	3	
Document No.				Made/Revised by				
Project No.		Sheet of		Checked by				
Client				Approved- Process				
Location				Approved				
Plant				Date	Date	Date	Date	

LINE NUMBER	FROM	TO	TEMPERATURE (°F) OPERATING / DESIGN	PRESSURE (PSIG) OPERATING / DESIGN	VELOCITY FT/SEC.	PRESSURE/100 FT PD/100 FT	FLOW$^{(1)}$	P & ID NUMBER

NOTES: 1. FLOW UNITS : G= gpm, C= acfm, P = lbs/hr.

Appendix D

PIPING PROCESS CONDITIONS SUMMARY

Job No.:		Sheet of		Issue No.:		1	Date	2	Date	3	Date
Job Name:				Description							
Document No.				Made/Revised by							
Project No.				Checked by							
Client				Approved - Process							
Location				Approved							
Plant											

INDEX NUMBER	MATERIAL AND COMPOSITION	FLUID SYMBOL	HAZARDOUS MATERIAL CLASSIFICATION	FLUID STATE LIQUID/GAS	MAXIMUM PIPE SIZE INCHES	MAXIMUM OPERATING TEMPERATURE, °F	CONTINUOUS RANGE PRESSURE, PSIA	PIPE SPECIFICATION	NOTES

	PLATE HEAT EXCHANGER DATA SHEET		Document No.		
			Sheet of		Rev.
Job			Item No.(s)		
Item Name			No. Working		Total No. Off

	PROCESS DESIGN	UNITS		FLUID 1	FLUID 2	FLUID 3
1						
2	Fluid Circulated	-	-			
3	Total Fluid Entering (Normal)	kg/h	lb/h			
4	Flow Margin	%	%			
5	Inlet Vapor and Gas (&MW)	kg/h	lb/h			
6	Inlet Liquid	kg/h	lb/h			
7	Fluid Vaporized/Condensed (&MW)	kg/h	lb/h			
8	Temperature In/Out	°C	°F			
9	Max. Pressure Drop	bar	psi			
10	Inlet Pressure (Operating)	bar g	psig			
11	Normal Heat Load	kW	Btu/h			
12	Heat Load Margin	%	%			
13	Fouling Resistance	$m^2\,°C/W$	$ft^2\,°Fh/Btu$			
14	FLUID PROPERTIES					
15	Specific Heat	kJ/kg°C	Btu/lb°F			
16	Thermal Conductivity	W/m°C	Btu/hr ft°F			
17	Density	kg/m^3	lb/ft^3			
18	Viscosity	cP	lb/hr.ft			
19	Liquid Viscosity (at Temp.)	cP	lb/hr.ft			
20	Latent Heat	kJ/kg	Btu/lb			
21	ENGINEERING DESIGN					
22	Process Design Pressure	bar g	psig			
23	Process Design Temperature	°C	°F			
24	Corrosion Allowance (Header)	mm	in.			
25	Line N.B. In/Out	in	in.			
26	Minimum Flow Passage	mm	in.			
27	Plate Material					
28	Frame Material					
29	Gasket Material					
30	Desgin Code					
31	NOTES					
32						
33	1. Unless otherwise stated fluid properties are for mean fluid temperatures.					
34	2. Sour Service Yes / No					

	1	Date	2	Date	3	Date
Description						
Made/Revised by						
Checked by						
Approved Process						
Approved by						

APPENDIX D

		PROCESS DATA SHEET	Job No.
Made	ΣΩZ A.K.C. TECHNOLOGY		Item No.
Checked			
Date		Job	Sheet of

CALCULATION OF PRESSURE DROP IN FIXED CATALYST BEDS

REQUIRED DATA

VESSEL NO:
SERVICE:
CATALYST: VOLUME REQUIRED .... V _____ ft³
             MANUFACTURER & NO. _____
             DIMENSIONS _____
             EQUIV. PARTICLE DIA. _____ D_p _____ ft.
             SHAPE/SIZE FACTOR _____ S_f _____

GAS DATA: FLOW RATE _____ W _____ lb/h
                                        W^2 _____ (lb/h)²
             MIN. MOLECULAR WEIGHT _____
             BED OUTLET PRESSURE _____ P _____ psia
             MAX. BED TEMPERATURE _____ °C
             ABS. TEMPERATURE (°K = °C + 273) _____ T _____ K

CALCULATION

$$k_1 = \frac{(T)(S_f)}{(M)(P)} = \underline{\hspace{3cm}} =$$

	SYMBOLS	UNITS	CALCULATION	CASE 1	CASE 2	CASE 3	CASE 4
BED DIAMETER	D	ft					
BED C.S.A.	A	ft²	$0.7854 D^2$				
BED DEPTH	L	ft.	V/A				
SUPERFICIAL MASS FLOW	G	lb/h.ft²	W/A				
MASS FLOW FACTOR	G_f	-					
$D_p G$		-					
REYNOLDS NO. FACTOR	Re_f	-					
k_2	k_2	-	$k_1 Re_f$				
CLEAN PRESSURE DROP/UNIT DEPTH	ΔP_c	psi/ft.	$k_2 G_f$				
CLEAN OVERALL PRESSURE DROP	ΔP_c	psi	$L \Delta P_c$				
SAFETY/FOULING FACTOR	f	-		1.25/			
FLOW SHEET PRESSURE DROP	ΔP	psi	$f \Delta P_c$				

* IF VALUE OF DpG > 200 (750 FOR RING CATALYSTS) PUT Re_f = 1.0
  IF VALUE OF DpG < 200 (750 FOR RING CATALYSTS) CALCULATE Re = DpG/μ
  (WHERE μ = DYNAMIC VISCOSITY OF GAS, lb/ft.hr)

	1	2	3	4	5	6
Description						
Made/Revised by						
Checked by						
Approved Process						
Approved by						

ΣΩΚ PROCESS ENGINEERING JOB ANALYSIS SUMMARY

Job Title			
Job No.	Charge No.	Date	
Based Upon Cost Estimated Dated		or Actual Construction Cost	
Summary Prepared By		Information Dated	
Production Basis (lbs/day, tons/day, lbs/month)			

	Service Requirements:	Unit Rate	Unit Rate/ Production Basis
1	Steam (30 lbs.)	lbs/hr.	
2	Steam (150 lbs.)	lbs/hr.	
3	Steam (400 lbs.)	lbs/hr.	
4	Steam (lbs.)	lbs/hr.	
5	Treated R.W.	gpm	
6	Untreated R.W.	gpm	
7	Fresh Water	gpm	
8	Sea Water	gpm	
9	Fuel Gas (psi)	cfm (60°F & 1 atm.)	
10	Air (psi)	cfm (60°F & 1 atm.)	
11	Power ( )		
12	Horsepower		
13	Condensate	lbs/hr.	
14			

	Raw Materials	Unit Rate	
1	Chlorine		
2	Hydrogen (%)		
3	Caustic (%)		
4	Salt		
5	Sat. Brine		
6	Natural Gas		
7	Air		
8	Ethylene		
9			
10			
11			

	Products and By-Products	Unit Rate	
1	Chlorine		
2	HCl (%)		
3	Salt (%)		
4	Caustic (%)		
5	Ammonia (%)		
6	H_2SO_4 (%)		
7	Gas ( )		
8			
9			
10			
11			

Process Engineering job analysis summary.

Appendix D

		UNITS		CASE I	CASE II
1					
2	Liquid Pumped				
3	Corrosion/Erosion				
4	Due To				
5	Operating Temp. (T)	°C	°F		
6	Specific Gravity at T				
7	Viscosity	cP	cP		
8	Vapor Pressure at T	bar a	psi a		
9	Normal mass Flowrate	kg/h	lb/h		
10	Normal Vol. Flowrate	m³/h	gpm		
11	Min. Vol. Flowrate				
12	Design Vol. Flowrate	m³/h	gpm		
13	SUCTION CONDITION				
14	Pressure at Equipment	barg	psi g	+	+
15	Static Head	bar	psi	+/-	+/-
16	Total - Lines 14 + 15	bar	psi	+	+
17	Suction Line ΔP	bar	psi	-	-
18	Filter/Strainer ΔP	bar	psi	-	-
19					
20	Total Suction Pressure	bar g	psi g	+	+
21	DISCHARGE CONDITION				
22	Pressure at Equipment	bar g	psi g	+	+
23	Static Head	bar	psi	+/-	+/-
24					
25	Exchanger ΔP	bar	psi	+	+
26					
27	Furnace ΔP	bar	psi	+	+
28	Orifice ΔP	bar	psi	+	+
29	Control Valve ΔP	bar	psi	+	+
30					
31	Line ΔP	bar	psi	+	+
32					
33	Total Discharge Press.	bar g	psi g	+	+
34	Differential Pressure	bar	psi		
35	Differential Head	bar	psi		
36	NPSH				
37	Total Suction Pressure	bar a	psi a		
38	Vapor Pressure	bar a	psi a		
39	NPSH - Lines 37 - 38	bar a	psi a		
40	=	m	ft.		
41	Safety Margin	m	ft.		
42	NPSH - Lines 40-41	m	ft.		
43	Hydraulic Power	kW	Hp		
44	Estimated Efficiency	%	%		
45	Estimated Abs. Power	kW	Hp		
46	Type of Pump				
47	Drive				
48					
49	Material - Casing				
50	- Impeller				
51	- Shaft				
52					
53	Sour Service		Yes/No		
54	HEAD m = 10.2 x bar /SG m = 10 x kg/cm2 / SG ft = 2.31 x psi / SG				
55	VOLUME m3/h x SG x 1000 = kg/h igpm x SG x 600 = lb/h				
56	POWER kW=m³/h x bar/36.0 kW=m³/h x kg/cm² /36.71 Hp = igpm x psi/1427				

PUMP CALCULATION SHEET

Document No.
Sheet   of   Rev.
Job
Item Name.
Item No. (s)
No. Working   Total No. off

SKETCH OF PUMP HOOK-UP

NOTES
1. Pump shut-ff head not to exceed.........
2. Relief valve on pump discharge to be set at ....
3. Pump case design pressure............
         design temperature........
4. Sealing/flushing fluid available:
5. Cooling medium available:
6. Insulation required:
7. Start-up/commissioning fluid SG.

	1	Date	2	Date	3	Date	4	Date	5	Date
58 Description										
59 Made/Revised by										
60 Checked by										
61 Approved Process										
62 Approved										

PUMP SCHEDULE

		Units				Sheet of		Issue No.:		Date	Date	Date
	Job No.:							Description				
	Job Name:							Made/Revised by				
	Document No.							Checked by				
	Project No.							Approved- Process				
	Client							Approved				
	Location									1	2	3
	Plant											
1	Item Number											
2	Item Name											
3	Number Off: No. Running / Total Installed											
4	Preferred Type											
5	Liquid Pumped											
6	Operating - Temperature	°C	°F									
7	Specific Gravity at Operating Temperature											
8	Viscosity at Operating Temperature	cP	lb/ft.h									
9	Vapor Pressure at Operating Temperature	bara	psia									
10	Design Capacity (per pump)	m³/h	ft³/h									
11	Normal Capacity (per pump)	m³/h	ft³/h									
12	Discharge Pressure	bar g	psig									
13	Suction Pressure	bar g	psig									
14	Differential Head	bar /m	psi/ft									
15	Minimum NPSH Available / Required	m	ft.									
16	Hydraulic Power (per pump)	kW	hp									
17	R.P.M											
18	Estimated Pump Efficiency	-	-									
19	Estimated Shaft Power (per pump)	kW	hp									
20	Type of Driver	-	-									
21	Driver Rated Power (per pump)	kW	hp									
22	Design Power Consumption (per pump)	kW	hp									
23	Normal Power Consumption (per pump)	kW	hp									
24	Max. Shut-Off Head	bar	psi									
25												
26	Material: Casing											
27	Impeller											
28	Shaft											
29	Sour Service											
30	Packing Type / Mechanical Seal Type											
31												
32	Services / Utilities											
33	Cooling Water											
34	Seal / Flush Fluid	m³/h	ft³/h									
35	Quench	m³/h	ft³/h									
36	Notes											
37												
38												
39												
40												
41												
42												
43												
44												
45												

APPENDIX D

ΣΩK A.R.C. TECHNOLOGY	RELIEF DEVICE PHILOSOPHY SHEET		DOCUMENT / ITEM REFERENCE		
	EQUIPMENT No.:				
	DATE:		SHEET No.:	OF	

CHECKED BY:		MADE BY:					1
							2
DESIGN CODES:	VESSELS		EXCHANGERS		LINES		3
							4
OTHER REQUIREMENTS							5
							6
BASIS FOR CALCULATION:							7
							8
SET PRESSURE, PSIG:	MAX. BACK PRESSURE:		(a) BEFORE RELIEVING		(b) WHILE RELIEVING		9
NORMAL CONDITIONS UNDER RELIEF DEVICE:					Calculated		10
STATE:	TEMPERATURE, °F:	PRESSURE, PSIG:	POSSIBLE CAUSE?	FLUID RELIEVED	RELIEF RATE, lb/h	ORIFICE AREA, in².	11 / 12
HAZARDS CONSIDERED							13
1. Outlets blocked							14
2. Control Valve malfunction							15
3. Machine trip/ overspeed/density change							16
4. Exchanger tube rupture							17
5. Power failure/ Voltage dip							18
6. Instrument air failure							19
7. Cooling failure							20
8. Reflux failure							21
9. Abnormal entry of volatile liquid							22
10. Loss of liquid level							23
11. Abnormal chemical reaction							24
12. Boxed in thermal expansion							25
13. External fire							26
14. (specify)							27
15. (specify)							28
16. (specifty)							29
SELECTED DESIGN CASE:							30
							31
RELIEVED FLUID: STATE	DENSITY / MW:		TEMPERATURE:		Cp/Cv:		32
COMPOSITION:					:FLASHING		33
RELIEF RATE REQUIRED, lb/h:	ORIFICE SELECTED:		AREA, in²:		TYPE:		34
ACTUAL CAPACITY, lb/h							35
REMARKS/SKETCH							36
							37–45

Issue No:	1	Date	2	Date	3	Date		46
Description								47
Made/Revised by								48
Checked by								49
Approved- Process								50
Approved								51

TANK & VESSEL AGITATOR DATA SHEET

Project Name:	Drawing No:
	Project No.:
	sheet .... Of ....
Equipment No: No. Off:	Associated Vessel/Tank* Item No:

PROCESS DEPT. INFORMATION

PROCESS
1. Largest and smallest charge:
2. Components added during mixing:
3. Agitator operating while vessel is being filled or product withdrawn? Yes / No*
4. If continuous, throughput per hour:
5. Process duty: Mixing liquids / Dissolving / Suspensions / Emulsions / Gas absorption / Homogenisation*
6. Mixing effect: Violent / Medium / Moderate*   Time available for mixing:
7. Working Pressure: Design Pressure:
8. Working Temperature: Design Temperature:
9. Special Remarks:

MIXING
10. Components % by weight:
11. Temperature during mixing:
12. Specific gravity of components:
13. Viscosity of mixing at mixing temperature:
14. Specific gravity of product at mixing temperature:
15. Size of solid particles:
16. Special Remarks:

DESIGN DEPT. INFORMATION

IMPELLER
17. Type of impeller:
18. Number of impellers on shaft:
19. Position of impellers:
20. Distance between shaft end & vessel: Bottom bearing:
21. Preferred impeller speed: Shaft diameter:
22. Type of drive: Direct / Vee belt / Fluid*
23. Type of seal: Vapor / Packed gland / Mechanical / Easy replacement*
24. Method of installation: Assembled in / Assembled out* of vessel
25. Entry position:
26. Materials of construction:
27. Absorbed HP / KW*:   Installed HP / KW*:
28. Other information:

TANK OR VESSEL
29. Dimensions: Capacity:
30. Coils, baffles, etc.:
31. Fixing agitator (beams, flanges, etc.):
32. If at atmospheric pressure: Closed / Open*   Can stuffing box be greased? Yes / No*
33. Headroom available above agitator:
34. Other information:

ELECT.
35. Agitator installed: Indoors/Outdoors   Motor enclosure:
36. Electrical specification:
37. Any other electrical information:

GENERAL
38. Motor to be included: Yes / No*
39. Motor Will / Will not* be sent to manufacturer for assembly and allignment.
40. Threads: Unified /*
41. All rotating parts must be strictly guarded to BS 1649/ASME
42. Fixing bolts supplied by:
43. Net weight including motor:
44. Witnessed run in air:
45. General notes:

NOTE * indicates delete as necessary.

	Date 1	Date 2	Date 3
Description			
Made/Revised by			
Checked by			
Approved Process			
Approved			

Appendix D

TANK PROCESS DATA SHEET

ΣΩΞ A.K.C. TECHNOLOGY

Doc. No:
Item No.
Job:
Sheet   of

Item Name:

NOTE: * indicates delete as necessary;   ** indicates for other than Code reason

5									
10									
15									
20									
25									
26	Shell Diameter (O.D. / I.D.):				Shell Length:			No. Required:	
27	Center Line: *Horizontal / Vertical								
28		Pressure: g		Temperature: °F / °C	Nozzles	Mark No.	Size	Number	
29	Item Number				Inlet	C-			
30	Operating								
31	Design				Vapor Out	C-			
32	Emergency Vacuum Design: * Yes / No								
33		Material		Corr. Allowance	Liquid Out:	C-			
34	Shell								
35	Heads								
36	Liner				Thermocouple	R-			
37	Type of Heads				Pressure Gauge	R-	by Inst. Op.		
38	Code:				Gauge Glass	R-			
39	Stress Relieve**:	Radiography**:			Level Control	R-			
40	Joint Efficiency:				Safety Valve	R-			
41	Density of Contents:		at	°F / °C					
42	Weight Empty:		Weight Full:						
43	Is vessel subject to mechanical vibration? *Yes / No				Vent	C-			
44	insulation:	Type: *Frost and Personnel Protection / Cold			Drain	C-			
45	*Yes / No	/ Anticondensation / Heat Conservation			Steam Out	C-			
46	REMARKS:								
47					Manhole	A-			
48									
49									
50									
51									
52									
53					min. Base Elev'n:		k- Skirt Length		
54					Material:				

Issue No.	1	Date	2	Date	3	Date	4	Date
Made/Revised by								
Checked by								
Approved - Process								
Approved by								

TANK SCHEDULE

			Sheet of		Issue No.:	1	2	3
Job No.:					Description			
Job Name:					Made/Revised by			
Document No.					Checked by			
Project No.					Approved- Process			
Client					Approved			
Location						Date	Date	Date
Plant								

		Units						
1	Item Number							
2	Item Name							
3	Number Required							
4	Type							
5	Length x Width	mm	in.					
6	Height	mm	in.					
7	Diameter							
8	Minimum Base Elevation	mm	in.					
9	Total Volume	m^3	ft^3					
10	Working Capacity	m^3	ft^3					
11	Internals / Fittings							
12								
13								
14	Operating Pressure	bar g	psig					
15	Operating Temperature	°C	°F					
16	Design Pressure +/−	bar g	psig					
17	Design Temperature	°C	°F					
18	Vacuum Design							
19	Material: Shell							
20	Liner							
21	Internals							
22								
23	Shell Corrosion Allowance	mm	in.					
24	Sour Service							
25	Stress Relieved for Process Reasons							
26	Insulation							
27								
28								
29								
30	Notes							
31								
32								
33								
34								
35								
36								
37								
38								
39								
40								
41								
42								
43								
44								
45								

APPENDIX D

TIE-IN-SCHEDULE

Job No.:		Sheet	of	Issue No.:		1	Date	2	Date	3	Date
Job Name:				Description							
Document No.				Made/Revised by							
Project No.				Checked by							
Client				Approved-Process							
Location				Approved							
Plant											

TIE-IN NUMBER	SERVICE	DESCRIPTION	EXISTING LINE NUMBER	NEW LINE NUMBER	EXISTING ELD* NUMBER	NEW ELD* NUMBER	REMARKS	TIE-IN NUMBER

ELD* = ENGINEERING LINE DRAWING

Made by:	ΣΩΚ A.K.C. TECHNOLOGY	TOWER PROCESS DATA SHEET		Document No.			
Date:							
		Job		Item No:(s)			
Checked by:		Item Name					
				Sheet of			

Tower Name:

NOTE: * indicates delete as necessary; ** indicates for other than code reason

				Top	Bottom				
1									
2	Shell Diameters O.D. - I.D.								
3	No. of Trays								
4	Pressure		Operating						
5	*psig		Design						
6	Temperature		Operating						
7	*°F / °C		Design						
8	Material		Shell						
9			Trays						
10			Caps						
11			Liner or Clad						
12	Corrosion -		Shell						
13	Allowance		Heads						
14			Trays						
15	Tray Spacing								
16	Type of Liquid Flow								
17	Type of Trays								
18	Joint Efficiency								
19	Code			Emergency Vac. Design *Yes / No					
20	Stress Relieved **Yes / No			Radiography **Yes / No					
21	Is vessel subject to mechanical vibration *Yes / No								
22	Insulation			Type: *Frost and Personnel Protection / Cold /					
23	*Yes / No			/anticondensation / Heat Conservation /					
24	Min. Base Elevation:				Skirt length:				
25	Weight Empty:				Full:				
26	Nozzles	Mark No.	Size	Number					
27	Feed	C-							
28									
29	Overhead Vpr.	C-							
30	Reflux In	C-							
31									
32	Bottoms	C-							
33	Reboiler Vpr	C-							
34	Reboiler liq	C-							
35									
36									
37	Thermocouple	R-	by inst. Gp.						
38	Level Glass	R-							
39	Press Gauge	R-							
40	Level Control	R-							
41	Safety Valve	R-							
42									
43									
44	Vent	C-							
45	Drain	C-							
46	Steam Out	C-							
47	Manholes	A-							
48	Handholes	A-							
49	Cap Type:			Tray Layout Ref:					
50	NOTES								
51									
52									
53									
54									
55			1	Date	2	Date	3	Date	4	Date
56	Description								
57	Made/Revised by								
58	Checked by								
59	Approved Process								
60	Approved by								

TRAY LOADING SUMMARY

Job No.
Item No.
Job:
Sheet of

Tower Name: Type of Tray

#	Item	Metric	British						
1	Manufacturer			Mfr. Ref:					
2	Pressure at top of Tower	barg	psia						
3	Max. ΔP over Tower	bar	psi	units Used: METRIC/BRITISH (delete one)					
4	No. of Trays			Aboved Feed			Below Feed		
5	Tray location								
6	Tray Number*								
7	Tray Spacing	mm	in.						
8	Tower Internal Diameter I.D.	mm	in.						
9	Vapor to Tray								
10	Temperature	°C	°F						
11	Compressibility								
12	Density	kg/m³	lb/ft³						
13	Molecular Weight								
14	Rate	kg/h	lb/h						
15									
16	Liquid from Tray								
17	Temperature	°C	°F						
18	Surface Tension	dynes/cm	lb/ft						
19	Viscosity	cps	lb ft/hr						
20	Density	kg/m³	lb/ft³						
21	Rate	kg/h	lb/h						
22	Foaming Tendency **								
23	Number of Passes								
24	Minimum Hole Diameter ***	mm	in.						
25	Minimum DC Residence Time***	secs	s						
26	Maximum Rate as % Design								
27	Minimum Rate as % Design								
28	Design Rate % Flood Rate***								
29									
30	Tray Material								
31	Valve or Cap Material								
32	Corrosion Allowance	mm	in.						
33	Tray Thickness***	mm	in.						
34									
35									
36									
37									

38 NOTES: * Trays are numbered from the bottom of the tower upwardstop of the tower downwards
39   ** Indicate whether 'non', 'moderate', 'high' or 'severe'
40   *** Data to be supplied by tray manufacturer unless special Process/Client requirement entered here.

41 REMARKS:

55			1	Date	2	Date	3	Date
56		Description						
57		Made/Revised by						
58		Checked by						
59		Approve - Process						
60		Approved by						

Trip Schedule

TRIP SCHEDULE				Sheet	of		Issue No.		Date		Date	
Job No.:							Description					
Job Name:							Made/Revised by					
Document No.							Checked by					
Project No.							Approved Process					
Client							Approved					
Location												
Plant												
1	2	3	4	5	6	7	8	9	10	11	12
INITIATING VARIABLE	BASIC	PRE-ALARM		CUT OUT		FIRST FAULT	INHIBIT	TRIP LINKAGE	ITEM	PRIMARY	REMARKS
	INST. No.	INST. No.	SETTING	INST. No.	SETTING	INDICATION	SWITCH		ACTUATED	PROCESS EFFECT	

Trip Schedule.

Utility summary sheet.

Vessel and tank schedule.

Vessel and tank summary

VESSEL SCHEDULE

			Sheet of	Issue No.:	1	Date	2	Date	3	Date
Job No.:				Description						
Job Name:				Made/Revised by						
Document No.				Checked by						
Project No.				Approved- Process						
Client				Approved						
Location										
Plant										

		Units								
1	Item Number									
2	Item Name									
3	Number Required									
4										
5	Shell Diameter (O.D/I.D.)	mm	in.							
6	Shell Length	mm	in.							
7	Center Line (H/V)									
8	Boot Size	mm	in.							
9	Base Elevation	mm	in.							
10	Internals									
11										
12										
13										
14	Operating Pressure	bar g	psig							
15	Operating Temperature	°C	°F							
16	Design Pressure	bar g	psig							
17	Design Temperature	°C	°F							
18	Vacuum Design									
19	Material: Shell									
20	Liner									
21	Internals									
22										
23	Shell Corrosion Allowance	mm	in.							
24	Sour Service									
25	Stress Relieved for Process Reasons									
26	Insulation									
27										
28										
29										
30	Notes									
31–45										

ΣΩZ A.R.C. TECHNOLOGY	**WATER ANALYSIS SHEET**	Submitted by: Address:		Location: Date:			
	** Analysis		No. Date:				
	pH Value						
	Suspended Solids	mg/l					
	Total Dissolved Solids	mg/l @ 110°C					
	Total Dissolved Solids	mg/l @ 180°C					
	Alkalinity to Pp	as mg/l CaCO$_3$					
	Alkalinity to M.O.	as mg/l CaCO$_3$					
	Sulphate	as mg/l SO$_4$					
	Chloride	as mg/l Cl					
	Nitrate	as mg/l NO$_3$					
	Silica	as mg/l SiO$_2$					
	Phosphate	as mg/l PO$_4$					
	Total Anions	as mg/l CaCO$_3$					
	Total Hardness	as mg/l CaCO$_3$					
	Calcium	as mg/l Ca					
	Magnesium	as mg/l Mg					
	*Sodium	as mg/l Na					
	*Potassium	as mg/l K					
	Iron	as mg/l Fe					
	Manganese	as mg/l Mn					
	Free and Saline Ammonia	as mg/l NH$_3$					
	Total Cations	as mg/l CaCO$_3$					
	Free Dissolved CO$_2$	as mg/l CO$_2$					
	Dissolved O$_2$	as mg/l O					
	Colour Hazaen Units						
	+Turbidity Formazin Units	FTU APHA					
	Lead	as mg/l Pb					
	Copper	as mg/l Cu					
	Residual Chlorine						
	Flouride	as mg/l F					
	Sulfite	as mg/l SO$_3$					

NOTES: 1. Please state if units other than milligrams per liter, or different conditions of test, are used.
2. * If Na and/or K form an appreciable amount of total cations, please state Alkalinity to Phenolphthalein (p), Alkalinity to Methyl Orange (M.O.), and Carbonate Hardness.
3. + DO NOT use Formazin Units as defined by British Standard BS. 2690 Pt. 9.
4. ** Tick if item definitely to be included in analysis.
5. 100 mg/l CaCO$_3$ = 2 milliequivalents/ (m val/l)

	WATER ANALYSIS SHEET	Submitted by: Address:		Location: R.W.A. Area: Date:			
PD12/2 **	Analysis	No. Date:					
	pH Value						
	Suspended Solids	mg/l					
	Total Dissolved Solids	mg/l @ 110°C					
	Total Dissolved Solids	mg/l @ 180°C					
	Alkalinity to Pp	as mg/l $CaCO_3$					
	Alkalinity to M.O.	as mg/l $CaCO_3$					
	Sulphate	as mg/l SO_4					
	Chloride	as mg/l Cl					
	Nitrate	as mg/l NO_3					
	Silica	as mg/l SiO_2					
	Phosphate	as mg/l PO_4					
	Total Anions	as mg/l $CaCO_3$					
	Total Hardness	as mg/l $CaCO_3$					
	Calcium	as mg/l Ca					
	Magnesium	as mg/l Mg					
	*Sodium	as mg/l Na					
	*Potassium	as mg/l K					
	Iron	as mg/l Fe					
	Manganese	as mg/l Mn					
	Free ans Saline Ammonia	as mg/l NH_3					
	Total Cations	as mg/l $CaCO_3$					
	Free Dissolved CO_2	as mg/l CO_2					
	Dissolved O_2	as mg/l O					
	Colour Hazaen Units						
	+Turbidity Formazin Units	FTU APHA					
	Lead	as mg/l Pb					
	Copper	as mg/l Cu					
	Residual Chlorine						
	Flouride	as mg/l F					
	Sulphite	as mg/l SO_3					

NOTES:
1. Please state if units other than milligrams per liter, or different conditions of test, are used.
2. * If Na and/or K form an appreciable amount of total cations, please state Alkalinity to Phenolphthalein (p), Alkalinity to Methyl Orange (M.O.), and Carbonate Hardness.
3. + DO NOT use Formazin Units as defined by British Standard BS. 2690 Pt. 9.
4. ** Tick if item definitely to be included in analysis.
5. 100 mg/l $CaCO_3$ = 2 milliequivalents (m val/l) = 10 French degrees (Frh) = 5.6 German degress (dH)

Glossary of Petroleum and Petrochemical Technical Terminologies

Abatement: **1**. The act or process of reducing the intensity of pollution. **2**. The use of some method of abating pollution. **3**. Putting an end to an undesirable or unlawful condition affecting the wastewater collection system.

Abrasion (Mechanical): Wearing away by friction.

Abrasive: Particles propelled at a velocity sufficient to cause cleaning or wearing away of a surface.

Absolute Porosity: The percentage of the total bulk volume, which is pore spaces, voids, or fractures.

Absolute Pressure: **1**. The reading of gauge pressure plus the atmospheric pressure. **2**. Gauge pressure plus barometric or atmospheric pressure. Absolute pressure can be zero in a perfect vacuum. Units, psia, bara. e.g., psia = psig + 14.7, bara = barg + 1.013.

Absolute Temperature: Temperature measurement starting at absolute zero. e.g., $°R = °F + 460$, $K = °C + 273.16$.

Absolute Viscosity: The measure of a fluid's ability to resist flow without regard to its density. It is defined as a fluid's kinematic viscosity multiplied by its density.

Absorbent: The material that can selectively remove a target constituent from another compound by dissolving it.

Absorption: A variation of fractionation. In a distillation column, the stream to be separated is introduced in vapor form near the bottom. An absorption liquid called lean oil is introduced at the top. The lean oil properties are such that as the two pass each other, the lean oil will selectively absorb components of the stream to be separated and exit the bottom of the fractionator as rich oil. The rich oil is then easily separated into the extra and lean oil in conventional fractionation.

Absorption Gasoline: Gasoline is extracted from wet natural gas by putting the gas in contact with oil.

Absorption Oil (Facilities): The wash oil is used to remove heavier hydrocarbons from the gas stream.

Accident: An event or sequence of events or occurrences, natural or man-made, that results in undesirable consequences and requires an emergency response to protect life and property.

Accumulator: A vessel that receives and temporarily stores a liquid used in the feedstock or the processing of a feed stream in a gas plant or other processing facility.

Acentric Factor: A correlating factor that gives a measure of the deviation in the behavior of a substance to that for an idealized simple fluid. It is a constant for each component and has been correlated with the component vapor pressure.

Acid Gas: **1**. A gas that contains compounds such as CO_2, H_2S, or mercaptans (RSH, where $R = C_nH_{2n+1}$, n= 1, 2) that can form an acid in solution with water. **2**. Group of gases that are found in raw natural gas and are usually considered pollutants. Among these are CO_2, H_2S, and mercaptans. **3**. Any produced gas primarily H_2S and CO_2 that forms an acid when produced in water.

Acid Inhibitor: Acid corrosion inhibitor. It slows the acid attack on metal.

Acid Number: A measure of the amount of potassium hydroxide (KOH) needed to neutralize all or part of the acidity of a petroleum product. Also referred to as neutralization number (NN) or value (NV) and total acid number (TAN).

Acid-Soluble Oil (ASO): 1. High boiling polymers are produced as an unwanted by-product in the alkylation processes. **2.** Polymers produced from side reactions in the alkylation process.

Acid Treating/Treatment: A process in which unfinished petroleum products, such as gasoline, naphthas, kerosene, diesel fuel, and lubricating oil stocks, are contacted with sulfuric acid to improve their color, odor, and other properties.

Acidity: The capacity of water or wastewater to neutralize bases. Acidity is expressed in milligrams per liter of equivalent calcium carbonate. Acidity is not the same as pH because water does not have to be strongly acidic (low pH) to have a high acidity. Acidity is a measure of how much base must be added to a liquid to raise the pH to 8.2.

AC Motor: Most of the pumps are driven by alternating current, three-phase motors. Such motors that drive pumps are usually fixed-speed drivers. DC motors are rarely used in process plants.

Actual Tray: A physical tray (contact device) in a distillation column, sometimes called a plate.

Activity of Catalyst: Activity generally means how well a catalyst performs with respect to reaction rate, temperature, or space velocity.

Adsorbents: Special materials like activated charcoal, alumina, or silica gel, used in an adsorption process that selectively cause some compounds, but not others to attach themselves mechanically as liquids.

Adsorption: 1. A process for removing target constituents from a stream by having them condense on an adsorbent, which is then taken offline so the target constituents can be recovered. **2.** The process by which gaseous components adhere to solids because of their molecular attraction to the solid surface.

Alarms: Process parameters (levels, temperatures, pressures, flows) are automatically controlled within a permissible range. If the parameter moves outside this range, it sometimes activates both an audible and a visual alarm. If the panel board operator fails to take corrective action, a trip may also then be activated.

Alcohol: The family name of a group of organic chemical compounds composed of carbon, hydrogen, and oxygen. The series of molecules vary in chain length and are composed of a hydrocarbon plus a hydroxyl group, $CH_3(CH_2)n - OH$ (e.g., methanol, ethanol, tertiary butyl alcohol).

Alkanolamine: An organic nitrogen bearing compound related to ammonia having at least one, two, or three of its hydrogen atoms substituted with at least one, two, or three linear or branched alkanol groups where only one or two could also be substituted with a linear or branched alkyl group (i.e., methyl diethanolamine MDEA). The number of hydrogen atoms substituted by alkanol or alkyl groups at the amino site determines whether the alkanolamine is primary, secondary, or tertiary.

Alkylate: 1. The gasoline produced by an alkylation process. It is made by combining the low boiling hydrocarbons catalytically to obtain a mixture of high-octane hydrocarbons boiling in the gasoline range. **2.** The product of an alkylation reaction. It usually refers to the high octane product from alkylation units. This alkylate is used in blending high octane gasoline.

Alkylation: 1. A refining process for chemically combining isobutane (iC_4H_{10}) with olefin hydrocarbons [e.g., propylene (C_3H_6), butylenes (C_4H_8)] through the control of temperature and pressure in the presence of an acid catalyst. 2. A refining process in which light olefins primarily a mixture of propylene (C_3H_6), butylenes (C_4H_8), and/or amylenes are combined with isobutane (iC_4H_{10}) over an acid catalyst to produce high octane gasoline (highly branched $C_5 - C_{12}$ i-paraffins), called alkylate. The commonly used catalysts are sulfuric acid (H_2SO_4) and hydrofluoric acid (HF). The major constituents of alkylate are isopentane and isooctane (2,2,4 – trimethyl pentane, TMP), the latter possessing an octane number of 100. The product, alkylate, is an isoparaffin, has high octane value, and is blended with motor and aviation gasoline to improve the antiknock value of the fuel.

Alkylate Bottoms: A thick, dark brown oil containing high molecular-weight polymerization products of alkylation reactions.

Aluminum Chloride Treating: A quality improvement process for steam cracked naphthas using aluminum chloride ($AlCl_3$) as a catalyst. The process improves the color and odor of the naphtha by the polymerization of undesirable olefins into resins. The process is also used when the production of resins is desirable.

American Petroleum Institute (API): An association, which among many things sets technical standards for measuring, testing, and other types of handling of petroleum.

Amine Treating: Contacting of a gas or light hydrocarbon liquid with an aqueous solution of an amine compound to remove the hydrogen sulfide (H_2S) and carbon dioxide (CO_2).

Anaerobic Digestion: This is a collection of processes by which microorganisms break down biodegradable material in the absence of oxygen. The process is used for industrial or domestic purposes to manage waste and/or to produce fuels. Much of the fermentation is used industrially to produce food and drink products, as well as home fermentation uses anaerobic digestion.

The digestion process begins with bacterial hydrolysis of the input materials. Insoluble organic polymers such as carbohydrates are broken down to soluble derivatives that become available for other bacteria. It is used as part of the process to treat biodegradable waste and sewage sludge. As part of an integrated waste management system, anaerobic digestion reduces the emission of landfill gas into the atmosphere. Anaerobic digestion is widely used as a source of renewable energy. The process produces biogas, consisting of methane, carbon dioxide, and traces of other "contaminant" gases. The biogas can be used directly as fuel in combined heat and power gas engines or upgraded to natural gas–quality biomethane. The nutrient-rich digestate also produced can be used as fertilizer.

Aniline Point: The minimum temperature for complete miscibility of equal volumes of aniline and the test sample. The test is considered an indication of the paraffinicity of the sample. The aniline point is used as a classification of the ignition quality of diesel fuels.

Antiknock Agent: 1. Is a gasoline additive used to reduce engine knocking and increase the fuel's octane rating by raising the temperature and pressure at which auto-ignition occurs. The mixture is gasoline or petrol, when used in high compression internal combustion engines, has a tendency to knock (also referred to as pinging, or pinking) and/or to ignite early before the correct time spark occurs (pre-ignition, refers to engine knocking). 2. The most wanted and widely used additives in gasoline are the antiknock compounds. They assist to enhance the octane number of gasoline. Lead in the form of tetraethyl lead (TEL) or tetramethyl lead (TML) is a good antiknock compound. TEL helps to increase the octane number of gasoline without affecting any other properties, including vapor pressure, but when used alone in gasoline gives rise to troublesome deposits.

Antiknock Index: The Research Octane Number (RON) test simulates driving under mild conditions while the Motor Octane Number (MON) test simulates driving under severe conditions, i.e., under load and at high speed.

The arithmetic average of RON and MON that gives an indication of the performance of the engine under the full range of conditions is referred to as Antiknock Index (AKI). It is determined by:

$$\text{Antiknock Index (AKI)} = \frac{RON + MON}{2} \tag{1}$$

Antiknock Quality (Octane Number): Knocking is a characteristic property of motor fuels that governs engine performance and is expressed in terms of octane number. It depends on the properties of hydrocarbon type and nature. Octane number is the percentage of isooctane in the reference fuel, which matches the knocking tendency of the fuel under test. Research octane number (RON) and motor octane number (MON) are two methods used and are measured with a standard single-cylinder, variable compression ratio engine. For both octane numbers, the same engine is used but operated at different conditions. The distinctions between two octane numbers (RON and MON) measurement procedures are engine speed, the temperature of admission, and spark advance. The motor method captures the gasoline at high engine speeds and loads, and the research octane method at a low speed depends on the fuel characteristics. The MON is normally 8–10 points lower than the RON. A high tendency to auto-ignite, or low octane rating, is undesirable in a gasoline engine, but desirable in a diesel engine.

$$\text{Antiknock index (AKI)} = (RON + MON)/2. \tag{2}$$

API Gravity: A method for reporting the density of petroleum streams. It is defined as:

$$°API = \left[\frac{141.5}{Sp.Gr @ 60/60°F} - 131.5 \right] \tag{3}$$

where Sp.Gr is the specific gravity relative to water. °API gravity is reported at a reference temperature of 60°F (15.9°C).

The scale allows representation of the gravity of oils, which on the specific gravity 60/60°F scale varies only over a range of 0.776 by a scale that ranges from less than 0 (heavy residual oil) to 340 (methane).

According to the expression, 10°API indicates a specific gravity of 1 (equivalent to water-specific gravity). Thus, higher values of API gravity indicate lower specific gravity and therefore lighter crude oils, or refinery products and vice versa. As far as crude oil is concerned, lighter API gravity value is desired as more amount of gas fraction, naphtha, and gas oils can be produced from the lighter crude oil than with the heavier crude oil. Therefore, crude oil with high values of API gravity is expensive to produce due to its quality.

Classification of crude oils

Crude category	°API gravity
Light crudes	°API > 38
Medium crudes	38 > °API > 29
Heavy crudes	29 > °API > 8.5
Very heavy crudes	°API < 8.5

The higher the API gravity, the lighter the compound. Light crudes generally exceed 38°API, and heavy crudes commonly are crudes with an °API of 22 or below. Intermediate crudes fall in the range of 22–38 °API. (See Figures 1a and 1b).

Aromatics: 1. A group of hydrocarbons is characterized by having at least one benzene ring-type structure of six carbon atoms with three double and three single bonds connecting them somewhere in the molecule. The general formula is C_nH_{2n-6} where n = 6, 7, 8, etc. The simplest is benzene, plus toluene and the xylenes. Aromatics in gas oils and residues can have many, even scores of rings. **2.** The three aromatic compounds—benzene (C_6H_6), toluene (C_7H_8), and xylene (C_8H_{10}).

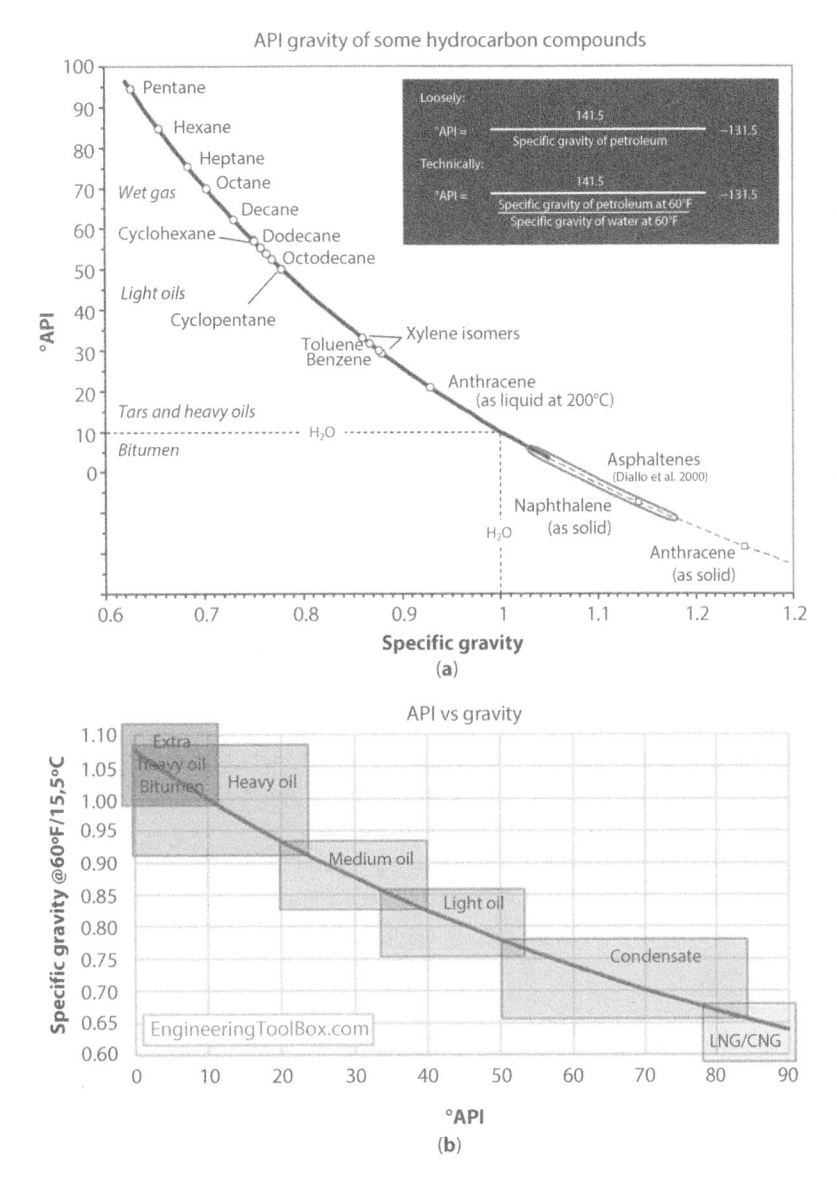

Figure 1 (a) A plot of °API vs. specific gravity of hydrocarbons compounds. (b) Specific gravity vs. °API of hydrocarbons compounds (Source: EngineeringToolBox.com)

As Low As Reasonably Practicable (ALARP): The principle that no industrial activity is entirely free from risk and that it is never possible to be sure that every eventuality has been covered by safety precautions, but that there would be a gross disproportion between the cost in (money, time, or trouble) of additional preventive or protective measures, and the reduction in risk in order to achieve such low risks (Figure 2).

Asphalt: 1. A heavy semi-solid petroleum product that gradually softens when heated and is used for surface cementing. Typically, brown or black in color, it is composed of high carbon to hydrogen hydrocarbons. It occurs naturally in crude oil or can be distilled or extracted. **2.** The end product used for area surfacing consists of refinery asphalt mixed with aggregation. **3.** Heavy tar-like residue from distillation of some types of crude oil. Asphalt components are high molecular weight derivatives of aromatic compounds. Not all asphalt materials are suitable for use as building agents in road pavement.

Asphaltenes: Highly condensed masses of high molecular weight aromatic compounds. They exist in petroleum residuum as the center of colloidal particles or micelles. The asphaltenes are kept in solution by an outer ring of

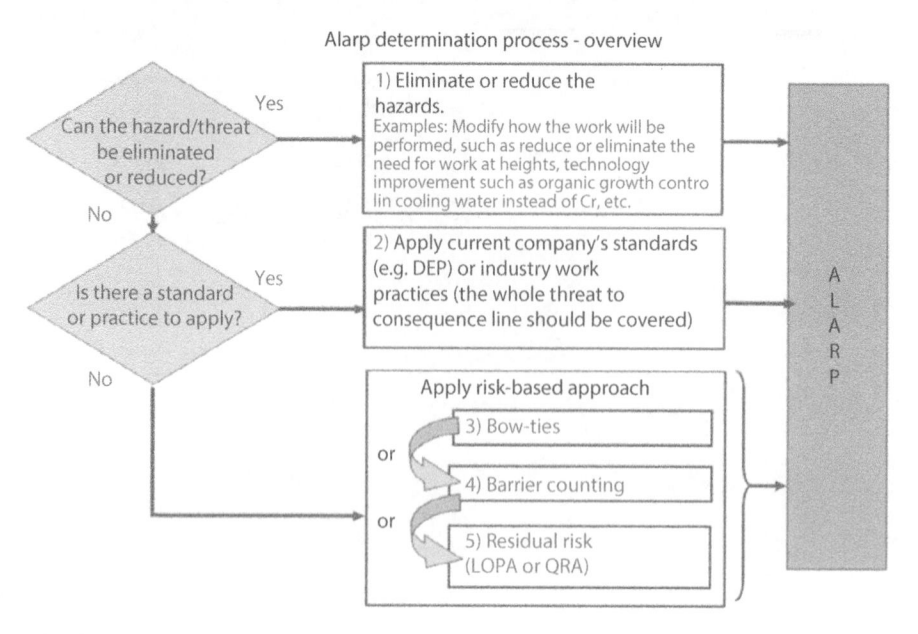

Figure 2 ALARP determination process overview. DEP = Design Engineering Practice.

aromatic compounds of lower molecular weight. They can precipitate when the continuous nature of the surrounding ring of aromatics is broken down by cracking processes.

Assay Data: Laboratory test data for a petroleum stream, including laboratory distillation, gravity, compositional breakdown, and other laboratory tests. Numerous important feed and product characterization properties in refinery engineering include:

1.	API gravity
2.	Watson Characterization factor
3.	Viscosity
4.	Sulfur content, wt %
5.	Nitrogen content, wt %
6.	Carbon residue, wt%
7.	Salt content
8.	Metal contents
9.	Asphaltene, %
10.	Naphthenes, %
11.	True boiling point (TBP) curve
12.	Pour point
13.	Cloud point
14.	Freeze point
15.	Aniline point

16.	Flash and fire point
17.	ASTM distillation curve
18.	Octane number
19.	Conradson carbon
21.	Reid vapor pressure
22.	Bottom sediment and water (BS &W)
23.	Light hydrocarbon yields ($C_1 - C_5$)

The crude quality is getting heavier worldwide. Existing refineries that are designed to handle normal crudes are being modified to handle heavy crude. New technology for upgrading is used to obtain clean and light products from lower cost feeds. The crude assay will determine the yields of different cuts and consequently the refinery configuration.

Associated Natural Gas: Natural gas that is dissolved in crude in the reservoir and is co-produced with the crude oil.

ASTM: American Society of Testing and Materials. Nearly all of the refinery product tests have been standardized by ASTM.

ASTM Distillation: A standardized laboratory batch distillation for naphthas and middle distillates carried out at atmospheric pressure without fractionation.

ASTM Distillation Range: Several distillation tests are commonly referred to as "ASTM distillations." These are usually used in product specifications. These ASTM distillations give results in terms of percentage distilled versus temperature for a sample laboratory distillation with no fractionation. The values do not correspond to those of refinery process distillations, where fractionation is significant.

ASTM D86 Distillation: Of an oil fraction takes place at laboratory room temperature and pressure. Note that the D86 distillation will end below an approximate temperature of 650°F (344°C), at which petroleum oils begin to crack at one atmospheric pressure.

ASTM D1160 Distillation: Of an oil fraction is applicable to high-boiling oil samples (e.g., heavy heating oil, cracker gas oil feed, residual oil, etc.) for which there is significant cracking at atmospheric pressures. The sample is distilled at reduced pressure, typically at 10 mm Hg, to inhibit cracking. In fact, at 10 mm Hg, we can distill an oil fraction up to temperatures of 950–1000°F (510–538°C), as reported on a 760 mm Hg basis. The reduced pressure used for D1160 distillation produces a separation of components that is more ideal than that for D86 distillation.

ASTM D2887 Distillation: Of oil fraction is a popular chromatographic procedure to "simulate" or predict the boiling point curve of an oil fraction. We determine the boiling point distribution by injecting the oil sample into a gas chromatograph that separates the hydrocarbons in a boiling-point order. We then relate the retention time inside the chromatograph to the boiling point through a calibration curve.

ASTM End Point of Distillates: End point is an important specification or way of describing gasolines, naphthas, or middle distillates. It is the approximate relationship between the end point of a fraction and its True Boiling Point (TBP) and other cut points.

Atmospheric distillation: 1. The refining process of separating crude oil components at atmospheric pressure by heating to temperatures of 600–750°F (316–400°C) (depending on the nature of the crude oil and desired products) and subsequent condensing of the fractions by cooling. **2.** Distillation/Fractionation of crude oil into various cuts/fractions under atmospheric conditions. The more volatile components (i.e., lower boiling points) rise through trays/bubble caps and are condensed at various temperatures and the least volatile components, short and long residues (i.e., higher boiler points), are removed as bottom products.

Atmospheric Crude Oil Distillation: The refining process of separating crude oil components at atmospheric pressure by heating to temperatures of about 600–750°F (316–400°C) (depending on the nature of the crude oil and desired products) and subsequent condensing of the fractions by cooling.

Atmospheric Gas Oil (AGO): A diesel fuel and No. 2 heating oil blending stock obtained from the crude oil as a side stream from the atmospheric distillation tower.

Atmospheric Reduced Crude (ARC): The bottoms stream from the atmospheric distillation tower.

Atmospheric Residuum: The heaviest material from the distillation of crude oil in a crude distillation column operating at a positive pressure.

Autoignition: The spontaneous ignition and resulting in a rapid reaction of a portion of or all the fuel–air mixture in the combustion chamber of an internal combustion engine. The flame speed is many times greater than that following normal ignition.

Autoignition Temperature (AIT): **1.** The lowest temperature at which a gas will ignite after an extended time of exposure. **2.** The lowest temperature at which a flammable gas or vapor air mixture will ignite from its own heat source or a contacted heat source without the necessity of a spark or a flame.

Aviation Gasoline Blending Components: Naphthas that will be used for blending or compounding into finished aviation gasoline (e.g., straight-run gasoline, alkylate, reformate, benzene, toluene, xylenes). Excludes oxygenates (alcohols, ethers), butanes, and pentanes. Oxygenates are reported as other hydrocarbons, hydrogen, and oxygenates.

Aviation Gasoline (Finished): A complex mixture of relatively volatile hydrocarbons with or without small quantities of additives, blended to form a fuel suitable for use in aviation reciprocating engines. Fuel specifications are provided in ASTM Specification D910 and Military Specification MIL-G-5572. Note: Data on blending components are not counted in data on finished aviation gasoline.

Azeotrope: A constant boiling point mixture for which the vapor and liquid have identical composition. Azeotropes cannot be separated from conventional distillation.

Backflow: **1.** A flow condition, caused by differential pressure, resulting in the flow of liquid into the potable water supply system from sources other than those intended; or the backing up of liquid, through a conduit or channel, in a direction opposite to normal flow. **2.** Return flow from an injection of a fluid into a formation.

Back Pressure: A pressure caused by a restriction or fluid head that exerts an opposing pressure to flow.

Barrel: A volumetric measure of refinery feedstocks and products equal to 42 US gal.

Barrels Per Calendar Day (BPCD or B/CD): Average flow rates based on operating 365 days per year. The amount of input that a distillation facility can process under usual operating conditions. The amount is expressed in terms of capacity during a 24-hour period and reduces the maximum process capability of all units at the facility under continuous operation to account for the following limitations that may delay, interrupt, or slow down production: The capability of downstream facilities to absorb the output of crude oil processing facilities of a given refinery. No reduction is made when a planned distribution of intermediate streams through other than downstream facilities are part of a refinery's normal operation; the types and grades of inputs to be processed; the types and grades of products expected to be manufactured; the environmental constraints associated with refinery operations; the reduction of capacity for scheduled downtime due to such conditions as routine inspection, maintenance, repairs, and turnaround, and the reduction of capacity for unscheduled downtime due to such conditions.

Barrels Per Stream Day (BPSD or B/SD): The maximum number of barrels of input that a distillation facility can process within a 24-hour period when running at full capacity under optimal crude and product slate conditions with no allowance for downtime. This notation equals barrels per calendar day divided by the service factor.

Basic Process Control System (BPCS): A system that responds to input signals from the process, its associated equipment, other programmable systems, and/or an operator and generates output signals causing the process and its associated equipment to operate in the desired manner but which does not perform any safety instrumented functions (SIF) with a claimed Safety Instrumented Level, SIL $\geq$ 1.

Battery Limits (BL): The periphery of the area surrounding any process unit, which includes the equipment for the particular process.

Baume gravity: Specific gravity of liquids expressed as degrees on the Baume scale. For liquids lighter than water,

$$\text{Sp.Gr} @ 15.6/15.6°C = \frac{140}{130 + \deg Be} \tag{4}$$

For liquids heavier than water

$$\text{Sp.Gr} @ 15.6/15.6°C = \frac{145}{145 - \deg Be} \tag{5}$$

Bbl: Abbreviation for a quantity of 42 US gal.

Benchmark crude: A reference crude oil with which the prices of other crudes are compared.

Benzene (C_6H_6): An aromatic hydrocarbon present in small proportion in some crude oils and made commercially from petroleum by the catalytic reforming of naphthenes in petroleum naphtha. It is also made from coal in the manufacture of coke. Used as a solvent, in manufacturing detergents, synthetic fibers, and petrochemicals and as a component of high-octane gasoline.

Bernoulli equation: A theorem in which the sum of the pressure-volume, potential, and kinetic energies of an incompressible and non-viscous fluid flowing in a pipe with the steady flow with no work or heat transfer is the same anywhere within a system. When expressed in head form, the total head is the sum of the pressure, velocity, and static head. It is applicable only for incompressible and non-viscous fluids as:

SI Units

$$\frac{P_1}{\rho_1} + \frac{v_1^2}{2g} + z_1 = \frac{P_2}{\rho_2} + \frac{v_2^2}{2g} + z_2 + h_f \tag{6}$$

where h_f is the pipe friction from point 1 to point 2 may be referred to as the head loss in meters of fluid.

English Engineering Units

$$\frac{144P_1}{\rho_1} + \frac{v_1^2}{2g_c} + z_1\frac{g}{g_c} = \frac{144P_2}{\rho_2} + \frac{v_2^2}{2g_c} + z_2\frac{g}{g_c} + h_f \qquad (7)$$

where h_f is the pipe friction from point 1 to point 2 in foot-pounds force per pound of flowing fluid; this is sometimes referred to as the head loss in feet of fluid.

where, P is pressure, ρ is density, g_c is a conversion factor $\left(32.174\frac{lb_m}{lb_f}\cdot\frac{ft}{s^2}\right)$, g is the acceleration due to gravity (32 ft/s²), v is velocity, z is elevation, and h_f is frictional head loss. It is a statement of the law of the conservation of energy, which was formulated by Daniel Bernoulli in 1738 (Figure 3).

Bioenergy with carbon capture and Storage (BECCS): 1. Is the process of extracting bioenergy from biomass and capturing and storing the carbon. 2. Is the process during which carbon is captured and stored. If carefully managed, using sustainable biomass, BECCS can generate "negative emissions" because while providing energy it also captures and stores the atmospheric carbon monoxide (CO) that is absorbed by plants as they grow.

Bitumen: That portion of petroleum, asphalt, and tar products that will dissolve completely in carbon disulfide (CS_2). This property permits a complete separation from foreign products not soluble in carbon disulfide.

Blast: A transient change in gas density, pressure (either positive or negative), and velocity of the air surrounding an explosion point.

Blending: One of the final operations in refining, in which two or more different components are mixed together to obtain the desired range of properties in the final product.

Blending Components: Modern gasoline is a blend of various refinery streams produced by distillation, cracking, reforming, and polymerization together with additives to achieve the specific fuel performance requirements.

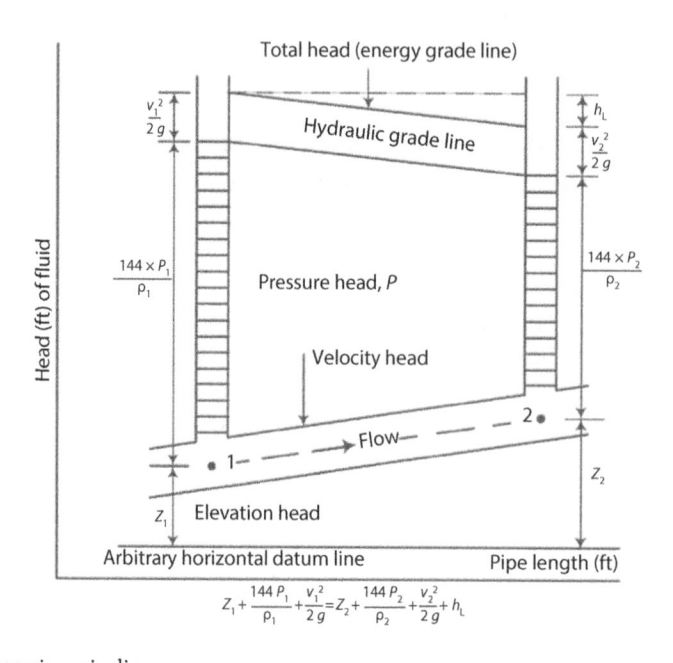

Figure 3 Distribution of fluid energy in a pipeline.

Blending Octane Number: When blended into gasoline in relatively small quantities, high-octane materials behave as though they had an octane number higher than shown by laboratory tests on the pure material. The effective octane number of the material in the blend is known as the blending octane number.

Blending Plant: A facility that has no refining capability but is either capable of producing finished motor gasoline through mechanical blending or blends oxygenates with motor gasoline.

Blending Value (Hydrocarbon): In octane ratings of a hydrocarbon made on blends of 20% hydrocarbon plus 80% of a 60:40 mixture of isooctane (iC_8H_{18}) and n-heptane (nC_7H_{16}), the blending octane number is a hypothetical value obtained by extrapolation of a rating of 100% concentration of the hydrocarbon.

Blocked Operation: A set of operating conditions and procedures that apply to particular feedstock and/or set of product specifications for a process.

Boiler: 1. A closed vessel in which a liquid is heated or heated and evaporated. Boilers are often classified as steam or hot water, low pressure or high pressure, and capable of burning one fuel or a number of fuels. 2. Vessel in which a liquid is heated with or without vaporization; boiling need not occur.

Boiler Feed Pump: A pump that returns condensed steam, makeup water, or both directly to the boiler.

Boiler Feed Water: Water supplied to a boiler by pumping.

Boiling Liquid Expanding Vapor Explosion (BLEVE): 1. The nearly instantaneous vaporization and corresponding release of energy of a liquid upon its sudden release from a containment under pressure than atmospheric pressure and at a temperature above its atmospheric boiling point. 2. A type of rapid phase transition in which a liquid contained above its atmospheric boiling point is rapidly depressurized, causing a nearly instantaneous transition from liquid to vapor with a corresponding energy release. A **BLEVE** is often accompanied by a large fireball if a flammable liquid is involved, since an external fire impinging on the vapor space of a pressure vessel is a common **BLEVE** scenario. However, it is not necessary for the liquid to be flammable to have a **BLEVE** to occur.

Blowdown: The disposal of voluntary discharges of liquids or condensable vapors from process and vessel drain valves, thermal relief or pressure relief valves.

Blowout: An uncontrolled flow of gas, oil, or other well fluids from a wellbore at the wellhead or into a ground formation, caused by the formation pressure exceeding the drilling fluid pressure. It usually occurs during drilling on unknown (exploratory) reservoirs.

Boiling Point: 1. Heat a liquid and its vapor pressure increases. When the liquid's vapor pressure equals the pressure in the vessel, the liquid starts to boil. The temperature at which this boiling starts is the liquid's boiling temperature. 2. Typically refers to the temperature at which a component or mixture of components starts to vaporize at a given pressure. When used in petroleum refining, it is usually synonymous with the normal boiling point (i.e., boiling point at one atmosphere). 3. The temperature at which the pressure exerted by molecules leaving a liquid equals the pressure exerted by the molecules in the air above it. A free-or-all of the molecules leaving the liquid then ensures. In a solution, the boiling point will be increased by a number that depends on the number of particles in the solution:

$$\text{delta (T)} = K_b \times (\text{number of solute molecules per liter})$$

where
    delta (T) = the rise in the boiling point.
    K_b = the ebullioscopic constant and varies from one solvent to another.

Boiling Range: 1. The spread of temperatures over which oil starts to boil or distill vapors and proceeds to complete evaporation. The boiling range is determined by ASTM test procedures for specific petroleum products. It is measured in °F or (°C). **2.** The lowest through to highest boiling temperatures for a petroleum stream when distilled. Boiling ranges are often reported on a TBP (true boiling point) basis, i.e., as normal boiling points.

Boiling Temperature: The temperature at which steam bubbles begin to appear within a liquid. When the fluid is a pure compound, the boiling point is unique for each pressure.

Boil Off: A small amount of LNG evaporates from the tank during storage, cooling the tank and keeping the pressure inside the tank constant and the LNG at its boiling point. A rise in temperature is encountered by the LNG being vented from the storage tank.

Boil Off Vapor: Usually refers to the gases generated during the storage or volatile liquefied gases such as LNG. Natural gas boils at slightly above -261°F (-163°C) at atmospheric pressure and is loaded, transported, and discharged at this temperature, which requires special materials, insulation, and handling equipment to deal with the low temperature and the boil-off vapor.

Boot, Boot Cooler: The section of a distillation column below the trays. For columns with very hot feeds, a portion of the bottom product is cooled and circulated through the boot or lowers the temperature of the liquid in the boot and prevents the depositing of coke. Many vacuum distillation columns have boot coolers.

Bottoms: 1. The heavy fractions or portions of crude oil that do not vaporize during fractionation/distillation. **2.** The accumulation of sediments, mud, and water in the bottoms of lease tanks. **3.** The product coming from the bottom of a fractionating column. In general, the higher-boiling residue is removed from the bottom of a fractionating tower. **4.** The liquid level is left in a tank after it has been pumped "empty" and the pump loses suction.

Bowtie analysis: 1. A qualitative risk analysis that portrays events and consequences on either side of a "bowtie." Barriers or safeguards are shown in between the two sides. It depicts the risks in ways that are readily understandable to all levels of operations and management. **2.** A type of qualitative safety review where cause scenarios are identified and depicted on the pre-event side (left side) of a bow-tie diagram. Credible consequences and scenarios outcomes are depicted on the post-event side (right side) of the diagram, and associated barrier safeguards are included (Figure 4).

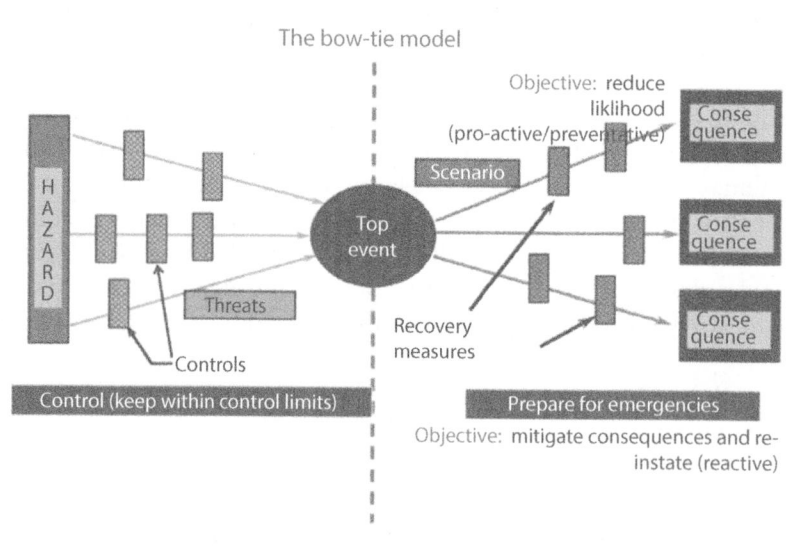

Figure 4 The bow-tie analysis.

Brackish Water: Indefinite term meaning water with small amounts of salt. Saltier than freshwater.

Brainstorming: A group problem-solving technique that involves the spontaneous contribution of ideas from all members of the group primarily based on their knowledge and experience.

Brent: A large oil field in the U.K. sector of the North Sea. Its name is used for a blend of crudes widely used as a price marker or benchmark for the international oil industry. Brent crude currently has an average quality of 38°API.

Brent Blend: A light sweet crude oil produced in the North Sea; a benchmark for pricing of many foreign crude oils.

Bright Stock: Heavy lube oils (frequently the vacuum still bottoms) from which asphaltic compounds, aromatics, and waxy paraffins have been removed. Bright stock is one of the feeds to a lube oil blending plant.

British thermal unit (Btu): A standard measure of energy; the quantity of heat required to raise the temperature of 1 pound of water by 1°F.

Bromine Index: Measure of the amount of bromine reactive material in a sample; ASTM D-2710.

Bromine Number: A test that indicates the degree of unsaturation in the sample (olefins and diolefins); ASTM D-1159.

BTX: The acronyms for the commercial petroleum aromatics benzene, toluene, and xylene.

Bubble Cap: **1.** It is an inverted cup with a notched or slotted periphery to disperse the vapor in small bubbles beneath the surface of the liquid on the bubble plate in a distillation column. The bubble caps cause the vapor coming from the bottom to come in intimate contact with the liquid sitting on the tray. **2.** A bubble cap tray has a riser or chimney fitted over each hole and a cap that covers the riser. The cap is mounted so that there is a space between the riser and cap to allow the passage of vapor. Vapor rises through the chimney and is directed downward by the cap, finally discharging through the slots in the cap, and finally bubbling through the liquid on the tray (Figure 5).

Bubble Point: **1.** This is the same as the boiling point. When a liquid is at its bubble point, it is said to be saturated liquid at temperature and pressure. If we raise the pressure, the liquid's bubble point temperature goes up. **2.** The temperature and pressure at which a liquid first begins to vaporize into a gas. **3.** The temperature at which the first bubbles appear when a liquid mixture is heated. **4.** The temperature at which a component or mixture of components begins to vaporize at a given pressure. It corresponds to the point of 0% vaporization or 100% condensation. The pressure should be specified, if not one atmosphere. **5.** The pressure at which gas begins to break out of undersaturated oil and form a free gas phase in the matrix or a gas cap.

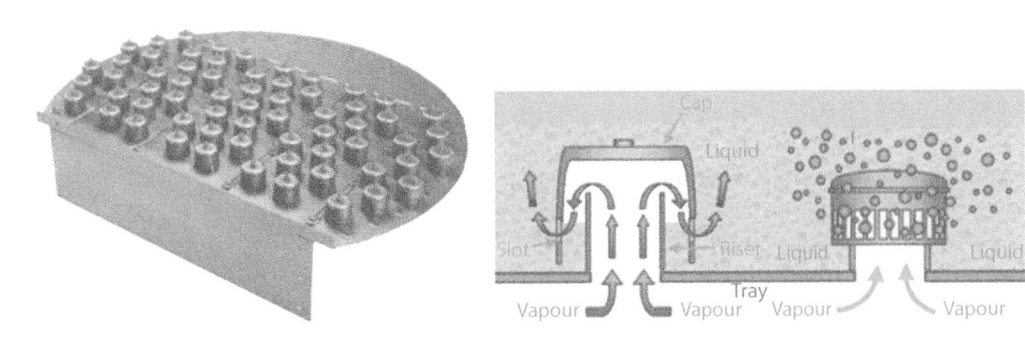

Figure 5 A bubble cap tray.

Bubble Tower or Column: A fractionating tower constructed in such a way that the vapors rising up pass through different layers of condensate on a series of plates. The less volatile portions of vapor condense in bubbling through the liquid on the plate, overflow to the next lower plate, and finally back to the boiler.

Bubble Tray: A horizontal tray fitted in the interior of a fractionating tower; meant to give intimate contact between rising vapors and falling liquid in the tower.

Bulk Properties: Provide a quick understanding of the type of the oil sample such as sweet or sour, light and heavy, etc. However, refineries require fractional properties of the oil sample that reflects the property and composition for specific boiling-point range to properly refine it into different end products such as gasoline, diesel, and raw materials for chemical process. Fractional properties usually contain paraffins, naphthenes, and aromatics (PNA) contents, sulfur content, nitrogen content for each boiling-point range, the octane number of gasoline, freezing point, cetane index, and smoke point for kerosene and diesel fuels.

Bulk Station: A facility used primarily for the storage and/or marketing of petroleum products, which has a total bulk storage capacity of less than 50,000 barrels and receives its petroleum products by tank car or truck.

Bulk Terminal: A facility used primarily for the storage and/or marketing of petroleum products, which has a total bulk storage capacity of 50,000 barrels or more and/or receives petroleum products by tanker, barge, or pipeline.

Bunker Fuel Oil: A heavy residual fuel oil used by ships, industry, and large-scale heating installations.

Butadiene (C_4H_6): A diolefin with two double bonds and two isomers. A colorless gas resulting from cracking processes. Traces result from the cat. cracking from catalytic dehydrogenation of butane (C_4H_{10}) or butylenes (C_4H_8) and in ethylene plants using butane, naphtha, or gas oil as feeds. Butadiene is principally used to make polymers like synthetic rubber and acrylonitrile butadiene styrene (ABS) plastics.

Butane (C_4H_{10}): A normally gaseous four-carbon straight-chain or branched-chain hydrocarbon extracted from natural gas or refinery gas streams. It includes normal butanes and refinery grade butanes and is designated in ASTM Specification D1835 and Gas Processors Association Specifications for commercial butane. Commercial butane is typically a mixture of normal and isobutene, predominantly normal. Hydrocarbons in the paraffin series with a general formula C_nH_{2n+2}, where n = 1, 2, 3, 4, 5, etc. To keep the liquid and economically stored, butane must be maintained under pressure or at low temperatures.

Butylene/Butene (C_4H_8): Hydrocarbons with several different isomers in the olefin series with a general formula C_nH_{2n}. Used in refining in an alkylation plant or in petrochemicals to make solvents and some polymers.

Carbon-to-Hydrogen Ratio: The carbon-to-hydrogen ratio is determined by the following:

$$\frac{C}{H} = \frac{74+15d}{26-15d} \tag{8}$$

where d is the specific gravity at 15°C

The carbon–hydrogen ratios of different products are:

LPG (d = 0.56)	= 4.68
Naphtha (d = 0.72)	= 5.57

Gasoline (d = 0.73)	= 5.64
ATF (d = 0.79)	= 6.067
SK (d = 0.795)	= 6.10
JP5 (d = 0.80)	= 6.14
HSD (d = 0.845)	= 6.50
LDO (d = 0.87)	= 6.72
LSHS (d = 0.98)	= 7.85
FO (d = 0.99)	=7.97

Calorific Value: 1. A measure of the amount of energy released as heat when a fuel is burned. **2.** The quantity of heat produced by the complete combustion of a fuel. This can be measured dry or saturated with water vapor, net or gross.

It is a measure of the heat-producing capacity of the fuel. It is determined by:

$$Q_v = 12400 - 2100d^2 \qquad (9)$$

where
Q_v = calorific value, gross cals/g
d = density at 15°C

Note: 1 cal = 4.184 Joules

Calorific value (average) of different fuels

Fuel	Calorific value, kcal/kg
Naphtha	11330
Kerosene	11070
HSD	10860
Fuel Oil	10219
Charcoal	6900
Hard coal	5000
Fire wood	4750
Lignite-Brown coal	2310

Carbon Number: The number of carbon atoms in one molecule of a given hydrocarbon, Figure 58 shows the names of hydrocarbons and their petroleum products.

Carbon Dioxide Equivalent (CO_{2eq}): US Environmental Protection Agency (EPA) defines carbon dioxide equivalent as the number of metric tons of CO_2 emissions with the same global warming potential as one metric ton of another greenhouse gas, and is calculated using Equation A-1 in 40 CFR Part 98.

Carbon Residue: Carbon residue is a measure of the coe-forming tendencies of oil. It is determined by destructive distillation in the absence of air of the sample to a coke residue. The coke residue is expressed as the weight percentage of the original sample. There are two standard ASTM tests, Conradson carbon residue (CCR) and Ramsbottom carbon residue (RCR).

Carbon capture and storage (CCS): CCS or carbon capture and sequestration is the process of capturing carbon dioxide (CO_2) before it enters the atmosphere.

Carbon intensity: The amount of CO_2 emitted when generating a unit of electricity, measured in gram of CO_2 per kWh of electricity produced.

Carbon Capture Utilization and Storage (CCUS): The process of capturing carbon dioxide from industrial processes, power generation, certain hydrogen production methods, and greenhouse gas removal technologies such as bioenergy with carbon capture and storage and direct air capture. The captured CO_2 is then either used, for example in chemical processes, or stored permanently in disused oil and gas fields or naturally occurring geological storage sites.

Carbon Footprint: While carbon footprints are usually reported in tons of emissions (CO_2-equivalent) per year, ecological footprints are usually reported in comparison to what the planet can renew. This assesses the number of "earths" that would be required if everyone on the planet consumed resources at the same level as the person calculating their ecological footprint. The carbon footprint is one part of the ecological footprint. Carbon footprints are more focused than ecological footprints since they measure merely the emissions of gases that cause climate change into the atmosphere.

Carbon Leakage: 1. Carbon leakage is the increase in CO_2 emissions outside the countries taking domestic mitigation action divided by the reduction in the emissions of these countries. It is expressed as a percentage and can be greater or less than 100%. **2.** Occurs when there is an increase in greenhouse gas emissions in one country as a result of emissions reduction by a second country with a strict climate policy.

Carbon leakage may occur for a number of reasons:

- If the emissions policy of a country raises local costs, then another country with a more relaxed policy may have a trading advantage. If demand for these goods remains the same, production may move offshore to the cheaper country with lower standards, and global emissions will not be reduced.
- If environmental policies in one country add a premium to certain fuels or commodities, then the demand may decline and their price may fall. Countries that do not place a premium on those items may then take up the demand and use the same supply, negating any benefit.

Carbon Price: A cost applied to carbon pollution to encourage polluters to reduce the amount of greenhouse gases they emit into the atmosphere.

Carbon Tax: 1. A carbon tax is a form of pollution tax. Its aim is to allow market forces to determine the most efficient way to reduce pollution. **2.** It is an indirect tax—i.e., a tax on a transaction as opposed to direct tax, which taxes income. Carbon taxes are price instruments since they set a price rather than an emission limit. In addition to creating incentives for energy conservation, a carbon tax puts renewable energy such as wind, solar, and geothermal on a more competitive footing. Although carbon tax covers only CO_2 emissions, it can also cover other greenhouse gases, such as methane (CH_4) or nitrous oxide (NO_x), by calculating their global warming potential (GWP) relative to CO_2. When a hydrocarbon fuel such as coal, petroleum, or natural gas is burnt, much of its carbon is converted to CO_2. Greenhouse gas emissions cause climate change, which damages the environment and human health. **3.** A carbon tax puts renewable energy such as wind, solar, and geothermal on a more competitive footing.

Catalyst: A substance present in a chemical reaction that will promote, accelerate, or selectively direct a reaction, but does not take part in it by changing chemically itself. Sometimes a catalyst is used to lower the temperature or pressure at which the reaction takes place.

Catalyst-to-Oil Ratio (C/O): The weight of circulating catalyst fed to the reactor of a fluid-bed catalytic cracking unit divided by the weight of hydrocarbons charged during the same interval.

Catalytic Cracking: 1. The refining process of breaking down the larger, heavier, and more complex hydrocarbon molecules into simpler and lighter molecules. Catalytic cracking is accomplished by the use of a catalyst and is an effective process for increasing the yield of gasoline from crude oil. Catalytic cracking processes fresh feeds and recycled feeds. 2. A central process in reforming in which heavy gas oil range feeds are subjected to heat in the presence of a catalyst and large molecules crack into smaller molecules in the gasoline, diesel, and surrounding ranges. 3. A petroleum refining process in which heavy hydrocarbon molecules are broken down (cracked) into lighter molecules by passing them over a suitable catalyst (generally heated). 4. A method of cracking that uses a catalyst to convert hydrocarbons to positively charged carbonations, which then break down into smaller molecules. This can be carried out at much lower temperatures than thermal cracking—still hot 932–1112°F (500–600°C) as compared to around 1292°F (700°C). But that difference adds up to a lot of dollars.

Catalytically Cracked Distillates: These are obtained when high boiling non-gasoline hydrocarbons are heated under pressure in the presence of a catalyst to obtain lower boiling gasoline components. Catalytically cracked distillates usually have high octane numbers than straight-run gasoline.

Catalytic Cycle Stock: That portion of a catalytic cracker reactor effluent that is not converted to naphtha and lighter products. This material, generally 340°F (170°C), either may be completely or partially recycled. In the latter case, the remainder will be blended to products or processed further.

Catalytic Hydrocracking: A refining process that uses hydrogen and catalysts with relatively low temperatures and high pressures for converting middle boiling or residual material to high-octane gasoline, reformer charge stock., jet fuel, and/or high-grade fuel oil. The process uses one or more catalysts, depending upon product output, and can handle high sulfur feedstocks without prior desulfurization.

Catalytic Hydrotreating: A refining process for treating petroleum fractions from atmospheric or vacuum distillation units (e.g., naphthas, middle distillates, reformer feeds, residual fuel oils, and heavy gas oil) and other petroleum (e.g., cat cracked naphtha, coker naphtha, gas oil, etc.) in the presence of catalysts and substantial quantities of hydrogen. Hydrotreating includes desulfurization, removal of substances (e.g., nitrogen compounds) that deactivate catalysts, conversion of olefins to paraffins to reduce gum formation in gasoline, and other processes to upgrade the quantity of the fractions.

Catalytic Polymerization (cat. poly): A process in which propylene and/or butylenes components are chemically joined to produce gasoline. A phosphoric acid (HPO_3) catalyst is usually employed in the process.

Catalyst Promoter: A substance added to a catalyst to increase the fraction of the total catalyst area, which is useful for a reaction.

Catalytic Reforming: 1. A refining process using controlled heat and pressure with catalysts to rearrange certain hydrocarbon molecules, thereby converting paraffinic and naphthenic hydrocarbons (e.g., low-octane gasoline boiling range fractions) into petrochemical feedstocks and higher octane stocks suitable for blending into finished gasoline. 2. A process where low octane straight-run naphthas are chemically changed into high octane gasoline, called reformate, and to produce aromatics (BTX: benzene, toluene, and xylene) for petrochemical plants over a platinum (Pt) catalyst. The reformate has higher aromatic and cyclic hydrocarbon contents. Catalytic reforming is reported into two categories, namely:

Low Pressure. A processing unit operating at less than 225 psig measured at the outlet separator.

High Pressure: A processing unit operating at either equal to or greater than 225 psig measured at the outlet separator.

Catalyst Selectivity: The relative activity of a catalyst with respect to a particular component or compound in a mixture.

Catalyst Stripping: The introduction of steam at a point where spent catalyst leaves the reactor, in order to remove or strip the hydrocarbons retained on the catalyst.

Catastrophic Incident: An incident involving a major uncontrolled emission, fire, or explosion with an outcome effect zone that extends offsite into the surrounding community.

Cause: The reasons why deviation might occur.

Caustic Soda: Name used for sodium hydroxide (NaOH); used in refineries to treat acidic hydrocarbon streams to neutralize them.

Cavitation: 1. The creating of high-speed, very low-pressure vapor bubbles that quickly and violently collapse. It is very detrimental to surfaces in the near proximity, and is often seen in a severe turbulent flow. **2.** Occurs during vaporization of a pumped fluid resulting in vibration, noise, and destruction of equipment. This is when the absolute pressure of the system equals the vapor pressure of the pumped fluid. In a centrifugal pump, it results in the damage of the impeller. **3.** When the pressure of liquid flowing into a centrifugal pump gets too low, the liquid boils inside the pump case and generates bubbles. The discharge pressure and flow become erratically low.

Centipoise (cP): A measure of viscosity related to centistokes by adjusting for density. **1.** Viscosity measurement, $1/1000^{th}$ of a poise. **2.** A centipoise (cP) is $1/1000^{th}$ of a poise (P), which is the fundamental unit of dynamic viscosity in the CGS system of units. In the SI system of units, the fundamental unit of dynamic viscosity is the Pascal second (Pa.s) and is equivalent to 10P.

Centistoke (cSt): Is $1/100^{th}$ of a Stoke (St), which is the fundamental unit of kinematic viscosity in the CSG system of units. In the SI system of units, the fundamental unit of kinematic viscosity is the millimeter squared per second (mm^2/s), which is equivalent to the centistokes.

Cetane (Hexadecane, $C_{16}H_{34}$): An alkane hydrocarbon with a chemical formula $C_{16}H_{34}$ used as a solvent and in cetane number determinations. **1.** A pure paraffin hydrocarbon used as standard reference fuel in determining the ignition qualities of diesel fuels. **2.** A number calculated from the API gravity and the D86 50% distilled for a petroleum stock. It is used to rate the performance of a fuel in diesel engines. It is arbitrarily given a cetane number of 100.

Cetane Index: 1. A number calculated from the average boiling point and gravity of a petroleum fraction in the diesel fuel boiling range, which estimates the cetane number of the fraction according to ASTM D976. An indication of the carbon–hydrogen ratio. **2.** An empirical method for determining the cetane number of diesel fuel by a formula based on API gravity and the mid-boiling point (ASTM D975) (see, for example, http://www.epa.gov/nvfel/testproc/121.pdf).

Cetane Number: 1. The percentage of pure cetane in a blend of cetane and alpha- methyl-naphthalene matches the ignition quality of a diesel fuel sample. This quality, specified for middle distillate fuel, is the opposite of the octane number of gasoline. It is an indication of ease of self-ignition. **2.** A term for expressing the ignition quality of diesel fuel. **3.** A measure of the ignition quality of diesel fuel, expressed as a percentage of cetane that must be mixed with methyl naphthalene to produce the same ignition performance as the diesel fuel being rated. The higher the number, the more easily the fuel is ignited under compression. It is an important factor in determining the quality of diesel fuel. In short, the higher the cetane number, the more easily the fuel will combust in a compression setting (such as a diesel engine). The characteristic diesel "knock" occurs when fuel that has been injected into the cylinder ignites after a delay causing a late shock wave. Minimizing this delay results in less unburned fuel in the cylinder and less intense knock. Therefore, higher-cetane fuel usually causes an engine to run more smoothly and quietly. This does not necessarily translate into greater efficiency, although it may in certain engines. The cetane number is determined in a single cylinder. Cooperative Fuel Research (CFR) engine by comparing its ignition quality with that of reference blends of known cetane number.

Cetane number = 0.72 diesel index (10)

Calculated cetane index (CCI) is determined by four variables:

$$\begin{aligned} CCI = &\ 45.2 + (0.0892)(T_{10}N) \\ &+ [0.131 + 0.901(B)][T_{50}N] \\ &+ [0.0523 - (0.420)B][T_{90}N] \\ &+ [0.00049][(T_{10}N)^2 - (T_{90}N)^2] \\ &+ 107B + 60B^2 \end{aligned} \qquad (10)$$

where T_{10} = 10 % distillation temperature, °C
T_{50} = 50 % distillation temperature, °C
T_{90} = 90% distillation temperature, °C
B = $e^{-3.5DN} - 1$
D = Density @ 15°C
$DN = D - 0.85$

CFR: Combined feed ratio. The ratio of total feed (including recycle) to fresh feed.

CGO: Coker gas oil.

Charge Capacity: The input (feed) capacity of the refinery processing facilities.

Characterization Factor (CF): 1. An index of feed quality, also useful for correlating data on physical properties. The Watson or Universal Oil Property (UOP) characterization factor K_W is defined as the cube root of the mean average boiling point in °R divided by the specific gravity. An indication of carbon-to-hydrogen ratio. K_W is expressed by

$$K_W = \frac{T_B^{1/3}}{Sp.Gr} \qquad (11)$$

where
T_B = mean average boiling point, °R [°F + 460]
Sp.Gr = Specific gravity at 60°F

T_B is the average boiling point in °R taken from five temperatures corresponding to 10, 30, 50, 70, and 90% volume vaporized.

2. A calculated factor used to correlate properties for petroleum streams. It is a measure of the paraffinicity of the stream and is defined as $CF = [MABP^{1/3}/Sp.Gr]$, where MABP = mean average boiling point temperature, °R (=460 + °F), and Sp.Gr. = specific gravity at 60°F (15.9°C) relative to water.

Typically, Watson characterization factor varies between 10.5 and 13 for various crude streams. Highly paraffinic crude typically possesses a K_w of 13. On the other hand, highly naphthenic crude possesses a K_w factor of 10.5. Therefore, the Watson characterization factor can be used to judge the quality of the crude oil in terms of the dominance of the paraffinic or naphthenic compounds.

Checklist: A detailed list of desired systems attributes for a facility. It is used to assess the acceptability of a facility compared to accepted norms.

Clarified Oil: The heaviest stream from a catalytic cracking process after settling to remove suspended catalyst particles.

Clear Treating: An elevated temperature and pressure process usually applied to thermally cracked naphthas to improve stability and color. The stability is increased by the adsorption and polymerization of reactive diolefins in the cracked naphtha. Clay treating is used for treating jet fuel to remove surface agents that adversely affect the water separator index specifications.

Clear: Without lead. Federal regulations require that fuels containing lead must be dyed.

Cloud Point: 1. The temperature at which solidifiable compounds (wax) present in the sample begin to crystallize or separate from the solution under a method of prescribed chilling. 2. The temperature at which a noticeable cloud of crystals or other solid materials appears when a sample is cooled under prescribed conditions. Cloud point is a typical specification of middle distillate fuels; ASTM D-2500.

Cold Filter Plugging: Is defined as the temperature at which a fuel suspension fails to flow through a standard filter when cooled as prescribed by the test method.

Coke Drum: A large upright drum used as a receptacle for coke formed in the delayed coking process.

Coke: 1. A product of the coking process in the form of mostly solid, densely packed carbon atoms. 2. Deposits of carbon that settle on catalysts in cat. crackers, cat. reformers, hydrocrackers, and hydrotreaters, and degrade their effectiveness. 3. A carbonaceous deposit is formed by the thermal decomposition of petroleum.

Coker: A refinery process in which heavy feed such as flasher bottoms, cycle oil from a catalytic cracker, or thermal cracked gas oil is cooked at high temperatures. Cracking creates light oils; coke forms in the reactors and needs to be removed after they fill up.

Coking: A refining process in which petroleum oil is heated destructively such that the heaviest materials are converted to coke. There are two processes: delayed coking and fluid coking, with delayed coking being the most widely used.

Coil: A series of pipes in a furnace through which oil flows and is heated.

Color: It is an indication of the thoroughness of the refining process. This is determined by the Saybolt Chromometer or by Lovibond Tintometer. The Saybolt color of petroleum products test is used for quality control and product identification purposes on refined products having an ASTM color of 0.5 or less.

The ASTM color of petroleum products applies to products having an ASTM color of 0.5 or darker, including lubricating oils, heating oils, and diesel fuel oils.

Pale	= 4.5 ASTM color or lighter
Red	= Darker than 4.5 ASTM
Dark	= Darker than 8.0 ASTM

Compressed Natural Gas: 1. Natural gas that has been compressed under high pressures (typically between 3000 and 3600 psi and held in a container; expands when released for use as a fuel. 2. Natural gas is compressed to a volume and density that is practical as a portable fuel supply (even when compressed, natural gas is not a liquid). 3. Natural gas in its gaseous state that has been compressed. 4. Natural gas that is under pressure. The pressure reduces the volume occupied for the gas so it can be contained in a smaller vessel.

Compressibility: The volume change of a material when pressure is applied.

Compressibility Factor (Z): 1. The fractional reduction in the volume of a substance with applied pressure. The compressibility factor is a measure of the compressibility of a gas, Z, and is used as a multiplier to adapt the ideal gas law for non-ideal gases. **2.** The ratio of the actual volume of a gas divided by the volume that would be predicted by the ideal gas law, usually referred to as the "Z" factor.

$$Z = \frac{pV}{RT} \qquad (12)$$

where p is the pressure, V is the volume, R is the universal gas constant, and T is the absolute temperature.

Compressible Fluid: A fluid in which the density changes with applied pressure. The compressibility of liquids is negligible in comparison with gases and vapors. The isothermal compressibility of a gas is the change in volume per unit volume or density for a unit change in applied pressure given as:

$$c = \frac{-1}{V}\left(\frac{\partial V}{\partial p}\right)_T = \frac{-1}{\rho}\left(\frac{\partial \rho}{\partial p}\right)_T \qquad (13)$$

Isothermal compressibility coefficients are frequently used in oil and gas engineering, transient fluid flow calculation, and the determination of the physical properties of substances.

Compression Ratio: Is a measure of the amount of compression that takes place in an engine's cylinder. The ratio of volumes in an internal combustion cylinder when the piston is at the bottom of the stroke to that when the piston is at the top of the stroke, giving a measure of how much the air or air/fuel mixture is compressed in the compression stroke.

$$CP = \frac{V_1}{V_2} = \frac{\text{Volume when piston is @ bottom of stroke}}{\text{Volume when piston is @ top of stroke}} \qquad (14)$$

Compressor: 1. A device that increases the pressure of the gas. Commonly used as a production rate increaser by increasing the gas pressure delivered from low-pressure gas wells to enter the pipeline. The intake into the compressor lowers the wellhead pressure, creating a larger drawdown. **2.** An engine used to increase the pressure of natural gas so that it will flow more easily through a pipeline. **3.** A thermodynamic machine that increases the pressure of a gas flow using mechanical energy. **4.** A mechanical device used to raise the pressure of a gas. Compressors can be of three types: *axial, centrifugal, or reciprocating*. The usual means of providing the required power are electrical motors, steam turbines, or gas turbines.

Compressor Station: 1. A booster station is associated with a gas pipeline that uses compressors to increase the gas pressure. When gas turbines are used to provide compressor power, stations can use some of the gas moving through the line as fuel. **2.** Stations located along natural gas pipelines that recompress gas to ensure an even flow.

Condensation: Reaction in which aromatic ring structures combine to form ring structures larger than the reactants.

Condensate: 1. The relatively small amount of liquid hydrocarbon, typically C_4, s, through naphtha or gas oil that gets produced in the oil patch with unassociated gas. **2.** The liquid formed when a vapor cools.

Conradson Carbon: A test used to determine the amount of carbon residue left after the evaporation and pyrolysis of oil under specified conditions. Expressed as weight percentage; ASTM D-189.

Conradson Carbon Residue (CCR): Results from ASTM test D189. It measures the coke-forming tendencies of oil. It is determined by destructive distillation of a sample to elemental carbon (coke residue), in the absence of air, expressed as the weight percentage of the original sample. A related measure of the carbon residue is called *Ramsbottom carbon residue*. A crude oil with a high CCR has a low value as refinery feedstock.

Conradson Carbon Residue (ASTM D 1289): ASTM D 4530 microcarbon residue: This procedure determines the carbon residue left after the evaporation and pyrolysis of an oil sample under prescribed conditions and is a rough indicator of oil's relative coke-forming tendency or the contamination of a lighter distillate fraction with a heavier distillate fraction or residue.

Carbon residue and atomic H-to-C ratio are correlated by:

$$H/C = 171 - 0.015CR \text{ (Conradson)} \tag{15}$$

Consequence: 1. Is the ultimate harm that may occur due to a credible hazard release scenario. **2.** The direct undesirable result of an accident sequence usually involving a fire, explosion, or the release of toxic material. Consequence description may include estimates of the effects of an accident in terms of factors such as health impacts, physical destruction, environmental damage, business interruption, and the public reaction of company prestige (see Figure 6).

Continuous Catalytic Reforming (CCR) process: Continuous catalytic reforming process occurs where the catalyst is circulated through the reactors and a regeneration step, analogous to catalytic cracking processes.

Continuous Stirred Tank Reactor (CSTR): 1. A type of idealized chemical reactor that is used to contain a chemical reaction in which liquid reactants continuously flow into the reactor and products are continuously removed such that there is no accumulation within the reactor. By assuming perfect mixing of the reactants within the reactor, by using a stirrer/mixer, the composition of the material is therefore assumed to be the same as the composition at all points within the reactor. **2.** Reactors are characterized by a continuous flow of reactants into and a continuous flow of products from the reaction system. Examples are the plug flow reactor and the continuous stirred flow reactor (Figure 7).

Control of Major Accident Hazards (COMAH): The legislation requires that businesses holding more than threshold quantities of named dangerous substances "Take all necessary measures to prevent major accidents involving dangerous substances. Limit the consequences to people and the environment of any major accidents which do occur." Plant designers need to consider whether their proposed plant will be covered by this legislation at the earliest stages.

Control of Substances Hazardous to Health (COSHH): The legislation that requires risk assessment and the control of hazards associated with all chemicals and used in a business that has potentially hazardous properties. A consideration of the properties of chemicals used as feedstock, intermediates, and products is a basic part of plant design. Inherently safe design requires us to consider these issues at the earliest stage.

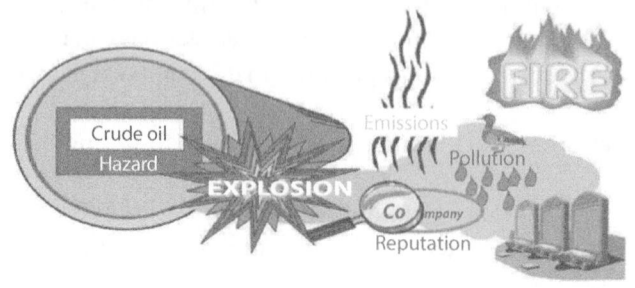

Resulting event or chain of events

Figure 6 A consequence.

Figure 7 (a) Continuous stirred tank reactor with different impeller types. (b) A series of continuous-flow stirred tank reactors.

Conversion: **1.** A measure of the completeness of a chemical reaction. It is often presented as the fraction of a particular reactant consumed by the chemical reaction. The *conversion per pass* is a measure of the limiting reactant that is converted in a chemical reactor and recycled in combination with fresh reactant feed. Not all reactions are complete within the reactor, and in many cases, unreacted reactants are separated from products and recycled for further action. **2.** Typically, the fraction of a feedstock is converted to gasoline and lighter components.

Correlation Index (CI): The US Bureau of Mines factor for evaluating individual fractions from crude oil. The CI scale is based upon straight chain hydrocarbons having a CI value of 0 and benzene having a value of 100. The lower the CI value, the greater the concentrations of paraffin hydrocarbons in the fraction, and the higher the CI value, the greater the concentrations of naphthenes and aromatics. CI is an indication of the hydrocarbon-to-carbon ratio and the aromaticity of the sample. CI is expressed by:

$$CI = \frac{87552}{T_B} + 473.7 Sp.Gr - 456.8 \tag{16}$$

$$CI = \frac{48640}{K} + 473.7 d - 456.8 \tag{17}$$

where
- d = specific gravity at 15°C/15°C
- K = average boiling point, (K = °C + 273.15)
- T_B = mean average boiling point, °R
- $Sp.Gr$ = specific gravity at 60°F

Corrosion: 1. The deteriorating chemical reaction of a metal with the fluids with which it is in contact. 2. The gradual decomposition or destruction of a material by chemical action, often due to an electrochemical reaction. Corrosion may be caused by (a) stray current electrolysis, (b) galvanic corrosion caused by dissimilar metals, and (c) differential-concentration cells. Corrosion starts at the surface of a material and proceeds inward.

Corrosion Inhibition: Corrosion can be defined as the unwanted production of a salt from a metal. Adding acid or oxygen is a good way for this to occur. The main way of slowing corrosion down (inhibition) is by providing an impermeable coating to stop the chemical reaction from occurring in the first place or by providing a more easily attacked metal that will be consumed first (a "sacrificial anode").

Corrosion Inhibitor: 1. A chemical substance or combination of substances that, when present in the environment, prevents or reduces corrosion. 2. A substance that slows the rate of corrosion.

Corrosive Gas: 1. A gas that attacks metal or other specified targets. Most commonly CO_2 and H_2S. Usually in association with water or water vapor. Oxygen can be described as a corrosive gas in some cases. 2. In water, dissolved oxygen reacts readily with metals at the anode of a corrosion cell, accelerating the rate of corrosion until a film of oxidation products such as rust forms. At the cathode where hydrogen gas may form a coating on it and therefore slows the corrosion rate, oxygen reacts rapidly with hydrogen gas, forming water, and again increases the rate of corrosion.

Cracking: The breaking down of higher-molecular-weight hydrocarbons to lighter components by the application of heat. Cracking in the presence of a suitable catalyst produces an improvement in yield and quality over simple thermal cracking.

Cracking Correction: Correction to a laboratory distillation to account for the lowering of the recorded temperatures because of the thermal cracking of the sample in the distillation flask. Cracking occurs for most petroleum stocks at temperatures greater than about 650°F (344°C) at atmospheric pressure.

Cracked Stock: A petroleum stock that has been produced in a cracking operation, either catalytic or thermal. Cracked stocks contain hydrogen-deficient compounds such as olefins (C_nH_{2n}) and aromatics (C_nH_{2n-6}).

Critical Point: The temperature and pressure at which a component or mixture of components enters a dense phase, being neither liquid nor vapor.

Critical Pressure: The vapor pressure at the critical temperature.

Critical Temperature: The temperature above which a component cannot be liquefied. For mixtures, the temperature above which all of the mixture cannot be liquefied.

Crude Assay Distillation: See Fifteen-Five (15/5) Distillation.

Crude Chemistry: Fundamentally, crude oil consists of 84–87 wt% carbon, 11–14% hydrogen, 0–3 wt% sulfur, 0–2 wt% oxygen, 0–0.6 wt% nitrogen, and metals ranging from 0 to 100 ppm. Understanding thoroughly the fundamentals of crude chemistry is very important in various refining processes. The existence of compounds with various functional groups and their dominance or reduction in various refinery products are what are essentially targeted in various chemical and physical processes in the refinery.

Based on chemical analysis and the existence of various functional groups, refinery crude can be broadly categorized into about nine categories:

1. Paraffins, C_nH_{2n+2}, CH_4, C_2H_6, C_3H_8	4. Aromatics, C_nH_{2n-6}, C_6H_6, C_7H_8, C_8H_{10}	7. Oxygen-containing compounds, R-OH, CH_3OH, C_6H_5OH
2. Olefins, C_nH_{2n}. C_2H_4, C_3H_6	5. Naphthalene	8. Resins
3. Naphthenes, C_nH_{2n}, C_6H_{12},	6. Organic sulfur compounds, RSH, CH_3SH, $R - S - R'$	9. Asphaltenes

Crude and Crude Oil: 1. A range of principally carbon–hydrogen chain compounds with generally straight carbon chain lengths of C_1 (methane) to C_{60+}, compounds boiling higher than 2000°F (1094°C). The straight-chain materials are alkanes. 2. Oil as it comes from the well; unrefined petroleum. 3. The petroleum liquids as they come from the ground; formed from animal and vegetable material that is collected at the bottom of ancient seas. 4. A tarry group consisting of mixed carbon compounds with a highly variable composition. 5. A mixture of hydrocarbons that exists in the liquid phase in natural underground reservoirs and remains liquid at atmospheric pressure after passing through surface-separating facilities. Depending upon the characteristics of the crude stream, it may also include the following:

- Small amounts of hydrocarbons that exist in the gaseous phase in natural underground reservoirs but are liquid at atmospheric pressure after being removed from oil well (casing head) gas in lease separators and are subsequently commingled with the crude stream without being separately measured. Lease condensate recovered as a liquid from natural gas wells in lease or field separation facilities and later mixed into the crude stream is also included.
- Small amounts of non-hydrocarbons produced from oil, such as sulfur and various metals.
- Drip gases and liquid hydrocarbons produced from tar sands, gilsonite, and oil shale.
- Liquid produced at natural gas processing plants is excluded. Crude oil is refined to produce a wide range of petroleum products, including heating oils; gasoline, diesel, and jet fuels; lubricants; asphalt; ethane, propane, and butane; and many other products used for their energy or chemical content.

The basic types of crudes are asphalt, naphthenic, or paraffinic depending on the relative proportion of these types of hydrocarbons present.

Crude Oil Assay: Is a precise and detailed analysis of carefully selected samples of crude thoroughly representative of average production quality. It helps to assess the potential sales value of new crude oil and to plan for its most effective utilization. Numerous important feed and product characterization properties in refinery engineering include:

1. API gravity
2. Watson characterization factor
3. Viscosity
4. Sulfur content
5. True boiling point (TBP) curve
6. Pour point
7. Flash and fire point
8. ASTM distillation curve
9. Octane number

Crude Oil Losses: This represents the volume of crude oil reported by petroleum refineries as being lost in their operations. These losses are due to spills, contamination, fires, etc., as opposed to refinery processing losses.

Crude Oil Production: The volume of crude oil produced from oil reservoirs during given periods of time. The amount of such production for a given period is measured as volumes delivered from lease storage tanks, (i.e., the point of custody transfer) to pipelines, trucks, or other media for transport to refineries or terminals with adjustments for (1) net differences between opening and closing lease inventories and (2) basic sediment and water (BS & W).

Crude Oil Qualities: This refers to two properties of crude oil, the sulfur content and API gravity, which affect processing complexity and product characteristics.

Cryogenics: The production and application of low-temperature phenomena. The cryogenic temperature range is usually from -238°F (-150°C) to absolute zero -460°F (-273°C), the temperature at which molecular motion essentially stops. The most important commercial application of the cryogenic gas liquefaction technique is the storage, transportation, and regasification of LNG.

Cryogenic Liquid or Cryogenics: Cryogenic liquids are liquefied gases that are kept in their liquid state at very low temperatures and have a normal boiling point below -238°F (-150°C). All cryogenic liquids are gases at normal temperatures and pressures. These liquids include methane (CH_4), oxygen (O_2), nitrogen (N_2), helium (He), and hydrogen (H_2). Cryogens are normally stored at low pressures.

Cryogenic Recovery: Cryogenic recovery processes are carried out at temperatures lower than -150°F (-101°C). The low temperatures allow the plant to recover over 90% of the ethane in the natural gas. Most new gas processing plants use cryogenic recovery technology.

CSB: An acronym for the US Chemical Safety and Hazard Investigation Board. An agency of the US government chartered to investigate chemical industry incidents, determine their root causes, and publish their findings to prevent similar incidents from occurring.

Cut: A portion of crude oil boiling within certain temperature limits. Usually, the limits are on a crude assay true boiling point (TBP) basis.

Cut Point Temperature, Cut Points: A temperature limit of a cut, usually on a true boiling point (TBP) basis, although ASTM distillation cut point is not uncommon. The boiling point curve most commonly used to define cut points is the TBP at one atmosphere of pressure.

Cut Point Ranges: A series of cut point temperatures are defined for a petroleum stock. The cut point ranges are the temperature differences between adjacent cut point temperatures. When developing petroleum pseudo-components for a petroleum stock, cut point ranges must be defined that include the total boiling point range of the stock.

Cutter Stock: Diluent added to residue to meet residual fuel specifications for viscosity and perhaps sulfur content. Typically cracked gas oil.

Cycloparaffin: A paraffin molecule with a ring structure.

Cycle Oil, Cycle Stock: An oil stock, containing a hydrogen-deficient compound that was produced in a thermal or catalytic cracking operation.

Cyclization: Chemical reaction in which a non-ring structure paraffin or olefins are converted into ring structures.

Cyclo-olefins: Unsaturated ring structure with one or two double bonds in the ring.

Darcy–Weisbach Equation: An equation used in fluid mechanics to determine the pressure or head loss due to friction within a straight length of pipe for a flowing fluid. The frictional pressure drop, Δp_f (psi, bar) is expressed by

$$\Delta p_f = f_D \left(\frac{L}{d}\right) \frac{\rho v^2}{2} \qquad (18)$$

where

$$f_D = \left(\frac{\tau_w}{\frac{\rho v^2}{2}}\right) \qquad (19)$$

In the form of frictional head loss, h_f (ft, m) is:

$$h_f = f_D \left(\frac{L}{d}\right) \frac{v^2}{2g} \qquad (20)$$

where τ_w is the shear stress, f_D is the Darcy friction factor, dimensionless ($f_D = 4f_F$), f_F is the fanning friction factor, L and d are the pipe length (ft, m) and inside diameter (ft., m), v is the average velocity of the fluid (ft), ρ is the fluid density (lb_m/ft^3, kg/m^3), and g is the acceleration due to gravity (ft/s^2, m/s^2). It is known as the Darcy–Weisbach or Moody friction fawctor, whose value depends on the nature of the flow and surface roughness of the pipe. This Darcy friction factor is four times the Fanning friction factor. The value of the friction factor can be determined from various empirical equations and published charts such as the Moody diagram (See Figure 8).

An empirical equation known as the Colebrook–White equation has been proposed for calculating the friction factor in the turbulent flow:

$$\frac{1}{\sqrt{f_D}} = -2\log_{10}\left(\frac{e}{3.7D} + \frac{2.51}{Re\sqrt{f_D}}\right) \qquad (21)$$

where
  D = pipe inside diameter, in
  e = absolute pipe roughness, in
  Re = Reynolds number, dimensionless

The term $f_D(L/d)$ may be substituted with a head loss coefficient K (also known as the resistance coefficient) and then becomes

$$h_f = K \frac{v^2}{2g} \qquad (22)$$

The head loss in a straight piece of pipe is represented as a multiple of the velocity head $v^2/2g$. Following a similar analysis, we can state that the pressure drop through a valve or fitting can be represented by $K(v^2/2g)$, where the coefficient K is specific to the valve and fitting. Note that this method is only applicable to turbulent flow through pipe fittings and valves. Recently, K is presented by Hooper's 2-K method and Darby's 3-K method.

DAO - Deasphalted oil: The raffinate product from the propane deasphalting unit.

D1160: ASTM laboratory distillation method for high-boiling streams. The D1160 is performed under vacuum conditions with 10 mm Hg being the most common pressure used for the test. D1160 data are normally reported at a 760 mm Hg basis.

D2887: ASTM-simulated distillation method for high-boiling streams. The D2887 has an upper limit of 1000°F (538°C), and the temperatures are reported versus weight percent distilled. A normal paraffin standard is used to convert the chromatographic results to a boiling point curve.

D3710: ASTM-simulated distillation method for gasoline and light naphthas. D3710 data are reported on a volume basis.

D86: ASTM laboratory distillation method conducted at atmospheric pressure for streams boiling below approximately 700°F (371°C). The D86 is the most commonly used laboratory distillation for petroleum stocks.

Deasphalting: Process for removing asphalt from petroleum fractions, such as reduced crude.

Debottlenecking: 1. Increasing the production capacity of existing facilities through the modification of existing equipment to remove throughput restrictions. Debottlenecking generally increases capacity for a fraction of the cost of building new facilities. **2.** The process of increasing the production capacity of existing facilities through the modification of existing equipment to remove throughput restrictions. **3.** A program, typically in surface facilities and lines, to remove pressure drop causing flow restrictions.

Debutanizer: A column that removes n-butanes (nC_4H_{10}) and lighter in the top product.

Decant Oil: The bottom stream from the FCC unit distillation tower after the catalyst has been separated from it.

Decanted Water: Insoluble water that is drawn from a drum containing condensed hydrocarbons and water.

Decarbonization: A process of reducing the amount of carbon dioxide released into the atmosphere.

Decoking: The process of removing coke from catalysts in a catalytic cracker, catalytic reformer, hydrocracker, or hydrotreaters. Usually, heated air will oxidize the coke to carbon monoxide or carbon dioxide.

Deethanizer: A column that removes ethane (C_2H_6) and lighter in the top product.

Deflagration (i.e., "to burn down"): Is a term describing subsonic combustion propagation through heat transfer; hot burning material heats the next layer of cold material and ignites it. Most "fire" found in daily life, from flames to explosions, is deflagration. Deflagration is different from detonation, which propagates supersonically through shock waves.

Delayed Coker: A process unit in which residue is cooked until it cracks to coke and light products.

Delayed Coking: 1. A semi-continuous thermal process for the conversion of heavy stock to lighter material. The method involves pre-heating the feedstock in a pipe still, discharging it into large insulated coke drums, and retaining it there for a particular length of time for cracking to occur. Gas, gasoline, and gas oil are recovered as overhead products, and finally, coke is removed. **2.** A process by which heavier crude oil fractions can be thermally decomposed under conditions of elevated temperatures and low pressure to produce a mixture of lighter oils and petroleum coke. The light oils can be processed further in other refinery units to meet product specifications. The coke can be used either as a fuel or in other applications such as the manufacturing of steel or aluminum.

Dehydrogenation: A chemical reaction in which a compound loses bonded hydrogen.

Deisobutanizer: A column that removes isobutane (iC_4H_{10}) and lighter in the top product.

Demethanizer: A column that removes methane (CH_4) and lighter in the top product.

Density: The density of crude oil and petroleum fractions is usually specified in °API, specific gravity, or kilograms per cubic meter (kg/m^3). The numerical values of specific gravity and kg/m^3 are equal; that is, a fraction with a specific gravity of 0.873 has a density of 873 kg/m^3. The API scale runs opposite to that of specific gravity, with larger values for less dense materials and smaller values for more dense fractions (water = 10°API). By definition, °API is always 60°F (15.6°C) for a liquid.

Depentanizer: A column that removes n-pentane (nC_5H_{12}) and lighter in the top product.

Depropanizer: A column that removes propane (C_3H_8) and lighter in the top product.

Desalting: A process that removes chlorides and other inorganic salts from crude oil by injecting water and applying an electrostatic field to force the salt into the aqueous phase.

Desiccant: An absorbent or adsorbent, liquid or solid, that removes water or water vapor from an air stream.

Desiccant Drying: The use of a drying agent to remove moisture from a stream of oil or gas. In certain product pipelines, great effort is made to remove all the water vapor before putting the line into service. To accomplish this, desiccant dried air or inert gas is pumped through the line to absorb the moisture that may be present even in the ambient air in the line.

Desiccation: The process of drying and removing the moisture within a material. It involves the use of a drying agent known as a desiccant. Desiccants that function by the adsorption of moisture include silica gel and activated alumina, while chemical desiccants that function by the reaction with water to form hydrates include calcium chloride and solid sodium hydroxide. A desiccator is a container used for drying substances or for keeping them dry and free of moisture. Laboratory desiccators are made of glass and contain a drying agent such as silica gel.

Design Codes (design standards): Published standards required for equipment and working practices within the chemical and process industries that represent good practice and define the level of standard of design. Developed and evolved over many years and based on tried and tested practices. There are a number of national standards organizations and institutions that provide published standards for design, materials, fabrication, the testing of processes, and equipment. These include the American Petroleum Institute (API), the American National Standards Institute (ANSI), the American Society of Mechanical Engineers (ASME), the American Society for Testing and Materials (ASTM), the American Iron and Steel Institute (AISI), and the British Standards Institute (BSI).

Desorption: The release of materials that have been absorbed or adsorbed in or onto a formation.

Desulfurization: The removal of sulfur, from molten metals, petroleum oil, or flue gases. Petroleum desulfurization is a process that removes sulfur and its compounds from various streams during the refining process. Desulfurization processes include catalytic hydrotreating and other chemical/physical processes such as adsorption. Desulfurization processes vary based on the type of stream treated (e.g., naphtha, distillate, heavy gas oil) and the amount of sulfur removed (e.g., sulfur reduction to 10 ppm). See Catalytic Hydrotreating.

Desuperheating zone: A section of a distillation/fractionating column where a superheated vapor is cooled and some liquid is condensed. FCC main fractionators have a desuperheating zone.

Detonation ("to thunder down"): Is a type of combustion involving a supersonic exothermic front accelerating through a medium that eventually drives a shock front propagating directly in front of it. Detonations occur in both conventional solid and liquid explosives, as well as in reactive gases. The velocity of detonation in solid and liquid explosives is much higher than that in gaseous ones, which allows the wave system to be observed with greater detail.

An extraordinary variety of fuels may occur as gases, droplet fogs, or dust suspensions. Oxidants include halogens, ozone, hydrogen peroxide, and oxides of nitrogen. Gaseous detonations are often associated with a mixture of fuel and oxidant in a composition somewhat below the conventional flammability ratio. They happen most often in confined systems, but they sometimes occur in large vapor clouds. Other materials, such as acetylene, ozone, and hydrogen peroxide are detonable in the absence of oxygen. See Knocking.

Dewaxing: The removal of wax from lubricating oils, either by chilling and filtering solvent extraction or selective hydrocracking.

Dew Point: 1. A vapor at its dew point temperature is on the verge of starting to condense to a liquid. Cool the vapor by 1°F, or raise its pressure by 1 psi and it will form drops of liquid. Air at 100% relative humidity is at its dew point temperature. Cool it by 1°F and it starts to rain. **2.** The temperature and pressure at which the first drop of liquid will condense for a component or a mixture of components. **3.** The temperature at a given pressure at which vapor will form the first drop of liquid on the subtraction of heat. Further cooling of the liquid at its dew point results in the condensation of a part or all of the vapors as a liquid. **4.** The temperature at which vaporized materials start to condense into liquid form. **5.** The temperature at which liquids begin to condense from the vapor phase in a gas stream.

Diene: Same as diolefin.

Diesel: 1. An internal combustion engine in which ignition occurs by injecting fuel in a cylinder where the air has been compressed and is at a very high temperature, causing self-ignition. **2.** Distillate fuel used in a diesel engine. (*See the Diesel engine*).

Diesel Fuel: A fuel produced for diesel engines with a typical ASTM 86 boiling point range of 450–675°F (233–358°C).

Diesel Index (DI): A measure of the ignition quality of diesel fuel. Diesel index is defined as

$$DI = \frac{(°API)(Aniline\ Point)}{100} \tag{23}$$

The higher the diesel index, the more satisfactory the ignition quality of the fuel. By means of correlations unique to each crude and manufacturing process, this quality can be used to predict the cetane number (if no standardized test for the latter is available).

Diolefin: C_nH_{2n}: Paraffin-type molecule except that it is missing hydrogen atoms, causing it to have two double bonds somewhere along the chains.

DIPE: Di-isopropyl ether. An oxygenate is used in motor fuels.

Disposition: The components of petroleum disposition are stock change, crude oil losses, refinery inputs, exports, and products supplied for domestic consumption.

Distillate Fuel Oil: A general classification for one of the petroleum fractions produced in conventional distillation operations. It includes diesel fuels and fuel oils. Products known as No. 1, No. 2, and No. 4 diesel fuels are used on highway diesel engines, such as those in trucks and automobiles, as well as off-highway engines, such as those in railroad locomotives and agricultural machinery. Products known as No.1, No. 2, and No. 4 fuel oils are used primarily for space heating and electric power generation.

No. 1 Distillate. A light petroleum distillate that can be used as either diesel fuel or fuel oil.

No. 1 Diesel Fuel. A light distillate fuel oil that has distillation temperatures of 550°F (288°C) at the 90% point and meets the specifications defined in ASTM Specification D 975. It is used in high-speed diesel engines generally operated under frequent speed and load changes, such as those in city buses and similar vehicles.

No. 1 Fuel Oil. A light distillate fuel oil that has distillation temperatures of 400°F (204°C) at the 10% recovery point and 550°F (288°C) at the 90% point and meets the specifications defined in ASTM Specifications D 396. It is used primarily as a fuel for portable outdoor stoves and portable outdoor heaters.

No. 2 Distillate: A petroleum distillate that can be used as either diesel fuel or fuel oil.

No. 2 Diesel Fuel: A fuel that has a distillation temperature of 500°F (260°C) at the 10% recovery point and 640°F (338°C) at the 90% recovery point and meets the specifications defined in ASTM Specification D 975. It is used in high-speed diesel engines that are generally operated under uniform speed and load conditions, such as those in railroad locomotives, trucks, and automobiles.

Low Sulfur No. 2 Diesel Fuel. No. 2 diesel fuel that has a sulfur level no higher than 0.05% by weight. It is used primarily in motor vehicle diesel engines for on-highway use.

High Sulfur No. 2. Diesel Fuel. No. 2 diesel fuel that has a sulfur level above 0.05% by weight.

No. 2 Fuel Oil (Heating Oil): A distillate fuel oil that has distillation temperatures of 400°F (204°C) at the 10% recovery point and 640°F (338°C) at the 90% recovery point and meets the specifications defined in ASTM Specification D 396. It is used in atomizing-type burners for domestic heating or for moderate-capacity commercial/industrial burner units.

No. 4 Fuel. A distillate fuel oil made by blending distillate fuel oil and residual fuel oil stocks. It conforms with ASTM Specification D 396 or Federal Specification VV–F–815C and is used extensively in industrial plants and in commercial burner installations that are not equipped with preheating facilities. It also includes No. 4 diesel fuel used for low- and medium-speed diesel engines and conforms to ASTM Specification D975.

No. 4 Diesel Fuel. See No. 4 Fuel

No. 4 Fuel Oil. See No. 4 Fuel.

Distillate: 1. The liquid obtained by condensing the vapor given off by a boiling liquid. **2.** Any stream except the bottoms coming from a fractionator. **3.** The products or streams in the light gas oil range such as straight-run light gas oil, cat. cracked light gas oil, heating oil, or diesel.

Distillation: Same as fractionation. A separation process that results in separated products with different boiling ranges. Distillation is carried out in a way that the materials being separated are not subjected to conditions that would cause them to crack or otherwise decompose or chemically change. It is a physical process.

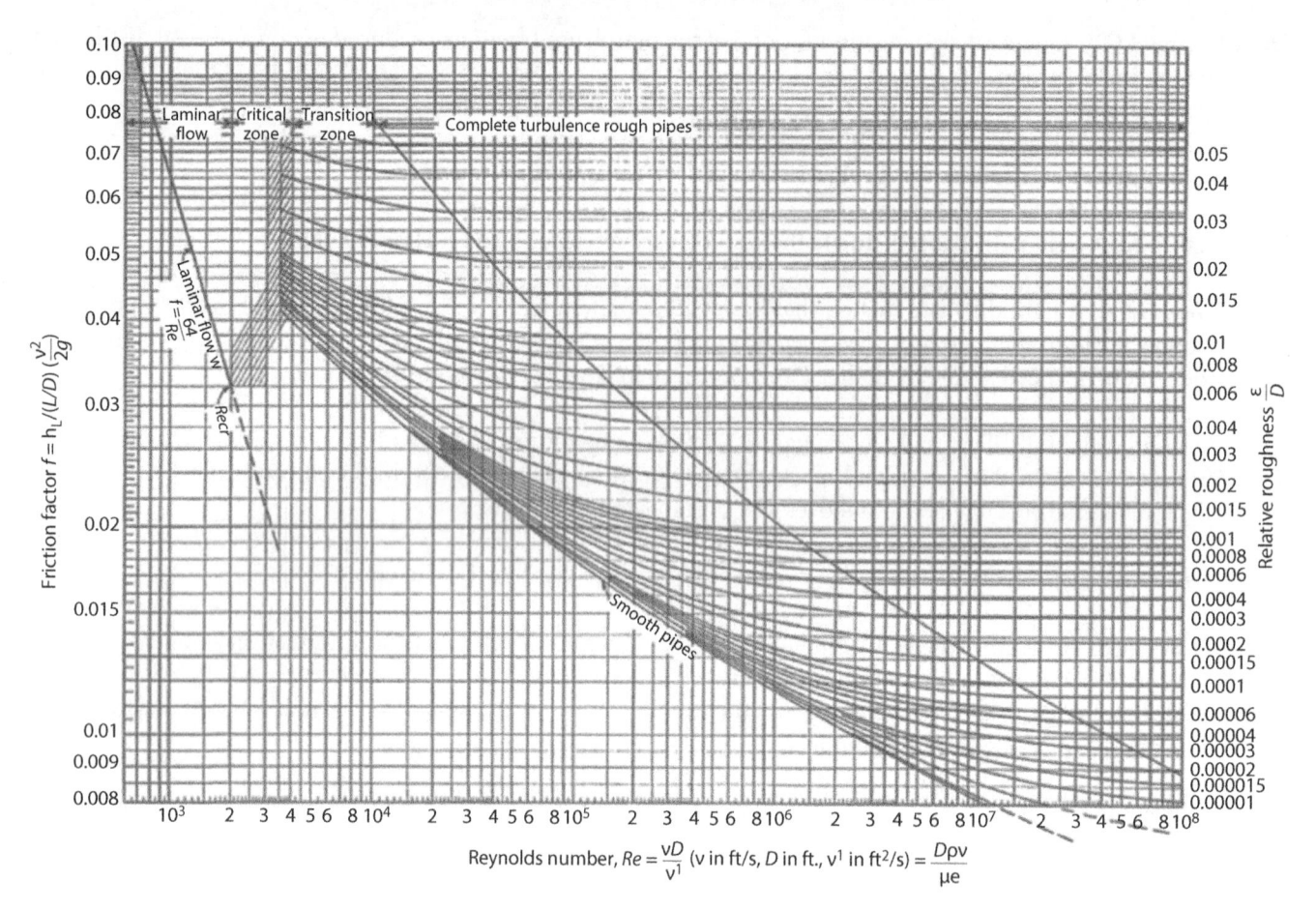

Figure 8 Moody diagram.

Distillation Column: A tall vertical cylindrical vessel used for the process of distillation. Hot vapor rises up the column, which is brought into intimate contact with cooled liquid descending on stages or trays for a sufficient period of time so as to reach equilibrium between the vapor and the liquid. The vapor rises up from the tray below through perforations in the tray, and the liquid on the tray flows over a weir to the tray below. In this way, the more volatile component increases in concentration progressively up the column. In continuous distillation, fresh feed is admitted at the tray corresponding to the same composition. Below the feed point, the section of a column is known as the stripping section, while the above is referred to as the rectifying section. A reboiler heat exchanger is used to boil the bottom product and produce vapor for the column. A condenser is used to condense some or all of the vapor from the top of the column. A small portion of liquid is returned to the column as reflux. The height of the column is an indication of the ease or difficulty of separation. For example, an ethylene splitter in a refinery used to separate ethylene from ethane, which has close boiling points, requires many trays, and the column is very tall. The width of the column is an indication of the internal vapor and liquid rates (See Figure 9).

Distillation Curves: In addition to True Boiling Point (TBP) or good fractionation distillations, there are at least three other major types of distillation curves or ways of relating vapor temperature and percentage vaporized: (a) equilibrium flash vaporization, (b) ASTM or non-fractionating distillations, and (c) Hempel or semi-fractionating distillations (See Figure 10).

Distillation Range: See boiling range.

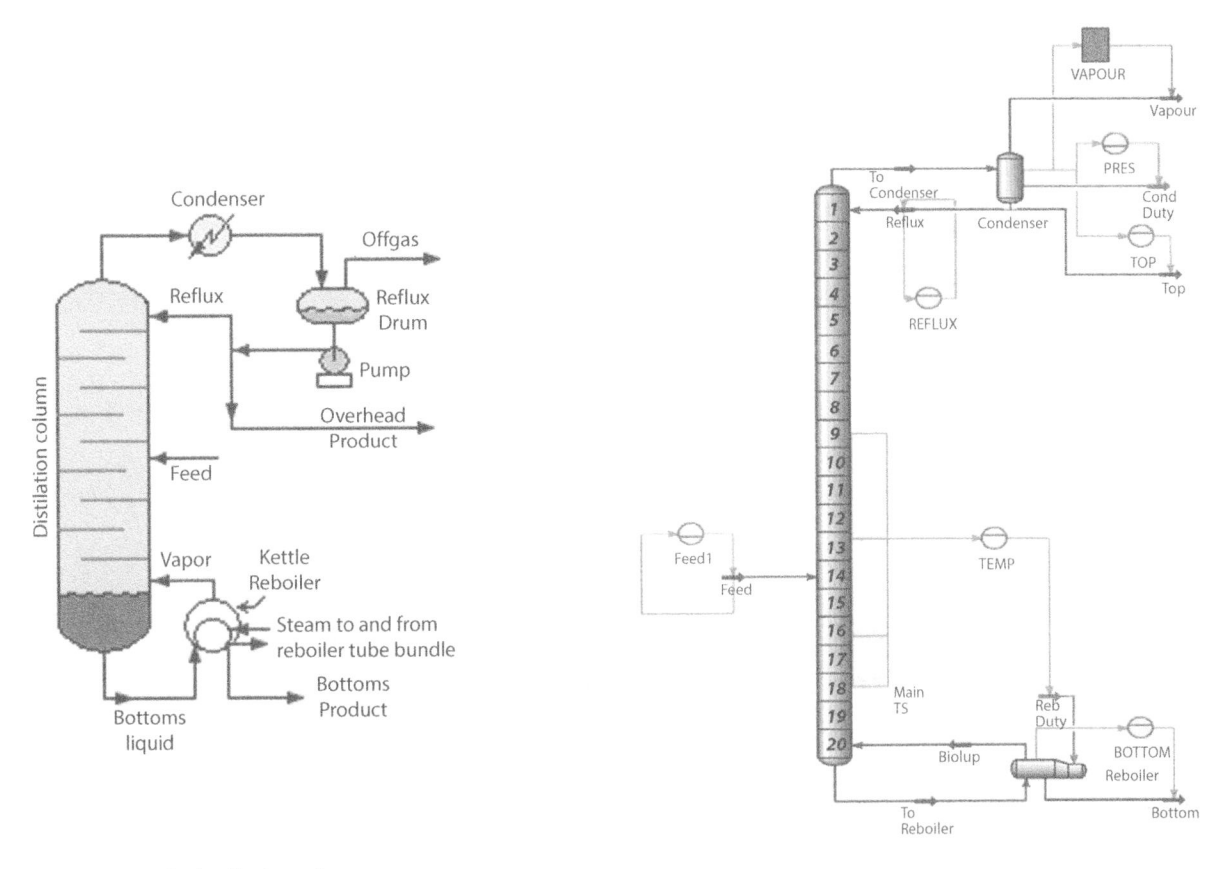

Figure 9 Diagrams of a distillation column.

Distillation Train: A sequence of distillation columns used to separate components from a multicomponent feed. Each column is required to perform a particular separation of either a pure component or a cut between two components. For example, in the separation of four components, ABCD in a mixture in which A is the most volatile and D is the least; then, the five possible separation sequences requiring three columns are:

Separation	Column 1	Column 2	Column 3
1	A:BCD	B:CD	C:D
2	A:BCD	BC:D	B:C
3	AB:CD	A:B	C:D
4	ABC:D	A:BC	B:C
5	ABC:D	AB:C	A:B

Where it is required to separate a larger number of components, the number of possible separation sequences becomes much larger according to the relationship

$$N = \frac{(2n-2)!}{n!(n-1)!} \qquad (24)$$

where N is the number of sequences and n is the number of components:

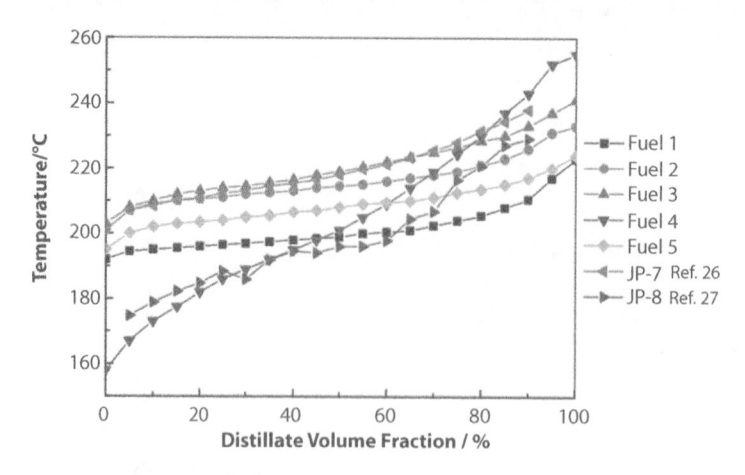

Figure 10 A typical distillation curve of petroleum products.

Components (n)	4	5	6	7	8	9	10
Sequences (N)	5	14	42	132	429	1430	4862

Distributed Component: A component that appears in both the top and bottom products from a distillation/fractionating column separating zone.

Distributed Control System: A system that divides process control functions into specific areas interconnected by communications (normally data highways) to form a single entity. It is characterized by digital controllers and typically by central operation interfaces.

Distributed control systems consist of subsystems that are functionally integrated but may be physically separated and remotely located from one another. Distributed control systems generally have at least one shared function within the system. This may be the controller, the communication link, or the display device. All three of these functions may be shared.

A system of dividing plant or process control into several areas of responsibility, each managed by its own Central Processing Unit, in which the whole are interconnected to form a single entity usually by communication buses of various kinds.

Distributor: A device in a vessel that disperses either liquid or vapor to promote better circulation.

Doctor Test: A method for determining the presence of mercaptan sulfur petroleum products. This test is used for products in which a "sweet" odor is desirable for commercial reasons, especially naphtha; ASTM D-484.

Dow Fire and Explosion Index (F & EI): A method (developed by Dow Chemical Company) for ranking the relative fire and explosion risk associated with a process. Analysts calculate various hazard and explosion indexes using material characteristics and process data.

Downcomer: A device to direct the liquid from a distillation column tray to the next lower distillation tray.

Draw, Side Draw: A product stream withdrawn from a distillation column at a location above the bottom tray and below the top tray. Draws may be vapor or liquid phase.

Dropping Point of Lubricating Greases: Dropping points are used for identification and quality control purposes and can be an identification of the highest temperature of utility for some applications. This is the temperature at which grease passes from a semisolid to a liquid state under prescribed conditions.

Dry Gas: All C_1 to C_3 material, whether associated with crude or produced as a by-product of refinery processing. Convention often includes hydrogen in dry gas yields.

Effective Cut Points: Cut points that can be considered a clean-cut, ignoring any tail ends.

Emergency: A condition of danger that requires immediate action.

Emergency Isolation Valve (EIV): A valve that, in event of a fire, rupture, or loss of containment, is used to stop the release of flammable or combustible liquids, combustible gas, or potentially toxic material. An EIV can be either hand-operated or power-operated (air, hydraulic, or electrical actuation).

Emergency Shutdown (ESD): A method to rapidly cease the operation of a process and isolate it from incoming and outgoing connections or flows to reduce the likelihood of an unwanted event from continuing or occurring. Critical valves shut to isolate sections of the process. Other valves may be opened to depressurize vessels or rapidly discharge the contents of reactors to quench tanks. Emergency shutdowns may occur due to changes in process conditions causing unstable or unsafe operating conditions, a failure in the control system, operator intervention causing unsafe conditions, plant and pipe failure, or some other external event such as an electrical storm or natural catastrophes as earthquakes or coastal flooding.

Emulsion: A colloidal suspension of one liquid dispersed within another. The dispersed phase has droplet sizes usually less than 1 mm. Surfactants or emulsifiers are surface-active agents and are used to stabilize emulsions. In the offshore oil industry, emulsions form at the interface of water and oil in crude oil gravity separators. Sufficient hold-up time is used to separate the emulsion, or, alternatively, surface-active agents are used to encourage separation.

Endothermic reaction: 1. A chemical reaction that absorbs heat from its surroundings in order for the reaction to proceed. Such reactions have a positive enthalpy change and therefore do not occur spontaneously. **2.** A reaction in which heat must be added to maintain reactants and products at a constant temperature.

E85: Fuel containing a blend of 70–85% ethanol.

End Point (final boiling point): 1. The highest boiling point recorded for a laboratory distillation. Usually, there is some residual material in the laboratory still, and the end point is not the highest boiling point material in the mixture being distilled. **2.** The lowest temperature at which virtually 100% of petroleum product will boil off to vapor form.

Energy: The capacity or ability of a system to do work. It may be identified by type as being kinetic, potential, internal, and flow or by a source such as electric, chemical, mechanical, nuclear, biological, solar, etc. Energy can be neither created nor destroyed but converted from one form to another. It can be stored as potential energy, nuclear, and chemical energy, whereas kinetic energy is the energy in motion of a body defined as the work that is done in bringing the body to rest. The internal energy is the sum of the potential energy and kinetic energy of the atoms and molecules in the body. Energy as the units Btu, cal, and J.

Energy Balance: 1. An accounting of the energy inputs and outputs to a process or part of a process, which is separated from the surroundings by an imaginary boundary. All energy forms are included in which the energy input across the boundary must equal the energy output plus any accumulation within the defined boundary. When the conditions are steady and unchanging with time, the energy input is equal to the energy output. The most important energy forms in most processes are kinetic energy, potential energy, enthalpy, heat, and work. Electrical energy is included in electrochemical processes and chemical energy is in processes involving chemical reactions that occur in various reactor types (e.g., batch, continuous stirred tank, plug flow, fixed and catalytic reactors). **2.** The summation of the energy entering a process and the summation of the energy leaving a process. They must be equal for a steady-state process.

Energy Management: Is the planning and operation of energy production and energy consumption units. Objectives are resource conservation, climate protection, and cost savings, while users have permanent access to the energy they need. Energy management is the proactive, organized, and systematic coordination of procurement, conversion, distribution, and the use of energy to meet the requirements, taking into account the environmental and economic objectives. It is also the solution for electric power producers to reduce emissions and improve efficiency and availability. Energy management requires reducing NO_x and greenhouse gas emissions; improving fuel efficiency and reducing SCR operating costs; and streamlining the detection, diagnosis, and remediation of plant reliability, capacity, and efficiency problems. Energy management programs incorporate energy policies, benchmarking, local and corporate goals, the types of energy audits and assessments, reporting systems, and the integration of energy efficiency elements into engineering procedures and purchasing protocols.

Pinch analysis is a tool that is employed in the energy management of chemical facilities and is a methodology for minimizing the energy consumption of chemical processes by calculating thermodynamically feasible energy targets (or minimum energy consumption) and achieving them by optimizing heat recovery systems, energy supply methods, and processing operating conditions. It is also known as process integration, heat integration, energy integration, or pinch technology (See Process Integration).

Engler Distillation: A standard test for determining the volatility characteristics by measuring the percent distilled at various specified temperatures (see ASTM D86).

Engine Knocking (knock, detonation, spark knocking, pinging, or pinking): Spark ignition in internal combustion engines occurs when the combustion of the air/fuel mixture in the cylinder does not start off correctly in response to ignition by the spark plug, but one or more pockets of air/fuel mixture explode outside the envelope of the normal combustion front.

The fuel–air charge is meant to be ignited by the spark plug only and at a precise point in the piston's stroke. Knock occurs when the peak of the combustion process no longer occurs at the optimum moment for the four-stroke cycle. The shock wave creates the characteristic metallic "pinging" sound, and the cylinder pressure increases dramatically. The effects of engine knocking range from inconsequential to completely destructive. See also Knocking.

Engineering line diagram (ELD): A diagrammatic representation of a process. Also referred to as ***engineering flow diagram***. It features all process equipment and piping that are required for the start-up and shutdown, emergency, and normal operation of the plant. It also includes insulation requirements; the direction of flows; the identification of the main process and start-up lines; all instrumentation, control, and interlock facilities; the key dimensions and duties of all equipment; the operating and design pressure and temperature for vessels; equipment elevations; set pressures for relief valves; and drainage requirements.

Engineering, Procurement, and Construction Contract: **1.** A legal agreement setting out the terms for all activities required to build a facility to the point that it is ready to undergo preparations for operations as designed. **2.** The final contracting phase in the development of the export portion of the LNG chain that defines the terms under which the detailed design, procurement, construction, and commissioning of the facilities will be conducted. Greenfield LNG project development involves a wide range of design, engineering, fabrication, and construction work far beyond the capabilities of a single contractor. Therefore, an LNG project developer divides the work into a number of segments, each one being the subject of an engineering, procurement, and construction (EPC) contract. **3.** Contract between the owner of a liquefaction plant and an engineering company for the project development and erection. See Front-End Engineering and Design Contract.

Enthalpy (H): The thermal energy of a substance or system with respect to an arbitrary reference point. The enthalpy of a substance is the sum of the internal energy and flow of energy, which is the product of the pressure and specific volume.

$$H = U + pV \tag{25}$$

The reference point for gases is 273 K and for chemical reactions is 298 K.

Enthalpy balance: A form of energy accounting for a process in which the stream energies to and from the process are expressed as enthalpies. At a steady state, the total enthalpy into a process is equal to the total enthalpy out. Where there is an inequality, there is either a loss or an accumulation of material with an associated loss or increase in enthalpy. An enthalpy balance is used to determine the amount of heat that will be generated in the process or that needs to be removed to ensure that the process operates safely and to specification.

Entrainment: A non-equilibrium process by which liquids are mechanically carried into a vapor leaving a process vessel or contacting device.

Entrance and exit losses: The irreversible energy loss caused when fluid enters or leaves an opening, such as into or out of a pipe into a vessel. Where there is a sudden enlargement, such as when a pipe enters a larger pipe or vessel, eddies form and there is a permanent energy loss expressible as a head loss as:

$$H_{exit} = \frac{v^2}{2g}\left(1 - \frac{a}{A}\right)^2 \tag{26}$$

where v is the velocity in the smaller pipe, a is the cross-sectional area of the smaller pipe, and A is the cross-sectional area of the larger pipe. For a considerable enlargement, the head loss tends to

$$H_{exit} = \frac{v^2}{2g} \tag{27}$$

With a rapid contraction, it has been found experimentally that the permanent head loss can be given by:

$$H_{exit} = K\frac{v^2}{2g} \tag{28}$$

where for very large contraction, K = 0.5

Entropy (dS): The extent to which the energy in a closed system is unavailable to do useful work. An increase in entropy occurs when the free energy decreases or when the disorder of molecules increases. For a reversible process, entropy remains constant such as in a friction-free adiabatic expansion or compression. The change in entropy is defined as:

$$dS = \frac{dQ}{T} \qquad (29)$$

where Q is the heat transferred to or from a system, and T is the absolute temperature. However, all real processes are irreversible, which means that in a closed system, there is a small increase in entropy.

Environmental Protection Agency (EPA), United States: **1.** Governmental agency, established in 1970. Its responsibilities include the regulation of fuel and fuel additives. **2.** The US federal agency that administers federal environmental policies, enforces environmental laws and regulations, performs research, and provides information on environmental subjects. The agency also acts as the chief advisor to the president on American environmental policy and issues. **3.** A federal agency created in 1970 to permit coordinated and effective government action, for the protection of the environment by the systematic abatement and control of pollution, through integration of research monitoring, standard setting, and enforcement activities. **4.** US pollution control enforcer. **5.** A regulatory agency established by the US Congress to administer the nation's environmental laws. Also called the US EPA.

Error: Discrepancy between a computed, observed, or measured value or condition and the true specified or theoretically correct value or condition.

Ethane (C_2H_6): A colorless gas; a minor constituent of natural gas and a component in refinery gas that, along with methane, is typically used as refinery fuel. An important feedstock for making ethylene.

Ether ($C_2H_5OC_2H_5$): **1.** A generic term applied to a group of organic chemical compounds composed of carbon, hydrogen, and oxygen, characterized by an oxygen atom attached to two carbon atoms (e.g., methyl tertiary butyl ether). **2.** Any carbon compound containing the functional group (C–O–C). A commonly used ether is diethyl ether, which is used as an anesthetic.

Ethylene (C_2H_4): A colorless gas created by cracking processes. In refineries, it is typically burned with the methane and ethane. In chemical plants, it is purposefully made in ethylene plants and it is a basic building block for a wide range of products including polyethylene and ethyl alcohol.

ETBE: Ethyl Tertiary Butyl Ether ($(CH_3)_3 CO C_2H_5$): **1.** A colorless, flammable, oxygenated hydrocarbon blend stock. It is produced by the catalytic etherification of ethanol (C_2H_5OH) with isobutylene (C_4H_8). **2.** An oxygenated gasoline blending compound to improve octane and reduce carbon monoxide emissions. It is commonly used as an oxygenate gasoline additive in the production of gasoline from crude oil.

Equation of state: A relationship that links the pressure, volume, and temperature of an amount of a substance. It is used to determine thermodynamic properties such as liquid and vapor densities, vapor pressures, fugacities, and deviations from ideality and enthalpies. Various equations of state have been developed to predict the properties of real substances. Commonly used equations of state include the ideal gas law; virial equation; van der Waals' equation; and Peng-Robinson, Soave–Redlich Kwong, and Lee–Kesler equations. Cubic equations are relatively easy to use and are fitted to experimental data. The van der Waals equation is comparatively poor at predicting state properties. The Lee–Kesler model, which is based on the theory of corresponding states, uses reduced temperature and pressure and covers a wide range of temperatures and pressures (see Table 1).

Equilibrium: A condition or state in which a balance exists within a system, which may be physical or chemical. A system is in equilibrium if it shows no tendency to change its properties with time. Static equilibrium occurs if there is no transfer of energy across the system boundary, whereas dynamic equilibrium is when a transfer occurs, but the net effect of the energy is zero. Thermodynamic equilibrium occurs when there is no heat or work exchange between a body and its surroundings. Chemical equilibrium occurs when a chemical reaction takes place in the forward direction, when reactants form products at exactly the same rate as the reverse reaction of products reverts to their original reactant form.

Equilibrium constant (K_c): A reversible process, chemical or physical, in a closed system will eventually reach a state of equilibrium. The equilibrium is dynamic and may be considered as a state at which the rate of the process in one direction exactly balances the rate in the opposite direction. For a chemical reaction, the equilibrium concentrations of the reactants and products will remain constant provided that the conditions remain unchanged for the homogeneous system:

$$wA + xB \leftrightarrow yC + zD \tag{30}$$

The ratio of the molar concentrations of products to reactants remain constant at a fixed temperature; the equilibrium constant, K_c, is:

$$K_c = \frac{[C]^y [D]^z}{[A]^w [B]^x} \tag{31}$$

Table 1 Useful Equations of State.

Name	Equation	Equation constants and functions
(1) ideal gas law	$P = \dfrac{RT}{V}$	None
(2) Generalized	$P = \dfrac{ZRT}{V}$	$Z = Z(p_r, T_r, Z_c \text{ or } \omega)$ as derived from data
(3) van-der-Waals	$P = \dfrac{RT}{(V-b)} - \dfrac{a}{V^2}$	a and b are species-dependent constants and estimated from the critical temperature and pressure
(4) Redlich-Kwong (R-K)	$P = \dfrac{RT}{(V-b)} - \dfrac{a}{V^2 + bV}$	$b = 0.08664\, RT_c/P_c$ $a = 0.42748\, R^2\, T_c^{2.5}/P_c\, T^{0.5}$
(5) Soave-Redlich-Kwong (S-R-K or R-K-S)	$P = \dfrac{RT}{(V-b)} - \dfrac{a}{V^2 + bV}$	$b = 0.08664\, RT_c/P_c$ $a = 0.42748\, R^2\, T_c^2 [1 + f_\omega(1 - T_r^{0.5})]^2/P_c$ $f_\omega = 0.48 + 1.574\omega - 0.176\omega^2$
(6) Peng-Robinson (P-R)	$P = \dfrac{RT}{V-b} - \dfrac{a}{V^2 + 2bV - b^2}$	$b = 0.07780R\, T_c/P_c$ $a = 0.45724\, R^2\, T_c^2 [1 + f_\omega(1 - T_r^{0.5})]^2/P_c$ $f_\omega = 0.37464 + 1.54226\omega - 0.26992\omega^2$

For the Haber process for the synthesis of ammonia, nitrogen is reacted with hydrogen as:

$$N_2 (g) + 3H_2 (g) \leftrightarrow 2NH_3 (g) \tag{32}$$

The equilibrium constant is expressed as partial pressure:

$$K_c = \frac{[NH_3]^2}{[N_2][H_2]^3} = \frac{p_{NH_3}^2}{p_{N_2} p_{H_2}^3} \tag{33}$$

Equilibrium K-value (K-value): The ratio of the mole fraction in the vapor divided by the mole fraction in the liquid for a component in the equilibrium state. Each K-value corresponds to a given temperature, pressure, and mixture composition.

Equilibrium ratio (K): the ratio of the mole fraction in the vapor phase of a component in a mixture, y, to the mole fraction in the liquid phase, x, at equilibrium:

$$K_A = \frac{y_A}{x_A} \tag{34}$$

It is a function of both temperature and pressure. The relative volatility, α, is less dependent on temperature and pressure than the equilibrium constant where for an ideal mixture of two components, A and B:

$$\alpha_{AB} = \frac{K_A}{K_B} \tag{35}$$

Equilibrium-Flash Vaporizer: When a mixture is heated without allowing the vapor to separate from the remaining liquid, the vapor assists in causing the high-boiling parts of the mixture to vaporize. Thus continuous-flash vaporization is used in almost all plant operations. The equipment is used to determine a flash vaporization curve, where a series of runs at different temperatures are conducted, and each run constitutes one point (of temperature and percentage vaporized) on the flash curve.

Equipment Reliability: The probability that, when operating under stated environment conditions, the process equipment will perform its intended function adequately for a specified exposure period.

Equivalent Length: A method used to determine the pressure drop across pipe fittings such as valves, bends, elbows, and T-pieces. The equivalent length of a fitting is the length of a pipe that would give the same pressure drop as the fitting. Since each size of a pipe or fitting requires a different equivalent length for any particular type of fitting, it is usual to express the equivalent length as so many pipe diameters and this number is independent of the pipe. For example, if a valve in a pipe of diameter, d, is said to have an equivalent length, n, of pipe diameters, then the pressure drop due to the valve is the same as that offered by a length, and of the pipe.

Ergun Equation: Sabri Ergun developed an equation in 1952 to determine the pressure drop per unit length of a fixed bed of particles such as a catalyst at incipient gas velocity, v:

$$\frac{-\Delta p}{L} = \frac{150(1-e)^2 \mu v}{\phi e^3 d^2} + \frac{1.75(1-e)\rho v^2}{\phi e^3 d} \tag{36}$$

where $-\Delta p/L$ is the pressure drop over the depth of bed, e is the bed voidage, d is the mean particle diameter, ρ is the fluid density, μ is the fluid viscosity, and ϕ is the sphericity. The incipient point of fluidization corresponds with the highest pressure drop at the minimum fluidization velocity.

Erosion: The physical removal of a material from a surface by mechanisms that exclude chemical attack. The usual phenomenon that causes erosion is impingement by either liquid droplets or entrained solid particles. If there are no corrosive substances present, then, in many cases, the most common mechanism for material damage due to erosion is impingement by solid particles.

Exothermic reaction: 1. A chemical reaction that gives out/liberates heat. No energy input is required for the reaction to proceed. It has a negative enthalpy change, and therefore, under the appropriate conditions, the reaction will occur spontaneously. Chemical reactors used to contain exothermic reactions, therefore, require cooling facilities to remove the excess heat that is generated and to maintain a constant temperature. 2. A reaction in which heat is evolved. Alkylation, polymerization, and hydrogenation reactions are exothermic reactions.

Expansion Loop: Piping thermally expands as it gets hot. Allowance must be made for the growth in pipe length; otherwise, the pipe will break by cracking at its welds. A fractionator at the Good Hope Refinery in the United States was burned down because of such an omission.

Explosion: 1. The sudden conversion of potential energy (chemical or mechanical) to kinetic energy with the production and release of gases under pressure or the release of gas under pressure. 2. A release of energy that causes a pressure discontinuity or blast wave.

Exports: Shipments of crude oil and petroleum products from countries, e.g., in the United States—shipments from the 50 states and the District of Columbia to foreign countries, Puerto Rico, the Virgin Islands, and other US possessions and territories.

Failure: 1. Termination of the ability of a functional unit to perform a required function. 2. An unacceptable difference between the expected and observed performance.

Fail Safe: A system design or condition such that the failure of a component, subsystem, or system or input to it will automatically revert to a predetermined safe static condition or state of least critical consequence for the component, subsystem, or system.

Fail Steady: A condition wherein the component stays in its last position when the actuating energy source fails. May also be called Fail in Place.

Failure Mode: The action of a device or system to revert to a specified state upon the failure of the utility power source that normally activates or controls the device or system. Failure modes are normally specified as fail open (FO), fail close (FC), or fail steady (FS), which will result in a fail-safe or fail-to-danger arrangement.

Failure Mode and Effects Analysis (FMEA): A systematic, tabular method for evaluating and documenting the causes and effects of known types of component failures.

Fault: Abnormal condition that may cause a reduction in, or loss of, the capability of a functional unit to perform a required function.

Fault Tree: A logic method that graphically portrays the combinations of failures that can lead to a specific main failure or accident of interest.

Field Production: Represents crude oil production on leases, natural gas liquids production at natural gas processing plants, a new supply of other hydrocarbons/oxygenates and motor gasoline blending components, and fuel ethanol blended into finished motor gasoline.

Final Boiling Point (FBP): The final boiling point of a cut, usually on an ASTM distillation basis.

Fifteen-five (15/5) distillation: A laboratory batch distillation performed in a 15-theoretical plate fractionating column with a 5:1 reflux ratio. A good fractionation results in accurate boiling temperatures. For this reason, this distillation is referred to as the true boiling point distillation. This distillation corresponds very closely to the type of fractionation obtained in a refinery.

Fire: 1. A combustible vapor or gas combined with an oxidizer in a combustion process manifested by the evolution of light, heat, and flame. **2.** The rapid thermal oxidation (combustion) of a fuel source, resulting in heat and light emission. There are various types of fire, classified by the type of fuel and associated hazards. In the United States, the National Fire Protection Association (NFPA) classifies fires and hazards by the types of fuels or combustibles in order to facilitate the control and extinguishing of fires:

Class A	Ordinary combustibles such as wood, cloth, paper, rubber, and certain plastics
Class B	Flammable or combustible liquids, flammable gases, greases, and similar materials
Class C	Energized electrical equipment
Class D	Combustible metals, such as magnesium, titanium, zirconium, sodium, and potassium

Fireball: The atmosphere burning of a fuel–air cloud in which the energy is mostly emitted in the form of radiant heat. The inner core of the fuel release consists of almost pure fuel, whereas the outer layer in which ignition first occurs is a flammable fuel–air mixture. As the buoyancy forces of the hot gases begin to dominate, the burning cloud rises and becomes more spherical in shape.

Fire Point: Is the temperature well above the flashpoint where the product could catch a fire. The fire point and flash point are always taken care of in the day-to-day operation of a refinery. (See also Flashpoint).

Fireproof: Resistant to specific fire exposure. Essentially nothing is absolutely fireproof, but some materials or building assemblies are resistant to damage or fire penetration at certain levels of fire exposures that may develop in the petroleum, chemical, or related industries.

Fireproofing: A common industry term used to denote the materials or methods of construction used to provide fire resistance for a defined fire exposure and specified time. Essentially nothing is fireproof if it is exposed to high temperatures for an extended period of time.

Fire Suppression System: A method, device, or system used to detect a fire or ignition source and to extinguish the fire in sufficient time so as to prevent structural damage and/or the debilitation of personnel.

Fire triangle: A way of illustrating the three factors necessary for the process of combustion, which are fuel, oxygen, and heat. All three are required for combustion to occur. A fire can therefore be prevented or extinguished by removing one of the factors. A fire is not able to occur without sufficient amounts of all three (See Figure 11).

First Law of Thermodynamics: A law that is applied to the conservation of energy in which the change in internal energy, ΔU, of a system is equal to the difference in the heat added, Q, to the system and the work done by the system:

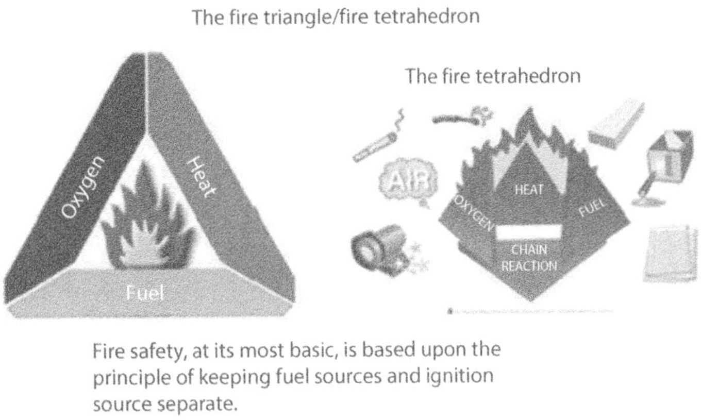

Figure 11 Diagram of a fire triangle.

$$\Delta U = Q - W \tag{37}$$

When considering chemical reactions and processes, it is more usual to consider situations where work is done on the system rather than by it.

Fittings: Connections and couplings used in pipework and tubing. The type of fittings used depends largely on the wall thickness as well as in part on the properties of the pipes and tubes including welds, flanges, and screw fittings. Fittings include elbow, bends, tees, reducers, and branches.

Fixed Bed: A place in a vessel for a catalyst through or by which feed can be passed for reactions, as opposed to a fluid bed, where the catalyst moves with the feed.

Fixed Bed Reactor: A reactor in which the catalyst is loaded into an immovable bed. The reactants enter the top of the bed, and the products exit from the bottom of the bed. The process must be taken offline, and hot gases are circulated through the catalyst bed to burn off coke deposits and restore the catalyst activity (See Figure 12).

Flame: The glowing gaseous part of a fire.

Flammable: 1. A substance or material that has the ability to support combustion and be capable of burning with a flame. It is easily ignited or highly combustible. The term is more widely used than inflammable as this is often confused with incombustible, which means an inability or lack of ability to combust. A flammable liquid is a liquid that has the capability of catching a fire. In the United States, the National Fire Protection Association defines a flammable liquid as a liquid that has a flashpoint below 100°F (37.8°C) and a vapor pressure not exceeding 40 psia (2.72 bara) at that temperature. 2. In general sense, refers to any material that is easily ignited and burns rapidly. It is synonymous with the term inflammable that is generally considered obsolete due to its prefix, which may be incorrectly misunderstood as not flammable (e.g., incomplete is not complete).

Flammable Liquid: 1. As defined by NFPA 30, a liquid having a flash point below 100°F (37.8°C) and having a vapor pressure not exceeding 2068 mm Hg (40 psia) at 100°F (37.8°C) as determined under specific conditions. 2. Any liquid having a flash point below 100°F (37.8°C), except any liquid mixture having one or more components with a flash point at or above the upper limit that makes up 99% or more of the total volume of the mixture. 3. Liquid with a flash point below 100°F (37.8°C). At that temperature, vapors from the substance can be ignited by a flame, spark, or other sources of ignition.

Flammability Limit: 1. The flammability limit of a fuel is the concentration of fuel (by volume) that must be present in the air for ignition to occur when an ignition source is present. 2. The range of gas or vapor amounts in air that

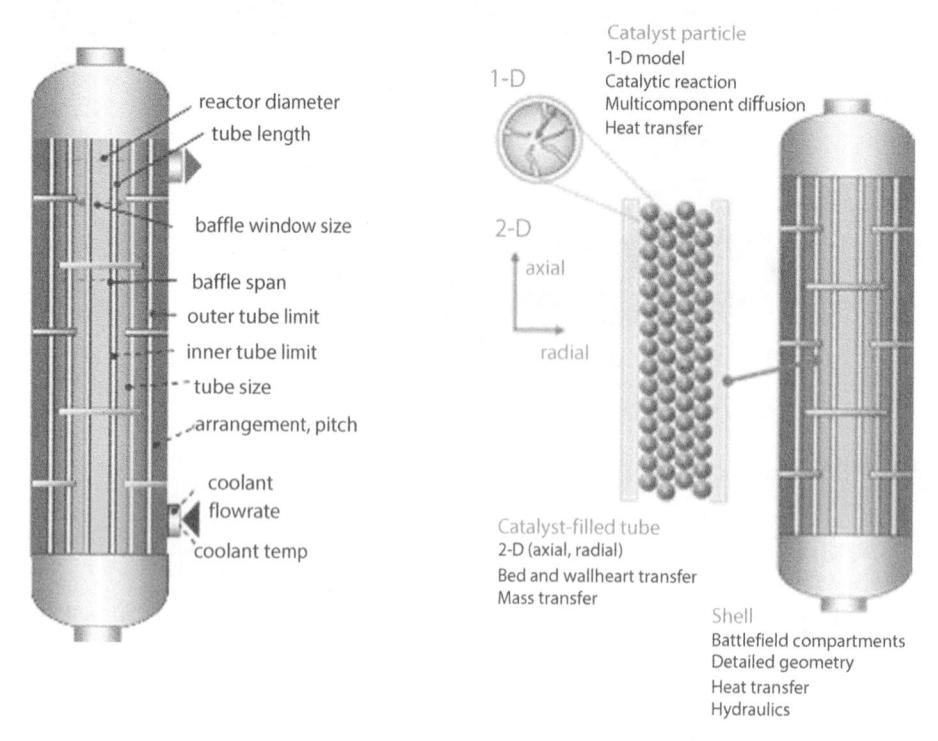

Figure 12 Diagrams of a fixed bed reactor.

will burn or explode if a flame or other ignition source is present. Importance: The range represents an unsafe gas or vapor mixture with air that may ignite or explode. Generally, the wider the range, the greater the fire potential.

Flange: It is a flat end of a pipe that is used to bolt up to a flange on another piece of piping. Bolts, with nuts at each end, are used to force the flanges together.

Flange Rating: Connections on vessels, spool pieces, and valves have a pressure rating called a flange rating. This rating can be confusing, e.g., a 150 psig flange rating is actually good for about 230 psi design.

Flare: **1.** A burner on a remote line used for the disposal of hydrocarbons during clean-up, for emergency, shut-downs, and the disposal of small-volume waste streams of mixed gases that cannot easily or safely be separated. **2.** A flame used to burn off unwanted natural gas; a "flare stack" is the steel structure on a processing facility from which gas is flared. **3.** An open flame used to burn off unwanted natural gas. **4.** To burn unwanted gas through a pipe or stack. **5.** The flame from a flare; the pipe or the stack itself.

Flared: Gas disposed of by burning in flares, usually at the production sites or at gas processing plants.

Flare Stack: The steel structure on an offshore rig or at a processing facility from which gas is flared.

Flare System: This is a piping network that runs through the plant to collect vents of gas so that they can be combusted at a safe location in the flare stack.

Flaring: Is the burning of a natural gas that cannot be processed or sold. Flaring disposes off the gas, and it releases emissions into the atmosphere.

Flaring/Venting: The controlled burning (flare) or release (vent) of natural gas that cannot be processed for sale or use because of technical or economic reasons.

Flashing: The vaporization of water or light ends as pressure is released during production or processing.

Flash Calculation: The determination of the compositions and quantities of liquid and vapor that co-exist in a mixture under equilibrium conditions.

Flash Chamber: A wide vessel in a vacuum flasher thermal cracking plant or similar operation into which a hot stream is introduced, causing the lighter fractions of that stream to vaporize and leave by the top.

Flash Fire: The combustion of flammable vapor and air mixture in which flame passes through that mixture at less than sonic velocity, such that negligible damaging overpressure is generated.

Flash Point: **1.** The minimum temperature at which a liquid, under specific test conditions, gives off sufficient flammable vapor to ignite momentarily on the application of an ignition source. **2.** The lowest temperature at which any combustible liquid will give off sufficient vapor to form an inflammable mixture with air (i.e., that can be readily ignited). Flash points are used to specify the volatility of fuel oils, mostly for safety reasons. They are generally an indication of the fire and explosion potential of a product; ASTM D-56, D-92, D-93, D-134, and D-1310. **3.** Hold a flame over a cup of diesel fuel; it will start to burn at its 160°F (71°C) flash temperature. Gasoline's flash point is below room temperature. For jet fuel, it is 110°F (43°C). The lighter the hydrocarbon, the lower the flash point.

Flash Tank: A container where the separation of gas and liquid phases is achieved after pressure reduction in flow fluid. Both phases appear when pressure is decreased as a consequence of the Joule–Thompson effect.

Flash Vapors: Vapors released from a stream of natural gas liquids as a result of an increase in temperature or a decrease in pressure.

Flash Zone: The section of a distillation/fractionating column containing the column feed nozzle(s). The column feed separates or "flashes" into liquid and vapor as it expands through the feed nozzle(s) and enters the column.

Flexicoking: A thermal cracking process that converts heavy hydrocarbons such as crude oil, oil sands bitumen, and distillation residues into light hydrocarbons. Feedstocks can be any pumpable hydrocarbons including those containing high concentrations of sulfur and metals.

Flooding: **1.** An excessive buildup of liquid in absorption columns or on the plate of a distillation column. It is due to high vapor flow rates up the column. In distillation columns, this is caused by high heating rates in the reboiler. **2.** An all-inclusive term that is given to non-equilibrium behavior in a distillation/fractionating column because of larger flows of liquid and/or vapor than the column can process. Flooding can be caused by liquid backing up in the column, vapor blowing through the column and lifting the liquid off the trays, etc. All columns are designed to handle about 80% of the flow before flooding occurs. Sometimes, flooding is caused by mechanical restrictions or damage to the internals in the column. The *flooding point* is a condition in a packed column such as an absorption column that receives a countercurrent flow of gas at the bottom and a liquid descending under gravity from the top where there is an insufficient liquid holdup in the packing for mass transfer to take place effectively. The liquid, therefore, descends to the bottom of the column without mass transfer. The rate of flow through the packing for effective mass transfer is controlled by the pressure drop across the packing material.

Flow: The movement of fluid under the influence of an external force such as gravity or a pump.

Flowline: A pipeline that carries materials from one place to another. In the offshore industry, a flowline is a pipeline that carries oil on the seabed from a well to a riser. On a process flow diagram, the flowline is indicated by a line entering and leaving a vessel or unit operation. An arrow indicates the direction of flow.

Flow Meter: A device used to measure the flow of process fluids. Flow meters are mainly classified into those that are intrusive and those that are non-intrusive to the flow of the fluid. Flow meters include differential pressure meters; positive displacement meters; and mechanical, acoustic, and electrically heated meters. The measurement of the flow of process fluids is essential not only for safe plant control but also for fiscal monitoring purposes. It is essential to select correctly the flow meter for a particular application, which requires a knowledge and comprehension of the nature of the fluid to be measured and an understanding of the operating principles of flow meters.

Flow Rate: The movement of material per unit time. The material may be gas, liquid, or solid particulates in suspension or a combination of all of these and expressed on a mass, volumetric, or molar basis. The volumetric flow of material moving through a pipe is the product of its average velocity and the cross-sectional area of the pipe.

Flow Regime: The behavior of a combined gas and liquid flow through a duct, channel, or pipe can take many forms, and there are descriptions used to define the possible flow patterns. Depending on the conditions of the flow of the two phases, one phase is considered to be the continuous phase, while the other is the discontinuous phase. An example is the flow of mist or fine dispersion of liquid droplets in a gas phase. The smaller the liquid droplets, the higher the surface tension effects. A distortion of the discontinuous phase results in the shape becoming non-spherical. Also, there is a tendency for the liquid phase to wet the wall of the pipe and for the gas phase to congregate at the center. An exception to this is in evaporation such as in refrigeration where nuclear boiling occurs on the pipe surface, resulting in a vapor film or bubbles forming at the surface with a central core of liquid. The flow of fluids through pipes and over surfaces can be described as:

1. Steady flow in which the flow parameters at any given point do not vary with time.
2. Unsteady flow in which the flow parameters at any given point vary with time.
3. Laminar flow in which the flow is generally considered to be smooth and streamlined.
4. Turbulent flow in which the flow is broken up into eddies and turbulence.
5. Transition flow, which is a condition lying between the laminar and turbulent flow regimes.

Flow regime maps are charts representing the various flow patterns that are possible for two-phase gas–liquid flow in both horizontal and vertical pipes and tubes. There are many types of flow regime maps that have been developed. The simplest form of the map involves a plot of superficial velocities or flow rates for the two phases, with the most widely used generalized flow regime map for horizontal flow as shown below (See Figure 13).

The maps are populated with experimental data in which lines are drawn to represent the boundaries between the various flow regimes. These include dispersed, bubble or froth, wavy, annular, stratified, slug, and plug flow. The boundaries between the various flow patterns are due to the regime becoming unstable as it approaches the boundary with the transition to another flow pattern. As with the transition between laminar and turbulent flow in a pipe, the transitions in a flow regime are unpredictable. The boundaries are therefore not distinct lines but loosely defined transition zones. A limitation of the maps is that they tend to be specific to a particular fluid and pipe. The seven types of flow regimes in the order of increasing gas rate at constant liquid flow rate are given below (See Figure 13).

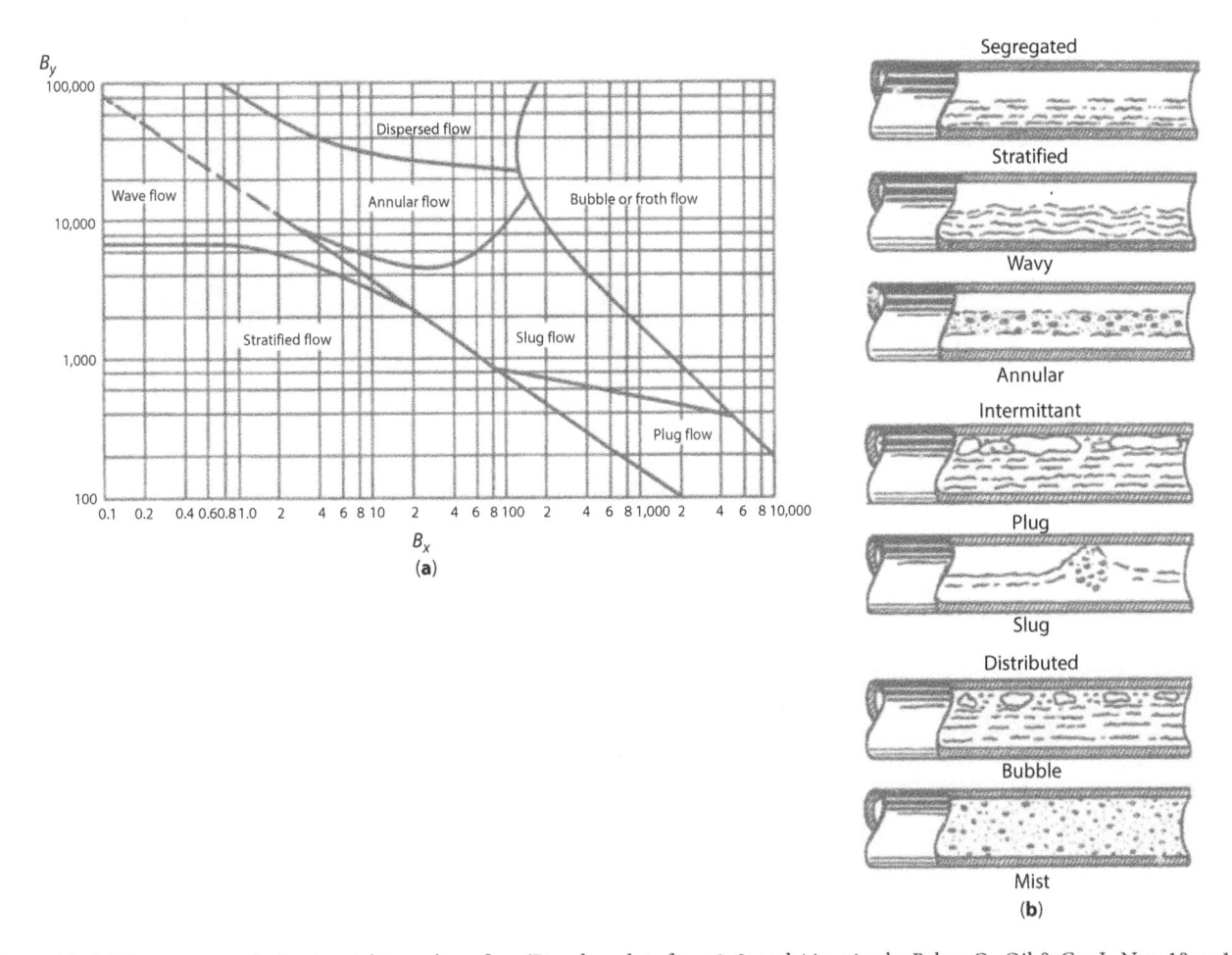

Figure 13 (a) Flow patterns for horizontal two-phase flow (Based on data from 1, 2, and 4 in. pipe by Baker, O., Oil & Gas J., Nov. 10, p. 156, 1958). (b) Representative forms of horizontal two-phase flow patterns as indicated in Figure 10a.

Bubble or Froth Flow: Bubbles of gas are dispersed throughout the liquid and are characterized by bubbles of gas moving along the upper part of the pipe at approximately the same velocity as the liquid. This type of flow can be expected when the gas content is less than about 30% of the total (weight) volume flow rate. (Note: About 30% gas by weight is over 99.9% by volume, normally.)

Plug Flow: Alternate plugs of liquid and gas move along the upper part of the pipe, and liquid moves along the bottom of the pipe.

Stratified Flow: Liquid flows along the bottom of the pipe, while gas flows over a smooth liquid–gas interface.

Wave Flow: Wave flow is similar to stratified flow, except that the gas is moving at a higher velocity and the gas–liquid interface is distributed by waves moving in the direction of flow.

Slug Flow: This pattern occurs when waves are picked up periodically by the more rapidly moving gas. These form frothy slugs that move along the pipeline at a much higher velocity than the average liquid velocity. This type of flow causes severe and, in most cases, dangerous vibrations in equipment because of the high-velocity slugs against fittings.

Annular Flow: In annular flow, liquid forms around the inside wall of the pipe and gas flows at a high velocity through the central core.

Dispersed, Spray, or Mist flow: Here, all of the liquid is entrained as fine droplets by the gas phase. This type of flow can be expected when the gas content is more than roughly 30% of the total weight flow rate. Some overhead condenser and reboiler-return lines have dispersed flow.

Flowsheet: A schematic diagram or representation of a process illustrating the layout of process units and their functions linked together by interconnecting process streams. The development of a flowsheet involves the process synthesis, analysis, and optimization. The heat and material balances are solved using thermodynamic properties and models. An economic analysis is also completed as well as a safety and environmental impact assessment. The choice of equipment and their interconnectivity are optimized along with the choice of operating parameters such as temperature, pressure, and flows. Steady-state flowsheet computer software packages are frequently used to develop flowsheets.

Flue: Passage through which flue gases pass from a combustion chamber to the outside atmosphere.

Flue Gas: **1**. A mixture of gases produced as a result of combustion that emerges from a stack or chimney. The gases contain smoke, particulates, carbon dioxide, water vapor, unburnt oxygen, nitrogen, etc. An Orsat analysis is a reliable device to determine the composition of the flue gas and the efficiency of combustion, although it has been replaced by other techniques. **2**. Gas from the various furnaces going up to the flue (stack).

Fluid Catalytic Cracking (FCC): A thermal process in which the oil is cracked in the presence of a finely divided catalyst that is maintained in an aerated or fluidized state by the oil vapors. The powder or fluid catalyst is continuously circulated between the reactor and the regenerator, using air, oil vapor, and steam as the conveying media. The most commonly used catalytic cracking process. The catalyst is a fine powder that is designed to form a fluidized bed in the reactor and regenerator (Figure 14).

Fluid Coking: **1**. A coking process in which the feed is preheated and sprayed into a reactor where it contacts a hot fluidized bed of coke returning from a burner vessel. The hot oil products are stripped from the coke, which is circulated back to the burner vessel. Coke not returned to the reactor from the burner vessel is withdrawn as a coke product. **2**. A thermal cracking process utilizing the fluidized solids technique to remove carbon (coke) for the continuous conversion of heavy, low-grade oils into lighter products.

Fossil Fuels: Fuels formed by natural processes such as anaerobic decomposition of buried dead organisms. The age of the organisms and their resulting fossil fuels is typically millions of years and sometimes exceeds 650 million years. Fossil fuels contain a high percentage of carbon and include coal, petroleum, and natural gas. Other more commonly used derivatives of fossil fuels are kerosene and propane. They range from volatile materials with low carbon-to-hydrogen ratios like methane to liquid petroleum, to non-volatile materials composed of almost pure carbon, like anthracite coal. Methane can be found in hydrocarbon fields, alone, associated with oil, or in the form of methane clathrates (See Figure 15).

Georg Agricola, in 1556, first introduced the theory that fossil fuels were formed from the fossilized remains of dead plants by exposure to heat and pressure in the earth's crust over millions of years and later expounded by Mikhail Lomonosov in the 18th century. Coal is one of the fossil fuels (See Figure 15).

The use of fossil fuels raises serious environmental concerns. The burning of fossil fuels produces around 21.3 billion tonnes (21.3 gigatonnes) of carbon dioxide (CO_2) per year, but it is estimated that natural processes can only absorb about half of that amount, so there is a net increase of 10.65 billion tonnes of atmospheric carbon dioxide per year (one tonne of atmospheric carbon is equivalent to 44/12 or 3.7 tonnes of carbon dioxide). Carbon dioxide is one of the

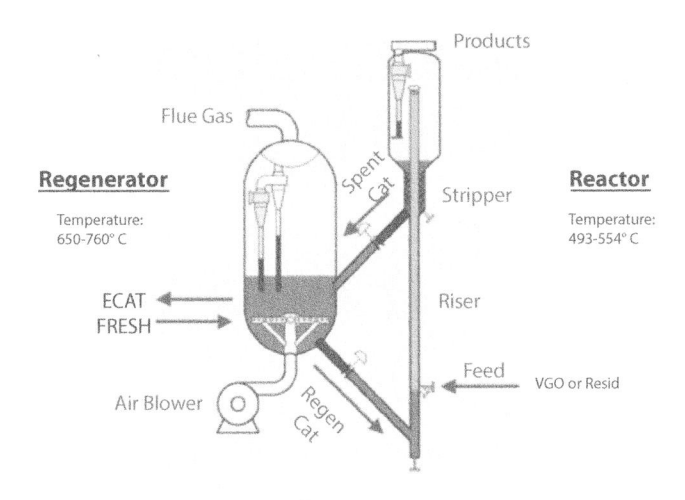

Figure 14 Fluid catalytic cracking unit.

Figure 15 Coal.

greenhouse gases that enhances the radiative forcing and contributes to global warming, causing the average surface temperature of the earth to rise with major adverse climatic effects. A global movement toward the generation of renewable energy is therefore essential to help reduce global greenhouse gas emissions.

The ratio of gross domestic product to kilograms of fossil fuel carbon consumed, for the world's 20 largest economies. The two countries with the highest GDP per kilogram carbon ratios, Brazil and France, produce large amounts of hydroelectric and nuclear power, respectively (See Figure 16).

Fractionation: The general name given to a process for separating mixtures of hydrocarbons or other chemicals into separate streams or cuts or fractions.

Free Carbon: The organic materials in tars that are insoluble in carbon disulfide.

Free Energy of Formation: The change in free energy when a compound is formed from its elements with each substance in its standard state at 77°F (25°C). The heat of reaction at 25°C may be calculated by subtracting the sum of the free energies of formation of the reactants from the sum of the free energies of formation of the products.

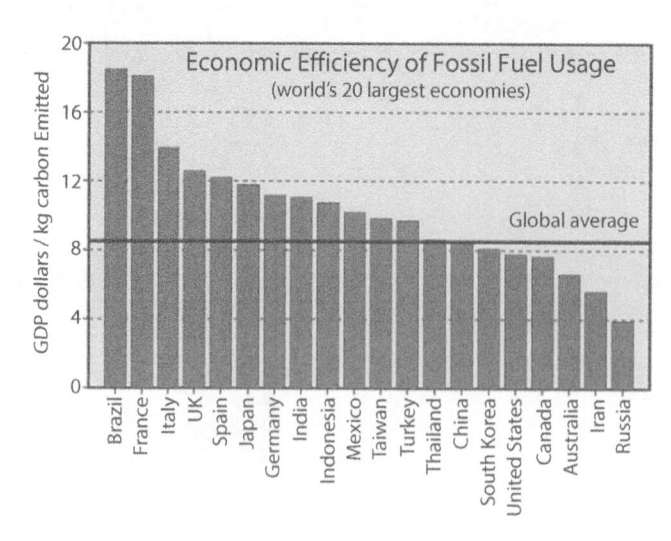

Figure 16 Economic efficiency of fossil fuel usage.

Free Water: Condensed water that exits as a separate liquid phase. Most refinery distillation columns are designed such that free water will not be present, since it can result in column upsets and promote the corrosion of the metal in the column.

Freeze Point: **1**. The temperature at which the hydrocarbon liquid solidifies at atmospheric pressure. It is important property for kerosene and jet fuels because of the very low temperatures encountered at high altitudes in jet planes. One of the standard test methods for the freeze point is ASTM D4790. **2**. The temperature at which a chilled petroleum product becomes solid and will no longer pour when a sample tube is tipped. Freeze point is a laboratory test. **3**. The temperature at which crystals first appear as a liquid is cooled, which is especially important in aviation fuels, diesel, and furnace oil.

Front End Engineering and Design (FEED) Contract: **1**. A legal agreement setting out the terms for all activities required to define the design of a facility to a level of definition necessary for the starting point of an EPC contract. **2**. Generally, the second contracting phase for the development of the export facilities in the LNG chain provides a greater definition than the prior conceptual design phase. In an LNG project, the single most important function of an FEED contract is to provide the maximum possible definition for the work to be ultimately performed by the engineering procurement and construction (EPC) contractor. **3**. A study used to analyze the various technical options for new field developments with the objective to define the facilities required. **4**. The stage of design between concept evaluation and detailed design during which the chosen concept is developed such that most key decisions can be taken. The output of FEED includes an estimate of the total installed cost and schedule. *See also Engineering, Procurement, and Construction Contract.*

Fuel Gas: **1**. A process stream internal to a facility is used to provide energy for operating the facility. **2**. Gas used as fuel in a liquefaction plant. It typically involves processing waste streams to LNG that are not profitable. It is used in gas turbines, boilers, and reaction furnaces.

Fuel Oil: Usually residual fuel but sometimes distillate fuel.

Fuel Oil Equivalent (FOE): The heating value of a standard barrel of fuel oil, equal to 6.05×10^6 Btu (LHV). On a yield chart, dry gas and refinery fuel gas are usually expressed in FOE barrels.

Furnace Oil: A distillate fuel made of cracked and straight-run light gas oils primarily for domestic heating because of its ease of handling and storing.

FVT: The final vapor temperature of a cut. The boiling ranges expressed in this manner are usually on a crude assay, true boiling point basis.

Gas/Liquid Chromatography (GC, GLC): Equipment used to determine the composition of a sample in the laboratory.

Gap: Gas is usually based on ASTM 86 distillation temperatures and is defined as the 5% distilled temperature of a distillation column product minus the 95% distilled temperature of the next higher product in the column. When the difference is positive, the difference is called a gap. When the difference is negative, the difference is sometimes called an overlap. The gap or overlap is a measure of the sharpness of the separation between adjacent products in a distillation column.

Gas Cap: An accumulation of natural gas at the top of a crude oil reservoir. The gas cap often provides the pressure to rapidly evacuate the crude oil from the reservoir.

Gasket: This is the softer material that is pressed between flanges to keep the fluid from leaking. Using the wrong gasket is a common cause of fires in process plants. Gaskets have different temperature and pressure ratings.

Gas Oil: 1. Any distillate stream having molecular weights and boiling points higher than heavy naphtha (>400°F or 205°C). The name gas oil probably traces its roots to "gasoline" bearing oil in the early days of refining. Early refiners used thermal cracking processes to produce more motor gasoline (MOG) from gas oil stocks. 2. The term is used for petroleum stocks with boiling ranges between approximately 650–1100°F (344–594°C). Unreacted gas oils are produced by distilling crude oil in crude and vacuum columns. Cracked gas oils are produced in refinery reaction processes, such as thermal and catalytic cracking, coking, visbreaking, and hydrocracking.

Gasoline: 1. A light petroleum product in the range of 80–400°F (27–204°C) for use in spark-ignition internal combustion engines. 2. An all-inclusive name for petroleum stocks that are used as fuel for internal combustion engines. Retail gasoline is a blend of several refinery gasolines and must meet the specifications of octane, Reid vapor pressure, distilling boiling range, sulfur content, and so on. Additives such as ethers or alcohols are used to improve the octane for the blended product.

Gasoline Blending Components: Naphthas that will be used for blending or compounding into finished aviation or motor gasoline (e.g., straight-run gasoline, alkylate, reformate benzene, toluene, and xylenes). Excludes oxygenates (alcohols, ethers), butane, and natural gasoline.

Gas sweetening: A process used to remove hydrogen sulfide and mercaptans from natural gas. Commonly used in petroleum refineries, the gas treatment uses amine solutions such as monoethanolamine. The process uses an absorber unit and regenerator. The amine solution flows down the scrubber and absorbs the hydrogen sulfide as well as carbon dioxide from the upflowing gases. The regenerator is used to strip the amine solution of the gases for reuse. It is known as gas sweetening as the foul smell is removed from the gas.

Gas Treating: The amine treating of light gases to remove such impurities as H_2S and CO_2. Molecular sieves are also used to concentrate hydrogen streams by removing inerts and light hydrocarbon contaminants.

Gas Turbine: An engine that uses internal combustion to convert the chemical energy of a fuel into mechanical energy and electrical energy. It uses air, which is compressed by a rotary compressor driven by the turbine and fed into a combustion chamber where it is mixed with the fuel, such as kerosene. The air and fuel are burnt under constant pressure conditions. The combustion gases are expanded through the turbine, causing the blades on the shaft to rotate. This is then converted to electrical energy. Gas turbines are used in the process industries and on offshore gas platforms for electrical generation.

Global Warming Potential: The global warming potential (GWP) of a greenhouse gas is its ability to trap extra heat in the atmosphere over time relative to carbon dioxide (CO_2). The GWP depends on the following factors:

- The absorption of infrared radiation by a given gas.
- The spectral location of its absorbing wavelengths
- The atmospheric lifetime of the gas

Grain: A unit of mass where one pound is equivalent to 7000 grains and a specification of 0.25 grain of H_2S per 100 scf is equivalent to an H_2S concentration of 4.0 ppmv.

Gravity: The specific gravity (Sp.Gr.) of a stream, often expressed as API Gravity by petroleum refiners. The basis is always the density of water.

Gravity Curve: The gravity of the material distilled from a petroleum stock in a laboratory still. The gravity curve is plotted against the percent distilled for the stock. Gravity curves are most commonly reported for true boiling point distillation (Figure 17).

Greenhouse Gas (GHG): A greenhouse gas is a gas in an atmosphere that absorbs and emits radiation within the thermal infrared range. This process is the fundamental cause of the greenhouse effect. The primary greenhouse gases in the Earth's atmosphere are water vapor, carbon dioxide, methane, nitrous oxide, and ozone. Without greenhouse gases, the average temperature of the Earth's surface would be about 15°C (27°F) colder than the present average of 14°C (57°F).

Greenhouse Gas Emissions: An addition to the atmosphere of gases that are a cause of global warming, including carbon dioxide, methane, and others.

Greenhouse Gas Removal Technologies (or negative emissions): Methods that actively remove greenhouse gases from the atmosphere, ranging from engineering to nature-based solutions.

Gross Heating Value: Is the total energy transferred as heat in an ideal combustion reaction at a standard temperature and pressure in which all water formed appears as a liquid. The gross heating is an ideal gas property in a hypothetical state (the water cannot all condense to the liquid because some of the water would saturate the CO_2 in the products).

Gross Heating Value of Fuels (GHV): The heat produced by the complete oxidation of material at 60°F (25°C) to carbon dioxide and liquid water at 60°F (25°C).

Gross Input to Atmospheric Crude Oil Distillation Units: Total inputs to atmospheric crude oil distillation units. Include all crude oil, lease condensate, natural gas plant liquids, and unfinished oils. Liquefied refinery gases, slop oils, and other liquid hydrocarbons produced from oil sands, gilsonite, and oil shale.

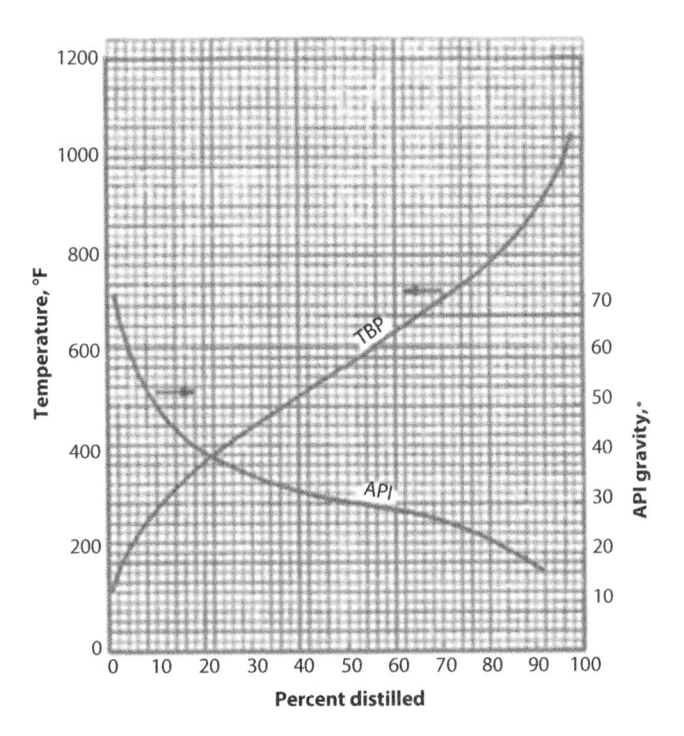

Figure 17 TBP and gravity—mid-percent curves.

Gum: A complex sticky substance that forms by the oxidation of gasoline, especially those stored over a long period of time. Gum fouls car engines, especially the fuel injection ports.

Harm: Physical injury or damage to the health of people, either directly or indirectly, as a result of the damage to property or to the environment.

Hazard: 1. A condition or object that has the potential to cause harm. 2. An unsafe condition, which, if not eliminated or controlled, may cause injury, illness, or death. 3. A physical or chemical characteristic that has the potential for causing harm to people, the environment, or property. Examples of hazards:

- *Combustible/Flammable substance. E.g., Ethylene is flammable*
- *Corrosive. E.g., Sulfuric acid is extremely corrosive to the skin.*
- *Explosive substance. E.g., Acrylic acid can polymerize, releasing large amounts of heat.*
- *Toxic fumes. E.g., Chlorine is toxic by inhalation.*
- *Substances kept at high pressure in containment (e.g., a vessel, tank)*
- *Objects or material with a high or low temperature.*
- *Radiation from heat source.*
- *Ionizing radiation source.*
- *Energy release during the decomposition of a substance. E.g., Steam confined in a drum at 600 psig contains a significant amount of potential energy (Figure 18).*

Hazard Analysis: Is the first step in a process used to assess risk. The result of a hazard analysis is the identification of the different types of hazards.

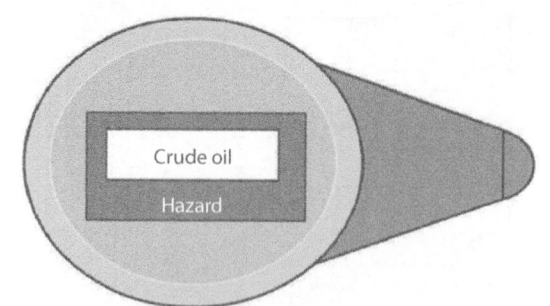

Consider pressure, temperature, composition, quantity, etc. into account.

Figure 18 A hazard.

It is assigned a classification, based on the worst-case severity of the end condition. Risk is the combination of probability and severity. Preliminary risk levels can be provided in the hazard analysis. The validation, more precise prediction (verification), and acceptance of risk is determined in the risk assessment (analysis). The main goal of both is to provide the best selection of the means of controlling or eliminating the risk.

Hazard Communication: Employees' "right-to-know" legislation requires the employers to inform employees (pretreatment inspectors) of the possible health effects resulting from contact with hazardous substances. At locations where this legislation is in force, employers must provide employees with information regarding any hazardous substances, which they might be exposed to under normal work conditions or reasonably foreseeable emergency conditions resulting from workplace conditions. OSHA's Hazard Communication Standard (HCS) (Title 29 CFR Part 1910.2100) is the Federal regulation and state statutes are called Worker Right-to- Know Laws.

Hazard Communication Program: A written plan to manage the hazards associated with the use of chemicals in the workplace.

Hazardous Chemical: A substance that may harm the worker either physically (e.g., fire, explosion) or chemically (e.g., toxic, corrosive).

Hazardous Events: Hazardous event is defined as a hazardous situation that results in harm. Each identified hazard could give a number of different hazardous events. For each identified hazardous event, it should also be described which factors contribute to it.

For example, the hazardous combustible substance could give the following hazardous events:

- Pool fire outside a tank, due to leakage, when an ignition source is present.
- Flash fire inside a tank when an ignition source is present.

The factors that could contribute to the leakage in the tank, for instance, could be:

- *Bad connection joint.*
- *Gasket damage.*
- *Tube damage.*
- *Pipe damage.*

Hazardous Situation: A circumstance in which a person is exposed to hazards.

HAZID/HAZOP: **1**. HAZard Identification / HAZard and Operability analysis systematic design review methods to identify and address hazards to ensure that the necessary safety measures to eliminate or mitigate hazards are incorporated in the design and operation of the unit. **2**. A qualitative process risk analysis tool used to identify hazards and evaluate if suitable protective arrangements are in place, if the process were not to perform as intended and unexpected consequences were to result.

HCGO: Heavy coker gas oil.

HCO: Heavy FCC cycle gas oil. See Heavy Cycle Oil.

Heart Cut Recycle: That unconverted portion of the catalytically cracked material is recycled to the catalytic cracker. This recycle is usually in the boiling range of the feed and, by definition, contains no bottoms. Recycling allows a less severed operation and suppresses the further cracking of desirable products.

Heat Balance: See Energy Balance.

Heat Exchanger: A pressure vessel for transferring heat from one liquid or vapor stream to another. A typical heat exchanger consists of a cylindrical vessel and nozzles through which one stream can flow and a set of pipes or tubes in series in the cylinder through which the other can flow. Heat transfer mechanisms are conduction and convection. See also Shell and Tube Heat Exchanger (See Figure 19).

Heat Flux: The rate of heat transfer per unit area normal to the direction of heat flow. It is the total heat transmitted by radiation, conduction, and convection.

Heat Pump: Thermodynamic heating/refrigerating system to transfer heat. The condenser and evaporator may change roles to transfer heat in either direction.

Heat Rate: The measure of efficiency in converting input fuel to electricity. Heat rate is expressed as the number of Btu of fuel (e.g., natural gas) per kilowatt-hour (Btu/kWh). The heat rate for power plants depends on the individual plant design, its operating conditions, and its level of electric power output. The lower the heat rate, the more efficient the plant.

Heat Recovery: Heat utilized that would otherwise be wasted from a heating system.

Heat Transfer Coefficient: Coefficient describing the total resistance to heat loss from a producing pipe to its surroundings. Includes heat loss by conduction, convection, and radiation.

Heating Oil: Any distillate or residual fuel. **1**. Oil used for residential heating. **2**. Trade term for the group of distillate fuel oils used in heating homes and buildings as distinguished from residual fuel oils used in heating and power installations. Both are burner fuel oils.

Heating Value: **1**. The average number of British thermal units per cubic foot of natural gas as determined from the tests of fuel samples. **2**. The amount of heat produced from the complete combustion of a unit quantity of fuel. **3**. The amount of energy or heat that is generated when a hydrocarbon is burned (chemically combined with oxygen). **4**. Energy released in the complete combustion of a unit of mass, matter, or volume of fuel in a stoichiometric mixture with air. **5**. The amount of heat produced by the complete combustion of a unit quantity of fuel.

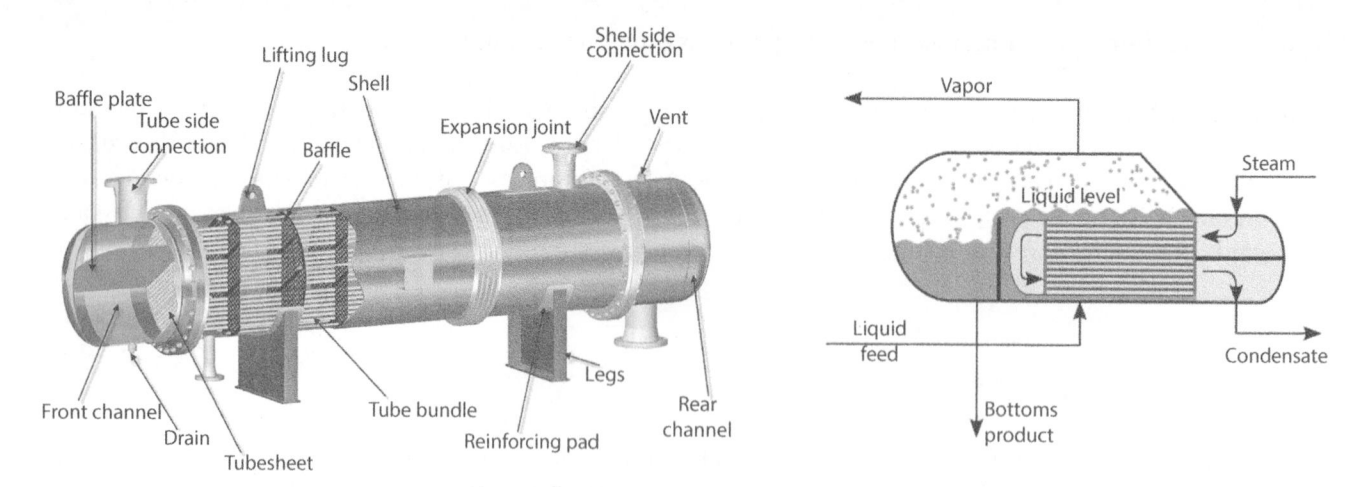

Figure 19 A shell and tube heat exchanger A reboiler.

Heat of Combustion: The amount of heat released in burning completely an amount of substance is its heat of combustion. The general formula for the combustion of a hydrocarbon compound is:

$$C_nH_{2n+2} + (3n+1)/2\ O_2 \rightarrow (n+1)\ H_2O + n\ CO_2 + \text{Energy} \tag{38}$$

Heat of Reaction: The heat release of heat absorbed when a chemical reaction takes place. The heat of reaction may be computed from the free energies of formation for the reacting components and the resultant products at the standard temperature of 77°F (25°C).

Heat of Vaporization: The amount of heat energy required to transform an amount of a substance from the liquid phase to the gas phase.

Heavy Crude: Crude oil of 20° API gravity or less; often very thick and viscous.

HCGO: Heavy coker gas oil.

Heavy Cycle Oil (HCO): Gas oil produced in an FCC operation that boils in the approximate TBP range of 400–1000°F (205–358°C). Heavy cycle oil is not generally withdrawn as a product, but it is recycled back to the reactor for further cracking to improve the overall conversion of the process.

Heavy Ends: The highest boiling portion of gasoline or other petroleum oil. The end point as determined by the distillate test reflects the amount and character of the heavy ends present in gasoline.

Heavy Gas Oil: Petroleum distillates with an approximate boiling range from 651 to 1000°F (344–538°C).

Heavy Key: A distributed component in a distillation section that is recovered in the (bottom) heavy product, with a small, specified amount leaving in the top product.

Heavy Oil: Lower gravity, often higher-viscosity oils. Normally less than 28° API gravity.

Hempel Distillation: The US Bureau of Mines' (now the Department of Energy, DOE) routine method of distillation. Results are frequently used interchangeably with TBP distillation.

Heptane (nC_7H_{16}): Normal heptane is a straight-chain alkane hydrocarbon with the chemical formula $H_3C(CH_2)_5CH_3$ or C_7H_{16}. Heptane (and its many isomers) is widely applied in laboratories as a totally non-polar solvent. As a liquid, it is ideal for transport and storage. Heptane is commercially available as mixed isomers for use in paints and coatings, as pure n- heptane for research and development and pharmaceutical manufacturing, and as a minor component of gasoline.

n-heptane is defined as the zero point of the octane rating scale. It is undesirable in gasoline because it burns explosively, causing engine knocking, as opposed to branched-chain octane isomers, which burn more slowly and give better performance. When used as a fuel component in antiknock test engines, a 100% heptane fuel is the zero point of the octane rating scale (the 100 point is 100% isooctane). Octane number equates to the antiknock qualities of a comparison mixture of heptane and isooctane, which is expressed as the percentage of isooctane in heptane and is listed in pumps for gasoline dispensed in the United States and internationally.

HF Alkylation: Alkylation using hydrofluoric acid as a catalyst.

High Pressure (HP): A processing unit operating at either equal to or greater than 225 psig measured at the outlet separator.

High Temperature Simulated Distillation (HTSD): Laboratory test designed for petroleum stocks boiling up to 1382°F (750°C).

HSR: Heavy Straight-Run. Usually a naphtha side stream from the atmospheric distillation tower.

HVGO: Heavy vacuum gas oil. A side stream from the vacuum distillation tower.

Hydrocarbon: Any organic compound that is comprised of hydrogen and carbon atoms, including crude oil, natural gas, and coal, Figure 58 provides names of hydrocarbons and their corresponding products.

Hydrocrackate: The gasoline range product from a hydrocracker.

Hydrocracking: 1. A process in which high or heavy gas oils or residue hydrocarbons are mixed with hydrogen under high pressure and temperature and in the presence of a catalyst to produce light oils. 2. A refining process in which a heavy oil fraction or wax is treated with hydrogen over a catalyst under relatively high pressure and temperature to give products of lower molecular mass.

Hydrocracked Naphtha: A high-quality blending stream obtained when high-boiling cracked distillates undergo a combination of processes like cracking, hydrogenation, and reforming in the presence of a catalyst and hydrogen.

Hydrocyclone: A cone-shaped device for separating fluids and the solids dispersed in fluids.

Hydrodesulfurization: A process in which sulfur is removed from the molecules in a refinery stream by reacting it with hydrogen in the presence of a catalyst.

Hydrodesulfurizing: A process for combining hydrogen with sulfur in refinery petroleum streams to make hydrogen sulfide, which is removed from the oil as a gas.

Hydrogen: The lightest of all gases, the element (hydrogen) occurs chiefly in combination with oxygen in the water. It also exists in acids, bases, alcohols, petroleum, and other hydrocarbons.

Hydrogen Consumption: The amount of hydrogen that is consumed in a hydrocracking or hydrotreating process, usually expressed on a per-unit-of-feed basis. Hydrogen may be consumed in chemical reactions and dissolved and lost from the process in liquid hydrocarbon products.

Hydrogen Embrittlement: A corrosion mechanism in which atomic hydrogen enters between the grains of the steel and causes the steel to become very brittle.

Hydrogen-Induced Cracking: Stepwise internal cracks that connect hydrogen blisters.

Hydrogen Sulfide: 1. "Rotten egg gas," H_2S. It is responsible for the distinctive odor of Rotorua. **2.** An objectionable impurity is present in some natural gas and crude oils and formed during the refining of sulfur-containing oils. It is removed from products by various treatment methods at the refining. **3.** Hydrogen sulfide is a gas with a rotten egg odor. This gas is produced under anaerobic conditions. Hydrogen sulfide gas is particularly dangerous because it dulls the sense of smell so that one does not notice it after one has been around it for a while. In high concentrations, hydrogen sulfide gas is only noticeable for a very short time before it dulls the sense of smell. The gas is very poisonous to the respiratory system, explosive, flammable, colorless, and heavier than air. **4.** A toxic, corrosive, colorless gas with the characteristic smell of rotten eggs in low concentration. An acid gas.

Hydrogen Sulfide Cracking: Minute cracking just under a metal's surface caused by exposure to hydrogen sulfide gas.

Hydrogenation: 1. Filling in with hydrogen the "free" places around the double bonds in an unsaturated hydrocarbon molecule. **2.** A refinery process in which hydrogen is added to the molecules of unsaturated hydrocarbon fractions.

Hydrofining: A process of treating petroleum fractions and unfinished oils in the presence of catalysts and substantial quantities of hydrogen to upgrade their quality.

Hydroforming: A process in which naphtha is passed over a solid catalyst at elevated temperature and moderate pressures in the presence of added hydrogen to obtain high-octane motor fuels.

Hydroskimming Refinery: A topping refinery with a catalytic reformer.

Hydrostatic Pressure: Pressure created by a column of fluid that expresses uniform pressure in all directions at a specific depth and fluid composition above the measurement point.

Hydrotreating: 1. A refinery process to remove sulfur and nitrogen from crude oil and other feedstocks. **2.** This is a term for a process by which product streams may be purified and otherwise be brought up to marketing specifications as to odor, color, stability, etc. **3.** A process in which a hydrocarbon is subjected to heat and pressure in the presence of a catalyst to remove sulfur and other contaminants such as nitrogen and metals and in which some hydrogenation can take place. Hydrotreating for the removal of sulfur is the major treating process in refineries. Cracked streams could be saturated and stabilized by converting olefins, albeit under more severe treating conditions. The process involves hydrogen under suitable temperature, pressure, and a catalyst.

Hyperforming: A catalytic hydrogenation process used for improving the octane number of naphtha by the removal of sulfur and nitrogen compounds.

H_2/Oil Ratio and Recycle Gas Rate: The H_2/oil ratio in standard cubic feet (scf) per barrel (bbl) is determined by:

$$\frac{H_2}{oil} = \frac{\text{total hydrogen gas to the reactor, scf/day}}{\text{total feed to the reactor, bbl/day}} [=] \frac{scf}{bbl} \qquad (39)$$

An H_2/oil ratio in m^3/bbl is obtained by multiplying the H_2/oil ratio in (scf/bbl) by a conversion factor 0.028317. A molar H_2/oil ratio can be calculated from the volumetric H_2/oil ratio by the following equation:

$$\frac{\text{molar } H_2}{oil} = 1.78093 \times 10^{-7} \left(\frac{H_2}{oil} \frac{scf}{bbl} \right) \frac{MW_{oil}}{MW_{H_2}} \frac{\rho_{H_2}}{\rho_{oil}} \qquad (40)$$

where MW_{oil} and MW_{H_2} are the molecular weights of the oil to be hydrotreated and of hydrogen, respectively, and ρ_{oil} and ρ_{H_2} are the densities of the oil and hydrogen (ρ_{H_2} at 15°C and 1 atm is 0.0898 kg/cm²).

Hypothetical State: Is defined as a fluid in a state that cannot actually exist, e.g., methane as a liquid at 60°F and 14.696 psia. Methane cannot be in its liquid phase at this temperature and pressure, but such a state, when defined, can be used in calculations.

Identification and Structural Group Analysis: The crude oil is a complex mixture of saturated hydrocarbons; saturated hetero-compounds; and aromatic hydrocarbons, olefinic hydrocarbons, and aromatic hetero-compounds. With the advancement of instrumental analysis techniques like chromatography and spectroscopic methods, now, it is possible to study in depth the identification and structural group analysis. Some of the major analytical instruments used are gas chromatography, ion-exchange chromatography, simulated distillation by gas chromatography, absorption chromatography, gel permeation chromatography, high-performance liquid chromatography, and supercritical fluid chromatography. The application of spectroscopy, mass spectroscopy, electron spin resonance, X-ray diffraction, inductively coupled plasma emission spectroscopy, X-ray absorption spectroscopy, and atomic absorption spectrophotometer.

Initial boiling point (IBP): The initial boiling point of a cut, usually on an ASTM basis. The lowest temperature at which a petroleum product will begin to boil. The boiling temperature in a laboratory still at which the first drop of distilled liquid is condensed. The initial boiling point may be higher than the boiling point for light components in the sample that are not condensed by the apparatus.

Ignition: The process of starting a combustion process through the input of energy. Ignition occurs when the temperature of a substance is raised to the point at which its molecules will react spontaneously with an oxidizer and combustion occurs.

Ignition Quality: Ignition quality is very important in the case of high-speed automotive diesel engines. The diesel engine knock, engine noise, smoke, gaseous emissions, and so on, all depend upon this factor. Ignition quality is measured in terms of cetane number using an ASTM standard test engine. The test method designated as D613 comprises a single-cylinder engine with a variable compression ratio combustion pre-chamber.

Incident: See Accident.

Independent Protection Layer (IPL): Protection measures that reduce the level of risk of a serious event to 100 times, which have a high degree of availability (greater than 0.99) or have specificity, independence, dependability, and auditability.

Inerting: The process of removing an oxidizer (usually air or oxygen) to prevent a combustion process from occurring, normally accomplished by purging.

Inflammable: Has an identical meaning as flammable; however, the prefix "in" indicates a negative in many words and can cause confusion. Therefore, the use of flammable is preferred over inflammable.

Inherently Safer: **1**. A chemical process is inherently safer if it reduces or eliminates the hazards associated with the materials and operations used in the process, and this reduction or elimination is permanent and inseparable. **2**. An essential character of a process, system, or equipment that makes it without or very low in hazard or risk. Inherent safety is a way of looking at processes in order to achieve this. There are four main keywords:

- *Minimize (Intensification)*: Reduce stocks of hazardous chemicals.
- *Substitute*: Replace hazardous chemicals with less hazardous ones.
- *Moderate (Attenuation)*: Reduce the energy of the system—lowering pressures and temperatures or adding stabilizing additives generally make for lower hazards.
- *Simplify*: Make the plant and process simpler to design, build, and operate, hence making them less prone to equipment control and human failings.

Note: The principles of inherent safety are applied at the conceptual design stage to the proposed process chemistry. In certain instances, these hazards cannot be avoided; they are the basic properties of the materials and the conditions of usage. The inherently safer approach is to reduce the hazard by reducing the quantity of hazardous material or energy or by completely eliminating the hazardous agent.

Inherently Safer Design: Is a fundamentally different way of thinking about the design of chemical processes and plants. It focuses on the elimination or reduction of the hazards, rather than on management and control. This approach should result in safer and more robust processes, and it is likely that these inherently safer processes will also be more economical in the due course.

Instrument: The apparatus used in performing an action (typically found in instrumented systems).

Note: Instrumented systems in the process sector are typically composed of sensors (e.g., pressure, flow, temperature transmitters), logic solvers or control systems (e.g., programmable controllers, distributed control systems), and final elements (e.g., control valves). In special cases, instrumented systems can be safety instrumented systems.

Internal Combustion Engine (ICE): Is a heat engine where the combustion of a fuel occurs with an oxidizer (usually air) in a combustion chamber that is an integral part of the working fluid flow circuit. In an internal combustion engine, the expansion of the high-temperature and high-pressure gases produced by combustion apply direct force to some components of the engine. The force is applied typically to pistons, turbine blades, or a nozzle. This force moves the component over a distance, transforming chemical energy into useful mechanical energy (See Figure 20).

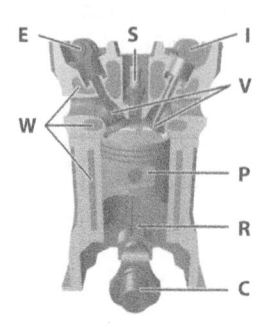

Figure 20 Diagram of a cylinder as found in four-stroke gasoline engines.

C	Crankshaft
E	Exhaust camshaft
I	Inlet camshaft
P	Piston
R	Connecting rod
S	Spark plug
V	Valves. Red: exhaust, blue: intake
W	Cooling water jacket
	Gray structure: Engine block.

Intrinsically Safe (IS): A circuit or device in which any spark or thermal effect is incapable of causing the ignition of a mixture of flammable or combustible materials in the air under prescribed test conditions.

IPTBE: Isopropyl tertiary butyl ether. An oxygenate used in motor fuels.

Isocracking: A hydrocracking process for the conversion of hydrocarbons to more valuable lower-boiling products by operation at relatively lower temperatures and pressures in the presence of hydrogen and a catalyst.

Isomerate: The product of an isomerization process.

Isomerization: 1. A refining process that alters the fundamental arrangement of atoms in the molecule without adding or removing anything from the original material. Used to convert normal butane into isobutane (iC_4H_{10}), an alkylation process feedstock, and normal pentane and hexane into isopentane (iC_5H_{12}) and isohexane (iC_6H_{14}), high-octane gasoline components. **2.** The rearrangement of straight-chain hydrocarbon molecules to form branched-chain products. Pentanes and hexanes, which are difficult to reform, are isomerized using precious metal catalysts to form the gasoline-blending components of a fairly high octane value. Normal butane may be isomerized to provide a portion of the isobutene feed needed for alkylation processes. The objective of isomerization is to convert low-octane n-paraffins to high-octane i-paraffins by using a chloride-promoted fixed bed reactor. **3.** Isomerization is the process by which one molecule is transformed into another molecule that has exactly the same atoms, but the atoms are rearranged. In some molecules and under some conditions, isomerization occurs spontaneously. Many isomers are equal or roughly equal in bond energy, and so they exist in roughly equal amounts, provided that they can interconvert relatively freely, that is, the energy barrier between the two isomers is not too high. When the isomerization occurs intermolecularly, it is considered a rearrangement reaction.

An example of an organometallic isomerization is the production of decaphenylferrocene [$(\eta^5\text{-}C_5Ph_5)_2Fe$] from its linkage isomer.

Isomers: Two compounds composed of identical atoms but with different structures/configurations, giving different physical properties. For example, hexane (C_6H_{14}) could be n-hexane, 2-methyl pentane, 3-methyl pentane, 2, 3-dimethyl butane, and 2, 2, - dimethylbutane.

A simple example of isomerism is given by propanol. It has the formula C_3H_8O (or C_3H_7OH) and occurs as two isomers: propanol-1-ol (n-propyl alcohol; II) and propanol-2-ol (isopropyl alcohol; III).

Note that the position of the oxygen atom differs between the two: It is attached to an end carbon in the first isomer and to the center carbon in the second (See Figures 21–23).

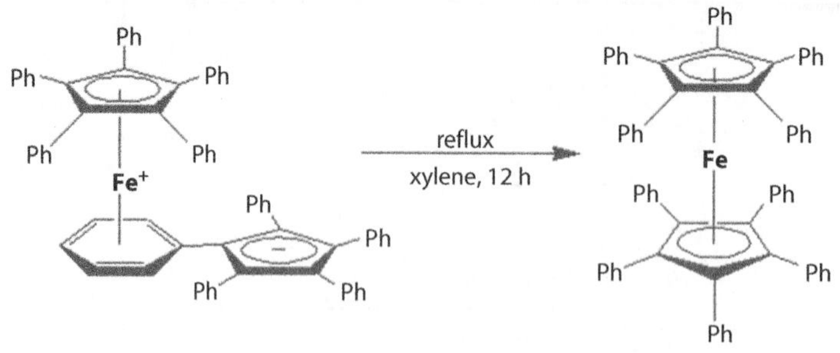

Figure 21

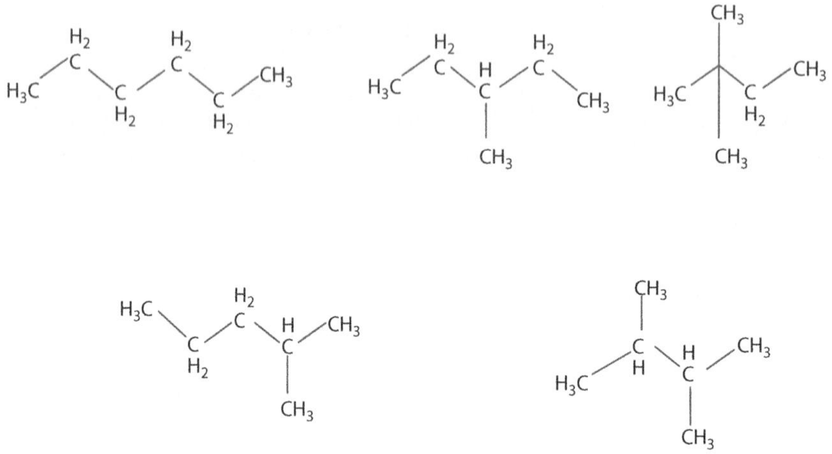

Figure 22

Isooctane – 2, 2, 4 – Trimethlypentane: Also known as isooctane or iso-octane, is an organic compound with the structure formula $(CH_3)_3CCH_2CH(CH_3)_2$.

It is one of several isomers of octane (C_8H_{18}). Engine knocking is an unwanted process that can occur during combustion in internal combustion engines.

Graham Edgar, in 1926, added different amounts of n-heptane and 2,2,4 – trimethylpentane to gasoline and discovered that the knocking stopped when 2,2,4 trimethlypentane was added. Test motors, using 2,2,4 trimethylpentane, gave a certain performance that was standardized as 100 octane. The same test motors, run in the same fashion, using heptane, gave a performance that was standardized as 0 octane. All other compounds and blends of components then were graded against these two standards and assigned octane numbers. 2,2,4 trimethylpentane is the liquid used with normal heptanes (nC_7H_{18}) to measure the octane number of gasoline. It is an important component of gasoline, frequently used in relatively large proportions to increase the knock resistance of the fuel (See Figure 24).

Isopentane: See Natural Gasoline.

IVT: The initial vaporization temperature of a cut, usually based on a crude assay distillation.

Jack: An oil well pumping unit that operates with an up-and-down, or seesawing motion; also called a pumping jack.

Jet fuel: A kerosene material of the typical ASTM D86 boiling point range 400–550°F (205–288°C) used as a fuel for commercial jet aircraft.

Figure 23

Joule–Thompson expansion: The pressure of a mixture is reduced with no heat transfer to or from the surroundings. A pressure decrease typically results in a temperature decrease, except for systems comprised largely of hydrogen gas.

Joule–Thompson Effect: **1.** The change in the temperature of a fluid that occurs when the fluid is allowed to expand in such a way that no external work is done and no heat transfer takes place. The case of most interest is the cooling of a compressed gas upon J-T expansion. NB: The J-T effect is not limited to gases; J-T expansion can, in some cases, produce an increase in temperature rather than a decrease, although this is not frequently encountered. **2.** The thermodynamic effect in a fluid whereby the reduction in its temperature is caused by pressure reduction without energy exchange with the environment. **3.** When a real (not ideal) gas expands, the temperature of the gas drops. During the passage of a gas through a choke, the internal energy is transferred to kinetic energy with a corresponding reduction in temperature as velocity increases. The effect for natural gas is approximately 7°F for every 100 psi pressure reduction.

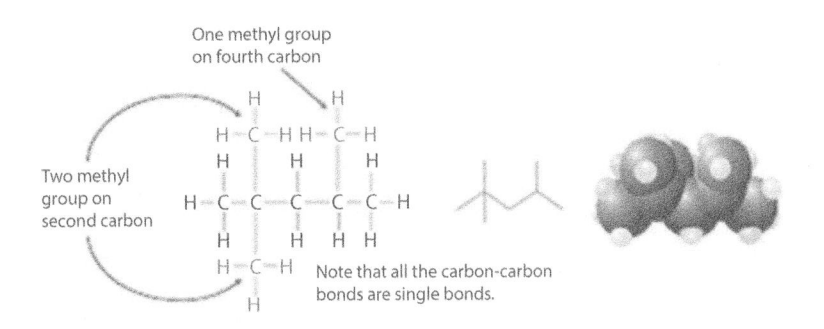

Figure 24

Joule–Thompson Valve: A device that, taking advantage of the Joule–Thompson effect, enables the cooling of fluid through throttling or the reduction of its pressure.

K factor: Sometimes used as a synonym for characterization factor.

K-value: Shortcut notation for the equilibrium K-value.

Kerogene: An initial stage of oil that never developed completely into crude. Typical of oil shales.

Kerosene/Kerosine: **1.** A medium range (C_9–C_{16}) straight-chain blend of hydrocarbons. The flashpoint is about 140°F (60°C), the boiling point is 345–550°F (174–288°C), and the density is 747–775 kg/m³. **2.** A medium–light distillate from the oil refining process; used for lighting and heating and for the manufacture of fuel for jet and turboprop aircraft engines. **3.** Any petroleum product with a boiling range between the approximate limits of 284 and 518°F (140 and 270°C), which satisfies specific quantity requirements. **4.** A middle distillate product material from the distillation of crude oil that boils in the approximate ASTM D86 range of 400–550°F (205–288°C) or from thermal and catalytic cracking operations (coker, visbreaker, FCC, hydrocracker, etc.). The exact cut is determined by various specifications of the finished kerosene. **5.** A light petroleum distillate that is used in space heaters, cook stoves, and water heaters and is suitable for use as a light source when burned in wick-fed lamps. Kerosene has a maximum distillation temperature of 400°F (204°C) at the 10% recovery point, a final boiling point of 572°F (300°C), and a minimum flash point of 100°F (38°C). Included are No. 1-K and No. 2-K, the two grades recognized by the American Society of Testing Materials (ASTM) Specification D3699 as well as all other grades of kerosene called range or stove oil, which have properties similar to those of No. 1 fuel oil. It is colorless and has a characteristic odor and taste. Kerosene is insoluble in water, moderately soluble in alcohol, and very soluble in ether, chloroform, or benzene.

Key Components: In a conventional distillation column with two products, two components or groups of components that define the separation. Both components must be distributed to the top and bottom products. The light key appears in the bottom product in a small significant quantity, and the heavy key appears in the top product in a small significant quantity.

Kinematic viscosity: Viscosity in centipoises (cP) divided by the liquid density at the same temperature gives kinematic viscosity in centistokes (cS) (100 cSt = 1 stoke). Water is the primary viscosity standard with an accepted viscosity at 20°C of 0.01002 poise. Kinematic viscosity is usually determined by the flow of a substance between two points in a capillary tube.

$$\text{Kinematic viscosity} = \frac{\text{Dynamic viscosity}}{\text{Density of fluid}}, \text{cSt} \tag{41}$$

$$\nu = \frac{\mu}{\rho}$$

Kinetic: The word "kinetic" is derived from the Greek word for "motion." In chemistry, kinetics is the study of how fast reactions occur. In many chemical reactions where there are a number of possible products, the first product formed may be the one that is formed most quickly, not necessarily the one that is most stable; if the reaction is left to proceed, eventually, a product is formed that involves the greatest change in bond energy—the thermodynamic product.

Knock: **1.** The sound associated with the autoignition in the combustion chamber of an automobile engine of a portion of the fuel–air mixture ahead of the advancing flame front. **2.** The noise is associated with premature ignition of the fuel–air mixture in the combustion chamber; also known as detonation or pinking.

Knocking (knock, detonation, spark knock, pinging, or pinking): In spark-ignition internal combustion engines, knocking occurs when the combustion of the air/fuel mixture in the cylinder does not start off correctly in response to ignition by the spark plug, but one or more pockets of air/fuel mixture explode outside the envelope of the normal combustion front. See also Engine Knocking.

Knocking is more or less unavoidable in diesel engines, where fuel is injected into highly compressed air toward the end of the compression stroke. There is a short lag between the fuel being injected and combustion starting. By this time, there is already a quantity of fuel in the combustion chamber that will ignite first in areas of greater oxygen density prior to the combustion of the complete charge. This sudden increase in pressure and temperature causes the distinctive diesel "knock" or "clatter," some of which must be allowed for in the engine design. A careful design of the injector pump, fuel injector, combustion chamber, piston crown, and cylinder head can reduce knocking greatly, and modern engines using electronic common rail injection have very low levels of knock. Engines using indirect injection generally have lower levels of knock than direct injection engines due to the greater dispersal of oxygen in the combustion chamber and lower injection pressures, providing a more complete mixing of fuel and air.

Knocking should not be confused with pre-ignition—they are two separate events. However, pre-ignition is usually followed by knocking. See Pre-ignition.

Knockout: A separator used to remove easily removed or excess gas or water from the produced fluid stream.

Knockout Drum: A vessel wherein suspended liquid is separated from gas or vapor.

Laminar flow: The streamlined flow of a fluid in which a fluid flows without fluctuations or turbulence. The velocities of fluid molecules are in the direction of flow with only minor movement across the streamlines caused by molecular diffusion. The existence was first demonstrated by Osborne Reynolds who injected a trace of colored fluid into a flow of water in a glass pipe. At low flow rates, the colored fluid was observed to remain as discrete filaments along the tube axis, indicating flow in parallel streams. At increased flow rates, oscillations were observed in the filaments, which eventually broke up and dispersed across the tube. There appeared to be a critical point for a particular tube and fluid above which the oscillations occurred. By varying the various parameters, Reynolds showed that the results could be correlated into the terms of a dimensionless number called the Reynolds number, Re. This is expressed by:

$$\text{Re} = \frac{\rho v d}{\mu} = \frac{4G}{\pi d \mu} = \frac{4\rho Q}{\pi d \mu} \tag{42}$$

where ρ is the density of the fluid, v is the velocity of the fluid, d is the inside diameter of the pipe, and μ is the fluid viscosity. The critical value of Re for the break-up of laminar flow in the pipes of circular cross-section is about 2000.

Leaded Gasoline: A gasoline that has TEL (tetraethyl lead) added to boost the octane number.

LCGO: Light coker gas oil

Lean Oil: The absorption oil entering the top tray of an absorber column.

Lease Condensate: A mixture consisting primarily of pentanes and heavier hydrocarbons that are recovered as a liquid from natural gas in lease separation facilities. This category excludes natural gas liquids, such as propane and butane, which are recovered at downstream natural gas processing plants or facilities. *See also Natural Gas Liquids.*

LHSV: Liquid hour space velocity, the volume of feed per hour per volume of a catalyst.

LHV: The lower heating value of fuels (net heat of combustion). The heat produced by complete oxidation of materials at 60°F (25°C) to carbon dioxide and water vapor at 60°F (25°C).

Light Cycle Oil (LCO): Gas oil produced in a catalytic cracking operation that boils in the approximate ASTM D86 range of 400–695°F (205–369°C).

Light Ends: Hydrocarbon fractions in the butane (C_4H_{10}) and lighter boiling range.

Light Gas Oils: Liquid petroleum distillates heavier than naphtha, with an approximate boiling range from 401 to 650°F (205–343°C).

Light key: A distributed component in a distillation section that is recovered in the top light product, with a small specified amount leaving the bottom product.

Light oil: Generally gasoline, kerosene, and distillate fuels.

Light Straight-Run (LSR): The low-boiling naphtha stream from the atmospheric distillation, usually composed of pentanes and hexanes.

Liquefaction: 1. The process by which gaseous natural gas is converted into liquid natural gas. **2.** The physical process from gas to liquid is condensation. For natural gas, this process requires cryogenic temperature since it is impossible to liquefy methane—the main component of natural gas—at a temperature above -117°F (-82.6°C), which is its critical temperature.

Liquefaction of Gases: Any process in which a gas is converted from a gaseous-to-liquid phase.

Liquefaction Plant: Industrial complex that processes natural gas into LNG by removing contaminants and cooling the natural gas into its condensation.

Liquefaction Unit or Liquefaction Train: Equipment that processes purified natural gas and brings it to a liquid state. Natural gas has been purified in the pretreatment unit before cooling and liquefying it.

Liquefied Natural Gas (LNG): 1. Natural gas that has been refrigerated to temperatures at which it exists in a liquid state. **2.** An odorless, colorless, non-corrosive, and non-toxic product of natural gas consisting primarily of methane (CH_4) that is in liquid form at near-atmospheric pressure. **3.** Natural gas liquefied either by refrigeration or by pressure to facilitate storage or transportation. **4.** A liquid composed chiefly of natural gas (e.g., mostly methane, CH_4). Natural gas is liquefied to make it easy to transport if a pipeline is not feasible (e.g., as across a body of water). LNG must be put under low temperature and high pressure or under extremely low (cryogenic) temperature and close to atmospheric pressure to liquefy. **5.** Natural gas, mainly methane, refrigerated to reach a liquid phase suitable for transportation in specialized vessels. **6.** Natural gas that has been cooled to -26°F (-32°C) and converted into a liquid so that its volume will be reduced for transportation. **7.** A hydrocarbon mixture, predominantly methane, kept in a liquid state at a temperature below its boiling point. **8.** Methane that has been compressed and cooled to the liquefaction point for shipping.

Liquefied Petroleum Gas (LPG): 1. Gaseous hydrocarbons at normal temperatures and pressures but that readily turns into liquids under moderate pressure at normal temperatures, i.e., propane, (C_3H_8) and butane (C_4H_{10}). **2.** Butane and propane mixture, separated from well fluid stream. LPG can be transported under pressure in refrigerated vessels (LPG carriers). **3.** A mixture of propane and butane and other light hydrocarbons derived from refining crude oil. At normal temperatures, it is a gas, but it can be cooled or subjected to pressure to facilitate storage and transportation. **4.** Of the gaseous hydrocarbons, propane and butane can be liquefied under relatively low pressure and at ambient temperature. Mixtures of these are known as LPG. **5.** A mixture of propane, propylene, butane, and butylenes. When compressed moderately at normal temperature, it becomes a liquid. It is obtained as light ends from the fractionation of crude oil. It has a good caloric value; it is used as cooking fuel. Because LPG has no natural odor, a distinctive odorant is added so that it will be noticeable should a leak occur. **6.** Light ends, usually C_3 and C_4 gases liquefied for storage and transport. 7. Propane, propylene, normal butane, butylenes, isobutane, and isobutylene produced at refineries or natural gas processing plants (includes plants that fractionate raw natural gas plant liquids). 8. A group of hydrocarbon-based gases derived from crude oil refining or natural gas fractionation. They include ethane, ethylene, propane, propylene, normal butane, butylenes, isobutane, and isobutylene. For the convenience of transportation, these gases are liquefied through pressurization.

Liquefied Refinery Gases: Liquefied petroleum gases fractionated from refinery or still gases. Through compression and/or refrigeration, they are retained in a liquid state. The reported categories are ethane/ethylene, propane/propylene, normal butane/butylenes, and isobutane/isobutylene.

Liquid Extraction: Light and heavy liquid phases are contacted in a column with contact surfaces and possibly mixing. Some components are transferred (extracted) from one liquid phase to the other.

Lockout–Tagout (LOTO): Refers to a program to control hazardous energy during the servicing and maintenance of machinery and equipment. Lockout refers to the placement of a locking mechanism on an energy-isolating device, such as a valve, so that the equipment cannot be operated until the mechanism is removed. Tagout refers to the secure placement of a tag on the energy-isolating device to indicate that the equipment cannot be operated until the tag is removed (See Figure 25).

Long Residue: The bottoms stream from the atmospheric distillation tower.

Long-Term Exposure Limit (LTEL): The time-weighted average concentration of a substance over an 8-h period thought not to be injurious to health.

Lower Explosive Limit (LEL): The minimum concentration of combustible gas or vapor in air below which the propagation of flame does not occur on contact with an ignition source. Also known as the lower flammable limit or the lower explosion limit.

Low Pressure (LP): A processing unit operating at less than 225 psig measured at the outlet separator.

Lubricants: Substances used to reduce the friction between bearing surfaces or as process materials either incorporated into other materials used as processing aids in the manufacture of other products or used as the carriers of other materials. Petroleum lubricants may be produced either from distillates or residues. Lubricants include all grades of lubricating oils, from spindle oil to cylinder oil and those used in greases.

Light Vacuum Gas Oil (LVGO): A side stream from the vacuum distillation tower.

Figure 25 An isolation tag on a piece of equipment (Source: hazardex, www.hazardexonthenet.net).

Make Up Stream: A feed to a process to replace a component that reacts or is otherwise depleted in a process.

Main Cryogenic Heat Exchanger: Main heat exchanger in the liquefaction unit where the cooling and liquefaction of natural gas take place by means of heat exchange with cooling fluids.

Main Fractionators: The first distillation column for an FCC or coking process.

Main Line: Branch or lateral sewers that collect wastewater from building sewers and service lines.

Main Sewers: A sewer that receives wastewater from many tributary branches and sewer lines and serves as an outlet for a large territory or is used to feed an intercepting sewer.

Management of Change (MOC): **1.** A process to understand all the implications of a change to a procedure. **2.** A process for evaluating and controlling hazards that may be introduced during modifications to the facility, equipment, operations, personnel, or activities; MOCs can also be used to identify, evaluate, and control unintended hazards introduced by modifying procedures or when developing a new plan or procedure.

Manhole: An opening in a sewer provided for the purpose of permitting operators or equipment to enter or leave a sewer. Sometimes called an "access hole" or a "maintenance hole."

Manifold(s): **1.** A junction or center for connecting several pipes and selectively routing the flow. **2.** A pipe spool in which a number of incoming pipes are combined to feed to a common output line.

Manometer: Instrument for measuring head or pressure; basically, a U-tube partially filled with a liquid, so constructed that the difference in the level of the liquid leg indicates the pressure exerted on the instrument.

MAOP: See Maximum Allowable Operating Pressure.

Mass Balance: The summation of the mass entering a process and the summation of the mass leaving a process. They must be equal for a steady-state process.

Material Safety Data Sheet: **1.** A description of the Health, Safety, and Environment (HSE) data for a marketed product. **2.** Printed information that describes the properties of a hazardous chemical and ways to control its hazards. **3.** A document that provides pertinent information and a profile of a particular hazardous substance or mixture. An MSDS is normally developed by the manufacturer or formulator of the hazardous substance or mixture. The MSDS is required to be made available to employees and operators whenever there is the likelihood of the hazardous substance or mixture being introduced into the workplace.

MAWP: *See Maximum Allowable Working Pressure.*

Maximum Allowable Operating Pressure (MAOP): The maximum gas pressure at which a pipeline system or process facility is allowed to operate.

Maximum Allowable Working Pressure (MAWP): **1.** This is a legal maximum pressure that a process vessel is allowed to experience. Above this pressure, a relief valve should be open to protect the vessel from catastrophic failure. **2.** The maximum pressure to which a surface vessel can be operated or the maximum pressure during treating to which a well should be exposed.

Mechanical Seal: This is the part of a centrifugal pump that keeps the liquid from squirting out along the shaft. It is often subject to leakage due to pump vibration and cavitation.

Melting Point: The temperature at which a solid turns into a liquid. As the temperature is a measure of the kinetic energy of molecules (i.e., how much they are moving around), this means that the molecules are moving too much to stay in one place.

Mercaptans: 1. Compounds of carbon, hydrogen, and sulfur (RSH, R=CH_3) found in sour crude and gas; the lower mercaptans have a strong, repulsive odor and are used, among other things, to odorize natural gas. 2. A class of compounds containing carbon, hydrogen, and sulfur. The shorter-chain materials are used as an odor marker in natural gas. 3. Organic sulfides of the formula RSH where R represents the organic radical and SH represents the thiol group.

Methane (CH_4): A light odorless flammable gas that is the principal component of natural gas.

Methanol (CH_3OH): Methyl alcohol from the general formula (ROH), where R = C_nH_{2n+1} is known as a radical and n = 1, 2, 3, etc. Methanol can be made by the destructive distillation of wood or through a process starting with methane or a heavier hydrocarbon, decomposing it to synthesis gas and recombining it to methanol.

Methyl Tertiary Butyl Ether (MTBE, - $(CH_3)_3COCH_3$): 1. Is manufactured by the etherification of methanol and isobutylene. Methanol is derived from natural gas, and isobutylene is derived from butane obtained from crude oil and natural gas 2. A gasoline additive used to increase the octane number. MTBE is produced by reacting methanol (CH_3OH) with isobutylene (iC_4H_8). 3. Blends up to 15.0% by volume of MTBE that must meet the ASTM D4814 specifications. Blenders must take precautions that the blends are not used as base gasoline for other oxygenated blends (commonly referred to as the "Sun waiver").

An ether intended for gasoline blending as described in oxygenate definition.

In the United States, it has been used in gasoline at low levels since 1979 to replace tetraethyl lead and to increase its octane rating, helping to prevent engine knocking. Oxygenates help gasoline burn more completely, reducing tailpipe emissions from pre-1984 motor vehicles; dilutes or displaces gasoline components such as aromatics (e.g., benzene) and sulfur; and optimizes the oxidation during combustion. Most refiners chose MTBE over other oxygenates primarily for its blending characteristics and low cost.

Middle Distillates: Atmospheric pipe still cuts boiling in the range of 300–700°F (149–371°C) vaporization temperature. The exact cut is determined by the specifications of the product. 1. A general classification of refined petroleum products that include distillate fuel oil and kerosene. 2. Medium-density refined petroleum products, including kerosene, stove oil, jet fuel, and light fuel oil. 3. Refinery products in the middle distillation range of refined products: kerosene, heating oil, and jet fuel.

Mid-Percent Point: The vapor temperature at which one-half of the material of a cut has been vaporized. Mid-percent point is used to characterize a cut in place of temperature limits.

Mixed Phase: More than one phase. Usually implies both vapor and liquid phase(s) present.

Molecular Sieve: A separation process that usually works by gaseous diffusion. A membrane is selected through which the compounds being removed or purified can pass, while the remaining compounds in the stream being processed cannot pass.

MONC: Motor octane number clear (unleaded).

Motor Octane Number (MON, ASTM ON F2): A measure of resistance to the self-ignition (knocking) of gasoline under laboratory conditions that correlate with road performance during highway driving conditions. The percentage by volume of isooctane in a mixture of isooctane and n-heptane that knocks with the same intensity as the fuel being tested. A standardized test engine operating under standardized conditions (900 rpm) is used. This test approximates the cruising conditions of an automobile; ASTM D-2723.

MPHC: Medium-pressure hydrocracking or partial conversion hydrocracking.

Motor Gasoline or Petrol: Gasoline is a volatile, flammable, complex petroleum fuel used mainly in internal combustion engines. It is used as fuel in specially designed heaters and lamps.

Motor Gasoline Blending: 1. Naphthas (e.g., straight-run gasoline, alkylate, reformate, benzene, toluene, xylenes) used for blending or compounding into finished motor gasoline. Includes receipts and inputs of Gasoline Treated as Blendstock (GTAB). Excludes conventional blendstock for oxygenate blending (CBOB), reformulated blendstock for oxygenate blending, oxygenates (e.g., fuel ethanol and methyl tertiary butyl ether), butane, and natural gasoline. 2. Mechanical mixing of motor gasoline blending components, and oxygenates when required, to produce finished motor gasoline. Finished motor gasoline may be further mixed with other motor gasoline blending components or oxygenates, resulting in increased volumes of finished motor gasoline and/or changes in the formulation of finished motor gasoline (e.g., conventional motor gasoline mixed with MTBE to produce oxygenated motor gasoline).

Motor Gasoline Blending Components: Naphthas (e.g., straight-run gasoline, alkylate, reformate, benzene, toluene, xylene) used for blending or compounding into finished motor gasoline. These components include reformulated gasoline blendstock for oxygenate blending (RBOB) but exclude oxygenates (alcohols, ethers), butane, and pentanes plus.

Motor gasoline (finished): A complex mixture of relatively volatile hydrocarbons with or without small quantities of additives, blended to form a fuel suitable for use in spark-ignition engines. Motor gasoline, as defined in ASTM Specification D4814 or Federal Specification VV – G – 1690C, is characterized as having a boiling range of 122–158°F (50–70°C) at the 10% recovery point to 365–374°F (185–190°C) at the 90% recovery point. "Motor gasoline" includes conventional gasoline, all types of oxygenated gasoline, including gasohol, and reformulated gasoline but excludes aviation gasoline.

Naphtha: 1. Straight-run gasoline distillate, below the boiling point of kerosene. Naphthas are generally unsuitable for blending as a component of premium gasoline; hence, they are used as a feedstock for catalytic reforming in hydrocarbon production processes or in chemical manufacturing processes. 2. A term that is applied to low-boiling mixtures of hydrocarbons with typical TBP boiling ranges between 150 and 450°F (66–233°C). Light and heavy naphthas are produced in the distillation of crude oils. Cracked naphthas are also produced by many of the refinery reaction processes.

Naphthas are subdivided according to the actual pipe still cuts—into light, intermediate and heavy, and very heavy virgin naphthas. A typical pipe still operation would be

C_5–160°F (C_5–71°C)	: light virgin naphtha
160–280°F (71–138°C)	: intermediate virgin naphtha
280–380°F (138–193°C)	: heavy virgin naphtha

Naphtha, the major constituent of gasoline, generally needs processing to make suitable quality gasoline.

Naphtha Less Than 401°F: A naphtha with a boiling range of less than 401°F (205°C) that is intended for use as a petrochemical feedstock.

Naphtha-Type Jet Fuel: A fuel in the heavy naphtha boiling range having an average gravity of 52.8 °API, 20–90% distillation temperature of 290–470°F (143–243°C), and meeting Military Specification MIL - T- 5624L (Grade JP-4). It is used primarily for military turbojet and turboprop aircraft engines because it has a lower freeze point than other aviation fuels and meets engine requirements at high altitudes and speeds.

Special Naphthas: All finished products within the naphtha boiling range that are used as paint thinners, cleaners, or solvents. These products are refined to a specified flash point. Special naphthas include all commercial hexane and cleaning solvents conforming to ASTM Specification D 1836 and D484, respectively. Naphthas to be blended or marketed as motor gasoline or aviation gasoline and synthetic natural gas (SNG) feedstocks are excluded.

Naphthenes: Hydrocarbons of the cyclane family, sometimes called cycloalkanes. Naphthenes have no double bonds and are saturated ring structures with the general formula C_nH_{2n}, where C = carbon atoms, H = hydrogen atoms, and n = 6, 7, 8, …

Naphthenic: Having the characteristics of naphthenes, saturated hydrocarbons whose molecules contain at least one closed ring of carbon atoms.

Naphthenic Acids: Organic acids occurring in petroleum that contain a naphthenic ring and one or more carboxylic acid groups. Naphthenic acids are used in the manufacture of paint driers and industrial soaps.

Naphthenic Crudes: A type of crude petroleum containing a relatively large proportion of naphthenic-type hydrocarbon.

Natural Gas: Naturally occurring gas consisting predominantly of methane, sometimes in conjunction with crude (associated gas) and sometimes alone (unassociated gas). **1.** A mixture of light hydrocarbons found naturally in the Earth's crust, often in association with oil (when it is known as associated gas). Methane is the most dominant component. It may also include some short-chain hydrocarbons (ethane, propane, butane) that may be in a gaseous state at standard conditions. **2.** A mixture of hydrocarbon compounds and small quantities of various non-hydrocarbons existing in the gaseous phase or in a solution with crude oil in natural underground reservoirs at reservoir conditions. The primary constituent compound is CH_4. Gas coming from wells also can contain significant amounts of ethane (C_2H_6), propane (C_3H_8), butane (C_4H_{10}), and pentanes (C_5H_{12}) and widely varying amounts of carbon dioxide (CO_2) and nitrogen (N_2).

Natural Gas Heating Value: The amount of thermal energy released by the complete combustion of one standard cubic foot of natural gas.

Natural Gas Liquids (NGL): **1.** Liquid hydrocarbons, such as ethane, propane, butane, pentane, and natural gasoline, extracted from field natural gas. **2.** Those hydrocarbons in natural gas that are separated from the gas as liquids through the process of absorption, condensation, adsorption, or other methods of gas processing or cycling plants. Generally, such liquids consist of propane and heavier hydrocarbons and are commonly referred to as lease condensate, natural gasoline, and liquefied petroleum gases. Natural gas liquids include natural gas plant liquids (primarily ethane, propane, butane, and isobutane). See Natural Gas Plant Liquids and Lease Condensate (primarily pentanes produced from natural gas and lease separators and field facilities). **3.** Liquids obtained during natural gas production include ethane, propane, butanes, and condensate.

Natural Gasoline: A gasoline range product separated at a location near the point of production from natural gas streams and used as a gasoline blending component.

Natural Gasoline and Isopentane: A mixture of hydrocarbons, mostly pentanes and heavier, extracted from natural gas, that meets vapor pressure, and endpoint and other specifications for natural gasoline set by the Gas Processors Association. Includes isopentane that is a saturated branch-chain hydrocarbon (iC_5H_{12}), obtained by the fractionation of natural gasoline or the isomerization of normal pentane (nC_5H_{12}).

Natural Gas Plant Liquids: Those hydrocarbons in natural gas that are separated as liquids at natural gas processing plants, fractionating and cycling plants, and, in some instances, field facilities. Lease condensate is excluded. Products obtained include ethane, liquefied petroleum gases (propane, butanes, propane–butane mixture, ethane–propane mixture), isopentane, and other small quantities of finished products, such as motor gasoline, special naphthas, jet fuel, kerosene, and distillate fuel oil.

Natural Gas Processing: 1. The purification of field gas at natural gas processing plants (or gas plants) or the fractionation of mixed NGLs to natural gas products to meet specifications for use of pipeline-quality gas. Gas processing includes removing liquids, solids, and vapors absorbing impurities and odorizing. 2. The process of separating natural gas liquids (NGLs) by absorption, adsorption, refrigeration, or cryogenics from the steam of natural gas.

Natural Gas Processing Plant: Facilities designed to recover natural gas liquids from a stream of natural gas that may or may not have passed through lease separators and/or field separation facilities. These facilities control the quality of the natural gas to be marketed. Cycling plants are classified as gas processing plants.

Net Heating Value: Is the total energy transferred as heat in an ideal combustion reaction at a standard temperature and pressure in which all water formed appears as vapor. The net heating is an ideal gas property in a hypothetical state (the water cannot all remain vapor because after the water saturates the CO_2 in the products, the rest would condense).

Net Positive Suction Head (NPSH): The net-positive suction head required to keep a centrifugal pump from cavitating. Cooling a liquid in a pump's suction line increases the pump's available NPSH, as does increasing the liquid level in the suction drum.

Net Zero: Refers to a point at which the amount of greenhouse gas being put into the atmosphere by human activity equals the amount of greenhouse gas that is being taken out of the atmosphere.

Non-associated Gas: Natural gas that exists in a reservoir alone and is produced without any crude oil.

Normal Boiling Point: *See Boiling Point.*

Nusselt Number (Nu): A dimensionless number Nu is used in heat transfer calculations characterizing the relation between the convective heat transfer of the boundary layer of a fluid and its thermal conductivity:

$$Nu = \frac{hd}{k} \tag{43}$$

where h is the surface heat transfer coefficient, d is the thickness of the fluid film, and k is the thermal conductivity.

Octane (C_8H_{18}): 1. Is a hydrocarbon and an alkane with the chemical formula C_8H_{18}, and the condensed structural formula $CH_3(CH_2)_6CH_3$. Octane has many structural isomers that differ by the amount and location of branching in the carbon chain. One of the isomers, 2, 2, 4 - trimethylpentane (isooctane) $(CH_3)_3CCH_2CH(CH_3)_2$

```
      H       H
      |       |
   H-C-H   H-C-H
    H  |    H  |    H
    |  |    |  |    |
H - C — C — C — C — C - H
    |  |    |  |    |
    H  |    H  H    H
      H-C-H
       |
       H
```
is used as one of the standard values in the octane rating scale. Octane is a component of gasoline (petrol). As with all low-molecular-weight hydrocarbons, octane is volatile and very flammable. **2.** A test used to measure the suitability of gasoline as motor fuel. The octane test determines the knocking characteristics of gasoline in a standard test engine relative to a standard of 2-2-4 trimethyl pentane (2 2 4 TMP). 2 2 4 TMP is assigned an octane number of 100.0. There are two octane tests. One is designated the research octane (F-1), and the second is the motor octane (F-2). Motor octane is determined in an engine more representative of actual operating conditions for automobiles and is lower than research octane for any gasoline stock.

Historically, gasoline was marketed based on the F-1 octane, but in recent years, the average of the F-1 and F-2 octane has been used. At the gasoline pump, this is reported as (R + M)/2.

The Research Octane Number (RON) test simulates driving mild conditions, while the Motor Octane Number (MON) test simulates driving under more severe conditions, i.e., under load and at high speeds. The arithmetic average of RON and MON, which gives an indication of the performance of the engine under the full range of conditions, is projected as Antiknock Index (AKI), i.e., Antiknock Index (AKI) = (RON + MON)/2.

Octane Number: 1. Is a measure of the knocking characteristics of fuel in a laboratory gasoline engine according to ASTM D2700. We determine the octane number of fuel by measuring its knocking value compared to the knocking of a mixture of n-heptane and isooctane of 2-2-4 trimethylpentane (224 TMP). **2.** An index measured by finding a blend of isooctane (iC_8H_{18}) and normal heptanes (nC_7H_{16}) that knocks under identical conditions as the gasoline being evaluated. It is a measure of the ease of self-ignition of fuel without the aid of a spark plug. **3.** The octane number is a measure of the antiknock resistance of gasoline. It is the percentage of isooctane in a mixture of isooctane and n-heptane, which gives a knock of the same intensity as the fuel being measured when compared in a standard engine. For example, if the fuel being tested matches in knocking to a blend of 90% isooctane and 10% n-heptane, then the test fuel is said to have an octane number of 90.

Iso-octane, which produces the least knocking or which knocks only at a much higher compression ratio, is given an octane number of 100, while n-heptane, which is very poor in its resistance to knocking or which knocks at a much lower compression ratio, is given an octane number of zero.

CH_3CH_2-CH_2-CH_2-CH_2-CH_2-CH_3 [Octane No. = 0]

n-heptane

```
        CH3    CH3
         |      |
CH3-CH-CH2-C-CH3        [Octane No. = 100]
                |
               CH3
```

Iso-octane
[2,2,4-trimethyl pentane]

Generally, the octane number increases as the degree of branching of the carbon chain increases, and thus iso-paraffins are found to give higher octane numbers than the corresponding normal isomers. Olefins are found to give higher octane numbers than the related paraffins. Naphthenes also give better octane numbers than the corresponding normal paraffins. Aromatics usually exhibit high octane numbers.

A single-cylinder test engine is made to obtain the antiknock characteristics of gasoline in terms of octane numbers. The octane numbers formed a scale ranging from 0 to 100; the higher the number, the greater the antiknock characteristics. The scale has been extended above 100 by comparing the knocking intensity with isooctane to which tetraethyl lead (TEL) is added. Numbers greater than 100 on the scale are referred to as performance numbers rather than octane numbers. *See also Motor Octane Number and Research Octane Number (Table 2).*

Octane numbers are very relevant in the reforming, isomerization, and alkylation processes in refining facilities. These processes enable the successful reactive transformations to yield long side-chain paraffin and aromatics that possess higher octane numbers than the feed constituents that do not consist of higher quantities of constituents possessing straight-chain paraffin and non-aromatics (naphthenes).

It is a measure of the ease of self-ignition of fuel without the aid of a spark plug.

Octane Scale: A series of arbitrary numbers from 0 to 120.3 used to rate the octane number of gasoline. Three reference materials define the scale; n-heptane (Octane number = 0), isooctane (Octane number = 100), and isooctane plus 6 ml tetramethyl lead (Octane number = 120.3). Above 100, the octane number of a fuel is based on the engine ratings, in terms of the ml of tetra ethyl lead in isooctane, which matches that of the unknown fuel.

Off Gas: The gas leaving a reflux drum or top tray of an absorber column.

Offline: When a process unit is shut down, it is said to be offline.

Oil: One of the various liquid, viscid, usually inflammable, chemically neutral substances that is lighter and insoluble in water but soluble in alcohol and ether and classified as non-volatile. Natural plant oils comprise terpenes and simple esters such as essential oils. Animal oils are glycerides of fatty acids. Mineral oils are mixtures of hydrocarbons. Oils have many uses and include fuels lubricants, soap constituents, vanishes, etc.

Oil and Gas: Refer to the industry associated with the recovery of liquid and gaseous hydrocarbons from underground deposits as reservoirs found both onshore and offshore around the world. A collection of localized deposits is known as an oil field or gas field. When they are drilled, they are known as oil and gas wells. Oil is mainly used as fuel for transportation purposes, whereas gas is primarily used as fuel for domestic and industrial purposes and for converting into other chemicals such as plastic. Oil is widely transported in ships. Gas is transported in underground, sub-sea, or overland pipelines covering large distances.

Table 2 Octane Numbers of Pure Hydrocarbons*.

Hydrocarbon	RON	MON	Hydrocarbon	RON	MON
n-Pentane	61.7	61.9	2,4 – Dimethyl hexane	62.5	69.9
n-Hexane	24.8	26.0	2,2,4 –Trimethyl pentane (isooctane)	100.0	100.0
n-Heptane	0.0	0.0	1 – Pentene	90.9	77.1
n-Octane	-19.5	-15.0	1 – Octane	28.7	34.7
n-Nonane	-17.0	-20.0	3 - Octene	72.5	68.1
2-Methyl butane (iso-pentane)	92.3	90.3	4 – Methyl – 1- Pentene	95.7	80.9
2 – Methyl hexane (iso-heptane)	42.4	46.4	Benzene	-	114.8
2 – Methyl heptane (iso-octane)	21.7	23.8	Toluene	120.1	103.5

*(Source: Speight, James G., The Chemistry & Technology of Petroleum, Marcel Dekker, Inc. 1991).

Oil Refinery: An industrial process plant where crude oil is converted into useful products such as naphtha, diesel fuel, kerosene, and LPG. Also known as petroleum refinery, the process involves the separation of the crude oil into fractions in the process of fractional distillation. By boiling the crude oil, the light or more volatile components with the lowest boiling point rise toward the top of the column, whereas the heavy fractions with the highest boiling points remain at the bottom. The heavy bottom fractions are then thermally cracked to form more useful light products. All the fractions are then processed further in other parts of the oil refinery, which may typically feature vacuum distillation used to distill the bottoms; hydrotreating, which is used to remove sulfur from naphtha, catalytic cracking, fluid catalytic cracking, hydrocracking, visbreaking, isomerization, steam reforming, alkylation, hydrodesulfurization, and the Claus process used to convert hydrogen sulfide into sulfur, solvent dewaxing, and water treatment (See Figure 26).

Olefins: Hydrocarbons of the alkenes family. Olefins have two carbon atoms in the molecular structure linked by a double bond to satisfy the absence of two hydrogen atoms that are present in the corresponding paraffin. This hydrogen deficiency is called unsaturation. The general formula for olefins is C_nH_{2n}, where C = carbon atoms, H = hydrogen atoms, and n = 2, 4, 6…

Olefins do not occur naturally in crude oil and are created in the thermal and catalytic cracking processes.

Online: When a process unit is in operation and the processing feed, it is said to be online.

OPEC: Organization of Petroleum Exporting Countries. These countries have organized for the purpose of negotiating with oil companies on matters of oil production, prices, and future concession rights. The current members are Algeria, Indonesia, Iran, Kuwait, Libya, Nigeria, Qatar, Saudi Arabia, United Arab Emirates, and Venezuela.

Operability Capacity: **1.** The amount of capacity that, at the beginning of the period, is in operation; not in operation and not under active repair but capable of being placed in operation within 30 days; or not in operation but under active repair that can be completed within 90 days. Operable capacity is the sum of the operating and idle capacity and is measured in barrels per calendar day or barrels per stream day. **2.** The component of operable capacity is operated at the beginning of the period.

Operating Pressure: Pressure indicated by a gauge when the system is in normal operation (working pressure).

Operation and Maintenance Manual: A manual that describes detailed procedures for operators to follow to operate and maintain specific water or wastewater treatment, pretreatment, or process plants and the equipment of the plants.

Figure 26 A photo of petroleum crude oil refinery.

Operator: 1. Term used to describe a company appointed by venture stakeholders to take primary responsibility for day-to-day operations for a specific plant or activity. 2. The company or individual responsible for managing an exploration, development, or production operation. 3. The company that has the legal authority to drill wells and undertake the production of hydrocarbons that are found. The operator is often part of a consortium and acts on behalf of this consortium. 4. The company that makes the decisions and is responsible for drilling, completing, operating, and repairing the well.

Operable Utilization Rate: Represents the utilization of the atmospheric crude oil distillation units. The rate is calculated by dividing the gross input to these units by the operable refining capacity of the units.

Organic Compounds: Compounds that include carbon and hydrogen atoms. Generally, organic compounds can be classified as either aliphatics (straight-chain compounds), cyclic (compounds with ring structures), and combinations of aliphatics and cyclic.

Orifice: An opening in a wall or plate used to control the rate of flow into or out of a tank or pipe.

Orifice Meter: A single-phase flow meter, primarily for liquid/gas, that measures the pressure drop created by the hole as gas is flowed (See Figure 27).

Other Hydrocarbons: Materials received by a refinery and consumed as a raw material. Includes hydrogen, coal tar derivatives, gilsonite, and natural gas received by the refinery for reforming into hydrogen. Natural gas to be used as fuel is excluded.

OSHA: 1. Occupational Safety and Health Administration: US government agency. 2. The Williams–Steiger Occupational Safety and Health Act of 1970 (OSHA) is a federal law designed to protect the health and safety of industrial workers, including the operators of water supply and treatment systems and wastewater treatment plants. The Act regulates the design, construction, operation, and maintenance of water supply systems, water treatment plants, wastewater collection systems, and wastewater treatment plants. OSHA also refers to the federal and state agencies that administer the OSHA regulations.

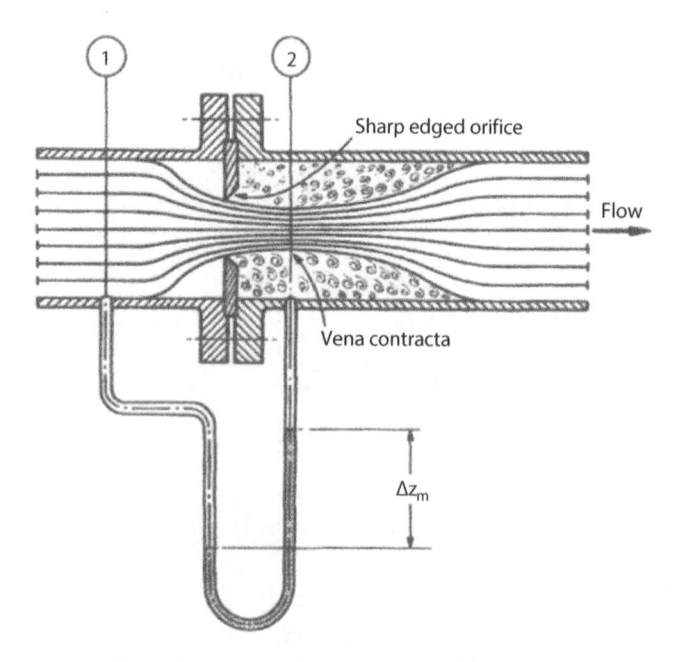

Figure 27 Orifice meter with Vena contracta formation.

Oxidation: Oxidation is the addition of oxygen, the removal of hydrogen, or the removal of electrons from an element or compound. In the environment, organic matter is oxidized to more stable substances. The opposite is reduction.

Oxidation Inhibitor: A substance added in small quantities to a petroleum product to increase its oxidation resistance, thereby lengthening its service or storage life; also called an antioxidant. The oxidation of fuels creates gums that become colloidal, then agglomerate and precipitate. Cracked distillates are found to be more prone to oxidation and deterioration than straight-run distillates. Oxidation fuels can also result in the formation of various acids, ketones, aldehydes, and esters from hydrocarbons. Amino guanidine derivatives when used in the range 3–30 ppm are found effective as antioxidants. Cyclic borates of polymers alkanolamines are effective anti-oxidants even in the 10 ppm range.

Oxidation Stability: 1. It is used for the evaluation of storage stability and resistance to oxidation as most of the oils, when exposed to air over time, react with oxygen, which are then degraded. Oil with poor oxidation stability, forms corrosive acids at high-temperature conditions in the engine. **2.** Gasoline contains cracked components having a tendency to form gum materials during storage and handling, which affect performance. Oxidation stability provides an indication of the tendency of gasoline and aviation fuels to form gum in storage. In this test, the sample is oxidized inside a stainless-steel pressure vessel initially charged with oxygen at 689 kPa and heated in a boiling water bath. The amount of time required for a specified drop in pressure (gasoline) or the amount of gum and precipitate formed after a specific aging period (aviation fuel) is determined.

Oxidizers: Reactants that oxidize, for example, bleach, chlorine, sodium hypochlorite, and sodium persulfate. Also, a compound that releases oxygen.

Oxidizing Agent: Any substance, such as oxygen (O_2) or chlorine (Cl_2), that will readily add (or take on) electrons. The opposite is a reducing agent.

Oxygen: A chemical element used by all known life forms for respiration.

Oxygenated Fuel: Any organic compound containing oxygen. Specifically for the petroleum industry, this term refers to oxygen-containing organic compounds, such as ethers, and alcohols added to fuels to reduce carbon monoxide in the engine exhausts. They are used as gasoline blending components. Oxygenated fuels tend to give a more complete combustion of its carbon into carbon dioxide (rather than monoxide), thereby reducing air pollution from exhaust emissions.

Oxygenated Fuels Program Reformulated Gasoline: A reformulated gasoline that is intended for use in an oxygenated fuels program control area during an oxygenated fuels program control period.

Oxygenated Gasoline: 1. Gasoline with an oxygen content of 1.8% or higher by weight that has been formulated for use in motor vehicles. **2.** Finished motor gasoline, other than reformulated gasoline, having an oxygen content 2.7% or higher by weight. It includes gasohol.

Oxygenates: Substances that, when added to gasoline, increase the amount of oxygen in that gasoline blend and thus boost the octane number of gasoline or petrol. Ethanol, methyl tertiary butyl ether (MTBE), ethyl tertiary butyl ether (ETBE), tertiary amyl methyl ether (TAME), and methanol are common oxygenates. MTBE, ETBE, and TAME have 1. Low water solubility, 2. Lower volatility, and 3. A compatibility with hydrocarbon fuels.

Overall Tray Efficiency: Overall tray efficiency can be defined as the number of theoretical trays in a distillation column section divided by the number of actual trays in the section and is reported as a percent. Overall tray efficiencies are less than 100% for all refinery distillation columns.

Overflash: The liquid that returns to the flash zone of a column.

Overhead: Usually refers to the vapor leaving the top tray of a distillation column. For an absorber column, the overhead and the top product are the same.

Overlap: See Gap.

Overpressure: Is any pressure relative to ambient pressure caused by an explosive blast, both positive and negative.

Ozone (O_3): An oxygen molecule with three oxygen atoms that occurs as a blue, harmful, pungent-smelling gas at room temperature. The stratosphere ozone layer, which is a concentration of ozone molecules located at 6–30 mi. above sea level, is in a state of dynamic equilibrium. Ultraviolet radiation not only forms the ozone from oxygen but can also reduce the ozone back to oxygen. The process absorbs most of the ultraviolet radiation from the Sun, shielding life from the harmful effects of radiation.

Packed Bed Scrubber: Vertical or horizontal vessels, partially filled with packing or devices of a large surface area, used for the continuous contact of liquid and gas such that absorption can take place. Frequently, the scrubber liquid or liquor has had chemicals added to react with the absorbed gas.

Packing (Seals): Seals around a moving shaft or other equipment.

Paraffins: **1**. Hydrocarbons of the alkanes family. Paraffins are saturated compounds, i.e., hydrogen atoms are appropriately attached to the carbon atoms such that the carbon atoms have only single bonds in the molecular structure. The general formula for paraffin is C_nH_{2n+2}, where C = carbon atoms, H = hydrogen atoms, and n = 1, 2, 3, 4, 5, … **2**. A white, odorless, tasteless, chemically inert, waxy substance derived from distilling petroleum; a crystalline, flammable substance composed of saturated hydrocarbons. **3**. Normal or straight carbon chain alkanes with carbon chain lengths of C_{18+}. The alkanes in this range solidify at temperatures from 80 to over 200°F (27–93°C).

Partial Pressure: In a gaseous mixture, the pressure contribution for a particular component of the mixture. The sum of the partial pressures of the components in the mixture is the total pressure. For example, in a mixture of two components A and B, with partial pressures as p_A, p_B respectively. The total pressure P_{Total} is: $P_{Total} = p_A + p_B$.

Penetration: A measure of the hardness and consistency of asphalt in terms of the depth that a special pointed device will penetrate the product in a set time and temperature.

Performance Rating: A method of expressing the quality of a high-octane gasoline relative to isooctane. This rating is used for fuels that are of better quality than isooctane.

Petroleum Administration for Defense Districts (PADD): Geographic aggregations of the 50 US states and the District of Columbia into five districts by the Petroleum Administration for Defense in 1950. These districts were originally defined during World War II for purposes of administering oil allocation.

Petroleum Coke: A residue high in carbon content and low in hydrogen that is the final product of thermal decomposition in the condensation process in cracking. This product is reported as marketable coke. The conversion is 5 bbl (of 42 US gal. each) per short ton. Coke from petroleum has a heating value of 6.024 million Btu per barrel.

Marketable coke: Those grades of coke produced in delayed or fluid cokers, which may be recovered as relatively pure carbon. This "green" coke may be sold as is or further purified by calcining.

Catalyst coke: The only catalytic coke used as a fuel is the coke on the catalyst in the FCC process. In other catalytic processes, there is coke deposited on the catalyst, but it is not regenerated in a way such that the heat of combustion is recovered.

Petrolatum: Microcrystalline wax or petroleum jelly.

Petroleum ether: A volatile fraction of petroleum consisting mainly of pentanes and hexanes.

Petrochemical Feedstocks: Chemical feedstocks derived from petroleum principally for the manufacture of chemicals, synthetic rubber, and a variety of plastics. These categories reported are "Naphthas less than 401°F and Other Oils Equal to or greater than 401°F."

Petroleum Products: Petroleum products are obtained from the processing of crude oil (including lease condensate), natural gas, and other hydrocarbon compounds. Petroleum products include unfinished oils, liquefied petroleum gases, pentanes plus, aviation gasoline, motor gasoline, naphtha-type jet fuel, kerosene-type jet fuel, kerosene, distillate fuel oil, residual fuel oil, petrochemical feedstocks, special naphthas, lubricants, waxes, petroleum coke, asphalt, road oil, still gas, and miscellaneous products.

pH: An important chemical property of an aqueous solution is its pH, which measures the acidity or basicity of the solution. In a neutral solution, such as pure water (H_2O), the hydrogen (H^+) and hydroxy (OH^-) ion concentrations are equal. At ordinary temperatures, this concentration is:

$$C_{H^+} = C_{OH^-} = 10^{-7} \left(\frac{g.ion}{L} \right) \tag{44}$$

where

C_{H^+} = hydrogen ion concentration

C_{OH^-} = hydroxyl ion concentration

The unit g. ion denotes gram ion, which represents an Avogadro number of ions. In all aqueous solutions, whether neutral basic, or acidic, a chemical equilibrium or balance is established between these two concentrations, so that

$$K_{eq} = C_{H^+} C_{OH^-} = 10^{-14} \tag{45}$$

where

K_{eq} = equilibrium constant

The numerical value for K_{eq} is given in Eq. (45) holds for room temperature and only when the concentrations are expressed in gram-ion per liter [(g. ion/l)]. In acid solutions, $C_{H^+} > C_{OH^-}$, in basic solutions, C_{OH^-} predominates.

The pH is a property that is a direct measure of the hydrogen ion concentration and is defined by

$$pH = -\log C_{H^+} \tag{46}$$

Thus, an acidic solution is characterized by a pH < 7 (i.e., the lower the pH, the higher the acidity), a basic solution by a pH > 7 and a neutral solubility by a pH = 7. It should be pointed out that Eq. (46) is not the exact definition of pH, but is a close approximation to it. It is strictly speaking the activity of the hydrogen ion a_{H^+} and not the ion concentration as in Eq. (46).

Phase Envelope: 1. The boundaries of an area on the P-T diagram for the material that encloses the region where both vapor and liquid coexist. **2.** Phase diagram or phase envelope is a relation between temperature and pressure that shows the condition of equilibria between the different phases of chemical compounds, the mixture of compounds, and solutions. Phase diagram is an important issue in chemical thermodynamics and hydrocarbon reservoir. It is very useful for process simulation, hydrocarbon reactor design, and petroleum engineering studies. It is constructed from the bubble line, dew line, and critical point. Bubble line and dew line are composed of bubble points and dew points, respectively. Bubble point is the first point at which the gas is formed when a liquid is heated. Meanwhile, dew point is the first point where the liquid is formed when the gas is cooled. The critical point is the point where all the properties of gases and liquids are equal, such as temperature, pressure, the amount of substance, and others. The critical point is very useful in fuel processing and the dissolution of certain chemicals.

According to the thermodynamic definition of the phase diagram, (phase envelope) is a graph showing the pressure at which the transition of different phases from a compound with respect to temperature. A bubble point that forms a bubble line is a point separating the liquid phase and the two-phase region, namely, the liquid phase and gaseous phase. The dew point that forms the dew line is a point separating the gaseous phase and two-phase region, namely, the liquid and gaseous phase. At the dew point, the following conditions must be satisfied (See Figure 28).

Physical Solvent: A liquid capable of absorbing selected gas components by solubility alone without associated chemical reactions.

Pig: 1. A cylindrical device that is inserted into a pipeline to clean the pipeline wall or monitor the internal condition of the pipeline. **2.** A device for cleaning a pipeline or separating two liquids being removed down the pipeline (intelligent pig—fitted with sensors to check for corrosion or defects in pipelines.). **3.** A flow line clearing device pumped through the line with normal flow. **4.** Refers to a poly pig, which is a bullet-shaped device made of hard rubber or similar material. This device is used to clean pipes. It is inserted in one end of a pipe, moves through the pipe under pressure, and is removed from the other end of the pipe.

Pinch Analysis: Bodo Linnhoff at the University of Leeds in 1977 developed a technique for minimizing energy usage in a process. It is based on calculating the minimum energy consumption by optimizing the heat recovery, energy supply, and process operating conditions. It uses process data represented as energy flows or streams as a function of heat load against temperature. These data are combined for all the hot and cold streams requiring heat. The point of closest approach between the hot and cold composite curves is called the pinch point and corresponds to the point where the design is most constrained. Using this point, the energy targets can be achieved using heat exchange to recover heat between the hot and cold streams in two separate systems, with one temperature above the pinch temperature and one for the temperature below the pinch temperatures.

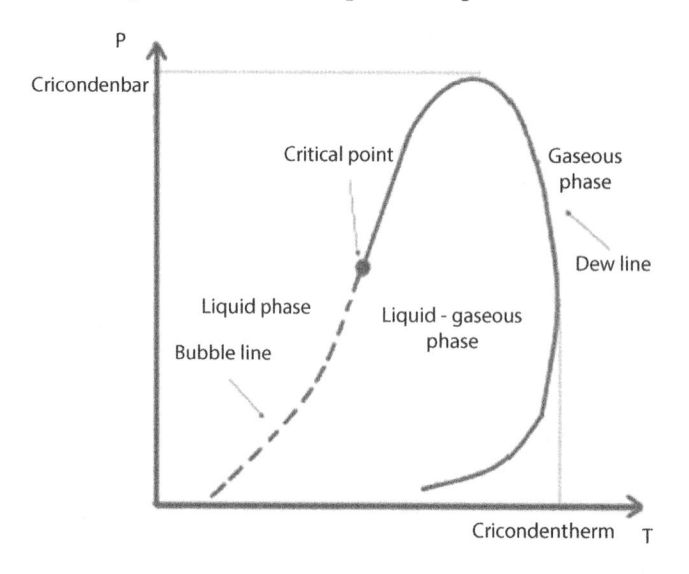

Figure 28 Phase diagram (Phase Envelope).

(a) The plus-minus principle.

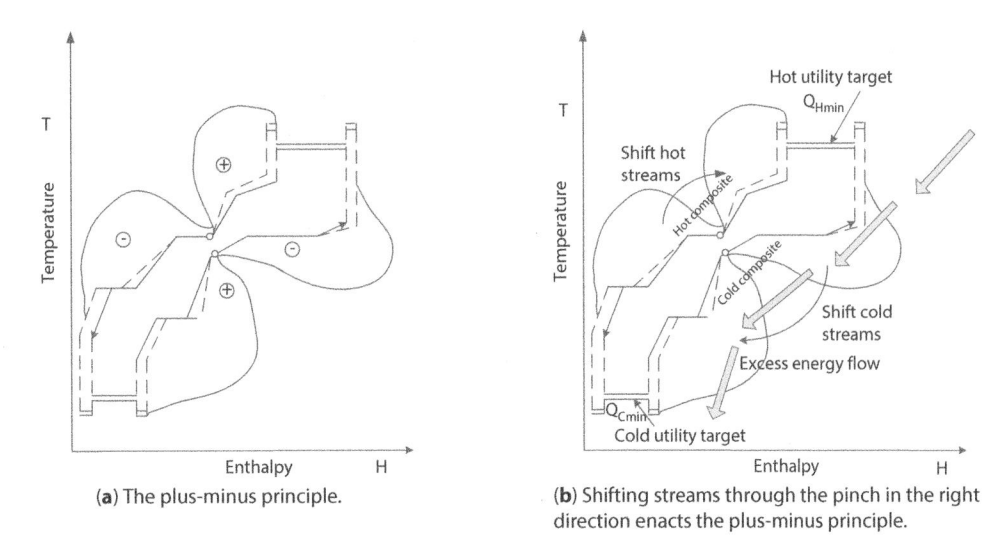

(b) Shifting streams through the pinch in the right direction enacts the plus-minus principle.

Figure 29 Shows the point in a pinch analysis that corresponds to the point where the hot and cold streams in an integrated process are most constrained.

Figure 29 shows the point in a pinch analysis that corresponds to the point where the hot and cold streams in an integrated process are most constrained.

Pipelines: Tubular arrangement for the transportation of crude oil, refined products, and natural gas from the well head, refinery, and storage facility to the consumer. The pipeline measures 14–42 in. (356–1067 mm) in diameter but is usually 20–36 in. (508–914 mm). It is often composed of 40 ft. (12 m) lengths, but the lengths may be as long as 60 or 80 ft. (18–24 m). The pipe is wrapped and coated for protection against corrosion, especially since it runs underground. About half of all gases and oils are moved by the pipeline.

Pipe Still: A heater or furnace containing tubes through which oil is pumped while being heated or vaporized. Pipe stills are fired with waste gas, natural gas, or heavy oils, and by providing for rapid heating under conditions of high pressure and temperature, are useful for thermal cracking as well as distillation operations.

Pipe size: Process piping comes in particular nominal sizes:

0.75 in.	-
1 in.	25 mm
2 in.	50 mm
2.5 in.	-
3 in.	80 mm
4 in.	100 mm
6 in.	150 mm
8 in.	200 mm
10 in.	250 mm

The nominal size does not refer either to the outside or inside diameter of the pipe. Pipe thickness affects the ID.

Piping and instrumentation diagram (P & ID): A schematic representation of the interconnecting pipelines and control systems for a process or part of a process (see Figure 30). Uses a standard set of symbols for process equipment

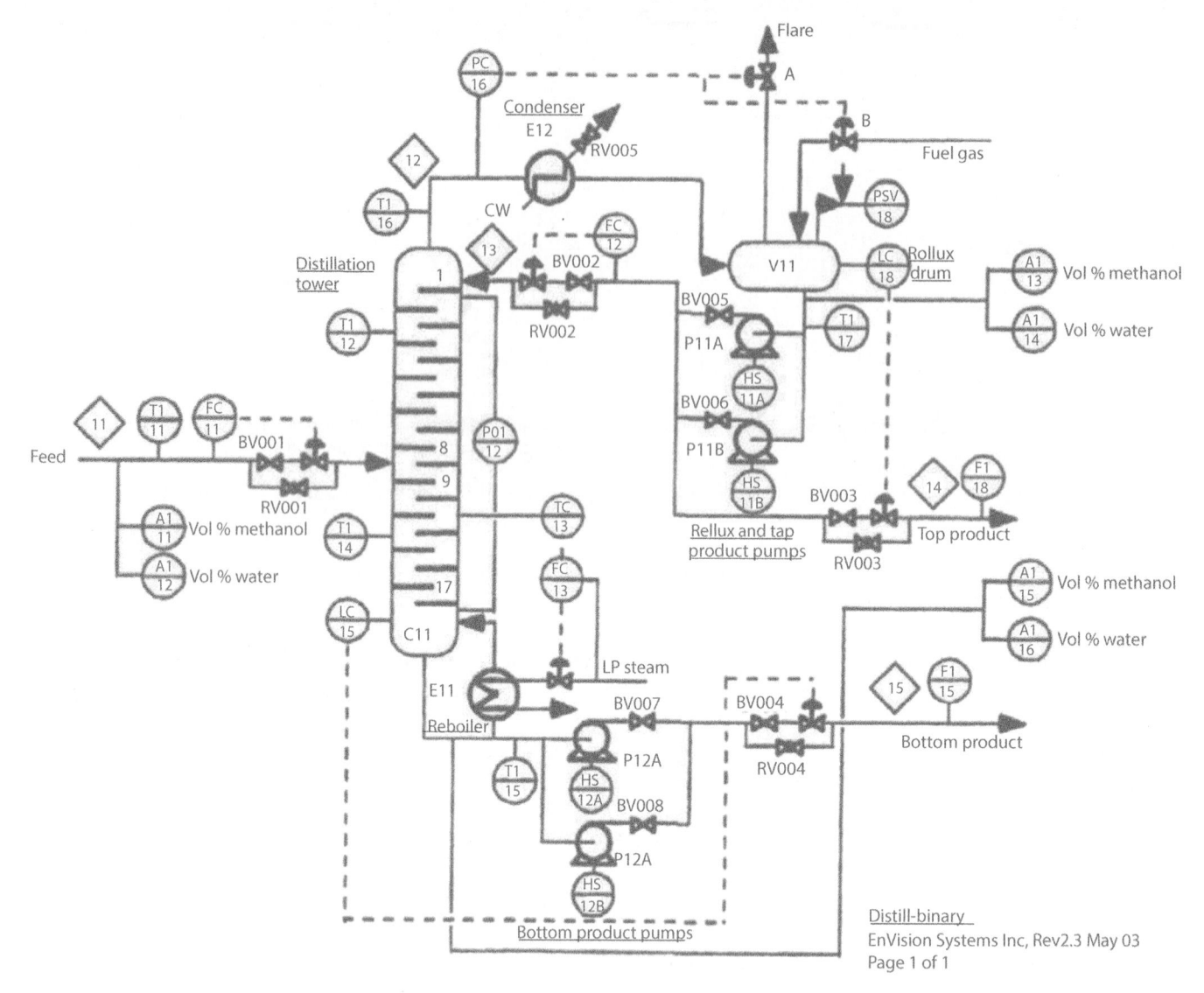

Figure 30 Piping and instrumentation diagram.

and controllers. It includes the layout of branches, reducers, valves, equipment, instrumentation, and control interlocks. It also includes process equipment names and numbers; process piping including sizes and identification; valves and their identification; flow directions, instrumentation, and designations; vents, drains, sampling lines, and flush lines. P & IDs are used to operate the process system, operators' trainings as well as being used in plant maintenance and process modifications. At the design stage, it is useful in carrying out safety and operations investigations such as Hazop. The list of P & ID items are:

- Instrumentation and designations
- Mechanical equipment with names and numbers
- All valves and their identifications
- Process piping, sizes, and identification
- Miscellanea—vents, drains, special fittings, sampling lines, reducers, enlargers, and swagers
- Permanent start-up and flush lines
- Flow directions.
- Interconnections references
- Control inputs and outputs and interlocks

- Interfaces for class changes
- Computer control systems
- Identification of components and subsystems

Plug Flow Reactor (PFR)/Continuous Tubular Reactor (CTR): Is a reactor tubular reactor where fluids enter continuously in an axial direction in a tube as the reaction takes place, and the products are withdrawn at the outlet. The fluid going through a plug flow reactor is modeled as flowing through the reactor as a series of infinitely thin coherent "plugs," each having a uniform composition. The plugs travel in the axial direction of the reactor, with each plug having a different composition from the ones before and after it. The key assumption is that as a plug flows through a PFR, the fluid is perfectly mixed in the radial direction but not mixed at all in the axial direction. Each plug is considered as a separate entity (i.e., effectively an infinitesimally small batch reactor with mixing approaching zero volume). As the plug flows downs the PRF, the residence time of the plug element is derived from its position in the reactor. In this description of the ideal plug flow reactor, the residence time distribution is therefore an impulse (a small narrow spike function). Although it is a powerful tool for estimating purposes, caution is required as a real flow system exhibits significant variability in residence times. Residence time distribution is one of the factors that need to be considered when scaling flow reactors (Figure 31).

Polymerization: A reaction in which like molecules are joined together to form dimer and trimer compounds, etc., of the reactant(s). This most often occurs with olefinic compounds in oil refineries. The objective of a polymerization unit is to combine or polymerize the light olefins propylene and butylenes into molecules two or three times their original molecular weight. The feed to this process consists of light gaseous hydrocarbons (C_3 and C_4) produced by catalytic cracking, which are highly unsaturated. The polymer gasoline produced has octane numbers above 90.

PONA Analysis: Analysis for paraffins (P), olefins (O), naphthenes (N), and aromatics (A). The method used is ASTM D 1319.

Pour Point: 1. Is a measure of how easy or difficult to pump the crude oil, especially in cold weather. Specifically, the pour point is the lowest temperature at which a crude oil will flow or pour when it is chilled without disturbance at a controlled rate. The pour point of the whole crude or oil fractions boiling above 450°F (232°C) is determined by the standard test ASTM D97. Both pour and cloud points are important properties of the product streams as far as heavier products are concerned. For heavier products, they are specified in the desired range, and this is achieved by blending appropriate amounts of lighter intermediate products. **2.** The temperature at which oil starts to solidify and no longer flows freely. Pour point usually occurs 40–42°F (4.5–5.5°C) below the cloud points. A sample tube of petroleum oil is chilled in the pour point test. The pour point is defined as the temperature at which the sample will

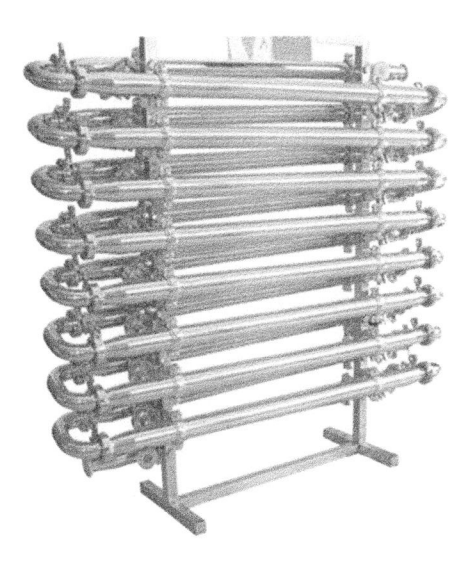

Figure 31 Plug flow reactor, 243' long, 8" S 40 CP pipe. A series of plug flow reactors.

still pour (move) when the sample tube is tipped. The pour temperature is typically about 5°F (2.8°C) lower than the cloud point.

Power Stroke: Is the downward motion of a piston that occurs after ignition as the fuel combusts and expands.

ppmv: A volume concentration of a species in bulk.

Prandtl number (Pr): A dimensionless number; Pr represents the ratio of the momentum of diffusivity to thermal diffusivity in fluid convection:

$$\Pr = \frac{C_p \mu}{k} \tag{47}$$

where C_p is the specific heat, μ is the viscosity, and k is the thermal conductivity.

Pre-ignition: Describes the event when the air/fuel mixture in the cylinder ignites before the spark plug fires. Pre-ignition is initiated by an ignition source other than the spark, such as hot spots in the combustion chamber, a spark plug that runs too hot for the application, or carbonaceous deposits in the combustion chamber heated to incandescence by previous engine combustion events. It is a technically different phenomenon from engine knocking.

The phenomenon is also referred to as "after-run" or "run-on" or sometimes diesel in when it causes the engine to carry on running after the ignition is shut off. This effect is more readily achieved on carbureted gasoline engines because the fuel supply to the carburetor is typically regulated by a passive mechanical float valve and fuel delivery can feasibly continue until fuel line pressure has been relieved, provided that the fuel can be somehow drawn past the throttle plate.

Pre-ignition and engine knock both sharply increase combustion chamber temperatures. Consequently, either effect increases the likelihood of the other effect occurring, and both can produce similar effects from the operator's perspective, such as rough engine operation or the loss of performance due to operational intervention by a computer. See Knocking.

Pre-Startup Safety Review (PSSR): Audit check performed prior to equipment operation to ensure adequate process safety management (PSM) activities have been performed. The check should verify (1) Construction and equipment is satisfactory, (2) Procedures are available and adequate, (3) A process hazard analysis (PHA) has been undertaken and recommendations resolved, and (4) The employees are trained.

Precursor: Compounds that are suitable or susceptible to specific conversion to another compound, e.g., methyl cyclopentane is a good precursor for making benzene in a catalytic reformer.

Preheat, Preheat Train: Heat exchanger or a network of heat exchangers in which the feed to a process (usually a distillation column) is heated by recovering heat from products being cooled.

Pressure, Absolute: 1. The force applied over a given area. The instrument gauges used to measure the pressure of fluids are either expressed as absolute pressure, which is measured above a vacuum, or 2. Gauge pressure plus barometric or atmospheric pressure. The absolute pressure can be zero only in a perfect vacuum. 3. The pressure is due to the weight of the atmosphere (air and water vapor) on the Earth's surface. The average atmospheric pressure at sea level has been defined as 16.69 lb_f/in^2 absolute.

Pressure, Atmospheric: 1. The pressure due to the weight of the atmosphere (air and water vapor) on the Earth's surface. The average atmospheric pressure at sea level is 14.696 lb_f/in^2. Absolute. 2. The pressure exerted by the atmosphere on a given point. It decreases as the elevation above sea level increases.

Pressuring Agent: The hydrocarbon, usually butane, used to bring gasoline blends up to acceptable vapor pressure.

Pressure drop: 1. The decrease in pressure between two points in a system caused by frictional losses of moving fluid in a pipe or duct or by some other resistance such as across a packed bed, filter, or catalyst, or due to the effects of hydrostatic head such as across the liquid on the tray of a distillation column. 2. Change in pressure with depth.

Pressure drop multiplier (φ^2): A parameter used in two-phase gas–liquid frictional pressure drop calculations where the overall pressure drop along a length of pipe is due to a combination from the flowing gas and liquid. This is expressed by:

$$\frac{dp_f}{dz} = \varphi_g^2 \left(\frac{dp_g}{dz}\right)_g = \varphi_L^2 \left(\frac{dp_L}{dz}\right)_L \tag{48}$$

where φ_g^2 and φ_L^2 are the pressure drop multipliers for the liquid and gas phases in which the parameter X^2 is defined as:

$$X = \sqrt{\left[\frac{\left(\frac{dp_L}{dz}\right)_L}{\left(\frac{dp_g}{dz}\right)_g}\right]} = \left(\frac{\varphi_g^2}{\varphi_L^2}\right)^{0.5} \tag{49}$$

Correlations have been developed to determine relationships for the multipliers for combinations of laminar and turbulent gas and liquid phases (See Figure 32).

Pressure, Hydrostatic: The pressure, volume per unit area, exerted by a body of water at rest.

Pressure Integrity Test: A pressure test of a vessel formed by the entire well or part of a well. It usually measures the ability of a pressure vessel to hold pressure without leaking at a given pressure.

Pressure, Negative: A pressure less than atmospheric.

Pressure Reducing Valve: Valve used to reduce high supply pressure to a usable level.

Pressure Relief Valve: A mechanical valve that opens at a preset pressure to relieve pressure in a vessel (See Figure 33).

Primary Absorber: The first absorber in an FCC gas plant.

Pretreatment: A group of processes that natural gas is subjected to prior to its liquefaction. Its purpose is to remove mainstream contaminants or compounds that may cause operational problems in the liquefaction unit.

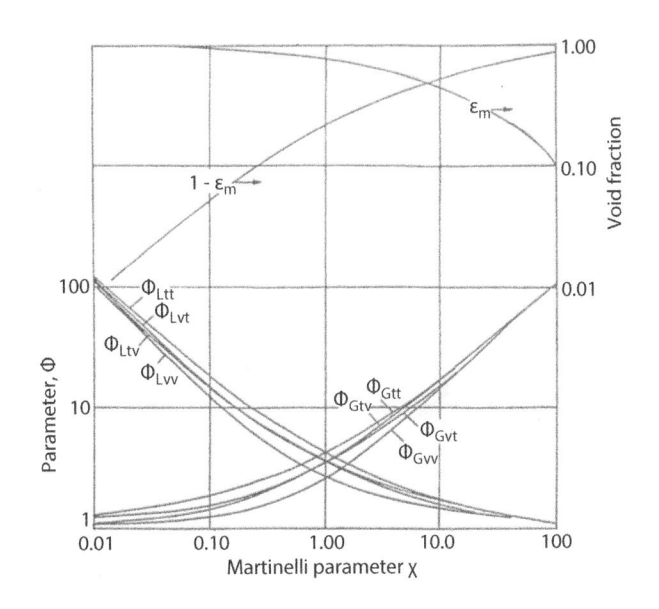

Figure 32 Lockhart–Martinelli two-phase multiplier.

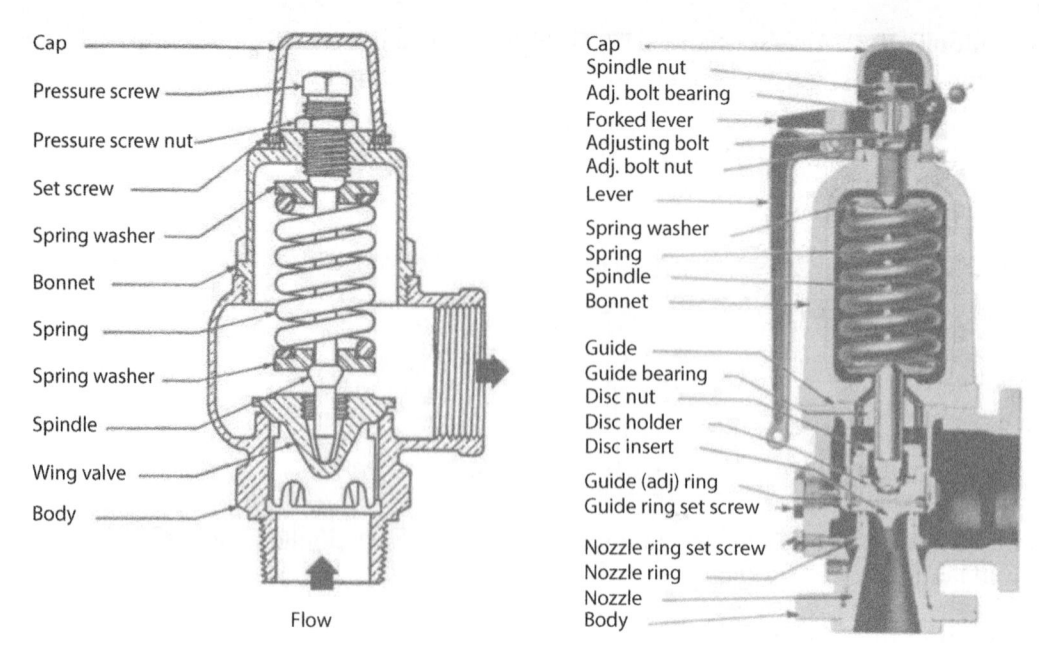

Figure 33 Relief valve and safety valve.

Pretreatment Facility: Industrial wastewater treatment plant consisting of one or more treatment devices designed to remove sufficient pollutants from wastewaters to allow an industry to comply with effluent limits established by the US EPA General and Categorical Pretreatment Regulations or locally derived prohibited discharge requirements and local effluent limits.

Preventative Maintenance: Maintenance carried out prior to unit or system failure.

Preventive Maintenance: Regularly scheduled servicing of machinery or other equipment using appropriate tools, tests, and lubricants. This type of maintenance can prolong the useful life of equipment and machinery and increase its efficiency by detecting and correcting problems before they cause a breakdown of the equipment.

Probability: The likelihood that the impact or event will occur. Impact (or consequence) is the effect on conditions or people if the hazard is realized (occurs) in practice, and probability is the likelihood that the impact will occur. Risk is a function of probability and impact (consequence). With these discrete data, it is determined by taking the number of occurrences for the particular type of event being considered and dividing that by the total number of outcomes for the event. Expressed as a deterministic value (quantitative single value or high, medium, low, etc.) or as a range of values—that is, uncertainty—that is represented by a probability distribution.

Probability Distribution (Risk): A mathematical relationship between the values and the associated probabilities for a variable across the entire range of possible values for that variable. Typically, probability distributions are displayed as a frequency or cumulative frequency plots.

Probability Distillation: The characteristic shape of laboratory distillation boiling curves tends to follow the shape of a normal distribution function, especially the TBP method. A probability distillation paper is constructed with a probability scale for the boiling point scale, and laboratory distillation curves may be plotted as straight lines on this paper. This provides a reasonable way to extrapolate partial laboratory distillation data.

Process: Any activity or operation leading to a particular event.

Process Flow Diagram (PFD): A schematic representation of a process or part of a process that converts raw materials to products through the various units' operations (Figure 34). It typically uses a symbolic representation for the major items of equipment such as storage vessels, reactors, separators, process piping to and from the equipment, as well as by-pass and recirculation lines, and the principal flow routes. Key temperatures and pressures corresponding to normal

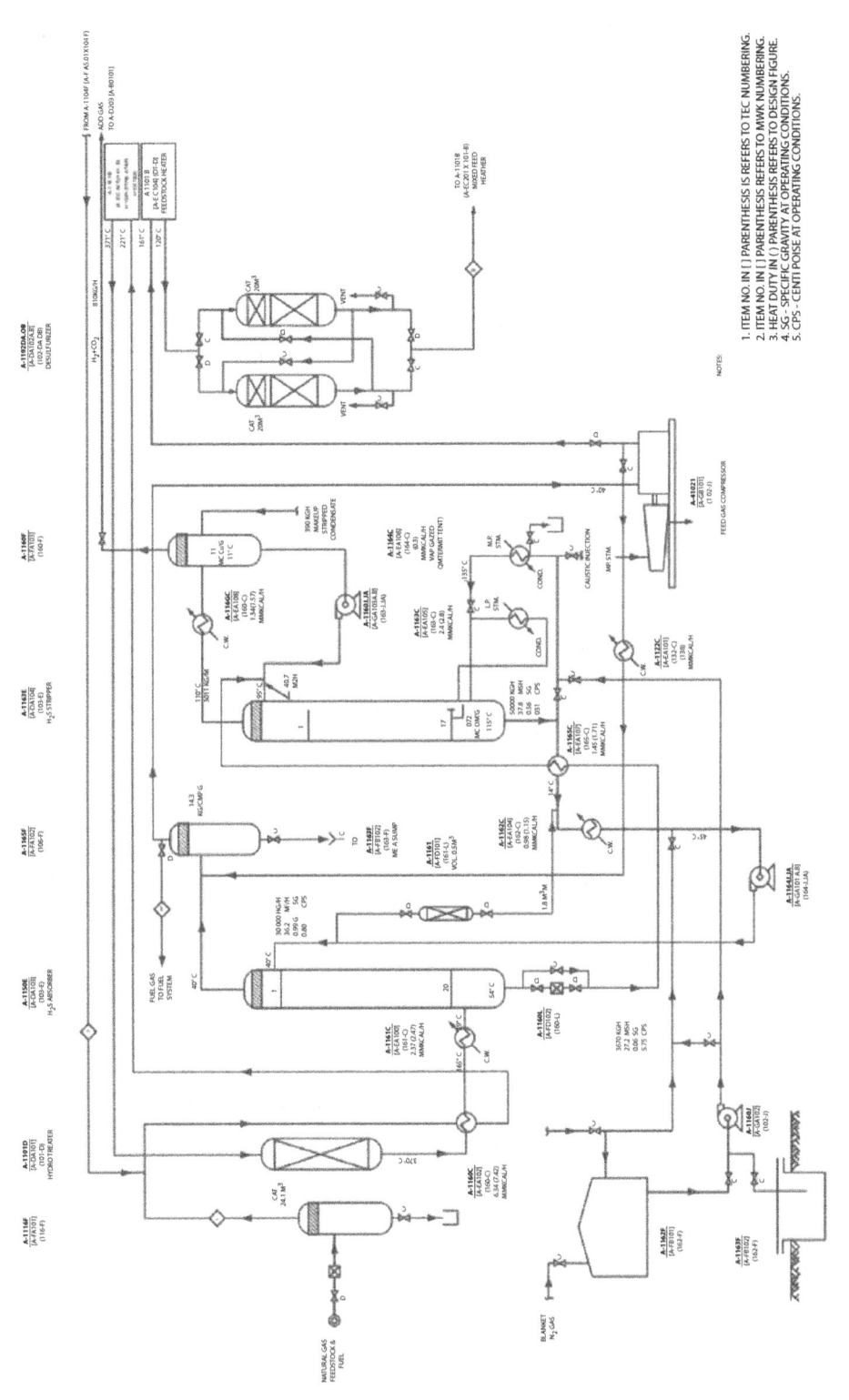

Figure 34 Process flow diagram (feed and fuel desulfurization section).

operation are included, as well as equipment ratings and minimum and maximum operational values. Material flows and compositions are included. It may also include important aspects of control and pumping, as well as any interaction with other process equipment or flows. The design duties or sizes of all the major equipment are also featured, which can collectively provide a comprehensive representation of the process. PFDs generally do not include the following:

- Pipe classes or piping line numbers
- Process control instrumentation (sensors and final elements)
- Minor bypass lines
- Isolation and shutoff valves
- Maintenance vents and drains
- Relief and safety valves
- Flanges

Programmable Logic Controller (PLC): A digital electronic controller that uses a computer-based programmable memory for implementing operating instructions through digital or analog inputs and outputs.

Process Hazard Analysis (PHA): An organized formal review to identify and evaluate hazards with industrial facilities and operations to enable their safe management. The review normally employs a qualitative technique to identify and access the importance of hazards as a result of identified consequences and risks. Conclusions and recommendations are provided for risks that are deemed at a level that is not acceptable to the organization. Quantitative methods may be also employed to embellish the understanding of the consequences and risks that have been identified.

Process Risk: Risk arising from the process conditions caused by abnormal events (including basic process control system (BPCS) malfunction).

Note: The risk in this context is that associated with the specific hazardous event in which Safety Instrument Systems (SIS) are to be used to provide the necessary risk reduction (i.e., the risk associated with functional safety).

Process Safety Management (PSM): A comprehensive set of plans, policies, procedures, practices, administrative, engineering, and operating controls designed to ensure that barriers to major incidents are in place, in use, and are effective.

Processing Gain: The volumetric amount by which the total output is greater than the input for a given period of time. This difference is due to the processing of crude oil into products that in total have lower specific gravity than the crude oil processed.

Processing Loss: The volumetric amount by which the total refinery output is less than the input for a given period of time. This difference is due to the process of crude oil into products that in total have higher specific gravity than the crude oil processed.

Production Capacity: The maximum amount of product that can be produced from processing facilities.

Products Supplied: 1. Crude Oil: Crude oil burned on leases and by pipelines as fuel. 2. Approximately represents the consumption of petroleum products because it measures the disappearance of these products from primary sources, i.e., refineries, natural gas processing plants, blending plants, pipelines, and bulk terminals. In general, the product supplied of each product in any given period is computed as follows: field production, plus refinery production, plus imports, plus unaccounted for crude oil (plus net receipts when calculated on a PAD District basis), minus stock change, minus crude oil losses, minus refinery inputs, minus exports.

Propane (C_3H_8): A hydrocarbon gas that is a principal constituent of the heating fuel, LPG. Propane is used extensively for domestic heating and as a feed to ethylene plants.

Propylene (C_3H_6): A hydrocarbon in the olefin series resulting from olefin plant operations and refinery cracking processes and used as an alkyl plant feed or chemical feedstock.

Propylene (C_3H_6) (non-fuel use): Propylene that is intended for use in non-fuel applications such as petrochemical manufacturing. Non-fuel use propylene includes chemical-grade propylene, polymer-grade propylene, and trace amounts of propane. Non-fuel use propylene also includes the propylene component of propane/propylene mixes where the propylene will be separated from the mix in a propane/propylene splitting process. Excluded is the propylene component of propane/propylene mixes where the propylene component of the mix is intended for sale into the fuel market.

Process design: The design of an industrial process that uses physical, chemical, or biochemical transformations for the production of useful products. It is used for the design of new processes, plant modifications, and revamps (Figure 35). It starts with conceptual and feasibility studies and includes detailed material and energy balances, the production of block flow diagrams (BFDs), process flow diagrams (PFDs), engineering line diagrams (ELDs), and piping and instrumentation diagrams (P & IDs). It also includes the production of reports and documents for plant construction, commissioning, start-up, operation, and shutdown. The reports and documents are used by vendors, regulatory bodies, operators, and other engineering disciplines.

Process economics: An evaluation of a process in terms of all the costs that are involved. It considers the cost of raw materials and how they are processed, as well as the costs associated with waste processing such as recycling or disposal. It also includes the optimization of a process to best utilize materials and energy. The fixed costs of a process are not dependent on the rate of production, but the variable costs are and must be met by the revenue generated by sales. Taxes are deducted, resulting in the net profit.

Process engineer: He/she uses the principles of heat and material balances, hydraulics, vapor–liquid equilibrium, and chemistry to solve plant operating problems and optimize operating variables.

Process Integration: 1. A holistic approach used in process design that considers the process as a whole with the interactions between unit operations in comparison with the optimization of unit operations separately and independently. It is known as process synthesis (See Figure 36). **2.** A technique used to minimize the energy consumption and heat recovery in a process. It is also known as process heat integration and pinch analysis (See Energy Management).

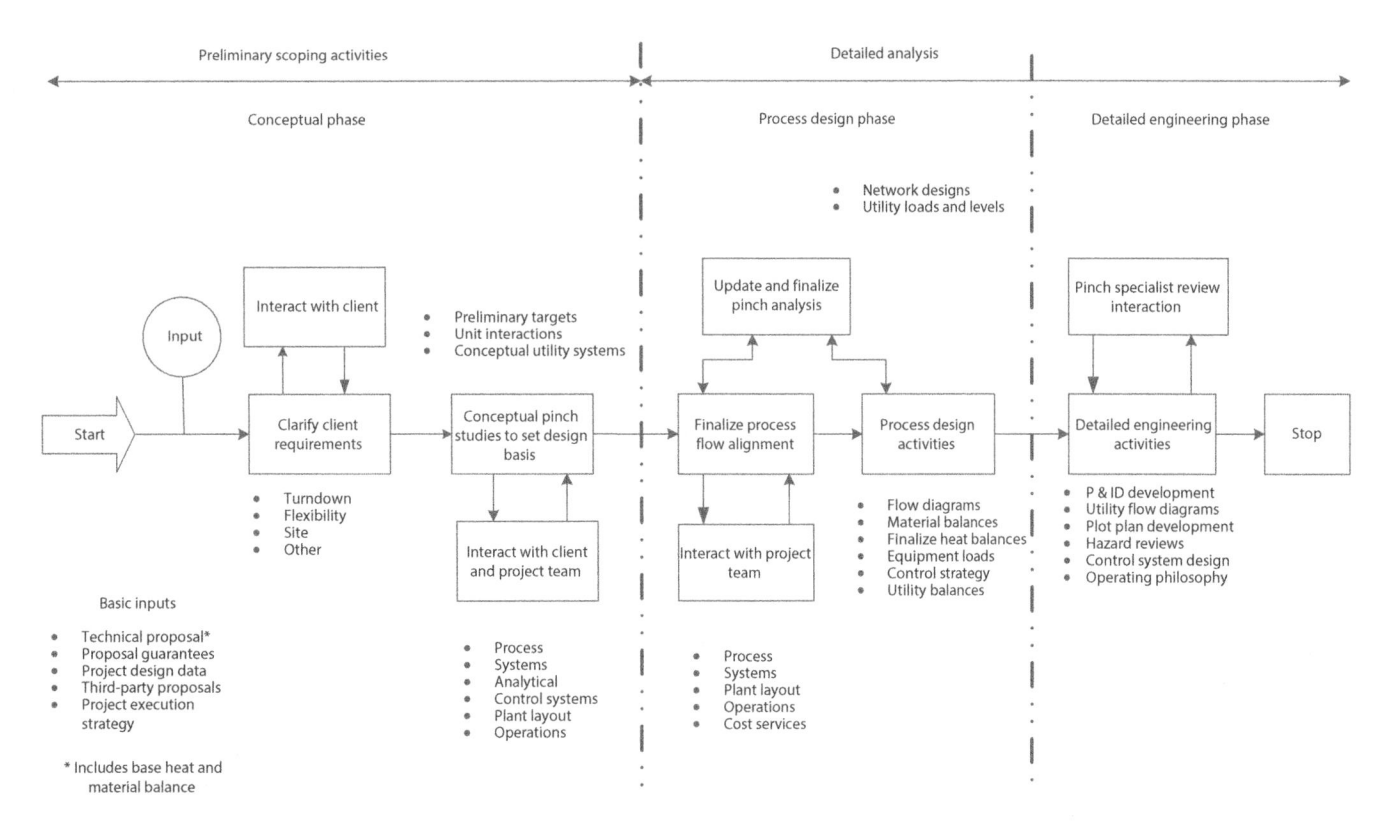

Figure 35 This new process design work process implements process integration effectively.

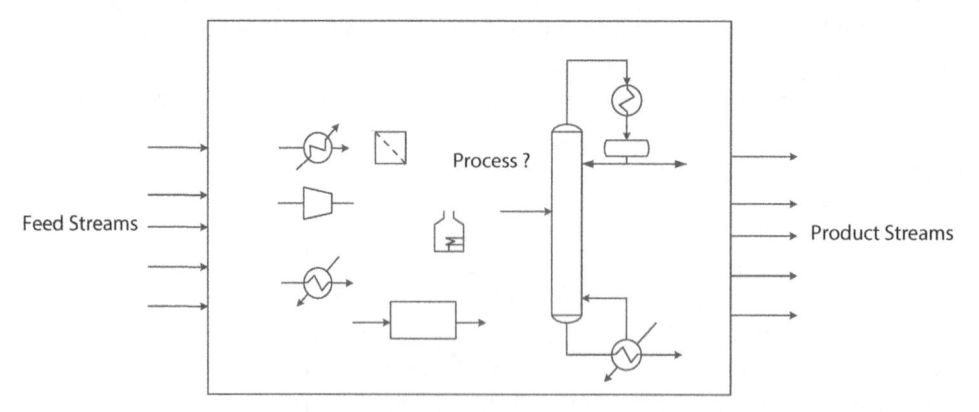

Figure 36 Process integration starts with the synthesis of a process to convert raw materials into desired products.

Process Intensification: An approach to the engineering design, manufacture, and operation of processes that aims to substantially improve process performance through energy efficiency, cost-effectiveness, reduction in waste, improvement in purification steps, reduction of equipment size, and increase in safety and operational simplicity. It involves a wide range of innovative reactor, mixing, and separation technologies that can result in dramatic improvements in process performance. It involves an integrative approach that considers overall process objectives rather than the separate performance of individual unit operations; process intensification can enable a process to achieve its maximal performance, leading to the development of cheaper, smaller, cleaner, safer, and sustainable technologies.

Any chemical engineering development that leads to a substantially smaller, cleaner, safer, sustainable, and more energy-efficient technology. Process Intensification (PI) can be employed in inherently safer plants and is grouped into four major strategies:

1. **Minimize:** Reduce quantities of hazardous substances.
2. **Substitute:** Replace a material with a less hazardous substance.
3. **Moderate:** Use less hazardous conditions, a less hazardous form of a material, or facilities that minimize the impact of a release of hazardous material or energy.
4. **Simplify:** Design facilities that eliminate unnecessary complexity and make operating errors less likely.

PI can allow one to moderate conditions to minimize the risk of explosions and to simplify processes by having fewer unit operations and less complex plants.

Process plant: A collective name for an industrial facility used to convert raw materials into useful products. It includes all the process equipment such as reactors, mixers and separating units, all the associated pipework and pumps, heat exchangers, and utilities such as steam and cooling water.

Process Safety: A comprehensive management system that focuses on the management and control of potential major hazards that arise from process operations. It aims at reducing risk to a level that is as low as is reasonably practicable by the prevention of fires, explosions, and accidental or unintended chemical releases that can cause harm to human life and to the environment. It includes the prevention of leaks, spills, equipment failure, over- and under-pressurization, over-temperatures, corrosion, and metal fatigue. It covers a range of tools and techniques required to ensure a safe operation of the plant and machinery to ensure the safety of personnel, the environment, and others through detailed design and engineering facilities, the maintenance of equipment, the use of effective alarms and control points, procedures, and training. It also includes risk assessment, layers of protection analysis, and the use of permit-to-work authorizations.

Process Simulation: The use of computers to model and predict the operational and thermodynamic behaviors of a process. Commercial software packages are used to simulate and model batch, continuous, steady-state, and transient processes. They require combined material and energy balances, the properties of the materials being processed, and sometimes combine the use of experimental data with mathematical descriptions of the process being simulated. Most software packages feature optimization capabilities involving the use of complex cost models and detailed process equipment size models. Some commercial software products are shown in the table below:

Software	Developer	Applications	Website
Aspen Plus/Aspen Hysys	Aspen Technology	Process simulation and optimization	www.aspentech.com
CHEMCAD	Chemstations	Software suite for process simulation	www.chemstations.com
Design II for Windows	WinSim Inc.	Process simulation	www.winsim.com
gPOMS	PSE Ltd.	Advanced process simulation and modeling	www.psenterprise.com
PRO II	SimSci	Process simulation	www.software.schneider-electric.com/simsci
ProSim Plus	ProSim	Process simulation and optimization	www.prosim.net
UniSim	Honeywell	Process simulation and optimization	www.honeywellprocess.com

Process synthesis: The conceptual design of a process that identifies the best process flowsheet structure, such as the conversion of raw materials into a product. This requires the consideration of many alternative designs. The complex structure of most processes is such that the flowsheet is split into smaller parts and each is reviewed in turn. Then, choices and decisions are made. Many techniques are used in arriving at the best flowsheet such as those based on total cost, which needs to be minimized. Use is made of graphical methods, heuristics, and various other forms of minimization such as the use of process integration.

Process upset: A sudden, gradual, or unintended change in the operational behavior of a process. It may be due to process equipment failure or malfunction, operator intervention, a surge or fall in pressure, flow, level, concentration, etc.

Process variable: A dynamic feature of a process or system that is required to be controlled to ensure that it operates according to design requirements and does not deviate to be unsafe or result in undesirable consequences. The commonly measured process variables include temperature, pressure, flow, level, and concentration.

Pseudo-component: For engineering calculation purposes, a component that represents a specified portion of the TBP distillation curve for a petroleum mixture. The pseudo-component is assigned a normal boiling point and gravity corresponding to the average for the boiling point range. The molecular weight and other properties are derived from the boiling point and gravity using literature correlations for hydrocarbons.

Pump: A mechanical device used to transport a fluid from one place or level to another by imparting energy to the fluid. The three bond groupings are centrifugal, reciprocating, and rotary-type pumps. The most commonly used is the centrifugal type, which has a rotating impeller used to increase the velocity of the fluid and where a part of the energy is converted to pressure energy. Rotary and reciprocating pumps are positive displacement pumps in which portions of fluid are moved in the pump between the teeth of gears and by the action of a piston in a cylinder. There are many variations of these types, and each has a particular application and suitability for fluid in terms of its properties, required flow rate, and delivery pressure (See Figure 37).

Pumparound: A liquid side-draw from a distillation/fractionating column that is pumped, cooled, and returned to a higher location in the column. Pumparounds recover usable heat that would be lost at the condenser. They also lower the vapor flow in a column and reduce the required column diameter for vapor-loaded columns such as crude and vacuum columns.

Pumpdown: A liquid side draw that is pumped down to a tray below the draw tray, usually the next tray lower. Pumpdowns are sometimes cooled prior to returning to the column.

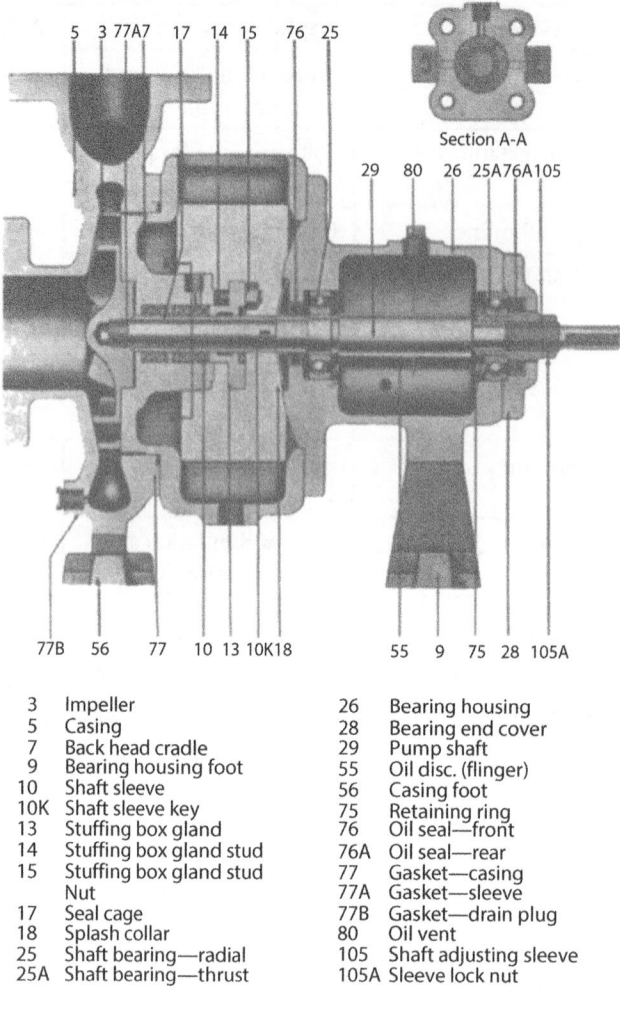

3	Impeller	26	Bearing housing
5	Casing	28	Bearing end cover
7	Back head cradle	29	Pump shaft
9	Bearing housing foot	55	Oil disc. (flinger)
10	Shaft sleeve	56	Casing foot
10K	Shaft sleeve key	75	Retaining ring
13	Stuffing box gland	76	Oil seal—front
14	Stuffing box gland stud	76A	Oil seal—rear
15	Stuffing box gland stud Nut	77	Gasket—casing
		77A	Gasket—sleeve
17	Seal cage	77B	Gasket—drain plug
18	Splash collar	80	Oil vent
25	Shaft bearing—radial	105	Shaft adjusting sleeve
25A	Shaft bearing—thrust	105A	Sleeve lock nut

Figure 37 General service centrifugal pump.

Pump priming: Used for the start-up and successful operation of centrifugal pumps in which the casing housing the "impeller" is first filled or primed with liquid before operation begins. Since the density of a liquid is many times greater than that of a gas, vapor, or air, the suction pressure is otherwise insufficient to draw in more liquid. Depending on the type of pump, priming can be achieved either manually or by drawing liquid using a vacuum pump. Valves can be used to prevent drainage and ensure that the pump does not require priming once the pump stops.

Purge: A stream that is removed from a recycling process to prevent the buildup of one or more components in the process streams.

Pyrolysis: **1**. Heating a feedstock to high temperature to promote cracking as in an ethylene plant. **2**. Destructive distillation that involves the decomposition of coal, woody materials, petroleum, and so on, by heating in the absence of air.

Pyrolysis Gasoline: The gasoline created in an ethylene plant cracking gas oil or naphtha feedstocks. Sometimes called pygas, it has a high content of aromatics and olefins and some diolefins.

Pyrophoric Iron Sulfide: A substance typically formed inside tanks and processing units by the corrosive interaction of sulfur compounds in the hydrocarbons and the iron and steel in the equipment. On exposure to air (oxygen), it ignites spontaneously.

Quality: The weight fraction of vapor in a vapor–liquid mixture.

Quench: Hitting a very hot stream coming out of a reactor with a cooler stream to stop immediately the reaction runaway.

Quench Crack: A crack in the steel resulting from stresses produced during the transformation from austenite to martensite.

Quench Hardening: Heat treating requiring austenitization followed by cooling, under conditions that austenite turns into martensite.

Quenching Oil: An oil introduced into high-temperature process streams during refining to cool them.

Quench Stream: A cooled stream that is used to cool another stream by direct contact. For example, hydrogen quench streams are used to quench the hot effluents from hydrocracker reactors.

Quench Zone: A section of a distillation column where a hot stream, usually vapor, is cooled by direct contact with a stream that has been cooled, usually a liquid.

Radiant Heat Transfer: Heat transfer without convection or conduction. Sunshine is radiant heat.

Radiation: Transmission of energy by means of electromagnetic waves emitted due to temperature.

Radical: A group of atoms that separate themselves from a compound momentarily and are highly reactive. For example, two methyl radicals *CH_3 can come from cracking an ethane compound, but they will rapidly attach themselves to some other atom or compound.

Raffinate: 1. The leftover from a solvent extraction process. 2. In solvent refining, that portion of the oil that remains undissolved and is not removed by the selective solvent.

Rating Calculations: Calculations in which a unit operation such as a column, heat exchanger, pump, etc., is checked for capacity restrictions.

Ratio of Specific Heats: 1. Thermodynamic comparison ($k = C_p/C_v$) of the ratio of specific heat (k) at constant pressure (C_p) to specific heat at constant volume (C_v). The ratio range of most gases is 1.2–1.4. 2. For gases, it is the ratio of the specific heat at constant pressure to the specific heat at constant volume. The ratio is important in thermodynamic equations as compressor horsepower calculations and is given the symbol k, where $k = C_p/C_v$. The ratio lies between 1.2 and 1.4 for most gases.

Reactor: The vessel in which chemical reactions take place (See Figure 38).

Reactive Distillation: A distillation column in which there is a section designed for chemical reaction, usually containing a catalyst bed. Some MTBE and TAME processes use a reactive distillation column in place of a second reactor prior to the product separation column.

Reactor Effluent: The outlet stream from a reactor.

Reboiler: 1. A heat exchanger used toward the bottom of a fractionator to reheat or oven-vaporize a liquid and introduce it several trays higher to help purify the incoming stream or get more heat into the column. 2. An auxiliary unit of a fractionating tower designed to supply additional heat to the lower portion of the tower (Figure 39).

Recovery: Usually refers to the fraction expressed as a percent of a component or group of components in the feed to a distillation column that is recovered in a given product stream.

Rectification Zone: The portion of a distillation column in which heavy components are washed down the column by contact with a liquid reflux stream. In conventional distillation columns, this is the portion of the column from the tray above the feed tray to the top tray.

Figure 38 Batch reactor types with (a) constant flux (Coflux) jacket, (b) half coil jacket, and (c) a single external cooling jacket.

Recycle: A process stream that is returned to an upstream operation.

Recycled Feeds: Streams that have been processed and are fed back to the reactors for additional processing.

Reduced Crude: A residual product remaining after the removal by distillation of an appreciable quantity of the more volatile components of crude oil.

Reduced Pressure: The ratio of the absolute pressure to the critical pressure.

Reduced Temperature: The ratio of the absolute temperature to the critical temperature.

Reducing Agent: Any substance, such as base metal (iron) or the sulfide ion, that will readily donate (give up) electrons. The opposite is an oxidizing agent.

Reduction: The addition of hydrogen, the removal of oxygen, or the addition of electrons to an element or compound. Under anaerobic conditions (no dissolved oxygen present), sulfur compounds are reduced to odor-producing hydrogen sulfide (HS) and other compounds.

Redwood Viscometer: Standard British viscometer. The number of seconds required for 50 ml of oil to flow out of a standard Redwood viscometer at a definite temperature is the Redwood viscosity.

Refinery Grade Butane (C_4H_{10}): A refinery-produced stream that is composed predominantly of normal butane and/or isobutane and may also contain propane and/or natural gasoline. These streams may also contain significant levels of olefin and/or fluoride contamination.

Refinery Input, Crude Oil: Total crude oil (domestic plus foreign) input to crude oil distillation units and other refinery processing units (cokers, etc.)

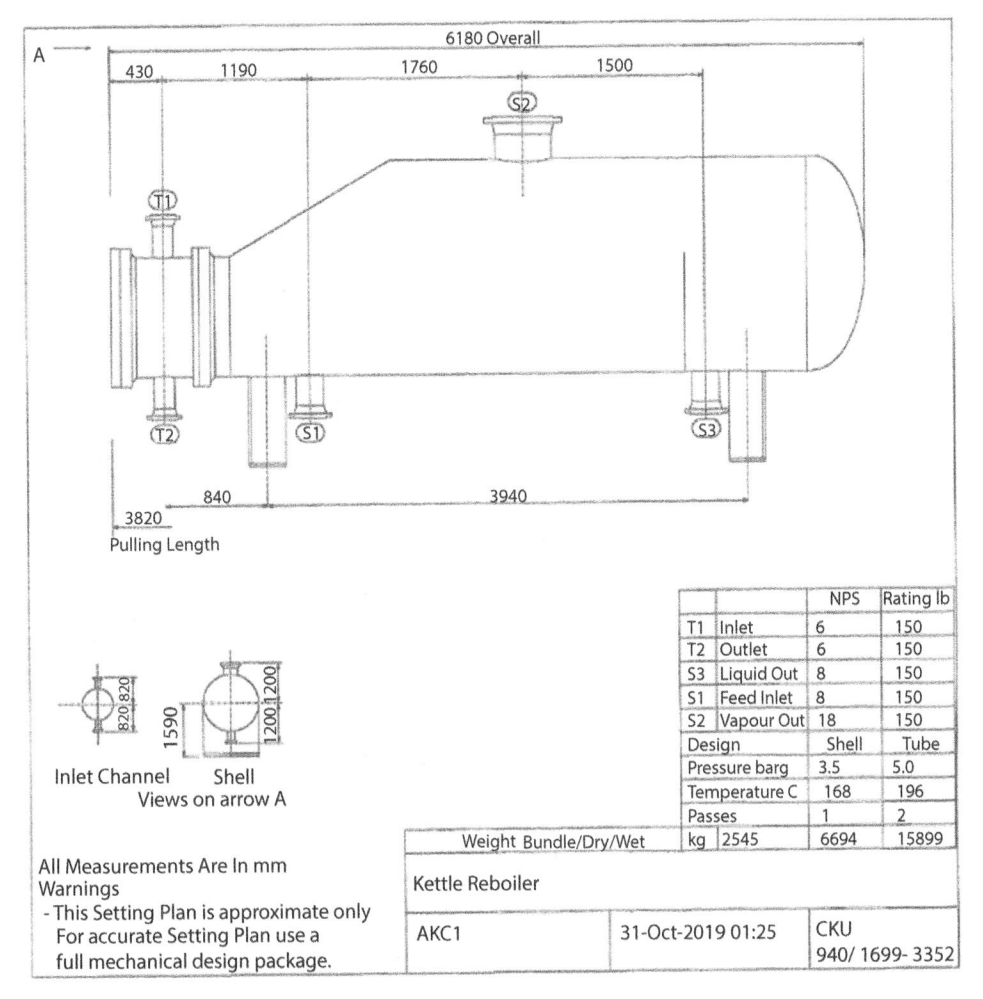

Figure 39 Kettle Reboiler diagram.

Refined Products: The various hydrocarbons obtained as a result of refining process separation from crude oil. Typical refined products are LPG, naphtha, gasoline, kerosene, jet fuel, home heating oil, diesel fuel, residual fuel oil, lubricants, and petroleum coke.

Refiner: A company involved in upgrading hydrocarbons to saleable products.

Refinery: 1. An installation that manufactures finished petroleum products from crude oil, unfinished oils, natural gas liquids, other hydrocarbons, and oxygenates. 2. A plant used to separate the various components present in crude oil and convert them into usable fuel products or feedstock for other processes. 3. A large plant composed of many different processing units that are used to convert crude oil into finished or refined products. These processes include heating, fractionating, reforming, cracking, and hydrotreating.

Refinery Gas: A non-condensable gas collected in petroleum refineries.

Refinery Input (Crude Oil): Total crude oil (domestic plus foreign) input to crude oil distillation units and other refinery processing units (cokers).

Refinery Input (total): The raw materials and intermediate materials processed at refineries to produce finished petroleum products. They include crude oil, products of natural gas processing plants, unfinished oils, other hydrocarbons and oxygenates, motor gasoline and aviation gasoline blending components, and finished petroleum products.

Refinery Margin: The difference in value between the products produced by a refinery and the value of the crude oil used to produce them. Refining margins will thus vary from refinery to refinery and depend on the price and characteristics of the crude used.

Refinery Production: Petroleum products produced at a refinery or blending plant. Published production of these products equals refinery production minus refinery input. Negative production occurs when the amount of a product produced during the month is less than the amount of that same product that is reprocessed (input) or reclassified to become another product during the same month. The refinery production of unfinished oils and motor and aviation gasoline blending components appear on a net basis under refinery input.

Refinery Yield: Represents the percentage of finished product produced from the input of crude oil and net input of unfinished oils (expressed as a percentage). It is calculated by dividing the sum of crude oil and net unfinished input into the individual net production of finished products. Before calculating the yield of finished motor gasoline, the input of natural gas liquids, other hydrocarbons and oxygenates, and the net input of motor gasoline blending components must be subtracted from the net production of finished motor gasoline. Before calculating the yield of finished aviation gasoline, the input of aviation gasoline blending components must be subtracted from the net production of the finished aviation gasoline.

Redwood Viscometer: Standard British viscometer. The number of seconds required for 50 ml of oil to flow out of a standard Redwood viscometer at a definite temperature is the Redwood viscosity.

Reflux: 1. Condensed liquid that is returned to the top tray of a distillation column. Reflux helps rectify the mixture being distilled by washing heavy components down the column. **2.** The portion of the distillate returned to the fractionating column to assist in attaining better separation into desired fractions.

Reflux drum: A drum that receives the outlet from the overhead condenser from a distillation column. The liquid and vapor portions are separated in the reflux drum.

Reformate: An upgraded naphtha resulting from catalytic or thermal reforming.

Reforming: 1. The mild thermal cracking of naphthas to obtain more volatile products such as olefins, of higher-octane values, or the catalytic conversion of naphthas' components to produce higher-octane aromatic compounds. **2.** A refining process used to change the molecular structure of a naphtha feed derived from crude oil by distillation. **3.** The gasoline produced in a catalytic reforming operation.

Reformulated Fuels: Gasoline, diesel, or other fuels that have been modified to reflect environmental concerns, performance standards, government regulations, customer preferences, or new technologies.

Reformed Gasoline: Gasoline made by a reformate process.

Reformulated Gasoline (RFG): 1. A gasoline whose composition has been changed (from that of gasoline sold in 1990) to (a) include oxygenates, (b) reduce the content of olefins and aromatics and volatile components, and (c) reduce the content of heavy hydrocarbons to meet performance specifications for ozone-forming tendency and for the release of toxic substances (benzene, formaldehyde, acetaldehyde, 1,3-butadiene, and polycyclic organic matter) into the air from both evaporation and tailpipe emissions. **2.** Is a cleaner-burning gasoline that reduces smog and other forms of air pollution. Federal law mandates the sale of reformulated gasoline in metropolitan areas with the worst ozone smog. **3.** Finished motor gasoline formulated for use in motor vehicles, the composition, and properties of which meet the requirements of the reformulated gasoline regulations promulgated by the US Environmental Protection Agency under Section 211 (k) of the Clean Air Act. NB: This category includes the oxygenated fuels program reformulated gasoline (OPRG) but excludes reformulated gasoline blendstock for oxygenate blending (RBOB). **4.** Gasoline that meets the requirements imposed by the Clean Air Act Amendment, passed by the United States Congress on November 15, 1990. Restrictions were placed on volatile organic compounds, nitrous oxides (NO_x) from combustion, and toxins primarily related to benzene (C_6H_6) and its derivatives.

Reformulated Gasoline Blendstock for Oxygenate Blending: A motor gasoline blending component that, when blended with a specified type and percentage of oxygenate, meets the definition of reformulated gasoline.

Refrigerant: 1. In a refrigerating system, the medium of heat transfer that picks up the heat by evaporating at a low temperature and pressure and gives up the heat on condensing at a higher temperature and pressure. 2. It is the fluid that performs an inverse thermodynamic cycle, generating the low temperature required for natural gas cooling and liquefaction.

Refrigerant Compressor: A component of a refrigerating system that increases the pressure of a compressible refrigerant fluid and simultaneously reduces its volume while moving the fluid through the device.

Refrigerating System: A system that, in operation between a heat source (evaporator) and a heat sink (condenser), at two different temperatures, is able to absorb heat from the heat source at the lower temperature and reject heat to the heat sink at the higher temperature.

Refrigeration: The process used to remove the natural gas liquids by cooling or refrigerating the natural gas until the liquids are condensed out. The plants use Freon or propane to cool the gas.

Refrigeration (or cooling cycle): 1. The process used to remove the natural gas liquids by cooling or refrigerating the natural gas until the liquids are condensed out. The plants use Freon or propane to cool the gas. 2. Inverse thermodynamic cycle whose purpose is to transfer heat from a medium at a low temperature to a medium at a higher temperature.

Regasification: The process by which LNG is heated, converting it into its gaseous state.

Regasification Plant: A plant that accepts deliveries of LNG and vaporizes it back to gaseous form by applying heat so that the gas can be delivered into a pipeline system.

Regenerator: The vessel in a catalytic process where a spent catalyst is cleaned up before being recycled back to the process. An example is the catalytic cracker regenerator where coke is burned off the catalyst.

Regeneration: 1. The process of burning off coke deposits on catalyst with an oxygen containing gas under carefully controlled conditions. 2. In a catalytic process, the reactivation of the catalyst, sometimes done by burning off the coke deposits under carefully controlled conditions of temperature and oxygen content of the regeneration gas stream (See Figure 40).

Reid Vapor Pressure (RVP): An ASTM test method to determine the vapor pressure of a light petroleum stream. The Reid vapor pressure is very nearly equal to the true vapor pressure for gasoline streams. There is also a Reid vapor pressure test for crude oil (See Figure 41 and Table 3).

Relative volatility (α): The ratio of the vapor pressure of one liquid component to another in a heterogeneous mixture and a measure of their separability. For a binary mixture, the relative volatility can be expressed in terms of the mole fraction of the more volatile component in the liquid and vapor phases, x and y, as:

$$\alpha = \frac{y(1-x)}{x(1-y)} \tag{50}$$

The greater the value of the relative volatility, the greater the degree of separation. If y = x, then no separation is possible.

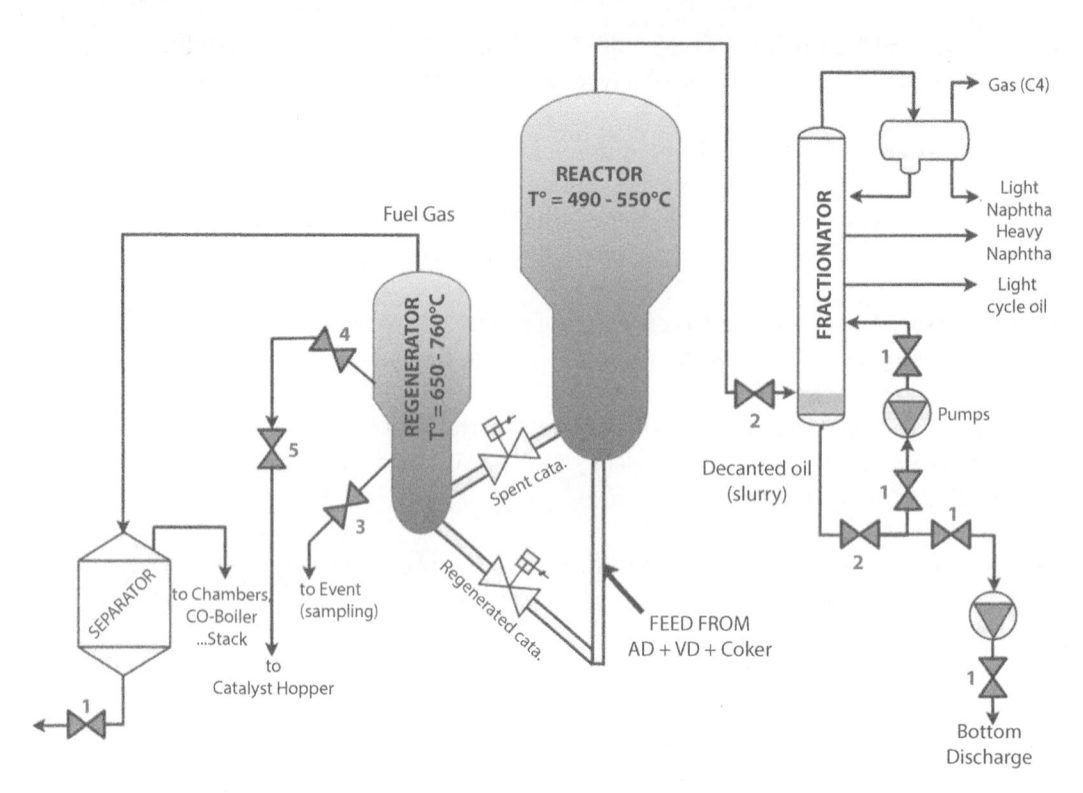

Figure 40 A diagram of an FCC showing the regenerator.

Reliability: The probability that a component or system will perform its defined logic functions under the stated conditions for a defined period of time.

Research Octane Number (RON): One of the two standard tests of gasoline knock; this one simulates less severe operating conditions like cruising. It is determined in a special laboratory test engine under mild "engine-severity" conditions, giving a measure of the low-speed knock properties of gasoline. In contrast with the Motor Octane Number.

Residence time: 1. The amount of time a hydrocarbon spends in a vessel where a reaction occurs. 2. The period of time in which a process stream will be contained within a certain volume or piece of equipment, seconds.

Residual Fuel: Heavy fuel oil made from long, short, or cracked residue plus whatever cutter stock is necessary to meet market specifications.

Residue: The bottoms from a crude oil distilling unit, vacuum flasher, thermal cracker, or visbreaker. See Long Residue and Short Residue.

Residuum: Residue from crude oil after distilling off all but the heaviest components with a boiling range greater than 1000°F (538°C).

Reynolds Number (Re): A dimensionless number, Re, expressing the ratio of inertial to viscous forces in a flowing fluid and can be used to determine the flow regime. For a fluid in a pipe of circular cross-section:

$$\text{Re} = \frac{\rho v d}{\mu} = \frac{4G}{\pi d \mu} = \frac{4\rho Q}{\pi d \mu} \tag{42}$$

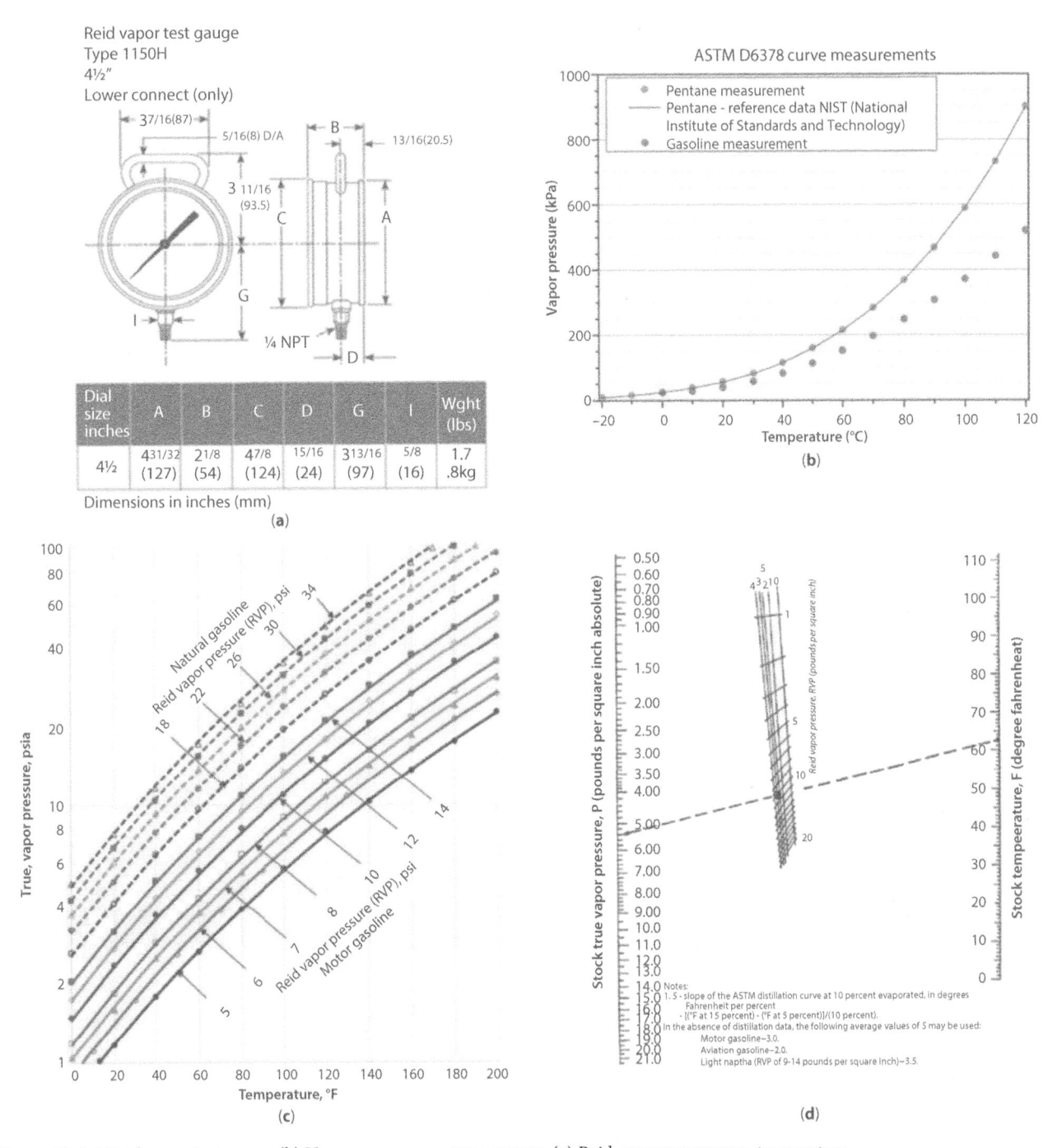

Figure 41 (a) Reid vapor test gauge. (b) Vapor pressure vs. temperature. (c) Reid vapor pressure vs. temperature.

where ρ is the density of the fluid, v is the velocity of the fluid, G is the mass flow rate, Q is the volumetric flow rate, d is the inside diameter of the pipe, and μ is the fluid viscosity. The critical value of Re for the break-up of laminar flow in the pipes of circular cross-section is about 2000.

Where the value for critical pipes falls below 2,000 the flow is laminar flow or streamline. For Reynolds number above 4,000, the flow is turbulent.

Rich oil: The absorption oil leaving the bottom tray of an absorption column. The rich oil contains the absorbed light components.

Ring Compounds: Hydrocarbon molecules in which the carbon atoms form at least one closed ring such as naphthenes or aromatics. Also called cyclic.

Table 3 RVP Blending Values.

RVP blending values		Vol% (aromatics)					
	rvp (pure HC)	0	10	20	30	40	50
Ethane	730.0	474.0	474.0	474.0	474.0	474.0	474.0
Propene	226.0	216.0	216.0	216.0	216.0	216.0	216.0
Propane	190.0	173.0	173.0	173.0	173.0	173.0	173.0
Isobutane	72.2	62.0	73.9	85.4	96.6	107.6	118.8
Isobutene	63.4	76.5	78.9	81.3	83.7	86.2	88.9
Butene-1	63.0	76.1	78.4	80.8	82.7	85.1	87.4
n-Butene	51.6	52.9	55.6	58.3	60.9	63.5	66.2
trans-2-Butene	49.8	62.1	64.0	66.0	68.0	70.0	72.0
cis-2-Butene	45.5	58.6	60.5	62.3	64.2	66.1	69.0
Isopentane	20.4	21.9	22.2	22.5	22.9	23.3	23.7
C_5 olefins*	16.5	17.9	18.1	18.4	18.6	18.8	19.0
n-Pentane	15.6	16.9	17.2	17.4	17.8	18.0	18.2

*C_5 olefins in FCC proportion.

Ring Structure: A compound in which some of the carbon atoms are linked with other carbon atoms to form a continuum. Carbon atoms attached to the ring carbon atoms are said to be "side chains."

Riser: 1. A pipe through which a fluid travels upwards (See Figure 40; a pipe to the reactor). 2. A steel or flexible pipe that transfer fluids well from the seabed to the surface.

Risk: 1. Is defined as a measure of economic loss, human injury, or environmental damage in terms of both the incident likelihood and the magnitude of the loss, injury, or damage. 2. The probability of an event happening times the impact of its occurrence on operations. (Impact is the effect on conditions or people if the hazard is realized (occurs) in practice and potentials are the likelihood that the impact will occur.

Risk Analysis: A decision-making tool that allows an examination of the level and significance of workplace risk for humans, equipment, weather, operations, or other conditions. Determines the probability of the risk occurring, the impact the risk will have, and how to mitigate the risk. *See Hazard Analysis.*

Risk Assessment: The process of identifying and evaluating the technical and non-technical risks associated with a project. It includes the amount or degree of potential danger perceived (by an assessor) when determining a course of action to accomplish a given task. Risk assessment may be qualitative or quantitative.

Risk Matrix: Is the common approach to risk assessment and hazard analysis. Its underlying idea is that acceptability of risk is a product of how likely a thing is to happen and how bad it would be if it happened. This is shown in the following tables.

Category	Definition	Range (failures per year)
Certain	Many times in system lifetime	$>10^{-3}$

(Continued)

(*Continued*)

Category	Definition	Range (failures per year)
Probable	Several times in system lifetime	10^{-3} to 10^{-4}
Occasional	Once in system lifetime	10^{-4} to 10^{-5}
Remote	Unlikely in system lifetime	10^{-5} to 10^{-6}
Improbable	Very unlikely to occur	10^{-6} to 10^{-7}
Inconceivable	Cannot believe that it could occur	$<10^{-7}$

Risk matrix categorization of severity of consequences

Category	Definition
Catastrophic	Multiple loss of life
Critical	Loss of a single life
Marginal	Major injuries to one or more persons
Negligible	Minor injuries to one or more persons

Risk matrix

Consequence

Likelihood	Catastrophic	Critical	Marginal	Negligible
Certain	Class I	Class I	Class I	Class II
Possible	Class I	Class I	Class II	Class III
Occasional	Class I	Class II	Class III	Class IV
Remote	Class II	Class III	Class III	Class IV
Improbable	Class III	Class III	Class IV	Class IV
Inconceivable	Class IV	Class IV	Class IV	Class IV

Key:
  Class I: Unacceptable
  Class II: Undesirable
  Class III: Tolerable
  Class IV: Acceptable

Road Oil: Any heavy petroleum oil, including residual asphaltic oil used as a dust palliative and surface treatment on roads and highways. It is generally produced in six grades from 0, the most liquid, to 5, the most viscous.

Rule of Thumb: Axioms based on practical experience and/or methods to approximate calculated results using simple formulae.

Runback: The liquid returning to the flash zone of a distillation column.

Safety: A general term denoting an acceptable level of risk of relative freedom from and low probability of harm.

Safeguard: A precautionary measure of stipulation. Usually equipment and/or procedures designed to interfere with incident propagation and/or prevent or reduce incident consequences.

Safety Integrity Level (SIL): 1. Is defined as a relative level of risk reduction provided by a safety function or to specify a target level of risk reduction. SIL is a measure of performance required for a safety instrumented function (SIF). **2.** The degree of redundancy and independence from the effects of inherent and operational failures and external conditions that may affect system performance.

The requirements for a given SIL are not consistent among all of the functional safety standards. In the European functional safety standards based on the IEC 61508 standard, four SILs are defined, with SIL 4 the most dependable and SIL 1 the least. An SIL is determined based on a number of quantitative factors in combination with qualitative factors such as development process and safety life cycle management.

The assignment of SIL is an exercise in risk analysis where the risk associated with a specific hazard, which is intended to be protected against by an SIF, is calculated without the beneficial risk reduction effect of the SIF. That "unmitigated" risk is then compared against a tolerable risk target. The difference between the "unmitigated" risk and the tolerable risk, if the "unmitigated" risk is higher than tolerable, must be addressed through the risk reduction of the SIF. This amount of required risk reduction is correlated with the SIL target. In essence, each order of magnitude of risk reduction that is required correlates with an increase in one of the required SIL numbers.

There are several methods used to assign an SIL. These are normally used in combination and may include:

- *Risk matrices*
- *Risk graphs*
- *Layers of Protection Analysis (LOPA)*

Of the methods presented above, LOPA is by far the most commonly used by large industrial facilities.

The assignment may be tested using both pragmatic and controllability approaches, applying guidance on the SIL assignment published by the UK HSE. SIL assignment processes that use the HSE guidance to ratify assignments developed from Risk Matrices have been certified to meet IEC EN 61508.

Safety Instrumented Function (SIF): A safety function with a specific safety integrity level that is necessary to achieve functional safety and that can either be a safety instrumented protection function or a safety instrumented control function.

Salt Content: Crude oil usually contains salts in solution in water that is emulsified with the crude. The salt content is expressed as the solution of sodium chloride (NaCl) equivalent in pounds per thousand barrels (PTB) of crude oil. Typical values range from 1 to 20 PTB. Although there is no simple conversion from PTB to parts per million by weight (ppm), 1 PTB is roughly equivalent to 3 ppm.

Saturated Compounds: Hydrocarbons in which there are no double bonds between carbon atoms. Saturated compounds contain the maximum number of hydrogen atoms that are possible.

Screwed Fittings: These are used to assemble screwed connections and field instruments on pipes. They are:

- Pipe thread fittings
- Instrument or tubing fittings
- Metric fittings

None of these will screw together.

Scrub: The removal of components (gas, liquids, or solids) from methane achieved by surface equipment (scrubbers).

Scrubber: 1. A reactor that removes various components from produced gas. **2.** Equipment that causes the separation of liquid and gaseous phases in a fluid system. The separation is usually based on the density differences of the two phases and can take place using gravity force, induced centrifugal force, and so on. **3.** A system to reduce noxious substances from a flowing stream of air, usually filled with plates or packing, through which the scrubbing fluid flows countercurrent or cross-current to the path of the contaminated air.

Scrubbing: The purification of a gas or liquid by washing it in a tower.

Secondary Absorber: The second absorber in an FCC gas plant. It is usually the last unit operation in the gas recovery plant and is also known as the sponge absorber.

Selectivity: The difference between the research octane number and the motor octane number of a given gasoline. Alkylate is an excellent low-sensitivity and reformates a high-sensitivity gasoline component. It is an indication of the sensitivity of the fuel to driving conditions (city vs. highway).

Selective Treating: The preferential removal of one acid gas component, leaving at least some of the other acid gas components in the treated stream.

Sensitivity: The difference in the research octane (F-1) and the motor octane (F-2) for a gasoline stream. Since the research octane is always larger, sensitivity is always a positive number.

Separation zone: A section of a distillation column in which a separation between two products occurs. The components that are found in both products are said to be distributed components.

Separator: Usually refers to a drum, in which the residence time is provided for a mixture of the liquid and vapor to separate into liquid and vapor streams. Also called a flash drum. The liquid and vapor leaving the separator are in phase equilibrium.

Severity: The degree of intensity of the operating conditions of a process unit. Severity may be indicated by the clear research octane number of the product (reformer), the percentage disappearance of the feed (catalytic cracking), or operating conditions alone (usually the temperature; the higher the temperature, the greater the severity).

Shale: **1.** A common sedimentary rock with porosity but little matrix permeability. Shales are one of the petroleum source rocks. Shales usually consist of particles finer than sand grade (less than 0.0625 mm) and include both silt and clay grade material. **2.** A very fine-grained sedimentary rock formed by the consolidation and compression of clay, silt, or mud. It has a finely laminated or layered structure. Shale breaks easily into thin parallel layers; a thinly laminated siltstone, mudstone, or claystone. Shale is soft but sufficiently hard-packed (indurated), so as not to disintegrate upon becoming wet. Some shales absorb water and swell considerably, causing problems in well drilling. Most shales are compacted and consequently do not contain commercial quantities of oil and gas. **3.** Rock formed from clay. **4.** Gas reserves found in unusually non-porous rock, requiring special drilling and completion techniques.

Shale Gas: Methane (CH_4) gas stored in shale. May be in the pore space, adsorbed to the mineral or rock surfaces, or as free gas in the natural fractures.

Shale Oil: **1.** Can be either an immature oil phase, often called kerogen or actual oil in the cracks or pores of shale. **2.** The liquid obtained from the destructive distillation of oil shale. Further processing is required to convert it into products similar to petroleum oils.

Shear force: An applied force to a material that acts in a direction that is parallel to a plane rather than perpendicular. A material such as a solid or fluid is deformed by the application of a shear force over a surface known as shear stress. The shear strain is the extent of the deformation defined as the ratio of the deformed distance with length. The shear modulus is the ratio of the shear stress to the shear strain.

Shear rate (γ): The deformation of fluid under the influence of an applied shear force presented as the change in velocity of the fluid perpendicular to flow:

$$\gamma = \frac{dv}{dz} \tag{51}$$

where dv/dz is referred to as the velocity gradient. The S.I. unit is s^{-1}.

Shear stress (τ): The shear force applied to a fluid that is applied over a surface. When the shear stress is proportional to the shear rate, the fluid exhibits Newtonian behavior and the viscosity is constant. The S.I. unit is Nm^{-2}.

$$\tau = \mu \frac{dv}{dz} \tag{52}$$

where μ is the viscosity.

Shell and Tube Heat Exchanger: A device used to transfer heat from one medium to another. It consists of a shell that contains tubes. One medium is contained within the shell and the other within the tubes, and heat is transferred from one to the other across the tubes. There are many designs commonly used, and the simplest is a single phase-type exchanger in which a cold liquid to be heated flows through the tubes from one side of the exchanger to the other. Steam is used as the heating medium and enters as vapor and leaves as condensate from the bottom. A kettle reboiler type is a type of shell and tube heat exchanger in which steam is admitted through the tubes. The choice of hot or cold fluid in the tubes or shell depends on the application and nature of the fluids, such as their susceptibility to fouling (See Figure 42).

Shell Side: The space between the outside of the tubes and the inside of the casing or shell of a shell and tube heat exchanger.

Sherwood Number (Sh): A dimensionless number that represents the relationship between mass diffusivity and molecular diffusivity:

$$Sh = \frac{kL}{D_{AB}} \tag{53}$$

where k is the mass transfer coefficient, L is the characteristic dimension, and D is the diffusivity of the solute A in solvent B. It corresponds to the Nusselt number used in heat transfer.

Shock wave: A pressure wave of very-high-pressure intensity and high temperature that is formed when a fluid flows supersonically or in which a projectile moves supersonically through a stationary fluid. It can be formed by a violent event such as a bomb blast or an explosion. A shock-wave compression is the non-isentropic adiabatic compression in a wave that is traveling above the speed of sound.

Short Residue: Flasher bottoms or residue from the vacuum tower bottoms.

Short-Term Exposure Limit (STEL): The time-weighted average concentration of a substance over a 15-min. period thought not to be injurious to health.

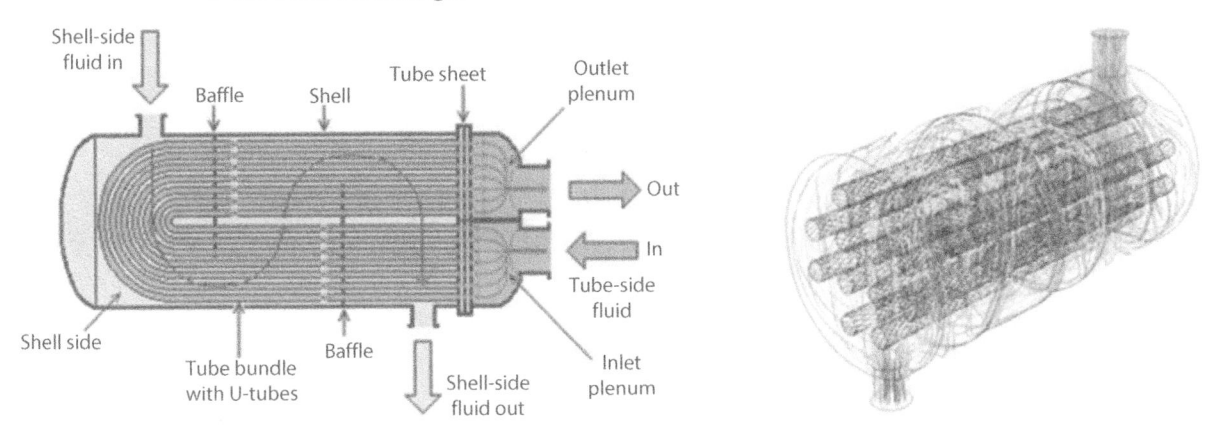

Figure 42 A shell and tube heat exchanger shows the direction of the flow of fluids in the shell and tube sides.

Shutdown: 1. The status of a process that is not currently in operation due to scheduled or unscheduled maintenance, cleaning, or failure. 2. A systematic sequence of action that is needed to stop a process safely.

Side Draw: See Draw

Side Heater (reboiler): A heat input to a distillation column that is located above the bottom tray of the column.

Side Reaction: A chemical reaction that takes place at the same time as a main reaction and produces unwanted products and therefore reduces the yield of the desired product. For example, in the high-temperature cracking reaction of propane (C_3H_8) to produce propylene (C_3H_6), $C_3H_8 \rightarrow C_3H_6 + H_2$, some of the hydrogen can react with the propane to produce methane and ethane as side reactions, $C_3H_8 + H_2 \rightarrow CH_4 + C_2H_6$. The conditions for the reaction must therefore be controlled to reduce this unwanted reaction.

Side Stream: The continuous removal of a liquid or a vapor from a process such as a distillation column that is not the main process flow. For example, drawing off vapor or liquid midway up the column can have an economic advantage in terms of the physical size of the column and the amount of boil-up energy required.

Side Stripper: A small auxiliary column that receives a liquid draw product from the main distillation column for the stripping of light components. Light components are stripped by stripping steam or reboiling and returned to the main column. Liquid products are sometimes stripped inside strippers to raise the flashpoint.

Sieve Plate Column: 1. A type of distillation column that uses a stack of perforated plates to enhance the distribution and intimate contact between vapor and liquid. The plates allow vapor to pass up and bubble through the liquid on the plates. The rate of flow of vapor is sufficient to prevent the liquid from draining down the sieve plates. Instead, the liquid flows over a weir and down a downcomer to the sieve plate below. 2. Sieve trays are metal plates with holes; vapor passes straight through the liquid on the plate. The arrangement, number, and size of the holes are designed parameters (See Figure 43).

Simulated distillation (Simdist): A relatively new laboratory technique in which a petroleum stream is separated into fractions with gas-phase chromatography. Carbon disulfide (C_2S) is used as the carrying agent to dissolve the petroleum stream. The component fractions elute from the chromatographic column in a time sequence, related to their boiling temperatures. Temperatures are assigned to the fractions based on the chromatographic separation

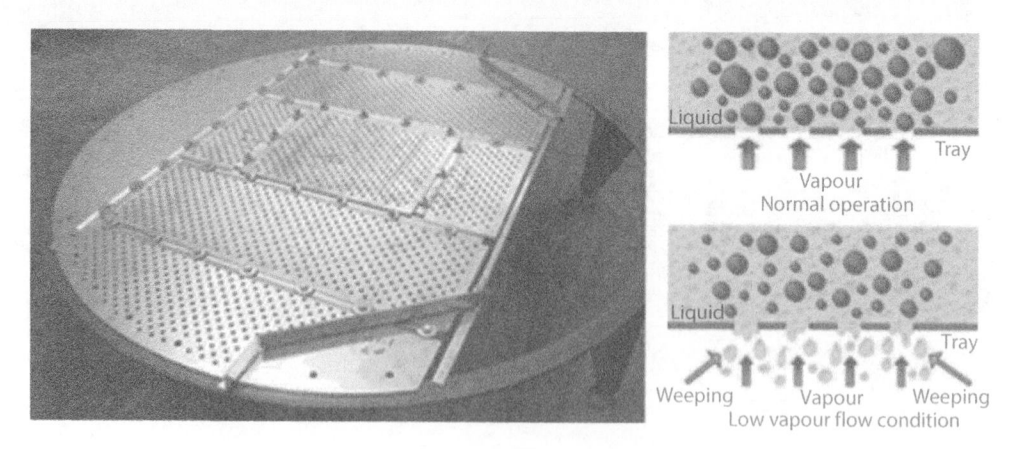

Figure 43 A sieve plate.

of a normal paraffin standard mixture. The simulated distillation approaches a true boiling point distillation and is reported on a mass basis for streams heavier than gasoline. Aromatic compounds elute from the column faster than the paraffin of similar boiling points. Therefore, simulated distillations must be corrected for aromatic content when stocks contain significant quantities of aromatic components.

Slack Wax: Wax produced in the dewaxing of lube oil base stocks. This wax still contains some oil and must be oiled to produce the finished wax product.

Slop Wax: The over flash from a vacuum column. The slop wax is usually withdrawn from the column and combined with the fresh charge to the vacuum furnace.

Slurry: The bottom stream from FCC main fractionators. It is termed slurry because it contains suspended catalyst particles.

Slurry Oil: The oil, from the bottoms of the FCC unit fractionating tower, containing FCC catalyst particles carried over by the vapor from the reactor cyclones. The remainder of the FCC bottoms is the decanted oil (See Figure 40).

Smoke: The gaseous products of the burning of carbonaceous materials made visible by the presence of small particles of carbon; the small particles that are of liquid and solid consistencies are produced as a byproduct of insufficient air supplies to a combustion process.

Smoke Point: **1**. Refers to the height of a smokeless flame of fuel in millimeters beyond which smoking takes place. It reflects the burning quality of kerosene and jet fuels. **2**. A test measuring the burning quality of jet fuels, kerosene, and illuminating oils. It is defined as the height of the flame in millimeters beyond which smoking takes place; ASTM D 1322.

Soaker, Soaking Drum: A soaker is a device that allows cracking time (soaking time) for heated oil in a thermal cracking operation. Furnace coils and/or drums are used for this purpose. Since some coke is deposited in the soaking device, it must be periodically taken offline and cleaned. Furnace coils are much easier to clean than drums.

Soave–Redlich–Kwong (SRK) equation of state: An equation of state widely used to predict the vapor–liquid equilibria of substances. It is a development of the "Redlich–Kwong" equation of state that correlated the vapor pressure of normal fluids:

$$p = \frac{RT}{v-b} - \frac{a\alpha(T)}{V(V+b)} \tag{54}$$

where a and b are constants and obtained from critical point data. It also involves a function that was developed to fit vapor pressure data using reduced temperature, T_r:

$$\alpha = \left[1 + \left(0.480 + 1.574\omega - 0.176\omega^2\right)\left(1 - T_r^{0.5}\right)\right]^2 \tag{55}$$

where ω is the acentric factor.

Solvent Extraction: A separation process based on selective solubility, where a liquid solvent is introduced at the top of a column. As it passes the feed, which enters near the bottom as a vapor, it selectively dissolves a target constituent. The solvent is then removed via the bottom of the column and put through an easy solvent/extract fractionation. From the top of the column comes a raffinate stream, the feed stripped out of the extract. Butadienes and aromatics are some products recovered by solvent extraction (Figure 44).

Sour Crude Oils: Crudes that contain sulfur in amounts greater than 0.5–1.0 wt % or that contain 0.05 ft.^{3} or more hydrogen sulfide (H_2S) per 100 gal. Such oils are dangerously toxic. Even 0.05 ft.^{3} per 100 gal can be present before severe corrosion tends to occur. Arabian crudes are high-sulfur crudes that are not always considered sour because they do not contain highly active sulfur compounds. The original definition was for any crude oil that smelled like rotten eggs.

Sour Gas: **1.** A light gas stream that contains acid gases, in particular sulfur compounds, ammonia compounds, and carbon dioxide. **2.** Gas rich in hydrogen sulfide (H_2S). **3.** Natural gas that contains a significant amount of hydrogen sulfide (usually greater than 16 ppm) and possibly other objectionable sulfur compounds (mercaptans, carbonyl sulfide). Also called "acid gas." **4.** Natural or associated gas with high sulfur content. **5.** Natural gas containing chemical

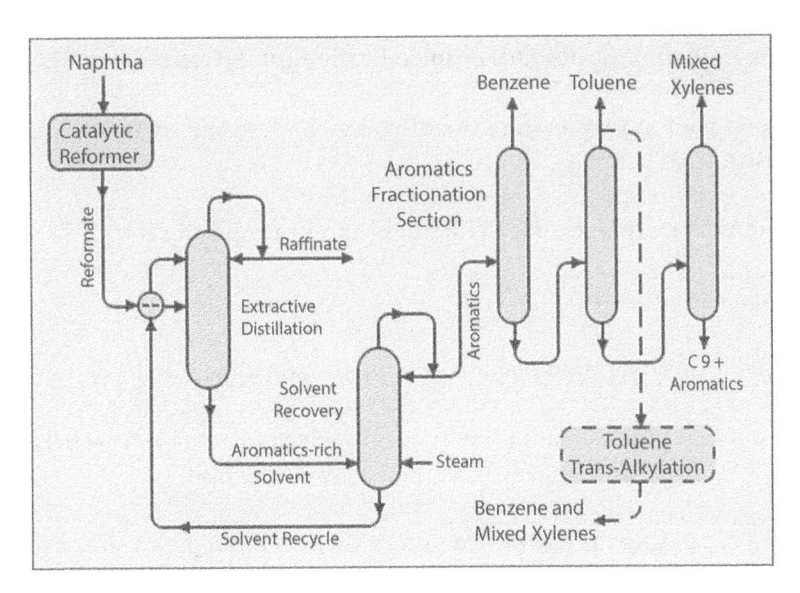

Figure 44 Solvent extraction involving aromatics (BTX).

impurities, a notable hydrogen sulfide (H_2S), or other sulfur compounds that make it extremely harmful to breathe even small amounts; a gas with a disagreeable odor resembling that of rotten eggs. **6.** A gas containing sulfur-bearing compounds such as hydrogen sulfide and mercaptans and that is usually corrosive. **7.** Raw natural gas to be processed, that is, gas received at the liquefaction plant before being subjected to any pretreatment.

Space Velocity: A unit generally used for expressing the relationship between the feed rate and reactor volume in a flow process. It is defined as the volume or weight of feed per unit time per unit volume of the reactor or per unit weight of the catalyst. Space velocity is normally expressed on a volume basis (LHSV: liquid hourly space velocity) or a weight basis (WHSV: weight hourly space velocity). The LHSV and WHSV are determined as follows:

$$\text{LHSV} = \frac{\text{total volumetric feed flow rate to the reactor}}{\text{total catalyst volume}} [=] h^{-1} \tag{56}$$

$$\text{WHSV} = \frac{\text{total mass feed flow rate to the reactor}}{\text{total catalyst weight}} [=] h^{-1} \tag{57}$$

LHSV and WHSV are related by the equation:

$$\text{WHSV} = \frac{\rho_{oil}}{\rho_{cat}} \text{LHSV} \tag{58}$$

where ρ_{oil} and ρ_{cat} are the densities of the hydrocarbon feed and the catalyst, respectively.

Specific gravity: By definition, is the ratio of gas density (at the temperature and pressure of the gas) to the density of dry air (at the air temperature and pressure).

Spent Catalyst: A catalyst that has been through a reaction and is no longer as active because of substances or other contaminants deposited on it (in the case of solid) or mixed with it (in the case of liquid).

Spillback: A spillback allows fluid to recycle from the discharge back to the suction of a machine. It is one way to stop a centrifugal compressor from surging.

Splitter: A distillation column that separates a feed into light and heavy products.

Sponge Absorber: See Secondary Absorber.

Sponge Oil: The liquid used in an absorption plant to soak up the constituent to be extracted.

Stability: Is the ability of a catalyst to maintain its activity and selectivity over a reasonable period. A catalyst with good stability has a long cycle life between regeneration in a commercial unit.

Stabilization: A process for separating the gaseous and more volatile liquid hydrocarbons from crude petroleum or gasoline and leaving a stable (less volatile) liquid so that it can be handled or stored with less change in composition.

Stabilizer: A distillation column that removes light components from a liquid product. This terminology is often used to describe debutanizer columns that remove C_4 hydrocarbons from gasoline to control the vapor pressure.

Standard cubic feet (scf): The volume of gas expressed as standard cubic feet. Standard conditions in petroleum and natural gas usage refer to a pressure base of 14.696 psia (101.5 kPa) and a temperature base of 60°F (15°C).

Static head: The potential energy of a liquid expressed in the head form:

$$h = \frac{p}{\rho g} \quad (59)$$

where p is the pressure, ρ is the density, and g is the acceleration due to gravity. It is used directly in the Bernoulli equation for which the other two head forms are velocity head and pressure head.

Steady State: Describes a process in which the mass and energy flowing both into and out the process are in perfect balance.

Steam: The gaseous form of water formed when water boils. At atmospheric pressure, steam is produced at 212°F (100°C) by boiling water. It is widely used in the chemical and process industries as a utility for heating processes such as a kettle-type reboiler for distillation columns. It is also used in power generation when steam is produced or raised from a thermal process and expanded through turbines. Other uses of steam at destroying harmful pathogens and is a harmless substance once cooled. Wet steam is water vapor that contains water droplets. With further heating, the water evaporates. The *dryness fraction* of steam is the ratio of the amount of water in steam to the total amount of water vapor. Superheated steam is produced by heating the steam above the boiling point of water. The thermodynamic properties of steam are presented in the published literature as steam tables.

Steam (purchased): Steam, purchased for use by a refinery that was not generated from within the refinery complex.

Steam Cracking: 1. The high-temperature reduction in length or cracking of long-chain hydrocarbons in the presence of steam to produce shorter-chain products such as ethylene (C_2H_4), propylene (C_3H_6), and other small-chain alkenes (C_nH_{2n}). 2. The same as catalytic cracking, but specifically refers to the steam injected with the catalyst and feed to give the mixture lift up the riser.

Steam Distillation: The separation of immiscible organic liquids by distillation using steam. It involves the injection of live steam into the bottom of the distillation column and into the heated mixture for separation. The steam reduces the partial pressure of the mixture and the temperature required for vaporization. When distilled, the components operate independently of one another, with each being in equilibrium with its own vapor. Steam distillation is used in the primary separation of crude distillation in a fractionating column.

Steam Injection: The use of live steam fed directly into a process to provide water and heat and to enhance either reaction or extraction. It is commonly used as an enhanced oil recovery method to recover oil from depleted reservoirs or from oil sands in which viscous heavy oil is recovered using steam injection to reduce the viscosity of the oil and aid transport and recovery. Steam is also directly used in the separation of crude oil and fed to the bottom of the fractionating/distillation column. This is the primary separation of crude oil into fractions that have different boiling points. Steam cracking uses steam for the thermal cracking and reforming of hydrocarbons.

Steam jet ejector: A type of fixed operating pump that uses high-pressure steam passed through a constriction to create a low pressure due to the venture effect and to which the equipment to be evacuated is connected such as a distillation column condenser. In spite of requiring high-pressure steam, the device has no moving parts and therefore has low maintenance costs. It can handle corrosive vapors.

Steam Methane Reformer: A primary source of hydrogen in a refinery, this operating unit converts methane (CH_4) and steam (H_2O) to hydrogen (H_2) with the byproducts of carbon monoxide (CO) and carbon dioxide (CO_2).

Steam Point: The temperature that corresponds to the maximum vapor pressure of water at standard atmospheric pressure (1.01325 bar). This corresponds to a temperature of 100°C.

Steam Reforming: The conversion of methane (CH_4) from natural gas into hydrogen (H_2). It is used in the production of ammonia (NH_3) in which the methane is first produced from desulfurized and scrubbed natural gas, mixed with steam, and passed over a nickel catalyst packed in tubes at a high temperature of around 1652°F (990°C)

$$CH_4 + H_2O \rightarrow CO + 3H_2$$
$$CH_4 + 2H_2O \rightarrow CO_2 + 4H_2$$
(60)

The reactions are endothermic (i.e., absorbing heat).

Steam Tables: Published tables that present thermodynamic data for enthalpy, entropy, and the specific volume of steam at various temperatures and pressures. Steam is a commonly encountered material in chemical processes, and its properties have been extensively tabulated.

Steam Tracing: An internal pipe or tube used in process vessels and pipelines carrying steam to provide adequate heating to a fluid to keep it at a controlled temperature. The amount of steam or heat supplied is sufficient to overcome losses. Steam tracing is typically used in pipelines carrying molten bitumen and other fluids prone to solidification on cooling, to ensure that they remain in a liquid state.

Steam Trap: A device used to automatically drain and remove condensate from steam lines to protect the steam main from condensate buildup. Various types of steam traps are used and generally consist of a valve that can be operated by a float, spring, or bellows arrangement. The discharge of the hot condensate may be either to the environment or into a collection pipe and returned to the boiler for reuse (See Figure 45).

Still Gas (refinery gas): Any form or mixture of gases produced in refineries by distillation, cracking, reforming, and other processes. The principal constituents are methane (CH_4), ethane (C_2H_6), ethylene (C_2H_4), normal butane (nC_4H_{10}), butylenes (C_4H_8), propane (C_3H_8), propylene (C_3H_6), and so on. Still gas is used as a refinery fuel and a petrochemical feedstock. The conversion factor is 6 million Btu per fuel oil-equivalent barrel.

Stoichiometric: Applied to reactors in which the reactants and products are defined in terms of the molar quantities reacting. For example, in the reaction: $3H_2 + 2N_2 \Leftrightarrow 2NH_3$, the stoichiometric coefficients are -3.0, -2.0, and 2.0 for the H_2, N_2, and NH_3, respectively.

Straight-Run Distillate or Natural Gasoline: 1. A fraction obtained on the simple distillation of crude oil without cracking. Its octane number is usually low and thus requires upgrading by catalytic reforming. 2. A product that has been distilled from crude oil but has not been through a process in which the composition has been chemically altered. 3. Gasoline produced by the primary distillation of crude oil. It contains no cracked, polymerized, alkylated, reformed, or visbroken stock.

Stress Relief: Coded vessels typically have a metal stamp attached that states "Do not weld, stress relieved." That means the vessel has been post-weld heat-treated to remove stresses in the vessel wall created by welding during fabrication.

Stripping: The removal (by steam-induced vaporization or flash evaporation) of the more volatile components from a cut or fraction.

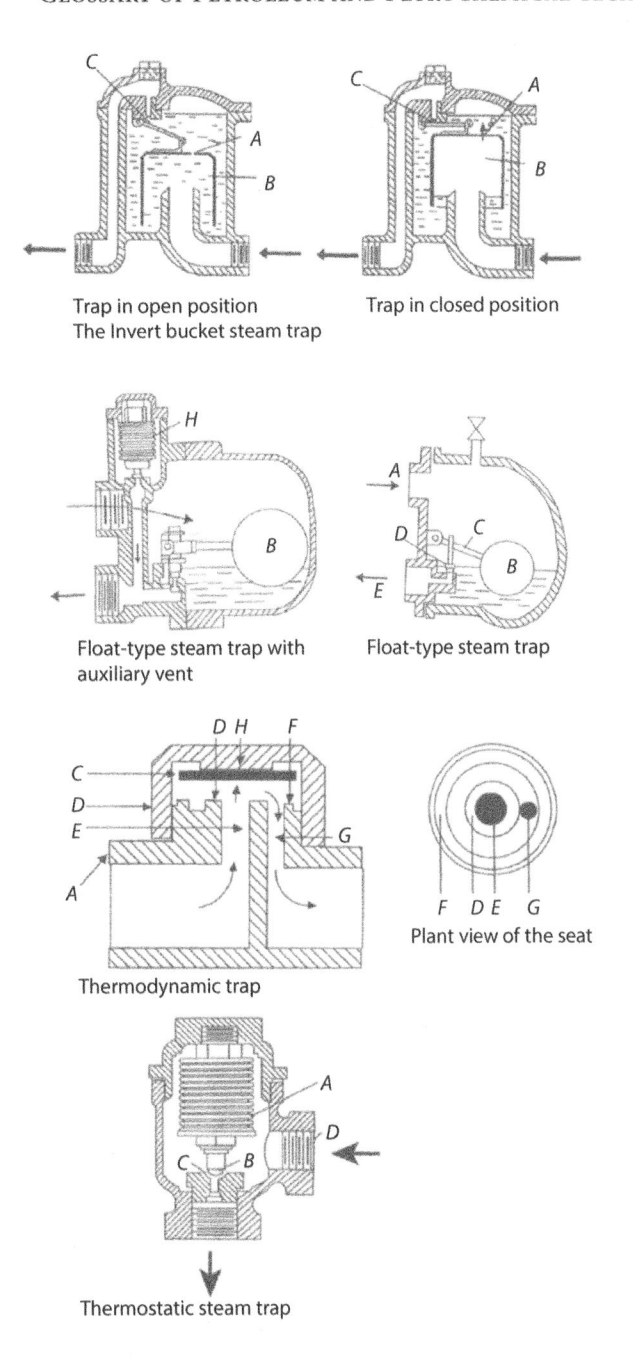

Figure 45 Steam traps.

Stripper Column: A loose designation applied to a distillation column in which light components are stripped from a heavier liquid product.

Stripping Steam: Steam that is injected into the bottom of a side-stripping column or used to strip oil from the catalyst in an FCC operation.

Stripping Zone: The section of the column in which light components are stripped from a heavier liquid product. In conventional distillation columns, this is the portion of the column from the reboiler to the feed tray.

Sulfolane $(CH_2)_4SO_2$: A chemical used as a solvent in extraction and extractive distillation processes (See Figure 44).

Sulfur: A yellowish non-metallic element, sometimes known as "brimstone." It is present at various levels of concentration in many fossil fuels whose combustion releases sulfur compounds that are considered harmful to the environment. Some of the most commonly used fossil fuels are categorized according to their sulfur content, with lower-sulfur fuels usually selling at a higher price. Note: No. 2 distillate fuel is currently reported as having either a 0.05% or lower sulfur level for on-highway vehicle use or a greater than 0.05% sulfur level for off-highway use, home heating oil, and commercial and industrial uses. This also includes Ultra-Low Sulfur Diesel (<15 ppm sulfur). Residual fuel, regardless of use, is classified as having either no more than 1% sulfur or greater than 1% sulfur. Coal is also classified as being low sulfur at a concentration of 1% or less or high sulfur at concentrations greater than 1%.

Sulfuric Acid Treating: A refining process in which unfinished petroleum products such as gasoline, kerosene, and lubricating oil stocks are treated with sulfuric acid to improve their color, odor, and other characteristics.

Sulfurization: Combining sulfur compounds with petroleum lubricants.

Sulfur Content: Is expressed as a percentage of sulfur by weight and varies from less than 0.1% to greater than 5%. Crude oils with less than 1 wt% sulfur are called low-sulfur content or sweet crude, and those with more than 1 wt% sulfur are called high-sulfur content or sour crude. The sulfur-containing constituents of the crude oil include simple mercaptans (also known as thiols), sulfides, and polycyclic sulfides. Mercaptan sulfur is simply an alkyl chain (R-) with an SH group attached to it at the end. The simplest form of R – SH is methyl mercaptan, CH_3SH.

Surface Area: The total area that a solid catalyst exposes to the feeds in a reaction. The surface area is enhanced in some catalysts like zeolites by extensive microscopic pores.

Supply: The components of petroleum supply are field production, refinery production, imports, and net receipts when calculated on a PADD basis.

Surge: This is a terrifying sound that centrifugal compressors make when they malfunction either due to low flow or excessive discharge pressure or low-molecular-weight gas.

Sweet Crude: **1**. Crude oil containing very little sulfur and having a good odor. **2**. Crude petroleum containing little sulfur with no offensive odor. **3**. Gets its name due to a pleasant and "sweet" smell. Sweet crude has a sulfur content of less than 1%. It is more valuable than sour crude because it costs less to process the crude into finished products. **4**. Oil containing little or no sulfur, especially little or no hydrogen sulfide. The original definition was for any crude oil that did not have a bad odor.

Sweetening: **1**. The removal or conversion to innocuous substances of sulfur compounds in a petroleum product by any of a number of processes (doctor treating, caustic and water washing, etc.). **2**. Processes that either remove obnoxious sulfur compounds (primarily hydrogen sulfide, mercaptans, and thiophenes) from petroleum fractions or streams or convert them, as in the case of mercaptans, to odorless disulfides to improve odor, color, and oxidation stability.

Sweet Gas: **1**. Gas sweetened. Gas processed in the acid gas removal unit that no longer contains gaseous pollutants. **2**. Natural gas that contains small amounts of hydrogen sulfide (and other sulfur compounds) and carbon dioxide that can be transported or used without purifying with no deleterious effect on piping and equipment. **3**. A gas stream from which the sulfur compounds have been removed. **4**. A gas containing no corrosive components such as hydrogen sulfide and mercaptans.

Symbols of Chemical Apparatus and Equipment: Listed below are some symbols of chemical apparatus and equipment normally used in a P & ID, according to ISO 10628 and ISO 14617 (See Figure 46).

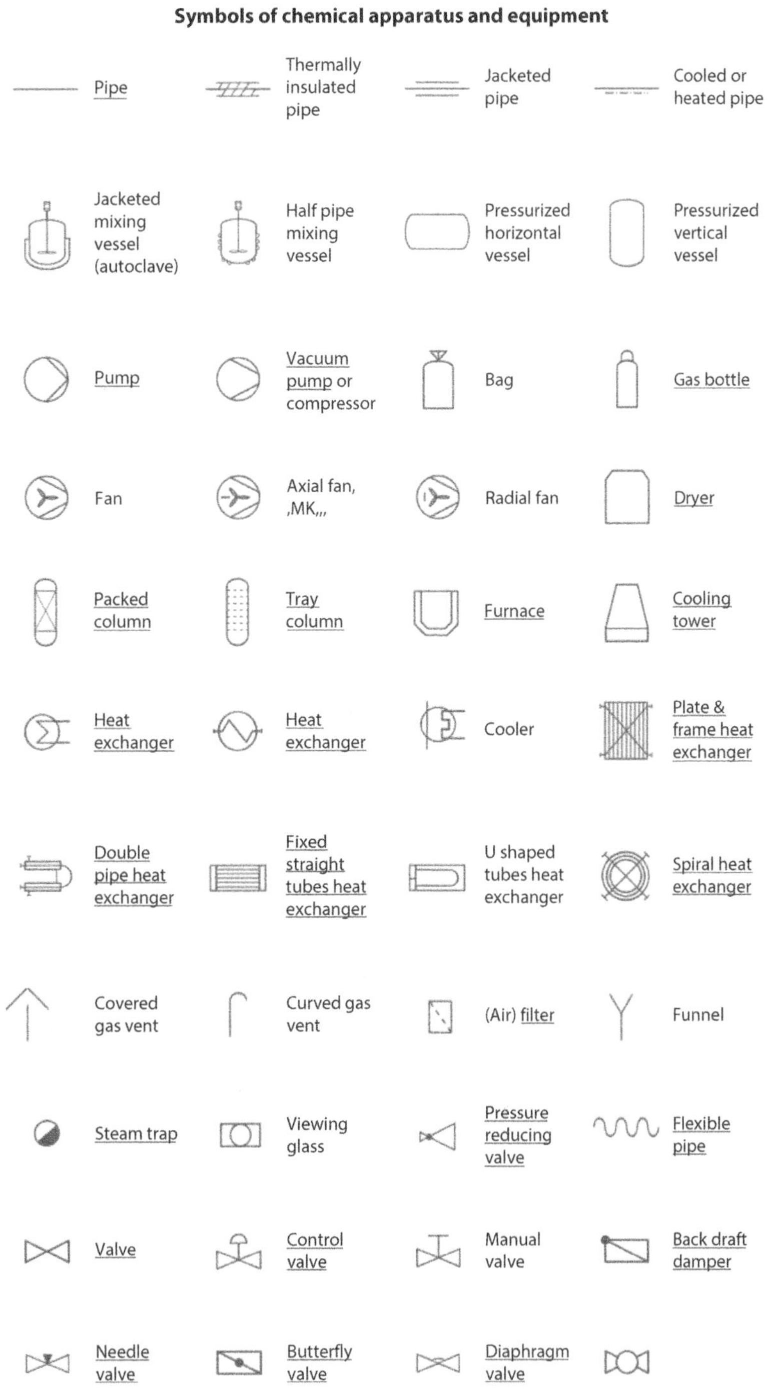

Figure 46 Symbols of chemical apparatus and equipment.

Synthetic Crude: A wide-boiling-range product of catalytic cracking, coking, hydrocracking, or some other chemical structure change operation.

Synthesis Gas: The product of a reforming operation in which hydrocarbons, usually methane and water, are chemically rearranged to produce carbon monoxide, carbon dioxide, and hydrogen. The composition of the product stream can be varied to fit the needs of hydrogen and carbon monoxide at refineries or a chemical plant. Also known as syn gas.

Tail Ends: Small amounts of hydrocarbon in a cut that vaporizes slightly outside the effective initial boiling point and the effective end point.

Tail Gas: Light gases C_1 to C_3 and H_2 produced as byproducts of refinery processing.

TAN: Total acid number.

Tank Farm: An installation used by gathering trunk pipeline companies, crude oil producers, and terminal operators (except refineries) to store crude oil.

Tanker and Barge: Vessels that transport crude oil or petroleum products. Data are reported for movements between PAD Districts; from a PAD District to the Panama Canal; or from the Panama Canal to a PAD District.

Tar: Complex, large molecules of predominantly carbon with some hydrogen and miscellaneous other elements that generally deteriorate the quality of processes and the apparatus.

TBP Distillation: See Fifteen-Five Distillation.

Tertiary Amyl Methyl Ether $(CH_3)_2(C_2H_5)COCH_3$ (TAME): A high-octane oxygenate blending stock produced by reacting isoamylene (isopentylene) produced in FCC processes with methanol. Used to enhance the octane of a motor gasoline pool.

Tertiary Butyl Alcohol – $(CH_3)_3COH$ (TBA): An alcohol primarily used as a chemical feedstock, solvent, or feedstock for isobutylene production for MTBE; produced as a co-product of propylene oxide production or by direct hydration of isobutylene.

Tetra Ethyl Lead (TEL): A compound added to gasoline to increase the octane. TEL has been superseded by other octane enhancers and is no longer used by refiners for motor gasoline.

Test Run: A time period during which the operating data and stream samples are collected for a process. During test runs, the operation of the processing unit is held as steady as possible. For good test runs, the average conditions and stream flows approximate a steady-state operation.

The Diesel Engine: The diesel engine is a reciprocating internal combustion engine. It is different from the petrol engine in that the air intake in the engine cylinder is unthrottled and not premixed with the fuel. Here, the ignition takes place spontaneously without the help of a spark plug. The air taken into the cylinder at atmospheric pressure is compressed to a volume ratio somewhere near to 1:16. At the end of the compression, fuel is injected into the cylinder. The quantity injected depends on the power of the engine, and the heat of compression heats the mass of the air compressed.

The Onion Model: The onion model depicts hazards, barriers, and recovery measures. It reflects the layers of protection and shows how the various measures fit together when viewed from the perspective of the hazard. The first layer is the basic containment of the feedstock, processes, and products (See Figure 47).

The Saybolt Universal Viscometer: Measures the time in seconds that would be required for the total flow of 60 cc of oil from a container tube at a given constant temperature through a calibrated orifice placed at the bottom of the tube. Lubricant viscosities are usually measured in Saybolt Universal seconds at 100°F (37.8°C), 130°F (54.4°C), or 210°F (98.9°C).

For example, the symbol SSU 100 represents the time in seconds that a fluid at 100°F (37.8°C) will take to flow through a given orifice of the Saybolt viscometer.

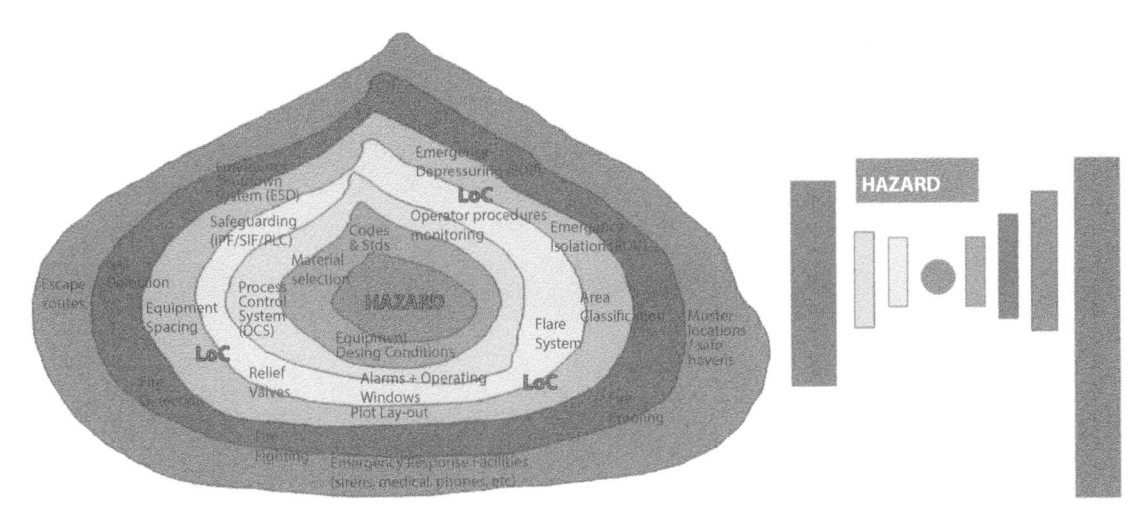

Figure 47 The onion model (LOC = loss of containment).

Kinematic viscosity can be converted to Saybolt viscosity SSU by the formula:

$$\text{Kinematic viscosity}, \nu = \frac{\mu}{\rho} = 0.219t - \frac{149.7}{t} \tag{61}$$

where
  μ = viscosity of fluid in centipoises, cP
  ρ = density of fluid, g/cc
  t = Saybolt Universal viscosity, sec

Theoretical Plate: **1.** A theoretical contacting unit useful in distillation, absorption, and extraction calculations. The vapor and liquid leaving any such unit are required to be in equilibrium under the conditions of temperature and pressure that apply. An actual fractionator tray or plate is generally less effective than a theoretical plate. The ratio of a number of theoretical plates required to perform a given distillation separation to the number of actual plates used, given the overall tray efficiency of the fractionator. **2.** Refers to a vapor/liquid contact device (e.g., distillation column) in which the liquid and vapor leaving the device are in perfect vapor/liquid-phase equilibrium. There are also perfect energy and mass balances for a theoretical tray.

Thermal Cracking: **1.** A refining process in which heat and pressure are used to break down, rearrange, or combine hydrocarbon molecules. Thermal cracking includes gas oil, visbreaking, fluid coking, delayed coking, and other thermal cracking processes (e.g., Flexicoking). **2.** The first cracking process in which the oil was cracked by heating only. Thermal cracking produces a lower-octane gasoline than catalytic processes.

Thermal Cracked Distillate: Is formed when a distillate heavier than gasoline is heated under pressure in order to break the heavy molecules into lighter ones that boil in the gasoline range. This is superseded by catalytic cracking, which gives a better distillate.

Thermal Conductivity: The ability of a material to let heat pass. Metals, water, and materials that are good conductors of electricity have a high thermal conductivity. Air, rubber, and materials that are bad conductors of electricity have a low thermal conductivity. High-viscosity hydrocarbons are bad conductors of heat.

Thermal Expansion: Railroad tracks grow longer in the heat of the sun. The hot tubes in an exchanger grow more than the cold shell. Hence, we have a floating head in the tube bundle to accommodate differential rates of thermal expansion between the tube bundle and the shell.

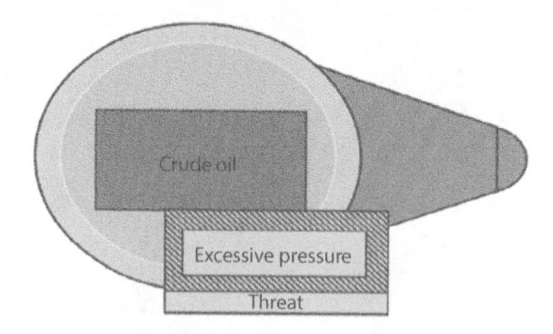

Figure 48 A threat.

Threat: A threat is something that can cause the release of a hazard and lead to a top event (See Figure 48).

Three Phase: A mixture consisting of one vapor in equilibrium with two mutually insoluble liquid phases.

Threshold Limit Value: The amount of a contaminant to which a person can have repeated exposure for an eight-hour day without adverse effects.

Toluene ($C_6H_5CH_3$): 1. A colorless liquid of the aromatic group of petroleum hydrocarbons, made by the catalytic reforming of petroleum naphthas containing methyl cyclohexane ($CH_3C_6H_{11}$). A high-octane gasoline blending agent, solvent, and chemical intermediate, base for TNT. **2.** One of the aromatic compounds used as a chemical feedstock most notoriously for the manufacture of TNT, trinitrotoluene.

Top Event: A top event is the "release" of the hazard, i.e., the first consequence, typically a loss of containment, a loss of control, or an exposure to something that may cause harm, such as the release of hydrocarbons, toxic substances, or energy (See Figure 49).

Top Product: For columns with condensers, the liquid and/or vapor streams from the reflux drum that exit the process.

Topping: Removal by distillation of the light products and transportation fuel products from crude oil, leaving in the still bottoms all of the components with boiling ranges greater than diesel fuel.

Topped Crude Oil: **1.** Crude that has been run through a distilling unit to remove the gas oil and lighter streams. The long residue is sold as residual fuel. **2.** The bottom product from a crude distillation column.

Toxic Compounds: NO_x, VOCs, and SO_x are toxic compounds such as formaldehyde, oxides of nitrogen, volatile organic compounds such as pentene, and oxides of sulfur.

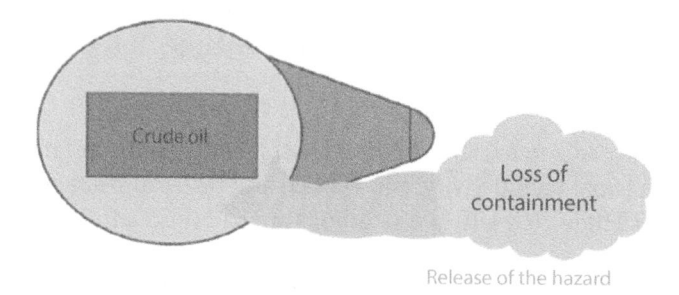

Figure 49 Top event.

Tray: A liquid/vapor contact device in a distillation column (Figure 43).

Tray Efficiency: See Overall Tray Efficiency.

Treat Gas: Light gases, usually high in hydrogen content, which are required for refinery hydrotreating processes such as hydrodesulfurization. The treat gas for hydrodesulfurization is usually the tail gas from catalytic reforming or the product from a hydrogen unit.

Trip: **1**. The fast shutdown of an item of chemical plant or process equipment such as a pump or compressor. The shutdown is the result of a process condition being exceeded such as an abnormal flow, pressure, temperature or concentration, etc. **2**. This a safety device that automatically shuts down a piece of equipment. It is a fail-safe mechanism often activated by unlatching a spring-operated valve, which then closes.

Troubleshooting: A form of problem-solving methodology used to identify, solve, and eliminate problems within a process that has failed or has the potential to fail. It is a logical and systematic search for the source or cause of the problem and solutions presented to ensure that the process is restored back to its full operability. Troubleshooting is applied once a problem has occurred and the process stops functioning. It can take the form of a systematic checklist and requires critical thinking. Computer techniques are employed for more complex systems where a sequential approach is either too lengthy or not practical or where the interaction between the elements in the system is not obvious.

True Boiling Point Distillation (TBP): **1**. Of a crude oil or petroleum fractions results from using the US Bureau of Mines Hempel method and the ASTM D-285 test procedures. Neither of these methods specifies the number of theoretical stages or the molar reflux ratio used in the distillation. Consequently, there is a trend toward applying a 15:5 distillation according to ASTM D2892, instead of the TBP. The 15:5 distillation uses 15 theoretical stages and a molar reflux ratio of 5. **2**. A laboratory test in which petroleum oil is distilled in a column having at least 15 theoretical trays and a reflux ratio of 5.0. The distillate is continually removed and further analyzed. The separation corresponds somewhat to a component-by-component separation and is a good measure of the true composition for the sample being distilled. As the temperatures in the still increase, the pressure of the still is lowered to suppress thermal cracking of the sample.

The minimum pressure for most TBP stills is about 38 mm Hg. This allows the distillation of petroleum components boiling up to about 900–950°F (483–510°C) at a pressure of 1 atm. Surprisingly, the TBP test has never been standardized and several different apparatuses are used for the test.

A key result from a distillation test is the boiling point curve, i.e., the boiling point of the oil fraction versus the fraction of oil vaporized. The initial boiling point (IBP) is defined as the temperature at which the first drop of liquid leaves the condenser tube of the distillation apparatus. The final boiling point or the end point (EP) is the highest temperature recorded in the test.

Additionally, oil fractions tend to decompose or crack at a temperature of approximately 650°F (344°C) at 1 atm. Thus, the pressure of TBP distillation is gradually reduced to as low as 40 mmHg, as this temperature is approached to avoid cracking the sample and distorting measurements of true components in the oil.

Tube Bundle: Pipes in a shell-and-tube heat exchanger that are packed into an arrangement to ensure effective heat transfer from the outer surface and good transport for fluids through the tubes. The tubes in the tube bundle are spaced and typically set with a rectangular or triangular pitch and held and sealed with a tube plate. Baffles also provide rigidity and encourage a turbulent flow of fluids through the shell side. The tubes can be a straight single-pass or hairpin double-pass arrangement. The tube bundle can be removed from the shell for periodic cleaning. Lugs are welded to the baffles for lifting purposes (Figures 19 and 42).

Tube Size: Tubing sizes are entirely different from pipe sizes. Tubing is often used in heat exchangers and fired equipment.

Tube Still: See Pipe Still.

Turbine: **1.** A machine used to generate electricity by the expansion of a gas or vapor at high pressure through a set of blades attached to a rotor. The blades rotate as the result of the expansion and conversion of energy. Gas turbines and steam turbines are commonly used to generate electricity. A nozzle is used to direct the high-speed gas or steam over a row of turbine blades. The fluid pushes the blades forward, causing them to rotate due to the change in momentum. A row of stationary blades within the turbine redirects the fluid in the correct direction again before it passes through another set of nozzles and expands to a lower pressure. A steam turbine may have several pressure sections and operate at high-pressure, medium-pressure, and, as the steam expands, a low-pressure section, all linked to the same shaft. The steam in the medium-pressure section may be returned to a boiler and reheated before doing further work to prevent the formation of water in the turbine. **2.** A turbine uses steam pressure or burning gas to drive pumps and compressors at variable speeds. Motor drives are usually fixed-speed machines. Variable speed is an energy-efficient way to control flows by eliminating the downstream control valve.

Turbulent Flow: A fluid flow regime characterized by the fluctuating motion and erratic paths of particles. In pipes of circular cross-section, this occurs at a Reynolds number in excess of 4000. Turbulent flow occurs when inertial forces predominate, resulting in a macroscopic mixing of the fluid.

Turnaround: A planned complete shutdown of an entire process or section of a refinery, or of an entire refinery, to perform major maintenance, overhaul, and repair operations and to inspect, test, and replace process materials and equipment.

Two Phase: A mixture consisting of one vapor in equilibrium with one homogeneous liquid phase.

ULSD: Ultra-low-sulfur diesel. Diesel fuel with <15 ppm sulfur.

Unaccounted for Crude Oil: Represents the arithmetic difference between the calculated supply and the calculated disposition of crude oil. The calculated supply is the sum of crude oil production plus imports minus changes in crude oil stocks. The calculated disposition of crude oil is the sum of crude oil input to refineries, crude oil exports, crude oil burned as fuel, and crude oil losses.

Undistributed Component: A component in a distillation column separation zone that is totally recovered in only one of the products.

Unfinished Oils: All oils requiring further processing, except those requiring only mechanical blending. Unfinished oils are produced by partial refining of crude oil and include naphthas and lighter oils, kerosene and light gas oils, heavy gas oils, and residuum.

Unfractionated Streams: Mixtures of unsegregated natural gas liquid components, excluding those plant condensates. This product is extracted from natural gas.

Unsaturated Compounds: Hydrocarbon compounds in which some of the carbon atoms have multiple bonds with other carbon atoms because of the lack of hydrogen atoms to satisfy the carbon atoms' valences.

Upper Explosive Limit (UEL): The maximum concentration of vapor in air above which the propagation of flame will not occur in the presence of an ignition source. Also referred to as the upper flammable limit or upper explosion limit.

Utilities: Most plants have some of the following utility systems connected to process units.

- Natural gas
- Nitrogen
- Plant air
- Instrument air
- Steam of various pressures
- Cooling water
- Service water
- Boiler feed water
- Fire water
- Fuel gas
- City water

Your company safety policy does not permit you to cross-connect these systems. Connecting natural gas to plant air killed 17 workers at a Louisiana refinery.

Vacuum Distillation: **1.** Distillation under reduced pressure (less than atmospheric), which lowers the boiling temperature of the liquid being distilled. This technique prevents the cracking or decomposition of the charge stock, which occurs above 1000°F (538°C). **2.** A distillation column that operates at sub-atmospheric pressure. Vacuum distillation permits the further distillation of heavy feedstocks at reduced temperatures that minimize cracking reactions.

Vacuum Gas Oil (VGO): A side stream from the vacuum distillation tower.

Vacuum Residuum: The heaviest product from a vacuum distillation column.

Valves: A valve is a device that regulates, directs, or controls the flow of a fluid (gases, liquids, fluidized solids, or slurries) by opening, closing, or partially obstructing various passageways.

Diaphragm Valve: A type of device in which a flexible membrane is used to restrict the rate of flow. The membrane is usually made from a flexible natural or synthetic rubber. Diaphragm valves are typically used for fluids that contain suspended solids (See Figure 50).

Gate Valve: This valve closes by sliding a plate or gate down between two grooves. Used to isolate different portions of the process equipment not used to control flow. The valve closes clockwise and takes about a dozen turns to close. Ninety percent of the valves used in process plants are gate valves (See Figure 51).

Globe Valve: A device that regulates the flow of a fluid in a pipe and consists of a flat disc that sits on a fixed ring seat. The disc is movable and allows flow through the valve (See Figure 52).

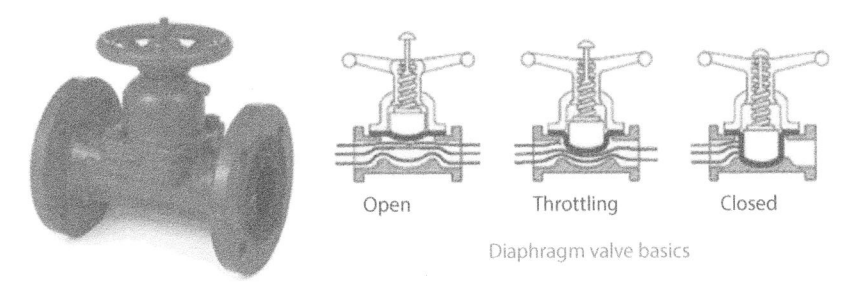

Figure 50 A diaphragm valve.

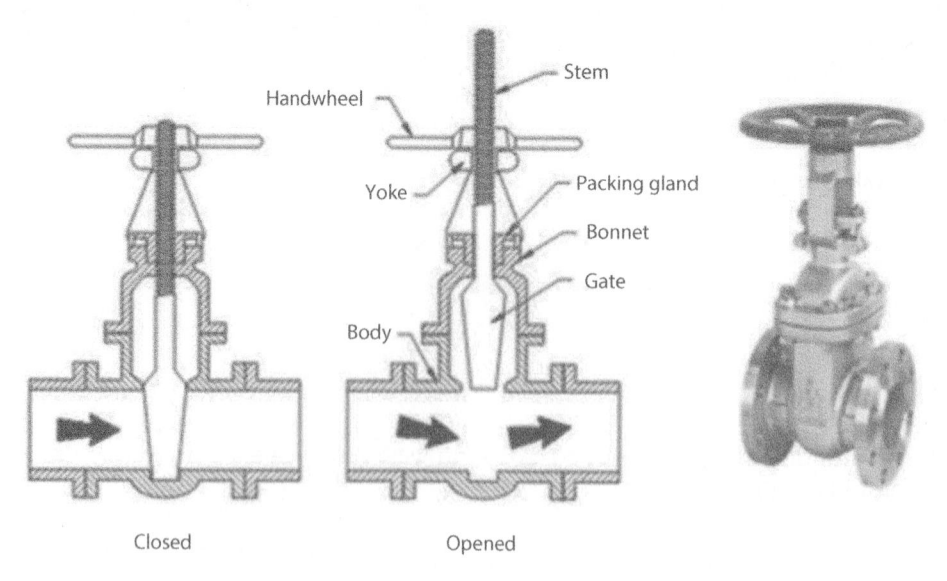

Figure 51 A gate valve.

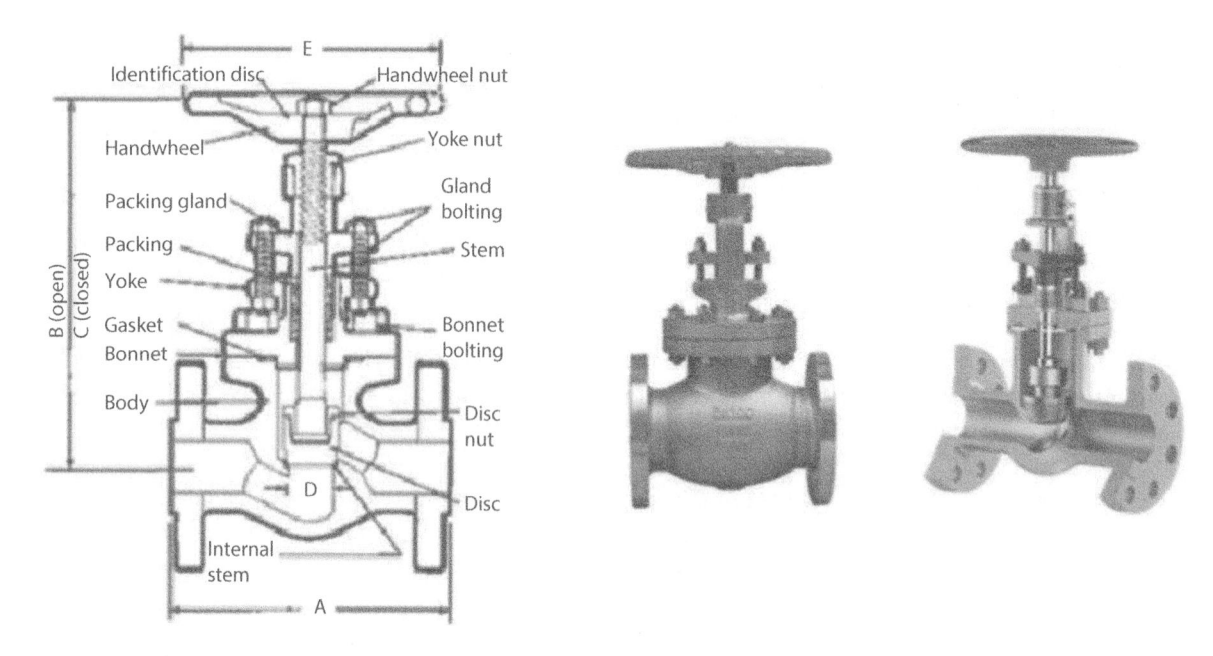

Figure 52 A globe valve away section of a globe valve.

Plug Valve: This valve goes from 100% open to shut by turning a valve 90°. The natural gas supply to your house is shut off with a plug valve (See Figure 53).

Control Valve: This valve is used to alter flows remotely. Normally, it is moved by air pressure. A gate valve is sometimes used to control flows locally, but this wears out the valve and is best avoided (See Figure 54).

Relief Valves: These valves open to relieve excess pressure to protect a vessel from failure. Also called safety or pop valves (See Figure 55).

Valve Trays: 1. Fractionator trays that have perforations covered by discs that operate as valves and allow the upward passage of vapor. **2.** In valve trays, perforations are covered by lift-able caps. Vapor flows lift the caps, thus

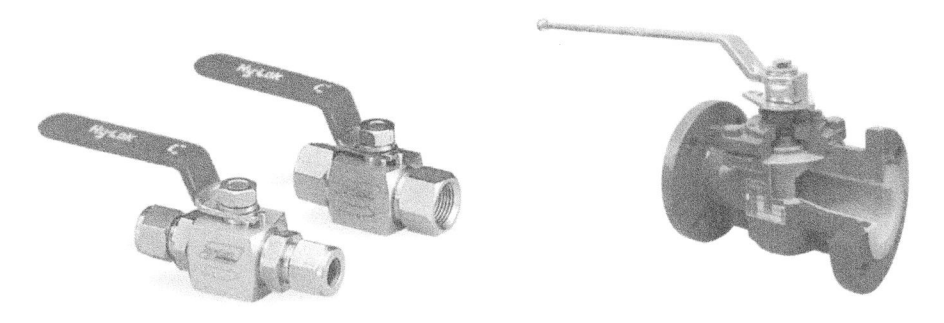

Figure 53 Plug valves. Cutaway section of a plug valve.

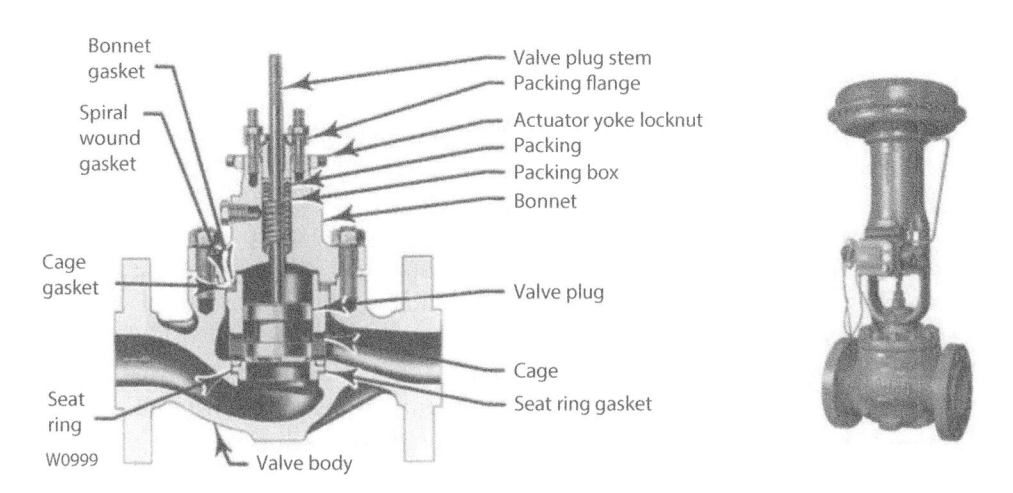

Figure 54 A control valve.

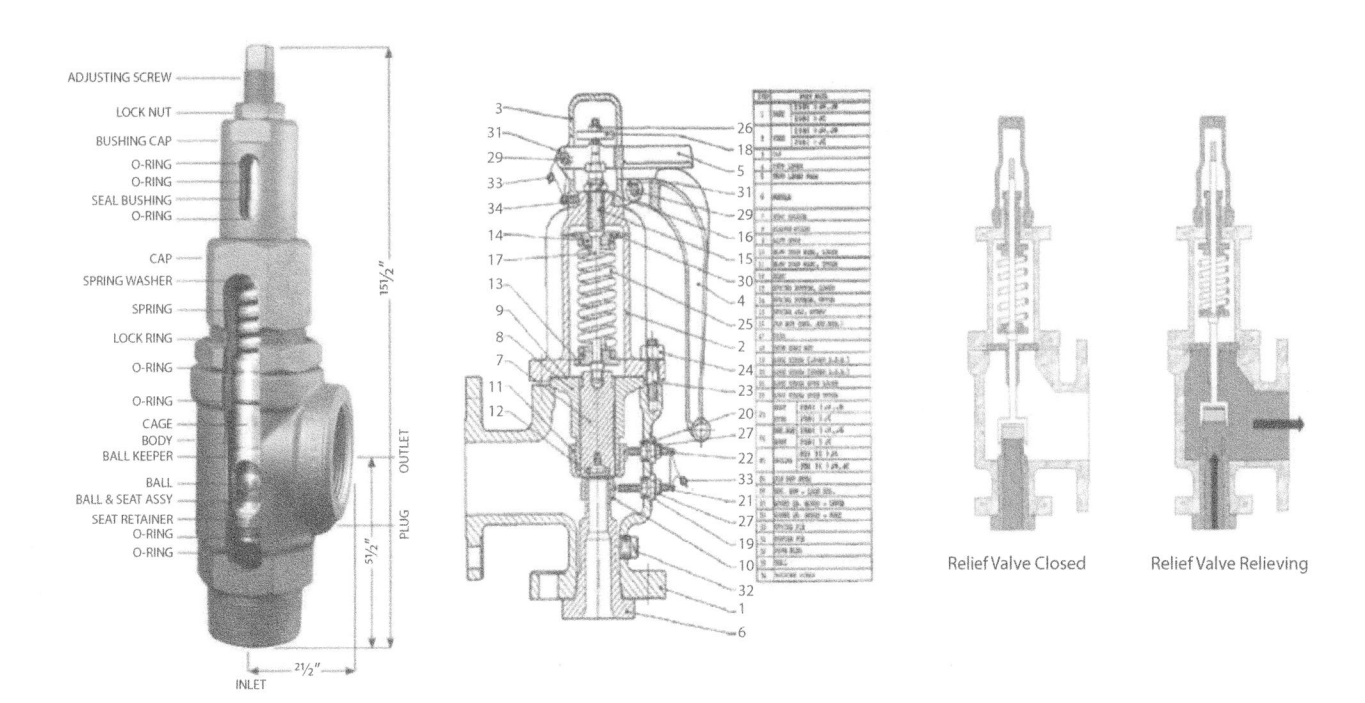

Figure 55 Relief valves.

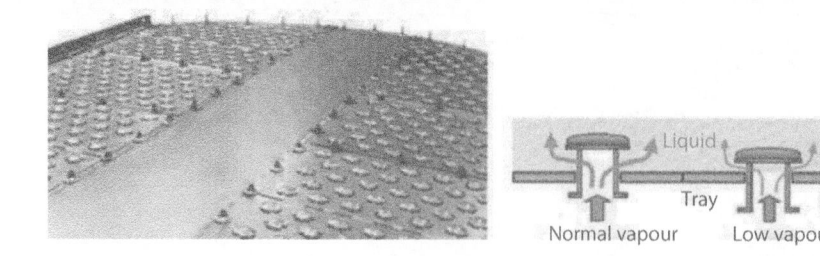

Figure 56 A valve tray.

self-creating a flow area for the passage of vapor. The lifting caps direct the vapor to flow horizontally into the liquid, thus providing better mixing than is possible in sieve trays (See Figure 56).

Vapor: The gaseous phase of a substance that is a liquid at normal temperature and pressure.

Vapor/Liquid Ratio (V/L): The vapor/liquid ratio (V/L) is the ratio of the volume of vapor formed at atmospheric pressure to the volume of gasoline in a standard test apparatus. The vapor-lock tendency of the gasoline sample can be measured more reliably in terms of its V/L ratio than in terms of its vapor pressure. The V/L ratio also increases with a rise in temperature.

Vapor Pressure: 1. As a liquid is heated, the molecules in the liquid try to escape into the vapor phase. The hotter the liquid, the harder they try to escape. The pressure that the molecules of liquid create as they push out into the vapor space is the liquid vapor pressure. More volatile liquids such as LPG have a higher vapor pressure than less volatile diesel oil. 2. The pressure exerted by a volatile liquid as determined by ASTM D-323, Standard Method of Test for Vapor Pressure of Petroleum Products (Reid Method). 3. Is a measure of the surface pressure necessary to keep a liquid from vaporizing. The vaporizing tendency of gasoline is measured in terms of its vapor pressure. It is related to vapor lock and engine starting. Vapor lock arises due to the vaporization of the fuel in fuel lines, fuel pump, carburetor, etc., making bubbles of vapor, which prevent the normal flow of fuel. This occurs if the gasoline contains a too-high percentage of low-boiling components as observed by a very high vapor pressure. Alternatively, if the gasoline contains only too few low-boiling components as indicated by a low vapor pressure, then the fuel will not vaporize readily, making it difficult in starting.

Vapor Lock: Is the phenomenon of insufficient gasoline flow from a fuel pump due to its inability to pump the mixture that results from low pressure or high temperature, which has high volatility.

Vapor Lock Index: A measure of the tendency of a gasoline to generate excessive vapors in the fuel line, thus causing the displacement of a liquid fuel and a subsequent interruption of normal engine operation. The vapor-lock index ia generally related to RVP and percentage distilled at 158°F (70°C).

Virgin material, gas oil, etc.: Virgin material is material distilled from crude oil but not subjected to processes that chemically alter its composition.

Virgin Stocks: Petroleum oils that have not been cracked or otherwise subjected to any treatment that would produce appreciable chemical change in their components.

Visbreaking: 1. A thermal cracking process in which heavy atmospheric or vacuum still bottoms are cracked at moderate temperatures to increase the production of distillate products and reduce the viscosity of the distillation residues. 2. A process in which heavy oil is thermally cracked just enough to lower or break the viscosity. A small quantity of gas oil and lighter products are formed in the process.

Viscosity ASTM D445: The internal resistance to the flow of liquids is expressed as viscosity. The property of liquids under flow conditions that causes them to resist instantaneous change of shape or instantaneous rearrangement of their parts due to internal friction. Viscosity is generally measured in seconds, at a definite temperature, required for a standard quantity of oil to flow through a standard apparatus. Common viscosity scales in use are Saybolt Universal, Saybolt Furol, poises, kinematic [stokes, or centistokes (cSt)]. Usually, the viscosity measurements are carried out at 100°F (38°C) and 210°F (99°C).

Viscosity is a very important property for the heavy products obtained from the crude oil. The viscosity acts as an important characterization property in the blending units associated to heavy products such as bunker fuel. Typically, the viscosity of these products is specified to be within a specified range, and this is achieved by adjusting the viscosities of the streams entering the blending unit.

Viscosity Index (VI): This index is a series of numbers ranging from 0 to 100 that indicate the rate of change of viscosity with temperature. A Viscosity Index of 100 indicates an oil that does not tend to become viscous at low temperatures or become thin at elevated temperatures.

Typically, paraffin-based lubricating oils exhibit a Viscosity Index of nearly 100, whereas naphthene-based oils on the market show about 40 Viscosity Index, and some naphthenic oils have a Viscosity Index of zero or lower. Paraffin wax has a VI of about 200, and hence its removal reduces the VI of raw lube stocks. By solvent extraction processes, lubricating oils of Viscosity Index higher than 100 can be produced.

Volatile: A hydrocarbon is volatile if it has a sufficient amount of butanes and higher material to noticeably give off vapors at atmospheric conditions.

Volatile Organic Compounds (VOCs): Organic chemicals that have a high vapor pressure at ordinary room temperature. Their high vapor pressure results from a low boiling point, which causes large numbers of molecules to evaporate or sublimate from the liquid or solid form of the compound and enter the surrounding air. For example, formaldehyde (HCHO), which evaporates from paint, has a boiling point of only -19°C (-2°F).

One VOC that is a known human carcinogen is benzene, which is a chemical found in environmental tobacco smoke, stored fuels, and exhaust from cars. Benzene also has natural sources such as volcanoes and forest fires. It is frequently used to make other chemicals in the production of plastics, resins, and synthetic fibers. Benzene evaporates into the air quickly, and the vapor of benzene is heavier than air, allowing the compound to sink into low-lying areas. Benzene has also been known to contaminate food and water and, if digested, can lead to vomiting, dizziness, sleepiness, and rapid heartbeat, and at high levels, even death may occur.

VOCs are many and varied, are dangerous to human health, or cause harm to the environment. Harmful VOCs typically are not acutely toxic but have compounding long-term health effects. Because the concentrations are usually low and the symptoms are slow to develop, the research into VOCs and their effects is difficult.

Volatility: As measured by the distillation characteristics, helps to determine the relative proportion of the various hydrocarbons throughout the boiling range of a gasoline. It is the distillation characteristics along with vapor pressure and vapor/liquid ratio that help to control the performance of the fuel with respect to starting, warm-up, acceleration, vapor-lock, evaporation losses, crankcase dilution, fuel economy, and carburetor icing.

Volatility Factor: An empirical quantity that indicates good gasoline performance with respect to volatility. It involves actual automobile operating conditions and climatic factors. The volatility factor is generally defined as a function of RVP (Reid vapor pressure), percentage distilled at 158°F (70°C), and percentage distilled at 212°F (100°C). This factor is an attempt to predict the vapor-lock tendency of a gasoline.

vppm: Parts per million by volume.

VRC: Vacuum-reduced crude; vacuum tower bottoms.

WABP: Weight average boiling point:

$$WABP = \sum_{i=1}^{n} X_{wi} T_{bi} \qquad (62)$$

where
X_{wi} = weight fraction of component i.
T_{bi} = average boiling point of component i.

Wash Zone: A section in a column where the column vapor is washed of entrained heavy materials by contact with a cooler injected liquid. A section of packed material is often used to promote good mixing of the liquid and vapor in the wash zone. All vacuum distillation columns have wash zones to remove heavy residual material that is carried up the column from the flash zone. If washing is inadequate, the heavy residual material forms petroleum coke and plugs the column above the flash zone.

Water Hammer: A violent and potentially damaging shock wave in a pipeline caused by the sudden change in flow rate, such as by the rapid closure of a valve. The effect is avoided by controlling the speed of valve closure, lowering the pressure of the fluid, or lowering the fluid flow rate.

Water vapor: The gaseous state of water dispersed with air at a temperature below the boiling point of the water. The amount present in air is designated by the humidity. The "relative humidity" is the amount of water vapor in a mixture of dry air. A relative humidity of 100% corresponds to the partial pressure of water vapor being equal to the equilibrium vapor pressure and depends on the temperature and pressure.

Watson Characterization Factor (K_w): See Characterization Factor.

Wax: A solid or semi-solid material consisting of a mixture of hydrocarbons obtained or derived from petroleum fractions, or through a Fischer–Tropsch type of process in which the straight-chained paraffins series predominates. This includes all marketable wax, whether crude or refined, with a congealing point (ASTM D 938) between 100 and 200°F (37.8–93°C) and a maximum oil content (ASTM D 3235) of 50% weight.

Weeping: A phenomenon that occurs in a distillation column in which the liquid on a sieve plate passes down through the perforations intended for the vapor to pass up. Weeping occurs when the velocity of the upward vapor is too low. This may be caused by insufficient boil-up.

Weir: A vertical obstruction across a channel carrying a liquid over which the liquid discharges. In a distillation column, a weir is used to retain an amount of liquid on a sieve tray or plate. While the vapor enriched with the more volatile component rises up through the perforations on the sieve tray or plate, the liquid cascades over the weir into the downcomer to the tray below. The weir crest is the top of the weir over which the liquid flows (See Figure 57).

Weighted Average Inlet Temperature (WAIT): [Weight of catalyst in reactor 1 x inlet temperature in reactor 1 + weight of catalyst in reactor 2 x inlet temperature in reactor 2 + weight of catalyst in reactor 3 x inlet temperature in reactors 3]/total weight of catalyst, i.e.,

$$[WCR_1 \times R_{1IT} + WCR_2 \times R_{2IT} + WCR_3 \times R_{3IT}]/(WCR_1 + WCR_2 + WCR_3). \qquad (63)$$

where WCR_1, WCR_2, and WCR_3 are the weights of catalysts in reactors 1, 2, and 3, and R_{1IT}, R_{2IT}, and R_{3IT} are the inlet temperatures for reactors 1, 2, and 3, respectively.

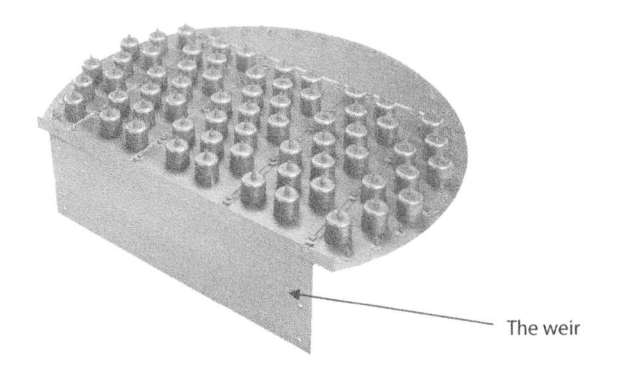

Figure 57 Bubble cap tray with the weir.

Weighted Average Bed Temperature (WABT):

$$[WCR_1 (R_{1IT} + R_{1OT})/2 + WCR_2 (R_{2IT} + R_{2OT})/2 + WCR_3 (R_{3IT} + R_{3OT})/2]/ \text{total weight of catalyst} \quad (64)$$

where WCR_1, WCR_2, and WCR_3 are the weights of the catalyst in reactors 1, 2, and 3; R_{1IT}, R_{2IT}, and R_{3IT} are the inlet temperatures for reactors 1, 2, and 3; and R_{1OT}, R_{2OT}, and R_{3OT} are the outlet temperatures for reactors 1, 2, and 3, respectively.

Well: 1. A natural oil or gas reservoir that exists below a layer of sedimentary rock. **2.** A hole bored or drilled into the earth for the purpose of obtaining water, oil, gas, or other natural resources.

West Texas Intermediate (WTI): A type of crude oil commonly used as a price benchmark.

Wet Gas: 1. Natural gas that has not had the butane, C_4, and natural gasoline removed. Also the equivalent refinery gas stream. **2.** A term used to describe light hydrocarbon gas dissolved in heavier hydrocarbons. Wet gas is an important source of LPG. **3.** Water that is present in natural gas in offshore platforms. It is necessary to remove the water from the gas for export through subsea pipelines. The pipeline is dosed with corrosion inhibitors to prevent hydrate formation.

White Oil: Sometimes kerosene, sometimes treated kerosene used for pharmaceutical purposes and in the food industry.

WHSV: Weight hour space velocity; weight of feed per hour per weight of a catalyst.

Wick Char: A test used as an indication of the burning quality of a kerosene or illuminating oil. It is defined as the weight of deposits remaining on the wick after a specified amount of sample is burned.

What-If Analysis (WIA): A safety review method, by which "What-If" investigative questions (i.e., brainstorming and/or checklist approach) are asked by an experienced and knowledgeable team of the system or component under review where there are concerns about possible undesired events. Recommendations for the mitigation of identified hazards are provided.

Wppm: Parts per million by weight.

Xylene, $C_6H_4(CH_3)_2$: 1. A colorless liquid of the aromatic group of hydrocarbons made from the catalytic reforming of certain naphthenic petroleum fractions. Used as high-octane motor and aviation gasoline blending agents,

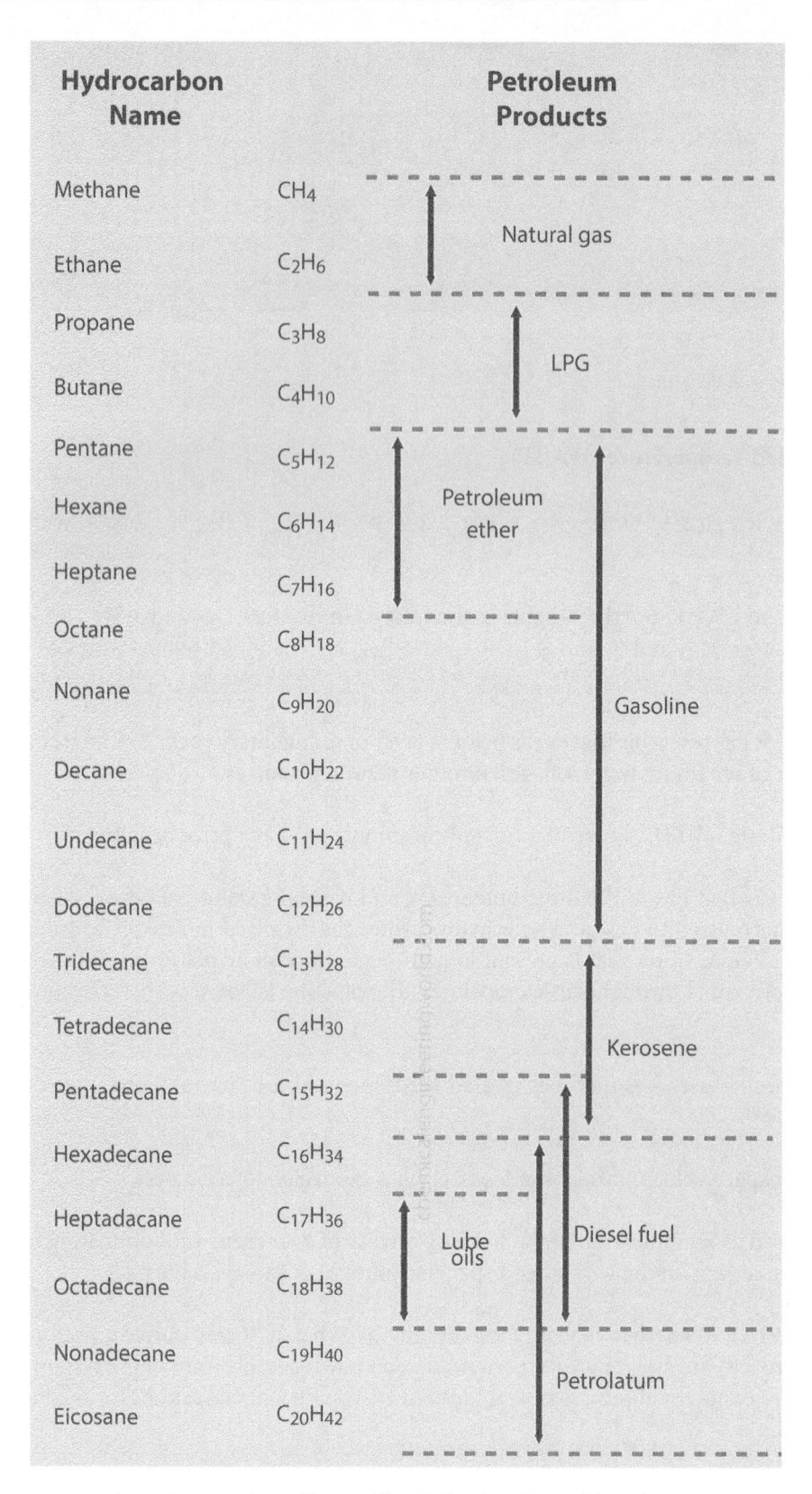

Hydrocarbon Name		Petroleum Products
Methane	CH_4	Natural gas
Ethane	C_2H_6	
Propane	C_3H_8	LPG
Butane	C_4H_{10}	
Pentane	C_5H_{12}	Petroleum ether
Hexane	C_6H_{14}	
Heptane	C_7H_{16}	
Octane	C_8H_{18}	Gasoline
Nonane	C_9H_{20}	
Decane	$C_{10}H_{22}$	
Undecane	$C_{11}H_{24}$	
Dodecane	$C_{12}H_{26}$	
Tridecane	$C_{13}H_{28}$	Kerosene
Tetradecane	$C_{14}H_{30}$	
Pentadecane	$C_{15}H_{32}$	
Hexadecane	$C_{16}H_{34}$	
Heptadacane	$C_{17}H_{36}$	Lube oils / Diesel fuel
Octadecane	$C_{18}H_{38}$	
Nonadecane	$C_{19}H_{40}$	Petrolatum
Eicosane	$C_{20}H_{42}$	

Figure 58 Hydrocarbon names and petroleum products. (Source: Chemicalengineeringworld.com).

solvents, and chemical intermediates. **2.** One of the aromatic compounds. Xylene has a benzene ring and two methyl radicals attached and three isomers, namely, ortho, para, and metaxylene. Used as a gasoline blending compound or chemical feedstock for making phthalic acids and resins.

Yield: Either the percent of a desired product or all the products resulting from a process involving chemical changes to the feed.

Zeolites: **1.** Compounds used extensively as catalysts, made of silica or aluminum as well as sodium or calcium and other compounds. Zeolites come in a variety of forms—porous and sand like or celatinous—and provide the platform for numerous catalysts. The solid zeolites have extensive pores that give very large surface areas. The precise control during the fabrication of the pore sizes enables selected access to different-sized molecules during reactions. **2.** A class of minerals that are hydrated aluminosilicates. An aluminosilicate is where some of the Si atoms in silica (SiO_4) are replaced with aluminum, giving an excess negative charge. Hydrated means that water is strongly associated with these materials by hydrogen bonding. A positively charged counter ion is required to balance the negative charge on the zeolite. Zeolites are extremely porous materials, with a regular internal structure of cavities of defined size and shape.

About the Author

A. Kayode Coker PhD, is Engineering Consultant for AKC Technology, an Honorary Research Fellow at the University of Wolverhampton, U.K., a former Engineering Coordinator at Saudi Aramco Shell Refinery Company (SASREF) and Chairman of the department of Chemical Engineering Technology at Jubail Industrial College, Saudi Arabia. He has been a chartered chemical/petroleum engineer for more than 30 years. He is a Fellow of the Institution of Chemical Engineers, U.K. and a senior member of the American Institute of Chemical Engineers. He holds a B.Sc. honors degree in Chemical Engineering, a Master of Science degree in Process Analysis and Development and Ph.D. in Chemical Engineering, all from Aston University, Birmingham, U.K. and a Teacher's Certificate in Education at the University of London, U.K. He has directed and conducted short courses extensively throughout the world and has been a lecturer at the university level. His articles have been published in several international journals. He is an author of ten books in chemical engineering, a contributor to the Encylopedia of Chemical Processing and Design, Vol. 61, and a certified train – the mentor trainer. A Technical Report Assessor and Interviewer for chartered chemical engineers (IChemE) in the U.K. He is a member of the International Biographical Centre in Cambridge, U.K. (IBC) as Leading Engineers of the World for 2008. Also, he is a member of International Who's Who for Professionals™ and Madison Who's Who in the U.S.

Index

absorbent, 118, 621, 632, 664, 701–702, 837
absorbers, 118, 344, 636, 693
Absorption, 272, 573, 632, 654, 692, 704, 723, 809
adsorbent, 664, 666, 672, 675, 690, 810, 837
Adsorption, xxii, 666–667, 672, 674, 690, 723, 810
amines, 281, 287, 289, 617–618, 620–621, 623–624, 626, 628–629, 632–633, 651–653, 681, 701
anaerobic, 811, 856, 866, 902
anthropogenic, 571, 576, 578, 654, 656
antiknock, 471, 811, 865, 881–882
Autocatalytic, 231, 346
Autoignition, 116, 231, 347, 816
autothermal, 669–671, 723

Bhopal, 121, 316
biochemical, 656, 897
biocides, 286–287, 360, 598
biodegradable, 342–343, 603, 607, 811
Biodiesel, xx–xxi, 556, 588–590, 592–598, 601–607, 729–731
Bioenergy, 724, 729, 818
bioethanol, 556, 600, 607
biofuels, xxiv, xxvi, 463, 554–556, 584, 587–588, 599, 607, 659, 675, 724, 730–731
Biomass, xxi, 554, 587–589, 598, 601, 610–611, 613, 676, 687, 724, 730, 732
biomethanol, 588
bioreactors, 591, 607
biorefinery, 556
BLENDING, 361, 429
blendstock, 464, 878, 904
BLEVE, 115–116, 214, 220, 223–224, 231, 240, 347, 354, 819
Blowdown, 8–9, 146, 208, 231, 258, 819
Bowtie, xv, 320–322, 820

CANSOLV, 653–655
carbamate, 627–629
carburetor, 892, 930–931
carcinogenic, 336, 455, 721
cavitation, 281, 286, 876
CCPS, 239, 247, 249–250, 264, 352, 354–355
Chatter, 232
choked, 53, 134, 228, 271
CSB, xiv, 217, 219–220, 223, 240, 260, 268, 272, 323, 354–355, 834
cyrogenic, 670

decarbonization, 661, 663, 676–678, 693, 705, 731–732
Decarburization, 282, 295
decoking, 334
deflagration, 118, 233, 242–243, 347, 836
dehydration, 598, 600, 612, 625
Dehydrogenation, 186, 837
Desalter, xiv, 267, 540, 545
desalting, 287, 289, 332–333, 338
desulfurization, 344, 542, 673, 825, 837, 895
detergents, 361, 817

electrochemical, 279, 282, 689, 701–702, 731, 832, 844
electrolysis, 661, 664, 676, 680, 704, 726, 832
enzymes, 589, 594, 681, 683–684, 728

FAME, 589, 593, 595, 598, 604–606, 660
FCCs, 338, 721
FCCU, 346, 361, 534, 709–711
Fermentation, 589, 598–599, 601
fertilizers, 682, 707
Flammability, 233, 348, 851
Flares, xiii, 182, 184, 234, 247, 338, 348
flarestack, 225–227
Flaring, xiii, xvi, xxv, 194, 336–337, 355, 577, 725, 852–853
FLASH, 361, 383
flashpoint, 850–851, 872, 913
Flexicoking, 853, 923
Flixborough, 316
flooding, 843, 853
flowsheet, 303, 471–472, 491, 591, 648, 856, 899
fluidity, xxv, 391, 415–416
fluidized, 358, 611, 655, 675, 721, 723, 856, 927
FMEA, xv, 263, 307, 317–318, 849
footprints, 558, 659, 824
Fractionation, 730, 816, 857, 915
fractionators, 271, 838, 914
fragmentation, 45
fugitive, 338, 344–345, 561, 659
Furnaces, 215, 331, 334

Gasification, 612, 687
gasifier, 687–688
gasohol, 556, 587, 878, 885
Glossary, xvi, xxiii, 346, 358, 569, 723, 809–935
Greenhouse, xx, 558, 562–566, 573–574, 576–577, 662, 726, 732, 824, 860

HAZAN, xv, 240, 301, 315, 317, 354
HAZARD, 310–314, 923
Hazardous, 234, 252, 323, 325–326, 349, 733, 830, 862–863
HAZID, 863
Hazop, xv, xxv, 301, 307, 354–355, 890
hydrocracker, 288, 371, 438, 442, 534, 836, 865, 872, 901
Hydrocracking, 265, 330, 430, 538, 717, 825, 865
hydrodesulfurization, 339, 345, 539, 621, 883, 925
Hydrofining, 866
Hydrofluoric, 273, 277, 294
Hydroforming, 866
HYDROTREATER, 734

inflammable, 851, 853, 868, 882
isobutane, 118, 271–273, 811, 837, 869, 875, 879, 902
isobutylene, 272, 846, 875, 877, 922
Isocracking, 869
isomer, 869
Isomerate, 360, 449, 579, 869
Isomerization, 257, 265, 330, 430, 598, 869, 871
isoparaffin, 811

Looping, xxiii, 688, 721, 732
LOPA, 349, 814, 910
LOTO, 322–325, 350, 875

MABP, 204, 209, 827
MDEA, xxvi, 617, 619, 623–625, 628, 651, 683, 691–692, 723, 730, 810
membranes, 578, 626, 681–682, 686, 690–692
mercaptan, 625, 842, 920
Methanation, xii, 117–118
Methanol, 130, 406, 474, 589–593, 595, 597, 601–603, 605–606, 612, 665, 877
microalgae, 587, 607, 729–730
microorganisms, 591, 661, 811
Monoethanolamine, 289, 617, 621

NFPA, 1, 114, 232, 240, 243, 245, 354, 850–851
nutrients, 343, 607

OVERPRESSURE, 62, 75–76
Oxygenates, 406, 816, 877, 885

Paraffin, 51, 366, 369, 838, 931
paraffinicity, 811, 827

permeance, 686, 691, 727
pinch, 345, 581–582, 585, 844, 888–889, 897
Plug, 3–5, 35, 855, 891, 928–929
Pollutants, 339
Pollution, xvi, 340, 355, 465, 568, 830
polycarbonate, 724
Polymerization, 117, 246, 265, 430, 474, 825, 891

reformates, 361, 911
reformers, 338, 663–664, 707, 828
Reforming, xiv, xxii, 265, 271, 330, 430, 465, 538, 678, 825, 830, 904, 918
regasification, 834
Regeneration, xxii, 654, 675, 692, 700, 704, 905
renewables, 663, 726, 731
Runaway, xii, 116, 118, 122–123, 128, 131, 133, 138, 191, 238, 246, 352
RV, xiv, 267, 890, 907

SAFETY, 7, 30, 306
Selexol, 617
SMR, xxii, 579, 666–672, 676, 678–680, 723, 732
SONIC, 210
SRU, 653
SRV, 22, 35–36, 46, 48
sulfides, 288, 342–343, 877, 920
Sulfinol, 617, 619
sulfonation, 119
sulfonic, 343, 605
sustainability, xxvi, 580–582, 584, 587, 657, 660–661, 663, 678, 720, 722–723
Sustainable, xxii, 580, 582, 584, 655–656, 659, 706, 729, 731
Sweetening, xxi, 265, 339, 430, 634, 636, 730, 920

Tagout, 323, 325, 875
tailpipes, 230, 583
TAN, 285–286, 616, 810, 922
Torrefaction, 588–589, 732
toxicity, 306, 316, 455, 464, 584, 603
toxins, 316, 904
Transalkylation, 708
transesterification, 594–595, 602, 605–606
triglycerides, 589–590, 593–595, 601–602, 607

Zeolites, 713, 920, 935

Also of Interest

Check out these other related titles from Scrivener Publishing

In the same set:

PETROLEUM REFINING DESIGN AND APPLICATIONS HANDBOOK VOLUME 1, by A. Kayode Coker, ISBN: 9781118233696. The most comprehensive and up-to-date coverage of the advances of petroleum refining designs and applications, written by one of the world's most well-known process engineers, this is a must-have for any chemical, process, or petroleum engineer.

PETROLEUM REFINING DESIGN AND APPLICATIONS HANDBOOK VOLUME 2: *Rules of Thumb, Process Planning, Scheduling, and Flowsheet Design, Process Piping Design, Pumps, Compressors, and Process Safety Incidents*, by A. Kayode Coker, ISBN: 9781119476412. The second of a three-volume set of the most comprehensive and up-to-date coverage of the advances of petroleum refining designs and applications, written by one of the world's most well-known process engineers, this is a must-have for any chemical, process, or petroleum engineer.

PETROLEUM REFINING DESIGN AND APPLICATIONS HANDBOOK VOLUME 3: *Mechanical Separation, Distillation, Packed Towers, Liquid-Liquid Extraction, Heat Transfer and Process Safety Incidents*, by A. Kayode Coker, ISBN: 9781119794868. The third volume of a four-volume set of the most comprehensive and up-to-date coverage of the advances of petroleum refining designs and applications, written by one of the world's most well-known process engineers, this is a must-have for any chemical, process, or petroleum engineer.

PETROLEUM REFINING DESIGN AND APPLICATIONS HANDBOOK VOLUME 4: *Heat Transfer, Pinch Analysis and Process Safety Incidents*, by A. Kayode Coker, ISBN: 9781119827528. The fourth volume of a multi-volume set of the most comprehensive and up-to-date coverage of the advances of petroleum refining designs and applications, written by one of the world's most well-known process engineers, this is a must-have for any chemical, process, or petroleum engineer.

By the same author:

CHEMICAL PROCESS ENGINEERING VOLUME 1: *Design, Analysis, Simulation, Integration, and Problem Solving with Microsoft Excel-UniSim Software for Chemical Engineers Computation, Physical Property, Fluid Flow, Equipment & Instrument Sizing, Pumps & Compressors, Mass Transfer*, by A. Kayode Coker and Rahmat Sotudeh-Gharebagh, ISBN 9781119510185. Written by one of the most prolific and respected chemical engineers in the world and his co-author, also a well-known and respected engineer, this two-volume set is the "new standard" in the industry, offering engineers and students alike the most up-do-date, comprehensive, and state-of-the-art coverage of processes and best practices in the field today.

CHEMICAL PROCESS ENGINEERING VOLUME 2: *Design, Analysis, Simulation, Integration, and Problem Solving with Microsoft Excel-UniSim Software for Chemical Engineers, Heat Transfer and Integration, Process Safety, Chemical Kinetics and Reactor Design, Engineering Economics, Optimization*, by A. Kayode Coker and Rahmat Sotudeh-Gharebagh, ISBN: 9781119853992. Written by one of the most prolific and respected chemical engineers in the world and his co-author, also a well-known and respected engineer, this two-volume set is the "new standard" in the industry, offering engineers and students alike the most up-do-date, comprehensive, and state-of-the-art coverage of processes and best practices in the field today.

Other Related Titles from Scrivener Publishing

Hydraulic Fracturing and Well Stimulation, edited by Fred Aminzadeh, ISBN 9781119555698. The first volume in the series, Sustainable Energy Engineering, written by some of the foremost authorities in the world on well stimulation, this groundbreaking new volume presents the advantages, drawbacks, and methods of one of the hottest topics in the energy industry: hydraulic fracturing ("fracking").

Energy Storage 2nd Edition, by Ralph Zito and Haleh Ardibili, ISBN 9781119083597. A revision of the groundbreaking study of methods for storing energy on a massive scale to be used in wind, solar, and other renewable energy systems.

The *Greening of Petroleum Operations*, by M. R. Islam *et al.*, ISBN 9780470625903. The state of the art in petroleum operations, from a "green" perspective.

Emergency Response Management for Offshore Oil Spills, by Nicholas P. Cheremisinoff, PhD, and Anton Davletshin, ISBN 9780470927120. The first book to examine the Deepwater Horizon disaster and offer processes for safety and environmental protection.

Bioremediation of Petroleum and Petroleum Products, by James Speight and Karuna Arjoon, ISBN 9780470938492. With petroleum-related spills, explosions, and health issues in the headlines almost every day, the issue of remediation of petroleum and petroleum products is taking on increasing importance, for the survival of our environment, our planet, and our future. This book is the first of its kind to explore this difficult issue from an engineering and scientific point of view and offer solutions and reasonable courses of action.